FOOD and FEED CROPS
of the
United States

Including all 50 states, U.S. territories, District of Columbia and Puerto Rico

Second Edition, Revised

A Descriptive List Classified According to Potentials for Pesticide Residues

A Regulatory Food Safety Focus

MEISTER**PRO**
reference guides *professional tools to grow with!*

Meister Publishing Co.
37733 Euclid Ave.
Willoughby, Ohio 44094

Library of Congress Cataloging-in-Publication Data.

Main entry under title:

Food and Feed Crops of the United States
Markle, George 1939-
Baron, Jerry, 1958-
Schneider, Bernard, 1944-

Includes index
Bibliography:
I. 1. Edible crops — common names, scientific names. 2. Feed crops – common names, scientific names. 3. EPA crop production regions. 4. Crop grouping schemes. 5. EPA residue chemistry guidelines – sampling, crop field trials, processed food/feed. 6. Germplasm Resources Information Network (GRIN). 7. Bayer plant codes. 8. Codex plant codes. 9. Canada's crops. 10. Mexico's crops. 11. Minor and major crops. 12. Taxonomy.
II. Title
III. Title: Food and Feed Crops of the United States

About the Authors

1. Professor G. M. Markle – Associate Director/Professor, IR-4 Project and Rutgers Cooperative Extension Department of Agricultural and Resource Management Agents, Rutgers University.

2. Dr. J. J. Baron – Assistant to the Director/Professor, IR-4 Project and Rutgers Cooperative Extension Department of Extension Specialists, Rutgers University.

3. Dr. B. A. Schneider – Senior Plant Physiologist, Health Effects Division, U.S. Environmental Protection Agency, Washington, D.C.

June 1998

Additional copies are available.
To order please write:
Meister Publishing Co.
Food and Feed Crops of the United States
37733 Euclid Ave.
Willoughby, OH 44094
440-942-2000
Fax: 440-942-0662

Special thanks to our immediate families for their support:
Wife, Nancy Markle and sons, John and David and daughter Nicole.
Wife, Brenda Baron and son, Liam.
Wife, Barbara Schneider and daughters, Eileen, Deborah and Susan.

This publication is Dedicated to:
Dr. Charles C. Compton
1898-1979
Project Leader, Friend and Counselor
Of the IR-4 Project

Preface

For over 25 years, the first edition of *Food and Feed Crops of the United States: A Descriptive List Classified According to Potentials for Pesticide Residues* has been an essential authority in the pesticide field for the U.S. and internationally, concerning the standardization of plant names and the development of extensive crop groupings, which included a large number of related minor crops, never before reported in such a concise manner. The book is known affectionately in the regulatory and academic arenas as the "Green Book" and it has become a classic publication for use by government agencies, reviewers, and as a reference for students, researchers, and other countries' agricultural departments.

The second edition updates its comprehensive coverage while maintaining the concise format. The primary revisions involve the regulatory agencies residue chemistry policies concerning crop groupings, commodity definitions, geographical location of residue field trials, updates on specific feed items, commodity vocabulary, changes to portion of commodity sampled for the raw agricultural commodity (RAC), international harmonization with Codex Alimentarius (Codex), and the development of many new food commodities. The majority of the regulatory changes were initiated in 1995. Throughout, the second edition reflects new nomenclature including plant names, botanical names, and regulations.

Food and Feed Crops of the United States, Second Edition, like its respected earlier edition, continues to be the most complete source of food and feed crop information as a significant part of the regulatory scheme for food safety. We are continuing the tradition of providing a basis for grouping minor crops. This continues to be important to research organizations, growers, industry, residue chemistry reviewers, and government agencies including international groups. Food and feed crop harmonization across these groups will provide a standardization on preferred or principal commodity names and common names that are used in the trade, thereby influencing FDA market inspectors.

Specific advantages of the second edition are:

• A reference for regulatory agencies dealing with commodity issues for petition reviews, inspections, etc.

• Standardized commodity nomenclature to include preferred, common and botanical names to meet the needs of a common vocabulary.

• Use by regulatory enforcement to determine commodity samples.

• Use as a training tool.

• Use by industry registrants for updated commodity information as it relates to the regulated community.

• Large step in international harmonization with Codex and North American Free Trade Agreement (NAFTA).

• Regulatory relationship of food and feed crops to include significant feed items.

• Assist emerging markets.

As noted in the first edition, this publication was the first of its kind to attempt to classify all commercial food and feed crops in the U.S. from the standpoint of degree of susceptibility to possible exposure to pesticides. The primary purpose of the groupings is for use in establishing safe pesticide tolerances and registering needed uses, especially for minor crops integrated pest management systems. The handy and useful glossary of edible crops is an additional benefit.

Pesticides, biological and conventional, are essential to the efficient production of commercially grown food and feed crops of high quality, wholesomeness and variety. The use of pesticides in the U.S. and internationally is normally regulated by the amount of residue permitted on the crop or the raw agricultural commodity (RAC) at harvest. The amount of any residue that is safe on the RAC is the tolerance level or maximum residue limit (MRL). A magnitude-of-residue study on a crop provides the actual amount of residue which is expressed in ppm by weight in or on the RAC or processed commodity. The level of residue that represents the use pattern under good agricultural practices is used as the tolerance level or MRL. Up-to-date tolerance level information on individual crops or crop groups for specific pesticides can be obtained from U.S. EPA publication Title 40, Code of Federal Regulations, Parts 180, 185 and 186. Eventually under the Food Quality Protection Act (FQPA) of 1996, Parts 185 and 186 will be merged into Part 180.

Instead of developing residue data on every food and feed crop for each pesticide use, data development on a smaller number of representative or indicator crops in crop groupings is more efficient, especially in regulatory research time, money and quality assurance. Using the botanical relationship does not always help in developing crop groups, unrelated crops may be better aligned for regulatory purposes. Therefore, botanical relationships, comparison of edible parts and their use, cultural practices, geographical distribution, production practices, feed items, and processed products are considered in crop group definition development.

Interregional Research Project No. 4 (IR-4) was established in 1963 to help the producers of minor crops obtain registrations for the conventional pesticides they need to successfully grow food and ornamental crops. A network of state and federal IR-4 cooperators develops field and laboratory data to support petitions for pesticide residue tolerances or exemptions for submission to US EPA. IR-4 has provided the national leadership in developing crop groups, subgroups, and definitions over the years to incorporate many minor crops in the regulatory scheme. The development of crop groups with the inclusion of additional minor crops will help facilitate the establishment of tolerances for needed uses in the U.S. and internationally. Therefore, this book will continue to assist in addressing IR-4's national objective of establishing safe tolerances for needed minor uses.

G.M. Markle
Senior Author

Introduction

The listing of crops from the first edition of *Food and Feed Crops of the United States* was utilized as the basis for this edition. As in the first edition new crops were included only if planted on farms or in home gardens or where we felt that there was a good potential for farm or home garden use, e.g., ethnic foods with an expanding market. No attempt was made to include varieties of these crops since differences in varieties are generally minor from the standpoint of pesticide exposure.

Crop Monograph Format

The majority of the book is established on a crop monograph template as follows:

Crop monograph format: For convenience in use, all crop monographs are listed in alphabetical order by principal commodity names. Each crop monograph is composed of 12 fields:

Crop Monograph Number
Definitive three digit number in bold for each crop monograph (001-691)

Field 1
Line one includes the commodity name or preferred name selected by the authors for standardization in bold print and other common names in parentheses. Commodity names are in the singular form except in a few cases, e.g., collards.
Line two includes the plant family name. In certain cases, the old family name is included in parentheses.
Line three includes the recognized scientific name. Additionally, synonyms will be added in parentheses for clarity with previous scientific identifier(s).
Note: In certain cases, multiple field 1's will be used for a crop monograph, e.g., mint.
Field 2
Includes a general paragraph on the plant descriptions which are written from the standpoint of potential surface exposure to pesticide residues that may occur in the plant part used for food or feed. Descriptions have been derived from the most authentic sources available.
Field 3
Includes information on crop season, as available, e.g., seeding to harvest time and bloom to harvest.
Field 4
Includes acreage of the crop in the U.S. Normally, 1994 USDA Agricultural Statistics and 1992 Census of Agriculture are used for the U.S. Additional references are noted in the monograph, as appropriate. (In certain cases, the USDA Plant Hardiness Zone Map was used for growing regions in fields 4 and 5).
Field 5
Includes other crop production regions outside the U.S., e.g., Canada, Mexico, West Indies, Central America, South America, etc.
Field 6
Includes the general use(s) of the edible part(s) of the crop.
Field 7
Includes the edible portion of the crop in commerce.
Field 8
Includes the portions of the plant that are sampled and recognized as the raw agricultural commodity and processed products from

EPA's Table 1 in OPPTS 860.1000 Guidelines dated August 1996 or as suggested by the Authors if the crop is not included in Table 1 (860.1000), Appendix V. The portion of the commodity to be analyzed is very important in establishing tolerances and MRL's. In determining residue levels, it is imperative that the raw agricultural commodity or processed commodity be adequately defined and consistently used by all persons conducting analyses for these purposes. Therefore, the U.S. EPA is proposing to amend 40 CFR 180.1(j) by establishing a new section, 40 CFR 180.45, with an expanded list of commodities including descriptions of the commodity portion(s) to be analyzed. This expanded list will reflect section 102 of the Pesticide Analytical Manual I which is published by Food and Drug Administration.

Field 9
Includes the various exposure classifications by crop groupings and individual crops.
Line 9a includes the Authors crop grouping scheme. Established U.S. EPA crop grouping terminology was used, as appropriate. New groups, e.g., tropical tree fruit and oilseed crops have been added. All commodities in the crop monograph were grouped. The authors used 31 crop groups (Appendix IV), to accomplish harmonization with crop grouping schemes – both national and international. Crop grouping names are in the plural form, except oilseed, and grass forage. Our primary interest is to propose an expansion of the presently accepted crop groupings for pesticide residue tolerance purposes while continuing to provide a working classification of crops for the possible grouping of minor crops within established major crop groups. As noted in the Introduction, there are 27 major food and feed crops which have been listed by U.S. EPA based on the FQPA criteria. Therefore, the remainder of the crops in the "Green Book" can be classified as minor based on acreage, but not minor in importance to the growers and consumers.
Line 9b includes the U.S. EPA crop grouping scheme as published in the 17 May 1995 *Federal Register* for the specific crop. If a crop is not included in the EPA crop grouping, the term "miscellaneous" is used. The present U.S. EPA crop grouping regulations enable the establishment of tolerances for a group of crops based on residue data from a subset of crops representative of the group. The established 19 groups include the following:
1. Root and tuber vegetables
2. Leaves of root and tuber vegetables (human food or animal feed)
3. Bulb vegetables
4. Leafy vegetables (except *Brassica* vegetables)
5. *Brassica* (Cole) leafy vegetables
6. Legume vegetables (succulent or dried)
7. Foliage of legume vegetables
8. Fruiting vegetables (except cucurbits)
9. Cucurbit vegetables

10. Citrus fruits
11. Pome fruits
12. Stone fruits
13. Berries
14. Tree nuts
15. Cereal grains
16. Forage, fodder, and straw of cereal grains
17. Grass forage, fodder, and hay
18. Nongrass animal feeds (forage, fodder, straw and hay)
19. Herbs and spices

The individual crops included in the 19 crop groups are exhibited in Appendix II.

Canada utilizes the same crop grouping scheme as the U.S. EPA, except for the addition of Group 20 (OILSEED) as follows:
(i) Representative commodities. Rapeseed and sunflower in Crop Group 20.
(ii) Commodities. The following is a list of all the commodities included in Crop Group 20:

Rapeseed (*Brassica napus*)
Rapeseed, Indian (*Brassica campestris*)
Mustard seed, Indian (*Brassica juncea*)
Mustard seed, Field (*Brassica campestris*)
Flax, Linseed (*Linum usitatissimum*)
Sunflower seed (*Helianthus annuus*)
Safflower (*Carthamus tinctorius*)

Line 9c includes the Codex Alimentarius Recommendations Concerning Pesticide Residue, dated 1993, for an international food grouping scheme. If a crop is not listed under Codex, the term "no entry" or "no specific entry" is used. The Codex classification of foods and animal feeds is intended to be as complete a listing of commodities in the trade as possible. EPA has recognized the importance of uniformity of commodity terms or standardization. One aspect of our effort is to provide a cross listing for eventual evaluation by the regulatory agencies to meet the needs of the growers. The Codex plant codes are exhibited in Appendix III.

Line 9d includes the U.S. EPA crop definition scheme of food groups as they relate to tolerances. The regulation is cited in 40 CFR 180.1(h). If a crop is not listed in the regulation, the term "none" is used. If a new entry has been proposed, the term "proposed" or "proposing" is used.

As listed in the specific crop monographs, certain "proposed" entries for the tropical/subtropical fruit definitions were recently accepted by the Chemistry Science Advisory Committee (CHEMSAC), Health Effects Division (HED), Office of Pesticide Programs and U.S. EPA. The accepted proposals are listed in the crop monographs for the representative commodities: avocado, grapefruit, guava, lychee, papaya, sugar apple (expansion) and citrus fruits (expansion) for white sapote, which see. The immediate impact of these U.S. EPA approvals is that researchers may develop residue data on the representative commodities in support of tolerances for the crops in the groups. Please note that HED approvals of these proposals does not constitute rulemaking. Therefore, the term "proposed" is retained for now. In the interim, researchers may utilize these new groupings to develop residue data under a recognized research program to support the new tolerance groups. The U.S. EPA Minor Use Officer plans to keep the regulated community informed on the eventual rulemaking under 40 CFR 180.1(h). These documents are available from the Minor Use Officer.

Field 10:
Includes the significant references used to develop the monograph. The monograph references are keyed to the Bibliography by means of its identifier. This concept is favorable for computer generated databases. Where the reference identifier is followed by the term "picture," the cited reference includes an image of the crop.

Field 11:
Includes reference to the EPA Crop Production Regional Map in Appendix I and identifies the region(s) for the crop, as appropriate.

Field 12:
Includes the Bayer plant code, recognized by the Weed Science Society of America (WSSA) and Japan (WSSJ). This international plant code system is utilized for crop standardization in computer models. Another international plant code system is Codex Alimentarius which is exhibited in field 9c of the crop monographs and Appendix III by code.

Taxonomy

All taxonomy used conforms to that of the Germplasm Resources Information Network (GRIN), the database of the USDA-Agricultural Research Service's National Plant Germplasm System. The taxonomic data provided includes the family name [e.g., Rosaceae], with acceptable alternative family names provided in parentheses for those families where they exist [e.g., Asteraceae (Compositae)]. This is followed by the currently accepted species name consisting of the Latin genus-species binomial and the authorship of the binomial (e.g., *Cryptotaenia canadensis* (L.) DC.). One or more frequently used taxonomic synonyms may follow the accepted species name to enable users of such names to locate the correct entry. The particular taxonomy accepted in this publication is based on the determinations of USDA taxonomists in GRIN as to the scientific accuracy and current usage of scientific names. All scientific nomenclature used conforms to the rules and recommendations of the International Code of Botanical Nomenclature (Tokyo Code) (W. Greuter *et al.*, 1994).

The authors in consultation with J.H. Wiersema of the Systematic Botany & Mycology laboratory USDA/ARS, decided to continue, for the present, the use of Liliaceae for the genus *Allium*. Alliaceae may be the eventual family name.

Food Quality Protection Act (FQPA)

FQPA (3 August 1996) amended both the Federal Food, Drug, and Cosmetic Act and Federal Insecticide, Fungicide, and Rodenticide Act. It provided the finite definition for a minor crop as one that is grown on less than 300,000 acres (Section 2 of the Act) and highlighted the IR-4 Minor Use Program in Section 32. Section 201 provides a definition for "food" as:

1. Articles used for food or drink for man or other animals.
2. Chewing gum.
3. Articles used for components of any such article. The term "food" shall mean a raw agricultural commodity or processed food.

The term "safe" with respect to a tolerance for a pesticide chemical residue, means that the U.S. EPA has determined that there is a reasonable certainty that no harm will result from aggregate exposure to the pesticide chemical residue, including all anticipated dietary exposures and all other exposures (Section 408). In our testing programs, we feel that it is important to develop real world residue data which include anticipated and actual (Market Basket) for potential exposure purposes.

For exposure purposes, it is best to examine the edible portion of the commodity for residues and discard the remainder. Anticipated residues can be generated for typical household preparations of food commodities, including peeling, washing, rinsing, cooking, etc. After the normal preparation of a food, the actual residues can be zero (less than the method sensitivity).

For international standardization or harmonization, FQPA requires the U.S. EPA to determine whether a maximum residue level for the pesticide chemical has been established by the Codex Alimentarius Commission (Section 408). This is required when U.S. EPA is establishing a tolerance for a pesticide chemical residue in or on food.

Major Crops

In the U.S., crops are considered major based on acreage. Major crops for the U.S., Canada and Europe are included as representative of the large acreage crops grown in the world.

United States

The Food Quality Protection Act (FQPA) of 1996 defines a minor crop as one grown on less than 300,000 acres. The major food and feed crops identified by the U.S. EPA based on the FQPA criteria are:
*Representative Commodities in the U.S. EPA 17 May 95 Crop Grouping Scheme

Almonds*	Peanuts
Apples*	Pecans*
Barley	Potato*
Bean (dry)*	Rice*
Bean (snap)*	Rye
Canola	Sorghum*
Corn (Sweet)*	Soybean*
Corn (Field)*	Sugar beet*
Corn (Pop)*	Sugarcane
Cottonseed	Sunflower
Grapes	Tobacco
Hay (alfalfa & other)*	Tomato*
Oats	Wheat*
Oranges*	

Canada

The large acreage Canadian crops grown on greater than 300,000 acres are:

Canadian Crop	Canadian area (in hectares)
Wheat (total)	12,625,600
Wheat, Spring	10,967,200
Hay, Tame	6,508,500
Barley	4,240,000
Canola	4,063,000
Alfalfa	3,229,119
Wheat, Durum	1,440,600
Oats	1,356,800
Corn, Field (total)	1,172,600
Corn, Grain	993,000
Soybeans	719,600
Flaxseed	501,700
Peas, Dry Field	467,400
Lentils	327,800
Mixed Grains	253,700
Wheat, Winter	217,800
Corn, Silage	198,902
Mustard Seed	186,200
Corn, Fodder	179,600
Rye (Total)	159,100
Rye, Fall	138,800
Potatoes	125,000
Canary Seed	123,400

Source: Rothwell 1996a, Pest Management Regulatory Agency, Health Canada. NOTE: Canada's Regulatory Directive for Residue Chemistry Guidelines is DIR 98-02, dated 1 June 98.

Europe

Major Crops by crop groupings grown in Europe are:

Citrus Fruits
Lemon
Mandarin
Orange

Pome Fruits
Apple
Pear

Stone Fruits
Apricot
Peach
Plum

Berries and Small Fruits
Grape (Wine)
Grape (Table)
Strawberry

Miscellaneous
Olive

Root and Tuber Vegetables
Carrot
Sugar beet
Fodder beet

Bulb Vegetables
Onion (dry)

Fruiting Vegetables
Tomato
Pepper
Cucumber
Melon

Source: PALLUTT 1996

Brassica **Vegetables**
Cauliflower
Brussels Sprouts
Cabbage (head)

Leafy Vegetables and Fresh Herbs
Lettuce

Legume Vegetables (Fresh)
Bean (green with pod)
Pea (green without pod)

Stem Vegetables
Leek

Pulses
Bean (dry (incl. broad bean)
Pea (dry (incl. chick pea)

Oilseed
Cottonseed
Rapeseed
Soybean
Sunflower

Cereals
Barley
Maize (Corn)
Oat
Rice
Rye
Sorghum
Triticale
Wheat

Others
Potato
Hops

Typical Development of Crop Definitions/Groups

The following 14 points are addressed in constructing a crop definition and/or group proposal:

1. Botany and nomenclature of crop.
2. U.S. Geographical distribution and production.
3. Cultural practices.
4. Commercial importance.
5. Possibility of genetic improvements.
6. Comparison of edible parts.
7. Livestock feed item(s) – Significant 250,000 tons.
8. Processed products.
9. Pest problems (especially if similar).
10. Comparison of potential residue levels.
11. Justification for crop group/subgroup/definition.
12. Codex classification(s).
13. Rotational crops.
14. References supporting the proposal.

Source: Authors

Common Crop Names

United States

Common names were captured from various sources including the first edition of *Food and Feed Crops of the United States*, GRIN, U.S. EPA 1995 Crop Grouping Index which includes 687 common commodity names for about 300 species, Hortus Third, Cornucopia, Fruits of Warm Climates, Codex, etc.

Puerto Rico

To exhibit the common names used in Puerto Rico and their relationship to English names as a cross reference, the 1992 U.S. Census provided the following:

Avocado = Aguacates
Banana = Guineos
Beans = Habichuelas
Cabbage = Repollo
Cantaloupes = Cantaloups
Cassava = Yuca
Celeriac = Apio
Citron = Cidra
Coconuts = Cocos
Coffee = Café
Corn = Maiz
Cucumber = Pepinillos
Dasheens = Malanga
Eggplant = Berenjenas
Cane = Caña
Grapefruit = Toronjas
Grasses = Yerbas
  Paragrass = Malojillo
  Star = Estrella
Lemons = Limones
Limes = Limas

Lettuce = Lechuga
Mangoes = Mangoes
Oranges = Chinas
Papayas = Papayas
Passionfruit = Parchas
Pasture = Pastos
Peppers = Pimientos
Pigeon Pea = Gandures
Pineapples = Piñas
Plantains = Plántanos
Pumpkins = Calabazas
Soursop = Guanábanas
Sugarcane = Caña de azucar
Sweet Corn = Maiz Tierno
Sweet Potatoes = Batatas
Taniers = Yautias
Tomatoes = Tomates
Watermelons = Sandias
Yams (True) = Ñames
Crops = Agricoles

Crop/Commodity Codes

This edition of *Food and Feed Crops of the United States* includes in the crop monographs standardized crop/commodity codes which are based on scientific and common names from
• Codex Alimentarius plant codes (letters and numbers).
• 1992 Bayer plant codes (letters).

Other plant codes that were reviewed but not included in the crop monographs are:
• USDA/Natural Resources Conservation Service extensive list which is primarily based on scientific names (letters and numbers).
• 1992 Census of Agriculture which is primarily based on common name (numbers).

The 1992 Census of Agriculture crop codes are as follows:

Noncitrus crops	Code
Apples	.123
Apricots	.129
Avocados	.135
Cherries, sweet	.345
Cherries, tart	.587
Grapes, dry	.171
Grapes, fresh	.177
Nectarines	.201
Olives	.207
Peaches	.225
Pears	.231
Plums and prunes, fresh	.243
Prunes, dry	.249
Other noncitrus - *Specify*	.261

Citrus crops	Code
Grapefruit	.267
Lemons	.279
Limes	.285
Oranges	.297
Tangelos	.303
Tangerines	.309
Other citrus - *Specify*	.315

Nut crops	Code
Almonds	.321
Filberts and hazelnuts	.327
Pecans	.339
Walnuts, English	.357
Other nut trees - *Specify*	.363

Vegetable Crops	Code
Asparagus	.379
Beans, snap (bush & pole)	.381
Beets	.383
Broccoli	.385
Cabbage, head	.391
Cantaloupes and muskmelons	.395
Carrots	.397
Cauliflower	.399
Celery	.401
Collards	.407
Cucumbers and pickles	.411

Vegetable Crops (continued)	Code	Crop name	Code
Eggplant	415	Alfalfa seed	542
Garlic	421	Beans, dry edible (DO NOT include dry limas.)	554
Herbs, fresh cut	455	Beans, dry lima	557
Honeydew melons	423	Buckwheat	575
Lettuce and romaine	427	Canola	614
Lima beans, green	429	Corn cut for dry fodder, hogged or grazed	581
Mustard greens	431	Dry southern peas (cowpeas)	584
Onions, dry	433	Emmer and spelt	599
Onions, green	435	Fescue seed	602
Okra	437	Flaxseed	605
Peas, English, green (DO NOT include southern peas.)	441	Industrial rapeseed	668
Peppers, sweet	443	Kentucky bluegrass seed	629
Peppers, hot	445	Lentils	635
Pumpkins	449	Lespedeza seed	638
Radishes	451	Mint for oil	644
Southern peas (cowpeas), green-blackeyed, crowder, etc.	409	Mustard seed	650
Spinach	457	Peanuts for nuts	656
Squash	459	Peas, dry edible	659
Sweet corn	461	Popcorn	662
Tomatoes	463	Proso millet	665
Turnips	465	Red clover seed	671
Turnip greens	467	Rice	677
Watermelons	473	Rye for grain	686
Other vegetables - *Specify*	475	Ryegrass seed	689
		Safflower	692
Bedding plants (include vegetable plants)	479	Sorghum cut for dry forage or hay	698
Bulbs, corms, rhizomes, or tubers - dry	482	Sorghum hogged or grazed	701
Cut flowers and cut florist greens	485	Sugar beets for sugar	719
Nursery crops - ornamentals fruit & nut trees, & vines	488	Sugarcane for sugar	722
Foliage plants	707	Sugarcane for seed	725
Potted flowering plants	710	Sunflower seed	734
Mushrooms	494	Timothy seed	746
		Other crops - *Specify*	752
Sod harvested	497		
Vegetable & flower seeds	500		
Greenhouse vegetables	503		
Other - *Specify*	506	Source: RICHE	

Abbreviations and Symbols Used In
Food and Feed Crops of the United States

A..acre
cm ....centimeter
cv. ....cultivar, cultivated variety
DM ....dry matter
Kg...kilogram
lb. ....pound
m ....meter
mm. ....millimeter
notho ....relating to (hybrid)
sp. ....one species
spp................................more than one species
ssp..............................refers to one subspecies
syn. ....synonym
T...ton
var. ....variety which is below ssp.
x ....lower case x, denotes a cross
between two spp.
/ ....per
% ....percent
( )......for scientific names, denotes synonyms
® ....registered trade mark

Note: Metric equivalents and approximate conversions are listed in Appendix VIII.

Acknowledgements

The authors recognize the generous input in revising or reviewing the general articles, advice and expertise for computerization of the data to better utilize the monographs as a database, and friends and colleagues for their support and assistance in writing the book.

American Mushroom Institute
Nilsa Acin
Drew Baker, Jr.
Brenda Baron
William Biehn
Robin Bellinder
Donald Bowen
Mark Burt
Raymond Choban
Jonathan Crane
John Cuddy
Joe Defrancisco
Keith Dorschner
Jere Downing
Debra Edwards
Ray Frank
Ann George
Mark Gnozzio
Jay Holmdal
Diane Infante
IR-4 Project Management Committee
Fred Ives
Hoyt Jamerson
Michael Kawate
Daniel Kunkel
Edward Kurtz
Robert Libby
Rick Loranger
Rocky Lundy
Edith Lurvey
Nancy Markle
John Martini
Robert McReynolds
Charles Meister
Rick Melnicoe
Lynn Miller
Satoru Miyazaki
Reed Olszack
Waltraud Pallutt
James Parochetti
Randy Perfetti
Ray Ratto
Margaret Reiff
Keith Richmond
Douglas Rothwell

Rutgers Cooperative Extension Departments (Agents & Specialists)
Chairs Bruce Barbour & Tom Orton
Kenneth Samoil
Pat Sarica
Alan Schreiber
Barbara Schneider
Elizabeth Schneider
Carroll Southard
Terry Spittler
David Thompson
Arthur Tucker
Melvin Tolliver
Brian Townshend
United States Department of Agriculture
United States Environmental Protection Agency
Mark Van Doren
Richard Van Vranken
Mark Wach
John Wiersema
James Willmott
Yukiko Yamada
Edward Zager
Rafael Montalvo-Zapata
Richard Zollinger

An Editorial Committee, consisting of Chair T. D. Spittler - Cornell University; J. R. Frank – USDA/ARS (retired); J. H. Wiersema - USDA/ARS; and G. M. Markle, was established to enhance the database. Also, scientists at the universities, including IR-4 Regional Coordinators, and regulatory experts provided state of the art data or grass root information on the very minor crops. The use of unpublished data has made this edition a very concise and complete book.

Since this is the 2nd Edition of *Food and Feed Crops of the United States*, we continue to recognize the contributors to the 1st Edition: C. Compton, G. Markle, J. Magness, V. Boswell, A. Kehr, S. Brooks, A. Hanson, A. Hovin, P. Reitz, W. Briggle, G. Sprague, C. Adair, R. Coleman, Dewey Stewart, Earl Smith, Paul Miller, N. Childers, P. Eck, Brad Johnson, Agripino Perez, Ed Swift, Stu Race, C. Smith, B. Pepper, United States Department of Agriculture Cooperative State Research Service (presently, Cooperative State Research, Education and Extension Service) and Agricultural Research Service, and the New Jersey Agricultural Experiment Station, College of Agriculture and Environmental Science (presently, Cook College), Rutgers University, the State University of New Jersey.

G.M. Markle
Senior Author

Table of Contents

Crop Monographs

001

1. Abiu

(Caimito, Madura verde, Luma, Temare, Yellow starapple)

Sapotaceae

Pouteria caimito (Ruiz & Pav.) Radlk. (syn: *Lucuma caimito* Roem. & Schult., *Achras caimito* Ruiz & Pav.)

2. The abiu tree is generally about 33 feet high. A gummy latex exudes from wounds in the bark. The leaves are 4 to 8 inches long, $1^1/_4$ to $2^3/_8$ inches wide and mostly smooth. The fruit, downy when young, is ovoid, elliptical or round. The fruit is $1^1/_2$ to 4 inches long with smooth, tough, pale-yellow skin when ripe with a white, mild-flavored pulp containing one to four seeds. Abiu can be picked while underripe and firm for transport to market.
3. Season, harvest: Fruit matures in October in Florida.
4. Production in U.S.: No data, limited. Few trees were planted in southern Florida (MORTON). In 1996, 9 trees reported in Hawaii (KAWATE 1996b).
5. Other production regions: South America tropics (MORTON 1987).
6. Use: Fruit eaten fresh or used to make ices and ice cream.
7. Part(s) of plant consumed: Fruit
8. Portion analyzed/sampled: Whole fruit
9. Classifications:
   a. Authors Class: Tropical and subtropical fruits – inedible peel
   b. EPA Crop Group (Group & Subgroup): Miscellaneous
   c. Codex Group: No specific entry
   d. EPA Crop Definition: None
10. References: GRIN, KAWATE 1996b, MORTON (picture)
11. Production Map: EPA Crop Production Region 13.
12. Plant Codes:
   a. Bayer Code: No specific entry

002

1. Abyssinian cabbage

(Abyssinian mustard, African cabbage, Karate, Ethiopian mustard, Mustard collard)

Brassicaceae (Cruciferae)

Brassica carinata A. Braun

2. Tender leaves and young stems up to 12 inches tall can be eaten raw in salads. Older leaves and stems are cooked. The cultivar, 'TexSel', developed by Texas A&M University is a green leafy vegetable with similar growth characteristics to collards and mustard. 'TexSel' was released in 1972. The greens are somewhat milder than collards and without the pungency of mustard greens.
3. Season, seeding to harvest: About 35-53 days. Can be harvested more than once if regrowth is permitted.
4. Production in U.S.: No data. Grown in Texas (FACCIOLA 1990). Produces well in northern Florida in the fall (STEPHENS 1988).
5. Other production regions: East Africa (FACCIOLA 1990).
6. Use: Leaves and stems eaten raw or cooked.
7. Part(s) of plant consumed: Leaves and stems
8. Portion analyzed/sampled: Leaves and stems
9. Classifications:
   a. Authors Class: *Brassica* (cole) leafy vegetables
   b. EPA Crop Group (Group & Subgroup): Miscellaneous
   c. Codex Group: No specific entry

d. EPA Crop Definition: None
10. References: GRIN, FACCIOLA, STEPHENS 1988 (picture)
11. Production Map: EPA Crop Production Regions 3 and 6.
12. Plant Codes:
   a. Bayer Code: No specific entry

003

1. Acerola

(West Indies cherry, Barbados cherry, Puerto Rican cherry, Cerza, Chereese, Jamaica cherry, Garden cherry, Native cherry, French cherry, Huesito, West Indian cherry, Cerise-de-cayenne)

Malpighiaceae

Malpighia emarginata DC.

M. glabra L. (syn: *M. punicifolia* L.)

2. The plant is a large evergreen shrub or small tree, up to 15 feet. It grows in hot tropical lowlands with medium to high rainfall. Acerola is propagated by seed, cutting, layering and grafting. The leaves are entire, elliptic and pointed, nearly glabrous. The plant produces waxy, bright cherry-like fruit during the wet season. Fruits are about an inch in diameter, with a single, rather large seed. The fruit skin is crimson red when mature with orange-yellow flesh. Its skin is smooth and thin with three ribs, or sutures. Acerola has come into prominence as having the highest known vitamin C content (up to 60 times higher than in orange juice).
3. Season: Fruits ripen in 3 to 4 months from bloom. Fruits in all stages of development are on the trees at the same time.
4. Production in U.S.: Some commercial plantings in Puerto Rico (500 acres, 1980) and Hawaii (no current data).
5. Other production regions: Other Caribbean countries and northern South America (MORTON).
6. Use: Mainly juice, also jelly, puree and powder.
7. Part(s) of plant consumed: Fruits are processed (extracted or canned)
8. Portion analyzed/sampled: Whole fruit
9. Classifications:
   a. Authors Class: Tropical and subtropical fruits – edible peel
   b. EPA Crop Group (Group & Subgroup): Miscellaneous
   c. Codex Group: 005 (FT 0287 and FT 4095) Assorted tropical and subtropical fruits – edible peel
   d. EPA Crop Definition: Proposing – Guava = Acerola
10. References: GRIN, ASENJO, CODEX, KNIGHT, LOGAN, MAGNESS, MARTIN 1987, US EPA 1994, MORTON (picture), PHILLIPS (d)
11. Production Map: EPA Crop Production Region 13.
12. Plant Codes:
   a. Bayer Code: MLPPU (*M. glabra*)

004

1. African horned cucumber

(Kiwano, Jelly melon, African horned melon)

Cucurbitaceae

Cucumis metuliferus E. Meyer ex Naudin

2. The fruit is yellowish-orange when ripe and studded with numerous spines. The flesh is green, translucent and slightly mucilaginous. The fruits are oblong and 2 to 4 inches long when mature and are bright orange on the outside with green inner pulp. The plants are grown like cucumbers, growing 5 to 10 feet long vines.
3. Season, seeding to harvest: No data.

4. Production in U.S.: Available year round from New Zealand. Eighteen acres in California at Los Banos (MYERS 1991).
5. Other production regions: New Zealand (MYERS 1991).
6. Use: Flesh used in fruit cups, cocktails, sundaes, parfaits and drinks. Marketed as a garnish or strained for juice in U.S.
7. Part(s) of plant consumed: Fruit
8. Portion analyzed/sampled: Whole fruit
9. Classifications:
   a. Authors Class: Cucurbit vegetables
   b. EPA Crop Group (Group & Subgroup): Miscellaneous
   c. Codex Group: No specific entry
   d. EPA Crop Definition: None
10. References: GRIN, FACCIOLA, MYERS (picture), REHM
11. Production Map: EPA Crop Production Region 10.
12. Plant Codes:
    a. Bayer Code: No specific entry

005

1. Agrimony
(Cocklebur)
Rosaceae
Agrimonia eupatoria L.

2. Erect perennial herb about 3 feet tall. Leaves largely in lower part with 7 to 13 leaflets, oblong to narrow-obovate and gray-hairy beneath to 1 inch wide. Bright yellow flowers. The whole plant is sweet-scented and the flowers have a spicy, apricot-like fragrance. Popular in France for making tea. A cultivar is called 'Sweet-scented'.
3. Season, planting: Propagated in the spring by seeds and leaves harvested just before flowers open.
4. Production in U.S.: No data. Introduced into the U.S. (GARLAND 1993).
5. Other production regions: Mostly in the north temperate zone (BAILEY 1976).
6. Use: Herbage has been used medicinally. Flowers, leaves and stems are harvested when the plant is in bloom and made into a tea. Especially, the dried leaves are used as a country tea.
7. Part(s) of plant consumed: Leaves
8. Portion analyzed/sampled: Leaves (fresh and dried)
9. Classifications:
   a. Authors Class: Herbs and spices
   b. EPA Crop Group (Group & Subgroup): Miscellaneous
   c. Codex Group: No specific entry
   d. EPA Crop Definition: None
10. References: GRIN, BAILEY 1976, FACCIOLA, GARLAND (picture)
11. Production Map: No entry
12. Plant Codes:
    a. Bayer Code: AGIEU

006

1. Airpotato
(Potato yam, Hoi, 'Ala 'ala, Aerial yam, Bulbil-bearing yam, Bitter yam, Acom)
Dioscoreaceae
Dioscorea bulbifera L.

2. This relative of the yam is a tall-climbing, twining herb to about 21 feet long with alternate, heart-shaped leaves. It forms large tubers of variable size and shape in the axils of the leaves. These may attain several pounds in weight and are palatable and potato-like in flavor. The plant is native to South Asia and is cultivated to a limited extent in tropical and subtropical areas.
3. Season, planting: Use bulbils or aerial tubers for propagation.
4. Production in U.S.: No data, Hawaii (NEAL 1965).
5. Other production regions: Tropical Asia (NEAL 1965).
6. Use: See Yam, True.
7. Part(s) of plant consumed: Tuber
8. Portion analyzed/sampled: Tuber
9. Classifications:
   a. Authors Class: Stalk and stem vegetables
   b. EPA Crop Group (Group & Subgroup): Miscellaneous
   c. Codex Group: No specific entry
   d. EPA Crop Definition: None
10. References: GRIN, MAGNESS, NEAL, YAMAGUCHI 1983, DUKE (a)(picture)
11. Production Map: EPA Crop Production Region 13.
12. Plant Codes:
    a. Bayer Code: DIUBU

007

1. Akee apple
(Akee, Aki, Seso vegetal)
Sapindaceae
Blighia sapida K.D. Koenig

2. Tropical evergreen tree of medium size (45 ft.). Grows in hot tropical lowlands where rainfall is well distributed or where seasonal dry periods occur. Tolerant of light frost. Propagated by seed. The flowers are produced laterally in racemes. The trivalued fruits turn yellow and red as they ripen. The mature fruit splits open along 3 sutures exposing the 3 large, shiny, black seeds attached to a white or milky-white aril.
3. Season: Bloom to harvest in about 90 days.
4. Production in U.S.: No data.
5. Other production regions: Africa, India, tropical America.
6. Use: The aril is consumed fresh or is cooked and used as a vegetable.
7. Part(s) of plant consumed: The firm and oily aril is the edible portion. Only naturally opened fruits can be eaten, immature fruits are very toxic
8. Portion analyzed/sampled: Aril
9. Classifications:
   a. Authors Class: Tropical and subtropical fruits – inedible peel
   b. EPA Crop Group (Group & Subgroup): Miscellaneous
   c. Codex Group: 006 (FI 0325 and FI 4147) Assorted tropical and subtropical fruits – inedible peel
   d. EPA Crop Definition: None
10. References: GRIN, CODEX, MAGNESS, MARTIN 1987
11. Production Map: EPA Crop Production Region 13.
12. Plant Codes:
    a. Bayer Code: No specific entry

008

1. Alfalfa
(Lucerne, Tame hay, Queen of forages, Mielga, Variegated alfalfa, Sickle alfalfa, Sickle medic, Yellow Lucerne, Yellow-flower alfalfa)
Fabaceae (Leguminosae)
Medicago sativa L. ssp. *sativa*

2. Alfalfa is probably native to Asia Minor and the Caucasus Mountain area but has been cultivated since antiquity. It was first established as a crop in the U.S. around 1850 in California. Its culture did not become widespread until the 20th century. At present alfalfa is grown for hay on about 25,000,000 acres in the U.S., with additional acreage grown for seed and pasture. This includes variegated alfalfas, hybrids between *M. sativa* L. and *M. falcata* L., a Siberian species, which presently is considered a subspecies of *M. sativa,* as *M. sativa* ssp. *falcata* (L.) Arcang. (Sickle alfalfa, Sickle medic, Yellow lucerne and Yellowflower alfalfa). Many cultivars of alfalfa have been selected for local adaptation, disease resistance and growth characteristics. These include spreading types, either by underground rhizomes or by adventitious shoots rising from lateral roots. The alfalfa plant is perennial, new growth rising each year from the crown. Stems reach to 3 feet and 5 to 25 stems may rise from a single crown. The pinnately trifoliate leaves are arranged alternately on the stems. Alfalfa pasture is excellent for hogs, but cattle on alfalfa pasture are subject to bloat unless carefully adapted to the pasture. When seeded with grasses, bloat is less of a problem. Much alfalfa is dehydrated and pelleted or ground as a meal and used in feed mixes. Alfalfa hay is highly palatable and nutritious with very high digestible protein, mineral and vitamin contents. About half of the harvested tame hay is alfalfa. Also cut for silage at late bud to one-tenth bloom stage. Alfalfa has about 227,000 seeds per pound. Alfalfa has the highest feeding value of all common legume hay crops.

3. Season, emergence to harvest: about 60 days with 2 or more cuttings per year (sometimes 4 to 9 cuttings per year). Stand life normally 3-5 years.

4. Production in U.S.: In 1995, 24,569,000 acres harvested for hay with 84,980,000 tons (USDA 1996c). In 1995, the top 10 states reported as South Dakota (2,600,000 acres harvested), Wisconsin (2,300,000), Montana (1,600,000), Minnesota (1,425,000), North Dakota (1,400,000), Iowa (1,350,000), Nebraska (1,350,000), Michigan (1,050,000), Idaho (1,100,000) and California (1,000,000) (USDA 1996c).

5. Other production regions: In 1995, Canada reported 7,979,339 acres of alfalfa and alfalfa mixtures. Alfalfa seed was produced on 148,965 acres. Canada produced 16,082,879 acres of tame hay in 1995 with alfalfa about half of it (ROTHWELL 1996a).

6. Use: See above. Seeds are also used as sprouts in salads. Pasture, hay, silage, haylage and honey crop.

7. Part(s) of plant consumed: Foliage (seeds for sprouts, leaves and stems). Also source of honey crop.

8. Portion analyzed/sampled: Primarily forage and hay. For alfalfa grown for seed, residue data should be provided on seed, forage and hay; for all other uses data should only be provided on forage and hay. No processed commodities are required. For alfalfa meal, residue data are not needed for meal; however, the meal should be included in the livestock diet, using the hay tolerance level. Hay should be field-dried to a moisture content of 10 to 20 percent. For alfalfa silage, residue data on silage are optional, but are desirable for assessment of dietary exposure. Cut at late bud to one-tenth bloom stage for alfalfa, allow to wilt to approximately 60 percent moisture, then chop fine, pack tight and allow to ferment for three weeks maximum in an air-tight environment

until it reaches pH 4. This applies to both silage and haylage. In the absence of silage data, residues in forage will be used for silage, with correction for dry matter.

9. Classifications:
   a. Authors Class: Nongrass animal feed
   b. EPA Crop Group (Group & Subgroup): Nongrass animal feed (18) (Representative crop)
   c. Codex Group: 050 (AL 1020, alfalfa fodder and 1021, alfalfa forage green) Legume animal feed
   d. EPA Crop Definition: Alfalfa = Sainfoin, Birdsfoot trefoil and varieties and/or hybrids of these

10. References: GRIN, CODEX, MAGNESS, ROTHWELL 1996a, USDA 1996c, US EPA 1996a, US EPA 1995a, US EPA 1994, BARNES, BALL (picture), FICK, SMALL

11. Production Map: EPA Crop Production Regions 1, 2, 5, 7, 9, 10 and 11.

12. Plant Codes:
    a. Bayer Code: MEDSA

009

1. Alkali sacaton
(Zacaton alcalino)
Poaceae (Gramineae)
Sporobolus airoides (Torr.) Torr.
1. Sacaton
(Giant sacaton)
   S. wrightii Munro ex Scribn.

2. Alkali sacaton is a native perennial bunchgrass found from South Dakota west to Washington and south into Mexico. It is densely tufted and long-lived, with erect, solid stems about 3 feet tall. Basal foliage is abundant, the leaves being up to 18 inches long and $^{1}/_{4}$ inch wide. Roots are fibrous and deep-penetrating. The grass will grow on moist, alkaline soils, hence the name, but also occurs on other soil types. It is palatable while succulent but becomes tough and unpalatable when ripe. Hay is of fair quality if cut early. Propagation is by seed, which is usually harvested from native stands. Sacaton is a more robust grower than alkali sacaton and is more southern in its range, being native from West Texas to Arizona and south into Mexico. Sacaton is not as widely distributed as alkali sacaton. The stems reach to 6 feet and are firm and hard. It is less drought resistant than alkali sacaton. It furnishes useful grazing both while succulent and in winter and is useful for hay if cut while young. It is rarely planted but is a useful grass in its native range. There are 80,000 seeds/lb.

3. Season: Growth midspring. Graze spring and summer. Seeded at 8 to 10 lb/acre, flower in June and mature in August.

4. Production in U.S.: No specific data, See Forage grass. South Dakota west to Washington and south.

5. Other production regions: Mexico.

6. Use: Pasture, hay, rangeland.

7. Part(s) of plant consumed: Foliage

8. Portion analyzed/sampled: Forage and hay

9. Classifications:
   a. Authors Class: See Forage grass
   b. EPA Crop Group (Group & Subgroup): See Forage grass
   c. Codex Group: See Forage grass
   d. EPA Crop Definition: None

10. References: GRIN, MAGNESS, ALDERSON, SKERMAN

11. Production Map: EPA Crop Production Regions 7, 8, 9, 11 and 12.
12. Plant Codes:
    a. Bayer Code: SPZAI (*S. airoides*)

010

1. Alkaligrass
Poaceae (Gramineae)
Puccinellia spp.
1. Nuttall's alkaligrass
P. nuttalliana (Schult.) Hitchc. (syn: *P. airoides* S. Watson & J.M. Coult.)
1. Slender alkaligrass
(European alkaligrass, Weeping alkaligrass, Reflexed saltgrass)
P. distans (Jacq.) Parl.
2. There are more than 40 species of cool season alkaligrass growing in North America with Nuttall alkaligrass and Slender alkaligrass being the most widely distributed and are members of the grass tribe, *Festuceae*. Nuttall alkaligrass occurs from Alaska to Newfoundland and south to California, New England and Kansas. It produces good to excellent forage for grazing animals. Slender alkaligrass is native to Europe and occur from Alaska to Newfoundland and south to California and Virginia. It is also being utilized for pasture and being developed for a salt tolerant turfgrass.
3. Season: Cool season grass that starts growing in early spring.
4. Production in U.S.: There is no production data for Alkaligrass, however, pasture and rangeland are produced on more than 410 million A (U.S. CENSUS, 1992).
5. Other production regions: Both species are widespread in Canada.
6. Use: Pasture grazing.
7. Part(s) of plant consumed: Leaves and stems
8. Portion analyzed/sampled: Forage and hay
9. Classifications:
    a. Authors Class: See Forage grass
    b. EPA Crop Group (Group & Subgroup): See Forage grass
    c. Codex Group: See Forage grass
    d. EPA Crop Definition: None
10. References: GRIN, US EPA 1994, US EPA 1995, CODEX, USDA NRCS, HOLZWORTH, HITCHCOCK, RICHE, MOSLER (picture)
11. Production Map: EPA Crop Production Regions for Nuttall alkaligrass are 1, 5, 7, 8, 9, 10, 11 and 12; and for Slender alkaligrass are 1, 5, 7, 9, 10, 11 and 12.
12. Plant Codes:
    a. Bayer Code: PUCDI (*P. distans*), PUCSS (*P. spp.*)

011

1. Allspice
(Clove pepper, Kubaba, Pimento, Jamaica pepper, Pimenta, Malaqueta)
Myrtaceae
Pimenta dioica (L.) Merr. (syn: *P. officinalis* Lindl.)
2. Allspice is the dried, unripe berries of a large evergreen tree, approximately 30 to 60 feet tall, native to the Caribbean area. The leaves are large and leathery, about 8 inches long by 2 inches wide. The fruits are about $1/3$-inch diameter, near globose, produced in clusters of a dozen or more at or near the terminals of branches. The fruit is harvested while unripe, as it is then most strongly flavored. It is a drupe, with 1 or 2 seeds. The whole dried fruit is ground to produce the allspice powder of commerce. Both pulp and seeds are aromatic and contain an oil with qualities similar to clove oil. An average of 1,070 tons were imported into the U.S. (1990-92).
3. Season, harvest: Harvest mature berries at unripened stage. Trees begin to bear in 5-8 years and are in full production at 20-25 years. Fruits ripen in 3-4 months after harvesting. Berries are sun-dried for 7-10 days.
4. Production in U.S.: No data.
5. Other production regions: Caribbean, especially Jamaica, and Mexico (FARRELL 1985).
6. Use: Spice known as allspice in the U.S. because its flavor resembles cinnamon, nutmeg and cloves put together.
7. Part(s) of plant consumed: Unriped fruits that are cured by fermenting and drying by sun or artifical heat with a moisture content of 10 percent. In some countries, leaves are used as tea or a spice
8. Portion analyzed/sampled: Fruit (dried)
9. Classifications:
    a. Authors Class: Herbs and spices
    b. EPA Crop Group (Group & Subgroup): Herbs and spices (19B)
    c. Codex Group: 028 (HS 0792) Spices (CODEX uses "Pimento, fruit" as the commodity name)
    d. EPA Crop Definition: None
10. References: GRIN, CODEX, FARRELL (picture), LEWIS, MAGNESS, PENZEY, USDA 1993a, US EPA 1995a, DUKE (a)
11. Production Map: EPA Crop Production Region 13.
12. Plant Codes:
    a. Bayer Code: PMTDI

012

1. Almond
(Almond oil, Amandier, Almendro)
Rosaceae
Prunus dulcis (Mill.) D.A. Webb (syn: *P. amygdalus* Batsch; *Amygdalus communis* L.)
2. The almond tree resembles peach, to which it is closely related. It grows to 25 feet, but under cultivation is usually held to under 20 feet by pruning. The leaves are simple, lanceolate and glabrous. The nuts are enclosed in a fleshy husk which becomes dry and fibrous and splits open allowing the nuts to drop out or be easily separated at maturity. The shell is porous to woody and encloses the oblong, flattened kernel. The latter is up to over an inch long and half as wide. Almonds are marketed both in the shell and as shelled kernels. Oil extracted from almond kernels is non-drying and is obtained from both the bitter almond, not edible as a nut, and from sweet almond kernels. The oil from the two kinds appears identical. The oil is edible, but is used largely in the manufacture of certain pharmaceuticals.
3. Season, bloom to harvest: 5 to 6 months. (Harvesting from August to October)
4. Production in U.S.: About 235,000 tons (shelled basis) and 402,000 acres (1993). 1994 Agricultural Statistics. In 1994, California reported 443,636 acres (MELNICOE 1996e). 1992 CENSUS reported California with about 100 percent of the acreage.

5. Other production regions: Mediterranean countries (WOODROOF 1967).
6. Use: Direct eating and in confections.
7. Part(s) of plant consumed: Inner kernels. Hulls as significant feed item
8. Portion analyzed/sampled: Nutmeat and hulls. Shells shall be removed and discarded from nuts before examination for residues
9. Classifications:
   a. Authors Class: Tree nuts
   b. EPA Crop Group (Group & Subgroup): Tree nuts (Representative crop)
   c. Codex Group: 022 (TN 0660) Tree nuts
   d. EPA Crop Definition: None
10. References: GRIN, CODEX, MAGNESS, RICHE, USDA 1994, US EPA 1995a, US EPA 1995b, US EPA 1994a, WOODROOF (b) (picture), MELNICOE 1996e, IVES, ROSENGARTEN
11. Production Map: EPA Crop Production Region 10
12. Plant Codes:
    a. Bayer Code: PRNDU

013

1. Almond, Tropical

(Badam, Kamani, Olive bark tree, Indian almond, Malabar almond, Almendro)

Combretaceae

Terminalia catappa L.

2. Large tree can grow 70 to 85 feet tall with egg shaped, leathery leaves. Grown for its edible nut. The fruit is the size of a plum with tender skin and juicy edible pulp. The nutmeat is edible, fresh or roasted. It also yields an edible oil and the meal is feed to pigs. The nuts yield about 50 percent oil.
3. Season, harvest: Ripens in January (WHEALY 1993).
4. Production in U.S.: No data. Grows in southern Florida (SCHNEIDER 1993a).
5. Other production regions: Native to southeast Asia (SCHNEIDER 1993a).
6. Use: Greenish red fruit with oil bearing edible seed. The nutmeat is edible fresh or roasted.
7. Part(s) of plant consumed: Fruit
8. Portion analyzed/sampled: Seed, meal and oil
9. Classifications:
   a. Authors Class: Tree nuts
   b. EPA Crop Group (Group & Subgroup): Miscellaneous
   c. Codex Group: 022 (TN 0677) Tree nuts
   d. EPA Crop Definition: None
10. References: GRIN, CODEX, SCHNEIDER 1993a, WHEALY
11. Production Map: EPA Crop Production Region 13.
12. Plant Codes:
    a. Bayer Code: TEMCA

014

1. Amaranth, Chinese

(Joseph's-coat, Amaranto, Caliloo, Calilu, Edible amaranth, Bush greens, Chinese amaranth, Amaranth, Leafy amaranth, Hon-toi-moi, Tampala, Chinese spinach, Aupa)

Amaranthaceae

Amaranthus tricolor L. (syn: *A. gangeticus* L., *A. melancholicus*)

2. Several species of *Amaranthus* are cultivated in Southeast Asia and to a limited extent by Chinese gardeners in the U.S. Plants are annuals, grown from seed and leaves and young stems are used as potherbs. Mature plants are 1 to 3 feet tall, with leaves 6 inches long. For use as potherbs, young plants may be pulled at 3 to 4 weeks; or the tops may be cut off at that stage and a second crop will be produced from lateral growth. When tender shoots are harvested, several cuttings can be made. Plant growth and parts used are similar to spinach. Some leaves are red, others green and variegated, usually with purplish patterns on a green background. The green leaf variety is 'tampala'. Callaloo is a leafy green vegetable from *Colocasia esculenta* (Dasheen) or *Amaranthus* spp. (Chinese spinach). Cultivated in southern Florida during the winter and a number of locations on the U.S. eastern seaboard near large metropolitan areas during the spring and summer. Adapted to all growing climates from the tropics northward. Canned callaloo (*Amaranthus*) is imported from Jamaica.
3. Season, seeding to harvest: 3 to 6 weeks. A second crop may be produced. In 3 to 4 weeks after transplanting when plants are 6 to 8 inches tall, the entire plant may be harvested. Seeding to transplant stage is 2 to 3 weeks. The young seedlings at thinning may be eaten. Frequently harvested at 7-10 day intervals.
4. Production in U.S.: No data. Mainly by Oriental gardeners for Chinese users. Eastern U.S. Seaboard (JANICK 1990) (STEPHENS 1988).
5. Other production regions: China (JANICK 1990).
6. Use: Cooked as a potherb.
7. Part(s) of plant consumed: Young stems and leaves
8. Portion analyzed/sampled: Tops (leaves and stems)
9. Classifications:
   a. Authors Class: Leafy vegetables (except *Brassica* vegetables)
   b. EPA Crop Group (Group & Subgroup): Leafy vegetables (except *Brassica* vegetables) (4A)
   c. Codex Group: 013 (VL 0460) Leafy vegetables (including *Brassica* leafy vegetables)
   d. EPA Crop Definition: None
10. References: GRIN, CODEX, JANICK 1990, MAGNESS, REHM, STEPHENS 1988, USDA NRCS, YAMAGUCHI 1983 (picture), SWIADER, RUBATZKY
11. Production Map: EPA Crop Production Regions 1, 2, 3 and 13.
12. Plant Codes:
    a. Bayer Code: AMATR

015

1. Amaranth, Grain

Amaranthaceae

Amaranthus spp.

1. Purple amaranth

(Red amaranth, Bush greens, Achita, Bledo, Red shank)

A. cruentus L. (syn: *A. hybridus* ssp. *cruentus* (L.) Thell)

1. Princess-feather

(Huantli, Alegria, Prince's feather, Amarante elegante)

A. hypochondriacus L. (syn: *A. hydridus* var. *erythrostachys* Moq.; *A. hydridus* auct., in part)

1. Inca wheat

(Love-lies-bleeding, Coimi, Kiwicha, Bledo frances, Amarante caudee, Amaranto, Achita, Chaquilla, Red cattail, Jataco, Kiwichi)

A. caudatus L. (syn: *A. edulis* Speg.; *A. mantegazzianus*)

2. The seeds of the above amaranths are used as grain. The seed protein is almost comparable to milk protein in nutritional quality. Crop is resistant to drought. The seeds are about the size of poppy seeds and plants have 100,000 or more seeds each. They are annuals to 7 feet tall with inflorescences to 3 feet, like a cattail. Young leaves of *A. cruentus* are used as a potherb (JANICK 1990). Amaranth, Leafy (which see).
3. Season, seeding to harvest: About 4 to 6 months. Direct combining is typical after killing frost. It can be broadcast or row seeded.
4. Production in U.S.: About 1,235 acres of the Mexican type in 1987 (NATIONAL RESEARCH COUNCIL 1989). In 1988, about 3,000 acres were planted in the Great Plains (JANICK 1990). Of particular interest to farmers growing dryland crops in areas of the Great Plains (NATIONAL RESEARCH COUNCIL 1989).
5. Other production regions: Native to Mexico, Central and South America (Andes Region) (JANICK 1990).
6. Use: Seeds are used like popcorn, cereal, breading and ground into flour or rolled into flakes. Forage and fodder are used as animal feed, as well as, grain and feed fractions.
7. Part(s) of plant consumed: Seed, forage and fodder
8. Portion analyzed/sampled: Grain and its processed commodity, flour and aspirated grain fractions; forage and fodder (stover)
9. Classifications:
   a. Authors Class: Cereal grains; Forage, Fodder and Straw of Cereal grains
   b. EPA Crop Group (Group & Subgroup): Miscellaneous
   c. Codex Group: No entry for grain use
   d. EPA Crop Definition: None
10. References: GRIN, JANICK 1990, NATIONAL RESEARCH COUNCIL 1989 (picture), REHM, BREENE PUTNAM 1992, KAUFFMAN, ROBINSON, SCARPA
11. Production Map: EPA Crop Production Regions 5 and 7.
12. Plant Codes:
   a. Bayer Code: AMACA (*A. caudatus*), AMACR (*A. cruentus*), AMACH (*A. hypochondriacus*)

016

1. Amaranth, Leafy
(Pigweed)
Amaranthaceae
Amaranthus spp.
1. Spiny amaranth
(Bledo, Pakai kuku, Blero, Thorny pigweed, Edlebur)
A. spinosus L.
1. Spleen amaranth
(Ibondwe, Bledo)
A. dubius C. Mart. ex. Thell.
1. Slim amaranth
(Vlete, Smooth pigweed)
A. hypochondriacus L. (syn: *A. hybridus* var. *erythostachys* Moq.; *A. hydridus* auct., in part)
1. Bush greens
(Bledo, Red amaranth, Purple amaranth, Achita)
A. cruentus L. (syn: *A. hybridus* ssp. *cruentus* (L.) Thell.)
1. Slender amaranth
(Green amaranth, Pakai, Cararu, Bledo)
A. viridis L. (syn: *A. gracilis* Desf.)

2. Tropical herbs which are similar to Amaranthus, Chinese (which see). Warm season crop growing 2 to 4 feet tall. Crop harvested by pulling tops with roots or by partial leaf removal for successive harvests by harvesting every 7-10 days to encourage new shoot/leaf growth. Plant tops are consumed like spinach (which see). Eaten as spinach statement is exhibited in JANICK 1990 and NEAL. *A. cruentus* is also a grain amaranth (JANICK 1990).
3. Season, seeding to harvest: Amaranth, Chinese (which see)
4. Production in U.S.: No data. *A. dubius* produced well in central Texas as summer greens (JANICK 1990).
5. Other production regions: *A. spinosus* used as spinach in India (NEAL 1965).
6. Use: Potherb.
7. Part(s) of plant consumed: Leaves and young stems
8. Portion analyzed/sampled: Tops (leaves and stems)
9. Classifications:
   a. Authors Class: Leafy vegetables (except *Brassica* vegetables)
   b. EPA Crop Group (Group & Subgroup): Leafy vegetables (except *Brassica* vegetables) (4A)
   c. Codex Group: 013 (VL 0460) Leafy vegetables (including *Brassica* leafy vegetables)
   d. EPA Crop Definition: None
10. References: GRIN, CODEX, JANICK 1990, MAGNESS (*A. spinosus*), NEAL (picture of Spiny amaranth), REHM, STEPHENS 1988, USDA NRCS, US EPA 1995a, MALKUS
11. Production Map: EPA Crop Production Region 13.
12. Plant Codes:
   a. Bayer Code: AMACR (*A. cruentus* and *A. hybridus*), AMADU (*A. dubius*), AMASP (*A. spinosa*), AMAVI (*A. viridis*), AMACH (*A. hypochondriacus*).

017

1. Ambarella
(Jobo de la India, Jew plum, Golden-apple, Otaheite apple, Great hog plum, Wi-tee, Yellow plum, Prunier de cythere, Purple mombin and Yellow mombin, ref in field 2)
Anacardiaceae
Spondias dulcis Sol. ex Parkinson (syn: *Spondias cytherea* Sonn.)
2. Tropical semi-deciduous tree to 60 feet tall. Small whitish flowers are borne in large, loose, terminal panicles. The fruits are 2 to 3 inches, globular to ovoid and somewhat irregular in outline with indistinct ridges on the sides. From 2 to 10 fruits are borne in a cluster. The tough skin is golden yellow when ripe. A single, large, spiny seed is embedded in the firm juicy flesh. Usually eaten fresh but may be preserved. Other *Spondias* spp. with edible fruit are Purple mombin (*S. purpurea* L.), Yellow mombin (*S. mombin* L.) and Imbu (which see).
3. Season, in Hawaii, fruit ripens from November to April. In Florida a single tree can provide a steady supply from fall to midwinter for a family. From bloom to harvest in 200 or more days.
4. Production in U.S.: No data. Home gardens. Hawaii and Florida (MORTON 1987).
5. Other production regions: Tropical areas and West Indies (MORTON 1987).
6. Use: Fresh, preserved, juice for cold beverages and sauce similar to apple sauce. Green fruit is pickled.

7. Part(s) of plant consumed: Fruit pulp
8. Portion analyzed/sampled: Whole fruit
9. Classifications:
   a. Authors Class: Tropical and subtropical fruits – edible peel
   b. EPA Crop Group (Group & Subgroup): Miscellaneous
   c. Codex Group: 005 (FT 0285) Assorted tropical and subtropical fruits – edible peel
   d. EPA Crop Definition: None
10. References: GRIN, MAGNESS, MARTIN 1987, MORTON (picture), CODEX
11. Production Map: EPA Crop Production Region 13.
12. Plant Codes:
    a. Bayer Code: SPXDU

018

1. Angelica
(Angelique, Garden angelica, Archangel, Wild Parsnip)
Apiaceae (Umbelliferae)
Angelica archangelica L.

2. The plants are perennial herbs to 6 feet tall with compound leaves and large umbels of flowers, related to carrots. The root is perennial, producing a new top each year. The odd flavor and odor are due to a volatile oil, contained in all parts of the plant. The roots, young stems, leaf petioles and midribs are steeped in syrups of increasing strength to produce candied angelica. Seeds are used for flavoring in beverages, cakes and candies. Oil distilled from the seeds is used in flavoring and the leaves can be cooked as a vegetable.
3. Season, start of growth to harvest: 3 to 4 months. Root usually harvested first year but if not, then grown for stems and/or seeds. Seeds mature in about 120 days after flowering.
4. Production in U.S.: No data. Naturalized in North America, mainly, Labrador to Delaware and west to Minnesota (STEPHENS 1988).
5. Other production regions: A northern plant naturalized in Europe and North America (GARLAND 1993).
6. Use: Flavoring in candies, beverages, soups, pastries, culinary and potherb
7. Part(s) of plant consumed: Roots, stems, seeds, leaves
8. Portion analyzed/ sampled: Primarily foliage and seeds; roots as required
9. Classifications:
   a. Authors Class: Herbs and spices
   b. EPA Crop Group (Group & Subgroup): Herbs and spices (19A)
   c. Codex Group: 027 (HH 0720) Herbs; 028 (HS 0720) Spices; 057 (DH 0720) Dried herbs
   d. EPA Crop Definition: None
10. References: GRIN, CODEX, GARLAND (picture), MAGNESS, STEPHENS 1988 (picture), US EPA 1995a, RUBATZKY
11. Production Map: EPA Crop Production Regions 1 and 5.
12. Plant Codes:
    a. Bayer Code: ANKAR

019

1. Anise
(Aniseed, Sweet Alice, Sweet cummin, Sweet cumin)
Apiaceae (Umbelliferae)
Pimpinella anisum L.

2. Anise is an annual herb related to carrot, which reaches a height of about 2 feet. Leaves are twice pinnate, resembling parsley. Flowers and seeds are produced in large, loose clusters. Seeds are oblong, about $1/_6$-inch long and curved. Fresh leaves are used for flavoring and garnishing, but the important articles of commerce are the seeds and oil obtained from them. Plants are cut or pulled when seeds begin to mature, stacked to dry in the field, then threshed. Seeds are used to flavor curry powder, pastries, confectionery, cheese and bread. Oil is used in beverages, to flavor drugs and in cosmetics.
3. Season, seeding to harvest: About 4 months.
4. Production in U.S.: Cultivated in Washington (SCHREIBER 1995). Grown in Florida gardens as a garnish green (STEPHENS 1988) and California has 474 acres of anise in 1992 (MELNICOE 1996a). Average yield in California range from 500 to 700 pounds of seed per acre (MYERS 1991)
5. Other production regions: Mediterranean region and China (USDA 1993a).
6. Use: Flavoring in pastries, cheese, bread; oil in beverages, drugs and cosmetics.
7. Part(s) of plant consumed: Mainly seeds and oil extracted from them. Fresh leaves to limited extent
8. Portion analyzed/sampled: Seeds
9. Classifications:
   a. Authors Class: Herbs and spices
   b. EPA Crop Group (Group & Subgroup): Herbs and spices (19B)
   c. Codex Group: 028 (HS 0771) Spices
   d. EPA Crop Definition: None
10. References: GRIN, CODEX, FARRELL (picture), FOSTER, JANICK 1990, LEWIS, MAGNESS, MELNICOE 1996a, MYERS, PENZEY, SCHREIBER, STEPHENS 1988, USDA 1993a, US EPA 1995a
11. Production Map: EPA Crop Production Regions 10 and 12.
12. Plant Codes:
    a. Bayer Code: PIMAN

020

1. Anise hyssop
(Licorice mint, Blue giant hyssop, Anise mint, Blue giant hissop)
Lamiaceae (Labiatae)
Agastache foeniculum (Pursh) Kuntze
1. Korean mint
(Huo-xiang, Agastache)
A. rugosa (Fisch. & C.A. Mey.) Kuntze

2. Hardy perennial to 4 feet tall with leaves generally oval shaped about 3 inches long and 2 inches wide and serrated. Flowers are borne on dense, terminal, cylinder-shaped spikes 2 to 4 inches long. Flowering begins from June to August and lasts into autumn.
3. Season, planting: Plant spacing of about 18 inches.
4. Production in U.S.: No data. Native to North America in the central to southwestern U.S.
5. Other production regions: Native to Asia.
6. Use: Flavoring as a spice in food and as an herbal tea.
7. Part(s) of plant consumed: Leaves and stems
8. Portion analyzed/sampled: Leaves (fresh and dried)
9. Classifications:
   a. Authors Class: Herbs and spices
   b. EPA Crop Group (Group & Subgroup): Miscellaneous

   c. Codex Group: No specific entry

   d. EPA Crop Definition: None

10. References: GRIN, FACCIOLA, FOSTER (picture), JANICK 1990, USDA NRCS (*A. foeniculum*)

11. Production Map: No entry.

12. Plant Codes:

   a. Bayer Code: No specific entry

021

1. Annatto

(Annato, Urucum, Roucou, Alaea, Arnotto, Achiote, Bija, Lipsticktree, Anatto, Lipstick pod, Annoto)

Bixaceae

Bixa orellana L.

2. Tropical evergreen shrub or small tree native to tropical America and the Caribbean. The fruits are heart-shaped, brown or reddish brown at maturity and are covered with short stiff hairs. When fully mature, the fruits split open exposing the numerous seeds. Although it does not produce an edible fruit, the annatto is widely grown for the yellow to red pulp that covers the seeds. The annatto seeds are a must for South American, Caribbean, Mexican and Spanish cooking. Imparts a red-yellow color and pungent flavor to rice and seafood dishes. To make achiote (annatto oil) cook seed in vegetable oil, strain out seeds and use for frying chicken or fish or coloring stews and casseroles. Also achiote dye can be prepared by stirring the seeds in water and used to color butter, cheese, rice and other foods.

3. Season, planting to harvest: About 2 years. Highest yields occur 3-4 years after planting and trees are productive for 10-15 years.

4. Production in U.S.: No data. Available from Florida (LOGAN 1996).

5. Other production regions: Tropical South American, Brazil (JANICK 1990).

6. Use: Seeds are used as spice and colorant. Achiote red dye is bixin and the yellow dye is orellin.

7. Part(s) of plant consumed: Seed (covered by sticky orange-red pulp or resin of bixin)

8. Portion analyzed/sampled: Seed

9. Classifications:

   a. Authors Class: Herbs and spices

   b. EPA Crop Group (Group & Subgroup): Herbs and spices (19B), seeds only

   c. Codex Group: 006 (FI 0030) Assorted tropical and subtropical fruits – inedible peel

   d. EPA Crop Definition: None

10. References: GRIN, CODEX, FACCIOLA, JANICK 1990, MAGNESS, NEAL (picture), PENZEY, LOGAN, KENNARD (picture), DUKE (a)

11. Production Map: EPA Crop Production Regions 3 and 13.

12. Plant Codes:

   a. Bayer Code: No specific entry

022

1. Apple

(Manzano, Pomme, Manzana)

Rosaceae

   Malus domestica Borkh. (syn: *Pyrus malus* L.)

2. Apples are the most important tree fruit of the temperate zone. The tree is of medium size, up to 30 to 40 feet in height, but usually held to under 20 feet by pruning. Propagated by budding or grafting onto rootstocks. Leaves are entire, up to 4 inches long, pubescent when young, near glabrous later. The fruits are oblate to slightly conic in shape, with depressions at both stem and calyx ends, $2^1/_4$ up to $3^1/_2$ inches in diameter. They consist of a thin outer peel, a thickened edible flesh and a central core of 5 carpels, in which the small seeds are borne. The peel is pubescent when young, later becoming smooth and waxy. Formerly some russeted varieties were grown. These have disappeared in commercial orchards, but russeted areas may occur on some varieties due to weather or other injury. Important commerical varieties/types include: 'Braeburn', 'Crispin'/'Mutsu', 'Empire', 'Fuji', 'Gala', 'Golden Delicious', 'Granny Smith', 'Ida Red', 'Jonagold', 'Jonathan', 'McIntosh', 'Red Delicious' and 'Rome'.

3. Season: Bloom to harvest in 70 to 170 days, depending on variety.

4. Production in U.S.: In 1995, 460,470 domestic acres produced about 5,368,000 fresh tons. Major production states are Washington (153,000 acres), New York (57,500 acres), Michigan (54,000 acres), California (35,000 acres), Pennsylvania (22,000 acres), Virginia (19,800 acres) and West Virginia (10,000 acres).

5. Other production regions: Significant imports from Canada, New Zealand, Chile, South Africa and Brazil. In 1995, Canada grew 86,136 acres (ROTHWELL 1996a).

6. Use: Fresh eating, canned sauce and slices, juice, dried, frozen, vinegar, culinary.

7. Part(s) of plant consumed: Mainly fruit flesh, but peel often eaten on fresh fruits. Peels and cores from processing plants may be used in vinegar, or as livestock feed. Pulp and peel from juice processing are used as livestock feed

8. Portion analyzed/sampled: Whole fruit, wet pomace and juice

9. Classifications:

   a. Authors Class: Pome fruits

   b. EPA Crop Group (Group & Subgroup): Crop Group 11: Pome Fruits Group (Representative Crop)

   c. Codex Group: 002 (FP 0226) Pome fruits

   d. EPA Crop Definition: None, but proposing – Apple = Crabapple

10. References: GRIN, CODEX, LOGAN, MAGNESS, SCHREIBER, US EPA 1995, US EPA 1994, ROTHWELL 1996a, IVES, REHM, USDA 1996f, CROCKER (d)

11. Production Map: EPA Crop Production Regions 1, 2, 5, 9, 10 and 11.

12. Plant Codes:

   a. Bayer Code: MABSD

023

1. Apricot

(Abricotier, Damasco, Abricot, Albaricoque)

Rosaceae

   Prunus armeniaca L.

2. The deciduous tree is of medium size, usually held to not over 18 feet by pruning. Plants are propagated by budding or grafting onto rootstocks. Leaves are entire, ovate to round ovate, deciduous. Apricots bloom very early, so production is limited to areas where blooms will not be injured by frost.

The fruit is generally globose to slightly oblong in shape, 1¼ to 2½ inches in diameter. Newer varieties produce larger fruit, with some fruit approaching a small peach in size. The fruit is pubescent when young, but nearly smooth when ripe. Its flesh is yellow, the skin yellow or blushed red. Fruit is subject to cracking in humid climates, so commercial production is largely in states west of the Rocky Mountains. Flavor ranges from mildly tart to sweet. Some important variety/types include 'Castlebrite', 'Patterson', 'Katy', 'Improved Flameing Gold', 'Royal', 'Tilton' and 'Blenheim'.

3. Season: Bloom to harvest in about 70 to 100 days.
4. Production in U.S.: In 1995, 21,290 domestic acres produced 58,500 fresh tons of fruit. California accounts for 19,800 acres and Washington 1,300 acres. Minor production in Idaho and Utah.
5. Other production regions: Some fruit are imported from Chile, New Zealand, Turkey, Italy, Greece, Spain and France. In 1995, Canada grew 917 acres (ROTHWELL 1996a).
6. Use: Fresh eating, canning, drying, preserves, juice. More than 84 percent of the apricots produced in the U.S. are canned, dried or frozen (GHORPADE).
7. Part(s) of plant consumed: Entire fruit except pit. However, "waste" pit can be processed to yield apricot kernel oil
8. Portion analyzed/sampled: Whole fruit, with pit and stem removed and discarded
9. Classifications:
   a. Authors Class: Stone fruits
   b. EPA Crop Group (Group & Subgroup): Crop Group 12: Stone Fruits Group
   c. Codex Group: 003 (FS 0240) Stone fruits
   d. EPA Crop Definition: None
10. References: GRIN, CODEX, LOGAN, MAGNESS, SCHREIBER, US EPA 1995, US EPA 1994, GHORPADE, ROTHWELL 1996a, USDA 1996f
11. Production Map: EPA Crop Production Regions 10 and 11.
12. Plant Codes:
    a. Bayer Code: PRNAR

024

1. Arizona cottontop
(Cottongrass, Zacate punta blanca)
Poaceae (Gramineae)
Digitaria californica (Benth.) Henrard
2. This is an annual warm season perennial grass native to the U.S. It is a member of the grass tribe *Paniceae*. It is an erect bunchgrass (0.4 to 1 m tall) with a panicle 8 to 15 cm long. It is adapted to the plains, open dry areas, chaparral and semi-desert grassland on sandy soils. It starts growing in late spring and early summer and reproduces readily from seed. It is a good forage for livestock and fair for wildlife and is palatable throughout the year.
3. Season: Warm season, growth starts in late spring and early August. It is palatable throughout the growing season.
4. Production in U.S.: No data for Arizona cottontop production, however, pasture and rangeland are produced on more than 410 million A (U.S. CENSUS, 1992).
5. Other production regions: Distributed throughout Mexico.
6. Use: Rangeland grazing and erosion control.
7. Part(s) of plant consumed: Leaves and stems
8. Portion analyzed/sampled: Forage and hay
9. Classifications:

a. Authors Class: See Forage grass
b. EPA Crop Group (Group & Subgroup): See Forage grass
c. Codex Group: See Forage grass
d. EPA Crop Definition: None
10. References: GRIN, US EPA 1994, CODEX, USDA NRCS, HOLZWORTH, RICHE, STUBBENDIECK (picture)
11. Production Map: EPA Crop Production Regions 8, 9 and 10.
12. Plant Codes:
    a. Bayer Code: TRCCA

025

1. Aronia berry
(Chokeberry, Aronia)
Rosaceae
Aronia spp.
1. Red chokeberry
A. arbutifolia (L.) Pers.
1. Black chokeberry
A. melanocarpa (Michx.) Elliott
2. Low deciduous shrubs of North America to 10 feet tall with simple short-petioled, finely serrated leaves. Fruit is a small berry-like pome. Fruits are red or black in color and about ¼ inch in diameter. 'Nero' is a cultivar for the Black chokeberry, fruits are shiny black and in large clusters.
3. Season: Harvest fruit in autumn.
4. Production in U.S.: No data. Native to eastern North America (BAILEY 1976).
5. Other production regions: Russia (FACCIOLA 1990).
6. Use: Extract of the fruit, Aronia juice, is used in beverages. Also eaten raw, stewed and made into jelly. Berries are astringent with much pectin.
7. Part(s) of plant consumed: Fruit
8. Portion analyzed/sampled: Whole fruit
9. Classifications:
   a. Authors Class: Berries
   b. EPA Crop Group (Group & Subgroup): Miscellaneous
   c. Codex Group: No specific entry
   d. EPA Crop Definition: None
10. References: GRIN, BAILEY 1976, FACCIOLA
11. Production Map: EPA Crop Production Regions 1, 2, 4 and 5.
12. Plant Codes:
    a. Bayer Code: ABOAR (*A. arbutifolia*), ABOME (*A. melanocarpa*)

026

1. Arracacha (ar-a-catch-a)
(Peruvian carrot, Fecula, Apio criollo, White carrot, Peruvian parsnip, Sancocho, Racacha, Mandioquina-salsa)
Apiaceae (Umbelliferae)
Arracacia xanthorrhiza Bancr. (syn: *A. esculenta* DC.)
2. Perennial herb, to 3 feet tall, is grown extensively in northern South America for its edible tuberous root. These tuberous roots have branches or lobes the size of carrots. Roots are parsnip-shaped with a flavor between parsnip and roasted chestnuts. Puerto Rico grows arracacha under the name 'Apio'. In Puerto Rico, arracacha is rotated with cabbage, dasheen and tanier.
3. Season, planting to harvest: About 300 to 400 days normally for roots. Immature roots may be harvested 120 to 240 days

after planting. At harvest time there may be up to 10 lateral roots (each the size of a carrot root) aggregated around the center rootstock. Stems are harvested like celery. In Puerto Rico, it is normally planted from July to September and harvested about a year later.

4. Production in U.S.: About 741 acres in Puerto Rico with normal yield about 6 tons per acre (NATIONAL RESEARCH COUNCIL 1989). The crop reported as celeriac or apio in the 1992 CENSUS is likely arracacha. The 1992 CENSUS reported about 756 acres of celeriac in Puerto Rico. In 1995, 600 acres were reported in Puerto Rico (IR-4 REQUEST DATABASE 1996).

5. Other production regions: Other Caribbean Islands, Peru and Brazil (NATIONAL RESEARCH COUNCIL 1989).

6. Use: Young, tender roots are eaten boiled, baked, fried or added to stews ("sancocho"). The young stems which look and taste like celery are used in salads or cooked as a vegetable. The central rootstock and foliage can be used as feed. The roots are not eaten raw like carrots.

7. Part(s) of plant consumed: Root and stems. Rootstock and foliage can be used as feed but are insignificant as feed items

8. Portion analyzed/sampled: Root and stems

9. Classifications:
   a. Authors Class: Root and tuber vegetable; Leaves of root and tuber vegetables
   b. EPA Crop Group (Group & Subgroup): Root tuber vegetables (1C and 1D)
   c. Codex Group: 016 (VR 0571) Root and tuber vegetables
   d. EPA Crop Definition: None

10. References: GRIN, CODEX, MAGNESS, NATIONAL RESEARCH COUNCIL 1989 (picture), US EPA 1995a, IR-4 REQUEST DATABASE 1996, RICHE, RUBATZKY

11. Production Map: EPA Crop Production Region 13.

12. Plant Codes:
    a. Bayer Code: No specific entry

027

1. Arrowhead
(Chinese arrowhead, Wapato, Duck potato, Chee-koo, Water-archer)
Alismataceae
Sagittaria sagittifolia ssp. *leucopelala* (Miq.) Hartoz (syn: *S. trifolia* L., *S. trifolia* var. *edulis* (Siebold ex. Miq.) Ohwi, *S. trifolia* var. *sinensis* (Sims.) Makino)
   (Chinese arrowhead, Chee-koo, Kuwai, Oldworld arrowhead, Swamppotato, Swan potato, Waterarcher, Flecha de agua)
S. latifolia Wild.
   (Common arrowhead, Wapato, Duck potato)

2. Grown under paddy conditions. Arrowhead is a perennial that grows to a height of 3 feet and has long-stemmed arrowhead-shaped leaves to 12 inches long. Corms are up to 2 inches in diameter with individual plants producing 6-10 corms. In Hawaii, the plant is grown for its yellow tubers (Chee-koo). Corms are similar to waterchestnut (*Eleocharis dulcis*).

3. Season, propagation: Division or by seed. Needs 6-7 months for corms to mature. Plantings start in the spring.

4. Production in U.S.: Grown in Hawaii (NEAL 1965).

5. Other production regions: Grows in North America (FACCIOLA 1990).

6. Use: Corms are cooked and not eaten raw. It is best to peel the tubers after cooking.

7. Part(s) of plant consumed: Corm
8. Portion analyzed/sampled: Corm
9. Classifications:
   a. Authors Class: Root and tuber vegetables
   b. EPA Crop Group (Group & Subgroup): Miscellaneous
   c. Codex Group: 016 (VR 0572) Root and tuber vegetables
   d. EPA Crop Definition: None

10. References: GRIN, CODEX, FACCIOLA, NEAL (picture), YAMAGUCHI 1983 (picture), RUBATZKY

11. Production Map: EPA Crop Production Region 13.

12. Plant Codes:
    a. Bayer Code: SAGLT (*S. latifolia*), SAGSA (*S. sagittifolia*)

028

1. Arrowleaf balsamroot
(Graydock, Breadroot)
Asteraceae (Compositae)
Balsamorhiza sagittata (Pursh) Nutt.

2. Arrowleaf balsamroot is a cool season perennial forb that grows 20 to 80 cm tall that is native to the U.S. and belongs to the tribe *Heliantheae*. It flowers in May through August and reproduces from seeds. Its habitat includes plains, open hillsides, valleys and on well drained soils. The Cheyenne Indians boiled the roots, stems and leaves for a headache remedy, and ground the ripe seeds into flour and the roots eaten raw or cooked. As a forage, arrowleaf balsamroot is most palatable late spring to early summer and is of good quality for sheep and fair for cattle. The green leaves and flowers are the most palatable parts of the plant.

3. Season: Cool season perennial forage that can be used for grazing in the spring and early summer.

4. Production in U.S.: No specific data for Arrowleaf balsamroot production, however, pasture and rangelands are produced on more than 410 million A (U.S. CENSUS, 1992).

5. Other production regions: British Columbia, Canada.

6. Use: Grazing.

7. Part(s) of plant consumed: Leaves, stems, seeds and flowers
8. Portion analyzed/sampled: Forage and hay
9. Classifications:
   a. Authors Class: Nongrass Animal Feeds
   b. EPA Crop Group (Group & Subgroup): Miscellaneous
   c. Codex Group: No specific citation although the general class 052 (AM 0165) Miscellaneous Fodder and Forage Crops could be used
   d. EPA Crop Definition: None

10. References: GRIN, US EPA 1994, US EPA 1995, CODEX, USDA NRCS, HOLZWORTH, RICHE, STUBBENDIECK (picture)

11. Production Map: EPA Crop Production Regions 9, 10, 11 and 12.

12. Plant Codes:
    a. Bayer Code: No specific entry

029

1. Arrowroot
(Arruruz, Common arrowroot, True arrowroot, Bamboo tuber, Yuquilla, West Indian arrowroot, Reed arrowroot, Bermuda arrowroot, Maranta, Jamachipeke)
Marantaceae
Maranta arundinacea L.

2. This arrowroot is the true arrowroot of the West Indies. Perennial to 3 feet in height. The underground tubers are a source of starch called arrowroot powder which is used in pastries, biscuits and thickening soups. A similar crop is Leren, which see.
3. Season, planting to harvest: About 11 months. Propagated by small rhizome pieces. Plantings productive for 5-7 years.
4. Production in U.S.: Grown to limited extent in south Florida (STEPHENS 1988) and Hawaii (NEAL 1965).
5. Other production regions: Brazil to Caribbean (YAMAGUCHI 1983).
6. Use: Tuber as a source of starch. Young arrowroots are prepared like sweet potatoes.
7. Part(s) of plant consumed: Tuber
8. Portion analyzed/sampled: Tuber
9. Classifications:
   a. Authors Class: Root and tuber vegetables
   b. EPA Crop Group (Group & Subgroup): Root and tuber vegetables (1C and 1D)
   c. Codex Group: 016 (VR 0573) Root and tuber vegetables
   d. EPA Crop Definition: None
10. References: GRIN, CODEX, FACCIOLA, MAGNESS, NEAL (picture), REHM, STEPHENS 1988 (picture), US EPA 1995a, YAMAGUCHI 1983
11. Production Map: EPA Crop Production Region 13.
12. Plant Codes:
    a. Bayer Code: MARAR

030

1. Artichoke, Chinese
(Chorogi, Knotroot, Ortiga japonesa, Cao shi can, Crosne, Japanese artichoke, Konloh, Artichoke betony, Crosnes du Japon, Tuberina, Alachofa tuberosa, Stachys tubereux, Chiyorogi)
Lamiaceae (Labiatae)
Stachys affinis Bunge (syn: *S. sieboldii* Miq; *S. tuberifera* Naudin)

2. The plant is mint-like, up to 18 inches, with ovate to lanceolate leaves. It produces numerous small, slender tubers, the edible part, just under the soil surface. These tubers are white, with crisp flesh which can be eaten raw or cooked. The plant and edible parts are comparable to potatoes, both in culture and exposure to pesticides. The tubers do not store well. Chinese artichoke is not grown commercially in the U.S., but may be found occasionally in home gardens. Tubers are 2 to 3 inches long and about $^3/_4$ inch in diameter.
3. Season, propagation: By tubers.
4. Production in U.S.: No data. Sometimes grown in U.S. in gardens (BAILEY 1976).
5. Other production regions: Much cultivated in Japan (BAILEY 1976).
6. Use: Mainly as cooked vegetable. In Japan, the tubers are pickled in a mixture of salt and red beefsteak leaves (Perilla, which see).
7. Part(s) of plant consumed: Tubers
8. Portion analyzed/sampled: Tubers
9. Classifications:
   a. Authors Class: Root and tuber vegetables
   b. EPA Crop Group (Group & Subgroup): Root and tuber vegetables (1C and 1D)
   c. Codex Group: 016 (VR 0584) Root and tuber vegetables
   d. EPA Crop Definition: None

10. References: GRIN, BAILEY 1976, CODEX, FACCIOLA 1990, KAYS, MAGNESS, US EPA 1995a, YAMAGUCHI 1983, RUBATZKY (picture)
11. Production Map: No entry
12. Plant Codes:
    a. Bayer Code: STASB

031

1. Artichoke, Globe
(French artichoke, Green artichoke, Alcachofa)
Asteraceae (Compositae)
Cynara cardunculus L. ssp. *cardunculus* (syn: *Cynara scolymus* L.)

2. The globe artichoke is an herbaceous perennial plant grown for its flower bud. The above-ground portion of the thistle-like plant dies down each year, but off-shoots rise from the rootstock. The flower buds are borne terminally on the main stem and on laterals. A plant may reach a height of 4 to 6 feet and bear several buds. The edible portions are the fleshy bases of the bracts, the thick, fleshy receptacle on which the bracts are borne and the flower primordia. Buds develop continuously on plants from September to May in coastal parts of California, where most commercial production occurs. In colder areas, buds may be harvested from mid to late summer. Buds are hand harvested with yields of 40 buds per plant. Buds can be stored for 3-4 weeks.
3. Season, planting to initial harvest: 3 months. In established fields, buds in various stages are on plants continuously during the growing period.
4. Production in U.S.: About 35,750 tons in 1994 (USDA 1996a). In 1992, 9,193 acres reported for U.S. with 9,128 acres in California (1992 CENSUS).
5. Other production regions: European countries in the Mediterranean region (RUBATZKY 1997).
6. Use: Mainly as salad after cooking. Canned, marinated, frozen commercially. Tops often silaged for livestock feed.
7. Part(s) of plant consumed: Interior base of bracts, receptacle and flower primordia. All are interior structures, not in direct contact with surface sprays. Preparation – artichokes should be washed under running water, pull off lower petals and cut stem to 1 inch or less and snip off tips of petals. Cut off top quarter of each artichoke. Small artichokes are better for pickling, stews and casseroles (LOGAN 1996)
8. Portion analyzed/sampled: Flower head
9. Classifications:
   a. Authors Class: Stalk and stem vegetables
   b. EPA Crop Group (Group & Subgroup): Miscellaneous
   c. Codex Group: 017 (VS 0620) Stalk and stem vegetables
   d. EPA Crop Definition: None
10. References: GRIN, CODEX, LOGAN, MAGNESS, RICHE, STEPHENS 1988 (picture), USDA 1996a, IVES, REHM, RUBATZKY, SWIADER
11. Production Map: EPA Crop Production Region 10.
12. Plant Codes:
    a. Bayer Code: CYUSC

032

1. Artichoke, Jerusalem
(Sunchoke, Sunflower artichoke, Sunroot, Topinambur, Pataca)
Asteraceae (Compositae)

Helianthus tuberosus L.

2. Jerusalem artichoke is an herbaceous perennial to 8 feet tall, arising from a fleshy rootstock that bears oblong tubers. In cultivation, however, it is grown as an annual, much like the potato. The tubers, used both for human food and livestock feed, are produced well underground. They are knobby, white, red or purple skinned and range in size up to 3 or 4 inches long and half as thick. They contain the carbohydrate inulin, rather than starch, which yields fructose on hydrolysis. Tubers are prepared by cooking and pickling. They do not store well. Jerusalem artichokes are native to North America.

3. Season, planting to harvest: 6 to 8 months. Tubers are dug in the fall after tops are killed by frost. Either a crop of tubers or forage can be harvested from a planting but only one use can occur.

4. Production in U.S.: Not available. Very limited for food or feed. North America, especially Northern U.S, (STEPHENS 1988 and YAMAGUCHI 1983).

5. Other production regions: Canada grows this crop (ROTHWELL 1996a).

6. Use: Fresh tubers cooked, mainly baked as potatoes, or pickled. Preparation – scrub to remove soil and peel before or after cooking or use unpeeled. Can be eaten raw in salads (SCHNEIDER 1985).

7. Part(s) of plant consumed: Tubers; Stalks and leaves can be used as feed but not a significant feed item

8. Portion analyzed/sampled: Tuber

9. Classifications:
   a. Authors Class: Root and tuber vegetables; Leaves of root and tuber vegetables
   b. EPA Crop Group (Group & Subgroup): Root and tuber vegetables (1C and 1D)
   c. Codex Group: 016 (VR 0585) Root tuber vegetables
   d. EPA Crop Definition: None

10. References: GRIN, CODEX, LOGAN (picture), MAGNESS, SCHNEIDER 1985 (picture), SCHREIBER, STEPHENS 1988 (picture), US EPA 1995a, YAMAGUCHI 1983, REHM, ROTHWELL 1996a, SPLITTSTOESSER

11. Production Map: No entry because of sporadic production especially with difficulty in harvesting, storing and marketing.

12. Plant Codes:
   a. Bayer Code: HELTU

033

1. Arugula
(Rocket salad, Tira, Arrugula, Gargeer, Roka, Roquette, Garden rocket, Roka, Rucola, Rugula, Salad rocket)

Brassicaceae (Cruciferae)

Eruca sativa Mill. (syn: *E. vesicaria* (L.) Cav. ssp. *sativa* (Mill.) Thell.)

2. This is a low growing annual 1 to 2 feet high. Leaves are compound. Flavor of young, tender leaves is pungent, resembling horseradish. Its leaves are harvested in spring or fall and additional growth and leaves are produced. The plant is grown more in Europe than the U.S. Exposure of leaves is comparable to that of spinach. Arugula seed crop production was on fewer than 50 acres in Washington state for 1993. The seed crop is direct seeded in early April and harvested early September by cutting, dried in the field for 10 to 14 days and combined.

3. Season, seeding to first harvest: About 2 to 3 months. Normally harvested when leaves are about 6 to 8 inches long and bunched for the trade.

4. Production in U.S.: No data; very minor. Arizona, California, Florida, New Jersey, New York and South Carolina (1996 IR-4 REQUEST DATABASE).

5. Other production regions: Old crop native to southern Europe and western Asia (RUBATZKY 1997).

6. Use: As potherb and in salads, mainly as flavoring or spice, fresh or cooked.

7. Part(s) of plant consumed: Young leaves only

8. Portion analyzed/sampled: Leaves (fresh)

9. Classifications:
   a. Authors Class: Leafy vegetables (except *Brassica* vegetables)
   b. EPA Crop Group (Group & Subgroup): Leafy vegetables (except *Brassica* vegetables) (4A)
   c. Codex Group: 013 (VL 0496) Leafy vegetables (including *Brassica* leafy vegetables)
   d. EPA Crop Definition: None

10. References: GRIN, CODEX, IR-4 REQUEST DATABASE 1996, MYERS (picture), SCHREIBER, STEPHENS 1988, US EPA 1995a, YAMAGUCHI 1983, RUBATZKY

11. Production Map: EPA Crop Production Regions 1, 2, 3 and 10.

12. Plant Codes:
   a. Bayer Code: ERUVE

034

1. Asafetida
(Hing, Ferula gum, Asafoetida, Galbanum, Kamol, Kavrak, Rochaek, Sassyr)

Apicaeae (Umbelliferae)

Ferula spp.
F. foetida (Bunge) Regel
F. assa-foetida L.
F. gummosa Boiss. (syn: *F. galbaniflua* Boiss. & Buhse)

2. The plant grows to a height of about 5 feet and has a large carrot-shaped root which is about 6 inches in diameter when the plant is 4-5 years old. The latex or ferula gum from the rootstock is harvested, dried and steam distilled for its volatile oil of asafoetida.

3. Season, harvest: Prior to flowering in the spring, the upper part of the root is laid bare and the stem removed close to the crown. A few days later the latex is collected. This continues for about 3 months after the first cut.

4. Production in U.S.: No data.

5. Other production regions: Asia.

6. Use: Spice used in flavoring curries, sauces, in Worcestershire sauce and French cookery. Also used as a medicine.

7. Part(s) of plant consumed: Root

8. Portion analyzed/sampled: Root latex (dried) and its processed commodity oil

9. Classifications:
   a. Authors Class: Herbs and spices
   b. EPA Crop Group (Group & Subgroup): Miscellaneous
   c. Codex Group: No specific entry
   d. EPA Crop Definition: None

10. References: GRIN, LEWIS, SCHNEIDER 1994c

11. Production Map: No entry

12. Plant Codes:
   a. Bayer Code: No specific entry

035

1. Asparagus
(Garden asparagus, Esparrago, Asperge)
Liliaceae

Asparagus officinalis L.

2. Asparagus is an important crop, both commercially and in home gardens. The plants are perennials. Plant growth is divided into three phases: crown development, spear development and fern development. Three methods are used to establish an asparagus bed: transplanting one-year-old crowns, direct-seeding in the field and transplanting 8-12-week-old seedling plants. The underground portion consists of stems or rhizomes and the edible aerial stems grow upward from them. Young "crowns" consisting of roots and rhizomes are grown from seed planted in beds and transplanted to the field. With good care, fields will produce for years. The tender, succulent aerial stems are cut for 2 to 3 months as they emerge in spring. In the California delta area, asparagus is harvested in the fall, 30 days after the ferns are mowed. Then cutting stops and they are allowed to grow to nourish the underground part for the following year's crop. Cutting may be deep in the soil with just the tip emerging for "white" or near the soil surface when spears are 6 to 8 inches high for "green" asparagus. Spears consist essentially of stem tissue only. New Jersey pioneered the male-only asparagus cultivars with the best sellers 'Jersey Giant' and 'Jersey Knight'. Another cultivar is 'Jersey Spartan'. The male only asparagus can produce longer (10 to 20 years if maintained) than the female asparagus, but the male asparagus still has the same fern (without fruit) pest problems.

3. Season, emergence of spears to harvest: About a week for green, none for white spears. Purple asparagus is harvested at 2-3 inches tall.

4. Production in U.S.: In 1993, 110,200 tons (1994 AGRICULTURAL STATISTICS). In 1993, California with 34,500 acres, Washington (25,500), Michigan (19,000), New Jersey (900) and Illinois (760) with 57 percent for fresh market (1994 AGRICULTURAL STATISTICS).

5. Other production regions: Canada for 1995 produced 3,304 tons from 3,793 acres (ROTHWELL 1996a). Mexico.

6. Use: As cooked vegetable,fresh, canned or frozen. Per capita consumption in the U.S. in 1994 is 0.5 pounds (PUTNAM 1996).

7. Part(s) of plant consumed: Young, succulent spears

8. Portion analyzed/sampled: Spears (stems)

9. Classifications:
   a. Authors Class: Stalk and stem vegetables
   b. EPA Crop Group (Group & Subgroup): Miscellaneous
   c. Codex Group: 017 (VS 0621) Stalk and stem vegetables
   d. EPA Crop Definition: None

10. References: GRIN, CODEX, MAGNESS, STEPHENS 1988, USDA 1994a, US EPA 1995b, US EPA 1994, IVES, GARRISON, SWIADER, RUBATZKY, MARR, ROTHWELL 1996a, PUTNAM 1996

11. Production Map: EPA Crop Production Regions 2, 5, 10 and 11.

12. Plant Codes:
   a. Bayer Code: ASPOF

036

1. Atemoya
(Annon, Annona, Hanumanphal)
Annonaceae

Annona cherimola Mill. x *A. squamosa* L. (syn: *A. atemoya* Hort.)

2. Atemoya is a hybrid of Cherimoya and Sugar apple with the first cross made in south Florida. The tree closely resembles the Cherimoya. It is fast-growing and can reach 25 to 30 feet tall and is short trunked. Leaves are deciduous, leathery and not as hairy as Cherimoya leaves. Fruit is conical or heart-shaped, about 4 inches long and 4 inches wide. The fruit rind is composed of fused areoles. The flesh is snowy-white, almost solid with fewer seeds than Sugar apple.

3. Season, bloom to harvest: Flowers in May-June and August-September. Fruits in August-October and November-January in Florida. About 150-180 days.

4. Production in U.S.: In 1994, 45 acres in Florida (CRANE 1995a). In 1992, 60 acres in Hawaii with about 16 tons (KAWATE 1996a).

5. Other production regions: Australia (MORTON 1987).

6. Use: Fruit flesh is snowy-white and eaten fresh from the shell or cubed for use in fruit cups. Also pulp is blended with citrus fruit and cream and frozen as ice cream.

7. Part(s) of plant consumed: Fruit pulp

8. Portion analyzed/sampled: Whole fruit

9. Classifications:
   a. Authors Class: Tropical and subtropical fruits – inedible peel
   b. EPA Crop Group (Group & Subgroup): Miscellaneous
   c. Codex Group: No specific entry. General entry is 006 (FI 0030) Assorted tropical and subtropical fruits – inedible peel. Sugar apple is 006 (FI 0368).
   d. EPA Crop Definition: Sugar apple = Atemoya

10. References: GRIN, CRANE 1995, KAWATE 1996a, MORTON (picture), CODEX, PUROHIT, CAMPELL(c)

11. Production Map: EPA Crop Production Region 13.

12. Plant Codes:
   a. Bayer Code: No specific entry for cross but sugar apple (ANUSQ) = Atemoya

037

1. Avocado
(Alligator pear, Aquacate)
Lauraceae

Persea americana Mill.

P. americana Mill. var. *americana*
(Avocado)

P. americana var. *drymifolia* (Schltdl. & Cham.) S.F. Blake
(Mexican avocado)

P. americana var. *nubigena* (L.O. Williams) L.E. Kopp
(Guatemalan avocado)

2. The tropical/subtropical evergreen avocado tree with large, leathery leaves may grow to 60 feet tall. The tree will tolerate only 2 to 8 degrees F below freezing. Plant is propagated by seed and grafting with fruit production within 3-4 years of grafting. There are 3 recognized races of avocado based on ecological origin: Mexican, Guatemalan and West Indian. The Mexican race is well adapted to the cool climates of tropical and subtropical areas. They produce small

type fruit (85 to 340 g) with a thin, smooth, soft skin. The West Indian race is best adapted to the lowland tropical conditions of high temperature and high humidity. Their fruit skin is smooth, thin and leathery. The Guatemalan type is intermediate between the Mexican and West Indian cultivars in climate adaption with their fruit ranging in size from 340 to 560 g with a warty, pebbly, brittle skin. All have a single, large seed. Pulp between peel and seed is soft and buttery when ripe. Oil content of pulp varies from 3 to 30 percent. The variation in oil content is mainly attributed to the racial origins of the fruit. Fruit is near round to oval or distinctly pyriform in shape, 2 to 4 inches in diameter.

3. Season, bloom to harvest: Flowers February-April and fruits June-March in South Florida. Fruit matures in 4 to 13 months, depending on race and variety. Popular cultivars include 'Fuerte' and 'Haas'.

4. Production in U.S.: In 1994-1995 season California produced 156,000 tons and Florida produced 20,000 tons on a total of 73,500 acres (USDA 1996a). In 1995, Puerto Rico with 200 acres (MONTAVO ZAPATA 1995). The 1992 CENSUS reported 637 acres for Hawaii. Minor production in Guam and Texas. California accounts for over 90 percent of production.

5. Other production regions: Mexico, Dominican Republic and other Caribbean Islands, Brazil, Columbia, South Africa, Venezuela and Israel.

6. Use: Mainly fresh eating, salads, "guacamole".

7. Part(s) of plant consumed: Pulp only

8. Portion analyzed/sampled: Whole fruit with pit and stem removed and discarded

9. Classifications:
   a. Authors Class: Tropical and subtropical fruits – inedible peel.
   b. EPA Crop Group (Group & Subgroup): Miscellaneous
   c. Codex Group: 006 (FI 0326) Assorted tropical and subtropical fruits – inedible peel
   d. EPA Crop Definition: Proposing avocado = Black sapote, Canistel, Mamey sapote, Mango, Papaya, Sapodilla and Star apple.

10. References: GRIN, CODEX, CRANE 1996a, CRANE 1995, FABER 1996a, MAGNESS, MARTIN 1987, MONTALVO-ZAPATA 1995, MORTON (picture), RICHE, USDA 1994a, US EPA 1994, USDA 1996a, IVES, KNIGHT, LOGAN, MOTT, USDA 1996f, KADAM, MALO(h), CRANE(a)

11. Production map: EPA Crop Production Regions 10 and 13 with some production in Region 6.

12. Plant Codes:
    a. Bayer Code: PEBAM

038

1. Bahiagrass
(Batatais, Jengibrillo, Tejona)
Poaceae (Gramineae)
Paspalum notatum Fluegge

2. This is a warm season perennial grass, first introduced from Brazil in 1914. It is low-growing, 12 to 20 inches tall and sod-forming, spreading by short rhizomes and also from seed. Leaves are rather short and broad. It is especially well adapted to droughty, sandy soils. It is not winter hardy so is adapted to the lower southeastern Coastal Plain, including all of Florida. In Florida, often seeded with annual legumes,

such as white and crimson clover. It is well suited for pastures and lawns, especially on sandy soils of low fertility. Propagation is by seed, which should be scarified or acid treated for prompt germination.

3. Season: Warm season pasture. April to October.

4. Production in U.S.: No specific data for forage use, See Forage grass. In 1992, 2,137 tons of seeds were produced from 38,510 acres (1992 CENSUS). In 1992, the top 4 states for seed production were Florida (34,452 acres), Texas (1,539), Alabama (1,218) and Mississippi (319) (1992 CENSUS). Southeast Coastal Plain, including all of Florida.

5. Other production regions: Native to Mexico and Central and South America (GOULD 1975).

6. Use: Pasture, hay and erosion control.

7. Part(s) of plant consumed: Foliage

8. Portion analyzed/sampled: Forage and hay

9. Classifications:
   a. Authors Class: See Forage grass
   b. EPA Crop Group (Group & Subgroup): See Forage grass
   c. Codex Group: See Forage grass
   d. EPA Crop Definition: None

10. References: GRIN, MAGNESS, RICHE, ALDERSON, BALL (picture), SKERMAN (picture), GOULD

11. Production Map: EPA Crop Production Regions 2, 3, 4, 6 and 13.

12. Plant Codes:
    a. Bayer Code: PASNO

039

1. Balm
(Lemon balm, Bee balm, Melissa)
Lamiaceae (Labiatae)
Melissa officinalis L.

2. Balm is of minor importance as a condiment and essential oil source. The plant is a perennial, erect herb, with broad, opposite, pubescent leaves. The dried leaves are used to flavor stews, soups and dressings. Fresh leaves are used in salads. Oil of balm, as well as leaves, has a lemon-like odor. The oil is used to flavor beverages. Culture is similar to that of mints, which see. No data are available on production in the U.S., but is very minor.

3. Season, harvest: As the plants come into bloom.

4. Production in U.S.: No data. Naturalized in U.S. (FOSTER 1993).

5. Other production regions: Native to southern Europe (FOSTER 1993).

6. Use: Flavoring.

7. Part(s) of plant consumed: Leaves

8. Portion analyzed/sampled: Leaves (fresh and dried)

9. Classifications:
   a. Authors Class: Herbs and spices
   b. EPA Crop Group (Group & Subgroup): Herbs and spices (19A)
   c. Codex Group: 027 (HH 0721) Herbs; 057 (DH 0721) Dried herbs
   d. EPA Crop Definition: None

10. References: GRIN, CODEX, FOSTER (picture), MAGNESS, US EPA 1995a, HYLTON

11. Production Map: No entry

12. Plant Codes:
    a. Bayer Code: MLSOF

040

1. Balsam pear

(Alligator pear, Bitter cucumber, Bitter gourd, Fuqua, Bitter-melon, Cundeamor, La-kwa, Margose, Fukwa, Lemon amargo, Balsamito, Paria, Karela, Carilla gourd)

Cucurbitaceae

Momordica charantia L.

1. Balsam apple

(Wonderapple, Balsamina, Pomme de merveille)

M. balsamina L.

1. Chinese cucumber

(Chinese bitter cucumber, cundeamor, Kakur, Spiny bitter cucumber, Sweet gourd)

M. cochinchinensis (Lour.) Spreng.

2. The balsam pear plant is a perennial grown as an annual running vine, to 12 feet or more, with near round, lobed leaves. The fruit is 4 to 12 inches long, oblong, pointed and furrowed lengthwise. When fully ripe it splits into 3 divisions. The immature fruit is boiled as a vegetable. A pulpy aril surrounds the seeds, which is esteemed by Orientals. In culture, similar to cucumber. The related balsam apple has a smaller to 3 inches long, egg-shaped fruit and is used in a similar manner. Chinese cucumber is 6-8 inches long and very warty on outside. About 100 percent of the Chinese cucumbers and bittermelons imported into the U.S. come from the Dominican Republic (LAMBERTS 1990).

3. Season, seed to harvest: 2 to 4 months. Propagated by seed and/or transplants and are usually grown on trellis support.

4. Production in U.S.: Mainly Oriental gardeners. In 1992, Guam reported 28 acres of bittermelons with about 40 tons (1992 CENSUS). Grows in Florida (STEPHENS 1988) and Guam. The 1996 IR-4 State Request Database exhibited needs from California, Florida and Hawaii for balsam pear.

5. Other production regions: Tropics and subtropics (YAMAGUCHI 1983).

6. Use: As boiled vegetable. Seed arils eaten out of hand. Leaves and shoots of balsam pear can be eaten as potherb. Fruit prepared as pickles and ingredient of curries.

7. Part(s) of plant consumed: Mainly whole immature fruits as summer squash. Seed aril when fruit is ripe and leaves of balsam pear. Fruits are peeled and parboiled in salty water to reduce bitterness

8. Portion analyzed/sampled: Whole fruit

9. Classifications:
   a. Authors Class: Cucurbit vegetables
   b. EPA Crop Group (Group & Subgroup): Cucurbit vegetables (9B)
   c. Codex Group: Balsam apple – 011 (VC 0420) Fruiting vegetables cucurbits; Balsam pear – 011 (VC 0421) Fruiting vegetables cucurbits; 013 (VL 0421) Leafy vegetables (including *Brassica* leafy vegetables)
   d. EPA Crop Definition: Summer squash = *Momordica* spp.

10. References: GRIN, CODEX, IR-4 REQUEST DATABASE 1996, LAMBERTS (a), MAGNESS (balsam pear and apple), MARTIN 1983, NEAL (picture), REHM, RICHE, STEPHENS 1988 (picture), US EPA 1995a, YAMAGUCHI 1983, RUBATZKY (picture)

11. Production Map: EPA Crop Production Regions 3, 10 and 13.

12. Plant Codes:
   a. Bayer Code: MOMBA (*M. balsamina*), MOMCH (*M. charantia*)

041

1. Bambara groundnut

(Indhlubu, Underground bean, Earth pea, Baffin pea, Voandzou, Groundbean, Madagascar peanut, Congo groundnut, Congo goober, Hog peanut)

Fabaceae (Leguminosae)

Vigna subterranea (L.) Verdc. (syn: *Voandzeia subterranea* (L.) DC.)

2. This species is cultivated extensively in central Africa for its underground peanut-like seeds. This plant is similar in nearly all respects to peanuts, being a prostrate annual with compound leaves consisting of 3 leaflets. The seeds are borne in roundish pods $1/2$ inch or more long, which lie buried in the ground, similar to peanuts, which see. Not grown commercially in the U.S.

3. Season, seeding to harvest: About $3^1/_2$ to $5^1/_2$ months for green shell. Dry shell seed is harvested about 4 to 7 months after seeding. Young pods are sometime harvested at 3 to 5 months after seeding. Flowering starts 40-50 days after seedling emergence. Seed matures 50-100 days after pollination.

4. Production in U.S.: No data. Can be grown in Florida gardens (STEPHENS 1988).

5. Other production regions: Tropical areas (YAMAGUCHI 1983).

6. Use: Seeds eaten raw, roasted or boiled like peanuts. Seeds also are crushed or ground into a meal.

7. Part(s) of plant consumed: Seeds which are usually harvested green. Young, tender pods are sometimes eaten raw

8. Portion analyzed/sampled: Seeds (succulent or dry)

9. Classifications:
   a. Authors Class: Legume vegetables (succulent or dried)
   b. EPA Crop Group (Group & Subgroup): Miscellaneous
   c. Codex Group: 014 (VP 0520) Legume vegetables (immature seeds); 015 (VD 0520) Pulses (dry seed)
   d. EPA Crop Definition: Bean = *Vigna* spp.

10. References: GRIN, CODEX, MAGNESS, MARTIN 1983, STEPHENS 1988 (picture), US EPA 1995a, YAMAGUCHI 1983, RUBATZKY

11. Production Map: EPA Crop Production Regions 3 and 13.

12. Plant Codes:
   a. Bayer Code: VOASU

042

1. Bamboo

(Bamboo shoots, Ohe, choke-sun, Take-noko)

Poaeae (Gramineae, Bambuseae)

Arundinaria spp.

Bambusa spp.

Dendrocalamus spp.

Gigantochloa spp.

Phyllostachys spp.

Schizostachyum spp.

1. Giant Cane

(Canebrake bamboo, Large cane, Southern cane)

A. gigantea ssp. *gigantea* (Walter) McClure

1. Common bamboo

(Golden bamboo)

B. vulgaris Schrad. ex. Wendl.

1. Beechey bamboo
(Silkball bamboo)
> B. beecheyana Munro

1. Giant bamboo
(Kyo-chiku)
> D. asper (Schult.f.) Backer ex Heyne
> D. latiflorus Munro

1. Male bamboo
(Calcutta bamboo, Solid bamboo)
> D. strictus (Roxb.) Nees

1. Gigantochloa
> G. verticillata (Willd.) Munro

1. Sweetshoot bamboo
> P. dulcis McClure

1. Pubescent bamboo
(Moso bamboo)
> P. edulis (Carriere) J. Houz. (syn: P. pubescens Mazel ex J.Houz.; P. heterocycla (Carriere) Mitf.)

1. Polynesian ohe
> S. glaucifolium (Rupr.) Munro

2. According to Bailey (HORTUS THIRD.), more than 200 species of bamboo are recognized, varying in size from a few feet to more than 100 feet in height. The tender, young shoot growth of many of these species is used as food, in the U.S. mainly in Chinese dishes. Sprouts harvested in the U.S. are mainly limited to Hawaii and Puerto Rico, but substantial quantities are imported. Since new sprout growth quickly becomes hard and woody, the maximum period of exposure of edible parts to direct pesticide application would be approximately a month. Also see Giant cane for forage and hay.

3. Season, start of growth to harvest: About 1 month.

4. Production in U.S.: Hawaii, Puerto Rico (MAGNESS 1971). Washington state (SCHREIBER 1995) and Florida and southern California (BAILEY 1976).

5. Other production regions: In 1985, about 62 percent of the total U.S. imports of bamboo shoots came from the Dominican Republic (LAMBERTS 1990).

6. Use: Tender, new bamboo shoots used in Oriental cuisine.

7. Part(s) of plant consumed: Tender shoots

8. Portion analyzed/sampled: Tender shoots

9. Classifications:
   a. Authors Class: Stalk and stem vegetables
   b. EPA Crop Group (Group & Subgroup): Miscellaneous
   c. Codex Group: 017 (VS 0622) Stalk and stem vegetables
   d. EPA Crop Definition: None

10. References: GRIN, BAILEY 1976, CODEX, FACCIOLA, JANICK 1990, LAMBERTS (a), MAGNESS, NEAL, SCHREIBER, STEPHENS 1988 (picture), US EPA 1994, US EPA 1995, STUDDENDIECK (picture)

11. Production Map: EPA Crop Production Regions 3, 10, 12 and 13.

12. Plant Codes:
   a. Bayer Code: BAMVU (B. vulgaris)

043

1. Banana
(Cavendish banana, Chinese banana, Banano, Guineo, Dwarf banana)

1. Plantain
(Plantano, Platano, Cooking banana, French plantain)

Musaceae
> Musa acuminata Colla , M. x paradisiaca L., M. acuminata Colla x M. balbisiana Colla hydrids (syn: M. cavendishii Lamb., M. nana Lour., M x paradisiaca ssp. sapientum (L.) Kuntze, M. x sapientum L.)

2. The plants are a perennial herb, which may grow to 20 feet or more. Grows in hot, wet tropical climates. Propagated by all types of division. Fruit production within 12 months of planting. The stalk, with great, broad leaves, produces a single bunch of fruit at the apex, then dies; new stalks rise from the rootstock. These are strictly tropical fruits that compose an extreme diverse group. The major bananas of commerical trade are all sterile triploids and are designated as AAA group. These bananas are almost wholly derived from M. acuminata. Plantains or cooking bananas are designated as AAB group with one-third of their genetic make-up from M. balbisiana. Each banana bunch may contain 100 or more individual seedless fruits, which range up to 10 inches or more in length and 1 inch or more in diameter. Peel is smooth and separates readily from edible pulp when ripe. The major type is cavendish. Fruit of the dwarf species averages smaller. Burro and Red Bananas becoming more popular.

3. Season: Bloom to harvest in 3 to 4 months. Plant flowers any time of year depending on planting time, nutrition and maturity. Fruits are harvested and shipped green. Bananas must be ripened with ethylene by the wholesaler prior to market.

4. Production in U.S.: Grown commercially in the U.S. only in Hawaii, Guam, Virgin Islands, Puerto Rico with some small plantings in south Florida and California. In 1992, Puerto Rico produced about 9,000 acres of banana and 19,500 acres of plantain (1992 CENSUS). EPA estimates 24,200 acres of bananas and 12,500 acres of plantain (US EPA 1994). In 1996, Hawaii harvested 930 acres of bananas with 14,000 pounds per acre and all used for the fresh market (USDA 1997a). In 1994, Florida harvested 400 acres of eating bananas and 200 acres of cooking plantain (CRANE 1995).

5. Other production regions: Grown in many tropical countries with significant US imports from Costa Rica, Ecuador, Honduras, Colombia, Mexico and Panama.

6. Use: Pulp eaten fresh or cooked in many ways. Bananas are mainly fresh eating, while plantain are often cooked. Ripe or unripe banana can be processed into several products, including pulp, juice, canned slices, deep fried chips, fruit bars and brandy.

7. Part(s) of plant consumed: Internal pulp. The whole green plantain fruit can be ground into flour

8. Portion analyzed/sampled: Whole fruit for both. Banana: Field residue data on both bagged and unbagged bananas should be provided. The required number of field trials may be split between bagged and unbagged bananas. Alternatively, one sample each of bagged and unbagged bananas may be taken from each site. For establishing tolerances, data are required on the whole commodity including peel after removing and discarding crown tissue and stalk. Residue data on just the banana pulp may be provided for purposes of dietary risk assessment: Banana tolerance will cover plantain

9. Classifications:
   a. Authors Class: Tropical and subtropical fruits – inedible peel
   b. EPA Crop Group (Group & Subgroup): Miscellaneous

c. Codex Group: 006 (FI 0327 and FI 0328 for banana and FI 0354 for plantain) Assorted tropical and subtropical fruits – inedible peel

d. EPA Crop Definition: Banana = Banana, Plantain

10. References: GRIN, CODEX, FORSYTH, LOGAN, MAG-NESS, MARTIN 1987, RICHE, US EPA 1995a, US EPA 1994, USDA 1997a, US EPA 1996a, CRANE 1995, KOTECHA, MALO (g)

11. Production Map: EPA Crop Production Regions 10 and 13 (primarily 13).

12. Plant Codes:

a. Bayer Code: MUBAC (*M. acuminata* = Dwarf banana tree, Chinese banana, Cavendish banana), MUBPA (*M.* x *paradisiaca* = Banana), MUBPK (*M.* x *paradisiaca* = Plantain, French plantain)

044

1. Barley

(Cebada, Orge)

Poaceae (Gramineae)

Hordeum vulgare L. ssp. *vulgare* (syn: *H. distichon* L., *H. irregulare* A.E. Aberg & Wiebe, *H. sativum* Jess.)

2. Barley is one of the top 5 important grain crops in the U.S. Acreage planted in 1966 and 1967 averaged about 10.6 million acres, with an average yield for the two years of 381.7 million bushels. Acreage planted in 1995 was 6,689,000 acres with a yield of about 360 million bushels. Barley is one of the most ancient of cultivated grains. Grains found in pits and pyramids in Egypt indicate that barley was cultivated there more than 5000 years ago. The most ancient glyph or pictograph found for barley is dated about 3000 B.C. Numerous references to barley and beer are found in the earliest Egyptian and Sumerian writings. Barley is native to the Middle East and was first grown in U.S. in 1602.

Types and Characteristics of Barley

Three types of *Hordeum vulgare* constitute the barleys under cultivation. They are derived types not known as such in nature. All have 14 chromosomes in the diploid stage and inter-cross readily. They are characterized as follows:

1) One is 6-rowed barleys with a tough rachis or spike stem. All florets are fertile and develop normal kernels. Within this type are two groups: (a) The typical group in which lateral kernels are only slightly smaller than the central one. (b) The intermediate group in which lateral kernels are distinctly smaller than central ones. This group may contain kinds with sterile or near sterile lateral spikelets. These two groups may overlap and are not fully distinct.

2) Another type is the 2-rowed barleys with a tough rachis. The central spikelets all contain a fertile flower, while flowers in the lateral spikelets are either male or sexless. Two groups of varieties occur: (a) The typical 2-row group, with lateral flowers containing lemma, palea and reduced sexual parts. (b) The deficiens group, with lateral flowers containing no sexual parts.

3) Last is an irregular barley with a tough rachis, but with lateral flowers reduced in some instances to a stem piece only; and others fertile, sterile or sexless. Central spikelets contain fertile flowers and set seeds.

Barley plants are annual grasses which may be either winter annuals or spring annuals. Winter annuals require a period of exposure to cold in order to produce flowers and set seeds, thus are planted in the fall. They form a rosette type of growth in fall and winter, developing elongated stems and flower heads in early summer. If seeded in the spring they fail to produce seed heads. Winter varieties form branch stems or tillers at the base so several stems rise from a single plant. The winter varieties of barley are more hardy than winter oats, but somewhat less hardy than winter wheat. Around a fourth of the barley grown in the U.S. is of winter varieties. Spring varieties do not require exposure to cold in order to develop seed heads. Also, they do not have a typical rosette stage and so develop fewer tillers than winter varieties. They are the only kind adapted to areas with very cold winters. For best production they should be seeded as early as land can be worked in the spring.

The stems of both winter and spring varieties may vary in length from 1 to 4 feet, depending on variety and growing conditions. Stems are round, hollow between nodes and develop 5 to 7 nodes below the head. At each node a clasping leaf develops. In most varieties the leaves are coated with a waxy chalk-like deposit. The density of this varies and in some varieties no waxiness is present and leaves are glossy. Shape and size of leaves vary with variety, growing conditions and position on the plant. The spike, which contains the flowers and later the mature seeds, consists of spikelets attached to the central stem or rachis. Stem intervals between spikelets are 2 mm or less in dense-headed varieties and up to 4 to 5 mm in lax or open-headed kinds. Three spikelets develop at each node on the rachis.

Barley varieties are classed as 2-row or 6-row. In 2-row varieties only the central spikelet develops a fertile flower and seed. In 6-row varieties all three of the spikelets at each node develop a seed. Each spikelet has two glumes rising from near the base. These are linear to lanceolate and flat and terminate in an awn. The glumes minus the awn are about half the length of the kernel in most varieties, but this varies from less than half to equal to the kernel in length. Glumes may be covered with hairs, weakly haired or hairless. The awns on the glumes may be shorter than the glume, equal in length or longer. The glumes are removed in threshing. The barley kernel consists of the caryopsis, or internal seed, the lemma and palea.

In most barley varieties the lemma and palea adhere to the caryopsis and are a part of the grain following threshing. However, naked or hulless varieties also occur. In these the caryopsis is free of the lemma and palea and threshes out free as in wheat. This type is mainly grown where barley is used for human food and is rarely found in the U.S. The lemmas in barley are usually awned. Awns vary from very short up to as much as 12 inches in length. Edges of awns may be rough or "barbed" (bearded) or nearly smooth. Awnless varieties are also known. In 6-row barley awns are usually more developed on the central spikelets than on the lateral ones. The barley kernel is generally spindle shaped. In commercial varieties grown in the U.S., length ranges from 7 to 12 mm. Kernels from 2-rowed varieties are symmetrical. In 6-rowed varieties the third of the kernels from the central spikelets are symmetrical, but the two-thirds from lateral spikelets are twisted. The twist is most apparent at the attachment end, less conspicuous at the terminal. The dorsal surface of kernels is smooth, the ventral surface grooved. The period from flowering until barley is ready for harvest may vary from 40 days to as long as 55 days, varying with varieties and climate conditions.

Cultivated Variety Groups

Some 150 varieties of barley are cultivated in the U.S., many on a minor scale. Varieties are constantly changing as new ones are developed and tested while others pass out of cultivation. These varieties fall into four general groups, as follows:

Manchuria, OAC 21, Aderbrucker Group. These are 6-rowed, awned, spring-type varieties with medium sized kernels. The type is believed to have come originally from Manchuria. Plants are tall with open or lax, nodding heads. They tend to shatter badly in dry climates. These are grown mainly in the upper Mississippi Valley and are extensively used for malting.

Coast Group. These varieties trace to North African ancestry and are grown in California and Arizona, also in the Inter-mountain Region. They are 6-rowed, awned, with large kernels and short to medium length stems. Spikes are medium to short, dense and generally held erect to semierect. They mature early and are not prone to shatter. They have a spring growth habit but may be fall or winter seeded in California and Arizona where winters are mild.

Tennessee Winter Group. Varieties of this group trace to the Balkan Caucasus Region or Korea. They are 6-rowed, awned, with mid-long lax spikes which tend to nod. Plants are medium tall, of winter habit. These varieties are fall seeded and are grown in the southeastern quarter of the U.S.

Two-rowed Group. This group includes types tracing to Europe and Turkey, the Turkish type being adapted to areas with marginal rainfall. Varieties in this group are grown principally in the Pacific and Inter-Mountain states and to some extent in the northern Great Plains. Varieties are mainly spring type though 2-rowed winter varieties are known. Some varieties are used mainly for malting, others for feed.

Barley cultivars in the U.S. are classified as malting or feed types. In the rest of the world, barley is a major human food grain. Most malting types are produced along the Red River Valley of Minnesota and North Dakota and irrigated Inter Mountain States of Idaho, Wyoming and Montana. Estimated that 25 percent of the total U.S. barley acreage is planted to winter types.

Uses of Barley

About half of the barley grown in the U.S. is used for livestock feed. As feed it is nearly equal in nutritive value to kernel corn. It is especially valuable as hog feed, giving desirable portions of firm fat and lean meat. The entire kernel is used in feed, generally after grinding or steam rolling. Malt sprouts from malting as well as brewers grain, by products of brewing, are also valuable livestock feeds. Around 42 percent of the barley crop is used for malting in the U.S. Of the malted barley some 80 percent is used for beer, around 14 percent for distilled alcohol products and 6 percent for malt syrup, malted milk and breakfast foods.

For malting, the barley is steeped in aerated water in large tanks for 45 to 65 hours, then transferred to germinating tanks or compartments where it is held with intermittent stirring for 5 to 7 days at temperatures of 60-70 degrees F. During this treatment root sprouts emerge, but not the stems. This "green" malt is then dried in hot air kilns. For making beer the dried malt is crushed between rollers, mixed in proper proportions with slightly warm water and held under rigidly controlled temperatures. The starch is converted by enzymatic reaction into maltose and dextrins. Proteins are also broken down by enzyme action. Upon completion of this process the solids settle out, the extract is filtered, then boiled with hops to add flavor, then cooled. Yeast is added to ferment the sugars into alcohol and carbon dioxide. The hop residue and proteins are then removed and the product (beer) is aged, chilled, filtered, pasteurized and bottled. Keg beer is similar but is not pasteurized or bottled. The solids from this process (brewer's grain) are a valuable livestock feed.

Barley for human food is made into pearl barley by using abrasive disks to grind the hulls and bran off the kernels. After three successive "pearlings" or grinding operations all the hull and most of the bran is removed. At this stage the remaining kernel part is known as pot barley. Two or three additional pearlings produce pearl barley, in which most of the embryo is removed. These later pearlings also produce barley flour. Pot and pearl barley are used in soups and dressings. The flour is used in baby foods and breakfast cereals, or mixed with wheat flour in baking. Barley is also grown as a hay crop in some areas. For hay, only smooth-awned varieties or awnless are used. Winter barley also may be pastured moderately before the stems start to elongate. It furnishes nutritive pasturage and grain yields are not seriously reduced.

3. Season, seeding to harvest: 75 to 100 days.
4. Production in U.S.: In 1995, 6,689,000 acres planted, producing 359,102,000 bushels. The top 6 states reported as North Dakota (2,300,000 acres), Montana (1,300,000), Idaho (780,000), Minnesota (610,000), Washington (300,000) and California (260,000) (USDA 1996c).
5. Other production regions: In 1995, Canada reported 10,477,285 acres (ROTHWELL 1996a).
6. Use: See above. Also barley bran is a significant food in the U.S. and is used in cooking, cake, cookie and bread. Bran is obtained from milling of brewers spent grain and not the traditional dry milling of barely grain. Barley flakes (rolled barley) are used in hot cereal or as a thickening agent.
7. Part(s) of plant consumed: Grain and foliage
8. Portion analyzed/sampled: Grain, hay and straw. Grain processed commodities pearled barley, flour and bran. Grain includes kernel plus hull. For barley hay, cut when the grain is in the milk to soft dough stage. Hay should be field-dried to a moisture content of 10 to 20 percent. For Barley straw, plant residue (dried stalks or stems with leaves) left after the grain has been harvested (threshed)
9. Classifications:
   a. Authors Class: Cereal grains; Forage, fodder and straw of cereal grains
   b. EPA Crop Group (Group & Subgroup): Cereal grains (15); Forage, fodder and straw of cereal grains (16)
   c. Codex Group: 020 (GC 0640) Cereal grains; 051 (AS 0640) Straw
   d. EPA Crop Definition: None
10. References: GRIN, CODEX, FACCIOLA, MAGNESS, ROTHWELL 1996a, USDA 1996c, US EPA 1996a, US EPA 1995a, US EPA 1994, IVES, REHM, SCHNEIDER 1996(a), MACGREGOR
11. Production Map: EPA Crop Production Regions 1, 2, 5, 7, 8, 9, 10 and 11 with most from 5, 7 and 11.
12. Plant Codes:
   a. Bayer Code: HORVX (*H. vulgare*), HORVS (spring barley), HORVW (winter barley)

045

1. Basil

(Albahaca, Bush basil, French basil, Garden basil, Herbe royal, Hsiangtsai, Italian basil, Purple basil, Sweet basil, Basilic, Pesto)

Lamiaceae (Labiatae)
Ocimum basilicum L.

1. Lemon basil

O. x *citriodorum* Vis.

1. Greek basil
(Bush basil)
> *O. minimum* L.

1. American basil
(Hoary basil)
> *O. americanum* L. (*O. canum* Sims)

1. Russian basil
> *O. gratissimum* L.

2. Basil is of the same plant family as the mints, but the plant is an annual, grown from seed. The plant is branched, 1 to 2 feet in height. The leaves are entire or toothed, opposite on the stem, ovate in shape. The dried leaves are used to flavor stews, dressings and soups. The essential oil, extracted from the leaves, like mint, is used in perfumes and to flavor liquors. 'Asian basil' is a cultivar that is anise scented with purple bracts. A typical set of ingredients for pesto is: fresh basil leaves, fresh parsley leaves, garlic, almonds and olive oil.
3. Season, seeding to harvest: About 60 days and repeated every two months. In the home gardens, fresh leaves can be continuously harvested during the season.
4. Production in U.S.: Hawaii produced basil on about 55 acres in 1994. Also California, and Indiana (SIMON 1985).
5. Other production regions: Tropical/subtropical areas of Asia and Europe, Cosmopolitan (BAYER 1992).
6. Use: Flavoring in culinary, liquors, commonly used in pesto.
7. Part(s) of plant consumed: Leaves and young stems or as source of essential oil
8. Portion analyzed/sampled: Leaves and young stems (fresh and dried); For basil oil – fresh hay and oil
9. Classifications:
   a. Authors Class: Herbs and spices
   b. EPA Crop Group (Group & Subgroup): Herbs and spices (19A) (Representative Crop)
   c. Codex Group: 027 (HH 0722) Herbs; 057 (DH 0722) Dried herbs
   d. EPA Crop Definition: None
10. References: GRIN, CODEX, FARRELL, FOSTER (picture), HAMASAKI, KAYS, KAWATE 1996c, KUEBEL, LEWIS (picture), MAGNESS, SIMON 1985, US EPA 1995a, BAYER
11. Production Map: EPA Crop Production Regions 5, 10 and 13.
12. Plant Codes:
   a. Bayer Code: OCIBA (*O. americanum* and *O. basilicum*)

046

1. Beachgrass
Poaceae (Gramineae)
> *Ammophila* spp.

1. American beachgrass
> *A. breviligulata* Fernald

1. European beachgrass
(Marram grass)
> *A. arenaria* (L.) Link

2. These cool season grasses are tough, coarse perennials with extensively creeping rhizomes that are native to the coastlines of North America. They produce good growth on sands of low fertility and are used primarily for erosion control in sandy areas and for sand dune stabilization. For this purpose, dividing and transplanting clumps of the grasses (3 to 5 culms/hill) is the most effective method of establishing stands. If planted not more than 3 feet apart, spaces between plants are rapidly filled in and they can spread up to ≥2 ft/season. American beachgrass is common to the Great Lakes region, the northern shore of eastern U.S. and Newfoundland, Canada. These grasses are resistant to the cutting effect of blowing sand. They are of little value as a pasture crop but excel in the stabilization of sandy soils. The European beachgrass averages 114,000 seeds/lb.
3. Season: Cool season and rapid growth up to 2 ft/season.
4. Production in U.S.: No data for beachgrass production, however, pasture and rangeland are produced on more than 410 million A (U.S. CENSUS, 1992).
5. Other production regions: Grown in Newfoundland, Canada (ROTHWELL 1996a).
6. Use: Sand and soil stabilization and erosion control, wildlife protection.
7. Part(s) of plant consumed: Leaves and stems, seeds
8. Portion analyzed/sampled: Forage and hay
9. Classifications:
   a. Authors Class: See Forage grass
   b. EPA Crop Group (Group & Subgroup): See Forage grass
   c. Codex Group: See Forage grass
   d. EPA Crop Definition: See Forage grass
10. References: GRIN, MAGNESS, US EPA 1994, US EPA 1995, CODEX, BARNES, ALDERSON, ROTHWELL 1996a, MOSLER (picture)
11. Production Map: EPA Production Regions 1, 2, 4, 6, 8 and 10.
12. Plant Codes:
   a. Bayer Code: AMOAR (*A. arenaria*), AMOBR (*A. breviligulata*)

047

1. Bean
(Chickpea, Garbanzo bean, Sweet lupine, White sweet lupine, White lupine, Grain lupin, Kidney bean, Lima bean, Mung bean, Navy bean, Pinto bean, Snap bean, Waxbean, Broad bean, Fava bean, Asparagus bean, Blackeyed pea, Cowpea, Pulse)
Fabaceae (Leguminosae)
> *Phaseolus* spp., *Vigna* spp., *Vicia* spp., *Glycine* spp.; *Cicer* spp. *Lupinus* spp.

2. Beans of several genera and species and numerous varieties, are important food crops and are of some importance as feed crops in many countries. All are annuals, grown from seeds. The fruits are pods in which the seeds are contained. In green or snap beans, also termed 'string', pods are harvested before ripening and both pods and the immature seeds are consumed, mainly as pot vegetables. In some kinds, the seeds when near full grown, but while still immature, are threshed from the pods and frozen or canned. In dry or field beans pods and seeds are allowed to ripen, then threshed and seeds only are consumed. Plants may be "bush", non-climbing and reaching a height of 15 to 30 inches; or "vine", "pole" or "runner", vining types reaching 10 or more feet in length or half-runner types that have short or medium length vines. Soybean (*Glycine* spp.) is not included in the regulatory definition of beans.

For bean, see crop group 6: Legume Vegetables under 40 CFR 180.41 for cultivars of beans. For bean seed: dried seed for uses on dried shelled beans; succulent seed without pod for uses on succulent shelled beans (e.g. lima beans); succulent seed with pod for edible-podded beans (e.g. snap beans). Cowpea is the only bean crop considered for livestock feeding. (See cowpea). Residue data for forage and hay are required only for cowpea.

For crop monograph fields 3-12 see specific species.

048

1. Bean, Adzuki
(Chinese red bean, Adanka bean, Azuki, Poroto arroz)
Fabaceae (Leguminosae)

Vigna angularis (Willd.) Ohwi & H. Ohashi var. *angularis* (syn: *Phaseolus angularis* (Willd.) W. Wight; *Dolichos angularis* Willd.)

2. This is a bush annual bean, with an erect plant, 1 to 2 feet high. Pods are small and cylindrical. Seeds are small and variously colored, commonly red, oblong to nearly round, often with flat ends. They are cultivated in Asian countries, but rarely in the U.S. Seeds are used as dry beans. Culture and exposure of edible parts are similar to dry or field beans.
3. Season, seeding to harvest: About 120 days. Seeds mature 40-50 days after pollination.
4. Production in U.S.: In 1994, about 4,500 acres with Washington reporting 900 acres (SCHREIBER 1995). North central region (IR-4 DATABASE 1996).
5. Other production regions: Grown in Canada (ROTHWELL 1996a). Asian countries.
6. Use: As dry beans and succulent beans. In Hawaii, eaten with ice cream.
7. Part(s) of plant consumed: Mainly for dried seed. Also the immature pods can be used as a vegetable, seeds for sprouts or green seed cooked
8. Portion analyzed/sampled: Seed (dried)
9. Classifications:
   a. Authors Class: Legume vegetables (succulent or dried)
   b. EPA Crop Group (Group & Subgroup): Legume vegetables (succulent or dried) (6C)
   c. Codex Group: 015 (VD 0560) Pulses for dry seed
   d. EPA Crop Definition: Beans = *Vigna* spp.
10. References: GRIN, CODEX, FACCIOLA, MAGNESS, ROTHWELL 1996a, SCHREIBER, STEPHENS 1988 (picture), US EPA 1996a, US EPA 1995a, YAMAGUCHI 1983, IR-4 DATABASE 1996, LUMPKIN, NELSON (a), RUBATZKY
11. Production Map: EPA Crop Production Regions 5 and 11.
12. Plant Codes:
    a. Bayer Code: PHSAN

049

1. Bean, Broad
(English bean, Fava bean, Horse bean, Windsor bean, Haba, Tick bean, Cold bean, Silkworm bean, Field bean, Pigeon bean, Faba bean, Fayot, Haba comun)
Fabaceae (Leguminosae)

Vicia faba L. var. *faba*

2. This type of bean is very important as a cool-season crop in Mediterranean areas and in cool regions of Europe, but is grown to only a limited extent in the U.S. Plants are erect annuals reaching 2 to 4 feet and very leafy. Pods are large and thick, 2 inches up to a foot or more in length. Seeds are

large and flat. They are used as green-shell, the seeds removed from the pod before maturity, or as dry beans. They are also used as feed for livestock.

3. Season, planting to harvest: 4 to 5 months. Pod set to harvest, 30 to 60 days.
4. Production in U.S.: No data; very limited. Grown in Washington (SCHREIBER 1995). Few are grown in Florida gardens (STEPHENS 1988). Commercially available from California, April through June (LOGAN 1996).
5. Other production regions: Grown as summer annual in northern climates and as a winter annual in southern climates.
6. Use: Mainly, cooked vegetables and stock feed. Seeds are used like lima beans, cooked or raw. This bean remains a specialty item in part because of its long cleaning process. The tough skin must be peeled, cut the tips and press open the seams and most of the bean discarded (LOGAN 1996).
7. Part(s) of plant consumed: Mainly seeds for food; immature pods as snap beans and whole plant for feed which is not a significant feed item. Also can be used as a cover crop
8. Portion analyzed/sampled: Seed (succulent and dry)
9. Classifications:
   a. Authors Class: Legume vegetables (succulent or dried)
   b. EPA Crop Group (Group & Subgroup): Legume vegetables (succulent or dried) (6B and 6C)
   c. Codex Group: 015 (VD 0523) Pulses; 014 (VP 0523) Legume vegetables (immature seed); 014 (VP 0522) Legume vegetables (green pod and immature seed)
   d. EPA Crop Definition: Beans = *Vicia* spp.
10. References: GRIN, CODEX, LOGAN (picture), MAGNESS, SCHREIBER, STEPHENS 1988 (picture), US EPA 1995a, YAMAGUCHI 1983, RUBATZKY
11. Production Map: EPA Crop Production Regions 10, 11 and 12.
12. Plant Codes:
    a. Bayer Code: VICFJ

050

1. Bean, Dry common
(Common bean, Kidney bean, Pea bean, Navy bean, Habichuela bean, Pink bean, Small red bean, Cranberry bean, Black bean, Soldier bean, Popbean, Nunas, Cannellini, Great northern bean, Small white bean, Dry bean, Pinto bean, Red kidney bean, Field bean, Frijoles comunes, Haricot commun, Frijol, Poroto, Judia comum)
Fabaceae (Leguminosae)

Phaseolus vulgaris L. var. *vulgaris*

2. These are beans that ripen prior to harvest and are threshed dry from the pods. Only the ripe seeds are marketed. Four main types are grown as follows: (1) Medium type includes Pinto, Great Northern, Sutter, Pink Bayo and Small Red or Mexican Red; (2) Pea or Navy; (3) Kidney; and (4) Marrow. Seeds vary in size from about $1/3$ inch long in the Pea or Navy bean to $3/4$ inch in the Kidney. All plants are of bush type. They are usually cut or pulled when most pods are ripe, then vines and pods are allowed to dry (7-14 days) before threshing. Popping beans called nunas or popbeans used as popcorn.
3. Season, seeding to harvest: 3 to 6 months.
4. Production in U.S.: In 1995, Dry edible beans reported on 2,069,300 acres with production of 1,551,600 tons, including Black-eyed peas (56,050 tons), Garbanzo (24,250 tons)

and Lima beans (49,250 tons), which see (USDA 1996c). In 1995, Dry edible beans from the top 12 states as North Dakota (600,000 acres), Michigan (390,000), Nebraska (225,000), Colorado (190,000), Minnesota (190,000), California (145,000), Idaho (110,000), Washington (41,000), Wyoming (35,000), New York (34,000), Kansas (34,000) and Texas (25,000). This included Limas (44,000 acres), Navy (487,100), Great northern (138,400), Small white (8,300), Pinto (843,000), Light red kidney (83,900), Dark red kidney (73,400), Pink (38,400), Small red (37,200), Cranberry (35,000), Black (130,900), Black-eyed pea (55,600), Garbanzo (30,600) and other (63,500) (USDA 1996c).

5. Other production regions: In 1995, Canada reported about 400,000 acres of dry beans (ROTHWELL 1996a).

6. Use: Commercially canned, soup, cooked in homes.

7. Part(s) of plant consumed: Seed only.

8. Portion analyzed/sampled: Seed (dry)

9. Classifications:
   a. Authors Class: Legume vegetables (succulent or dried)
   b. EPA Crop Group (Group & Subgroup): Legume vegetables (succulent or dried) (6C) (Representative crop)
   c. Codex Group: 015 (VD 0526) Pulses for dry
   d. EPA Crop Definition: Beans = *Phaseolus* spp.

10. References: GRIN, CODEX, FACCIOLA, MAGNESS, ROTHWELL 1996a, USDA 1996c, US EPA 1996a, US EPA 1995a, US EPA 1994, REHM, IVES, SMITH 1989, UEBERSAX, WILLIAMS, MARTIN 1983

11. Production Map: EPA Crop Production Regions, Dried beans: 1, 5, 7, 8, 9, 10 and 11 with 99 percent of the total U.S. acreage.

12. Plant Codes:
    a. Bayer Code: PHSVV

051

1. Bean, Goa

(Princess bean, Winged bean, Asparagus pea, Yi dou, Pois aile, Frijol alado, Four-angled bean, Manila bean, Kok tau)

Fabaceae (Leguminosae)

Psophocarpus tetragonolobus (L.) DC.

2. The Goa bean is a perennial grown as an annual plant with entire ovate leaves 3 to 6 inches long, on weak, vining stems. It produces a large, tuberous root which is eaten both cooked and raw in the Orient. The pods are 6 to 9 inches long and 1 inch broad When young, pods are cooked and eaten like green beans. Growth habit and exposure for the pods are like pole beans as grown in the U.S. Root production is entirely underground. Goa beans are grown commercially in the U.S. and in gardens of Oriental vegetables.

3. Season, emergence to first harvest: Young pods about 65 days; Plant (seed) matures in about 110 days. Pods contain 4 to 17 seeds. Propagation is usually by seed and stem cuttings can be used.

4. Production in U.S.: No data. Grown commercially in south Florida during winter (STEPHENS 1988).

5. Other production regions: Southeast Asia (RUBATZKY 1997).

6. Use: Young pods used as snap beans; roots are eaten cooked or raw. Preparation is by dunking the pods in a sink full of water, drain and cook or pickle (SCHNEIDER 1985).

7. Part(s) of plant consumed: Primarily roots and young pods. The entire plant can be used as food and feed

8. Portion analyzed/sampled: Roots; young pods

9. Classifications:
   a. Authors Class: Legume vegetables (succulent or dried); Root and tuber vegetables
   b. EPA Crop Group (Group & Subgroup): Miscellaneous
   c. Codex Group: 014 (VP 0530) Legume vegetables for immature pod; 016 (VR 0530) Root and tuber vegetables for the root
   d. EPA Crop Definition: None

10. References: GRIN, CODEX, KAYS, MAGNESS, SCHNEIDER 1985 (picture), STEPHENS 1988 (picture), YAMAGUCHI 1983 (picture), RUBATZKY, DUKE(a)

11. Production Map: EPA Crop Production Region 13.

12. Plant Codes:
    a. Bayer Code: PSHTE

052

1. Bean, Hyacinth

(Lablab bean, Bonivist bean, Bonavist bean, Chicharos, Frijoles caballero, Chinese flowering bean, Egyptian bean, Indian bean, Pharao bean, Seem, Shink, Val, Wild field bean)

Fabaceae (Leguminosae)

Lablab purpureus (L.) Sweet ssp. *purpureus* (syn: *Dolichos lablab* L., *L. niger* Medik.; *L. vulgaris* Savi)

2. This bean, known under many names, is a tall viner, up to 20 feet, with broad, ovate leaves. In the U.S., it is cultivated mainly as an ornamental, but in the tropics the pods and seeds are eaten. The pods are small, 2 to 3 inches long, flat and smooth. Seeds are from $1/4$ to $1/2$ inch long, nearly as wide and flattened. Exposure of plant parts is similar to that of pole snap beans. No data are available on U.S. production, but it is insignificant in continental U.S.

3. Season, seeding to first harvest: Minimum of $1^1/2$ months. Seeds mature 3-5 months.

4. Production in U.S.: No data. Not cultivated much in Florida or the rest of the U.S. (STEPHENS 1988).

5. Other production regions: Widely distributed in the tropics eg. Caribbean (MARTIN 1983).

6. Use: Green pod used as snap beans. Seeds as dried beans.

7. Part(s) of plant consumed: Ripe seed and green pod; Forage crop

8. Portion analyzed/sampled: Seed and Pod; Dried seed. Forage not significant feed item

9. Classifications:
   a. Authors Class: Legume vegetables (succulent or dried)
   b. EPA Crop Group (Group & Subgroup): Legume vegetables (succulent or dried) (6C)
   c. Codex Group: 014 (VP 0531) Legume vegetables (immature pod); 015 (VD 0531) Pulses (Dried)
   d. EPA Crop Definition: None

10. References: GRIN, CODEX, MAGNESS, MARTIN 1983, STEPHENS 1988 (picture), US EPA 1996a, US EPA 1995a, RUBATZKY

11. Production Map: EPA Crop Production Region 13.

12. Plant Codes:
    a. Bayer Code: DOLLA

053

1. Bean, Lima
(Butter bean, Haba, Burma bean, Guffin bean, Hibbert bean, Java bean, Sieva bean, Rangood bean, Madagascar bean, Paiga, Paigya, Prolific bean, Civet bean, Sugar bean, Dried lima, Potato bean, Green lima, Succulent lima, Haba lima, Frijol de luna, Towe, Judia de lima, Butterpea, Fordhook)

Fabaceae (Leguminosae)

Phaseolus lunatus L. var. *lunatus* (syn: *P. limensis* Macfad.; *P. inamoenus* L.; *P. tunkinensis* Lour.)

2. While both vining and bush type lima beans are available, the bush types are mainly grown commercially for green limas. There are two main types, the baby limas with small, rather thin seeds and those with large, thick seeds. "Green" lima beans are harvested when seeds are near full grown but the pods are still green in color. Dry or ripe limas are mainly of vine type and grown in California. Vines are allowed to run on the ground. In harvesting green limas for processing, plants are cut and the beans are shelled by machine, as are peas. Dry limas are harvested much as other dry beans. Limited quantities of green limas are hand picked and sold in the pods. There are 20-70 lima beans per ounce. Bush forms grow from 20-35 inches tall while climbing types reach heights of 9-12 feet. Pods range from 2-6 inches in length.

3. Season, seeding to green-shell stage: 2 to 3 months; dry stage, 4 to 5 months.

4. Production in U.S.: In 1995, Processing limas 52,100 acres with 70,160 tons which included Fordhooks and Baby lima; Fresh market limas 6,500 acres with 8,250 tons; Dry limas (large lima) 21,000 acres with 22,500 tons; Dry limas (baby lima) 23,000 acres with 26,750 tons (USDA 1996b and c). In 1995, Processed limas reported from California, Delaware, Illinois, Maryland, Minnesota, New Jersey, Oregon, Tennessee, Washington and Wisconsin; Fresh market limas reported from Georgia only; and Dry limas (large limas) reported from California only, as were baby limas (USDA 1996b and c). In 1977, Green processed limas were reported for California (26,900 acres), Delaware (9,000), Wisconsin (5,900) and Maryland (2,800) with total U.S. acreage at 60,310 acres (WARE 1980). The bush limas cultivars include 'Butterpea', 'Fordhook', 'Henderson', 'Nemagreen', 'Red calico' and 'Thornogreen'. Pole limas include 'Butterbean', 'Black', 'Christmas', 'Sieva' and 'Indian red'.

5. Other production regions: In 1995, Canada reported 190 acres of limas (ROTHWELL 1996a).

6. Use: Green limas, processed as frozen or canned; some few sold fresh in pods. Dry limas, sold as seeds, canned.

7. Part(s) of plant consumed: Seeds only

8. Portion analyzed/sampled: Seed (succulent or dried). 40 CFR 180.1(J)(8) dated 7-1-96 states that the term lima beans means the beans and the pod which is out of date as per EPA's Table 1 (US EPA 1996a). Limas are used as green shell *Phaseolus* for EPA crop group

9. Classifications:
   a. Authors Class: Legume vegetables (succulent or dried)
   b. EPA Crop Group (Group & Subgroup): Legume vegetables (succulent or dried) (6B and 6C)
   c. Codex Group: 014 (VP 0534) Legume vegetables for young pods and/or immature beans; 015 (VD 0534) Pulses for dry
   d. EPA Crop Definition: Beans = *Phaseolus* spp.

10. References: GRIN, CODEX, FACCIOLA, LOGAN (picture), MAGNESS, ROTHWELL 1996a, USDA 1996b and c, US EPA 1996a, US EPA 1995a, US EPA 1994, WARE, IVES, REHM, SWIADER

11. Production Map: EPA Crop Production Regions: Dried lima: 10 and 11 with 99 percent of the total U.S. acreage; Green or Succulent lima: 2, 5, 10 and 11 with 97 percent of the total U.S. acreage.

12. Plant Codes:
    a. Bayer Code: PHSLU

054

1. Bean, Moth
(Mat bean, Dew gram, Haricot papillon, Matki bean, Mout bean)

Fabaceae (Leguminosae)

Vigna aconitifolia (Jacq.) Marechal (syn: *Phaseolus aconitifolius* Jacq.)

2. This bean is a trailing plant, with stems up to 2 feet, covered with stiff hairs. Pods are small, about 2 inches long, nearly round and glabrous. Seeds are small. Moth bean is cultivated for food in South Asia and for forage, but is rarely grown in the U.S. Seeds are used as dry beans. In India the green pods are used as a vegetable.

3. Season, seeding to harvest: About 3 months.

4. Production in U.S.: No data. Tried for cattle feed in Texas and California (STEPHENS 1988).

5. Other production regions: Native to India or Southeast Asia (MARTIN 1983).

6. Use: Green pod as vegetable and dry seed as dry beans.

7. Part(s) of plant consumed: Seeds as dry beans; Green pods as snap bean; and fodder as feed

8. Portion analyzed/sampled: Succulent seed with pod; and dry seed

9. Classifications:
   a. Authors Class: Legume vegetables (succulent or dried)
   b. EPA Crop Group (Group & Subgroup): legume vegetables (succulent or dried) (6A and 6C)
   c. Codex Group: 014 (VP 0535) Legume vegetables for green pods, mature, fresh seeds; 015 (VD 0535) Pulses for dry seed as Mat bean
   d. EPA Crop Definition: Beans = *Vigna* spp.

10. References: GRIN, CODEX, MAGNESS, MARTIN 1983, STEPHENS 1988 (picture), US EPA 1996a, US EPA 1995a, YAMAGUCHI 1983

11. Production Map: No entry

12. Plant Codes:
    a. Bayer Code: PHSAC

055

1. Bean, Mung
(Mungo bean, Habichuela mungo, Oorud bean, Bundo, Mongo, Mash bean, Golden gram, Green gram, Chinese bean sprouts, Amberique)

Fabaceae (Leguminosae)

Vigna radiata (L.) Wilczek var. *radiata* (syn: *Phaseolus aureus* Roxb.; *P. radiatus* L.)

2. Mung beans for food are used mainly as the sprouts. The dry beans are soaked overnight, then placed in containers in a warm, dark room and kept thoroughly sprinkled. Sprouts are

harvested after about a week, then in the U.S. they are mostly canned. They are extensively used in Oriental dishes. Mung beans are annuals grown much as other dry beans. Plants are 12 to 24 inches tall with pods borne near the top. Pods are about 4 inches long. Mung beans are also grown for hay in the southwestern states of the U.S.

3. Season, seeding to harvest: 3 to 5 months.
4. Production in U.S.: In 1992, 22,770 acres with 5,631 tons (1992 CENUS). In 1992, Oklahoma with 18,114 acres (1992 CENSUS). In 1997, Texas reported 15,000 acres in the Texas Panhandle region (SMITH 1997).
5. Other production regions: Grown in Canada (ROTHWELL 1996a). Southeast Asia.
6. Use: Sprout production, sprouts used mainly cooked, but some raw in salads.
7. Part(s) of plant consumed: Seeds as food, mainly sprouted. Whole plant as feed which is not a significant feed item. Also, green pods can be used as a vegetable. The green grain is used for sprouts. One gram of seeds will produce 6 to 8 grams of sprouts
8. Portion analyzed/sampled: Bean (seed). Data on mung bean covers sprouts except when pesticide is used on sprouts *per se*
9. Classifications:
   a. Authors Class: Legume vegetables (succulent or dried)
   b. EPA Crop Group (Group & Subgroup): Legume vegetables (succulent or dried) (6C)
   c. Codex Group: 014 (VP 0536) Legume vegetables for green pods; 015 (VD 0536) Pulses for dry
   d. EPA Crop Definition: Beans = *Vigna* spp.
10. References: GRIN, CODEX, MAGNESS, RICHE, ROTHWELL 1996a, US EPA 1996a, US EPA 1995a, US EPA 1994, YAMAGUCHI, SMITH 1997, OPLINGER, SCHLAR, RUBATZKY, MARTIN 1983
11. Production Map: EPA Crop Production Region 6, with 95 percent of total U.S. acreage, and 8.
12. Plant Codes:
    a. Bayer Code: PHSAU

056

1. Bean, Rice
(Red bean, Frijol arroz, Haricot de riz, Oriental bean, Mambi bean, Climbing mountain bean)
Fabaceae (Leguminosae)
Vigna umbellata (Thunb.) Ohwi & H. Ohashi (syn: *Phaseolus calcaratus* Roxb.; *V. calcarata* (Roxb.) Kurz)

2. A twining annual to 3 to 10 feet tall. Boiled rice bean is used instead of boiled rice in Southeast Asia. The immature pods and leaves are used as vegetables and the whole plant used as fodder. The long slender pods bear 6 to 12 seeds which range in color from red to speckled. Usually sown as a second crop, after rice, for example.
3. Season, seeding to first harvest: About 2 to 6 months.
4. Production in U.S.: No data. Grows in the Caribbean and southern U.S. (MARTIN 1983).
5. Other production regions: Southeast Asia.
6. Use: Whole seeds are cooked, young pods can be a green vegetable. Also green shelled seeds are cooked, the leaves are cooked as spinach and the seeds are sprouted as a vegetable.
7. Part(s) of plant consumed: Mainly seed; Also young pods for food and plant residue for feed
8. Portion analyzed/sampled: Seed (dry)

9. Classifications:
   a. Authors Class: Legume vegetables (succulent or dried)
   b. EPA Crop Group (Group & Subgroup): Legume vegetables (succulent or dried) (6C)
   c. Codex Group: 014 (VP 0539) Legume vegetables for young pods; 015 (VD 0539) Pulses for dry.
   d. EPA Crop Definition: Beans = *Vigna*
10. References: GRIN, CODEX, MARTIN 1983 (picture), US EPA 1996a, US EPA 1995a, YAMAGUCHI 1983, RUBATZKY
11. Production Map: EPA Crop Production Region 13.
12. Plant Codes:
    a. Bayer Code: PHSPU

057

1. Bean, Scarlet runner
(White dutch runner bean, Astec bean, Oregon lima, Multiflora bean, Runner bean, Harecot despagna, Ayocote, Pilay, Cuba, Judia pinta, Dutch case-knife bean)
Fabaceae (Leguminosae)
Phaseolus coccineus L.
   P. coccineus L. ssp. *coccineus* (syn: *P. multiflorus* Lam.)
(Ayocote)

2. The Scarlet runner bean is grown mainly as an ornamental for its showy red flowers, but the White Dutch varieties are grown for the edible pods and the white mature seeds. Plants are strong growing, slender and vining, with large leaves. They are perennials often grown as an annual up to 10 feet long. Pods are 3 to 6 inches long and curved. Seeds are large, near 1 inch long, flattened or near cylindrical. In exposure and general culture, edible varieties resemble pole snap beans. The edible types are sparingly grown in the U.S., but are occasionally seen in markets, particularly in the Southwest. No data are available on total production. In the tropics, it is used for its dry seeds. However, in the temperate zone the green pods and green shell beans are the principal use.
3. Season, seeding to first harvest: About 2½ months. Up to 4 months for dried seeds.
4. Production in U.S.: No data.
5. Other production regions: Native to Mexico (MARTIN 1983).
6. Use: Mainly immature pods used as snap beans. Also green and dry shell beans are used as food.
7. Part(s) of plant consumed: Immature pods, seeds and tuberous roots
8. Portion analyzed/sampled: Succulent seed with pod and mature seed
9. Classifications:
   a. Authors Class: Legume vegetables (succulent or dried)
   b. EPA Crop Group (Group & Subgroup): Legume vegetables (succulent or dried) (6A)
   c. Codex Group: 014 (VP 0540) Legume vegetables for pods and seeds; 015 (VD 4517) Pulses
   d. EPA Crop Definition: Beans = *Phaseolus* spp.
10. References: GRIN, CODEX, FACCIOLA, MAGNESS, MARTIN 1983, US EPA 1996a, US EPA 1995a, YAMAGUCHI 1983, REHM, SWIADER
11. Production Map: No entry
12. Plant Codes:
    a. Bayer Code: PHSCO

058

1. Bean, Succulent common
(Common bean, Wax bean, Green bean, Snap bean, Blue lake bean, Kentucky wonder bean, Magic bean, Filet bean, Haricot bean, String bean, Flageolet bean, Garden bean, French bean, Romano bean, Frijoles comunes, Haricot commun, Judia comum, Italian bean, French horticultural bean, Pole bean, Bush bean, Vainica, Edible-podded bean)
Fabaceae (Leguminosae)
Phaseolus vulgaris L. var. *vulgaris*

2. Beans harvested as snap or green beans are of two types, bush and pole, or climbers. Bush beans are non-climbing plants 15 to 20 inches tall and set the pods and reach harvest in a short season. Pole beans are trained on stakes or trellis, grow to 10 feet and blossom and set pods over a long season. Pods on both types are 5 to 10 inches long and are harvested while seeds are immature. Some varieties are termed wax beans. Pole beans are more adapted to home garden use. Green shell beans are called Buttergreen, Flageolet or French bean.
3. Season, bloom to harvest: About 30 days. Bush beans set fruit and mature over a short time. Pole beans may continue to bloom and set over several weeks. Bush beans can produce a crop in 50-70 days after seeding while pole beans in 70-90 days.
4. Production in U.S.: In 1995, Snap beans for fresh market reported on 96,300 acres with 219,100 tons; Snap beans for processing reported on 233,040 acres with 708,170 tons (USDA 1996b). In 1995, Snap beans for fresh market reported from the top 10 states as Florida (31,800 acres), Georgia (16,500), California (8,800), Tennessee (8,700), North Carolina (7,000), Virginia (5,500), New York (5,100), New Jersey (4,100), Maryland (2,500) and Michigan (2,300); Snap beans for processing reported from the top 8 states as Wisconsin (74,400 acres), Oregon (23,600), Illinois (23,600), Michigan (23,000), New York (22,300), Pennsylvania (8,500), Indiana (4,400) and Florida (1,000) (USDA 1996b).
5. Other production regions: In 1995, Canada reported 17,500 acres of green and wax beans and 1,119 acres of snap and baking beans (ROTHWELL 1996a).
6. Use: Marketed as fresh, canned, frozen. U.S. per capita consumption of snap beans in 1994 was 7.4 pounds (PUTNAM 1996).
7. Part(s) of plant consumed: Mainly whole pods; Also green shelled seed (Limas are used as green shell *Phaseolus* for EPA crop group)
8. Portion analyzed/sampled: Succulent seed with pod; and/or succulent seed without pod
9. Classifications:
   a. Authors Class: Legume vegetables (succulent or dried)
   b. EPA Crop Group (Group & Subgroup): Legume vegetables (succulent or dried) (6A and 6B) (Representative crop)
   c. Codex Group: 014 (VP 0526) Legume vegetables for pods and/or immature seeds
   d. EPA Crop Definition: Beans = *Phaseolus* spp.
10. References: GRIN, CODEX, FACCIOLA, MAGNESS, ROTHWELL 1996a, USDA 1996b, US EPA 1996a, US EPA 1995a, US EPA 1994, REHM, PUTNAM 1996
11. Production Map: EPA Crop Production Regions; Snap beans: 1, 2, 3, 5, 10 and 11 with 97 percent of the total U.S. acreage.
12. Plant Codes:
   a. Bayer Code: PHSVV

059

1. Bean, Tepary
(Tepari bean, Yori mui, Pavi, Texas bean, Escomite)
Fabaceae (Leguminosae)
Phaseolus acutifolius A. Gray var. *acutifolius* (syn: *P. acutifolius* var. *latifolius* G.F. Freeman)

2. This bean is native to southwestern U.S. and Mexico and long grown by the Native Americans there. It is highly heat and drought resistant, but eating quality is less desirable than *P. vulgaris*. Culture is similar to that of other dry or field beans. Cultivars are black-seeded, brown/yellow-seeded, mottled and white-seeded. The cultivated varieties are bush type to about 30 inches tall.
3. Season, seeding to harvest: About 2 to 3 months. Pods are about 3 inches.
4. Production in U.S.: No data. Southwestern North America (FACCIOLA 1990).
5. Other production regions: In northern Mexico (FACCIOLA 1990).
6. Use: Dried seeds are eaten boiled or baked. Also, ground into meal which is added to boiling water for instant beans.
7. Part(s) of plant consumed: Seed
8. Portion analyzed/sampled: Seed (dry)
9. Classifications:
   a. Authors Class: Legume vegetables (succulent or dried)
   b. EPA Crop Group (Group & Subgroup): Legume vegetables (succulent or dried) (6C)
   c. Codex Group: 015 (VD 0564) Pulses for dry
   d. EPA Crop Definition: Bean = *Phaseolus* spp.
10. References: GRIN, CODEX, FACCIOLA, MAGNESS, STEPHENS 1988 (picture), US EPA 1995a, US EPA 1996a, REHM
11. Production Map: No entry
12. Plant Codes:
   a. Bayer Code: PHSAF

060

1. Bean, Urd
(Urud bean, Black gram, Urid, Amberique)
Fabaceae (Leguminosae)
Vigna mungo (L.) Hepper (syn: *Phaseolus mungo* L.)

2. This bean is similar to Mung bean, which see, except plants are more prostrate, pods are long and hairy and seeds are oblong and black. Use is similar to that of Mung bean. Urd beans are frequently planted after rice. In the West Indies, the cultivar, 'Wooly pyrol', is grown.
3. Season, seeding to first harvest: $2^1/_2$ to $4^1/_2$ months. Pods mature 20 days after flowering and have 6-10 seeds.
4. Production in U.S.: No data. Grows in Caribbean and U.S. (MARTIN 1983).
5. Other production regions: India (YAMAGUCHI 1983).
6. Use: Seeds are cooked.
7. Part(s) of plant consumed: Mainly seeds; Also green pods
8. Portion analyzed/sampled: Seed (dry)
9. Classifications:
   a. Authors Class: Legume vegetables (succulent or dried)
   b. EPA Crop Group (Group & Subgroup): Legume vegetables (succulent or dried) (6C)
   c. Codex Group: 014 (VP 0521) Legume vegetables for green pods; 015 (VD 0521) Pulses for dry

d. EPA Crop Definition: Beans = *Vigna* spp.
10. References: GRIN, CODEX, MAGNESS, MARTIN 1983, US EPA 1996a, US EPA 1995a, YAMAGUCHI 1983
11. Production Map: EPA Crop Production Region 13.
12. Plant Codes:
   a. Bayer Code: PHSMU

061

1. Bean, Yardlong

(Asparagus bean, Tau kok, Peru bean, Snake bean, Bodi bean, Chinese longbean, Peabean, Stringbean, Judia esparago)
Fabaceae (Leguminosae)
   Vigna unguiculata ssp. *sesquipedalis* (L.) Verdc. (syn: *V. sesquipedalis* (L.) Fruwirth; *Dolichos sesquipedalis* L.)
2. This bean is closely related to southern pea, which see. The pods are greatly elongated, up to 3 feet, fleshy and brittle and can be cooked as a snap bean. It is little grown as a vegetable in the U.S., since it sets few pods. The plant is used somewhat for forage. The plant is a trailing vine, similar in all respects to southern pea except for longer and more tender pods. Cultivars are 'Bulacan dark purple'; 'Bush sitao No. 1, No. 2'; 'Economic garden'; and 'Singapore light green' (MARTIN 1983).
3. Season, seeding to first harvest: About 2 to 3 months.
4. Production in U.S.: No data. Grown in Florida primarily in home gardens (STEPHENS 1988). Grown in Hawaii (NEAL 1965) and California (LOGAN 1996).
5. Other production regions: Available year-round from Mexico (LOGAN 1996).
6. Use: As snap beans, cooked vegetable.
7. Part(s) of plant consumed: Pods (fruit)
8. Portion analyzed/sampled: Pods
9. Classifications:
   a. Authors Class: Legume vegetables (succulent or dried)
   b. EPA Crop Group (Group & Subgroup): Legume vegetables (succulent or dried) (6A)
   c. Codex Group: 014 (VP 0544) Legume vegetables (pods)
   d. EPA Crop Definition: Bean = *Vigna* spp.
10. References: GRIN, CODEX, LOGAN, MAGNESS, MARTIN 1983, NEAL, STEPHENS 1988 (picture), US EPA 1995a, YAMAGUCHI 1983 (picture)
11. Production Map: EPA Crop Production Regions 3, 10 and 13.
12. Plant Codes:
   a. Bayer Code: VIGSQ

062

1. Bearberry

(Kinnikinik, Kutai tea, Manzanita, Mealberry, Hog cranberry, Sandberry, Mountain box, Bear's grape, Creashak, Gayuba)
Ericaceae
   Arctostaphylos uva-ursi (L.) Spreng.
2. It is an evergreen, prostrate creeping shrub. Leaves obovate to spatulate, to 1 inch long and glabrous. Fruit is red colored, a berry-like drupe and smooth.
3. Season, harvest: Usually harvested in late summer.
4. Production in U.S.: No data.
5. Other production regions: Northern temperate region (FACCIOLA 1990). USDA Hardiness Zone 2 (BAILEY 1976).

6. Use: Fruits are usually cooked, preserved or made into jellies. The dried leaves are used as tea in parts of Russia.
7. Part(s) of plant consumed: Fruit
8. Portion analyzed/sampled: Whole fruit
9. Classifications:
   a. Authors Class: Berries
   b. EPA Crop Group (Group & Subgroup): Miscellaneous
   c. Codex Group: 004 (FB 0260) Berries and other small fruits
   d. EPA Crop Definition: None
10. References: GRIN, BAILEY 1976, CODEX, FACCIOLA, KUHNLEIN (picture)
11. Production Map: No entry
12. Plant Codes:
   a. Bayer Code: ARYUU

063

1. Beechnut

Fagaceae
   Fagus spp.
1. American beech

   F. grandifolia Ehrh. (syn: *F. americana* Sweet; *F. ferruginea* Aiton)
1. European beech

   F. sylvatica L.
2. The seeds of these beech species are sometimes gathered from native trees, or from trees planted for other purposes and used as food. The seeds are formed in prickly burrs, about ³/₄ to 1 inch in diameter, which remain closed until ripe, then partially open. The angular seeds or nuts are up to ³/₄ inch long. The seed coat must be removed from the kernel before eating. In general, beechnuts are similar to small chestnuts. The trees are not cultivated for the purpose of nut production.
3. Season, harvest: Mature in late summer and fall to the ground, over a period of several weeks.
4. Production in U.S.: No data. Not cultivated but grows from New Jersey to Florida and west to south Illinois and Texas.
5. Other production regions: Western Europe (WOODROOF 1967).
6. Use: Direct eating.
7. Part(s) of plant consumed: Nutmeat and oil extracted for cooking purposes and as a salad oil in Europe
8. Portion analyzed/sampled: Nutmeat
9. Classifications:
   a. Authors Class: Tree nuts
   b. EPA Crop Group (Group & Subgroup): Tree nuts
   c. Codex Group: 022 (TN 0661) Tree nuts
   d. EPA Crop Definition: None
10. References: GRIN, CODEX, MAGNESS, US EPA 1995a, US EPA 1995b, WOODROOF(b) (picture)
11. Production Map: EPA Crop Production Regions 2, 4 and 5.
12. Plant Codes:
   a. Bayer Code: FAUGR (*F. grandifolia*), FAUSY (*F. sylvatica*)

064

1. Beet, Fodder

(Mangold, Mangels, Mangoldwurzel, Acelga)
Chenopodiaceae
   Beta vulgaris L. ssp. *vulgaris*

2. For general information, see Beet, Garden. Mangels are large rooted beets generally used as animal feed due to their coarseness. Roots are frequently 2 feet long, weighing up to 15 pounds each with about half of the root growing above the surface of the ground.
3. Season, seeding to harvest: About 100 to 110 days.
4. Production in U.S.: No data.
5. Other production regions: Adapted to Beet, Garden climatic zone (which see).
6. Use: Roots and tops used as animal feed.
7. Part(s) of plant consumed: Mainly the root; tops
8. Portion analyzed/sampled: Root and tops
9. Classifications:
   a. Authors Class: Root and tuber vegetables; Leaves of root and tuber vegetables
   b. EPA Crop Group (Group & Subgroup): As *B. vulgaris*: Root and tuber vegetables (1A); and leaves of root and tuber vegetables
   c. Codex Group: Fodder beet; 052 (AM 1051) Miscellaneous fodder and forage crops; Fodder beet leaves or tops: 052 (AV 1051) Miscellaneous fodder and forage crops
   d. EPA Crop Definition: None
10. References: GRIN, CODEX, FACCIOLA, US EPA 1995a, YAMAGUCHI 1983, IVES, REHM
11. Production Map: See Garden beet
12. Plant Codes:
   a. Bayer Code: BEAVC

065

1. Beet, Garden
(Table beet, Ramolacha, Red beet, Remolacha, Betterave, Betabel)

Chenopodiaceae

Beta vulgaris L. ssp. *vulgaris*

2. Beets are grown primarily for the enlarged bulbous root which forms with the top of the enlarged root near or somewhat above the soil surface. The plant is normally a biennial, producing a rosette of leaves and the bulbous root one year and a seed stalk the following year. Except for seed production, however, it is grown from seed as an annual. Plants are usually harvested for fresh market or processing when the near globular or oblate enlarged root is not more than 2 inches in diameter. At that stage the root is tender, but becomes harder and tougher with greater age. The beet develops best under cool conditions, so may be grown in winter in the far South, or in summer in the North. Baby beets are about 1 inch in diameter and harvested about 40 to 54 days after seeding.
3. Season, seeding to harvest: About 2 to 4 months. Fresh market beets are generally harvested 55-80 days after planting and processing beets 90-110 days.
4. Production in U.S.: In 1993, fresh market 3,200 tons. Canned beets not reported in 1993 (1994 Agricultural Statistics). In 1992 for the U.S., 10,523 acres reported with 82 percent of the acreage in 6 states; Wisconsin (3,251 acres), New York (1,856 acres), Oregon (1,560 acres), Texas (1,209 acres), California (395 acres) and New Jersey (338 acres) (1992 CENSUS). In 1994, California harvested 503 acres (MELNICOE 1996e).
5. Other production regions: In 1995, Canada reported 2,088 acres (ROTHWELL 1996a).

6. Use: Canned, pickled, used in soups and as cooked vegetable and leafy greens.
7. Part(s) of plant consumed: Enlarged root. Leaves sometimes used as green pot vegetable
8. Portion analyzed/sampled: Root and tops (leaves); Analyzed separately
9. Classifications:
   a. Authors Class: Root and tuber vegetables; Leaves of root and tuber vegetables
   b. EPA Crop Group (Group & Subgroup): Root and tuber vegetables (1A and 1B); Leaves of root and tuber vegetables. (Representative Crop)
   c. Codex Group: 016 (VR 05074) Root and tuber vegetables; Beet leaves, see Swiss Chard (VL 0474) Leafy vegetables (including *Brassica* leafy vegetables)
   d. EPA Crop Definition: None
10. References: GRIN, CODEX, FACCIOLA, MAGNESS, RICHE, ROTHWELL 1996a, SCHREIBER, USDA 1994a, US EPA 1995a, US EPA 1995b, US EPA 1994, YAMAGUCHI 1983, MELNICOE 1996e, IVES, REHM, RUBATZKY (picture), SWIADER
11. Production Map: EPA Crop Production Regions 1, 5, 6, 10 and 12.
12. Plant Codes:
   a. Bayer Code: BEVD

066

1. Ben moringa seed
(Ben, Ben moringa seed oil, Coatli, Horseradish tree, Reseda, West Indian ben, Benzolive tree, Drumstick-tree, Moringa, Murung kai, Sohnja)

Moringaceae

Moringa oleifera Lam. (syn. *M. pterygosperma* Gaertn.)

2. This oil is obtained from the seed of the above species, called ben or horseradish tree. The tree is small, up to 25 feet, with deciduous, pinnate leaves. While native to India, the tree is now naturalized in the West Indies. The fruit is an angled pod, 6 to 18 inches in length, which contains about 20 small, angled seeds. The seeds contain 25 to 35 percent of non-drying oil. The oil is used for cooking and in cosmetics.
3. Season, planting to harvest: Leaves after 8 weeks, young fruit during the first year. Immature fruit harvested 50-75 days after flowering. Usually propagated by cuttings and productive within one year of planting.
4. Production in U.S.: No data. Sometimes planted in southern Florida as a backyard crop (STEPHENS 1988).
5. Other production regions: Hot tropics.
6. Use: Roots used as condiment; Flowers, shoots and foliage are edible as greens and the unripe pods known as "susumber" or "drumsticks" are cooked and boiled like beans. The seeds yield the ben oil.
7. Part(s) of plant consumed: Seeds, leaves, immature pods and roots
8. Portion analyzed/sampled: Seed and its oil
9. Classifications:
   a. Authors Class: Oilseed
   b. EPA Crop Group (Group & Subgroup): Miscellaneous
   c. Codex Group: 023 (SO 0690) Oilseed
   d. EPA Crop Definition: None
10. References: GRIN, CODEX, MAGNESS, MARTIN 1984 (picture), STEPHENS 1988

11. Production Map: EPA Crop Production Region 13.
12. Plant Codes:
    a. Bayer Code: No specific entry

067

1. Bentgrass
Poaceae (Gramineae)
Agrostis L. spp.
1. Colonial bentgrass
(Common bentgrass, Browntop, Rhode Island bentgrass, New Zealand bentgrass, Prince Edward Island bentgrass, Yerba fina)
A. capillaris L. (syn: *A. tenuis* Sibthorp).
1. Creeping bentgrass
A. stolonifera var. *palustris* (Huds.) Farw. (syn: *A. palustris* Huds.)
1. Velvet bentgrass
A. canina L.

2. The bentgrasses are cool season, perennial, rather fine-leaved, creeping grasses, used mainly in lawns and golf putting greens and for soil erosion control over the northern half of the U.S. These species were all introduced into this country from Europe and belong to the grass tribe *Aveneae*. Colonial bent is most commonly used in lawns and golf course fairways. Included here are the cultivars 'Astoria' bent and 'Highland' bent. The creeping bents generally do not produce seed and are propagated by stolons, which can grow up to 5 ft/season. 'Seaside' bent and 'Penncross' cultivars do produce seed and are propagated in that way. None of these bentgrasses is used for pasture or hay, except Colonial bentgrass can be used for erosion control, whereas other bentgrasses such as spikebentgrass and redtop bentgrass have agricultural value and are discussed elsewhere. Seeding rates are 40 to 60 lb/A. Seed yields can vary from 420 to 530 lb/A. Bentgrasses depending upon the species vary from 7,800,000 to 10,800,000 seeds/lb.
3. Season: Cool season grass with spring and fall growth.
4. Production in U.S.: There were 6,904,896 lb of bentgrass seed produced in OR, WA and NC in 1992 (U.S. CENSUS, 1992). In 1995, approximately 3,478,000 lb of creeping bentgrass was produced on 6,540 A and 2,715,000 lb of Colonial bentgrass was produced on 6,400 A.
5. Other production regions: Adapted to plant hardiness zones 1-10. Also see above.
6. Use: Turfgrass, golf courses, soil erosion control.
7. Part(s) of plant consumed: Leaves and stems
8. Portion analyzed/sampled: Forage and hay
9. Classifications:
    a. Authors Class: See Forage grass
    b. EPA Crop Group (Group & Subgroup): See Forage grass
    c. Codex Group: See Forage grass
    d. EPA Crop Definition: None
10. References: GRIN, MAGNESS, US EPA 1994, US EPA 1995, CODEX, BARNES, ALDERSON, STUDDEND-IECK, ANON(d), MOSLER (picture), RICHE
11. Production Map: EPA Production Regions 1, 5, 6, 7, 8, 9, 10 and 11.
12. Plant Codes:
    a. Bayer Code: AGSTE (*A. capillaris*), AGSCA (*A. canina*), AGSPL (*A. stolonifera*)

068

1. Bermudagrass
(Pata de gallo, Agarista, couchgrass, Devilgrass, Wiregrass)
(Stargrass and Giant bermudagrass – see field 2 below)
Gramineae
Cynodon dactylon (L.) Pers. var. *dactylon*

2. Bermudagrass is believed native to India but has now spread throughout the tropical, subtropical and mild-temperate regions of the world. It became established throughout the southern half of the U.S. during the 18th century. Bermudagrass is a long-lived perennial which propagates by runners, underground rootstocks and seed. Runners may reach many feet in length and it is difficult to eradicate where not wanted. The flowering branches reach only 6 to 12 inches high. The leaves are up to 4 inches long, flat and somewhat hairy at the base. It thrives during warm weather and gives nutritious pasturage even after frost in the fall. Improved varieties have been developed, adapted to the Southern states. Propagation by planting stolons is used for the improved varieties. Principal use is for pastures and for lawns. 'Giant bermudagrass' is var. *aridus* Harlan & de Wet. 'Stargrass', *C. plectostachyus* (K. Schum.) Pilg. grows in Puerto Rico with 44,635 acres (1992 CENSUS). 'Coastal bermudagrass' is an improved cultivar and 'Midland' is a hybrid that is more winter hardy than coastal.
3. Season: warm season sod forming. Harvest hay at 4-6 week intervals. Cuttings per season are 4 to 5.
4. Production in U.S.: No specific forage data, See Forage grass. In 1992, about 25,000 acres of seed crop with 4,109 tons of seed which is about the same as 1987 production (1992 CENSUS). In 1992, the seed crop was produced in California (18,747 acres harvested) and Arizona (5,779) (1992 CENSUS). Southern U.S.
5. Other production regions: See above.
6. Use: Pasture, hay, pellets, silage, erosion control, turfgrass.
7. Part(s) of plant consumed: Foliage
8. Portion analyzed/sampled: Forage and hay
9. Classifications:
    a. Authors Class: See Forage grass
    b. EPA Crop Group (Group & Subgroup): See Forage grass (Representative Crop)
    c. Codex Group: 051 (AS 5241) Straw, fodder and forage of grasses
    d. EPA Crop Definition: None
10. References: GRIN, MAGNESS, RICHE, BALL (picture)
11. Production Map: EPA Crop Production Regions 2, 3, 4, 5, 6, 8, 9, 10, and 13. Seed crop production primarily in 10 with some in 2 and 6.
12. Plant Codes:
    a. Bayer Code: CYNDA

069

1. Betelnut
(Betel palm, Pan, Nioi-Pekela, Indian-nut, Areca, Areca-nut, Areca-nut palm, Catechu, Pinang)
Arecaceae (Palmae)
Areca catechu L. (syn: *A. cathecu* L.)

2. Tall palm tree with a straight and slender trunk, about 6 inches in diameter and grows to 30 to 100 feet tall. The many leaf divisions are 1 to 2 feet long. The fruit is egg-shaped 1.5

to 2.5 inches long with a seed within a thick fibrous husk surrounded by a fleshy covering. Leaves are 4 to 6 feet long.

3. Season: No data.
4. Production in U.S.: 4,990 pounds harvested in Guam, 1992 CENSUS. The 1987 CENSUS exhibited 14,612 pounds harvested in Guam.
5. Other production regions: Tropical areas mainly Southeast Asia (NEAL 1965).
6. Use: Nuts are used in a preparation of fresh betel pepper leaves (*Piper betle* L.) smeared with lime and wrapped around the nut which is chewed for enjoyment.
7. Part(s) of plant consumed: Nutmeat
8. Portion analyzed/sampled: Nutmeat
9. Classifications:
   a. Authors Class: Tree nuts
   b. EPA Crop Group (Group & Subgroup): Miscellaneous
   c. Codex Group: No specific entry
   d. EPA Crop Definition: None
10. References: GRIN, FACCIOLA, NEAL, RICHE
11. Production Map: EPA Crop Production Region: Guam.
12. Plant Codes:
   a. Bayer Code: ARMCA

070

1. Bilberry
(Whortleberry, Velvet-leaf blueberry, Sourtop, Blaeberry, Dwarf bilberry, Whinberry, Mirtillo)

Ericaceae

Vaccinium myrtillus L.

V. myrtilloides Michx. **(Sourtop blueberry, Velvetleaf blueberry)**

2. The plant is a tiny deciduous shrub which rarely grows over 18 inches tall. Similar to lowbush blueberries except berries are only 4 or 5-celled and bilberry plant being more shade tolerant. Fruit is a glabrous berry, blue to black-colored, sweet and $1/3$ inch diameter.
3. Season, bloom to harvest. No entry.
4. Production in U.S.: Hardy in USDA plant zones 4-9.
5. Other production regions: Favorite fruit in Scotland.
6. Use: Fruit used for preserves and wine. Also used medicinally for cystitis.
7. Part(s) of plant consumed: Fruit
8. Portion analyzed/sampled: Whole fruit
9. Classifications:
   a. Authors Class: Berries
   b. EPA Crop Group (Group & Subgroup): Crop Group 13: Berries Group and this group's Bushberry Subgroup based on blueberry entry being denoted as *Vaccinium* spp.
   c. Codex Group: 004 (FB 0261) Berries and other Small Fruit
   d. EPA Crop Definition: None
10. References: GRIN, CODEX, MAGNESS, REICH, US EPA 1994, WHEALYS
11. Production Map: No entry.
12. Plant Codes:
   a. Bayer Code: VACMY (*V. myrtillus*)

071

1. Biriba
(Corosol, Wild sugarapple, Zambo, Anona, Wild sweetsop)

Annonaceae

Rollinia mucosa (Jacq.) Baill. (syn: *Rollinia pulchrinervia* A. DC.; *R. deliciosa* Saff.)

2. The tree is fast-growing from 13 to 50 feet tall and has brown hairy twigs and deciduous leaves, 4 to 10 inches long. The fruit is ovate, rough and about 3 to 5 inches in diameter. Similar to Sugar apple, which see.
3. Season, bloom to harvest: About 90+ days, July to August in Florida for continuous harvest.
4. Production in U.S.: No data. Grows in Puerto Rico (MORTON, 1987).
5. Other production regions: Tropical South America, Central America, southern Mexico and West Indies (MORTON).
6. Use: Eaten fresh.
7. Part(s) of plant consumed: Fruit flesh
8. Portion analyzed/sampled: Whole fruit
9. Classifications:
   a. Authors Class: Tropical and subtropical fruits – inedible peel
   b. EPA Crop Group (Group & Subgroup): Miscellaneous
   c. Codex Group: No specific entry. General entry is 006 (FI 0030) Assorted tropical and subtropical fruits – inedible peel
   d. EPA Crop Definition: Proposing – Sugar apple = Biriba
10. References: GRIN, CRANE 1995, MORTON (picture)
11. Production Map: EPA Crop Production Region 13.
12. Plant Codes:
   a. Bayer Code: No specific entry

072

1. Black bread weed
(Wild fennel)

Ranunculaceae

Nigella arvensis L.

2. Annual herb to $1\frac{1}{2}$ feet tall. Leaves 1 inch across or more.
3. Season: No data.
4. Production in U.S.: No data.
5. Other production regions: Eurasia (FACCIOLA 1990).
6. Use: Seeds used for flavoring breads and other foods.
7. Part(s) of plant consumed: Seeds
8. Portion analyzed/sampled: Seeds
9. Classifications:
   a. Authors Class: Herbs and spices
   b. EPA Crop Group (Group & Subgroup): Miscellaneous
   c. Codex Group: No specific entry
   d. EPA Crop Definition: None
10. References: GRIN, BAILEY 1976, FACCIOLA
11. Production Map: No entry
12. Plant Codes:
   a. Bayer Code: NIGAR

073

1. Black wattle
(Tan wattle, Black bay)

Fabaceae (Leguminosae)

Acacia mearnsii DeWild.

2. Black wattle is a perennial warm season tropical tree up to 7 feet tall that is native to the tropics and is grown in the Caribbean and Hawaii. This tree is common to areas receiving >40 inches rainfall and at altitudes of ≤5000 ft. It has been used in rotations with vegetable crops and tobacco in Indonesia

and has produced 8.5 to 11.3 T/A/season (wet weight) and it is used as a green manure crop. Black wattle is also the principal source of tanbark in the world, the bark contains 40 percent tannins and is used in leathermaking. It reproduces by seeds and there are 33,000 to 40,000 seeds/lb.

3. Season: Warm season perennial shrubs, leaves are readily grazed and it is used as a green manure crop.
4. Production in U.S.: No specific data for Black wattle production, however, pasture and rangelands are produced on more than 410 million A (U.S. CENSUS, 1992).
5. Other production regions: See above.
6. Use: Grazing, green maure crop, tanbark and a source of tannins.
7. Part(s) of plant consumed: Leaves and stems
8. Portion analyzed/sampled: Forage and hay
9. Classifications:
   a. Authors Class: Nongrass Animal Feeds Group
   b. EPA Crop Group (Group & Subgroup): Miscellaneous
   c. Codex Group: No specific citation although the general class 050 (AL 0157) Legume Animal Feeds could be used
   d. EPA Crop Definition: None
10. References: GRIN, US EPA 1994, US EPA 1995, CODEX, RICHE, HOLZWORTH, SKERMAN(a) (picture)
11. Production Map: EPA Crop Production Region 13.
12. Plant Codes:
   a. Bayer Code: ACAMR

074

1. Blackberry
(Caneberry, Bingleberry, Black satin berry, Boysenberry, Cherokee blackberry, Chesterberry, Cheyene blackberry, Coryberry, Darrowberry, Dewberry, Dirksen thornless berry, Himalayaberry, Hullberry, Lavacaberry, Lowberry, Lucretiaberry, Mammoth blackberry, Marionberry, Nectarberry, Olallieberry, Oregon evergreen berry, Phenomenalberry, Rangeberry, Ravenberry, Rossberry, Shawnee blackberry, Moras, Brombeere, Zarzamora, Tayberry, Mures de ronce)
Rosaceae
Rubus spp.

1. Common blackberry
(Allegheny blackberry)
R. allegheniensis Porter (syn: *R. nigrobaccus* L.H. Bailey)

1. Arctic blackberry
(Crimsonberry, Arctic bramble, Crimson bramble, Nectarberry)
R. arcticus L. ssp. *arcticus*

1. Northern dewberry
(American dewberry)
R. flagellaris Willd.

1. Evergreen blackberry
(Cutleaf blackberry, Parsleyleaved blackberry)
R. laciniatus Willd.

1. Southern dewberry
R. trivialis Michx.

1. California blackberry
(Pacific blackberry, Pacific dewberry)
R. ursinus Cham. & Schltdl. var. *ursinus*

1. Loganberry
(Boysenberry, Veitchberry, Ronce framboise)
R. loganbaccus L.H. Bailey (syn: *R. ursinus* var. *loganobaccus* (L.H. Bailey) L.H. Bailey)

1. Youngberry
(European dewberry, Ronce bleuatre)
R. caesius L.

2. Blackberries are an extremely diverse group of cultivated plants. The group consists of numerous species of *Rubus*. Some varieties are heavily thorned while others are thornless, some varieties grow erect while others are trailing. The trailing types are commonly referred as dewberry (see dewberry for additional information). Blackberry plants are propagated by vegetative systems, i.e. root cuttings, root suckers, tip layering, softwood stem cuttings and tissue culture. The canes grow from the crown one year, fruit the following season, then die. The roots and crown are perennial. Canes are usually supported on a trellis. They may grow to 15 or more feet in length. The aggregate fruit are borne in loose clusters on laterals that grow from the canes. They consist of numerous small seeds each imbedded in a juicy pulp and all adhering to a fleshy base. The base separates from the plant when the fruit is harvested, in contrast to raspberries, in which the base or receptacle is retained on the plant. Fruits are near globose to cylindrical in shape, $^1/_2$ to $^3/_4$ inch in diameter and $^3/_4$ to $1^1/_2$ inches long. Important cultivars of erect blackberry include: 'Cherokee', 'Darrow' and 'Shawnee'. 'Chester' is the most important semi-erect blackberry variety. Also see dewberry.
3. Season: typical bloom to harvest is 60 to 90 days.
4. Production in U.S.: Approx. 8000 acres cultivated in U.S. with Oregon cultivating 4900 acres in 1995.
5. Other production regions: Imported fruit: Guatemala and Costa Rica. Also grown in Europe.
6. Use: Fruit are used fresh, frozen, or processed into desserts, canned, fruit salads, preserves and jams. Some use as wines and brandies.
7. Part(s) of plant consumed: Fruit
8. Portion analyzed/sampled: Whole fruit
9. Classifications:
   a. Authors Class: Berries
   b. EPA Crop Group (Group & Subgroup): Crop Group 13: Berries Group and its Caneberry Subgroup (Representative Crop)
   c. Codex Group: 004 (FB 0278) Berries and other Small Fruit
   d. EPA Crop Definition: Caneberry = *Rubus* spp.; Blackberry = *Rubus* subgenus *Eubatus*
10. References: GRIN, CODEX, LOGAN, MAGNESS, MOORE, SCHREIBER, SHOEMAKER, US EPA 1994, US EPA 1995, WHEALY, IVES, USDA 1996f
11. Production Map: EPA Crop Production Regions 2, 6, 10 and 12 (Primarily 12).
12. Plant Codes:
   a. Bayer Code: RUBAL (*R. allegheniensis*), RUBCA (*R. caesius*), RUBFL (*R. flagellaris*), RUBLA (*R. laciniatus*), RUBLO (*R. loganbaccus*), RUBTV (*R. trivialis*), RUBSS (*R. spp.*)

075

1. Blimbe
(Carambolier bilimbi, Tree sorrel, Mimbro, Grosella china, Cucumber tree, Bilimbe)
1. Bilimbi
Oxalidaceae
Averrhoa bilimbi L.

2. The blimbe tree attains a height from 20 to 60 feet. The mature fruit resembles small cucumbers and range from 2 to 3 inches in length. They are smooth, thin, green, rind sometimes faintly 5 angles. Fruits of blimbe are candied as a preserve. The pulp also is used to make a refreshing drink. It is closely related to starfruit.
3. Season, bloom to harvest: In Florida, tree begins to flower in February, then bloom and fruit more or less continuously until December.
4. Production in U.S.: No data, limited in southern Florida and Hawaii (MORTON 1987).
5. Other production regions: Tropical Asia, Central America, West Indies, South America (MORTON 1987).
6. Use: Fruits used to make chutney, preserves, pickled.
7. Part(s) of plant consumed: Fruit
8. Portion analyzed/sampled: Whole fruit
9. Classifications:
   a. Authors Class: Tropical and subtropical fruits – edible peel
   b. EPA Crop Group (Group & Subgroup): Miscellaneous
   c. Codex Group: 005 (FT 0288) Assorted tropical and subtropical fruits – edible peel
   d. EPA Crop Definition: None
10. References: GRIN, CODEX, MAGNESS, MORTON (picture)
11. Production Map: EPA Crop Production Region 13.
12. Plant Codes:
   a. Bayer Code: No specific entry

076

1. Blowoutgrass
Poaceae (Gramineae)
Redfieldia flexuosa (Thurb.) Vasey
2. This is a warm season perennial native to the U.S. and is a member of the grass tribe *Eragrosteae*. It is an erect to ascending bunchgrass from 0.5 to 1.3 m tall that spreads by long rhizomes, with blades 10 to 75 cm long and has a panicle seedhead (22 to 40 cm long). Blowoutgrass is adapted to sandy soils on plains and rolling sandhills areas. It has fair forage quality for cattle and horses in the summer and provides limited amounts of forage in the fall and winter. It reproduces from rhizomes and seeds and is planted vegetatively by sprigs. There is 579,915 seed/kg.
3. Season: Warm season, spring growth, flowers in July to October.
4. Production in U.S.: There is no production data for Blowoutgrass, however, pasture and rangeland are produced on more than 410 million A (U.S. CENSUS, 1992).
5. Other production regions: A monotypic North American genus (GOULD).
6. Use: Grazing, hay and soil erosion control.
7. Part(s) of plant consumed: Leaves and stems
8. Portion analyzed/sampled: Forage and hay
9. Classifications:
   a. Authors Class: See Forage grass
   b. EPA Crop Group (Group & Subgroup): See Forage grass
   c. Codex Group: See Forage grass
   d. EPA Crop Definition: None
10. References: GRIN, US EPA 1994, US EPA 1995, CODEX, RICHE, GOULD, HOLZWORTH, STUDDENDIECK (picture)
11. Production Map: EPA Crop Production Regions 7, 8 and 9.

12. Plant Codes:
   a. Bayer Code: No specific entry

077

1. Blueberry
(Huckleberry, Airelle, Arandanos americanos)
Ericaceae
Vaccinium spp.
1. Highbush blueberry
(American blueberry, Swamp blueberry)
V. corymbosum L.
1. Rabbiteye blueberry
(Southern black blueberry)
V. virgatum Aiton (syn: *V. ashei* J.M. Reade)
2. There are three species that are of commerical importance, highbush blueberry, lowbush blueberry and rabbiteye blueberry. The lowbush is not commercially planted, but thousands of acres of wild stands are managed in Maine. The highbush and rabbiteye varieties, produced by breeding are widely grown. The production and cultural practices of highbush and rabbiteye are very similar and will be discussed here. See the entry for **Blueberry, Lowbush** for a discussion on *V. angustifolium*. Blueberries are long lived perennials that are native in the U.S. Highbush blueberry require well drained acid soils. Adequate moisture is extremely important. Rabbiteyes are much more tolerant to their soil and environmental conditions. Plants of all are woody shrubs growing to 10 or more feet naturally, but usually held to 5 to 6 feet by pruning. The plants are propagated by hardwood and softwood cuttings. Blueberry plants can produce a light crop during the third year from planting with full production in 6-8 years. Bloom is often very early in spring, thus fruit are susceptible to frost and freeze injury. Fruit is smooth skinned with a waxy coating or bloom. Individual berries are borne in clusters, are round to oblate in shape, with fruit size up to $^3/_4$ inch diameter. Rabbiteye berries are generally larger-seeded and less finely flavored than highbush types. Pruning is required to regulate current year's crop and fruiting potential for the following year. Important highbush varieties include 'Bluecop', 'Blueray', 'Bluetta', 'Duke', 'Elliot', 'Jersey' and 'Wymouth'. 'Tifblue' is the most important rabbiteye variety. The names huckleberry and blueberry are used interchangeably although they are not the same species. Blueberry has a much smaller seed.
3. Season: Bloom to harvest is in 2 to 4 months. Variety dependent, however, fruit of even the same variety does not ripen uniformly. Rabbiteyes tend to have longer bloom to harvest interval than highbush.
4. Production in U.S.: About 59,000 total blueberry acres with major highbush production in New Jersey, Michigan, North Carolina, Oregon and Washington. Rabbiteye production is typically in the Southeast, with major production states being North Carolina, Georgia, Florida and Arkansas.
5. Other production regions: Chile, New Zealand. In 1995, Canada grew 40,649 acres (ROTHWELL 1996a).
6. Use: Fresh, frozen, canned, jam, culinary.
7. Part(s) of plant consumed: Fruit
8. Portion analyzed/sampled: Whole fruit
9. Classifications:
   a. Authors Class: Berries

b. EPA Crop Group (Group & Subgroup): Representative commodity of Crop Group 13: Berries Group. Is the representative commodity of this group's Bushberry Subgroup (13B)

c. Codex Group: 004 (FB 0020) Berries and other Small Fruit. Highbush is FB 4073 and Rabbiteye is FB 4077

d. EPA Crop Definition: None

10. References: GRIN, CODEX, LOGAN, MAGNESS, SCHREIBER, SHOEMAKER, US EPA 1994, US EPA 1995, ROTHWELL 1996a, DAVIS, LYRENE

11. Production Map: EPA Crop Production Regions 1, 2, 5 and 12.

12. Plant Codes:
    a. Bayer Code: No specific entry

078

1. Blueberry, Lowbush

(Wild blueberry, Sweet hurts, Lowsweet blueberry, Late sweet blueberry)

Ericaceae

Vaccinium angustifolium Aiton

2. There are three species that are of commerical importance, *V. corymbosum* or highbush blueberry, *V. angustifolium* the lowbush blueberry and rabbiteye blueberry or *V. virgatum*. The lowbush is not commercially planted, but thousands of acres of wild stands are managed in Maine. The highbush and rabbiteye varieties, produced by breeding are widely grown. The production and cultural practices of lowbush will be discussed here. See the entry for **Blueberry** for a discussion on highbush blueberry and rabbiteye blueberry. Lowbush blueberries are long lived perennials that are native in the U.S. Plants of all are woody shrubs, varying from 1 to 2 feet. The plants are not propagated, natural stands are managed in the wild. Bloom is often very early in spring, thus fruit are susceptible to frost and freeze injury. Fruit is smooth skinned with a waxy coating or bloom. Individual berries are borne in clusters, are round to oblate in shape. Size ranges from $1/4$ to $1/2$ inch in diameter. Fruit is harvested by raking the fields. Burning or mowing of fields is required to regulate crop fruiting.

3. Season: Bloom to harvest: 2 to 4 months.

4. Production in U.S.: Mostly in Maine; almost 33,000 tons valued over 21 million dollars in 1995 (DAVIS 1996). Less than 1 percent of the crop is sold in the fresh market.

5. Other production regions: In 1995, Canada grew 14,295 acres (ROTHWELL 1996a).

6. Use: Lowbush fruit mostly used for processing (frozen, canned, jam, culinary).

7. Part(s) of plant consumed: Fruit

8. Portion analyzed/sampled: Whole fruit

9. Classifications:
   a. Authors Class: Berries
   b. EPA Crop Group (Group & Subgroup): Crop Group 13: Berries Group. Bushberry Subgroup (13B)
   c. Codex Group: 004 (FB 0020) Berries and other Small Fruit for Blueberries. The specific Lowbush blueberry is FB 4075
   d. EPA Crop Definition: None

10. References: GRIN, CODEX, DAVIS, LOGAN, MAGNESS, SHOEMAKER, US EPA 1994, US EPA 1995, ROTHWELL 1996a

11. Production Map: EPA Crop Production Region 1.

12. Plant Codes:
    a. Bayer Code: VACAN

079

1. Bluegrass
(Meadowgrass)
Poaceae (Gramineae)
Poa spp.

1. Texas bluegrass
P. arachnifera Torr.

1. Bulbous bluegrass
(Bulbous meadowgrass)
P. bulbosa L.

1. Canada bluegrass
(Flattened meadowgrass)
P. compressa L.

1. Mutton bluegrass
(Muttongrass)
P. fendleriana (Steud.) Vasey

1. Kentucky bluegrass
(Smooth meadowgrass, Smooth-stalked meadowgrass)
P. pratensis L.

1. Sandberg bluegrass
(Big bluegrass, Nevada bluegrass, Alkali bluegrass)
P. secunda J. Presl (*P. ampla* Merr.)

1. Roughstalk bluegrass
(Rough bluegrass, Rough meadowgrass, Rough-stalk bluegrass)
P. trivialis L.

2. Some 300 species of *Poa* are distributed throughout the cool, temperate regions of the world, with 69 species recognized in the U.S. They are valuable for pasturage, hay and lawns. They are considered the most palatable of range and pasture grasses. The more important agricultural species follow.

Texas bluegrass is a vigorous sod-forming perennial native in the Southeastern and Southern Plains states. Plants grow up to 3 feet on good soil, with numerous leaves 6 to 12 inches long and $1/4$ inch wide. The grass grows throughout the winter producing abundant, nutritious pasture which is highly palatable. This is a valuable species where native, but seeding is difficult. The species is dioecious, with male and female plants. It produces only limited quantities of seed which is covered with woolly hairs that are difficult to remove. Consequently, establishment of stands for agricultural use is limited.

Bulbous bluegrass is a cool-season bunchgrass believed native to Europe. It is now found in nearly all temperate and subtropical regions. In this country it is grown mainly in the Pacific area, especially southern Oregon and northern California. Its distinctive feature is that it forms small, true bulbs at the base and small bulblets in the panicle. It rarely forms true seeds but is propagated by planting the bulblets, which can be handled in harvesting and planting much as though they were seed. The stems reach to 18 inches high. Growth starts in the fall and continues through winter and spring in mild climates. It ceases when the bulblets form about mid-May.

Canada bluegrass, like Kentucky bluegrass, is native to Europe but is extensively naturalized in Northern states. The foliage is a distinctive bluegreen in color. It spreads by underground rhizomes. It is adapted to open, rather poor, dry soils and under these conditions maybe better than Kentucky bluegrass for pastures or lawns. Forage is highly palatable and nutritious. Seed production is good and propagation is by seeding.

Mutton bluegrass is native from the Great Lakes westward to the Cascade Mountains and south into Mexico. It is a perennial bunchgrass with erect stems up to 24 inches tall. It develops tillers at the base and rarely produces short rhizomes. Leaves are mainly basal, are rather firm and stiff. They are folded or inrolled, rarely flat. The species grows under a wide range of conditions including elevations to near the top of the Rocky Mountains. It is also found among sagebrush and in open timber stands. It is well adapted to dry slopes and is found on clay loam as well as sandy or gravelly soils. It is drought resistant, palatable and nutritious and starts growth very early. Even the dry growth is grazed well. These characteristics make it a valuable range grass. The name reflects the value sheepmen place on the grass for sheep feed.

Kentucky bluegrass occurs over much of Europe and Asia. where it is believed native. It probably was brought to this country in early Colonial days. It is now so widely distributed throughout the Northern and Central states that its origin is questionable. The grass is long-lived, with underground rhizomes, resulting in dense sods. It becomes semidormant during hot, dry periods. The seed stems reach up to 24 inches high. Leaves are abundant, long, medium in width and blunt at the terminal. It is a highly palatable pasture grass and is also extensively used for lawns and turf. It is readily established by seeding. Numerous varieties are in commercial use. It is the most extensively used bluegrass and has 2,200,000 seeds per pound.

Sandberg bluegrass is a native bluegrass occurring generally throughout the Northern Plains and Western states. It is a bunchgrass reaching to 24 inches under optimum conditions. Leaves are rather sparse. The grass starts growth early in spring and matures and dries in midsummer. While green and even after drying, the foliage is quite palatable, so it is a valuable range grass. Seed germination is usually low. Sandberg is usually seeded in combination with later growing grasses to obtain maximum pasturage over a long season.

Roughstalk bluegrass generally resembles Kentucky bluegrass but differs in that it does not have rhizomes. It is native to Europe and is more important there as a pasture grass than in this country. It is used in this country for pasture on wet soils and for lawns in shaded areas. It is not adapted to droughty conditions. Although highly palatable, its restricted areas of adaptation limit its general agricultural importance.

3. Season: Cool season grass.
4. Production in U.S.: No specific forage data, See Forage grass. In 1992, 22,544 tons of Kentucky bluegrass seed were produced on 87,610 acres (1992 CENSUS). In 1992, the top 4 Kentucky bluegrass seed crop producing states are Idaho (29,197 acres), Washington (28,338), Oregon (21,163) and Minnesota (8,332) (1992 CENSUS). In 1993, Washington grew 35,000 acres for seed crop use (SCHREIBER 1995). Cool temperate regions including Northern and Central states, Southern Plains, Great Lakes west to the Pacific and south into Mexico.

5. Other production regions: Bluegrasses are native to Europe and widespread in temperate and cold regions of the world (GOULD 1975).
6. Use: Pasture, hay, turf, rangeland, erosion control.
7. Part(s) of plant consumed: Foliage
8. Portion analyzed/sampled: Forage and hay
9. Classifications:
   a. Authors Class: See Forage grass
   b. EPA Crop Group (Group & Subgroup): See Forage grass; Bluegrass is a representative grass
   c. Codex Group: See Forage grass; 051 (AS 5243) Bluegrass
   d. EPA Crop Definition: None
10. References: GRIN, MAGNESS, RICHE, SCHREIBER, MOSLER, BARNES, BALL (picture), GOULD
11. Production Map: EPA Crop Production Regions for seed crop by order of importance 11, 12 and 5. For forage 1, 2, 3, 4, 5, 6, 7, 8, 11 and 12.
12. Plant Codes:
   a. Bayer Code: POABU (*P. bulbosa*), POACO (*P. compressa*), POAPR (*P. pratensis*), POATR (*P. trivialis*), POASS (*P.* spp.)

080

1. Bluestem, Australian
(Forest bluegrass, Lautoka grass, Caucasian bluestem, Australian beardgrass)
Poaceae (Gramineae)
Bothriochloa bladhii (Retz.) S.T. Blake (syn: *B. intermedia* (R.Br.) A. Camus; *A. intermedius* R.Br.; *B. caucasica* (Trin.) C.E. Hubb.)

2. This is a warm season, perennial grass native to Africa, India and Pacific Islands that has been introduced into the U.S. It is adapted to areas receiving 700 to 800 mm rainfall and can tolerate short periods of flooding. In India it is propagated by rooted slips and it roots readily from its stems. It grows erect up to 1 m in height and is used in reseeding rangeland areas in the southern U.S. It produces growth in the summer and can be grazed or cut for hay for beef cattle. When grazed for 2 weeks Australian bluestem should be rested for an additional 6 weeks before it is regrazed.
3. Season: Warm season, growth starts in late June and is completed in fall and can be grazed in the wet season.
4. Production in U.S.: No data for Australian bluestem production, however, pasture and rangeland are produced on more than 410 million A (U.S. CENSUS, 1992). Southern U.S.
5. Other production regions: Native to tropical and subtropical Asia, Australia and islands of the Pacific (GOULD 1975).
6. Use: Rangeland grazing, hay, revegetation of rangelands and for erosion control.
7. Part(s) of plant consumed: Leaves and stems
8. Portion analyzed/sampled: Forage and hay
9. Classifications:
   a. Authors Class: See Forage grass
   b. EPA Crop Group (Group & Subgroup): See Forage grass
   c. Codex Group: See Forage grass
   d. EPA Crop Definition: None
10. References: GRIN, US EPA 1994, US EPA 1995, CODEX, HOLZWORTH, RICHE, GOULD, SKERMAN (picture), MAGNESS, US EPA 1996a, US EPA 1995a, USDA NRCS
11. Production Map: EPA Crop Production Regions 2, 4, 6, 7, 8 and 9.

12. Plant Codes:
   a. Bayer Code: No specific entry

081

1. Bluestem, Big
(Turkeyfoot bluestem, Popotillo gigante)
Poaceae (Gramineae)
Andropogon gerardii Vitman

2. This is a vigorous, rather coarse perennial, warm season, bunchgrass with short rhizomes, belongs to the grass tribe *Andropogoneae* and is native over most of the U.S.; but of major importance in the Central States and the eastern edge of the Great Plains. It is also becoming important in the southeast U.S. The stems, which may reach to 6 feet, are solid between nodes. The leaves reach 12 inches or more in length and are $1/2$ inch or less in width and hairy near the base. Big bluestem is more drought tolerant than most other warm season grasses and has been reported to survive up to 50 years in the Great Plains. Big bluestem was a dominant species of the native tall grassland species making up to 80 percent of the vegetation on certain areas. Growth starts late in the spring and continues throughout the summer, providing good grazing for all kinds of livestock. It is seeded in April and May at a rate 5 to 10 lb/A. Good quality hay is produced if mowed before seed heads have formed. Roots penetrate deeply; but the grass thrives on all types of soils except for sands but is best on moist, well drained soils of good quality and can be used to control erosion. Propagation is by seeds and seed yields range from 150 to 200 lb/A. There are 165,000 seeds/lb.
3. Season: Warm season forage and grows rapidly from mid-spring through the early fall. Harvest for hay before seedheads are formed. Seeds mature in late September and October.
4. Production in U.S.: No data for Big bluestem production, however, pasture and rangeland are produced on more than 410 million A (U.S. CENSUS, 1992). Adapted to plant hardiness zones 3 through 5.
5. Other production regions: Grown in prairie regions of Canada.
6. Use: Pasture and rangeland, hay and for erosion control.
7. Part(s) of plant consumed: Leaves and stems
8. Portion analyzed/sampled: Forage and hay
9. Classifications:
   a. Authors Class: See Forage grass
   b. EPA Crop Group (Group & Subgroup): See Forage grass
   c. Codex Group: See Forage grass
   d. EPA Crop Definition: None
10. References: GRIN, MAGNESS, US EPA 1994, US EPA 1995, CODEX, BARNES, ALDERSON, STUBBENDIECK (picture), BALL, RICHE
11. Production Map: EPA Production Regions 1, 2, 4, 5, 7, 8 and 11.
12. Plant Codes:
   a. Bayer Code: ANOGE

082

1. Bluestem, Caucasian
(Old World bluestem)
Poaceae (Gramineae)
Bothriochloa bladhii (Retz.) S.T. Blake (syn: *B. caucasica*

(Trin.) C.E. Hubb.; *Andropogon caucasicus* Trin.)
2. This perennial bunchgrass bluestem from Russia shows promise as a pasture and hay grass for the central and southern Great Plains. It grows in areas with 500 to 700 mm of rainfall. The plant is leafy with fine stems and reseeds readily. It is palatable while succulent but ripens early so is less useful for late pasture than the native bluestems. It is especially useful for erosion control and is also used for pasture and hay. It thrives best on medium to fine textured soils but will also grow on sandy soil. There are 4.9 million seeds/lb. The authors have retained this monograph as a separate entry from Australian bluestem, which see, since some differences exist.
3. Season: Seeds planted May to June. It is foraged late May to August. Seeded at a rate of 1.8 to 2.7 lb/acre and grows 2 to 4 feet tall.
4. Production in U.S.: No specific data. See Forage grass. Central and southern Great Plains.
5. Other production regions: Russia.
6. Use: Pasture, hay and soil stabilization.
7. Part(s) of plant consumed: Foliage
8. Portion analyzed/sampled: Forage and hay
9. Classifications:
   a. Authors Class: See Forage grass
   b. EPA Crop Group (Group & Subgroup): See Forage grass
   c. Codex Group: See Forage grass
   d. EPA Crop Definition: None
10. References: GRIN, CODEX, MAGNESS, US EPA 1996a, US EPA 1995a, USDA NRCS, ANDERSON, BALL (picture)
11. Production Map: EPA Crop Production Regions 4, 5 and 6.
12. Plant Codes:
   a. Bayer Code: No specific entry

083

1. Bluestem, Diaz
(Kleberg bluestem, Hindigrass, Shedagrass, Marvalgrass)
Poacea (Gramineae)
Dichanthium annulatum (Forssk.) Stapf var. *annulatum*

2. This is a warm season bunchgrass from South Africa, useful for pasture in South Texas. Stems are slender, erect to semi-decumbent, up to 5 feet in height. In one variety stems are leafy and have stiff leaves at the nodes. Plants reseed aggressively. The grass is palatable both as pasture and hay. Plants are drought tolerant, somewhat alkali tolerant and well adapted to range seeding on heavy soils.
3. Season: Warm season pasture. Palatable even after flowering.
4. Production in U.S.: No specific data, See Forage grass. Hay yields are 10 tons/hectare. South Texas.
5. Other production regions: South Africa.
6. Use: Pasture, rangegrass, hay.
7. Part(s) of plant consumed: Foliage
8. Portion analyzed/sampled: Forage and hay
9. Classifications:
   a. Authors Class: See Forage grass
   b. EPA Crop Group (Group & Subgroup): See Forage grass
   c. Codex Group: See Forage grass
   d. EPA Crop Definition: None
10. References: GRIN, MAGNESS, SKERMAN
11. Production Map: EPA Crop Production Regions 6 and 8.

12. Plant Codes:
    a. Bayer Code: DIHAN

084

1. Bluestem, Yellow
(Turkestan bluestem, Angleton bluestem, Yellow bluestem, King Ranch bluestem, Plain's bluestem)
Poaceae (Gramineae)
Bothriochloa ischaemum (L.) Keng. var. *ischaemum*
(Turkestan bluestem, Plain's bluestem)
Dichanthium aristatum (Poir.) C. E. Hubb.
(Angleton bluestem)
Bothriochloa ischaemum var. *songarica* (Rupr. & Fisch. & C.A. Mey.) Celarier & Harlan
(King Ranch bluestem, Texas yellow bluestem)
2. These bluestems are both native to central or southern Asia. *B. ischaemum* is used for pasture through the southern Great Plains. *D. aristatum* cultivars are known as 'Angleton' and 'Gordo'. They are used as pasture and hay in the humid Gulf Coastal Plain in Texas. Both of these species are warm season, semiprostrate perennial bunchgrasses, leafy or medium leafy. In palatability *D. aristatum* appears superior to *B. ischaemum* but both furnish good pasturage and good hay if cut early.
3. Season: Warm season pasture. Flowers June to October.
4. Production in U.S.: No specific data, See Forage grass. southern Great Plains and Gulf coast of Texas.
5. Other production regions: Native to central or southern Asia.
6. Use: Pasture, hay and range reseeding.
7. Part(s) of plant consumed: Foliage
8. Portion analyzed/sampled: Forage and hay
9. Classifications:
    a. Authors Class: See Forage grass
    b. EPA Crop Group (Group & Subgroup): See Forage grass
    c. Codex Group: See Forage grass
    d. EPA Crop Definition: None
10. References: GRIN, MAGNESS, SKERMAN, ANDERSON
11. Production Map: EPA Crop Production Regions 5, 6, 7, 8, 9 and 13.
12. Plant Codes:
    a. Bayer Code: DIHIS (*B. ischaemum*), DIHAR (*D. aristatum*)

085

1. Bok choy
1. Pak choi
1. Mustard cabbage
(Chongee, Shanghai, Baak choi, Pai tsai white stalk, Taisai, Lei choy, Bok choi, Paksoi, Bok choy Chinese cabbage, Paktsoi, White mustard cabbage, Spoon cabbage, Japanese white celery mustard, Pak choy sum, Chinese chard, Pak choy, Pak toy, Celery mustard, Chinese mustard cabbage, Joi choy, Tak tsai, Chinese savoy, Celery cabbage, Tatsoi, Osaka-na, Chinese cabbage bok choy)
Brassicaceae (Cruciferae)
Brassica rapa spp. *chinensis* (L.) Hanelt (syn: *Brassica chinensis* L., *B. campestris* L. (Chinese group), *B. sinensis* L.)
1. Choy sum
1. Choisum

(Flowering bok choy, Yu choy, U-choy, Tsai shim, Tsoi sum, False pak choi, Chinese cabbage choy sum, Mock pakchoi, Choisum, Yow choy, Edible rape)
Brassica rapa var. *parachinensis* (L.H. Bailey) Hanelt (syn: *B. parachinensis* L.H. Bailey; *B. chinensis* var. *parachinensis*)
2. Bok choy is quite closely related to Chinese Cabbage, which see. Bok choy resembles miniature Swiss chard. As compared to Chinese Cabbage, the head is shorter and more loose or non-heading. The mustard green, kai choy, which see, has broader petioles as compared to Bok choy. Leaves are cupped or spoon-shaped. In culture and use Bok choy is similar to Chinese Cabbage, but exposure of edible parts to pesticides is somewhat greater. Hawaii produced 655 tons on 80 acres in 1968. White stem Choy sum is the bok choy heart which is available year-round from California (LOGAN 1996). The Shanghai variety, commonly called baby bok choy, has green midribs and leaf bases. Also young bok choy at thinning is used as baby bok choy. This Oriental vegetable is very similar in growth habit and culture to Chinese cabbage, which see. However, it is grown for the flowering stalk. The entire plant with the flowering stalk may be harvested and marketed. In exposure of edible parts during growth the plant is comparable to sprouting broccoli. "Sum" in Chinese means flower stalk.
3. Season, seeding to harvest: 40 to 80 days.
4. Production in U.S.: In 1992, 132 acres (1992 CENSUS) as mustard cabbage. In 1992, Hawaii with 79 acres and New Jersey reported some (1992 CENSUS). Also grown in Washington, Regions 11 and 12 (SCHREIBER 1995). Baby bok choy available from California in fall and winter (LOGAN 1996).
5. Other production regions: In 1995, Canada grew some bok choy (ROTHWELL 1996a).
6. Use: Dark green leaves and white ribs can be eaten raw, but cooking enhances its flavor. Preparation: Trim off the heavy base (can be used with stalks if not too tough), separate stalks from the base, like celery, slice leaves from stalks and cook both leaves and stalks (SCHNEIDER 1985).
7. Part(s) of plant consumed: Leaves and petioles cut at the ground level
8. Portion analyzed/sampled: Leaves and petioles
9. Classifications:
    a. Authors Class: *Brassica* (cole) leafy vegetables
    b. EPA Crop Group (Group & Subgroup): *Brassica* (cole) leafy vegetables (5B)
    c. Codex Group: Pakchoi: 013 (VL 0466) Leafy vegetables (including *Brassica* leafy vegetables; Choisum: 013 (VL 0468) Leafy vegetables (including *Brassica* leafy vegetables); 013 (VL 0054) *Brassica* leafy vegetables
    d. EPA Crop Definition: None
10. References: GRIN, CODEX, LOGAN (picture), MAGNESS, MYERS (picture), RICHE, ROTHWELL 1996a, SCHNEIDER 1985, SCHREIBER, US EPA 1995a, US EPA 1994, SCHNEIDER 1994
11. Production Map: EPA Crop Production Regions: Bok choy: 2, 3, 10 and 13.
12. Plant Codes:
    a. Bayer Code: BRSCH (*B. rapa* ssp. *chinensis*), BRSPA (*B. rapa* var. *parachinensis*)

086

1. Borage
(Beebread, Beeplant, Talewort, Borraja, Common borage)
Boraginaceae

Borago officinalis L.

2. Borage is a rather coarse annual plant grown for culinary use in Europe, but little grown in the U.S. Plant is hairy, $1^1/_2$ to 2 feet high. Leaves are oval to oblong, 4 to 5 inches long. Only the young leaves are palatable. In the U.S. borage also is grown as an ornamental. In addition to its occasional use as a pot-herb, borage flowers are used for flavoring, mainly in drinks, as the English drink "cool tankard" and in lemonade and other fruit juice drinks and teas. The flowers are blue or purple, borne in large racemes. Also borage oil from its seed is used as a nutritional supplement and for cosmetics.

3. Season, seeding to harvest: About 2 months. Harvest as it begins to flower up to 2 or 3 times during growing season or a couple of plantings. Borage blooms from midsummer. For seed crop about 4 months.

4. Production in U.S.: No data. About 1,000 acres for seed in North Dakota.

5. Other production regions: Native to Mediterranean region (FACCIOLA 1990)

6. Use: Young leaves as potherb or occasionally as salad. Flowers as flavoring or garnish. Seeds for oil.

7. Part(s) of plant consumed: Young leaves, stems and flowers; Seeds

8. Portion analyzed/sampled: Tops (Fresh and dried); Seeds and oil (meal if used for food or feed)

9. Classifications:
   a. Authors Class: Herbs and spices; Oilseed
   b. EPA Crop Group (Group & Subgroup): Herbs and spices group (19A)
   c. Codex Group: 027 (HH 0724) Herbs; 057 (DH 0724) Dried herbs
   d. EPA Crop Definition: None

10. References: GRIN, CODEX, FOSTER (picture), GARLAND (picture), LEWIS, MAGNESS, SEPTHENS 1988 (picture), US EPA 1995a, USDA NRCS, FACCIOLA, JANICK 1996

11. Production Map: EPA Crop Production Region 5 or 7 for seed.

12. Plant Codes:
    a. Bayer Code: BOROF

087

1. Brazil nut
(Brazilnut, Butternut, Creamnut, Paranut)
Lecythidaceae (Myrtaceae)

Bertholletia excelsa Humb. & Bonpl.

2. The tree is very large, up to 150 feet. It is a tropical evergreen, native to northern Brazil, with large leathery leaves, 2 feet long and 6 inches wide. The tree forms forests along the Amazon and Rio Nigro Rivers. Large quantities of nuts are gathered from such trees, but they are little cultivated. The fruits are round, about 6 inches in diameter, with a hard shell near $^1/_2$ inch thick, which contains 18 to 24 of the 3-sided angular nuts. The shell of the individual nut is woody, rather thin and completely filled with the white, creamy kernel.

Brazil nuts are not produced commercially in the U.S., but large quantities are imported.

3. Season, bloom to harvest: About one year when pod (nut) matures and drops to the ground, starting in late November and continue into early June.

4. Production in U.S.: No Data.

5. Other production regions: Tropical regions.

6. Use: Direct eating.

7. Part(s) of plant consumed: Nutmeat

8. Portion analyzed/sampled: Nutmeat

9. Classifications:
   a. Authors Class: Tree nuts
   b. EPA Crop Group (Group & Subgroup): Tree nuts
   c. Codex Group: 022 (TN 0662) Tree nuts
   d. EPA Crop Definition: None

10. References: GRIN, CODEX, MAGNESS, US EPA 1995a, WOODROOF(a)

11. Production Map: EPA Crop Production Region 13.

12. Plant Codes:
    a. Bayer Code: BTHEX

088

1. Breadfruit
(Panapen, Breadnut, Fruta de pan, Pana de pepita, Ulu)
Moraceae

Artocarpus altilis (Parkinson) Fosberg (syn: *A. communis* Forster, *A. incisus* L.f.)

2. These are large (90 feet), tropical evergreen trees, important food sources in the tropics. The tree likes hot, wet tropical lowlands and is tolerant to a variety of soils if they are well drained. Plants are propagated by cuttings and layering. Fruits are generally oblong and large, often 10 pounds or more. Skin is thin but roughened by projections of carpels. Ripe breadfruit has a brown-speckled peel. The raw pulp is hard and starchy. When cooked, the pulp is slightly musky, but is extremely bland. **Breadnut** (Pana de pepita) closely resembles the seedless breadfruit but has numerous 1 inch or more seeds.

3. Season: Bloom to mature fruit is about 90 days. Generally two crops of fruit mature each year. Fruit in all stages of development being present on the tree the year round. In Hawaii, fruit is most abundant July to February.

4. Production in U.S.: Scattered dooryard trees in Hawaii and South Florida. Guam and the US Virgin Islands harvested 3,720 and 5,535 lbs in 1992, respectively.

5. Other production regions: Caribbean and tropical Pacific areas (MORTON 1987).

6. Use: Breadfruit is usually consumed only after cooking (bake, boil, roast, or fry). Seeds of breadnut are boiled or roasted with flavor resembling chestnuts. Dried leaves used for tea.

7. Part(s) of plant consumed: Primarily the edible portion is the interior starchy pulp of the fruit

8. Portion analyzed/sampled: Whole fruit

9. Classifications:
   a. Authors Class: Tropical and subtropical fruits – inedible peel
   b. EPA Crop Group (Group & Subgroup): Miscellaneous
   c. Codex Group: 006 (FI 0329) Assorted tropical and subtropical fruits – inedible peel
   d. EPA Crop Definition: None

10. References: GRIN, CODEX, LOGAN, MAGNESS, MARTIN 1987, RICHE, US EPA 1994, NEAL, MORTON (picture)
11. Production Map: EPA Crop Production Region 13.
12. Plant Codes:
    a. Bayer Code: ABFAL

089

1. Broadleaf carpetgrass
(Savannah grass, Nudillo, Bes-chaitgras, Tropical carpetgrass, Carpetgrass-broadleaf)
Poaceae (Gramineae)
Axonopus compressus (Sw.) P. Beauv.

2. Carpetgrass is native to Southern U.S., Mexico and Brazil and has been introduced into tropical and subtropical areas. Broadleaf carpetgrass is similar to carpetgrass, but is more stoloniferous and vigorous in growth. It is a perennial, warm season creeping grass spreading quickly by stolons, which makes a dense coarse sod. Broadleaf carpetgrass is adapted to moist sandy soils and rainfall ≥775 mm/year. Its greatest value is for permanent pastures in Florida and Louisana. It can be grazed closely in the summer without being damaged. It is also useful for lawns and is widely used to prevent erosion and stabilize slopes. Its nutrient value is low but cattle readily consumed it. It forms seeds abundantly and is readily established by seeding at 6 kg/ha. Broadleaf carpetgrass has 2,970,000 seeds/kg.
3. Season: Warm season forage and grows in the summer to fall. It can be grazed May through August and withstands heavy defoliation. It is not used for hay since quality at maturity is poor, but it is palatable until heading stage.
4. Production in U.S.: No data for Broadleaf carpetgrass, however, pasture and rangeland are produced on more than 410 million A (U.S. CENSUS, 1992).
5. Other production regions: See above.
6. Use: Grazing for pasture, turfgrass, slope stabilization and for erosion control.
7. Part(s) of plant consumed: Leaves and stems
8. Portion analyzed/sampled: Forage and hay
9. Classifications:
    a. Authors Class: See Forage grass
    b. EPA Crop Group (Group & Subgroup): See Forage grass
    c. Codex Group: See Forage grass
    d. EPA Crop Definition: None
10. References: GRIN, US EPA 1994, US EPA 1995, CODEX, SKERMAN (PICTURE), RICHE
11. Production Map: EPA Crop Production Regions 3, 4 and 8.
12. Plant Codes:
    a. Bayer Code: AXOCO

090

1. Broccoli
(Sprouting broccoli, Cavolo broccoli, Italian broccoli, Asparagus broccoli, Calabrese, Purple cauliflower, Cape broccoli, Heading broccoli, Brecoles, Winter broccoli, Brocoli, Broccolini)
Brassicaceae (Cruciferae)
Brassica oleracea var. *italica* Plenck

2. The broccoli plant is a strong growing, upright annual, up to 3 feet in height, with large spreading leaves. The edible portion of commerce is the flower heads, cut when the buds are relatively small and green. The flowers do not condense into a solid head, like heading broccoli or cauliflower, which see. The central head may be 3 to 6 inches across and rather flat and is ready for harvest first. It is cut along with 6 to 8 inches of the thick, tender stem. Later lateral stems and flower heads grow from leaf axils. These heads are generally 1 to 3 inches across and are also cut with stems. Plants may continue to produce from laterals for several weeks. Broccolini is a cross between broccoli and Chinese broccoli.
3. Season, planting to first cutting: 60 to over 100 days, depending on variety and growing area.
4. Production in U.S.: As dual usage (fresh and processing) crop, broccoli in 1995 planted on 104,100 acres with fresh market production at 544,400 tons and processing at 64,590 tons (USDA 1996b). In 1995, broccoli reported in the top 4 states as California (89,500 acres), Arizona (8,600), Texas (3,300) and Oregon (2,700). California ranked first for both fresh market and processing production. For fresh market, California and Arizona produced about 99 percent of the production and California produced about 74 percent of the processing crop (USDA 1996b). Fresh market and processing winter season production comes from California (USDA 1996b).
5. Other production regions: In 1995, Canada reported 8,780 acres of broccoli (ROTHWELL 1996a).
6. Use: Fresh market, frozen. Generally cooked, but limited use as raw salad ingredient. Preparation: For cooking broccoli, wash and trim the main stem lightly. Also, can be used fresh in salads (LOGAN 1996). Per capita consumption in U.S. in 1994 is 5.7 pounds (PUTNAM 1996).
7. Part(s) of plant consumed: Green flower head and adjoining tender stem.
8. Portion analyzed/sampled: Flower head and stem (stalk)
9. Classifications:
    a. Authors Class: *Brassica* (cole) leafy vegetables
    b. EPA Crop Group (Group & Subgroup): *Brassica* (cole) leafy vegetables (5A) (Representative Crop)
    c. Codex Group: See Cauliflower
    d. EPA Crop Definition: Broccoli = Chinese broccoli
10. References: GRIN, BAILEY 1976, CODEX, FACCIOLA, LOGAN (picture), MAGNESS, ROTHWELL 1996a, STEPHENS 1988 (picture), USDA 1996b, US EPA 1995a, US EPA 1994, WARE, SWIADER, PUTNAM 1996
11. Production Map: EPA Crop Production Regions 6, 10 and 12.
12. Plant Codes:
    a. Bayer Code: BSROK

091

1. Broccoli, Chinese
(White flowering Chinese broccoli, Gay Lon, Gai Lan, Kailaan, Gai Iohn, Chinese kale, Gai Ion, Kailan, Yellow flowering Chinese broccoli, Kaai-laan, White flowered broccoli, Tsai shim)
Brassicaceae (Cruciferae)
Brassica oleracea var. *alboglabra* (L.H. Bailey) Musil (syn: *Brassica alboglabra* L.H. Bailey)

2. The plant and edible portions are similar to sprouting broccoli, which see, except for the lighter color of the edible flower stalk and head. Like sprouting broccoli, these are harvested at the bud stage and used as potherbs or in salads. At harvest time, cut 8 inch stalks including a few leaves just before the flowers open. Each plant can be harvested several times.
3. Season, seeding to first harvest: 40 to 60 days.

4. Production in U.S.: No data. Limited. Grows in California, Florida and Washington (MYERS, STEPHENS 1988, SCHREIBER). New Jersey.
5. Other production regions: In 1995, Canada grew some Chinese broccoli (ROTHWELL 1996a).
6. Use: Used stir-fried and salads.
7. Part(s) of plant consumed: Immature flowers and stalk
8. Portion analyzed/sampled: Flowerhead and stalk
9. Classifications:
   a. Authors Class: *Brassica* (cole) leafy vegetables
   b. EPA Crop Group (Group & Subgroup): *Brassica* (cole) leafy vegetables (5B)
   c. Codex Group: 010 (VB 0401) *Brassica* (cole or cabbage) vegetables, Head cabbages, Flowerhead *Brassicas*; 010 (VB 0042) Flowerhead *Brassicas*
   d. EPA Crop Definition: Broccoli = Chinese broccoli (gai lon, white flowering broccoli)
10. References: GRIN, CODEX, LOGAN, MAGNESS, MYERS (picture), ROTHWELL 1996a, SCHREIBER, STEPHENS 1988 (picture), US EPA 1995a, RUBATZKY
11. Production Map: EPA Crop Production Regions 2, 3, 10 and 12.
12. Plant Codes:
    a. Bayer Code: BRSAG

092

1. Broccoli raab
(Rapa, Rapini, Taitcat, Italian turnip, Broccoli turnip, Chinese flowering cabbage, Choy sum, Spring broccoli, Italian mustard, Turnip rape, Rappone, Italian turnip broccoli, Nabana, Ruvo kale, Saishin, Tsai-hsin, Tsai-tai, Cima-de-rapa)
Brassicaceae (Cruciferae)
Brassica ruvo L.H. Bailey (syn: *B. campestris* L. (Ruvo group); *B. rapa* L. (Ruvo group))
2. This plant is grown for its tender leaves and flowers shoots which are used as greens or potherbs. Plants develop rather rapidly and are harvested before the flower buds open. General growth habit and exposure of edible parts are similar to spinach and Chinese broccoli. The dark green chard-like leaves that grow on a stalk are grown for fall or spring harvests.
3. Season, seeding to harvest: About 60 days.
4. Production in U.S.: No data, very limited. Grown in New Jersey (STEPHENS 1988). Available from California, August through March (LOGAN 1996).
5. Other production regions: Grown in temperate Europe and Italy (RUBATZKY 1997).
6. Use: As potherb. Can be cooked like broccoli but less cooking time required (LOGAN 1996).
7. Part(s) of plant consumed: Entire top, leaves, stems and flower buds
8. Portion analyzed/sampled: Tops
9. Classifications:
   a. Authors Class: *Brassica* (cole) leafy vegetables
   b. EPA Crop Group (Group & Subgroup): *Brassica* (cole) leafy vegetables (5B)
   c. Codex Group: See Turnip – similar to turnip greens
   d. EPA Crop Definition: See Turnip
10. References: GRIN, BAILEY 1976, CODEX, LOGAN, MAGNESS, STEPHENS 1988 (picture), US EPA 1995a, RUBATZKY

11. Production Map: EPA Crop Production Regions 2 and 10.
12. Plant Codes:
    a. Bayer Code: No specific entry

093

1. Bromegrass
(Brome)
Poaceae (Gramineae)
Bromus spp.
1. Field brome
B. arvensis L.
1. California brome
(Sweet brome, Bromo)
B. carinatus Hook. & Arn.
1. Rescuegrass
(Prairie grass)
B. catharticus Vahl. (syn: *B. willenowii* Kunth)
1. Smooth brome
(Awnless brome, Russian brome, Hungarian brome)
B. inermis Leyss. ssp. *inermis*
1. Mountain brome
B. marginatus Nees ex. Steud.
2. Some 42 species of *Bromus* are native cool season grasses in the U.S. Some of these are important forage sources, others are troublesome weeds. Most bromes are highly palatable during succulent growth, even the ones classed as weeds. The leaf blades are flat and the seed heads are open, spreading panicles. Species of major value in agriculture are described as follows.
   Field bromegrass is a winter annual grass introduced from Europe in the late 1920's and now grown in the Cornbelt and eastward. When seeded in late summer it develops an extensive fibrous root system, making it excellent for erosion control. It is winter hardy and grows rapidly the following spring. The seeds ripen in midsummer and the plants die. Although most used for erosion control and soil improvement, it furnishes palatable pasturage during the spring.
   California brome is a bunchgrass native in the Rocky Mountain and Pacific Coast regions. The plant grows up to 4 feet. Leaf blades are up to 8 inches long, about $1/2$ inch wide. It produces large quantities of leafy forage, relished by all kinds of livestock while immature. The mature foliage is less palatable, but the seed heads are palatable and nutritious. The plant is relatively short-lived.
   Rescuegrass is native to Argentina, but was introduced into the Southern states around 1850. It is now naturalized there in many areas. It is a short-lived perennial, adapted to humid areas with mild winters. Plants reach up to 3 feet, with leaves up to a foot long and about $1/4$ inch wide. Young plants are pubescent, but mature plants are sparingly so. Growth occurs throughout the winter, the plants becoming mature in early summer. On strong soils, a good amount of forage palatable to livestock is produced. Seed is produced in quantity and stands are readily obtained by seeding.
   Smooth brome is native to northern Europe and Asia. It was introduced into the U.S. in 1884 and is now widely distributed. It is best adapted to regions of moderate

rainfall and moderate to cool summer temperatures. The plant is a long-lived perennial with creeping rhizomes. It forms a dense sod. Leaf blades are smooth, up to 12 inches long and $1/2$ inch wide. The flower head is a panicle, 6 to 8 inches long. The root system is strong and interlaced, making the plant excellent for erosion control. Smooth brome is among the best of the pasture and hay grasses, being both highly palatable and nutritious. Two types of smooth brome are recognized. "Southern" type came originally from Europe and is best suited to the corn belt and adjacent plains areas. "Northern" type, originally from Siberia, is adapted to the Northern Plains and adjacent Canada areas.

Mountain brome is a short-lived perennial bunchgrass native to the Rocky Mountain and Pacific Coast regions. Plants grow to 4 feet, with leaves up to 12 inches long and about $1/4$ inch wide. Leaf blades are flat and hairy underneath. Growth starts early in the spring, producing much leafy forage relished by livestock. Because of rapid seedling growth and a well-branched, deep root system, mountain brome is excellent where a rapid cover development is needed. This grass is frequently seeded with alfalfa or sweet clover in the Pacific Northwest. The mixture is ideal both for prevention of erosion and as a well balanced animal diet. Stands of mountain brome grass are readily established by seeding.

3. Season: Seeded in late summer. Harvest in full bloom stage.
4. Production in U.S.: No specific data for forage use. See Forage grass. In 1992, 10,518 acres of seed crop with 1,227 tons of seeds produced (1992 CENSUS). In 1992, the top 3 states in seed crop production were Kansas (6,638 acres), Washington (1,041) and South Dakota (900) (1992 CENSUS). In Washington, bromegrass can produce for 10 years. It is winter hardy and adapted to a range of moisture conditions. In eastern Washington, meadow and mountain brome are common. In western Washington, mature prairie grass is most common (SCHREIBER 1995). Rocky Mountains and Pacific Coast, Southern states, corn belt and east and Northern plains (including Canada).
5. Other production regions: In 1995, Canada reported 23,221 acres of bromegrass and 1,764 acres of meadow brome (ROTHWELL 1996a).
6. Use: Erosion control, pasture, hay, rangeland, silage.
7. Part(s) of plant consumed: Foliage
8. Portion analyzed/sampled: Forage and hay
9. Classifications:
   a. Authors Class: See Forage grass
   b. EPA Crop Group (Group & Subgroup): See Forage grass; Bromegrass or fescue is a representative crop
   c. Codex Group: See Forage grass; 051 (AS 5245) Bromegrass
   d. EPA Crop Definition: None
10. References: GRIN, MAGNESS, RICHE, ROTHWELL 1996a, SCHREIBER, BALL, STUBBENDIECK, ALDERSON, MOSLER (picture), BARNES
11. Production Map: Seed crop production in EPA Crop Production Regions 5, 11 and 12. Forage production in 1, 2, 4, 5, 6, 7, 8, 9, 11 and 12.
12. Plant Codes:

a. Bayer Code: BROAV (*B. arvensis*), BROCN (*B. carinatus*), BROCA (*B. catharticus*), BROIN (*B. inermis*), BROMG (*B. marginatus*)

094

1. Bromegrass, Minor annual

Poaceae (Gramineae)

Bromus spp.

1. Cheatgrass

(Chess, Rye brome, Cheat)

B. secalinus L.

1. Downy brome

(Broncograss, Drooping brome, Downy chess, Cheatgrass, Junegrass)

B. tectorum L.

1. Japanese chess

(Japanese brome)

B. japonicus Thunb.

1. Hairy chess

(Meadowbrome, Meadow brome)

B. commutatus Schrad.

2. These are all winter annual grasses that are widely distributed in pasture and rangelands and may be troublesome weeds in grain fields. They are prolific seed producers and may become dominant in overgrazed perennial pastures. For a short period in spring they furnish good pasturage before flowering. After flowering, they become poor quality. Downy brome (*B. tectorum*) has 208,000 seeds per pound and Cheatgrass (*B. secalinus*) has 71,000 seeds per pound.
3. Season: Cool season crop. Seeds germinate late fall or early spring.
4. Production in U.S.: No specific data, See Forage grass. Adapted to cool season U.S.
5. Other production regions: Adapted to cool western Canada.
6. Use: Pasture, rangeland, hay, erosion control.
7. Part(s) of plant consumed: Foliage
8. Portion analyzed/sampled: Forage and hay
9. Classifications:
   a. Authors Class: See Forage grass
   b. EPA Crop Group (Group & Subgroup): See Forage grass
   c. Codex Group: See Forage grass; 051 (AS 5245) Bromegrass
   d. EPA Crop Definition: None
10. References: GRIN, MAGNESS, SCHREIBER, ALDERSON, STUBBENDIECK
11. Production Map: EPA Crop Production Regions 9, 11 and 12.
12. Plant Codes:
   a. Bayer Code: BROCO (*B. commutatus*), BROJA (*B. japonicus*), BROSE (*B. secalinus*), BROTE (*B. tectorum*)

095

1. Broomsedge

(Broomsedge bluestem, Popotillo, Whiskey grass, Bluestem broomsedge)

Poaceae (Gramineae)

Andropogon virginicus L.

2. Broomsedge is a perennial, warm season, native bunchgrass, widely distributed over the U.S., closely related to big and little bluestem grasses and also a member of the grass tribe

Andropogoneae. It grows to a height of 1.4 m. The palatability of broomsedge is poor, so it ranks low as livestock feed. However, it thrives on soils of low fertility and is often the dominant growth on worn out, unproductive soils, affording protection from erosion and some forage. It has poor forage quality except early growth in the spring. Because of its low palatability it is not used in seeded pastures.

3. Season: Warm season poor quality forage and grows from mid-spring through the early fall. Seeds mature in September to late October.
4. Production in U.S.: No data for Broomsedge production, however, pasture and rangeland are produced on more than 410 million A (U.S. CENSUS, 1992).
5. Other production regions: Bluestem grasses in warm temperature and tropical regions of the world (BAILEY 1976).
6. Use: Worn out pastureland and for erosion control.
7. Part(s) of plant consumed: Leaves and stems
8. Portion analyzed/sampled: Forage and hay
9. Classifications:
   a. Authors Class: See Forage grass
   b. EPA Crop Group (Group & Subgroup): See Forage grass
   c. Codex Group: See Forage grass
   d. EPA Crop Definition: None
10. References: GRIN, MAGNESS, US EPA 1994, US EPA 1995, CODEX, BARNES, STUBBENDIECK (picture), RICHE, BAILEY 1976
11. Production Map: EPA Crop Production Regions 1, 4, 5, 6 and 8.
12. Plant Codes:
    a. Bayer Code: ANOVI

096

1. Brussels sprouts
(Toy cabbage, Breton, Col de Bruselas)
Brassicacae (Cruciferae)
Brassica oleracea var. *gemmifera* Zenker

2. This biennial plant of the cabbage family attains a height of up to 3 feet. The edible parts are small, very compact heads or "sprouts" formed in the axils of the leaves along the stem. Leaves are large, with long petioles, forming a canopy above and around the edible heads. The latter are 1 to 2 inches across and are comprised of tightly packed leaves and core, resembling miniature cabbage heads. As the main plant stem elongates additional small heads are formed, so harvest is extended with heads ready to harvest and others forming at the same time. Plants are usually started in outdoor beds and transplanted to the field. Transplants are spaced 2 feet apart in rows 2-3 feet apart. Each stem bears 20-40 heads.
3. Season, from transplanting to first harvest: 3 to 3½ months.
4. Production in U.S.: In 1995, 32,300 tons (USDA 1996b). Also grown in New Jersey and New York (STEPHENS 1988). In 1995, 3,800 acres reported in California with 32,300 tons produced (USDA 1996b).
5. Other production regions: In 1995, Canada reported 1,443 acres (ROTHWELL 1996a).
6. Use: Fresh market and frozen. Potherb. Preparation: Trim stem and outer leaves before freezing or cooking (LOGAN 1996). Crop can be held in storage for 3-5 weeks. Per capita consumption in U.S. in 1994 is 0.4 pounds (PUTNAM 1996).
7. Part(s) of plant consumed: Small heads or "sprouts" produced in leaf axils.

8. Portion analyzed/sampled: Leaf sprouts
9. Classifications:
   a. Authors Class: *Brassica* (cole) leafy vegetables
   b. EPA Crop Group (Group & Subgroup): *Brassica* (cole) leafy vegetables (5A)
   c. Codex Group: 010 (VB 0402) *Brassica* (cole or cabbage), head cabbages, flowerhead *Brassicas*
   d. EPA Crop Definition: None
10. References: GRIN, CODEX, LOGAN (picture), MAGNESS, ROTHWELL 1996a, STEPHENS 1988 (picture), USDA 1996b, US EPA 1996a, US EPA 1995a, US EPA 1994, REHM, PUTNAM 1996
11. Production Map: EPA Crop Production Regions 1, 2 and 10.
12. Plant Codes:
    a. Bayer Code: BRSOF

097

1. Buckwheat
(Soba, Ble-noir, Grano-sarraceno, Beechwheat, Bouquette, Sarrasin)
Polygonaceae
Fagopyrum spp.

1. Japanese buckwheat
F. esculentum Moench

1. Silverhull buckwheat
(Common gray buckwheat)
F. esculentum Moench (syn: *F. sagittatum* Gilib., *F. vulgare* Hill, *F. vulgare* T. Nees)

1. Tartary buckwheat
(Ku-chiao-mai, Sarrasin-de-tartarie, Tartarian buckwheat, Greenbuckshot)
F. tataricum (L.) Gaertn.

2. Buckwheat is believed to have originated in central and western China. It is now a relatively minor crop in the U.S. Average acreage grown, 1960-64 inclusive, was about 57,000. This contrasts with near 1 million acres 50 years earlier. Present production is under 1 million bushels. In 1992, acreage was about 65,000. The buckwheat plant is entirely different from other grains and is not a grass. It is a summer annual with rather coarse, branched stems and large, broadly arrow-shaped leaves. Flower panicles and leaves rise from the nodes, both on the main stem and branches. Growth habit is indeterminate with flowers opening throughout a long season so the seed crop does not mature at one time. The seed is partially but not entirely enclosed by adhering flower parts during development. Buckwheat is usually seeded only after the ground is thoroughly warm in early summer. Plants will begin blooming in about 40 days from seeding and first seeds mature about 35 days later. Harvesting is done when a substantial part of the seed is ripe. Fields are then mowed and plants are stacked to dry before they can be threshed. Seeds are pointed, broad at the base and triangular to nearly round in cross section. They vary in size in different kinds from about 4 mm at maximum width and 6 mm long to 7 mm wide and 4 mm long. The seed consists of an outer layer or hull, an inner layer, the seed coat proper and within this a starchy endosperm and the germ. In milling, the hull, which comprises 18 to 20 percent of the whole grain weight, is first removed. A second milling removes most of the seed coat or "middlings" which comprise 4 to 18 percent of the whole grain weight, depending on how completely the seed coat

tissues are removed. In most buckwheat flour some of the seed coat particles remain, resulting in a light brown color. More complete milling results in a white flour. The Japanese buckwheat is most widely grown in this country. The seeds are large, brown in color and triangular in cross section. Plants are about 40 inches tall, rather coarse growers with large, broad arrow-shaped leaves. Silverhull plants are smaller than Japanese buckwheat with smaller leaves. The seeds are small, nearly round in cross section, glossy and gray in color. The hulls are thinner than in Japanese. A kind termed Common Gray is probably identical with Silverhull. Tartary buckwheat has small seeds, nearly round in cross section. Color varies from gray to black. The hull may be smooth to rough and spiny. Leaves are relatively small, narrow and arrow-shaped. Plants may be almost viny in habit. The flour from Tartary buckwheat is inferior to that from Japanese or Silverhull and is not used to make pancake flour. Tartary buckwheat has a lower feeding value than the common variety and is used mostly for poultry and cattle feed.

Buckwheat hulls have little feeding value, sold as feed or bran, or used as soil mulch or poultry litter in U.S. or for stuffing pillows in Japan. Buckwheat is used in commercial birdseed mixtures.

Uses of Buckwheat

Most of the buckwheat grown in the U.S. is milled into flour which is used largely in pancakes. For pancakes the flour is usually blended with that from other grains. Approximately 100 lb of clean dry buckwheat seed yields 60 to 75 lb flour (52 lb pure white flour), 4 to 18 lb middlings and 18 to 26 lb hulls. Whole buckwheat grain may be used in poultry scratch feed mixtures. The middlings from milling make good livestock feed as they are high in protein. The straw is higher in protein but lower in digestible carbohydrates than grass grain straw. The buckwheat plant is an excellent honey source as the blossoms are rich in nectar and blooming continues into the fall months. Some beekeepers plant buckwheat primarily for such use. Buckwheat is a short-season crop that is double-cropped following wheat, potato, sweet corn or peas in Washington.

3. Season, seeding to harvest: About 75 days.
4. Production in U.S.: In 1992, 64,554 acres with 899,632 bushels (1992 CENSUS). In 1992, the top 7 states reported as North Dakota (21,847 acres), South Dakota (17,726), Minnesota (10,114), Washington (3,865), New York (2,520), Montana (2,222) and Pennsylvania (2,087) (1992 CENSUS). In 1994, Washington grew 20,000 acres and 250 acres for seed. Most of the buckwheat raised in Washington is exported to Japan in 1994 where it is made into soba noodles (SCHREIBER 1995).
5. Other production regions: In 1995 Canada reported 22,487 acres (ROTHWELL 1996a).
6. Use: See above. Plus green manure crop; extracted for pharmaceutical drug, Rutin; soups; thicking agent; Kasha; and gravies. Once a major livestock feed, now its main uses are for human food.
7. Part(s) of plant consumed: Grain and straw
8. Portion analyzed/sampled: Grain and its processed commodity flour. Grain is the seed plus hull
9. Classifications:
   a. Authors Class: Cereal grains; Forage, fodder and straw of cereal grains
   b. EPA Crop Group (Group & Subgroup): Cereal grains (15); Forage, Fodder and Straw of Cereal grains

c. Codex Group: 020 (GC 0614) Cereal grain; 051 (AS 0641) Straw
   d. EPA Crop Definition: None
10. References: GRIN, CODEX, MAGNESS, RICHE, ROTHWELL 1996a, SCHREIBER, US EPA 1996a, US EPA 1994, OPLINGER(b), REHM
11. Production Map: EPA Crop Production Regions 1, 5, 7 and 11.
12. Plant Codes:
   a. Bayer Code: FAGES (*F. esculentum*), FAGTA (*F. tataricum*)

098

1. Buffalo gourd
(Calabazilla, Missouri gourd, Fetid wild pumpkin, Chilicote)
Cucurbitaceae
Cucurbita foetidissima Kunth
2. Long-running perennial with stiff cordate-triangular leaves. The fruit is striped green and yellow and the size and shape of an orange. It is a potential new source of vegetable oil in arid lands. The edible seeds which are harvested at the end of the second year can be used for vegetable oil. The vine growth can be used for fodder. Also, in the third year, roots can be harvested for starch. The seeds contain 30-35 percent protein and up to 34 percent oil.
3. Season, seeding to harvest: Seeds harvested the second year.
4. Production in U.S.: No data. Limited. Grown in California (MELNICOE 1997b). Southwestern North America (FACCIOLA 1990).
5. Other production regions: Mexico (BAILEY 1976).
6. Use: Seeds for edible oil and protein (roasted/meal); roots are a source of starch used as a sweetener and vines for animal fodder. The young fruits are said to be edible, fresh or dried, but are bitter.
7. Part(s) of plant consumed: Seeds, roots and vines. Primarily the seeds
8. Portion analyzed/sampled: Seeds, oil and meal fractions
9. Classifications:
   a. Authors Class: Oilseed
   b. EPA Crop Group (Group & Subgroup): Miscellaneous
   c. Codex Group: No specific entry
   d. EPA Crop Definition: None
10. References: GRIN, BAILEY 1976, BAYER, FACCIOLA, GRIN, JANICK 1990, MELNICOE 1997b
11. Production Map: EPA Crop Production Region 10
12. Plant Codes:
   a. Bayer Code: CUUFO

099

1. Buffalograss
Poaceae (Gramineae)
Buchloe dactyloides (Nutt.) Engelm.
2. Buffalograss, a native warm season species, is the dominant grass in parts of the Great Plains. It is a fine leaved, sod forming perennial, generally only 6 to 8 inches high, with leaves 3 to 6 inches long and less than $1/_8$ inch wide. It spreads by surface runners or stolons to form a dense sod. Growth starts in late spring and continues through the summer. It is very palatable and stands grazing well. It is readily established either by seeding or by sod pieces and is valuable both as pasture

and for erosion control. The species is unisexual, about half the plants being female and producing seed; and half male, the flowers of which produce only pollen. There are 280,000 seeds/lb. It is seeded at a rate of 1-2 lb/A.

3. Season, growth starts in late spring and continues through the summer.
4. Production in U.S.: No specific data, See Forage grass. Great Plains. Adapted to plant hardiness zones 3 to 9.
5. Other production regions: Northern Mexico (GOULD 1975).
6. Use: Pasture, rangeland, hay, erosion control.
7. Part(s) of plant consumed: Foliage
8. Portion analyzed/sampled: Forage and hay
9. Classifications:
   a. Authors Class: See Forage grass
   b. EPA Crop Group (Group & Subgroup): See Forage grass
   c. Codex Group: See Forage grass
   d. EPA Crop Definition: None
10. References: GRIN, MAGNESS, ALDERSON, GOULD
11. Production Map: EPA Crop Production Regions 6, 7, 8, 9 and 11. Also See Forage grass.
12. Plant Codes:
   a. Bayer Code: BUCDA

100

1. Buffelgrass
(Zacate buffel, African foxtailgrass)
Poaceae (Gramineae)
   Cenchrus ciliaris L. (syn: *Pennisetum ciliare* (L.) Link)
2. This warm season grass from South Africa includes both bunch and spreading perennial types. Stems are generally under 4 feet in height. It does well on light, sandy soil and is drought resistant but not cold tolerant. It is nutritious and palatable and stands grazing well. It is grown mainly in south Texas for pasture. It is considered one of the best forage grasses in semi-arid subtropical and tropical areas. Seeding rate is 0.5 to 4 kg/ha. There are 450,000 to 703,000 seeds/kg.
3. Season: Warm season grass, drought resistant.
4. Production in U.S.: No specific data, See Forage grass. In 1997, Texas reported 2.1 million acres of Buffel (SMITH 1997).
5. Other production regions: Northern Mexico.
6. Use: Pasture, rangeland reseeding.
7. Part(s) of plant consumed: Foliage
8. Portion analyzed/sampled: Forage and hay
9. Classifications:
   a. Authors Class: See Forage grass
   b. EPA Crop Group (Group & Subgroup): See Forage grass
   c. Codex Group: See Forage grass
   d. EPA Crop Definition: None
10. References: GRIN, MAGNESS, SMITH 1997
11. Production Map: EPA Crop Production Regions 6 and 8.
12. Plant Codes:
   a. Bayer Code: PESCI

101

1. Burclover
(Medic)
Fabaceae (Leguminosae)
   Medicago spp.

1. California burclover
(Toothed burclover, Toothed medic)
   M. polymorpha L.
1. Spotted burclover
(Southern burclover, Spotted medic)
   M. arabica (L.) Huds. (*M. maculata* Sibth.)
1. Black medic
(Hop clover, Nonesuch, Yellow trefoil)
   M. lupulina L.
2. The burclovers are so-called because the seed pods are spiny and coiled. The three species listed are naturalized in the U.S. These annuals are native to the Mediterranean area and they became established in this country by chance. They are adapted only to areas where winter temperatures do not go much below freezing. California burclover is found mainly in the far west while spotted burclover is found in the Southeast. Although annuals, they commonly reseed, starting growth in the fall and furnishing winter and early summer pasture under mild conditions. Stems are rather weak, usually semiprostrate unless supported. Leaves are trifoliate with 3-heart-shaped leaflets. The burclovers combine well with grasses as pasturage. While nutritious, either as hay or pasture, they are less palatable to livestock than alfalfa or most other clovers. Excessive grazing may cause bloating. They are valuable as green-manure crops, especially in orchards where volunteer growth can be maintained if plants are allowed to mature a seed crop at intervals before turning under. There are 20 annual *Medicago* called 'Medic'.
3. Season: Winter and early summer pasture. About 65 days to flower and flowers March to June.
4. Production in U.S.: No data. Western and Southern U.S.
5. Other production regions: Native to Mediterranean area.
6. Use: See above. Also leaves can be used in salads.
7. Part(s) of plant consumed: Foliage. Leaves also used as a salad garnish
8. Portion analyzed/sampled: Forage and hay
9. Classifications:
   a. Authors Class: Nongrass animal feeds
   b. EPA Crop Group (Group & Subgroup): Miscellanous
   c. Codex Group: No specific entry
   d. EPA Crop Definition: None
10. References: GRIN, FACCIOLA, MAGNESS, STUBBENDIECK, BARNES, DUKE(h), BALL
11. Production Map: EPA Crop Production Regions 2, 4, 10 and 12.
12. Plant Codes:
   a. Bayer Code: MEDAB (*M. arabica*), MEDLU (*M. lupulina*), MEDPO (*M. polymorpha*), MEDSS (*M.* spp.)

102

1. Burdock, Edible
(Gobo, Harlock, Clotbur, Ngau pong, Great burdock, Cuckold, Lappa)
Asteraceae (Compositae)
   Arctium lappa L. (syn: *Lappa officinalis* All.; *L. major* Gaertn.; *A. edule* (Sieb.) Nakai; *L. edulis* Sieb.)
2. Burdock plants are rather coarse perennials to 10 feet tall which are weeds in many temperate areas, including the U.S. Tops die down in winter. New sprouts rising from roots in spring are peeled and eaten raw or after cooking like spinach. Long slender roots are also eaten like radish when

young or cooked when mature. Sparingly grown as a vegetable in Japan and possibly other countries.

3. Season, start of growth from old roots to harvest of sprouts: 2 to 4 weeks. Mature roots can be harvested from about $2^1/_2$ to 10 months after seeding, depending on climate and size of root.
4. Production in U.S.: No recent data. Naturalized in North America (BAILEY 1976). Edible burdock seed is produced in Washington state on about 5 acres every other year in EPA Crop Production Region 11 (SCHREIBER 1995).
5. Other production regions: Japan. Native to China and Siberia.
6. Use: Sprouts as salad or potherb. Roots cooked after peeling. Very young roots can be eaten raw like radish.
7. Part(s) of plant consumed: Tender spring sprouts, roots which can grow to 4 feet long but normally 2 feet long
8. Portion analyzed/sampled: Young leaves and stems; roots
9. Classifications:
   a. Authors Class: Root and tuber vegetables; Leaves of root and tuber vegetables
   b. EPA Crop Group (Group & Subgroup): Root and tuber vegetables (1A and 1B); Leaves of root and tuber vegetables
   c. Codex Group: 016 (VR 0575) Root and tuber vegetables
   d. EPA Crop Definition: None
10. References: GRIN, BAILEY 1976, CODEX, MAGNESS, REHM, SCHREIBER, STEPHENS 1988 (picture), US EPA 1995a, SPLITTSTOESSER, FORTIN, RUBATZKY (picture)
11. Production Map: EPA Crop Production Region 10.
12. Plant Codes:
    a. Bayer Code: ARFLA

103

1. Burnet, Salad
(Small burnet, Lesser burnet, Petite pimprenelle, Pimpinela menor)
Rosaceae
Sanguisorba minor Scop.
1. Garden burnet
(Great burnet, Grande pimprenelle, Pimpinela major, Burnet bloodwort)
S. officinalis L.
2. The plant is a hardy perennial 1 to $2^1/_2$ feet in height, with long compound leaves, each with 6 to 10 pairs of small leaflets. It is sometimes grown in gardens for the fresh young leaves which are used in salads. While it has long been listed in gardening manuals, it is not normally grown commercially in U.S.
3. Season, planting: Sow seed in the fall or spring.
4. Production in U.S.: No data. Mainly gardens.
5. Other production regions: Native to Britain and Europe (GARLAND 1993).
6. Use: In salads for their cucumber-like flavor. Lesser burnet can be grown in pastures for sheep feed.
7. Part(s) of plant consumed: Leaves
8. Portion analyzed/sampled: Leaves
9. Classifications:
   a. Authors Class: Herbs and spices
   b. EPA Crop Group (Group & Subgroup): Herbs and spices (19A)
   c. Codex Group: 027 (HH 0725) Herbs for both *S. minor* and *S. officinalis*

d. EPA Crop Definition: None
10. References: GRIN, CODEX, GARLAND (picture), MAGNESS, NEAL, US EPA 1995a
11. Production Map: EPA Crop Production Region 12.
12. Plant Codes:
    a. Bayer Code: SANMI (*S. minor*), SANOF (*S. officinalis*)

104

1. Butternut
(White Walnut)
Juglandaceae
Juglans cinerea L.
2. The butternut tree is large, up to 100 feet, with long compound leaves, with up to 20 oblong-lanceolate, pubescent leaflets. Fruits are up to 5 in clusters, about 3 inches long and ribbed. The nut is oblong, ribbed, with a hard, woody, thick shell. The kernel is separated from the shell with difficulty. The fleshy husk which surrounds the nut is difficult to remove. Butternuts are not cultivated in the U.S., but some are gathered from native trees, found from southern Canada to Georgia and west to the Great Plains.
3. Season, harvest: See Black walnut.
4. Production in U.S.: No data, see above.
5. Other production regions: See above.
6. Use: Primarily used in candy, ice cream and bakery products.
7. Part(s) of plant consumed: Nutmeat and rarely its oil for cooking
8. Portion analyzed/sampled: Nutmeat
9. Classifications:
   a. Authors Class: Tree nuts
   b. EPA Crop Group (Group & Subgroup): Tree nuts
   c. Codex Group: 022 (TN 0663) Tree nuts
   d. EPA Crop Definition: None
10. References: GRIN, CODEX, FACCIOLA, MAGNESS, US EPA 1995a, US EPA 1995b, WOODROOF(a) (picture)
11. Production Map: EPA Crop Production Regions 1, 2, 4 and 5.
12. Plant Codes:
    a. Bayer Code: IUGCI

105

1. Cabbage
(Repollo, Head Cabbage, Savoy cabbage, Red cabbage, Oxhead cabbage, Pointed cabbage, Yellow cabbage, Green cabbage, White cabbage, Col, Purple cabbage)
Brassicaceae (Cruciferae)
Brassica oleracea var. *capitata* L.
B. oleracea var. *sabauda* L.
(Savoy cabbage)
2. The cabbage plant, except for seed production, is grown as an annual. Plants are usually started in beds, then set in the field 30 to 40 days after seeding. Basal leaves become large, up to a foot or more long and wide and spreading. Later the "head" consisting of the "core" or stem terminal and densely packed leaves develops. This increases in size by growth of inside leaves. The outermost leaves spread away from the head. Outermost leaves that encircle the head and adhere closely to it are rather coarse and are usually removed in preparation for eating, so the part consumed is largely protected during growth by enveloping leaves which are not

consumed. Heads at harvest vary from 5 to 12 or more inches in diameter and in shape from conic to oblate.

3. Season, field setting to harvest: About 2 to 4 months. Early cultivars require 80-95 days and late cultivars require 100-115 days. Heads weigh 2-12 pounds with preference for 3-5 pound heads.

4. Production in U.S.: In 1995, 1,374,120 tons, of which about 1,200,250 tons is marketed fresh and 173,870 tons is made into sauerkraut (USDA 1996b). In 1995, fresh market cabbage reported on 80,300 acres with the top 7 states as New York (13,800 acres), California (13,000), Georgia (10,500), Texas (10,300), Florida (8,400), North Carolina (5,700) and Wisconsin (4,200) with 82 percent of the acreage. Processed cabbage grown for kraut reported on 7,770 acres with the top 2 states as Wisconsin (3,300 acres) and New York (3,000) with 81 percent of the acreage and 77 percent of the production. Fresh market winter season production from Florida and Texas (USDA 1996b).

5. Other production regions: In 1995, Canada reported 11,110 acres (ROTHWELL 1996a).

6. Use: As potherb, salad, sauerkraut. Eaten raw or cooked. Per capita consumption in 1994 for U.S. is 10.2 pounds (PUTNAM 1996).

7. Part(s) of plant consumed: Compact inner leaves. Core and outer leaves generally discarded. Outer leaves after removal from head may be fed to livestock

8. Portion analyzed/sampled: Fresh leaves with wrapper leaves. Entire cabbage head with obviously decomposed or withered leaves removed. Also, residue data on cabbage heads without wrapper leaves is desirable

9. Classifications:
   a. Authors Class: *Brassica* (cole) leafy vegetables
   b. EPA Crop Group (Group & Subgroup): *Brassica* (cole) leafy vegetables (5A) (Representative Crop)
   c. Codex Group: 010 (VB 0041) *Brassica* (cole or cabbage) vegetables, head cabbages, flowerhead cabbage for head cabbage; 010 (VB 0403) *Brassica* (cole or cabbage) vegetables, head cabbages, flowerhead cabbage for savory cabbage or green cabbage
   d. EPA Crop Definition: Cabbage = Chinese cabbage (tight-heading varieties only)

10. References: GRIN, CODEX, MAGNESS, ROTHWELL 1996a, USDA 1996b, US EPA 1996a, US EPA 1995a, IVES, SWIADER, PUTNAM 1996

11. Production Map: EPA Crop Production Regions 1, 2, 3, 5, 6, 8 and 10.

12. Plant Codes:
   a. Bayer Code: BRSOL

106

1. Cabbage, Chinese

(Celery cabbage, Wong Bok, Pe-Tsai, Repollo chino, Peking cabbage, Kim chee, Chinese white cabbage, Nappa cabbage, Che-foo, Chihili, Michihli, Napa cabbage, Hakusai, Pao, Chinese cabbage napa)

Brassicaceae (Cruciferae)

Brassica rapa ssp. *pekinensis* (Lour.) Hanelt (syn: *Brassica pekinensis* (Lour.) Rupr., *B. campestris* L. (Pekinensis group); *B. petsai* L.H. Bailey)

2. The general growth habit of Chinese cabbage is similar to cabbage, but both leaves and heads are elongated and rela-

tively narrow. Heads are less densely packed and leaves are much thinner than in cabbage. Heads range from densely packed (Wong Bok or Chee-foo) to semi- or loose-heading (Shantung or Santo) varieties. As with cabbage, initial leaves are somewhat spreading. These, as well as leaves immediately surrounding the edible portion, are usually discarded. The Wong Bok type head is up to 20 inches long and 4 inches wide and is quite dense. The Chinese word "Bok Choy" refers to a non-heading form of Chinese cabbage, Bok choy which see. The Chee-foo or Napa type is tight-heading and includes Wong bok, Wintertime, Tropical pride, Spring gaint. The Chihili type is loose-heading and includes Michihili, Jade pagoda, Shantung and Shako tsai (MYERS 1991). Data generated for both Napa and Bok choy Chinese cabbages provide research for all Chinese cabbages including tight-, semi- and non-heading forms under the regulatory term "Chinese cabbage".

3. Season, seeding to harvest: 70 to 80 days. Transplanting to harvest: 55 to 70 days.

4. Production in U.S.: 1,200 acres reported 1959 CENSUS. 8,824 acres (1992 CENSUS). In 1992, the top 4 states reported as California (4,540 acres), Florida (3,053), Hawaii (399) and Michigan (83) with about 92 percent of the U.S. acreage. Other states reporting were Massachusetts, Washington, Arkansas, Arizona, Minnesota and New Jersey. Also, Guam reported 18 acres (1992 CENSUS).

5. Other production regions: In 1995, Canada reported 828 acres (ROTHWELL 1996a).

6. Use: Fresh, as salad or as potherb.

7. Part(s) of plant consumed: Inner leaves

8. Portion analyzed/sampled: See cabbage and Bok choy for Chinese cabbage all

9. Classifications:
   a. Authors Class: *Brassica* (cole) leafy vegetables
   b. EPA Crop Group (Group & Subgroup): *Brassica* (cole) leafy vegetables(5A)
   c. Codex Group: 013 (VL 0467) Leafy vegetables (including *Brassica* leafy vegetables)
   d. EPA Crop Definition: Cabbage = Chinese cabbage (tight-heading varieties only)

10. References: GRIN, CODEX, LOGAN, LORENZ, MAGNESS, MYERS (picture), RICHE, ROTHWELL 1996a, STEPHENS 1988 (picture), US EPA 1995a, KRAUS, STERRETT, VAVRINA

11. Production Map: EPA Crop Production Regions 1, 2, 3, 10 and 13.

12. Plant Codes:
   a. Bayer Code: BRSPK

107

1. Cabbage, Seakale

(Couve tronchuda, Butter cabbage, Braganza cabbage, Portugal cabbage, Tronchuda cabbage, Tronchuda kale, Bedford cabbage, Portuguese cole)

Brassicaceae (Cruciferae)

Brassica oleracea var. *costata* DC. (syn: *B. oleracea* (Tronchuda group))

2. Cabbage-like plant, classified as one of the savoy cabbages, meaning its leaves are crinkled or wrinkled rather than smooth. The plant resembles a thick-stemmed collard with

large floppy leaves. Hardy perennial with large leaves and emerging leaf shoots harvested before they unfold.

3. Season, seeding to harvest: Plants are propagated from seeds or cuttings.
4. Production in U.S.: No data. Grown in Florida (STEPHENS 1988).
5. Other production regions: No entry.
6. Use: Cooking greens.
7. Part(s) of plant consumed: Leaves and petiole
8. Portion analyzed/sampled: Leaves and petiole
9. Classifications:
   a. Authors Class: *Brassica* (cole) leafy vegetables
   b. EPA Crop Group (Group & Subgroup): Miscellaneous
   c. Codex Group: 013 (VL 0499) Leafy vegetables (including *Brassica* leafy vegetables)
   d. EPA Crop Definition: None
10. References: GRIN, FACCIOLA, STEPHENS 1988 (picture), CODEX, RUBATZKY
11. Production Map: No entry
12. Plant Codes:
   a. Bayer Code: BRSOT

108

1. Cacao bean
(Cacao, Chocolate, Cocoa, Cacao butter, Cacaotier)
Sterculiaceae
   Theobroma cacao L. ssp. *cacao*
2. The tree is a small, tropical evergreen, up to 25 feet, with thick, oblong-oval, entire leaves. Propagation is by cuttings, budding, grafting and seedings. Fields are shaded for three years and floral buds are removed until trees are 5 years. Maximum yields are obtained by 8-10 years and can produce pods up to 100 years.
   The cacao "beans" are produced in large, angular capsules, up to a foot long and 4 inches in diameter. The capsule rind is thick, hard and leathery. The capsules are borne along the trunk and main branches. The beans, or seeds, about an inch across, range from 20 to 60 per capsule and are imbedded in an acid, fleshy pulp. After removal from the capsule, the beans are washed or fermented to remove the mucilaginous pulp. Every 1000 pounds of dried cocoa beans produce 930 pounds of dried pod husk which can be used as a source of potash or ground and added to livestock feed. Chocolate is the sweetened or unsweetened product of the roasted and ground beans, with most of the fat retained. Cocoa of commerce is the finely ground product, with most of the fat removed. Both forms are very widely used in confections, ice cream, cookery and drinks.
   Cacao is not grown in continental U.S. Over 453,130 tons of the beans are imported annually (1993), mainly from Africa, Asia and South America. Mexico exported 18,348 tons in 1992. Cacao butter is obtained from cacao seeds, which contain 50 percent or more of a non-drying fat. The butter is a byproduct from the manufacture of beverage cacao. The butter may be extracted by pressing or with solvents. The chief use of the butter is in confections. It is also used in pharmaceutical preparations. U.S. imported 107,096 tons of this vegetable oil in 1993.
3. Season, harvest: Fruit (capsule or pod) is cut from the tree with a knife. From fertilization to harvest, fruit requires 5-6 months.

4. Production in U.S.: No data. Few acres in Hawaii (IR-4 REQUEST DATABASE 1996).
5. Other production regions: Best production in hot tropical climate.
6. Use: In confections, ice cream, cookery, beverages.
7. Part(s) of plant consumed: Seed (bean)
8. Portion analyzed/sampled: Seed (dried); and processed commodities roasted bean, cocoa powder and chocolate
9. Classifications:
   a. Authors Class: Tropical and subtropical trees with edible seeds for beverages and sweets
   b. EPA Crop Group (Group & Subgroup): Miscellaneous
   c. Codex Group: 024 (SB 0715) Seed for beverages and sweets
   d. EPA Crop Definition: None
10. References: GRIN, CODEX, MAGNESS, MARTIN 1984, USDA 1994a, US EPA 1995b, DUKE (a), IR-4 REQUEST DATABASE 1996
11. Production Map: EPA Crop Production Region 13.
12. Plant Codes:
   a. Bayer Code: THOCA

109

1. Calamondin
(Golden lime, Scarlet lime, Panama orange, Chinese orange)
Rutaceae
   Citrus madurensis Lour. (syn: x *Citrofortunella* mitis (Blanco) J. Ingram & H. Moore, *Citrus reticulata* var. *austera* Swingle x *Fortunella* sp., *C. mitis* Blanco)
2. The calamondin is a small, slightly oblate fruit, generally not over $1^{1}/_{2}$ inches in diameter, with a deep orange color. The tree is relatively hardy for a citrus and may reach 20 feet in height. It is rather widely grown as an ornamental. Fruit is somewhat intermediate between the kumquat and tangerine. Peel is tender and edible like the kumquat and separates from the flesh like a tangerine when fully mature. The fruit makes an excellent marmalade and an acceptable acid drink.
3. Season, bloom to maturity: 8 to 12 months.
4. Production in U.S.: No data, limited. California and Arizona (LOGAN 1996). Also popular in Florida and Texas (MORTON).
5. Other production regions: Widely grown in India and throughout southern Asia and Malaysia. Important *Citrus* juice source in the Philippine Islands (MORTON 1987).
6. Use: Fruit may be made into marmalade or juice.
7. Part(s) of plant consumed: Fruit except seeds
8. Portion analyzed/sampled: Whole fruit
9. Classifications:
   a. Authors Class: See citrus hybrids
   b. EPA Crop Group (Group & Subgroup): See citrus hybrids
   c. Codex Group: 001 (FC 0201) Citrus fruits
   d. EPA Crop Definition: See citrus hybrids
10. References: GRIN, CODEX, LOGAN, MAGNESS, MORTON (picture)
11. Production Map: EPA Crop Production Region 10.
12. Plant Codes:
   a. Bayer Code: No specific entry, genus entry is CIDSS

110

1. Calamus-root
(Sweet flag, Flagroot, Calamus, Myrtle flag, Sweet calamus, Sweetroot, Acore odorant)
Araceae
Acorus calamus L.
2. The plant is a hardy, herbaceous perennial, adapted to moist soil. The root is pungent and aromatic. In south Asia it is valued chiefly as a drug, but is sometimes used for flavoring beer, cordials and other drinks, or is coated with sugar as candy. Only the roots are used. There are no data on production in the U.S.
3. Season, planting to harvest: Horizontal rhizomes are harvested for use as spice and propagation. The rhizome is harvested about a year after planting.
4. Production in U.S.: No data.
5. Other production regions: Grows in Europe, Asia and North America.
6. Use: Spice.
7. Part(s) of plant consumed: Root
8. Portion analyzed/sampled: Root (dried)
9. Classifications:
   a. Authors Class: Herbs and spices
   b. EPA Crop Group (Group & Subgroup): Miscellaneous
   c. Codex Group: 028 (HS 0772) Spices
   d. EPA Crop Definition: None
10. References: GRIN, CODEX, FOSTER (picture), GARLAND (picture), MAGNESS
11. Production Map: No entry
12. Plant Codes:
    a. Bayer Code: ACSCA

111

1. Calopo
(Frisolilla, Rabo de iguana, False oro)
Fabaceae (Leguminosae)
Calopogonium mucunoides Desv. (syn: *C. orthocarpum* Urb.)
2. Calopo is an annual to short-lived warm season perennial herbaceous vine that is native to tropical South America and the West Indies and is also widespread in tropical Asia and Africa. It forms a dense matted cover up to 60 cm deep and spreads up to 1 m. It is adapted to areas receiving >1000 mm rainfall and ≤1500 m elevation and soils with pH 4.3 to 8.0. Alone, it is not readily grazed by cattle, but it is readily grazed when mixed with grasses, yielding 4 tons DM/ha. It also makes an excellent green manure and cover crop. It should be rotationally grazed every 8 to 12 weeks. Its uses are similar to Tropical kudzu. Calopo flowers in December and fruit (pods linear 2.5 to 4 cm with 3 to 8 seeds) matures in December through January. It makes a good ground cover within 5 months and yields up to 60 MT green manure/ha. There are 65,000 to 70,000 seeds/kg and its seed yields are 200 to 300 kg/ha.
3. Season: Warm season perennial forage that can be used for grazing, but is not used for hay or silage.
4. Production in U.S.: No specific data for Calopo production, however, pasture and rangelands are produced on more than 410 million A (U.S. CENSUS, 1992).
5. Other production regions: See above.

6. Use: Grazing when mixed with grasses, green manure crop and cover crop.
7. Part(s) of plant consumed: Leaves and stems
8. Portion analyzed/sampled: Forage and hay
9. Classifications:
   a. Authors Class: Nongrass Animal Feeds
   b. EPA Crop Group (Group & Subgroup): Miscellaneous
   c. Codex Group: No specific citation although the general class 050 (AL 0157) Legume Animal Feeds could be used
   d. EPA Crop Definition: None
10. References: GRIN, US EPA 1994, US EPA 1995, CODEX, HOLZWORTH, DUKE(h)(picture), SKERMAN(a), RICHE
11. Production Map: EPA Crop Production Region 13.
12. Plant Codes:
    a. Bayer Code: CLOMU

112

1. Camomile
(Chamomile)
Asteraceae (Compositae)
Chamaemelum spp.
Matricaria spp.
1. Garden dogfennel
(Garden camomile, Noble camomile, Roman camomile, Russian camomile)
C. nobile (L.) All. (syn: *Anthemis nobilis* L.)
1. Mayweed
(German camomile, Scented camomile, Wild camomile, Sweet false camomile)
M. recutita L. (syn: *M. chamomilla* L., *Chamomilla chamomilla* (L.) Rydb., *Chamomilla recutita* (L.) Rauschert)
2. The *Chamaemelum* plant is a much-branched perennial herb with finely cut leaves, bearing flowers about an inch in diameter. For medicinal or condiment purposes, only the flowers are used. They are cut as soon as fully expanded and dried. The main use is in medicine. No data are available on production in the U.S. The *Matricaria* plant is a glabrous annual to 2¹/₂ ft in height. Flower heads 1 inch across.
3. Season, seeding to harvest: About 2 months with 2 to 3 cuttings of flowers every 2 weeks.
4. Production in U.S.: No data. Naturalized in U.S.
5. Other production regions: Europe.
6. Use: Herb used for preparation of infusion (Tea) and to flavor sherry which gives it an apple-like scent and termed, manzanilla.
7. Part(s) of plant consumed: Flowers
8. Portion analyzed/sampled: Flowers
9. Classifications:
   a. Authors Class: Herbs and spices (Both species)
   b. EPA Crop Group (Group & Subgroup): Herbs and spices (19A), Only *Chamaemelum nobile*
   c. Codex Group: 066 (DT 1110) Teas (Both species)
   d. EPA Crop Definition: None
10. References: GRIN, CODEX, FOSTER (picture), GARLAND (picture), MAGNESS, US EPA 1995a, USDA NRCS, HYLTON
11. Production Map: EPA Crop Production Region 10.
12. Plant Codes:
    a. Bayer Code: ANTNO (*C. nobile*), MATCH (*M. recutita*)

113

1. Camwood
(African sandalwood, Barwood, Bois de cam)
Fabaceae (Leguminosae)
Baphia nitida Lodd. (syn: *B. haematoxylon* Hook.; *Carpolobia versicolor* G. Don)

2. Camwood is a large erect evergreen shrub 2.6 to 3.3 m tall with legume pods 10 cm long. It is native to western tropical coastal areas of Africa. It may grow in Florida and eastern Texas and is adapted to areas receiving 13.6 to 40.3 dm rainfall and for soils with a pH 5.0 to 5.3. It was formerly planted as a dye source for wool but has been replaced by the American logwood. Camwood is used to make ax handles, walking sticks, and it's mainly used now for a potential stockfeed because its leaves are high in protein and it can be browsed at anytime. It reproduces by seeds or from cuttings.
3. Season: Warm season perennial shrub to small tree whose nutritious leaves can be utilized at anytime.
4. Production in U.S.: No specific data for Camwood production, however, pasture and rangelands are produced on more than 410 million A (U.S. CENSUS, 1992). See above.
5. Other production regions: No entry.
6. Use: Grazing and for soil erosion control.
7. Part(s) of plant consumed: Leaves and stems
8. Portion analyzed/sampled: Forage and hay
9. Classifications:
   a. Authors Class: Nongrass Animal Feeds
   b. EPA Crop Group (Group & Subgroup): Miscellaneous
   c. Codex Group: No specific citation although the general class 050 (AL 0157) Legume Animal Feeds could be used
   d. EPA Crop Definition: None
10. References: GRIN, US EPA 1994, US EPA 1995, CODEX, HOLZWORTH, DUKE(h)(picture), SKERMAN(a), US EPA 1995, RICHE
11. Production Map: EPA Crop Production Region are 6 and 13.
12. Plant Codes:
    a. Bayer Code: No specific entry

114

1. Canarygrass, Annual
(Birdseed grass, Canaryseed, Common canarygrass, Alpiste, Canarygrass, Birdseedgrass)
Poaceae (Gramineae)
Phalaris canariensis L.

2. This is a cool season bunchgrass native to the Mediterranean region and has been introduced into the U.S. There is some production in the northern Red River valley of North Dakota and western Canada provinces of Alberta, Saskatchewan and Manitoba. Annual canarygrass is often classified as a grain crop, and are often confused with Reed canarygrass which is a forage. Both species have inflorescences called panicles, but annual canarygrass is more spike-like and resembles club wheat. The plant grows to 36 inches tall, and readily tillers with it heading in 65 days and maturing in 104 to 107 days which is similar to spring wheat. Annual canarygrass is adapted to the same climatic areas as spring wheat, and performs better on heavier clay loam soils. It is planted early in the spring as soon as the ground can be worked, which is late March or April in southern Minnesota or central Wisconsin. It can replace wheat as part of a crop rotation system. Seed-

ing rates are 30 lb/A. One bushel of annual canarygrass weighs 50 lbs. Yields in Minnesota average from 1,093 to 1,459 lb/A, while in Saskatchewan yields averaged 1,180 lb/A. There are 149,940 seeds/kg. The primary market for annual canarygrass is currently birdseed, but several other uses include flour and it must be dehulled before it is milled.
3. Season: Cool season, spring growth through early summer. There are 63 to 66 days from planting to heading. Seeding to harvest for grain: About 100 days.
4. Production in U.S.: There were 3,000 acres contracted in 1987 for North Dakota and Minnesota (Putnam 1990). Introduced into Canada and states south to Virginia, Arizona and California.
5. Other production regions: In Canada it is grown in Saskatchewan and western provinces. In 1995, 304,929 acres in Canada that produced 127,800 tons of seed (ROTHWELL 1996a).
6. Use: Forage, birdseed, and potential human grain crop.
7. Part(s) of plant consumed: Leaves, stems and seed
8. Portion analyzed/sampled: Seed. Straw optional (insignificant feed item). As forage, need forage and hay samples
9. Classifications:
   a. Authors Class: Cereal grains; Forage, Fodder and Straw of Cereal grains
   b. EPA Crop Group (Group & Subgroup): Grass, Forage, Fodder, and Hay Group (17). Also see Forage grass
   c. Codex Group: 051 (AS 0162) Straw, Fodder, and Forage Cereal Grasses. (Also see Forage grass)
   d. EPA Crop Definition: None
10. References: GRIN, US EPA 1994, US EPA 1995, CODEX, BARNES, HOLZWORTH, PUTNAM 1990, ROTHWELL 1996a, BAILEY 1976, IVES, REHM, RICHE
11. Production Map: EPA Crop Production Regions 5 and 7.
12. Plant Codes:
    a. Bayer Code: PHACA

115

1. Canarygrass, Reed
(Perennial canarygrass)
Poaceae (Gramineae)
Phalaris arundinacea L.

2. Reed canarygrass is native to North America, Europe and Asia. It has been introduced to all continents except Antartica. Its seed has long been used as canary bird feed, hence the name. It is 2 to 6 feet tall and spreads by short rhizomes. It grows in broad clumps that may be 3 feet across. It is one of the best grasses for wet soils, but also is adapted to uplands. It is hardy and grows rapidly during spring and early summer, persists well and has a long grazing season. It produces heavy yields of nutritious, palatable forage, silage or hay. It is important in the North Central states and Pacific states from northern California northward to Alaska. Propagation is by seed. Improved varieties are available. It can be harvested for silage and hay before heading stage. Up to 4 (2 to 4) cuttings for the hay crop can be made with each cutting 40 days apart.
3. Season: Cool season perennial pasture. Seeded early spring. Grazing begins when grass is 6 to 12 inches tall.
4. Production in U.S.: No specific data, See Forage grass. North central states and Pacific states from northern California north.

5. Other production regions: In 1995, Canada listed reed canarygrass as grown in Canada. The annual canarygrass grain *P. canariensis* crop reported at 304,929 acres (ROTHWELL 1996a).
6. Use: Pasture, hay, silage and erosion control.
7. Part(s) of plant consumed: Foliage
8. Portion analyzed/sampled: Forage and hay
9. Classifications:
   a. Authors Class: See Forage grass
   b. EPA Crop Group (Group & Subgroup): See Forage grass
   c. Codex Group: See Forage grass
   d. EPA Crop Definition: None
10. References: GRIN, MAGNESS, ROTHWELL 1996a, SCHREIBER, MOSLER(picture), BALL (picture)
11. Production Map: EPA Crop Production Regions 1, 2, 5, 7, 9, 11 and 12.
12. Plant Codes:
    a. Bayer Code: TYPAR

116

1. **Caneberries**
(Blackberry, Youngberry, Loganberry, Raspberry, Brambles, Zarza)
Roaceae
   Rubus spp.
2. See blackberry, raspberry, dewberry and cloudberry crop monographs.
3. Season: See above crop monographs.
4. Production in U.S.: See above.
5. Other production regions: See above.
6. Use: See above.
7. Part(s) of plant consumed: Fruit
8. Portion analyzed/sampled: Whole fruit
9. Classifications:
   a. Authors Class: Berries
   b. EPA Crop Group (Group & Subgroup): Berries (13)
   c. Codex Group: *Rubus* spp. listed with 004 (FB 0018) Berries and other small fruits
   d. EPA Crop Definition: Caneberries = *Rubus* spp.
10. References: GRIN, CODEX, US EPA 1995a
11. Production Map: See specific crop monographs.
12. Plant Codes:
    a. Bayer Code: RUBSS

117

1. **Canistel**
(Eggfruit, Fruta huevo, Sapote amarillo, Ti-es, Yellow sapote, Zapote mante, Huevo vegetal, Zubul, Eggfruit tree)
Sapotaceae
   Pouteria campechiana (Kunth) Baehni (syn: *Lucuma campechiana* HBK.)
2. The tree is a rather small, up to 25 feet, tropical evergreen native to Central America. Grows in hot tropical lowlands with medium to high rainfall. Propagated by seed and grafting. Leaves are oblong-lanceolate, up to 8 inches long, and glabrous. Fruit is globose to ovoid, 2 to 4 inches long, 2 to 4 seeded, with a fairly smooth skin. Flesh is yellow and mealy. The fruit is edible, but not highly regarded in spite of the fact it contains a fair source of protein and provitamin A.

3. Season: Bloom to harvest in approximately 180 days. Flowers June-August in Florida.
4. Production in U.S.: Very minor production in South Florida. In 1994, 1 acre (CRANE 1995a).
5. Other production regions: Southern Mexico, Central America and West Indies (MORTON 1987).
6. Use: Pulp eaten fresh or used in beverages and desserts.
7. Part(s) of plant consumed: Pulp only
8. Portion analyzed/sampled: Whole fruit
9. Classifications:
   a. Authors Class: Tropical and subtropical fruits – inedible peel
   b. EPA Crop Group (Group & Subgroup): Miscellaneous
   c. Codex Group: 006 (FI 0330 and FI 4129) Assorted tropical and subtropical fruits – inedible peel
   d. EPA Crop Definition: See Papaya and Avocado
10. References: GRIN, CODEX, MAGNESS, MARTIN 1987, US EPA 1994, CRANE 1995a, MORTON (picture), SAULS(c)
11. Production map: EPA Crop Production Region 13.
12. Plant Codes:
    a. Bayer Code: No specific entry

118

1. **Canna, Edible**
(Gruya, Australian arrowroot, Queensland arrowroot, Tous-les-mois, Canna, Poloke, Indian shot, Aliipoe, Purple arrowroot, Doug rieng, Achira, Spanish arrowroot)
Cannaceae
   Canna indica L.
   C. edulis Ker Gawl.
2. Tropical perennial reed or cane-like herb to 10 feet in height. Rhizomes are tuber bearing and consumed as potatoes. Leaves are oblong, 2 feet long and 1 foot wide, and may be used as a potherb.
3. Season, planting to harvest: About 8 months for young tubers or corms and may be allowed to grow for 18 or more months for more starch content of mature tubers before harvesting.
4. Production in U.S.: Grown in Hawaii for tubers and tops for cattle and swine feed (NEAL 1965).
5. Other production regions: Native to Central and South America (NEAL 1965) and West Indies (YAMAGUCHI 1983).
6. Use: Tuber cooked as potatoes.
7. Part(s) of plant consumed: Mainly tuber
8. Portion analyzed/sampled: Tuber
9. Classifications:
   a. Authors Class: Root and tuber vegetables
   b. EPA Crop Group (Group & Subgroup): Root and tuber vegetables (1C and 1D) for *C. indica*
   c. Codex Group: 016 (VR 0576) Root and tuber vegetables
   d. EPA Crop Definition: None
10. References: GRIN, CODEX, MAGNESS, NATIONAL RESEARCH COUNCIL 1989 (picture), NEAL (picture), STEPHENS 1988, US EPA 1995a, YAMAGUCHI 1983, RUBATZKY
11. Production Map: EPA Crop Production Region 13.
12. Plant Codes:
    a. Bayer Code: CNNIN (*C. indica*)

119

1. Cantaloupe
(Nutmeg melon, Netted melon, Muskmelon, Cantaloup, Cantaloupe, Persian melon, Honey ball)
Cucurbitaceae
Cucumis melo var. *cantalupensis* Naudin (syn: *C. melo* var. *reticulatus* Naudin)

2. Cantaloupes are the most widely grown type of muskmelon, which see. They are produced on long-running annual plants, or non-climbing "vines" that are prostrate on the soil. On healthy plants there is a canopy of large, soft-hairy leaves, generally heart shaped and somewhat lobed. Fruits may be orange fleshed or green fleshed. Fruit surface is generally netted and roughened, some somewhat sutured. Fruit shape is generally round to oval. Size in different kinds ranges from 5 to 8 or more inches long, and near same diameter.
3. Season, seeding to harvest: 3 to 4 months.
4. Production in U.S.: 1,053,950 tons in 1995 on planted acreage of 104,890 (USDA 1996b). In 1995, California (59,300 acres), Arizona (16,000), Texas (12,900), Georgia (6,500), Indiana (3,500), Colorado (2,000), Maryland (1,800), Pennsylvania (1,300), Michigan (1,100). (USDA 1996b).
5. Other production regions: In 1995, Canada reported 628 acres for cantaloupe and melons (ROTHWELL 1996a).
6. Use: Fresh dessert, salads.
7. Part(s) of plant consumed: Internal pulp only
8. Portion analyzed/sampled: Whole fruit (discard stem)
9. Classifications:
   a. Authors Class: Cucurbit vegetables
   b. EPA Crop Group (Group & Subgroup): Cucurbit vegetables (9A) (Representative Crop)
   c. Codex Group: 011 (VC 0046) Cucurbits fruiting vegetables
   d. EPA Crop Definition: See muskmelon
10. References: GRIN, CODEX, MAGNESS, USDA 1996b, US EPA 1994, US EPA 1995a, ROTHWELL 1996a
11. Production Map: EPA Crop Production Regions 1, 2, 5, 6 and 10.
12. Plant Codes:
    a. Bayer Code: CUMMC

120

1. Caper
(Caperbush, Caper buds)
Capparidaceae
Capparis spinosa L.

2. The caper plant is a spiny, straggling vine-like shrub, up to 5 feet tall, with round to ovate, deciduous leaves. The capers of commerce are the unopened flower buds, which are picked daily. The youngest buds make the finest product. The buds are pickled in strong vinegar, and used as pickles or in sauce. Capers are produced commercially in Mediterranean countries, in the climate of the olive tree. A mature plant can produce 17-19 pounds of fresh buds annually. Seeds are used for propagation but one-year-old cuttings are commonly used. It takes up to 4 years to reach maximum production.
3. Season, harvest: Flower buds are harvested before opening and picked daily.
4. Production in U.S.: No data. Limited scale in California (FARRELL 1985).
5. Other production regions: Mediterranean region.

6. Use: Spice for salads, meat, fish, poultry, and flavoring relishes, sauces and pickles.
7. Part(s) of plant consumed: Buds (unopened)
8. Portion analyzed/sampled: Buds
9. Classifications:
   a. Authors Class: Herbs and spices
   b. EPA Crop Group (Group & Subgroup): Herbs and spices (19B)
   c. Codex Group: 028 (HS 0773) Spices
   d. EPA Crop Definition: None
10. References: GRIN, BAILEY 1976, CODEX, FARRELL (picture), MAGNESS, STEPHENS 1988 (picture), US EPA 1995a, RUBATZKY
11. Production Map: EPA Crop Production Region 10.
12. Plant Codes:
    a. Bayer Code: CPPSP

121

1. Caraway
(Caraway seed, Carum, Fang-feng, Se-lu-zi, Siyah-jira)
Apiaceae (Umbelliferae)
Carum carvi L.

2. Caraway is an annual or biennial plant of the carrot family, grown for the seeds, which are widely used in cookery. The leaves are greatly compound, with thread-like divisions. The flower stalk is 2 feet or more in height. Flowers and seeds are borne in clusters. Seeds are ovate to oblong, somewhat ribbed. Like carrots, the plant forms a rosette of leaves and a tuberous root the first season, and the seed stalk the second year.
3. Season, start of growth to harvest, second year: About 4 months.
4. Production in U.S.: Cultivated in U.S. (FOSTER 1993).
5. Other production regions: Canadian acreage reported as 1,658 in 1995 (ROTHWELL 1996a).
6. Use: Seeds mainly in cookery as flavoring. Oil from seeds in cosmetics and to flavor meat and confectionery.
7. Part(s) of plant consumed: Seed
8. Portion analyzed/sampled: Seed
9. Classifications:
   a. Authors Class: Herbs and spices
   b. EPA Crop Group (Group & Subgroup): Herbs and spices (19B)
   c. Codex Group: 028 (HS 0774) Spices
   d. EPA Crop Definition: None
10. References: GRIN, CODEX, FARRELL (picture), FOSTER, LEWIS, MAGNESS, ROTHWELL 1996a, US EPA 1995a
11. Production Map: No entry
12. Plant Codes:
    a. Bayer Code: CRYCA

122

1. Caraway, Black
(Black cumin, Charnushka, Fennelflower, Nutmeg flower, Roman coriander, Russian caraway)
Ranunculaceae
Nigella sativa L.

2. Black caraway is an annual herb of the buttercup family, grown for the aromatic seeds which are used as seasoning.

The plant attains a height of $1^{1}/_{2}$ feet with segmented leaves and solitary flowers to $1^{1}/_{2}$ inches across. The seeds are enclosed in a capsule during the season of growth.

3. Season, seeding to harvest: Grown as annual.
4. Production in U.S.: No data.
5. Other production regions: Native to Mediterranean region (BAILEY 1976).
6. Use: Seeds are used for seasoning/flavoring. Tiny black seeds on top of Jewish rye bread.
7. Part(s) of plant consumed: Seed
8. Portion analyzed/sampled: Seed
9. Classifications:
   a. Authors Class: Herbs and spices
   b. EPA Crop Group (Group & Subgroup): Herbs and spices (19B)
   c. Codex Group: No specific entry
   d. EPA Crop Definition: None
10. References: GRIN, BAILEY 1976, FACCIOLA, PENZEY, US EPA 1995a
11. Production map: No entry.
12. Plant Codes:
    a. Bayer Code: No specific entry

123

1. Cardamom
(Cardamon seed, True cardamom, Small cardamom)
Zingiberaceae
Elettaria cardamomum (L.) Maton (syn: *Amomum cardamomum* L.)

2. The above species of tropical plant produces the cardamom seeds of commerce. They are tropical, perennial herbs, the tops growing each year from underground rhizomes. *E. cardamomum* reaches 5 to 10 feet, with lanceolate leaves up to 2 feet long. The capsules are oblong to globular, ribbed and indehiscent. These dried capsules are the principal cardamoms of commerce. The important grades of cardamom in the trade are: Greens (green pods or capsules artifically dried); Sun dried capsules; Bleached capsules; and Hulled seeds. The U.S. imported 175 tons from Guatemala in 1992. The seeds and plant of *Amomum* spp. are very similar to "grains of paradise" which see. These false cardamon seeds (*Amomum* spp.) are also sold as cardamom but of local importance only because of their inferior quality.
3. Season, planting to harvest: Third year for up to 15 years. Harvest starts in August with peak in October to November. Bloom to harvest is 3 to 4 months.
4. Production in U.S.: No data.
5. Other production regions: Grown in tropical Central America and Asia.
6. Use: Flavoring for coffee, meat, baked goods, curries and chewing gums.
7. Part(s) of plant consumed: Seed only without hull, may be extracted for its oil
8. Portion analyzed/sampled: Seed
9. Classifications:
   a. Authors Class: Herbs and spices
   b. EPA Crop Group (Group & Subgroup): Herbs and spices (19B)
   c. Codex Group: 028 (HS 0775) Spices
   d. EPA Crop Definition: None

10. References: GRIN, CODEX, FACCIOLA, LEWIS (picture), MAGNESS, NEAL, USDA 1993a, US EPA 1995a, DUKE(a)
11. Production Map: No entry.
12. Plant Codes:
    a. Bayer Code: ETACA

124

1. Cardamon-amomum
(False cardamom, Round cardamom, Chester cardamon, Siam cardamon)
Zingiberaceae
Amomum compactum Sol. ex Maton

2. Plant is a rhizomatic herb to 12 feet tall with a leafy stem. Leaves are oblong lanceolate and glabrous. The fruits with numerous seeds are produced on pendent spikes. Seeds are used locally as a spice. Cheap substitute for true cardamoms.
3. Season, seeding to harvest: No data.
4. Production in U.S.: No data.
5. Other production regions: India.
6. Use: Seeds are used as a spice for flavoring curries, meats and beverages.
7. Part(s) of plant consumed: Seeds
8. Portion analyzed/sampled: Seeds
9. Classifications:
   a. Authors Class: Herbs and spices
   b. EPA Crop Group (Group & Subgroup): Miscellaneous
   c. Codex Group: No specific entry
   d. EPA Crop Definition: None
10. References: GRIN, FACCIOLA, LEWIS (picture), NEAL
11. Production Map: No entry
12. Plant Codes:
    a. Bayer Code: No specific entry

125

1. Cardoon
(Cardoni, Cardi, Cardo)
Asteraceae (Compositae)
Cynara cardunculus L. ssp. *cardunculus*

2. Cardoon is closely related to globe artichoke. Growth is up to 6 feet, normally 3-5 feet, with large pinnate, prickly leaves. In cultivation, plants are grown from seed. When leaves are near fully grown they are tied together near the top, and plants are banked with straw and soil, or other material, to blanch the leaf stalks or petioles. These are edible parts, but are inedible unless blanched. In exposure of edible parts during growth, the plant is comparable to celery.
3. Season, seeding to harvest: 4 to 5 months.
4. Production in U.S.: No data. California (LOGAN 1996).
5. Other production regions: Italy and temperate regions (RUBATZKY 1997).
6. Use: Fresh, as potherb.
7. Part(s) of plant consumed: Petioles and main ribs (Leaves and tender stalks). Trimmed leaf stalks are normally 12 to 16 inches long, like celery. Preparation – the fleshly silver-gray stalks grow in bunches, like celery, but are flat. Recommended to trim and precook to remove bitterness before use (SCHNEIDER 1985). Also cardoon can be marinated and eaten uncooked in salads (MYERS 1991)
8. Portion analyzed/sampled: Untrimmed leaf stalk (petiole)

9. Classifications:
   a. Authors Class: Leafy vegetables (except *Brassica* vegetables)
   b. EPA Crop Group (Group & Subgroup): Leafy vegetables (except *Brassica* vegetables) (4B)
   c. Codex Group: 017 (VS 0623) Stalk and stem vegetables
   d. EPA Crop Definition: None
10. References: GRIN, CODEX, LOGAN, MAGNESS, MYERS (picture), SCHNEIDER 1985 (picture), STEPHENS 1988 (picture), US EPA 1995a, REHM, RUBATZKY
11. Production Map: EPA Crop Production Region 10.
12. Plant Codes:
    a. Bayer Code: CYUCA

126

1. **Caribgrass**
(Janeiro, Malojilla)
Poaceae (Gramineae)
  Eriochloa polystachya Kunth
2. This is a perennial warm season grass native to Caribbean region, West Indies and Brazil, and was introduced into the Southern U.S. and Puerto Rico. It is an erect grass up to 1 m tall with leaves 10 to 15 mm wide that spreads by stolons. It prefers sandy loam soils. It is more palatable than Paragrass, and blooms throughout the year in the Southern states. It is adapted to areas from sea level to 1800 m that receive 1500 to 2000 mm rainfall. Caribgrass is propagated vegetatively by stem cuttings or rootstocks placed 1 m apart. It can be grazed once per season at a height of 0.5 m and is also used for greenchop or hay with yields ≤1760 kg/ha.
3. Season: Warm season tropical grass that starts growing in the spring.
4. Production in U.S.: There is no production data for Caribgrass, however, pasture and rangeland are produced on more than 410 million A (U.S. CENSUS, 1992). See Forage grass.
5. Other production regions: Eastern Mexico and the Caribbean to South America (GOULD 1975).
6. Use: Pasture grazing, hay, and greenchop.
7. Part(s) of plant consumed: Leaves and stems
8. Portion analyzed/sampled: Forage and hay
9. Classifications:
   a. Authors Class: See Forage grass
   b. EPA Crop Group (Group & Subgroup): See Forage grass
   c. Codex Group: See Forage grass
   d. EPA Crop Definition: None
10. References: GRIN, US EPA 1994, US EPA 1995, CODEX, HOLZWORTH, SKERMAN (picture), RICHE, GOULD
11. Production Map: EPA Crop Production Regions 3, 6 and 13.
12. Plant Codes:
    a. Bayer Code: ERBPO

127

1. **Carob bean**
(St. Johns bread, Algarroba, Locust bean, Pasteli, Carob)
Fabaceae (Leguminosae)
  Ceratonia siliqua L.
2. The carob tree is of medium size, occasionally up to 50 feet, with compound, glossy evergreen leaves. Grown in dry, seasonally hot subtropical climate. Does not fruit well in high rainfall areas. The plant is hardier than citrus, drought resistant and long lived. It is propagated by seed and grafting. It has tiny red flower buds that expand into greenish to cream colored flowers. The fruits are large dried flat brown fleshy pods, up to a foot long and about an inch wide, containing 5 to 15 seeds.
3. Season: Fruit production is 10 to 12 years from seeding and 5 to 6 years from grafting. Bloom to maturity is in 6 to 8 months.
4. Production in U.S.: No data, numerous scattered trees in semi-tropical areas. The cultivars 'Casuda', 'Santa Fe', 'Sfax', and 'Tylliria' are adapted to southern California in the foothills but also in the desert (FACCIOLA 1990).
5. Other production regions: Mediterranean region (FACCIOLA 1990).
6. Use: Pulp is eaten fresh or pods and seeds are ground and processed as chocolate substitute. Can also be used for livestock feed.
7. Part(s) of plant consumed: All of pod and seeds
8. Portion analyzed/sampled: Whole pod with seed. In U.S. only the bean is sampled
9. Classifications:
   a. Authors Class: Tropical and subtropical fruits – edible peel
   b. EPA Crop Group (Group & Subgroup): Miscellaneous
   c. Codex Group: 005 (FT 0291 and FT 4121) Assorted tropical and subtropical fruits – edible peel
   d. EPA Crop Definition: None
10. References: GRIN, CODEX, LOGAN, MARTIN 1987, FACCIOLA, MAGNESS, ROY
11. Production Map: EPA Crop Production Region 10.
12. Plant Codes:
    a. Bayer Code: CEQSI

128

1. **Carpetgrass**
(Common carpetgrass, Mat grass, Compressum, Durrington grass, Narrow-leaved carpetgrass)
Poaceae (Gramineae)
  Axonopus fissifolius (Raddi) Kuhlm. (syn: *A. affinis* Chase)
2. Carpetgrass is native to Central America and the West Indies. It was introduced into the U.S. early in the last century and is now spread over the Southeastern Coastal Plain region from Virginia south to Florida to Texas. It is a perennial, warm season creeping grass spreading by stolons, which makes a dense sod. The creeping stems are compressed, and root at each joint. The numerous leaves are blunt, up to about 6 to 8 inches long, rather broad, and the plant can grow 6 to 10 inches tall. Carpetgrass is adapted to sandy and sandy loam soils, with moisture near the surface, and rainfall ≥750 mm/ year. Its greatest value is for permanent pastures that can be grazed from May through August and can be grazed closely without being damaged. It is also useful for lawns and is widely used to prevent erosion and stabilize roadbanks. Its nutrient value is low but cattle readily consume it. Yields in southeast U.S. range from 812 to 5,197 kg DM/ha. It forms seeds abundantly, and is readily established by seeding at rates of 5 to 12 lb/A. Carpetgrass has 1,350,000 seeds/lb.
3. Season: Warm season forage and grows in the summer to fall. It can be grazed May through August and withstands heavy defoliation. Hay quality at maturity is poor, but it is palatable until heading stage.

4. Production in U.S.: No data for Carpetgrass, however, pasture and rangeland are produced on more than 410 million A (U.S. CENSUS, 1992). See Forage grass.

5. Other production regions: Native to Central America and the West Indies.

6. Use: Pasture, hay, turfgrass, roadbank stabilization and for erosion control.

7. Part(s) of plant consumed: Leaves and stems

8. Portion analyzed/sampled: Forage and hay

9. Classifications:
   a. Authors Class: See Forage grass
   b. EPA Crop Group (Group & Subgroup): See Forage grass
   c. Codex Group: See Forage grass
   d. EPA Crop Definition: None

10. References: GRIN, US EPA 1994, US EPA 1995, CODEX, SKERMAN (picture), BALL, RICHE

11. Production Map: EPA Crop Production Regions 2, 3, 4 and 6.

12. Plant Codes:
    a. Bayer Code: AXOAF

129

1. Carrot
(Peen, Zanahoria, Carotte)
Apiaceae (Umbelliferae)
  Daucus carota ssp. *sativus* (Hoffm.) Arcang.

2. Carrots, except for seed production, are grown as annuals. The plant consists of a crown of greatly compound leaves which rise directly from the top of the root, the edible portion. The root top or crown is about even with the soil surface. Roots of most varieties are broadest at top and tapering, from 4 or 5 to 10 or 12 inches long. For fresh market they are usually harvested when 1 to 1 1/2 inches maximum diameter. Since plants are closely spaced, 1 to 2 or 3 inches in the row, and grow slowly at the start, weed control is a major problem. Grouped by shape and length, carrot cultivars are either 'Danvers', 'Chantenery', 'Nantes', or 'Imperator' with the first three making up most of the processing cultivars. Nearly all carrots for fresh market and processing are machine-harvested except those marketed with their tops attached and are called bunch carrots. Those without are called bulk carrots. Mature topped carrots can be stored for 7-9 months, while immature bunched carrots can usually be held no more than 2-3 weeks.

3. Season, seeding to harvest: 2 to 3 months for fresh market, longer for processing.

4. Production in U.S.: In 1995, **Fresh**: 95,050 acres and 1,314,600 tons. **Processed**: 29,140 acres and 588,650 tons. **Total Acres**: 124,190 acres (USDA 1996b). In 1995, the top 10 states for fresh market were: California (63,500 acres), Florida (7,800), Michigan (6,200), Texas (5,200), Colorado (4,000), Minnesota (2,100), Arizona (1,900), Washington (1,900), Oregon (1,400) and New York (750) for a total of 94,750 acres. Processing carrots were planted to 29,140 acres with a production of 588,650 tons from the top 8 states; Washington (7,500 acres), Texas (5,300), California (5,100), Wisconsin (5,000), Michigan (1,700), Minnesota (1,600), New York (850), and Oregon (780) for a total of 27,830 acres (USDA 1996b).

5. Other production regions: In 1995, Canada reported 19,015 acres (ROTHWELL 1996a). Grown in most temperate regions (RUBATZKY 1997).

6. Use: Fresh market, canned, frozen, for use in culinary and salads. Culls are used as livestock feed. Per capita consumption in 1994 for U.S. is 13.7 pounds (PUTNAM 1996).

7. Part(s) of plant consumed: Fleshy root

8. Portion analyzed/sampled: Root

9. Classifications:
   a. Authors Class: Root and tuber vegetables
   b. EPA Crop Group (Group & Subgroup): Root and tuber vegetables (1A and 1B); Leaves of root and tuber vegetables (2) (Representative Crop)
   c. Codex Group: 016 (VR 0577) Root and tuber vegetables
   d. EPA Crop Definition: None

10. References: GRIN, CODEX, LOGAN (PICTURE), MAGNESS, ROTHWELL 1996a, USDA 1996b, US EPA 1995a, US EPA 1996a, US EPA 1994, RUBATZKY, SWIADER, PUTNAM 1996, MCCALLUM

11. Production Map: EPA Crop Production Regions 3, 5, 6, 10, 11 and 12 which account for 98 percent of U.S. acreage.

12. Plant Codes:
    a. Bayer Code: DAUCS

130

1. Cashew
(Cajuil, Cashew apple, Cashew nut)
Anacardiaceae
  Anacardium occidentale L.

2. The tree is a large, spreading tropical evergreen, 30 to 40 feet high, and spreading up to 60 feet. The leaves are oval or obovate, leathery, 4 to 8 inches long and half as wide. The nuts are borne on a fleshy receptacle, 2 to 4 inches long, called the apple, which is edible and used for eating fresh or in jams or jellies in the tropics. The whole nuts are removed from the apple and sun dried for two days. The nuts with surrounding tissues are 1 to 1 1/2 inches long and kidney shaped. The shell has an oily outer layer, and a thin hard inner one. Between these layers is a honeycombed tissue, from which cashew oil is obtained. The kernels are difficult to separate from the shell because of the curved shape. After roasting the whole nut to extract the oil, shells are removed by hand or machine. Most of the cashews come from India, but production is increasing somewhat in other tropical countries. It is a native of Brazil.

3. Season, bloom to harvest: About 3-4 months for main crop which is gathered off the ground.

4. Production in U.S.: No data. Grown in southern Florida gardens (MORTON 1987).

5. Other production regions: Tropical areas, especially Brazil (MORTON 1987).

6. Use: Nutmeat for direct eating. "Apple" used fresh or in jams or jellies. Oil is used for industrial purpose.

7. Part(s) of plant consumed: Nutmeat and "apples" which are consumed locally

8. Portion analyzed/sampled: Nutmeat

9. Classifications:
   a. Authors Class: Tree nuts; Tropical and subtropical fruits – edible peel
   b. EPA Crop Group (Group & Subgroup): Tree nuts only

c. Codex Group: 005 (FT 0292) Assorted tropical and sub-tropical fruits – edible peel; 022 (TN 0295) Tree nuts

d. EPA Crop Definition: None

10. References: GRIN, CODEX, MAGNESS, US EPA 1995a, US EPA 1995b, WOODROOF(a)(picture), MORTON (picture)

11. Production Map: EPA Crop Production Region 13.

12. Plant Codes:

a. Bayer Code: ANAOC

131

1. Cassava

(Manioc, Mandioca, Tapioca plant, Sweet potato tree, Yuca, Miami fries, Manihot)

Euphorbiaceae

Manihot esculenta Crantz (syn: *M. aipi* Pohl; *M. dulcis* (J.F. Gmel.) Pax; *M. melanobasis* Muell., Arg.; *M. utilissima* Pohl)

2. Cassava is a highly important food crop of the tropics and is grown to some extent in the Southern states, mainly for stock feed. The plant is a large herbaceous shrub up to 10 feet, resembling castor bean in appearance, with large, compound leaves. It is cultivated for the large, tuberous roots which are rich in starch and are the source of tapioca, Brazilian arrow root and other foods. The tuberous roots form in a cluster at the stem base and are 8 to 15 inches long with some to 3 feet long. Plants are propagated by stem pieces laid horizontally in furrows, somewhat like sugarcane. The roots used as food sources are formed entirely underground. About 800 acres were reported for continental U.S., South Florida, in 1984 (STEPHENS 1988), but there is substantial production in the tropical areas. The bitter type is used to manufacture starch (tapioca).

3. Season, planting to harvest: Usable roots in 8 to 16 months.

4. Production in U.S.: About 500 tons in 1992 (1992 CENSUS). In 1992, 723 acres in Puerto Rico, 8 acres in Guam and 7 acres in Virgin Islands (1992 CENSUS).

5. Other production regions: Africa, South America (mainly Brazil) and Asia. Mainly used for food in Africa, whereas South America and Asia also use it for feed (RUBATZKY 1997).

6. Use: Leaves and dried roots for animal feed. Roots are used for tapioca or as potatoes. Young leaves as potherb. Roots are eaten raw or cooked but must be peeled before used.

7. Part(s) of plant consumed: Roots and young leaves (sweet type)

8. Portion analyzed/sampled: Roots and leaves

9. Classifications:

a. Authors Class: Root and tuber vegetables; Leaves of root and tuber vegetables

b. EPA Crop Group (Group & Subgroup): Root and tuber vegetables (1C and 1D); Leaves of root and tuber vegetables

c. Codex Group: 016 (VR 0463) Root and tuber vegetables; 013 (VL 0463) Leafy vegetables (including *Brassica* leafy vegetables)

d. EPA Crop Definition: None

10. References: GRIN, CODEX, JANICK 1990, MAGNESS, MARTIN 1983, STEPHENS 1988 (picture), US EPA 1995a, YAMAGUCHI 1983 (picture), RUBATZKY, DUKE(a)

11. Production Map: EPA Crop Production Region 13.

12. Plant Codes:

a. Bayer Code: MANES

132

1. Cassia

(Cassia buds, Chinese cassia, Chinese cinnamon, Tej pat, Cassia bark)

Lauraceae

Cinnamomum aromaticum Nees (syn: *C. cassia* auct.)

C. burmanni (Nees) Blume **(Batavia cassia, Padang cassia)**

C. loureirii Nees **(Saigon cassia)**

2. The tree is a tropical evergreen, reaching up to 50 feet, with thick, oblong leaves 3 to 6 inches long. The trees are hardy in the Gulf States in the U.S., but commercial cassia and cassia buds are not produced. The cassia of commerce, quite similar to cinnamon, is the ground, dried bark (stem and branches) of the tree; while cassia buds are the dried, immature fruits harvested when about one-fourth their full size. They resemble cloves, but are smaller. The buds are used as a spice, mainly in confections, while the powdered bark is used in cookery, often as a substitute for cinnamon, which is more expensive. Cassia powder is reddish brown in color.

3. Season: Propagated by seed or cuttings and after 6-7 years the stems are cut close to the ground. Dark green cassia buds are gathered from trees 10 years or older.

4. Production in U.S.: Not grown commercially in U.S.

5. Other production regions: China, Vietnam and Indonesia.

6. Use: Bark and buds are used as spice. Also dried leaves used as a flavoring, Tej pat, in Indian cookery.

7. Part(s) of plant consumed: Bark and buds

8. Portion analyzed/sampled: Bark and buds

9. Classifications:

a. Authors Class: Herbs and spices

b. EPA Crop Group (Group & Subgroup): Herbs and spices (19B)

c. Codex Group: 028 (HS 0776) Spices for Cassia buds; Cassia bark is included with Cinnamon bark 028 (HS 0777) Spices

d. EPA Crop Definition: None

10. References: GRIN, CODEX, GARLAND (picture), LEWIS (picture), MAGNESS, REHM, US EPA 1995a, DUKE(a)

11. Production Map: No specific entry.

12. Plant Codes:

a. Bayer Code: No entry

133

1. Castor oil plant

(Castorbean, Koli, Pa'aila, la'au-'aila, Palma christi, Wonder tree, Higuerilla, Ricin, Herba mora, Castor bean)

Euphorbiaceae

Ricinus communis L.

2. The plant is herbaceous to 15 feet perennial that acts as an annual in temperate regions, but may become a small tree in the tropics. The leaves are large and palmately pinnate. The seeds are borne in capsules, generally covered with soft spines. The capsules split, sometimes explosively, at ripening. The seeds are ovoid in shape, variable in size, generally about a gram in weight. The shelled beans contain 35 to 55 percent of oil, which is extracted by pressure or by solvents. The non-drying oil is used as a medicine and extensively in industry, rarely in cookery. In Asia the leaves are fed to silkworms and cattle.

3. Season, planting: Transplanted about the middle of May since it is susceptible to frost in the early growing season. Fruits are harvested in about 95-180 days depending on cultivar. U.S. harvest begins in October.
4. Production in U.S.: No data. About 459 tons of beans and 46,437 tons as vegetable oil were imported into U.S. in 1993.
5. Other production regions: Naturalized in the tropics and warm regions. Main producing countries – India, Brazil and China in 1985.
6. Use: See above.
7. Part(s) of plant consumed: Oil from seed. Food grade oil is used as an antistick agent in candy molds, and a flavor component in baked goods, beverages and candies. The untreated meal is used as fertilizer or as a fish feed. It needs to be detoxified to be fed to livestock
8. Portion analyzed/sampled: Seed and its oil
9. Classifications:
   a. Authors Class: Oilseed
   b. EPA Crop Group (Group & Subgroup): Miscellaneous
   c. Codex Group: No specific entry
   d. EPA Crop Definition: None
10. References: GRIN, BAILEY 1976, MAGNESS, NEAL (picture), ROBBELEN (picture), USDA 1994a, DUKE(a)
11. Production Map: No entry
12. Plant Codes:
    a. Bayer Code: RIICO

134

1. Catjang
(Catjung, Bombay cowpea, Dolique de Chine, Catjang cowpea, Sowpea, Loubia, Jiang dou)
Fabaceae (Leguminosae)
Vigna unguiculata ssp. *cylindrica* (L.) Verdc. (syn: *V. catjang* (Burm. f.) Walp.; *V. cylindrica* (L.) Skeels; *Dolichos catjang* Burm.; *D. biflorus* L.)
2. The catjang is closely related to southern pea, which see. The plant and culture are similar. The pod of catjang is smaller, 3 to 5 inches long, and the seeds also are smaller. Catjang is grown mostly for feed, but is also grown for the seeds, used as food. The seeds are harvested at the green shell stage or when ripe, as are southern peas. A cultivar is 'MITA 58'. Catjangs have short, compact cylindrical pods and are climbers.
3. Season, bloom to harvest: Green shell, 15 to 20 days. Dry beans, 30 or more days.
4. Production in U.S.: No separate data – included with southern peas.
5. Other production regions: Grown in tropical, subtropical regions, India and Africa (RUBATZKY 1997).
6. Use: Green shell canned or frozen as cooked vegetable. Dry beans as cooked vegetable. Also used as a forage, hay and green manure crop.
7. Part(s) of plant consumed: Seeds only as food; whole plant for feed as cowpeas
8. Portion analyzed/sampled: Seed (succulent or dry); Forage and hay
9. Classifications:
   a. Authors Class: Legume vegetables (succulent or dried); Foliage of legume vegetables
   b. EPA Crop Group (Group & Subgroup): Legume vegetables (succulent or dried) (6B and 6C); Foliage of legume vegetables
   c. Codex Group: 014 (VP 0527) Legume vegetables (immature pods and green seeds)
   d. EPA Crop Definition: Beans = *Vigna* spp.
10. References: GRIN, CODEX, MAGNESS, MARTIN 1983 (picture), US EPA 1995a, YAMAGUCHI 1983, DUKE(h), RUBATZKY
11. Production Map: No entry.
12. Plant Codes:
    a. Bayer Code: VIGSC

135

1. Catnip
(Catmint)
Lamiaceae (Labiatae)
Nepeta cataria L.
2. Catnip is a perennial plant of the mint family, native to Europe and the Orient, but now naturalized in most temperate regions. The plant is densely downy and grows to 4 feet in height with heart-shaped leaves. The leaves were formerly much used as a condiment, especially in sauces, but are now little utilized in foods. The name indicates the attractiveness of the plant to cats. Catnip is by weight 60 to 80 percent stem material.
3. Season, production: About 3 years. Commercial plantings can produce about a ton per acre of dried herb.
4. Production in U.S.: Limited data, grown in home gardens. North Carolina (FOSTER 1993), Florida (STEPHENS 1988) as herbal tea.
5. Other production regions: Limited acres in Alberta, Canada (ROTHWELL 1997).
6. Use: Fresh leaves chewed and dried catnip used as herbal tea. Also used as an ornamental.
7. Part(s) of plant consumed: Top
8. Portion analyzed/sampled: Top (fresh and dried)
9. Classifications:
   a. Authors Class: Herbs and spices
   b. EPA Crop Group (Group & Subgroup): Herbs and spices (19A)
   c. Codex Group: 027 (HH 0726) Herbs; 057 (DH 0726) Dried herbs. Codex uses "catmint" as the commodity name
   d. EPA Crop Definition: None
10. References: GRIN, CODEX, FOSTER (picture), MAGNESS, STEPHENS 1988 (picture), US EPA 1995a, ROTHWELL 1997
11. Production Map: EPA Crop Production Regions 2 and 3.
12. Plant Codes:
    a. Bayer Code: NEPCA

136

1. Cauliflower
(Broccoli, Coliflor, Cauliflower broccoli, Heading broccoli, Broccoflower, Cavalo broccolo, Green cauliflower)
Brassicaceae (Cruciferae)
Brassica oleracea var. *botrytis* L.
2. Cauliflower is grown for the edible, partially developed, multiple flower head. For food, the plant is grown as an annual and is usually seeded in beds, then transplanted to the

field at 4 to 5 weeks of age. It develops large basal leaves which surround but do not completely cover the head which is borne terminally on the stem. Since a white head is desired it is necessary to blanch the head by bringing the large outer leaves up over it and fastening them with twine or rubber bands. This is done 3 to 10 days before harvest, depending on prevailing temperatures. Some kinds have sufficient leaf cover to make tying unnecessary. Heads are harvested while still compact, white and firm. The white heads are called curds, the outer leaves and stalk are removed before cooking and the head washed.

3. Season, field setting to harvest: 2 months in summer. Winter varieties grown where winters are mild, mainly California, 4 to 5 months. Cauliflower is a cool season crop.

4. Production in U.S.: As dual usage (fresh and processing) crop, cauliflower in 1995 planted on 51,500 acres with fresh market production at 282,750 tons and processing at 43,510 tons (USDA 1996b). In 1995, cauliflower reported in the top 6 states as California (40,700 acres), Arizona (4,500), Oregon (3,300), New York (1,300), Michigan (900) and Texas (800). California and Arizona account for 90 percent of the production for fresh and process. Fresh market and processing winter season production for cauliflower comes from California (USDA 1996b).

5. Other production regions: In 1995, Canada reported 6,719 acres of cauliflower including broccoflower (ROTHWELL 1996a).

6. Use: Fresh, frozen, mostly as pot vegetable, some as salad or pickles. Preparation: For cooking, wash and trim the main stem lightly. Also, can be used fresh in salads (LOGAN 1996).

7. Part(s) of plant consumed: Undeveloped flower head, consisting of succulent stems (stalks) and flower initials

8. Portion analyzed/sampled: Flower head and stem (stalk)

9. Classifications:
   a. Authors Class: *Brassica* (cole) leafy vegetables
   b. EPA Crop Group (Group & Subgroup): *Brassica* (cole) leafy vegetables (5A) (Representative Crop)
   c. Codex Group: Broccoli: 010 (VB 0400) *Brassica* (cole or cabbage) vegetables, head cabbage, flowerhead *Brassica*; Cauliflower: 010 (VB 0404) *Brassica* (cole or cabbage) vegetables, head cabbages, flowerhead *Brassica*; Broccoli and cauliflower: 010 (VB 0042) Flowerhead *Brassica*
   d. EPA Crop Definition: See Broccoli

10. References: GRIN, CODEX, LOGAN (picture), MAGNESS, ROTHWELL 1996a, STEPHENS 1988 (picture), USDA 1996b, US EPA 1996a, US EPA 1995a, SWIADER

11. Production Map: EPA Crop Production Regions: Caulifower: 1, 3, 5, 10 and 12.

12. Plant Codes:
    a. Bayer Code: BRSOB

137

1. Celeriac
(Apio nabo, Gen qin cai, Celery root, Turnip-rooted celery, Knob celery)
Apiaceae (Umbelliferae)

Apium graveolens var. *rapaceum* (Mill.) Gaudin (syn: *A. rapaceum* Miller)

2. Celeriac to 3 feet tall is a plant closely related to celery and is like celery in growth habit and general appearance. How-

ever, it develops a thick, tuberous base and root, which is used as a salad and cooked vegetable. The leaves rise directly from this thickened base which may reach 3 to 4 inches in diameter and similar length. Plants are usually started in seed beds and transplanted to the field. The celery-like leaves and stalks that are hollow are not very palatable. The bulbous root is covered with soil during growth but one-half exposed when ready to dig. Celeriac is of minor importance in the U.S.

3. Season, field setting to harvest: 3½ to 4 months. Seeding to harvest: About 4 to 6 months (normally 6 months).

4. Production in U.S.: About 756 acres with 1,034 tons reported as celeriac or apio from Puerto Rico (1992 CENSUS) but likely Arracacha, which see (NATIONAL RESEARCH COUNCIL 1989). In California, transplant 3 times a year starting June to September. Direct seed crop could be harvested in 120 days but normally 180 days. June transplants harvested in October and November. September transplants harvested in early April. The types of packs include root and stem stubs and whole plant. About 50 acres grown in California around the San Francisco area (RATTO 1988). Similar growing requirements to celery (which see) (STEPHENS 1988).

5. Other production regions: Grown in temperate Europe (RUBATZKY 1997).

6. Use: Pot vegetable, salad; not processed, largely fresh market. Wash root before use, eaten raw, cooked or pickled. When it is marketed with its celery-like leaves, they can be trimmed from the root and used as flavoring in soups and stews.

7. Part(s) of plant consumed: Tuberous root primarily but sometimes the whole plant is shipped to market

8. Portion analyzed/sampled: Root and top (leaves), analyze separately

9. Classifications:
   a. Authors Class: Root and tuber vegetables; Leaves of root and tuber vegetables
   b. EPA Crop Group (Group & Subgroup): Root and tuber vegetables (1A and 1B); Leaves of root and tuber vegetables
   c. Codex Group: 016 (VR 0578) Root and tuber vegetables
   d. EPA Crop Definition: None

10. References: GRIN, CODEX, MAGNESS, RATTO 1988, REHM, RICHE, STEPHENS 1988 (picture), US EPA 1995a, RUBATZKY (picture), NATIONAL RESEARCH COUNCIL 1989, SCHNEIDER 1985

11. Production Map: EPA Crop Production Region: celery (which see).

12. Plant Codes:
    a. Bayer Code: APUGR

138

1. Celery
(Celery seed, Apio, Quin cai, Stalk celery)
Apiaceae (Umbelliferae)

Apium graveolens var. *dulce* (Mill.) Pers.

2. Celery is grown for the thick, succulent leaf stalks or petioles the first year which are esteemed as a salad and to a lesser extent as a cooked vegetable. The leaves rise from a crown at ground level. Leaf stems are up to a foot long, with the greatly compound leaf blades extending an additional foot during growth. Plants are usually started in beds, but

seed may be field sown in cool climates. Outer leaves develop first and to largest size, inner leaf stems are smaller and more succulent. Formerly most of the celery was blanched by placing heavy paper or boards against each side of the rows to exclude most of the light. At present, green or non-blanched celery is mainly marketed. Outermost leaves, particularly if the stems are very coarse or scarred, are frequently removed before packing. Plants must be grown rapidly for most succulent and desirable quality. As a condiment, the seeds of celery are used either as the whole seeds or ground and mixed with salt in celery salt. For seed production, celery plants are exposed to cold, mainly by overwintering in mild climates. They then develop the seed stalks, 2 to 3 feet tall. Seeds are numerous in compound umbels. The small seeds are flattened and broader than long. Harvest by pulling the plants, allowing them to dry, then threshing. It is in late summer of the second growing season, 5 to 6 months from the start of growth the second year.

3. Season, seeding to field planting: 2½ to 3 months; field planting to harvest, 3 to 4 months for fresh celery.
4. Production in U.S.: 35,580 acres in 1992 from California, Florida, Michigan, Texas, New York and Ohio (1994 AGRICULTURAL STATISTICS).
5. Other production regions: Grows in temperate regions (RUBATZKY 1997).
6. Use: Mainly salad, also flavoring in soups, stews, juices and as a cooked vegetable. Celery seed as condiment. Small amount dehydrated as a flavoring. Per capita consumption in 1994 for U.S. is 6.8 pounds (PUTNAM 1996).
7. Part(s) of plant consumed: Mainly petioles. Leaf blades used sparingly as flavoring. Celery seed
8. Portion analyzed/sampled: Untrimmed leaf stalk (petiole) for fresh. For seed: celery seed
9. Classifications:
   a. Authors Class: Leafy vegetables (except *Brassica* vegetable); Herbs and spices
   b. EPA Crop Group (Group & Subgroup): Leafy vegetables (except *Brassica* vegetables) (4B); Herbs and spices (19B). (Representative Commodities)
   c. Codex Group: 017 (VS 0624) Stalk and stem vegetables; 057 (DH 0624) Dried herbs; 028 (HS 0624) Spices (seed)
   d. EPA Crop Definition: Celery = Florence fennel (sweet anise, sweet fennel, finochio) (fresh leaves and stalks only)
10. References: GRIN, CODEX, FARRELL, MAGNESS, RICHE, USDA 1994a, US EPA 1995a, US EPA 1995b, US EPA 1994, IVES, REHM, RUBATZKY, SWIADER, PUTNAM 1996
11. Production Map: EPA Crop Production Regions 3, 5, 6 and 10.
12. Plant Codes:
   a. Bayer Code: APUGD

139

1. Celery, Chinese
(Smallage, Leaf celery, Soup celery, Cutting celery, Celeri a couper, Kintsai, Heung kunn)
Apiaceae (Umbelliferae)
Apium graveolens L. var. *secalinum*
2. Plant closely resembles the wild celery (slender celery, smallage) *Cyclospermum leptophyllum* (Pers.) Sprague ex Britton & P. Wilson (syn: *A. leptophyllum* (Pers.) F. Muell ex

Benth.). Chinese celery has a branching rather than bunching habit, which makes it easier to harvest individual leafstalks. The stalks are hollow, rather thin, tender and brittle. Grown for their strongly aromatic leaves and stalks which are cut like parsley.
3. Season, seeding to harvest: About 90 days.
4. Production in U.S.: No data. Limited. Grown in Florida (SHULER 1994).
5. Other production regions: Grows in Asia and Mediterranean region (RUBATZKY 1997).
6. Use: Leaves and stalks used in stir-fried dishes or used as a garnish in bowls of soups.
7. Part(s) of plant consumed: Leaves and stalks
8. Portion analyzed/sampled: Leaves and stalks
9. Classifications:
   a. Authors Class: Leafy vegetables (except *Brassica* vegetables)
   b. EPA Crop Group (Group & Subgroup): Leafy vegetables (except *Brassica* vegetables) (4B)
   c. Codex Group: No specific entry
   d. EPA Crop Definition: None, but proposing – Celery = Chinese celery
10. References: GRIN, FACCIOLA, SHULER (picture) (in part), US EPA 1995a, RUBATZKY (picture)
11. Production Map: EPA Crop Production Regions 3 and 13.
12. Plant Codes:
   a. Bayer Code: APUGS, APULE (*C. leptophyllum*)

140

1. Centipedegrass
(Lazy-man's-grass)
Poaceae (Gramineae)
Eremochloa ophiuroides (Munro) Hack.
2. This warm season grass, native to Asia, was introduced into the U.S. in 1919 from China. It is a low growing perennial, spreading by stolons. It makes a dense sod mat of stems and leaves. It is of very low nutritive value, so is of little value for pastures, but is useful for lawns and erosion control. It is not hardy, but is grown in the Coastal Plain from North Carolina to Texas. Propagation is by planting stolons or from seed.
3. Season: Warm season pasture. Flowers mostly September to November.
4. Production in U.S.: No specific data, See Forage grass. Coastal Plain from North Carolina to Texas.
5. Other production regions: Native to southeastern Asia (GOULD 1975).
6. Use: Erosion control, pasture, turfgrass.
7. Part(s) of plant consumed: Foliage
8. Portion analyzed/sampled: Forage and hay
9. Classifications:
   a. Authors Class: See Forage grass
   b. EPA Crop Group (Group & Subgroup): See Forage grass
   c. Codex Group: See Forage grass
   d. EPA Crop Definition: None
10. References: GRIN, MAGNESS, ALDERSON, GOULD
11. Production Map: EPA Crop Production Regions 2, 3, 4, 6 and 8.
12. Plant Codes:
   a. Bayer Code: ERLOP

141

1. Chaya
(Tree spinach, Vegetable chaya)

Euphorbiaceae

Cnidoscolus chayamansa McVaugh

2. Chaya is a large, leafy shrub to 8 feet tall. The dark green leaves are 6 to 8 inches across. 'Pig chaya' is one of best eating varieties. It somewhat resembles the cassava plant with foliage similar to okra.

3. Season, planting to harvest: Stem cuttings about 6-12 inches long are used and the leaves can be harvested after the first year of planting in Florida. In Puerto Rico, 3 to 4 months from planting to first harvest.

4. Production in U.S.: No data. Southern Florida (STEPHENS 1988) and Puerto Rico (MARTIN 1979).

5. Other production regions: Grows in Mexico, Central and South America (RUBATZKY 1997).

6. Use: Potherb. Used like cooked spinach, must be cooked because raw leaves contain hydrocyanic acid.

7. Part(s) of plant consumed: Leaves

8. Portion analyzed/sampled: Leaves (fresh)

9. Classifications:
   a. Authors Class: Herbs and spices
   b. EPA Crop Group (Group & Subgroup): Miscellaneous
   c. Codex Group: No specific entry
   d. EPA Crop Definition: None

10. References: GRIN, MARTIN 1979, STEPHENS 1988, RUBATZKY

11. Production Map: EPA Crop Production Region 13.

12. Plant Codes:
    a. Bayer Code: No specific entry

142

1. Chayote (Chi-o-tay)
(Choco, Xuxu, Alligator pear, Kajot, Choke, Merliton, Christophine, Vegetable pear, Pepineca, Cristophine, Chayotli, Mirliton, Mango squash, Custard marrow, Cho-cho, Talote, Chayotte, Brionne, Choko, Chuchu, Pipinela, Guispui, Chou-Chou, Tallote)

Cucurbitaceae

Sechium edule (Jacq.) Sw. (syn: *S. edulis* Jacq.)

2. This is an important food crop in tropical regions, including the West Indies. In the U.S., it can be grown in areas having mild winters. The plant is a perennial-rooted vine, which bears fruit that is green to white and varying in size from a few ounces up to 3 pounds. It resembles summer squash both in appearance and use. Fruit surface may be smooth, wrinkled or prickly. In tropical regions the plants are often trained on trellis or other supports and produce fruit almost continuously. Fruits contain a single large seed. In the U.S., fruits are usually eaten, after boiling or frying. In the tropics, roots and vine tops as well as fruits are used as food.

3. Season, bloom to maturity: About 2 months, but bloom and fruit setting continuous. Seeding to first harvest is about 6 months. The entire fruit is planted in the spring and spaced 10 feet apart with each plant producing 30 to 35 fruits.

4. Production in U.S.: No data. Minor in continental U.S. Grown in Florida, Louisana and Puerto Rico (SCHNEIDER 1985). California and Florida are main suppliers (LOGAN 1996). Planted in all areas of Florida (STEPHENS 1988).

5. Other production regions: Costa Rica and Mexico are main importers (LOGAN 1996). Mediterranean region.

6. Use: Fresh market. Cooked vegetable. In preparation, the fruit is normally peeled, cubed and the flesh is used boiled, fried, stuffed or baked. Also the tips of the vines are cut periodically and used as a vegetable, and the tuberous root is harvested when the plant is to be destroyed and used as a valuable farinaceous crop.

7. Part(s) of plant consumed: Fruit including seed; tuberous root and vine tips are pimarily consumed locally

8. Portion analyzed/sampled: Whole fruit

9. Classifications:
   a. Authors Class: Root and tuber vegetables; Cucurbit vegetables
   b. EPA Crop Group (Group & Subgroup): Root and tuber vegetables (1C and 1D); Cucurbit vegetables (9B)
   c. Codex Group: 016 (VR 0423) Root and tuber vegetables; 011 (VC 0423) Cucurbits fruiting vegetables
   d. EPA Crop Definition: Summer squash = Chayote

10. References: GRIN, CODEX, LOGAN, MAGNESS, MARTIN 1983, SCHNEIDER 1985, STEPHENS 1988 (picture), US EPA 1995a, US EPA 1995c, REHM

11. Production Map: EPA Crop Production Regions 3, 4, 10 and 13.

12. Plant Codes:
    a. Bayer Code: SEHED

143

1. Cherimoya
(Chirimoya, Anon, Cherimolier, Chirimolia, Custard apple, Cherimoyar, Lakshmanphal)

Annonaceae

Annona cherimola Mill.

2. The cherimoya, native to the Northern Andes in South America, is a small spreading tree, up to 25 feet, with highly pubescent deciduous leaves up to 10 inches long. Propagated by seed and grafting. The plant will not set fruit well in hot, humid tropics. It thrives best in tropical highlands. The fruits are large, up to 4 to 6 pounds, generally heart shaped or conical. The surface may be nearly smooth, or irregular due to points on the carpel terminals. The rather thin rind encloses a juicy, white, custard-like pulp in which numerous seeds are embedded. Fruit is sensitive to extremes of heat and cold.

3. Season: Bloom to maturity in about 5 months.

4. Production in U.S.: California and Hawaii, exact acreage unknown.

5. Other production regions: South America and Spain.

6. Use: The fruit is generally consumed fresh or made into ice cream.

7. Part(s) of plant consumed: Fruit pulp, the pulp's flavor is of the finest among tropical fruits, with a taste of pineapple, mango, papaya and vanilla custard

8. Portion analyzed/sampled: Whole fruit

9. Classifications:
   a. Authors Class: Tropical and subtropical fruits – inedible peel
   b. EPA Crop Group (Group & Subgroup): Miscellaneous
   c. Codex Group: 006 (FI 0331) Assorted tropical and subtropical fruits – inedible peel
   d. EPA Crop Definition: Proposing – Sugar apple = Cherimoya

10. References: GRIN, CODEX, LOGAN, MAGNESS, MARTIN 1987, US EPA 1994, PUROHIT
11. Production Map: EPA Crop Production Regions 10 and 13.
12. Plant Codes:
    a. Bayer Code: ANUCH

144

1. Cherry, Black
(Wild black cherry, Rum cherry, Black rum cherry)
Rosaceae
Prunus serotina Ehrh. ssp. *serotina*
2. Native black cherry tree of eastern North America, grows rapidly to 50 feet or more tall. Dense foliage with peach-shaped leaves. Fruits are globose, red to purple-black and sweet or bitter.
3. Season: Flowers in May.
4. Production in U.S.: No data. Grows from Nova Scotia to North Dakota, south to Florida and Texas (BAILEY 1976). Hardy in USDA hardiness Zones 3-9 (WHEALY 1993).
5. Other production regions: North America (FACCIOLA 1990).
6. Use: Eaten fresh, made into preserves, pies, cherry bounce, and flavoring cider, brandy, rum and liqueurs. The bark is used in medicine.
7. Part(s) of plant consumed: Fruit
8. Portion analyzed/sampled: Whole fruit with pit and stem removed and discarded
9. Classifications:
    a. Authors Class: Stone fruits
    b. EPA Crop Group (Group & Subgroup): Miscellaneous
    c. Codex Group: No specific entry
    d. EPA Crop Definition: None
10. References: GRIN, BAILEY 1976, FACCIOLA, WHEALY
11. Production Map: EPA Crop Production Regions 1, 2, 4 and 5.
12. Plant Codes:
    a. Bayer Code: PRNSO

145

1. Cherry, Nanking
(Manchu cherry, Downy cherry, Mongolian cherry, Chinese bush cherry, Hansens bush cherry)
Rosaceae
Prunus tomentosa Thunb.
2. This hardy deciduous bush is native to central Asia. The plant is a spreading shrub or small tree that grows 10 to 15 feet high. The plant flowers very early in spring. The flowers are profuse and somewhat frost tolerant and yield a heavy crop of short stemmed half-inch red fruit. The fruit has a tangy taste similar to tart cherry, however fruit are extremely soft limiting the ability to ship the fruit any distance. In addition, the fruit has a short shelf-life. The plant can hybridize easily with apricot, plum and other cherries. Some improved cultivars include 'Drilea', 'Eileen' and 'Baton Rouge'.
3. Season: Fruit ripens last half of July.
4. Production in U.S.: Northern tier states, no statistical data available.
5. Other production regions: Canada and Russia.
6. Use: Fruit is used for pies, jams and jellies as well as eaten fresh out of hand. The plant is a beautiful ornamental specimen.

7. Part(s) of plant consumed: Mainly the fruit pulp and fruit peel
8. Portion analyzed/sampled: Whole fruit with pit and stem removed and discarded
9. Classifications:
    a. Authors Class: Stone fruits
    b. EPA Crop Group (Group & Subgroup): Could include in Crop Group 12: Stone Fruits Group based on "Plum (*Prunus domestica, Prunus* spp.)". But considered miscellaneous for now
    c. Codex Group: 003 (FS 0012) Stone fruits; (FS 0013) Cherries
    d. EPA Crop Definition: None
10. References: GRIN, CODEX, REICH, US EPA 1994, WHEALY, USDA NRCS, FACCIOLA
11. Production Map: EPA Crop Production Regions 1, 5 and 12.
12. Plant Codes:
    a. Bayer Code: PRNTO

146

1. Cherry, Sweet
(Mazzard cherry, Cerezo, Cerise, Bird cherry, Duke cherry, Maraschino cherry, Gean)
Rosaceae
Prunus avium (L.) L.
2. The sweet cherry tree is an upright grower, medium in size, up to 40 or more feet, but usually held to 20 feet or less by pruning. The deciduous leaves are oblong. The sweet cherries are separated into two types, the Hearts and the Bigarreaus. The Hearts are characterized by soft tender flesh and heart shaped fruit. The Bigarreaus are smooth round shaped fruit. All types have smooth, thin skins which adhere to the fleshy pulp and a single round pit. The fruit is $^3/_4$ to 1 inch in diameter. There is a depression in the skin where the stem attaches. Fruits are borne on 1 to 3 inch stems, in groups of 1 to 5. Fruit color varies in different varieties from light red to near black, with a few yellow. Prominent varieties include 'Bing', 'Black Tartarian', 'Van', 'Emperior Francis', 'Lambert', 'Napoleon', and 'Rainier'. To produce fruit, sweet cherries must be cross pollinated. Some cultivars are incompatable with others. The plant is sensitive to extreme heat and cold conditions which limits its production areas. In addition, the plant is subjected to early season frost injury and fruit cracking, factors which reduce the quantity and quality of fruit. 'Duke' cherry is a hybrid between sweet and tart cherry.
3. Season: Bloom to harvest in 80 to 100 days.
4. Production in U.S.: In 1995, there were 47,380 domestic bearing acres which produced 165,250 tons. Key producing states include Washington (14,800 acres), California (12,500 acres), Oregon (10,400 acres) and Michigan (7,300 acres) (USDA 1996f).
5. Other production regions: Significant imports from Canada and Chile. In 1995, Canada reported 2,659 acres (ROTHWELL 1996a).
6. Use: Fruit is used for fresh eating out of hand, cooked or canned into sauces for desserts or salads. Also fresh sweet cherries are used for juice, wine, brandy and freezing. In the U.S., sweet cherries are primarily used fresh, approximately 50 percent. The processed sweet cherries are primarily brined and used for maraschino cherries.

7. Part(s) of plant consumed: Fruit skin and pulp without the pit
8. Portion analyzed/sampled: Whole fruit with the pit and stem removed and discarded
9. Classifications:
   a. Authors Class: Stone fruits
   b. EPA Crop Group (Group & Subgroup): Crop Group 12: Stone Fruits Group (Representative Crop)
   c. Codex Group: Stone fruits 003 (FS 0013) Cherry; (FS 0244) Sweet cherry
   d. EPA Crop Definition: Cherry = Sour cherry, Sweet cherry
10. References: GRIN, CODEX, LOGAN, MAGNESS, TESKEY, US EPA 1995, US EPA 1994, ROTHWELL 1996a, IVES, USDA 1996f, DESAI
11. Production Map: EPA Crop Production Regions 5, 10, 11 and 12.
12. Plant Codes:
   a. Bayer Code: PRNAV

147

1. Cherry, Tart
(Sour cherry, Pie cherry, Red cherry, Pitted cherry, Cerisier aigre, Cerezo, Dwarf cherry, Duke cherry)
Rosaceae
Prunus cerasus L.

2. The tart cherry tree is rather small, to 20 feet, spreading and much branched. It is much hardier than the sweet cherry tree with production usually in northern areas (not south of Ohio and Potomac rivers). The leaves are ovate, obovate in shape, rather stiff and glossy. There are two types, the Amarelles, with light red flesh and juice and the Morellos. Fruits are in groups of 1 to 4 on 1- to 2-inch stems. They are near round, slightly depressed at the stem, about $3/4$ inch diameter. The peel is smooth and thin, and adheres to the soft, juicy, tart pulp. The single pit is near round. The main varieties are 'Montmorency' and 'Early Richmond'. 'Duke' cherry is a hybrid between tart and sweet cherry. Almost all tart cherries are mechanically harvested and flushed with water.
3. Season: Bloom to harvest in 80 to 100 days.
4. Production in U.S.: In 1995, there were 45,625 domestic bearing acres which produced 197,800 tons. Key producing states include Michigan (30,000 acres), New York (4,000 acres), Wisconsin (3,100 acres), Oregon (1,600 acres) and Pennsylvania (1,500 acres) (USDA 1996f).
5. Other production regions: In 1995, Canada grew 2,256 acres (ROTHWELL 1996a).
6. Use: Mostly frozen or canned for culinary use (pies, preserves) or processed into juice. Approximately 10 percent of all tart cherries are dried, infused with sugar and made into a snack food. Tart varieties are primarily processed, about 94 percent. Small quantities are brined or used for juice, wine, preserves and candied cherries.
7. Part(s) of plant consumed: All except pit. Some oil extracted from seed kernels
8. Portion analyzed/sampled: Whole fruit with the pit and stem removed and discarded (unwashed fruit)
9. Classifications:
   a. Authors Class: Stone fruits
   b. EPA Crop Group (Group & Subgroup): Crop Group 12: Stone Fruits Group (Representative Crop)
   c. Codex Group: Stone fruits 003 (FS 0013) Cherry; (FS 0243) Sour cherry

d. EPA Crop Definition: Cherry = Sour cherry, Sweet cherry
10. References: GRIN, CODEX, MAGNESS, McCALLUM, TESKEY, US EPA 1995, US EPA 1994, DESAI, ROTHWELL 1996a, USDA 1996f
11. Production Map: EPA Crop Production Regions 1, 5, 9, 11 and 12.
12. Plant Codes:
   a. Bayer Code: PRNCE

148

1. Chervil
(Salad chervil, Sweet cicely, Leaf chervil, Garden chervil)
Apiaceae (Umbelliferae)
Anthriscus cerefolium (L.) Hoffm. (syn: *A. longirostris* Bertol.)

2. Chervil is an annual plant, reaching a height of 1 to 2 feet. The leaves are much compounded, in general resembling parsley. Garden forms with curled leaves are available. Chervil is grown for its pungent, aromatic and decorative leaves, which are used for garnishing and flavoring. One of the "fines herbes" of French cookery including basil, chives, parsley, sage, savory and tarragon.
3. Season, seed to first harvest: 6 to 8 weeks. Several harvests can be made.
4. Production in U.S.: No data. Very limited. Cultivated in U.S. (FARRELL 1985).
5. Other production regions: Cultivated in Mediterranean region and Russia (FARRELL 1985).
6. Use: Decorative and as flavoring in culinary cookery, salads.
7. Part(s) of plant consumed: Leaves
8. Portion analyzed/sampled: Leaves (fresh and dried)
9. Classifications:
   a. Authors Class: Leafy vegetables (except *Brassica* vegetables); Herbs and spices
   b. EPA Crop Group (Group & Subgroup): Leafy vegetables (except *Brassica* vegetables) (4A); Herbs and spices (19A)
   c. Codex Group: 013 (VL 0465) Leafy vegetables (including *Brassica* leafy vegetables); 027 (HH 0465) Herbs
   d. EPA Crop Definition: None
10. References: GRIN, CODEX, FARRELL (picture), FOSTER (picture), LEWIS, MAGNESS, US EPA 1995a, FORTIN
11. Production Map: EPA Crop Production Region 10.
12. Plant Codes:
   a. Bayer Code: ANRCE

149

1. Chervil, Turnip-rooted
(Sham, Tuberous chervil, Parsnip chervil)
Apiaceae (Umbelliferae)
Chaerophyllum bulbosum L.

2. This is a root-crop vegetable, grown to some extent in Europe, but little known in the U.S. Except for seed, it is grown as an annual, like carrots. The edible roots are 4 to 5 inches long, tapered and grey or near black. Interior flesh is yellowish white. Roots are used as cooked vegetables, and in stews and soups. Conditions of growth in general are similar to those for carrots. Grows to a height of 3 feet.
3. Season, seeding to harvest: 3 to 5 months for roots. Leaves can be utilized and can be harvested 6-10 weeks after planting.
4. Production in U.S.: No data.

5. Other production regions: France (JANICK 1990).
6. Use: Culinary, as boiled vegetable and in soups and stews.
7. Part(s) of plant consumed: Primarily tuberous root. Secondarily the leaves
8. Portion analyzed/sampled: Root; Leaves
9. Classifications:
   a. Authors Class: Root and tuber vegetables; Leaves of root and tuber vegetables
   b. EPA Crop Group (Group & Subgroup): Root and tuber vegetables (1A and 1B); Leaves of root and tuber vegetables
   c. Codex Group: 016 (VR 0579) Root and tuber vegetables
   d. EPA Crop Definition: None
10. References: GRIN, CODEX, FACCIOLA, JANICK 1990, MAGNESS, US EPA 1995a, RUBATZKY
11. Production Map: No entry.
12. Plant Codes:
   a. Bayer Code: CHPBU

150

1. Chestnut
Fagaceae
Castanea spp.
1. American chestnut
(Castana)
C. dentata (Marshall) Borkh. (syn: *C. americana* (Michaux) Raf.)
1. Chinese chestnut
(Chinese hairy chestnut)
C. mollissima Blume
1. European chestnut
(Spanish chestnut, Sweet chestnut, Eurasian chestnut)
C. sativa Mill. (syn: *C. vesca* Gaertn.; *C. vulgaris* Lam.)
1. Japanese chestnut
C. crenata Siebold & Zucc. (syn: *C. japonica* Blume; *C. pubinervis* (Hassk.) C.K. Schneid.; *C. stricta* Siebold & Zucc.)
1. Chinquapin
(Allegheny chinkapin, Bush chestnut)
C. pumila (L.) Mill.
2. The various chestnut species vary in height to 100 feet for the American and about 30 feet for the Chinese. All have elongated, oval to lanceolate entire leaves. The fruits are spined burrs, fibrous as they mature, which partially open at maturity. They usually contain up to 5 angular nuts. The "shell" or covering of the kernel is rather thin and tough, leathery rather than woody. A fungus disease, chestnut blight, has killed practically all large American chestnut trees in the U.S. The Chinese chestnut, resistant to the blight, is now planted somewhat, and considerable quantities of European nuts are imported.
3. Season, bloom to harvest: About 100 to 120 days with nut harvested from the ground in August to October.
4. Production in U.S.: No data. Eastern U.S., California and maritime northwest U.S. Almost 400 acres in California (1992) and over 2000 acres have been established nationwide (1992) with most in the southeast (WALLACE 1992).
5. Other production regions: Japan and China (FACCIOLA 1990).

6. Use: Served raw, boiled, roasted, cooked with meat, puree and ground into flour.
7. Part(s) of plant consumed: Nutmeat
8. Portion analyzed/sampled: Nutmeat
9. Classifications:
   a. Authors Class: Tree nuts
   b. EPA Crop Group (Group & Subgroup): Tree nuts
   c. Codex Group: 022 (TN 0664) Tree nuts
   d. EPA Crop Definition: None
10. References: GRIN, CODEX, MAGNESS, US EPA 1995a, US EPA 1995b, WALLACE 1992, WININGER 1991, WOODROOF(a) (picture), FACCIOLA
11. Production Map: EPA Crop Production Regions 1, 2, 3, 4, 5, 10 and 12.
12. Plant Codes:
   a. Bayer Code: CSNCR (*C. crenata*), CSNDE (*C. dentata*), CSNMO (*C. mollissima*), CSNPU (*C. pumila*), CSNSA (*C. sativa*)

151

1. Chia
(Chia pet, Mexican chia)
Lamiaceae (Labiatae)
Salvia hispanica L.
2. Summer annual to 2 feet tall with glaborous leaves ovate and 2 to 3 inches long. Cultivated varieties are grown in central Mexico and Guatemala. The winter annuals are *S. columbariae* Benth. (**California** and **Golden chia**) and *S. carduacea* Benth. that are grown in southern California with its Mediterranean type of climate. The summer annuals do not grow well in California. The winter annuals germinate best in the fall and grow slowly through December and January and harvest early summer. Another summer annual chia is *Hyptis suaveolens* (L.) Poit called **Chia grande**, **Cham**, **Conivari**, **Horehound** and **Wild spikenard** which is grown in Mexico.
3. Season, seeding to harvest: Summer annuals germinate in the warm days of late spring and flower in the short days of September for seed harvest in fall.
4. Production in U.S.: No data. See above.
5. Other production regions: Central America and Mexico (FACCIOLA 1990). Naturalized in West Indies (BAILEY 1976).
6. Use: Seeds sprouted and eaten in salads. Seeds used for making drinks in Mexico. Seed also ground into meal for bread and cakes. Seeds are eaten raw or parched and used in sauces and as a thickening agent. Mixed with orange juice the gel-like seeds make a nutritious breakfast.
7. Part(s) of plant consumed: Seeds
8. Portion analyzed/sampled: Seeds
9. Classifications:
   a. Authors Class: Herbs and spices
   b. EPA Crop Group (Group & Subgroup): Miscellaneous
   c. Codex Group: No specific entry
   d. EPA Crop Definition: None
10. References: GRIN, BAILEY 1976, FACCIOLA, GRIN, JANICK 1990 (picture of *Salvia* spp.), GENTRY
11. Production Map: EPA Crop Production Region 13.
12. Plant Codes:
   a. Bayer Code: No specific entry

152

1. Chickpea
(Garbanzo bean, Pea bean, Garbs, Bengal gram, Common gram, Cecci, Indian gram, Gram pea, Yellow gram, Jierdou, Egyptian pea, Cece bean, Homos, Chana dal, Besan, Dhokla, Chana dahl)

Fabaceae (Leguminosae)

Cicer arietinum L.

2. Chickpea is a warm-season pea-like, bushy, hairy, annual plant about 2 feet in height. Leaves are compound pinnate. There are 2-3 seeds per pod. Chickpea culture is similar to snap beans and harvesting is similar to dry beans. The edible seeds borne in pods are near round, but flattened on the sides about $^1/_3$-inch diameter. There are two types of chickpea – kabuli and desi which is more common. In India it is called Bengal gram; and when split, Chana dal or dhal; as a flour it is called Besan; and the fermented product is called Dhokla.

3. Season, seeding to harvest: $3^1/_2$ to 5 months.

4. Production in U.S.: 43,900 acres in U.S. for 1996 (USDA NASS 1996). Western U.S. (SCHREIBER 1995 and STEPHENS 1988). In 1993, Washington state reported 4,805 acres and 1994 as 2,466 acres. In Washington, most chickpeas are used for canning. In 1996, California reported 25,000 acres, Washington (8,600), Idaho (7,500) and Oregon (2,800 acres). (USDA NASS 1996).

5. Other production regions: Grows in warm temperate, subtropical regions, India and Middle Eastern countries (RUBATZKY 1997).

6. Use: Cooked in soup, salad, flour, canned, frozen, fresh or fermented.

7. Part(s) of plant consumed: Seeds. Foliage is not a significant feed item when chickpeas are combined. The pod and straw are deposited in the field

8. Portion analyzed/sampled: Seeds

9. Classifications:
   a. Authors Class: Legume vegetables (succulent or dried)
   b. EPA Crop Group (Group & Subgroup): legume vegetables (succulent or dried) (6C); Foliage of legume vegetables
   c. Codex Group: 014 (VP 0524) Legume vegetables; 015 (VD 0524) Pulses; 050 (AL 0524) Legume animal feeds
   d. EPA Crop Definition: Beans = Chickpeas; Peas = Chickpeas

10. References: GRIN, CODEX, MAGNESS, SCHREIBER, STEPHENS 1988, USDA NASS 1996, US EPA 1995a, RUBATZKY, SPLITTSTOESSER, FACCIOLA

11. Production Map: EPA Crop Production Regions 10 and 11.

12. Plant Codes:
    a. Bayer Code: CIEAR

153

1. Chicory
(Common chicory, Italian dandelion, Asparagus chicory, Catalogua chicory, Succory, Achicoria, Frisee, Radichetta, Green chicory)
1. Chicory root
(Coffee chicory, Brunswick chicory, Magdeburg chicory)
1. Radicchio
(Red chicory)
1. Belgium endive

(White endive, Red endive, French endive, Ondeev, Chicon, Witloof chicory, Brussels chicory)

Asteraceae (Compositae)

Cichorium intybus L.

2. Chicory produces a large, tapered root, and both leaves and roots are utilized. Asparagus or Catalogua chicory is grown for the tender leaves and flower shoots, and used as potherbs or greens. Radicchio is red and resembles a small romaine or head lettuce with a zesty flavor. Cultural practices are similar to endive and lettuce with 80 to 85 days to harvest from seeding. Roots of chicory are dried, ground and used as a coffee substitute or supplement. Roots are also grown during the summer, dug and buried upright in damp sand or other material for year round forcing in a dark, warm place for blanched tops. This produces the witloof, Belgium, white, red or French endive, used as a salad. Roots are held in cold storage for 6 to 11 months before forced. In general growth characteristics, chicory is very similar to carrots.

3. Season, seeding to harvest: For greens, 2 to 3 months. For roots for coffee or forcing, 5 to 6 months. Roots are forced for 3 to 4 weeks for tops.

4. Production in U.S.: About 847 acres for chicory. No data for coffee roots (1992 CENSUS). About 200 acres forcing roots grown in California for Belgium endive (COLLINS 1996). In 1992, California reported 469 acres of chicory, New Jersey (86) and New York with 25 acres (1992 CENSUS).

5. Other production regions: Belgium endive in grown in temperate Europe. Chicory is native to the Mediterranean region (RUBATZKY 1997).

6. Use: Green tops as potherbs. Roots as coffee supplement or food extract. The pulp from extracted roots may be used as feed but would be an insignificant feed item. Forced tops as salad. Radicchio tops as salad and potherb.

7. Part(s) of plant consumed: Green, red or forced tops, roots for coffee substitute or food extract

8. Portion analyzed/sampled: Tops (leaves) for chicory, radicchio (fresh leaves) and Belgium endive; Roots for chicory coffee substitute. Roots for forcing are discarded after tops are harvested for Belgium endive and some radicchio cultivars

9. Classifications:
   a. Authors Class: Root and tuber vegetables; Leaves of root and tuber vegetables; Leafy vegetables (except *Brassica* vegetables)
   b. EPA Crop Group (Group & Subgroup): Chicory: Root and tuber vegetables (1A and 1B); Leaves of root and tuber vegetables (2). Radicchio: Leafy vegetables (except *Brassica* vegetables (4A)
   c. Codex Group: Chicory: 013 (VL 0469) Leafy vegetables (including *Brassica* leafy vegetables); 016 (VR 0469) Root and tuber vegetables. Witloof chicory (sprouts): 017 (VS 0469) Stalk and stem vegetables
   d. EPA Crop Definition: None

10. References: GRIN, CODEX, COLLINS 1996, MAGNESS, MYERS (picture), RICHE, STEPHENS 1988 (picture), US EPA 1995a, US EPA 1995b, LOGAN, RUBATZKY, HANSON(a), SWIADER

11. Production Map: EPA Crop Production Regions 1, 2 and 10 which represent about 70 percent of the harvested chicory acreage.

12. Plant Codes:
    a. Bayer Code: CICIF (Belgium endive), CICIN (Chicory), CICIS (Chicory root)

154

1. Chive
(Cive, Schnittlauch, Ciboulette, Cebollin, Cebollino, Siuheung, Tsung, He, Ezo-negi, Chives)
Liliaceae (Amaryllidaceae)
Allium schoenoprasum L. (syn: *A. sibiricum* L.)

2. The chive plant, an onion relative, is grown in thick clumps from small, oval bulbs. It may be grown from seed, or by separating the clumps of bulbs. It is a perennial, grown for the leaves, which are used as seasoning in salads, omelettes and other dishes. The leaves are slender, tubular and about 6 inches long. They may be cut several times in one season, after the clumps of bulbs are established.
3. Season: Leaves may be cut every 6 to 8 weeks and plantings cultivated for 3-4 years. Seeding to harvest is about 3 months.
4. Production in U.S.: In 1988, about 20 acres in California (MYERS 1991).
5. Other production regions: Widely distributed across Northern Hemiphere (RUBATZKY 1997).
6. Use: Leaves, finely chopped, used in salads; culinary cookery. Some dehydrated or freeze dried.
7. Part(s) of plant consumed: Leaves only
8. Portion analyzed/sampled: Leaves (Herbs are generally fresh and dried)
9. Classifications:
   a. Authors Class: Herbs and spices; Bulb vegetables
   b. EPA Crop Group (Group & Subgroup): Herbs and spices (19A) (Representative Crop)
   c. Codex Group: 027 (HH 0727) Herbs
   d. EPA Crop Definition: None
10. References: GRIN, CODEX, FACCIOLA, KAYS, MAGNESS, MYERS (picture), US EPA 1995a, RUBATZKY
11. Production Map: EPA Crop Production Region 10.
12. Plant Codes:
    a. Bayer Code: ALLSC

155

1. Chive, Chinese
(Cebollino de la China, Alko chines, Kau tsai, Gau choi, Gau tsoi, Gow choy, Gil choy, Garlic chive, Oriental garlic, Kui tsai, Nira, su-en, Shu-an, Kau tsai, Jiu cai, Chinese leek, Flowering leek)
Liliaceae (Amaryllidaceae)
Allium tuberosum Rottler ex Spreng. (syn: *A. odoratum* L.)

2. This onion relative is an important crop in the Orient, and is grown to some extent by Oriental gardeners in the U.S. Bulbs, if any, are poorly developed but rhizomes are thick and conspicuous. The plant spreads from these rhizomes to form dense clumps. The edible portion is the long grass-like leaves, which are flat and bend down at the tips, and the young, tender inflorescences. Flowers are borne on a solid stalk, harvested at the bud stage and bunched for market. Leaves are sometimes blanched by excluding light. The flavor is strong and garlic like. In established beds, leaves may be cut repeatedly. The same planting can produce for 10 years or more. The plant grows to 12 to 18 inches tall. The bulbs can also be harvested and used like garlic cloves. Mainly harvested for green leaves alternated with blanched leaves.

3. Season, from seed to first cutting: 4 to 5 months. (Limited cuttings the first year). Can be harvested at 15-20 day intervals of green/blanched leaves and conclude annual production by harvesting the scapes.
4. Production in U.S.: No data. Very limited. California (MYERS 1991).
5. Other production regions: China, Southeast Asia (RUBATZKY 1997).
6. Use: Mainly as flavoring and mainly used fresh.
7. Part(s) of plant consumed: Mainly leaves, and young inflorescences (buds and stalks).
8. Portion analyzed/sampled: Leaves
9. Classifications:
   a. Authors Class: Herbs and spices; Bulb vegetables
   b. EPA Crop Group (Group & Subgroup): Herbs and spices (19A)
   c. Codex Group: 027 (HH 0727) Herbs
   d. EPA Crop Definition: None
10. References: GRIN, CODEX, FACCIOLA, KAYS, MAGNESS, MYERS (picture), US EPA 1995a, YAMAGUCHI 1983 (picture), MCCLURE, JANICK 1996, RUBATZKY, FORTIN
11. Production Map: EPA Crop Production Region 10.
12. Plant Codes:
    a. Bayer Code: ALLTU

156

1. Chrysanthemum, Edible-leaved
(Pua-Pake, Margarita, Crisantemo)
Asteraceae (Compositae)
Chrysanthemum spp.

1. Japanese greens
(Shungiku, Tan O, Kor Tongho, Chopsuey greens, Chopsuey, Crowndaisy, Garland chrysanthemum, Tongho)
C. coronarium var. *spatiosum* L.H. Bailey (syn: *C. spatiosum* (L.H. Bailey))

1. Corn chrysanthemum
(Tangho, Tongho, Corn marigold)
C. segetum L.

2. These species are used sparingly as potherbs. For that purpose, young plants are harvested when 4 to 8 inches high. Leaves are glabrous, lobed and grow as a rosette. They are aromatic and are cooked like spinach. In exposure of edible parts they are comparable to spinach.
3. Season, seeding to harvest: About 30 to 60 days.
4. Production in U.S.: No data. Grows well in Florida (STEPHENS 1988). Climatically adapted to the eastern U.S. (YAMAGUCHI 1983).
5. Other production regions: Probably native to the Mediterranean region and popular in Japan, China and other Asian countries (RUBATZKY 1997).
6. Use: Young plant tops used as potherb. Also see flower, edible.
7. Part(s) of plant consumed: Leaves and tender shoots
8. Portion analyzed/sampled: Tops
9. Classifications:
   a. Authors Class: Leafy vegetables (except *Brassica* leafy vegetables)
   b. EPA Crop Group (Group & Subgroup): Leafy vegetables (except *Brassica* leafy vegetables) (4A)

c. Codex Group: 013 (VL 0479) Leafy vegetables (including *Brassica* leafy vegetables)

d. EPA Crop Definition: None

10. References: GRIN, BAILEY 1976, CODEX, FACCIOLA, KUEBEL, KAYS, MAGNESS, NEAL, STEPHENS 1988 (picture), USDA NRCS, US EPA 1995a, YAMAGUCHI 1983 (picture), RUBATZKY

11. Production Map: EPA Crop Production Regions 1, 2, 3, 4, 5 and 6.

12. Plant Codes:

a. Bayer Code: CHYCO (*C. coronarium*), CHYSE (*C. segetum*)

157

1. Chufa

(Ground almond, Edible rush, Tigernut, Yellow nut grass, Chufa oil, Yellow nutsedge, Rush nut, Earth almond, Watergrass)

Cyperaceae

Cyperus esculentus L.

2. Chufa plants are perennials grown for the edible tubers. The top is grass-like, with simple leaves and flower stalk rising from the ground surface, up to 3 feet. Tubers are small, $1/2$ to $3/4$ of an inch long, cylindrical and hard, produced entirely underground. They are eaten raw or baked. Plants are propagated by planting the tubers, similar to potatoes. They are a minor crop, grown mainly in the Southern states. The tubers of Chufa contain 20 to 28 percent of a non-drying oil. The oil is obtained by pressing the cleaned tubers and has a mild, pleasant flavor. The separated emulsion of oil and juice, obtained by pressing, is consumed in quantity in Europe, especially Spain, as a drink (Horchata de Chufas).

3. Season, planting to harvest: 3 to 6 months.

4. Production in U.S.: Limited, mainly in Southern states, mainly Florida.

5. Other production regions: North Africa and Spain (RUBATZKY 1997).

6. Use: Tubers eaten raw or baked. Sometimes used as coffee substitute and edible oil. Also planted as wild game food.

7. Part(s) of plant consumed: Tuber

8. Portion analyzed/sampled: Tuber

9. Classifications:

a. Authors Class: Root and tuber vegetables

b. EPA Crop Group (Group & Subgroup): Root and tuber vegetables (1C and 1D)

c. Codex Group: 016 (VR 0580) Root and tuber vegetables

d. EPA Crop Definition: None

10. References: GRIN, CODEX, MAGNESS, NEAL, US EPA 1995a, FRANK, RUBATZKY

11. Production Map: EPA Crop Production Regions 2 and 3.

12. Plant Codes:

a. Bayer Code: CYPES

158

1. Cinnamon

(Canela, Canelero, Ceylon cinnamon)

Lauraceae

Cinnamomum verum J. Presl (syn: *C. zeylanicum* Blume)

2. This is a small semi-tropical tree, 20 to 30 feet high, with thick ovate to lanceolate leaves 4 to 7 inches long. It is hardy in the Gulf States of the U.S. Most of the cinnamon of commerce comes from Sri Lanka (Ceylon). It is the ground bark of the tree. The best quality of bark is from branches at least 2 years old. Cinnamon is widely used in cookery and confections. The flavor is due to a volatile oil contained in the bark. The cinnamon oil of commerce is extracted from inferior bark, not suitable for grinding. Cinnamon powder is tan in color.

3. Season: 10 month old seedlings are transplanted to the field. Every 2-3 years the shoots are cut back and the bark removed. The outer bark is scraped off and the inner bark is peeled in two strips. The strips are dried and curl into quills or sticks.

4. Production in U.S.: Not grown commercially in U.S.

5. Other production regions: Cultivated in Mexico, Grenada and South Vietnam.

6. Use: Dried bark used as spice.

7. Part(s) of plant consumed: Bark

8. Portion analyzed/sampled: Bark (dried)

9. Classifications:

a. Authors Class: Herbs and spices

b. EPA Crop Group (Group & Subgroup): Herbs and spices (19B)

c. Codex Group: 028 (HS 0777) Spices

d. EPA Crop Definition: None

10. References: GRIN, CODEX, FARRELL (picture), LEWIS (picture), MAGNESS, REHM, US EPA 1995a, FRANK

11. Production Map: No entry.

12. Plant Codes:

a. Bayer Code: CINZE

159

1. Citron, Citrus

(Cidra, Limon cidra, Etrog, Citron citrus)

Rutaceae

Citrus medica L. (syn: *C. cedra* Link; *C. cedratus* Raf.)

2. This citron is a citrus fruit in which the peel or rind makes up most of the fruit volume. The sweet or Corsican citron fruit is large and elliptical in shape. Diameter 3 to 5 inches, length 4 to 7 inches. Rind about 1 inch thick. The acid citrons are similar in size, shape and rind thickness. The Etrog, used by Jewish people in rites of the Feast of the Tabernacles, is a small citron a little larger than lemons.

3. Season, bloom to harvest: 8 to 10 months.

4. Production in U.S.: In 1992, 1,018 acres with 1,514 tons from Puerto Rico (1992 CENSUS).

5. Other production regions: Mediterranean region (MORTON 1987).

6. Use: Peel candied, used in confections and culinary.

7. Part(s) of plant consumed: Mainly thick peel or rind; the entire Etrog fruit is eaten

8. Portion analyzed/sampled: Whole fruit

9. Classifications:

a. Authors Class: See citrus fruits

b. EPA Crop Group (Group & Subgroup): See citrus fruits

c. Codex Group: 001 (FC 0202) Citrus fruits

d. EPA Crop Definition: See citrus fruits

10. References: GRIN, CODEX, MAGNESS, MORTON (picture), RICHE, FACCIOLA

11. Production Map: EPA Crop Production Region 13.

12. Plant Codes:

a. Bayer Code: CIDME

160

1. Citron melon
(Citron, Stock melon, Preserving melon)
Cucurbitaceae
Citrullus lanatus var. *citroides* (L.H. Bailey) Mansf. (syn: *C. vulgaris* var. *citroides* L. H. Bailey)

2. Citron melon is a plant of the same species as watermelon, but the fruit flesh is white, hard and inedible in the raw state. The plant is prostrate growing. The fruits are round to oval, up to 6 inches long, with smooth surface, resembling small watermelons. The flesh is used for making conserves and pickles. Citrons are sometimes used as feed for hogs. Citron melon grows as a weed in California (MYERS 1991).
3. Season, planting to harvest: 4 to 5 months.
4. Production in U.S.: No Data. Very Minor. Similar to watermelon which see.
5. Other production regions: Grows in temperate and subtropical regions (RUBATZKY 1997).
6. Use: Conserves, pickling, stock feed.
7. Part(s) of plant consumed: White flesh of fruit
8. Portion analyzed/sampled: Whole fruit
9. Classifications:
   a. Authors Class: Cucurbit vegetables
   b. EPA Crop Group (Group & Subgroup): Cucurbit vegetables (9A)
   c. Codex Group: 011 (VC 0432) Fruiting vegetables, Cucurbits
   d. EPA Crop Definition: Melons = *Citrullus* spp.
10. References: GRIN, CODEX, MAGNESS, MYERS, STEPHENS 1988 (picture), US EPA 1995a, RUBATZKY
11. Production Map: No entry.
12. Plant Codes:
    a. Bayer Code: CITLC

161

1. Citrus fruits
(Zest, Zeste)
Rutaceae
Citrus spp.

2. Citrus fruits produced commercially in the U.S. are of 5 major and 3 minor kinds. Major kinds are *C. limon*, lemon; *C. sinensis*, sweet orange; *C. reticulata*, Mandarin orange; *C. xparadisi*, grapefruit; and *C. aurantiifolia*, lime. Minor kinds are *C. maxima*, pummelo; *C. aurantium*, sour or Seville orange; and *C. medica*, citron. All are produced on relatively small, evergreen trees. Citrus trees normally set fruit in the fourth year.

All are injured by winter temperatures below about 25 degrees F. Citrus can grow well in a wide range of soils, however they are highly sensitive to overly moist conditions. Leaves and fruit peel contain oil vesicles. Most fruit peels are relatively thick, consisting of a white, spongy endocarp, and with the surface epidermis containing numerous oil vesicles. Interior of fruit consists of 8 to 18 segments, each of which may contain seeds near the inner angle. Segments are composed largely of rather large vesicles filled with juice, which may vary from sweet to very acid. All citrus fruits will cross with each other, producing literally hundreds of cultivars. Many such crosses have been made by breeders and some have originated by chance. As a result, we now have in commerce Mandarin orange x grapefruit crosses called tangelos, sweet orange x Mandarin crosses called tangors, and others. Specialty citrus includes tangelos, tangerines, mandarins and tangors (LOGAN 1996).

3. Season, bloom to harvest: See specific citrus fruit.
4. Production in U.S.: See specific citrus fruit. In 1995-96, total citrus fruits reported as 1,103,620 bearing acres with 16,000,000 tons. About 72 percent of the productions was processed (USDA 1996d). Florida, Texas, Arizona and California (US EPA 1994).
5. Other production regions: Mexico, Central America, South America, West Indies, southern Europe, North Africa, Middle East, India and the Orient (MORTON 1987).
6. Use: Fresh, drinks, flavoring, marmalade, juice or confection. Fruit preparation for citrus fruits: Navel oranges, tangerines, tangelos, temple oranges and clementines are peeled easily by simply pulling off the skin. Other citrus fruits can be peeled by cutting as close to the flesh as possible with a sharp paring knife. To segment grapefruit: Cut in half horizontally, scoop out seeds with the point of a knife and use a small knife, preferably a curved grapefruit knife, to cut around each segment and free it from the surrounding membrane. Lightly roll whole lemons or limes on a countertop or warm them slightly before cutting to release more juice. Squeeze through a piece of cheesecloth to avoid pits in juice. (LOGAN 1996). Zest (zeste) is a piece of the peel or of the thin oily, outer skin of normally an orange or lemon used for flavoring foods.
7. Part(s) of plant consumed: Fruit
8. Portion analyzed/sampled: Whole fruit. For sweet orange, whole fruit and its processed commodities dried pulp, oil and juice
9. Classifications:
   a. Authors Class: Citrus fruit
   b. EPA Crop Group (Group & Subgroup): Citrus fruits (10)
   c. Codex Group: 001 (FC 0001) Citrus fruits
   d. EPA Crop Definition: Citrus fruits = Grapefruit, lemons, limes, oranges, tangelos, tangerines, citrus citron, kumquats, and hybrids of these. Also Tangerines = Tangerines (mandarins or mandarin oranges); tangelos, tangors, and other hybrids of tangerine with other citrus. Also proposing to add White Sapote to the Citrus fruits and a new entry for Grapefruit = Pummelo, Ugli fruit, Pompelmousse and Chinese Grapefruit and its hybrids and/or cultivars
10. References: GRIN, CODEX, LOGAN (picture), MAGNESS, USDA 1996d, US EPA 1996a, US EPA 1995a, US EPA 1994, KALE, MORTON, JACKSON, REUTHER, SAUNT
11. Production Map: No entry, see specific citrus.
12. Plant Codes:
    a. Bayer Code: CIDSS

162

1. Citrus hybrids
(Tangor, Citrange, Citrangequat, Citrangedin, Limequat, Uniq fruit, Tangelolo, Tangelo, Chironja, Ugli fruit, Calamondin, Panama orange, Oro Blanco, Orangelo, King orange, Temple orange, Lemandarin, Rangpur lime, Mandarin lime, Otaheite orange, Uniq fruit)
Rutaceae
Citrus spp. and *Fortunella* spp.

2. A number of citrus hybrids, in addition to tangelo and cala-mondin, listed separately, have been developed by breeding. These include:

(1) Tangors (King orange, Tankan mandarin) (Lour.) are Mandarin orange x sweet orange, and are generally interme-diate in characteristics. Several varieties are grown to a lim-ited extent. Temple, generally listed as a sweet orange, is probably of such parentage.

(2) The citranges (x *Citroncircus webberi* J.W. Ingram & H.E. Moore) are crosses of trifoliate orange and sweet orange. Several varieties are in cultivation to a limited extent. The juice is tart, and makes an acceptable ade drink. Trees are hardier than lemons or limes. Fruit $2^1/_2$ inches or more in diameter, rather thin skinned.

(3) Citrangequats are crosses of citranges and kumquats. Trees are relatively hardy. Fruits useful for ade drinks and marmalade.

(4) Citrangedins are hybrids of citrange and calamondin. The one named variety is hardy, and suitable for both ade and culinary use.

(5) Limequats (x *citrofortunella* sp.) are crosses of lime and kumquat, and resemble limes except they are much hardier.

(6) Lemandarin (*Citrus limonia* Osbeck) (Rangpur lime, Mandarin lime, Otaheite orange) are crosses of lemon and tangerine.

(7) Natsudaidai and small or medium sized tangelos are pos-sibly a cross of mandarin and shaddock.

(8) Tangelolo is a cross of grapefruit and tangelo.

(9) Chironja is a cross of sweet orange and grapefruit as an orangelo which is sweeter than a grapefruit and grown in Puerto Rico.

(10) Oro blanco is a cross of pummelo and grapefruit.

(11) Calamondin is a cross of kumquat and tangerine.

(12) Tangelos (*C.* x *tangelo* J.W. Ingram & H.E. Moore) (Uniq fruit, Ugli fruit) are a cross of grapefruit and tangerine. Generally, none of the above hybrids, except tangors, tange-los, calamondin, oro blanco, and chironja, are in commerce, but are useful for home gardens in areas too cold for other citrus. Citranges are important as rootstocks for other kinds of citrus.

3. Season, bloom to harvest: See specific hybrids listed.
4. Production in U.S.: See Citrus fruits.
5. Other production regions: See citrus fruits.
6. Use: Fresh, drinks, flavoring, marmalade or juice.
7. Part(s) of plant consumed: Fruit
8. Portion analyzed/sampled: Whole fruit
9. Classifications:
   a. Authors Class: See citrus fruits
   b. EPA Crop Group (Group & Subgroup): See citrus fruits
   c. Codex Group: 001 (FC 0001, 0002, 0003, 0004, and 0005) Citrus fruits and hybrids. For additional codex numbers see specific hybrids listed
   d. EPA Crop Definition: See citrus fruits
10. References: GRIN, CODEX, LOGAN, MAGNESS, MOR-TON
11. Production Map: No entry, see specific hybrids listed.
12. Plant Codes:
   a. Bayer Code: CIDSS (*Citrus* spp.), FOLSS (*Fortunella* spp.)

163

1. Clary
(Clary sage, Garden clary, Sauge sclaree)
Lamiaceae (Labiatae)
Salvia sclarea L.

2. The plant is an erect, biennial herb, 2 to 3 feet high. The leaves are large, up to 9 inches long by 5 inches wide, pubes-cent. The dried leaves are used sparingly for flavoring in cookery. Its use in this country has been largely replaced by sage.
3. Season, seeding to harvest: See Sage.
4. Production in U.S.: Grows in USDA Hardiness Zones 7 to 9 (GARLAND 1993)
5. Other production regions: Native to Eurasia (FACCIOLA 1990).
6. Use: Flavoring in sauces, soups and added to Rhine wine, vermouth and ale.
7. Part(s) of plant consumed: Leaves
8. Portion analyzed/sampled: Leaves (fresh and dried)
9. Classifications:
   a. Authors Class: Herbs and spices
   b. EPA Crop Group (Group & Subgroup): Herbs and spices (19A)
   c. Codex Group: 027 (HH 0743) Herbs; 057 (DH 0743) Dried herbs. Also sage
   d. EPA Crop Definition: None
10. References: GRIN, CODEX, FACCIOLA, GARLAND (pic-ture), MAGNESS, US EPA 1995a
11. Production Map: No entry.
12. Plant Codes:
   a. Bayer Code: SALSC

164

1. Cloudberry
(Malka, Salmonberry, Yellowberry, Bakeapple, Baked-apple, Baked-apple berry, Foxberry, Mars apple)
Rosaceae
Rubus chamaemorus L.

2. Short-stemmed herbaceous perennial, 3 to 10 inches high growing from a creeping rhizome. Leaves are rounded in outline. Fruit is red or yellow and a bramble type fruit with an aggregate of small drupes. The fruits are like raspberries. Bakeapples are important to Newfoundlanders and are har-vested commercially.
3. Season: Ripen from July to September.
4. Production in U.S.: No data
5. Other production regions: Arctic and subarctic, south to Maine (BAILEY 1976). Occur in peat bogs throughout Can-ada from British Columbia to Maritimes and Newfoundland, also found in Alaska (KUHNLEIN 1991).
6. Use: Berries are consumed raw, stewed, preserved or pro-cessed into confections, jellies, vinegar and wine.
7. Part(s) of plant consumed: Fruit
8. Portion analyzed/sampled: Whole fruit
9. Classifications:
   a. Authors Class: Berries
   b. EPA Crop Group (Group & Subgroup): Miscellaneous
   c. Codex Group: 004 (FB 0277) Berries and other small fruit
   d. EPA Crop Definition: Caneberries = *Rubus* spp.

10. References: GRIN, BAILEY 1976, CODEX, FACCIOLA, KUHNLEIN, US EPA 1995a
11. Production Map: No entry
12. Plant Codes:
    a. Bayer Code: No specific entry

165

1. Clove
(Clove buds, Giroflier, Clove tree, Clavo-de-olor, Cravo-da-India)
Myrtaceae

Syzygium aromaticum (L.) Merr. & L.M. Perry (syn: *Eugenia caryophyllus* (Spreng.) Bullock & S.G. Harrison; *E. caryophyllata* Thunb.; *E. aromaticum* (L.) Baill.; *Caryophyllus aromaticus* L.)

2. The clove tree is a small, tropical evergreen, up to 20 feet tall, with oblong leaves, 5 to 10 inches long and 2 to 4 inches wide. It is native to the Moluccas, but has been introduced into all tropical countries. The tree is not hardy in continental U.S. The cloves of commerce are the dried flower buds, which grow in clusters at the ends of branches and are harvested and dried before they open. Cloves are widely used in cookery and confections. The extracted clove oil is widely used in cosmetics and confections. By oxidation, it produces vanillin. U.S. imported about 1,140 tons mostly from Madagascar, Indonesia and Brazil in 1992.
3. Season, planting to harvest: Produces crops when 6-8 years of age. Full bearing age when 20 years old and may be productive for 100 years. Can be harvested twice a year, July-October and December-January. Cloves are propagated by seeds and transplanted when 15 to 24 months old.
4. Production in U.S.: No data.
5. Other production regions: Madagascar, Indonesia, and Brazil.
6. Use: To flavor meats, fruits and vegetables dishes, and tobacco.
7. Part(s) of plant consumed: Unopened flower buds, may be extracted for its oil
8. Portion analyzed/sampled: Flower buds (Dried)
9. Classifications:
    a. Authors Class: Herbs and spices
    b. EPA Crop Group (Group & Subgroup): Herbs and spices group (19B)
    c. Codex Group: 028 (HS 0778) Spices
    d. EPA Crop Definition: None
10. References: GRIN, CODEX, FARRELL (picture), LEWIS (picture), MAGNESS, USDA 1993a, US EPA 1995a, DUKE(a)
11. Production Map: No entry.
12. Plant Codes:
    a. Bayer Code: SYZAR

166

1. Clover, Alsike
(Swedish clover, Trefol hibrido, Trefle hybride, Hybrid clover)
Fabaceae (Leguminosae)
Trifolium hybridum L.

2. Alsike clover is believed native to northern Europe. It was introduced into the U.S. about 1839 and is now mainly grown in states bordering on the Great Lakes, northern U.S. west to Kansas, Idaho and in northern California and Oregon. Its special merit is its adaptation to wet soils, even tolerating some flooding. It is more tolerant to both acid and alkaline soils than other clovers. The stems are quite slender, up to 3 feet long, and tend to be prostrate except in dense stands. The trifoliate leaves are long stemmed with obovate leaflets. Flower heads are not at the terminal of the main stem but are at the terminals of branch stems. They are white, yellow or pink in color. Alsike clover is a short-lived perennial species that is utilized as a forage and hay crop, often in mixtures with red clover and grasses, such as timothy. Alsike is better adapted to colder and wetter climates, and poor fertility than red clover. Both stems and leaves are smooth. Alsike clover is valuable both for pasture and hay on soils too wet or too acid for other clovers. In 1995, Canada reported 19,867 acres. There are 728,000 seeds/lb.
3. Season, See Clovers, True. Harvest hay at full bloom stage. Only one harvest per season.
4. Production in U.S.: See Clovers, True.
5. Other production regions: See Clovers, True.
6. Use: See Clovers, True. Cover crop, hay and pasture.
7. Part(s) of plant consumed: See Clovers, True
8. Portion analyzed/sampled: See Clovers, True
9. Classifications:
    a. Authors Class: See Clovers, True
    b. EPA Crop Group (Group & Subgroup): See Clovers, True
    c. Codex Group: See Clovers, True
    d. EPA Crop Definition: See Clovers, True
10. References: GRIN, See Clovers, True
11. Production Map: EPA Crop Production Region, See Clovers, True.
12. Plant Codes:
    a. Bayer Code: TRFHY

167

1. Clover, Alyce
(Oneleaf clover, Buffalo clover)
Fabaceae (Leguminosae)
Alysicarpus vaginalis (L.) DC.

2. This is a summer annual native to tropical Asia. It was introduced into the U.S. in 1910 and has proven adapted to areas near the Gulf of Mexico. It is best adapted to sandy soil with pH ≥4.5. In thin stands the stems branch, but in thick stands stems reach 3 feet in height with little branching. Leaves are oval, unifoliate, and borne on short petioles along the entire stem. There are two distinct phenological types, erect and prostrate. The erect type is more productive. Alyce clover is grown mainly for hay and soil improvement, but also makes good pasturage. The hay appears about equal to other legume hays in feeding value. The crop is seeded in late spring and matures seed the same season. If seed matures and shatters, a volunteer crop is produced in succeeding years. Seed was harvested from about 5000 acres (1954, 1959 CENSUSes), sufficient to plant about 80,000 acres. There are 301,000 seeds per pound. It is seeded as a cover crop at 10 lb/A and at 15 to 20 lb/A for pasture. It often reseeds itself.
3. Season: Pasture and hay, seeded in the spring. Hay cut at 18-24 inches with second cutting possible. Production season is July to September. Grazing begins when 12 to 15 inches tall.
4. Production in U.S.: No recent data. Gulf of Mexico area.

5. Other production regions: Native to Old World tropics, India (BAILEY 1976).
6. Use: Hay, pasture, soil improvement.
7. Part(s) of plant consumed: Foliage
8. Portion analyzed/sampled: Forage and hay
9. Classifications:
   a. Authors Class: Nongrass animal feeds
   b. EPA Crop Group (Group & Subgroup): Miscellaneous
   c. Codex Group: No specific entry, general class 050 (AL 0157) Legume animal feeds
   d. EPA Crop Definition: None
10. References: GRIN, MAGNESS, BALL (picture), CODEX, BARNES, BAILEY 1976
11. Production Map: EPA Crop Production Regions 2, 3 and 13.
12. Plant Codes:
    a. Bayer Code: ALZVA

168

1. Clover, Arrowleaf

Fabaceae (Leguminosae)
Trifolium vesiculosum Savi

2. Annual true clover which was introduced from Italy in 1963. Three U.S. releases are 'Amclo', 'Meecher', and 'Yuchi' cultivars.
3. Season: See Clovers, True.
4. Production in U.S.: In 1997, Texas reported 140, 000 acres primarily in the eastern part (SMITH 1997).
5. Other production regions: Italy (TAYLOR 1990). See Clovers, True.
6. Use: See Clovers, True.
7. Part(s) of plant consumed: See Clovers, True
8. Portion analyzed/sampled: See Clovers, True
9. Classifications:
   a. Authors Class: See Clovers, True
   b. EPA Crop Group (Group & Subgroup): See Clovers, True
   c. Codex Group: See Clovers, True
   d. EPA Crop Definition: See Clovers, True
10. References: GRIN, SMITH 1997, TAYLOR, See Clovers, True
11. Production Map: EPA Crop Production Region 6.
12. Plant Codes:
    a. Bayer Code: TRFVE

169

1. Clover, Ball
(Small white clover)

Fabaceae (Leguminosae)
Trifolium nigrescens Viv.

2. This is a winter annual clover introduced from Turkey in 1953. It is now established in many southern locations. It is largely grown from eastern Texas across the Gulf Coast to Alabama and tolerates heavy grazing. It thrives well on rather thin, sandy soils and reseeds well in heavy sods. Its use is primarily as a legume in grass pastures. There are 1 million seeds/lb.
3. Season: See Clovers, True.
4. Production in U.S.: See Clovers, True.
5. Other production regions: See Clovers, True.
6. Use: See Clovers, True. Pasture, Hay, green manure, roadside beautification and honey crop.
7. Part(s) of plant consumed: See Clovers, True

8. Portion analyzed/sampled: See Clovers, True
9. Classifications:
   a. Authors Class: See Clovers, True
   b. EPA Crop Group (Group & Subgroup): See Clovers, True
   c. Codex Group: See Clovers, True
   d. EPA Crop Definition: See Clovers, True
10. References: GRIN, See Clovers, True
11. Production Map: EPA Crop Production Region, See Clovers, True.
12. Plant Codes:
    a. Bayer Code: TRFNI

170

1. Clover, Berseem
(Egyptian clover, Moscino)

Fabaceae (Leguminosae)
Trifolium alexandrinum L.

2. Annual true clover which was introduced from Syria and Egypt. The U.S. release is 'Bigbee' cultivar.
3. Season: See Clovers, True.
4. Production in U.S.: In 1997, Texas reported 10,000 acres of Berseem clovers (SMITH 1997).
5. Other production regions: Syria and Egypt (TAYLOR 1990). See Clovers, True.
6. Use: See Clovers, True.
7. Part(s) of plant consumed: See Clovers, True
8. Portion analyzed/sampled: See Clovers, True
9. Classifications:
   a. Authors Class: See Clovers, True
   b. EPA Crop Group (Group & Subgroup): See Clovers, True
   c. Codex Group: See Clovers, True
   d. EPA Crop Definition: See Clovers, True
10. References: GRIN, SMITH 1997, TAYLOR, See Clovers, True
11. Production Map: No entry.
12. Plant Codes:
    a. Bayer Code: TRFAL

171

1. Clover, Bigflower
(Mike's clover)

Fabaceae (Leguminosae)
Trifolium michelianum Savi

2. This is a weak-stemmed, tall-growing coarse annual clover recently introduced from Turkey and apparently well adapted to the Deep South coastal region. In general appearance it resembles Alsike Clover, *T. hybridum*. It is useful both for pasturage and hay.
3. Season: See Clovers, True.
4. Production in U.S.: See Clovers, True.
5. Other production regions: See Clovers, True.
6. Use: See Clovers, True.
7. Part(s) of plant consumed: See Clovers, True
8. Portion analyzed/sampled: See Clovers, True
9. Classifications:
   a. Authors Class: See Clovers, True
   b. EPA Crop Group (Group & Subgroup): See Clovers, True
   c. Codex Group: See Clovers, True
   d. EPA Crop Definition: See Clovers, True
10. References: GRIN, See Clovers, True

11. Production Map: EPA Crop Production Region, See Clovers, True.
12. Plant Codes:
   a. Bayer Code: No specific entry

172

1. Clover, Crimson
(Italian clover, Scarlet clover, Carnation clover, Trebol rosado, Trefle incarnat)
Fabaceae (Leguminosae)
Trifolium incarnatum L.

2. Crimson clover, native to the eastern Mediterranean area, was brought to the U.S. from Italy in 1819. It was near the turn of this century that it became recognized as a useful crop, especially in the Southeastern states and in the western parts of the Pacific states. It is grown mainly as an overwintering annual and grows to a height of 1 to 3 feet. Seeded in late summer, it forms a dense rosette of leaves. In mild winter climates growth continues in winter, followed in early spring by the development of leafy flower stalks. Leaves are trifoliate with leaflets narrow at the base and broad at the terminals. Both leaves and stems are hairy. The flower heads are elongated and pointed, bright crimson in color, and contain up to more than 100 florets. Crimson clover is excellent for winter and spring grazing in mild winter areas. It also yields good hay crops. Both as pasture and hay it is highly nutritious and palatable. Crimson clover is one of the easiest crop to established. It makes a good cool season annual for no tillage crop rotations in southeastern U.S. with corn and grain sorghum. Planted mid-August through October, seasonal production in the southeast U.S. is late November-December. It can be grazed throughout the winter. If hay is desired, the cattle must be removed in mid-March. In 1992, crimson clover seed was produced on 7,216 acres with 4,936,846 pounds. The primary seed state was Oregon (6,462 acres), next Tennessee (447 acres) (1992 CENSUS). There are 150,000 seeds/lb.
3. Season: See Clovers, True. Flowers in the spring, with seeds produced 24 to 30 days after flowering.
4. Production in U.S.: See Clovers, True. In 1993, there were about 500,000 acres used for pasture and hay (BARNES 1995). In 1997, Texas reported 160,000 acres.
5. Other production regions: See Clovers, True.
6. Use: See Clovers, True. Cover crop, pasture, hay, silage, green manure crop, and road stabilization.
7. Part(s) of plant consumed: See Clovers, True
8. Portion analyzed/sampled: See Clovers, True
9. Classifications:
   a. Authors Class: See Clovers, True
   b. EPA Crop Group (Group & Subgroup): See Clovers, True
   c. Codex Group: See Clovers, True
   d. EPA Crop Definition: See Clovers, True
10. References: GRIN, DUKE(h) (picture), BARNES, BALL, SMITH 1997, See Clovers, True
11. Production Map: EPA Crop Production Region, See Clovers, True.
12. Plant Codes:
   a. Bayer Code: TRFIN

173

1. Clover, Hop
Fabaceae (Leguminosae)
Trifolium spp.
1. Small hop clover
(Trefle douteux, Trebol filifore, Shamrock, Low hop clover, Lesser yellow trefoil, Suckling clover, Yellow clover)
T. dubium Sibth.
1. Large hop clover
(Low clover, Low hop clover, Rabbitfoot clover, Hop trefoil, Trebol campreste, Trefle jaune, Field clover)
T. campestre Schreb.
1. Hop clover
(Trefle dore, Golden clover)
T. aureum Pollich (syn: *T. agrarium* L.)

2. All three species of hop clovers are of European and Asia Minor origin and apparently came to the U.S. in mixtures of other seeds. The small and large hop clovers are adapted as winter annuals in the Southern and Pacific States. They occur in pastures and lawns and are valuable for winter and early spring pasture, being highly nutritious. Stems are small, reaching up to 18 inches. Two species are quite similar but the Large hop clover is somewhat larger and higher yielding. Hop clover, *T. aureum* occurs somewhat sparingly, mainly in Northern states. It is not seeded but is valuable in grassland pastures where present. Hop clovers are similar in appearance with small round heads and yellow flowers. Also, the hop clovers are adapted to droughty, infertile eroded soils of the Southern U.S. There are 884,000 seeds/lb.
3. Season: See Clovers, True. Forage production for short period in the spring.
4. Production in U.S.: See Clovers, True.
5. Other production regions: See Clovers, True.
6. Use: See Clovers, True. Pasture.
7. Part(s) of plant consumed: See Clovers, True
8. Portion analyzed/sampled: See Clovers, True
9. Classifications:
   a. Authors Class: See Clovers, True
   b. EPA Crop Group (Group & Subgroup): See Clovers, True
   c. Codex Group: See Clovers, True
   d. EPA Crop Definition: See Clovers, True
10. References: GRIN, See Clovers, True, BARNES, BALL
11. Production Map: EPA Crop Production Region, See Clovers, True.
12. Plant Codes:
   a. Bayer Code: TRFSS (*T.* spp.), TRFAU (*T. aureum*), TRFCA (*T. campestre*), TRFDU (*T. dubium*)

174

1. Clover, Lappa
(Burdock clover)
Fabaceae (Leguminosae)
Trifolium lappaceum L.
1. Cluster clover
(Clustered clover)
T. glomeratum L.

2. These two cool season clovers were both introduced from the Mediterranean region as small-growing winter annuals. Lappa clover is now naturalized on the black soils of Ala-

bama and Mississippi, while cluster clover does well in southern Mississippi. The value of both is for pasturage. There are 180,000 seeds/lb.

3. Season: See Clovers, True.
4. Production in U.S.: See Clovers, True. Seasonal production April-May, produce dense growth up to 2 feet tall. Seeded at a rate of 10 to 15 lb/acre in September-October.
5. Other production regions: See Clovers, True.
6. Use: See Clovers, True; Pasture.
7. Part(s) of plant consumed: See Clovers, True
8. Portion analyzed/sampled: See Clovers, True
9. Classifications:
   a. Authors Class: See Clovers, True
   b. EPA Crop Group (Group & Subgroup): See Clovers, True
   c. Codex Group: See Clovers, True
   d. EPA Crop Definition: See Clovers, True
10. References: GRIN, See Clovers, True
11. Production Map: EPA Crop Production Region, See Clovers, True.
12. Plant Codes:
    a. Bayer Code: TRFGL (*T. glomeratum*), TRFLA (*T. lappaceum*)

175

1. Clover, Persian
(Reversed clover, Birdeye clover, Shaftal clover, Birdseye clover)
Fabaceae (Leguminosae)
Trifolium resupinatum L.

2. Persian clover is native to Asia Minor and is widely used in Eastern Mediterranean countries and India as a winter annual. It was found in Louisiana in 1928 and is now established, either through planting or natural reseeding, from Texas eastward to the Atlantic. It is adapted to heavy, moist soils but is not tolerant to low winter temperatures. Under cultivation it is seeded in the fall and grows rapidly during late winter and early spring. Stems are soft and hollow, reaching to 3 feet under the best growing conditions. Seed pods are inflated and light in weight, subject to distribution by wind or floating on water. Natural reseeding generally occurs. Although used mainly for grazing, Persian clover is also excellent for silage and hay. Seeded at the rate of 3 to 5 lbs/acre in September to October and productive March to April. Withstands close grazing. There are 909,091 seeds per pound.
3. Season: See Clovers, True; Harvested for hay at one-fourth to full bloom stage.
4. Production in U.S.: See Clovers, True.
5. Other production regions: See Clovers, True.
6. Use: See Clovers, True; Grazing, silage, hay, green manure and ornamental.
7. Part(s) of plant consumed: See Clovers, True
8. Portion analyzed/sampled: See Clovers, True
9. Classifications:
   a. Authors Class: See Clovers, True
   b. EPA Crop Group (Group & Subgroup): See Clovers, True
   c. Codex Group: See Clovers, True
   d. EPA Crop Definition: See Clovers, True
10. References: GRIN, See Clovers, True; BARNES, DUKE(h)
11. Production Map: EPA Crop Production Region, See Clovers, True.

12. Plant Codes:
    a. Bayer Code: TRFRS

176

1. Clover, Red
(Purple clover, Peavine clover)
Fabaceae (Leguminosae)
Trifolium pratense L.

2. Red clover is the most widely grown of the true clovers and is often grown as a biennial. It is believed native to the Eastern Mediterranean region. It was widely cultivated in Europe before the settlement of America by Europeans and was brought here by the early colonists. It is now most extensively grown from the edge of the Great Plains east to the Atlantic and west of the Rocky Mountains in irrigated areas. In the Southeast it is grown somewhat as a winter annual. The plant is an herbaceous short-lived perennial with a number of leafy stems rising from a crown. Leaves and stems are pubescent. Leaves are trifoliate with leaflets near round to oblong and $1/2$-inch or more across. Stems reach 2 to 3 feet under favorable conditions and bear at the terminals. The rose-to-magenta flower heads comprised of 100 or more individual flowers. The plants are highly nutritious and palatable both as pasture and hay. The crop is grouped into 3 divisions or types. These are (1) early-flowering which includes most North American red clovers and known collectively as Medium red clover and produces two crops per season; (2) late-flowering which includes the American mammoth red types that produce one crop per season; and (3) Wild red clover type which is found in England, not in North America. Red clover is cross pollinated. It is usually grown with small grain crops, such as oats or barley. For seed production, the first crop is harvested for hay and the second crop for seed. Acreage in red clover in the U.S. is estimated at 8 to 10 million in the 1960s. In 1992, 49,600 acres of red clover were grown for seed with 10,249,841 pounds of seed produced. The top six seed producing states are Oregon (12,067 acres), Missouri (11,003), Illinois (6,823), Minnesota (5,913), Ohio (2,721) and Washington (2,013) (1992 CENSUS). In 1995, Canada reported 59,189 acres. There are 1,334,025 seeds per pound.
3. Season: See Clovers, True. Hay can be harvested 2 times per season. Harvested 60 to 70 days after seeding. Harvest at prebloom or early bloom stage, 6 to 7 week intervals.
4. Production in U.S.: See Clovers, True.
5. Other production regions: See Clovers, True.
6. Use: See Clovers, True; Hay, pasture, green manure, silage, cover crop.
7. Part(s) of plant consumed: See Clovers, True
8. Portion analyzed/sampled: See Clovers, True
9. Classifications:
   a. Authors Class: See Clovers, True
   b. EPA Crop Group (Group & Subgroup): See Clovers, True
   c. Codex Group: See Clovers, True
   d. EPA Crop Definition: See Clovers, True
10. References: GRIN, See Clovers, True, BALL (picture)
11. Production Map: EPA Crop Production Region, See Clovers, True.
12. Plant Codes:
    a. Bayer Code: TRFPR

177

1. Clover, Rose
Fabaceae (Leguminosae)
Trifolium hirtum All.

2. This is a winter annual from the Mediterranean region, now of some importance in the Southeastern Coastal Plain and California. It is not tolerant to low winter temperatures. It reseeds well where adapted. Plants are hairy, with rose colored flower heads, and are relished by all kinds of livestock. Use is primarily for pasturage. Rose clover grows semierect to height of 16 to 20 inches on well drained soils. It is also adapted to central Texas and Oklahoma, and used as a cover crop in orchards and vineyards in California. It can be grazed continuosly until maturity and with proper grazing management, the clover will reseed itself. There are 164,000 seeds/lb.
3. Season: See Clovers, True. Flowers May-June.
4. Production in U.S.: See Clovers, True. About 50,000 acres grown in U.S. (BARNES 1995). In 1997, about 4,000 acres in Texas.
5. Other production regions: See Clovers, True.
6. Use: See Clovers, True; Pasture, rangeland, cover crop and road revegetation.
7. Part(s) of plant consumed: See Clovers, True
8. Portion analyzed/sampled: See Clovers, True
9. Classifications:
   a. Authors Class: See Clovers, True
   b. EPA Crop Group (Group & Subgroup): See Clovers, True
   c. Codex Group: See Clovers, True
   d. EPA Crop Definition: See Clovers, True
10. References: GRIN, . DUKE(h) (picture), BARNES, SMITH 1997. See Clovers, True.
11. Production Map: EPA Crop Production Region, See Clovers, True.
12. Plant Codes:
   a. Bayer Code: No specific entry

178

1. Clover, Seaside
(Cow clover)
Fabaceae (Leguminosae)
Trifolium wormskioldii Lehm. (syn: *T. willdenovii* auct.)
1. Tomcat clover
T. willdenovii Sprengel
2. Perennial clovers with creeping rootstocks. Seaside clover is native in the Western states where it often occurs in dense stands. It is somewhat tolerant to saline soils and competes well with salt grasses on such soils. It is palatable to livestock both as hay and pasture. It may well become more important than at present for coastal or other saline locations.
3. Season: See Clovers, True.
4. Production in U.S.: See Clovers, True.
5. Other production regions: See Clovers, True.
6. Use: See Clovers, True.
7. Part(s) of plant consumed: See Clovers, True
8. Portion analyzed/sampled: See Clovers, True
9. Classifications:
   a. Authors Class: See Clovers, True
   b. EPA Crop Group (Group & Subgroup): See Clovers, True

   c. Codex Group: See Clovers, True
   d. EPA Crop Definition: See Clovers, True
10. References: GRIN, See Clovers, True
11. Production Map: EPA Crop Production Region, See Clovers, True.
12. Plant Codes:
   a. Bayer Code: No specific entry

179

1. Clover, Strawberry
(Trefle fraise, Trebol fresa, Stawberry-headed clover)
Fabaceae (Leguminosae)
Trifolium fragiferum L.

2. This is a perennial stoloniferous plant, in general similar to white clover, *T. repens*, but spreads slower. Has pink to white flower heads somewhat resembling strawberries. It is adapted to wet, saline or alkaline soils in mild climates. It will tolerate flooding up to several weeks. It is grown to a limited extent in the wetter parts of the Pacific states, mainly for pastures on wet or saline soils. It seldom grows enough for a hay crop. One acre of pasture would carry 1 to 2 cows/season. There are 300,680 seeds/lb.
3. Season: See Clovers, True. Flowers May-June. Graze in spring until fall.
4. Production in U.S.: See Clovers, True. Used in California as cover crop.
5. Other production regions: See Clovers, True.
6. Use: See Clovers, True; Grazing pasture, green manure, cover crop, turf.
7. Part(s) of plant consumed: See Clovers, True
8. Portion analyzed/sampled: See Clovers, True
9. Classifications:
   a. Authors Class: See Clovers, True
   b. EPA Crop Group (Group & Subgroup): See Clovers, True
   c. Codex Group: See Clovers, True
   d. EPA Crop Definition: See Clovers, True
10. References: GRIN, DUKE(h). See Clovers, True.
11. Production Map: EPA Crop Production Region, See Clovers, True.
12. Plant Codes:
   a. Bayer Code: TRFFR

180

1. Clover, Striate
(Knotted clover)
Fabaceae (Leguminosae)
Trifolium striatum L.

2. This is a winter annual clover, apparently introduced by chance and now established in the South. It is adapted to heavy clay, wet soils. It appears less palatable to livestock than most other winter annual clovers, especially at the flowering stage and later when it becomes harsh and tough.
3. Season: See Clovers, True.
4. Production in U.S.: See Clovers, True.
5. Other production regions: See Clovers, True.
6. Use: See Clovers, True.
7. Part(s) of plant consumed: See Clovers, True
8. Portion analyzed/sampled: See Clovers, True
9. Classifications:
   a. Authors Class: See Clovers, True

b. EPA Crop Group (Group & Subgroup): See Clovers, True
c. Codex Group: See Clovers, True
d. EPA Crop Definition: See Clovers, True
10. References: GRIN, See Clovers, True
11. Production Map: EPA Crop Production Region, See Clovers, True.
12. Plant Codes:
a. Bayer Code: TRFST

181

1. Clover, Sub
(Subterranean clover, Subclover)
Fabaceae (Leguminosae)
Trifolium subterraneum L.
2. This is a self-reseeding winter-annual clover important in New Zealand and Australia. It is now of some importance in the Pacific Northwest and of possible value for the South. Plant stems are decumbent. The ripening flower heads turn downward and bury themselves in the soil. In areas with dry summers, seeds remain dormant until fall rains occur, then germinate. Its value is primarily for pasture and cover crop. Seasonal production: is grazed in early April, and flowers in April to May. Planted in the fall at a broadcast rate of 10 to 20 lbs per acre. As a cover crop, it is seeded at 20 to 30 lbs per acre in orchards and vineyards. It has persisted in some areas for 25 years and can be cut four to eight times per season for hay. It tolerates shade better than most clovers. There are 68,182 seeds/lb.
3. Season: See Clover, True.
4. Production in U.S.: See Clovers, True. In 1997, Texas reported 30,000 acres.
5. Other production regions: See Clovers, True.
6. Use: See Clovers, True; Pasture and cover crop.
7. Part(s) of plant consumed: See Clovers, True
8. Portion analyzed/sampled: See Clovers, True
9. Classifications:
a. Authors Class: See Clovers, True
b. EPA Crop Group (Group & Subgroup): See Clovers, True
c. Codex Group: See Clovers, True
d. EPA Crop Definition: See Clovers, True
10. References: GRIN, See Clovers, True; BALL
11. Production Map: EPA Crop Production Region, See Clovers, True.
12. Plant Codes:
a. Bayer Code: TRFSU

182

1. Clover, White
(Ladino clover, Common white clover, White Dutch clover, White clover, Dutch clover)
Fabaceae (Leguminosae)
Trifolium repens L.
2. White clover, like red, is believed native to the eastern Mediterranean region. It was widely grown in Europe before America was colonized and was brought here by the earliest settlers. It is now grown in all areas of the U.S. except the Great Plains and the extreme South. The plants are perennials with a prostrate growth habit. They develop stolons which root at the nodes, resulting in thickening of the stands. They do not develop upright stems. Leaves grow at the crown and at the nodes of stolons. They are trifoliate and the leaflets vary in shape from broadly elliptical to obovate, generally about $1/2$ inch across. Flower heads are generally white, sometimes pink-tinted, and contain up to 100 florets. Among many strains, Ladino white clover is now most widely grown. It differs from the common kinds in that it grows two to four times as large, otherwise, it is similar. White clovers are used primarily as pasturage, often in combination with grasses. They are highly nutritious and palatable. White clover is one of the most important and widely distributed forage legumes in the world. Over 230 white clover cultivars have been developed over the last 60 years. Usually grown in mixture with a grass such as Tall fescue, Orchardgrass, Kentucky bluegrass, Centipedegrass, Dallisgrass and Timothy. As pasture, the clover is managed rather than the grass. It is grazed to about 2 inches and allowed to regrow to about 10 inches. The stand life is 3 years or more. Seed crops are harvested 25 to 30 days after flowering. In 1992, ladino clover seed crop was grown in California on 6,137 acres with 2,218,032 pounds of seeds. White clover seed was produced in Idaho (2,261 acres) and Oregon (2,588). The total white clover seed production reported as 4,952 acres with 1,355,723 pounds of seed (1992 CENSUS). Ladino and white clovers reported separately in 1992 Census. There are over 3,022,800 seeds/lb.
3. Season: See Clover, True.
4. Production in U.S.: See Clovers,True; About half of the acreage in humid or irrigated pastures contains some white cover.
5. Other production regions: See Clovers, True; Canada (BARNES 1995).
6. Use: See Clover, True; Pasture, hay.
7. Part(s) of plant consumed: See Clovers, True
8. Portion analyzed/sampled: See Clovers, True
9. Classifications:
a. Authors Class: See Clovers, True
b. EPA Crop Group (Group & Subgroup): See Clovers, True
c. Codex Group: See Clovers, True
d. EPA Crop Definition: See Clovers, True
10. References: GRIN, See Clovers, True. BALL (picture), DUKE(h), BARNES
11. Production Map: EPA Crop Production Region, See Clovers, True.
12. Plant Codes:
a. Bayer Code: TRFRE

183

1. Clover, Whitetip
Fabaceae (Leguminosae)
Trifolium variegatum Nutt.
2. This clover is a native in the Pacific States where it occurs frequently in thick stands to 2 feet tall. It is an annual and may be either a summer-annual or winter-annual, depending on the local climate. Flower heads are purple with a white tip. It is relished by livestock both as pasture and hay. Grows in mountain meadows of western U.S. and Canada. It is rarely cultivated as a crop.
3. Season: See Clovers, True.
4. Production in U.S.: See Clovers, True.
5. Other production regions:See Clovers, True.
6. Use: See Clovers, True.
7. Part(s) of plant consumed: See Clovers, True

8. Portion analyzed/sampled: See Clovers, True
9. Classifications:
   a. Authors Class: See Clovers, True
   b. EPA Crop Group (Group & Subgroup): See Clovers, True
   c. Codex Group: See Clovers, True
   d. EPA Crop Definition: See Clovers, True
10. References: GRIN, See Clovers, True; DUKE(h)(picture)
11. Production Map: EPA Crop Production Region, See Clovers, True.
12. Plant Codes:
    a. Bayer Code: No specific entry

184

1. Clovers, True
(Klee, Trebol, Trefle)
Fabaceae (Leguminosae)
Trifolium spp.

2. Some 250 species of *Trifolium* are recognized throughout the world, with about 50 indigenous in the U.S. None of these native species is cultivated, although they contribute to grazing and may be an important part of wild hay crops. They contribute nitrogen and thus promote the growth of associated grass. The clovers may be annual or perennial. Leaves are mostly trifoliate, rarely five to seven leaflets. Flower heads are usually short spikes or umbels with numerous small individual flowers in the head. Annual clovers are used where perennial clovers do not persist and are used extensively in the southeast U.S. by overseeding with warm season perennial grasses and with winter annual grasses to reduce costs and improve overall performance. Clovers provide high quality forage and extend the grazing season. In the west coast where summers are dry, annual clovers are also grown with cool season annual grasses. Use of forage legumes as cover crops a year before corn planting can help reduce the amount of nitrogen fertilizer used in corn production. The species important agriculturally in the U.S. are described, which see.
3. Season: Winter and early spring pasture. Initial harvest for hay, about 60-70 days after seeding for red clover.
4. Production in U.S.: EPA estimated 37,300,000 acres of clover (US EPA 1994). For seed production, see individual species, as appropriate..
5. Other production regions: Worldwide.
6. Use: Pasture, hay, silage, roadside beautification and stabilization, cover crops and honey crops. Also, seeds are sprouted and used in salads. Sprouted clover seeds are grown in greenhouses and take about one week to produce and production is year-round (SCHREIBER 1995).
7. Part(s) of plant consumed: Foliage and seed
8. Portion analyzed/sampled: Forage and hay. For clover forage, cut sample at 4 to 8 inch to prebloom stage, at approximately 30 percent dry matter. For clover hay, cut at early to full bloom stage. Hay should be field-dried to a moisture content of 10 to 20 percent. Residue data for clover seed are not needed. For clover silage, residue data on silage are optional, but are desirable for assessment of dietary exposure. Cut sample at early to one-fourth bloom for clover, allow to wilt to approximately 60 percent moisture, then chop fine, pack tight, and allow to ferment for 3 weeks maximum in an air-tight environment until it reaches pH 4. This applies to both silage and haylage. In the absence of silage

data, residues in forage will be used for silage, with correction for dry matter
9. Classifications:
   a. Authors Class: Nongrass animal feed
   b. EPA Crop Group (Group & Subgroup): Nongrass animal feed (18) (Representative crop)
   c. Codex Group: 050 (AL 1023) Legume animal feed
   d. EPA Crop Definition: None
10. References: GRIN, CODEX, MAGNESS, SCHREIBER, US EPA 1996a, US EPA 1995a, US EPA 1994, IVES, BALL, BARNES, BARNES(a), TAYLOR
11. Production Map: EPA Crop Production Regions 1, 2, 4, 5, 6, 7, 8, 9, 10 and 11.
12. Plant Codes:
    a. Bayer Code: TRFSS

185

1. Coconut
(Coconut oil, Coco palm, Coconut palm, Water coconut, Copra, Khopra, Nariyal)
Arecaceae (Palmae)
Cocos nucifera L.

2. The coconut palm may reach to 100 feet or more. It can be grown in a wide variety of soils. Coconut plants are commercially propagated by seeds. The leaves are very large, up to 18 feet long, with lanceolate leaflets up to 3 feet long. The fruits are produced in clusters near the growing tip. They vary in shape, but are generally near globose to oblong, up to a foot or more in length. The nut is encased in a thick, fibrous husk which is persistent and must be cut away to expose the nut. The shell is very hard and woody, near $1/_4$-inch thick. When mature, the edible, oily flesh or kernel adheres to the shell, and is about $1/_2$-inch thick, with a hollow center which contains a liquid during growth. The immature coconut or "water coconut" is used for its edible liquid and gelatinous flesh. The commerical coconut is 1 to 3 pounds in weight. The dried flesh or meat of the mature coconut is the copra of commerce, produced in great quantities mainly for its oil. In preparing copra, the dried meat of the coconut, the nuts are cut in half, the milk drained off, and the nuts are exposed to sun. The partially dried meat contracts and can be readily removed from the shell. Further drying reduces the moisture to under 8 percent, necessary to prevent mold growth. Artificial heat is often used for this. The fresh meats contain 30 to 40 percent of oil, the dried copra 60 to 70 percent. The oil is extracted from the copra by heating and pressing in various types of expellers. The oil is used for margarine and vegetable shortenings. The press cake is used for livestock feed. Crude oil is made into soap. World production of coconut oil is near 3,320,900 tons. Around 436 tons of copra and over 487,846 tons of coconut oil are imported into the U.S. annually (1993).
3. Season, fruit set to maturity: 11 to 12 months. One mature bunch of coconuts can be harvested each month.
4. Production in U.S.: About 662,013 pounds in shell, 1992 CENSUS for Puerto Rico, Virgin Islands and Guam. U.S. exported about 7,000 tons of coconut oil (1993). Also Florida and Hawaii (CRANE 1994).
5. Other production regions: Indonesia and Philippines account for 60 percent of the world's coconut production (CHAVAN).

6. Use: Mainly oil used in cooking fats, soaps, etc. Also used in confections, cookery. A principal item of diet in tropical countries. All parts of the plant are useful in some parts of the world.

7. Part(s) of plant consumed: Internal kernel and oil from it

8. Portion analyzed/sampled: Meat and liquid combined; Copra (dried meat); and oil

9. Classifications:
   a. Authors Class: Tree nuts; Oilseed
   b. EPA Crop Group (Group & Subgroup): Miscellaneous
   c. Codex Group: 022 (TN 0665) Tree nuts; 023 (SO 0665) Oilseed; 067 (OC 0665) Vegetable crude oils; 068 (OR 0665) Vegetable oil edible or refined
   d. EPA Crop Definition: None

10. References: GRIN, CODEX, CRANE 1994, MAGNESS, OLSZACK 1995, RICHE, ROBBELEN 1989 (picture), USDA 1994a, US EPA 1995b, US EPA 1994, CHAVAN, PHILLIPS(c)

11. Production Map: EPA Crop Production Region 13.

12. Plant Codes:
   a. Bayer Code: CCNNU

186

1. Coffee
(Cafe, Cafeto, Arabian Coffee, Common Coffee)
Rubiaceae
Coffea arabica L. (Arabica coffee)
C. liberica W. Bull ex Hiern (Liberica coffee)

2. These two species constitute most of the coffee of commerce. The plants are woody, tropical, evergreen shrubs up to 15 feet in height. The leaves are elliptical, glossy, up to 6 inches long and $^1/_3$ as wide. The fruit is a fleshy berry, in which 2 seeds are imbedded. Blossoming and fruit setting occur mainly 2 to 3 times per year. About 6 to 7 months are required to ripen the fruit, so fruits at various stages of maturity are on the plants at the same time. Ripe berries are picked at about 2 week intervals. The pulp is removed by machines. Seeds are dried in the sun or in dehydrators and roasted before being marketed. U.S. production was given as 1,450 tons in Hawaii only, (1994 AGRICULTURAL STATISTICS). Imports total about 1,191,212 tons annually with 200,000 tons from Mexico (1993).

3. Season, bloom to harvest: 6 to 7 months. Peak harvest in Puerto Rico is October and November.

4. Production in U.S.: 7,783 acres in Hawaii (1992 CENSUS); About 75,000 acres in Puerto Rico (1996).

5. Other production regions: Native to Africa (FACCIOLA 1990).

6. Use: Beverage.

7. Part(s) of plant consumed: Seed (bean)

8. Portion analyzed/sampled: Green bean (dried seed of coffee bean) and processed commodities roasted bean and instant

9. Classifications:
   a. Authors Class: Tropical and subtropical trees with edible seed for beverages and sweets
   b. EPA Crop Group (Group & Subgroup): Miscellaneous
   c. Codex Group: 024 (SB 0716) Seed for beverages and sweets; 059 (SM 0716) Miscellaneous secondary food commodities of plant origin as coffee beans roasted
   d. EPA Crop Definition: None

10. References: GRIN, CODEX, MAGNESS, MONTALVO-ZAPATA 1996a, USDA 1994a, US EPA 1995b, US EPA 1994, IVES, FACCIOLA

11. Production Map: EPA Crop Production Region 13.

12. Plant Codes:
   a. Bayer Code: COFAR (*C. arabica*); COFLI (*C. liberica*)

187

1. Cola
(Cola nut, Kola)
Sterculiaceae
Cola acuminata (P. Beauv.) Schott & Endl.
(Abata cola, Cola tree, Guru)
C. nitida (Vent.) Schott & Endl.
(Ghanja kola, Bitter cola)
C. anomala K. Schum.
(Bamenda cola)
C. verticillata (Thonn.) Stapf ex A. Chev.
(Owe cola)

2. The cola seeds of commerce come from medium-size spreading trees, up to 40 feet, native to tropical Africa, but now cultivated in the West Indies and other tropical countries. The seeds or nuts are borne in leathery or woody oblong capsules, about 6 inches long. Seeds are near globose, about an inch in diameter. They, or preparations from them, are widely used in tropical countries as a stimulant and appetizer. Preparations from the seeds are widely used for flavoring the cola drinks, very popular in the U.S. and other countries.

3. Season, seeding to harvest: 7 to 10 years and may bear fruit for about 75 years.

4. Production in U.S.: No data.

5. Other production regions: Hot, wet tropical lowlands.

6. Use: Beverage.

7. Part(s) of plant consumed: Seed used as source of extract for beverages

8. Portion analyzed/sampled: Seed

9. Classifications:
   a. Authors Class: Tropical and subtropical trees with edible seeds for beverages and sweets
   b. EPA Crop Group (Group & Subgroup): Miscellaneous
   c. Codex Group: 024 (SB 0717) Seed for beverages and sweets
   d. EPA Crop Definition: None

10. References: GRIN, CODEX, KENNARD (picture), MAGNESS, MARTIN 1987, MARTIN 1983

11. Production Map: EPA Crop Production Region 13.

12. Plant Codes:
   a. Bayer Code: COHAC (*C. acuminata*)

188

1. Collards
(Winter greens, Cow cabbage, Spring heading cabbage, Tall kale, Tree kale)
Brassicaceae (Cruciferae)
Brassica oleracea var. *viridis* L. (syn: *B. oleracea* var. *acephala* DC. in part)

2. Collards are closely related to kale and cabbage, and might be described as a non-heading cabbage. They are grown for the smooth, rather thick, tender leaves which are used as

greens or potherbs. Plants may be started in beds, or direct seeded in the field. They produce a rosette of leaves. The whole rosette may be cut off and marketed, the usual commercial practice, or leaves may be stripped off and the central axis will continue to grow and produce. Leaf surface is generally exposed, as in spinach. Used as a winter green in the South. The plants are biennials grown as annuals and grow to a height of 2-4 feet.

3. Season, seeding to harvest: About 2 to 3 months.
4. Production in U.S.: 10,050 acres (1959 CENSUS). 16,062 acres (1992 CENSUS). In 1992, the top 10 collard-producing states were Georgia (4,354 acres), South Carolina (1,610), North Carolina (1,244), Texas (1,029), New Jersey (778), Maryland (777), California (775), Alabama (761), Arizona (729), and Tennessee (691) with about 80 percent of the U.S. acreage (1992 CENSUS).
5. Other production regions: In 1995, Canada grew collards (ROTHWELL 1996a).
6. Use: Mainly as greens or potherbs.
7. Part(s) of plant consumed: Leaves
8. Portion analyzed/sampled: Greens (leaves)
9. Classifications:
   a. Authors Class: *Brassica* (cole) leafy vegetables
   b. EPA Crop Group (Group & Subgroup): *Brassica* (cole) leafy vegetables (5B)
   c. Codex Group: Kale: 013 (VL 0480) Leafy vegetables (including *Brassica* leafy vegetables)
   d. EPA Crop Definition: None
10. References: GRIN, CODEX, MAGNESS, MYERS, RICHE, ROTHWELL 1996a, US EPA 1996a, US EPA 1995a, US EPA 1994, SWIADER
11. Production Map: EPA Crop Production Regions 1, 2, 3, 4, 5, 6 and 10.
12. Plant Codes:
    a. Bayer Code: BRSOA

189

1. Comfrey
(Boneset, Knitbone, Common comfrey, Bonset, Blackwort, Consuelda major, Grande consoude, Healing herb)
Boraginaceae
Symphytum officinale L.
1. Russian comfrey
(Quaker comfrey, Blue comfrey, Blackwort, Wallwort, Gumplant, Prickly comfrey)
S. x *uplandicum* Nyman (syn: *S. peregrinum* auct.)
2. Hardy, herbaceous, perennial that grows to 4 feet tall. Leaves are 5 inches wide and 8 to 12 inches long and are covered on the upper surface by many short, bristly hairs. The leaves appear to be stacked. Cultivated comfrey is called Russian comfrey. Tap root may stretch 6 feet into ground.
3. Season, planting: Use root or crown cuttings that are 1-6 inches long. Once established, harvest as needed as the flowers bud up. Many cuttings are possible.
4. Production in U.S.: Grows well in Florida gardens where it grows year round (STEPHENS 1988). Commonly grown in North America (FOSTER 1993).
5. Other production regions: Eurasia (FACCIOLA 1990).
6. Use: Cooking green and herb. Dried leaves used as tea.
7. Part(s) of plant consumed: Leaves
8. Portion analyzed/sampled: Leaves (fresh and dried)

9. Classifications:
   a. Authors Class: Leafy vegetables (except *Brassica* vegetables); Teas
   b. EPA Crop Group (Group & Subgroup): Miscellaneous
   c. Codex Group: No specific entry
   d. EPA Crop Definition: None
10. References: GRIN, FACCIOLA, FOSTER (picture), GARLAND (picture), SCHNEIDER 1993a, STEPHENS 1988 (picture)
11. Production Map: EPA Crop Production Region 3.
12. Plant Codes:
    a. Bayer Code: SYMOF (*S. officinale*)

190

1. Coneflower, Purple
(Echinacea, Droops, Nutraceutical)
Asteraceae (Compositae)
Echinacea spp.
1. Eastern Purple coneflower
E. purpurea (L.) Moench (syn: *E. purpurea* var. *arkansana* Steyermark, *Rudbeckia purpurea* L., *Brauneria purpurea* (L.) Britt.)
1. Blacksamson echinacea
(Narrowleaf purple coneflower, Kansas snakeroot)
E. angustifolia DC.
1. Pale purple coneflower
(Paleflower echinacea)
E. pallida (Nutt.) Nutt. (syn: *Brauneria pallida* (Nutt.) Britt., *Rudbeckia pallida* Nutt.)
1. Bush's purple coneflower
E. paradoxa (Norton) Britton
1. Wavyleaf purple coneflower
E. simulata McGregor
2. Perennial herb, 1 to 4 feet tall, with erect stems, native to North America and drought resistant. This genus is represented by nine species and at least two varieties which are native to North America. The common species in cultivation are *E. angustifolia*, *pallida* and *purpurea*. The cone-shaped flower head is a hemispherical receptacle with radiating ray florets which droop. Flowering begins as early as mid-May in the South, extending into October in the northern limits of its range, southern Canada. Several rosettes of leaf and flower stalks may arise from a single root. The fibrous or tap rootstock is pungently aromatic and 6-24 inches in length. Only *E. purpurea* has a fibrous root, also it is the one most widely distributed and cultivated. Cultivars of *E. purpurea* in international trade include 'Schleissheim', 'Hybrida', 'Leuchtstern', 'Magnus', and 'Rubinstern' which are suitable as medicinal plants. Polysaccharides participate in the immunostimulatory activity in the expressed juice of *E. purpurea*, its water extracts and its whole powdered leaves or root. More information in "Echinacea – Nature's Immune Enhancer" by S. Foster. 1991. (Rochester, VT, Healing Arts Press). Foods or food components that confer medical benefits are termed "Nutraceuticals". These include both raw and processed foods for medical benefits, including health promotion and disease prevention (Hannon, J. (Editor). 1997. *Connexions*. Cook College/NJAES, Office of Communications and Public Affairs. Rutgers University, New Brunswick, NJ 20 pp.). Other crops with nutraceutical properties

are: Asparagus, Blueberry, Broccoli, Cauliflower, Cranberry, Feverfew, Hot pepper, Potato, Soybean, Raspberry, Strawberry and Tomato.

3. Season, seeding to root harvest: About 3 to 4 years for sizable root, after plant has gone to seed.
   Season, planting divided crowns to root harvest: About 2 years.
   Season, seeding to top or herbage harvest: Late in the 1st through 3rd year around flowering.
4. Production in U.S.: Produced in Grant County, Washington (SCHREIBER 1995). Mainly harvested from wild (FOSTER 1993). Cultivated *E. purpurea* has produced 1200 pounds of dry root per acre (FOSTER 1993). Plant distribution centered in Arkansas, Kansas, Missouri and Oklahoma.
5. Other production regions: Europe and Australia (FOSTER 1993).
6. Use: Medicinal herb. Extracts of roots and tops used in topical and injected products for humans, especially in Germany. Root preparations used as antidote for all types of venomous bites or stings and also used to support and stimulate the immune system. Also grown as an ornamental.
7. Part(s) of plant consumed: Roots and tops
8. Portion analyzed/sampled: Roots (dry) and tops (fresh)
9. Classifications:
   a. Authors Class: Herbs and spices
   b. EPA Crop Group (Group & Subgroup): Miscellaneous
   c. Codex Group: No specific entry
   d. EPA Crop Definition: None
10. References: GRIN, BAILEY 1976, FOSTER (picture), HANNON, SCHREIBER, USDA NRCS 1995
11. Production Map: EPA Crop Production Region 11.
12. Plant Codes:
    a. Bayer Code: ECEPA (*E. pallida*), RUDPU (*E. purpurea*)

191

1. Coriander
(Coriandre, Coriandro, Coriender seed)
1. Cilantro
(Chinese parsley, Mexican parsley, Yun tsai, Culantro, Celantor, Ngo, Cilantro leaves, Yan-sui, Culantrillo, Dhaniya, Coriander leaves, Yuensai, Kasturi)
Apiaceae (Umbelliferae)
Coriandrum sativum L.

2. Coriander is a hardy annual herb of the carrot family, native to the Mediterranean area. The plant is 1 to 3 feet in height with glabrous, greatly divided strong smelling leaves, primarily termed Cilantro, Chinese parsley or Mexican parsley. Similar cultural practices for growing the leaves as parsley and exposure of leaves is similar. Fresh leaves and stems are harvested 1 inch above ground level. Dried cilantro leaves are generally rehydrated prior to consumption. Foliage is utilized in Indian, Chinese and Mexican meals and the roots are eaten in China and Thailand. The seeds mature about 6 months after planting and are used for flavoring candies, beverages, tobacco products and in cookery. Oil distilled from the seeds is used in perfumes and soaps. The seeds, sold as the spice coriander, are harvested when the entire plant is dry and the seed pods have not opened. The whole plant is cut, threshed and the seed is further dried. Undried seed has a bitter taste. U.S. imported about 2,314 metric tons

of coriander seed with about half from Canada and exported about 39 metric tons in 1992.

3. Season, seeding to first harvest: Seeds: 6 months; Leaves: 80 days with the potential for one to three additional harvests or cuttings every 30 days. In tropical/subtropical conditions, leaves can be produced in approximately 40 days.
4. Production in U.S.: Less than 100 acres in Washington state, 375 acres in Texas, 1,733 acres in California, 175 acres in Florida and minor acreage in Hawaii. Primarily grown for its leaves in U.S. for the fresh market. Washington State averages 200 acres of coriander grown for seed as a seed crop which represents 33 percent of the U.S. acreage. Primary commercial production is in Southwest U.S.
5. Other production regions: Native to Mediterranean region (FACCIOLA 1990). Canada, Mexico, Asia.
6. Use: Mainly for garnishing and flavoring of foods or cooked.
7. Part(s) of plant consumed: Fresh leaves and stems; may be dehydrated for dried leaves; Dried seeds
8. Portion analyzed/sampled: Fresh leaves and stems, dried leaves, dried seeds
9. Classifications:
   a. Authors Class: Herbs and spices group; leafy vegetables (except *Brassica* vegetables)
   b. EPA Crop Group (Group & Subgroup): Herbs and spice (19A and 19B)
   c. Codex Group: Seed only: 028 (HS 0779) Spices
   d. EPA Crop Definition: Proposed (Parsley = Cilantro)
10. References: GRIN, CHAMBERS 1996, CODEX, KAWATE 1996c, KAYS, KUEBEL, MAGNESS, MELNICOE 1996a, MONTALVO ZAPATA 1996b, MYERS (picture), PENZEY, RATTO 1996a, REHM, SCHREIBER, USDA 1993a, US EPA 1995a, US EPA 1994, FACCIOLA, TUCKER, RUBATZKY
11. Production map: EPA Crop Production Regions 2, 3, 6, 8, 10, 11, 12 and 13.
12. Plant Codes:
    a. Bayer Code: CORSA

192

1. Coriander, False
(Puerto Rican coriander, Spiny coriander, Acopate, Culantro, Ngo Gai, Recaito, Recao, Thoryn coriander, Stinkweed, Azier la fievre, Culantrillo, Fitweed, Culantro del monte, Cilantrillo, Spiritweed, Wideleaf coriander, Shadow beni)
Apiaceae (Umbelliferae)
Eryngium foetidum L.

2. False Coriander is cultivated in tropical areas and is a member of the carrot family. This plant may be treated as a biennial or short-lived perennial herb. The leaves are caulescent spinosely serrated. The use is similar to that of cilantro (coriander) leaves.
3. Season, seeding to harvest: No data.
4. Production in U.S.: Limited data. Grown in Puerto Rico (MARTIN 1979).
5. Other production regions: Native to Central America (KUEBEL 1988).
6. Use: Flavoring. Young leaves are eaten raw, steamed with rice, or used as garnish for fish. The seeds can be roasted and are also used in West Indies as part of seasoning called "sofrito". Leaves used in a seasoning sauce called "Recaito".
7. Part(s) of plant consumed: Leaves (fresh) and seeds

8. Portion analyzed/sampled: Leaves (fresh) and seeds
9. Classifications:
    a. Authors Class: Herbs and spices; Leafy vegetables (except *Brassica* vegetables)
    b. EPA Crop Group (Group & Subgroup): Herbs and spices (19A and 19B) as Culantro
    c. Codex Group: No specific entry
    d. EPA Crop Definition: None
10. References: GRIN, SCHNEIDER 1993a, MARTIN 1979 (picture), KUEBEL, US EPA 1995a, REHM, JANICK, MONTALVO ZAPATA 1996b, USDA NRCS
11. Production Map: EPA Crop Production Region 13.
12. Plant Codes:
    a. Bayer Code: No specific entry

193

1. Corn

(Maize, Maiz, Field corn, Popcorn, Pod corn, Dent corn, Flint corn, Sweet corn, Maiz tierno, Mais doux, Maiz dulce, Chulo)
Subsets of common names:
- **Field corn**
(Maiz, Maize, Dent corn, Flint corn, Corn oil, Pod corn, Chulo)
- **Popcorn**
(–)
- **Sweet corn**
(Maiz tierno, Mais doux, Maiz dulce)
Poaceae (Gramineae)
Zea mays L. ssp. *mays*

1. Teosinte

Z. mays L. ssp. *mexicana* (Schrad.) Iltis (syn: *Z. mexicana* (Schrad.) Kuntze; *Euchleaena mexicana* Schrad.)

2. Native Americans were growing corn extensively long before the discovery of these continents by Europeans. Archaeological studies indicate that corn was cultivated in the Americas at least 5600 years ago. The exact origin of corn is unknown as the plant is found only under cultivation. The probable center of origin is Central America and Mexico. The corn plant is a warm-weather annual, deep-rooted but requiring abundant moisture for best development.

From a seed a single stalk arises that will reach from 2 to near 20 feet depending on kind and growing conditions. This stalk terminates in the tassel or staminate flowers. At the stem nodes are attached the large, smooth leaves which may be more than 2 feet long and 2 inches wide along the mid-point of the stem. At the base of the main stem, side shoots or suckers commonly rise, which may produce seeds. The female flowers are borne on a receptacle, termed ear, which arises at a leaf axil near the mid-point along the stem. Normally one to three or more such ears develop. The flower organs, and later the grain kernels, are enclosed in several layers of papery tissue, termed husks. Strands of "silk", actually the stigmas from the flowers, emerge from the terminals of the ears and husks at the same time the pollen from the terminal tassels is shed. The pollen is wind-blown and comes in contact with the emerged silk or stigma. The pollen then germinates and a pollen tube grows down through the silk to the egg cell of the female flower. The male gamete fuses with the egg, and from the fertilized egg the corn seed or kernel develops.

Most varieties of corn require 100 to 140 days from seeding to full ripeness of the kernels though some kinds will ripen in as little as 80 days. The time of pollen shed and fertilization of the egg is a little after the midpoint of this period. Corn kernels or seeds vary in size and shape in different kinds and varieties. They may be only $1/8$ inch long and near round in popcorn to $1/2$ inch long and a flattened cylinder shape in some other kinds. The kernel consists of the following: (1) An outer thin covering which is made up of two layers, an outer pericarp and an inner testa or true seed coat; (2) The endosperm which makes up nearly two-thirds of the total volume. This consists almost entirely of starch, except in sweet corn; (3) The embryo, the miniature plant structure that develops into a new plant if the seed is planted and grows.

The embryo is near one side of the kernel in most kinds rather than in the middle. It contains most of the oil in corn. The oil is obtained from the germ of the corn seed. In the preparation of certain corn products, as hominy, starch, and glucose, the germ is separated from the rest of the seed mechanically. The germ contains near 50 percent of oil, which is separated with expellers or with solvents. About $1/2$ pound of oil is obtained from a bushel of corn. Most of the oil is used as a salad or cooking oil or in the manufacture of lard substitutes. There is some use in industry.

Three major types of corn are grown in the U.S.:

Grain or field corn is grown annually for grain on 60 to 70 million acres, with seed production in excess of 7 billion bushels. In addition, around 6 million acres of this type are harvested for silage.

Sweet corn, used mainly as food, is grown on around 767,710 acres. Sweet corn is distinguished from field corn by the high sugar content of the kernels at the early "dough" stage, and by wrinkled, translucent kernels when dry. The plant is a single-stemmed annual, grown from one seed, though sucker shoots rise from the base. The stem produces one to three ears, each consisting of a base or cob on which the seed is embedded and completely encased in several layers of thin, papery husks. Plants attain a height up to 6 to 8 feet, with grass-like leaves up to 2 feet long and 2 inches wide. For food, sweet corn must be harvested when kernels are fully developed but still in an immature or "dough" stage. Otherwise they lose sweetness and become tough.

Popcorn, also used mainly for food, is grown on about 321,000 acres.

Grain corn is further classified commercially into four main types: (1) **Dent corn**, when fully ripe, has a pronounced depression or dent at the crown of the kernels. The kernels contain a hard form of starch at the sides and a soft type in the center. This latter starch shrinks as the kernel ripens resulting in the terminal depression. Dent varieties vary in kernel shape from long and narrow to wide and shallow. It is the type mainly grown in the U.S. (2) **Flint corn** has the hard starch layer entirely surrounding the outer part of the kernel. Consequently, on drying, the kernel shrinks uniformly and does not develop a depressed area and are round in shape. (3) **Flour** or **soft corn kernels** contain almost entirely soft starch, with only a very thin layer of hard starch. This type is little grown commercially. (4) **Waxy corn** is so-called because the endosperm when cut or broken is wax-like in appearance. The starch consists almost entirely of amylopectin, while in ordinary corn the starch is near 30 percent amylose and the remaining 70 percent is amylopectin. Waxy corn is largely used industrially although it is also suitable for food or feed. Acreage of waxy corn is small as compared with dent or flint.

Popcorn is characterized by a very high proportion of hard starch. Under heat the moisture in the starch grains expands rapidly resulting in an explosive rupture of the epidermis and the starch grains. Increase in volume after "popping" is 15 to 35 fold, depending on variety. Insufficient moisture in the kernels results in poor popping. Both the plant and ears of popcorn are smaller than the grain corns but are otherwise similar.

Pod corn is a curiosity in which each individual kernel is covered by a pod-like growth in addition to the husks enclosing the ear as a whole. It is not grown commercially.

Nearly all the corn now grown in the U.S. is of hybrid varieties. Seed is obtained by crossing inbred lines which are obtained by self pollination through several generations. This results in reduced vigor and yield but increased uniformity in the inbreds. To produce hybrid seed, two inbreds are planted together and the tassels removed from one before any pollen is shed. Thus kernels on the detasselled variety are from pollen produced on the other inbred line. This restores and increases vigor and is known as a single cross. Two single crosses may be similarly crossed producing what is termed double cross seed. Properly selected and adapted hybrid corn varieties produce higher yields and more uniform plants and ears than the open pollinated varieties formerly used.

About 75 percent of the grain corn produced in the U.S. is fed to livestock. In hog feeding, whole ears may be used. For other livestock and poultry, and some hog feeding, kernels are removed from the cob by machinery and are often partially ground before feeding. About half of the feeding is done on the farms where the corn is produced. Over 10 percent of grain corn grown in the U.S. is exported either as grain or corn products. From 12 to 15 percent of the crop is processed for starch, corn sugar, syrup, corn oil, corn-oil meal, gluten feed and meal, whiskey, alcohol, and for direct human food in the form of corn flakes, corn meal, hominy and grits.

For silage, the corn seed is planted more closely spaced than for grain, to obtain maximum yields. The corn is harvested before the kernels ripen from the late dough to early dent stage. At this stage the plant is still green and somewhat succulent. The whole plant is cut near the ground and passed through silage cutters either in the field or at the silo or pit. Thus silage is a mixture of all the above-ground plant parts. Corn silage is mainly used for feeding cattle, especially dairy cows. Corn stover is mature dried stalks from which the whole ear is removed.

There is strong evidence that teosinte is the wild ancestor of corn and not wild pod corn. Teosinte originated in Mexico (WATSON 1987). 'Chulo' is a corn cultivar released by USDA-ARS in Puerto Rico that can be grown year round in the Tropics. Makes good poultry feed because of its smaller size kernels. Some popcorn cultivars are grown for baby corn. Baby corn production is similar to sweet corn. Ears are normally 10 cm long and 1.2 cm in diameter (KOTCH).

3. Season, seeding to harvest: Most require 100 to 140 days and some as little as 80 days for field corn. Sweet corn 2½ to 4 months. Popcorn 90 days to more than 110 days. Baby corn 60 days.

4. Production in U.S.: In 1992, popcorn was reported on 321,485 acres with production of 557,404 tons shelled (1992 CENSUS). In 1995, corn for grain was reported on 64,995,000 acres with production of 206,035,588 tons. Corn for silage was reported on 5,295,000 acres with production of 77,867,000 tons (USDA 1996a and c). In 1995, sweet corn was reported on 767,710 acres with 4,399,300 tons of production and about 70 percent of the acreage for processing sweet corn (USDA 1996b). In 1992, popcorn was reported in Indiana (75,853 acres), Nebraska (60,040), Illinois (58,156) and Ohio (28,978). These top 4 states represent about 70 percent of the total U.S. acreage (1992 CENSUS). In 1995, the top 7 states for grain corn were Iowa (11,400,000 acres), Illinois (10,000,000), Nebraska (7,700,000), Minnesota (6,150,000), Indiana (5,300,000), Ohio (3,100,000) and Wisconsin (3,050,000). The top 7 states for corn silage were Wisconsin (580,000 acres), New York (485,000), Minnesota (450,000), Pennsylvania (390,000), South Dakota (320,000), California (280,000) and Michigan (260,000) (USDA 1996a and c). In 1995, sweet corn's top 5 states for processing were Wisconsin (140,400 acres), Minnesota (133,900), Washington (84,700), Oregon (49,400) and New York (40,300). In 1995, the top 8 fresh market states were Florida (38,100 acres), New York (25,800), California (22,300), Georgia (21,000), Pennsylvania (19,500), Ohio (15,900), Michigan (14,000) and New Jersey (9,000). Fresh market sweet corn was grown on a total of 236,300 acres with 1,075,150 tons of production. Whereas, processing sweet corn was grown on a total of 531,410 acres with 3,324,150 tons of production (USDA 1996b). In 1992, Puerto Rico reported about 640 acres of field corn (Maiz) with a production of 942,700 pounds. About 60 acres of sweet corn (Maiz tierno) were reported with a production of 65,390 pounds (1992 CENSUS).

5. Other production regions: In 1995, Canada reported 2,453,760 acres of corn grain, about 1 million acres of fodder and silage corn and 83,136 acres of sweet corn (ROTHWELL 1996a).

6. Use: Sweet corn is used as a vegetable for fresh market, canned and frozen. Corn grain is either dry or wet milled. Dry milling is primarily concerned with separation of the parts of the grain. Wet milling provides the same separation but further separates some of those parts into their chemical constituents, primarily starch, protein, oil and fiber. Dry milling provides bran, germ and endosperm (HOSENEY 1986). Corn grains, silage, stover, forage, aspirated grain fractions, cannery waste and milled byproducts are used for livestock feed. The popcorn seed is used for popping and direct consumption. Corn bran is also used commercially in breakfast cereals and comprise 5.3 percent of the kernel. Corn extracts are also fermented to produce an alcoholic beer product in several countries. Blue corn has been traditionally used for tortilla and chips in New Mexico and Arizona. Corn for "Cornuts" was developed from the Cuzco Gigante race from Peru. Yearly, the U.S. dry milling industry processes 163 million bushels of corn. In 1995, the per capita consumption of corn by-products are: flour and meal 15.8 lbs, hominy and grits 4.0 lbs, syrup 77.3 lbs, sugar 3.9 lbs and starch 3.9 lbs.

7. Part(s) of plant consumed: Whole plant. Seed only for food. All of the plant may be fed to livestock.

8. Portion analyzed/sampled: **Field corn** – RAC's are the grain, forage, stover, aspirated grain fractions. The processed commodities are for wet milling-starch and refined oil; for dry milling-grits, meal, flour and refined oil. **Popcorn** – RAC's

are grain and stover. **Sweet corn** – RAC's are sweet corn (kernel plus cob with husk removed), forage and stover. Cannery waste includes husks, leaves, cobs and kernels. Residue data for forage can be used for sweet corn cannery waste. For field corn forage, cut sample (whole aerial portion of the plant) at late dough/early dent stage (black ring/layer stage for corn only). Forage samples should be analyzed as is, or may be analyzed after ensiling for 3 weeks maximum, and reaching pH 5 or less, with correction for dry matter. For corn stover, mature dried stalks from which the grain or whole ear (cob + grain) have been removed; containing 80 to 85 percent dry matter. For aspirated grain fractions (previously called grain dust), dust collected at grain elevators for environmental and safety reasons. Residue data should be provided for any postharvest use on corn. For a preharvest use after the reproduction stage begins and seed heads are formed, data are needed unless residues in the grain are less than the limit of quantitation of the analytical method. For a preharvest use during the vegetative stage (before the reproduction stage begins), data will not normally be needed unless the plant metabolism or processing study shows a concentration of residues of regulatory concern in a grain fraction. For corn starch, residue data for starch will be used for corn syrup. Petitioners may also provide data on syrup for a more accurate assessment of dietary exposure. For corn milled byproducts, use residue data for corn dry-milled processed commodities having the highest residues, excluding oils. For sweet corn, residue data on early sampled field corn should suffice to provide residue data on sweet corn, provided the residue data are generated at the milk stage on kernel plus cob with husk removed and there are adequate numbers of trials and geographical representation from the sweet corn growing regions. For sweet corn (K + CWHR), kernels plus cob with husks removed. For sweet corn forage, samples should be taken when sweet corn is normally harvested for fresh market, and may not include the ears. Petitioners may analyze the freshly cut samples, or may analyze the ensiled samples after ensiling for 3 weeks maximum, and reaching pH 5 or less, with correction for percent dry matter. For sweet corn cannery waste, includes husks, leaves, cobs, and kernels. Residue data for forage will be used for sweet corn cannery waste.

9. Classifications:
   a. Authors Class: Cereal grains: Forage, fodder and straw of cereal grains
   b. EPA Crop Group (Group & Subgroup): Cereal grains (15); Forage fodder and straw of cereal grains (16) (Representative crop)
   c. Codex Group: Codex does not include popcorn and sweet corn under Maize. **Maize:** 020 (GC 0645) Cereal grains; 078 (CP 0645) Manufactured multi-ingredient cereal products; 065 (CF 0645 and CF 1255) Cereal grain milling fractions; 051 (AS 0645) (AF 0645) Straw, fodder and forage of cereal grains and grasses, except grasses for sugar production (including buckwheat fodder); 067 (OC 0645) Crude vegetable oil; 068 (OR 0645) Edible or refined vegetable oils. **Popcorn:** 020 (GC 0656) Cereal grains. **Teosinte:** 020 (GC 0657) Cereal grains; 051 (AS 0657) Straw, fodder and forage of cereal grains and grasses. **Sweet corn:** 012 (VO 0447) Corn-on-the-cob, 012 (VO 1275) Kernels, Fruiting vegetables other than cucurbits
   d. EPA Crop Definition: None

10. References: GRIN, CODEX, HOSENEY, MAGNESS, RICHE, ROBBELEN 1989, USDA 1996a,b,c, US EPA 1994, US EPA 1995a, US EPA 1996a, WATSON, CARTER 1989, SERNA-SALDIVAR, KOTCH, JOHNSON(b)
11. Production Map: EPA Crop Production Regions: **Field corn:** 1, 2, 5 and 6 (86 percent of the acreage in 5); **Popcorn:** 5 and 8 (91 percent of the acreage in 5); **Sweet corn:** 1,2, 3, 5, 10, 11 and 12 (50 percent of the acreage in 5).
12. Plant Codes:
   a. Bayer Code: ZEAMX (*Z. mays*), ZEAMA (Soft or flour corn), ZEAMD (Dent corn), ZEAME (popcorn), ZEAMI (Flint corn), ZEAMS (Sweet corn), ZEAMT (Podcorn)

194

1. Corn salad
(Lambs lettuce, Fetticus, European corn salad, Mache, Doucette, Lechuga de campo)
Valerianaceae
   Valerianella locusta (L.) Laterr. (syn: *V. olitoria* (L.) Pollich)

1. Italian Corn salad
(Hairy-fruited cornsalad)
   V. eriocarpa Desv.

2. Corn salad is extensively grown in Europe as a cool season salad vegetable, but little grown in the U.S. The plant is a biennial, forming a rosette of leaves the first year, and a seed stalk the second. Leaves are spoon shaped to round, up to 6 inches long. Exposure of leaves is comparable to that of spinach. Leaves are used both as raw salad and as potherbs. Italian corn salad is similar in growth and use. Leaves are slightly smaller, somewhat pubescent and toothed near the base. The plant is more Southern in adaptation than corn salad.
3. Season, seeding to usable leaves: 2 to 3 months.
4. Production in U.S.: No data. Very limited. Corn salad is grown in New York. Grown in gardens and cold frames in Maine. The vegetable plant is grown in Florida similarly to endive or lettuce (STEPHENS 1988).
5. Other production regions: Available year-round from Holland (LOGAN 1996).
6. Use: Salad, potherb.
7. Part(s) of plant consumed: Leaves
8. Portion analyzed/sampled: Leaves
9. Classifications:
   a. Authors Class: Leafy vegetables (except *Brassica* vegetables)
   b. EPA Crop Group (Group & Subgroup): Leafy vegetables (except *Brassica* vegetables) (4A)
   c. Codex Group: 013 (VL 0470) Leafy vegetables (including *Brassica* leafy vegetables)
   d. EPA Crop Definition: None
10. References: GRIN, CODEX, LOGAN, MAGNESS, STEPHENS 1988 (picture), US EPA 1995a
11. Production Map: EPA Crop Production Regions 1, 3 and 10.
12. Plant Codes:
   a. Bayer Code: VLLER (*V. eriocarpa*), VLLLO (*V. locusta*)

195

1. Costmary
(Mint geranium, Alecost)
Asteraceae (Compositae)

Tanacetum balsamita L. ssp. *balsamita* (syn: *Chrysanthemum balsamita* L.; *Balsamita major* Desf.)

2. The costmary plant is a tall and erect perennial chrysanthemum, of minor importance as a condiment. The sweet-scented leaves are oval or oblong, up to almost 1-foot long. They are used somewhat in salads and cookery.
3. Season, seeding to harvest: About 1-2 months for young leaves.
4. Production in U.S.: Limited acreage, mainly herb gardens (STEPHENS 1988). Produced in Florida.
5. Other production regions: Native to Southwest Asia (FACCIOLA 1990).
6. Use: Flavor meats and salads.
7. Part(s) of plant consumed: Leaves
8. Portion analyzed/sampled: Leaves (Fresh and dried)
9. Classifications:
   a. Authors Class: Herbs and spices
   b. EPA Crop Group (Group & Subgroup): Herbs and spices (19A)
   c. Codex Group: 027 (HH 0748) Herbs; 057 (DH 0748) Dried herbs
   d. EPA Crop Definition: None
10. References: GRIN, CODEX, FOSTER (picture), LEWIS, MAGNESS, STEPHENS 1988, US EPA 1995a, FACCIOLA, HYLTON
11. Production Map: EPA Crop Production Region 3.
12. Plant Codes:
    a. Bayer Code: No specific entry

196

1. Cottonseed
(Cotton, Algodonera, Cottonseed oil)
Malvaceae
Gossypium spp.

1. American-Egyptian cotton
(American pima cotton, Brazilian cotton, Egyptian cotton, Gallini cotton, Kidney cotton, Peruvian cotton, Pima cotton, Sea island cotton, Extra-long staple cotton, Long-staple cotton)
G. barbadense L. (Syn: *G. peruvianum* Cav.; *G. vitifolium* Lam.)

1. Upland cotton
(American cotton, American upland cotton, Bourbon cotton)
G. hirsutum L. (syn: *G. mexicanum* Tod.)

1. Hawaiian cotton
G. tomentosum Nutt. ex Seem.

1. Short-staple cotton
(Arabian cotton, Lecant cotton, Maltese cotton, Syrian cotton)
G. herbaceum L.

2. Cotton is grown primarily for the fibers or lint, but the oil-containing seeds are highly important. World production of cotton seed oil averaged 3,857,150 tons, 1992-94. The cotton plant is a stiff-growing herbaceous annual outside the tropics, with fairly large, lobed leaves. The fruits are capsules which dehisce as they ripen. Each capsule contains up to 40 or 50 obovate, rounded or angular seeds, to which are attached the fibers or lint. The lint and seeds are harvested from the dehisced bolls, partly by hand but now largely by machine in the U.S. Harvested commercially by stripper or mechanical picker. The longer lint is removed from the seeds mechanically at cotton gins, then baled. The seeds of most varieties are still covered with short fibers or linters after the ginning.

The seeds consist about half of hull and half of kernel. The kernels contain 28 to 40 percent oil. In extracting the oil the seeds are cleaned, delinted, and pressed or put through expellers either whole or after dehulling. A ton of seeds yields around 300 pounds of oil. The meal or press cake is a valuable high-protein livestock feed. After harvest, the cotton fields may be used for livestock pasturage. The oil is used mainly for shortenings. Smaller quantities are used for cooking and salad oils, margarines and soap manufacture.

3. Season, planting to harvest: Begins to flower about 60 days after planting, flowering continues for about 45 days and individual bolls mature in about 55 days. Cotton emerges 5 to 20 days after planting.
4. Production in U.S.: 6,343,200 tons of cottonseed from 12,783,300 acres of cotton in 1993. 79 percent of the total production is reported from 6 states – Texas, California, Mississippi, Arkansas, Louisiana and Arizona in 1993 (USDA 1994a).
5. Other production regions: Tropical and subtropical regions of the world (BAILEY 1976).
6. Use: See above, and fiber used to produce yarn for weaving baby diapers to NASA space suits.
7. Part(s) of plant consumed: Seed and gin trash
8. Portion analyzed/sampled: Undelinted seed, cotton gin byproducts and seed processed commodities – meal, hulls and refined oil. For cotton gin byproducts (commonly called gin trash), include the plant residues from ginning cotton, and consist of burrs, leaves, stems, lint, immature seeds, and sand and/or dirt. Cotton must be harvested by commercial equipment (stripper and mechanical picker) to provide an adequate representation of plant residue for the ginning process. At least three field trials for each type of harvesting (stripper and picker) are needed
9. Classifications:
   a. Authors Class: Oilseed
   b. EPA Crop Group (Group & Subgroup): Miscellaneous
   c. Codex Group: 023 (SO 0691) Oilseed; 052 (AM 0691) Miscellaneous fodder and forage crops; 067 (OC 0691) Vegetable crude oil; 068 (OR 0691) Vegetable edible oil
   d. EPA Crop Definition: None
10. References: GRIN, CODEX, MAGNESS, ROBBELEN 1989 (picture), USDA 1994a, US EPA 1995a, US EPA 1995b, US EPA 1994, IVES, SMITH 1995, BAILEY 1976, EDMINSTEN
11. Production Map: EPA Crop Production Regions 2, 4, 6, 8 and 10.
12. Plant Codes:
    a. Bayer Code: GOSBA (*G. barbadense*), GOSHI (*G. hirsutum*), GOSHE (*G. herbaceum*), GOSTO (*G. tomentosum*)

197

1. Crabapple
(Applecrab)
Rosaceae
Malus spp.

1. Southern crabapple
(American crab, Southern wild crabapple, Wild crab)
M. augustifolia (Aiton) Michx.

1. Siberian crabapple
(Manzano)
M. baccata (L.) Borkh. (syn: *Pyrus baccata* L.)

1. Sweet crabapple
(Wild sweet crab, Sweet scented crab)
M. coronaria (L.) Mill.

1. Paradise crabapple
(Paradise apple)
M. pumila Mill.

1. Wild crabapple
M. sylvestris Mill.

1. Japanese flowering crabapple
(Purple chokeberry, Showy crabapple)
M. floribunda Siebold ex Van Houtte

1. Prairie crabapple
(Wild crab)

M. ioensis (A.W. Wood) Britton

2. Crabapples are native species of apple-like fruits, that are grown mostly for pollinators in apple orchards or as a beautiful ornamental specimen. There are varieties, notably 'Hyslop' and 'Red Siberian', that produce relatively large fruit that can be used as a food crop and are processed for cinnamon apple rings. These are hybrids of native species of apples. In general, the trees are hardier than most apple varieties. These are round yellow fruits 1 to 2 inches in diameter, with a red to maroon blush. Crabapples are considered to be apples except for being less than 2 inches in diameter, but some crabapples are greater than 2 inches in diameter and some dessert apples are small. For additional information see Apple.
3. Season: Bloom to harvest in approximately 4 months.
4. Production in U.S.: Pollinator and ornamental varieties are grown everywhere apples are grown. Crabapple crop grown in California, New York, and Washington. No statistics available.
5. Other production regions: Canada (British Columbia)
6. Use: Fruit is processed and used for apple jelly, tarts and apple butter. Fruit can also be used as a garnish. Crabapple trees are used for pollinators in apple orchards and as a showy ornamental crop.
7. Part(s) of plant consumed: The whole fruit is processed in pickling and preserving. The fruit can be cooked, with core tissue and peel sieved out
8. Portion analyzed/sampled: Whole fruit
9. Classifications:
   a. Authors Class: Pome fruits
   b. EPA Crop Group (Group & Subgroup): Crop Group 11: Pome Fruits Group
   c. Codex Group: 002 (FP 0227) Pome fruits
   d. EPA Crop Definition: None, but proposing – Apple = Crabapple
10. References: GRIN, CODEX, LOGAN, MAGNESS, SCHREIBER, US EPA 1995, USDA NRCS, IVES, REHM, FACCIOLA
11. Production Map: See Apple
12. Plant Codes:
   a. Bayer Code: MABAN (*M. angustifolia*), MABCO (*M. coronaria*), MABIO (*M. ioensis*), MABSY (*M. sylvestris*), MABPM (*M. pumila*)

198

1. Crabgrass
Poaceae (Gramineae)
Digitaria spp.

1. Hairy crabgrass
(Large crabgrass)
D. sanguinalis (L.) Scop.

1. Wild crabgrass
(Summer grass, Pata de gallo, Pata de gallina, Southern crabgrass)
D. ciliaris (Retz.) Koeler

2. Hairy and wild crabgrasses are annual warm season grasses introduced from southern Africa to the U.S. They are members of the grass tribe *Paniceae*. Hairy crabgrass is a decumbent creeping growing grass 2 to 4 ft. tall that readily roots by stolons at its nodes. It is frequently grazed when necessary and makes a palatable forage. It can support grazing in late spring until early fall for up to 122 days. Hairy crabgrass are most productive in the Gulf Coast area. In most farming situations, crabgrasses is considered weeds, but they are sometimes useful for pasture and as a hay crop, since their forage quality is often superior to many perennial warm season grasses. Wild crabgrass is common to sandy loam soils and rather poor quality soils and prefer tropical climates. It is adapted to areas from sea level to 1800 m. It has been used as a forage in Oklahoma and has produced up to 9520 kg DM/ha. Crabgrasses readily reseed.
3. Season: Warm season annual, growth starts in late spring, and continues until fall. It is palatable throughout the growing season, and makes good quality forage where it is grown.
4. Production in U.S.: No data for either Hairy or Wild crabgrass production, however, pasture and rangeland are produced on more than 410 million A (U.S. CENSUS, 1992).
5. Other production regions: Mexico.
6. Use: Pasture grazing and for erosion control.
7. Part(s) of plant consumed: Leaves and stems
8. Portion analyzed/sampled: Forage and hay
9. Classifications:
   a. Authors Class: See Forage grass
   b. EPA Crop Group (Group & Subgroup): See Forage grass
   c. Codex Group: See Forage grass
   d. EPA Crop Definition: None
10. References: GRIN, US EPA 1994, US EPA 1995, CODEX, HOLZWORTH, RICHE, SKERMAN (picture), BALL
11. Production Map: EPA Crop Production Regions 2, 3, 4, and 6.
12. Plant Codes:
   a. Bayer Code: DIGAD (*D. ciliaris*), DIGSA (*D. sanguinalis*)

199

1. Crambe
(Colewort)
Brassicaceae (Cruciferae)
Crambe abyssinica Hochst. ex R.E. Fr.

2. Crambe is an erect annual herb with numerous branches growing to 3 feet tall with leaves large, oval and smooth. Crambe is seeded in the spring and produces white blossoms and light-colored seeds. The seeds yield about 35 percent oil (dry matter basis). Seeds containing more than 25 percent oil, such as crambe and rapeseed, typically warrant use of pre-press/solvent extraction for processing. Dehulling the seed near the point of production would reduce volume and ease transportation costs. Presently, the hull is considered part of the harvested product. The seed pod protects the seed from direct contact with pesticide residues.

3. Season, seeding to harvest: About 90 days. To prevent shatter losses, harvest should be done while lower stem is still green.
4. Production in U.S.: In 1991, 2,200 acres in North Dakota and 200 acres in Iowa (ROETHELI 1991a). In 1986, Iowa raised crambe in excess of 2,000 pounds per acre (VAN DYNE 1990). Midwest U.S. to include Missouri, Iowa, Nebraska and North Dakota (VANDYNE 1990). Crambe is best grown in areas of Midwest, upper Plains states and Northwest (COOKE 1991). In 1997, 50,000 acres in North Dakota.
5. Other production regions: Mediterranean region (BAILEY 1976).
6. Use: Crambe meal at a concentration not to exceed 4.2 percent of the ration can be used as beef cattle feed. With pre-cooking during processing or other methods to reduce glucosinolate, the meal can be used for all livestock and poultry. Oil is used for industrial purposes.
7. Part(s) of plant consumed: Seed
8. Portion analyzed/sampled: Seed and its meal
9. Classifications:
   a. Authors Class: Oilseed
   b. EPA Crop Group (Group & Subgroup): Miscellaneous
   c. Codex Group: No specific entry
   d. EPA Crop Definition: Rapeseed = *Crambe abyssinica*
10. References: GRIN, COOKE 1991, ROBBELEN 1989, ROETHELI 1991a, US EPA 1995a, US EPA 1995b, VANDYNE 1990 (picture), WOOLLEY 1986, BAILEY 1976
11. Production Map: EPA Crop Production Region 5. Rapeseed (canola) field trials conducted in regions 2, 5, 7 and 11.
12. Plant Codes:
   a. Bayer Code: CRMAB

200

1. Cranberry
(Large cranberry, American cranberry, Canneberge agrands, Krannbeere, Arandano agrios, Grosse canneberge d'Amerique)
Ericaceae
Vaccinium macrocarpon Aiton
2. This native American fruit grows on a prostrate evergreen "vine", though not a climber. Propagated from cuttings, the plant requires at least 4 years to produce the first crop. The stems are actually rather tender to cold, but stand winter covering with water well. Thus in commercial culture, where most are grown, they are planted on peat bogs prepared so they can be covered with water in winter. The berries are borne on short uprights 6 to 8 inches in length, rising from the dense mass of stems prostrate on the soil surface. Fruit has a smooth skin, is generally round, elliptical, or bell shaped and about $^1/_3$ inch in diameter and $^1/_2$ to 1 inch long. Inconspicuous seeds are attached at the center of the fruit and surrounded by the tart pulp. Plantings persist for many years if properly managed. Harvested by flooding or floating the berries or by dry raking.
3. Season: Bloom to harvest: 100 to 130 days.
4. Production in U.S.: Domestic production reported at 31,700 harvested acres in 1995. Key production states are Massachusetts (13,500 acres), Wisconsin (11,500 acres), New Jersey (3,500 acres), Oregon (1,700 acres) and Washington (1,500 acres). About 200,000 tons. Also minimal acreage in Delaware, Maine, Michigan, Minnesota, New Hamphire,

New York and Rhode Island (DOWNING 1997). A barrel of cranberries is reported as 100 pounds.
5. Other production regions: Some domestic imports from Canada and Chile. In 1995, Canada grew 2,869 acres (ROTHWELL 1996a).
6. Use: Fresh, canned, frozen, juice and jellied.
7. Part(s) of plant consumed: Whole fruit or interior pulp and juice. Skins and seeds are screened out in juice and jellied products, after heating
8. Portion analyzed/sampled: Whole fruit
9. Classifications:
   a. Authors Class: Small fruits
   b. EPA Crop Group (Group & Subgroup): Miscellaneous, used to be part of Small Fruits and berry group
   c. Codex Group: 004 (FB 0265) Small Fruits and Berries
   d. EPA Crop Definition: None
10. References: GRIN, CODEX, LOGAN, MAGNESS, SCHREIBER, US EPA 1994, ROTHWELL 1996a, DOWNING, USDA 1996f
11. Production Map: EPA Crop Production Region 1, 5 and 12.
12. Plant Codes:
   a. Bayer Code: VACMA

201

1. Creeping foxtail
(Creeping meadow foxtail, Foxtail-creeping)
Poaceae (Gramineae)
Alopecurus arundinaceus Poir.
2. This is a cool season sod-forming grass by rhizomes native to Eurasia that is adapted to wet meadows of Central and Pacific U.S., and it is a member of the grass tribe *Aveneae*. It resembles meadow foxtail but has more vigorous rhizomes and broader leaves and grows 40 to 120 cm tall. It is adapted to poorly drained acidic soils and is winter hardy. It is used to a limited extent for hay, pasture and erosion control on moist sites in the northern Great Plains, Intermountain area, Pacific Northwest and Canada. Forage is palatable and nutritious. It is seeded at 15 to 25 lb/A. There are 576,000 seeds/lb.
3. Season: Cool season, spring and fall growth. Excellent early season growth and regrowth after cut. Seeded in the fall or spring. Good for pasture, supporting a high stocking rate.
4. Production in U.S.: No data for Creeping foxtail production, however, pasture and rangeland are produced on more than 410 million A (U.S. CENSUS, 1992). Adapted to plant hardiness zones 2 through 5.
5. Other production regions: Grown in Newfoundland, Canada.
6. Use: Pasture, hay, silage and for erosion control.
7. Part(s) of plant consumed: Leaves and stems
8. Portion analyzed/sampled: Forage and hay
9. Classifications:
   a. Authors Class: See Forage grass
   b. EPA Crop Group (Group & Subgroup): See Forage grass
   c. Codex Group: See Forage grass
   d. EPA Crop Definition: None
10. References: GRIN, MAGNESS, US EPA 1994, US EPA 1995, CODEX, BARNES, ALDERSON, ROTHWELL 1996a, MOSLER (picture), RICHE
11. Production Map: EPA Production Regions 1, 5, 6, 7, 9, 11 and 12.
12. Plant Codes:
   a. Bayer Code: ALOAR

202

1. Cress, Garden
Brassicaceae (Cruciferae)
Lepidium sativum L.
1. Peppergrass
(Virginia peppergrass)
L. virginicum L.

2. Cress plants are grown for the leaves, which are used as salads and garnishes. The plant is an cool season annual, attaining a height of 1 to 2 feet. The leaves are aromatic and peppery in flavor. They rise from the root crown at the soil surface, and are variable in shape, some forms being greatly divided like parsley, others curled. If lower leaves are the only ones removed, new leaves will continue to be formed on the central stalk. Growth habit and leaf exposure are similar to spinach. Peppergrass is a closely related plant, native to the U.S. It is not cultivated, but often gathered and used as with garden cress. A relative, *L. meyenii* Walp., is called Maca or Peruvian ginseng and it is grown for its turnip-like root in Peru.
3. Season, seeding to harvest: 6 to 8 weeks.
4. Production in U.S.: No data. Minor importance. It can be grown in the South and Pacific Southwest during the winter and in other areas in the spring until fall.
5. Other production regions: Native to Egypt and western Asia (BAILEY 1976).
6. Use: Fresh for salads, garnishes. Seeds can be sprouted for salad use.
7. Part(s) of plant consumed: Leaves
8. Portion analyzed/sampled: Leaves
9. Classifications:
   a. Authors Class: Leafy vegetables (except *Brassica* vegetables)
   b. EPA Crop Group (Group & Subgroup): Garden cress – Leafy vegetables (except *Brassica* vegetables) (4A)
   c. Codex Group: 013 (VL 0472) Leafy vegetables (including *Brassica* leafy vegetables) for Garden cress
   d. EPA Crop Definition: None
10. References: GRIN, CODEX, MAGNESS, NATIONAL RESEARCH COUNCIL 1989, US EPA 1995a, YAMAGUCHI 1983, BAILEY 1976
11. Production Map: No entry
12. Plant Codes:
   a. Bayer Code: LEPSA (*L. sativum*), LEPVI (*L. virginicum*)

203

1. Cress, Upland
(Dryland cress, Wintercress, Scurvy grass, Belle Isle cress, Spring cress, Yellow rocket, Toi, Early wintercress, Landcress, Garden yellowrocket, Early yellowrocket)
Brassicaceae (Cruciferae)
Barbarea vulgaris R. Br.
B. verna (Mill.) Asch. (syn: *B. praecox* (Sm.) R.Br.)

2. These cresses, related to watercress and horseradish, are cultivated sparingly for winter salads and potherbs. Plants are naturalized in many parts of the U.S. Plants are hardy biennials. Leaves are generally entire but notched, and smooth with 6 to 8 inch long leafstems. As grown under cultivation, leaf exposure and general culture are similar to those of spinach and turnips for greens. Cress seed is produced on about 10 acres in Washington state in EPA Crop Production Region 12. *B. verna* is termed early wintercress or landcress.
3. Season, seeding to harvest: Leaves are picked when plant is about 4 inches tall with repeated pickings. In Tennessee, it is planted in mid-to-late August with its first cutting in late November to early December. Its primary harvest is in March. The primary harvest is $7^1/_2$ months from seeding while the first cutting harvest is $3^1/_2$ months from seeding.
4. Production in U.S.: About 150 acres in Tennessee for processing and a limited acreage in Tennesse, Virginia, North Carolina and South Carolina for fresh market (MULLINS 1997).
5. Other production regions: Native to Europe (BAILEY 1976).
6. Use: Cooked as greens or used raw in salads. Seeds sprouted for salad use.
7. Part(s) of plant consumed: Leaves
8. Portion analyzed/sampled: Leaves
9. Classifications:
   a. Authors Class: Leafy vegetables (except *Brassica* vegetables)
   b. EPA Crop Group (Group & Subgroup): *B. vulgaris*: Leafy vegetables (except *Brassica* vegetables) (4A)
   c. Codex Group: No specific entry
   d. EPA Crop Definition: None
10. References: GRIN, MAGNESS, SCHREIBER, STEPHENS 1988 (picture), US EPA 1995a, USDA NRCS, MULLINS, BAILEY 1976
11. Production Map: EPA Crop Production Region 2.
12. Plant Codes:
   a. Bayer Code: BARVE (*B. verna*), BARVU (*B. vulgaris*)

204

1. Crotalaria
(Rattlebox)
Fabaceae (Leguminosae)
Crotalaria spp.
1. Striped crotalaria
(Smooth crotalaria, Smooth rattlebox)
C. pallida Aiton (syn: *C. mucronata* Desv.)
1. Showy crotalaria
C. spectabilis Roth
1. Slenderleaf crotalaria
(Ethiopian rattlebox)
C. brevidens Benth. (syn: C. *intermedia* Kotschy)
1. Lanceleaf crotalaria
(Lance crotalaria)
C. lanceolata E. Mey.
1. Sunnhemp
(Indian hemp, Madras hemp, Brown hemp)
C. juncea L.

2. All are upright-growing summer annuals. Stems are coarse and well branched except in thick stands. Leaves are trifoliate, the leaflets varying in shape from linear to ovate. Crotalarias are adapted only to warm climates with a long growing season. They are resistant to rootknot nematode and are valuable to reduce the nematode population in infested soils. They do well on soils of low fertility and are most used to turn into the soil for soil improvement. *C. spectabilis* contains an alkaloid poisonous to livestock. The other four species can be used for pasture or silage though palatability is

low. According to the 1960 CENSUS, seed was harvested in 1959 from 2,657 acres in this country, and sufficient seed was produced to plant around 35,000 acres.

Sunnhemp is widely grown in tropical areas as a cover crop. The dried stalks and hay are used as livestock forage. It is an annual legume to 6 feet tall. Also a fiber crop that grows rapidly in California.

3. Season: Long growing season. Flowers July to August.
4. Production in U.S.: No recent data.
5. Other production regions: Warm climate.
6. Use: Forage, green manure, soil erosion control.
7. Part(s) of plant consumed: Foliage
8. Portion analyzed/sampled: Forage and hay
9. Classifications:
   a. Authors Class: Nongrass animal feeds
   b. EPA Crop Group (Group & Subgroup): Miscellanous
   c. Codex Group: O50 (AL 0157) Legume animal feed
   d. EPA Crop Definition: None
10. References: GRIN, MAGNESS, USDA NRCS, DUKE(h) (picture)
11. Production Map: EPA Crop Production Regions 2, 10 and 13.
12. Plant Codes:
    a. Bayer Code: CVTBD (*C. brevidens*), CVTLA (*C. lanceolata*), CVTMU (*C. pallida*), CVTSP (*C. spectabilis*), CVTJU (*C. juncea*)

205

1. Crown vetch
(Trailing crownvetch,Coronille variee, Kronwicke, Purple crownvetch, Vetch-crown, Crownvetch)
Fabaceae (Leguminosae)
Securigera varia (L.) Lassen (syn: *Coronilla varia* L.)

2. This is not a true vetch but in general appearance resembles the vetches. It is a long-lived perennial legume, native to the Mediterranean area but now widely distributed in local areas throughout the Central and Northern states. Stems are weak and hollow, up to 4 feet, procumbent unless supported. The plant spreads by creeping underground roots and is excellent for erosion control. Seed should be scarified and inoculated before planting. Palatability of crownvetch is said to be low except while growth is young and tender. It is a nonbloating forage. The crop has increased in importance for erosion control, pasture and hay in the northeastern quarter of the U.S. Used along highways for erosion control.
3. Season, cuttings: 1-2 cuttings per year. Cut hay in the full bloom stage. Flowers May-August.
4. Production in U.S.: No specific crownvetch data. See Vetch. Central and Northern states.
5. Other production regions: Native to Europe (BAILEY 1976).
6. Use: Pasture, hay, roadbank erosion control and as ornamentals in Europe.
7. Part(s) of plant consumed: Foliage
8. Portion analyzed/sampled: Forage and hay. For crownvetch forage, cut sample at 6 inch to prebloom stage, at approximately 30 percent dry matter. For crownvetch hay, cut at full bloom stage. Hay should be field-dried to a moisture content of 10 to 20 percent
9. Classifications:
   a. Authors Class: Nongrass animal feeds

b. EPA Crop Group (Group & Subgroup): Nongrass animal feeds (18)
c. Codex Group: 050 (AL 1029) Legume animal feed
d. EPA Crop Definition: None
10. References: GRIN, CODEX, MAGNESS, US EPA 1996a, US EPA 1995a, USDA NRCS, BARNES, DUKE(h), BAILEY 1976
11. Production Map: EPA Crop Production Regions 1, 2, 5 and 13.
12. Plant Codes:
    a. Bayer Code: CZRVA

206

1. Cucumber
(Pepino, Ka'ukama, Cornichons, Kiuri, Gherkin, Pepinillo, Concombre, Khira)
Cucurbitaceae
Cucumis sativus L. var. *sativus*

2. Cucumber is one of the most widely grown vegetables, common in home gardens, on truck farms and as a greenhouse crop. The plant is trailing, usually on the ground in the open, but on trellises in greenhouses. Leaves are large, with long petioles and form a canopy over the stems and fruits in well-grown plants. Fruits are greatly elongated and cylindrical, with a non-pubescent but somewhat wrinkled surface, sometimes very lightly spined. Large forcing varieties may be more than a foot long and 2 inches in diameter. Fruits are commonly harvested while still green and for small whole pickles when very immature, and of a size determined by the desire of the processor. Plants will continue to bloom and set fruits for months if healthy. Gherkins are a small pickling cucumber. Also see West Indian Gherkin. Types of cucumbers are pickling (common and cornichons) and slicing (common; hothouse, forcing, greenhouse or European; Middle Eastern or beit alpha; and Oriental).
3. Season, seeding to first harvest: About 2 to 3 months. Pickling type picked 5-12 days after flowering and fresh market picked 15-18 days after flowering. Of the pickling crop, about 50 percent is brined and the rest fresh packed.
4. Production in U.S.: Commercial, in 1995 about 500,000 tons (fresh market); 600,000 tons (processing) (USDA 1996b). In 1995, the top 10 states for fresh market cucumbers were: Florida (13,100 acres), Georgia (12,500), North Carolina (6,500), Virginia (6,400), Michigan (5,900), California (5,200), New York (3,500), South Carolina (2,600), Texas (2,400), New Jersey (2,400). In 1995, the top 10 states for processing cucumbers were: Michigan (27,000 acres), North Carolina (18,800), Texas (13,000), South Carolina (7,000), Wisconsin (6,200), Florida (6,000), California (5,300), Ohio (3,000), Indiana (2,800) and Colorado (950) (USDA 1996b).
5. Other production regions: In 1995, Canada reported 6,405 acres of field cucumbers, 282 acres of hothouse cucumbers, 156 acres of pickling cucumbers and 82 acres of slicing cucumbers (ROTHWELL 1996a).
6. Use: Pickles, fresh in salads. In Indonesia, young leaves and stems are eaten as potherb. U.S. per capita consumption in 1994 for pickles is 4.8 pounds and fresh cucumbers is 5.6 pounds (PUTNAM 1996).
7. Part(s) of plant consumed: Whole fruit
8. Portion analyzed/sampled: Whole fruit
9. Classifications:

a. Authors Class: Cucurbit vegetables
b. EPA Crop Group (Group & Subgroup): Cucurbit vegetables (9B) (Representative Crop)
c. Codex Group: 011 (VC 0424) Cucurbits fruiting vegetables
d. EPA Crop Definition: None
10. References: GRIN, CODEX, FACCIOLA, MAGNESS, NEAL, REHM, ROTHWELL 1996a, USDA 1996b, US EPA 1994, US EPA 1996a, US EPA 1995a, SWIADER, PUTNAM 1996
11. Production Map: EPA Crop Production Regions 1, 2, 3, 5, 6, 10 and 12 which accounts for 94 percent of acreage.
12. Plant Codes:
a. Bayer Code: CUMSA

207

1. Cucumber, Armenian
(Japanese cucumber, Snake melon, Snake cucumber, Tarra, Cohombre serpentino, Serpent melon, Kakri, Yard long cucumber)
Cucurbitaceae
Cucumis melo var. *flexuosus* (L.) Naudin
2. The plant on which this melon or cucumber is grown is closely related and similar to muskmelon, which see. Fruits are very long and slender, up to 36 inches long and 3 inches in diameter. They are grown mainly as curiosities, but are used to some extent for preserves. It is a warm season crop.
3. Season, seeding to harvest: About 3 or more months.
4. Production in U.S.: No data. Limited. California (LOGAN 1996). Grown in Florida (STEPHENS 1988).
5. Other production regions: India (FACCIOLA 1990).
6. Use: Preserves, fresh in salads or cooked. Taste like cucumber. Small fruit utilized as summer squash.
7. Part(s) of plant consumed: Whole fruit
8. Portion analyzed/sampled: Whole fruit
9. Classifications:
a. Authors Class: Cucurbit vegetables
b. EPA Crop Group (Group & Subgroup): Cucurbit vegetables (9A)
c. Codex Group: 011 (VC 0046) Cucurbits fruiting vegetables
d. EPA Crop Definition: see muskmelon
10. References: GRIN, CODEX, FACCIOLA, LOGAN, MAGNESS, REHM, STEPHENS 1988 (picture), US EPA 1995a
11. Production Map: EPA Crop Production Regions 3, 10 and 13.
12. Plant Codes:
a. Bayer Code: CUMMF

208

1. Cucumber, Wild
(Caihua, Achoccha, Korila, Stuffing cucumber, Caygua, Korilla)
Cucurbitaceae
Cyclanthera pedata (L.) Schrad.
2. Annual crawling vine branched at the lower nodes and spreading as much as 16 feet in length. Leaves are 3 to 5 inches wide and lobed. Fruits are soft, flattened, oblong, 4 to 6 inches long and 2 to $2^3/_4$ inches wide. The interior of the fruit is hollow.
3. Season, seeding to first harvest: About 3 months. Can continue to harvest for an additional 3 months. Per crop year, 4-6 harvests are possible.

4. Production in U.S.: No data. Grown in Florida, (STEPHENS 1988).
5. Other production regions: Caribbean, and Mexico to Bolivia (YAMAGUCHI 1983).
6. Use: Fully grown fruits are used either raw or cooked. Fruits are often stuffed like bell peppers after removing the seeds and pulp.
7. Part(s) of plant consumed: Fruit
8. Portion analyzed/sampled: Whole fruit
9. Classifications:
a. Authors Class: Cucurbit vegetables
b. EPA Crop Group (Group & Subgroup): Miscellaneous
c. Codex Group: No specific entry. There is a general code of 011 (VC 0045) cucurbits fruiting vegetables
d. EPA Crop Definition: None
10. References: GRIN, CODEX, MARTIN 1983, STEPHENS 1988 (picture), YAMAGUCHI 1983 (picture), RUBATZKY
11. Production Map: EPA Crop Production Region 13.
12. Plant Codes:
a. Bayer Code: No specific entry

209

1. Cumin
(Cumin seed, Safed-zira, Comino)
Apiaceae (Umbelliferae)
Cuminum cyminum L.
2. Cumin is grown for the seeds, which are used as an ingredient in curry powder, and for flavoring pickles, soups, meats, pastry and cheese. The plant is a low-growing annual herb to 10 inches tall, of the carrot family, cultivated in southern Europe and Asia, but rare in America. The plant is grown from seed, with a mature crop in about 4 months. At harvest, plants are cut and dried, then threshed. About 14 million pounds of the seed are imported into the U.S. annually with Pakistan supplying 68 percent (1993).
3. Season, seeding to harvest: About 4 months.
4. Production in U.S.: No commercial production in U.S.
5. Other production regions: Mediterranean region origin (RUBATZKY 1997). Pakistan.
6. Use: Ground cumin seed is an essential ingredient of curry and chili powders. Seed and oil used to flavor various foods.
7. Part(s) of plant consumed: Seeds, may be crushed and steam distilled for its oil
8. Portion analyzed/sampled: Seeds
9. Classifications:
a. Authors Class: Herbs and spices
b. EPA Crop Group (Group & Subgroup): Herbs and spices group (19B)
c. Codex Group: 028 (HS 0780) Spices
d. EPA Crop Definition: None
10. References: GRIN, CODEX, GARLAND (picture), FARRELL (picture), LEWIS, MAGNESS, USDA 1993a, US EPA 1995a, IVES, REHM, RUBATZKY
11. Production Map: No entry.
12. Plant Codes:
a. Bayer Code: CVUCY

210

1. Cuphea
(Waxweed)

Lythraceae

Cuphea spp.

C. carthagenensis (Jacq.) J.F. Macbr.

(Colombian waxweed, Sete-sangrias)

C. wrightii A. Gray

(Wrights' waxweed)

2. Annual plant to about 3¹/₂ feet tall. Leaves are simple, entire and opposite or whorled.
3. Season, seeding to harvest: No data.
4. Production in U.S.: No data. Could be adapted to midwest U.S. and western U.S. (ROBBELEN 1989).
5. Other production regions: Widely distributed in Mexico and Brazil (ROBBELEN 1989).
6. Use: Use like coconut or palm kernel oil. For lauric and capric fatty acids. The feed value of the defatted meal is being studied.
7. Part(s) of plant consumed: Seed
8. Portion analyzed/sampled: Seed and its oil and meal
9. Classifications:
   a. Authors Class: Oilseed
   b. EPA Crop Group (Group & Subgroup): Miscellaneous
   c. Codex Group: No specific entry
   d. EPA Crop Definition: None
10. References: GRIN, JANICK 1996, NEAL, ROBBELEN 1989 (picture), USDA NRCS, YEARBOOK OF AGRICULTURE 1992 (picture)
11. Production Map: No entry
12. Plant Codes:
   a. Bayer Code: CPHCA (*C. carthagenensis*), CPHWR (*C. wrightii*)

211

1. Curly mesquite

Poaceae (Gramineae)

Hilaria belangeri (Steud.) Nash

2. This is a warm season perennial grass native from Central Texas and Arizona south throughout Mexico. It is stolon forming and grows on dry soils that may range in texture from clay to gravelly. The plant grows in tufts, new tufts developing at the nodes of the stolons. Stems are slender, up to a foot tall, with short narrow leaves. It is highly drought resistant, palatable both green and when dry, thus is a valuable range grass for warm dry areas.
3. Season, grass for warm dry areas.
4. Production in U.S.: No specific data, See Forage grass. Native from Central Texas and Arizona south.
5. Other production regions: Mexico.
6. Use: Rangegrass.
7. Part(s) of plant consumed: Foliage
8. Portion analyzed/sampled: Forage and hay
9. Classifications:
   a. Authors Class: See Forage grass
   b. EPA Crop Group (Group & Subgroup): See Forage grass
   c. Codex Group: See Forage grass
   d. EPA Crop Definition: None
10. References: GRIN, MAGNESS, ALDERSON, STUBBENDIECK(a)
11. Production Map: EPA Crop Production Regions 6, 7, 8 and 10.
12. Plant Codes:
   a. Bayer Code: HILBE

212

1. Currant, Black

(Cassis, European black currant)

Grossulariaceae (Saxifragaceae)

Ribes nigrum L.

2. This crop is propagated by hardwood cuttings. Black currant bushes are small shrubs, up to 4 feet high. Stems are nearly free of thorns. Flowers and fruits are produced on racemes, resulting in loose clusters of a dozen or more fruits. Individual fruits have a thin, smooth skin, are globose and about ¹/₃ inch diameter. Highly flavored tart pulp surrounds several inconspicuous seeds. Dried petals and parts of pedicle adhere to fruit at harvest.
3. Season: Bloom to harvest in about 60 days.
4. Production in U.S.: Mostly California and northwestern U.S., no statistics available.
5. Other production regions: New Zealand and Europe.
6. Use: Mainly jelly, but can be used in desserts.
7. Part(s) of plant consumed: Mainly pulp and juice after pressing
8. Portion analyzed/sampled: Whole fruit (including stem)
9. Classifications:
   a. Authors Class: Berries
   b. EPA Crop Group (Group & Subgroup): Crop Group 13: Berries Group as well as this group's Bushberry Subgroup
   c. Codex Group: 004 (FB 0278) Berries and other Small Fruit
   d. EPA Crop Definition: None
10. References: GRIN, CODEX, LOGAN, MAGNESS, SCHREIBER, SHOEMAKER, US EPA 1994, US EPA 1995
11. Production Map: EPA Crop Production Regions 10 and 12.
12. Plant Codes:
   a. Bayer Code: RIBNI

213

1. Currant, Red

(White currant, Common currant, Garden currant, Grosellero rojo, Groseillier rouge)

Grossulariaceae (Saxifragaceae)

Ribes rubrum L. (syn: *R. sativum* (Rchb.) Syme)

1. Northern red currant

(Nordic currant, Red currant)

R. spicatum E. Robson

2. Currants grown in the U.S. are mainly red or black, with a few white. Red currant bushes are small shrubs, up to 4 feet high. They are propagated by hardwood cuttings. Stems are nearly free of thorns. Flowers and fruits are produced on racemes, resulting in loose clusters of a dozen or more fruits. Individual fruits have a thin, smooth skin, are globose and about ¹/₃ inch diameter. Highly flavored pulp surrounds several inconspicuous seeds. Dried petals and parts of pedicle adhere to fruit at harvest.
3. Season: Bloom to harvest in about 60 days.
4. Production in U.S.: Washington and California. Approx. 140 acres with 75 percent produced by three growers in Washington.
5. Other production regions: New Zealand.
6. Use: Mainly jelly, some culinary.
7. Part(s) of plant consumed: Mainly pulp and juice after pressing
8. Portion analyzed/sampled: Whole fruit (including stem)
9. Classifications:

a. Authors Class: Berries
b. EPA Crop Group (Group & Subgroup): Crop Group 13: Berries Group as well as this group's Bushberry Subgroup
c. Codex Group: 004 (FB 0279) Berries and other Small Fruit
d. EPA Crop Definition: None
10. References: GRIN, CODEX, LOGAN, MAGNESS, SCHREIBER, SHOEMAKER, US EPA 1994, US EPA 1995
11. Production Map: EPA Crop Production Regions 10 & 11.
12. Plant Codes:
a. Bayer Code: RIBRU (*R. rubrum*)

214

1. Curry
(Curry leaf, Curry – leaf tree, Mock orange)
Rutaceae
Murraya koenigii (L.) Spreng.
2. A small deciduous tree of the citrus family with fresh leaves 1-2 inches long, used as seasoning. Dried leaves are also used.
3. Season, harvest: Leaves picked as needed.
4. Production in U.S.: No data. Limited number of curry trees are grown in south Florida and Southern California. Localized use in U.S.
5. Other production regions: Native to Southeast Asia (RUBATZKY 1997).
6. Use: Pungent aromatic leaves used as seasoning. Dried leaves used in curry powders.
7. Part(s) of plant consumed: Leaves
8. Portion analyzed/sampled: Leaves, fresh and dried
9. Classifications:
a. Authors Class: Herbs and spices
b. EPA Crop Group (Group & Subgroup): Herbs and spices group (19A)
c. Codex Group: 027 (HH 0729) Herbs
d. EPA Crop Definition: None
10. References: GRIN, BAILEY 1976, CODEX, LEWIS, MARTIN 1987, US EPA 1995a, RUBATZKY
11. Production Map: EPA Crop Production Region 10 and 13.
12. Plant Codes:
a. Bayer Code: No specific entry

215

1. Custard apple
(True custard apple, Bullock's heart, Anona, Corazon, Ramphal)
Annonaceae
Annona reticulata L.
2. Tropical tree from 15 to 35 feet tall. Ill-smelling leaves are deciduous, 4 to 8 inches long and 2 inches wide. The compound fruit is smooth, pale yellow color, with a red blush, 3 to 6$\frac{1}{2}$ inches in diameter and may be heart-shaped. The custard apple is believed to be a native of the West Indies. Pond apple, *Annona glabra* L., is used as a dwarf rootstock for other cultivated *Annona* spp. Pond apple fruit is edible but rarely cultivated outside the wild.
3. Season, bloom to harvest: About 4 to 7 months.
4. Production in U.S.: No data. Occasionally grown in south Florida (MORTON 1987).
5. Other production regions: West Indies, Central America, southern Mexico, and tropical South America (MORTON 1987). It is of commercial importance in Egypt and Israel.
6. Use: Eaten fresh or added to milk shakes.

7. Part(s) of plant consumed: Fruit pulp
8. Portion analyzed/sampled: Whole fruit
9. Classifications:
a. Authors Class: Tropical and subtropical fruits – inedible peel
b. EPA Crop Group (Group & Subgroup): Miscellaneous
c. Codex Group: 006 (FI 0332) Assorted tropical and subtropical fruits – inedible peel
d. EPA Crop Definition: Sugar apple = Custard apple
10. References: GRIN, CODEX, CRANE 1995, MORTON (picture), US EPA 1995a, PUROHIT
11. Production Map: EPA Crop Production Region 13.
12. Plant Codes:
a. Bayer Code: ANURE

216

1. Dahlia, Garden
(Dacopa, Pinnate dahlia, Common dahlia, Waiohinu)
Asteraceae (Compositae)
Dahlia pinnata Cav. x *D. coccinea* Cav.
2. Perennial herb to 6 feet tall with unbranched stem, and leaves are simple to pinnately dissected. The sugar extract in the tuberous root is also in chicory and Jerusalem artichoke roots. The polysaccharide is inulin. The roots are processed like sugar beets for the extract.
3. Season, seeding to harvest: No data.
4. Production in U.S.: No data.
5. Other production regions: Mexico (FACCIOLA 1990). Native to mountains of Mexico, Central America and Columbia (BAILEY 1976).
6. Use: Tuberous roots are eaten as a vegetable in parts of Mexico. A sweet extract of the root is marketed as "Dacopa" which is used as a beverage or flavoring. Also grown as an ornamental.
7. Part(s) of plant consumed: Tuberous root
8. Portion analyzed/sampled: Tuberous root and its extract
9. Classifications:
a. Authors Class: Root and tuber vegetables
b. EPA Crop Group (Group & Subgroup): Miscellaneous
c. Codex Group: No specific entry
d. EPA Crop Definition: None
10. References: GRIN, BAILEY 1976, FACCIOLA, NEAL, USDA NRCS
11. Production Map: No entry.
12. Plant Codes:
a. Bayer Code: DAHHY (*D. hybrids*)

217

1. Dallisgrass
(Dallis, Watergrass, Common dallisgrass)
Poaceae (Gramineae)
Paspalum dilatatum Poir.
2. This is a major warm season perennial pasture grass in the Cotton Belt. It was introduced from South America around 1875. It is an upright-growing bunchgrass, which requires moist soil. It is not hardy north of the Cotton Belt. It does not form a dense sod, so is well suited to mixed planting with legumes or other grasses. It is often seeded into rice stubble in Louisiana and Texas. Its palatability is superior to bahiagrass. It is cut for hay before flowering, will withstand close grazing and should be grazed to prevent accumulation of

dead leaves and stems. It is palatable and nutritious. It's susceptible to ergot fungus which is poisonous to livestock. There are 340,000 seeds/pound.

3. Season: Warm season pasture. Flowers April to November.
4. Production in U.S.: About 20,000 acres of pasture seeded with dallisgrass (BALL 1991). Cotton Belt.
5. Other production regions: Native to South America. Introduced into the Southern U.S. from Uruguay or Argentina (GOULD 1975).
6. Use: Pasture, hay, silage, erosion control.
7. Part(s) of plant consumed: Foliage
8. Portion analyzed/sampled: Forage and hay
9. Classifications:
   a. Authors Class: See Forage grass
   b. EPA Crop Group (Group & Subgroup): See Forage grass
   c. Codex Group: See Forage grass
   d. EPA Crop Definition: None
10. References: GRIN, MAGNESS, ALDERSON, STUBBENDIECK, BALL, SKERMAN, BARNES, GOULD
11. Production Map: EPA Crop Production Regions 2, 3, 4, 6 and 13.
12. Plant Codes:
    a. Bayer Code: PASDI

218

1. Dandelion

(Common dandelion, Pissenlit, Seiyotanpopo, Chinese dandelion, Diente-de-lion, Russian dandelion, Lion's tooth, Gow gay)

Asteraceae (Compositae)

Taraxacum officinale F.H. Wigg. (syn: *T. vulgare* Schrank)

2. Dandelions have become naturalized throughout northern temperate zones, and leaves from such plants are widely gathered and used as potherbs. Cultivated forms are grown to a limited extent in the U.S., more in Europe. The plants are perennials, forming a rosette of leaves from the root crown. Leaves are highly variable in shape. Cultivated forms may have rosettes of leaves spreading to 18 to 24 inches on established plants. Leaf exposure is similar to that of spinach. Leaves are sometimes tied together or covered for blanching. Italian dandelion is *Cichorium intybus*, Chicory, which see.
3. Season, seeding to harvest: About 3 months.
4. Production in U.S.: Grown on a large scale in Florida (STEPHENS 1988). Grows wild in lawns and gardens throughout the U.S. (STEPHENS 1988).
5. Other production regions: Weedy pest of the Northern Hemisphere, naturalized in Europe and central temperate Asia (RUBATZKY 1997).
6. Use: Mainly as potherbs and in salads. Preparation: Wash throughly, trim roots and any torn or wilted leaves (LOGAN 1996).
7. Part(s) of plant consumed: Leaves for food. Dried roots are used for medicinal purposes under name Taraxacum. Flowers are used in wine making.
8. Portion analyzed/sampled: Leaves
9. Classifications:
   a. Authors Class: Leafy vegetables (except *Brassica* vegetables)
   b. EPA Crop Group (Group & Subgroup): Leafy vegetables (except *Brassica* vegetables) (4A)
   c. Codex Group: 013 (VL 0474) Leafy vegetables (including *Brassica* leafy vegetables)

   d. EPA Crop Definition: None
10. References: GRIN, CODEX, LOGAN, MAGNESS, STEPHENS 1988 (picture), US EPA 1995a, MYERS, FRANK, RUBATZKY
11. Production Map: EPA Crop Production Region 3.
12. Plant Codes:
    a. Bayer Code: TAROF

219

1. Dasheen

(Malanga, Chinese potato, Old cocoyam, Calalou, Elephants'-ear, Upland taro, Nampi, Chamol)

Araceae

Colocasia esculenta (L.) Schott (syn: *C. antiquorum* Schott)

2. Dasheen is a variety of taro, an important food crop in tropical countries, that is grown to a limited extent in the U.S. for its edible corms. A corm from the previous season is planted in early spring and produces a vigorous plant with very large cordate leaves on long petioles. A new crop of corms forms about the original "mother" corm. These are harvested in late fall. They are starchy foods, similar to potatoes, but a little sweeter. The edible corms develop entirely underground. The dasheen "mother" corm which is about 2 pounds produces about 20 or more side tubers or cormels up to 2 inches in diameter and usually 2 to 4 ounces. Corms and cormels are used as food. Little grown in continental U.S. Dasheens require higher soil moisture than Yautia, which see, for the development of the "mother" corm. At the time the tubers begin to form, the older leaves begin to die. Dasheen is not suitable for poi.
3. Season, planting to harvest: About 7 to 12 months.
4. Production in U.S.: In 1992, Puerto Rico reported 1,382 acres and 1,850 tons (1992 CENSUS). Grows in Florida (STEPHENS 1988). Mainly Florida and the Caribbean area.
5. Other production regions: Caribbean areas. Old World origin (RUBATZKY 1997).
6. Use: As cooked vegetable, similar to potato. Leaves used as greens and blanched young shoots from corms forced in dark. A stew called calalou is prepared from the leaves.
7. Part(s) of plant consumed: Corms and young leaves
8. Portion analyzed/sampled: Corm and foliage
9. Classifications:
   a. Authors Class: Root and tuber vegetables; Leaves of root and tuber vegetables
   b. EPA Crop Group (Group & Subgroup): Root and tuber vegetables (1C and 1D); Leaves of root and tuber vegetables
   c. Codex Group: 016 (VR 0505) Root and tuber vegetables
   d. EPA Crop Definition: None
10. References: GRIN, CODEX, MAGNESS, NEAL, RICHE, STEPHENS 1988 (picture), US EPA 1995a, YAMAGUCHI 1983, RUBATZKY
11. Production Map: EPA Crop Production Regions 3 and 13.
12. Plant Codes:
    a. Bayer Code: CXSES

220

1. Date

(Date palm, Datil, Palmera-datilera, Desert date)

Arecaceae (Palmae)

Phoenix dactylifera L.

2. Grows in the dry subtropical and tropical regions. Known as the "Tree of Life" in Genesis. Needs hot dry climate for proper fruit ripening. Propagated by seeds or suckers. First fruit harvest 5-8 years after suckering. Date fruits are borne on palm trees that grow only from a terminal bud. Palms may begin to fruit at 5 to 6 feet in height, but increase in height each year. Thus in older plantings the fruit, which is produced only near the top, may be 20 or more feet in the air. Fruit is produced in large bunches on strands. A bunch may have 40 or more strands, each carrying up to 20 or more fruits. In commercial production, strands are usually pruned off to about half length to improve fruit size and quality. Bunches are often covered with weather resistant material, open at the bottom, to protect from rain, birds and to some extent insects. Individual fruits vary in shape from nearly round to cylindrical and from $^3/_4$ to 1 inch diameter. Each contains a single rather large seed. Surface is smooth in the growing fruit, becoming wrinkled as the fruit ripens and loses moisture. Fruit of most varieties loses most of its moisture before harvest. Key varieties include 'Deglet noor', 'Zahidi', 'Medjool', 'Halawy' and 'Khadrawy'.

The **desert date (soapberry tree)** is *Balanites aegyptiaca* (L.) Delile (Balanitaceae), the fruits are eaten fresh or dried, and the seeds are used for oil (zachun oil). It grows in drier Africa to Southwest Asia. The Codex crop group code for desert date is 005 (FL 0296).

3. Season: Bloom to harvest in about 7 to 8 months. Dates ripen in late September to December with 6-8 pickings per tree in the Coachella Valley of California.
4. Production in U.S.: Production in U.S. (5500 acres) is mainly in Coachella Valley, California, with some in Arizona.
5. Other production regions: Most U.S. imports come from Middle East countries.
6. Use: Mainly dried and sold as whole, pitted or chopped for confections or to be eaten out of hand.
7. Part(s) of plant consumed: All except seed (pit)
8. Portion analyzed/sampled: Whole dried fruit without pit which is discarded
9. Classifications:
   a. Authors Class: Tropical and subtropical fruits – edible peel
   b. EPA Crop Group (Group & Subgroup): Miscellaneous
   c. Codex Group: 005 (FT 0295) Assorted tropical and subtropical fruits – edible peel
   d. EPA Crop Definition: None
10. References: GRIN, CODEX, KNIGHT(a), LOGAN, MAGNESS, MARTIN 1987, FACCIOLA, USDA 1996f
11. Production Map: Primarily, EPA Crop Production Region 10.
12. Plant Codes:
    a. Bayer Code: PHXDA

221

1. Daylily
(Kanzou, Gum jum, Huang hau tsai, Golden needles, Gum tsoy, Tawny daylily, Skina-kanzo, Fulvous daylily, Orange daylily)
Liliaceae
Hemerocallis fulva (L.) L.

2. Clump forming perennial herb to 6 feet tall native to Eurasia. Hardy summer blooming herbs with lilylike flowers which are born in a more or less open or branched cluster terminating the scape. Leaves to 2 feet long and $1^3/_8$ inches wide. Flowers to 5 inches long and $3^1/_2$ inches across when fully opened. Also other *Hemerocallis* spp. are edible, reference Facciola (1990).
3. Season, seeding to harvest: No data.
4. Production in U.S.: No data. Naturalized in eastern U.S.
5. Other production regions: Native to Eurasia (BAILEY 1976).
6. Use: Young shoots eaten cooked. Flower buds used in salads or cooked. Dried flowers called golden needles, gum-tsoy, or gum-jum are used in soups and stews. Bulbs are eaten raw or cooked. Also grown as an ornamental.
7. Part(s) of plant consumed: Young shoots, flowers and bulbs
8. Portion analyzed/sampled: Whole plant including bulb without roots for vegetable use. Flowers (fresh and dried) for herb use
9. Classifications:
   a. Authors Class: Bulb vegetables; Herbs and spices for flowers
   b. EPA Crop Group (Group & Subgroup): Miscellaneous
   c. Codex Group: No specific entry
   d. EPA Crop Definition: None
10. References: GRIN, BAILEY 1976, BAYER, FACCIOLA, MYERS
11. Production Map: No entry
12. Plant Codes:
    a. Bayer Code: HEGFU

222

1. Dewberry
(Trailing blackberry, Boysenberry, Loganberry, Youngberry, Caneberry, Zarzamoras, Mures des haies)
Rosaceae
Rubus spp.

2. Dewberries are very similar to blackberries (see blackberry for additional information). Dewberry is considered the proper name for blackberry varieties that have a trailing growth habit. With these, the canes are slender and trailing, and generally are less thorny than blackberries. Fruits and fruiting habit are similar to blackberry. Fruits generally are cylindrical in shape, $^1/_2$ inch in diameter and 1 to $1^1/_2$ inches in length. Some dewberries varieties, such as 'Logan', 'Young' and 'Boysen' are derived from crossing blackberry and raspberry. Leading varieties are 'Boysen', 'Logan', 'Marion', 'Olallie', 'Evergreen' and 'Young'.
3. Season: Bloom to maturity in 50 to 70 days.
4. Production in U.S.: 1,470 acres of Boysenberry in 1995, with 1,200 acres in Oregon. Oregon also produced 80 acres of Loganberry in 1995. California.
5. Other production regions: Some imports from New Zealand and Australia.
6. Use: Fruit is utilized fresh, frozen, or canned. The juice can be extracted and used for wine and preserves.
7. Part(s) of plant consumed: Fruit
8. Portion analyzed/sampled: Whole fruit
9. Classifications:
   a. Authors Class: Berries
   b. EPA Crop Group (Group & Subgroup): Crop Group 13: Berries Group as well as this group's Bushberry Subgroup
   c. Codex Group: 004 (FB 0266) Berries and other Small Fruit
   d. EPA Crop Definition: Blackberry = Dewberry

10. References: GRIN, CODEX, LOGAN, MAGNESS, MOORE, US EPA 1995, USDA 1996f
11. Production Map: Same as blackberry, EPA Crop Production Regions 2, 6, 10 and 12.
12. Plant Codes:
    a. Bayer Code: RUBSS

223

1. Dill

(Dillweed, Dill seed, Dill oil, Babydill, Sowa, Shi-luo, Garden dill)

Apiaceae (Umbelliferae)

Anethum graveolens L.

2. The dill plant is an upright annual of the carrot family, grown for its foliage, seeds and oil. The plant is 2 to 3 feet tall, with finely divided leaves, being near thread-like. The flowers and seeds are produced terminally in flat umbels. Plants are grown from seed. Most of the limited U.S. production is in the north central and Northwestern states. Both the herb and its oil are used in flavoring foods, especially pickles. Primarily, oil is obtained by distilling from the whole plant like mint. The whole plant with immature seeds is called dillweed or babydill. The dried ripe fruits are dill seeds.
3. Season, seeding to harvest: 56-65 days for fresh and about 4 months for seed or seed oil. Dill weed oil is distilled from the whole plant when the seed has just started to ripen.
4. Production in U.S.: For oil, 1,071 acres in Washington State (1992). No data on fresh flowers or seed. North central and Northwestern states.
5. Other production regions: Native to Eurasia (RUBATZKY 1997).
6. Use: Fresh herb, dried herb, oil and seed as spice. Dill seeds are used to flavor salads, soups, pickles, breads and vegetables. Dill seed and oil are used in processed meat. Dillweed and its oil are used for pickles. Dillweed in salads and cream cheese.
7. Part(s) of plant consumed: Whole plant for fresh or distillation of oil. Seeds for dill seed or crushed for oil. Flowers (fresh or dried) used for flavoring
8. Portion analyzed/sampled: Foliage (fresh, oil, dried), seeds (spice and oil) and flowers (fresh and dried). Recommendation is to only sample the dillweed as fresh and dried and the dill seed
9. Classifications:
    a. Authors Class: Herbs and spices
    b. EPA Crop Group (Group & Subgroup): Herbs and spices (19A, 19B) (Dill seed is representative commodity)
    c. Codex Group: 027 (HH 0730) Herbs; 028 (HS 0730) Spices
    d. EPA Crop Definition: None
10. References: GRIN, CODEX, FARRELL (picture), GARLAND (picture), LEWIS, MAGNESS, RICHE, SCHREIBER, US EPA 1995a, FACCIOLA, RUBATZKY, SPLITTSTOESSER
11. Production Map: EPA Crop Production Regions 5, 11 and 12.
12. Plant Codes:
    a. Bayer Code: AFEGR

224

1. Dock

(Spinach dock, Herb patience, Patience dock, Hierba de la paciencia, Ba tian suan mo, Oseille epinard, Romaza hortense)

Polygonaceae

Rumex patientia L.

2. Spinach dock is a strong growing perennial, reaching 5 feet when in flower. Rosette leaves are 8 to 12 inches long, tapering at both ends. Stem leaves are rounded at the base. Leaves are used as greens, especially leaves which develop in early spring. Leaf exposure is similar to that of spinach.
3. Season, seed to harvest: 4 to 5 months. Leaves available in early spring from established roots.
4. Production in U.S.: No data. Negligible. Naturalized in North America (FACCIOLA 1990).
5. Other production regions: Native to Southern Europe (FACCIOLA 1990).
6. Use: Potherb greens.
7. Part(s) of plant consumed: Leaves only
8. Portion analyzed/sampled: Leaves
9. Classifications:
    a. Authors Class: Leafy vegetables (except *Brassica* vegetables); Herbs and spices
    b. EPA Crop Group (Group & Subgroup): Leafy vegetables (except *Brassica* vegetables) (4A)
    c. Codex Group: 027 (HH 0746) Herbs (Common and related *Rumex*)
    d. EPA Crop Definition: None
10. References: GRIN, CODEX, FACCIOLA, KAYS, MAGNESS, US EPA 1995a
11. Production Map: No entry
12. Plant Codes:
    a. Bayer Code: RUMPA

225

1. Dokudami

(Chi, Tsi, Yuxing, Giap Ca)

Saururaceae

Houttuynia cordata Thunb.

2. Herbaceous perennial herb to 15 inches tall, rhizome creeping. Leaves cordate, 2 to 3 inches long and gland-dotted. Grown in moist locations and propagated by division and seeds.
3. Season: No data.
4. Production in U.S.: Hardy to at least USDA plant hardiness zone 5 (KUEBEL 1988).
5. Other production regions: Southeast Asia (KUEBEL 1988) and Japan (BAILEY 1976).
6. Use: Leaves used in salads and to garnish fish stew.
7. Part(s) of plant consumed: Leaves
8. Portion analyzed/sampled: Leaves (fresh)
9. Classifications:
    a. Authors Class: Herbs and spices
    b. EPA Crop Group (Group & Subgroup): Miscellaneous
    c. Codex Group: No specific entry
    d. EPA Crop Definition: None
10. References: GRIN, BAILEY 1976, FACCIOLA, KUEBEL, USDA NRCS
11. Production Map: No entry
12. Plant Codes:
    a. Bayer Code: HOTCO

226

1. Durian
(Durion, Civetfruit, Stinkvrucht)
Bombacaceae
Durio zibethinus L.

2. The durian is a fruit native to Southeast Asia, where it is important in native diets. Plant requires a warm, humid atmosphere and a rich soil. The tree is large (120 ft.), long-lived (80 to 150 years) and strictly tropical, with leathery leaves 6 to 7 inches long. Fruits are large, up to 5 pounds or more, with a hard rind covered with sharp woody protuberances. The edible pulp has a strong reeking odor similar to the smell of a rotting onion. However, it has a rich, butter-like, bland taste, much esteemed where the tree is native. It provides a good source of iron and niacin. It flowers in clusters from branches or tree trunks with 1 to 45 flowers per cluster. The fruit is an aril that forms 4 weeks after pollination and covers the whole seed. The fruit matures 90 to 150 days after fruit set. The edible pulp is 19 to 32 percent of the whole fruit. Tree bears 4 to 8 years after planting, depending on cultivar and reach maximum production after 15 to 20 years. Mature fruits are 12 inches long by 8 inches wide. U.S. and Canada prefer 2.5 to 3.5 kg fruits for imports. Best cultivars have small seeds and large arils. Seedless varieties are known.
3. Season: Plant flowers throughout the year. In the tropics, harvest is April to September.
4. Production in U.S.: Small plantings are developed in Hawaii (approx 10 acres) and Puerto Rico.
5. Other production regions: Thailand, Malaysia, Indonesia and Philippines.
6. Use: Mostly fresh eating, however it is also preserved as a paste or frozen, or dried. Pulp can be added to ice cream, cake or made into a jam. Seeds can be boiled or roasted.
7. Part(s) of plant consumed: The pulp is the item of commerce. Three kg of fresh pulp = 1 kg of dried powder.
8. Portion analyzed/sampled: Whole fruit
9. Classifications:
   a. Authors Class: Tropical and subtropical fruits – inedible peel.
   b. EPA Crop Group (Group & Subgroup): Miscellaneous
   c. Codex Group: 006 (FI 0334) Assorted tropical and subtropical fruits – inedible peel
   d. EPA Crop Definition: None
10. References: GRIN, ANON(b), CODEX, MAGNESS, MARTIN 1980, MARTIN 1987, KNIGHT(a), US EPA 1994, MORTON, NANTACHAI, ROY
11. Production Map: EPA Crop Production Region 13.
12. Plant Codes:
    a. Bayer Code: DURZI

227

1. Eastern gamagrass
(Macillo, Pasto Guatemala, Zacate maicero, Gamagrass, Eastern gammagrass)
Poaceae (Gramineae)
Tripsacum dactyloides (L.) L. (syn: *T. dactyloides* var. *occidentale* Cutler & Andes)

2. This is a warm season perennial bunchgrass native to the U.S. and it is a member of the grass tribe *Andropogoneae*. It is a erect bunchgrass from 1.5 to 3 m tall that spreads by rhizomes, and has a panicle seedhead (12 to 25 cm long). Eastern gamagrass is adapted to river banks and well-drained grasslands, and is occasionally seeded into pastures in areas which receive 1000 to 1500 mm annual rainfall. It has excellent forage quality for all livestock throughout the growing season. It reproduces from seeds but mostly spreads by rhizomes. It is usually planted by vegetative sprigs. There are 16,052 seeds/kg.
3. Season: Warm season, spring growth, flowers in late spring. Seeds mature July through September, and grass is grazed spring and summer. Also an excellent hay crop if cut at 15 to 25 cm when seedheads appear. Forage can be grazed until late summer to fall.
4. Production in U.S.: There is no production data for Eastern gamagrass, however, pasture and rangeland are produced on more than 410 million A (U.S. CENSUS, 1992).
5. Other production regions: West Indies and northern Mexico (GOULD 1975).
6. Use: Grazing on pasture and for hay.
7. Part(s) of plant consumed: Leaves, stems, and seed
8. Portion analyzed/sampled: Forage and hay
9. Classifications:
   a. Authors Class: See Forage grass
   b. EPA Crop Group (Group & Subgroup): See Forage grass
   c. Codex Group: See Forage grass
   d. EPA Crop Definition: None
10. References: GRIN, US EPA 1994, US EPA 1995, CODEX, BARNES, USDA 1995, HOLZWORTH, SKERMAN, STUBBENDIECK (picture), RICHE, GOULD
11. Production Map: EPA Crop Production Regions 1, 2, 3, 4, 5, 6 and 13.
12. Plant Codes:
    a. Bayer Code: No specific entry

228

1. Eggplant
(Guinea squash, Berenjena, Boulanger, Aubergine, White eggplant, Japanese eggplant, Italian eggplant, American eggplant, Brinjal eggplant, Oriental eggplant, Grape eggplant)
Solanaceae
Solanum melongena L.

2. The plant, as grown commercially, is a warm season annual, much branched, 2 to 4 feet tall, with large ovate or oblong leaves, densely pubescent on the lower surface. Fruits are normally large, 3 inches to as much as 10 inches long, and smooth skinned, with calyx retained at base when marketed. They are near globular to pyriform, or markedly elongated in shape. Plants are usually started in beds and moved to the field at 8 to 10 weeks of age. Cultural practices are similar to those for tomatoes. The Japanese eggplant is small, slender and purple in color. Marketable eggplant fruit can range in size from thumb or grape size to almost football size and are purple, yellow, white or green color.
3. Season, field setting to harvest. About 45 to 80 days.
4. Production in U.S.: 8,097 acres (1992 CENSUS). In 1992, top 8 states with more than 200 acres are Florida (2,544 acres), New Jersey (1,117), California (1,050), Georgia (717), Oregon (386), Washington (273), North Carolina (267) and Texas (256) (1992 CENSUS).
5. Other production regions: Baby Japanese and Japanese eggplants (variety 'Millionare') are provide by California and

Mexico from February to October (LOGAN 1996). Canada grows eggplant (ROTHWELL 1996a).

6. Use: Mainly culinary, as cooked vegetable. U.S. per capita consumption of eggplants in 1994 was 0.4 pounds (PUTNAM 1996).
7. Part(s) of plant consumed: Whole fruit after calyx removed
8. Portion analyzed/sampled: Whole fruit
9. Classifications:
   a. Authors Class: Fruiting vegetables (except cucurbits)
   b. EPA Crop Group (Group & Subgroup): Fruiting vegetables (except cucurbits) (8)
   c. Codex Group: 012 (VO 0440) Fruiting vegetables (other than cucurbits)
   d. EPA Crop Definition: None
10. References: GRIN, CODEX, LOGAN (picture), MAGNESS, RICHE, ROTHWELL 1996a, US EPA 1996a, US EPA 1995a, US EPA 1994, SWIADER, PUTNAM 1996
11. Production Map: EPA Crop Production Regions 1, 2, 3, 4, 5, 10 and 13.
12. Plant Codes:
    a. Bayer Code: SOLME

229

1. Elderberry
(American elder, Blueberry elder, Common elder, Elder, European elderberry, Mexican elder, Red elder, Southern elder, Sureau, Holunder)
Caprifoliaceae
Sambucus spp.
1. American elderberry
(American elder, Mexican elder, Sweet elder)
S. canadensis L.
1. Blueberry elder
(Blue elderberry, Western elderberry, Blue elder)
S. cerulea Raf. (syn: *S. glauca* Nutt.)
1. European elder
(European elderberry, Black elderberry, Common elder, Black elder, Elder)
S. nigra L.
1. European red elder
(American red elder, Redberried elder, Scarlet elder, Stinking elder)
S. racemosa ssp. *pubens* (Michx.) House

2. Cultivated elderberry plants are deciduous large shrubs or small trees, sometimes up to 20 feet, grown to a limited acreage in the U.S. for the fruit. Plants are adopted from the temperate Northern to mid-Southern regions. Plants are propagated from hardwood or softwood cuttings, root cuttings or suckers. Leaves are large and compound. Stems are hollow or pithy. Fruits are produced in large flat clusters 6 to 9 inches across. Individual berries are small, about $1/6$ to $1/4$ inch diameter, globose, with prominent seeds. Colors vary from red to blue-black. Important commerical cultivars of the species *S. canadensis* include 'Adams 1', 'Adams 2', 'Johns', 'Scotia', 'York', 'Nova', 'Kent', 'Victoria' and 'Ezyoff'. Fruits of elderberry are often found in the wild in considerable quantities.
3. Season: Bloom to harvest in 60 to 90 days. Depending on cultivar and location, harvest is mid-August to mid-September.
4. Production in U.S.: About 500 tons.
5. Other production regions: Canada and Eurasia (FACCIOLA 1990).

6. Use: Most important use is the fruit for jelly, wine and sauces. Some less important uses include leaves, flowers and fruit for dyes, oil from seed as a flavorant, and use as medicine.
7. Part(s) of plant consumed: Fruit
8. Portion analyzed/sampled: Whole Fruit
9. Classifications:
   a. Authors Class: Berries
   b. EPA Crop Group (Group & Subgroup): Crop Group 13: Berries Group as well as this group's Bushberry Subgroup
   c. Codex Group: 004 (FB 0267) Berries and other Small Fruit
   d. EPA Crop Definition: None
10. References: GRIN, CODEX, MAGNESS, STANG, US EPA 1995, FACCIOLA
11. Production Map: EPA Crop Production Region 5.
12. Plant Codes:
    a. Bayer Code: SAMCN (*S. canadensis*), SAMGL (*S. cerulea*), SAMNI (*S. nigra*), SAMRA (*S. racemosa*), SAMSS (*S.* spp.)

230

1. Elecampane
(Elf dock, Horse-heal)
Asteraceae (Compositae)
Inula helenium L.

2. The plant is a coarse growing perennial herb to 6 feet tall, cultivated for the roots. Leaves are up to a foot long, elliptic-oblong and hairy. Roots should be dug the second year. A preparation of the mucilaginous roots is used as a medicine, and roots are also used in sweetmeats.
3. Season, seeding to harvest: Fall of second year.
4. Production in U.S.: No data, very limited if at all. Naturalized in eastern North America.
5. Other production regions: Eurasia (FACCIOLA 1990).
6. Use: Flavoring for candy.
7. Part(s) of plant consumed: Root
8. Portion analyzed/sampled: Root (dried)
9. Classifications:
   a. Authors: Class: Herbs and spices
   b. EPA Crop Group (Group & Subgroup): Miscellaneous
   c. Codex Group: 028 (HS 0781) Spices
   d. EPA Crop Definition: None
10. References: GRIN, BAILEY 1976, CODEX, FOSTER (picture), MAGNESS, FACCIOLA
11. Production Map: No data, but naturalized in North America (BAILEY 1976).
12. Plant Codes:
    a. Bayer Code: INUHE

231

1. Endive
(Chicory)
1. Escarole
Asteraceae (Compositae)
Cichorium endivia L. ssp. *endivia*

2. These are cool season annual plants as produced commercially, grown from seed for the loose-headed leaves, which are used in salads and to some extent as potherbs. The leaves rise from near the soil surface, and are 6 to 8 inches long. Two leaf types are grown. In one, commonly termed endive, the numerous leaves are oblong, curled and fringed. In the

Florida vegetable trade, curly endive is often referred to as chicory. The second type, termed escarole, has broad, generally nearly flat leaves. Method of growing endive and escarole is similar to that for lettuce. Plants may be started in beds, but more generally seeds are field sown. Leaf exposure during growth is similar to that of spinach.

3. Season, seeding to harvest: 3 to 3½ months.
4. Production in U.S.: In 1992, about 2,000 acres of endive and about 2,000 acres of escarole (1992 CENSUS). Endive: Florida (1,004 acres), New Jersey (407), California (247) and New York 70 acres (1992 CENSUS). Escarole: Florida (1,264 acres), New Jersey (496), California (105) and New York 100 acres (1992 CENSUS). In 1994, California reported 1,239 acres of endive (MELNICOE 1996e). In 1996, Ohio reported about 50 acres of endive and escarole (IR-4 REQUEST DATABASE 1996).
5. Other production regions: Native to Mediterranean region (RUBATZKY 1997).
6. Use: Mainly in salads, some as potherb. U.S. per capita consumption in 1994 was 0.2 pounds (PUTNAM 1996).
7. Part(s) of plant consumed: Leaves only
8. Portion analyzed/sampled: Leaves
9. Classifications:
   a. Authors Class: Leafy vegetables (except *Brassica* vegetables)
   b. EPA Crop Group (Group & Subgroup): Leafy vegetables (except *Brassica* vegetables) (4A)
   c. Codex Group: 013 (VL 0476) Leafy vegetables (including *Brassica* leafy vegetables)
   d. EPA Crop Definition: Endive = Escarole
10. References: GRIN, CODEX, MAGNESS, RICHE, US EPA 1994, US EPA 1995b, MELNICOE 1996e, STEPHENS 1988, RUBATZKY (picture), IR-4 REQUEST DATABASE 1996, BEATTIE, SWIADER, PUTNAM 1996
11. Production Map: EPA Crop Production Regions 1, 2, 3, 5 and 10.
12. Plant Codes:
   a. Bayer Code: CICEC

232

1. Epazote
(American wormseed, Epasote, Mexican tea, Spanish tea, Wormseed, Wormwood, Indian goosefoot)
Chenopodiaceae
   Chenopodium ambrosioides L.
2. Annual or perennial herb to 3½ feet, with strong odor. Pungent herb with a flavor similar to coriander that is naturalized in Mexico and the U.S. It has flat-pointed leaves to 5 inches long and available fresh or usually dried. It is used as a tea substitute or to flavor bean, corn or shellfish dishes. Historically, it has been used to reduce the gassiness that occurs with beans while adding a sweet flavor.
3. Season, seeding to harvest: No data.
4. Production in U.S.: Grows in backyards in California with up to 200 plants each. U.S. mainland and Hawaii.
5. Other production regions: It is found in Mexico.
6. Use: Tender leaves used as potherb. Dried leaves also used as herb. Can be cultivated for its essential oils which have medicinal properties to treat for worms.
7. Part(s) of plant consumed: Leaves
8. Portion analyzed/sampled: Leaves (Fresh and dried)

9. Classifications:
   a. Authors Class: Herbs and spices
   b. EPA Crop Group (Group & Subgroup): Miscellaneous
   c. Codex Group: No specific entry
   d. EPA Crop Definition: None
10. References: GRIN, BAILEY 1976, DUKE(c), FACCIOLA, MELNICOE 1996b, NEAL, PENZEY, SCHNEIDER 1994c
11. Production map: EPA Crop Production Region 10.
12. Plant Codes:
   a. Bayer Code: CHEAM

233

1. Euphorbia
(Caper spurge, Moleplant, Myrtle spurge, Gopher plant)
Euphorbiaceae
   Euphorbia lathyris L. (syn: *E. lathyrus* L., *Galarhoeus lathyris* (L.) Haw.)
2. Annual or biennial latex-bearing plant to 3 feet tall. Stem leaves are green 4-ranked, oblong-linear to lanceolate, 2 to 6 inches long. The seed has an oil content of about 50 percent, of which oleic acid makes up 80 to 90 percent.
3. Season: Winter or spring crop. It requires a long growing period for seed set.
4. Production in U.S.: No data. Naturalized in North America (BAILEY 1976).
5. Other production regions: Native to Europe (BAILEY 1976).
6. Use: Used like lard and tallow, in soaps, detergents, cosmetics, paints and lubricants. No data available on the palatability of the meal for livestock.
7. Part(s) of plant consumed: Seed
8. Portion analyzed/sampled: Seed and its oil (meal, if eventually used for feed)
9. Classifications:
   a. Authors Class: Oilseed
   b. EPA Crop Group (Group & Subgroup): Miscellaneous
   c. Codex Group: No specific entry
   d. EPA Crop Definition: None
10. References: GRIN, JANICK 1996, NEAL, ROBBELEN 1989 (picture), BAILEY 1976
11. Production Map: No entry
12. Plant Codes:
   a. Bayer Code: EPHLA

234

1. Evening primrose, Common
(Onagre, Onagra, Nachtkerze, German rampion)
Onagraceae
   Oenothera biennis L. (syn: *O. muricata* L.)
1. Tufted evening primrose
   O. caespitosa Nutt.
2. Common evening primrose is an evening flowering winter biennial herb 1 to 6 feet tall which normally germinates the first year, making a taproot for the winter, then flowers the second season. The stems are branched with lance-shaped leaves 4 to 12 inches long. Roots may be eaten as a vegetable, and the shoots in salads. Blooms from June to September. In Canada the seed oil is an approved dietary supplement. The oil contains gamma linolenic acid. The tufted evening primrose is a stemless perennial with leaves 1 to 4 inches long.

3. Season, seeding to harvest: Fall or spring seeded. Fall seedings in August are harvested for seeds the following August (12 months). Spring seedings in late March to early April are harvested for seed in November.
4. Production in U.S.: In 1994, 150 acres grown for oil in Washington state which accounted for about 70 percent of the acreage in U.S. (SCHREIBER 1995). Native to North America (FOSTER 1993).
5. Other production regions: Canada (ROTHWELL 1996a).
6. Use: Roots are sweet, somewhat resembling salsify or parsnips and eaten boiled or added to soups and stews. Young shoots used in salads or as potherb. Seed for oil or confectionary as substitute for poppyseed.
7. Part(s) of plant consumed: Roots, tops, and seeds
8. Portion analyzed/sampled: Roots and tops; seeds and its oil
9. Classifications:
   a. Authors Class: Oilseed; Root and tuber vegetables; Herbs and spices
   b. EPA Crop Group (Group & Subgroup): Miscellaneous
   c. Codex Group: No specific entry
   d. EPA Crop Definition: None
10. References: GRIN, BAILEY 1976, FACCIOLA, FOSTER (picture), REHM, SCHREIBER, USDA NRCS, ROTHWELL 1996a
11. Production Map: EPA Crop Production Region 11.
12. Plant Codes:
    a. Bayer Code: OEOBI (*O. biennis*)

235

1. Fameflower
(Waterleaf, Surinam purslane, Surinam spinach, Pourpier grand bois, Pink purslane, Verdolaga de playa)
Portulacaceae
Talinum triangulare (Jacq.) Willd.
2. Slightly succulent herb cultivated widely throughout the tropics. Small upright herb that branches readily with glabrous leaves. It is best to let the plants reach about 12 inches before harvesting the leaves. A new crop of leaves are produced every 2 weeks for about a year.
3. Season, planting to harvest: Cuttings can root in about 2 weeks and shortly thereafter produce new edible growth.
4. Production in U.S.: No data. Reseeds itself in the garden (MARTIN 1979). Naturalized in Florida (FACCIOLA 1990).
5. Other production regions: West Indies (NEAL 1965). Tropical America.
6. Use: As greens used in salads or cooked as potherb.
7. Part(s) of plant consumed: Young leaves and tender stems
8. Portion analyzed/sampled: Leaves
9. Classifications:
   a. Authors Class: Leafy vegetables (except *Brassica* vegetables)
   b. EPA Crop Group (Group & Subgroup): Miscellaneous
   c. Codex Group: No specific entry
   d. EPA Crop Definition: None
10. References: GRIN, FACCIOLA, MARTIN 1979 (picture), NEAL
11. Production Map: EPA Crop Production Region 13.
12. Plant Codes:
    a. Bayer Code: TALTR

236

1. Feather fingergrass
(Woolly-tip,Feathertop Rhodesgrass, Blackseed, Feather-top chloris, Windmill grass, Showy chloris)
Poaceae (Gramineae)
Chloris virgata Sw.
2. This is an annual warm season grass native to the tropics and introduced into the U.S. It is a member of the grass tribe *Chlorideae*. It is an erect bunchgrass (15 to 90 cm tall) that sometimes spreads by short stolons, and is adapted to areas receiving 500 to 750 mm rainfall and altitudes from sea level to 2000 m. It can be used as a hay crop, but is best utilized to reseed denuded rangelands. Feather fingergrass is seeded at a rate of 0.5 kg/ha, and seed yields range from 100 to 650 kg/ha.
3. Season: Warm season, growth starts in late spring, and matures in August. It can be cut for hay at flowering.
4. Production in U.S.: No data for Feather fingergrass production, however, pasture and rangeland are produced on more than 410 million A (U.S. CENSUS, 1992).
5. Other production regions: Worldwide in tropics to warm temperate regions (GOULD 1975).
6. Use: Rangeland grazing, hay and reseeding rangeland for erosion control.
7. Part(s) of plant consumed: Leaves and stems
8. Portion analyzed/sampled: Forage and hay
9. Classifications:
   a. Authors Class: See Forage grass
   b. EPA Crop Group (Group & Subgroup): See Forage grass
   c. Codex Group: See Forage grass
   d. EPA Crop Definition: None
10. References: GRIN, US EPA 1994, US EPA 1995, CODEX, HOLZWORTH, SKERMAN (picture), RICHE, GOULD
11. Production Map: EPA Crop Production Regions 3, 6, 8 and 9.
12. Plant Codes:
    a. Bayer Code: CHRVI

237

1. Feijoa
(Pineapple guava, Guayaba, Chilena)
Myrtaceae
Acca sellowiana (O. Berg) Burret (syn: *Feijoa sellowiana* (O. Berg)
2. Feijoa trees are best grown in cool subtropical or highland tropical climates. Plants are usually vegetatively propagated with first crop in 2-3 years. The small evergreen tree or shrub grows to 18 feet, with leaves 2 to 3 inches long. The plant can endure winter temperatures down to about 15 degrees F. Fruit is generally oval in shape, 1.5-3 inches long. The prominent calyx is persistent. Seeds are edible. Skin is waxy, dark green to yellow.
3. Season: Bloom to harvest is 150-180 days. Most domestic production harvested September to January.
4. Production in U.S.: Commercial production in California. Scattered trees have been planted in Florida and Hawaii.
5. Other production regions: New Zealand.
6. Use: Fresh eating, used in salads or processed in to jelly, marmalade and other preserves.
7. Part(s) of plant consumed: Interior pulp of fruit
8. Portion analyzed/sampled: Whole fruit

9. Classifications:
   a. Authors Class: Tropical and subtropical fruits – inedible peel
   b. EPA Crop Group (Group & Subgroup): Miscellaneous
   c. Codex Group: 006 (FI 0335 and FI 4143) Assorted tropical and subtropical fruits – inedible peel
   d. EPA Crop Definition: Proposing – Guava = Feijoa
10. References: GRIN, CODEX, LOGAN, MAGNESS, MARTIN 1987, US EPA 1994
11. Production map: EPA Crop Production Region 10.
12. Plant Codes:
   a. Bayer Code: FEJSE

238

1. Fennel
(Wild fennel, Common fennel, Dry fennel, Fennel seed, Hinojo)
Apiaceae (Umbelliferae)
Foeniculum vulgare Mill. var. *vulgare* (syn: *F. officinale* All.)

2. The condiment fennel is closely related to the vegetable Florence fennel *F. vulgare* Mill. var. *azoricum*, which see. Leaves are not thickened into a bulb at the base like Florence fennel. The plant is an herbaceous perennial 2 to 5 feet tall, which may be grown as an annual. However, it yields more seed the second and succeeding years. The leaves are pinnate, greatly compounded and glabrous, with ultimate segments threadlike. The flower head is large. Fresh young leaves and stems are minced and added to sauces and used as flavoring in puddings, soups and with fish. Stems and leaves used before bloom are tender. The young stems are enclosed in a sheath formed by the bases of the petioles or leaf stems. Stems and young leaves may be eaten raw or used for flavoring. Resembles celery in general habit and use. The seeds are the main item of commerce. They are used in cookery, confections and liquors. Volatile oil from the seeds is used in toilet articles. Seeds are small and oblong. The seed heads are cut and dried before threshing. U.S. imported about 3,418 tons from Egypt, India and Turkey in 1992.
3. Season, start of growth to harvest: 3 to 4 months (for seed).
4. Production in U.S.: Limited acreage. Cultivated in the U.S. on a limited basis (SCHRIEBER 1995).
5. Other production regions: Southern Europe (FACCIOLA 1990).
6. Use: Flavoring in cookery, liquors and bakery products.
7. Part(s) of plant consumed: Seed and oil from seed of commerce. Leaves, flower heads and young stems locally
8. Portion analyzed/sampled: Seed (dried)
9. Classifications:
   a. Authors Class: Herbs and spices
   b. EPA Crop Group (Group & Subgroup): Herbs and spices (19B)
   c. Codex Group: 028 (HS 0731) Spices; 027 (HH 0731) Herbs; 057 (DH 0731) Dried herbs
   d. EPA Crop Definition: None
10. References: GRIN, BAILEY 1976, CODEX, FACCIOLA, FARRELL (picture), GARLAND (picture), LEWIS, MAGNESS, SCHRIEBER, USDA 1993a, US EPA 1995a
11. Production Map: No entry
12. Plant Codes:
   a. Bayer Code: FOEVU

239

1. Fennel, Florence
(Bulb fennel, Fino, Italian fennel, Bulb anise, Sweet fennel, Sweet anise, Fetticus, Finocchio, Finochio, Carosella, Roman fennel)
Apiaceae (Umbelliferae)
Foeniculum vulgare Mill. var. *azoricum* (Mill.) Thell. (syn: *F. azoricum* Mill.)

2. This plant is grown as an annual, grown for its thickened bulb-like leaf-stem bases. These make a bulb-like structure just above the ground, up to 3 or 4 inches long, and oval in cross section. By covering with soil, these are sometimes blanched. They have an aromatic and distinctive flavor, and are generally used as a boiled vegetable. Plants attain a height of 2 to 3 feet. In general habit, it is somewhat similar to celery.
3. Season, seeding to harvest: About 3 months for edible leaf base.
4. Production in U.S.: In 1988, 3,220 tons on 274 acres in California (MYERS 1991). Grown in Florida (STEPHENS 1988).
5. Other production regions: Europe (FACCIOLA 1990).
6. Use: As cooked vegetable and raw as a salad. Has been sold in American market as anise. Preparation – cut off stems and leaves. The stems can be used in stews for flavoring; the leaves for sprinkling like dill. The bulb is prepared by cutting out the knob-like core. For raw use remove first heavy wrapping of bulbs or at least de-string it like celery (SCHNEIDER 1985).
7. Part(s) of plant consumed: Thickened leaf-stem bases and fresh leaves and flower heads. Seeds for spice
8. Portion analyzed/sampled: Enlarged bulbous leaf base and leaves; seeds as appropriate
9. Classifications:
   a. Authors Class: Leafy vegetables (except *Brassica* vegetables); Herbs and spices
   b. EPA Crop Group (Group & Subgroup): Leafy vegetables (except *Brassica* vegetables) (4B); Herbs and spices (19B)
   c. Codex Group: 009 (VA 0380) Bulb vegetables; (VA 4159 – Italian fennel); (VA 4161 – Roman fennel); (VA 4163 – Sweet fennel); 013 (VL 4347 and VL 4345) Leafy vegetables; 027 (HH 4747) Herbs
   d. EPA Crop Definition: Celery = Florence fennel (Sweet anise, Sweet fennel, Finochio) (fresh leaves and stalks only)
10. References: GRIN, BAILEY 1976, FACCIOLA, FOSTER, MAGNESS, MARTIN 1979, MYERS, SCHNEIDER 1985, STEPHENS 1988 (picture), US EPA 1995a, SPITTSTOESSER
11. Production Map: EPA Crop Production Regions 3 and 10.
12. Plant Codes:
   a. Bayer Code: FOEVA

240

1. Fenugreek
(Alholva, Bockshornklee, Common methi, Helbeh, Hilbah, Methi, Fenugreco, Greek hay, Greek clover)
Fabaceae (Leguminosae)
Trigonella foenum-graecum L.

2. Fenugreek is widely but not extensively grown in many temperate climates, but sparingly in the U.S. The plant is an erect, unbranched, annual legume up to 30 inches high, with 3-pinnate leaves. Seeds are produced in pods, several per pod. The plant is grown for fodder in Mediterranean countries. Seeds are used as condiments in various foods, and in curries. Seeds are also used in stock feeds and human and livestock medicines. In Pennsylvania about 1 acre cultivated for its fresh leaves and stems as a specialty herb. As fresh herbage, it grew from seeding to harvest in about 1 month (1991).

3. Season, seeding to harvest: 3 to 4 months for seed: About 1 month for greens.

4. Production in U.S.: 50 to 100 acres in New Jersey for greens (VANVRANKEN 1996a).

5. Other production regions: Native to eastern Mediterranean region.

6. Use: In food as condiment, livestock feed, and in medicines.

7. Part(s) of plant consumed: Mainly seed for food. Whole plant for feed. Leaves and stems for potherb or flavoring

8. Portion analyzed/sampled: Seed (if used as potherb, fresh leaves and stems)

9. Classifications:
   a. Authors Class: Herbs and spices; Leafy vegetables (except *Brassica* vegetables)
   b. EPA Crop Group (Group & Subgroup): Herbs and spices (19B) (Seed only)
   c. Codex Group: 028 (HS 0782) Spices (Seed only)
   d. EPA Crop Definition: None

10. References: GRIN, CODEX, FACCIOLA, FARRELL (picture), GARLAND, JANICK 1990, LEWIS, MAGNESS, PENZEY, REHM, US EPA 1995a, VANVRANKEN 1996a

11. Production Map: EPA Crop Production Region 1 and 2 for greens.

12. Plant Codes:
    a. Bayer Code: TRKFG

241

1. Fern, Edible
(Fiddlehead, Croziers, Buckhorns)
1. Brackenfern, Western
(Brakenfern, Pasture brakefern, Eagle fern, Warabi, Brackenfern, Brake)
Dennstaediaceae
Pteridium aquilinum (L.) Kuhn (syn: *Pteris aquilina* L.)
1. Zenmai
(Cinnamon fern, Buckhorn, Woolly cinnamon, Flowering fern)
Osmundaceae
Osmunda japonica Thunb.

2. Ferns are flowerless plants normally producing true roots from a rhizome. The leaves are termed "fronds". The uncoiled fronds are boiled until tender and eaten.

3. Season, harvest: Perennial, harvest immature fronds in the early spring.

4. Production in U.S.: No data. In home gardens (BAILEY 1976).

5. Other production regions: Grown in temperature regions (YAMAGUCHI 1983).

6. Use: Immature fronds, called croziers, buckhorns or fiddleheads are used in soups or boiled and served on toast.

7. Part(s) of plant consumed: Immature frond. The food safety of brackenfern must be considered as to possible health consequences

8. Portion analyzed/sampled: Immature frond

9. Classifications:
   a. Authors Class: Leafy vegetables (except *Brassica* vegetables)
   b. EPA Crop Group (Group & Subgroup): Miscellaneous
   c. Codex Group: No specific entry
   d. EPA Crop Definition: None

10. References: GRIN, BAILEY 1976, FACCIOLA, USDA NRCS, YAMAGUCHI 1983, RUBATZKY

11. Production Map: No entry, cosmopolitan.

12. Plant Codes:
    a. Bayer Code: PTEAP (*P. aquilinum*), OSMJA (*O. japonica*)

242

1. Fescue grass
Poaceae (Gramineae)
Festuca spp.
1. Tall fescue
(Reed fescue, Alta fescue)
F. arundinacea Schreb.
1. Meadow fescue
(English bluegrass)
F. pratensis Huds. (syn: *F. elatior* auct. Amer.)
1. Idaho fescue
(Blue bunchgrass)
F. idahoensis Elmer
1. Sheep fescue
(Sheep's fescue)
F. ovina L. var. *ovina*
1. Red fescue
(Fine fescue)
F. rubra L. ssp. *rubra*
1. Chewing's fescue
F. rubra ssp. *fallax* (Thuill.) Nyman (syn: *F. rubra* var. *comutata* Gaudin)
1. Rough fescue
F. scabrella Torr. ex Hook.

2. Some 80 species of *Festuca* are known, generally adapted to temperate or cool climates. They vary widely in texture and growth habit. Some are annuals, others perennial. Annuals may be troublesome weeds, while perennials are excellent for forage, pasturage and turf. The most valuable kinds follow.

Tall fescue is a perennial bunchgrass, introduced from Europe, which reaches up to 4 feet. It is used for pasture, hay, turf and erosion control throughout the Northern states, and is also adapted southward except in the Coastal Plain. Basal leaves are numerous, broad and flat. Plants are vigorous and grow well on both wet and dry sites. They thrive best on heavy soils. Palatability for livestock is lower than that of some other grasses, and pastures should be grazed close for best acceptance and nutritive value. A number of improved varieties are available in the trade for both forage and turf uses. Commonly used in the South.

Meadow fescue is a hardy perennial bunchgrass, introduced from England and adapted to cool climates. On rich soils it reaches up to 30 inches in height. Leaves are long and slender, bright green and succulent. It is now grown less than in the past for pasturage and hay because of the general superiority of tall fescue. It is not as heavy yielding nor as persistent as tall fescue.

Idaho fescue is a native bunchgrass found from Washington and Montana south to California and Colorado. The large bunches reach to 3 feet with numerous smooth leaves. This is a valuable range grass, palatable while green, and also curing well for fall and winter forage. It has potential for seeding on range lands.

Sheep fescue is a bunchgrass that forms dense clumps, with numerous stiff, sharp, bluish grey leaves. It is cold and drought tolerant and is better than most grasses on sandy or gravelly soils in Northern states. It is useful for grazing in early spring, but its greatest value is providing a durable turf on sandy or gravelly soils.

Red fescue was introduced from Europe. It differs from sheep fescue in that it creeps by underground stems and forms a sod rather than growing in tufts. It is extensively used for lawns and erosion control in northern parts of the U.S. Plants are hardy and vigorous. It is not highly palatable and is not generally used for pastures or hay. Chewing's fescue is a closely related kind but grows in clumps instead of forming a dense sod. It is also used for lawns and general purpose turf in shaded areas.

Rough fescue is a perennial cool season grass native to the U.S. that is a member of the grass tribe *Festuceae*. It is an erect bunchgrass 30 to 90 cm tall growing in tufts up to 60 cm in diameter. It is adapted to sandy loam soils in mountain foothills and valleys in the western U.S., and provides excellent forage for cattle and horses and is also good for sheep and wildlife. It has been productive in areas for up to 15 years and can withstand heavy grazing pressure as it matures and can produce from 1900 to 3800 kg/ha. Rough fescue reproduces by seed, tillers and rhizomes.

3. Season: Long-lived perennials but slow to establish.
4. Production in U.S.: No specific data, See Forage grass. Fescue seed is produced in eastern Washington (SCHREIBER 1995). In 1992, 320,334 acres of seed were harvested with 67,000 tons (1992 CENSUS). In 1992, the top 6 seed crop states reported as Missouri (171,652 acres), Oregon (117,044), Arkansas (7,628), Oklahoma (5,594), Kansas (4,267) and Washington (3,354) (1992 CENSUS).
5. Other production regions: Primarily the northern hemisphere (GOULD 1975).
6. Use: Pasture, hay, rangegrass, turf, erosion control.
7. Part(s) of plant consumed: Foliage
8. Portion analyzed/sampled: Forage and hay
9. Classifications:
   a. Authors Class: See Forage grass
   b. EPA Crop Group (Group & Subgroup): See Forage grass; Bermudagrass or Fescue, representative crop
   c. Codex Group: See Forage grass; 051 (AS 5253) Fescue
   d. EPA Crop Definition: None
10. References: GRIN, MAGNESS, RICHE, SCHREIBER, CODEX, HOLZWORTH, GOULD, STUBBENDIECK (picture), ALDERSON, BALL, MOSLER
11. Production Map: EPA Crop Production Regions 1, 2, 4, 5, 6, 7, 9, 10, 11 and 12. Seed crop regions 4, 5, 6, 11 and 12.
12. Plant Codes:
   a. Bayer Code: FESAR (*F. arundinacea*) FESOV (*F. ovina*), FESPR (*F. pratensis*), FESRU (*F. rubra*), FESNI (*F. rubra* ssp. *fallax*)

243

1. Fig
(Common fig, Higo)
Moraceae
   Ficus carica L.

2. The tree grows in warm temperate conditions with mild wet winters and hot dry summers. Usually propagated from cuttings. The fig tree is deciduous and usually held to a maximum size of 12 to 15 feet by pruning. The fruit is generally pear shaped, 1 to 2.5 inches in diameter, and has a tough, rather rough, ridged surface. The interior consists of numerous, very small seeds imbedded in a sweet pulp. Most of the fruits are borne on new growth in the axils of the leaves; but some large, early ripening fruits are set on old wood. Fruit color varies from yellowish green with white flecks to black. Important varieties include 'Calimyrna', 'Black Mission', 'Kadota' and 'White Adriatic'. Fruits are rich in vitamin A, calcium and iron. Primarily, dried figs are normally harvested daily by picking them off the ground after the ripe fruits become dry and fall onto a clean orchard floor. These are normally dehydrated to remove surplus moisture to approximately 16 percent. Of the four commercial figs grown in California, hand picked fresh fruit harvested from the 'Kadota' variety is the favorite for canning and preserving. The standard dried fig is the 'White Adriatic' variety which is used as fig paste in fig bars. The self-pollinating varieties ('Black Mission', 'White Adriatic' and 'Kadota') are termed common figs, while the 'Calimyrna' figs require an outside pollination source and that is a tiny wasp.
3. Season: Fruit set to harvest for individual fruits occur in 3 to 4 months. Fruits in various stages of development on tree from midsummer until frost occurs.
4. Production in U.S.: Almost all of the 15,000 acres of commercial production is in California. Figs are also grown in home gardens throughout the Southern states.
5. Other production regions: Turkey, Greece, Italy, Algeria, Morocco.
6. Use: Mostly dried, some canned and preserves. Some fruit eaten fresh (in recent years less than 5 percent). Fruits and leaves have potential for medical use.
7. Part(s) of plant consumed: Whole fruit
8. Portion analyzed/sampled: Whole fruit, fresh and dried
9. Classifications:
   a. Authors Class: Tropical and subtropical fruits – edible peel
   b. EPA Crop Group (Group & Subgroup): Miscellaneous
   c. Codex Group: 005 (FT 0297) Assorted tropical and subtropical fruits – edible peel
   d. EPA Crop Definition: None
10. References: GRIN, CODEX, KNIGHT, LOGAN, MAGNESS, US EPA 1994, IVES, DESAI(a), USDA 1996(f), KREZDORN, USDA 1996f
11. Production map: EPA Crop Production Region 10.
12. Plant Codes:
   a. Bayer Code: FIUCA

244

1. Flax
(Solin, Linseed oil, Flaxseed, Lino)
Linaceae
   Linum usitatissimum L.

2. Linseed oil, obtained from seed of the flax plant, is primarily used in industry; but some is used for edible purposes in eastern Europe. The flax plant is erect, growing to 3 feet, with narrow, entire leaves. The fruit is a pod or capsule, which is indehiscent. The seeds contain around 35 to 44 percent of drying oil. In eastern Europe, the seed is generally first cold pressed, the cold-press oil being used in foods. A later hot press yields additional industrial oil. In the U.S., oil extraction is generally hot press, followed by solvent extraction, and the oil is not used as food. The press cake from hot pressing is a valuable livestock feed. The flax seed contains a cyanogenic glucoside which forms hydrocyanic acid by enzyme action unless the enzyme is inactivated by heat. Flax seed for oil was grown in the U.S. on an average of about 2.7 million acres, 1964-66 but down in 1993 to 206,000 acres. World production of linseed oil is down and last reported 1991-92 at 569,800 tons. The downward production trend for flax may be reversed with the recent introduction of solin (so-lin) which is a low (<5 percent) linolenic acid flax. Solin oil is good for salad dressing and cooking oil. These varieties are called ‘Linola’. The GRAS petition (5G0416) was filed with the U.S. FDA by the Flax Council of Canada proposing that the low linolenic acid flaxseed oil be generally recognized as safe (GRAS) for use as a food oil (27 MAR 96 *Federal Register*: 61:13505-6).

3. Season, seeding to harvest: About 110-120 days.

4. Production in U.S.: 206,000 acres planted in 1993. 97 percent of the acreage in 1993 reported in North Dakota, South Dakota and Minnesota.

5. Other production regions: Recently, about 60 percent of the total flax seed utilized was imported. Canada reported 1,239,730 acres for 1995.

6. Use: Seed oil is used primarily for industrial purposes. The meal which remains after oil extraction is fed to livestock. The seed is used as a condiment in some breads and a feed supplement for laying hens. The straw is used for fiber in garments and linen paper, e.g. cigarette paper and bibles.

7. Part(s) of plant consumed: Seed

8. Portion analyzed/sampled: Seed, and seed processed commodity meal which is the present flax entry in EPA’s Table 1. But with solin, the oil will also be a required processed fraction sample

9. Classifications:
   a. Authors Class: Oilseed
   b. EPA Crop Group (Group & Subgroup): Miscellaneous
   c. Codex Group: 023 (SO 0693) Oilseed
   d. EPA Crop Definition: None

10. References: GRIN, CODEX, FLAX COUNCIL OF CANADA 1995, MAGNESS, NALAWAJA 1996, ROBBELEN 1989 (picture), ROTHWELL 1996a, USDA 1994a, US EPA 1995b

11. Production Map: EPA Crop Production Regions 5 and 7.

12. Plant Codes:
    a. Bayer Code: LIUUT

245

1. Flower, Edible
(Edible flowers, Herb flowers, Zatar)
Various, see below
Various spp., see below

2. Common edible flowers to include the ones in commercial production are listed below:
Scientific name
Family name
(Common name(s) or Principal vernacular referencing a specific crop monograph in this book)
Comments

Allium tuberosum Rottler ex Spreng.
Liliaceae
(Chive, Chinese – which see)
The flowers are used in eggs, cheese and fish dishes or used as garnish. Normally, the terms garlic chives or chive garlic, are used

Aloysia citrodora Palau
Verbenaceae
(Lemon verbena – which see)
Flowers are edible.

Anethum graveolens L.
Apiaceae (Umbelliferae)
(Dill – which see)
Flowers are used fresh or dried for flavoring.

Borago officinalis L.
Boraginaceae
(Borage – which see)
The blue flowers are eaten in salads, preserves, candied, made into syrup or used as garnish. Both flowers and leaves are brewed into tea.

Calendula officinalis L.
Asteraceae (Compositae)
(Marigold, Pot – which see)
Fresh flower petals are chopped and sprinkled on tossed salads. When dried, used for flavoring soups and stews. Both flowers and petals can be used for tea. The commercial fresh flowers are called Calendula.

Calochortus spp.
Liliaceae
(Tulip, Mariposa)
Flowers are eaten raw in tossed salads. The two species of interest are *C. gunnisonii* and *C. nuttallii*.

Centaurea cyanus L.
Asteraceae (Compositae)
(Bachelor's button, Blue bottle, Cornflower)
Flowers are eaten as a vegetable, used as garnish or added to beer. Flowers also yield blue dye for coloring sugar and gelatin.

Chamaemelum nobile (L.) All.
Asteraceae (Compositae)
(Camomile – which see)
Dried flower heads brewed into tea or added to fine liqueurs.

Chrysanthemum coronarium L.
Asteraceae (Compositae)
(Mum, Crowndaisy, Garland chrysanthemum)
Flowers can be used in pickling and salads.

Cucurbita spp.
Cucurbitaceae
(Squash)
Flowers are cooked and eaten, flowers used fresh or dried.

Dendranthema x *grandiflorum* (Ramat.) Kitam.
Asteraceae (Compositae)
(Florist's chrysanthemum, Mum)
Flowers and petals brewed into tea.

Dianthus barbatus L.
Caryophyllaceae
(Sweet william)
The flowers have a mild flavor and are used as garnish for vegetables and fruit salads, cakes, soups, cold drinks and deviled eggs.

Dianthus caryophyllus L.
Caryophyllaceae
(Carnation, Clove pink)
Flower petals smell of cloves and are candied, used as garnish in salads or flavoring fruits, ice cream and vinegars.

Foeniculum vulgare Mill.
Apiaceae (Umbelliferae)
(Fennel – which see)
The stems and flower heads are used as a vegetable.

Fushsia spp.
Onagraceae
(Fuchsia)
Flowers are rather fleshy.

Hedychium coronarium J. Konig
Zingiberaceae
(Ginger, White – which see)
Flowers are used for flavoring.

Hemerocallis fulva (L.) L.
Liliaceae
(Daylily – which see)
Flowerbuds are used in salads or cooked as a vegetable. Dried flowers, called Golden needles, gum-tsoy, or gum-jum, are used in soups and stews.

Lavandula angustifolia Mill.
Lamiaceae (Labiatae)
(Lavender – which see)
Flower petals and flowering tips are added to salads, soups, stews, vinegars, jellies, wine and soft drinks. The flowers can be candied or used as garnish. Fresh and dried flowers are brewed into tea.

Matricaria recutita L.
Asteraceae (Compositae)
(Camomile – which see)
Dried flower heads brewed into tea or added to fine liqueurs.

Matthiola incana (L.) R.Br.
Brassicaceae (Cruciferae)
(Stock, Gillyflower, Brompton stock)
Flowers are eaten as a vegetable or used as a garnish.

Mentha spp.
Lamiaceae (Labiatae)
(Mint – which see)
Spearmint (*M. spicata* L.) leaves and flowers used as flavoring or garnish in salads, etc.

Monarda didyma L.
Lamiaceae (Labiatae)
(Monarda – which see)
Fresh and dried leaves and flower heads used for tea. Fresh flowers used in tossed salads. One cultivar is 'Lavender'. Other common names are **Bee balm**, **Bergamont**, **Indian plume**, **Mountain mint**, **Fragrant balm**, **Oswego tea**.

Origanum spp.
Lamiaceae (Labiatae)
(Marjoram – which see)
Flowering tops used in drinks as flavoring. Flowering tops are dried and brewed into tea. Some dried marjoram is mixed with sumac to form a spice blend called Zatar. Flowers are called Oregano.

Pelargonium graveolens L'Her.
Geraniaceae
(Geranium, Sweet scented geranium, Rose geranium)
Leaves and flowers are eaten in salads.

Perilla frustescens (L.) Britton
Lamiaceae (Labiatae)
(Perilla – which see)
Immature flower clusters used as garnish for soup and chilled tofu. Mature flowers are fried.

Pimpinella saxifraga L.
Apiaceae (Umbelliferae)
(Burnet saxifrage)
Flower heads made into wine.

Rosa spp.
Rosaceae
(Rose)
Hybrid flowers used for flavoring honey, confections and sorbets. Also flavor or garnish teas and salads.

Salvia officinalis L.
Lamiaceae (Labiatae)
(Sage – which see)
Purple and red flowers eaten raw in salads or as garnish.

Tagetes spp.
Asteraceae (Compositae)
(Marigold)
– **Saffron marigold** (*T. erecta* L.) is used as a dried flower source of yellow dye for coloring butter and cheese. Flowers are edible. (Also **marigold** – which see).
– **Sweet marigold** (*T. lucida* Cav.) is used for its dried leaves and flowering tops which are brewed into anise-flavored tea.
– **French marigold** (*T. patula* L.) is used as a dried flower source for saffron.
– **Signet marigold** or **Slenderleaf marigold** (*T. tenuifolia* Cav.) flowers can be used as garnish in salads and sandwiches or added to desserts and wine.

Thymus vulgaris L.

Lamiaceae (Labiatae)

(Thyme – which see)

Leaves and flowering tops (fresh and dried) used for flavoring stuffings, soups, cheeses and meats.

....

Tropaeolum spp.

Tropaeolaceae

(Nasturtium – which see)

(Mashua – which see)

– **Garden nasturtium** (*T. majus* L.) flowers are used in salads, vegetable dishes and as garnish. Flower buds and young fruit can be used as a substitute for capers.

– **Bush nasturtium** (*T. minus* L.) leaves and flowers are eaten in salads or as garnish. Flowers are more abundant then *T. majus*.

– **Mashua** (*T. tuberosum* Ruiz & Pav.) flowers are eaten.

....

Viola spp.

Violaceae

(Violet – which see)

– **Sweet violet** (*V. odorata* L.) flowers are candied, used as garnish, made into syrup and jellies or added to salad dressings.

– **Downy blue violet** (*V. sororia* Willd.) flowers are eaten in salads.

– **Johnny-jump-up** (*V. tricolor* L.) flowers are eaten in salads.

– **Pansy** (*V.* x *wittrockiana* Gams) flowers are eaten in salads.

....

Zingiber mioga (Thunb.) Roscoe

Zingiberaceae

(Mioga – which see)

Flowers are used for flavoring.

3. Season, harvest: At flowering.

4. Production in U.S.: Limited for flower production. California (QUAIL MOUNTAIN 1996).

5. Other production regions: Cosmopolitan.

6. Use: Edible fresh flowers are used for flavoring in salads, cooked as a vegetable, added to drinks or as garnishes. Edible dried flowers are used for seasoning of soups, puddings, cakes and cookies; coloring; or brewed into tea.

7. Part(s) of plant consumed: Flowers (fresh and dried)

8. Portion analyzed/sampled: Flowers (fresh and dried); Primarily fresh but some are dried, see above

9. Classifications:
   a. Authors Class: Herbs and Spices
   b. EPA Crop Group (Group & Subgroup): Miscellaneous
   c. Codex Group: No specific entry
   d. EPA Crop Definition: None

10. References: GRIN, BAILEY 1976, FACCIOLA, MC-CLURE, QUAIL MOUNTAIN (picture), SISSON

11. Production Map: EPA Crop Production Region 10.

12. Plant Codes:
   a. Bayer Code: ALLTU (*A. tuberosum*), ALYTR (*A. citrodora*), AFEGR (*A. graveolens*), BOROF (*B. officinalis*), CLDOF (*C. officinalis*), CENCY (*C. cyanus*), ANTNO (*C. nobile*), CHYCO (*C. cornarium*), CUUSS (*Cucurbita* spp.), DINCA (*D. caryophullus*), DINBA (*D. barbatus*), FOEVU (*F. vulgare*), HEGFU (*H. fulva*), HEYCO (*H. cornarium*), MATCH (*M. recutita*), MTLIN (*M. incana*), MENSS (*Mentha* spp.), PELGV (*P. graveolens*), PRJFR (*P. frustescens*), PIMSA (*p. saxifraga*), ROSSS (*Rosa* spp.), SALOF (*S. officinalis*), TAGER (*T. erecta*), TAGPA

(*T. patula*), THYVU (*T. vulgaris*), TOPMA (*T. majus*), VIOOD (*V. odorata*), VIOTR (*V. tricolor*), VIOWH (*V.* x *wittrockiana*), ZINMI (*Z. mioga*)

246

1. Forage grass
(Tame hay, Yerbas, Rangegrass, Pasture grass, Grass, Pasture-grass)

Poaceae (Gramineae)

Various spp.

2. See specific forage grasses. Rangegrasses include wheatgrass, fescue, bluegrass and wildrye.

3. Season: Warm to cool season grasses.

4. Production in U.S.: Pastureland and rangeland other than cropland and woodland pastured were 410,834,565 acres in 1992 (1992 CENSUS). Harvested cropland was about 300 million acres (1992 CENSUS). In 1992, 4,257,569 acres harvested with over 24 million tons of grass silage, haylage, and green chop hay (tons, green) (1992 CENSUS). In 1992, tame hay other than alfalfa, small grain, and wild hay which included clover, lespedeza, timothy, Bermudagrass, Sudangrass and other types of legumes and tame grasses were grown on about 20 million acres with about 37 million tons of dry hay (1992 CENSUS). The states with a million or more harvested acres of tame hay were Texas (2,832,033 acres), Missouri (2,248,566), Kentucky (1,306,116) Tennessee (1,093,736) and Oklahoma (1,069,655) (1992 CENSUS). In 1992, the top 10 states reported grass silage, haylage and green chop hay as Wisconsin (1,166,659 acres harvested), New York (446,649), Pennsylvania (289,130), Michigan (218,274), Minnesota (185,896), California (141,014), Vermont (136,000), Iowa (119,707), Texas (118,504) and Ohio (113,877) (1992 CENSUS).

5. Other production regions: Canada and Mexico.

6. Use: Pasture, rangeland, hay, silage, fodder, erosion control.

7. Part(s) of plant consumed: Foliage (leaves and stems)

8. Portion analyzed/sampled: Forage and hay

Grass: Zero day crop field residue data for grasses cut for forage should be provided unless it is not feasible, e.g., preplant/preemergent pesticide uses. A reasonable interval before cutting for hay is allowed.

Grass forage: Cut sample at 6-8 inch to boot stage, at approximately 25 percent DM.

Grass hay: Cut in boot to early head stage. Hay should be field-dried to a moisture content of 10 to 20 percent. Grasses include barnyardgrass, bentgrass, bermudagrass, Kentucky bluegrass, big bluestem, smooth bromegrass, buffalograss, reed canarygrass, crabgrass, cupgrass, dallisgrass, sand dropseed, meadow foxtail, eastern gamagrass, side-oats grama, guineagrass, Indiangrass, Johnsongrass, lovegrass, napiergrass, oatgrass, orchardgrass, pangolagrass, redtop, Italian ryegrass, sprangletop, squirreltailgrass, stargrass, switchgrass, timothy, crested wheatgrass and wildryegrass. Also included are sudangrass and sorghum forages and their hybrids. For grasses grown for seed only: Residue data for grass straw (plant material remaining in field after harvest of seeds) and seed screenings should be provided only for uses on grass grown for seed. A label restriction against the feeding or grazing of directly treated forage is considered practical for uses on grass grown for seed. In such cases, PGIs (pregrazing intervals) and PHIs (preharvest intervals) should be included in the use directions and the residue data for forage and hay may be based on the regrowth after the seed crop has been harvested. If a pesticide is to be used on pasture/range grass in addition to grass grown for seed, the forage (usually zero day) and hay data for the former will cover these two commodities for grass grown for seed (provided the application rates on the latter are not higher). In such cases, only residue data for straw and seed screenings should be provided for the use on grass grown for seed.

Grass silage: Residue data on silage are optional, but are desirable for assessment of dietary exposure. Cut sample at boot to early head stage, allow to wilt to 55 to 65 percent moisture, then chop fine, pack tight, and allow to ferment for three weeks maximum in an air-tight environment until it reaches pH 4. In the absence of silage data, residues in forage will be used for silage, with correction for dry matter.

9. Classifications:
   a. Authors Class: Grass forage, fodder and hay
   b. EPA Crop Group (Group & Subgroup): Grass forage, fodder and hay (17). Representative crops are Bermudagrass, Bluegrass; and Bromegrass or Fescue
   c. Codex Group: 051 (AS 0162) Hay or fodder (dry) of grasses
   d. EPA Crop Definition: None, except sorghum, which see
10. References: GRIN, CODEX, RICHE, SCHREIBER, US EPA 1996a, US EPA 1995a, US EPA 1994, BARNES, BARNES (a)
11. Production Map: EPA Crop Production Regions: Grasses – field trial distribution for all areas across country. Regions with significant forage grass harvested of greater than 100,000 acres are 1, 2, 5, 6 and 10. Regions with over a million acres of harvested tame hay except alfalfa are 2, 4, 5 and 6.
12. Plant Codes:
    a. Bayer Code: No specific entry

247

1. Galangal
(Greater galangal, Lesser galangal, Rieng, Gieng, Laos powder, China root)
Zingiberaceae
Alpinia officinarum Hance
(Lesser galangal)
A. galanga (L.) Sw.
(Greater galangal, Languas)

2. These two species, respectively known as lesser and greater galangal are tropical perennials, cultivated mainly for the underground rhizomes. The plants and rhizomes are similar to ginger, which see. The tuberous rhizomes are the spice used in making vinegar and beer and in liquors, especially in Russia. The spice is also used in curries. The ground dried root is termed "Laos powder".
3. Season: Propagated by division in the spring.
4. Production in U.S.: No data.
5. Other production regions: Asia.
6. Use: Spice.
7. Part(s) of plant consumed: Root
8. Portion analyzed/sampled: Root (fresh and dried)
9. Classifications:
   a. Authors Class: Herbs and spices
   b. EPA Crop Group (Group & Subgroup): Miscellaneous

c. Codex Group: 016 (VR 0581)(VR 0582) Root and tuber vegetables; 028 (HS 0783) Spices
   d. EPA Crop Definition: None
10. References: GRIN, CODEX, GARLAND (picture), KUEBEL, MAGNESS, NEAL, PENZEY
11. Production Map: No entry.
12. Plant Codes:
    a. Bayer Code: No specific entry

248

1. Galleta grass
(Galleta)
Poaceae (Gramineae)
Hilaria jamesii (Torr.) Benth.
1. Tobosa grass
H. mutica (Buckley) Benth.

2. Both of these native species are slightly spreading range grasses. Galleta grass occurs from Wyoming to California and west Texas while Tobosa grass is in west Texas and Arizona and south into Mexico. The bases of the stems of both are rhizome-like. Stems may reach to 2 feet, with small, narrow leaves. Both are found under arid conditions and are highly drought resistant. They are moderately palatable while succulent, but not when dry. Produces good to fair quality forage for cattle and horses.
3. Season: Grows under arid conditions as perennial grass in late spring to summer.
4. Production in U.S.: No specific data, See Forage grass. Wyoming to California and west Texas.
5. Other production regions: South into Mexico.
6. Use: Rangegrass.
7. Part(s) of plant consumed: Foliage
8. Portion analyzed/sampled: Forage and hay
9. Classifications:
   a. Authors Class: See Forage grass
   b. EPA Crop Group (Group & Subgroup): See Forage grass
   c. Codex Group: See Forage grass
   d. EPA Crop Definition: None
10. References: GRIN, MAGNESS, STUBBENDIECK
11. Production Map: EPA Crop Production Regions 8, 9 and 10.
12. Plant Codes:
    a. Bayer Code: HILMU (*H. mutica*)

249

1. Garlic
(Ajo, Alho)
Liliaceae (Amaryllidaceae)
Allium sativum L. var. *sativum*
1. Serpent garlic
(Rocambole)
A. sativum var. *ophioscorodon* (Link) Doll

2. The garlic plant is similar to onion, except it produces a group of small bulbs, called cloves, all enclosed in thin papery scales, instead of a single bulb. The leaves reach about 12 inches in height, and are narrow, but not hollow. Plants are usually produced by planting a clove, or a bulblet that forms in the flower head. All commercial plantings in the U.S. is in areas of mild-winter climate, mainly in California. Cloves are planted in October to January, and harvest is in mid- to late-summer. Bulb development is below the soil

surface. The strongly scented and flavored bulbs are used mainly for flavoring meats, stews and soups. The common names Rocambole and Serpent garlic are applied to garlic varieties having coiled or twisted scapes, the flower stalks.

3. Season, planting to harvest: About 8 months.
4. Production in U.S.: About 21,179 acres in U.S. (1992 CENSUS). California with about 88 percent of production, 7 percent in Oregon and 3 percent in Nevada (1992 CENSUS). In 1994, California harvested 27,594 acres of garlic (MELNICOE 1996e).
5. Other production regions: Origin – central Asia with Asia leading the world in production. China produces over 50 percent of the world supply (RUBATZKY 1997).
6. Use: Dried and ground as powder, fresh, used mainly as flavoring in other foods. U.S. per capita consumption in 1994 was 2.0 pounds (PUTNAM 1996).
7. Part(s) of plant consumed: Cloves – small bulbs enclosed in scales during growth
8. Portion analyzed/sampled: Bulb. 40 CFR 180.1(J)(5) states that roots, stems and outer sheaths (or husks) shall be removed and discarded from garlic bulbs and dry bulb onions, and only the garlic cloves and onion bulbs shall be examined for pesticide residues
9. Classifications:
   a. Authors Class: Bulb vegetables
   b. EPA Crop Group (Group & Subgroup): Bulb vegetables
   c. Codex Group: 009 (VA 0381) Bulb vegetables
   d. EPA Crop Definition: Onions = Garlic; Dry bulb onions = Garlic
10. References: GRIN, CODEX, KAYS, MAGNESS, RICHE, US EPA 1995b, US EPA 1995a, US EPA 1994, MELNICOE 1996e, RUBATZKY, PUTNAM 1996
11. Production Map: EPA Crop Production Regions 9, 10 and 11.
12. Plant Codes:
    a. Bayer Code: ALLSA

250

1. Garlic, Great-headed
(Elephant garlic, Suan, Suen, Suahn, Puerro agreste, Levant garlic, Wild leek, Da tou suan)
Liliaceae (Amaryllidaceae)
Allium ampeloprasum L. var. *ampeloprasum*

2. Plants have appearance of very robust garlic plants with very flat leaves that resemble leeks. Great-headed garlic forms large flower heads which usually lack bulblets. It may produce a cluster of several cloves similar to garlic, or a single massive bulb, with small bulblets around its main bulb. The bulb flavor is intermediate between onion and garlic. Cultural practices are similar to those for garlic. The bulbs are developed entirely underground. Great-headed garlic is grown in many home gardens, but rarely as a commercial crop.
3. Season, setting to harvest: About 8 months.
4. Production in U.S.: No data separate from garlic. Mainly in home gardens.
5. Other production regions: Grown in temperate regions (RUBATZKY 1997).
6. Use: Culinary, and as part of stews and soups.
7. Part(s) of plant consumed: Bulbs and leaves
8. Portion analyzed/sampled: Bulb
9. Classifications:

   a. Authors Class: Bulb vegetables
   b. EPA Crop Group (Group & Subgroup): Bulb vegetables
   c. Codex Group: 009 (VA 0382) Bulb vegetables
   d. EPA Crop Definition: None
10. References: GRIN, CODEX, KAYS, MAGNESS, MYERS, US EPA 1995a, RUBATZKY
11. Production Map: No entry.
12. Plant Codes:
    a. Bayer Code: ALLAM

251

1. Geranium, Scented
Geraniaceae
Pelargonium spp.
1. Lemon geranium
P. crispum (P. J. Bergius) L'Her.
1. Rose geranium
(Sweetscented geranium, Rober's lemon rose)
P. graveolens L'Her.

2. They are tender woody perennials that are native to South Africa, and can grow to 3 feet tall in full sun. The crushed leaves of *P. crispum* are used to flavor soups, poultry, fish, sauces, fruit dishes and vinegar. Cultivars for *P. crispum* are 'Lemon Crispum', 'Fingerbowl Geranium', 'Limoneum', 'Orange', 'Citronella', 'Prince of Orange', and 'Prince Rupert'. Leaves and flowers of *P. graveolens* are used to flavor soups, sugar, and desserts. Flowers can be eaten in salads, and leaves are brewed into tea. Some *P. graveolens* cultivars are 'Dr. Livingston' and 'Rober's Lemon Rose'.
3. Season, propagation: Take cuttings in spring or fall to grow overwinter indoors. Flowers in the summer and fall.
4. Production in U.S.: No data.
5. Other production regions: Native to South Africa (MCCLURE 1995).
6. Use: Leaves are used in herb sugars, herb butters, teas, cakes and soups. Flowers used in salads.
7. Part(s) of plant consumed: Leaves and flowers
8. Portion analyzed/sampled: Leaves (fresh and dried)
9. Classifications:
   a. Authors Class: Herbs and spices
   b. EPA Crop Group (Group & Subgroup): Miscellaneous
   c. Codex Group: No specific entry
   d. EPA Crop Definition: None
10. References: GRIN, BAILEY 1976, BAYER, FACCIOLA, MCCLURE (picture)
11. Production Map: EPA Crop Production Region: No entry
12. Plant Codes:
    a. Bayer Code: PELGV (*P. graveolens*), PELSS (*P.* spp.)

252

1. Gherkin, West Indian
(Gherkin, Cohombro, Pepinillo, Burr cucumber, Pepino, Jerusalem cucumber, Horned cucumber, Gooseberry gourd, Burr gherkin, Xi yin du huang gua)
Cucurbitaceae
Cucumis anguria L. var. *anguria*

2. This annual plant is similar to its close relative, the cucumber. The fruit is generally oval in shape, 1 to 3 inches long, much more warty than cucumber. It is used mainly for pickling. Plants continue to blossom and set fruits throughout the

summer. The common "gherkins" used in pickles are immature cucumbers, *C. sativus*.

3. Season, seeding to first harvest: 2 to 2¹/₂ months.
4. Production in U.S.: No data. Grown in home gardens in Florida (STEPHENS 1988).
5. Other production regions: Introduced from Africa and grown in warm temperate, subtropical regions (RUBATZKY 1997).
6. Use: Mainly pickling. Can be used in salads or cooked.
7. Part(s) of plant consumed: Whole fruit
8. Portion analyzed/sampled: Whole fruit
9. Classifications:
   a. Authors Class: Cucurbit vegetables
   b. EPA Crop Group (Group & Subgroup): Cucurbit vegetables (9B)
   c. Codex Group: 011 (VC 0426) Cucurbits fruiting vegetable
   d. EPA Crop Definition: None
10. References: GRIN, BAILEY 1976, CODEX, MAGNESS, REHM, STEPHENS 1988 (picture), US EPA 1995a, YAMAGUCHI 1983 (picture), RUBATZKY
11. Production Map: EPA Crop Production Regions 3 and 13.
12. Plant Codes:
    a. Bayer Code: CUMAN

253

1. Giant cane
(Bamboo, Canebrake bamboo, Southern cane)
Poaceae (Gramineae)
Arundinaria gigantea (Walter) Muhl.

2. This is a perennial, cool season native grass belonging to the grass tribe *Bambuseae*. It occurs on moist soils along streams and woodlands. Giant cane grows erect to a height of 8 m, with blades 2-4 cm wide with reproduction from rhizomes which form dense colonies. Giant cane begins growth in March, flowers from a panicle in April and May, every 4 to 6 years. Propagation by stem pieces or rhizomes. Young rhizomes are used as a vegetable in Asia. It produces valuable forage for cattle and wildlife, especially in winter and spring and can be grazed all year long. Also see Bamboo for food uses.
3. Season: Cool season forage and grows rapidly from early spring through the early fall. It can be grazed all year.
4. Production in U.S.: No data for Giant cane production, however, pasture and rangeland are produced on more than 410 million A (U.S. CENSUS, 1992). Generally southeastern U.S. from Ohio and Missouri to Florida and eastern Texas (GOULD 1975).
5. Other production regions: Southeastern Asia (GOULD 1975).
6. Use: Grazing and browsing and for erosion control.
7. Part(s) of plant consumed: Leaves and young stems
8. Portion analyzed/sampled: Forage and hay
9. Classifications:
   a. Authors Class: See Forage grass
   b. EPA Crop Group (Group & Subgroup): See Forage grass
   c. Codex Group: See Forage grass
   d. EPA Crop Definition: None
10. References: GRIN, US EPA 1994, US EPA 1995, CODEX, STUBBENDIECK (picture), RICHE, GOULD
11. Production Map: EPA Production Regions 2, 3, 4, 5 and 6.
12. Plant Codes:
    a. Bayer Code: No specific entry

254

1. Ginger
(Chinese ginger, Japanese ginger, Jengibre, Gingembre, Amome, Jiang, Shioga, True ginger, Common ginger, Canton ginger, Commercial ginger, Jamaica ginger, Keong, Green ginger)
Zingiberaceae
Zingiber officinale Roscoe

2. Ginger is a biennial or perennial reed-like herb, grown for the pungent, spicy underground stems or rhizomes. The stems reach a height of 3 feet, with lanceolate, smooth leaves up to 8 inches long. The plants are propagated by small divisions of the rhizomes. Two varieties are grown in Hawaii, the small root is usually referred to as Japanese ginger and the large root which is predominant is usually referred to as Chinese ginger. A crop of rhizomes can be harvested approximately a year after planting. After harvesting, the rhizomes may be cleaned, washed and dried directly, or they may be peeled before drying. Preserved ginger is prepared from immature rhizomes (young ginger or green ginger with low fiber content) by washing, boiling successively in sugar and water and placed in containers in syrup, or dried and rolled in sugar. Ginger is used as a spice or condiment especially in carbonated beverages. Oil of ginger is also extracted from the rhizomes. Ginger is produced in many tropical countries, and is grown in Florida; produced commercially in Hawaii with 4,439 tons on 253 acres (1987).
3. Season, planting to harvest: Plant in early spring with mature roots harvested in approximately a year (usually 8-10 months).
4. Production in U.S.: 1992 CENSUS reported 325 acres harvested in U.S. Ginger is produced in Florida and Hawaii with the majority in Hawaii and concentrated in the Hilo area.
5. Other production regions: Grown in subtropical and tropical regions (RUBATZKY 1997).
6. Use: Spice or condiment, especially in carbonated beverages.
7. Part(s) of plant consumed: Root (Rhizome). Harvested roots are washed and dried (cured for 3-5 days)
8. Portion analyzed/sampled: Dried root
9. Classifications:
   a. Authors Class: Root and Tuber Vegetables
   b. EPA Crop Group (Group & Subgroup): Root and Tuber Vegetables (Subgroups 1C and 1D)
   c. Codex Group: Ginger root: 028 (HS 0784)
   d. EPA Crop Definition: None
10. References: GRIN, CODEX, MAGNESS, NISHINA (picture), US EPA 1995a, RUBATZKY, RICHE, NEAL
11. Production Map: EPA Crop Production Region 13.
12. Plant Codes:
    a. Bayer Code: ZINOF

255

1. Ginger, White
(Garlandflower, Butterfly ginger, Ginger lily, Cinnamon jasmine, Butterfly lily, Longouze, Mariposa, Perlas de Orient)
Zingiberacea
Hedychium coronarium J. Konig

2. Perennial herb to 6 feet tall with leaves to 2 feet long and 5 inches wide. Flower bracts large and firm. White flowers are very fragrant. Prized for showy flowers and fragrance. White

ginger requires rich soil and plenty of water. It was introduced to Hawaii in the late 1800s.

3. Season, propagation: By division of rhizomes.

4. Production in U.S.: No data. Grows in Hawaii (USDI 1997).

5. Other production regions: Tropical Asia (FACCIOLA 1990) and South America (BAYER 1992).

6. Use: Young buds and flowers are eaten or used for flavoring.

7. Part(s) of plant consumed: Flowers

8. Portion analyzed/sampled: Flowers (fresh)

9. Classifications:
   a. Authors Class: Herbs and spices
   b. EPA Crop Group (Group & Subgroup): Miscellaneous
   c. Codex Group: None
   d. EPA Crop Definition: None

10. References: GRIN, BAILEY 1976, BAYER, FACCIOLA, USDI 1997

11. Production Map: EPA Crop Production Region 13.

12. Plant Codes:
   a. Bayer Code: HEYCO

256

1. Ginkgo

(Maidenhair tree, Icho, Ya-chiao-tzu, Ginkgo nut, Ginnan, Pai-kua)

Ginkgoaceae

Ginkgo biloba L.

2. Ginkgo is an ancient, deciduous, dioecious tree to 120 feet tall which produces fruits after they are 30 years old. The fruits resemble plums up to 1 inch in diameter. The pulp is foul smelling. The fruit contain one seed which is about $^3/_4$ inch in diameter. After the pulp is removed, the seeds or nuts are washed and sun-dried. In U.S., ginkgos are mostly grown as ornamentals, whereas, in Japan and China, the nutmeat is enjoyed as an appetizer or cooked with meat. The fan-shaped leaves are used for medicinal purposes.

3. Season, harvest: Seeds harvested in the fall, roasted and eaten as a seasonal delicacy.

4. Production in U.S.: No data. Hardy through USDA Hardiness Zone 5 (BAILEY 1976).

5. Other production regions: Southeast China and Japan. (BAILEY 1976, SCHNEIDER 1993a).

6. Use: Canned or dried seeds are boiled and used in soups. Seeds are boiled or roasted. The kernels are low in fat (2.6 percent) and taste like swiss cheese. Leaves are used for medicinal purposes.

7. Part(s) of plant consumed: Seed (nut) and leaves

8. Portion analyzed/sampled: Seed (fresh and dried); Leaves (dried)

9. Classifications:
   a. Authors Class: Tree nuts; Herbs and spices
   b. EPA Crop Group (Group & Subgroup): Miscellaneous
   c. Codex Group: No specific entry
   d. EPA Crop Definition: None

10. References: GRIN, BAILEY 1976, FACCIOLA, SCHNEIDER 1993a

11. Production Map: EPA Crop Production Region: U.S. except northern Great Plains.

12. Plant Codes:
   a. Bayer Code: GIKBI

257

1. Ginseng

(American ginseng, Sang)

Araliaceae

Panax quinquefolius L.

2. Ginseng is a perennial herbaceous plant that is cultivated for its odd-shaped root. The root is a fleshy light-yellow, taproot, often with 2 to 5 lateral branches. It is grown either in the woods or cultivated under artificial shade. Commercially cultivated ginseng is harvested after 3 or 4 years. Harvesting may start as early as July with the most active harvest period as September to October. Roots are harvested by hand or with a potato digger. The fresh roots (approximately 30 percent dry matter) are dried to approximately 90 percent dry matter. Drying is accomplished by air or heat. *P. ginseng* C.A. Mey. is the Chinese or Korean ginseng; *P. vietnamensis* Ha & Grushv. is the Vietnamese ginseng; and *Eleutherococcus senticosis* is the Siberian ginseng. American ginseng is highly favored by the Chinese.

3. Season, seeding to harvest: Cultivated under artificial shade 3-4 years. To reach a desirable size, wild roots may need 20 years. The seeds alone need 18 months to germinate. Can also be propagated by seedlings and roots.

4. Production in U.S.: 1992 CENSUS listed 1,505 acres harvested in U.S. with 1,428 in Wisconsin. Wisconsin growers cultivate approximately 5000 acres of ginseng each year with approximately one-third harvested each year. Washington State grew 20 acres in 1994. The remainder in Kentucky, Tennessee, North Carolina, Virginia and West Virginia. In 1995, 973,160 kg of cultivated ginseng and 104,329 kg of wild ginseng exported from U.S. to Hong Kong.

5. Other production regions: In 1995 Canada listed 237 acres, mainly in Ontario and British Columbia (ROTHWELL 1996a).

6. Use: The root can be sold as dried, powder, sliced or extract and are used as flavoring in teas, candy and drinks. It is normally mixed with other ingredients before consumed.

7. Part(s) of plant consumed: Dried Root

8. Portion analyzed/sampled: Dried Root Only

9. Classifications:
   a. Authors Class: Root and tuber vegetables
   b. EPA Crop Group (Group & Subgroup): Root and Tuber Vegetables (Subgroups 1A and 1B)
   c. Codex Group: No specific entry
   d. EPA Crop Definition: None

10. References: GRIN, DUKE(d), PARKE, PROCTOR, RICHE, ROTHWELL 1996a, SCHREIBER, USDA 1995, USDA 1994b, US EPA 1995a, US EPA 1994, WILLIAMS(a) (picture), FOSTER

11. Production map: EPA Crop Production Regions 1, 2, 4 and 5 (Major cultivated ginseng region is 5).

12. Plant Codes:
   a. Bayer Code: PNXGI (*P. ginseng*), PNXQU (*P. quinquefolius*)

258

1. Globemallow

Malvaceae

Sphaeralcea spp.

1. Munroe globemallow

(Munro's globemallow)

S. munroana (Douglas ex Lindl.) Spach ex A. Gray

1. Scarlet globemallow

(Prairie mallow)

S. coccinea (Nutt.) Rydb.

2. Perennial herbs from 4 to 20 inches tall with leaves palmately lobed and native to the western U.S. They are drought resistant. Munroe globemallow when included with adapted grasses can be used for range revegetation and is native to southern British Columbia, southwestern Montana and Wyoming, Utah, Nevada and California. Scarlet globemallow is widely distributed in the Rocky Mountains and Great Plains rangelands of western U.S. It is an important rangeland dietary component of small mammals, pronghorn, sheep and cattle.
3. Season, harvest: Rangeland grazing by sheep and cattle.
4. Production in U.S.: No data. Native to the western U.S. from Arizona to California.
5. Other production regions: Arid North and South America (BAILEY 1976).
6. Use: Tops for rangeland forage as sheep and cattle feed.
7. Part(s) of plant consumed: Tops
8. Portion analyzed/sampled: Tops
9. Classifications:
   a. Authors Class: Nongrass animal feeds
   b. EPA Crop Group (Group & Subgroup): Miscellaneous
   c. Codex Group: No specific entry
   d. EPA Crop Definition: None
10. References: GRIN, CHATTERTON 1996, USDA NRCS, BAILEY 1976
11. Production Map: EPA Crop Production Regions 9 and 10.
12. Plant Codes:
    a. Bayer Code: SPHCO (*S. coccinea*)

259

1. Good-King-Henry

(Mercury, Markery, Fathen, Wild spinach, Perennial goosefoot, Allgood, Lincolnshire asparagus)

Chenopodiaceae

Chenopodium bonus-henricus L.

2. The plant is a stout, erect herb, up to 2¹/₂ feet, with broad, triangular or ovate leaves. Leaves have wide spreading basal points, and are entire or undulate. Plants are sparingly cultivated for the leaves, used as potherbs. The plant is similar to spinach both in general growth habit and use.
3. Season, seeding to first harvest: about 2 months.
4. Production in U.S.: No data. Apparently not commercial.
5. Other production regions: Grown in temperate regions (RUBATZKY 1997).
6. Use: As potherb for leaves, young shoots and flowers. Young shoots used as asparagus.
7. Part(s) of plant consumed: Leaves, young stems and tender flower clusters
8. Portion analyzed/sampled: Tops
9. Classifications:
   a. Authors Class: Leafy vegetables (except *Brassica* leafy vegetables)
   b. EPA Crop Group (Group & Subgroup): Miscellaneous
   c. Codex Group: No specific entry
   d. EPA Crop Definition: None
10. References: GRIN, FACCIOLA, MAGNESS, RUBATZKY
11. Production Map: No entry

12. Plant Codes:
    a. Bayer Code: CHEBH

260

1. Gooseberry

(Crossberry)

Grossulariaceae (Saxifragaceae)

Ribes spp.

1. European gooseberry

(Groseillier epineux, Grosellero espinoso, English gooseberry)

R. uva-crispa L. (syn: *R. grossularia* L.)

1. Hairy gooseberry

R. hirtellum Michx.

2. Most domestic cultivars of gooseberry are hybrids of the American and European types (European types susceptible to mildew, native types yielded small fruit). Plants are small, deciduous, woody shrubs, up to 4 to 5 feet in height, with prominent thorns at the nodes. They are very hardy to cold. The plant blooms early in the spring and flowers are borne on one-year-old wood and fruit spurs. The fruits are produced along the stems singly or in small groups of 2 to 4. Fruits generally are near globose, with dried flower parts adhering to maturity. Fruit surface is somewhat pubescent, green to red in color. The pulp encloses several small seeds. Fruit size is ¹/₂ to near 1 inch diameter. Flavor is generally tart.
3. Season: Bloom to maturity in 60 to 80 days.
4. Production in U.S.: Oregon, with limited production in New York, Minnesota, North Dakota and Washington. No statistics available.
5. Other production regions: New Zealand.
6. Use: Frozen, canned for culinary. Green berries must be cooked. Ripe berries can be used for desserts.
7. Part(s) of plant consumed: Fruit
8. Portion analyzed/sampled: Whole fruit
9. Classifications:
   a. Authors Class: Berries
   b. EPA Crop Group (Group & Subgroup): Crop Group 13: Berries Group as well as this group's Bushberry Subgroup
   c. Codex Group: 004 (FB 0268) Berries and other Small Fruit
   d. EPA Crop Definition: None
10. References: GRIN, CODEX, LOGAN, MAGNESS, REICH, SCHREIBER, SHOEMAKER, US EPA 1994, US EPA 1995
11. Production Map: EPA Crop Production Region 12.
12. Plant Codes:
    a. Bayer Code: RIBHI (*R. hirtellum*), RIBUC (*R. uva-crispa*)

261

1. Gourd, Edible

Cucurbitaceae

Lagenaria spp.

Luffa spp.

1. Gourd, Bottle

(Bau, Calabash gourd, Yugao, Pogua, Upo, Trumpet gourd, Calabash gourd, White-flowered gourd, Cucuzzi, Suzza melon, Zucca, Italian edible gourd, Tasmania bean, Guinea bean, New Guinea bean, Hyotan, Mokwa, Kashi, Botella, Calebasse, Cucuzza squash)

Lagenaria siceraria (Molina) Standl. (syn: *L. leucantha* (Duchesne) Rusby; *L. vulgaris* Ser.; *Cucurbita lagenaria*)

1. Loofah, Angled
(Chinese okra, California okra, Sinkwa towel gourd, Sinkwa, Sing gwa, Togado Hechima, Calabaza de aristas, Lufa, Purpengaye, Seequa, Vine okra, Sinqua, Ribbed gourd, Ribbed luffa, Silky gourd, Muop khia, Tatsu kua)
Luffa acutangula (L.) Roxb.

1. Loofah, Smooth
(Vegetable sponge gourd, Sponge gourd, Dishcloth gourd, Rag gourd, Hechima, Calabaza de aristas, Esponja, Eponge vegetable, Sze kwa, Dishrag gourd, Strainer vine, Vegetable sponge, Smooth luffa, Bark qua, Muop huong)
Luffa aegyptiaca Mill. (syn: *L. cylindrica* M. Roem.)

2. Plants are running vines, similar to cucumber and melon. Leaves are rounded. Fruits are strongly ribbed, elongated, pyriform or cylindrical in shape, 1 foot or more in length, gourd-like. Sometimes young fruits are consumed as cooked vegetables, like summer squash. When ripe, shell is hard and interior is fibrous, sponge-like and inedible. Bottle gourd is an edible gourd, often classed as a summer squash, but not a true squash. The plant is a vining annual, with large, pubescent leaves on long petioles which form a canopy over the stems and fruits. Fruits are up to 2 to 3 feet long and 3 inches in diameter, light green in color, with a smooth skin. They are harvested when the skin and flesh are tender. Similar to vining type summer squash in growth habit. *Lagenaria* and *Luffa* are included together because similar in form, culture and uses. Chinese call the immature fruit of the bottle gourd, moqua, as are immature waxgourd fruit.
3. Season, seeding to harvest as vegetable: About 2 to 3 months. (Mature fruits require about 4 to 5 months).
4. Production in U.S.: No data. Very limited. Grown in Florida (STEPHENS 1988), Louisiana (BLACK 1988) and California (LOGAN 1996).
5. Other production regions: Tropics (MARTIN 1983).
6. Use: Immature fruits as cooked vegetable like summer squash. A vegetable curd similar to soybean tofu can be made from the seeds.
7. Part(s) of plant consumed: All of young fruits
8. Portion analyzed/sampled: Whole fruit
9. Classifications:
   a. Authors Class: Cucurbit vegetables
   b. EPA Crop Group (Group & Subgroup): Cucurbit vegetables (9B). In crop group 9 edible gourd = *Lagenaria* spp., *Luffa acutangula* and *Luffa cylindrica*
   c. Codex Group: **Bottle gourd** 011 (VC 0422); **Angled loofah** 011 (VC 0427); **Smooth loofah** 011 (VC 0428) Fruiting vegetables, Cucurbits
   d. EPA Crop Definition: Summer squash = *Lagenaria* spp. and *Luffa* spp. Squash = summer squash
10. References: GRIN, BAILEY 1976, BLACK, CODEX, KAYS, LOGAN, MAGNESS, MARTIN 1983, NEAL, STEPHENS 1988 (picture), YAMAGUCHI 1983 (picture), MYERS
11. Production Map: EPA Crop Production Regions 3, 4, 10 and 13.
12. Plant Codes:
    a. Bayer Code: LGNSI (*Lagenaria siceraria*), LUFAC (*Luffa acutangula*), LUFAE (*Luffa aegyptiaca*)

262

1. Governor's plum
(Ciruela de Madagascar, Ramontchi)

Flacourtiaceae
Flacourtia indica (Burm. f.) Merr. (syn: *F. ramontchi* L'Her.)
2. Tropical shrub or small tree, sometimes sparsely armed with sharp, stout thorns. Fruits are subglobose and may be slightly over 1 inch in diameter. The fruits are purplish red or blackish, and are surmounted by the remains of the pistils. The reddish juicy pulp encloses 8 to 10 small seeds. The fruits may be eaten fresh or used in jams and jellies.
3. Season, bloom to harvest: 60-90 days.
4. Production in U.S.: No data. Home gardens (MARTIN 1987). Florida (MORTON 1987).
5. Other production regions: West Indies (MORTON 1987).
6. Use: Jelly, jams and fresh.
7. Part(s) of plant consumed: Fruit pulp
8. Portion analyzed/sampled: Whole fruit
9. Classifications:
   a. Authors Class: Tropical and subtropical fruits – inedible peel
   b. EPA Crop Group (Group & Subgroup): Miscellaneous
   c. Codex Group: No specific entry
   d. EPA Crop Definition: None
10. References: GRIN, MAGNESS, MARTIN 1987, MORTON (picture)
11. Production Map: EPA Crop Production Region 13.
12. Plant Codes:
    a. Bayer Code: No specific entry

263

1. Gow Kee
(Chinese wolfberry, Chinese matrimony vine, Chu Chi, Matrimony vine, Box thorn, Kichi, Kau-kei, Chinese boxthorn)
Solanaceae
Lycium chinense Mill.
2. This plant is listed under vegetables because the tender leaves and shoots are used as potherbs. The plant is a shrub with slender, arching branches which may reach up to 12 feet in length. Leaves are ovate to lanceolate, up to 3 inches long, bright green in color. Fruits are oblong, up to near 1 inch in length. They are edible but rather tasteless. They are said to be eaten in Asia. In the U.S., the plant is grown to a limited extent by Oriental gardeners for the leaves and tender stems.
3. Season, harvest: Cut as needed.
4. Production in U.S.: No data.
5. Other production regions: Native to eastern Asia (NEAL 1965).
6. Use: Potherb. Leaves used for flavorings, especially in soup with pork.
7. Part(s) of plant consumed: Leaves and tender stems
8. Portion analyzed/sampled: Leaves and tender stems
9. Classifications:
   a. Authors Class: Leafy vegetables (except *Brassica* vegetables)
   b. EPA Crop Group (Group & Subgroup): Miscellaneous
   c. Codex Group: 013 (VL 0462) Leafy vegetables (including *Brassica* leafy vegetables)
   d. EPA Crop Definition: None
10. References: GRIN, CODEX, MAGNESS, MARTIN 1979, NEAL
11. Production Map: No entry
12. Plant Codes:
    a. Bayer Code: LYUCN

264

1. Grains of Paradise
(Guinea grains, Melegueta pepper, Alligator pepper)
Zingiberaceae

Aframomum melegueta K. Schum.

2. Grains of paradise are the aromatic, pungent seeds of this species, native to western Africa. The plants are tropical perennials, spreading by rhizomes and forming dense clumps. Plants attain heights of 4 to 5 feet, with lanceolate, glabrous leaves up to 9 inches long. Seeds are borne in ovoid or flask-shaped, pubescent capsules 2 to 3 inches long. Seeds were formerly used as a substitute or adulterant of pepper. Now they are sometimes used in wine, beer, spirits and vinegars. They are not produced, so far as known, in the U.S.
3. Season: Propagated by division in the spring.
4. Production in U.S.: No data.
5. Other production regions: Tropics.
6. Use: Spice.
7. Part(s) of plant consumed: Aromatic seeds
8. Portion analyzed/sampled: Seeds
9. Classifications:
   a. Authors Class: Herbs and spices
   b. EPA Crop Group (Group & Subgroup): Herbs and spices (19B)
   c. Codex Group: 028 (HS 0785) Spices
   d. EPA Crop Definition: None
10. References: GRIN, CODEX, MAGNESS, NEAL, US EPA 1995a
11. Production Map: No entry
12. Plant Codes:
    a. Bayer Code: No specific entry

265

1. Grama grass
Poaceae (Gramineae)

Bouteloua spp.

1. Side-oats grama
B. curtipendula (Michx.) Torr.

1. Black grama
(Navajita negra)
B. eriopoda (Torr.) Torr.

1. Blue grama
(Navagita)
B. gracilis (Kunth) Lag. ex. Griffiths (syn: *B. oligostachya* (Nutt.) Torr. ex. A. Gray)

2. Some 18 species of *Bouteloua*, the grama grasses, are native in the U.S., mainly throughout the Great Plains and Western states and north to Canada. They are summer growers, and the amount of growth produced is dependent on available moisture. Most species cure naturally, so growth from a previous season is palatable for livestock. They are prized as forage producers on range and pastureland. The three most valuable species are discussed as follows.

Side-oats grama is a long-lived warm-season native bunchgrass, widely distributed, but most abundant in the central and southern Great Plains. It produces short rhizomes and tends to a bunch-type growth. The leaves are about 6 inches long and under $1/4$ inch wide. Flower stems may reach to 3 feet. It produces an abundance of leafy forage well liked by all classes of livestock. Hay of good quality is produced if mowed sufficiently early. It is adapted to wide ranges of soil and climate. Seedling vigor is good, and stands are readily established by seeding. Generally, side-oats grama is seeded in mixtures with other grasses. Several varieties of superior local adaptation are in the trade.

Black grama is a major native grass of the semiarid to arid areas from west Texas westward to California. The stems, when in contact with soil, form roots at the nodes to form other nearby plants. Stems are slender and wiry, up to 2 feet. Plants are leafy and highly palatable with good feeding value both summer and winter. They are highly drought resistant.

Blue grama is a long-lived native perennial grass that grows throughout the Great Plains. It is low growing, up to 18 inches, with small leaves, not over 6 inches long and $1/8$ inch or less in width. It is found on all soil types, but thrives best on upland, rather heavy soils. It is drought resistant. Growth is late starting in spring. It is relished as pasture by all classes of livestock. It is one of the more important range grass species, standing heavy grazing well. It is readily established by seeding.

3. Season: Growth starts early April and flowers May-November. Warm season grasses.
4. Production in U.S.: No specific data, See Forage grass. Great Plains and Western states.
5. Other production regions: Canada.
6. Use: Rangeland, pasture, hay and soil conservation.
7. Part(s) of plant consumed: Foliage
8. Portion analyzed/sampled: Forage and hay
9. Classifications:
   a. Authors Class: See Forage grass
   b. EPA Crop Group (Group & Subgroup): See Forage grass
   c. Codex Group: See Forage grass
   d. EPA Crop Definition: None
10. References: GRIN, MAGNESS, BARNES
11. Production Map: EPA Crop Production Regions 4, 5, 6, 7, 8, 9 and 10.
12. Plant Codes:
    a. Bayer Code: BOBCU (*B. curtipendula*), BOBGR (*B. gracilis*)

266

1. Grape
(Raisin, Uva pasa, Wine grape, Foxgrape, Slipskin grape, Vigne, Weinrebe, European grape, Concord grape, Vid, Parra, Skunk grape, Parson, Moscada, Vite moscata, Corinthian grape, Currant grape, Zante currant, Champagne grape, Staphis, Passonilla, Sultanas, Scuppernong, Vine leaf)
Vitaceae

Vitis spp.

2. Grapes are long-lived woody perennial vines. They are propagated by rootstocks. There are three main types of grapes grown commerially in North America, the vinifera or Old World European, the American bunch grapes and the muscadine. Characteristics are noted below:

....

Vinifera
V. vinifera L.

(Wine grape, European grape, Vid, Parra, Corinthian grape, Currant grape, Zante currant, Champagne grape, Staphis, Passonilla, Sultanas)

This is the grape type of major world production with over 5,000 named cultivars. Requires a climate with a long season, high summer/mild winter temperatures. Fruit is borne in clusters. Individual fruits or berries vary in size from $^1/_3$ inch diameter to 1 inch or more; in shape from round to oval or cylindrical; and in color from light green to red or black. Skin is smooth and waxy, enclosing a juicy pulp to which it adheres. Both seedless and seeded kinds are among the varieties grown commercially. Some pea-sized vinifera are marketed as Champagne grape when fresh, and Zante currants after drying as small raisins.

American Bunch

V. labrusca L.

(Fox grape, Slipskin grape, Concord grape, Skunk grape, Parson)

These are grape varieties developed in whole or in part from species indigenous in this country. They are hardier and more disease resistant than the Old World grape. Can tolerate cooler summer and winter temperatures than vinifera types. In fact, there is some of this type grown in all the mainland states. Fruit is produced in bunches. Individual berries, from $^1/_2$ inch to near 1 inch diameter, vary in shape from oval to slightly oblate and in color from green to red or black. Skin is thin and waxy, and separates readily from pulp, hence the name "slip skin". Fruits generally contain 2 to 4 seeds.

Muscadine

V. rotundifolia Michx.

(Moscada, Vite moscata, Scuppernong, Bullace, Southern fox grape)

These are a distinctive type of grape, native in the Southeastern states, and largely grown there. Vines are characteristically strong growers, quite disease resistant. Does best in Cotton Belt climate, will not tolerate temperatures below 0 degree F. Fruit is borne singly or in small clusters, usually not more than a dozen berries. Fruit skin is very tough, and separates from the pulp. Berries are generally nearly round, $^3/_4$ to 1 inch or more in diameter, and are harvested without the stem adhering. Plants supported on trellises.

3. Season: Bloom to harvest in about 2.5 to 6 months depending on type and variety.
4. Production in U.S.: Domestic grape acreage was 752,620 acres in 1995 that produced 5.93 million tons. Major production states include California (645,200 acres), Washington (34,000 acres), New York (33,000 acres), Michigan (11,800 acres) and Pennsylvania (11,000 acres).
5. Other production regions: A major worldwide crop with Italy, France, Russia, and Spain major grape producing countries. In 1995, Canada grew 15,560 acres (ROTHWELL 1996a).
6. Use: Grapes are used for wine, fresh eating (table grapes), dried into raisins, unfermented juice, and preserves. Grape seed oil is extracted from bunching grape seeds. Some use of grape leaves for ethnic dishes.
7. Part(s) of plant consumed: Whole fruit with or without skin and/or seeds. Leaves for ethnic dishes
8. Portion analyzed/sampled: Whole fruit and processed commodities raisin and juice
9. Classifications:
   a. Authors Class: Small Fruits
   b. EPA Crop Group (Group & Subgroup): Recently reclassified as Miscellaneous, previously a member of Crop Group 13: Small Fruits and Berries Group
   c. Codex Group: 004 (FB 0269) Grapes, (FB 1235) Table Grapes: Berries and other Small Fruit; 055 (DF 0269) Dried grapes (= Currants, Raisins, Sultanas); 070 (JF 0269) Grape juice; 013 (VL 0269) Grape leaves; 071 (AB 0269) Dry grape pomace
   d. EPA Crop Definition: None
10. References: GRIN, AHMEDULLAH, CODEX, LOGAN, MAGNESS, SHOEMAKER, US EPA 1995, US EPA 1994, ROTHWELL 1996a, REHM, IVES, FACCIOLA, USDA 1996f, PATIL, US EPA 1996a, CROCKER(c and e), HALBROOKS
11. Production Map: EPA Crop Production Regions 1, 10, 11 and 12.
12. Plant Codes:
    a. Bayer Code: VITSS (*V.* spp.), VITLA (*V. labrusca*), VITRF (*V. rotundifolia*), VITVI (*V. vinifera*)

267

1. Grapefruit

(Toronja, Pomelo, Toronjo)

Rutaceae

Citrus x *paradisi* Macfad. (syn: *C. maxima uvacarpa* Merr. & Lee)

2. Grapefruit is one of the important citrus fruits (see citrus fruits). The tree tends to be more vigorous than orange, especially when young, with larger and thicker leaves. Fruit is near round to oblate, 3 to 4$^1/_2$ inches in diameter with a rather thick rind. Varieties may be seeded or seedless, and flesh may be white, red or pink. Fruit sets in small clusters on vigorous trees. Cultivars are Red type ('Henderson', 'Ray', 'Rio red', 'Flame', 'Star ruby', 'Marsh ruby') and White type ('Marsh seedless').
3. Season, bloom to harvest: 8 to 14 months, depending on location.
4. Production in U.S.: In 1995-96, about 2,718,000 tons from 174,570 bearing acres (USDA 1996d). In 1995-96, the top 4 states reported Florida (132,800 bearing acres), California (18,800), Texas (17,670) and Arizona (5,300) (USDA 1996d). In 1992, Puerto Rico reported 605 acres (1992 CENUS).
5. Other production regions: Jamaica, Trinidad, Argentina, Brazil, Israel, Mexico, India, Cyprus and Morocco (MORTON 1987).
6. Use: Fresh, canned juice and segments, marmalade. About equal amounts fresh and processed.
7. Part(s) of plant consumed: Interior segments or juice. Rind from processing plants is dried for livestock feed
8. Portion analyzed/sampled: Whole fruit
9. Classifications:
   a. Authors Class: See citrus fruits
   b. EPA Crop Group (Group & Subgroup): See citrus fruits (Grapefruit is a representative crop).
   c. Codex Group: 001 (FC 0203) Citrus fruits
   d. EPA Crop Definition: See citrus fruits
10. References: GRIN, CODEX, LOGAN (picture), MAGNESS, RICHE, USDA 1996d, US EPA 1995a, US EPA 1994, IVES, JACKSON(b), MORTON

11. Production Map: EPA Crop Production Regions 3, 6, 10 and 13.
12. Plant Codes:
    a. Bayer Code: CIDPA

268

1. Green sprangletop
(Zacate gigante)
Poaceae (Gramineae)
Leptochloa dubia (Kunth) Nees

2. This is a perennial warm season grass native to the U.S. that is a member of the grass tribe *Festuceae*, and is distributed in well drained and sandy soils in rocky hills, canyons and prairies. It is an erect bunchgrass 0.3 to 1.1 m tall growing with a branched-panicle inflorescence (4 to 12 cm long). It produces a good quality forage for livestock and fair for wildlife when green, however it has a short life span. Green sprangletop reproduces by seeds and tillers and is often found in mixed forage stands.
3. Season: Warm season grass that starts growing in early April and flowers most of the season.
4. Production in U.S.: There is no production data for Green sprangletop, however, pasture and rangeland are produced on more than 410 million A (U.S. CENSUS, 1992).
5. Other production regions: Mexico.
6. Use: Pasture grazing, hay, and soil erosion.
7. Part(s) of plant consumed: Leaves and stems
8. Portion analyzed/sampled: Forage and hay
9. Classifications:
    a. Authors Class: See Forage grass
    b. EPA Crop Group (Group & Subgroup): See Forage grass
    c. Codex Group: See Forage grass
    d. EPA Crop Definition: None
10. References: GRIN, US EPA 1994, CODEX, HOLZWORTH, STUBBENDIECK(a) (picture), RICHE, US EPA 1995
11. Production Map: EPA Crop Production Regions are 3, 6, 8, 9 and 13.
12. Plant Codes:
    a. Bayer Code: LEFDU

269

1. Groundcherry
(Cherry tomato, Chinese lantern plant, Husk tomato, Bladder cherry, Strawberry tomato, Tomatillo ground cherry, Peruvian ground cherry, Poha, Goldenberry, Jamberry, Alkekengi, Miltomate, Uchuba)
Solanaceae
Physalis spp.

1. Chinese lantern plant
(Alkekengi, Winter cherry, Strawberry ground cherry, Japanese lantern, Strawberry tomato)
P. alkekengi L.

1. Tomatillo
(Husk tomato, Tomatillo, Groundcherry, Tomate de casarna, Miltomate, Jamberry, Tomatoverde, Fresadilla, Mexican green tomato, Tomatillo groundcherry)
P. philadephica Lam. (syn: *P. ixocarpa* auct.)

1. Cape gooseberry
(Peruvian cherry, Peruvian groundcherry, Poha, Goldenberry, Uchuba, Gooseberry tomato)
P. peruviana L. (syn: *P. edulis* Sims)

1. Hairy groundcherry
(Dwarf cape gooseberry, Downy groundcherry)
P. grisea (Waterf.) M. Martinez (syn: *P. pruinosa* auct.)

1. Downy groundcherry
P. pubescens L.

2. Plants are annuals in the north, some forms are perennial in the tropics. Both upright and trailing forms occur. Fruits are generally globose, red or yellow in color, an inch or less in diameter, with many small, inconspicuous seeds. Fruits are smooth skinned and completely enclosed in a thin papery husk, which is free and easily removed. Groundcherries are widely grown in home gardens in many parts of the world, including the U.S., but enter commerce only to a very limited extent. The tomatillo plant is an annual, 3 to 4 feet high, with thin ovate or elongated notched leaves. The fruit is smooth, sticky, purplish or green in color, and entirely enclosed in a thin husk. Fruit is 3 inches in diameter, near globose which is larger than the other Groundcherries. It is used in making chili sauce and for flavoring meat by Mexicans and Latin Americans; but is little grown in the U.S.
3. Season, bloom to maturity: 2 to 3 months. Seeding to first harvest: About 3 to 4 months. Harvest can proceed for 2 months.
4. Production in U.S.: Limited data. Often in home gardens, but rarely in commerce. California for commercial tomatillos year round (LOGAN 1996). Poha is grown in Hawaii on less than 10 acres (KAWATE 1995a). Grown in south Florida (STEPHENS 1988). In Washington, grown for local use (SCHREIBER 1995).
5. Other production regions: Tomatillos were introduced from Mexico (STEPHENS 1988). Tomatillos available year round from Mexico (LOGAN 1996). Other groundcherries are commercially available from New Zealand from April to June (LOGAN 1996).
6. Use: Tomatillos are used mainly as seasoning, in chili sauce, salsas and meats. The other groundcherries, which are yellow in color, are mainly used for preserves, some culinary. Preparation is by removing the husk, washing the fruit and removing stem (SCHNEIDER 1985).
7. Part(s) of plant consumed: Whole fruit, after husk removed
8. Portion analyzed/sampled: Whole fruit, after husk removed
9. Classifications:
    a. Authors Class: Fruiting vegetables (except cucurbits)
    b. EPA Crop Group (Group & Subgroup): Fruiting vegetables (except cucurbits) (8)
    c. Codex Group: 012 (VO 0441) Fruiting vegetables other than cucurbits
    d. EPA Crop Definition: Tomato = Tomatillo
10. References: GRIN, CODEX, KAWATE 1995a, LOGAN, MAGNESS, MYERS, SCHNEIDER 1985, SCHREIBER, STEPHENS 1988 (picture), US EPA 1995a, YAMAGUCHI 1983 (picture)
11. Production Map: Tomatillos, EPA Crop Production Regions 10, 11 and 13. Other groundcherries in gardens including Regions 3 and 13.
12. Plant Codes:
    a. Bayer Code: PHYAL (*P. alkekengi*), PHYIX (*P. ixocarpa*), PHYPE (*P. peruviana*), PHYPU (*P. pubescens*)

270

1. Guajillo
(Berlandier acacia)
Fabaceae (Leguminosae)
Acacia berlandieri Benth.

2. Guajillo is a perennial warm season shrub native to the U.S. This shrub is up to 4 m tall and is common to limestone hills and sandy soils in southern Texas and Mexico. It flowers in November to March, with its fruits (a pod, oblong 8 to 15 cm long and 1.5 to 2.5 cm wide) maturing June to July. It reproduces by seed. It has been used as an important honey plant, a source of gums and dyes, as an ornamental shrub, and as a fair quality forage for livestock and wildlife that browse its leaves and stems. If consumed in large amounts it may cause hydrocyanic acid poisoning to livestock. The seeds are used by birds as a food source.
3. Season: Warm season perennial shrubs, grazed when young to mature.
4. Production in U.S.: No specific data for Guajillo production, however, pasture and rangelands are produced on more than 410 million A (U.S. CENSUS, 1992).
5. Other production regions: Mexico.
6. Use: Grazing, honey crop, and as an ornamental crop.
7. Part(s) of plant consumed: Leaves and stems
8. Portion analyzed/sampled: Forage, hay is not required
9. Classifications:
   a. Authors Class: Nongrass Animal Feeds
   b. EPA Crop Group (Group & Subgroup): Miscellaneous
   c. Codex Group: No specific citation although the general class 050 (AL 0157) Legume Animal Feeds could be used
   d. EPA Crop Definition: None
10. References: GRIN, US EPA 1994, CODEX, HOLZ-WORTH, STUBBENDIECK (a)(picture), RICHE, US EPA 1995
11. Production Map: EPA Crop Production Regions 6 and 8.
12. Plant Codes:
   a. Bayer Code: ACABE

271

1. Guar
(Clusterbean, Guarbean, Calcutta lucerne, Siambean, Guar bean)
Fabaceae (Leguminosae)
Cyamopsis tetragonoloba (L.) Taub. (Syn: *C. psoraleoides* (Lam.) DC.)

2. Guar is a branched warm season annual legume, native to India and Pakistan. In the 1960s it was grown on more than 100,000 acres in this country, mainly in west Texas and Oklahoma. It is an excellent soil improvement crop that grows to 9 feet tall, but is produced mainly for the seeds or "beans" which are enclosed in pods until harvested by combine threshers. The seeds contain a mannogalactan gum that is extracted and widely used as a stabilizer and smoother in ice cream and other frozen desserts. It also has industrial uses, as in paper manufacture and color printing on fabrics. The meal following gum extraction is excellent high-protein feed. For hay or forage, guar appears inferior to many other legumes, but the straw and stubble following combining furnish acceptable livestock pasturage. Guar is grown in rotation with cotton, grain sorghum, corn or vegetable crops.

Green pods can be harvested 45 to 55 days after planting. If used for hay, cut when lower pods turn brown. India and Pakistan provide about 90 percent of the U.S. guar gum needs. In India, there are three main types: (1) Deshi, as a seed crop; (2) Papdeshi, mostly as vegetable pod crop and (3) Sotiaguvar, mostly as green manure or forage crop.
3. Season, seeding to harvest: 123-135 days to 160-175 days.
4. Production in U.S.: In 1992, 6,836 acres with 2,760 tons (1992 CENSUS). Texas with 5,738 acres. Arizona and Oklahoma with the remainder (1992 CENSUS).
5. Other production regions: India and Pakistan.
6. Use: See above, and the young pods can be eaten like green beans, and leaves like spinach.
7. Part(s) of plant consumed: Dried shelled seed, young pods and forage
8. Portion analyzed/sampled: Seed, straw and meal. If used at immature stage: pods and forage. In the U.S., the primary use is for the seed. With the straw being an insignificant feed item, only the seed and meal should be analyzed
9. Classifications:
   a. Authors Class: Legume vegetables (succulent or dried); Foliage of legume vegetables. Primarily dried
   b. EPA Crop Group (Group & Subgroup): Legume vegetables (succulent or dried) (6C); Foliage of legume vegetables (7)
   c. Codex Group: 014 (VP 0525) Legume vegetables for young pods
   d. EPA Crop Definition: None
10. References: GRIN, CODEX, MAGNESS, RICHE, US EPA 1995a, US EPA 1994, DUKE(h) (picture)
11. Production Map: EPA Crop Production Regions 7, 8 and 9.
12. Plant Codes:
   a. Bayer Code: CMOTE

272

1. Guarana (Guar-a-na)
Sapindaceae
Paullinia cupana Kunth (syn: *P. sorbilis*)

2. Bushy vine 9 to 15 feet tall in cultivation. It has pinnate leaves and hard, small triangular fruit capsules with black seeds which are the source of guarana paste. Fruits are picked by hand from October to December in the tropics. Racemes are 2 to 6 inches long.
3. Season, planting: Vegetative of selected plants or by seed.
4. Production in U.S.: No data.
5. Other production regions: Grown in Brazil (JANICK 1990) and other tropical and subtropical areas in America but not Hawaii (NEAL 1965) (FACCIOLA 1990).
6. Use: Drink resembling coffee prepared from seeds. Marketed as stimulant cola-type beverage in U.S. Roasted, pounded seeds are pressed into a paste which is used in carbonated beverages and a tea, containing about 5 percent caffeine. The sweetened paste, called "Brazilian chocolate" is used in soft drinks, candy, and for flavoring liqueurs.
7. Part(s) of plant consumed: Seeds
8. Portion analyzed/sampled: Seeds
9. Classifications:
   a. Authors Class: Tropical and subtropical trees with edible seed for beverages and sweets
   b. EPA Crop Group (Group & Subgroup): Miscellaneous
   c. Codex Group: No specific entry
   d. EPA Crop Definition: None

10. References: GRIN, FACCIOLA, JANICK 1990, MORTON, NEAL, SCHERY 1972
11. Production Map: EPA Crop Production Region 13.
12. Plant Codes:
    a. Bayer Code: No specific entry

273

1. Guava
(Guayaba, Sand plum)
Myrtaceae
Psidium guajava L.
(Common guava, Guayaba grande, Goyavier commun)
P. cattleianum Sabine
(Strawberry guava, Guayaba fresca, Goyavier fraise)
P. cattleianum Sabine var. *cattleianum*
(Purple guava, Red strawberry guava)
P. cattleianum var. *littorale* (Raddi) Fosberg
(Strawberry guava, Waiawi, Yellow strawberry guava)
2. Grows in warm climate with medium rainfall. First crop within 2-3 years after vegetative propagation. The evergreen trees can attain heights of 20 to 25 feet. Plants can be propagated from seed and root cuttings, and by grafting, budding or air layering. The common guava fruits vary in shape from spherical to pyriform, and in diameter from 1 to 4 inches, commonly 2 inches. Its skin is light yellow, somewhat rough, free of pubescence. The fruit has many small hard seeds and is considered a berry. Seedless (triploid) varieties have been developed. The fruit color varies from white to deep pink to salmon red. Pulp within the peel is soft when ripe, sweet to slightly acid, musky. Strawberry guava has a much smaller fruit than common guava. The fruit is usually round with a purplish-red or light yellow skin and a sweet aromatic flavor reminiscent of the strawberry.
3. Season: Fruit mature in 90 to 150 days from flowering. Flowers and mature fruits may be on a plant at the same time.
4. Production in U.S.: Guava orchards were fairly extensive around 1900, but commercial production has declined. Current domestic production is mainly in Florida (150 acres – 1995) with some acreage in Guam (292 lbs – 1992), Hawaii (750 acres – 1995) and Puerto Rico.
5. Other production regions: India, Egypt, Brazil, Mexico, South Africa.
6. Use: Jelly, jam, juice, and sometimes eaten out of hand. Fruit is high in vitamin C.
7. Part(s) of plant consumed: Inner pulp, primarily
8. Portion analyzed/sampled: Whole fruit
9. Classifications:
    a. Authors Class: Tropical and subtropical fruits – edible peel
    b. EPA Crop Group (Group & Subgroup): Miscellaneous
    c. Codex Group: 006 (FI 0336) Assorted tropical and subtropical fruits – inedible peel
    d. EPA Crop Definition: Proposing – Guava = Feijoa, Acerola, Jaboticaba, Passionfruit, Starfruit, Wax jambu
10. References: GRIN, CODEX, KNIGHT, MAGNESS, MARTIN 1987, RICHE, US EPA 1994, WILSON, ADSULE, USDA 1996f, MALO(a), CRANE 1995a
11. Production Map: EPA Crop Production Region 13.
12. Plant Codes:
    a. Bayer Code: PSICA (*P. cattleianum*), PSIGU (*P. guajava*)

274

1. Gumweed, Curlycup
(Resinweed, Gum plant, Pitchweed)
Asteraceae (Compositae)
Grindelia squarrosa (Pursh) Dunal
2. Erect annual, biennial or short-lived perennial to $3^1/_2$ feet tall. Middle and upper leaves mostly oblong or ovate, 2 to 4 times as long as wide and glandular-dotted.
3. Season, seeding to harvest: No data.
4. Production in U.S.: No data. Western North America (FACCIOLA 1990).
5. Other production regions: No entry.
6. Use: Leaves used to make an aromatic tea. Also used as a chewing gum substitute.
7. Part(s) of plant consumed: Leaves
8. Portion analyzed/sampled: Leaves (fresh and dried)
9. Classifications:
    a. Authors Class: Herbs and spices
    b. EPA Crop Group (Group & Subgroup): Miscellaneous
    c. Codex Group: No specific entry
    d. EPA Crop Definition: None
10. References: GRIN, BAILEY 1976, FACCIOLA, USDA NRCS
11. Production Map: No entry
12. Plant Codes:
    a. Bayer Code: GRNSQ

275

1. Gumweed, Great valley
(Grindelia, Gumweed)
Asteraceae (Compositae)
Grindelia camporum Greene
2. Arid-adapted, herbaceous perennial to $4^1/_2$ feet tall, found in the Central Valley of California. The plant produces significant quantities of extractable diterpene resin acids. The resins are produced in glands on the surfaces of stems, leaves and involucres. The crude resins are solvent extracted, the remaining bagasse or material remaining after resin extraction may have potential as animal feed or fuel.
3. Season, harvest: The biomass is harvested two times a year.
4. Production in U.S.: No data. Native to Central Valley of California (JANICK 1990).
5. Other production regions: No entry.
6. Use: Diterpene resins used in Naval stores as tar, etc. for building and repairing wooden ships. Potential for bagasse as feed.
7. Part(s) of plant consumed: Potential for bagasse as animal feed
8. Portion analyzed/sampled: Bagasse, if used as feed
9. Classifications:
    a. Authors Class: Nongrass animal feed
    b. EPA Crop Group (Group & Subgroup): Miscellaneous
    c. Codex Group: No specific entry
    d. EPA Crop Definition: None
10. References: GRIN, BAILEY 1976, JANICK 1990
11. Production Map: EPA Crop Production Region 10.
12. Plant Codes:
    a. Bayer Code: No specific entry

276

1. Hanover salad
(Hanover kale, Spring kale, Hanover turnip, Siberian kale, Winter rape, Curled kitchen kale)
Brassicaceae (Cruciferae)
Brassica napus var. *pabularia* (DC.) Rchb.

2. This member of the cabbage family is grown for the tender leaves used as potherbs and in salads. In growth, it is much like turnip, but the root is non-tuberous. The leaves form a rosette and are smooth and generally scalloped. The stems vary from purple to white. General culture and exposure of leaves are comparable to spinach. Siberian kale differs from Common or Scotch (which see) in that the foliage is bluish-green in color and is less curled. Cultural conditions, season and use are similar to Common Kale.
3. Season, seeding to harvest: 2 to 3 months.
4. Production in U.S.: No data, limited. In Florida, plant statewide from September through March (STEPHENS 1988). In the Southern U.S., it also is planted as animal fodder (FACCIOLA 1990).
5. Other production regions: Grown in temperate regions (RUBATZKY 1997).
6. Use: As salad, potherb and fodder.
7. Part(s) of plant consumed: Leaves only
8. Portion analyzed/sampled: Leaves
9. Classifications:
   a. Authors Class: *Brassica* (cole) leafy vegetables
   b. EPA Crop Group (Group & Subgroup): Miscellaneous
   c. Codex Group: 013 (VL 0480) Leafy vegetables (including *Brassica* leafy vegetables) as Kale
   d. EPA Crop Definition: See Turnip
10. References: GRIN, FACCIOLA, MAGNESS, STEPHENS 1988 (picture), US EPA 1995a, YAMAGUCHI 1983, CODEX, RUBATZKY
11. Production Map: See Kale.
12. Plant Codes:
    a. Bayer Code: No specific entry

277

1. Hardinggrass
(Toowoomba canarygrass, Bulbous canarygrass, Perla)
Poaceae (Gramineae)
Phalaris aquatica L. (syn: *Phalaris stenoptera* Hack.; *P. tuberosa* L.; *P. tuberosa* var. *stenoptera* (Hack.) Hitchc.)

2. This is a cool-season perennial grass, native to Africa, but brought to the U.S. from Australia in 1914. It is a long-lived bunchgrass with short, stout rhizomes. It is adapted to mild climates with winter rainfall and thrives best on heavy soils. It is the most widely adapted rangegrass in California but is grown only sparingly in other areas of the Southwest. Where adapted, forage yields are high and quality is good. It is distinguished from Reed canarygrass by its bulbous culm bases.
3. Season: Cool season grass. Productive October to May and dormant in the summer.
4. Production in U.S.: No specific data, See Forage grass. California.
5. Other production regions: Africa, Australia.
6. Use: Rangeland, pasture, erosion control.
7. Part(s) of plant consumed: Foliage

8. Portion analyzed/sampled: Forage and hay
9. Classifications:
   a. Authors Class: See Forage grass
   b. EPA Crop Group (Group & Subgroup): See Forage grass
   c. Codex Group: See Forage grass
   d. EPA Crop Definition: None
10. References: GRIN, MAGNESS, STUBBENDIECK, BALL (picture), MOSLER, BALL(a)
11. Production Map: EPA Crop Production Regions 2, 3, 4, 6 and 10 (Primarily 10).
12. Plant Codes:
    a. Bayer Code: PHATU

278

1. Hazelnut
(Cobnut, Filbert)
Betulaceae
Corylus spp.
1. European filbert
(European hazel, Avellana)
C. avellana L.
1. American hazelnut
(American filbert)
C. americana Marshall
1. California hazelnut
(Western hazel, Beaked filbert, Beaked hazelnut)
C. cornuta var. *californica* (A. DC.) W.M. Sharp
1. Giant filbert
C. maxima Mill.

2. The hazelnut tree is small, about 15 feet high, spreading and much branched. The leaves are deciduous, roundish oval, serrate, near glabrous above, somewhat pubescent on the lower veins. The nuts are nearly enclosed in a leafy involucre, but the apex is partially exposed. The shell is hard and woody. The kernel is free inside the shell, and separates freely when cracked. The inconspicuous female blossoms are exposed and pollinated in advance of the leafing-out of the trees. In addition to cultivated kinds, 2 species of *Corylus*, namely *C. americana* and *C. cornuta* are native in the U.S., and nuts from them are often harvested locally. Mature hazelnuts (without hulls) fall to the ground and are harvested.
3. Season, tree leafing to harvest: About 4 months.
4. Production in U.S.: About 27,148 tons in shell on 32,674 acres (1992 CENSUS). Oregon (about 99 percent of the U.S. acreage) and Washington (1992 CENSUS).
5. Other production regions: Europe.
6. Use: Direct eating, confections.
7. Part(s) of plant consumed: Internal kernels only
8. Portion analyzed/sampled: Nutmeat
9. Classifications:
   a. Authors Class: Tree nuts
   b. EPA Crop Group (Group & Subgroup): Tree nuts
   c. Codex Group: 022 (TN 0666) Tree nuts
   d. EPA Crop Definition: None
10. References: GRIN, CODEX, MAGNESS, RICHE, US EPA 1995a, US EPA 1995b, US EPA 1994, WOODROOF(a) (picture)
11. Production Map: EPA Crop Production Region 12.
12. Plant Codes:

a. Bayer Code: CYLAM (*C. americana*), CYLAV (*C. avellana*), CYLCC (*C. cornuta*), CYLMA (*C. maxima*)

279

1. Heartnut
(Hime-gurumi, Japanese walnut, Siebold walnut)
Juglandaceae
Juglans ailantifolia var. *cordiformis* (Makino) Rehder (syn: *J. sieboldiana* Maxim. var. *cordiformis* Makino)

2. Related to butternut, which see. Kernel is heart-shaped.
3. Season, harvest: See Black walnut.
4. Production in U.S.: No data. See Butternut.
5. Other production regions: See Butternut. Japan.
6. Use: Seeds eaten raw and used in cooking.
7. Part(s) of plant consumed: Nutmeat
8. Portion analyzed/sampled: Nutmeat
9. Classifications:
   a. Authors Class: Tree nuts
   b. EPA Crop Group (Group & Subgroup): Tree nuts under Black and English Walnut entry as *Juglans* spp., but no specific entry for Heartnut
   c. Codex Group: No specific entry
   d. EPA Crop Definition: None
10. References: GRIN, BAILEY 1976, FACCIOLA, WOODROOF(a)
11. Production Map: See Butternut
12. Plant Codes:
    a. Bayer Code: No specific entry

280

1. Hickory nut
Juglandaceae
Carya spp.
1. Pignut hickory
(Pignut, Red hickory, Redheart hickory, Small-fruited hickory, Sweet pignut, Broom hickory)
C. glabra (Mill.) Sweet
1. Shagbark hickory
(Scalybark hickory, Shellbark hickory, Upland hickory)
C. ovata (Mill.) K. Koch
1. Shellbark hickory
(Big shellbark, Kingnut)
C. laciniosa (F. Michx.) Loudon
1. Mockernut hickory
(Bigbud hickory, Mockernut, Squarenut, White-heart hickory)
C. tomentosa (Poir.) Nutt.
1. Nutmeg hickory
C. myristiciformis (F. Michx.) Nutt.

2. The above species are all native to parts of the U.S. and produce nuts with edible kernels. They are not grown commercially for the nuts, but some quantities are harvested from native or ornamental trees. The trees become large, up to 100 feet or more, with compound, pinnate leaves. Fruits are generally near globose, glabrous, and somewhat ridged, and 1 to 1¹/₂ inches long. The nut is encased in a fleshy husk which becomes fibrous and opens as the nuts mature. The shells are hard and woody. The kernels do not separate from the shells readily. Limited quantities either in shell or as kernels are marketed.

3. Season, harvest: Mature nuts fall out of hulls while on the tree from September to December. Leaf bud opening to nut maturity is almost 150 days.
4. Production in U.S.: No data. Central and Eastern U.S.
5. Other production regions: No entry.
6. Use: Direct eating, and used in ice cream, cookies and candies.
7. Part(s) of plant consumed: Nutmeat
8. Portion analyzed/sampled: Nutmeat
9. Classifications:
   a. Authors Class: Tree nuts
   b. EPA Crop Group (Group & Subgroup): Tree nuts
   c. Codex Group: 022 (TN 0667) Tree nuts
   d. EPA Crop Definition: None
10. References: GRIN, BAILEY 1976, CODEX, MAGNESS, US EPA 1995a, US EPA 1995b, WOODROOF(a) (picture)
11. Production Map: EPA Crop Production Regions 1, 2, 3, 4 and 5
12. Plant Codes:
    a. Bayer Code: CYAGL (*C. galbra*), CYALA (*C. laciniosa*), CYAOV (*C. ovata*), CYATO (*C. tomentosa*)

281

1. Highbush cranberry
(American Cranberrybush)
Caprifoliaceae
Viburnum opulus L. var. *Americanum* Aiton (syn: *Viburnum trilobum* Marshall)
1. European cranberry bush
(Crampbark, Guelder rose, Snowball bush)
V. opulus L. var. *opulus*
1. Nannyberry
(Blackhaw, Sheepberry, Sweet viburnum, Wild raisin, Nanny plum, Sweetberry, Tea plant, Cowberry)
V. lentago L.

2. **Highbush cranberry** is a deciduous shrub which can grow up to 12 feet with maple-like leaves. It has a open spreading growth pattern. The aggregate fruits are borne in loose clusters on laterals that grow from the canes. Propagated by hardwood and softwood cuttings. Plant produces showy flowers in early summer. Bright scarlet fruit by the end of July. Fruits are ¹/₃ inch long. Limited fruit production occurs during the third season after planting, with full production in the fifth. Important cultivars include: 'Wentworth', 'Hahs', 'Andrews', 'Compactum', 'Manitou' and 'Phillips'.
Nannyberry is a deciduous upright shrub to 30 feet with edible sweet blue-black fruits which are 1-seeded drupes. The fruits are borne in loose clusters and are ¹/₂ inch long.
3. Season: Typical bloom to harvest is in 90 to 110 days.
4. Production in U.S.: No production statistics, some acreage in New Hampshire, Massachusetts and other northern states.
5. Other production regions: Canada and Europe.
6. Use: Highbush cranberry fruits are mostly used to make jelly, pies or as a substitute for cranberry sauce. Nannyberry fruits are very sweet, pulpy, and somewhat juicy with the best being ¹/₂ inch long. Viburnums are among the most popular ornamental shrubs (BAILEY 1976).
7. Part(s) of plant consumed: Fruit. However the large seed of these fruit preclude their use in whole-fruit products
8. Portion analyzed/sampled: Whole Fruit
9. Classifications:

a. Authors Class: Berries

b. EPA Crop Group (Group & Subgroup): Miscellaneous

c. Codex Group: No specific entry

d. EPA Crop Definition: None

10. References: GRIN, STANG, USDA NRCS, FACCIOLA, BAILEY 1976

11. Production Map: EPA Crop Production Region 1 is logical production region.

12. Plant Codes:

a. Bayer Code: VIBLE (*V. lentago*), VIBOP (*V. opulus*)

282

1. Honewort

(Wild chervil, White chervil, Canadian honewort)

Apiaceae (Umbelliferae)

Cryptotaenia canadensis (L.) DC.

2. Slender, hairless, branching perennial 3 to 4 feet tall. Leaves are three leaflets with each leaflet 2 to 6 inches long and 1 to 3 inches wide. Flowers are white.

3. Season, flowering: May to June.

4. Production in U.S.: No data. Japanese honewort was cultivated in Oregon (DUKE 1992). Distribution mostly in the piedmont and lower mountains, Minnesota-Maine-Oregon-Georgia, USDA Zones 3 to 8 (DUKE 1992).

5. Other production regions: No entry.

6. Use: Roots are cooked like parsnips. Young leaves or tops are cooked as a spicy potherb or added to soups. Seeds used like aniseed.

7. Part(s) of plant consumed: Mainly young tops (primarily leaves and stems), and roots

8. Portion analyzed/sampled: Young tops and roots

9. Classifications:

a. Authors Class: Herbs and spices; Root and tuber vegetables; Leaves of root and tuber vegetables

b. EPA Crop Group (Group & Subgroup): Miscellaneous

c. Codex Group: No specific entry

d. EPA Crop Definition: None

10. References: GRIN, DUKE(c) (picture), NEAL, USDA NRCS

11. Production Map: EPA Crop Production Regions 1, 2, 5 and 12.

12. Plant Codes:

a. Bayer Code: CPBCA

283

1. Hooded windmillgrass

(Pata de gallo arensa, Windmillgrass)

Poaceae (Gramineae)

Chloris cucullata Bisch.

2. This is a perennial warm season grass native to North America, and is a member of the grass tribe *Chlorideae*. It is an erect bunchgrass 15 to 60 cm tall that sometimes spreads by short rhizomes and flowers with a panicle 5 to 20 cm long. Hooded windmillgrass used in pastures, plains, and road rights-of-ways. It produces fair-to-good forage starting early in the spring and is still available as forage in the winter.

3. Season: Warm season, growth starts in early spring and is completed in fall and can be grazed in the winter.

4. Production in U.S.: No data for Hooded windmillgrass production, however, pasture and rangeland are produced on more than 410 million A (U.S. CENSUS, 1992).

5. Other production regions: Mexico (GOULD 1975).

6. Use: Pasture, hay, and erosion control.

7. Part(s) of plant consumed: Leaves and stems

8. Portion analyzed/sampled: Forage and hay

9. Classifications:

a. Authors Class: See Forage grass

b. EPA Crop Group (Group & Subgroup): See Forage grass

c. Codex Group: See Forage grass

d. EPA Crop Definition: None

10. References: GRIN, US EPA 1994, US EPA 1995, CODEX, HOLZWORTH, STUBBENDIECK (picture), RICHE, GOULD

11. Production Map: EPA Crop Production Regions 6, 8 and 9.

12. Plant Codes:

a. Bayer Code: No specific entry, genus entry is CHRSS

284

1. Hop

(Hops, Hopfen, Lupulo, Houblon)

Cannabinaceae

Humulus lupulus L.

2. Hop is a dioecious perennial climbing plant lacking tendrils. The bines (erroneously called vines) climb by twining around any available support in a clockwise direction with the aid of hooked hairs. Hop plants may attain heights of 20 feet or more. The aerial portion of the plant dies back to the subterranean perennial crown at the end of the growing season. Large leaves with a toothed margin are borne in pairs at each node. Leaf shape varies from cordate to 7-lobed, with 3 to 5 lobes being commonly found in most commercial cultivars. The inflorescence of the female plant, known as hop cones, are the hops of commerce. The hop cones are formed in clusters, mainly on side branches (arms). Cones are usually 1 to 3 inches long and up to 1 inch in diameter. Hops have a characteristic aroma and bitter flavor and lend unique character to fermented malt beverages. In commercial production, hops are trained to climb a rough twine supported by a trellis structure approximately 18 feet high. The hops reach the top of the trellis and begin to flower by midsummer. Hop harvest in the northwestern U.S. begins in mid-August and sometimes continues until mid-September. Bines are normally cut and hauled to a stationary picking machine. The machine strips the plant and separates the valuable hop cones from leaves and other debris. The cleaned hops are transferred to a kiln where they are dried to a moisture content of approximately 10 percent. The dried hops are then compressed into 200 lb bales and transported off the farm. Dried hops may be ground into pellets, extracted for their essential oils and resins or used unmodified in the brewing of beer. About 0.25 to 1 pound of dried hops are needed to flavor 30 to 50 gallons of beer. The hops are added to a malt-water mixture and boiled to extract flavor. After the hops have been removed and the mixture cooled, yeast is added for fermentation. In 1966 to 1967, an average of 23,730 tons of hops were produced on about 30,000 acres in the U.S., almost all in Washington, Oregon, Idaho, and California. Significant production ceased in Cali-

fornia by 1992, but increased nationally to over 36,000 tons of hops from 40,549 acres.

3. Season, start of fruiting body formation to harvest: About 2 months.

4. Production in U.S.: 40,549 acres with production over 36,000 tons, 1992 U.S. CENSUS of Agriculture. EPA Crop Production Region 11 which accounts for 94 percent of the U.S. hops. The major states are Idaho, Oregon and Washington.

5. Other production regions: Europe.

6. Use: Flavoring, mainly for beer.

7. Part(s) of plant consumed: Hop cones; may be extracted for its hop extract. Spent hops from the extracting process is not a significant feed item

8. Portion analyzed/sampled: Hop Cones (Dried)

9. Classifications:
   a. Authors: Class: Herbs and spices
   b. EPA Crop Group (Group & Subgroup): Miscellaneous
   c. Codex Group: 057 (DH 1100) Dried herbs
   d. EPA Crop Definition: None

10. References: GRIN, CODEX, FARRELL (picture), MAGNESS, NEVE, RICHE, US EPA 1994, US EPA 1995a, IVES, DORSCHNER

11. Production Map: EPA Crop Production Region 11.

12. Plant Codes:
   a. Bayer Code: HUMLU

285

1. Horehound
(Hoarhound, White horehound)
Lamiaceae (Labiatae)

Marrubium vulgare L.

2. Horehound is an aromatic perennial herb, 1 to 3 feet in height, with hairy, ovate to near round leaves. It has escaped from gardens and become naturalized in many parts of the world, including the U.S. It was formerly much esteemed in cookery, but is now used mainly in candy, where it is believed to relieve tickling of the throat and coughing. Leaves and tender stems are extracted for use in candies. No data are available on commercial production in the U.S.

3. Season, harvest: At peak bloom. Flowers June to September.

4. Production in U.S.: No data. Naturalized in U.S. (FOSTER 1993).

5. Other production regions: Native to Europe (FOSTER 1993).

6. Use: Cookery and in candy and tea.

7. Part(s) of plant consumed: Leaves and flowering tops

8. Portion analyzed/sampled: Leaves and flowers (fresh and dried)

9. Classifications:
   a. Authors Class: Herbs and spices
   b. EPA Crop Group (Group & Subgroup): Herbs and spices (19A)
   c. Codex Group: 027 (HH 0732) Herbs; 057 (DH 0732) Dried herbs
   d. EPA Crop Definition: None

10. References: GRIN, CODEX, FOSTER (picture), MAGNESS, US EPA 1995a

11. Production Map: No entry.

12. Plant Codes:
   a. Bayer Code: MAQVU

286

1. Horseradish
Brassicaceae (Cruciferae)

Armoracia rusticana Gaertn., *et al.* (syn: *A. lapathifolia* Gilib. ex Usteri)

2. Horseradish is a root crop, grown for the very pungent roots, which contain an oil with a strong pungent odor and hot, biting taste. The plant attains a height of 2 to 3 feet when in flower. It is propagated by planting pieces of side roots, which are taken from the main root when the latter is harvested. The top of the plant consists of a rosette of large leaves, and a flower stalk; it rarely produces seeds. The roots develop entirely underground.

3. Season, planting to harvest: 6 to 8 months.

4. Production in U.S.: Historically, Illinois is the major producing state. In 1990, East St. Louis – Collinsville, Illinois region was a major horseradish growing area with about 1000 acres. In 1975, reported as about 2000 acres (IR-4 DATABASE 1996); and in 1995; reported 1360 acres (McMILLIN 1997). One grower in the Eau Claire, Wisconsin area harvests about 700 acres a year, Huntsinger Farms. Also with growers in Tulelake, California and New Jersey, eastern Maryland and Virginia about 3,000 acres are grown. Normally, horseradish is grown in the same field only once every 3 to 5 years and rotated with corn and soybean (GREAT LAKES VEGETABLE GROWERS NEWS 1990 and McMILLIN 1997). In 1994, California harvested 1,151 acres with 2,678 tons (MELNICOE 1996e). The longer the roots are left in the ground the better the production. Growers in California may leave the roots in the ground for up to 2 years (MCMILLIN 1997).

5. Other production regions: Native of southeastern Europe (RUBATZKY 1997).

6. Use: As a condiment. Preparation: Harvested roots are cleaned and lateral roots trimmed. The main roots (called "sticks") are trimmed around the crown and lower end, removing small rootlets. Wash roots and pack them for market. Prepare horseradish commercially for table use by peeling or scraping the roots and removing all defects, then the root is grated in vinegar. Refrigerate to keep it "hot" (USDA 1970).

7. Part(s) of plant consumed: Root. Leaves can be eaten raw or cooked but usually only grown for its root

8. Portion analyzed/sampled: Root

9. Classifications:
   a. Authors Class: Root and tuber vegetables; Herbs and spices
   b. EPA Crop Group (Group & Subgroup): Root and tuber vegetables (1A, 1B)
   c. Codex Group: 016 (VR 0583) Root and tuber vegetables; 028 (HS 0583) Spices
   d. EPA Crop Definition: None

10. References: GRIN, CODEX, MAGNESS, STEPHENS 1988 (picture), US EPA 1995a, US EPA 1995b, MELNICOE 1996e,GREAT LAKES VEGETABLES GROWERS NEWS 1990, IR-4 DATABASE 1996, USDA 1970, MCMILLIN, RUBATZKY

11. Production Map: EPA Crop Production Regions 2, 5 and 10.

12. Plant Codes:
   a. Bayer Code: ARWLA

287

1. Huckleberry

Ericaceae

Gaylussacia spp. (syn: *Buxella* Small, *Decachaena* Torr. & A. Gray, *Lasiococcus* Small)

1. Black huckleberry

G. baccata (Wangenh.) K. Koch

1. Dwarf huckleberry

G. dumosa (Andrews) Torr. & A. Gray

1. Dangleberry

(Blue tangle, Blue huckleberry, Dwarf huckleberry)

G. frondosa (L.) Torr. & A. Gray

1. Box huckleberry

G. brachycera (Michx.) A. Gray

1. Bear huckleberry

G. ursina (M.A. Curtis) Torr. & A. Gray

2. In the U.S., the name huckleberry is often used for blueberry, which see. While the two fruits are similar in appearance and flavor, the huckleberry, which is a drupe, has a 10-celled ovary, each cell (drupelet) normally containing a seed large enough to be conspicuously noticeable when the whole fruit is eaten. Blueberries, in contrast, contain many seeds so small as not to be noticeable when the fruit, which is a berry, is consumed. Normally only blueberries are a cultivated crop, but quantities of huckleberries are harvested from native plants as well as some cultivated plants. The huckleberry plant is a shrub, to 6 feet, with small, entire oval leaves. Fruits are borne in small clusters. Individual fruits are generally $1/3$ inch or less in diameter, mainly blue to black in color, sweet or slightly tart when ripe. Several species, mainly in eastern North America, produce fruits valued locally.

3. Season, harvested: In spring.

4. Production in U.S.: No data. Harvested from wild for local consumption. Also domesticated huckleberries are grown in Washington for commercial production with fewer than 50 growers (SCHREIBER 1995). Mainly eastern North America and northwestern U.S. (BAILEY 1976, FACCIOLA, SCHREIBER).

5. Other production regions: No entry.

6. Use: Primarily pies and preserves, also fresh. Secondary product is huckleberry honey. Huckleberries can be kept frozen for years.

7. Part(s) of plant consumed: Fruit

8. Portion analyzed/sampled: Whole fruit

9. Classifications:
   a. Authors Class: Berries
   b. EPA Crop Group (Group & Subgroup): Crop Group 13; Berries Group as well as this group's Bushberry Subgroup
   c. Codex Group: 004 (FB 4083) Berries and other Small Fruits
   d. EPA Crop Definition: None

10. References: GRIN, SCHREIBER, US EPA 1996a, FACCIOLA, USDA NRCS, BAILEY 1976, CODEX, MAGNESS, STANG, US EPA 1995a

11. Production Map: EPA Crop Production Regions 1, 2, 5 and 11.

12. Plant Codes:
   a. Bayer Code: GAYBA (*G. baccata*), GAYFR (*G. frondosa*)

288

1. Huckleberry, Garden

(Petty morel, Solanberry, Quonderberry, Moralle, Houndsberry, Wonderberry, Sunberry, Black berried nightshade, Morella)

Solanaceae

Solanum scabrum Mill. (syn: *S. melanocerasum* All.)

2. This is an herbaceous annual, up to 2 or more feet in height, having simple, ovate leaves about 6 inches long, pointed at both ends. Fruits are globular, smooth skinned, black, and borne in small clusters. The plant is widely distributed in the wild, and leaves are sometimes gathered and used as potherbs. Fruits are also used for preserves and pies. A garden form, with fruits about $1/2$-inch in diameter, is occasionally cultivated in home gardens. It is a close relative to the poisonous nightshades.

3. Season, seeding to harvest: For leaves, 2 to 3 months. For fruit, about 3 months.

4. Production in U.S.: None commercial. Rare in home gardens. Native to North America.

5. Other production regions: Wide distribution in temperate to tropical regions (YAMAGUCHI 1983).

6. Use: Leaves as potherbs; fruits culinary (preserves, pies, or cooked dishes).

7. Part(s) of plant consumed: Leaves and whole fruits. Leaves are used locally

8. Portion analyzed/sampled: Whole fruit

9. Classifications:
   a. Authors Class: Fruiting vegetables (except cucurbits)
   b. EPA Crop Group (Group & Subgroup): Miscellaneous
   c. Codex Group: No specific entry
   d. EPA Crop Definition: None

10. References: GRIN, MAGNESS, STEPHENS 1988 (picture), YAMAGUCHI 1983

11. Production Map: EPA Crop Production Region: Many regions for garden use including 3.

12. Plant Codes:
   a. Bayer Code: No specific entry, genus entry is SOLSS

289

1. Huisache

(Mimosa bush, Sweet acacia, Huizache, Aromo, Espino ruco, Oponax Needle bush, Klu, Ellington, Ellington curve, Cassie, Opopanax, Popinac)

Fabaceae (Leguminosae)

Acacia farnesiana (L.) Willd. (syn: *Mimosa farnesiana* L.)

2. Huisache is a perennial warm season tropical shrub or small tree up to 9 m tall that is native to the U.S. It is grown in Hawaii and is called Klu. This shrub is common to brushy areas and open woods on dry, sandy soils in the southern coastal U.S. and Mexico. It is common to tropical growing areas including Australia and Africa. It flowers from February to March and produces legume pods 2 to 8 cm long. Huisache has been a source of perfume oils, fence posts, honey plant and an ornamental plant. While its forage quality is usually poor, it can provide good quality forage for cattle and sheep in dry seasons. The young leaves are commonly eaten, and its seeds are readily eaten by wildlife. It reproduces by seeds.

3. Season: Warm season perennial shrubs, leaves are grazed when young.

4. Production in U.S.: No specific data for Huisache production, however, pasture and rangelands are produced on more than 410 million A (U.S. CENSUS, 1992).
5. Other production regions: Mexico, Australia and Africa.
6. Use: Grazing, honey crop, perfume oils, and as an ornamental crop.
7. Part(s) of plant consumed: Leaves
8. Portion analyzed/sampled: Forage, hay is not required
9. Classifications:
   a. Authors Class: Nongrass Animal Feeds
   b. EPA Crop Group (Group & Subgroup): Miscellaneous
   c. Codex Group: No specific citation although the general class 050 (AL 0157) Legume Animal Feeds could be used
   d. EPA Crop Definition: None
10. References: GRIN, US EPA 1994, US EPA 1995, CODEX, HOLZWORTH, SKERMAN(a), STUBBENDIECK (picture), RICHE
11. Production Map: EPA Crop Production Regions 3, 4, 6, 9 and 13.
12. Plant Codes:
   a. Bayer Code: ACAFA

290

1. Hyssop
(Hisopo, Hysope)
Lamiaceae (Labiatae)
Hyssopus officinalis L.

2. The hyssop plant is a perennial subshrub, 1 to 1¹/₂ feet high, with a woody stem-base, from which herbaceous shoots bearing flowers grow each year. The leaves are entire, up to 3 inches long, and slender. The whole plant has a pungent, bitter taste. The green parts are used in making absinthe and occasionally in flavoring salads. Dried flower parts are sometimes used in flavoring soups, and to some extent in medicine; but hyssop is now less grown and utilized, both in cookery and medicine, than in the past.
3. Season, start of growth to harvest: About 2 months for green parts just before blooming.
4. Production in U.S.: No data.
5. Other production regions: Native to southern Europe.
6. Use: Flavoring in tomato sauce and tea.
7. Part(s) of plant consumed: Tops; Flowers may be used
8. Portion analyzed/sampled: Tops (fresh and dried)
9. Classifications:
   a. Authors Class: Herbs and spices
   b. EPA Crop Group (Group & Subgroup): Herbs and spices group (19A)
   c. Codex Group: 027 (HH 0733) Herbs; 057 (DH 0733) Dried herbs
   d. EPA Crop Definition: None
10. References: GRIN, BAILEY 1976, CODEX, FOSTER (picture), MAGNESS, US EPA 1995a
11. Production Map: No entry.
12. Plant Codes:
   a. Bayer Code: HYSOF

291

1. Iceplant
(Fig marigold, Frost plant, Diamond plant, Midday flowers, Dew plant, Binghua, Algazul, Ficoide, Crystalline iceplant)
Aizoaceae
Mesembryanthemum crystallinum L.

2. Introduced into warm areas of the U.S. The plant is a perennial, but grown in gardens as an annual. It is small, about the size of bibb lettuce, with numerous 12- to 15-inch stems. Iceplant grows best in hot and dry climates. Same family as New Zealand spinach, which see.
3. Season, seeding to harvest: Pick the leaves once plant is well established.
4. Production in U.S.: No data. Grown in California.
5. Other production regions: Its source is India (MARTIN 1979).
6. Use: Fleshy parts of the leaves are boiled and served like spinach.
7. Part(s) of plant consumed: Leaves. Also, edible fruit
8. Portion analyzed/sampled: Leaves
9. Classifications:
   a. Authors Class: Leafy vegetables (except *Brassica* vegetables)
   b. EPA Crop Group (Group & Subgroup): Miscellaneous
   c. Codex Group: No specific entry
   d. EPA Crop Definition: None
10. References: GRIN, KAYS, MARTIN 1979, STEPHENS 1988
11. Production Map: EPA Crop Production Region 10.
12. Plant Codes:
   a. Bayer Code: MEKCR

292

1. Ilama
(Papauce, Izlama, Anona banana, Anona blanca)
Annonaceae
Annona diversifolia Saff.

2. The ilama plant is an erect tree which can grow to 25 feet tall. It is usually propagated by seeding or grafting. It thrives best in hot tropical lowlands with medium rainfall. The deciduous leaves are 2 to 6 inches long, glossy and thin. The flowers are maroon and are solitary and long stalked. The plant is quite similar in fruit and tree characters of other Annons. The fruits are spherical to ovoid, 4 to 6 inches in diameter and usually weighing 0.75 to 1.5 lbs. The fruit skin is studded with triangular protuberances and colored green to lavender. Each fruit contains 25 to 80 smooth seeds that are imbedded in the white to lavender pulp. There is one named variety, 'Imery'.
3. Season, bloom to fruit maturity in 150 days. September to October is approximate period of continuous harvest.
4. Production in U.S.: No statistical data available. It can be grown in California, Florida, Puerto Rico and Hawaii. Ilama production is rare in the U.S
5. Other production regions: South America, Mexico and Chile (MORTON 1987).
6. Use: Pulp eaten fresh in sherbets and other desserts. Some consider the fruit less flavorful than Cherimoya.
7. Part(s) of plant consumed: Fruit pulp
8. Portion analyzed/sampled: Whole fruit
9. Classifications:
   a. Authors Class: Tropical and subtropical fruits – inedible peel
   b. EPA Crop Group (Group & Subgroup): Miscellaneous

c. Codex Group: 006 (FI 0337) Assorted tropical and subtropical fruits – inedible peel

d. EPA Crop Definition: Proposing – Sugar apple = Ilama

10. References: GRIN, CODEX, CRANE 1995, MAGNESS, MARTIN 1987, MORTON (picture)

11. Production Map: EPA Crop Production Regions 10 and 13.

12. Plant Codes:

a. Bayer Code: No specific entry

293

1. Imbe
(African mangosteen)

Clusiaceae (Guttiferae)

Garcinia livingstonei T. Anderson

2. The Imbe is a small tree up to 20 feet, native to East Africa. Leaves are leathery, dark green, oblong, up to 6 inches in length. Fruits are up to 2 inches long, nearly as broad, and ripen in mid-summer. The skin is tender and encloses a thin, tart, watery pulp, with generally a single seed. There is no commercial production in the U.S., but occasional trees may be found in tropical areas.

3. Season, bloom to harvest: 180 to 200 days.

4. Production in U.S.: No data.

5. Other production regions: Tropical (MARTIN 1987).

6. Use: Pulp used in the preparation of a wine and eaten fresh.

7. Part(s) of plant consumed: Fruit pulp

8. Portion analyzed/sampled: Whole fruit

9. Classifications:

a. Authors Class: Tropical and subtropical fruits – inedible peel

b. EPA Crop Group (Group & Subgroup): Miscellaneous

c. Codex Group: No specific entry

d. EPA Crop Definition: None

10. References: GRIN, FACCIOLA, MAGNESS, MARTIN 1987

11. Production Map: EPA Crop Production Region 13.

12. Plant Codes:

a. Bayer Code: No specific entry

294

1. Imbu
(Umbu)

Anacardiaceae

Spondias tuberosa Arruda ex Kost.

2. Imbu grows wild in northeastern Brazil and is cultivated sparingly in other relatively dry, tropical areas. Propagated by seed and cuttings. The tree is spreading and low growing (20 ft.), with compound leaves up to 6 inches long, each with 5 to 9 leaflets. Fruits are produced near the ends of the branches. They are oval in shape, about 1.5 inches long, with a single seed. The fruit generally resembles plums, but with a tougher, thicker skin. The flesh is soft, almost watery. It is eaten fresh or made into jelly.

3. Season: Fruits are gather from the ground.

4. Production in U.S.: Minor production in South Florida

5. Other production regions: Northeastern Brazil

6. Use: Pulp eaten fresh, made into beverages, desserts or preserves.

7. Part(s) of plant consumed: Fruit

8. Portion analyzed/sampled: Whole fruit

9. Classifications:

a. Authors Class: Tropical and subtropical fruits – edible peel

b. EPA Crop Group (Group & Subgroup): Miscellaneous

c. Codex Group: No specific entry

d. EPA Crop Definition: None

10. References: GRIN, MAGNESS, MARTIN 1987, MORTON (picture)

11. Production Map: EPA Crop Production Region 13.

12. Plant Codes:

a. Bayer Code: No specific entry

295

1. Indian borage
(Puerto Rico oregano, Cuban oregano, Spanish thyme, Can day la, West Indies oregano, Country borage, Indian mint, Soupmint, French thyme, Mexican mint)

Lamiaceae (Labiatae)

Plectranthus amboinicus (Lour.) Spreng. (syn: *Coleus amboinicus* Lour.)

2. Semi-shrubby tender perennial to 3 feet tall bearing fleshy leaves that are hirsute above and 2 inches long.

3. Season, seeding to harvest: No data.

4. Production in U.S.: No data.

5. Other production regions: Cultivated throughout the tropics (KUEBEL 1988).

6. Use: In meat dishes and stews.

7. Part(s) of plant consumed: Leaves

8. Portion analyzed/sampled: Leaves (fresh and dried)

9. Classifications:

a. Authors Class: Herbs and spices

b. EPA Crop Group (Group & Subgroup): Miscellaneous

c. Codex Group: No specific entry

d. EPA Crop Definition: None

10. References: GRIN, MARTIN 1979, NEAL, KUEBEL

11. Production Map: EPA Crop Production Region 13.

12. Plant Codes:

a. Bayer Code: No specific entry

296

1. Indian ricegrass
(Silkgrass, Indian millet)

Poaceae (Gramineae)

Achnatherum hymenoides (Roem. & Schult.) Barkworth (syn: *Oryzopsis hymenoides* (Roem. & Schult.) Ricker ex Piper)

2. This is a native cool season short-lived perennial bunchgrass distributed from the Dakotas south to Texas and west to the Pacific Ocean. It is drought-resistant, adapted to dry, sandy soils. The plant grows in dense clumps, up to 2 feet tall. The leaves are slender and nearly as long as the stems. Panicle resembles a rice inflorescence. It is highly palatable to livestock, both while green in summer and dried in winter. Natural stands in many areas have been greatly depleted by overgrazing. This is an important species for reseeding range lands. Seed produced 10 to 11 weeks after initiation of spring growth.

3. Season: Primarily winter grazing.

4. Production in U.S.: No specific data. Dakotas south to Texas and west to the Pacific Ocean.

5. Other production regions: Canada and Mexico (BAILEY 1976).

6. Use: Pasture, rangegrass.
7. Part(s) of plant consumed: Foliage. Seeds were formerly used by Native Americans for grinding into meal and making bread
8. Portion analyzed/sampled: Forage and hay. Possible market potential for flour made from its seed
9. Classifications:
   a. Authors Class: See Forage grasses
   b. EPA Crop Group (Group & Subgroup): See Forage grasses
   c. Codex Group: See Forage grasses
   d. EPA Crop Definition: None
10. References: GRIN, MAGNESS, MOSLER, BAILEY 1976
11. Production Map: EPA Crop Production Regions 7, 8, 9, 10, 11 and 12.
12. Plant Codes:
    a. Bayer Code: ORZHY

297

1. Indiangrass
(Woodgrass, Yellow indiangrass, Slender indiangrass)
Poaceae (Gramineae)
Sorghastrum nutans (L.) Nash

2. This is a native warm season bunchgrass with short rhizomes, widely distributed east of the Rocky Mountains from Canada south to the Gulf of Mexico and into Mexico. It tillers and grows as a biennial with the first year vegetative and the second year reproductive. It is the only forage outside of sorghum that contains cyanogenic glucoside (dhurrin). Under the best conditions stems may reach to 10 feet. Leaves are smooth and flat, near $1/2$-inch wide, elongated, narrow at the base. Indiangrass thrives best on fertile bottom soils but also occurs on sandy soils and dry slopes. It is palatable while succulent but only fairly so when dry. It must be grazed before heading stage for best quality. It is most useful in the central and southern Great Plains. Some selected varieties are in commerce. There are 120,000 seeds/lb.
3. Season: Starts growth mid-spring from short rhizomes. It matures September to October.
4. Production in U.S.: No specific data, See Forage grass. East of Rocky Mountains, especially central and southern Great Plains.
5. Other production regions: Canada and Mexico.
6. Use: Pasture, hay.
7. Part(s) of plant consumed: Foliage
8. Portion analyzed/sampled: Forage and hay
9. Classifications:
   a. Authors Class: See Forage grass
   b. EPA Crop Group (Group & Subgroup): See Forage grass
   c. Codex Group: See Forage grass
   d. EPA Crop Definition: None
10. References: GRIN, MAGNESS, BALL
11. Production Map: EPA Crop Production Regions 2, 5, 6, 7, 8 and 9.
12. Plant Codes:
    a. Bayer Code: SOSNU

298

1. Jaboticaba (Ja-boat-e-ka-ba)
(Brazilian grapetree)
Myrtaceae

Myrciaria cauliflora (C. Mart.) O. Berg
2. Grows in cool tropical and subtropical areas with medium to high rainfall. Propagated by seed or grafting. The tree or large shrub is a slow-growing tropical evergreen, reaching to 30 feet. Flowers and fruits are borne in clusters along the trunk and main branches. Individual fruits are about $1/2$ inch in diameter. The skin is rather tough, thick and maroon in color. The pulp has a flavor similar to muscadine grape. Each fruit contains 1 to 5 seeds.
3. Season: Fruits mature in 30-40 days after flowering. Flowering and fruiting are nearly continuous, thus flowers and fruits (unripe and mature) can be on the plant at any given time.
4. Production in U.S.: Some production in South Florida.
5. Other production regions: South America (MORTON 1987).
6. Use: Pulp eaten fresh out of hand or in salads or desserts. Fruit can also be processed into jellies or wine.
7. Part(s) of plant consumed: Whole fruit
8. Portion analyzed/sampled: Whole fruit
9. Classifications:
   a. Authors Class: Tropical and subtropical fruits – edible peel
   b. EPA Crop Group (Group & Subgroup): Miscellaneous
   c. Codex Group: 006 (FT 0300) Assorted tropical and subtropical fruits – edible peel
   d. EPA Crop Definition: Proposing Guava = Jaboticaba
10. References: GRIN, CODEX, LOGAN, MAGNESS, MARTIN 1987, US EPA 1994, MORTON (picture), PHILLIPS(b)
11. Production map: EPA Crop Production Region 13.
12. Plant Codes:
    a. Bayer Code: No specific entry

299

1. Jackbean
(Haba de Caballo, Chickasaw lima bean, Brazilian broad bean, Coffee bean, Ensiform bean, Horsebean, Mole bean, Go-Ta-Ki, Overlook bean, Pearson bean, Watanka, Swordbean, Wonderbean, Giant stockbean, Gotani bean)
Fabaceae (Leguminosae)
Canavalia ensiformis (L.) DC.

2. This bean is grown in the Southern U.S. mainly for stock feed, but young pods can be used as snap beans. The plant is bushy. Pods reach 10 to 14 inches in length, but are harvested at half that size for eating. Pods may contain 3 to 18 white seeds. Seeds are large, $1/2$ to $3/4$ inch long, and nearly as broad, and are sometimes used as coffee substitute. Jackbeans are not grown as a commercial food crop in this country.
3. Season, seeding to first harvest: 8 to 12 weeks.
4. Production in U.S.: Grown in Florida gardens (STEPHENS 1988). Southwest U.S.
5. Other production regions: Tropical and subtropical plant (YAMAGUCHI 1983). Mexico and southwest U.S. (JANICK 1990) as secondary grain legume.
6. Use: Immature pods and seeds used as snap beans. Fodder for feed or green manure crop.
7. Part(s) of plant consumed: Young pods and immature seeds
8. Portion analyzed/sampled: Young pods and immature seeds (edible-podded)
9. Classifications:
   a. Authors Class: Legume vegetables (succulent or dried)
   b. EPA Crop Group (Group & Subgroup): Legume vegetables (succulent or dried) (6A)
   c. Codex Group: 014 (VP 0532) Legume vegetables

d. EPA Crop Definition: None
10. References: GRIN, CODEX, JANICK 1990, MAGNESS, MARTIN 1983, STEPHENS 1988 (picture), US EPA 1995a, YAMAGUCHI 1983, USDA NRCS, RUBATZKY
11. Production Map: EPA Crop Production Regions 3 and 13.
12. Plant Codes:
   a. Bayer Code: CNAEN

300

1. Jackfruit
(Jaca, Jak-fruit, Jack, Kathal)
Moraceae
Artocarpus heterophyllus Lam. (syn: *A. integrifolia* L.f., *A. integer* auct. *A. integrifolius* auct., *Rademachia integra* Thunb)

2. These are large, tropical evergreen trees that are native to the rain forests of Southern India. The tree grows to 70 feet tall. The leaves are 9 inches long and oval. Both male and female flowers are borne in clusters. The aggregate fruit grows 8 to 36 inches long and 6 to 20 inches in diameter. A fruit can weigh over 100 pounds but usually in the 10 to 60 lb range. The fruit skin varies in color from green to orange-yellow in color. The interior consists of numerous bulbs of yellow flesh and a central core. Each bulb covers a seed. Unopened ripe fruit has a strong odor like that of decaying onions. The pulp of open ripe fruit are much more agreeable, smelling like pineapple and banana. There are two main types, koozha chakka which are the smaller soft fruits and koozha pazham, the large crisp fruits. Some prominent varieties include 'Singapore' or 'Ceylon', 'Safeda', 'Khaja' and 'T Nagar Jack'.
3. Season, bloom to harvest: In 3 to 6 months. Flowers and fruits all year.
4. Production in U.S.: In 1994, 5 acres in Florida (CRANE 1995a). Plant is likely to be grown in South Florida, Puerto Rico and Hawaii.
5. Other production regions: Important food sources in several tropical areas, including India where there are over 14,000 acres. U.S. imports from Thailand. Important in Brazil.
6. Use: Unripe fruit can be cut into chunks and cooked or pickled. Bulbs must be removed from ripe fruit. The bulbs can be eaten raw, cooked or processed into ice cream, jams/jellies, liquor or syrup. There are also some canned and dried products. Young leaves and flowers can be used a vegetable. Tree used for wood.
7. Part(s) of plant consumed: Whole fruit; pectin extracted from rind, bulbs and seeds for food products. Sometimes leaves and flowers
8. Portion analyzed/sampled: Whole fruit
9. Classifications:
   a. Authors Class: Tropical and subtropical fruits – inedible peel
   b. EPA Crop Group (Group & Subgroup): Miscellaneous
   c. Codex Group: 006 (FI 0338) Assorted tropical and subtropical fruits – inedible peel
   d. EPA Crop Definition: None
10. References: GRIN, CODEX, CRANE 1995a, CRANE 1995, LOGAN, MAGNESS, MARTIN 1987, MORTON (picture), ROY
11. Production Map: EPA Crop Production Region 13.
12. Plant Codes:
   a. Bayer Code: ABFHE

301

1. Japanese honewort
(Mitsuba, Honewort, Japanese parsley, Ya-er-qin)
Apicaeae (Umbelliferae)
Cryptotaenia japonica Hassk.

2. An important minor crop in Japan. A cool season crop, the main production is March through May, but is grown throughout the year in Japan. The petioles are 4 to 6 inches long. Leaves with three broad leaflets used much like celery.
3. Season, seeding to harvest: Propagated by seed or by rhizome section. It can be propagated hydroponically.
4. Production in U.S.: No data. Japanese honewort was cultivated in Oregon (DUKE 1992).
5. Other production regions: Honewort (*C. canadensis*), which see, in North America and Japan (NEAL 1965).
6. Use: Used in soups, raw in salads and flavoring (raw or cooked).
7. Part(s) of plant consumed: Leaves and petioles
8. Portion analyzed/sampled: Tops (fresh)
9. Classifications:
   a. Authors Class: Leafy vegetables (except *Brassica* vegetables)
   b. EPA Crop Group (Group & Subgroup): Miscellaneous
   c. Codex Group: No specific entry
   d. EPA Crop Definition: None
10. References: GRIN, MYERS, NEAL, YAMAGUCHI, DUKE(c), RUBATZKY
11. Production Map: EPA Crop Production Region 12.
12. Plant Codes:
   a. Bayer Code: CPBCA

302

1. Japanese knotweed
(Jointweed, Mexican Bamboo, Ong Toy, Japanese bamboo, Itadori)
Polygonaceae
Polygonum cuspidatum Siebold and Zucc. (syn: *Reynoutria japonica* Houtt.)

2. Erect unarmed glabrous perennial with round ovate leaves. The immature stems are edible and used especially by the Chinese in salads or cooked as a potherb. The long succulent young stems can be produced in the field or grown under glass. A crop grown under glass is under high humidity in water-saturated soil and harvested continuously. Plants root very rapidly. This is a close relative to Smartweed.
3. Season, harvest: The sprouts are harvested in the spring (March to April) in the wild. Continuously under glass.
4. Production in U.S.: Limited. New Jersey. Midwest; east and south to North Carolina and Tennessee (DUKE 1992).
5. Other production regions: Japan, China.
6. Use: Mainly in salads and cooked like asparagus.
7. Part(s) of plant consumed: Immature stems up to 1 foot long are cut
8. Portion analyzed/sampled: Immature stems
9. Classifications:
   a. Authors Class: Stalk and stem vegetables
   b. EPA Crop Group (Group & Subgroup): Miscellaneous
   c. Codex Group: No specific entry
   d. EPA Crop Definition: None

10. References: GRIN, BAILEY 1976, DUKE(c) (picture), MAGNESS
11. Production Map: EPA Crop Production Regions 1, 2 and 5.
12. Plant Codes:
    a. Bayer Code: POLCU

303

1. Jojoba (Ho-ho-ba)
(Goatnut, Deernut)
Simmondsiaceae
Simmondsia chinensis (Link) C.K. Schneid. (syn: *S. californica* Nutt.)

2. Slow growing evergreen perennial desert shrub native to Sonora Desert and Arizona. May grow to 10 feet tall with a natural life span of over 100 years. The plant produces fruit capsules that contain 1-3 oily brown seeds about the size of a hazelnut or peanut. A mature plant averages 5 pounds of seeds (dry weight) per year. The seeds are extracted for its oil which is 50 percent of the seeds dry weight, and is used for shampoo, in cosmetics, and as a sperm oil substitute. The extracted oil seed meal has potential as an animal feed. The seed can be roasted and eaten or used as a coffee substitute. Collecting seeds from the ground by vacuum appears the most promising for mechanical harvesting.
3. Season, flowering to harvest: In southern California, flower from November to Feburary with the first ripe fruit on the plant in May. The growers wait until all mature fruits fall and vacuum them up in August. Planting to first commerical harvest is about 5 years.
4. Production in U.S.: 1992 CENSUS reported 15,010 acres harvested in U.S. with 11,311 acres in Arizona and 3,699 acres in California with most farms irrigated. Some production in Texas (FRANK 1997). Desert southwest U.S. (SCHNEIDER 1993a).
5. Other production regions: Mexico (ROBBBELEN 1989).
6. Use: Oil used in cosmetics. Meal as a feed supplement in lamb rations at 5 percent. The treatment of the meal with *Lactobacillus* increases its palatability for livestock (MANOS 1986).
7. Part(s) of plant consumed: Seed. Meal has potential as animal feed. Not significant feed item to date
8. Portion analyzed/sampled: Seed and its meal
9. Classifications:
   a. Authors Class: Oilseed
   b. EPA Crop Group (Group & Subgroup): Miscellaneous
   c. Codex Group: No specific entry
   d. EPA Crop Definition: None
10. References: GRIN, MANOS 1986, MELNICOE 1996c, NATIONAL RESEARCH COUNCIL 1975, RICHE, ROBBELEN 1989 (picture), SCHNEIDER 1993a, FRANK
11. Production Map: EPA Crop Production Regions 9 and 10 (Region 9, since some plants grown just north of Phoenix, Arizona).
12. Plant Codes:
    a. Bayer Code: SMMCH

304

1. Jostaberry
(Pruterberry, Yostaberry))
Grossulariaceae (Saxifragaceae)

Ribes x *nidigrolaria* Rud. Bauer & A. Bauer (*R. nigrum* L. x *R. divaricatum* x *R. uva-crispa* L.)

2. Jostaberries are a cross between gooseberry and black currants. They are a nonthorny, woody plant; they do not have primocanes. The berry is deep black when ripe but, unlike currants, the berries do not grow in clusters. The berries are difficult to harvest, because the fruit ripens unevenly. The fruit has a tart taste.
3. Season, harvest: Picked by hand one to two times per week during harvest season.
4. Production in U.S.: Some production in Eastern Washington state and Idaho (SCHREIBER 1995). No available production statistics.
5. Other production regions: Grown in Canada (ROTHWELL 1996a).
6. Use: Fresh often processed into jams, jellies and fruit spreads.
7. Part(s) of plant consumed: Berry
8. Portion analyzed/sampled: Whole fruit
9. Classifications:
   a. Authors Class: Berries
   b. EPA Crop Group (Group & Subgroup): Miscellaneous
   c. Codex Group: No specific entry
   d. EPA Crop Definition: None
10. References: GRIN, SCHREIBER, ROTHWELL 1996a
11. Production Map: EPA Crop Production Region 11.
12. Plant Codes:
    a. Bayer Code: No specific entry

305

1. Jujube (Ju-ju-be)
(Ber, Chinese date, Gingeolier, Common jujube, Tsao, Chinese jujube, Tsa, Azufaifo)
Rhamnaceae
Ziziphus jujuba Mill. (syn: *Z. sativa* Gaertn.; *Z. spinosa* (Bunge) Hu ex F.H. Chen; *Z. zizyphus* (L.) Meikle)

2. The jujube or Chinese date, introduced into this country from China, is a medium size decidous tree, 25 feet or taller, with glossy green, deciduous foliage. It thrives best in warm, dry climates; but will withstand winter temperatures down to -20 degrees F. Small flowers grow in clusters in some of the leaf axils. Plants have an extended blooming period. Fruit is generally dark, mahogany brown when ripe, oval to pyroform in shape and range in size from a cherry to a plum. The fruit have a single stone. It is very sweet, with sugar content up to 22 percent and fruit will dry if left on tree, similar to figs. Fruit skin is smooth and thin until drying of fruit occurs, then becomes wrinkled. Pulp is dryer than in most fruits.
3. Season: Bloom to maturity in 2 to 4 months, depending on kind and climate.
4. Production in U.S.: Not commercial. Scattered trees, mostly home gardens in Southern tier of states.
5. Other production regions: China.
6. Use: Some fresh eating, dried, smoked, pickled, or candied.
7. Part(s) of plant consumed: Fruit pulp and fruit skin
8. Portion analyzed/sampled: Whole fruit with stone removed
9. Classifications:
   a. Authors Class: Tropical and subtropical fruits – edible peel
   b. EPA Crop Group (Group & Subgroup): Miscellaneous

c. Codex Group: 005 (FT 0302) Assorted tropical and subtropical fruit – edible peel

d. EPA Crop Definition: None

10. References: GRIN, CODEX, MAGNESS, REICH, MARTIN 1987, GUPTA

11. Production Map: No entry

12. Plant Codes:
   a. Bayer Code: ZIPJU

306

1. **Jujube, Indian**

(Aprin, Dunks, Beri, Chinese date, Indian cherry, Indian plum, Ber, Bor, Cottony jujube)

Rhamnaceae

Ziziphus mauritiana Lam. (syn: *Z. jujuba* (L.) Gaertn.)

2. Tropical evergreen shrub or tree 10 to 40 feet in height. The orange to brown fruit is rounded or oblong in outline, and $1/2$ to 1 inch in length. A layer of edible pulp surrounds the seed.

3. Season, bloom to harvest: About 180 days.

4. Production in U.S.: No data.

5. Other production regions: Subtropical (MARTIN 1987).

6. Use: Fresh, dried, stewed, candied, preserved.

7. Part(s) of plant consumed: Fruit pulp

8. Portion analyzed/sampled: Whole fruit

9. Classifications:
   a. Authors Class: Tropical and subtropical fruit – edible peel
   b. EPA Crop Group (Group & Subgroup): Miscellaneous
   c. Codex Group: 005 (FT 0301) Assorted tropical and subtropical fruit – edible peel
   d. EPA Crop Definition: None

10. References: GRIN, CODEX, MAGNESS, MARTIN 1987

11. Production Map: EPA Crop Production Region 13.

12. Plant Codes:
   a. Bayer Code: ZIPMA

307

1. **Juneberry**

(Serviceberry, Sarvisberry, Saskatoon, Alleghney serviceberry, Chinese serviceberry, Sarvistree, Shadblow, Shadbush, Swamp sugar pear, Currant tree, Snowy mespilus, Indian pear, Shad, Sugarplum)

Rosaceae

Amelanchier spp.

1. **Saskatoon serviceberry**

(Western serviceberry)

A. alnifolia (Nutt.) Nutt. ex M. Roem.

1. **Downy serviceberry**

(Woodland saskatoon)

A. arborea (F. Michx.) Fernald

1. **Allegheny serviceberry**

A. laevis Wiegand

2. Some species of *Amelanchier* occur over most parts of the U.S. and produce edible fruits. Most are shrubs or small slender trees usually 6 to 15 feet high but can grow to 20 feet. They are propagated readily from seed, root sprouts, root cuttings or softwood cuttings. The plants bear fruit 2 to 4 years after planting. The fruits are borne on previous year's growth and older wood. The flowers are in open clusters or racemes. Individual fruits are under $1/3$ inch diameter, near round, generally black with a waxy bloom. They are sweet with little acid when ripe. Cultivated forms are known, but serviceberries are but sparingly planted for fruit. Some may be found in Northern Plains gardens, where adapted fruits are limited in number. Substantial quantities may be harvested from native plants. 'Regent' and 'Saskatoon' are cultivars of *A. alnifolia* which normally bears extra sweet, high quality, fruit the second year after transplanting (WHEALY 1993).

3. Season: In northern states they bloom early in the season, making them susceptible to spring frosts. Fruit ripens in late June/early July. The season from bloom to harvest is 2 to 3 months.

4. Production in U.S.: Very minor, no data available.

5. Other production regions: In 1995, Canada reported 850 acres (ROTHWELL 1996a).

6. Use: The fruit may be eaten fresh, in pies or other baked desserts. It may also be canned, frozen or made into wine, jellies, jams, preserves and syrup. The fruit is used by Native Americans in making pemmican, a semidry mixture of fruit and meat. Can be used as an ornamental for its showy flowers.

7. Part(s) of plant consumed: Fruit

8. Portion analyzed/sampled: Whole fruit

9. Classifications:
   a. Authors Class: Berries
   b. EPA Crop Group (Group & Subgroup): Miscellaneous
   c. Codex Group: 004 (FB 0270) Berries and other Small Fruit
   d. EPA Crop Definition: None

10. References: GRIN, CODEX, MAGNESS, STANG, US EPA 1995, ROTHWELL 1996a, WHEALY, FRANK

11. Production Map: No entry.

12. Plant Codes:
   a. Bayer Code: AMEAL (*A. alnifolia*), AMEAR (*A. arborea*), AMELA (*A. laevis*)

308

1. **Junegrass**

(Prairie junegrass, Koelersgrass, Zacate de Junio, Crested hairgrass)

Poaceae (Gramineae)

Koeleria macrantha (Ledeb.) Schult. (syn: *Koeleria pyramidata* auct. Amer.; *K. nitia* Nutt.)

2. This is a perennial cool season grass native to the U.S. that is a member of the grass tribe *Aveneae*, and is distributed over most of Canada, northern and central Mexico and the U.S. except for the Southeastern states. It is an erect bunchgrass 20 to 60 cm tall growing with a panicle inflorescence (3 to 18 cm long) in prairie and open woodland areas. Junegrass provides an excellent quality forage for all livestock, and has a fair to good hay quality. Seeding rates for Junegrass are 9 to 13 kg/ha. Low maintenance orchard grass.

3. Season: Cool season grass that starts growing in early spring, flowers in June and July and seed matures in September.

4. Production in U.S.: There is no production data for Junegrass, however, pasture and rangeland are produced on more than 410 million A (U.S. CENSUS, 1992). Throughout U.S. except in the Southeastern states (GOULD 1975).

5. Other production regions: Widespread in temperate regions of Northern Hemisphere (GOULD 1975).

6. Use: Pasture grazing, hay, soil erosion and turfgrass.
7. Part(s) of plant consumed: Leaves and stems
8. Portion analyzed/sampled: Forage and hay
9. Classifications:
   a. Authors Class: See Forage grass
   b. EPA Crop Group (Group & Subgroup): See Forage grass
   c. Codex Group: See Forage grass
   d. EPA Crop Definition: None
10. References: GRIN, US EPA 1994, US EPA 1995, CODEX, HOLZWORTH, ALDERSON, BARNES, MOSLER, STUBBENDIECK (picture), RICHE, GOULD
11. Production Map: EPA Crop Production Regions 1, 4, 5, 6, 7, 8, 9, 10, 11 and 12.
12. Plant Codes:
   a. Bayer Code: No specific entry

309

1. Juniper
(Common juniper, Genevrier, Enebro)
Cupressaceae
   Juniperus communis L.
2. Juniper berries are the fruits of the small, evergreen juniper tree. The berries are about $1/3$ inch in diameter, dark blue, nearly globular, smooth-skinned, with a sweet, bitter flavor. They become near black when dried, and are used in the manufacture of "gin" drinks. The characteristics of the berries are due to a volatile oil.
3. Season, bud swell to harvest: About 3 months.
4. Production in U.S.: No data, very limited. Hardy in Northern U.S.
5. Other production regions: It is hardy in Southern Canada (BAILEY 1976). Native to Eurasia (FACCIOLA 1990).
6. Use: Flavoring of gin and seasoning of game. The dried berries are ground to add flavor to meat. Also grown as an ornamental.
7. Part(s) of plant consumed: Fruit
8. Portion analyzed/sampled: Fruit (Dried)
9. Classifications:
   a. Authors: Class: Herbs and spices
   b. EPA Crop Group (Group & Subgroup): Herbs and spices group (19B)
   c. Codex Group: 028 (HS 0786) Spices
   d. EPA Crop Definition: None
10. References: GRIN, BAILEY 1976, CODEX, MAGNESS, PENZEY, US EPA 1995a, FACCIOLA
11. Production Map: No data, but can grow in USDA Zone 3 (BAILEY 1976)
12. Plant Codes:
   a. Bayer Code: IUPCO

310

1. Jute, Nalta
(Jews mallow, Tussa jute, Tossa jute, Bush okra, West African sorrel, Jute mallow)
Tiliaceae
   Corchorus olitorius L.
2. Tropical herb, grown both for the jute fiber or gunny obtained from the stem bark and for the young shoots, used as potherbs. Plants reach 10 to 12 feet in height with slender stems branched only at the top. It is grown in many tropical and subtropical areas as a garden vegetable. Leaves are used much as in spinach. Leaves are narrow and serrate, about 2 to 5 inches long.
3. Season, planting to first harvest for food: About 3 months. Depending on conditions, the first edible crop may be harvested in as little as 1 to $2^{1}/_{2}$ months after planting with subsequent cuttings each month.
4. Production in U.S.: No data. Limited to Puerto Rico and Hawaii.
5. Other production regions: Tropics (MARTIN 1979).
6. Use: As potherb. The leaves may be dried and used either as a tea or as a cooked vegetable (fresh or dried).
7. Part(s) of plant consumed: Young leaves
8. Portion analyzed/sampled: Leaves
9. Classifications:
   a. Authors Class: Leafy vegetables (except *Brassica* vegetables)
   b. EPA Crop Group (Group & Subgroup): Miscellaneous
   c. Codex Group: No specific entry
   d. EPA Crop Definition: None
10. References: GRIN, MAGNESS, MARTIN 1979, NEAL, RUBATZKY
11. Production Map: EPA Crop Production Region 13.
12. Plant Codes:
   a. Bayer Code: CRGOL

311

1. Kale, Common
(Scotch kale, Borecole, Flowering kale, Curly kale, Kitchen kale, Dwarf siberian kale)
Brassicaceae (Cruciferae)
   Brassica oleracea var. *sabellica* L. (syn: *Brassica oleracea* var. *acephala* DC., in part)
2. Kale grown for food is handled as an annual, although the plant is biennial, producing a seed crop the second year. The Scotch varieties have very curled, grayish-green thick smooth leaves. In the north they may be seeded in early spring for summer production, or in midsummer for fall harvest. In the Southern states, kale is planted in the fall and harvested throughout the winter. For market, the whole young plant is cut off and trimmed. For home use, leaves are stripped off, and the plant continues to produce. Kale as food is used mainly as a potherb, although sometimes in salads. Kale is also used as feed for livestock. Kale growth is similar to collards.
3. Season, seeding to harvest: About 2 to 3 months.
4. Production in U.S.: All kale – 7,950 acres (1992 CENSUS). No separate data for Scotch kale. In 1992, all kale reported from the top 10 states as Georgia (1,481 acres), California (1,268), Maryland (870), Texas (701), New Jersey (482), Pennsylvania (411), North Carolina (330), Virginia (315), Illinois (300) and Ohio (284) with about 81 percent of the U.S. acreage (1992 CENSUS). Flowering kale is available year-round from California. It has large wrinkled leaves of either purple, cream or spruce color laced with green trim (LOGAN 1996).
5. Other production regions: Grown in the temperate regions (RUBATZKY 1997).
6. Use: As potherb. Frozen commercially. Used in salads, washed before use.
7. Part(s) of plant consumed: Leaves

8. Portion analyzed/sampled: Leaves
9. Classifications:
   a. Authors Class: *Brassica* (cole) leafy vegetables
   b. EPA Crop Group (Group & Subgroup): *Brassica* (cole) leafy vegetables (5B)
   c. Codex Group: 013 (VL 0480) Leafy vegetables (including *Brassica* leafy vegetables) for food; 052 (AV 1052) Miscellaneous fodder and forage crop for feed
   d. EPA Crop Definition: None
10. References: GRIN, CODEX, LOGAN (picture), MAGNESS, RICHE, US EPA 1996a, US EPA 1995a, RUBATZKY, SWIADER, SPLITTSTOESSER
11. Production Map: EPA Crop Production Regions 1, 2, 5, 6 and 10.
12. Plant Codes:
    a. Bayer Code: BRSOA (Kale), BRSOC (Curly kale)

312

1. Kale, Sea

(Colewort, Scurvygrass, Halmyrides, Abyssinian kale)

Brassicaceae (Cruciferae)

Crambe maritima L.

2. Sea kale is a large-leaved perennial to 2 feet tall, the young shoots are eaten in the spring, usually after blanching. Plants may be grown from seed or root cuttings, and usually are given a year to become firmly established before any harvesting. The plant for food is grown much as is rhubarb. The second year and thereafter the young, etiolated shoots are harvested when 4 to 8 inches high. Usually they are blanched by covering the crown deeply with soil, through which the young shoots grow, or by otherwise darkening. They are prepared for eating like asparagus, mainly cooked, but also as salad. Large growth of sea kale may also be used as feed for livestock.
3. Season: Harvested early in spring, as shoots emerge. The roots may be forced like Belgium endive. Shoots are harvested in 4 to 5 weeks.
4. Production in U.S.: No data. Little grown. Rare in North America (RYDER 1979). Production requires a cool, moist climate (STEPHENS 1988).
5. Other production regions: Native along shores of western Europe (YAMAGUCHI 1983).
6. Use: Potherb, or occasionally in salad. Leaves are insignificant feed item.
7. Part(s) of plant consumed: Young shoots and leaves before expanding
8. Portion analyzed/sampled: Young shoots and small leaves
9. Classifications:
   a. Authors Class: Leafy vegetables (except *Brassica* vegetables)
   b. EPA Crop Group (Group & Subgroup): Miscellaneous
   c. Codex Group: 013 (VL 0499) Leafy vegetables (including *Brassica* leafy vegetables)
   d. EPA Crop Definition: None
10. References: GRIN, CODEX, MAGNESS, RYDER 1979, STEPHENS 1988 (picture), YAMAGUCHI 1983
11. Production Map: No entry
12. Plant Codes:
    a. Bayer Code: CRMMA

313

1. Kapok oil

(Ceiba, Kapok, Silk-Cotton tree, Kapoktree, White silk-cotton tree)

Bombacaceae

Ceiba pentandra (L.) Gaertn.

2. Kapok oil is obtained from the seeds of the Kapok tree, which is a very large, deciduous tropical or semi-tropical tree, now grown in many tropical areas. The fruit is a pod, about 6 inches long and 2 inches in diameter, which is lined inside with hairs or lint, the kapok fiber of commerce, for which the tree is mainly grown. The seeds are free of lint, and are a byproduct of lint production. Seeds may be crushed for oil locally or exported. The oil is suitable for the same purposes as cotton seed oil. The press cake and meal are used for cattle feed.
3. Season, flower fall to harvest: About 2 months.
4. Production in U.S.: No data.
5. Other production regions: Tropical areas. Native of tropical America and grows in Mexico.
6. Use: Cooking oil.
7. Part(s) of plant consumed: Seed: Also seeds and flowers are used as feed
8. Portion analyzed/sampled: Seed, and its oil and meal
9. Classifications:
   a. Authors Class: Oilseed
   b. EPA Crop Group (Group & Subgroup): Miscellaneous
   c. Codex Group: 023 (SO 0692) Oilseed
   d. EPA Crop Definition: None
10. References: GRIN, CODEX, FACCIOLA, MAGNESS, NEAL, USDA NRCS
11. Production Map: No entry
12. Plant Codes:
    a. Bayer Code: CEIPE

314

1. Kenaf (Ka-naf)

(Indian hemp, Deccan hemp, Brown indian hemp, Deckaner hemp, Bimli jute, Bastard jute, Bimlipatum jute)

Malvaceae

Hibiscus cannabinus L.

2. Kenaf is an annual, nonwood fiber plant native to Africa and related to cotton. It is grown in rows, reaches a height of 12-18 feet, and yields 6-10 tons of dry fiber per acre. The bamboo-like kenaf can be grown as a high-protein livestock feed (dried leaves and stems) which is harvested 60-80 days after planting that could afford double cropping. Kenaf is normally separated into fiber and core. The fiber is used primarily in making paper and the core is used in horse and chicken bedding which is very absorbent.
3. Season, seeding to harvest: 150 days for maximum dry matter harvest for paper fiber and core. For feed 60-80 days.
4. Production in U.S.: Limited data but increasing importance. Southern U.S.
5. Other production regions: Africa (BAILEY 1976).
6. Use: Fodder, forage, animal bedding, and pulp, paper and other fiber products.
7. Part(s) of plant consumed: Dried leaves and stems as animal feed
8. Portion analyzed/sampled: Dried leaves and stems

9. Classifications:
   a. Authors Class: Nongrass animal feeds
   b. EPA Crop Group (Group & Subgroup): Miscellaneous
   c. Codex Group: No specific entry
   d. EPA Crop Definition: None
10. References: GRIN, GOFORTH, KUGLER (picture), USDA 1993b, USDA NRCS, BAILEY 1976, DUKE(a)
11. Production map: EPA Crop Production Regions 2, 4, 6, 8 and 10 (same as Cotton).
12. Plant Codes:
    a. Bayer Code: HIBCA

315

1. Kidney vetch
(Kidneyvetch)
Fabaceae (Leguminosae)
Anthyllis vulneraria L.
2. Kidney vetch is a cultivated cool season forage legume of temperate regions such as Europe along the Atlantic Ocean coast and can be grown as an annual, biennial or perennial, depending how it is managed. It grows up to 90 cm tall, on soils with pH varying from 4.8 to 8.0 and from sea level to 2230 m elevation and annual precipatation 4.4 to 13.6 dm. Kidney vetch is cultivated like clovers and cut for hay with yields varying from 4 to 20 mt/ha. It flowers June to August and is in production until frost. There are 396,900 seeds/kg.
3. Season: Cool season annual, biennial or perennial forage that starts growth in the spring to the first killing frost. It is preferred as a forage for sheep and goats.
4. Production in U.S.: No specific data for Kidney vetch production, however, pasture and rangelands are produced on more than 410 million A (U.S. CENSUS, 1992).
5. Other production regions: Europe.
6. Use: Grazing and hay.
7. Part(s) of plant consumed: Leaves and stems
8. Portion analyzed/sampled: Forage and hay
9. Classifications:
   a. Authors Class: Nongrass Animal Feeds
   b. EPA Crop Group (Group & Subgroup): Miscellaneous
   c. Codex Group: No specific citation although the general class 050 (AL 0157) Legume Animal Feeds could be used
   d. EPA Crop Definition: None
10. References: GRIN, US EPA 1994, US EPA 1995, CODEX, HOLZWORTH, DUKE(h) (picture), RICHE
11. Production Map: EPA Crop Production Region are 1, 2 and 3.
12. Plant Codes:
    a. Bayer Code: AYLVU

316

1. Kiwifruit
(Chinese gooseberry, Kiwi, Yang-t'ao, Hardy kiwi, Strawberry peach, Fuzzy kiwi, Kiwi fruit, Babykiwi)
Actinidaceae
Actinidia deliciosa (A. Chev.) C. F. Liang & A.R. Ferguson (syn: *A. chinensis* var. *deliciosa* (A. Chev.) A. Chev.)
2. Grows in subtropical and warm temperate areas where there is significant prolonged cold weather during dormancy for successful development. Propagated from seed, cuttings and grafting with first fruit production in one year from vegeta-tive propagation. The kiwifruit plant is a deciduous vine which requires trellises for support. The vines lack tendrils, however terminal shoots will entwine the supporting wire. Flowers are borne in axils of leaves, either singly or in small inflorescences. The brown fuzzy fruit are eggshaped and about 3-5 inches long. The internal pulp is green with many seeds. 'Hayward' is the most recognized kiwifruit variety marketed in the U.S. Hardy kiwis (Babykiwi) (*Actinidia* spp.) are grown farther north, hardy to -25 degrees F and lack the fuzzy fruit surface. They are sweeter when mature and grape-like in size.
3. Season: Bloom to fruit maturity: 180 to 200 days.
4. Production in U.S.: 1995 California production reported at 6,600 bearing acres. Minor production in coastal South Carolina and Georgia. Hardy kiwi is grown in Oregon on small acreage.
5. Other production regions: Worldwide production up to 170,000 acres. U.S. imports significant amount of kiwifruit from Chile and New Zealand. In 1995, Canada grew 40 acres (ROTHWELL 1996a) of a cold hardy type.
6. Use: Almost all used as fresh fruit or juice. Some used in cooking and puree. Also grown as an ornamental.
7. Part(s) of plant consumed: Pulp
8. Portion analyzed/sampled: Whole fruit
9. Classifications:
   a. Authors Class: Small fruits
   b. EPA Crop Group (Group & Subgroup): Miscellaneous
   c. Codex Group: 006 (FI 0341) Assorted tropical and subtropical fruits – inedible peel
   d. EPA Crop Definition: None
10. References: GRIN, CODEX, HASSY, LOGAN, ROTHWELL 1996a, MARTIN 1987, US EPA 1994, USDA 1996f
11. Production Map: EPA Crop Production Region 10.
12. Plant Codes:
    a. Bayer Code: ATICH

317

1. Koa
(Koa acacia)
Fabaceae (Leguminosae)
Acacia koa A. Gray
2. Koa is a perennial warm season tropical spreading tree 15 to 30 m tall that is native to the Asian tropics and is grown in Hawaii. This tree occurs at elevations from 230 to 850 m, and grows rapidly and within 6 years it will reach a height of 3 m. It also produces pods 7 to 12 cm long and 1.3 cm wide. It is used as a forage and must not be heavily grazed to avoid eliminating it. It reproduces by seeds and suckers.
3. Season: Warm season perennial tree, leaves and stems are readily grazed.
4. Production in U.S.: No specific data for Koa production, however, pasture and rangelands are produced on more than 410 million A (U.S. CENSUS, 1992). Hawaii.
5. Other production regions: Asian tropics.
6. Use: For grazing as a forage crop.
7. Part(s) of plant consumed: Leaves and stems
8. Portion analyzed/sampled: Forage and hay.
9. Classifications:
   a. Authors Class: Nongrass Animal Feeds
   b. EPA Crop Group (Group & Subgroup): Miscellaneous

c. Codex Group: No specific citation although the general class 050 (AL 0157) Legume Animal Feeds could be used

d. EPA Crop Definition: None

10. References: GRIN, US EPA 1994, US EPA 1995, CODEX, HOLZWORTH, SKERMAN(a) (picture), RICHE

11. Production Map: EPA Crop Production Region 13.

12. Plant Codes:
   a. Bayer Code: No specific entry

318

1. Kochia

Chenopodiaceae
Bassia spp.

1. Kochia

(Fireweed, Burningbush, Summer cypress, Belvedere, Mexican firebush)

Bassia scoparia (L.) A.J. Scott (syn: *Kochia scoparia* (L.) Schrad.; *Chenopodium scoparia* L.)

1. Greenmolly summercypress

(Perennial summercypress, Red sage, Green molly)

Bassia americana (S. Watson) A.J. Scott (syn: *Kochia americana* S. Watson)

2. Kochia is an annual warm season forb, native to Eurasia. It was introduced in the U.S. around 1900 as an ornamental, since it turns fiery red in the fall. Kochia is an erect and spreading bushy forb growing from 1 to 7 feet tall, and has taproots. It grows wild throughout the U.S. and Canada, except in the Pacific Northwest. It is good forage quality for livestock and wildlife when immature, but quality declines rapidly with maturity, and it can become a weed. Kochia can break off at the stem and be wind carried as a tumbleweed. Its habitat includes roadsides, pastures, wastelands and fields. Since it has a low moisture requirement it is being researched as a drought resistant forage in dry areas. Its palatability is better than grasses such as bromegrass, but a little lower than alfalfa. It is also called the "poor man's alfalfa", and can be cut for hay. All kochias may cause nitrate and/or oxalate poisoning if not carefully managed. They have been used for erosion control and revegetation programs.

Greenmolly summercypress is a perennial, warm season small shrub growing up to 20 inches tall that flowers in late summer and reproduces from seed. It produces excellent forage for sheep, cattle and deer. It is often used as a winter forage for sheep, and is high in protein in the fall. Its habitat includes desert valleys, marshes, roadsides and foothills having alkaline soils.

3. Season: Warm season shrubs grow from April through the fall. Provides forage in winter. Livestock should not graze for more than 90 to 120 days to prevent oxalate poisoning. The entire plant is edible. For hay or silage it should be cut when it is 18 to 26 in. tall. In the southwest 3 to 4 cuttings may be made.

4. Production in U.S.: There are no production data for Kochias, however, pasture and rangeland are produced on more than 410 million A (U.S. CENSUS, 1992).

5. Other production regions: Widely distributed in Canada.

6. Use: Grazing, pasture, hay, silage, revegetation and erosion control.

7. Part(s) of plant consumed: Leaves and stems

8. Portion analyzed/sampled: Forage and hay

9. Classifications:
   a. Authors Class: Nongrass Animal Feeds
   b. EPA Crop Group (Group & Subgroup): Miscellaneous
   c. Codex Group: 052 (AM 0165) Miscellaneous Fodder and Forage Crops
   d. EPA Crop Definition: None

10. References: GRIN, US EPA 1994, US EPA 1995, CODEX, HOLZWORTH, UNDERSANDER(a), STUBBENDIECK (picture), RICHE

11. Production Map: EPA Crop Production Regions for Kochia are 1, 5, 7, 8, 9, 10 and 11. EPA Crop Production Regions for Greenmolly summercypress are 8, 9 and 11.

12. Plant Codes:
   a. Bayer Code: KCHAM (*K. americana*), KCHSC (*B. scoparia*)

319

1. Kohlrabi

(Stem turnip, Colinabo, Cabbage turnip)

Brassicaceae (Cruciferae)
Brassica oleracea var. *gongylodes* L.

2. Kohlrabi is a cabbage relative, grown for the turnip-like enlargement of the stem just above ground level. Plants are grown from seed, and must be grown rapidly for good quality "bulbs". Leaves rise from the enlarged stem. The enlargement is globose or flattened, and 2 to 4 inches across at harvest. The "bulbs" are tender and succulent if rapidly grown and harvested, but become tough and fibrous with age. For eating, the peel is removed and the interior sliced or diced and cooked.

3. Season, seeding to harvest: About 2 months.

4. Production in U.S.: 147 acres reported in 1954 CENSUS. No later CENSUS data. Mainly in home gardens. In 1989, one county in California produced 12 acres (MYERS 1991). Grown in Washington, Regions 11 and 12 (SCHREIBER 1995).

5. Other production regions: In 1995, Canada reported limited kohlrabi production (ROTHWELL 1996a).

6. Use: Cooked vegetable but can be eaten raw.

7. Part(s) of plant consumed: Enlarged, bulb-like stem, after peeling, and young leaves. Baby kohlrabi can be served whole. The swollen stem of the baby kohlrabi can be eaten raw in salads. A 1992 study provided weight-to-weight ratio data between the tops and the enlarged stem as 57 percent bulbous stem and 43 percent tops (MARTINI 1992)

8. Portion analyzed/sampled: Bulbous stem and leaves

9. Classifications:
   a. Authors Class: *Brassica* (cole) leafy vegetables
   b. EPA Crop Group (Group & Subgroup): *Brassica* (cole) leafy vegetables (5A)
   c. Codex Group: 010 (VB 0405) *Brassica* (cole or cabbage) vegetables, head cabbages, flowerhead *Brassicas*
   d. EPA Crop Definition: None

10. References: GRIN, CODEX, MAGNESS, MARTINI, MYERS (picture), ROTHWELL 1996a, SCHREIBER, US EPA 1996a, US EPA 1995a, SPITTSTOESSER

11. Production Map: EPA Crop Production Regions 10, 11 and 12.

12. Plant Codes:
   a. Bayer Code: BRSOG

320

1. Kudzu (Kud-zu)
(Japanese arrowroot, Kuzu, Tropical kudzu, Puero, Ko, Ko-hemp, Kudsu, Kuzu)
Fabaceae (Leguminosae)
Pueraria montana var. *lobata* (Willd.) Maesen & S.M. Almeida (syn: *Pueraria lobata* (Willd.) Ohwi, *P. thunbergiana* (Siebold & Zucc.) Benth., *P. hirsuta* (Thunb.) Matsum.)
2. Kudzu is a long-lived perennial adapted to the southeastern quarter of the U.S. It was introduced from Japan in 1876 but its widespread use for pasture, hay and erosion control has been during the last half century. The plant is a coarse-growing vine that forms long runners which root and form crowns at the nodes. Leaves are trifoliate and large with leaflets up to 6 inches long and equally broad. Above-ground parts are killed by temperatures much below freezing, and deep freezing of the soil will kill the entire plant. In warm winter areas, stems become woody. Kudzu is nearly equal to alfalfa in nutritive value and palatability, both as pasture and cured as hay. It is not particularly valuable for erosion control. It continues growth until frost and gives a long season for pasturing. Propagation is by planting sections of the rooted runners with the crowns. Under good growing conditions a complete ground cover can be attained in 2 to 3 years and will persist indefinitely if not overgrazed or mowed too often. Commonly used for soil conservation uses in 1920 to 30s, but it is now a major weed problem in many areas of the South and as far north as Washington, DC. Tropical kudzu (Puero) is *P. phaseoloides* (Roxb.) Benth.
3. Season, rapid growth and flowers July-September. Production is May to October with one hay cutting per season.
4. Production in U.S.: No data. Southeastern U.S.
5. Other production regions: China.
6. Use: Hay, silage, pasture. Also roots, tender leaves, shoots and flowers are edible food.
7. Part(s) of plant consumed: Foliage
8. Portion analyzed/sampled: Forage and hay
9. Classifications:
   a. Authors Class: Nongrass animal feeds
   b. EPA Crop Group (Group & Subgroup): Nongrass animal feed (18)
   c. Codex Group: 050 (AL 1024) Legume animal feeds
   d. EPA Crop Definition: None
10. References: GRIN, CODEX, FACCIOLA, MAGNESS, US EPA 1995a, FRANK, BALL (picture), DUKE(h)
11. Production Map: EPA Crop Production Regions 2, 3, 4, 5 and 13.
12. Plant Codes:
    a. Bayer Code: PUELO

321

1. Kumquat
(Kunquat, Comquot, Kin kan)
Rutaceae
Fortunella spp.
1. Hongkong kumquat
F. hindsii (Champ. ex Benth.) Swingle
1. Round kumquat
(Meiwa, Marumi kumquat)

F. japonica (Thunb.) Swingle
1. Oval kumquat
(Nagami, Nagami kumquat)
F. margarita (Lour.) Swingle
1. Meiwa kumquat
F. crassifolia Swingle
2. Kumquats are small, citrus-like fruits, that will hybridize readily with citrus, but are not now classed botanically as citrus. The fruits are small and deeply colored, produced on small evergreen trees that are somewhat hardier than citrus. Fruit shape is round or distinctly oval, the latter being about 1 inch in diameter by 2 inches long. Rind is thin and edible, so the whole fruit may be eaten out of hand. In the U.S., kumquats are grown mainly in home gardens as ornamentals. Clusters of the highly colored, attractive fruits are frequently placed in gift packages for the ornamental effect. A few trees may be found in citrus orchards, especially of growers catering to gift package trade. Whole fruits, except for seeds, are sometimes made into marmalade.
3. Season, bloom to harvest: 8 to 10 months.
4. Production in U.S.: No data, limited. Available November to March. Southern California and Florida (LOGAN 1996).
5. Other production regions: Native to China, also grown in Central and South America and South India (MORTON 1987).
6. Use: Mainly decorative, but whole fruits may be eaten out of hand, or made into marmalade.
7. Part(s) of plant consumed: Fruit, entirely edible
8. Portion analyzed/sampled: Whole fruit
9. Classifications:
   a. Authors Class: Citrus fruits
   b. EPA Crop Group (Group & Subgroup): Citrus fruits (10)
   c. Codex Group: 005 (FT 0303) Assorted tropical and subtropical fruits – edible peel
   d. EPA Crop Definition: See citrus fruits
10. References: GRIN, CODEX, LOGAN (picture), MAGNESS, US EPA 1996a, US EPA 1995a, US EPA 1994, MORTON (picture)
11. Production Map: EPA Crop Production Regions 3 and 10.
12. Plant Codes:
    a. Bayer Code: FOLJA (*F. japonica*), FOLMA (*F. margarita*), FOLSS (*F.* spp.)

322

1. Laurel, Cherry
(English laurel, Lauroceraso, Skip laurel)
Rosaceae
Prunus laurocerasus L. (syn: *Laurocerasus officinalis* Roem.)
2. The plant is an evergreen bush, up to 10 feet, with thick, glossy, oval to lanceolate leaves, 2-7 inches long, widely grown as an ornamental. There are many horticultural varieties. The leaves have a taste and flavor resembling bitter almond, due to a glucoside, laurocerasin. The leaves are used in cookery, particularly to flavor puddings and custards. A distillate from the leaves is used medicinally.
3. Season, harvest: Harvest leaves as needed.
4. Production in U.S.: No data. USDA Plant Hardiness Zone 7 (Southern Region) (FACCIOLA 1990).
5. Other production regions: Eastern Mediterranean region (FACCIOLA 1990).
6. Use: Almond-flavoring distilled from leaves.
7. Part(s) of plant consumed: Leaves

8. Portion analyzed/sampled: Leaves (fresh)
9. Classifications:
   a. Authors Class: Herbs and spices
   b. EPA Crop Group (Group & Subgroup): Miscellaneous
   c. Codex Group: No specific entry
   d. EPA Crop Definition: None
10. References: GRIN, BAILEY 1976, FACCIOLA, MAGNESS, FRANK
11. Production Map: No entry
12. Plant Codes:
    a. Bayer Code: PRNLR

323

1. **Lavender**

(English lavender, True lavender, Lavandula, Fragrant lavender)
Lamiaceae (Labiatae)

Lavandula angustifolia Mill. (syn: *L. officinalis* Chaix; *L. vera* DC.)

2. Lavender is a perennial shrub of the mint family, which under good culture may reach 5 feet in height. The leaves are oblong-linear to lanceolate, up to 1 1/2 inches long, somewhat hairy. Flower spikes may reach up to 3 feet, and the oil distilled from the flowers only is of the highest quality. The oil is used mainly in perfumes, but may be used in aromatic vinegar. Dried lavender is sometimes used to flavor salads, dressings, etc., and dried flowers are widely used in sachet bags to perfume clothes. French or Spanish lavender is *L. stoechas* L.
3. Season, planting to harvest: From cuttings or seeds, the plant grows very slowly the first year, overwinters and normally flowers in June.
4. Production in U.S.: No data. Flowers in USDA Zone 7 from mid-to-late June (BAILEY 1976).
5. Other production regions: Europe.
6. Use: Flavoring, e.g. tea and meat dishes. Herbes De Provence includes lavender. Also flavors ice cream.
7. Part(s) of plant consumed: Dried flowers
8. Portion analyzed/sampled: Dried flowers
9. Classifications:
   a. Authors Class: Herbs and spices
   b. EPA Crop Group (Group & Subgroup): Herbs and spices (Subgroup 19A)
   c. Codex Group: 27 (HH 0734) Herbs; 057 (DH 0734) Dried herbs
   d. EPA Crop Definition: None
10. References: GRIN, BAILEY 1976, CODEX, FOSTER (picture), MAGNESS, PENZEY, US EPA 1995a
11. Production Map: USDA Zone 7 which mainly includes EPA Crop Production Regions 2, 4, 6, 8 and 9
12. Plant Codes:
    a. Bayer Code: No specific entry

324

1. **Leadplant**

(Prairie shoestring, False indigo)
Fabaceae (Leguminosae)

Amorpha canescens Pursh

2. Leadplant is a perennial warm season shrub that grows from 1.2 to 2 m tall and is native to the U.S. It flowers in July to early August and has a one-seeded legume pod 3 to 5 mm long. Its habitat includes prairies, dry plains and hills and is adapted to a wide range of soil types. It is abundant on over-grazed rangeland. Native Americans dried the leaves to make a tea and it has been cultivated as an ornamental shrub. Leadplant is grazed and produces an excellent palatable feed for livestock and wildlife. It can reproduce by seeds or rhizomes.
3. Season: Warm season perennial shrub that flowers in July to early August and is grazed to heights less than 2 m for its nutritious leaves.
4. Production in U.S.: No specific data for Leadplant production, however, pasture and rangelands are produced on more than 410 million A (U.S. CENSUS, 1992).
5. Other production regions: Distributed in middle Canadian provinces.
6. Use: Grazing, forage, ornamental, and as a human tea.
7. Part(s) of plant consumed: Leaves and stems
8. Portion analyzed/sampled: Forage and hay
9. Classifications:
   a. Authors Class: Nongrass Animal Feeds
   b. EPA Crop Group (Group & Subgroup): Miscellaneous
   c. Codex Group: No specific citation although the general class 050 (AL 0157) Legume Animal Feeds could be used
   d. EPA Crop Definition: None
10. References: GRIN, US EPA 1994, US EPA 1995, CODEX, HOLZWORTH, STUBBIENDIECK (picture), RICHE
11. Production Map: EPA Crop Production Region are 5, 6, 7, 8 and 11.
12. Plant Codes:
    a. Bayer Code: AMHCN

325

1. **Leek**

(Ajo porro, Porro, Cebollin, Garden leek, Pearl onion, Purret)
Liliaceae (Amaryllidaceae)

Allium porrum L. (syn: *A. ampeloprasum* var. *porrum* (L.) J. Gay)

1. **Kurrat**

A. kurrat Schweinf. ex Krause (syn: *A. ampeloprasum* var. *kurrat* auct.)

2. These onion relatives differ from onion in having flat leaves instead of tubular and relatively little bulb development. Leaves are smooth, about 1/2-inch wide and 10 inches long. Plants are grown from seed and usually are started in beds for later field planting. Plants are usually blanched by building up soil against them as they grow in the field, but are also grown without blanching. The thick leaf bases and slightly developed bulb, appearing similar to "green" onions, are eaten as a cooked vegetable with or without attached leaves. The green leaves are also eaten and have a pungent odor and acrid taste. They are used as flavoring in culinary, cookery and in salads. Kurrat is leek-like in stature, but is much smaller than leek. It is used mainly fresh and for seasoning.
3. Season, seeding to harvest: About 5 to 15 months; transplant to harvest: About 5 to 6 months.
4. Production in U.S.: In 1988, California harvested 62 acres (MYERS). In 1994, California harvested 420 acres of leeks (MELNICOE 1996e). California, New Jersey, Michigan and Virginia (MYERS 1991).
5. Other production regions: Domesticated in the eastern Mediterranean and now are cultivated throughout the world (RUBATZKY 1997).
6. Use: For flavoring raw and cooked dishes, and in salads wash before use.

7. Part(s) of plant consumed: Leaves and bulb
8. Portion analyzed/sampled: Whole plant without roots
9. Classifications:
   a. Authors Class: Bulb vegetables
   b. EPA Crop Group (Group & Subgroup): Bulb vegetables
   c. Codex Group: Kurrat: 009 (VA 0383) Bulb vegetables; Leek: 009 (VA 0384) Bulb vegetables
   d. EPA Crop Definition: Onions = green onions; Green onions = Leeks
10. References: GRIN, CODEX, KAYS, MAGNESS, MYERS (picture), SCHREIBER, STEPHENS 1988, US EPA 1995a, US EPA 1995b, MELNICOE 1996e, USDA NRCS, YAMAGUCHI 1983, RUBATZKY
11. Production Map: EPA Crop Production Regions 2, 3, 5 and 10.
12. Plant Codes:
   a. Bayer Code: ALLPO

326

1. Lemon
(Limon, Limon amarillo)
Rutaceae
Citrus limon (L.) Burm. f. (syn: *C. medica limon* L., *C. limonum* Risso, *C. medica limonum* Hook.f.)
1. Lemon, Rough
(Jambhiri orange)
C. jambhiri Lush.

2. Lemons are one of the important citrus fruits (see citrus fruits). Young lemon trees grow rapidly, and tend to be more spreading than oranges or grapefruit. Trees are slightly less hardy than oranges. Fruits are generally oval in shape, about $2\frac{1}{4}$ inches in diameter and $2\frac{1}{2}$ to 3 inches long. Most are picked when they reach a certain size and are still green in color. They may be stored up to several months prior to marketing. Major harvest is during the spring months, while major demand is during mid and late summer. Acidic lemons are the only type grown for commercial purposes in the U.S., along with 'Eureka' and 'Lisbon' cultivars in California and Arizona. Florida grows the 'Sicilian' types ('Avon', 'Harney' and 'Villofranco').
3. Season, bloom to harvest: 9 to 14 months, depending on temperatures.
4. Production in U.S.: In 1995-96, 992,000 tons from 60,500 acres (USDA 1996d). In 1995-96, the top 2 states reported as California (46,300 bearing acres) and Arizona (14,200) (USDA 1996d). In 1992, Puerto Rico reported 210 acres of lemons and limes (1992 CENSUS). California, Arizona and Florida (LOGAN 1996).
5. Other production regions: The largest importers were Bahamas and Chile for 1994 and 1995 (LOGAN 1996).
6. Use: Fresh fruit, canned juice, ade drinks, culinary source of pectin. Mainly fresh, some processed. Lemon oil is used for flavoring purposes in soft drinks and baked foods.
7. Part(s) of plant consumed: Interior juice vesicles. Rind is candied or used in marmalade. Dried rind from processing plants used as livestock feed
8. Portion analyzed/sampled: Whole fruit
9. Classifications:
   a. Authors Class: See citrus fruits
   b. EPA Crop Group (Group & Subgroup): See citrus fruits (Lemon is a representative crop)
   c. Codex Group: 001 (FC 0204) Citrus fruits
   d. EPA Crop Definition: See citrus fruits. Also proposing lemon=lime
10. References: GRIN, CODEX, LOGAN (picture), MAGNESS, RICHE, USDA 1996d, US EPA 1995a, US EPA 1994, KALE, SAULS(d)
11. Production Map: EPA Crop Production Regions 3 and 10 with 97 percent of the acreage in Region 10.
12. Plant Codes:
   a. Bayer Code: CIDLI (*C. limon*); No specific entry for *C. jambhiri*.

327

1. Lemon verbena
(Wapine, Limonetto, Cidron, Verbena, Sweet-scented verbena, Hierba luisa, Verveine citronelle)
Verbenaceae
Aloysia citrodora Palau (syn: *Aloysia triphylla* (L'Her.) Britton; *Lippia citriodora* (Lam.) Kunth)

2. Perennial deciduous shrubs that grow to about 10 feet tall in temperate climates, taller in its native areas in South America. The leaves are 3 to 4 inches long, lance-shaped, entire, lemon fragrance, and arranged in whorls of 3 or 4 leaves. Plant can be cut back to the roots before winter and covered with protective straw or leaves until the spring. Plants can grow up to 4 feet high in a year after being cut back. Flowers appear in July to September.
3. Season, planting to harvest: Normally propagated from cuttings. Leaves are best when harvested as the flowers come into bloom.
4. Production in U.S.: No data. Is grown in gardens (FOSTER 1993). Was a favorite plant in Hawaiian gardens (NEAL 1965).
5. Other production regions: Native to South America and is widely grown in Mediterranean region (FOSTER 1993).
6. Use: Leaves used for flavoring of fruit salads, melons, jellies, beverages and desserts. Used in herbal tea. The essential oil is traded as "true verbena oil". Flowers are edible.
7. Part(s) of plant consumed: Leaves primarily; flowers
8. Portion analyzed/sampled: Leaves (fresh and dried)
9. Classifications:
   a. Authors Class: Herbs and spices
   b. EPA Crop Group (Group & Subgroup): Miscellaneous
   c. Codex Group: 066 (DT 1111) Teas (Dried leaves)
   d. EPA Crop Definition: None
10. References: GRIN, CODEX, FOSTER (picture), GARLAND, NEAL
11. Production Map: EPA Crop Production Region 13.
12. Plant Codes:
   a. Bayer Code: ALYTR

328

1. Lemongrass
(X'a, S'a, West Indian lemongrass, Lapine, Takrai, Sereh, Fever grass, Lukini)
Poaceae (Gramineae)
Cymbopogon citratus (DC. ex Nees) Stapf
2. Lemongrass is a tufted perennial oil grass. The plant has fragrant, rough-edged leaf blades normally to 3 feet by 0.5 inches. Leaves are used for flavoring. In Hawaii leaves are

dried and used as tea. Citronella grass is similar and it is widely cultivated in the tropics, also southern California and Florida. Lemongrass can grow in thick clumps up to 6 feet in height and width.

3. Season, planting to harvest: Propagated using root cuttings or plant divisions. Plant cuttings in furrows 3-4 feet apart on beds 4-5 feet wide. Usually harvested once a year by chopping the entire plant clump at the base.

4. Production in U.S.: In California, Stanislaus county, a marketable bunch consists of 4 to 8 stems (tillers) (MYERS 1991). Grows in Florida (SCHNEIDER 1985). Hawaii. In California, lemongrass is usually harvested once but in the tropics it is harvested up to 4 times per year.

5. Other production regions: Southeast Asia, Mexico.

6. Use: Flavoring. Best if used fresh. Best dried substitute is dried cross cuts of the lower stem. Used in soups and sauces or ground with other spices to make the traditional Thai style curry paste.

7. Part(s) of plant consumed: Leaves. The basal portions of the leafy shoot are chopped and used as flavoring. The heart of the young shoot is eaten as a vegetable with rice and the outer leaves are tied and cooked with food but removed before serving. Lemongrass oil (distilled from leaves) is used in food and drink

8. Portion analyzed/sampled: Leaves (fresh and dried)

9. Classifications:
   a. Authors Class: Herbs and spices
   b. EPA Crop Group (Group & Subgroup): Herbs and spices (19A)
   c. Codex Group: No specific entry
   d. EPA Crop Definition: None

10. References: GRIN, BAILEY 1976, FACCIOLA, KUEBEL, MYERS, NEAL, PENZEY, SCHNEIDER 1985, US EPA 1995a, US EPA 1995b, DUKE(a)(picture)

11. Production Map: EPA Crop Production Regions 3 and 10.

12. Plant Codes:
    a. Bayer Code: CYGCI

329

1. Lentil

(Lenteja, Lentille, Masur dhal, Gram)

Fabaceae (Leguminosae)

Lens culinaris Medik. (syn: *L. esculenta* Moench; *Ervum lens* L.)

2. The lentil plant is a much branched annual, 1 to 2 feet high, with compound leaves having numerous oval leaflets. The edible rounded lens-shaped seeds are borne in short pods, 2 per pod. In culture, lentils are similar to dry beans or peas. Seeds only are used as food. The herbage is sometimes used as fodder. Lentils are an important food in many countries.

3. Season, seeding to harvest: 3 to 5 months.

4. Production in U.S.: 148,000 acres reported in 1995 with 98,400 tons (USDA 1996c). In 1995, Washington reported 79,000 acres and Idaho with 69,000 acres (USDA 1996c).

5. Other production regions: In 1995, Canada reported 810,013 acres (ROTHWELL 1996a).

6. Use: As cooked vegetable, mainly soup and stews.

7. Part(s) of plant consumed: In U.S. seed only as food; whole plant as feed but not a significant feed item. In India, the immature pods are used as a vegetable

8. Portion analyzed/sampled: Seed (dried)

9. Classifications:
   a. Authors Class: Legume vegetables (succulent or dried)
   b. EPA Crop Group (Group & Subgroup): Legume vegetables (succulent or dried) (6C)
   c. Codex Group: 014 (VP 0533) Legume vegetables (young pods); 015 (VD 0533) Pulses
   d. EPA Crop Definition: Peas = Lentils

10. References: GRIN, CODEX, MAGNESS, MARTIN 1983, ROTHWELL 1996a, SCHREIBER, USDA 1996c, US EPA 1995a, US EPA 1996a, YAMAGUCHI 1983, IVES 1997a, MUHLBAUER

11. Production Map: EPA Crop Production Regions 7 and 11 which comprise 99 percent of the U.S. acreage, of which 95 percent is in Region 11.

12. Plant Codes:
    a. Bayer Code: LENCU

330

1. Leren

(Allouia, Sweet cornroot, Allouya, Llerenes, Topee-tambu, Guinea arrowroot, Sweet cormroot, Sweet corn tuber, Bamboo tuber, Topi, Tambo, Aria)

Marantaceae

Calathea allouia (Aubl.) Lindl.

2. This plant is grown for the small potato-like tubers (³/₄ to 3 inches long) which remain crisp after boiling. Growth is erect. Four to ten leaves, each 2 to 4 feet long, with blades up to 2 feet, rise from the crown. The tubers are produced entirely under ground. The crop is of local importance in Puerto Rico and other Caribbean areas.

3. Season, transplanting to harvest: About 6 to 10 months.

4. Production in U.S.: No data. Grown in Puerto Rico (YAMAGUCHI 1983).

5. Other production regions: Native to Caribbean region and parts of South America (YAMAGUCHI 1983)

6. Use: Tubers are cooked as a vegetable like white potatoes.

7. Part(s) of plant consumed: Tuber. Also leaves in the Caribbean to wrap tamales

8. Portion analyzed/sampled: Tuber

9. Classifications:
   a. Authors Class: Root and tuber vegetables
   b. EPA Crop Group (Group & Subgroup): Root and tuber vegetables (1C and 1D)
   c. Codex Group: 016 (VR 0598) Root and tuber vegetables
   d. EPA Crop Definition: None

10. References: GRIN, CODEX, FACCIOLA, MAGNESS, US EPA 1995a, YAMAGUCHI 1983, DUKE(a), RUBATZKY

11. Production Map: EPA Crop Production Region 13.

12. Plant Codes:
    a. Bayer Code: No specific entry, genus entry is CBASS

331

1. Lespedeza

Fabaceae (Leguminosae)

Lespedeza spp.

   Kummerowia spp.

1. Sericea lespedeza

(Chinese bush clover, Chinese lespedeza, Perennial lespedeza)

   L. cuneata (Dum. Cours.) G.Don (syn: *L. sericea* Miq., *L. juncea* var. *sericea* Maxim.)

1. Korean lespedeza

(Annual lespedeza, Annual lespedaza, Japanese bushclover, Japanese clover)

K. stipulacea (Maxim.) Makino (syn: *L. stipulacea* Maxim.)

1. Striate lespedeza

(Common lespedeza, Japanese lespedeza, Annual lespedeza, Japanese bushclover, Japanese clover)

K. striata (Thunb.) Schindl. (syn: *L. striata* (Thunb.) Hook & Arn.)

2. Lespedezas are of major importance for pasture and, to a lesser extent, for hay production in the southeastern quarter of the U.S. Average seed production, 1966-67, was about 46 million pounds, sufficient to seed 2 million acres. Since pastures are generally self reseeding, acreage in lespedeza is probably near 40 million. In 1992, only about 3 million pounds of seed were produced on 11,156 acres (1992 CENSUS). Although some 140 species of lespedeza have been described, mostly native to Eastern Asia, only three are of importance in American agriculture. The species grown in the U.S. are described as follows.

Sericea is the only perennial lespedeza of importance agriculturally in the U.S. Seed from Japan was first tested in North Carolina in 1896. From this and later introductions, seed of vigorous, productive lines was widely distributed. The plants are long-lived, leafy, erect, with rather coarse stems reaching 2 to 4 feet high. Leaflets are long, narrow and blunt at the terminals. Annual growth dies in winter, new growth rising from crown buds. Range of adaptation is roughly from the Atlantic Coast west to Texas and Kansas and north to the Ohio River. Where adapted the crop is useful for pasturage, hay and soil improvement.

Korean is an annual lespedeza introduced into the U.S. in 1919. It is adapted to a more northern area than striate varieties. It is mainly grown from Eastern Oklahoma and Kansas eastward to the Atlantic Seaboard. Leaves are broader and somewhat larger than striate leaves, and seed is borne somewhat differently. Otherwise, the two annual lespedezas are quite similar. Korean lespedeza is excellent both for hay and pasture, especially on soils that are acid or of low fertility. It reseeds readily to give semipermanent pasture.

Striate lespedeza is an annual of which one variety, 'Common', was established in Georgia by 1850. Its mode of introduction from Asia is unknown. The plant has slender, branched stems up to a foot or more in height and small, trifoliate leaves. An introduction from Kobe, Japan, called 'Kobe', was made in 1919. It is taller and larger growing than 'Common' and is useful both for pasture and hay. Although plants of both kinds are annuals they produce seed in late summer and reseed readily, so stands are long lived if properly handled. Striate lespedezas are palatable and nutritious both as pasture and hay. They are best adapted to the area suitable for cotton from east Texas to the Eastern Seaboard.

3. Season: Hay harvested 1 to 3 times per year, first cut early bloom. Mainly pasture.

4. Production in U.S.: In 1992, 11,156 acres for seed with 1,330 tons (1992 CENSUS). In 1992, the top 5 states for seed production reported as Missouri (2,353 acres), South Carolina (1,743), Alabama (1,530), Arkansas (1,517), and Kansas (1,261) (1992 CENSUS). For pasture and hay, primarily Southern U.S.

5. Other production regions: Asia

6. Use: Pasture, hay, green manure, erosion control and soil improvement.

7. Part(s) of plant consumed: Foliage

8. Portion analyzed/sampled: Forage and hay. For lespedeza forage, cut sample at 4 to 6 inch to prebloom stage, at 20 to 25 percent dry matter. For lespedeza hay, Annual/Korean, cut at early blossom to full bloom stage. For Sericea, cut when 12 to 15 inches tall. Hay should be field-dried to a moisture content of 10 to 20 percent.

9. Classifications:

   a. Authors Class: Nongrass animal feeds

   b. EPA Crop Group (Group & Subgroup): Nongrass animal feeds (18)

   c. Codex Group: 050 (AL 1025) Legume animal feeds (only *L. cuneata*); 050 (AL 5229) Sericea

   d. EPA Crop Definition: None

10. References: GRIN, CODEX, MAGNESS, RICHE, US EPA 1996a, US EPA 1995a, BARNES, DUKE(h), BALL (picture),

11. Production Map: EPA Crop Production Regions for seed 2, 4 and 5; for forage 1, 2, 4, 5 and 6.

12. Plant Codes:

   a. Bayer Code: LESCU (*L. cuneata*), LESSL (*K. stipulacea*), LESST (*K. striata*)

332

1. Lesquerella

(Bladderpod, Fendler's bladderpod)

Brassicaceae (Cruciferae)

Lesquerella fendleri (A.Gray) S. Watson

2. Perennial desert plant native to the Southwest (west Texas, New Mexico, Arizona, Colorado and Utah). It grows to a height of about 14 to 16 inches with several stems. Capsules develop along the stem and contain numerous small, flat seeds. Similar cropping system to winter wheat or other small grains. Oil is very similar to castor oil. Protein content in the meal runs to 35 percent and can be used for cattle feed when thioglucosidase enzyme is inactivated. Oil is a substitute for castor oil. The seed contains 25 percent oil by weight. Oil is extracted by extrusion followed by solvent extraction.

3. Season, seeding to harvest: About 7 to 8 months.

4. Production in U.S.: Limited. Fall plantings in Texas and Arizona. Spring plantings in Oregon.

5. Other production regions: Canada.

6. Use: Oil for industrial use and meal as cattle feed.

7. Part(s) of plant consumed: Seed

8. Portion analyzed/sampled: Seed and meal

9. Classifications:

   a. Authors Class: Oilseed

   b. EPA Crop Group (Group & Subgroup): Miscellaneous

   c. Codex Group: No specific entry

   d. EPA Crop Definition: None

10. References: GRIN, ROBBELEN 1989, ROETHELI 1991 (picture)

11. Production Map: EPA Crop Production Regions 8, 9 and 11.

12. Plant Codes:

   a. Bayer Code: No specific entry

333

1. Lettuce
(Lechuga, Garden lettuce, Mesclun, Laituc, Lekuke, Salad)
Asteraceae (Compositae)

Lactuca sativa L.

2. Lettuce is the most important of the leafy vegetable crops grown in the U.S., both commercially and in home gardens. The plant is an annual, grown from seed, and reaches a usable stage in 40 to 90 days, depending on kind and climate. Three main types are grown: Heading varieties, Cutting or leaf varieties, and Cos or Romaine. A fourth, very minor type, is the so-called celery lettuce. These are discussed individually, as they differ substantially in exposure of the edible parts during development. There are 26,000 seeds per ounce.

3. Season, seeding to harvest: About 40 to 90 days.

4. Production in U.S.: Commercial 3,898,000 tons, grown on about 261,360 acres as reported (USDA 1996b). Mainly California and Arizona with 95 percent of the acreage (248,000) (USDA 1996b). Also very widely planted in home gardens.

5. Other production regions: Native to eastern Mediterranean (RUBATZKY 1997). Canada, Mexico and Peru (LOGAN 1996). In 1995, Canada reported 6,781 acres of lettuce (ROTHWELL 1996a).

6. Use: Mainly as raw salads. Lettuce can be prepared in a salad mix with other vegetables, including carrots, red cabbage and romaine lettuce. Mesclun is a salad mixture of lettuces, herbs, edible flowers, etc. The types of salad mixes include Iceberg base or Garden salad European base, Gourmet Mesclun, Coleslaw base and Spinach (flat and curly leaf). The mesclun salads can consist of mixtures of 7 to 15 commodities.

7. Part(s) of plant consumed: Leaves except for celery lettuce, which see

8. Portion analyzed/sampled: Leaves except for celery lettuce, which see

9. Classifications:
   a. Authors Class: Leafy vegetables (except *Brassica* vegetables)
   b. EPA Crop Group (Group & Subgroup): Leafy vegetables (except *Brassica* vegetables) (4). (Head and Leaf Lettuce are representative crops)
   c. Codex Group: See Leaf lettuce, Head lettuce, and Cos lettuce for CODEX classification
   d. EPA Crop Definition: Lettuce = Leaf lettuce and head lettuce

10. References: GRIN, LOGAN (picture), MAGNESS, MARTIN 1983, NEAL, USDA 1996b, US EPA 1996a, US EPA 1995a, RUBATZKY, ROTHWELL 1996a, SCOTT, RYDER 1984

11. Production Map: EPA Crop Production Regions for lettuce (head & leaf) 1, 2, 3 and 10.

12. Plant Codes:
   a. Bayer Code: LACSA

334

1. Lettuce, Celery
(Stem lettuce, Celtuce, Asparagus lettuce, Nen jing wo ju, Chinese lettuce)
Asteraceae (Compositae)

Lactuca sativa var. *angustana* L.H. Bailey (syn: *L. sativa* var. *asparagina* L.H. Bailey)

2. This type of lettuce is grown for the edible, enlarged stem or seed stalk. Stem portions are used in many Chinese dishes, but it is a minor crop in the U.S. The leaves are edible, but of inferior quality compared to other lettuce types, so usually are discarded. Peeled stems may be eaten raw but are more frequently boiled or stewed. It looks like a cross between celery and leaf lettuce. The stem or flower stalk is cut about 12 to 18 inches long. The stems are peeled to remove the bitter sap. The central core is edible.

3. Season, propagation: Seed.

4. Production in U.S.: No data. Grown in Florida Gardens on a limited basis (STEPHENS 1988).

5. Other production regions: Popular in China and Egypt (RUBATZKY 1997).

6. Use: Mainly cooked, stem sometimes eaten raw. Young leaves in salad.

7. Part(s) of plant consumed: Stem and leaves

8. Portion analyzed/sampled: Untrimmed stem (see Celery)

9. Classifications:
   a. Authors Class: Leafy vegetables (except *Brassica* vegetables)
   b. EPA Crop Group (Group & Subgroup): Leafy vegetables (except *Brassica* vegetables) (4B)
   c. Codex Group: 017 (VS 0625) Stalk and stem vegetables
   d. EPA Crop Definition: None

10. References: GRIN, CODEX, KAYS, MAGNESS, MYERS (picture), US EPA 1995a, YAMAGUCHI 1983, RUBATZKY, SPLITTSTOESSER

11. Production Map: No entry.

12. Plant Codes:
   a. Bayer Code: No specific entry, *L.* spp. is LACSS

335

1. Lettuce, Head
(Great lakes lettuce, Imperial lettuce, Iceberg lettuce, Crisphead lettuce, Woju, Cabbage lettuce)
Compositae

Lactuca sativa var. *capitata* L.

2. Head lettuce varieties form a definite head of leaves closely packed about the very short stem or core. Heads are near round to oblate at harvest. There are two types, crisphead and butterhead. For butterhead see Leaf lettuce. The crisphead varieties are more closely packed, with large wrapper leaves substantially covering the part usually consumed. Thus the edible portion is largely protected from spray materials applied late in the growing season. This is the type mainly grown for distant markets. Also see lettuce.

3. Season, seeding to harvest: 60 to 90 days or more.

4. Production in U.S.: In 1995, 190,460 acres with about 3 million tons (USDA 1996b). In 1995, the top 7 states reported as California (136,000 acres), Arizona (44,100 with 41,700 acres in the western part of Arizona), Colorado (4,100), New Mexico (2,000), New Jersey (1,700), Washington (1,400) and New York (1,000) (USDA 1996b).

5. Other production regions: Canada and Mexico (LOGAN 1996).

6. Use: Leaves in salad; Also sold processed or chopped. U.S. per capita consumption in 1994 is 24.3 pounds (PUTNAM 1996).

7. Part(s) of plant consumed: Entire head of leaves with obviously decomposed or withered leaves removed

8. Portion analyzed/sampled: Leaves fresh with wrapper leaves. Entire lettuce head with obviously decomposed or withered leaves removed. In addition, residue data on head lettuce without wrapper leaves are desirable for dietary exposure

9. Classifications:
   a. Authors Class: Leafy vegetables (except *Brassica* vegetables)
   b. EPA Crop Group (Group & Subgroup): Leafy vegetables (except *Brassica* vegetables) (4A) (Representative crop)
   c. Codex Group: 013 (VL 0482) Leafy vegetables (including *Brassica* vegetables)
   d. EPA Crop Definition: Head lettuce = Crisphead varieties only. Lettuce = Head lettuce and Leaf lettuce

10. References: GRIN, CODEX, LOGAN (picture), MAGNESS, USDA 1996b, US EPA 1996a, US EPA 1995a, US EPA 1994, SCOTT, PUTNAM 1996

11. Production Map: EPA Crop Production Regions for Lettuce (head and leaf) 1, 2, 3 and 10. About 95 percent of the head lettuce production is in Region 10.

12. Plant Codes:
   a. Bayer Code: LACSC

336

1. **Lettuce, Leaf**

Asteraceae (Compositae)

Lactuca sativa L.

1. Cos Lettuce

(Romaine, Roman lettuce, Avignon lettuce)

L. sativa var. *longifolia* Lam. (syn: *L. sativa* L. var. *romana* Gars.)

1. Leaf lettuce

(Green oakleaf lettuce, Ye woju, Red oakleaf lettuce, Lolla Rossa lettuce, Greenhouse lettuce, Bunch lettuce, Simpson lettuce, Prayhead lettuce, Grand rapid lettuce, Salad bowl lettuce, Cutting lettuce, Looseleaf lettuce, Redleaf lettuce, Greenleaf lettuce, Curled lettuce, Tango lettuce)

L. sativa var. *crispa* L. (syn: *L. sativa* L. var. *foliosa*)

1. Butterhead lettuce

(Boston lettuce, Bibb lettuce, May king lettuce, Greenhouse lettuce, Red perella lettuce, Tom thumb lettuce, Eaten lettuce, Batavian lettuce)

L. sativa var. *capitata* L.

2. **Leaf lettuce** leaves rise from short stems and tend to roll outward, so the leaves are largely separated during development and at harvest. Leaves may be of different shapes, from spatulate to deeply lobed or cut leaf. This is the most easily grown type of lettuce and is most popular in home gardens. It is grown somewhat for local markets. Leaf lettuce varieties are ready for harvest in about 40 days from seeding. Exposure of edible parts is comparable to that of spinach. May be grown in greenhouses. The **butterhead types**, some of which are called Boston or Bibb lettuce, have a looser head, and most of the leaves in the head are partially exposed, even at late stages of growth. This type does not ship and handle as well as the crisphead varieties, so is produced mainly for nearby markets. **Cos lettuce** develops into an elongated, somewhat oval-shaped head. Leaves are elongated, with thick stems and mid-ribs. Heads are only medium firm, and up to 9 or 10 inches in length. Outer leaves, which are largely discarded, enfold the head during late stages of growth. Cos lettuce is grown much less extensively than the head type.

3. Season, seeding to harvest: About 40 to 50 days for leaf lettuce, 70 days for cos and 55 to 70 days for butterhead.

4. Production in U.S.: In 1995, leaf lettuce reported on 39,200 acres with 446,100 tons. Romaine reported on 31,750 acres with 451,950 tons. Total acres for leaf lettuce, including romaine reported as 70,900 (USDA 1996b). In 1995, leaf lettuce reported for the top 4 states as California (35,000 acres), Arizona (3,200), Ohio (550), and Florida (450). Romaine reported for the top 4 states as California (24,000 acres), Arizona (5,700), Florida (1,600), and Ohio (400) (USDA 1996b).

5. Other production regions: See Lettuce.

6. Use: Mainly as raw salad.

7. Part(s) of plant consumed: Leaves

8. Portion analyzed/sampled: Leaves with obviously decomposed or withered leaves removed

9. Classifications:
   a. Authors Class: Leafy vegetables (except *Brassica* vegetables)
   b. EPA Crop Group (Group & Subgroup): Leafy vegetables (except *Brassica* vegetables) (4A) (Representative crop)
   c. Codex Group: 013 (VL 0483) Leafy vegetables (including *Brassica* leafy vegetables) for leaf lettuce; 013 (VL 0510) Leafy vegetables (including *Brassica* leafy vegetables) for cos lettuce
   d. EPA Crop Definition: Leaf lettuce = Cos (Romaine) and Butterhead varieties. Lettuce = Leaf lettuce and Head lettuce

10. References: GRIN, CODEX, LOGAN (picture), MAGNESS, MYERS (picture), SCHREIBER, STEPHENS 1988 (picture), USDA 1996b, US EPA 1996a, US EPA 1995a, US EPA 1994

11. Production Map: EPA Crop Production Regions for Lettuce (head and leaf) 1, 2, 3 and 10. About 95 percent of the commercial leaf lettuce production is in Region 10.

12. Plant Codes:
   a. Bayer Code: LACSA (*L. sativa*), LACSC (*L. sativa* var *capitata*), LACSP (*L. sativa* var *crispa*), LACSO (*L. sativa* var *longifolia*), LASSR (romaine and cos)

337

1. **Leucaena (Lu-seen-a)**

(Koa-haole, Leadtree, Jumbie bean, Ipil-ipil, White popinac, Tantan)

Fabaceae (Leguminosae)

Leucaena leucocephala (Lam.) deWit (syn: *L. glauca* of authors)

2. It is a subtropical-tropical small thornless evergreen legume tree or shrub that grows to 18 feet in its first year. Leaves are twice divided similar to mimosa. Leucaena is a perennial crop normally propagated by seeds but in Texas leucaena can be transplanted when there is a lack of irrigation. In Texas, seedlings planted in mid-March are about 4 feet tall by early July for the first harvest. It is cut above 20 inches for green chop, dried and cubed. For established plantings more than 1 year old, the first harvest for the season is early May and additional cuttings at 4 to 6 week intervals. Leucaena is planted in 38 inch rows and managed as a hedge.

3. Season, seeding to first harvest: About 4-5 months with additional cuttings at 4-6 week intervals.
4. Production in U.S.: Limited in Texas (FELKER 1993) and Hawaii (NEAL 1965). About 300 acres in Texas (FELKER 1993).
5. Other production regions: Primarily tropical America (BAILEY 1976).
6. Use: In U.S. used as animal feed but it is not a significant feed item (US EPA 1995b). Forage used as cattle feed.
7. Part(s) of plant consumed: Forage, including mainly young stems and leaves
8. Portion analyzed/sampled: Forage and hay
9. Classifications:
   a. Authors Class: Nongrass animal feeds
   b. EPA Crop Group (Group & Subgroup): Miscellaneous
   c. Codex Group: No specific entry
   d. EPA Crop Definition: None
10. References: GRIN, FACCIOLA, FELKER 1993, NEAL, US EPA 1995b, BAILEY 1976, SCHNEIDER 1993b, NATIONAL RESEARCH COUNCIL 1984, THOMPSON
11. Production Map: EPA Crop Production Regions 6 and 13.
12. Plant Codes:
   a. Bayer Code: LUAGL

338

1. **Licorice**
(Common licorice, Liquorice roots, Sweetwood)
Fabaceae (Leguminosae)
Glycyrrhiza glabra L.
2. The plant is a perennial herb, grown from seed or from divisions of the roots, for the licorice of commerce obtained from the rhizomes or roots. The leaves are compound-pinnate, with entire leaflets. Three to four years of growth are required for rhizomes to reach sufficient size for harvest. The plant top dies down each year. In harvesting, rhizomes are dug, partially dried and extracted. For use in medicine, candies and tobacco products. Licorice is little grown in the U.S., as production is generally uneconomical; but large quantities of the roots are imported.
3. Season, root propagation or seeding to harvest: The root is harvested in the fall of the third or fourth year.
4. Production in U.S.: Licorice is little grown in the U.S. (FOSTER 1993).
5. Other production regions: Mediterranean region and southwestern Asia (BAILEY 1976).
6. Use: Flavoring in medicine, candies, tobacco products, soft drinks, baked goods and chewing gum. Its extract is used in beer which increases foam.
7. Part(s) of plant consumed: Root
8. Portion analyzed/sampled: Root
9. Classifications:
   a. Authors: Class: Herbs and spices
   b. EPA Crop Group (Group & Subgroup): Miscellaneous
   c. Codex Group: 028 (HS 0787) Spices with the commodity name "Liquorice"
   d. EPA Crop Definition: None
10. References: GRIN, BAILEY 1976, CODEX, FOSTER (picture), GARLAND (picture), MAGNESS
11. Production Map: No entry
12. Plant Codes:
   a. Bayer Code: GYCGL

339

1. **Lime**
(Lima, West Indian lime, Limette, Key lime, Bartender Lime, Mexican lime, Sour lime)
Rutaceae
Citrus aurantiifolia (Christm.) Swingle (syn: *C. aurantifolia* Swingle, *C. lima* Lunan., *C. acida* Roxb., *C. limonellus* Hassk., *Limonia aurantifolia* Christm., *L. acidissima* Houtt.)

1. **Lime, Tahiti**
(Persian lime, Green lemon, Bearss seedless lime)
C. latifolia (Yu. Tanaka) Tanaka
2. Limes grow on relatively small, much branched citrus trees (See Citrus fruits). Fruits are of several types. The Mexican or Key lime is near round 1 to 2 inches diameter, with thin rind and acid pulp. Tahiti limes are larger, 2 to $2^{1}/_{2}$ inches diameter and usually seedless. Low acid or sweet limes are available and grown in some countries, but rarely in the U.S. Limes have been crossed with other types of citrus (see Citrus hybrids). Cultivars of Tahiti lime are 'Bearss', 'Idemor', and 'Pond'. Tahiti limes are harvested 8 to 12 times per year with the peak period July to September. The Tahiti lime is presumed to be a hybrid of the Mexican lime and citron (MORTON).
3. Season, bloom to harvest: 9 months for Mexican; 12 months for Tahiti type.
4. Production in U.S.: In 1995-96, 14,000 tons from 2,000 bearing acres (USDA 1996d). In 1995-96, Florida reported all the acreage (USDA 1996d). In 1992, Puerto Rico reported 210 acres of lemon and lime (1992 CENSUS).
5. Other production regions: Largest importer is Mexico in 1995 (LOGAN 1996).
6. Use: Ade drinks, culinary, flavoring, in ice cream, confections. Mainly fresh, some processed.
7. Part(s) of plant consumed: Juice mainly. Peel may be used in flavoring
8. Portion analyzed/sampled: Whole fruit
9. Classifications:
   a. Authors Class: See citrus fruits
   b. EPA Crop Group (Group & Subgroup): See citrus fruits
   c. Codex Group: 001 (FC 0205) Citrus fruits; 001 (FC 0002) Lemons and limes
   d. EPA Crop Definition: See citrus fruits. Proposing Lemon = Lime
10. References: GRIN, CODEX, LOGAN (picture), MAGNESS, MORTON, RICHE, USDA 1996d, US EPA 1994, MALO(d), PHILLIPS(g), SAULS(d)
11. Production Map: EPA Crop Production Regions 3 and 10 with most of the acreage in 3.
12. Plant Codes:
   a. Bayer Code: CIDAF (*C. aurantiifolia*); No entry for *C. latifolia*

340

1. **Lime, Sweet**
(Limettier doux, Lima dulce, Mitha limbu, Sweet lemon, Lumia)
Rutaceae
Citrus limetta Risso (syn: *C. limettioides* Tan., *C. lumia* Risso)
2. The tree, its foliage, and the form and size of the fruit resemble the Tahiti lime. Sweet lime is thought to be a hybrid

between lime (Mexican-type), and a sweet lemon or sweet citron (MORTON 1987).

3. Season, planting: Propagated from cuttings.
4. Production in U.S.: No data. Limited culture in California (MORTON 1987).
5. Other production regions: West Indies, Central America, India, northern Vietnam, Mediterranean Region and tropical America (MORTON 1987).
6. Use: In the West Indies and Central America, the fruit are commonly used fresh.
7. Part(s) of plant consumed: Fruit
8. Portion analyzed/sampled: Whole fruit
9. Classifications:
   a. Authors Class: Citrus fruits
   b. EPA Crop Group (Group & Subgroup): Miscellaneous
   c. Codex Group: 001 (FC 0002) Lemons and limes
   d. EPA Crop Definition: See citrus fruits. Proposing Lemon = Lime
10. References: GRIN, CODEX, MORTON, US EPA 1995a
11. Production Map: EPA Crop Production Region 10.
12. Plant Codes:
   a. Bayer Code: CIDLM

341

1. Lime blossom
(Linden, Lime tree, Basswood, Whitewood, Tilo)
Tiliaceae
Tilia spp.
1. Littleleaf linden
(Small-leaved lime, Small leaf linden, Small-leaved European linden)
T. cordata Mill. (syn: *T. ulmifolia* Scop., *T. parvifolia* Ehrh.)
1. Bigleaf linden
(Largeleaf lime, Largeleaf linden, Large-leaved linden)
T. platyphyllos Scop. (syn: *T. grandifolia* Ehrh.)
1. American linden
(American basswood, American lime, Whitewood)
T. americana L. (syn: *T. glabra* Vent.)
1. European linden
(Common linden, Lime)
T. x europaea L. (*T. cordata* x *T. platyphyllos* – natural hybrid)
2. Large deciduous trees to about 100 feet tall in temperate Northern Hemisphere. Leaves are to 2$\frac{1}{2}$ inches long for the Littleleaf linden and to 5 inches long for the Bigleaf linden. Cymes are 5 to 7 flowered for the Littleleaf linden to 3 flowered for Bigleaf linden. Flowers have a honey-like fragance and are made into tea in France and sold under the name "tilleul". According to Facciola, this tea comes from *Tilia* x *europaea*. Flowers and leaves are sold in commerce for tea.
3. Season, planting: Propagated by seeds in the spring or autumn.
4. Production in U.S.: No data. Sold in U.S. stores as "Linden leaves tilo" from a spice company in Miami, Florida.
5. Other production regions: Northern hemisphere (BAILEY 1976).
6. Use: Flowers and leaves are used to make tea. Also grown as an ornamental shade tree.
7. Part(s) of plant consumed: Flowers and leaves
8. Portion analyzed/sampled: Flowers and leaves
9. Classifications:
   a. Authors Class: Teas
   b. EPA Crop Group (Group & Subgroup): Miscellaneous
   c. Codex Group: 066 (DT 1112) Teas (except *T. americana*)
   d. EPA Crop Definition: None
10. References: GRIN, BAILEY 1976, CODEX, FACCIOLA, FRANK
11. Production Map: No entry
12. Plant Codes:
   a. Bayer Code: TILSS (*T. spp.*), TILAM (*T. americana*), TILCO (*T. cordata*), TILPL (*T. platyphyllos*)

342

1. Limpograss
(African jointgrass, Haltgrass, Red veltgrass)
Poaceae (Gramineae)
Hemarthria altissima (Poir.) Stapf & C.E. Hubb.
2. This is a hardy perennial warm season grass native to southern Africa, and was introduced into Florida in 1964. Is a member of the grass tribe *Andropogoneae*. It is also grown in Central America and Hawaii. It is an erect bunchgrass up to 1.5 m tall that spreads by stolons and its inflorescence is a spike-like raceme. It is more winter hardy than either digitgrass or stargrass, and more frost tolerant than bahiagrass. It is adapted to wet soils with a pH 5.5 to 6.5, especially seasonally flooded flatland soils of the southeast U.S. Limpograss is usually planted with legumes and it is readily grazed for up to 3 to 4 months during late spring through the summer, and can be grown throughout the mild winters in southern Florida. It should be grazed at a height of ≥ 30 cm for best persistence. Dry matter yields up to 29,000 kg/ha have been reported, but most yields vary from 12,000 to 20,000 kg/ha. It is propagated vegetatively, since its seeds are of low viability.
3. Season: Warm season tropical grass that starts growing in the spring and it can be utilized all winter in the southernmost U.S. It can be utilized as hay or silage.
4. Production in U.S.: There are greater than 50,000 ha of limpograss reported in 1978. Pasture and rangeland are produced on more than 410 million A (U.S. CENSUS, 1992). See above.
5. Other production regions: Africa and Central America.
6. Use: Pasture grazing, hay and silage.
7. Part(s) of plant consumed: Leaves and stems
8. Portion analyzed/sampled: Forage and hay.
9. Classifications:
   a. Authors Class: See Forage grass
   b. EPA Crop Group (Group & Subgroup): See Forage grass
   c. Codex Group: See Forage grass
   d. EPA Crop Definition: None
10. References: GRIN, US EPA 1994, US EPA 1995, CODEX, HOLZWORTH,SKERMAN (picture), RICHE
11. Production Map: EPA Crop Production Regions 2, 3, 4, 6 and 13.
12. Plant Codes:
   a. Bayer Code: HEMAL

343

1. Lingonberry
(Cowberry, Mountain cranberry, Lingberry, Rock cranberry, Foxberry, Whimberry, Partridge berry, Alpine cranberry, Red

whortleberry, Lingen, Moss cranberry, Arandano encarnado, Lingenberry, Kronsbeere, Airella rouge)

Ericaceae

Vaccinium vitis-idaea L.

2. The lingonberry is native throughout northern portions of North America and Europe. It is a woody, evergreen dwarf shrub. It is akin to cranberry, the plants being creeping and low growing up to 10 inches. The fruits grow on short uprights and are about $1/3$ inch in diameter. They are acid and bitter raw, but make excellent jellies and preserves. They are grown to a limited extent under cultivation, but substantial quantities are gathered from the wild and are marketed to some extent. They are an important native food source in areas where other fruits are not available.

3. Season, fruit can hold on the plant all winter if not picked.

4. Production in U.S.: No data. Harvested from the wild. Research at the University of Wisconsin since 1984 suggest lingonberries are suited for cultivation (JANICK 1990). About 1 acre is cultivated in both Wisconsin and Maine, and about 10-15 acres in North America.

5. Other production regions: Northern temperate region (FACCIOLA 1990). Canada is a major source of fresh fruit for the U.S. market. Germany with about 50 acres under cultivation and about 200 acres worldwide.

6. Use: Jellies and preserves. Cranberry-like sauce served with pancakes.

7. Part(s) of plant consumed: Fruit

8. Portion analyzed/sampled: Whole fruit

9. Classifications:
   a. Authors Class: Berries
   b. EPA Crop Group (Group & Subgroup): Miscellaneous
   c. Codex Group: No specific entry
   d. EPA Crop Definition: None

10. References: GRIN, FACCIOLA, JANICK 1990 (picture), MAGNESS

11. Production Map: EPA Crop Production Regions 1 and 5.

12. Plant Codes:
    a. Bayer Code: VACVI, VACVM

344

1. Little bluestem
(Prairie beardgrass, Popotillo colorado)

Poaceae (Gramineae)

Schizachyrium scoparium (Michx.) Nash (syn: *Andropogon scoparius* Michx.)

2. This is a warm season perennial, sod-forming grass native to the Great Plains area. It belongs to the grass tribe *Andropogoneae*. It is smaller than big bluestem and is more drought tolerant than the other bluestems. It has rhizomes and grows 1 to 3 ft tall. It is found mainly on sandy soils and is a valuable rangegrass under limited moisture conditions. After seeds begin to form, its palatability declines. Little bluestem is seeded in the spring at a rate of 15 lb/A and contains 260,000 seeds/lb. Seed yields average 200 lb/A.

3. Season: Warm season forage. Grows from mid-spring through early fall. Seeds mature in late September-November.

4. Production in U.S.: No data for Little bluestem production, however, pasture and rangeland are produced on more than 410 million A (U.S. CENSUS, 1992).

5. Other production regions: Canada (BAILEY 1976).

6. Use: Pasture and rangeland, hay, and for erosion control.

7. Part(s) of plant consumed: Leaves and stems

8. Portion analyzed/sampled: Forage and hay

9. Classifications:
   a. Authors Class: See Forage grass
   b. EPA Crop Group (Group & Subgroup): See Forage grass
   c. Codex Group: See Forage grass
   d. EPA Crop Definition: None

10. References: GRIN, US EPA 1994, US EPA 1995, CODEX, BARNES, ALDERSON, RICHE, BAILEY 1976

11. Production Map: EPA Production Regions 5, 6, 7, 8, 9 and 10.

12. Plant Codes:
    a. Bayer Code: ANOSC

345

1. Longan
(Lungan, Mamoncillo chino, Dragons eye)

Sapindaceae

Dimocarpus longan Lour. (syn: *Euphoria longan* (Lour.) Steud., *E. longana* Lam.)

2. The tree and fruit are closely related to lychee. The tree is a tropical and sub-tropical evergreen, up to 40 feet, with pinnate leaves. The tan-colored fruits are borne in panicles or loose clusters. They are globular, about an inch in diameter, with a nearly smooth rind and a single, rather large seed. The pulp, between the rind and the seed, is gelatinous and of pleasant flavor. The rind separates from the pulp readily. Only the pulp is consumed. The fruit ripens a little later than lychee, and the tree is a little more cold resistant.

3. Season, harvest: August and September in Florida. Bloom to harvest: 120-150 days.

4. Production in U.S.: Home garden (MARTIN 1987). In 1994, Florida reported 200 acres (CRANE 1995a). Also grown in Hawaii.

5. Other production regions: West Indies (MORTON 1987). Longan is substantially commercialized in China, Taiwan and Thailand.

6. Use: Fresh or cooked. Also dried after seeds and rind removed. Used in Oriental medicine.

7. Part(s) of plant consumed: Fruit pulp

8. Portion analyzed/sampled: Whole fruit

9. Classifications:
   a. Authors Class: Tropical and subtropical fruits – inedible peel
   b. EPA Crop Group (Group & Subgroup): Miscellaneous
   c. Codex Group: 006 (FI 0342) Assorted tropical and subtropical fruits – inedible peel
   d. EPA Crop Definition: None but proposing – Lychee = Longan

10. References: GRIN, CODEX, MAGNESS, MARTIN 1987, MORTON (picture), CRANE 1995a, ROY, MENZEL

11. Production Map: EPA Crop Production Region 13.

12. Plant Codes:
    a. Bayer Code: No specific entry

346

1. Loquat
(Japanese plum, Japanese medlar, Nispero, Nispero-del-Japon)

Rosaceae

Eriobotrya japonica (Thunb.) Lindl.

2. Grows in subtropical areas or tropical highlands. Propagated by seed or grafting. First crop 2-3 years after grafting. Loquat trees are evergreen, rounded in shape, up to 25 feet high. Dark green leaves are up to a foot long. Trees will stand winter temperatures down to about 12 degrees F. Flowers open and set fruit in autumn, and if temperatures reach 25 degrees F or lower in winter, fruit crop will be lost. Flowers and fruits are produced on large panicles. Individual fruits are pyriform or oval, $1^{1}/_{2}$ to near 3 inches long, and covered with a tough, pubescent skin which separates readily from the pulp when ripe. The fruit is a pome, like apple and pear, generally with 3 to 5 seeds, and with the calyx persistent. Flesh is firm and creamy, mild, sub-acid in flavor. Domestic cultivars include 'Champagne', 'Thales', 'Advance', 'Fletcher' and 'Miller'.

3. Season: Bloom to maturity in 120-180 days.

4. Production in U.S.: California, Hawaii, and Florida.

5. Other production regions: Japan, India, Israel.

6. Use: Fresh eating out of hand or fruit salads. Also cooked for desserts, sauces or preserves. Also grown as an ornamental.

7. Part(s) of plant consumed: Inner pulp only

8. Portion analyzed/sampled: Whole fruit

9. Classifications:
   a. Authors Class: Pome fruits
   b. EPA Crop Group (Group & Subgroup): Crop Group 11: Pome Fruits Group
   c. Codex Group: 002 (FP 0228, FP 4044) Pome fruits
   d. EPA Crop Definition: None

10. References: GRIN, CODEX, KAWATE 1995, KNIGHT(a), LOGAN, MAGNESS, MARTIN 1987, US EPA 1994, ROY, CAMPBELL (d)

11. Production Map: EPA Crop Production Regions 10 and 13.

12. Plant Codes:
    a. Bayer Code: EIOJA

347

1. Lotus root

(Loto sagrado, Renkon, Linngau, Lenkon, Ho, Lian, Oriental lotus, Sacred lotus, East Indian lotus, Egyptian lotus, Nelumbium, Hasu)

Nelumbonaceae (Nymphaeaceae)

Nelumbo nucifera Gaertn. (syn: *Nelumbuim nelumbo* L. (Druce); *N. speciosum* Willd.)

2. This is a perennial water plant, long grown in Oriental countries both as an ornamental and for its edible rhizomes and seeds. Rhizomes are thick and tuberous and creep in the earth at the bottoms of ponds or slow moving streams. The leaves are nearly round, large and long stemmed. They may float on deep water, but rise high above shallow water. The fruit is an enlarged receptacle, in which the many seeds are embedded. In the U.S., this plant is grown almost entirely as an ornamental. The starchy rhizomes are of some importance as foods in the Orient. Rhizomes are harvested from the mud in the pond or paddy while still flooded or drained. Rhizomes are 2 to 4 feet long. The acorn-like seeds are also an Oriental delicacy.

3. Season: planting to harvest: About 4 to 8 months. Or after leaves start turning brown.

4. Production in U.S.: Mainly as ornamental, negligible for food. Hawaii produced about 32 tons of root from about 10 acres in 1992. Also Nebraska reported one farm (1992 CENSUS). It was grown in California (JANICK 1990).

5. Other production regions: Cultivated in China, Egypt, India and Japan (RUBATZKY 1997).

6. Use: Rhizomes as cooked food. Seeds as nut-like delicacy in Orient.

7. Part(s) of plant consumed: Underwater rhizomes; seeds. Mainly the terminal rhizomes with 3 to 4 swollen internodes

8. Portion analyzed/sampled: Rhizomes

9. Classifications:
   a. Authors Class: Root and tuber vegetables
   b. EPA Crop Group (Group & Subgroup): Miscellaneous
   c. Codex Group: No specific entry
   d. EPA Crop Definition: None

10. References: GRIN, JANICK 1990, MAGNESS, NEAL (picture), REHM, RICHE, YAMAGUCHI 1983 (picture), RUBATZKY (picture)

11. Production Map: EPA Crop Production Region 13.

12. Plant Codes:
    a. Bayer Code: NELNU

348

1. Lovage
(Levistica, Liveche)

Apiaceae (Umbelliferae)

Levisticum officinale W.D.J. Koch

2. Lovage is a perennial plant of the carrot family, grown somewhat for the aromatic seeds used in confectionery, and for the young stems which are preserved in sugar like angelica. The leaves are dark green, shiny, and much compounded like carrot. The plant is fast growing. While long used in Europe, lovage now is of minor importance as a condiment.

3. Season, seeding to harvest: Leaves in second or third year before bloom. Roots in the fall of the third year and seeds in autumn.

4. Production in U.S.: No data. Naturalized in North America (FOSTER 1993).

5. Other production regions: Native to southern Europe.

6. Use: Fresh leaves used in salads; stems used in drinks; seed in confectionery and roots to flavor tobacco products.

7. Part(s) of plant consumed: Leaves, stems, roots and seeds

8. Portion analyzed/sampled: Leaves (fresh and dried); Roots (fresh and dried); Seeds

9. Classifications:
   a. Authors Class: Herbs and spices
   b. EPA Crop Group (Group & Subgroup): Herbs and spices (Subgroup 19A and 19B)
   c. Codex Group: 027 (HH 0735) Herbs, 057 (DH 0735) Dried Herbs, 028 (HS 0735) Spices
   d. EPA Crop Definition: None

10. References: GRIN, CODEX, FOSTER (picture), MAGNESS, US EPA 1995a

11. Production Map: No entry

12. Plant Codes:
    a. Bayer Code: LEWOF

349

1. Lovegrass

Poaceae (Gramineae)

Eragrostis spp.

1. Weeping lovegrass
(Boer lovegrass, Zacate del amor)

E. curvula (Schrad.) Nees (syn: *E. chloromelas* Steud.)

1. Lehmann lovegrass

E. lehmanniana Nees

1. Sand lovegrass

E. trichodes (Nutt.) A.W. Wood

2. Some 250 species of *Eragrostis*, the lovegrasses, are known, with about 40 native in the U.S. Only a few have agricultural value. Some species produce abundant growth on soils of low fertility and are valuable for protection of eroding sites. The species of most value in the U.S. are described as follows.

Weeping lovegrass is a vigorous warm season perennial bunch-grass, introduced from Tanzania, Africa, in 1927. It has proved well adapted in the Southern states, particularly the Southern Plains. Seed stalks are numerous and slender, reaching to 5 feet under the best conditions. Basal leaves are numerous, slender, 10 to 20 inches long, palatable while succulent, but of low palatability when ripe. Plants are semi-hardy, enduring temperatures to about 10 degrees F. They are readily established by seeding and make a quick ground cover.

Lehmann lovegrass is a warm season, slightly spreading grass introduced from South Africa in 1932. It is used for range reseeding in the warm semidesert areas of the Southwestern U.S. It forms prostrate stems which root at the nodes and is readily established by seeding. The plants are smaller and less cold-tolerant than Weeping lovegrass.

Sand lovegrass is a vigorous, long-lived, native bunchgrass occurring on sandy soils of the central and southern Great Plains. It is drought resistant, with a deep root system. Stems may reach to 6 feet. Leaves are abundant, slightly hairy, 12 inches long and $1/4$ inch wide. Plants start growth early in spring and continue through the summer. This grass is highly palatable and nutritious, and often is overgrazed. It is easily established from seed. Most important lovegrass in U.S.

3. Season: Start growth early in spring. Most palatable in the spring, declines as it matures. Matures in July.
4. Production in U.S.: No specific data, See Forage grass. Central and southern Great Plains and Southwestern states.
5. Other production regions: Africa.
6. Use: Rangegrass, erosion control, hay.
7. Part(s) of plant consumed: Foliage
8. Portion analyzed/sampled: Forage and hay
9. Classifications:
   a. Authors Class: See Forage grass
   b. EPA Crop Group (Group & Subgroup): See Forage grass
   c. Codex Group: See Forage grass
   d. EPA Crop Definition: None
10. References: GRIN, MAGNESS, ALDERSON, SKERMAN
11. Production Map: EPA Crop Production Regions 4, 5, 6, 7, 8, 9 and 10.
12. Plant Codes:
   a. Bayer Code: ERACU (*E. curvula*), ERALE (*E. lehmanniana*)

350

1. Lupin
(Lupine, Forage lupine, Grain lupin)

Fabaceae (Leguminosae)

Lupinus spp.

1. Blue lupin
(European blue lupine, Narrow-leafed lupin, Blue lupine, New Zealand blue lupin)

L. augustifolius L.

1. Yellow lupin
(European yellow lupine, Yellow lupine)

L. luteus L.

1. White lupin
(Egyptian lupine, White lupine, Sweet lupine, White sweet lupine)

L. albus L. var *albus* (syn: *L. termis* Forssk.)

1. Tarwi
(Pearl lupin, Andean lupin, Sweet lupine)

L. mutabilis Sweet

2. Several hundred species of *Lupinus* are known, mostly native to America. Many are grown in gardens as ornamentals. The species important agriculturally, however, are winter annuals introduced from Europe, as listed above. They are not winter-hardy and are grown mainly as green manure crops for turning under for soil improvement in the Southeastern states. The use of lupines for feed is limited since they contain a poisonous alkaloid. Kinds lacking or low in this alkaloid (termed "sweet" lupines) are now available, so use as pasture is increasing. The plants are bushy unless in dense stands, with coarse stems up to 3 or more feet high. Leaves are palmate with 6 to 8 leaflets. Seeding is in the fall and growth is heavy during early spring. The primary species of grain lupin or sweet lupin are *L. albus* and *L. mutabilis* in the Americas. Grain lupins are not processed. White lupins are the most winter hardy lupin. White lupins are used late winter and early spring for grazing. Blue lupins are the most widely grown and are used for forage, silage and soil improvement. The yellow lupin grows 10 to 32 inches tall, flowers March to July, and is widely cultivated in the Southern U.S.

3. Season, seeding to harvest: Grain: about 6 to 8 months, depending on variety. Flowers May to June.
4. Production in U.S.: No data. Limited. Forage in southeastern U.S. and eastern Washington (SCHREIBER 1995). Grain in Michigan and California (IR-4 REQUEST DATABASE 1996).
5. Other production regions: Europe (BAILEY 1976).
6. Use: Grain for livestock feed, forage, pasture, some seed for food, green manure crop.
7. Part(s) of plant consumed: Grain and foliage
8. Portion analyzed/sampled: Seed for grain; forage and hay for forage crops
9. Classifications:
   a. Authors Class: Nongrass animal feeds; Legume vegetable (succulent or dried); Foliage of legume vegetables
   b. EPA Crop Group (Group & Subgroup): Nongrass animal feed (18); Legume vegetable (dried or succulent) (6C); Foliage of legume vegetables (7)
   c. Codex Group: 050 (AL 0545) Legume animal feed for forage; 015 (VD 0545) Pulses for grain lupin; 014 (VP 0545) Legume vegetables
   d. EPA Crop Definition: Beans = *Lupinus* spp.
10. References: GRIN, CODEX, IR-4 REQUEST DATABASE 1996, MAGNESS, SCHREIBER, US EPA 1996a, US EPA 1995a, DUKE(h), BAILEY 1976, MERONUCK
11. Production Map: EPA Crop Production Regions 2 and 11 as forage legume. Also grain lupin in 5 and 10.

12. Plant Codes:
  a. Bayer Code: LUPAL (*L. albus*), LUPAN (*L. augustifolius*), LUPLU (*L. luteus*), LUPSS (*L.* spp.)

351

1. Lychee (Lee-chee)
(Litchi, Lychee nut, Lichi, Chinese nut)
Sapindaceae
  Litchi chinensis Sonn.
2. The lychee tree is an evergreen, medium to large, up to 45 feet, with pinnate, leathery leaves. Grows in cool tropical or warm subtropical climate with well distributed rainfall. Does not fruit in hot lowland tropics. Young trees are very sensetive to frost. Established trees are somewhat tolerant to frost injury (somewhat more so than sweet orange, and a little less so than mango and avocado). Propagated by seed, layering and grafting. Fruits are generally bright red, round to ovate in shape, and 1 to 1½ inches in diameter, with a single rather large seed. They are borne on large, loose panicles. The "shell" or peel is thin and leathery, with a rough surface due to numerous small protuberances and separates from the pulp readily. The edible pulp, between the shell and seed is juicy, pearl white in color and translucent.
3. Season: Fruit matures in 60 to 90 days of flowering.
4. Production in U.S.: In 1994, 300 acres in Florida (CRANE 1995a) and 55 acres in Hawaii (KAWATE 1995). In Hawaii, 680 bearing trees produced about 43 tons of fruit (KAWATE 1995).
5. Other production regions: China, India, South Africa, Taiwan; Mexico and Israel (LOGAN 1996).
6. Use: Mainly as fresh fruit in U.S. In China, canned and dried.
7. Part(s) of plant consumed: Interior pulp only
8. Portion analyzed/sampled: Whole fruit (pit can be removed and discarded)
9. Classifications:
  a. Authors Class: Tropical and subtropical fruits – inedible peel
  b. EPA Crop Group (Group & Subgroup): Miscellaneous
  c. Codex Group: 006 (FI 0343) Assorted tropical and subtropical fruits – inedible peel
  d. EPA Crop Definition: None, but proposing: Lychee = Longan, Spanish lime, Rambutan and Pulasan
10. References: GRIN, CODEX, KNIGHT(a), LOGAN, MAGNESS, MARTIN 1987, US EPA 1994, CRANE 1995a, KAWATE 1995, KADAM(a), CAMPBELL(g)
11. Production Map: EPA Crop Production Region 13.
12. Plant Codes:
  a. Bayer Code: LIHCH

352

1. Maca (mak-kah)
(Maka, Peruvian ginseng)
Brassicaceae (Cruciferae)
  Lepidium meyenii Walp.
2. Maca is a mat-like perennial with a tuberous root (3 inches in diameter) which resembles the radish. The dried root can be stored for years. It is a unique crop because it grows at very high elevations. Maca is used as cocoa, pudding and jam.

3. Season, seeding to harvest: About 6 to 11 months for the roots.
4. Production in U.S.: No data.
5. Other production regions: South America (NATIONAL RESEARCH COUNCIL 1989).
6. Use: Fresh roots are baked or roasted. The sun dried roots are boiled in milk or water to create a porridge. Leaves can be used in salads.
7. Part(s) of plant consumed: Root and top
8. Portion analyzed/sampled: Root (fresh and dried) and tops
9. Classifications:
  a. Authors Class: Root and tuber vegetables; Leaves of root and tuber vegetables
  b. EPA Crop Group (Group & Subgroup): Miscellaneous
  c. Codex Group: No specific entry
  d. EPA Crop Definition: None
10. References: GRIN, NATIONAL RESEARCH COUNCIL 1989 (picture), RUBATZKY
11. Production Map: No entry
12. Plant Codes:
  a. Bayer Code: No specific entry

353

1. Macadamia nut
(Queensland nut, Bushnut, Australian nut, Bopplenut, Bauplenut)
Proteaceae
  Macadamia integrifolia Maiden & Betche
  (Smooth-shell Queensland nut, Nuez-de-macadamia)
  M. tetraphylla L. A. S. Johnson
  (Rough-shell Queensland nut)
2. The tree is a tall, tropical evergreen, native to Australia, with oblong or lanceolate leaves up to a foot long and glabrous. Fruits are near globular, up to 1½ inches diameter. The outer covering or husk is leathery and separates readily from the shell of the nut. The shell is very hard, woody and thick. When the shell is cracked, the round kernel is free. Occasionally, two hemispherical kernels occur in a nut. The single kernels are about ½ inch in diameter, and rich in flavor.
3. Season, bloom to maturity: 8 to 10 months. Peak harvest is in October as the ripe nuts drop to the ground.
4. Production in U.S.: 18,500 acres with 24,250 tons (in-shell) for 1993 in Hawaii (USDA 1994). Hawaii with 22,634 acres and California with 521 acres in 1992 CENSUS.
5. Other production regions: Australia (FACCIOLA 1990), Costa Rica.
6. Use: Mainly eaten directly. Some use in confections.
7. Part(s) of plant consumed: Internal kernels only
8. Portion analyzed/sampled: Nutmeat
9. Classifications:
  a. Authors Class: Tree nuts
  b. EPA Crop Group (Group & Subgroup): Tree nuts
  c. Codex Group: 022 (TN 0669) Tree nuts
  d. EPA Crop Definition: None
10. References: GRIN, CODEX, MAGNESS, RICHE, REHM, USDA 1994, US EPA 1995a, US EPA 1995b, US EPA 1994, WOODROOF(a) (picture), FACCIOLA, MALO(b)
11. Production Map: EPA Crop Production Region 10 and 13 (primarily 13).
12. Plant Codes:
  a. Bayer Code: MCDTE

354

1. Maidencane
(Paille fine)

Poaceae (Gramineae)

Panicum hemitomon Schult.

2. This is a warm season perennial native to the U.S. and is a member of the grass tribe *Paniceae*. It is an erect bunchgrass from 0.5 to 1.5 m tall that spreads by creeping rhizomes and occasionally stolons, with flat leaf blades 12 to 30 cm long and has a panicle seedhead (10 to 30 cm long). Maidencane is adapted from aquatic to semi-aquatic moist soils along river banks, ditches, lakes and ponds. It is an excellent forage for livestock and it rapidly declines in quality as it matures. It reproduces from rhizomes with shoots arising during June and July and by seeds. Also see Forage grass.
3. Season: Cool season, spring growth with shoots from rhizomes developing in January to March. Seeds mature in June and July. It is occasionally used as hay.
4. Production in U.S.: There is no production data for Maidencane, however, pasture and rangeland are produced on more than 410 million A (U.S. CENSUS, 1992). Distribution is on the U.S. Coastal Plain, New Jersey to Florida and Texas and in Tennessee (GOULD 1975).
5. Other production regions: Brazil (GOULD 1975).
6. Use: Grazing and hay.
7. Part(s) of plant consumed: Leaves and stems
8. Portion analyzed/sampled: Forage and hay
9. Classifications:
   a. Authors Class: See Forage grass
   b. EPA Crop Group (Group & Subgroup): See Forage grass
   c. Codex Group: See Forage grass
   d. EPA Crop Definition: None.
10. References: GRIN, US EPA 1994, US EPA 1995, CODEX, HOLZWORTH, STUBBENDIECK(a) (picture), GOULD, RICHE
11. Production Map: EPA Crop Production Regions 2, 3, 4, 6 and 13.
12. Plant Codes:
    a. Bayer Code: PANHE

355

1. Mamey apple
(Mamey, Mammee apple, Mammy apple)

Clusiaceae (Guttiferae)

Mammea americana L. (syn: *M. emarginata* Moc. & Sesse ex DC.)

2. The mamey tree grows in hot tropical climates. The plant is propagated by seed and grafting. First fruit crop within 4-5 years of grafting. The tree reaches to 60 feet in height. The leaves are oblong-obovate, up to 8 inches long, thick and glossy. Fruit is round to oblate, up to 6 inches in diameter. The skin is about $1/8$ inch thick, leathery and rather rough and russetted. The juicy but firm flesh surrounds 1 to 4 seeds. The flesh is yellow to orange in color.
3. Season: Harvest in southern Florida in late June through August. In Puerto Rico, some trees produce two crops a year.
4. Production in U.S.: Minor crop production in South Florida and Puerto Rico.

5. Other production regions: Native to West Indies and northern South America. Also grown in Central America and southern Mexico (MORTON 1987).
6. Use: The fruit may be eaten fresh, but more commonly is cooked, as a sauce, preserves or jam.
7. Part(s) of plant consumed: Fruit pulp
8. Portion analyzed/sampled: Whole fruit
9. Classifications:
   a. Authors Class: Tropical and subtropical fruits – inedible peel
   b. EPA Crop Group (Group & Subgroup): Miscellaneous
   c. Codex Group: 006 (FI 0344) Assorted tropical and subtropical fruits – inedible peel
   d. EPA Crop Definition: None
10. References: GRIN, CODEX, MAGNESS, MARTIN 1987, US EPA 1994, MORTON (picture)
11. Production map: EPA Crop Production Region 13.
12. Plant Codes:
    a. Bayer Code: No specific entry

356

1. Mango
(Mangga, Manguier)

Anacardiaceae

Mangifera indica L.

1. Saipan mango
(Kwini, Bembem, Kuwini, Kweni)

M. odorata Griff.

2. Mangoes are produced on evergreen trees that may attain great size unless restricted by pruning. They grow in hot tropical lowlands and are frost-sensitive. Mango plants grow well in any type of soil. Propagated by seed and grafts with earliest fruit crop within 3 years of graft. Fruits hang on long stems and their skin is smooth. The fruit is flattened and generally kidney shaped and may reach 7 to 8 inches in length. Each fruit contains one large flattened seed. The edible pulp surrounding the seed is fibrous and varies greatly in flavor. Fruit is usually harvested in a physiologically mature but unripe stage 15 to 16 weeks after fruit setting. The fruit is an excellent source of vitamin A. Important commercial varieties include 'Tommy Atkins', 'Haden', 'Keitt', 'Kent', 'Pierie' and 'Francisque'. Guam also grows a close relative to mangos, *M. odorata* Griff. (**Saipan mango**).
3. Season: Bloom to harvest: 3 to 6 months. Flowering December-March and fruits May-October in Florida.
4. Production in U.S.: In 1995, 1880 acres in South Florida (CRANE 1996a) and 2500 acres in Puerto Rico (MONTALVO-ZAPATA 1995). Minor production in Hawaii with 50 acres in 1994 and 40 acres in South California.
5. Other production regions: This is a major fruit in all tropical countries, with significant U.S. imports from Mexico, Haiti, Brazil, Guatemala, Peru and Venezuela. India produces about 64 percent of the world's production (KALRA 1995).
6. Use: Usually the fruit is consumed fresh. Ripe mango can be processed into mango slices in syrup, juice, jam, pickles and chutney.
7. Part(s) of plant consumed: Pulp only, skin contains allerigin
8. Portion analyzed/sampled: Whole fruit with stem and pit removed and discarded
9. Classifications:

a. Authors Class: Tropical and subtropical fruits – inedible peel

b. EPA Crop Group (Group & Subgroup): Miscellaneous

c. Codex Group: 006 (FI 0345) Assorted tropical and subtropical fruits – inedible peel

d. EPA Crop Definition: See Avocado and Papaya

10. References: GRIN, ANON(a), CODEX, KAWATE 1996b, LAKSHMINARAYANA, LOGAN, MAGNESS, MORTON, MARTIN 1987, MELNICOE, US EPA 1994, KALRA, CRANE 1996a, MONTALVO-ZAPATA 1995, CAMPBELL(f), RUELE, SAULS(b), SWEET, TEMPLETON

11. Production map: EPA Crop Production Region 13 with some production in Region 10.

12. Plant Codes:

a. Bayer Code: MNGIN (*M. indica*)

357

1. Mangosteen
(Manggis, Mangostan, King's fruit)
Clusiaceae (Guttiferae)

Garcinia mangostana L.

2. Grows in hot, wet tropical lowlands. The plant requires abundant moisture. It is propagated by seed and grafting with first harvest in 4-5 years after the graft. The mangosteen tree is a small, slow growing tropical evergreen with leathery, glabrous leaves up to 10 inches long. The fruit is the size of a small peach, brown to purple in color when ripe. Fruits are borne on lateral branches. They have a thick rind, which encloses 4 to 8 fleshy segments, in which the pulp and seeds are imbedded. The pulp has excellent flavor, proclaimed by many as the best among tropical fruits. The proportion of edible pulp is rather small. The trees often bear sparingly.

3. Season: Fruit matures in 150-180 days from flowering. Often two crops per year, one in the autumn, and one in early summer.

4. Production in U.S.: Few small plantings in Puerto Rico, South Florida, and Hawaii (12 acres – 1996)

5. Other production regions: Honduras, Panama, Southeast Asia.

6. Use: Fresh fruit, preserve (halwa manggis), juice, jelly, syrup, and canned fruit segments.

7. Part(s) of plant consumed: The pulp is the only part consumed.

8. Portion analyzed/sampled: Whole fruit

9. Classifications:

a. Authors Class: Tropical and subtropical fruits – inedible peel

b. EPA Crop Group (Group & Subgroup): Miscellaneous

c. Codex Group: 006 (FI 0346 and FI 4137) Assorted tropical and subtropical fruits – inedible peel

d. EPA Crop Definition: None

10. References: GRIN, CODEX, MAGNESS, MARTIN 1980, MARTIN 1987, KNIGHT(a), KAWATE 1996b, US EPA 1994, MORTON (picture), ROY

11. Production Map: EPA Crop Production Region 13.

12. Plant Codes:

a. Bayer Code: GANMA

358

1. Mannagrass
Poaceae (Gramineae)

Glyceria spp.

1. American mannagrass
Glyceria grandis S.Watson

1. Fowl mannagrass
G. striata (Lam.) Hitchc.

1. Tall mannagrass
G. elata (Nash.) Hitchc.

2. Mannagrasses are cool season perennial bunchgrasses that spread by rhizomes, and are members of the grass tribe *Meliceae*. Most are palatable to livestock, but are located in marshes and wetland areas. There are 20 species of mannagrasses distributed in North America, and the following three are the most important. **American mannagrass** is distributed along lakes and streams from Alaska and throughout the U.S. It grows tall (50 to 200 cm) with a large panicle (15 to 40 cm long). **Fowl mannagrass** is distributed from Newfoundland to British Columbia and south to New Mexico and California, and grow to a height of 30 to 100 cm. **Tall mannagrass** is distributed from Montana to British Columbia and south to Florida and northern Mexico. It grows to a height of 1 to 2 m.

3. Season: Cool season. Growth is in the spring and matures in fall. It is a very palatable forage.

4. Production in U.S.: No production data for any of the Mannagrasses is available, however, pasture and rangeland are produced on more than 410 million A (U.S. CENSUS, 1992). See above.

5. Other production regions: Canada.

6. Use: Pasture, hay and for erosion control.

7. Part(s) of plant consumed: Leaves and stems

8. Portion analyzed/sampled: Forage and hay

9. Classifications:

a. Authors Class: See Forage grass

b. EPA Crop Group (Group & Subgroup): See Forage grass

c. Codex Group: See Forage grass

d. EPA Crop Definition: None

10. References: GRIN, US EPA 1994, US EPA 1995, CODEX, HOLZWORTH, HITCHCOCK, STUBBENDIECK (picture), MOSLER, RICHE

11. Production Map: EPA Production Regions for American mannagrass are 1, 2, 5, 7, 8, 9, 10, 11 and 12; for Fowl mannagrass are 1, 7, 8, 10, 11 and 12; and for Tall mannagrass are 1, 2, 3, 5, 7, 8, 11 and 12.

12. Plant Codes:

a. Bayer Code: No specific entry

359

1. Marigold
(African marigold, Big marigold, Aztec marigold, Okole-oi-oi, Flor de muerto, Tagete etalee, Saffron marigold, Inca marigold, American marigold)
Asteraceae (Compositae)

Tagetes erecta L.

2. Stout glabrous annual to 3 feet tall. Leaves are pinnate, and flower heads solitary, 2 to 5 inches across, light yellow to orange color. The harvested flower heads are dried to 10 per-

cent moisture. The dried flower heads and its pigment extract are used as a feed supplement for poultry. With a similarity to hop processing, in the future, the fresh sample use may be eliminated with the raw agricultural commodity being the dried marigold flowers.
3. Season, seeding to first harvest: About 90 days with the second and third flower harvests at intervals of 3 to 5 weeks. Normally planted in early April.
4. Production in U.S.: Limited acreage in California, about 300 acres (SARRACINO 1988) and more recently 750 acres (SCHNEIDER 1992a).
5. Other production regions: Native to Mexico and Central America (BAILEY 1976).
6. Use: Dried flowers are ground, then added to poultry feed. The xanthophyll pigment can also be extracted by solvent, dried to powder and added to poultry feed to enhance the yellow color of the skin or egg yolk color. Also grown as an ornamental.
7. Part(s) of plant consumed: Flowers (dried) and extract as feed supplement
8. Portion analyzed/sampled: Flowers (fresh and dried) and its pigment extract. With mechanical harvesting, some stems and leaves are included with fresh samples but the goal is to harvest only the flower heads
9. Classifications:
   a. Authors Class: Herbs and spices
   b. EPA Crop Group (Group & Subgroup): Miscellaneous
   c. Codex Group: No specific entry
   d. EPA Crop Definition: None
10. References: GRIN, BAILEY 1976, FOHNER 1986, NEAL, SARRACINO 1988, SCHNEIDER 1992a
11. Production Map: EPA Crop Production Region 10.
12. Plant Codes:
   a. Bayer Code: TAGER

360

1. Marigold, Pot
(Calendula, Marigold flowers, Scotch marigold, Ruddles)
Asteraceae (Compositae)
Calendula officinalis L.
2. The plant is an annual related to sunflower, 1 to 2 feet high and hairy, with oblong, entire, thick leaves. The flowers are large heads with yellow or orange rays. The dried flower heads are used to flavor soups and stews. Such use is more general in Europe than in the U.S.
3. Season, harvest: Blooms continuously for the season. Mainly propagated by seed with first blossoms about 6 weeks after planting.
4. Production in U.S.: No data. Grows in gardens in California (BAILEY 1976).
5. Other production regions: Europe.
6. Use: Flavoring soups and stews as herb used like saffron. Also grown as an ornamental.
7. Part(s) of plant consumed: Flower heads
8. Portion analyzed/sampled: Flower heads (fresh and dried)
9. Classifications:
   a. Authors Class: Herbs and spices
   b. EPA Crop Group (Group & Subgroup): Herbs and spices (19A)
   c. Codex Group: 027 (HH 0737) Herbs
   d. EPA Crop Definition: None

10. References: GRIN, BAILEY 1976, CODEX, FACCIOLA, FOSTER (picture), MAGNESS, US EPA 1995a
11. Production Map: EPA Crop Production Region 10.
12. Plant Codes:
   a. Bayer Code: CLDOF

361

1. Marjoram
Lamiaceae (Labiatae)
Origanum spp.
1. Marjoram, Sweet
(Annual marjoram)
O. majorana L. (syn: *O. hortensis* Moench; *Majorana hortensis* Moench)
1. Marjoram, Pot
O. onites L.
1. Oregano
(Wild marjoram, Greek oregano, Turkish oregano)
O. vulgare L.
2. The marjorams are perennial plants, although sweet or annual marjoram, the most important kind, is usually grown as an annual. The plants are 1 to 2 feet in height, with oblong to ovate, somewhat hairy leaves. Leaves and tender stems are fragrant and spicy. They may be cut at intervals during the summer, or the whole young plants may be cut near the ground before the flowers open. Commercially, the cut plants are dried and pulverized. Marjoram is used in flavoring soups, stews, dressings, etc. In the U.S., marjoram is often found in home gardens, with some grown commercially. An average of 5,574 tons of dried marjoram was imported annually for the years 1990-92, mainly from the Mediterranean area and Mexico.
3. Season, harvest: Just before bloom. Marjoram is direct seeded. An established planting can be cut 2-6 times during season.
4. Production in U.S.: Limited data. About 200 acres in California (1980). Produced in California and Florida (STEPHENS 1988).
5. Other production regions: Mediterranean region and Mexico.
6. Use: Flavoring of vegetables, meats, soups, stews and sauces. Flowering tops used in drinks.
7. Part(s) of plant consumed: Leaves, stems and flowers
8. Portion analyzed/sampled: Leaves and stems (fresh and dried)
9. Classifications:
   a. Authors Class: Herbs and spices
   b. EPA Crop Group (Group & Subgroup): Herbs and spices (19A)
   c. Codex Group: 027 (HH 0736) Herbs; 057 (DH 0736) Dried herbs
   d. EPA Crop Definition: Marjoram = *Origanum* spp.
10. References: GRIN, CODEX, ELMORE 1980, FOSTER, MAGNESS, MARTIN 1979, SCHREIBER, STEPHENS 1988, USDA 1993a, US EPA 1995a
11. Production Map: EPA Crop Production Regions 3 and 10.
12. Plant Codes:
   a. Bayer Code: MAJHO (*O. majorana*), ORIVU (*O. vulgare*)

362

1. Marmaladebox
(Genipap, Genip, Maluco, Huito)
Rubiaceae

Genipa americana L.

2. This fruit is borne on a medium-size tree, native to the Caribbean area. Leaves are up to a foot long and are clustered at branch tips. Fruits are ovoid in shape, up to 3 inches broad and 4 inches long. The skin is thin, enclosing a granular pulp and a seed cavity with numerous small seeds. The fruit is popular in Puerto Rico, especially for macerating the pulp in water to make a refreshing drink. The fruit is rarely eaten out of hand.
3. Season, in Puerto Rico, flowers and fruits appear continuous from spring to fall.
4. Production in U.S.: No data. Puerto Rico.
5. Other production regions: West Indies and South America (MARTIN 1987).
6. Use: Fresh and in beverages.
7. Part(s) of plant consumed: Fruit pulp
8. Portion analyzed/sampled: Whole fruit
9. Classifications:
   a. Authors Class: Tropical and subtropical fruits – inedible peel
   b. EPA Crop Group (Group & Subgroup): Miscellaneous
   c. Codex Group: 006 (FI 0347) Assorted tropical and subtropical fruits – inedible peel
   d. EPA Crop Definition: None
10. References: GRIN, CODEX, MAGNESS, MARTIN 1987, MORTON (picture)
11. Production Map: EPA Crop Production Region 13.
12. Plant Codes:
    a. Bayer Code: No specific entry

363

1. Marshhay cordgrass
(Saltmeadow cordgrass)
Poaceae (Gramineae)

Spartina patens (Aiton) H.L. Muhl.

2. This is a perennial warm season grass native to the U.S. and is a member of the grass tribe *Chlorideae*. It is an erect bunchgrass up to 1.5 m tall that spreads by creeping rhizomes with leaf blades 15 to 40 cm long, and its inflorescence is a panicle (9 to 20 cm long). It is adapted to salt marshes, sandy meadows, beaches and other saline sites. Marshhay cordgrass produces fair to good forage for cattle and for wild geese, but poor for sheep. It can be managed for winter grazing. It reproduces by seeds or rhizomes.
3. Season: Warm season grass that starts growing in the late spring and it flowers from May to September, and produces seed in October. It can be utilized all winter in the southernmost U.S.
4. Production in U.S.: There are no production data for Marshhay cordgrass, however, pasture and rangeland are produced on more than 410 million A (U.S. CENSUS, 1992). Eastern U.S. coast to Mexico (GOULD 1975).
5. Other production regions: Mediterranean region, Mexico and Canada (GOULD 1975).
6. Use: Grazing, hay, and soil stabilization.
7. Part(s) of plant consumed: Leaves and stems

8. Portion analyzed/sampled: Forage and hay
9. Classifications:
   a. Authors Class: See Forage grass
   b. EPA Crop Group (Group & Subgroup): See Forage grass
   c. Codex Group: See Forage grass
   d. EPA Crop Definition: None
10. References: GRIN, US EPA 1994, US EPA 1995, CODEX, HOLZWORTH, STUBBENDIECK (picture), RICHE, GOULD
11. Production Map: EPA Crop Production Regions 2, 3, 4 and 6.
12. Plant Codes:
    a. Bayer Code: SPTPA

364

1. Marshmarigold
(Cowslip greens, Kingcup, Mayblob, Meadowbright, Cowslip, American cowslip, Yellow marshmarigold)
Ranunculaceae

Caltha palustris L. (syn: *C. radicans* J.R. Forst., *C. zetlandica* (Beeby) Dorfl.)

2. This plant is a perennial native in eastern U.S., from the Carolinas to Canada, growing in marshy areas. Stems are hollow, 1 to 2 feet high. Leaves are cordate or rounded. They are gathered in spring before flowering, and used as potherbs.
3. Season: Leaves gathered within 2 or 3 weeks of growth start in spring.
4. Production in U.S.: None commercial, harvested from wild plants.
5. Other production regions: No data.
6. Use: Potherbs. Young leaves eaten like spinach. Also grown as an ornamental.
7. Part(s) of plant consumed: Leaves
8. Portion analyzed/sampled: Leaves
9. Classifications:
   a. Authors Class: Leafy vegetables (except *Brassica* vegetables)
   b. EPA Crop Group (Group & Subgroup): Miscellaneous
   c. Codex Group: 013 (VL 0471) Leafy vegetables (including *Brassica* leafy vegetables)
   d. EPA Crop Definition: None
10. References: GRIN, BAILEY 1976, CODEX, FACCIOLA, MAGNESS, NEAL, USDA NRCS
11. Production Map: No entry
12. Plant Codes:
    a. Bayer Code: CTAPA

365

1. Martynia
(Unicorn plant, Proboscis flower, Ram's-horn, Devil's-claw, Purple flower devil's claw)
Pedaliaceae (Martyniaceae)

Proboscidea louisianica (Mill.) Thell.

P. louisianica (Mill.) Thell. ssp. *louisianica* (syn: *P. jussieui* Medik.)

2. This is an annual plant, up to 2 to 3 feet in height, with entire leaves, round to cordate, 4 to 12 inches wide. The fruit is hairy, about 1 inch thick and 4 to 6 inches long at maturity, about half the length consisting of a slender curved beak. The small, immature pods are made into pickles, like cucum-

bers. The plant is grown as an ornamental, and sparingly as a pickling vegetable. Culturally similar to okra.

3. Season, seeding to first harvest: 3 to 4 months. Continue picking pods as they develop, to encourage more pod set.
4. Production in U.S.: No data. Of minor importance. Native to southwestern U.S. and occasionally grown in home gardens throughout the U.S. (STEPHENS 1988).
5. Other production regions: Mexico.
6. Use: Pickles or use like okra.
7. Part(s) of plant consumed: Entire young pods
8. Portion analyzed/sampled: Pod
9. Classifications:
   a. Authors Class: Fruiting vegetables (except cucurbits)
   b. EPA Crop Group (Group & Subgroup): Miscellaneous
   c. Codex Group: No specific entry
   d. EPA Crop Definition: None
10. References: GRIN, MAGNESS, STEPHENS 1988 (picture), FACCIOLA
11. Production Map: See Okra
12. Plant Codes:
   a. Bayer Code: PROLO

366

1. **Mashua (mah-shoo-ah)**
(Anu, Isano, Navo, Ysano, Capucine)
Tropaeolaceae
   Tropaeolum tuberosum Ruiz & Pav.
2. Mashua is closely related to the garden nasturtium with tubers the size of small potatoes. The hot taste is due to isothiocyanates (mustard oils). Mashua is a perennial, herbaceous, semiprostrate climber to 6 feet tall. Leaves are 3 to 5 lobed and glabrous. Stems are twining.
3. Season, planting to harvest: About 6 to 8 months using small tubers as seed.
4. Production in U.S.: No data.
5. Other production regions: Northern South America (NATIONAL RESEARCH COUNCIL 1989).
6. Use: Tubers are normally cooked. The young tender leaves are used as a boiled green vegetable. The flowers are also eaten.
7. Part(s) of plant consumed: Tuber and leaves
8. Portion analyzed/sampled: Tuber and leaves
9. Classifications:
   a. Authors Class: Root and tuber vegetables; Leaves of root and tuber vegetables
   b. EPA Crop Group (Group & Subgroup): Miscellaneous
   c. Codex Group: No specific entry
   d. EPA Crop Definition: None
10. References: GRIN, NATIONAL RESEARCH COUNCIL 1989, FACCIOLA
11. Production Map: No entry
12. Plant Codes:
   a. Bayer Code: No specific entry

367

1. **Mauka (mah-oo-kah)**
(Yuca inca, Chago, Arracacha de toro)
Nyctaginaceae
   Mirabilis expansa (Ruiz & Pav.) Standl.

2. Mauka is a low, compact perennial plant to 3 feet tall which is grown as an annual. The edible parts are the upper part of the root and the lower part of the stem which are below ground level.
3. Season, planting to harvest: About 12 months. Propagated by portions of stem or root.
4. Production in U.S.: No data
5. Other production regions: Northern South America (NATIONAL RESEARCH COUNCIL 1989).
6. Use: Root and swollen stem are used boiled or fried like cassava. Leaves can be used in salads.
7. Part(s) of plant consumed: Leaves, roots and swollen stem
8. Portion analyzed/sampled: Roots and underground swollen stem, and leaves
9. Classifications:
   a. Authors Class: Root and tuber vegetables; Leaves of root and tuber vegetables
   b. EPA Crop Group (Group & Subgroup): Miscellaneous
   c. Codex Group: No specific entry
   d. EPA Crop Definition: None
10. References: GRIN, NATIONAL RESEARCH COUNCIL 1989, RUBATZKY
11. Production Map: No entry
12. Plant Codes:
   a. Bayer Code: No specific entry

368

1. **Mayhaw**
Rosaceae
   Crataegus spp.
1. **Mayhawthorn**
(Lori mayhaw, Lindsey mayhaw, Eastern mayhaw)
   C. aestivalis (Walter) Torr. & A. Gray (syn: *C. maloides* Sarg.)
1. **Riverflat hawthorn**
(Texas super berry, Big red mayhaw, Applehaw, Western mayhaw)
   C. opaca Hook. & Arn.
1. **Rusty hawthorn**
(Rufous mayhaw)
   C. rufula Sarg.
2. Arborescent shrub or round-topped small tree to 30 feet tall. Mayhaws flower early (late February to mid-March in southern Georgia) and the fruit ripens mostly in early May but some ripen through June. The fruit is a small pome, yellow to red, acid and juicy, resembling cranberries in appearance.
3. Season, bloom to harvest: About 3 months and some can be harvested for about 30 days.
4. Production in U.S.: No data. Limited. North Carolina south to northern Florida and west to east Texas (PAYNE 1990).
5. Other production regions: Native to north temperate zone (BAILEY 1976).
6. Use: Mainly in jellies, preserves. Also in wines and syrup.
7. Part(s) of plant consumed: Fruit
8. Portion analyzed/sampled: Whole fruit
9. Classifications:
   a. Authors Class: Pome fruita
   b. EPA Crop Group (Group & Subgroup): Pome fruits (11)
   c. Codex Group: No entry for Mayhaw. Pome fruits in 002 (FP 0009).
   d. EPA Crop Definition: None

10. References: GRIN, PAYNE (picture), USDA NRCS, US EPA 1995a, CODEX, BAILEY 1976
11. Production Map: EPA Crop Production Regions 2, 3, 4 and 6.
12. Plant Codes:
    a. Bayer Code: No specific entries, genus entry is CSCSS

369

1. Maypop
(Apricot vine, Wild passionflower, Maypop passionflower)
Passifloraceae
Passiflora incarnata L.

2. The plant has been termed the passionflower for the North as it is a cold-hardy species of *Passiflora*. Captain John Smith of the Jamestown settlement noted that the Native Americans cultivated Maypop. The plant is a herbaceous perennial climbing vine, that clings to available support by tendrils. It dies to the ground each winter and starts regrowth in early summer. Blossoms appear about one month after the shoots first emerge. The large colorful flower lasts only one day, however the plant continues to produce new flowers well into autumn. Maypop fruits are yellow to yellow green, oval, and 1-2 inches across. The inside of the fruit is filled with seeds surrounded by a tasty gelatinous pulp. It grows from Virgina to Florida and west to Missouri and Texas (BAILEY 1976), especially as an ornamental. Its showy flowers are colored white and purple (FRANK 1997).
3. Season: Bloom to ripe fruit: one month, but flowers and mature fruits may be on a plant at the same time.
4. Production in U.S.: No data. See above.
5. Other production regions: No data.
6. Use: Fresh eating of fruit. Dried maypop vines are a mild sedative drug.
7. Part(s) of plant consumed: Internal pulp only
8. Portion analyzed/sampled: Whole fruit
9. Classifications:
    a. Authors Class: Small fruits
    b. EPA Crop Group (Group & Subgroup): Miscellaneous
    c. Codex Group: No specific entry
    d. EPA Crop Definition: None
10. References: GRIN, REICH, BAILEY 1976, FRANK
11. Production Map: EPA Crop Production Regions 2, 3, 4, 5 and 6.
12. Plant Codes:
    a. Bayer Code: PAQIN

370

1. Meadow foxtail
(Golden foxtail grass, Yellow foxtail grass, Foxtail-meadow)
Poaceae (Gramineae)
Alopecurus pratensis L.

2. This grass is native to temperate Europe and Asia, and was introduced into the U.S. around the middle of the last century, and it is a member of the grass tribe *Aveneae*. It is a long-lived perennial, which forms a few short rootstocks and underground rhizomes. It produces a medium dense sod. Leaves are medium in width, dark green and numerous. Stems generally reach about 3 feet. The species is especially adapted to cool, moist climates, such as west of the Cascade Mountains in Oregon and Washington; but is useful in other Northern states, and in southern areas of Canada. It is espe-

cially useful as pasture, and in low-lying areas. Grows throughout a long season, with its regrowth superior to creeping foxtail. In combination with legumes such as alsike clover and birdsfoot trefoil it is seeded at a rate of 3 to 5 lb/A and is made into silage in the Pacific Northwest, but is rarely harvested for hay. Seed yields range from 300 to 450 kg/ha. Seeds vary in size from 500,000 to 1,000,000/lb.
3. Season: Cool season, good early spring growth. Seeded in the fall or spring. Good for pasture, flowers June or July.
4. Production in U.S.: No data for Meadow foxtail production, however, pasture and rangeland are produced on more than 410 million A (U.S. CENSUS, 1992). See above.
5. Other production regions: Grown in Ontario, Canada.
6. Use: Pasture, silage and hay.
7. Part(s) of plant consumed: Leaves and stems.
8. Portion analyzed/sampled: Forage and hay.
9. Classifications:
    a. Authors Class: See Forage grass
    b. EPA Crop Group (Group & Subgroup): See Forage grass
    c. Codex Group: See Forage grass
    d. EPA Crop Definition: None
10. References: GRIN, MAGNESS, US EPA 1994, US EPA 1995, CODEX, BARNES, ANDERSON, ROTHWELL 1996a MOSLER (picture), RICHE
11. Production Map: EPA Production Regions 2, 5, 11 and 12.
12. Plant Codes:
    a. Bayer Code: ALOPR

371

1. Meadowfoam
Limnanthaceae
Limnathes alba Hartw. ex Benth.

2. Meadowfoam can be grown as a winter annual crop. It is a small erect herbaceous crop, 10 to 18 inches tall, native to northern California. Leaves are pinnately dissected, and flowers are solitary, 1 inch across. The seeds (nutlets) are pear-shaped, striated and tiny, and yield about 1,000 pounds per acre and produce only 20 to 30 percent oil. It has been grown to a limited extent in parts of the eastern U.S. and Alaska. Meadowfoam is well adapted to cool wet Mediterranean climate of the Pacific Northwest. The seeds can be harvested with the same equipment used in grass seed production and is cut when 90 percent of the seeds are mature. It is allowed to dry in windrows for 7 to 10 days to a seed moisture content of 12 to 16 percent. The meal can be fed to beef cattle. The oil is in direct competition with rapeseed oil as an industrial source. It also can be a substitute for sperm whale oil and jojoba oil.
3. Season, seeding to harvest: Planted in the fall about October and harvested in early summer, June-July.
4. Production in U.S.: Willamette Valley in Oregon with about 400 acres in 1986 (ROBBELEN 1989). In 1993, about 1,000 acres with a goal of about 3,000 acres in 1994 (EHRENSING 1994).
5. Other production regions: West coast of North America (BAILEY 1976).
6. Use: Oil is useful for cosmetics. Meal as livestock feed.
7. Part(s) of plant consumed: Seed
8. Portion analyzed/sampled: Seed and its meal and oil
9. Classifications:
    a. Authors Class: Oilseed
    b. EPA Crop Group (Group & Subgroup): Miscellaneous

c. Codex Group: No specific entry
d. EPA Crop Definition: None
10. References: GRIN, BAILEY 1976, BOSISIO 1989 (picture), EHRENSING 1994, ROBBELEN 1989, OELKE (e)
11. Production Map: EPA Crop Production Region 12.
12. Plant Codes:
a. Bayer Code: No specific entry

372

1. Medlar
(Nispero, Northern Loquat)
Rosaceae
Mespilus germanica L.
2. This plant is a small, flat-topped, deciduous tree that grows to 20 feet high. A close relative of hawthorn which lacks thorns. In late spring, large white or slightly pink blossoms are borne singly on ends of short shoots. Almost every flower sets fruit. The fruit is 1 to $2^1/_2$ inches in diameter and resembles a small russetted apple, tinged dull yellow or brown color with the calyx end flared open. The flesh is as soft as a baked apple. Embedded in the pulp are five large stone-like seeds. Fruits are best picked at early leaf drop prior to heavy frost. At harvest, fruits are rock hard and must soften prior to consumption. Ripening, or bletting takes two weeks to one month after harvest.
3. Season: Ripens in October and is ready by Christmas.
4. Production in U.S.: No statistics available.
5. Other production regions: Once a major fruit crop in Europe.
6. Use: Fresh eating out of hand or used for beverage, dessert or preserves. Can be baked whole or roasted. The plant is a beautiful ornamental.
7. Part(s) of plant consumed: Inner pulp only, skin and seeds discarded
8. Portion analyzed/sampled: Whole fruit with stem removed
9. Classifications:
a. Authors Class: Pome fruits
b. EPA Crop Group (Group & Subgroup): Miscellaneous
c. Codex Group: 002 (FP 0229) Pome fruits
d. EPA Crop Definition: None
10. References: GRIN, CODEX, REICH, FACCIOLA, WEISS
11. Production Map: No entry
12. Plant Codes:
a. Bayer Code: MSPGE

373

1. Melon, Garden
(Orange melon, Mango melon, Melon apple, Vine peach, Vegetable orange, Garden lemon, Lemon cucumber, Queen Annes' pocket melon, Dudaim melon, Garden melon, Pomegranate melon, Plum granny, Stink melon)
Cucurbitaceae
Cucumis melo var. *chito* (C. Morren) Naudin (syn: *C. melo* var. *dudaim* (L.) Naudin)
2. This is a close relative of the muskmelon which it resembles but has a less vigorous vine and smaller leaves. Leaves and vines are hairy. Fruit is of size, shape and color of an orange or lemon, with a smooth surface. Fruit flesh is white or yellow, resembling cucumber.
3. Season, seed to first harvest: 2 to 3 months.
4. Production in U.S.: No data. Very limited.

5. Other production regions: Asia Minor and northern Africa (RUBATZKY 1997).
6. Use: Preserves, pickles. Mature fruits are peeled and eaten fresh or in preserves. Unripe fruits are pickled whole.
7. Part(s) of plant consumed: Whole fruit
8. Portion analyzed/sampled: Whole fruit
9. Classifications:
a. Authors Class: Cucurbit vegetables
b. EPA Crop Group (Group & Subgroup): Cucurbit vegetables (9A)
c. Codex Group: 011 (VC 0046) Cucurbits fruiting vegetables
d. EPA Crop Definition: See muskmelon
10. References: GRIN, CODEX, FACCIOLA, MAGNESS, US EPA 1995a, RUBATZKY
11. Production Map: No entry
12. Plant Codes:
a. Bayer Code: CUMMH

374

1. Melon, Oriental pickling
(Yuetkwa, Uri, Yueh kua, Chinese white cucumber, Tsukemono, Koko, Pickling melon, Tea melon, Tsa gwa, Sweet melon)
Cucurbitaceae
Cucumis melo var. *conomon* (Thunb.) Makino
2. Culture is similar to cucumber and melons. The fruits are usually cylindrical, 8 to 12 inches long and $2^1/_2$ to $3^1/_2$ inches in diameter. For pickling, fruits are normally harvested when full size but still immature.
3. Season, planting to harvest: About 40 days. Seeding to harvest about 50 to 85 days, depending on cultivar.
4. Production in U.S.: No data. Grown in Hawaii (NEAL 1965).
5. Other production regions: Orient (YAMAGUCHI 1983).
6. Use: Pickling and eaten raw. Both immature and mature fruits are made into sweet or sour pickles. Those pickled in sake are called "Nara-Zuke".
7. Part(s) of plant consumed: The seed cavity is normally removed and only the rind is used for pickling. In Asia, also the fruits are used like summer squash
8. Portion analyzed/sampled: Whole fruit
9. Classifications:
a. Authors Class: Cucurbit vegetables
b. EPA Crop Group (Group & Subgroup): Miscellaneous
c. Codex Group: 011 (VC 0046) Cucurbits fruiting vegetables
d. EPA Crop Definition: See muskmelon
10. References: GRIN, CODEX, FACCIOLA, NEAL, REHM, YAMAGUCHI 1983
11. Production Map: No entry
12. Plant Codes:
a. Bayer Code: CUMMO

375

1. Melon, Winter
(Casaba melon, Honeydew melon, White-skinned melon, Crenshaw melon, Canary melon, Pineapple melon, Santa Claus melon, Sungold, Golden pershaw, Yellow canary, Juan canarymelon, Gold king, Golden honeymoon, Honeyloupe, Green fleshed, Morgan, Venus, White antibes, Winter muskmelon)
Cucurbitaceae
Cucumis melo var. *inodorus* H. Jacq.

2. These are melons grown on plants similar to cantaloupes, but fruits are without distinctive odor and late in ripening. The fruits are medium to large with generally smooth surface, casabas are rough. In all cultural aspects they resemble cantaloupes except for a longer growing season. Included in the group are the Honeydew, Casaba, Crenshaw and Canary. Casaba melons include Santa claus melon, Sungold and Winter pineapple melon. Crenshaw melons include golden pershaw. Canary melon include yellow canary (Juan canarymelon) and Gold king. Honeydew melons include Golden honeymoon, Honeyloupe, Green fleshed, Morgan, Venus and White antibes. Honeydews have best flavor weighing 4 to 5 pounds (LOGAN 1996).
3. Season, seeding to maturity: About 4 months.
4. Production in U.S.: In 1995, honeydew melons were 282,000 tons on 29,300 acres (USDA 1996b). In 1995, California (21,100 acres), Texas (4,600), Arizona (3,600) (USDA 1996b).
5. Other production regions: Mexico (LOGAN 1996).
6. Use: Fresh eating, salads.
7. Part(s) of plant consumed: Internal pulp
8. Portion analyzed/sampled: Fruit
9. Classifications:
   a. Authors Class: Cucurbit vegetables
   b. EPA Crop Group (Group & Subgroup): Cucurbit vegetables (9A)
   c. Codex Group: 011 (VC 0046) Cucurbits fruiting vegetables
   d. EPA Crop Definition: See muskmelon
10. References: GRIN, CODEX, FACCIOLA, LOGAN (picture), MAGNESS, STEPHENS 1988 (picture), USDA 1996b, US EPA 1995a, US EPA 1994
11. Production Map: EPA Crop Production Regions for honeydew melons 6 and 10.
12. Plant Codes:
   a. Bayer Code: CUMMI

376

1. Mexican mint marigold
(Sweet-scent marigold, Mexican tarragon, Sweet marigold, Sweet mace, Pericon, Sweet-scented marigold)
Asteraceae (Compositae)
   Tagetes lucida Cav.
1. Signet marigold
(Lemon marigold, Slender-leaf marigold)
   T. tenuifolia Cav. (syn: *T. signata* Bartl.)
2. **Mexican mint marigold** is native to Mexico. It grows to 30 inches tall. As a tender perennial, it thrives in hot weather and needs a long, warm growing season to be most productive.

   The **signet marigold** is a glabrous annual to 2 feet tall. It is related, and includes, citrus-scented 'Lemon Gem' and 'Orange Gem' cultivars. Both leaves and flowers have a lemon-like flavor.
3. Season, propagation: By seed or cutting.
4. Production in U.S.: No data
5. Other production regions: Mexico (MCCLURE 1995).
6. Use: Leaves and sprigs used for tea and food flavoring. Leaves are licorice-flavored and used as substitute for tarragon. Also grown as an ornamental.
7. Part(s) of plant consumed: Leaves and sprigs
8. Portion analyzed/sampled: Leaves and flowering tops (fresh and dried).

9. Classifications:
   a. Authors Class: Herbs and spices
   b. EPA Crop Group (Group & Subgroup): Miscellaenous
   c. Codex Group: No specific entry
   d. EPA Crop Definition: None
10. References: GRIN, BAILEY 1976, FACCIOLA, MCCLURE (picture)
11. Production Map: EPA Crop Production Regions 10 and 13.
12. Plant Codes:
   a. Bayer Code: No specific entry

377

1. Mexican oregano
(Mexican sage, Te de pais, Scented matgrass, Oregano)
Verbenaceae
   Lippia graveolens Kunth
1. Puerto Rico oregano
(Heller's false thyme, Spanish thyme)
   L. micromera Schauer
2. Small shrub, woody and erect. Mexican oregano is principally gathered from *L. graveolens*.
3. Season, harvest: Multiple harvests, as needed.
4. Production in U.S.: No data. Puerto Rico (FACCIOLA 1990).
5. Other production regions: Grown on limited scale in tropics (MARTIN 1979). Mexico (JANICK 1990).
6. Use: To flavor Mexican foods, pizza, barbecue sauces, pozole, and dried leaves may be brewed into an herbal tea.
7. Part(s) of plant consumed: Leaves
8. Portion analyzed/sampled: Leaves (fresh and dried)
9. Classifications:
   a. Authors Class: Herbs and spices
   b. EPA Crop Group (Group & Subgroup): Miscellaneous
   c. Codex Group: No specific entry
   d. EPA Crop Definition: None
10. References: GRIN, FACCIOLA, JANICK 1990, LEWIS, MARTIN 1979 (picture), NEAL, USDA NRCS
11. Production Map: EPA Crop Production Region 13.
12. Plant Codes:
   a. Bayer Code: No specific entry

378

1. Millet, Foxtail
(Italian millet, German millet, Hungarian millet, Mijo menor, Foxtail millet, Dwarf setaria, Foxtail bristlegrass, Moha, Panizo)
Poaceae (Gramineae)
   Setaria italica (L.) P. Beauv.
2. This warm season annual grass was cultivated in China more than 4,000 years ago. It was introduced into this country from Europe in 1849 and is now grown north from Texas through the Plains and Central States. It is used as hay, pasture and green fodder. The seed is used as bird feed. Foxtail millet is an annual grass growing 3 to 5 feet under the best conditions. It is a warm weather crop, usually seeded after the soil becomes warm in late spring. Flowering stems are leafy throughout their length, but the hay or fodder is less nutritious than a number of other grasses and legumes. For this reason it is now grown less than in the past. A number of

varieties, differing slightly in characteristics, are available. A weed grass called foxtail is a close relative.

3. Season: Seeded in late spring, 60 to 70 days to harvest. Cut for hay at late boot to late bloom stage.
4. Production in U.S.: No specific data for forage use, See Forage grass. In 1992, 1,664 tons of seed were produced from 3,935 acres. In 1992, the top two seed crop states were Colorado (3,348 acres) and Nebraska (45) (1992 CENSUS). Plains and Central states north from Texas.
5. Other production regions: China and Europe.
6. Use: Silage, hay, pasture. Seeds used as bird feed.
7. Part(s) of plant consumed: Foliage; seed which is an insignificant feed item
8. Portion analyzed/sampled: Forage and hay
9. Classifications:
   a. Authors Class: Grass forage, fodder and hay; Cereal grains
   b. EPA Crop Group (Group & Subgroup): See Forage grass
   c. Codex Group: See Forage grass. 051 (AS 0646) Hay or fodder (dry) of grasses; 020 (GC 4633 and GC 4653) Cereal grains
   d. EPA Crop Definition: None
10. References: GRIN, MAGNESS, RICHE, BALL
11. Production Map: EPA Crop Production Regions 2, 3, 4, 5, 7 and 8.
12. Plant Codes:
    a. Bayer Code: SETIT

379

1. Millet, Japanese
(Japanese barnyardgrass, Mijo japones)
Poaceae (Gramineae)
Echinochloa esculenta (A. Braun) H. Scholz (syn: *E. utilis* Ohwi & Yabuno)

1. Barnyard millet
(Indian barnyard millet, Billion dollar grass)
E. frumentacea Link (syn: *E. crus-galli* var. *frumentacea* (Link) W.F. Wright)

2. Warm season annual grasses grown to a limited extent in the northeastern states for green feed, silage and hay. They are superior under cool summers to sudangrass or foxtail millet. Stems reach to 4 feet or more. Leaves are large, more than $1/_2$ inch broad. Seed heads are dense and drooping. Once exploited as "a billion dollar grass" it is now grown much less than formerly. It produces good tonnage but is coarse and only fair in feed value. It is a weed in paddy rice.
3. Season: Warm season pasture.
4. Production in U.S.: No specific data, See Forage grass. Northeastern states.
5. Other production regions: Asia
6. Use: Hay, silage, pasture.
7. Part(s) of plant consumed: Foliage
8. Portion analyzed/sampled: Forage and hay
9. Classifications:
   a. Authors Class: See Forage grass
   b. EPA Crop Group (Group & Subgroup): See Forage grass
   c. Codex Group: 051 (AS 0646) Straw, fodder and forage of cereal grains and grasses; 020 (GC 0646 and GC 4645) Cereal grains
   d. EPA Crop Definition: None
10. References: GRIN, MAGNESS, SKERMAN
11. Production Map: EPA Crop Production Regions 1, 2, 4 and 5.

12. Plant Codes:
    a. Bayer Code: ECHCF (*E. frumentacea*)

380

1. Millet, Pearl
(Cattail millet, Bulrush millet, Spiked millet, Kambu, Dukhn, Bajra, Mijo, American fountain grass)
Poaceae (Gramineae)
Pennisetum glaucum (L.) R. Br. (syn: *P. typhoideum* L.C. Rich., *P. americanum* (L.) Schumann, *P. americanum* (L.) Leeke, *P. spicatum* (L.) Koern., *P. typhoides* (Burm. f.) Stapf & C.E. Hubb.)

2. Pearl millet is one of five millet species of commercial importance, the others are proso, foxtail, Japanese and browntop. Pearl millet is native to India but came to the U.S. via the West Indies. It is a tall upright annual grass, up to 10 feet, with coarse stems which grow in dense clumps. Leaves are coarse, 2 to 3 feet long and an inch wide and numerous. In fertile soil it produces great amounts of green fodder, which is palatable and nutritious, and can be cut repeatedly during a season. It is grown both for pasture and silage. It can be grown as far north as Maryland, but grows better farther south. The seed is planted directly in the field, generally in rows about 4 feet apart. Pearl millet is also a food grain crop. It was introduced into the U.S. but seldom grown until 1875. It is primarily grown in Southern U.S. as a temporary pasture. Millets are annual grasses that have seedbed preparation similar to spring seeded small grains. Outside the U.S., more than 95 percent of the pearl millet crop is grown for grain. U.S. farmers regard it as a high yielding summer annual forage crop. Excellent dry season grass for beef cattle. Also, there are up to 3 cuttings per season at 6 to 7 week intervals for hay, silage or green chop.
3. Season, seeding to harvest: 75 to 85 days. For pasture as a warm season annual.
4. Production in U.S.: No data, limited. Also see Forage grass. Coastal Plains region from Maryland south for pasture.
5. Other production regions: Native to India, West Indies. A food crop of tropical Asia and Africa (BAILEY 1976).
6. Use: Seed use for bird seed and food grain similar to proso millet. Also hay, pasture, silage and green chop.
7. Part(s) of plant consumed: Grain and foliage
8. Portion analyzed/sampled: Grain and straw; Forage and hay. Pearl millet grain is kernel with hull removed. Also see Millet, Proso
9. Classifications:
   a. Authors Class: Cereal grain; Forage, fodder and straw of cereal grains. Also see Forage grass
   b. EPA Crop Group (Group & Subgroup): Cereal grains (15); Forage, fodder, and straw of cereal grains (16). Also see Forage grass
   c. Codex Group: 020 (GC 0646) Cereal grains. Also see Forage grass
   d. EPA Crop Definition: None
10. References: GRIN, CODEX, MAGNESS, OELKE(g), IVES, ALDERSON, BALL (picture), SKERMAN, SCHNEIDER 1970, BAILEY 1976, ANDREWS
11. Production Map: EPA Crop Production Regions 2, 3, 4, 6 and 13.
12. Plant Codes:
    a. Bayer Code: PESGL

381

1. Millet, Proso
(Common millet, Broomcorn millet, Hog millet, Hershey millet, Mijo, Brown top millet)
Poaceae (Gramineae)
Panicum miliaceum L. ssp. *miliaceum*

2. Proso millet is the only millet grown as a grain crop in the U.S. Other millets, which see, as foxtail, Japanese or barnyard, and pearl millet are grown mainly for forage or pasture. Most production is in the Northern Plains and other short-growing season areas. In Asia, Africa and Russia, grain millet is an important food crop, but is less important than formerly as other adapted grains are more desirable. Since proso millet will mature a grain crop 50 to 75 days after seeding, and is low in moisture requirement, it will produce some food or feed where other grain crops would fail. Millets have been grown in Asia and North Africa since prehistoric times, and little is known of their origin. They probably came originally from Eastern or Central Asia. They were important in Europe during the Middle Ages before corn and potatoes were known there. Today they are of minor importance in Western Europe.

Proso millet grows up to 4 feet with stout, erect stems which may spread at the base. Stems and leaves are hairy. The panicle or flower head is rather open, like oats, and drooping. In different varieties it may be spreading, one-sided, or erect. The branches in the panicle bear spikelets only toward the tips. Each spikelet has two unequal glumes and a single flower. The flower consists of the lemma and palea, enclosing the stamens and pistil. As in oats, the lemma and palea adhere to and are a part of the threshed grain. The ripened seed is small (about 2 mm wide and 2.5 mm long), ovate and rounded on the dorsal side. Seeds range in color from white or cream to yellow, brown or nearly black. The seeds do not mature uniformly and shattering of those ripening first often occurs before others are mature. For this reason the crop is usually mowed and cured in the swath or windrow prior to combining.

As food in Old World countries, millet is used as a meal for making baked foods, as a paste from pounded wet seeds or as boiled gruel. As feed the grain is eaten readily by livestock, and is equal to or superior to oats in feed value. It should be ground for livestock feed. It is also used in poultry and bird seed mixes. A related species is browntop millet *Urochloa ramosa* (L.) R.D. Webster (syn: *P. ramosum* L., *Brachiaria ramosa* (L.) Stapf.), which is sometimes seeded for game bird pasturage in the Southeastern states.

3. Season, seeding to harvest: 50 to 75 days. Yields range 2,500 to 2,800 lb/A in Minnesota.
4. Production in U.S.: In 1992, 239,761 acres with 6,619,230 bushels (1992 CENSUS). In 1992, the top six states were South Dakota (91,071 acres), Colorado (65,501), Nebraska (43,383), North Dakota (32,936), Kansas (3,904) and Minnesota (1,097) (1992 CENSUS).
5. Other production regions: In 1995, Canada reported 39,814 acres (ROTHWELL 1996a).
6. Use: Grain crop produced for food or feed, or forage. Mainly livestock feed. See above.
7. Part(s) of plant consumed: Grain and foliage
8. Portion analyzed/sampled: Grain, forage, hay and straw. The grain processed commodity flour. Residue data are not required for flour since it is not produced significantly in U.S. for human consumption. For millet forage, cut sample at 10 inches to early boot stage, at approximately 30 percent dry matter. For millet hay, cut at early boot stage or approximately 40 inches tall, whichever is reached first. Hay should be field dried to a moisture content of 10 to 20 percent. Millet includes pearl millet. For millet grain, kernel plus hull (lemma and palea). For pearl millet grain, kernel with hull (lemma and palea) removed. For millet straw, data are required for proso millet only. For proso millet straw, plant residue (dried stalks or stem with leaves) left after the grain has been harvested
9. Classifications:
   a. Authors Class: Cereal grains; Forage, fodder and straw of cereal grains
   b. EPA Crop Group (Group & Subgroup): Cereal grains (15); Forage, fodder and straw of cereal grains (16)
   c. Codex Group: 020 (GC 0646) Cereal grains; 051 (AS 0646) Dry fodder
   d. EPA Crop Definition: None
10. References: GRIN, CODEX, MAGNESS, RICHE, ROTHWELL 1996a, SCHREIBER, US EPA 1996a, US EPA 1995a, US EPA 1994, IVES, OELKE (g)
11. Production Map: EPA Crop Production Regions 5, 7 and 8.
12. Plant Codes:
    a. Bayer Code: PANMI

382

1. Mint
(Minze, Menthe, Mint Tops, Mint Hay)
Lamiaceae (Labiatae)
Mentha spp.
1. Applemint
(Pineapple mint, Roundleaf mint)
M. suaveolens Ehrh. (syn: *M. rotundifolia* auct.)
1. Peppermint
(Black mint, White mint)
M. x *piperita* L. nothossp. *piperita* (*M. aquatica* x *M. spicata*)
1. Spearmint
(Native spearmint, Garden mint)
M. spicata L. (syn: *M. viridis* (L.) L.)
1. Scotch Spearmint
(Red mint, Scotch mint, Ginger mint)
M. x *gracilis* Sole (*M. arvensis* x *M. spicata*; *M.* x *cardiaca* J. Gerard ex Baker)
1. Lemon mint
(Bergamot mint, Orangemint, Eau-de-cologne mint)
M. x *piperita* nothossp. *citrata* (Ehrh.) Briq.
1. Watermint
(Limemint)
M. aquatica L.
1. Horsemint
(Rossminze)
M. longifolia (L.) Huds.

2. The various kinds of mint are so similar in plant and culture that they can be discussed together. They differ in the flavor of the oil. The three major types of mint grown in the U.S. are peppermint, scotch spearmint and native spearmint. All are perennial herbs, with square stems and opposite, simple leaves. Leaves are 2 to 3 inches in length, entire and near

glabrous in the more important peppermint and spearmint. Plants are semi-prostrate except flower stems, which reach 2 feet. They are grown for the volatile aromatic oil, present in all parts but mainly the leaves. They are propagated by planting underground stems or rhizomes, and form a complete ground cover by the second season. Tops are mowed and the oil distilled off immediately. Plantings last for about 6 years. First year or baby mint produces a different quality oil. The leaves, either fresh or dried, are sometimes used as flavoring. Spearmint is the usual type for seasoning lamb or used in jellies, salads and marinated vegetables. Peppermint is used mostly for candy making and chocolate sauces.

3. Season, between cuttings: 1½ to 2½ months. Midwest U.S. mint is harvested once in mid-August and no irrigation is used. In the Northwest U.S., spearmint is cut twice in a season with the first cutting in June and the second cutting in late August. Peppermint is primarily cut once around mid-August (LUNDY 1997).

4. Production in U.S.: 1992 CENSUS reported about 158,000 acres all mints and about 5,307 tons of oil. The top-producing states in 1994 were Oregon (47,160 acres), Indiana (34,487), Washington (31,731), and Idaho and Wisconsin (about 16,000 each) with the remainder from Michigan, Montana and South Dakota. 1994 Agricultural Statistics reported 130,800 acres and 4,375 tons of oil. Normal yields are 60 to 100 pounds of oil per acre. In 1995, U.S. harvested 164,100 acres of mint for oil. Of that total, 135,300 acres were for peppermint and 28,800 acres represented spearmint. Average oil production for 3 years (1994-1996) was about 80 percent peppermint (8,777,000 pounds) and 20 percent spearmint (2,212,000 pounds). In the U.S., the top three commercial peppermint growing states by rank are Oregon, Washington and Idaho. The top three spearmint states by rank are Washington, Wisconsin and Idaho/Oregon (USDA 1997b). About 2,500 acres of fresh leaves used for tea (LUNDY 1997). Some fresh market mint, primarily spearmint, is produced in Arizona, California and Florida (IR-4 REQUEST DATABASE 1996).

5. Other production regions: Eurasia (FACCIOLA 1990). In 1996, Canada reported about 300 acres of peppermint in Alberta and 2,840 acres of spearmint in Alberta and Saskatchewan (ROTHWELL 1997).

6. Use: In candies, chewing gum, toothpaste, cookery, drinks and garnish. The percent of oil used in mint flavored products ranges from 0.1 to 1.0 percent. Chewing gum (45 percent) and toothpaste (45 percent) account for 90 percent of the mint oil production with the remainder used for flavoring in confections, liqueurs and pharmaceuticals (LUNDY 1997).

7. Part(s) of plant consumed: Tops distilled for oil; leaves (fresh or dried) used for flavoring. Fresh spearmint plant parts also used as flavoring or garnish in salads. Both peppermint and spearmint dried leaves are used for tea

8. Portion analyzed/sampled: Tops (leaves and stems) and its oil. Tops also termed mint hay. Spent mint hay is an insignificant feed item

9. Classifications:
   a. Authors Class: Herbs and spices
   b. EPA Crop Group (Group & Subgroup): Miscellaneous
   c. Codex Group: 052 (AM 0738) Miscellaneous Fodder and Forage for mint hay; 027 (HH 0738) Herbs; 057 (DH 0738) Dried herbs for mints. Peppermint (HH 4761); Spearmint (HH 4765)
   d. EPA Crop Definition: None

10. References: GRIN, CODEX, FARRELL, FOSTER (picture), LEWIS, MAGNESS, PENZEY, RICHE, USDA 1994a, US EPA 1995a, US EPA 1995b, US EPA 1994, LUNDY, FACCIOLA, ROTHWELL 1997, USDA 1997b, IR-4 REQUEST DATABASE 1996, GREEN

11. Production Map: EPA Crop Production Regions 5 and 11, representing 99 percent of the commercial mint production. Some fresh market mint is produced in regions 3 and 10.

12. Plant Codes:
   a. Bayer Code: MENSS (*M.* spp.), MENGE (*M. gracilis*), MENPI (*M. piperita*), MENRO (*M. rotundifolia*), MENSP (*M. spicata*), MENSU (*M. suaveolens*), MENPC (*M. piperita citrata*), MENAQ (*M. aquatica*), MENLO (*M. longifolia*)

383

1. Mioga
(Japanese wild ginger, Mioga ginger, Japanese ginger, Ginger)
Zingiberaceae
Zingiber mioga (Thunb.) Roscoe

2. A native ginger of Japan, less than 3 feet high with smooth leaves about a foot long. Mioga grows as a perennial. Rhizome "seeds" are planted late winter to early spring in Hawaii. Conical-shaped flowers, 2-4 inches long, are usually borne below the soil surface with soil mound around the plants so the flower bracts are blanched at harvest. If soil is not mounded around plant, the stalked flower heads rise less than 2 inches above the ground.

3. Season, seeding to first harvest: Approximately 5 months with a harvest period of about 4 weeks from the mature plants.

4. Production in U.S.: Limited acreage in Hawaii.

5. Other production regions: Japan.

6. Use: Flavoring in soups or sliced as a relish.

7. Part(s) of plant consumed: Flower heads (flower bracts)

8. Portion analyzed/sampled: Flower heads

9. Classifications:
   a. Authors Class: Herbs and spices
   b. EPA Crop Group (Group & Subgroup): Miscellaneous
   c. Codex Group: No specific entry
   d. EPA Crop Definition: None

10. References: GRIN, FUKUDA, KAYS

11. Production map: EPA Crop Production Region 13 (Hawaii)

12. Plant Codes:
   a. Bayer Code: ZINMI

384

1. Molassesgrass
(Yerba melao, Efwatakala grass, Venezuelagrass, Gordura, Calinguero)
Poaceae (Gramineae)
Melinis minutiflora P. Beauv.

2. Molassesgrass is native to Africa but was introduced into Puerto Rico from Brazil. It is a tropical grass, reaching to 3 feet, excellent for grazing, but poorly adapted for hay or silage as it does not tolerate repeated mowing. It is valuable for pastures in Puerto Rico. It produces seed, in contrast to most tropical grasses, and is seed propagated. There are 6 to 15 million seeds/kg.

3. Season, tropical pasture.

4. Production in U.S.: No specific data. See above.

5. Other production regions: See above.
6. Use: Pasture, silage and erosion control.
7. Part(s) of plant consumed: Foliage
8. Portion analyzed/sampled: Forage and hay
9. Classifications:
   a. Authors Class: See Forage grass
   b. EPA Crop Group (Group & Subgroup): See Forage grass
   c. Codex Group: See Forage grass
   d. EPA Crop Definition: None
10. References: GRIN, MAGNESS, SKERMAN (picture)
11. Production Map: EPA Crop Production Regions 3 and 13.
12. Plant Codes:
   a. Bayer Code: MILMI

385

1. Monarda
(Bee balm, Bergamot)
Lamiaceae (Labiatae)
Monarda spp.
1. Oswego tea
(Red bergamont, Indian plume, Mountain mint, Fragrant balm)
M. didyma L.
1. Wild bergamot
M. fistulosa L.
1. Spotted beebalm
(Horsemint, Dotted mint)
M. punctata L.

2. The above species range from annual to perennial to 4 feet tall. Leaves about 3 inches long. Also grown as ornamentals.
3. Season: Propagation by root division in spring. Harvest leaves before bloom which occurs July into September.
4. Production in U.S.: No data. Eastern North America (FOSTER 1993).
5. Other production regions: No entry.
6. Use: Dried leaves used as herbal tea and meat seasoning. Fresh leaves as herb in salad. Fresh and dried flowers can be used in salads and teas, respectively.
7. Part(s) of plant consumed: Leaves and flowers
8. Portion analyzed/sampled: Leaves (fresh and dried) primarily; but if the whole top of the plant is used, whole top (fresh and dried)
9. Classifications:
   a. Authors Class: Herbs and spices
   b. EPA Crop Group (Group & Subgroup): Miscellaneous
   c. Codex Group: No specific entry
   d. EPA Crop Definition: None
10. References: GRIN, BAILEY 1976, FACCIOLA, FOSTER (picture), GARLAND (picture)
11. Production Map: No entry
12. Plant Codes:
   a. Bayer Code: MOAFI (*M. fistulosa*), MOAPU (*M. punctata*)

386

1. Monstera
(Monstera deliciosa, Ceriman, Balazo, Pina anona, Japanese pineapple, Pinanona, Cutleaf philodendron, Swiss cheese plant, Hurricane plant, Breadfruit vine, Splitleaf philodendron, Mexican breadfruit, Fruit salad plant, Window plant)

Araceae
Monstera deliciosa Liebm.

2. The monstera grows in hot, humid tropical lowland. Mostly propagated by root cuttings with first fruit production in 3-4 years. The plant is a large, creeping tropical vine with thick stems which bear very large cut and perforated leaves, 2 to 3 feet long and equally broad. The fruit is shaped like an ear of corn, up to 8 inches long, slender, with the rind consisting of hexagonal plates, the terminals of the individual berries in the multiple fruit. The platelets fall off when the product is ripe. The fruit flavor is intermediate between pineapple and banana.
3. Season: The plant flowers all year with mature fruit in 12 to 14 months after opening of the inflorescence. Therefore, there are flowers, immature fruits and mature fruits together on the same plant.
4. Production in U.S.: Hawaii and South Florida.
5. Other production regions: Mexico, Latin America and tropical South America.
6. Use: Pulp is eaten fresh or made into beverages and preserves. Also grown as an ornamental.
7. Part(s) of plant consumed: Pulp only. Many parts of this plant are toxic due to the high concentration of oxalic acid. The green thick, hard rind or scales and core of the fruit are discarded
8. Portion analyzed/sampled: Whole fruit
9. Classifications:
   a. Authors Class: Tropical and subtropical fruits – inedible peel
   b. EPA Crop Group (Group & Subgroup): Miscellaneous
   c. Codex Group: No specific entry
   d. EPA Crop Definition: None
10. References: GRIN, LOGAN, MAGNESS, MARTIN 1987, US EPA 1994, MORTON (picture)
11. Production Map: EPA Crop Production Region 13.
12. Plant Codes:
   a. Bayer Code: MOSDE

387

1. Mountainmint
Lamiaceae (Labiatae)
Pycnanthemum spp.
1. Hoary mountainmint
P. incanum Michx.
1. Clustered mountainmint
(Short-toothed mountainmint)
P. muticum (Michx.) Pers.
1. Whorled mountainmint
(Hairy mountainmint)
P. verticillatum (Michx.) Pers.
1. Virginia mountainmint
(Hairy mountainmint, Wild basil)
P. virginianum (L.) T. Durand & B.D. Jacks. ex B.L. Rob. & Fernald

2. Erect perennial herbs to about 3 feet tall in North America with pungent, mint-like odor. Stems are square in cross section and leaves are opposite. Blooms in summer or autumn.
3. Season, planting: Easily propagated by seeds.
4. Production in U.S.: No data. Native to North America (FACCIOLA 1990).
5. Other production regions: No entry.
6. Use: Leaves used as tea and condiment.

7. Part(s) of plant consumed: Leaves
8. Portion analyzed/sampled: Leaves (fresh and dried)
9. Classifications:
   a. Authors Class: Herbs and spices
   b. EPA Crop Group (Group & Subgroup): Miscellaneous
   c. Codex Group: No specific entry
   d. EPA Crop Definition: None
10. References: GRIN, BAILEY 1976, FACCIOLA, KUHN-LEIN, USDA NRCS
11. Production Map: No entry
12. Plant Codes:
    a. Bayer Code: No specific entry

388

1. **Muhly grass**
Poaceae (Gramineae)
Muhlenbergia spp.
1. **Bush muhly**
(Mesquite grass, Zacate arana)
M. porteri Scribn.
1. **Green muhly**
(Marsh muhly)
M. racemosa (Michx.) Britton
1. **Mountain muhly**
M. montana (Nutt.) Hitchc.
1. **Ring muhly**
(Ringgrass, Ringgrass muhly)
M. torreyi (Kunth) Hitchc.
2. The muhly grasses are perennial warm season bunchgrasses native to the western U.S. and Canada, and are members of the grass tribe, *Eragrosteae*.
Bush muhly is an erect bunchgrass growing 10 to 30 cm tall, with leaf blades 2 to 8 cm long and 2 mm wide, and a panicle seedhead (5 to 10 cm long). It is common to dry mesas, canyons, and rocky deserts on calcereous soils. It produces excellent forage for all livestock and will remain green all season if moisture is adequate. It reproduces from seeds and tillers.
Green muhly is an erect bunchgrass growing 0.3 to 1.3 m tall, with leaf blades 4 to 18 cm long and 2 to 7 cm wide, and a panicle seedhead (2 to 15 cm long). It is common to meadows, prairies, along rivers and rocky slopes on moist and dry soils. It produces fair quality forage for cattle when immature and is poor for sheep, with its value declining rapidly with maturity. It reproduces from seeds and rhizomes. There are 4,943,610 seeds/kg.
Mountain muhly is an erect bunchgrass growing 15 to 80 cm tall, with flat leaf blades 2 to 8 cm long and 1 to 3 mm wide, and a panicle seedhead (5 to 20 cm long). It is common to open woodlands, hillsides and canyons up to elevations ≤3000 m on sandy and gravelly soils. It produces good forage for cattle and horses, and is fair for sheep and deer when immature, and declines rapidly upon maturity. It starts growth in late spring, matures in August and reproduces by seeds and tillers.
Ring muhly is also an erect bunchgrass spreading by rhizomes and growing 10 to 30 cm tall, with leaf blades 1 to 4 cm long and up to 1.5 mm wide, and has a panicle seedhead (7 to 25 cm long). It is common to canyons, mesas, rocky slopes, woodlands and plains on sandy to clay loam soils. It produces fair to good quality forage for cattle when immature.

Its quality declines rapidly with maturity. The plant reproduces from seeds, tillers and short rhizomes. It is the lowest quality of the muhly grasses.
3. Season: Warm season grasses that start growing in early spring and are varying in palatability from high to poor upon maturity. Bush muhly flowers from June to November. Most start growth in late spring and early summer.
4. Production in U.S.: There are no production data for the muhly grasses, however, pasture and rangeland are produced on more than 410 million A (U.S. CENSUS, 1992). See above.
5. Other production regions: Grows in Mexico and Canada.
6. Use: Pasture grazing, hay, and for erosion control.
7. Part(s) of plant consumed: Leaves and stems
8. Portion analyzed/sampled: Forage and hay
9. Classifications:
   a. Authors Class: See Forage grass
   b. EPA Crop Group (Group & Subgroup): See Forage grass
   c. Codex Group: See Forage grass
   d. EPA Crop Definition: None
10. References: GRIN, US EPA 1994, US EPA 1995, CODEX, HOLZWORTH, SKERMAN, STUBBENDIECK (picture), RICHE
11. Production Map: EPA Crop Production Regions for Bush muhly are 8, 9, and 10. EPA Crop Production Regions for Green muhly are 5, 6, 7, 8, 9, 11 and 12. EPA Crop Production Regions for Mountain muhly are 8, 9, 10 and 11. EPA Crop Production Regions for Ring muhly are 8 and 9.
12. Plant Codes:
    a. Bayer Code: No specific entry

389

1. **Mulberry**
Moraceae
Morus spp.
1. **Mulberry, White**
(Chinese mulberry)
M. alba L.
1. **Mulberry, Black**
M. nigra L.
1. **Mulberry, Red**
M. rubra L.
2. Mulberry plants are large (up to 60 feet) trees with large leaves that bear fruits resembling blackberries in size and shape. Fruit quality is variable, and too soft for shipment to market, so rarely enters into commerce. Trees are often planted as ornamentals, and fruit may be used locally. *M. alba* L., the white or Chinese mulberry was used for its leaves to feed silkworms. This is the most hardy mulberry. In fact, it is the second most common weed in New York City. Generally, the plants bear heavy reliable crops. Foliage and unripe fruit may be poisonous. Some of the more prominent varieties include 'Black Persian', 'Downing' and 'Illinois Everbearing'.
3. Season: Bloom to ripening in 2 to 3 months.
4. Production: In U.S., none commercially, some fruit consumed locally.
5. Other production regions: Asia and Europe (BAILEY 1976).
6. Use: Mainly fresh. Can be used to make wine. Once used for murrey, a puree in medieval England to spice meats.
7. Part(s) of plant consumed: Fruit
8. Portion analyzed/sampled: Whole fruit

9. Classifications:
   a. Authors Class: Berries
   b. EPA Crop Group (Group & Subgroup): Currently Miscellaneous, potential addition to Crop Group 13: Berries Group as well as this group's Bushberry Subgroup
   c. Codex Group: 004 (FB 0271) Berries and other Small Fruit
   d. EPA Crop Definition: None
10. References: GRIN, CODEX, MAGNESS, REICH, US EPA 1994, BAILEY 1976
11. Production Map: No entry
12. Plant Codes:
    a. Bayer Code: MORAL (*M. alba*), MORNI (*M. nigra*), MORRU (*M. rubra*)

390

1. Mulga

Fabaceae (Leguminosae)
Acacia aneura F. Muell. ex Benth.

2. Mulga is a perennial warm season tropical tree up to 7 feet tall that is native to the tropics and is grown mostly in Australia. There are four forms of Mulga depending on its height: Low mulga that is a short shrub kept trimmed on top for grazing by sheep and cattle; whipstick mulga that are immature stands of trees with leaves only on the top and it is often bulldozed for feed; umbrella mulga that are mature trees and very leafy with 175 to 200 trees/ha that are best utilized in drought feeding; and tall mulga that are mature old trees with only the top leaves palatable. This one is utilized during drought by bulldozing the trees so they can be grazed. It is grown in areas receiving 300 to 450 mm rainfall. Approximately 250 trees/ha are retained as seed trees.
3. Season: Warm season perennial trees, leaves are readily grazed and it is used as as a drought forage by bulldozing the more mature trees.
4. Production in U.S.: No specific data for Mulga production, however, pasture and rangelands are produced on more than 410 million A (U.S. CENSUS, 1992).
5. Other production regions: Australia.
6. Use: Grazing and erosion control.
7. Part(s) of plant consumed: Leaves
8. Portion analyzed/sampled: Forage. Hay will not be needed
9. Classifications:
   a. Authors Class: Nongrass Animal Feeds
   b. EPA Crop Group (Group & Subgroup): Miscellaneous
   c. Codex Group: No specific citation although the general class 050 (AL 0157) Legume Animal Feeds could be used
   d. EPA Crop Definition: None
10. References: GRIN, US EPA 1994, US EPA 1995, CODEX, HOLZWORTH, SKERMAN(a) (picture), RICHE
11. Production Map: EPA Crop Production Region is 13.
12. Plant Codes:
    a. Bayer Code: No specific entry

391

1. Multiflower false-rhodesgrass
(Zacate pelillo, Windmill grass)

Poaceae (Gramineae)
Trichloris pluriflora E. Fourn. (syn: *Chloris plurifora* (E. Fourn.) Clayton)

2. This is a perennial warm season grass native to the Southern U.S., and grows south into Mexico. It is a member of the grass tribe *Chlorideae*. It is an erect bunchgrass (0.5 to 1.5 m tall) that sometimes spreads by short stolons and flowers with a panicle 7 to 20 cm long. It is used in prairies, plains and dry shady wooded areas, and produces good to excellent forage for livestock and wildlife. It cures well and is stored for winter forage. Growth is started in late spring, flowers in July to September and can produce more than one seed crop/season.
3. Season: Warm season, growth starts in late spring, flowers July to August and completes growth in the fall. It is stored for winter feed.
4. Production in U.S.: No data for Multiflower false-rhodesgrass production, however, pasture and rangeland are produced on more than 410 million A (U.S. CENSUS, 1992). See above.
5. Other production regions: Mexico
6. Use: Pasture, hay, and erosion control.
7. Part(s) of plant consumed: Leaves and stems
8. Portion analyzed/sampled: Forage and hay
9. Classifications:
   a. Authors Class: See Forage grass
   b. EPA Crop Group (Group & Subgroup): See Forage grass
   c. Codex Group: See Forage grass
   d. EPA Crop Definition: None
10. References: GRIN, US EPA 1994, US EPA 1995, CODEX, HOLZWORTH, STUBBENDIECK (picture), RICHE
11. Production Map: EPA Crop Production Regions 6 and 8.
12. Plant Codes:
    a. Bayer Code: TCLPL

392

1. Mushroom, Agaricus
1. Button mushroom
(Common mushroom, Cremini, Portobello, White mushroom, Champignon, Champinon, Masherum, Seta, Yang gu, Haratake)
Agaricaceae
Agaricus bisporus (Lange) Imbach (syn: *A. brunnescens* Peck; *Psalliota* spp.)

1. Rodman's Agaricus
(Meadow mushroom)
A. bitorquis (Quel.) Saccardo (syn: *A. rodmanii* Peck; *A. campestris* var. *edulis* Vitt.; *Psalliota edulis* (Vitt.) Moller and Schaeff.)

2. The term mushroom applies to edible, fleshy fungi, either gathered from the wild or grown in cultivation. Under cultivation, mushrooms are grown mainly in the dark, in caves or light-tight buildings, with temperature, moisture and ventilation control. The "spawn" or mycelium is seeded in specially prepared compost in beds or suitable containers. After the mycelia have spread through the compost, a layer of soil or "casing" is applied.

The mushrooms, the fruiting bodies of the fungi, first appear at the soil surface about 5 or more weeks after seeding with spawn, and continue to appear in flushes. They are usually harvested by cutting off the cap with a small portion of the stem before the caps have become fully expanded. Beds produce the main crops in the first 40 days after fruiting starts, but may be retained with light production for several months.

Agaricus bisporus includes both brown and white skinned types. The brown skinned cultivars includes the Italian forms as

Cremini (button stage) and the Portobello (cap fully open to 8 inches in diameter).

3. Season, seeding spawn to first harvest: About 5 to 7 weeks with flushing intervals at 7-10 days.

4. Production: In U.S. for 1994-95, about 390,489 tons of *Agaricus* with 33,718,000 sq. ft. of growing area with total fillings amounting to 139,594 sq. ft. (1995 Agricultural Statistics for Mushrooms). In 1994-95 sq. ft. of growing area for the leading states reported as Pennsylvania (19,334,000), California (3,585,000), Florida (1,600,000), Ohio (359,000), Delaware (332,000), and Michigan (326,000) with total fillings averaging about 4 times during the year. These six states are about 76 percent of the total growing area for U.S. In the U.S. for 1994-95, 68 percent of the mushroom production reported as fresh market and 32 percent as processed. Mushroom can be sold fresh, canned or dried. (1995 Agricultural Statistics for Mushrooms).

5. Other production regions: In Canada for 1995, 60,137 tons with 73,430 sq. ft. (Rothwell 1996a).

6. Use: Food flavoring, soups, sometimes as pot vegetable, or raw in salads. Fresh, canned or dried. U.S. per capita consumption in 1994 is 3.0 pounds (PUTNAM 1996).

7. Part(s) of crop consumed: Fruiting "cap" and stem. (About 8 percent dry matter)

8. Portion analyzed/sampled: Whole commodity which includes cap and stem.

9. Classifications:
   a. Authors Class: Edible fungi
   b. EPA Crop Group (Group & Subgroup): Miscellaneous
   c. Codex Group: 012 (VO 0450) Fruiting vegetables other than cucurbits
   d. EPA Crop Definition: None

10. References: GRIN, BAILEY 1976, CODEX, FACCIOLA, MAGNESS, MUSHROOMS 1995 STATS, ROTHWELL 1996a, SCHREIBER, STAMETS 1983 (picture), USDA 1994a, US EPA 1995b, IVES, REHM, RUBATZKY, PUTNAM 1996

11. Production Map: EPA Crop Production Regions 1, 2, 3, 5 and 10.

12. Plant Codes:
    a. Bayer Code: AGCBI (*A. bisporus*)

393

1. Mushroom, Specialty

1. Shiitake mushroom

(Japanese black mushroom, Chinese black mushroom, Black forest mushroom, Shan ku, Xiang gu)

Polyporaceae

  Lentinula edodes (Berk.) Pegl. (syn: *Lentinus edodes* (Berk.) Singer)

1. Oyster mushroom

(Hiratake, Tree oyster, Florida oyster, Blue oyster)

Tricholomataceae

  Pleurotus ostreatus (Jacq.) Kummer
  P. ostreatus (Jacq.) Kummer, Florida variety (Pleurotus Florida) (syn: *P. ostreatus* var. *florida* nom. prov. Eger; *P. floridanus* Singer).

1. Enoke

(Enokitake, Velvet stew, Winter mushroom, Jin Tsen gu, Enoki, Velvet shank)

Tricholomataceae

  Flammulina velutipes (Curt.) Singer (syn: *Collybia velutipes* (Curt.) Kumm.)

1. Chinese mushroom

(Paddy-straw mushroom, Cao gu, Fukurotake, Paddy mushroom, Straw mushroom)

Pluteaceae

  Volvariella volvacea (Bull.) Singer

1. Morel

(Yellow morel, Morille)

Morchellaceae

  Morchella spp.

1. Truffle

(Oregon white truffle, Black truffle, Perigord truffle, White truffle)

Tuberaceae

  Tuber spp.

2. Specialty mushrooms can be grown indoor and outdoor under shade. They grow from various substrates as compared to *Agaricus* which is grown normally from enriched wheat straw and/or horse manure based compost. The specialty mushrooms are grown on logs, cereal straw fruiting substrates and underground. Spawn run durations vary from 6 to 12 months for cut logs (normally oak) to 30-60 days for artifical logs of sawdust blocks or bags for shiitake to less than 1 week for Chinese mushrooms.

Normally soil casing is not required and pinhead initiation is 7-14 days. Cropping duration on oak logs can be 3-5 years and flushing intervals can be 2 per year outdoor or 4 per year indoor for shiitake. Oyster mushroom cropping duration is 5-7 weeks with a flushing interval of 10 days. Enoke cropping duration is 2-3 weeks with a flushing interval of 10 days. The fruit bodies are edible and grow above the substrate except for truffles which grow underground and are harvested commercially in the Pacific Northwest (FACCIOLA 1990). Virtually all specialty mushrooms are sold fresh but some are sold dried, like shiitake.

Other specialty mushrooms include Matsutake (*Tricholoma matsutake*); Iwatake (*Umbilicaria* sp. (syn: *Gyrophora esculenta*); Nameko (*Pholiota nameko*); Shimeji (*Hypsizygus marmoreus*); Maitake (*Grifola frondosa*); Shaggy mane (*Coprinus comatus*); Wine cap (*Stropharia rugoso-annulata*); Teonanacatl, Nize, Pajaritos (*Psilocybe mexicana*); San Isidro (*Psilocybe cubensis*); Chanterelle (*Cantharellus cibarius*); Corn smut, Cuitlacoche, Huitlacoche, Mexican truffle, Maize mushroom, Caviar azteca (*Ustilago maydis*).

3. Season, seeding spawn to first harvest:
   Shittake on natural logs: About 8 to 14 months.
   Shiitake on artificial logs: About 9 to 14 weeks.
   Oyster: About 4 to 6 weeks.
   Enoke: About 5 to 6 weeks.
   Chinese mushroom: About 2 to 3 weeks.

4. Production in U.S.: For 1994-95, total production for specialty mushrooms reported at 8,626,000 pounds with shiitake and oyster mushrooms 87 percent of the total (Mushrooms 1995 Stats.). States reporting to the Agricultural Statistics Board for specialty mushrooms for 1994-95 included Arkansas, California, Colorado, Connecticut, Florida, Hawaii, Iowa, Idaho, Illinois, Kentucky, Massachusetts, Michigan, Minnesota, Missouri, Mississippi, North Carolina, New York, Ohio, Oregon, Pennsylvania, South Carolina, Tennessee, Virginia, Washington, Wiscon-

sin and West Virginia (Mushrooms 1995 Stats.). Since 1993, black Perigord truffle mushrooms have been cultivated in the Chapel Hill area of North Carolina on filbert tree roots with a grower's return of $300 to $500 per pound (DEAN 1997).

5. Other production regions: Southeast Asia, Japan, Europe, Argentina, Australia, New Zealand and Canada (RUBATZKY 1997). Mexico.

6. Use: Food flavoring, soups, pot vegetables, or in salads. Raw, cooked or dried.

7. Part(s) of crop consumed: Fruiting bodies

8. Portion analyzed/sampled: Whole commodity or fruit body which includes the cap and stem except truffles

9. Classifications:
   a. Authors Class: Edible fungi
   b. EPA Crop Group (Group & Subgroup): Miscellaneous
   c. Codex Group: Generally no specific entry; Specialty types 012 (VO 0449) and Chanterelle is 012 (VO 4287)
   d. EPA Crop Definition: None

10. References: GRIN, FACCIOLA, MUSHROOMS 1995 STATS., ROYSE 1995, STAMETS 1983 (pictures), DEAN, RUBATZKY, PATAKY, VALVERDE, FORTIN

11. Production Map: No entry

12. Plant Codes:
   a. Bayer Code: PUWOS (*P. ostreatus*), no entries for remaining spp.

394

1. Muskmelon

(Melon, Dudaim melon, Stink melon, Persian melon, Netted muskmelon, Nutmeg muskmelon, Winter melon, Santa Claus melon, Oriental pickling melon, Cantaloupe, Honeydew melon, Casaba, Armenian cucumber, Snake melon, Serpent melon, Melofon, Mango melon, Lemon cucumber, Garden melon, Plum granny, Pomegranate melon,, Queen Anne's pocket melon, Smellmelon)

Cucurbitaceae
   Cucumis melo L.
   C. melo L. ssp. *melo*

2. Muskmelon or melon is a generic name applied to annual plants of the above species, with trailing, vine-like stems, and producing fleshy fruits with interior seed cavities. Other type plants also may have the term melon in the common name. Also the term melons applies to both muskmelon and watermelon (40 CFR 180.1h, EPA regulation). The principal types, the first five of which are treated separately, are as follows:

C. melo var. *cantalupensis* Naudin (syn: *C. melo* var. *reticulatus* Naudin): **Cantaloupe, Muskmelon, Persian melon, Netted** or **Nutmeg melon**. Fruits medium, scented, with sutured and netted surface. Variety *cantalupensis* also includes the true cantaloupe of Europe, including the names **Ipu-ala**, and **Rock melon**.

C. melo var. *chito* (C. Morren) Naudin (syn: *C. melo* var. *dudaim* (L.) Naudin): **Queen Anne's pocket melon, Dudaim melon, Garden melon, Garden lemon, Lemon cucumber, Mango melon, Melon apple, Orange melon, Smell melon, Pomegranate melon, Plum granny, Stink melon, Vegetable orange, Vine peach**: Fruit small, very fragrant and used for preserves.

C. melo var. *conomon* (Thunb.) Makino: **Oriental pickling melon, Pickling melon, Sweet melon**.

C. melo var. *flexuosus* (L.) Naudin: **Armenian cucumber, Serpent melon, Snake melon**. Fruit very long, slender and curved.

C. melo var. *inodorus* H. Jacq.: **Casaba melon, Crenshaw, Honeydew melon, Winter melon**. Fruit late ripening, little odor, surface smooth for Honeydew and Crenshaw. Casaba are usually rough-skinned (wrinkled).

C. melo var. *momordica* (Roxb.) Duthie & J.B. Fuller: **Snapmelon, Phut**.

3. Season, seeding to harvest: 85 to 120 days depending on cultivar and growing conditions.

4. Production in U.S.: In 1995, 104,890 acres of cantaloupes were planted with a production of 1,053,950 tons. Also in 1995, 29,300 acres of honeydew melons were planted with a production of 282,800 tons (USDA 1996b). Neal 1965 reported 50 acres in Hawaii devoted to true cantaloupes. In 1995, the top 5 states (94 percent) for cantaloupe acreage were California (59,300 acres), Arizona (16,000), Texas (12,900), Georgia (6,500), Indiana (3,500). The 3 states (100 percent) that reported honeydew acreage in 1995 were California (21,100 acres), Texas (4,600) and Arizona (3,600) (USDA 1996b).

5. Other production regions: In 1995, Canada reported 628 acres of cantaloupe and melons (ROTHWELL 1996a).

6. Use: Fresh eating and pickling. Cantaloupes do not require ethylene treatment to ripen like some honeydews. Cantaloupes need to be rapidly cooled at harvest whereas honeydews do not. Cantaloupes and honeydews are harvested before eating ripe. Melofon is a pickling melon which can be mechanically harvested.

7. Part(s) of plant consumed: Fruit

8. Portion analyzed/sampled: Fruit. 40 CFR 180.1(J)(4) states that stems shall be removed and discarded from melons before examination for pesticide residue

9. Classifications:
   a. Authors Class: Cucurbit vegetables
   b. EPA Crop Group (Group & Subgroup): Cucurbit vegetables (9A) (Representative crop)
   c. Codex Group: 011 (VC 0046) Cucurbits fruiting vegetables
   d. EPA Crop Definition: Melons = Muskmelons and watermelon; and Muskmelons = *Cucumis melo* including its hybrids and/or varieties.

10. References: GRIN, BAILEY 1976, CODEX, FACCIOLA, JANICK 1990, MAGNESS, NEAL (In Part), USDA 1996b, US EPA 1994, US EPA 1995a, US EPA 1996a, YAMAGUCHI 1983.

11. Production Map: EPA Crop Production Regions: Melons (Honeydew), 98 percent produced in 6 and 10. Cantaloupes, 95 percent produced in 1, 2, 5, 6 and 10.

12. Plant Codes:
   a. Bayer Code: CUMME (*C. melo* ssp. *melo*), CUMHY (*C. melo* hybrids)

395

1. Mustard, Tuberous rooted Chinese

(Chinese turnip, Tuberous root mustard, Dai tou jie, Root mustard, Turnip-root mustard, Large-root mustard)

Brassicacea (Cruciferae)
   Brassica juncea var. *napiformis* (Pallieux & Bois) Kitam.

2. This mustard relative develops a tuberous root, much like a globular turnip in appearance. Roots reach a diameter of 3 to 4 inches, and are similar to turnips in appearance, texture, flavor and culture. It is rarely grown in the U.S.
3. Season, seeding to harvest: See Turnip.
4. Production in U.S.: No data.
5. Other production regions: No data.
6. Use: Root and leaves used, as turnips.
7. Part(s) of plant consumed: The root, leaves and petioles
8. Portion analyzed/sampled: Root and leaves
9. Classifications:
   a. Authors Class: Root and tuber vegetables; Leaves of root and tuber vegetables
   b. EPA Crop Group (Group & Subgroup): Miscellaneous
   c. Codex Group: No specific entry
   d. EPA Crop Definition: None
10. References: GRIN, BAILEY 1976, FACCIOLA, KAYS, MAGNESS
11. Production Map: No entry
12. Plant Codes:
    a. Bayer Code: BRSJU (*B. juncea*)

396

1. Mustard, Wild
(Weed mustard)
Brassicaceae (Cruciferae)
Brassica spp. and *Sinapis* spp.
1. Black mustard
B. nigra (L.) W.D.J. Koch
1. White mustard
S. alba L. ssp. *alba* (syn: *B. hirta* Moench)
1. Indian mustard
(India mustard)
B. juncea (L.) Czern. var. *juncea*
1. Charlock
(California rape)
S. arvensis L. ssp. *arvensis* (syn: *B. kaber* (DC.) L.C. Wheeler)
1. Birdsrape mustard
B. rapa L.
1. African mustard
(Asian mustard)
B. tournefortii Gouan

2. Black mustard is in general similar to white mustard in growth and appearance. The leaves are compound pinnate, and hairy. The seed pods are short and glabrous. Black Mustard is cultivated in the U.S., but also is a common weed. Its young leaves are sometimes gathered and used as potherbs. The condiment mustard is mainly the ground seed or "flour" made from the seeds of this species. Other *Brassica* species on the WSSA list are *B. hirta* (white mustard), *B. juncea* (Indian mustard), *B. kaber* (Wild mustard), *B. rapa* (Birdsrape mustard), and *B. tournefortii* (African mustard) (DUKE 1992).
3. Season, mainly as spring greens.
4. Production in U.S.: No data. In garden and the wild. All the American species are introduced, frequent in gardens and fields in U.S. (DUKE 1992).
5. Other production regions: No entry.
6. Use: Mainly potherb.
7. Part(s) of plant consumed: Tops

8. Portion analyzed/sampled: Tops
9. Classifications:
   a. Authors Class: *Brassica* (cole) leafy vegetables
   b. EPA Crop Group (Group & Subgroup): Mustard greens: *Brassica* (cole) leafy vegetables (5B)
   c. Codex Group: 013 (VL 0485) Leafy vegetables (including *Brassica* leafy vegetables).
   d. EPA Crop Definition: None
10. References: GRIN, CODEX, DUKE(c) (picture), US EPA 1995a, USDA NRCS
11. Production Map: No entry except U.S.
12. Plant Codes:
    a. Bayer Code: BRSNI (*B. nigra*), BRSJU (*B. juncea*), SINAL (*S. alba*), SINAR (*S. arvensis*), BRSRO (*B. rapa*), BRSTO (*B. tournefortii*)

397

1. Mustard greens
(Gui choy, Kai choy, Chinese mustard, Gai chow, Karashina, Japanese greens, Mizuna, Potherb mustard, Chinese green mustard, Specialty mustards, Kyona, Indian mustard, Chinese mustard, Mostaza, Takana, Tendergreen, Brown mustard, Leaf mustard, Prong, Gar choy, Mustard spinach, California peppergrass, Gai choy, Komatsuma, Cabbage leaf mustard, Komatsuna, Curled mustard, Cutleaf mustard, Dissected-leaf mustard, Yau choi, Ostrich plume, Mibuna, Raya, Moutarde Chinese, Kabuna, Raapstecltjes, Southern curled mustard, Head mustard, Spinach mustard, Zairainatane)
Brassicaceae (Cruciferae)
Brassica rapa ssp. *nipposinica* (L.H. Bailey) Hanelt
   (Mibuna, Mizuna)
B. rapa var. *perviridis* L.H. Bailey (syn: *B. perviridis* (L.H. Bailey) L.H. Bailey; *B. camprestis* L. (Perviridis group); *B. rapa* (Perviridis group))
   (Kabuna, Komatsuna, Raapstecltjes, Spinach mustard, Tendergreen, Yau choi, Zairainatane)
B. juncea var. *crispifolia* L.H. Bailey
   (Curled mustard, Dissected-leaf mustard, Ostrich plume, Southern curled mustard)
B. juncea var. *folisa* L.H. Bailey
   (Leaf mustard)
B. juncea var. *rugosa* (Roxb.) N. Tsen & S.N. Lee
   (Cabbage leaf mustard, Head mustard)
B. juncea var. *japonica* (Thunb.) L.H. Bailey (syn: *B. rapa* (Japonica group))
   (Cutleaf mustard, Dissected-leaf mustard)

2. These, and possibly other species of *Brassica*, are grown under the name mustard for the young green leaves which are used as potherbs or "greens". All are annuals grown from seed, and form clusters of leaves, the edible portion, prior to forming a seed stalk. Leaves are rather large, with blades 6 inches or more in length, smooth or curled or notched, somewhat pubescent when young, later glabrous. They are harvested early, while the leaves are tender. Culture and exposure of leaves are comparable to spinach. The condiment "mustard" is the ground seed of another species.
3. Season, seeding to harvest: 35 to 60 days.
4. Production in U.S.: 9,493 acres reported (1959 CENSUS). 12,775 acres (1992 CENSUS). In 1992, mustard greens reported for the top 11 states as Georgia (2,330 acres), Cali-

fornia (2,218), Florida (1,301), Texas (1,218), Arkansas (670), North Carolina (561), Tennessee (490), South Carolina (387), Ohio (319), Mississippi (308) and Arizona (295) with about 80 percent of the U.S. acreage (1992 CENSUS). Kai choy leaves are lime green and available year-round from California (LOGAN 1996).

5. Other production regions: Mexico (LOGAN 1996).
6. Use: Fresh, canned, frozen for potherbs and to limited extent in salads or cooked.
7. Part(s) of plant consumed: Leaves, including stems
8. Portion analyzed/sampled: Greens (leaves)
9. Classifications:
   a. Authors Class: *Brassica* (cole) leafy vegetables
   b. EPA Crop Group (Group & Subgroup): *Brassica* (cole) leafy vegetables (5B) (Representative crop)
   c. Codex Group: 013 (VL 0485) Leafy vegetables (including *Brassica* leafy vegetables) for Mustard greens; 013 (VL 0481) Leafy vegetables (including *Brassica* leafy vegetables) for Komatsuma; 013 (VL 0478) for Indian mustard
   d. EPA Crop Definition: None
10. References: GRIN, CODEX, FACCIOLA, LOGAN, LORENZ 1988, MAGNESS, MYERS (picture), RICHE, STEPHENS 1988 (picture), US EPA 1996a, US EPA 1995a, YAMAGUCHI 1983
11. Production Map: EPA Crop Production Regions 2, 3, 4, 5, 6 and 10.
12. Plant Codes:
    a. Bayer Code: BRSJU (*B. juncea*), BRSPE (*B. rapa* var *perviridis*)

398

1. Mustard seed
Brassicaceae (Cruciferae)
Brassica spp. and *Sinapis* spp.
1. Mustard, Black
Brassica nigra (L.) W.D.J. Koch
1. Mustard, Brown
(Indian mustard, Oriental mustard)
B. juncea (L.) Czern. var. *juncea*
1. Mustard, White
(Yellow mustard)
Sinapis alba L. ssp. *alba* (syn: *B. hirta* Moench)

2. The condiment mustards of commerce are the ground seeds of these three plants which grow to about 3 feet tall, mainly with the seed coats removed. The seeds are near round, $1/10$ inch or less in diameter, and produced in pods, which are removed by threshing. Black mustard seed is yellow inside and is the kind used for commercial production. The ground brown or black mustard seeds exhibit a strong odor while the white or yellow ones do not. The oilseed is protected by the pod or shell from pesticides applied during the growing season. Oil may be extracted by three methods. Fixed oil is extracted by cold press, the remaining press cake is steam distilled or solvent extracted for the essential oil. There are basically four types of condiment mustard-whole mustard seed; ground mustard mainly for the meat industry; mustard flour for sauces, salad dressings, baked beans and Chinese or hot English mustard; and oleoresin mustard.
Codex Alimentarius has included *B. rapa* ssp. *sarson* (Prain) Denford (**Brown sarson**) and var. *dichotoma* (Roxb. ex

Fleming) Kitam. (**Toria**) as Field mustard seed (SO 0694). See Rapeseed oil.

3. Season, seeding to harvest: About 5 months.
4. Production in U.S.: U.S. produced 5,875 tons of seed on 12,636 acres as reported in 1992 CENSUS. 96 percent of the U.S. production is in North Dakota and Montana.
5. Other production regions: Canada produces about 250,000 tons of seed on about 540,109 acres.
6. Use: As flavoring for meats, pickles, etc., see above.
7. Part(s) of plant consumed: Ground seed, mainly with seed coat removed; and edible oil
8. Portion analyzed/sampled: Seed only
9. Classifications:
   a. Authors Class: Herbs and spices group; Oilseed group.
   b. EPA Crop Group (Group & Subgroup): Herbs and spices group (19B)
   c. Codex Group: 023 (SO 0485, SO 0090, SO 0478) Oilseed
   d. EPA Crop Definition: None
10. References: GRIN, CODEX, FARRELL (picture), LEWIS, MAGNESS, RICHE, ROTHWELL 1996a, US EPA 1995a
11. Production Map: EPA Crop Production Regions 7 and 11.
12. Plant Codes:
    a. Bayer Code: BRSNI (*B. nigra*), BRSJU (*B. juncea*), SINAL (*S. alba*)

399

1. Napiergrass
(Elephantgrass, Merkergrass)
Poaceae (Gramineae)
Pennisetum purpureum Schumach.

2. This is a perennial warm season, slightly spreading bunchgrass, introduced from Africa in 1913. In growth habit it is similar to pearl millet, forming clumps up to 10 feet tall of coarse cane-like stems and large leaves. Leaves are 2 to 3 feet long, and about 1 inch wide. Napiergrass requires a rich soil for best growth. It does not tolerate much frost, so culture is limited to warmest parts of mainland U.S. and Hawaii. It is usually propagated vegetatively; if grown from seed, it is started in a nursery and transplanted. It is grown in rows and cultivated for highest yield. In areas where adapted, it is grown for green feed, silage and rotational grazing. Hybrids between pearl millet and napiergrass are being evaluated to provide a hay crop and fall forage crop for fall use in the Gulf Coast region of the U.S.
3. Season: Warm season pasture. Late summer production, grazed once every 8 to 9 weeks.
4. Production in U.S.: No specific data for forage use. In 1992, 1,020 acres of Merkergrass grown in Puerto Rico (1992 CENSUS). Warmer parts of mainland U.S. and Hawaii and Puerto Rico.
5. Other production regions: Africa.
6. Use: Pasture, hay, silage, green chop and erosion control.
7. Part(s) of plant consumed: Foliage
8. Portion analyzed/sampled: Forage and hay
9. Classifications:
   a. Authors Class: See Forage grass
   b. EPA Crop Group (Group & Subgroup): See Forage grass
   c. Codex Group: See Forage grass
   d. EPA Crop Definition: None
10. References: GRIN, MAGNESS, RICHE, CUOMO, ALDERSON, SKERMAN, BARNES

11. Production Map: EPA Crop Production Regions 3, 4, 6 and 13.
12. Plant Codes:
    a. Bayer Code: PESPU

400

1. Naranjilla
(Lulo, Lulun, Narangillo)
Solanaceae
Solanum quitoense Lam.
2. This fruit is native to the Northern Andes in South America, where it is extensively cultivated at elevations of 3,000 to 7,000 feet. The plant is a strong-growing herb, up to 10 feet. Leaves are large, ovate, up to 18 inches long. Plants are thorny and most parts are pubescent. Fruits are borne at the leaf axils. They are globular, up to 2^1/$_2$ inches diameter, and covered with brittle hairs which are readily rubbed off. The rather thin skin encloses a green colored acid pulp in which seeds are embedded, as in tomatoes. Fruits are used for juice, preserves and pies. Plants produce a crop in less than a year from seed, and continue to produce for 3 or more years. Most of the crop ripens in early spring. It is not grown commercially in the U.S. at present, but will thrive at higher elevations in Puerto Rico and probably in Hawaii and south Florida.
3. Season, seeding to harvest: Less than a year. Fruits mainly in winter in Florida and fruit must be collected every 7-10 days because of continuous bearing.
4. Production in U.S.: No data. Hawaii. South Florida (MORTON 1987).
5. Other production regions: West Indies.
6. Use: See above.
7. Part(s) of plant consumed: Fruit pulp
8. Portion analyzed/sampled: Whole fruit
9. Classifications:
    a. Authors Class: Tropical and subtropical fruits – inedible peel
    b. EPA Crop Group (Group & Subgroup): Miscellaneous
    c. Codex Group: 006 (FI 0349) Assorted tropical and subtropical fruit – inedible peel
    d. EPA Crop Definition: None
10. References: GRIN, CODEX, MAGNESS, MORTON (picture), RUBATZKY.
11. Production Map: EPA Crop Production Region 13.
12. Plant Codes:
    a. Bayer Code: SOLQU

401

1. Nasturtium
(Garden nasturtium, Indian cress, Mexican cress, Peruvian cress, Grande capucine, Capuchina)
Tropaeolaceae
Tropaeolum majus L.
1. Bush nasturtium
(Dwarf nasturtium)
T. minus L.
2. Nasturium is a somewhat succulent, climbing annual, with nearly round smooth leaves, 3 to 5 inches in diameter. The pods are near globular, ridged, about ^{1}/$_3$ inch in diameter. Nasturtiums are widely grown as ornamentals, but the pep-

pery flavored leaves are sometimes used in salads, like cress. The young pods are made into pickles. Plants will develop pods about 4 months after seeding. While one of the most common garden flowers, nasturtium leaves and pods are rarely used as condiments, and appear not to be commercially grown for this purpose.
3. Season, seeding to harvest (pods): About 4 months.
4. Production in U.S.: Backyard gardens. Grows well in Florida (STEPHENS 1988).
5. Other production regions: Native to South America and are grown worldwide (STEPHENS 1988).
6. Use: Condiments and used in salads.
7. Part(s) of plant consumed: Mainly leaves; some use of buds, flowers and pods
8. Portion analyzed/sampled: Leaves (fresh and dried)
9. Classifications:
    a. Authors Class: Herbs and spices
    b. EPA Crop Group (Group & Subgroup): Herbs and spices (19A)
    c. Codex Group: 027 (HH 0739) Herbs; 028 (HS 0739) Spices
    d. EPA Crop Definition: None
10. References: GRIN, CODEX, FACCIOLA, GARLAND, MAGNESS, STEPHENS 1988 (picture)
11. Production Map: EPA Crop Production Regions 3 and 13.
12. Plant Codes:
    a. Bayer Code: TOPMA (*T. majus*)

402

1. Natal plum
(Amatungula, Carissa)
Apocynaceae
Carissa macrocarpa (Eckl.) A.DC. (syn: *Carissa grandiflora* (E. Mey.) A.DC.)
2. The natal plum grows well in a variety of tropical and subtropical climates and soil conditions. The plant is tolerant to salt water spray. It is propagated by seed, cutting and layering with first fruit production within 2-3 years after vegetative propagation. The natal plum plant is a small tree or shrub growing to about 15 feet tall. Branches are heavily thorned. Leaves are glossy-green and thick. Fruits are ovoid to elliptical, up to 2 inches long. The red skin is thin and papery, enclosing a pinkish pulp in which several seeds are imbedded. Pulp is quite tart, but makes a sauce resembling cranberry.
3. Season: Bloom to harvest is about 60 days. The main ripening season is midsummer, following spring bloom; but some fruits may ripen throughout the year.
4. Production in U.S.: While not grown commercially, it is found in home plantings in Florida.
5. Other production regions: South Africa.
6. Use: Pulp eaten fresh or made into desserts or preserves. Plant is important as an ornamental.
7. Part(s) of plant consumed: Fruit
8. Portion analyzed/sampled: Whole fruit
9. Classifications:
    a. Authors Class: Tropical and subtropical fruit – edible peel
    b. EPA Crop Group (Group & Subgroup): Miscellaneous
    c. Codex Group: 005 (FT 0304) Assorted tropical and subtropical fruits – edible peel
    d. EPA Crop Definition: None

10. References: GRIN, CODEX, MAGNESS, MARTIN 1987, US EPA 1994, USDA NRCS, MORTON (picture)
11. Production Map: EPA Crop Production Region 13.
12. Plant Codes:
    a. Bayer Code: CISMA

403

1. Needlegrass
Poaceae (Gramineae)
1. Needle-and-thread grass
Heterostipa comata ssp. comata (Trin. & Rupr.) Barkworth (syn: *Stipa comata* Trin. & Rupr.)
1. Green needlegrass
(Feather bunchgrass, Green stipagrass)
Nassella viridula (Trin.) Barkworth (syn: *Stipa viridula* Trin.)

2. Many species are indigenous to the Western states. They are long-lived cool season bunchgrasses. Each spikelet has a single flower which terminates in a prominent awn, accounting for the name needlegrasses. These grasses are abundant, widely distributed, cure well on the ground, and rank high for forage. The needle-like awns, however, cause injury to animals, especially sheep, and greatly detract from their value. The two most valuable species are described as follows.

Needle-and-thread grass is a native perennial bunchgrass widely distributed over the Western states and common on dry, sandy, gravelly soils of the Northern Plains. Seed stalks may reach to 4 feet. Leaves are up to 12 inches long and $1/8$ inch wide. Growth starts early in spring and continues through the summer when moisture is available. The seeds are sharp-pointed and have long, twisted thread-like awns. The shape of the seed and awn account for the name. The grass is palatable and readily grazed except when in seed. This period begins in June, and the seed is shed in July. During this period the sharp seeds penetrate mouth parts and hides of stock. Although the grass is widely adapted and has many useful characteristics, it is little used for reseeding because of injury to livestock while in seed.

Green needlegrass, a native bunchgrass, is most abundant on the upland prairie and ranges of the Northern Plains. It frequently invades abandoned cropland. It grows on most soil types but thrives best on sandy soils. It grows up to 3 feet. Leaves are mostly basal, up to 12 inches long and $1/2$ inch wide. Seed spikes have bent awns about an inch long but are less troublesome to livestock than those of Needle-and-thread grass. Growth starts early in the spring and continues through summer when moisture is available. The grass is palatable and nutritious and makes excellent hay. It also cures well on the ground so is useful for winter grazing. It is useful for revegetation as seedlings are vigorous and drought resistant. Seed of some improved varieties is available.

3. Season, not grazed June and July. Cool season perennial.
4. Production in U.S.: No specific data, See Forage grass. Western states and Northern Plains of U.S.
5. Other production regions: Canada and Mexico.
6. Use: Pasture.
7. Part(s) of plant consumed: Foliage
8. Portion analyzed/sampled: Forage and hay
9. Classifications:
    a. Authors Class: See Forage grass
    b. EPA Crop Group (Group & Subgroup): See Forage grass

c. Codex Group: See Forage grass
d. EPA Crop Definition: None
10. References: GRIN, MAGNESS, MOSLER, STUBBENDIECK
11. Production Map: EPA Crop Production Regions 6, 7, 8, 9 and 10.
12. Plant Codes:
    a. Bayer Code: STDCO (*H. comata*), STDVI (*N. viridula*)

404

1. Niger seed
(Nug, Noog, Noug, Ramtil, Niger thistle)
Asteraceae (Compositae)
Guizotia abyssinica (L.f.) Cass.

2. Niger is an important oil crop in Ethiopia and parts of India. The crop is tolerant to high salinity, high boron and low soil oxygen levels. Oil content of the seeds range from 36 to 42 percent. Niger is an annual herb to $6^1/2$ feet tall. A single plant often bears 20 to 40 flower heads. Flowering begins 50 to 110 days after sowing.
3. Season, seeding to harvest: In Ethiopia, generally 150 to 180 days are required from emergence to maturity. The crop is normally planted in July, flowers in early October and is harvested in December or January. In India, 75 to 150 days.
4. Production in U.S.: No data.
5. Other production regions: Native to Ethiopia (ROBBELEN 1989). See above.
6. Use: Meal for livestock feed. The fatty acid composition of the Niger oil is similar to safflower and sunflower oil. Used as edible oil. Also the seed can be used as snack food or ground into flour for bread.
7. Part(s) of plant consumed: Seed
8. Portion analyzed/sampled: Seed and its oil and meal
9. Classifications:
    a. Authors Class: Oilseed
    b. EPA Crop Group (Group & Subgroup): Miscellaneous
    c. Codex Group: 023 (SO 0695) Oilseed
    d. EPA Crop Definition: None
10. References: GRIN, CODEX, ROBBELEN 1989 (picture), ROBINSON
11. Production Map: No entry
12. Plant Codes:
    a. Bayer Code: GUIAB

405

1. Nutmeg
(Nuez moscada)
1. Mace
Myristicaceae
Myristica fragrans Houtt. (syn: *M. officinalis* L.f.)

2. While other species of *Myristica* produce nutmegs, the principal nutmeg of commerce is *M. fragrans*. The tree is a tropical evergreen, reaching up to 40 feet, with entire oval leaves up to 10 inches long. The fruit resembles a small peach and is oval or pyriform about 2 inches long, and consists of an outer fleshy husk and inner seed. The husk splits when ripe, exposing the seed. Immediately surrounding the seed coat is a crimson network of tissue or aril, the mace. The mace is rather leathery in texture. Nutmeg is the inner seed or kernel. Both nutmeg and mace are used mainly as spices. The dis-

tinctive flavors are due to volatile oils, present in both tissues. The oil of nutmeg, used as a nutmeg butter, is obtained by crushing and pressing the seed.

3. Season: Mature trees can bear fruit for 60 years or more. Seedling trees begin bearing at 5-8 years old.
4. Production in U.S.: No data
5. Other production regions: In the Americas, widely cultivated in Brazil and Grenada. Grenada with about 40 percent of world production (LEWIS 1984). About 1,854 tons of nutmeg were imported into the U.S. in 1992 (USDA 1993a).
6. Use: Spice used in flavoring food.
7. Part(s) of plant consumed: Dried aril and inner seed of the nut
8. Portion analyzed/sampled: Dried aril and dried inner seed of the nut with shell discarded
9. Classifications:
   a. Authors Class: Herbs and spices
   b. EPA Crop Group (Group & Subgroup): Herbs and spices (19B)
   c. Codex Group: Nutmeg 028 (HS 0789) Spices; Mace 028 (HS 0788) Spices
   d. EPA Crop Definition: None
10. References: GRIN, CODEX, FARRELL, GARLAND (picture), LEWIS (picture), MAGNESS, NEAL, US EPA 1995a, USDA 1993a
11. Production Map: As appropriate, EPA Crop Production Region 13, since the tree thrives best in tropical climate, especially islands, at an elevation of about 1,000 feet above sea level.
12. Plant Codes:
    a. Bayer Code: MYIFR

406

1. Oak

(Acorn, Bur oak, Mossy-cup oak)

Fagaceae

Quercus macrocarpa Michx.

2. Bur oak has the largest leaves and acorns of any oak and most palatable nut of all the oaks. Cultivars are 'Ashworth', 'Krieder' and 'Sweet Idaho'. Deciduous trees are up to 80 feet or more tall with obovate leaves to 10 inches long. As with acorn nuts the cup encloses the bottom half of the nut.
3. Season, harvest: in the fall of the year.
4. Production in U.S.: No data. Primarily northern U.S. (BAILEY 1976).
5. Other production regions: Canada.
6. Use: Large, edible nuts (acorns) are low in tannic acid. Nutmeat used in breads, muffins, cakes, soups and dumplings.
7. Part(s) of plant consumed: Nutmeat
8. Portion analyzed/sampled: Nutmeat
9. Classifications:
   a. Authors Class: Tree nuts
   b. EPA Crop Group (Group & Subgroup): Miscellaneous
   c. Codex Group: No specific entry
   d. EPA Crop Definition: None
10. References: GRIN, BAILEY 1976, FACCIOLA, WHEALY
11. Production Map: EPA Crop Production Regions 1, 5, 7 and 11.
12. Plant Codes:
    a. Bayer Code: QUEMC

407

1. Oat
1. Oats
(Avena, Avoine)

Poaceae (Gramineae)

Avena spp.

2. Oats are one of the top 5 grain crops in the U.S. They are grown to some extent in most continental states. However, the importance of oats as a farm crop continues to decrease but food uses are increasing. Acreage planted during the 3-year period 1954-56 averaged near 46 million. Average acreage planted to oats, 1993-95 was about 7 million. Production during that period averaged about 200,000,000 bushels. Most oats are used for livestock feed. About 23 percent of the total production in this country is used as food, mainly in the form of breakfast food, oat bran and oat flour. Some acreage is also used for pasture, hay, green chop and silage. Spring oats are less likely to be grown for forage then winter oats.

Oats are believed to be mainly Asiatic in origin. Different kinds of oats probably came from different parts of that continent or Europe. As a cultivated crop oats appear to be substantially later in origin than wheat. Early use of oats appears to have been medicinal. Not until about the beginning of the Christian era are references to oats as a cultivated crop found in literature. Cultivation of oats was extensive in Europe prior to the discovery of America, and the earliest settlers brought seed to the new world. They are now an important crop in all temperate zone countries.

The oat plant, like wheat, is an annual grass with kinds and varieties adapted either to fall planting and midsummer harvest or spring planting and late summer harvest. In general, overwintering kinds are grown where winter climates are mild, as throughout the Cotton Belt and in the western portions of the Pacific states. Spring seeding is generally practiced in other areas.

The early growth of the plant consists of leaves and a greatly shortened stem, giving a rosette type of plant. The early habit may be prostrate, semi-prostrate or upright. Tiller or branch buds under the soil surface grow into additional "branch plants" or tillers. The number of tillers formed depends on density of seeding, variety and growing conditions. In general, varieties adapted for spring planting form relatively few tillers, while up to 30 tillers may form on fall varieties under favorable conditions. However, all may not produce panicles. With fall seeding, plants generally remain in the rosette stage until spring. With spring seeding this stage is relatively short. The main stem and tiller stems then push upward reaching to 2 feet or more, again depending on variety and growing condition. These stems terminate in a large, generally loose panicle on which the flowers and seeds or kernels are borne. The panicle consists of a central stem or rachis, side or rachis branches which rise in whorls at the nodes and spikelets in which the flowers and seeds are borne. Each main and lateral stem or rachis terminates in a spikelet, but spikelets also are produced at the nodes of the branch stems. The panicles may be spreading in equilateral fashion, or one-sided. From 20 up to 150 spikelets may be produced on one panicle. Spikelets often droop or hang downward but may be upright. The spikelets are subtended by two loose membranous glumes, which generally are longer than the spikelet and largely cover it during flower and seed develop-ment. These are removed in threshing. In hulled varieties the spikelets usually contain three florets or flowers, one of which is rudimentary and nonfunctional.

Generally two kernels or seeds are produced per spikelet, but sometimes only one develops. In so-called naked oats one to three flowers may be produced per spikelet. The oat flower consists of the palea and lemma which enclose the sex organs, the stamens and single ovary. The palea may be awned or awnless and the lemma enclosed and adhered to the developing ovary or seed, except in so-called hull-less oats. Thus they are a part of the seed following threshing in most kinds. The oat kernel, also termed caryopsis or groat, is the part remaining after removal of the palea and lemma. It is elongated, spindle in shape, up to about $1/2$ inch in length and $1/8$ inch or less in width. It is generally thinly covered with fine, silky hairs and includes the seedcoat layers of cells, the starchy endosperm and the embryo. The oat kernel consists of 25 to 41 percent pericarp, 55 to 70 percent endosperm and 2 to 4 percent germ. When used for feed the whole oat is fed either whole or after grinding. This includes the palea and lemma but not the glumes. In products used for food the lemma and palea are also removed.

Uses of Oats

Oats are a nutritious feed for all classes of livestock. The hull, composed of lemma and palea, comprises on the average about 23 percent of the weight of the whole grain. Oats are high in mineral content and also in several vitamins. Formerly largely fed to horses, oats are now used as feed for dairy cattle and poultry as well. Oat hulls from milling are used in poultry mash.

For food use, the groat or inner kernel is rolled into flakes and used as oat meal in breakfast foods and baking. Oat flour contains an antioxidant which is used to preserve quality by delaying rancidity. Oat flour may also be mixed with wheat flour for multi-grain baked products. Oat food products include rolled oats, oat flakes, oat flour, oatmeal, oat bran, steel-cut oats (Scottish or Irish oats) and oat groats. Oat straw is more nutritious and palatable than wheat straw and is important as a supplementary feed on many farms. Fall sown oats furnish nutritious and palatable winter grazing in areas having mild winters.

3. Season, seeding to harvest: About 85 to 100 days.
4. Production in U.S.: In 1995, 6,336,000 acres planted with 161,847,000 bushels (USDA 1996c). In 1995, the top 10 states reported as Iowa (750,000 acres), North Dakota (650,000), Texas (650,000), Minnesota (625,000), Wisconsin (590,000), Illinois (500,000), South Dakota (350,000), California (350,000), Pennsylvania (190,000) and Nebraska (155,000) (USDA 1996c).
5. Other production regions: In 1995 Canada reported 3,352,731 acres (ROTHWELL 1996a).
6. Use: Grain, forage, hay, silage and straw used as feed. Grain also as food, see above. Also used as companion crop.
7. Part(s) of plant consumed: Grain and foliage
8. Portion analyzed/sampled: Grain, forage, hay, straw and the grain processed commodities: flour and groats/rolled oats. Oat grain is kernel plus hull. For oats forage, cut sample between tillering to stem elongation (jointing) stage. For oats hay, cut sample from early flower to soft dough stage. Hay should be field-dried to a moisture content of 10 to 20 percent. For oats straw, cut plant residue (dried stalks or stems with leaves) left after the grain has been harvested (threshed)

9. Classifications:
   a. Authors Class: Cereal grains; Forage, fodder and straw of cereal grains
   b. EPA Crop Group (Group & Subgroup): Cereal grains (15); Forage, fodder and straw of cereal grains (16)
   c. Codex Group: 020 (GC 0647) Cereal grains; 051 (AS 0647) Straw; 051 (AF 0647) Forage; 020 (GC 0080) Cereal grains; Codex lists *A. fatua*, *A. abyssinica* and *A. byzantina* in 020
   d. EPA Crop Definition: None
10. References: GRIN, CODEX, FACCIOLA, MAGNESS, ROTHWELL 1996a, USDA NRCS, USDA 1996c, US EPA 1996a, US EPA 1995a, US EPA 1994, MARSHALL 1992, SCHNEIDER 1996a, COFFMAN, WEBSTER
11. Production Map: EPA Crop Production Regions 1, 2, 5, 6, 7, 8 and 10 with most production in 5 and 7.
12. Plant Codes:
   a. Bayer Code: AVESS

408

1. Oat, Abyssinian
(Ethiopian oat)
Poaceae (Gramineae)
   Avena abyssinica Hochst.
2. This oat is similar in habit to the desert oat, *A. wiestii*, but differs in chromosome number. It is adapted to high elevations and seldom grown alone, usually grown with barley. Stems are erect 40 to 90 cm tall, rather small and fairly stiff. Panicles are equilateral, medium sized, very drooping. This species is grown to some extent in Ethiopia but not in the U.S. It is often considered a weedy species.
3. Season: See Oat.
4. Production in U.S.: See Oat.
5. Other production regions: See Oat.
6. Use: See Oat.
7. Part(s) of plant consumed: See Oat
8. Portion analyzed/sampled: See Oat
9. Classifications:
   a. Authors Class: See Oat
   b. EPA Crop Group (Group & Subgroup): See Oat
   c. Codex Group: See Oat
   d. EPA Crop Definition: See Oat
10. References: GRIN, See Oat
11. Production Map: EPA Crop Production Region, See Oat.
12. Plant Codes:
   a. Bayer Code: No specific entry

409

1. Oat, Animated
(Wild red oat, Winter wild oat, Avoine animmee, Avoine sterile, Sterile oat, Wildoat)
Poacea (Gramineae)
   Avena sterilis L.
2. This species is believed to be that from which cultivated red oats developed. It is characterized by large, hairy lemmas, which have strong twisted awns and adhere tightly to the kernel. Three botanical varieties are recognized. This species has become naturalized in some areas in this country but is not in cultivation. Animated oats grow prostrate to erect when young and grow erect 50 to 120 cm tall.

3. Season: See Oat.
4. Production in U.S.: See Oat.
5. Other production regions: See Oat. This species is distributed in the Middle East, Europe and Africa.
6. Use: See Oat.
7. Part(s) of plant consumed: See Oat
8. Portion analyzed/sampled: See Oat
9. Classifications:
   a. Authors Class: See Oat
   b. EPA Crop Group (Group & Subgroup): See Oat
   c. Codex Group: See Oat
   d. EPA Crop Definition: See Oat
10. References: GRIN, See Oat
11. Production Map: EPA Crop Production Regions 1, 2, 5, 7 and 8.
12. Plant Codes:
   a. Bayer Code: AVEST

410

1. Oat, Common
(Tree oat, Common side oat, Oat, Red oat, Side oat)
Poaceae (Gramineae)
   Avena sativa L. (syn: *A. byzantina* C. Koch, *A. sativa* var. *orientalis* (Schreb.) Hook. f.)
2. This is the most important of the cultivated oats. Magness *et al* noted 146 varieties of common oats, most of which were grown commercially in some part of the U.S. although production of many of them was very limited. Numerous new varieties have been released since that time, while a good many listed are no longer grown. Included are many varieties of winter and spring habit. The plants of different varieties differ in vigor and height. All are characterized by a panicle that is roughly pyramidal in shape with equilateral branches that spread outward. Many varieties are awnless, and in awned varieties, usually only the first flower is awned. Lemmas and paleas adhere to the kernel and may be white, gray, yellow or black in color. Shape and size of grains also are highly variable. Basal hairs are few, under the lemma. A number of important cultivated varieties of both winter and spring habit are included in this species. Red oat varieties are the kinds generally grown in the southern half of the U.S. Presently, red oat is considered to be the same species as common oats. Side oat is of much less economic importance in the U.S. than the common or tree oat since varieties of it are generally lower yielding. All the side oat varieties are of spring type. The principal distinguishing characteristic of this sub-species is the panicle which is almost entirely at one side of the stem axis. Side branches on the panicle arise from all sides of the stem, but branches definitely turn to one side of the stem or rachis. They also tend to turn upward more than in common oats. Like in the common oat, lemmas and paleas adhere to the kernel and may be black, gray, yellow or white in color.
3. Season: See Oat.
4. Production in U.S.: See Oat.
5. Other production regions: See Oat.
6. Use: See Oat.
7. Part(s) of plant consumed: See Oat
8. Portion analyzed/sampled: See Oat
9. Classifications:
   a. Authors Class: See Oat

b. EPA Crop Group (Group & Subgroup): See Oat
   c. Codex Group: See Oat
   d. EPA Crop Definition: See Oat
10. References: GRIN, See Oat
11. Production Map: EPA Crop Production Regions 1, 2, 5, 6, 7 and 8.
12. Plant Codes:
   a. Bayer Code: AVESA

411

1. Oat, Naked
(Hull-less oat, Hulless oat, Large naked oat, Sandoat, Naked-seeded oat, Naked oat)
Poaceae (Gramineae)
   Avena nuda L.
2. In this species the kernel or caryopsis is loose within the palea as in the major kinds of wheat. The stems grow erect to 60-80 cm tall. The origin of the species appears to have been Central and Eastern Asia. Several varieties have been introduced or have been developed in this country, but they are grown only to a very limited extent.
3. Season: See Oat.
4. Production in U.S.: See Oat.
5. Other production regions: See Oat.
6. Use: See Oat.
7. Part(s) of plant consumed: See Oat
8. Portion analyzed/sampled: See Oat
9. Classifications:
   a. Authors Class: See Oat
   b. EPA Crop Group (Group & Subgroup): See Oat
   c. Codex Group: See Oat
   d. EPA Crop Definition: See Oat
10. References: GRIN, See Oat
11. Production Map: EPA Crop Production Region, See Oat.
12. Plant Codes:
   a. Bayer Code: AVESG

412

1. Oat, Sand
(Bristle oat, Small oat, Black oat, Lopsided oat)
Poaceae (Gramineae)
   Avena strigosa Schreb.
2. In this species the lemmas are lance-like, extending to two distinct points. The plant has small, erect stems that vary widely in height from 80 to 200 cm. Panicles are near equilateral. The species is widely distributed in Europe and has become naturalized in California. It is not grown as a grain crop.
3. Season: See Oat.
4. Production in U.S.: See Oat.
5. Other production regions: See Oat.
6. Use: See Oat.
7. Part(s) of plant consumed: See Oat
8. Portion analyzed/sampled: See Oat
9. Classifications:
   a. Authors Class: See Oat
   b. EPA Crop Group (Group & Subgroup): See Oat
   c. Codex Group: See Oat
   d. EPA Crop Definition: See Oat
10. References: GRIN, See Oat

11. Production Map: EPA Crop Production Region 10.
12. Plant Codes:
    a. Bayer Code: AVESG

413

1. Oat, Slender
(Avenilla, Barbed oat)
Poaceae (Gramineae)
  Avena barbata Pott ex Link
2. This is an annual, cool season bunchgrass, native to Europe and introduced to the U.S. It belongs to the grass tribe *Aveneae*. Slender oat and wild oat are the main two forage species of the oat (*Avena*) family. It occurs on foothill rangeland areas, fields and roadside right-of-ways, and is most abundant on coarse-textured dry soils. It grows erect up to a height of 1.2 m with leaf blades 10 to 40 cm long and 5 to 10 cm wide, and has a panicle (20 to 40 cm long) typical of the oat family. It is adapted to a narrow area along the Pacific coast from southern Canada to Baja California. Slender oat produces good quality forage when young (winter and spring), but is unpalatable after it reaches maturity. Germinates in late fall or early winter and flowers in March to June. This species has small, weak stems, resulting in a decumbent growth habit. Panicles are equilateral, rather large and drooping. Seeds fall to the ground at maturity. The species is now widely distributed throughout the world. The species is not grown as a grain crop but has become naturalized widely. In California it is a reseeding range grass. Also see Oat.
3. Season: Cool season forage. It grows early spring and matures in summer. It can be grazed winter and spring and is unpalatable at maturity.
4. Production in U.S.: No data for slender oat production, however, pasture and rangeland are produced on more than 410 million A (U.S. CENSUS, 1992). See above.
5. Other production regions: Grows in western Canada.
6. Use: Grazing , hay, and erosion control.
7. Part(s) of plant consumed: Leaves and stems
8. Portion analyzed/sampled: Forage and hay
9. Classifications:
    a. Authors Class: See Forage grass
    b. EPA Crop Group (Group & Subgroup): See Forage grass
    c. Codex Group: See Forage grass
    d. EPA Crop Definition: None
10. References: GRIN, US EPA 1994, US EPA 1995, CODEX, STUBBENDIECK (picture), MOSLER, RICHE
11. Production Map: EPA Production Regions 10, 11 and 12.
12. Plant Codes:
    a. Bayer Code: AVEBA

414

1. Oat, Wild
(Avena loca, Spring wild oat)
Poaceae (Gramineae)
  Avena fatua L.
2. This is a annual, cool season bunchgrass, native to Europe and introduced to the U.S. It belongs to the grass tribe *Aveneae*. Wild oat and slender oat are the main two forage species of the oat (*Avena*) family. It occurs on valleys and slopes of foothill rangeland areas, fields and roadside right-of-

ways, and is most abundant on coarse-textured dry soils. It grows erect to a height ranging from 0.3 to 1.3 m with leaf blades 10 to 30 cm long and 5 to 10 cm wide, and has a panicle (10 to 30 cm long) typical of the oat family. It grows from Canada to Mexico except for the southern U.S. Wild oat makes up a majority of the early forage in its adaptable sites in the annual rangeland areas of the western U.S. It is good to exellent quality forage for grazing and hay by all classes of livestock. It germinates in early winter with most growth in the early spring and it flowers in March to May with seeds maturing in June. Wild oat reproduces from seed. This species is now widely distributed throughout temperate regions and is a troublesome weed in Northern states and in Canada. The plant resembles the common oat, *A. sativa* L., but is of greater vigor. Flowering stems are erect, 80 to 160 cm tall. Panicles are very large and drooping. The spikelets carry long twisted awns and separate from their pedicels by abscission, dropping when ripe. Thus the plant is self-seeding with a wide range of habitats. The species is of some value for hay or pasturage but is not grown for grain. Also see Oat.
3. Season: Cool season forage and grows early spring and matures in summer. It can be grazed winter and spring until it matures in June.
4. Production in U.S.: No data for wild oat production, however, pasture and rangeland are produced on more than 410 million A (U.S. CENSUS, 1992). See above
5. Other production regions: Grows in Canada.
6. Use: Grazing , hay, and rangeland revegetation.
7. Part(s) of plant consumed: Leaves and stems
8. Portion analyzed/sampled: Forage and hay
9. Classifications:
    a. Authors Class: See Forage grass
    b. EPA Crop Group (Group & Subgroup): See Forage grass
    c. Codex Group: See Forage grass
    d. EPA Crop Definition: None
10. References: GRIN, US EPA 1994, US EPA 1995, CODEX, STUBBENDIECK (picture), MOSLER
11. Production Map: EPA Production Regions 1, 5, 6, 7, 8, 10, 11 and 12.
12. Plant Codes:
    a. Bayer Code: AVEFA

415

1. Oatgrass
(Wild oatgrass)
Poaceae (Gramineae)
  Danthonia spp.
1. California oatgrass
  Danthonia californica Bol.
1. Flatstem oatgrass
(Flatstem danthonia, Mountain oatgrass)
  D. compressa Austin
1. Parry oatgrass
(Parry danthonia, Parry's danthonian)
  D. parryi Scribn.
1. Timber oatgrass
(Timber danthonia, Intermediate oatgrass)
  D. intermedia Vasey
2. Most oatgrasses (*Danthonia* spp.) are cool season perennial bunchgrasses and members of the grass tribe *Danthonieae*.

Oatgrasses are important forage rangeland grasses native to the Rocky Mountain and Western states, and Canadian provinces. Flatstem oatgrass is an exception, since it is important to the eastern areas from Nova Scotia south to Georgia. California oatgrass is distributed in mountain meadows and coastal prairies from the Rocky Mountains to the Pacific Coast and from southern Canada to central California. It grows from 30 to 60 cm tall with an open panicle 2 to 6 cm long. It grows early in the spring, flowers May through June, and seed matures in August. It reproduces from seed and tillers. California oatgrass produces good to excellent forage for cattle and sheep but are somewhat less palatable for sheep and goats. Parry oatgrass is found in the Rocky Mountain area at high altitudes (≤3000 m) and open woods. It grows erect (20 to 60 cm tall) with a panicle 5 to 20 cm long, and starts growing in the spring and matures in September. Forage palatability of Parry oatgrass declines rapidly upon maturity, but is fair to good for cattle and horses, and is important because it is so abundant. Timber oatgrass is primarily found in mountain meadows and forest understory areas of northeastern U.S. and western Canada at altitudes >1800 m. It grows 10 to 50 cm tall with a panicle 2 to 6 cm long. It begins growth early in the spring and reproduces from seeds and tillers. It produces good to excellent forage for livestock and deer and withstands heavy grazing.

3. Season: Cool season, growth is in the spring and it matures in September. Most palatable in the spring. Timber oatgrass can withstand heavier grazing pressure than the other oatgrasses.
4. Production in U.S.: No production data for any of the oatgrasses, however, pasture and rangeland are produced on more than 410 million A (U.S. CENSUS, 1992). See above.
5. Other production regions: Canada.
6. Use: Grazing on pasture, hay, also for erosion control.
7. Part(s) of plant consumed: Leaves and stems
8. Portion analyzed/sampled: Forage and hay
9. Classifications:
   a. Authors Class: See Forage grass
   b. EPA Crop Group (Group & Subgroup): See Forage grass
   c. Codex Group: See Forage grass
   d. EPA Crop Definition: None
10. References: GRIN, US EPA 1994, US EPA 1995, CODEX, HOLZWORTH, HITCHCOCK, STUBBENDIECK (picture), MOSLER, RICHE
11. Production Map: EPA Production Regions for Flatstem oatgrass are 1 and 2; for California oatgrass are 8, 9, 10, 11 and 12; for Parry oatgrass is 9; and for Timber oatgrass are 1, 9, 10, 11 and 12.
12. Plant Codes:
   a. Bayer Code: No specific entry

416

1. Oca (oh-kah)
(Sorrel, Kao, New Zealand yam, Truffette acide, Papa roja, Red potato, Cavi, Caya, Ibias, Apilla, Wood sorrel)
Oxalidaceae
Oxalis tuberosa Molina (syn: *O. crenata*)
2. Bushy plant that grows 8 to 12 inches in height with clover-like leaves. Oca tubers resemble stubby, wrinkled carrots. The tubers are firm, white flesh and shiny skin from white to red. Harvested like standard potatoes. The sour or bitter

tubers contain some oxalic acid, while the sweet tuber contains trace amounts. For over 100 years, oca has been grown in Britain and continental Europe as a home garden ornamental.
3. Season, planting to harvest: About 6 months. Planting the whole tuber.
4. Production in U.S.: No data.
5. Other production regions: Grown in Mexico, Central America and South America and New Zealand (NATIONAL RESEARCH COUNCIL 1989).
6. Use: Tubers are boiled, baked, fried, or canned or mixed fresh with salads or pickled in vinegar.
7. Part(s) of plant consumed: Primarily tuber but leaves can be eaten in salads or as a cooked vegetable
8. Portion analyzed/sampled: Tuber
9. Classifications:
   a. Authors Class: Root and tuber vegetables
   b. EPA Crop Group (Group & Subgroup): Miscellaneous
   c. Codex Group: 016 (VR 0586) Root and tuber vegetables
   d. EPA Crop Definition: None
10. References: GRIN, CODEX, LOGAN, NATIONAL RESEARCH COUNCIL 1989, YAMAGUCHI 1983 (picture)
11. Production Map: No entry.
12. Plant Codes:
   a. Bayer Code: No specific entry, genus entry is OXASS

417

1. Okra
(Bhindi, Gumbo, Quimgombo, Lady's fingers, Okro, Ochro, Bamia, Quiabo, Gombo, Quinbombo, Bandakai, Gobbo, Okra leaves, Bumbo, Bombo)
Malvaceae
Abelmoschus esculentus (L.) Moench (syn: *Hibiscus esculentus* L.)
2. The okra plant is an annual, requiring warm growing conditions. It attains heights of from 3 feet in dwarf varieties to 7 or 10 feet in others. Leaves are cordate in shape, and lobed or divided. The fruit is a long pod, generally ribbed and spineless in cultivated kinds. Pods, the edible portion, are harvested while still tender and immature. They attain lengths of 4 to 10 inches, and up to 1 inch or more in diameter. First pods are ready for harvest about 2 months after planting, but plants continue to bloom and set if all pods are harvested at the proper early stage. Okra is generally similar in culture and exposure to summer squash, except the plants are tall and upright, with many short branches. Chinese okra is *Luffa acutangula*, edible gourd, which see.
3. Season, planting to first harvest: 2 months.
4. Production in U.S.: 19,804 acres reported in 1959 CENSUS. In 1992, 4,336 acres reported (1992 CENSUS). Top 6 states are: Texas (734 acres), Florida (661), Georgia (528), Alabama (437), California (206) and Louisiana (178) for a total of 2,744 acres or 63 percent of total acres reported for okra.
5. Other production regions: In 1994, U.S. imported 46,450 tons from Mexico (LOGAN 1996). Okra is grown in Canada (ROTHWELL 1996a).
6. Use: Fresh, canned, frozen, dried, for use in culinary – mainly soups, stews or as cooked vegetable. Preparation – rinse and cut off cap but do not expose the seeds. Fuzzy varieties need to be rubbed with a towel prior to washing and trimming (SCHNEIDER 1985).

7. Part(s) of plant consumed: All of immature pods. Also okra leaves and shoots can be eaten like spinach
8. Portion analyzed/sampled: Fruit (pod)
9. Classifications:
   a. Authors Class: Fruiting vegetables (except cucurbits)
   b. EPA Crop Group (Group & Subgroup): Miscellaneous
   c. Codex Group: 012 (VO 0442) Fruiting vegetables other than cucurbits
   d. EPA Crop Definition: None
10. References: GRIN, CODEX, LOGAN (picture), MAGNESS, RICHE, ROTHWELL 1996a, SCHNEIDER 1985, US EPA 1996a, US EPA 1994, YAMAGUCHI 1983 (picture), SWIADER, DUKE(a), RUBATZKY
11. Production Map: EPA Crop Production Regions 2, 3, 4, 6 and 10 which accounts for 97 percent of U.S. acreage.
12. Plant Codes:
    a. Bayer Code: ABMES

418

1. Olive
(Olive oil, Aceituna, Olivo, Common olive)
Oleaceae

Olea europaea L. ssp. *europaea*

2. Olive trees are evergreens; small or medium size, sometimes up to 25 feet, resistant to drought, and generally very long lived. They will withstand winter temperatures to 15 degrees F. Leaves are small, $1^1/_2$ to 3 inches long, and thick. Fruit is borne on panicles rising from leaf axils. Fruits have a thin, smooth skin, green when immature, through red to nearly black when ripe. Shape is generally oval, $^3/_4$ by 1 inch in small-fruited kinds to 1 inch diameter and $1^1/_2$ inches long in large-fruited varieties. Each fruit has a single, elongated seed. Pulp is extremely bitter, due to tannin in raw fruit, and contains up to 20 percent oil. Olive oil is obtained from the fleshy portion of the fruit, which contains from 14 percent to as high as 40 percent of non-drying oil, depending on variety and growing conditions.

   World production of oil totals around 1,800,000 metric tons annually (1993-94), mainly in Mediterranean countries. For best oil, fruit is harvested just before full maturity. The fruit is crushed then pressed in various ways, depending on available facilities. Solvents are generally used for final extraction of the press cake, to obtain maximum oil yield. The oil is used as salad and cooking oil. Low-grade oils are used mainly for making soap. In the U.S., only around 5,000 tons are crushed for oil annually. Annual imports of olive oil average near 120,000 metric tons, 1992-93. Generally, 90 percent of the U.S. crop is canned. Olives too ripe or culls are crushed for oil. Cold pressed oil is the highest quality.

3. Season, bloom to harvest: 6 to 8 months for green olives; 8 to 10 months for ripe olives or oil.
4. Production in U.S.: 30,100 acres with 122,000 tons in 1993 with only 5,300 tons being crushed for oil (USDA 1994a). In 1994, California harvested 33,630 acres with 92,486 tons (MELNICOE 1996e).
5. Other production regions: Mediterranean area, especially Spain, Italy and Greece.
6. Use: In U.S. grown mainly for its fresh fruit. Pickled for green olives, brined and canned for ripe, crushed for oil. Also grown as an ornamental.

7. Part(s) of plant consumed: Fruit. Seed sometimes crushed for oil extraction
8. Portion analyzed/sampled: Fruit and its processed commodity oil. In EPA Table 1, EPA notes that the fruit should be analyzed after removing and discarding stem and pit. Olives processed for oil normally include the whole olive with its pit in the grinding process
9. Classifications:
   a. Authors Class: Tropical and subtropical fruits – edible peel; Oilseed
   b. EPA Crop Group (Group & Subgroup): Miscellaneous
   c. Codex Group: 005 (FT 0305) Assorted tropical and subtropical fruits (edible peel); 023 (SO 0305) Oilseed; 067 (OC 0305) Vegetable crude oil; 068 (OR 0305) Vegetable refined oil; 069 (DM 0305) Miscellaneous derived edible products of plant origin.
   d. EPA Crop Definition: None
10. References: GRIN, CODEX, MAGNESS, ROBBELEN 1989 (picture), USDA 1994a, US EPA 1995a, US EPA 1995b, US EPA 1994, MELNICOE 1996e, IVES, FERGUSON
11. Production Map: EPA Crop Production Region 10.
12. Plant Codes:
    a. Bayer Code: OLVEU

419

1. Onion
(Green onion, Dry bulb onion, Cebolla, Cipolla, Silverskin onion, Spring onion, Cipollini, Spanish onion, Tama-negi, Pearl onion, Pickling onion, Boilers, Mini onion, Baby onion)
Liliaceae (Amaryllidaceae)

Allium cepa L. var. *cepa*

2. Onions, grown mainly for the dry bulbs, are the most important of the vegetable bulb crops. They are also grown less extensively for the green leaves and succulent leaf bases and young bulbs. They may be grown from seed planted in place, or from small bulb sets. The leaves are tubular, up to 18 inches in height, $^1/_3$- to $^1/_2$-inch in diameter, generally smooth. In the north, seeds, young plants or sets are field-planted in early spring; and bulbs are harvested in the late summer or fall. In the South, seed or sets are planted in the fall for early summer harvest. Bulbs range from oblate to oval in shape and from 1 inch to over 3 inches diameter. They vary in pungency from mild to very strong. The top of the mature bulb commonly extends slightly above the soil, so may be partially exposed. The outer layers, or scales, are generally removed prior to use. Cipollini is a name used for small bulbs. Pearl onion (Pickling onion, Boilers, Mini onion, Baby onion) are bulbing onions that are planted thickly and harvested when very small about 2 to 3 months after planting.
3. Season, planting to harvest: About 3-6 months for dry bulb, varies. Overwintering bulb onions can take almost a year to mature for harvest. Green onions are about 2 months from planting to harvest.
4. Production: In U.S. about 2,853,100 tons of dry bulb onions (1994 Agricultural Statistics). In the 1992 CENSUS for dry bulb onion production: California (31,153 acres), Texas (16,949), Colorado (15,742), Oregon (15,602), New York (12,066), Idaho (9,198), Georgia (7,437), New Mexico (6,839), Michigan (6,771) and Washington (6,419) reported 93 percent of the total U.S. acreage of 138,060 with storage

onions accounting for about 50 percent of the acreage. Also in the 1992 CENSUS for green onion production; California (5,298 acres), Texas (1,023), Arizona (968), New York (885), Colorado (330), Michigan (379), New Jersey (327), New Mexico (311), Florida (231), and Georgia (226) reported 81 percent of the total U.S. acreage of 12,395. The 1994 Agricultural Statistics listed onions with total U.S. acreage of 150,680 and the top 10 states for dry bulb onion production in the 1992 CENSUS accounted for 96 percent of the total U.S. acreage.

5. Other production regions: In 1995, Canada produced about 104,642 tons of dry bulb on 10,606 acres and 3,653 tons of green onion on 1,337 acres (ROTHWELL 1996a).

6. Use: Culinary, salads, fresh, canned, pickled, frozen, dehydrated. U.S. per capita consumption in 1994 is 15.4 pounds (PUTNAM 1996).

7. Part(s) of plant consumed: Bulbs, after outer papery scales are removed. Also less extensively as green onions from sets

8. Portion analyzed/sampled: Bulb for dry bulb; whole plant without roots for green onions. 40 CFR 180.1 (J)(5) states that roots, stems and outer sheaths (or husks) shall be removed and discarded from garlic bulbs and dry bulb onions, and only the garlic cloves and onion bulbs shall be examined for pesticide residues

9. Classifications:
   a. Authors Class: Bulb vegetables
   b. EPA Crop Group (Group & Subgroup): Bulb vegetables (Dry bulb and green onion are representative crops)
   c. Codex Group: Bulb onion: 009 (VA 0385) Bulb vegetables; Spring onion: 009 (VA 0389) Bulb vegetables; Silverskin onion: 009 (VA 0390) Bulb vegetables
   d. EPA Crop Definition: Onions = Dry bulb onions, green onions and garlic; Dry bulb onions = Garlic and shallots (dry); Green Onions = Leeks, spring onions or scallions, Japanese bunching onions, green shallots, or green eschalots

10. References: GRIN, CODEX, KAYS, MAGNESS, ROTHWELL 1996a, SCHREIBER, USDA 1994a, US EPA 1995b, US EPA 1995a, US EPA 1994, RABINOWITCH, SWIADER, PUTNAM 1996

11. Production Map: EPA Crop Production Regions: Dry bulb onion – 1, 2, 5, 6, 8, 9 10, 11 and 12. Green onion – 1, 2, 5, 6, 8, 10, 11 and 12.

12. Plant Codes:
    a. Bayer Code: ALLCE, ALLXP (transplanted onions), ALLXS (direct-seeded)

420

1. Onion, Beltsville bunching

Liliaceae (Amaryllidaceae)

Allium x *proliferum* (Moench) Schrad. (syn: *Allium cepa* L. x *A. fistulosum* L.)

2. This is a multiplier onion, produced by breeding, similar in culture to the Welsh or Japanese bunching, which see. It is the bunching onion that is mainly grown in the U.S., and is marketed as a green onion.

3. Season, bulb setting to first harvest: 2 to 3 months.

4. Production in U.S.: No separate data. 12,071 acres all green onions reported (1959 CENSUS) and 12,395 acres all green onions reported (1992 CENSUS).

5. Other production regions: No entry.

6. Use: Salads, and as flavoring in culinary cookery. Some dehydrated.

7. Part(s) of plant consumed: Leaves and young bulblets

8. Portion analyzed/sampled: Whole plant without roots

9. Classifications:
   a. Authors Class: Bulb vegetables
   b. EPA Crop Group (Group & Subgroup): Bulb vegetables, both taxonomic synonyms are included, along with *Allium* spp.
   c. Codex Group: Similar to Welsh onion, which see
   d. EPA Crop Definition: Onions = Green onions

10. References: GRIN, MAGNESS, US EPA 1995a, US EPA 1995b

11. Production Map: No entry.

12. Plant Codes:
    a. Bayer Code: No specific entry, genus entry ALLSS

421

1. Onion, Chinese
(Ch'iao t'ou, Chalote chinesa, Japanese scallions, Oriental onion, Kiltow, Chinese scallion)
1. Rakkyo

Liliaceae (Amaryllidaceae)

Allium chinense G. Don (syn: *A. bakeri* Regel)

2. Rakkyo is an onion relative. It is an important vegetable in the Orient. In the U.S. it is grown and used mainly by Orientals. The plants do not produce seeds and are propagated by bulb division. In mild climates, bulbs are planted in late summer, and the crop is harvested in midsummer of the following year. Several small bulbs are obtained from each bulb planted. Rakkyo bulbs are mainly pickled, some are canned. Also, they are used as a cooked vegetable. The leaves have hollow blades. Culture and exposure of plant parts is similar to that of bulb-set onions.

3. Season, planting to harvest: About 10 months.

4. Production in U.S.: No data. Grown to limited extent by Oriental gardeners. Grown in Florida (STEPHENS 1988).

5. Other production regions: Primarily the Orient.

6. Use: Mainly pickles, some canned and some used as fresh cooked vegetable. Canned as pickled scallions or rakkyo-zuke.

7. Part(s) of plant consumed: Bulbs

8. Portion analyzed/sampled: Bulbs

9. Classifications:
   a. Authors Class: Bulb vegetables
   b. EPA Crop Group (Group & Subgroup): Miscellaneous
   c. Codex Group: 009 (VA 0386) Bulb vegetables
   d. EPA Crop Definition: None

10. References: GRIN, CODEX, FACCIOLA, KAYS, MAGNESS, MYERS, STEPHENS 1988, RUBATZKY

11. Production Map: No entry

12. Plant Codes:
    a. Bayer Code: ALLCH

422

1. Onion, Potato
(Multiplier onion, Hill onion, Pregnant onion, Nest onion, Mother onion)

Liliaceae (Amaryllidaceae)

Allium cepa var. *aggregatum* G. Don

2. When large bulbs of the potato onion are planted, they form a number of small bulblets, each with a leafy top. These can be harvested while green as green onions. If allowed to

ripen, they are small bulbs suitable for planting the following year. The potato onion varieties rarely set seeds, so are propagated by the small bulbs. When grown for the bulbs, culture is in all ways similar to that for common onion. The leaves are broad and hollow.

3. Season, planted bulbs to harvest: For green onions, 3 to 4 months.
4. Production in U.S.: No separate data. It is grown in Florida, planted from September through March (STEPHENS 1988).
5. Other production regions: No entry.
6. Use: Dry bulbs as cooked vegetable, and in culinary and salad. Green onion mainly as salad.
7. Part(s) of plant consumed: Bulb after outer scales removed; or leaves and immature bulbs as green onions
8. Portion analyzed/sampled: Bulb for dry bulb; whole plant without roots for green onions
9. Classifications:
   a. Authors Class: Bulb vegetables
   b. EPA Crop Group (Group & Subgroup): Bulb vegetables
   c. Codex Group: Welsh onion: 009 (VA 0387) Bulb vegetables
   d. EPA Crop Definition: Onions = Dry bulb onions and green onions
10. References: GRIN, CODEX, MAGNESS, STEPHENS 1988 (picture), US EPA 1995a, US EPA 1995b, YAMAGUCHI 1983, RUBATZKY, FACCIOLA
11. Production Map: EPA Crop Production Region 3.
12. Plant Codes:
   a. Bayer Code: No specific entry, genus entry ALLSS

423

1. Onion, Tree
(Egyptian onion, Top onion, Topset onion, Catawissa onion, Perennial onion, Walking onion, Kitsune negi, Egyptian topset onion)
Liliaceae (Amaryllidaceae)
Allium x *proliferum* (Moench) Schrad. ex Willd. (syn: *A. cepa* var. *proliferum* (Moench) Regel; *A. cepa* L. var. *bulbiferum* L.H. Bailey; *A. cepa* L. var. *viviparum* (Metz.) Alef.)

2. Green onion plant without any large bulb formation at the base of the stem. The top bears bulbils, or bulblets, instead of flowers and seeds. Several offsets form at the base of the stem. It resembles a green shallot. The leaves are round and hollow. In Florida, they are primarily started in the fall. The bulblets (topsets) or lower offsets are used for propagation. Popular in the Orient for its foliage.
3. Season, propagation is by topset or lower offsets in the fall in Florida.
4. Production in U.S.: No data. Grown in Florida (STEPHENS 1988).
5. Other production regions: Orient (RUBATZKY 1997).
6. Use: Top bulblets are used as flavoring in pickle cucumbers and green leaves for green onions.
7. Part(s) of plant consumed: Top bulblets and leaves
8. Portion analyzed/sampled: Top bulblets; whole plant without roots for green onion
9. Classifications:
   a. Authors Class: Bulb vegetables
   b. EPA Crop Group (Group & Subgroup): Bulb vegetables, taxonomic synonym for tree onions is *A. cepa* which is included, along with *Allium* spp.
   c. Codex Group: 009 (VA 0391) Bulb vegetables

d. EPA Crop Definition: None
10. References: GRIN, BAILEY 1976, CODEX, FACCIOLA, STEPHENS 1988, RUBATZKY (picture)
11. Production Map: EPA Crop Production Region 3.
12. Plant Codes:
   a. Bayer Code: ALLCV

424

1. Onion, Welsh
(Cebolinha, Cebolleta, Chung, Cong, Multiplier onion, Japanese bunching onion, Negi, Cibal, Spring onion, Nebuka, Cebollin, Hanh-ta, Zwiebel, Bunching onion, Salad onion, Chinese small onion, Scallion)
Liliaceae (Amaryllidaceae)
Allium fistulosum L.

2. This is the principle onion of Japan and China, but of limited importance in the U.S. Leaves are rigid and tubular and inflated or swollen in appearance. The bulbs become only slightly enlarged. Plants multiply by tillers from a mother plant, so clusters of plants result from planting a single one. They may also be grown from seed or seed propagated transplants. In the Orient the leaves and leaf bases are often blanched by covering with soil. In Asia and the U.S., they are also marketed as green onions. The thick, swollen leaves and leaf bases are harvested. The plant grows 6 to 24 inches tall.
3. Season, seed to first harvest: 4 to 5 months for green onions, a year or more for blanched. After transplanting, green onions can be harvested in 2-3 months.
4. Production in U.S.: No separate data. Total green onions 12,071 acres (1959 CENSUS) and total green onions 12,395 acres (1992 CENSUS) as a comparsion. Grown in California (HEATON 1997).
5. Other production regions: See above.
6. Use: As potherbs, flavoring in culinary cookery, salads. Also dehydrated.
7. Part(s) of plant consumed: Thick leaves and leaf bases
8. Portion analyzed/sampled: Whole plant without roots
9. Classifications:
   a. Authors Class: Bulb vegetables
   b. EPA Crop Group (Group & Subgroup): Bulb vegetables
   c. Codex Group: 009 (VA 0387) Bulb vegetables
   d. EPA Crop Definition: Onions = Green onions; Green onions = Japanese bunching onion and Spring onion
10. References: GRIN, CODEX, KAYS, MAGNESS, MYERS (picture), RICHE, STEPHENS 1988, US EPA 1995a, HEATON
11. Production Map: EPA Crop Production Region 10.
12. Plant Codes:
   a. Bayer Code: ALLFI

425

1. Oniongrass
(Melic)
Poaceae (Gramineae)
Melica spp.
1. Bulbous oniongrass
(Onion melic, Cebollin)
M. bulbosa Geyer ex Porter & J.M. Coult.
1. Alaska oniongrass

M. subulata (Griseb.) Scribn.

1. California oniongrass
(Smallflower melicgrass, California melic)

M. imperfecta Trin.

1. Porter oniongrass

M. porteri Scribn.

2. Oniongrasses are cool season perennial bunchgrasses that contain more than 60 species with 20 naturally occurring in the U.S., and they are the most widely distributed western grass species. Some of the species are available for ornamental uses. They occur in the Rocky Mountains states and areas along the Pacific Coast north through British Columbia and Alaska. Bulbous oniongrass is erect growing 30 to 90 cm tall and the blades are flat 10 to 30 cm long and 2 to 5 cm wide and it has a panicle (8 to 18 cm long) as its inflorescence. They are located along streambanks, dry woods, rocky hillsides and in meadows. Produces an excellent forage for all livestock.

3. Season: Cool season, growth is in the early spring, flowers in late spring and early summer and seeds mature in July. They are a very palatable forage but available in limited amounts.

4. Production in U.S.: No production data for any of the oniongrasses is available, however, pasture and rangeland are produced on more than 410 million A (U.S. CENSUS, 1992). See above.

5. Other production regions: Canada.

6. Use: Pasture, ornamental use, and for erosion control.

7. Part(s) of plant consumed: Leaves, stems and seeds

8. Portion analyzed/sampled: Forage and hay

9. Classifications:
   a. Authors Class: See Forage grass
   b. EPA Crop Group (Group & Subgroup): See Forage grass
   c. Codex Group: See Forage grass
   d. EPA Crop Definition: None

10. References: GRIN, US EPA 1994, US EPA 1995, CODEX, HOLZWORTH, MOSLER, STUBBENDIECK (picture), RICHE

11. Production Map: EPA Production Regions for Oniongrasses are 9 and 11.

12. Plant Codes:
   a. Bayer Code: No specific entry

426

1. Orach
(Arroche, Garden orach, Orache, Mountain spinach, French spinach, Sea purslane, Butter leaves)

Chenopodiaceae

Atriplex hortensis L.

2. Orach is grown as a substitute for spinach in Europe and in the Northern Plains in the U.S. The plant is drought resistant, and slower to form a seed stalk than spinach. Leaves are cordate or triangular oblong, 4 to 5 inches long, 2 to 3 inches wide. A rosette of leaves first develops, followed by the seed stalk, which may reach a height up to 8 feet. Leaves are used as potherbs. In culture (exposure of edible parts and use), orach is comparable to spinach. Orach is a cool season annual with tender leaves and stems that can tolerate higher temperatures than spinach.

3. Season, seeding to harvest: 40 to 60 days.

4. Production in U.S.: No data. Mostly in home gardens. Northern Plains of U.S.

5. Other production regions: Europe.

6. Use: As potherb, like spinach.

7. Part(s) of plant consumed: Leaves and young stems

8. Portion analyzed/sampled: Leaves and young stems

9. Classifications:
   a. Authors Class: Leafy vegetables (except *Brassica* vegetables)
   b. EPA Crop Group (Group & Subgroup): Leafy vegetables (except *Brassica* vegetables) (4A)
   c. Codex Group: 013 (VL 0488) Leafy vegetables (including *Brassica* leafy vegetables)
   d. EPA Crop Definition: None

10. References: GRIN, CODEX, MAGNESS, MARTIN 1979, STEPHENS 1988 (picture), US EPA 1995a, SWIADER, RUBATZKY

11. Production Map: EPA Crop Production Region 5.

12. Plant Codes:
   a. Bayer Code: ATXHO

427

1. Orange, Sour
(Bitter orange, Seville orange, Bigarade orange, Naranja agria, Chinotto, Myrtleleaf orange, Moli, Soap orange)

Rutaceae

Citrus aurantium L. (syn: *C. myrtifolia* Raf., *C. vulgaris* Risso, *C. bigarradia* Loisel., *C. communis* Le Maout & Dec.)

1. Orange, Bergamot
(Bergamoto, Bergamotier)

C. bergamia Risso & Poit. (syn: *C. aurantium* ssp. *bergamia* (Risso & Poit.) Wight & Arn. ex. Engl.; *C. vulgaris* Risso)

2. These oranges (See also citrus fruits) are similar to the sweet orange in tree and fruit appearance, but are characterized by a very acid pulp and by a hollow axis or core in the fruit. Cut fruits and crushed leaves have a characteristic, very strong odor. Fruit is too acid for fresh use but is used for marmalade. The sour orange was widely used as a rootstock for other varieties prior to the appearance of the Tristizia virus disease, which affects trees on this stock. It is still used in areas where Tristizia is not important. The Bergamot oranges, grown in Mediterranean areas for oil, belong in this group. They are not grown commercially in the U.S.

3. Season, bloom to harvest: 8 to 12 months.

4. Production in U.S.: Florida (LOGAN 1996) for Seville orange. Also Guam and Samoa (MORTON 1987).

5. Other production regions: Orient, Mediterranean Region, Haiti, Dominican Republic, Brazil and Paraquay (MORTON 1987).

6. Use: Used for marmalade, flavoring and in drinks.

7. Part(s) of plant consumed: Whole fruit when made into marmalade

8. Portion analyzed/sampled: Whole fruit

9. Classifications:
   a. Authors Class: See citrus fruits
   b. EPA Crop Group (Group & Subgroup): See citrus fruits
   c. Codex Group: 001 (FC 0207) Citrus fruits
   d. EPA Crop Definition: See citrus fruits

10. References: GRIN, CODEX, LOGAN, MAGNESS, US EPA 1994, MORTON (picture)

11. Production Map: EPA Crop Production Regions 3, 6 and 10.

12. Plant Codes:

a. Bayer Code: CIDAU (*C. aurantium*), CIDAB (*C. bergamia*)

428

1. Orange, Sweet

(China sweet orange, Pigmented orange, Sanguine orange, Chinese navel orange, Chinas, Blood orange, Malta orange, Oranger, Valencia orange, Naranjo dulce, Shamouti)

Rutaceae

Citrus sinensis (L.) Osbeck (syn: *C. aurantium sinensis* L., *C. dulcis* Pers., *C. aurantium vulgare* Risso & Poit., *C. aurantium dulce* Hayne).

2. Sweet oranges are the most important of the citrus fruits (See citrus fruits). Varieties vary from early ripening, about 8 months from bloom to late, up to 16 months from bloom. There are three main groups: The normal fruited, without navels and with light orange colored flesh; the navel oranges, with a distinct navel development at the styler end; and blood oranges, with red flesh and juice. The latter are little grown commercially in the U.S. Sweet oranges vary in size from 2 inches upward in diameter, are generally round to oblong in shape and have a medium thick and tough rind. Although 73 varieties are listed by Webber (THE CITRUS INDUSTRY) major production in U.S. is of 6 varieties: 'Valencia', 'Washington Navel', 'Hamlin', 'Parson Brown', 'Pineapple' and 'Temple'. 'Valencia' has few seeds and available from Florida, California, Arizona and Texas. 'Navel' is seedless and primarily available from Arizona, California and Texas, 'Hamlin' is nearly seedless and 'Pineapple' is seedy. Blood orange (Pigmented orange, Sanguine orange) has few seeds with deep burgundy interior. It's available from California. Chinese navel oranges are grown in California. The Shamouti (Jaffa) is a popular eating sweet orange which is easy to peel and has no navel. It is grown in the western U.S. (LOGAN 1996).

3. Season, bloom to harvest: 8 to 16 months depending on variety and area of production.

4. Production in U.S.: In 1995-96, 11,723,000 tons from 808,250 bearing acres (USDA 1996d). In 1995-96, the top 4 states for Navel and Valenica reported as Florida (594,800 bearing acres), California (196,000), Arizona (9,600) and Texas (7,850). Temples were grown on 6,600 acres with 97,000 tons all from Florida (USDA 1996d). In 1992, Puerto Rico grown 8,279 acres (1992 CENSUS).

5. Other production regions: The top importers in 1994 and '95 were Australia and Mexico (LOGAN 1996). Brazil and Central America contribute significantly to the world's orange production.

6. Use: Fresh, canned and frozen juice, canned segments, marmalade. Mainly processed, some fresh.

7. Part(s) of plant consumed: Interior segments. Some rind is used in marmalade; dried rind from processing plants is used as livestock feed

8. Portion analyzed/sampled: Whole fruit and its processed commodities dried pulp, oil and juice

9. Classifications:
   a. Authors Class: See citrus fruits
   b. EPA Crop Group (Group & Subgroup): See citrus fruits (Orange is a representative crop)
   c. Codex Group: 001 (FC 0208) Citrus fruits
   d. EPA Crop Definition: See citrus fruits

10. References: GRIN, CODEX, LOGAN (picture), MAGNESS, RICHE, USDA 1996d, US EPA 1996a, US EPA 1995a, US EPA 1994, KALE, WILLIAMSON

11. Production Map: EPA Crop Production Regions 3, 6, 10, and 13.

12. Plant Codes:
    a. Bayer Code: CIDSI

429

1. Orange, Trifoliate

Rutaceae

Poncirus trifoliata (L.) Raf.

2. The genus *Poncirus* is closely related to citrus, but the small, thorny trees are deciduous. Trees are hardier than citrus, in mid-winter enduring temperatures to near 0 degrees F. Fruits are globose, up to $1^1/_2$ inches in diameter, and very acid and bitter. *Poncirus* will hybridize with citrus, and such hybrids have been made in efforts to increase hardiness in edible citrus fruits. *Poncirus* seedlings are used as rootstocks for citrus, particularly for Mandarin oranges. Hybrids of *Poncirus* with sweet orange, called citranges, are now widely used as rootstocks. *Poncirus* is grown as an ornamental dooryard tree in areas too cold for citrus, and occasional trees may be maintained by nurserymen as seed sources. Fruit appears valueless in the U.S., but may be used for flavoring or marmalade in other countries.

3. Season, bloom to maturity: 8 to 10 months.

4. Production in U.S.: No data. The commercial rootstock group variety is 'Rubidoux'; hybrid cultivated variety is 'Carrizo' and is very widespread. The ornamental variety is 'Flying Dragon'.

5. Other production regions: See citrus fruits.

6. Use: Flavoring and marmalade.

7. Part(s) of plant consumed: Fruit

8. Portion analyzed/sampled: Whole fruit

9. Classifications:
   a. Authors Class: Citrus fruit
   b. EPA Crop Group (Group & Subgroup): Miscellaneous
   c. Codex Group: No specific entry
   d. EPA Crop Definition: None

10. References: GRIN, MAGNESS, MARTIN 1987 (picture)

11. Production Map: EPA Crop Production Region, See citrus fruits.

12. Plant Codes:
    a. Bayer Code: PMITR

430

1. Orchardgrass

(Cocksfoot, Zacate ovillo, Knaulgras)

Poaceae (Gramineae)

Dactylis glomerata L.

2. Orchardgrass is a cool season grass native to Europe, but has been in cultivation in the U.S. for over 200 years. It is a long-lived perennial, distinctly bunch-type, with folded leaf blades. Since it does not produce stolons, it does not produce a dense sod and is especially suitable for planting in mixtures with legumes. It grows erect 2 to 3 feet tall. It is most extensively grown from southern New York to Virginia and westward to Kansas. Grows throughout the U.S. except Flor-

ida, Louisiana and Texas. It is more tolerant to shade than most grasses. It starts growth early in the spring and produces excellent pasturage throughout the summer. If cut early it makes excellent silage or hay. Following such cutting, abundant high-quality pasturage is produced under favorable moisture and fertility conditions. Numerous selected varieties are in the trade.

3. Season: Rapid regrowth permits 2-3 cuttings a year. Harvest late June to early July for seed.
4. Production in U.S.: No specific data, See Forage grass. In 1992, the seed crop was produced on 18,642 acres with 7,448 tons of seeds (1992 CENSUS). In 1992, seed crop was produced in 3 primary states: Oregon (18,334 acres), Virginia (144) and Tennessee (56) (1992 CENSUS). It yields 1 to 2 tons/acre of hay. Southern New York to Virginia and west to Kansas and West coast (MAGNESS and SCHREIBER 1995).
5. Other production regions: Europe.
6. Use: Pasture, cover crop, hay and silage. Also grown as an ornamental.
7. Part(s) of plant consumed: Foliage
8. Portion analyzed/sampled: Forage and hay
9. Classifications:
   a. Authors Class: See Forage grass
   b. EPA Crop Group (Group & Subgroup): See Forage grass
   c. Codex Group: See Forage grass
   d. EPA Crop Definition: None
10. References: GRIN, MAGNESS, RICHE, SCHREIBER, BALL (picture), MOSLER
11. Production Map: EPA Crop Production Regions 1, 2, 4, 5, 9, 11 and 12.
12. Plant Codes:
    a. Bayer Code: DACGL

431

1. Otaheite gooseberry
(Grosella, Gooseberry tree, Cerezo, Star gooseberry, Indian gooseberry, Otaheite gooseberry-tree)
Euphorbiaceae
Phyllanthus acidus (L.) Skeels (syn: *P. distichus* (L.) Muell. – Arg.)

2. This is a small tropical tree, to 20 feet, native to South Asia but now naturalized in the West Indies and south Florida. Leaves are compound with numerous pinnate leaflets. Fruits are borne in loose clusters, which hang from the tree trunk and main branches. Individual fruits are up to $^3/_4$ inch in diameter, near globose and angled. Fruits are generally too tart for fresh eating but are esteemed for jellies, preserves and pastries. They ripen in midsummer. There are no commercial plantings; but trees are in home gardens, and fruit is harvested from naturalized trees.
3. Season, ripens in midsummer. Bloom to harvest: 90-100 days.
4. Production in U.S.: No data. Florida home gardens. Hawaii and south Florida (MORTON 1987).
5. Other production regions: West Indies, southern Mexico.
6. Use: Cooked whole and served as relish. Juice is used in drinks. Also, see above.
7. Part(s) of plant consumed: Fruit pulp
8. Portion analyzed/sampled: Whole fruit
9. Classifications:

a. Authors Class: Tropical and subtropical fruits – edible peel
b. EPA Crop Group (Group & Subgroup): Miscellaneous
c. Codex Group: 005 (FI 0306) Assorted tropical and subtropical fruits – edible peel
d. EPA Crop Definition: None
10. References: GRIN, CODEX, MAGNESS, MARTIN 1987, MORTON (picture), IVES
11. Production Map: EPA Crop Production Region 13.
12. Plant Codes:
    a. Bayer Code: PYLAC

432

1. Palm heart
(Palm cabbage, Millionaire's salad, Palmillo, Chou-palmeta, Hearts of palm, Palmeto, Spiny club palm, Palmito, Palmita)
Arecaceae (Palmae)
Various spp.

1. Pejibaye (Pay-e-by-e)
(Peach palm, Parepou, Pupunha, Pewa, Peachnut, Pijibaye, Palmito, Palmita)
Bactris gasipaes Kunth

1. Palmyra palm
(Borassus palm, African fan palm, Toddy palm, Wine palm, Tala palm, Doub palm, Deleb palm)
Borassus aethiopum C. Mart.
B. flabellifera L.

1. Coconut
(Coconut palm, Coco, Nariyal)
Cocos nucifera L.

1. Cabbage palm
(Assai palm)
Euterpe oleracea C. Mart.

1. Wine palm
Raphia spp.

1. Royal palm
(Cabbage palm, Caribbee royal palm)
Roystonea oleracea (Jacq.) O.F. Cook (syn: *Oreodoxa oleraceae* (Jacq.) Mart.)

1. Salac palm
(Salak)
Salacca zalacca (Gaertn.) Voss (syn: *Salacca edulis* Reinw.)

1. Saw palmetto
(Scrub palmetto)
Serenoa repens (W. Bartram) Small

1. Cabbage palmetto
(Florida cabbage palm, Swamp cabbage tree, Thatch palm, Palmetto palm, Cabbage palm, Blue palmetto)
Sabal palmetto (Walter) Schult. & Schult. f.

2. The Florida cabbage palm reaches a height of up to 90 feet at maturity, most are 10-20 feet tall, but it is the 8-10 feet palms that are cut for the heart or cabbage. The heart is the growing bud at the apex within and protected by the stem. The palm offers three potential products: Palm heart, the tender leaves above the palm growing point that are wrapped within the tender sheath of an older leaf; Palm stem, the tender stem portion immediately below the growing point; and Palm leaf, the tender leaves above the palm heart that are not wrapped within the tender sheath of the older leaf; of which only the heart is currently commercialized. "Sago" also comes from

trunks of certain palms. Sago is the powdered starch prepared from the pith and used as thickening agent in foods.

3. Season, seeding to harvest: All palms are propagated by seed. Time to harvest is 5 to 10 years. The tree is cut down to harvest the heart. Under orchard cultivation with about 2000 plants per acre, the pejibaye is a multi-shoot palm with each clump yielding a harvestable stem to 4 feet from the base up to the lowest petiole every 9 to 15 months and can weigh about 20 pounds. Each stem can produce 0.5 pound of edible palm heart and 1 pound of edible stem (KAWATE 1997b).

4. Production in U.S.: The cabbage palm grows wild in Florida in great abundance but authorization to cut it is required (STEPHENS 1988). Pejibaye in Hawaii (CLEMENT 1991). Palmetto is native from North Carolina to Florida (NEAL 1965), especially south Florida (MARTIN 1979).

5. Other production regions: Brazil (CLEMENT 1991). Costa Rica.

6. Use: The heart is cut into thin slices and used fresh like cole slaw or in a tossed salad. "Hearts of palm" salad includes sliced palm heart, and dates or guava paste into a tossed salad. Palm heart is canned in Brazil. They can be pickled or canned.

7. Part(s) of plant consumed: The center core or cabbage of the plant is edible and creamy white. The edible part consists of the base of the young developing fronds or tender leaves from the meristem before they expand and green. Also the young unfolding leaves or fronds of *Arenga* (Sugar palms) are used as a green vegetable. The fruit (drupe) of the Saw-palmetto is edible, which is dried and used as a pharmaceutical extract (TANNER). Pejibaye normally refers to the fruit, and palmito refers to the stem in Costa Rica.

8. Portion analyzed/sampled: Palm heart; stem and leaf (see above for definitions). Fruit (Saw palmetto)

9. Classifications:
   a. Authors Class: Stalk and stem vegetables
   b. EPA Crop Group (Group & Subgroup): Miscellaneous
   c. Codex Group: 017 (VS 0626) Stalk and stem vegetables, for *Cocos, Borassus, Raphia, Salocca* and others
   d. EPA Crop Definition: None

10. References: GRIN, CLEMENT, CODEX, MARTIN 1979, MARTIN 1983, NEAL, STEPHENS 1988 (picture), TANNER, KAWATE 1997b, RUBATZKY, DUKE(a)

11. Production Map: EPA Crop Production Regions 2, 3 and 13.

12. Plant Codes:
   a. Bayer Code: BASFL (*B. flabellifer*), CCNNU (*C. nucifera*), SABPA (*Sabal palmetto*), SERRE (*Serenoa repens*)

433

1. Palm oil
(Palm nut, Oil palm)
Arecaceae (Palmae)
Elaeis spp.
1. African oil palm
E. guineensis Jacq.
1. American oil palm
(Corozo, Elaeocarpus)
E. oleifera (Kunth) Cortes

2. The source of palm oil is the fleshy fruit of tropical, spineless palm trees, native to western Africa but extensively cultivated in other tropical countries. The fruits are borne in large bunches, each of which may carry up to 20 pounds of fruit. The fruits are oval, 1 to 2 inches long and an inch or more in diameter. The flesh contains 30 to 70 percent of non-drying oil. The seed also contains oil, the palm kernel oil of commerce. Various methods are used to extract the oil from the pulp, including pressing or centrifuging or macerating the pulp and boiling in water, the oil floating and skimmed off. The oil is used in margarine and vegetable shortening as well as in soap manufacture industry. The kernel oil is extracted by crushing and pressing, or with solvents. The oil is similar to coconut oil, and is used mainly in margarine and soap. Several other species of palm, some indigenous to South America, also yield pulp and kernel oils that are generally similar to African palm oil, but production of these oils is much less extensive. About 273,288 tons of palm oil were imported into the U.S. in 1993 about half of which was kernel oil.

3. Season, seeding to harvest: 3 to 4 years; Flowering to harvest: 5 to 6 months.

4. Production in U.S.: No data.

5. Other production regions: Widely distributed in tropics including Central and South America.

6. Use: Oil from pulp and kernel is used as food and for cooking.

7. Part(s) of plant consumed: Oil from pulp (mesocarp) and seed kernel ("red oil") of fruit

8. Portion analyzed/sampled: Fruit and processed commodity, oil

9. Classifications:
   a. Authors Class: Oilseed
   b. EPA Crop Group (Group & Subgroup): Miscellaneous
   c. Codex Group: 023 (SO 0696) Oilseed; 067 (OC 0696 and OC 1240) Vegetable crude oil; 068 (OR 0696 and OR 1240) Vegetable edible oils
   d. EPA Crop Definition: None

10. References: GRIN, CODEX, MAGNESS, MARTIN 1984, ROBBELEN 1989 (picture), USDA 1994a

11. Production Map: No entry.

12. Plant Codes:
   a. Bayer Code: EAIGU (*E. guineensis*), no specific entry for *E. oleifera*

434

1. Pangolagrass
(Woolly fingergrass, Pangola digitgrass, Pasto pangola)
Poaceae (Gramineae)
Digitaria eriantha Steud. (syn: *Digitaria eriantha* ssp. *pentzii* (Stent) Kok; *D. decumbens* Stent, *D. valida* Stent, *D. pentzii* Stent)

2. Pangolagrass is a warm season sod-forming grass introduced from South Africa in 1935. It resembles crabgrass in appearance but is a long-lived, creeping to decumbent, perennial with strong-growing stolons. It is used mainly for pasture in central and southern Florida and other Caribbean and Gulf Coast areas. It may winter kill in northern Florida. Seedstalks may grow to 4 feet in height, but few seeds are developed. Propagation is by planting rooted stems and runners. The grass is nutritious and palatable as pasture and also useful as hay or silage.

3. Season, warm season pasture.

4. Production in U.S.: In 1992, Puerto Rico with 85,146 acres (1992 CENSUS). Southern Florida and Gulf Coast areas.
5. Other production regions: Other Caribbean areas.
6. Use: Pasture, hay, silage.
7. Part(s) of plant consumed: Foliage
8. Portion analyzed/sampled: Forage and hay
9. Classifications:
   a. Authors Class: See Forage grass
   b. EPA Crop Group (Group & Subgroup): See Forage grass
   c. Codex Group: See Forage grass
   d. EPA Crop Definition: None
10. References: GRIN, MAGNESS, RICHE
11. Production Map: EPA Crop Production Regions 2, 3 and 13.
12. Plant Codes:
   a. Bayer Code: DIGER

435

1. Panicgrass
Poaceae (Gramineae)
Panicum spp.; *Urochloa* spp.
1. Vine mesquitegrass
(Zacate guia, Obtuse panicgrass)
P. obtusum Kunth
1. Paragrass
(Malojillo, Pasto para, Parana, Mauritus, Signalgrass)
U. mutica (Forssk.) T.O. Nguyen (syn: *P. purpurascens* Raddi)
1. Torpedograss
(Victoriagrass, Cheno, Creeping panicum, Couch panicum)
P. repens L.
1. Switchgrass
(Blackwell switchgrass)
P. virgatum L.
1. Browntop millet
U. ramosa (L.) R.D. Webster (syn: *P. ramosum* L.)

2. Vine mesquitegrass is a sod-forming grass, native from Missouri west to Colorado and south into Mexico, useful both for grazing and erosion control. Leaves are 4 to 6 inches long and $^1/_2$ inch wide. Plants form stolons which may grow to 15 feet in a season. Stands can be established by seeding or sod pieces. Vine mesquitegrass is a warm season perennial native to the U.S. and is a member of the grass tribe *Paniceae*. It is an erect bunchgrass from 1 to $2^1/_2$ feet tall that spreads by stolons and rhizomes, and has a panicle seedhead (3 to 14 cm long and 5 to 13 mm wide). It is adapted to sandy soils in moist areas on banks of irrigation ditches and rivers and lowland pastures. It has fair to good forage quality for livestock grazing and withstands heavy grazing and produces fair quality hay. It reproduces from rhizomes, stolons and seeds, and is seeded at rates of 6 to 11 kg/ha. There are 313,315 seeds/kg. Warm season, spring growth, flowers in May and matures in October. EPA Crop Production Regions 6, 8 and 9.

Paragrass is a straggling tropical perennial important in Puerto Rico. It forms wide-creeping stolons which root at the nodes. It grows to 3 feet in height and is valuable both for pasturage and hay. Propagation is by planting pieces of the stolons or sod pieces.

Torpedograss is native along the Gulf Coast from Florida to Texas. It is a sod-forming grass with creeping rhizomes and upright stems reaching 2 to 3 feet in height. Leaves are flat or folded. It thrives on coarse sands and wet muck soils and grows so aggressively that it may become a serious weed. Palatability is good but it is less nutritious than many grasses. It is used to a limited extent for pasture and erosion control along the Gulf Coast. It's reportedly no longer planted in the U.S. but cultivated in Asia.

Switchgrass is an important native perennial sod-forming grass which occurs over much of the U.S. It is especially valuable in the central and southern Plains. It grows up to 5 feet tall, with short rhizomes. Leaves are 6 to 18 inches long, $^1/_4$ to $^1/_2$ inch wide. While it grows on most soil types, it thrives best on low, moist areas of good fertility. Growth starts rather late but continues throughout the summer if moisture is available. Forage yield is high, with quality fair while green; but poor as standing winter feed. Propagation is by seeds. Also, switchgrass grows in the eastern two-thirds of the U.S. and eastern Canada.

Browntop millet is an annual warm season grass to 3 feet tall. It is adapted to Oklahoma and Texas and used as pasture or hay. Also used for game bird pasturage in the southeastern states; see Millet, Proso.
3. Season: Warm season pasture.
4. Production in U.S.: In 1992, Paragrass grown on 13,462 acres acres in Puero Rico (1992 CENSUS). Central and southern plains, and south to Mexico.
5. Other production regions: Canada and Mexico.
6. Use: Erosion control, pasture, hay and rangeland grass.
7. Part(s) of plant consumed: Foliage
8. Portion analyzed/sampled: Forage and hay
9. Classifications:
   a. Authors Class: See Forage grass
   b. EPA Crop Group (Group & Subgroup): See Forage grass
   c. Codex Group: See Forage grass
   d. EPA Crop Definition: None
10. References: GRIN, BARNES, HOLZWORTH, MAGNESS, RICHE, STUBBENDIECK (picture), SKERMAN (picture), ALDERSON, BALL
11. Production Map: EPA Crop Production Regions 2, 3, 4, 5, 6, 8, 9 and 13.
12. Plant Codes:
   a. Bayer Code: PANOB (*P. obtusum*), PANRE (*P. repens*), PANRA (*P. ramosum*), PANVI (*P. virgatum*), UROSS (*U.* spp.)

436

1. Panicgrass, Introduced
Poaceae (Gramineae)
Panicum spp.
1. Blue panicgrass
P. antidotale Retz.
1. Kleingrass
(Keriagrass, Small panicum, Small buffalograss)
P. coloratum L. var. *coloratum*
1. Guineagrass
(Zania, Pasto guinea, Gramalote)
P. maximum Jacq.

2. Three species of *Panicum*, one introduced from India and two from Africa, are of some importance in parts of the U.S. Blue panicgrass, *P. antidotale* Retz., native to India, was introduced via Australia in 1912. It has a coarse, vigorous root system and is sodforming. Forage yields are high on fer-

tile, well drained soils. It is important in parts of the south-western U.S., both for dry-land and irrigated pastures. It is not winter hardy in northern locations. Kleingrass, *P. colora-tum* L., is a complex of grasses which includes both bunch and sod-forming types. It was introduced from Africa. Adapted to moist, heavy soil, it is used for pasture, hay and silage, mainly in south Texas and Oklahoma. Plants have slender stems up to 4 feet, with abundant dark green leaves. It tillers throughout the season, has excellent regrowth and adapted to sandy loam and clay soils. Guineagrass, *P. maxi-mum* Jacq., is a warm season, spreading grass from Africa used to a limited extent for pastures and silage in Hawaii, Florida and other southern areas. It is a tall, coarse grower with high nutritive value when leafy and green. It is not cold hardy. Propagation is by sod pieces.

3. Season: warm season pasture.
4. Production in U.S.: In 1992, Guineagrass grown on 38,313 acres in Puerto Rico (1992 CENSUS). Also see above.
5. Other production regions: Africa, India and Australia.
6. Use: Pasture, hay, erosion control and silage.
7. Part(s) of plant consumed: Foliage
8. Portion analyzed/sampled: Forage and hay
9. Classifications:
   a. Authors Class: See Forage grass
   b. EPA Crop Group (Group & Subgroup): See Forage grass
   c. Codex Group: See Forage grass
   d. EPA Crop Definition: None
10. References: GRIN, MAGNESS, RICHE, STUBBEND-IECK(a) (picture), BARNES, SKERMAN (picture)
11. Production Map: EPA Crop Production Regions 2, 3, 4, 8 and 13 for guineagrass; 6, 7 and 8 for kleingrass; 9 for blue panicgrass.
12. Plant Codes:
   a. Bayer Code: PANAN (*P. antidotale*), PANCO (*P. colora-tum*), PANMA (*P. maximum*), PANSS (*P. spp.*)

437

1. Papaya
(Papaw, Pawpay, Melon pawpaw, Tree melon, Papaya lechosa, Fruta bomba, Papayo, Lechosa, Mamao, Melon tree)
Caricaceae
Carica papaya L.

2. The papaya, a native of tropical America, grows best in warm tropical or subtropical climate. Not tolerant to frost or wind. Plants are grown from seed and the first fruits will mature within a year after planting. Sometimes plants are grown from cuttings or graftings. The tropical tree is non-woody with a large stem hollow between the nodes. The trunk may persist for several years, but best fruit production is on plants not over four years of age. Leaves are very large and lobed. Fruits are borne along the new growth of the largely unbranched trunk or main stem. New flowers and fruits are produced continuously, so fruits in all stages of development will be on the plant at one time. Fruits are very large, melon-like, averaging 3 pounds with some kinds up to 15 or 20 pounds. Fruit rind is thin, smooth and rather tender. The pulp is smooth textured and mild to strongly flavored. At the center is a cavity, along the walls of which numerous seeds the size of small peas are borne. Papaya fruit are an economically important source of the enzyme papain which is widely used as a meat tenderizer. Papain is obtained by drying and col-

lecting the latex contained in developing fruit. 'Kaoho', 'Sunrise' and 'Waimanalo' are the three main varieties grown.

3. Season: Seeding to first harvest: 6 to 11 months. Normally, flowers all year and fruit matures 60 days from bloom. Fruit available year round.
4. Production in U.S.: Domestic production of 2,435 acres in 1995. Hawaii reported production of 31,000 tons from 2,200 acres in 1994. Florida had 300 acres in 1995. Guam, U.S. Virgin Islands and Puerto Rico reported production of 5,220 lbs, 11,825 lbs, and 25,383 cwt, respectively in 1992.
5. Other production regions: Mexico, Dominican Republic, Jamaica, Brazil, Thailand.
6. Use: Fresh eating, juice and jam; unripe fruits may be used in cooking. Production of enzyme, papain, from fruit.
7. Part(s) of plant consumed: Pulp only
8. Portion analyzed/sampled: Whole fruit
9. Classifications:
   a. Authors Class: Tropical and subtropical fruits – inedible peel
   b. EPA Crop Group (Group & Subgroup): Miscellaneous
   c. Codex Group: 006 (FI 0350 and FI 4139) Assorted tropical and subtropical fruits – inedible peel
   d. EPA Crop Definition: Proposing – Papaya = Star apple, Black sapote, Canistel, Mamey sapote, Mango, Sapodilla
10. References: GRIN, ANON(a), ARRIOLA, CODEX, KNIGHT(a), LOGAN, MAGNESS, MARTIN 1987, RICHE, US EPA 1994, CRANE 1996, IVES, DESAI, USDA 1996f, MALO(c)
11. Production Map: EPA Crop Production Region 13.
12. Plant Codes:
   a. Bayer Code: CIAPA

438

1. Papaya, Mountain
(Chamburo, Toronchi)
1. Babaco
Caricaceae
Carica x *heilbornii* var. *pentagona* (Heilborn) V.M. Badillo (syn: *C. pentagona* Heilborn)
(Babaco)
C. pubescens (A.DC.) Solms
(Chamburo)

2. The small plant is papaya-like, slender, 3 to 10 feet tall and occasionally branched. The 5-angled fruits reach a foot in length. In Ecuador, the fruits are locally eaten only after cooking. The fruits can weigh about 4 pounds and are normally seedless. Total yields are better than those of a good papaya plantation. Other mountain papaya cultivars are 'Tor-onchi' and 'Lemon Creme'.
3. Season, planting: Propagated by cuttings. Reported to bear commerically a year or two after planting.
4. Production in U.S.: Limited. Grown in California (MELNI-COE 1996d).
5. Other production regions: Commonly cultivated in mountain valleys of Ecuador (MORTON 1987). Also New Zealand, Australia, Israel, Italy and Guernsey (NATIONAL RESEARCH COUNCIL 1989).
6. Use: Fruit is eaten cooked or fresh. It is easily prepared because it can be eaten skin and all.
7. Part(s) of plant consumed: Fruit

8. Portion analyzed/sampled: Whole fruit
9. Classifications:
   a. Authors Class: Tropical and subtropical fruits – edible peel
   b. EPA Crop Group (Group & Subgroup): Miscellaneous
   c. Codex Group: No specific entry
   d. EPA Crop Definition: None
10. References: GRIN, MELNICOE 1996d, MORTON, NATIONAL RESEARCH COUNCIL 1989 (picture), FACCIOLA
11. Production Map: EPA Crop Production Region 10.
12. Plant Codes:
    a. Bayer Code: No specific entry

439

1. Parsley
(American parsley, Perejil, Persil, Italian parsley)
Apiaceae (Umbelliferae)

Petroselinum crispum (Mill.) Nyman ex A.W. Hill (syn: *P. hortense* Hoffm.; *P. sativum* Hoffm.)

P. crispum var. *neapolitanum* Danert **(Italian parsley)**

2. Parsley is a leafy plant, the leaves being used mainly for garnishing meats, fish and other dishes. The finely chopped leaves are also used as flavoring. Leaf shape is generally triangular, and varies from three-leaflet to greatly curled and cut. The two main foliage types are plain (Italian) and curled leaf. For market, the plant is grown as an annual. Seed may be sown in beds for field transplanting, or direct in the field. In harvesting, the outer leaves may be removed for fresh market as they attain suitable size, and the plant continues to produce. Normally, the fresh market parsley is cut one to two times, while dehydrated acreage is cut three to five times during the production cycle on a 30 day schedule. The dehydrated parsley consists of dry leaves only (parsley flakes). Dried parsley is generally rehydrated prior to consumption. U.S. imported about 227 metric tons of dried parsley in 1992. Canada contributed about 3 metric tons and Mexico about 29 metric tons in 1992. Fresh market crop is often hand harvested and tied in bunches or it can be mechanically harvested and handled as loose leaves. The processed crop is usually mechanically harvested.
3. Season, seeding to first harvest: 70 to 90 days.
4. Production in U.S.: 5,100 acres in U.S. reported by EPA in 1994, with approximately half in California. The majority of the acreage in California is for dehydration. Minor acreage in Hawaii. Washington State grows about 70 acres of parsley as a seed crop which represents about 50 percent of the U.S. acreage.
5. Other production regions: 294 acres in Canada. Mexico.
6. Use: Mainly for garnishing and flavoring of foods. Fresh (herb/leafy green); Dried (herb).
7. Part(s) of plant consumed: Fresh (Leaves and stems); May be dehydrated (dried leaves only)
8. Portion analyzed/sampled: Leaves and stems (fresh); Leaves only (dried). Parsley: Fresh parsley is included in Crop Group 04: Leafy Vegetables under 40 CFR 180.41. Dried parsley is included in Crop Subgroup 19A: Herbs under 40 CFR 180.41
9. Classifications:
   a. Authors Class: Leafy vegetables (except *Brassica* vegetables); Herbs and spices
   b. EPA Crop Group (Group & Subgroup): Leafy vegetables (except *Brassica* vegetables) group (4A); Herbs and spices group (19A)
   c. Codex Group: 027 (HH 0740) Herbs
   d. EPA Crop Definition: None, but proposing Parsley = Cilantro
10. References: GRIN, CHAMBERS 1996, CODEX, KAWATE 1996c, KAYS, KURTZ 1995, MAGNESS, MELNICOE 1996a, PENZEY, ROTHWELL 1996a, SCHREIBER, USDA 1993a, US EPA 1995a, US EPA 1994, RUBATZKY
11. Production Map: EPA Crop Production Regions 2, 3, 6 and 10, accounting for 88 percent of the U.S. production.
12. Plant Codes:
    a. Bayer Code: PARCR

440

1. Parsley, Turnip-rooted
(Dutch parsley, Hamburg parsley, Heimischer, Turnip-parsley, Parsley root, Petrouska)
Apiaceae (Umbelliferae)

Petroselinum crispum (Mill.) Nyman ex A.W. Hill var. *tuberosum* (Benth.) Mart. Crov.

2. This is a type of parsley which forms an edible root. The culture is similar to that for carrot, and exposure of leaves and roots is similar. The edible root flesh is white, firm and celery-like in flavor. In shape and appearance, the root resembles a slender carrot.
3. Season, seeding to harvest: 3 to 4 months.
4. Production in U.S.: No data, minor. Available year round from California and Texas (LOGAN 1996). Grown in New Jersey (SCHNEIDER 1985).
5. Other production regions: Europe.
6. Use: As cooked vegetable, like carrot or parsnip. Leaves used as garnish.
7. Part(s) of plant consumed: Roots and leaves
8. Portion analyzed/sampled: Roots and leaves
9. Classifications:
   a. Authors Class: Root and tuber vegetables; Leaves of root and tuber vegetables
   b. EPA Crop Group (Group & Subgroup): Root and tuber vegetables (1A and 1B)
   c. Codex Group: 016 (VR 0587) Root and tuber vegetables
   d. EPA Crop Definition: None
10. References: GRIN, CODEX, LOGAN, MAGNESS, SCHNEIDER 1985, STEPHENS 1988 (picture), US EPA 1995a, YAMAGUCHI 1983, RUBATZKY
11. Production Map: EPA Crop Production Regions 2, 6 and 10.
12. Plant Codes:
    a. Bayer Code: PARCT

441

1. Parsnip
(Panais, Chirivias)
Apiaceae (Umbelliferae)

Pastinaca sativa L. ssp. *sativa*

2. Parsnips, except for seed production, are grown as an annual, for the edible roots. Seed is sown in place. The plant forms a rosette of compound leaves, with ovate or oblong, notched leaflets. The edible root is large, up to 3 inches diameter at the top, tapering from 6 to 15 inches long, depending on

variety. Culture and exposure of the edible root are similar to those of carrot, except the growing season is longer. Foliage is normally removed before root is harvested.

3. Season, seeding to harvest: 3½ to 4 or more months.
4. Production in U.S.: No data. Limited. Grown in Pennsylvania, Illinois, California and New York (WARE 1980). In 1993, 50 percent of the parsnip seed in the U.S. was produced in Washington on 70 to 100 acres (SCHREIBER 1995).
5. Other production regions: In 1995, Canada reported 1,013 acres (ROTHWELL 1996a).
6. Use: As cooked vegetable.
7. Part(s) of plant consumed: Root only
8. Portion analyzed/sampled: Root with top discarded
9. Classifications:
   a. Authors Class: Root and tuber vegetables
   b. EPA Crop Group (Group & Subgroup): Root and tuber vegetables (1A and 1B); Leaves of root and tuber vegetables (2)
   c. Codex Group: 016 (VR 0588) Root and tuber vegetables
   d. EPA Crop Definition: None
10. References: GRIN, CODEX, MAGNESS, ROTHWELL 1996a, SCHREIBER, STEPHENS 1988, US EPA 1995a, US EPA 1996a, WARE, YAMAGUCHI 1983, SWIADER
11. Production Map: EPA Crop Production Regions 1, 5 and 10.
12. Plant Codes:
    a. Bayer Code: PAVSA

442

1. Partridgeberry
(Squawberry, Twinberry, Squaw vine, Two-eyed berry, Running box)
Rubiaceae
Mitchella repens L.

2. Evergreen perennial herbs, native to North America and east Asia, with trailing rooting stems to 1 foot long. The fruits are twin berries and normally red with 8 seeds and to ³/₈ inches in diameter. Leaves are ³/₄ inches long and dark green and glossy above. Grown from Nova Scotia to Ontario and Minnesota, south to Florida and east Texas.
3. Season, seeding to havest: No data.
4. Production in U.S.: No data. USDA Hardiness Zone 4, native to eastern and central U.S. (WHEALY 1993).
5. Other production regions: Canada.
6. Use: Eaten fresh.
7. Part(s) of plant consumed: Fruit
8. Portion analyzed/sampled: Whole fruit
9. Classifications:
   a. Authors Class: Berries
   b. EPA Crop Group (Group & Subgroup): Miscellaneous
   c. Codex Group: No specific entry
   d. EPA Crop Definition: None
10. References: GRIN, FACCIOLA, WHEALY, USDA NRCS
11. Production Map: EPA Crop Production Regions 1, 5 and 7.
12. Plant Codes:
    a. Bayer Code: No specific entry

443

1. Paspalum
Poaceae (Graminae)

Paspalum spp.

1. Knotgrass
(Creeping perennial jointgrass, Mercergrass, Grama de agua, Salaillo, Seashore paspalum, Saltwater couch, Gramabobo, Water coughgrass, Couch paspalum, Gingergrass, Eternity grass)
P. distichum L. (*P. paspaloides = paspalodes* (Michx.) Scribn.)

1. Brunswick grass
P. nicroae Parodi

1. Brownseed paspalum
(Pasto gallito, Zacaton)
P. plicatulum Michx.

2. There are over 350 *Paspalum* warm season perennial grass spp. that are mostly native to tropical and subtropical areas of South America. Some *Paspalums* are native to southern U.S.

Brunswick grass is adapted to moist sandy soils. It is an erect sod-forming grass growing to 40 cm tall and spreads by rhizomes. In Georgia, it begins growth in early spring through November with most forage growth April to September. It is shade tolerant and regenerates well after grazing.

Brownseed paspalum is adapted to marshy areas such as in Florida. It grows to 1.2 meters tall and is used as a hay or pasture crop. It spreads by seeds or short rhizomes and flowers in early summer. There are 780,000 to 1,000,000 seeds/kg. It can be grazed October to April in Georgia and Texas with a 30 day regeneration period. Seeds mature 18-22 days after flowering. Seed yields are 50-150 kg/ha.

Knotgrass is a warm season sod forming grass. It is an erect perennial to 2 feet tall that spreads by rhizomes and stolons. It is adapted to marshy areas and withstands heavy grazing.

3. Season: Warm season perennial.
4. Production in U.S.: No specific data, See Forage grass. In U.S. southern coastal areas and Hawaii (SKERMAN 1989).
5. Other production regions: Tropical and subtropical South and Central America (GOULD 1975).
6. Use: Pasture, hay, erosion control.
7. Part(s) of plant consumed: Foliage (leaves and stems)
8. Portion analyzed/sampled: Forage and hay, See Forage grass
9. Classifications:
   a. Authors Class: See Forage grass
   b. EPA Crop Group (Group & Subgroup): See Forage grass
   c. Codex Group: See Forage grass
   d. EPA Crop Definition: None
10. References: GRIN, SKERMAN (picture), BAYER, GOULD
11. Production Map: EPA Crop Production Regions 2, 3, 4, 6, 10, 12 and 13.
12. Plant Codes:
    a. Bayer Code: PASDA (*P. distichum*), PASPL (*P. plicatulum*)

444

1. Passionfruit
(Granadilla, Maracuya, Maypop, Parcha, Parcha cimarrona, Passionflower, Ceibey, Percha, Lilikoi, Passion fruit)
Passifloraceae
Passiflora edulis Sims
P. edulis f. *flavicarpa*. O. Deg.
   (Yellow passionfruit)
P. edulis f. *edulis* Sims
   (Purple passionfruit, Purple granadilla)

2. The plant is native to tropical America. Passionfruit plants grow in cool tropical or subtropical climate with well distributed rainfall. They are not tolerant to frost or high winds nor will they set fruit if temperatures are too high. The plant is propagated by seed, cuttings or grafting. Plants produce fruits within a year of grafting. Fruits are produced on evergreen vines. Vines are trained on trellises similar to grapes and pruned annually either to stubs or canes. Flowers are produced on new growth. Leaves are large and three-lobed. Fruits are generally globose in shape, purple or yellow in color, about 2 inches diameter. They have a tough outer skin which encloses a mass of seeds, each imbedded in a juicy pulp, the edible portion. It is an excellent source of provitamin A and ascorbic acid. Key cultivar in Hawaii is 'Flavicarp'.
3. Season: Bloom to ripe fruit: 2 to 3 months, but flowers and mature fruits may be on a plant at the same time.
4. Production in U.S.: Hawaii, Florida, Puerto Rico and California. In 1992, Florida reported 37 acres, Hawaii 15 and California 12. U.S. harvested 256,972 pounds with about half from California (1992 CENSUS).
5. Other production regions: South Africa, Australia, Kenya, New Zealand.
6. Use: Beverages, desserts, jellies and some eating of fresh fruit.
7. Part(s) of plant consumed: Internal pulp only
8. Portion analyzed/sampled: Whole fruit
9. Classifications:
   a. Authors Class: Tropical and subtropical fruits – inedible peel
   b. EPA Crop Group (Group & Subgroup): Miscellaneous
   c. Codex Group: 006 (FI 0351) Assorted and subtropical fruits – inedible peel
   d. EPA Crop Definition: Proposing – Guava = Passionfruit
10. References: GRIN, CODEX, KNIGHT, LOGAN, MAGNESS, MARTIN 1987, MOTT, RICHE, US EPA 1987, CHAVAN(a)
11. Production Map: EPA Crop Production Region 13.
12. Plant Codes:
   a. Bayer Code: PAQED (*P. edulis*), PAQEF (*P. edulis f. flavicarpa*)

445

1. Pawpaw
(Hooiser banana, Poorman's banana, Custard banana, Banana tree, Michigan banana, Nebraska banana, Pawpaw tree, Indiana banana)
Annonaceae
Asimina triloba (L.) Dunal

1. Smallflower pawpaw
(Dwarf pawpaw)
A. parviflora (Michx.) Dunal
2. The pawpaw is the North American native plant of the family Annonaceae. It is the only plant of this family which grows in the temperate zone. In 1916, its growing zone was deliniated as "from the Gulf of Mexico to the Atlantic, west to Oklahoma and as far north as New York and Michigan" (ANON(c)). The pawpaw is a small pyramidal, deciduous tree, with long drooping leaves. It usually grows 10 to 25 feet in height, however it occasionally grows to 40 feet. It is mainly propagated via seedlings in deep containers that are

transplanted into the field when plants are 2-3 feet tall. Sprouts commonly shoot up from horizontal roots. Pawpaw flowers are borne singly, but each flower contains separate ovaries, with the potential for fruit clusters. The fruits are 2 to 6 inches long, half as wide, generally oval in shape, dark brown when mature. Up to several large seeds are surrounded by a soft pulp that is strongly aromatic, esteemed by some, but objectionable to others. 'Davis', 'Overleese' and 'Sunflower' are the most widely grown varieties. Tree lives 30-80 years. Fruits look like elongated potatoes.
3. Season: Fruit need about 150 days to ripen.
4. Production in U.S.: Very limited, only one commerical planting in 1991. Normally collected from the wild and from home gardens, experimental orchards are growing in Kentucky and Maryland (CALLAWAY 1992).
5. Other production regions: Japan and Italy (CALLAWAY 1992).
6. Use: Primarily consumed as fresh fruit, however it can be processed into desserts. The pawpaw has been shown to have high nutritional quality, especially compared to typical temperate fruits (PETERSON). Certain vegetative parts (bark) of pawpaw plant may contain compounds that exhibit highly effective pesticidal and anticancer properties (ALKOFAHI, RUPPRECHT).
7. Part(s) of plant consumed: Fruit pulp
8. Portion analyzed/sampled: Whole fruit
9. Classifications:
   a. Authors Class: Tropical and subtropical fruits – inedible peel
   b. EPA Crop Group (Group & Subgroup): Miscellaneous
   c. Codex Group: No specific entry
   d. EPA Crop Definition: None currently established. There is potential to classify pawpaw as banana, but proposing – Sugar apple = Pawpaw
10. References: GRIN, ALKOFAHI, ANON(c), CALLAWAY, MAGNESS, REICH, RUPPRECHT, PETERSON (picture), USDA NRCS, WHEALY
11. Production Map: EPA Crop Production Regions 2, 4 and 5.
12. Plant Codes:
   a. Bayer Code: ASITR (*A. triloba*)

446

1. Pea, Edible-podded
(Sugar pea, Chinese pea, Pois mange tout, Snap pea, Ming pea, Podded pea, Snow pea, China pea, Chicharo, Shi hia wandou, Saya-endo, Sugar snap pea)
Fabaceae (Leguminosae)
Pisum sativum var. *macrocarpon* Ser. (syn: *P. sativum* ssp. *sativum* var. *saccharatum*)
2. Varieties of edible-podded peas are similar to the garden pea in plant and growth characteristics, except that the pods are more fleshy and less fibrous than in garden peas. They are harvested while the seeds are quite immature, and the pods are cooked as with snap beans. Such varieties are little grown in the U.S. but are grown frequently in Europe and the Orient. In exposure of edible parts they are similar to snap beans. Snap peas are round-shaped pods as compared to snow peas (Sugar pea, Wan tou, Sic Kap woon dou, Hawlaan tau, Saya-endo, china pea) which are flat pods.
3. Season, seeding to harvest: About 60 to 75 days. Edible pods, 3-5 inches long, are hand harvested.

4. Production in U.S.: In 1992, 5,233 acres (1992 CENSUS). In 1992, California (4,145 acres), Washington (839) and Hawaii (6) (1992 CENSUS). In 1995, California produced 5,300 tons, Texas (250 tons), and New Jersey and Washington each with 50 tons (LOGAN 1996).
5. Other production regions: Europe and the Orient.
6. Use: Eaten raw in salads or cooked.
7. Part(s) of plant consumed: Succulent seed with pod
8. Portion analyzed/sampled: Succulent seed with pod
9. Classifications:
   a. Authors Class: Legume vegetables (succulent or dried)
   b. EPA Crop Group (Group & Subgroup): Legume vegetables (Succulent or Dried) (6A)
   c. Codex Group: 014 (VP 0538) Legume vegetables for podded pea (young pods)
   d. EPA Crop Definition: Peas (succulent) = Edible-podded pea
10. References: GRIN, CODEX, KAYS, LOGAN, MAGNESS, MYERS, RICHE, STEPHENS 1988 (picture), US EPA 1995a
11. Production Map: EPA Crop Production Regions 10 and 12.
12. Plant Codes:
    a. Bayer Code: No specific entry

only such as Austrian winter pea. Field pea vines, cut sample anytime after pods begin to form, at approximatley 25 percent dry matter. For field pea hay, succulent plant cut from full bloom through pod formation. Hay should be field-dried to a moisture content of 10 to 20 percent. For pea, field, silage, use field pea vine residue data for field pea silage with correction for dry matter
9. Classifications:
   a. Authors Class: Legume vegetables (succulent or dried); Foliage of legume vegetables
   b. EPA Crop Group (Group & Subgroup): Legume vegetables (succulent or dried) (6C) for dried peas; Foliage of legume vegetables (7) – Representative commodity for (6c and7)
   c. Codex Group: 015 (VD 0561) Pulses; 050 (AL 0072) Legume animal feeds for pea hay or pea fodder (dry)
   d. EPA Crop Definition: See garden pea
10. References: GRIN, CODEX, MAGNESS, USDA 1996c, US EPA 1996a, US EPA 1995a, BARNES, BALL, BAILEY 1976
11. Production Map: EPA Crop Production Region 11 for Austrian winter pea. See garden pea for dried edible pea.
12. Plant Codes:
    a. Bayer Code: PIBSA

447

1. Pea, Field

(Austrian winter pea, Pois gris, Bisalto, Grey pea, Winter pea)
Fabaceae (Leguminosae)

Pisum sativum var. *arvense* (L.) Poir. (syn: *P. arvense* L.)

2. Similar to garden pea in plant and growth characteristics. Field peas do not include the canning field pea cultivars used for human food. Field peas do include cultivars grown for livestock feed only, such as Austrian winter pea. Field pea vines are sampled anytime after pods begin to form at approximately 25 percent dry matter. Field pea hay is the succulent plant cut from full bloom thru pod formation. Hay should be field dried to a moisture content of 10 to 20 percent. In the past, it was grown primarily as a soil improvement crop but now there is renewed interest in its forage quality. It is also used as a winter annual in Palouse region of the Pacific Northwest and is grown as a seed crop for export. Field peas are seeded in September to October in the South at 30 to 40 pounds per acre or with small grain at 20 to 30 pounds per acre. They are not well adapted for pasture since they can be tramped easily and are best used as a silage crop. The food dried pea is under garden pea, which see.
3. Season, bloom to harvest: See garden pea. Field peas are winter annuals in the South, summer annuals in the North, and grow 2 to 4 feet tall.
4. Production in U.S.: In 1995, Austrian winter peas were planted on 10,900 acres and produced 5,800 tons (USDA 1996c). In 1995, Idaho with 10,000 acres and Oregon with 900 acres (USDA 1996c). For dried peas, see garden pea.
5. Other production regions: Native to the Old World (BAILEY 1976).
6. Use: Feed only. Hay, silage and also cover crop and soil improvement.
7. Part(s) of plant consumed: Seed, vines, hay as feed only
8. Portion analyzed/sampled: Seed, vines, hay. For pea, field, it does not include the canning field pea cultivars used for human food. Includes cultivars grown for livestock feeding

---

448

1. Pea, Garden

(English pea, Guisante, Pois potager, Dry pea, Wan dou, Green pea, Chicharo, Arvejas, Common pea, Wrinkled pea, Little marvel)
Fabaceae (Leguminosae)

Pisum sativum L. var. *sativum*

2. Garden peas are a major vegetable crop in the U.S. and in all temperate zone countries. Field peas, which see, grown for feed are also important. There are many varieties, varying in height of plant, size, tenderness and sweetness of the peas, season of maturity and other properties. All garden peas grown commercially are on vining type plants, generally 2 to 3 feet in length. Seeds are enclosed in pods, from which they are removed by machine for commercial processing or by hand for home use. Edible podded kinds are known, which see. For commercial processing, vines are cut and passed through threshers or "viners" which remove the peas from vines and pods. Garden peas and dry peas are included together to harmonize with the EPA sampling guide. Succulent pea seed without pod for uses on succulent shelled peas, including English pea, Green pea, Garden pea and Guisante. Mature dried seed for uses on dried shelled peas, including dry pea (field peas).
3. Season, bloom to harvest: 20 to 30 or more days, depending on variety and climate. Seeding to harvest about 56-75 days. Planting to harvest for dry pea production is about 90 days.
4. Production in U.S.: About 500,000 tons commercially for green peas for processing for 1995 (USDA 1996b) (USDA 1996c). 187,450 tons dried peas for 1995 (USDA 1996c). In 1995, green peas for processing planted to 320,300 acres with the top 5 states as Minnesota (92,900 acres), Wisconsin (68,200), Washington (59,200), Oregon (36,600) and New York (18,400) (USDA 1996b). In 1995, dried edible peas planted to 166,000 acres with Idaho (70,000 acres) and Washington (96,000) (USDA 1996c). In 1995, the wrinkled seed peas production was 52,400 tons from Washington and

Idaho with almost equal production (USDA 1996c). The green pea wrinkled cultivar, also called little marvel types, are used for canning or freezing.

5. Other production regions: In 1995, Canada reported 48,727 acres of green pea, 1,154,972 acres of dry peas and 44,002 acres of peas (ROTHWELL 1996a).
6. Use: Processed canned and frozen, canned soup, dried split peas for soup. U.S. per capita consumption in 1994 is 3.7 pounds (PUTNAM 1996).
7. Part(s) of plant consumed: Seeds only for food. Vines and pods from threshers for livestock feed are not significant feed items
8. Portion analyzed/sampled: Seeds (succulent) or Seeds (dried), all shelled (without pod)
9. Classifications:
   a. Authors Class: Legume vegetables (succulent or dried)
   b. EPA Crop Group (Group & Subgroup): Legume vegetables (succulent or dried) (6B and 6C). (Representative crop)
   c. Codex Group: 014 (VP 0528) Legume vegetables for succulent immature seed (young pods); 014 (VP 0529) Legume vegetables for succulent seeds shelled; 050 (AL 0528) Legume animal feed for pea vines (green); 015 (VD 0561) Pulses for dried pea; 050 (AL 0072) Legume animal feeds for pea hay or pea fodder (dry)
   d. EPA Crop Definition: Peas (succulent) = Garden Pea; Peas (dry) = Dry pea (field pea), Chickpea, Pigeon pea and Lentils
10. References: GRIN, CODEX, KAYS, MAGNESS, ROTHWELL 1996a, USDA 1996b, USDA 1996c, US EPA 1996a, US EPA 1995a, IVES, SWIADER, SPITTSTOESSER, RUBATZKY, PUTNAM 1996, SCHREIBER
11. Production Map: EPA Crop Production Regions, Succulent 1, 2, 5, 10, 11 and 12; Dried 11.
12. Plant Codes:
   a. Bayer Code: PIBST

449

1. Pea, Pigeon

(Congo pea, No-eye pea, Red gram, Arhur, Grandul, Gandules, Dhal, Toor, Gunds pea, Porto Rico pea, Urhur, Gandul, Guandu, Pois-d'angole, Gungo pea)

Fabaceae (Leguminosae)

Cajanus cajan (L.) Millsp. (syn: *C. indicus* Spreng.)

2. These pea-like legumes are erect, short-lived perennial plants 3 to 10 feet high, much propagated as annuals in the tropics for the edible seeds and pods. They are important food plants in the West Indies and tropical areas. In exposure of edible parts they are comparable to garden peas. The ripe seeds are a source of flour and used split (dhal) in soups or with rice. Tender leaves are rarely used as a potherb.
3. Season, seeding to harvest: Pod set in about 3 to 4$\frac{1}{2}$ months; mature seeds 5 to 6 months with late types requiring almost a year.
4. Production in U.S.: In 1992, about 5,000 acres with 2,336 tons reported in Puerto Rico (1992 CENSUS).
5. Other production regions: India.
6. Use: Immature seeds cooked like green peas; Dried seeds used in soups and curries; Unripe pods used in curries; Split dried seeds are available as Toovar, Toor or Arhar dol. The canned fresh peas are known as Gandules in Latin America. Also can be used as forage crop or green manure crop in the south.

7. Part(s) of plant consumed: Mainly the mature dried seeds. Sometimes the immature seeds and pods are used as vegetables
8. Portion analyzed/sampled: Seed (dried); Immature seed and pod
9. Classifications:
   a. Authors Class: Legume vegetables (succulent or dried)
   b. EPA Crop Group (Group & Subgroup): Legume vegetables (succulent or dried) (6A, 6B, 6C)
   c. Codex Group: 014 (VP 0537) Legume vegetables for green pods and/or young green seeds; 015 (VD 0537) Pulses for pigeon pea (dry)
   d. EPA Crop Definition: Peas = Pigeon peas
10. References: GRIN, CODEX, FACCIOLA, MAGNESS, US EPA 1995a, YAMAGUCHI 1983, MORTON(a), NENE, DUKE(a), LONGBRAKE
11. Production Map: EPA Crop Production Region 13.
12. Plant Codes:
   a. Bayer Code: CAJCA

450

1. Pea, Southern

(Cowpea, Black-eyed pea, Callivance, Cherry bean, Indian pea, Cornfield pea, Crowder pea, Pois a vache, Frijol de costa, Niebe, Caupi, Costeno, Rabizo)

Fabaceae (Leguminosae)

Vigna unguiculata (L.) Walp. ssp. *unguiculata* (syn: *V. sinensis* (L.) Savi ex Hassk.; *Dolichos sinensis* L.)

2. This so-called pea (*Vigna*) is more closely related to the beans (*Phaseolus*) than to the peas (*Pisum*). It is an important food and stock-feed crop in the Southern states. Several slightly differing types are grown, as blackeye, brown eye, cream and cream crowder, differing in the flower color and color markings on the seeds. The plant needs warm weather, and is injured by any frost. It is bushy, or procumbent, twining but not climbing. Pods are 3 to 12 inches long and slender. Seeds are small, $\frac{1}{6}$-$\frac{1}{4}$ inch in length. For food, they are harvested either at the green-shell state, while pods are still green, or as dry-shells, when ripe. Harvesting is usually by mowing and threshing in a viner, as with green peas. The pods may also be hand picked with multiple pickings. For stock feed, the whole plant is harvested as hay, or pastured. In this book and by general usage, we are using the terms Southern pea for green shell and green pod; Cowpea and Black-eyed pea for dry shell; and feed use as Cowpea. Black-eyed peas are used in Hoppin' John.
3. Season, bloom to harvest: Green-shell stage, 15 to 20 days. Dry-shell, 30 or more days. For Southern pea, seeding to harvest is 65 to 85 days, depending on variety.
4. Production in U.S.: 17,676 tons shelled dry; no data on green shell (1992 CENSUS). Green-shell southern pea (green cowpea) 32,329 reported acreage with the top 10 states as Texas (5,486 acres), Oklahoma (3,927), Georgia (3,638), Tennessee (3,565), Florida (2,731), Mississippi (2,429), Alabama (2,410), Missouri (1,929), Arkansas (1,914) and South Carolina (1,424) for 25,888 acres or 80 percent of the U.S. total (1992 CENSUS). Dry-shell cowpea (dry southern pea or black-eyed pea) 36,757 reported acreage with the top 5 states as Texas (18,696 acres), Florida (6,563), Oklahoma (3,907), Missouri (3,525) and Arkansas (974) for 33,665 acres or 92 percent of the U.S. total (1992 CENSUS). Black-eyed peas

harvested as dry edible beans were planted to 55,600 acres in 1995 with 44,500 acres from California and 11,100 from Texas accounting for 56,050 tons of beans (USDA 1996c).

5. Other production regions: Mexico, Africa.
6. Use: Commercially, green-shell canned or frozen; dry-shell marketed dry. Normally for green and dry shell, the threshings are left in the field. Cowpeas are used as forage, green manure, silage and pasture crop.
7. Part(s) of plant consumed: Seed only for food; whole plant for feed which is a significant feed item for cowpeas. Immature pods and seed may be used as snapbeans
8. Portion analyzed/sampled: Southern pea: Seed (succulent); Black-eyed pea: Seed (dry); Cowpea: Seed, forage (approximately 30 percent dry matter) and hay (approximately 80 to 90 percent dry matter). For cowpea forage, cut sample at 6 inch to prebloom stage, at approximately 30 percent dry matter. For cowpea hay, cut when pods are one-half to fully mature. Hay should be field-dried to a moisture content of 10 to 20 percent
9. Classifications:
   a. Authors Class: Legume vegetables (succulent or dried); Foliage of legume vegetables
   b. EPA Crop Group (Group & Subgroup): Legume vegetables (succulent or dried) (6B and 6C); Cowpea = Foliage of legume vegetables (7)
   c. Codex Group: 015 (VD 0527) Pulses (dry); 014 (VP 0527) Legume vegetables (immature pod)
   d. EPA Crop Definition: Beans = *Vigna* spp.
10. References: GRIN, CODEX, LOGAN (picture), LORENZ, MAGNESS, RICHE, USDA 1996C, US EPA 1996A, US EPA 1995a, REHM, IVES, DUKE(h), GRANBERRY, DAVIS(a)
11. Production Map: EPA Crop Production Regions for Green shell: 2, 3, 4, 5 and 6; Dry shell: 3, 4, 5, 6 and 10.
12. Plant Codes:
   a. Bayer Code: VIGSI

451

1. Pea, Winged
(Bin dow, Asparagus pea, Guisantillo rojo, Pois asperge, Lotier rouge)
Fabaceae (Leguminosae)
Lotus tetragonolobus L. (syn: *Tetragonolobus purpureus* Moench)

2. The plant is an annual trailer, with broad, ovate bean-like leaves. Pods are 2 to 3 inches long, somewhat rectangular in cross section, fleshy when immature. Winged pea is grown for the edible pods, and for the seeds, which are roasted as a substitute for coffee. Exposure is similar to that of snap or field beans. There are no data on production in U.S., but it is very slight.
3. Season, propagation: Seed.
4. Production in U.S.: No data.
5. Other production regions: Cultivated in Mediterranean region (FACCIOLA 1990).
6. Use: Young pods used as snap beans; seeds used as coffee bean substitute.
7. Part(s) of plant consumed: Mainly young pod; seed
8. Portion analyzed/sampled: Young pods
9. Classifications:
   a. Authors Class: Legume vegetables (succulent or dried)

b. EPA Crop Group (Group & Subgroup): Miscellaneous
c. Codex Group: 014 (VP 0543) Legume vegetables for young pods
d. EPA Crop Definition: None
10. References: GRIN, CODEX, FACCIOLA, KAYS, MAGNESS
11. Production Map: No entry
12. Plant Codes:
   a. Bayer Code: TTGPU

452

1. Peach
(Pecher, Duraznero, Duraznos, Melocotonero, Peche)
Rosaceae
Prunus persica (L.) Batsch var. *persica*

1. Nectarine
(Brugnon, Duraznero nectarin, Nectarinas)
P. persica var. *nucipersica* (Suckow) C.K. Schneid. (syn: *P.p.* var. *nectarina* (Aiton) Maxim.)

2. Trees of the two fruits are indistinguishable and are relatively small, usually held to under 15 feet by pruning. Peaches and nectarines are relatively large fruits with large, deeply ridged stones. Fruit 2 to $3^1/_2$ inches diameter. Peach fruits are pubescent throughout the growing season, and are usually brushed by machine prior to marketing to remove most of the pubescence. Nectarines have a smooth, plum-like peel. Nectarines have apparently originated from peaches by mutation. Both peaches and nectarines may be freestone, pit relatively free of the flesh, or clingstone, pit adheres to flesh. In some areas, late frost can significantly damage the crop. Winter chilling is required to break dormancy.
3. Season: Bloom to harvest range is 75-150 days with most crops being 100-120 days.
4. Production in U.S.: In 1995 there were 172,710 acres of peaches and 32,400 acres of nectarines reported. Domestic production was 1,150,700 and 176,000 fresh tons for peaches and nectarines, respectively. Major peach production states include California (32,500 acres freestone/28,100 acres clingstone), South Carolina (23,000 acres), Georgia (21,000 acres), Texas (12,000 acres) and New Jersey (10,000 acres). California was the only state to report nectarine acreage in 1995.
5. Other production regions: Significant imports of fruit come from Chile. In 1995, Canada grew 341 acres of nectarines and 10,300 acres of peaches (ROTHWELL 1996a). Also grown in Italy, Spain, China and France.
6. Use: Fresh fruit, and processed (i.e. canned, dried, marmalade, baby food).
7. Part(s) of plant consumed: Fruit for fresh market utilize entire fruit minus the pit. Processed fruit uses mainly the fruit pulp. The peel (skin) is often included with dried fruit
8. Portion analyzed/sampled: Whole fruit with pit and stem removed and discarded
9. Classifications:
   a. Authors Class: Stone fruits
   b. EPA Crop Group (Group & Subgroup): Representative commodity of Crop Group 12: Stone Fruits Group
   c. Codex Group: Stone fruits 003 (FS 0245) Nectarine; (FS 0247) Peach
   d. EPA Crop Definition: Peach = Peach & Nectarine

10. References: GRIN, ANNON(c), CODEX, LOGAN, MAG-NESS, ROTHWELL 1996a, SCHREIBER, US EPA 1995, US EPA 1994, US EPA 1996a, IVES, REHM, JOSHI, CROCKER(a)
11. Production Map:
Peach: EPA Crop Production Regions 1, 2, 4, 5, 6 and 10.
Nectarine: EPA Crop Production Regions 1, 2, 10 and 11.
12. Plant Codes:
a. Bayer Code: Peach: PRNPS; Nectarine: PRNPN

453

1. Peanut

(Groundnut, Goober, Mani, Peanut Oil, Grassnut, Earthnut, Pindar, Monkeynut, Cacahuate, Manillanut, Common peanut)
Fabaceae (Leguminosae)
Arachis hypogaea L.
2. The peanut plant is procumbent or semi-erect, with rather small compound-pinnate, smooth leaves. Three market types are in commercial production: (1) runner and (2) Virginia types with a longer growing season of 120-160 days, usually 2 seeds per pod; and (3) Spanish and Valencia types are bunch types which are erect with a short growing season of 90-150 days. The Spanish type has 2-3 seeds per pod and the Valencia type has 3 to 4 seeds. Valenica type is grown primarily in New Mexico. The runner type accounts for 78 percent of the total U.S. production, while the Virgina type accounts for 16 percent and the remainder for Spanish and Valancia. The Virginia type accounts for most of the roasted, salted and inshell peanuts while the runners are mostly used for peanut butter. The seeds are enclosed in a rather fibrous pod. After the flowers are pollinated, a short, thick stem at the flower base, termed gynophore or peg, grows downward and penetrates into the soil, so the fruiting body (pod) develops entirely underground. Seeds, the edible part, are 1 to 4 per pod, $1/4$ to $3/4$ inch long and vary from near globose to elongated. In harvesting, the entire plant with adhering seed pods is mechanically lifted from the soil with a digger-shaker-windrower, dried in windrows, then threshed to remove the seeds. Among edible vegetables oils, peanut oil is exceeded in world production by soybean, rapeseed (canola) and sunflower. World production of peanut oil, 1992-94, averaged 3,930,300 tons. In extracting the oil in the U.S., the cleaned nuts are passed through hullers or shellers to separate the kernels. The kernels, which contain 48 to 56 percent of oil, are then crushed, heated and pressed hot in hydraulic presses. The oil is used in the manufacture of margarines and shortenings, and as a salad and cooking oil. The press cake is used for cattle food. Shelled peanuts crushed for oil annually in the U.S. averaged near 668 million pounds, 1991-93. Oil production averaged near 285 million pounds for the 3 years. Peanuts are native to South America (Brazil). Normally rotated with grasses.
3. Season, seeding to harvest: About 3 to 5 months.
4. Production in U.S.: About 1,690,00 acres grown, almost 1,696,208 tons of nuts in 1993. Georgia, Texas, Alabama, North Carolina, Oklahoma, Virginia and Florida. (1994 Agricultural Statistics).
5. Other production regions: South America, Mexico, Africa and Asia (USDA 1994a).
6. Use: Seeds consumed roasted, boiled, steamed or raw and eaten directly or in confections and baked goods; ground for

peanut butter; crushed for oil. Tops of plants, hulls and cake from oil extraction used for feed. Also tender shoots and leaves are used as vegetables.
7. Part(s) of plant consumed: Nutmeat and hay
8. Portion analyzed/sampled: Nutmeat and peanut hay (dried vines and leaves), and processed commodities meal and refined oil. Peanut hay consists of the dried vines and leaves left after the mechanical harvesting of peanuts from vines that have sun-dried to a moisture content of 10 to 20 percent. Label restrictions against feeding may be allowed; e.g. *Do not feed green immature growing plants to livestock,* or *Do not harvest for livestock feed*
9. Classifications:
a. Authors Class: Oilseed: Legume vegetables (succulent or dried); Foliage of legume vegetables
b. EPA Crop Group (Group & Subgroup): Miscellaneous
c. Codex Group: 023 (SO 0697 and SO 0703) Oilseed; 050 (AL 0697 and AL 1270) legume animal feed; 067 (OC 0697) Crude vegetable oil; 068 (OR 0697) Refined vegetable oil
d. EPA Crop Definition: None
10. References: GRIN, CODEX, MAGNESS, ROBBELEN 1989 (picture), USDA 1994a, US EPA 1995b, US EPA 1994, IVES, YAMAGUCHI 1983, BAILEY 1976, SMITH 1995, BOOTE, PATTEE, SULLIVAN
11. Production Map: EPA Crop Production Regions 2, 3, 6 and 8 which account for 100 percent.
12. Plant Codes:
a. Bayer Code: ARHHY

454

1. Pear

(Common pear, European pear, Pera, Poirier, Peral, Birnbaum, Poire)
Rosaceae
Pyrus communis L.

1. Oriental pear

(Asian pear, Sand pear, Nashi, Mizunaski, Salad pear, Pear apple, Water pear, Apple pear, Japanese pear, Chinese pear, Nihonnaski, Shalea pear)
P. pyrifolia (Burm. f.) Nakai
P. pyrifolia var. *culta* (Makino) Nakai
(Asian pear, Chinese pear, Japanese pear, Apple pear, Nashi)
P. ussuriensis Maxim.
(Asian pear, Harbin pear)
2. Pears are temperate zone fruits that are grown on a wide variety of soils. They are similar to apples in many respects. The tree is of medium size, up to 30 to 40 feet in height, but usually held to under 20 feet by pruning. Propagated by budding or grafting onto rootstocks. They differ from apples in having "grit" cells in the flesh of the fruit. In general, pears are pyriform in shape, tapering toward the stem, although some varieties are nearly round. Normally fruit size varies in varieties from less than 2 inches in diameter up to 3 inches. Fruit surface in some varieties is russeted but in others is free of russet and covered with a thin layer of wax. Trees tend to be more upright than apples. Pears can be classified by maturity date; summer, autumn and winter. They are the only temperate tree fruit that cannot be left on the tree to ripen. Major production in U.S. is of the common or European

pear. Major commerical varieties/types of this pear include: 'Anjou', 'Bartlett', 'Bosc', 'Comice' and 'Seckel'. Most popular Oriental pears are '20th Century' ('Nijisseiki'), 'Kosui', 'Kikusui', 'Hosui', 'Shinseiki', 'Shinko' and 'Niitaki'. There are a few cultivars of hybrid origin (KADAM).

3. Season: Bloom to harvest in 100 to 170 days.

4. Production in U.S.: In 1995, 70,550 domestic acres produced about 948,300 fresh tons. Major production states include Washington (24,200 acres), California (23,900 acres), Oregon (17,000 acres), New York (2,500 acres), Michigan (1,000 acres) and Pennsylvania (1,000 acres). Pears are available year-round (LOGAN 1996).

5. Other production regions: Significant imports from Chile and Argentina. Other growing regions of pears include Australia, Canada, Japan, New Zealand, South Korea, South Africa and Europe. In 1995, Canada grew 5,419 acres (ROTHWELL 1996a).

6. Use: Pears are used primarily as a fresh fruit. Also processed as canned, dried, baby food and beverages.

7. Part(s) of plant consumed: Inner flesh only in canned or baby food preparations. Peel may be consumed when eaten fresh and is retained on dried fruit

8. Portion analyzed/sampled: Whole fruit

9. Classifications:
   a. Authors Class: Pome fruits
   b. EPA Crop Group (Group & Subgroup): Crop Group 11: Pome Fruits Group (Representative crop)
   c. Codex Group: Pome fruits 002 (FP 0230) Pear
   d. EPA Crop Definition: Pear = Oriental Pear (US EPA letter, 25 Nov. 86, to Professor Markle with Pears = Asian or Oriental Pear

10. References: GRIN, CODEX, LOGAN, MAGNESS, SCHREIBER, US EPA 1995, US EPA 1994, ROTHWELL 1996a, KADAM(b), IVES, USDA 1996f, CROCKER(g)

11. Production Map: EPA Crop Production Regions 1, 10 and 11.

12. Plant Codes:
    a. Bayer Code: PYUCO (*P. communis*), PYUPC (*P. pyrifolia*)

455

1. Pecan
(Black hickory, Pecana, Nogal americano, Nogal pecanero)
Juglandaceae
Carya illinoinensis (Wangenh.) K.Koch (syn: *Carya illinoensis* (Wangenh.) K. Koch; *C. oliviformis* (Michx.) Nutt.; *C. pecan* (Marshall) Engl. & Graebn.)

2. The pecan is a large tree, up to 100 feet in height, and with trunk diameter up to 6 feet. It is native in the lower Mississippi Valley and westward through Texas, into California and in northern Mexico. Leaves are large and compound, with a dozen or more long-oval, near glabrous leaflets. The fruits are generally oval, up to 2¹/₂ inches long, and fairly smooth. The outer husk is fleshy early, becoming fibrous and splitting open at maturity. The shells are relatively thin, hard and woody. The kernel separates rather readily. Improved varieties are widely cultivated. In addition, large quantities are harvested from native trees.

3. Season, bloom to harvest: 5 to 6 months.

4. Production in U.S.: About 182,500 tons, in shell (1993). In order of production, Georgia, Texas, New Mexico, Alabama, Oklahoma and Louisiana which account for about 88 percent

of total production (1993). The 1992 CENSUS reported 473,426 acres in U.S. In 1994, California reported 1,907 acres (MELNICOE 1996e).

5. Other production regions: Mexico.

6. Use: Direct eating, confections, ice cream, cookery.

7. Part(s) of plant consumed: Internal kernels only

8. Portion analyzed/sampled: Nutmeat

9. Classifications:
   a. Authors Class: Tree nuts
   b. EPA Crop Group (Group & Subgroup): Tree nuts (Representative crop)
   c. Codex Group: 022 (TN 0672) Tree nuts
   d. EPA Crop Definition: None

10. References: GRIN, CODEX, FACCIOLA, MAGNESS, RICHE, USDA 1994a, US EPA 1995a, US EPA 1995b, US EPA 1994, WOODROOF(a) (picture), IVES, REHM, MELNICOE 1996e, CROCKER(b), PONCAVAGE, RIOTTE

11. Production Map: EPA Crop Production Regions 2, 3, 4, 5, 6, 8, 9 and 10, with 87 percent in Regions 2, 4, 6 and 8.

12. Plant Codes:
    a. Bayer Code: CYAIL

456

1. Pennyroyal
(European pennyroyal)
Lamiaceae (Labiatae)
Mentha pulegium L. (syn: *Pulegium vulgare* Mill.)

2. Pennyroyal is very similar in plant habit and culture to the other mints, which see. The plant is prostrate and much branched, with small, round-oval leaves about an inch long. The leaves, either dried or fresh, are sometimes used for flavoring. The flavor of pennyroyal is more pungent and less agreeable than peppermint or spearmint. Oils of pennyroyal are obtained by distillation of the entire plant, as with the mints. Pennyroyal is grown more in Europe than in the U.S. No data are available on U.S. production, as it is included with other mints.

3. Season, planting: Propagated by dividing root runners in spring or early September and by summer stem cuttings.

4. Production in U.S.: An acre will produce about ¹/₂ ton of dried leaves or 10-30 pounds of oil. Naturalized in California (FOSTER 1993).

5. Other production regions: Europe.

6. Use: Flavoring in beverages, ice cream, candy, baked goods and tea.

7. Part(s) of plant consumed: Leaves; oil from tops

8. Portion analyzed/sampled: Leaves (fresh and dried)

9. Classifications:
   a. Authors Class: Herbs and spices
   b. EPA Crop Group (Group & Subgroup): Herbs and spices (19A)
   c. Codex Group: 027 (HH 0738) Herbs; 057 (DH 0738) Dried herbs
   d. EPA Crop Definition: None

10. References: GRIN, CODEX, FOSTER (picture), MAGNESS, US EPA 1995a

11. Production Map: EPA Crop Production Region 10.

12. Plant Codes:
    a. Bayer Code: MENPU

457

1. Pennyroyal, American
(Squaw mint, American falsepennyroyal)
Lamiaceae (Labiatae)
Hedeoma pulegioides (L.) Pers.

1. New Mexican pennyroyal
(Toronjil, Poleo)
H. drumondii Benth.

2. Annual plant to 15 inches tall with leaves $3/4$ inch wide and about 1 inch long. American pennyroyal leaves yield about 2 percent oil.
3. Season, seeding: Seeds sown in the spring or fall and spaced 8 to 12 inches apart.
4. Production in U.S.: No data. Native to eastern North America. Primarily found in the southwestern U.S. (FOSTER 1993).
5. Other production regions: South America (FOSTER 1993).
6. Use: Used as flavoring in tea, ice cream, candy and baked goods.
7. Part(s) of plant consumed: Leaves; oil from tops
8. Portion analyzed/sampled: Leaves (fresh and dried)
9. Classifications:
   a. Authors Class: Herbs and spices
   b. EPA Crop Group (Group & Subgroup): Miscellaneous
   c. Codex Group: No specific entry
   d. EPA Crop Definition: None
10. References: GRIN, FACCIOLA, FOSTER
11. Production Map: No entry
12. Plant Codes:
    a. Bayer Code: HEDPU (*H. pulegioides*)

458

1. Pepino
(Melon pear, Pepino melon, Melon shrub, Pepino dulce, Apple of the Inca, Mellowfruit, Peruvian pepino, Manguena)
Solanaceae
Solanum muricatum Aiton

2. The plant is an erect, spineless, bushy herb, reaching 2 to 3 feet in height, with oblong to lanceolate, entire leaves. The edible fruit is ovoid to oval, up to 6 inches long, borne on a long stem. It is yellow or light green to white fleshed and generally seedless, tender, highly aromatic and juicy. It is cultivated in the tropics and subtropics, and can be grown in most parts of the U.S., but is little grown. In general culture it resembles eggplant, and the fruit is somewhat similar to eggplant, but melon-like. Immature fruits are eaten cooked while mature fruits are eaten as a dessert food.
3. Season, seeding to first harvest: 4 to 5 months. Commonly propagated by cutting.
4. Production in U.S.: 10 acres in Washington (SCHREIBER 1995). California and Hawaii (NATIONAL RESEARCH COUNCIL 1989).
5. Other production regions: Native to Ecuador and Peru (RUBATZKY 1997). New Zealand (LOGAN 1996).
6. Use: Mainly used as fresh dessert like a melon. Pepinos are often peeled because the edible skin of some varieties has a disagreeable flavor. The seeds are edible and located in a center cavity. The seeds can easily be removed as a honeydew melon. Can be used as a garnish for meats, fish or soup.

7. Part(s) of plant consumed: Large fleshy fruits only. Range in size and shape from plum to small football
8. Portion analyzed/sampled: Whole fruit
9. Classifications:
   a. Authors Class: Fruiting vegetables (except cucurbits)
   b. EPA Crop Group (Group & Subgroup): Fruiting vegetables (except cucurbits) (8)
   c. Codex Group: 012 (VO 0443) Fruiting vegetables (other than cucurbits)
   d. EPA Crop Definition: None
10. References: GRIN, CODEX, LOGAN (picture), MAGNESS, MYERS, NATIONAL RESEARCH COUNCIL 1989 (picture), SCHNEIDER 1985, SCHREIBER, US EPA 1995a, RUBATZKY, PROHENS
11. Production Map: EPA Crop Production Regions 10, 12 and 13.
12. Plant Codes:
    a. Bayer Code: SOLMU

459

1. Pepper
(Green peppercorn, Peppercorns, Pimienta, Common pepper)
1. Pepper, Black
1. Pepper, White
Piperaceae
Piper nigrum L.

2. Black pepper is the mature whole dried, unripe berry fruit of a perennial, tropical, woody evergreen vine. It may climb to 20 feet. The leaves are rather thick, broadly ovate-oblong or nearly round and evergreen. The fruit is a small, globular drupe or berry, red in color. Powdered black pepper is the ground, entire fruit. White pepper is prepared by removing the outer skin of the more mature fruit and grinding. It is less pungent than the black. Immature green peppercorns are used for flavoring pickles, vegetables and meats. Pepper is widely used in many forms of cookery. Desire for it was a strong motivation for searching for sea routes to India. Main commercial sources are now southern Asia and Brazil.
3. Season, transplant to harvest: About 3 years and continue to produce for 10 to 20 years.
4. Production in U.S.: None.
5. Other production regions: In 1991, Indonesia, Brazil, Malaysia, India and Vietnam are major producers of black and white pepper, with Indonesia the largest producer of white pepper (USDA 1993a). Most popular is Lampong pepper from Indonesia, and second is Tellicherry from India. U.S. imported 176 tons and 22,845 tons of black pepper from Mexico and Indonesia, respectively, in 1992 (USDA 1993a).
6. Use: Flavoring and garnishing in many forms of cookery.
7. Part(s) of plant consumed: Fruit (dried)
8. Portion analyzed/sampled: Fruit (dried)
9. Classifications:
   a. Authors Class: Herbs and spices
   b. EPA Crop Group (Group & Subgroup): Herbs and spices (19B) (Representative commodity)
   c. Codex Group: 028 (HS 0790) Spices
   d. EPA Crop Definition: None
10. References: GRIN, CODEX, FARRELL (picture), GARLAND (picture), LEWIS (picture), MAGNESS, PENSEY, USDA 1993a, US EPA 1995a, DUKE(a)
11. Production Map: No entry.

12. Plant Codes:
   a. Bayer Code: PIPNI

460

1. Pepper, Long
(Indian long pepper, Java long pepper, Jaborandi pepper)
Piperaceae

Piper longum L.
P. retrofractum Vahl. (syn: *P. officinarum* (Miq.) C.DC.)

2. These two species are similar in growth habit and culture to black pepper, which see. The spikes are gathered when they begin to color red or yellow, and dried rapidly. They may be used directly in pickling and are also ground and used in preserves or curries. They are grown mainly in South Asia, with some exported to Europe.
3. Season, transplant to harvest: See Pepper.
4. Production in U.S.: No data.
5. Other production regions: South Asia.
6. Use: mainly as a spice for pickling and curries.
7. Part(s) of plant consumed: Unripe fruit
8. Portion analyzed/sampled: Fruit (dried)
9. Classifications:
   a. Authors Class: Herbs and spices
   b. EPA Crop Group (Group & Subgroup): Miscellaneous
   c. Codex Group: 028 (HS 0791) Spices
   d. EPA Crop Definition: None
10. References: GRIN, CODEX, FACCIOLA, MAGNESS
11. Production Map: No entry.
12. Plant Codes:
   a. Bayer Code: No specific entry

461

1. Pepper leaf
(Acuyo, Boombo, La Lot, Makulan)
Piperaceae

Piper lolot C.DC
Piper sanctum (Miq.) Schltdl.
Piper auritum Kunth
Piper umbellatum L. (syn: *Pothomorphe umbellata* (L.) Miq.)

2. Perennial tropical, woody vines of the pepper family. Leaves are glossy and cordate. Pepper leaf is grown in Asia and Central and South America. The leaves, having the delicate flavor and aroma of sarsaparilla, are used for seasoning tamales; and flavoring soup, fish and beef.
3. Season: No data.
4. Production in U.S.: No data. Grown in southern U.S. near Vietnamese communities (KUEBEL 1988).
5. Other production regions: La Lot can be easily carried over into the greenhouse. Native of southeastern Asia (KUEBEL 1988).
6. Use: Leaves used as condiment for meat.
7. Part(s) of plant consumed: Leaves
8. Portion analyzed/sampled: Leaves (fresh)
9. Classifications:
   a. Authors Class: Herbs and spices
   b. EPA Crop Group (Group & Subgroup): Miscellaneous
   c. Codex Group: 013 (VL 0489) Leafy vegetables (including *Brassica* leafy vegetables)
   d. EPA Crop Definition: None
10. References: GRIN, BAILEY 1976, CODEX, FACCIOLA, KUEBEL, MARTIN 1979

11. Production Map: No entry
12. Plant Codes:
   a. Bayer Code: No specific entry

462

1. Peppers
(Bell pepper, Non-bell pepper, Aji, Pimiento, Piment, Pfeffer, Paprika, Pimento)
1. Pepper
Solanaceae

Capsicum spp.

2. Pepper plants are warm season annuals, ranging in various types from 1 to 6 feet in height, much branched. Leaves are elliptical to lanceolate. Plants are usually started in beds and moved to fields after 6 to 8 weeks. Fruits vary greatly in size, color, shape and degree of pungency or "hotness". See statements for main types below. Plants continue to flower and set fruits from midsummer until frost, so fruits in all stages of development are on plants at the same time. For regulatory food safety purposes, EPA divided peppers into bell and non-bell. CODEX divided peppers into Sweet and Chili. We have followed both themes with the addition of two super vernaculars: Pepper, Sweet or Mild type and Pepper, Hot or Pungent type, which see. In Puerto Rico, plants usually start to bloom 35 days after transplanting. From blooming to fruit set, there are about 10 to 12 days, and 20 days from fruit set to harvest. Transplanting to first harvest for non-bell peppers (usually Cubanelle type) is 60 to 70 days, and Bell peppers 70 days. The number of pickings for Non-bell peppers are about 6 to 8, and Bell peppers about 3 to 5. In Puerto Rico, peppers are called Pimientos. In Spanish, peppers are Ajis (MONTALVO-ZAPATA 1988). There are 4,500 seeds per ounce.
3. Season, plants setting to harvest: About 2 to 3 months.
4. Production in U.S.: See (Pepper, sweet) and (Pepper, hot) for production data.
5. Other production regions: Peppers are native to America. See (Pepper, sweet) and (Pepper, hot) for production data.
6. Use: Cooked, raw, dried, or pickled. U.S. per capita consumption in 1994 of bell peppers is 6.1 pounds (PUTNAM 1996).
7. Part(s) of plant consumed: Fruit
8. Portion analyzed/sampled: Whole fruit
9. Classifications:
   a. Authors Class: Fruiting vegetables (except cucurbits).
   b. EPA Crop Group (Group & Subgroup): Fruiting vegetables (except cucurbits) (8) (For all *Capsicum*) (Bell pepper and one cultivar of non-bell pepper are representative commodities)
   c. Codex Group: 012 (VO 0051) Fruiting vegetables other than cucurbits (For all peppers)
   d. EPA Crop Definition: Peppers = All varieties of peppers including pimentos and bell, hot, and sweet peppers. A regulatory crop definition is needed for bell peppers
10. References: GRIN, CODEX, ESHBAUGH, MAGNESS, MONTALVO-ZAPATA 1988, RICHE, US EPA 1996a, US EPA 1995a, US EPA 1994, BAYER, USDA NRCS, PUTNAM 1996, NAJ, SWIADER
11. Production Map: EPA Crop Production Regions: **Bell peppers:** 1, 2, 3, 5, 6, 8 and 10; **Non-bell peppers:** 2, 3, 5, 8, 9 and 10; **Pimentos:** 2 and 9.
12. Plant Codes:
   a. Bayer Code: Sweet and hot peppers, which see; CPSSS (*Capsicum* spp.)

463

1. Peppers, Hot

(Chile pepper, Aji picante, Chili pepper, Cayenne pepper, Peruvian pepper, Rocoto, Habanero pepper, Ajibravo, Chilli)

1. Non-bell pepper (hot type)

Solanaceae

Capsicum spp.

1. Chili pepper

(Tabasco pepper, Bird pepper, Red Chili, Cluster pepper, Spur pepper, Chilli, Chile, Bird chilli, Cone pepper)

C. frutescens L.

1. Peruvian pepper

(Aji amarillo, Uchu, Cusqueno, Piris)

C. baccatum L. var. *pendulum* (Willd.) Eshbaugh (syn: *C. pendulum* Willd.)

1. Rocoto pepper

(Chile manzana, Chamburoto, Apple chili, Locoto, Cuzco, Manzano, Horse chili)

C. pubescens Ruiz & Pav.

1. Habanero pepper

(Bohemian chili pepper, Bonnet pepper, Squash pepper, Rocotillo, Scotch bonnet, Piri-Piri pepper, Datil pepper, Yellow squash pepper)

C. chinense Jacq. (syn: *C. sinense* Jacq.)

1. Bird pepper

(Chilipiquin, Chiltepine, Bird's-eye pepper, American bird pepper, Chiltepe)

C. annuum var. *glabriusculum* (Dunal) Heiser & Pickersgill (syn: *C. annuum* var. *aviculare* auct.; *C. annuum* var. *minimum* (Mill.) Heiser)

1. Chili pepper

(Chile, Cayenne pepper, Cherry pepper, Jalapeno, Hungarian wax, Italian wax, Goat pepper)

C. annuum L. var. *annuum*

2. All the hot peppers are non-bell peppers which include Chile peppers but not all non-bell peppers are hot. In Spanish, the name chili refers to any kind of pepper, but in the U.S. the term is limited to mildly pungent varieties that are mainly dried and used for flavoring other foods and for pickling. Fruits are smooth skinned, up to 8 inches long and $1^{1}/_{2}$ inches broad at the top in largest-fruited kinds. Various types of relatively small-fruited hot peppers are grown for drying and grinding to produce the cayenne, and "red pepper" condiments. The very small Tabasco and Tonka types are crushed without drying in preparing very hot sauces. Some types are also pickled. Fruits are generally small, 1 inch or less in width. Length varies in different types from under 2 inches up to 6 inches or more. The fresh hot peppers include 'Anaheim', 'Fresno chile', 'Habanero', 'Jalapeno', 'Peperoni', 'Poblano', 'Serrano', 'Scotch bonnet', and 'Yellow chile' ('Yellowwax', 'Caribe', 'Banana pepper', 'Hungarian wax'). The dried hot peppers include 'Anaheim red chile', 'Ancho chile' ('Pasilla', 'Chile de arbol'. Cayenne fruits are slender, rounder and more wrinkled then Anaheim chiles. Jalapenos are small cylindrical shape and smooth. Tabasco peppers are small and tapered to a point. Hot peppers can be mechanically harvested since they do not bruise as easily as other non-bells. The pepper varieties with the hottest Scoville Heat Units are the Habanero and Scotch bonnet peppers with about 200,000 to 400,000 Scoville units. In comparison, the Jalapenos have 3,500 to 4,500 units with Tabasco 30,000 to 50,000 units (LOGAN 1996). In the *Capsicum baccatum* L. var. *pendulum* species, most are very hot but a few are mild. Only in this species and the common pepper are nonpungent cultivars known (NATIONAL RESEARCH COUNCIL 1989).

3. Season, plant setting to first harvest: About $2^{1}/_{2}$ to 3 months. Chili peppers usually are direct-seeded.

4. Production in U.S.: 50,851 acres reported for Hot peppers (1992 CENSUS). In 1992, Hot peppers reported for the top 8 states as New Mexico (29,698 acres), California (7,463), Texas (6,464), Arizona (1,859), Florida (1,049), New Jersey (612), Colorado (572) and Oklahoma (427) (1992 CENSUS).

5. Other production regions: Tropical and subtropical America (RUBATZKY 1997).

6. Use: Mainly dried, used for flavoring foods, some pickled, fresh, processed into sauces and canned. U.S. per capita consumption in 1994 for hot peppers is 6.2 pounds (PUTNAM 1996).

7. Part(s) of plant consumed: Fruit

8. Portion analyzed/sampled: See Peppers

9. Classifications:

   a. Authors Class: See Peppers

   b. EPA Crop Group (Group & Subgroup): See Peppers. (Non-bell pepper is representative crop)

   c. Codex Group: Chili peppers: 012 (VO 0444 and VO 4277) Fruiting vegetables other than cucurbits; Cherry pepper (VO 4273); Cluster pepper (VO 4281); Cone pepper (VO 4283)

   d. EPA Crop Definition: See Peppers

10. References: GRIN, CODEX, LOGAN (picture), MAGNESS, NATIONAL RESEARCH COUNCIL 1989, RICHE, BAYER, RUBATZKY, DEWITT, DEWITT(a), PUTNAM 1996

11. Production Map: See Peppers.

12. Plant Codes:

   a. Bayer Code: CPSFR (*C. frutescens*), CPSFC (*C. frutescens* – cherry pepper)

464

1. Peppers, Sweet

(Bell pepper, Long pepper, Pimento pepper, Non-bell pepper (sweet type), Chili dulce, Pimiento pepper)

1. Bell pepper

(Green pepper, Mango pepper, Garden pepper, Common pepper)

1. Non-bell pepper (sweet type)

(Cubanelle pepper, Bull's horn pepper, Hungarian long pepper, Yellow banana pepper, Italian sweet pepper, Long John pepper, Red cherry pepper, Sweet cherry pepper, Paprika pepper, Pimento pepper, Squash pepper, Cheese pepper, Cooking pepper, Aji dulce, Long pepper)

Solanaceae

Capsicum annuum L. var *annuum*

2. Bell peppers are the peppers generally marketed as "green" peppers, although many kinds become red when ripe. Fruits are large, thick walled and nonpungent. Largest varieties have fruits up to 4 inches long by 3 to $3^{1}/_{2}$ inches broad, but more generally a little smaller. Skin is smooth, but fruits are lobed and depressed at both stem and styler ends. Shape varies from conic to near rectangular. This group comprises

most of the peppers sold fresh in the U.S. Sweet peppers can be green, yellow, orange, red, brown or purple in color. The LaRouge Royale® is a sweet red pepper 10 inches long by 4 inches wide and grown in California (LOGAN).

Pimento peppers are generally similar to the "bell" type, but have distinctive shape and flavor. Fruit is up to 3 inches long and 2 to 2¹/₂ inches at the shoulder, generally heart shaped. This type is used for canning, harvested when deep red in color. For canning, the peel, seeds and placenta are removed, so only the interior fleshy walls are utilized. Pimentos are called Squash peppers and Cheese peppers.

Paprika peppers, especially the Hungarian type have some strains that are more pungent than others, but most are mild. The Spanish paprika has larger fruits. In the past, paprika peppers were grown commercially in South Carolina and Louisiana (STEPHENS 1988).

3. Season, field setting to first harvest: 2¹/₂ months for Bells; and 2¹/₂ to 3 months for Pimentos. Usually harvested every 7-10 days for fresh market.

4. Production in U.S.: Bells, in 1995, 66,300 acres with 658,200 tons (USDA 1996b). No separate data for Pimentos or other Non-bell sweet type in 1995 report. In 1995, the top 10 states for bells reported as Florida (21,500 acres), California (20,000), North Carolina (7,600), New Jersey (5,400), Texas (4,900), Michigan (2,700), Virginia (1,900), Louisana (1,100), Ohio (1,000) and Hawaii (200) (USDA 1996b). In 1992, Sweet peppers and Pimentos reported on 75,202 acres with the top 10 states for sweet peppers as California, Florida, Texas, New Jersey, North Carolina, Georgia, Michigan, Ohio, Virginia and New York. Pimentos reported on 1,236 acres in Tennessee, Alabama, Kentucky, Mississippi and Georgia (1992 CENSUS).

5. Other production regions: In 1995, Canada reported 4,722 acres of peppers and 62 acres of greenhouse peppers (ROTHWELL 1996a). Mexico is the largest Bell importer to the U.S. (LOGAN 1996).

6. Use: Bells: Fresh market, for cooking and in salads. Pimentos: Mostly commercially canned. U.S. per capita consumption in 1994 for sweet peppers is 6.1 pounds (PUTNAM 1996).

7. Part(s) of plant consumed: Bells: Walls, including peel, after seeds and placenta are removed. Pimentos: Thick walls, after surface peel, seeds and placenta removed

8. Portion analyzed/sampled: See Peppers

9. Classifications:
   a. Authors Class: See Peppers
   b. EPA Crop Group (Group & Subgroup): See Peppers. (Bell pepper is representative crop)
   c. Codex Group: Sweet peppers: 012 (VO 0445) Fruiting vegetables other than cucubits
   d. EPA Crop Definition: See Peppers

10. References: GRIN, LOGAN (picture), MAGNESS, RICHE, ROTHWELL 1996a, STEPHENS 1988 (picture), USDA 1996b, US EPA 1996b, US EPA 1995a, BAYER, PUTNAM 1996

11. Production Map: See Peppers.

12. Plant Codes:
   a. Bayer Code: CPSAN

465

1. Perennial peanut
(Rhizoma peanut, Arb, Peanut-perennial)

Fabaceae (Leguminosae)
Arachis glabrata Benth. var. *glabrata*

1. Pinto peanut
   A. pintoi Krapov. & W.C. Greg.

2. Perennial peanuts are native to Argentina, Brazil and Paraguay and were introduced into the U.S. Pinto peanut is native to Brazil and is cultivated in Australia and was introduced into the U.S. in 1954. Perennial peanuts are long-lived perennial legumes that are 12 to 16 inches tall and spread by rhizomes and stems up to 35 cm long. They produce very small amounts of seed (10 mm long and 5 to 6 mm wide) and propagation is usually vegetatively by using rhizomes at a rate of 60 to 80 bu of rhizomes/A in December through early March. The cultivar 'Florigraze' has been commercially developed for use as grazing and hay production crop in sandy soils of Florida and Georgia. It is slow to develop and takes 2 to 3 years to develop a thick stand after planting. They are usually interplanted with grasses and production will occur between April through October. It is maintained at a grazing height of 4 inches and livestock should be rotated to another field after 10 days or less and the field allowed 3 weeks rest. Hay can be cut two to three times/season and should not be cut 5 to 6 weeks before the first killing frost. The pinto peanut has prostrate to erect stems up to 20 cm long and spreads by stolons. It tolerates heavy grazing pressure and competes well in association with tall grasses. It can produce a nut with seeds 8 to 10 mm long and 4 to 6 mm wide.

3. Season: Warm season perennial, most productive after it is well established in the spring and summer. It can be rotationally grazed or cut for hay two to three times per season.

4. Production in U.S.: No specific data for Perennial peanut production, however, pasture and rangelands are produced on more than 410 million A (U.S. CENSUS, 1992). See above.

5. Other production regions: See above.

6. Use: Pasture grazing, forage, and hay crop.

7. Part(s) of plant consumed: Leaves and stems

8. Portion analyzed/sampled: Forage and hay

9. Classifications:
   a. Authors Class: Nongrass Animal Feeds
   b. EPA Crop Group (Group & Subgroup): Miscellaneous
   c. Codex Group: No specific citation although the general class 050 (AL 0157) Legume Animal Feeds could be used
   d. EPA Crop Definition: None

10. References: GRIN, US EPA 1994, US EPA 1995, CODEX, BARNES, HOLZWORTH, SKERMAN (a), BALL (picture), SCHNEIDER 1993(a)

11. Production Map: EPA Crop Production Regions 2, 3 and 13.

12. Plant Codes:
   a. Bayer Code: No specific entry

466

1. Perennial veldtgrass

Poaceae (Gramineae)
Ehrharta calycina Sm.

2. This is a cool season bunchgrass native to South Africa and introduced via Australia in 1929. It is a highly palatable, drought resistant grass adapted to light soils and nonirrigated rangelands. The leafy stems reach to 3 feet in height. It has proved valuable on sandy, coastal soils in California where it is used for range reseeding.

3. Season: Drought resistant rangegrass.

4. Production in U.S.: No specific data, See Forage grasses. Coastal California.
5. Other production regions: See above.
6. Use: Rangeland, pasture.
7. Part(s) of plant consumed: Foliage
8. Portion analyzed/sampled: Forage and hay
9. Classifications:
   a. Authors Class: See Forage grass
   b. EPA Crop Group (Group & Subgroup): See Forage grass
   c. Codex Group: See Forage grass
   d. EPA Crop Definition: None
10. References: GRIN, MAGNESS
11. Production Map: EPA Crop Production Region 10.
12. Plant Codes:
   a. Bayer Code: EHRCA

467

1. Perilla
(Tia To, Shiso, Baisuzi, Beefsteak plant, Rattlesnake weed, Red shiso)
Lamiaceae (Labiatae)
Perilla frutescens (L.) Britton (syn: *P. arguta* Benth.; *P. ocymoides* L.)

2. Bears large flat leaves that are green above and purple below. Used as a salad herb or garnish. It is an erect, branching, tender annual herb to 4 feet tall. Leaves are about 3 inches broad and up to 5 inches long and toothed. In the U.S., it blooms from August to October. It is a common weed in the Southern states.
3. Season, seeding: Annual that is very easy to grow, sown in the spring.
4. Production in U.S.: No data. Naturalized from Illinois to New York and south to Georgia (FOSTER 1993).
5. Other production regions: Asia (KUEBEL 1988)
6. Use: Leaves used as spice in bean curd or as a garnish for tempura and as a salad herb. The flower spikes are used in soups and the seedlings as a spice for raw fish. The purple leaves impart color and fragrance to pickled apricots. Pickled whole leaves are used to flavor rice dishes.
7. Part(s) of plant consumed: Leaves, flower spikes, seedlings and seeds
8. Portion analyzed/sampled: Tops (fresh and dried)
9. Classifications:
   a. Authors Class: Herbs and spices
   b. EPA Crop Group (Group & Subgroup): Miscellaneous
   c. Codex Group: No specific entry
   d. EPA Crop Definition: None
10. References: GRIN, FOSTER (picture), KUEBEL
11. Production Map: No entry
12. Plant Codes:
   a. Bayer Code: PRJFR

468

1. Persimmon
Ebenaceae
Diospyros spp.
1. Japanese Persimmon
(Kaki persimmon, Chinese persimmon, Kaki, Oriental persimmon)
D. kaki Thunb. (syn: *D. chinensis* Blume)

1. American persimmon
(Common persimmon, Caqui silvestre)
D. virginiana L. (syn: *D. mosieri* Small)
1. Black persimmon
(Texas persimmon, Mexican persimmon, Chapote)
D. texana Scheele

2. Almost all of the persimmons that are grown commercially in the U.S are *D. kaki* or Japanese persimmon. The American persimmon and black persimmon are native to North America. The fruit from the native species is often gathered in the wild, however some improved varieties have been commercialized but they are planted sparingly in home gardens. The Japanese varieties are semi-hardy, enduring temperatures to 15 degrees F, and are roughly adapted to the Cotton Belt. They require seasonal cool period for successful flowering and fruiting. Trees attain medium size, up to 25 feet unless pruned and have large, shiny leaves which are shed in winter. Fruit roughly resembles tomatoes in size and shape, with a similar smooth skin. They are 2 to 4 inches in diameter, oblate to conic in shape, yellow to deep red in color, seedless in some varieties, 8 to 10 large seeds in others. Hundred of varieties exist but only two are sold commerically, the 'Fuyu' and 'Hachiya'. The Black persimmon fruit is 1 inch in diameter and black.
3. Season: Bloom to harvest in approximately 200 days.
4. Production in U.S.: Production included Hawaii (20 acres), Florida, California, Alabama, and Texas. In 1994, California harvested 2,208 acres with 10,729 tons (MELNICOE 1996e).
5. Other production regions: Japan, Israel, China, Brazil, Australia.
6. Use: Mainly fresh eating, whole fruit, salads and desserts. Some fruits are dried and preserved.
7. Part(s) of plant consumed: Edible peel and inner pulp
8. Portion analyzed/sampled: Whole fruit
9. Classifications:
   a. Authors Class: Tropical and subtropical fruits – edible peel
   b. EPA Crop Group (Group & Subgroup): Miscellaneous
   c. Codex Group: 005 (FT 0307, FT 4113, and FT 4105) Assorted tropical and subtropical fruits – edible peel
   d. EPA Crop Definition: None; but proposing – Guava = Persimmon
10. References: GRIN, ANON (b), CODEX, KNIGHT(a), MAGNESS, MARTIN 1987, US EPA 1994, ROY, WHEALY, KAWATE 1997a, MELNICOE 1996e, CROCKER (f)
11. Production Map: EPA Crop Production Regions 3, 4, 6, 10 and 13.
12. Plant Codes:
   a. Bayer Code: DOSKA (*D. khaki*), DOSTE (*D. texana*), DOSVI (*D. virginiana*)

469

1. Pili nut
(Canary tree, Java almond, Kanari, Kenari)
Burseraceae
Canarium ovatum Engl.
C. vulgare Leenh. (syn: *C. commune* auct.)

2. Native to tropical areas of S.E. Asia. The tree grows to 60 feet tall, produces oblong, black fruit in clusters, 2½ inches long, and each fruit is a small nut with a thin hard shell. The

leaves, about 1 foot long, have 3 to 5 pairs of oval leaflets. *C. pimela* Leenh. is Chinese black olive.

3. Season, seeding to harvest: 7 to 10 years.
4. Production in U.S.: No data.
5. Other production regions: Tropical areas (NEAL 1965). Hot wet tropics (MARTIN 1983).
6. Use: The nut, after removing the seedcoat, can be eaten raw, or can be roasted and salted. Flavor of almonds. An edible cooking oil is also extracted from seed and pulp.
7. Part(s) of plant consumed: Seed and its oil primarily; Also the pulp is edible when cooked and yields a cooking oil
8. Portion analyzed/sampled: Seed
9. Classifications:
   a. Authors Class: Tree nuts
   b. EPA Crop Group (Group & Subgroup): Miscellaneous
   c. Codex Group: 022 (TN 0674) Tree nuts
   d. EPA Crop Definition: None
10. References: GRIN, CODEX, MARTIN 1983, NEAL (picture), SCHNEIDER 1993a
11. Production Map: No entry
12. Plant Codes:
    a. Bayer Code: No specific entry

470

1. Pine dropseed
(Hairy dropseed, Popotillo del pinar, Dropseed)
Poaceae (Gramineae)
Blepharoneuron tricholepis (Torr.) Nash

2. This is a native bunchgrass that is a member of the grass tribe *Eragrosteae* and is generally distributed in the southwest U.S. into Mexico. It is a warm season perennial that grows from 20 to 70 cm tall and has a panicle 5 to 20 cm long, and is adapted to areas from 2000 to 3500 m in elevation. Its habitat includes slopes and woodlands including rocky soils. Pine dropseed is palatable for all classes of livestock and is one of the highest quality grasses in timbered areas, but drops rapidly in quality as it matures. It reproduces from seeds and tillers.
3. Season: Warm season, growth starts in late June and is completed in September. Forage quality is good until plant matures.
4. Production in U.S.: No data for Pine dropseed production, however, pasture and rangeland are produced on more than 410 million A (U.S. CENSUS, 1992). See above.
5. Other production regions: Mexico.
6. Use: Rangeland grazing, forage, and hay, and cover crop.
7. Part(s) of plant consumed: Leaves and stems
8. Portion analyzed/sampled: Forage and hay
9. Classifications:
   a. Authors Class: See Forage grass
   b. EPA Crop Group (Group & Subgroup): See Forage grass
   c. Codex Group: See Forage grass
   d. EPA Crop Definition: None
10. References: GRIN, US EPA 1994, US EPA 1995, CODEX, HOLZWORTH, STUBBENDIECK (picture), RICHE
11. Production Map: EPA Crop Production Regions 8 and 9.
12. Plant Codes:
    a. Bayer Code: No specific entry

471

1. Pine nut
(Pinyon, Pinyon pine, Pinocchi, Pignoli, Stone nut, Pinon, Pignolia, Pino, American pinon, Colorado pinyon)
Pinaceae
Pinus edulis Engelm.
P. quadrifolia Parl. ex Sudw.
(Parry pinyon)

2. The seeds of these species of *Pinus* are gathered from native trees for food by the Native Americans in the southwestern U.S. and occasionally appear in markets. The trees are up to 40 feet, spreading, with stout branches and short needles not over 1½ inches long. Cones are near globose or broad ovate, compact until mature. Seeds are about ½ inch long, elongated and angular. They are eaten directly, also used in confections under the name pignolia. The trees are not in cultivation. Other edible ones are *P. pinea* L., *P. lamertiana* Doug., *P. cembra* L.
3. Season, harvest: After fall frost, mature cones open and the nuts fall to ground.
4. Production in U.S.: No data. Southwestern U.S., most abundant in New Mexico.
5. Other production regions: No entry.
6. Use: Eaten directly and used in confections.
7. Part(s) of plant consumed: Nutmeat
8. Portion analyzed/sampled: Nutmeat
9. Classifications:
   a. Authors Class: Tree nuts
   b. EPA Crop Group (Group & Subgroup): Miscellaneous
   c. Codex Group: 022 (TN 0673) Tree nuts
   d. EPA Crop Definition: None
10. References: GRIN, CODEX, MAGNESS, WOODROOF(a), IVES
11. Production Map: EPA Crop Production Region 9.
12. Plant Codes:
    a. Bayer Code: PIUED (*P. edulis*)

472

1. Pineapple
(Pina, Ananas)
Bromeliaceae
Ananas comosus (L.) Merr. (syn: *A. ananas* (L.) Voss; *A. sativas* Schult. & Schult.f.)

2. Pineapple is a tropical American fruit that grew wild in Brazil. It was originally named "anana" meaning fragrance or excellent fruit by the Native Americans. The plants are herbaceous with long, stiff sword-shaped leaves with rough edges. They are vegetatively propagated by crowns, slips and suckers. The plants produce multiple crops with the newly planted plants known as the main crop, and subsequent crops produced from auxillary plants are called ratoon crops. Commercially, one to two ratoon crops are harvested. The central stalk, at the terminal of which the single fruit is borne, attains a height of 2 to 4 feet. Commercial varieties are seedless and are usually propagated by suckers which develop near the base or terminal of the fruiting stalk. The fruit is composed of the thickened rachis or stalk, in which the numerous fleshy fruitlets, botanically berries, are imbedded. The fleshy, persistent bracts make the surface of the composite fruit greatly roughened and tough. Plants are tender to frost. They blossom 12 to 14 months after setting and fruit matures about 6 months later. Time of blossoming can be controlled in part by use of plant growth regulators. Fruits

are large, 2 to 6 pounds or more. The most widely planted variety is 'Smooth Cayenne'. It has a high acid and sugar content. Red pineapple (Wild pineapple) is *A. bracteatus* (Lindley) Schultes f.

3. Season: Bloom to maturity in about 6 months. Fruits do not all ripen simultaneously.
4. Production in U.S.: Domestic production in Hawaii (20,800 acres, 1995), Guam (50 tons), Virgin Islands (1437 lbs), and Puerto Rico (58,764 tons). U.S. production in 1995, 345,000 tons (USDA 1996a). In 1986, U.S. production was 646,000 tons (USDA 1996a).
5. Other production regions: Costa Rica, Dominican Republic, Philippines, Mexico, Brazil, Taiwan.
6. Use: Pulp eaten fresh, canned, frozen, dried, or made into juice. Fresh whole fruit has become more available in retail outlets. Good grade fruit is used for tibbits, slices and chunks; offgrade fruit is used for crushed pineapple; and the cores, trimmings and juice from crushing are used for beverage juice.
7. Part(s) of plant consumed: Tender inner flesh, after removal of rough surface and tough fibrous central cylinder. Residue from processing can be used as livestock feed
8. Portion analyzed/sampled: Whole fruit after crown has been removed, processed residue and juice. Pineapple process residue (also known as wet bran) is a wet waste byproduct from the fresh-cut product line that includes pineapple tops (minus crown), bottoms, peels, any trimmings with peel cut up, and the pulp (left after squeezing for juice); it can include culls
9. Classifications:
   a. Authors Class: Tropical and subtropical fruits – inedible peel
   b. EPA Crop Group (Group & Subgroup): Miscellaneous
   c. Codex Group: 006 (FI 0353) Assorted tropical and subtropical fruits – inedible peel
   d. EPA Crop Definition: None
10. References: GRIN, CODEX, COLLINS, FLATH, LOGAN, MAGNESS, MARTIN 1987, RICHE, US EPA 1994, USDA NRCS, USDA 1996a, USDA 1996f, MALO(e), PURSEG-LOVE(a), SALVI
11. Production Map: EPA Crop Production Region 13.
12. Plant Codes:
    a. Bayer Code: ANHCO

473

1. Pistachio
(Pistache nut, Pistachio nut, Pistacia nut, Green almond)
Anacardiaceae
Pistacia vera L.
2. The pistachio tree is small, up to 30 feet, and spreading. Leaves are compound-pinnate, hairy when young, later glabrous. The fruit is ovoid to oblong in shape, an inch or less long. It has an external, fleshy hull which loosens from the nut at maturity, but must be removed either by hand or mechanically. The nut has a thin, woody shell which often partially splits when ripe. The kernel is small and smooth, rich flavored and green in internal color. Most of the imported nuts are from Mediterranean countries.
3. Season, bloom to maturity: About 5 months.
4. Production in U.S.: 1992 CENSUS reported 69,344 acres with 72,200 tons, in shell. 1992 CENSUS reported Califor-

nia with 96 percent of the U.S. acreage. Arizona, New Mexico, Nevada and Texas were reported with the remainder. In 1994, California reported 56,923 acres (MELNICOE 1996e).
5. Other production regions: Mediterranean countries.
6. Use: Mainly for flavoring and coloring confections and ice cream. Some direct eating.
7. Part(s) of plant consumed: Internal kernel only
8. Portion analyzed/sampled: Nutmeat
9. Classifications:
   a. Authors Class: Tree nuts
   b. EPA Crop Group (Group & Subgroup): Tree nuts (Proposed)
   c. Codex Group: 022 (TN 0675) Tree nuts
   d. EPA Crop Definition: None
10. References: GRIN, CODEX, MAGNESS, RICHE, US EPA 1995b, US EPA 1994, WOODROOF(a) (picture), MELNICOE 1996e, NORTON, BAILEY 1976
11. Production Map: EPA Crop Production Region 10.
12. Plant Codes:
    a. Bayer Code: PIAVE

474

1. Plains bristlegrass
(Zacate tempranero)
Poaceae (Gramineae)
Setaria leucopila (Scribn. & Merr.) Schum.
S. macrostachya Kunth
2. This is a perennial warm season grass native to the U.S. and is a member of the grass tribe *Paniceae*. It is an erect bunchgrass from 0.2 to 1.2 m tall with leaf blades flat 8 to 25 cm long and 2 to 6 mm wide, and its inflorescence is a panicle (6 to 25 cm long). It is adapted to prairies, dry woods and rocky slopes and on alkaline soils along gullies or streams. Plains bristlegrass produces good forage for cattle and horses and fair to good for sheep and wildlife. It reproduces by seeds and tillers.
3. Season: Warm season grass that starts growing midspring and it flowers from May to September. Its cannot withstand heavy grazing and does not grow in dense stands.
4. Production in U.S.: There are no production data for Plains bristlegrass, however, pasture and rangeland are produced on more than 410 million A (U.S. CENSUS, 1992). Grown in southern Colorado, Texas, New Mexico, Arizona and south to central Mexico (GOULD 1975).
5. Other production regions: Mexico.
6. Use: Grazing and to prevent soil erosion.
7. Part(s) of plant consumed: Leaves and stems
8. Portion analyzed/sampled: Forage and hay
9. Classifications:
   a. Authors Class: See Forage grass
   b. EPA Crop Group (Group & Subgroup): See Forage grass
   c. Codex Group: See Forage grass
   d. EPA Crop Definition: None
10. References: GRIN, US EPA 1994, US EPA 1995, CODEX, HOLZWORTH, STUBBENDIECK (picture), RICHE, GOULD
11. Production Map: EPA Crop Production Regions 8 and 9.
12. Plant Codes:
    a. Bayer Code: No specific entry, genus entry is SETSS

475

1. Plaintain
Plantaginaceae
Plantago spp.

1. Buckhorn plaintain
(Chou qi zhuang che qian, Estrellamer, Cuerno de ciervo, Minutina, Herba stella, Capuchins beard, Misticanza, English plantain)
P. lanceolata L.

1. Common plaintain
(Greater plantain, Broadleaf plantain, Whitemans foot, Carttrack plant)
P. major L.

2. Perennial herbs are low growing with a rosette of basal leaves. Leaves are 2 to 6 inches long. It's normally growing in grassy and well-used areas as a weed.
3. Season, harvest: Fresh leaves as needed.
4. Production in U.S.: No data. Grows wild.
5. Other production regions: Eurasia (FACCIOLA 1990). Naturalized in North America and most parts of the world (BAILEY 1976).
6. Use: Young leaves used in salads and as potherb. In Italy, *P. coronopus* leaves are one of the ingredients in "misticanze".
7. Part(s) of plant consumed: Leaves primarily. Roots and seeds are also edible
8. Portion analyzed/sampled: Leaves (fresh)
9. Classifications:
   a. Authors Class: Leafy vegetables (except *Brassica* vegetables)
   b. EPA Crop Group (Group & Subgroup): Miscellaneous
   c. Codex Group: 013 (VL 0490) Leafy vegetables including *Brassica* leafy vegetables) for *P. major*
   d. EPA Crop Definition: None
10. References: GRIN, BAILEY 1976, CODEX, DUKE(c)(picture), KAYS, USDA NRCS, FACCIOLA, GARLAND (picture)
11. Production Map: EPA Crop Production Region: U.S.
12. Plant Codes:
   a. Bayer Code: PLALA (*P. lanceolata*), PLAMA (*P. major*)

476

1. Plum
(European plum, Garden plum, Common plum, Ciruelo, Plum prune, Gages, Prunier, Greengage plum)
1. Prune
(Pruneaux, Pasa)
Rosaceae
Prunus domestica L. ssp. *domestica*

2. According to Childers "*a prune is a plum which, because of higher sugar content, can be dried whole without fermentation at the pit. Growers in California who ship their fruit to canneries are known as plum growers, while those producing drying varieties are known as prune growers. Commercially, the prune is largely dried, but it is also canned and sold fresh. Plums are sold fresh, canned, or split and dried with the pit removed.*" The species *P. domestica* is the most important plums/prunes grown in the U.S. It includes all the prunes grown for drying and most of those canned, as well as a number of varieties mainly marketed fresh. There are 125 prune varieties. Trees are medium sized, usually held to 15-18 feet by pruning. Trees are of medium hardiness. The plant is characterized by large thick leaves with an upper dark green surface and pale green lower surface. The leaf margins are coarsely notched. The fruits are borne on spurs. There is extreme variation in size, shape and color of fruit; 1 to 1½ inches diameter, globose to oval, and with a firm, meaty flesh. Peel can be purple, red, yellow or green and is smooth, with a waxy surface and adheres to flesh. Important varieties include 'Italian', 'Agen', 'French', 'Imperial', 'Epineuse', 'Stanley', 'Lombard', and 'Reine Claude'.
3. Season: Bloom to harvest in 130-160 days.
4. Production in U.S.: In 1995, domestic production was 127,200 acres which yielded 744,000 tons. Most of commercial U.S. production is in the Pacific states, but some plantings occur in all areas of U.S. except the south and the coldest areas. Key production states include California (about 120,000 acres), Oregon (2,500 acres), Michigan (1,700 acres), Washington (1,500 acres) and Idaho (700 acres). In 1996, California produced about 80,000 acres of dried processed prunes and about 42,000 acres of plums for a total of 122,000; and Idaho, Michigan, Oregon and Washington produced about 6,000 acres of plum/prunes with about half processed (USDA 1997a).
5. Other production regions: Significant imports from Chile. Many European countries, Canada, New Zealand and Australia produce plums/prunes. In 1995, Canada grew 2,550 acres (ROTHWELL 1996a).
6. Use: Fresh market and processing. Most processed plums are dehydrated into prunes. Some are canned, or used for preserves and beverages. Prunes are dried within 18-24 hours after harvest. In Europe, plum kernel oil is extracted and used as a salad oil or in cosmetics.
7. Part(s) of plant consumed: Usually just the fruit pulp and fruit skin. Undersized prunes can be used for livestock feed
8. Portion analyzed/sampled: Whole fruit with pit and stem removed and discarded; and its dried processed commodity, prune
9. Classifications:
   a. Authors Class: Stone fruits
   b. EPA Crop Group (Group & Subgroup): Crop Group 12:Stone Fruits Group (Representative crop)
   c. Codex Group: Stone fruits 003 (FS 0014) Plume (Prune); (FS 4065) Greengage plum
   d. EPA Crop Definition: None, but proposing – Plum = Prune (fresh)
10. References: GRIN, CHILDERS, CODEX, MAGNESS, SCHREIBER, TESKEY, US EPA 1995, US EPA 1994, USDA 1997a, LOGAN, ROTHWELL 1996a, IVES, USDA 1996f, WESTPORT, CHILDERS(a)
11. Production Map: EPA Crop Production Regions 5, 10, 11 and 12.
12. Plant Codes:
   a. Bayer Code: PRNDO

477

1. Plum, American
(Native plum, Sloe, Blackthorn, Prunier americain)
Rosaceae
Prunus spp.
1. American plum
(American red plum, River plum)
P. americana Marshall

1. Sloe
(Blackthorn, Prunellier, Espino negro)
P. spinosa L.
1. Beach plum
(Shore plum)
P. maritima Marshall
1. Klamath plum
(Sierra plum, Pacific plum)
P. subcordata Benth.

2. There are at least 14 species of native *Prunus* that are considered American plums. Some of these native species have been crossed with traditional cultivated plums or used as a rootstock to yield commerical varieties. Some of the more prominent species includes *P. americana*, a small graceful tree that yields abundant 1 inch red and yellow fruit that are excellent for wildlife and out-of-hand fresh eating. The fruit is especially good for preserves. Leading cultivars are 'Desoto', 'Hawkeye', 'Wyant', 'Weaver' and 'Terry'. Beach plum or shore plum (*P. maritima*) is grown in the coastal areas from New Brunswick, Canada to Virginia. The 1-inch diameter fruit was used widely by early colonial settlers for fresh eating, jams, jellies and sauces. It is a straggling stiff, thorny bush which grows 6-10 feet high. The fruit color ranges from bluish purple, to red to yellow. The Sierra, Pacific or Klamath plum (*P. subcordata*) grows in California and Oregon. The fruit are yellow to red, growing to 1 inch in diameter. The plant is a medium shrub or small tree with fragrant white flowers. The fruit makes valued wild plum preserves. There are very limited commercial plantings of these.
3. Season: Bloom to harvest varies with species, generally 70-120 days.
4. Production in U.S.: No statistical data available.
5. Other production regions: Northern hemisphere (BAILEY 1976).
6. Use: Fruit of most species is gathered from the wild for making jams, and jellies. Some eaten fresh.
7. Part(s) of plant consumed: Mainly the fruit pulp, pit and peel usually removed during processing into preserves
8. Portion analyzed/sampled: Whole fruit with pit and stem removed and discarded
9. Classifications:
   a. Authors Class: Stone fruits
   b. EPA Crop Group (Group & Subgroup): Crop Group 12:Stone Fruits Group as "Plum (*Prunus domestica, Prunus* spp.)"
   c. Codex Group: Stone fruits 003 (FS 0249) Wild *Prunus* spp. and Sloe (*Prunus spinosa*) and (FS 4061) American plum
   d. EPA Crop Definition: None
10. References: GRIN, CHILDERS, CODEX, MAGNESS, TESKEY, US EPA 1994, WHEALY, BAILEY 1976, US EPA 1995a, BHUTANI
11. Production Map: EPA Crop Production Regions 1, 2, 4, 5, 6, 9 and 12.
12. Plant Codes:
    a. Bayer Code: PRNSS (*P.* spp.), PRNAM (*P. americana*), PRNSN (*P. spinosa*)

478
1. Plum, Cherry
(Myrobalan, Purple leaf plum, Japanese cherry plum, Ciruela mirobalana)
Rosaceae

Prunus cerasifera Ehrh.
2. Childers noted that this plant is mainly used for a rootstock for European and Japanese varieties of plums. Vigorous, upright tree grows rapidly up to 20 feet. Dark, reddish purple leaves with bright red tips hold their color all season. Small fragrant pink flowers in spring yield small fruit (round or oval) that are yellow or red in color. Fruiting is sporadic. 'Thundercloud' and 'All Red' are most prominent varieties.
3. Season: Bloom to harvest: 70-120 days.
4. Production in U.S.: No statistical data available. Naturalized in USDA Hardiness Zone 4 (BAILEY 1976).
5. Other production regions: No entry.
6. Use: Fruit is used for preserves, fresh out of hand eating and as a seed source from which the seedling rootstocks are grown. Also a beautiful ornamental shade tree.
7. Part(s) of plant consumed: Mainly the fruit pulp, pit and peel usually removed during processing into preserves
8. Portion analyzed/sampled: Whole fruit with pit and stem removed and discarded
9. Classifications:
   a. Authors Class: Stone fruits
   b. EPA Crop Group (Group & Subgroup): Included in Crop Group 12:Stone Fruits Group based on "Plum (*Prunus domestica, Prunus* spp.)"
   c. Codex Group: Stone fruits 003 (FS 0242) Cherry plum
   d. EPA Crop Definition: None
10. References: GRIN, CHILDERS, CODEX, TESKEY, US EPA 1994, WHEALY, BAILEY 1976
11. Production Map: EPA Crop Production Regions 1, 5 and 7.
12. Plant Codes:
    a. Bayer Code: PRNCF

479
1. Plum, Chickasaw
(Sand plum, Mountain cherry, Chabacano)
Rosaceae
Prunus angustifolia Marshall (syn: *P. chicasaw* Michx.)
2. Tree to 16 feet tall with leaves 1 to 3 inches long. Fruit is $1/2$ inch in diameter, red or yellow, thinned skin with soft, juicy sweet pulp. Grows from New Jersey to Missouri, south to Florida and Texas.
3. Season: Bloom to harvest: 70-120 days.
4. Production in U.S.: No data. Hardy in USDA Hardiness Zone 4-9 (WHEALY 1993).
5. Other production regions: No entry.
6. Use: Fruit of most species is gathered from the wild for making jams, and jellies. Some eaten fresh.
7. Part(s) of plant consumed: Mainly the fruit pulp, pit and peel usually removed during processing into preserves
8. Portion analyzed/sampled: Whole fruit with pit and stem removed and discarded
9. Classifications:
   a. Authors Class: Stone fruits
   b. EPA Crop Group (Group & Subgroup): Crop Group 12: Stone Fruits Group as "Plum, Chickasaw (*Prunus angustifolia*)"
   c. Codex Group: Stone fruits 003 (FS 0248, FS 4053) Chickasaw plum
   d. EPA Crop Definition: None
10. References: GRIN, BAILEY 1976, CODEX, TESKEY, US EPA 1994, US EPA 1995a, WHEALY

11. Production Map: EPA Crop Production Regions 2, 4 and 5.
12. Plant Codes:
    a. Bayer Code: PRNAN

480

1. Plum, Damson
(Bullace, Damson plum, Mirabelle, Bullace plum)
Rosaceae
Prunus domestica spp. *insititia* (L.) C.K. Schneid. (syn: *Prunus insititia* L.)

2. These are small, oval or round, firm fruited plums. Grown in Europe before the Christian era and brought to America by the earliest settlers. The small, compact trees are hardy and disease resistant. They are distinguished from European plums *(P. domestica)* by having a more compact and dwarfed habit of growth. They thrive better in eastern U.S. than other European plum types. The fruits are less than 1 inch in diameter, round to oval and purple (Damson), yellow (Mirabelle) or white/black (Bullace). The skin is very acid, making the fruits rather unsuitable for eating out of hand. However, they make highly esteemed jellies and jams, and are grown commercially for that purpose.
3. Season: Bloom to harvest in 100-125 days.
4. Production in U.S.: No data available, possibly 2,000 tons Western U.S.
5. Other production regions: Europe (FACCIOLA 1990).
6. Use: Used mainly for jelly/jam or other culinary purposes.
7. Part(s) of plant consumed: Pulp in making jelly and jam, but whole fruit cooked prior to separating out seeds and peel
8. Portion analyzed/sampled: Whole fruit with pit and stem removed and discarded
9. Classifications:
   a. Authors Class: Stone fruits
   b. EPA Crop Group (Group & Subgroup): Crop Group 12: Stone Fruits Group as "Plum, Damson (*Prunus domestica* ssp. *insititia*)"
   c. Codex Group: Stone fruits 003 (FS 0241) Bullace; (FS 4055) Damson; and (FS 4063) Bullace and Damson; (FS 4057) Mirabelle; (FS 4071) Mirabelle plum
   d. EPA Crop Definition: None
10. References: GRIN, CHILDERS, CODEX, MAGNESS, TESKEY, US EPA 1994, BHUTANI, FACCIOLA
11. Production Map: EPA Crop Production Region 11.
12. Plant Codes:
    a. Bayer Code: No specific entry

481

1. Plum, Japanese
(Oriental plum)
Rosaceae
Prunus salicina Lindl. (syn: *P. triflora* Roxb.)

2. The Japanese plums were introduced into the U.S. during the latter part of the 19th century, and have become popular in our market. Trees of Japanese plums are of medium size, and not very hardy to winter cold. In addition, they are early blooming and susceptible to frost. Hence commercial production is mainly in California. The fruit is relatively large for plums, $1^{1}/_{2}$ to over 2 inches in diameter. Shape is variable, but mostly conic to oval. Color generally light to dark red, never blue. Flesh generally soft and very juicy when ripe. Peel smooth, with waxy surface. Because these plums have high quality and will cross with American species they have been used in breeding to improve the quality of the hardy native types. Important cultivars include 'Santa Rosa', 'Burbank' and 'Kelsey'.
3. Season: Bloom to harvest in 60-160 days, depending on variety.
4. Production in U.S.: No statistical data available. Mainly California. Also grown in southeast and north central U.S. (FACCIOLA 1990).
5. Other production regions: Japan, Canada.
6. Use: Mainly fresh market, eating out of hand.
7. Part(s) of plant consumed: Generally all except pit. Peel may be separated from pulp in mouth and not eaten
8. Portion analyzed/sampled: Whole fruit with pit and stem removed and discarded
9. Classifications:
   a. Authors Class: Stone fruits
   b. EPA Crop Group (Group & Subgroup): Crop Group 12: Stone Fruits Group as "Plum, Japanese (*Prunus salicina*)"
   c. Codex Group: Stone fruits 003 (FS 0014) Plums; (FS 4069) Japanese plum
   d. EPA Crop Definition: None
10. References: GRIN, CHILDERS, CODEX, MAGNESS, TESKEY, US EPA 1994, WESTPORT, FACCIOLA
11. Production Map: EPA Crop Production Region 10.
12. Plant Codes:
    a. Bayer Code: PRNSC

482

1. Plumcot
(Stone fruit, Drupe)
Rosaceae
Prunus domestica L. x *P. armeniaca* L.
1. Peachcot
P. domestica x *P. persica*. (L.) Batsch
1. Other Stone Fruit Crosses
(Aprium, Plum cherry, Pluot®)
P. domestica L. or *P. salicina* Lindl. x *P.* spp.

2. These hybrids are made by crossing members of the stone fruit group. The most common are the plumcots which are crosses of European type plums (prunes) with apricots. These plumcots are not the Apricot Plum *(P. simonii)* that are common in China. Plants and fruit can take on many of their parent's characteristics. For example some varieties of plumcot can yield relatively large yellow fruit with golden yellow flesh and apricot flavor. Others are much more like their plum parents. Peachcots have a large yellow fruit with a peach-like flavor, however the tree has all apricot characteristics. Some unique crosses include 'Aprium' (plumcot backcrossed to an apricot), Plum-Cherries (Sand cherry x Japanese Plum), and Pluot® [(*P. domestica* L. x *P. armeniaca* L.) x (*P. domestica* L.)] which is a hybrid [(plum x apricot) x plum] and is termed an "ISP plum" or "interspecific plum".
3. Season: Bloom to harvest: Varies with crosses.
4. Production in U.S.: No statistical data available.
5. Other production regions: See Peach.
6. Use: Fruit is used for preserves or fresh out of hand eating. Some are beautiful ornamental specimens.
7. Part(s) of plant consumed: Mainly the fruit pulp and fruit peel. The pit is removed and discarded

8. Portion analyzed/sampled: Whole fruit with pit and stem removed and discarded
9. Classifications:
   a. Authors Class: Stone fruits
   b. EPA Crop Group (Group & Subgroup): Except for plumcot there is no mention of these crosses in Crop Group 12: Stone Fruits Group. These crosses logically fit into the crop group
   c. Codex Group: Stone fruits 003 (FS 0012) Stone fruits
   d. EPA Crop Definition: None
10. References:CODEX, US EPA 1994, WHEALY
11. Production Map: EPA Crop Production Region, See Peach.
12. Plant Codes:
    a. Bayer Code: No specific entry, genus entry is PRNSS

483

1. **Pokeweed**
(**Poke, Scoke, Garget, Inkberry, Pigeonberry, Coakun, Pocan bush, Pokeberry, American nightshade, Poke salad**)
Phytolaccaceae
Phytolacca americana L. (syn: *P. decandra* L.)
2. Pokeweed is a native plant throughout eastern North America. It is a large-rooted perennial, with a strong growing top, up to 10 or more feet in height, which dies down in winter. The young, asparagus-like shoots formed in the spring are sometimes used as potherbs. They are sometimes grown from lifted roots in winter, similar to forced rhubarb. The roots and seeds are poisonous. The fruit was used for ink in colonial times.
3. Season: New shoots harvested 3 to 5 days after start of growth. Propagation is by seed.
4. Production in U.S.: No data. Little cultivation, but some gathered from wild. Native plant to eastern North America.
5. Other production regions: No entry.
6. Use: As potherb and greens. Before use as greens, must be boiled to remove bitterness.
7. Part(s) of plant consumed: Young tender shoots only
8. Portion analyzed/sampled: Young tender shoots
9. Classifications:
   a. Authors Class: Leafy vegetables (except *Brassica* vegetables)
   b. EPA Crop Group (Group & Subgroup): Miscellaneous
   c. Codex Group: 013 (VL 0491) Leafy vegetables (including *Brassica* leafy vegetables)
   d. EPA Crop Definition: None
10. References: GRIN, CODEX, MAGNESS, STEPHENS 1988 (picture)
11. Production Map: Native to EPA Crop Production Regions 1, 2, 3, 4 and 5.
12. Plant Codes:
    a. Bayer Code: PHTAM

484

1. **Polargrass**
(**Articgrass, Arctagrostis**)
Poaceae (Gramineae)
Arctagrostis latifolia (R. Br.) Griseb.
2. Polargrass is a cool season perennial grass native to North America, Europe, and Asia, that grows throughout Alaska. It is also grown throughout Europe and Asia for hay and pas-

ture. It is adapted to a wide range of sites from dry areas to boggy lowlands, but it is most common to the moist soils of tundras and marshes. It is a member of the grass tribe *Agrostideae*, and grows erect (30 to 150 cm) with rhizomes and a open panicle (3-30 cm long). to a height of 1.4 m. Polargrass is valuable for grazing and revegetation. Yields are less than smooth bromegrass, but it tolerates acid soils and icing conditions better than smooth bromegrass or timothy. Seed production varies from 125 to 280 kg/ha.
3. Season: Cool season forage and grows from mid-spring through the early fall.
4. Production in U.S.: No data for Polargrass production, however, pasture and rangeland are produced on more than 410 million A (U.S. CENSUS, 1992). Alaska.
5. Other production regions: Adapted to northern Canada, Europe and Asia.
6. Use: Pasture, hay, revegetation and erosion control.
7. Part(s) of plant consumed: Leaves and stems
8. Portion analyzed/sampled: Forage and hay
9. Classifications:
   a. Authors Class: See Forage grass
   b. EPA Crop Group (Group & Subgroup): See Forage grass
   c. Codex Group: See Forage grass
   d. EPA Crop Definition: None
10. References: GRIN, US EPA 1994, US EPA 1995, CODEX, MOSLER, RICHE
11. Production Map: EPA Crop Production Region, Alaska.
12. Plant Codes:
    a. Bayer Code: ARSLA

485

1. **Polynesian arrowroot**
(**East Indian arrowroot, Tahiti arrowroot, Fiji arrowroot, Tacca, Pia, Salep, Hawaii arrowroot, Southsea arrowroot, Mokmok**)
Taccaceae
Tacca leontopetaloides (L.) Kuntze (syn: *T. involucrata* Schumach. & Thonn.; *T. pinnatifida* J.R. Forst. & G. Forst. F.; *T. hawaiiensis* Limpr.)
2. Similar to arrowroot, which see. Starch is obtained from this arrowroot. A perennial grown for its starchy underground tubers which are 6-8 inches in diameter.
3. Season, seeding to harvest: Matures in 8-10 months.
4. Production in U.S.: Rarely cultivated in Hawaii (NEAL 1965).
5. Other production regions: Grown in tropics (NEAL 1965).
6. Use: Potato-like tubers are prepared by grating them, then soaking and repeatedly washing the starch. The starch is added to coconut milk and baked or boiled. Used like cornstarch. Used in breads and soups. Can be cooked into poi.
7. Part(s) of plant consumed: Tubers
8. Portion analyzed/sampled: Tubers
9. Classifications:
   a. Authors Class: Root and tuber vegetables
   b. EPA Crop Group (Group & Subgroup): Miscellaneous
   c. Codex Group: No specific entry
   d. EPA Crop Definition: None
10. References: GRIN, FACCIOLA, NEAL, STEPHENS 1988, RUBATZKY
11. Production Map: EPA Crop Production Region 13.
12. Plant Codes:
    a. Bayer Code: No specific entry

486

1. Pomegranate
(Granada, Chinese apple, Indian apple, Grenadier, Grandado)
Punicaceae
Punica granatum L.

2. Grows in subtropical climates. The plant is well adapted to both hot and moderate-cool conditions. Pomegranates are small deciduous, semi-hardy trees or large shrubs. The fruit is globular to oblate to 4 inches diameter and weighing up to 1 pound. All fruit, even when ripe, has the calyx persistent at the terminal. The outer red colored rind is smooth and rather thin. Within, the fruit is filled with seeds, each imbedded in juicy pulp, and this pulp is commonly the edible portion. Seeds and pulp may be scooped out and pressed for juice, or the pulp and seeds may be separated in the mouth. The rind, boiled, has long been used as a remedy for tapeworm. The plant is highly ornamental, both in flower and fruit, and is found in many southern gardens, and also as a potted plant.
3. Season: Bloom to harvest in 4 to 10 months depending on variety.
4. Production in U.S.: Hawaii. In 1994, California reported 2,889 acres with 12,597 tons (MELNICOE 1996e).
5. Other production regions: Native of Iran and India. Cultivated in the Mediterranean region and Mexico (MORTON 1987).
6. Use: Mainly beverages with some fresh. The aggregate of berry-like pulp is consumed raw in salads.
7. Part(s) of plant consumed: Interior pulp (flesh and seed)
8. Portion analyzed/sampled: Whole fruit
9. Classifications:
   a. Authors Class: Tropical and subtropical fruits – inedible peel
   b. EPA Crop Group (Group & Subgroup): Miscellaneous
   c. Codex Group: 006 (FI 0355) Assorted tropical and subtropical fruits – inedible peel
   d. EPA Crop Definition: None
10. References: GRIN, CODEX, LOGAN, MAGNESS, MARTIN 1987, US EPA 1994, MORTON (picture), MELNICOE 1996e, ADSULE(a), SHEETS
11. Production map: EPA Crop Production Regions 10 and 13.
12. Plant Codes:
    a. Bayer Code: PUNGR

487

1. Pomerac
(Mountain apple, Pomarrosa malay, Malay apple, Pomarrosa Americana, Longfruited roseapple, Rose-apple, Ohia)
Myrtaceae
Syzygium malaccense (L.) Merr. & L.M. Perry (syn: *Eugenia malaccensis* L.)

2. Tropical evergreen tree to 35 feet in height. The fruits are pear-shaped, 4 inches long, 3 inches wide and are reddish pink with longitudinal stripes and splashes of darker red. The fruit flesh is thick, white, dry and surrounds a large seed. The fruits may be eaten fresh or used to make jelly, but are usually stewed with some flavoring material such as cloves.
3. Season, bloom to harvest: About 60 days.
4. Production in U.S.: No data. Hawaii (MORTON 1987).
5. Other production regions: Tropics – West Indies (MORTON 1987).

6. Use: Fresh, jelly and stewed. Table wine in Puerto Rico.
7. Part(s) of plant consumed: Fruit
8. Portion analyzed/sampled: Whole fruit
9. Classifications:
   a. Authors Class: Tropical and subtropical fruits – edible peel
   b. EPA Crop Group (Group & Subgroup): Miscellaneous
   c. Codex Group: 005 (FT 0308) Assorted tropical and subtropical fruits – edible peel
   d. EPA Crop Definition: None
10. References: GRIN, CODEX, MAGNESS, MARTIN 1987, MORTON (picture)
11. Production Map: EPA Crop Production Region 13.
12. Plant Codes:
    a. Bayer Code: SYZMA

488

1. Poppy seed
(Opium poppy)
Papaveraceae
Papaver somniferum L. ssp. *somniferum*

2. Annual herb native to Mediterranean region and Central Asia. Poppy seeds are not narcotic. Generally, blue-seeded cultivars are superior in seed oil yield to white-seeded ones. White-seeded ones are used more in confectionery trade. U.S. imported about 5,284 tons from Australia, Netherlands and Turkey in 1992 (USDA 1993a).
3. Season, seeding to harvest: About 7 months.
4. Production in U.S.: No data.
5. Other production regions: Australia, Netherlands and Turkey (USDA 1993a).
6. Use: Dried seeds as a topping for baked goods, fill for pastry, seed oil for use in salads and cooking oil and seed cake is also used as food. Seeds also added to curry paste, pasta, salads and vegetables.
7. Part(s) of plant consumed: Seed
8. Portion analyzed/sampled: Seed (dried) and its oil
9. Classifications:
   a. Authors Class: Herbs and spices, Oilseed
   b. EPA Crop Group (Group & Subgroup): Herbs and spices (19B)
   c. Codex Group: 023 (SO 0698) Oilseed; 028 (HS 0698) Spices
   d. EPA Crop Definition: None
10. References: GRIN, CODEX, LEWIS (picture), ROBBELEN 1989 (picture), USDA 1993a, US EPA 1995a
11. Production Map: No entry
12. Plant Codes:
    a. Bayer Code: PAPSO

489

1. Potato
(Irish potato, White potato, Pomme de terre, Papa, Patata, Kartoffel, New potato)
Solanaceae
Solanum tuberosum L. ssp. *tuberosum*

2. The potato is the most important vegetable in the world, being a staple food in all temperate climates. The plant is an annual, and is propagated by planting pieces of the underground tubers or enlarged stems. The top attains heights up to 2 feet or more, with fibrous, fleshy stems carrying a can-

opy of compound leaves up to 10 inches in length. The edible, starchy tubers are produced entirely underground. In cultivated varieties they may be smooth-skinned or russeted, oblong to near round, but usually somewhat flattened. Marketable size is generally 1^1/$_2$ inch diameter and larger, and up to 6 inches or more in length. Potatoes are grown year round in the U.S. with the bulk of the harvest (89 percent) occurring in the fall. The spring, winter and summer crops are normally shipped directly to market or used for processing. The fall crop is normally stored for later use. Vines are killed before harvest and most cultivars require 10-14 days after vine kill to have their skin firm.

3. Season, planting to harvest: Normally about 4 months but can range from 90 to 160 days.

4. Production in U.S.: About 22,115,450 tons and 1,396,900 acres in 1995 (USDA 1996c). In 1995, the top 11 states for acreage planted: Idaho (400,000 acres), Washington (147,000), North Dakota (125,000), Colorado (86,000), Minnesota (83,000), Wisconsin (83,000), Maine (78,000), Michigan (55,000), Oregon (52,000), Florida (46,800) and California (41,500) which accounts for about 86 percent of total U.S. acreage. The winter crop comes out of California and Florida, the spring crop comes out of Alabama, Arizona, California, Florida, North Carolina and Texas. The summer crop from 14 states, eg., New Jersey. The fall crop is the largest from 23 states, eg., Idaho (USDA 1996c).

5. Other production regions: In 1995, Canada reported 308,882 acres (ROTHWELL 1996a).

6. Use: As cooked vegetable and processed as chips and other snack products. Also a source of edible starch and flour. Culls or surplus and processed potato waste fed to livestock. U.S. per capita consumption in 1994 is 140.7 pounds (PUTNAM 1996).

7. Part(s) of plant consumed: Tubers only, mainly with peel removed

8. Portion analyzed/sampled: Tuber and its processed commodities, granules or flakes, chips and wet peel. For potato granules/flakes, residue data may be provided for either. For processed potato waste, tolerance levels for wet peel should be used for dietary burden calculations. Residue data may be provided from actual processed potato waste generated using a pilot or commercial scale process that gives the highest percentage of wet peel in the waste

9. Classifications:
   a. Authors Class: Root and tuber vegetables
   b. EPA Crop Group (Group & Subgroup): Root and tuber vegetables (1C) (Representative crop)
   c. Codex Group: 016 (VR 0589) Root and tuber vegetables
   d. EPA Crop Definition: None

10. References: GRIN, CODEX, MAGNESS, ROTHWELL 1996a, USDA 1996c, US EPA 1996a, US EPA 1995a, US EPA 1994, HENNINGER, IVES, SWIADER, PUTNAM 1996

11. Production Map: EPA Crop Production Regions 1, 2, 3, 5, 9, 10 and 11.

12. Plant Codes:
   a. Bayer Code: SOLTU

490

1. Potato, Specialty
(Papa)

Solanaceae
Solanum spp.
1. Pitiquina (pee-tee-keen-ya)
S. stenotomum Juz. & Bukasov ssp. *stenotomum*
1. Limena (lie-main-ya)
(Papa amarilla, Yellow potato)
S. stenotomum ssp. *goniocalyx* (Juz. & Bukasov) Hawkes (syn: *S. goniocalyx* Juz. & Bukasov)
1. Phureja (foo-ray-ha)
S. phureja Juz. & Bukasov
1. Andigena (an-di-je-na)
S. tuberosum ssp. *andigena* Hawkes (syn: S. *andigenum* Juz. & Bukasov)
1. Chaucha (chow-cha)
S. x *chaucha* Juz. & Bukasov (syn: *S. stenotomum* x *S. andigenum*)
1. Ajankuiri (a-han-hwee-ri)
(Ajawiri)
S. x *ajanhuiri* Juz. & Bukasov
1. Rucki (rue-kee)
S. x *juzepczukii* Bukasov and *S.* x *curtilobum* Juz. & Bukasov

2. During approximately 8,000 years that potatoes have been cultivated in the Andes, farmers have selected types to meet their needs. Specialty potatoes differ from standard potatoes in color, size and growing conditions.

3. Season: Propagated by tuber.

4. Production in U.S.: No data.

5. Other production regions: Andes region of South America (NATIONAL RESEARCH COUNCIL 1989).

6. Use: Tuber used like standard potato.

7. Part(s) of plant consumed: Tuber

8. Portion analyzed/sampled: Tuber

9. Classifications:
   a. Authors Class: Root and tuber vegetables
   b. EPA Crop Group (Group & Subgroup): Miscellaneous
   c. Codex Group: No specific entry
   d. EPA Crop Definition: None

10. References: GRIN, NATIONAL RESEARCH COUNCIL 1989, HENNINGER

11. Production Map: No entry.

12. Plant Codes:
   a. Bayer Code: No specific entry, genus entry is SOLSS

491

1. Prairie sandreed
Poaceae (Gramineae)
Calamovilfa longifolia (Hook.) Scribn.

2. This is a native warm season, perennial bunchgrass that may form dense colonies and it is a member of the grass tribe *Eragrosteae* and is generally distributed in the northern Great Plains areas of the U.S. and adjacent Canadian areas. Prairie sandreed is considered to be the most important grass of the genus *Calamovilfa*. It grows erect from 0.5 to 1.8 m tall and has a panicle that is 15 to 40 cm long. Its habitat includes irrigated or dryland prairies, plains and open woods areas, and is most abundant on sandy soils. It is grazed, but sensitive to repeated defoliation. Forage quality is less than sand bluestem. Prairie sandreed is palatable for all classes of livestock and is one of the highest quality grasses in timbered areas, but drops rapidly in quality as it matures. It

reproduces from seeds and rhizomes and can be difficult to establish. It is drought tolerant, and provides good forage for cattle, horses and wildlife, as well as providing good standing winter feed. Prairie sandreed has 274,000 seeds/lb.

3. Season: Warm season, growth starts rapidly in the late spring and continues throughout the summer and matures in October. It is cut for hay up to the hard dough stage. Forage quality is good until the plant matures.
4. Production in U.S.: No data for Prairie sandreed production, however, pasture and rangeland are produced on more than 410 million A (U.S. CENSUS, 1992). Northern Great Plains.
5. Other production regions: Canada.
6. Use: Prairie and plain grazing, hay, erosion control and cover crop.
7. Part(s) of plant consumed: Leaves and stems
8. Portion analyzed/sampled: Forage and hay
9. Classifications:
   a. Authors Class: See Forage grass
   b. EPA Crop Group (Group & Subgroup): See Forage grass
   c. Codex Group: See Forage grass
   d. EPA Crop Definition: None.
10. References: GRIN, US EPA 1994, US EPA 1995, CODEX, HOLZWORTH, STUBBENDIECK(a) (picture), RICHE
11. Production Map: EPA Crop Production Regions 5, 7, 8 and 9.
12. Plant Codes:
    a. Bayer Code: No specific entry

492

1. Prickly pear
(Indian fig, Cactus pear, Tuna, Chumbo, Higo, Nopalitos, Spineless cactus, Prickly pear cactus)
Cactaceae
Opuntia ficus-indica (L.) Mill.
1. Texas prickly pear
(Lindheimer prickly pear)
O. lindheimeri Engelm.
2. Grown in cool semiarid climate. Propagated by cuttings with first crop in 2-3 years. Plants attain a height of 10 feet or more, with woody, cylindrical trunks, and joints up to 18 inches. Fruits are egg-shaped 1.5 to 2 inches in diameter, covered with a medium-thick, spiny rind (some spineless). The exterior fruit color is medium-green to dark magenta. Spines are usually rubbed off before picking. Peel separates from the flesh readily, and is removed before the edible pulp is consumed. Small seeds are edible.
3. Season: Bloom to harvest is no less than 90 days, commonly 4 to 8 months or more.
4. Production in U.S.: California and Texas. Grown widely in home gardens and small commercial plantings in southwestern U.S.
5. Other production regions: Chile and Mexico supply most of the U.S. imports (LOGAN 1996).
6. Use: Fruit used for mostly fresh eating, however, it can be made into preserves, candies or dried. Young cacti leaves (pads, nopalitos, cladophylls or cladodes) can be mixed with other ingredients to make relish (salsa) or used as livestock feed. Also the young pads are boiled and used like snap beans in salads and soups.
7. Part(s) of plant consumed: Fruit with spines and skin removed and cacti pads

8. Portion analyzed/sampled: Whole fruit and young cacti leaves
9. Classifications:
   a. Authors Class: Tropical and subtropical fruits – inedible peel
   b. EPA Crop Group (Group & Subgroup): Miscellaneous
   c. Codex Group: 006 (FI 0356 and FI 4133) Assorted tropical and subtropical fruits – inedible peel
   d. EPA Crop Definition: None
10. References: GRIN, CODEX, KNIGHT, LAKSHMINARA-YANA, LOGAN, MAGNESS, MARTIN 1987, US EPA 1994, FACCIOLA
11. Production Map: EPA Crop Production Regions 6 and 10.
12. Plant Codes:
    a. Bayer Code: OPUFI (*O. ficus-indica*), OPULI (*O. lindheimeri*)

493

1. Psyllium
(Plantago)
Plantaginaceae
Plantago spp.
1. Psyllium seed
(Whorled plantain, Llanten, Poormans bran flakes, Fleawort, Fleaseed, Spanish psyllium, French psyllium, Black psyllium)
P. arenaria Waldst. & Kit. (syn: *P. scabra* Moench)
P. afra L. (syn: *P. psyllium* auct.)
1. Blond psyllium
(Desert Indian wheat, Indian plantago, Ispaghul, Ispaghula, Spogel, Isabgol, White psyllium)
P. ovata Forssk. (syn: *P. brunnea* Greene)
2. Annual herbaceous plants to 18 inches tall cultivated for its seed which is used for its mucilage content. The mucilage layer (husk) is the outer seed layer which accounts for about 25 percent of the total weight of the seed. The mucilage is obtained by mechanical grinding of the outer seed layer. It is a true dietary fiber in diets, medicines and cereals. As a thickener, it is used in ice cream and frozen desserts. Seed kernel is grown and fed to cattle and horses. Its primary use is for the husk as a dietary fiber for humans. Psyllium is the common name used for several members of the plant genus *Plantago* and the genus contains over 200 species. The U.S. is the world's largest importer with over 50 percent of the total imports to pharmaceutical companies (about 8,800 tons per year). The seed mucilage is referred to as husk or psyllium husk. In India, the dehusked seed is used for chicken and cattle feed.
3. Season, seeding to harvest: About 107-130 days. Plants flower 60 days after planting.
4. Production in U.S.: No data. Grown in Arizona in late 1970s and Washington in 1985 (1996 IR-4 DATABASE). Also reported in the Carolinas (DUKE 1992). Naturalized in eastern U.S. (BAILEY 1976).
5. Other production regions: Mainly grown in India (SCHNEIDER 1993a). Eurasia (FACCIOLA 1990).
6. Use: Seed husk used for its mucilage content in ice cream and frozen desserts, and as dietary fiber in cereal and diet.
7. Part(s) of plant consumed: Seeds
8. Portion analyzed/sampled: Seed and its processed husk and threshings
9. Classifications:

a. Authors Class: Cereal grains
b. EPA Crop Group (Group & Subgroup): Miscellaneous
c. Codex Group: No specific entry
d. EPA Crop Definition: None
10. References: GRIN, BAILEY 1976, DUKE(b & c), FACCI-OLA, IR-4 REQUEST DATABASE 1996, SCHNEIDER 1993a, USDA NRCS, HANSON
11. Production Map: EPA Crop Production Regions 10 and 11.
12. Plant Codes:
a. Bayer Code: PLAAF (*P. afra*), PLAIN (*P. arenaria*)

494

1. Pulasan
(Pooasan, Kapoolasan)
Sapindaceae
Nephelium ramboutan-ake (Labill.) Leenh. (syn: *Nephelium mutabile* Blume)
2. Grows in hot, wet tropical lowlands. Pulasan trees, a handsome ornamental, are a close relative to rambutan. In fact, some confuse the two. It reaches 50 feet tall. The plant has small green petalless flowers that are borne singly or in clusters. The globose fruits are up to 3 inches long, yellow to red in external color with a thick leathery rind. The white flesh contains seeds that are 0.75 to 1.33 inches long. There are two main types, 'Seebabat' or 'Kapoolasan'. The flavor is tart and pleasant. Pulasan fruit is without the hairlike covering like Rambutan fruit. "Rambut" means "hair" in Malay.
3. Season, bloom to harvest: See Rambutan.
4. Production in U.S.: No data available. The plant requires ultra-tropical conditions. This limits production to extreme south Florida, Hawaii and Puerto Rico.
5. Other production regions: Malaya, Thailand, and Costa Rica (MORTON 1987) and Java (ROY 1995).
6. Use: Pulp eaten fresh or stewed, canned or preserved into jams and jellies. Seeds can be boiled or roasted and used to make beverages or extracted to yield oil.
7. Part(s) of plant consumed: Fruit pulp and seeds
8. Portion analyzed/sampled: Whole fruit
9. Classifications:
a. Authors Class: Tropical and subtropical fruits – inedible peel
b. EPA Crop Group (Group & Subgroup): Miscellaneous
c. Codex Group: 006 (FI 0357) Assorted tropical and subtropical fruits – inedible peel
d. EPA Crop Definition: Proposing Lychee = Pulasan
10. References: GRIN, CODEX, MARTIN 1987, MORTON, ROY
11. Production Map: EPA Crop Production Region 13.
12. Plant Codes:
a. Bayer Code: No specific entry

495

1. Pummelo
(Shaddock, Pompelmous, Oro blanco, Pumelo, Pomelo, Large-sized tangelo, Pompelmouse, Chinese grapefruit)
Rutaceae
Citrus maxima (Burm.) Merr. (syn: *C. grandis* (L.)Osbeck, *C. decumana* (L.) L., *C. aurantium decumana* L.)
2. Pummelos (see citrus fruits) are important fruits in the Orient, but only scattered trees are found in the U.S. Quality of

the fruit has been disappointing in this country as compared to that of grapefruit, which the pummelo resembles. The fruit is oblate to globose in shape and of large size, 4 to 7 inches, or more, in diameter. Rind generally thick $^1/_2$ inch or more. Juice vesicles separate from peel and segment fibers readily. The variety 'Tresca' is grown in Florida, where it is marketed as grapefruit. Oro blanco is a cross between pummelo and grapefruit.
3. Season, bloom to harvest: Similar to grapefruit.
4. Production in U.S.: Available from California mid-January through mid-February (LOGAN 1996). In 1995, almost 500 acres in the San Joaquin Valley of California (ENGLER 1995). Florida.
5. Other production regions: Orient.
6. Use: Mainly eaten as fresh fruit sweeter than grapefruit. It is also good for jams, jellies, marmalades and syrups.
7. Part(s) of plant consumed: Interior juice vesicles
8. Portion analyzed/sampled: Whole fruit
9. Classifications:
a. Authors Class: See citrus fruits
b. EPA Crop Group (Group & Subgroup): See citrus fruits
c. Codex Group: 001 (FC 0209) Citrus fruits
d. EPA Crop Definition: See citrus fruits
10. References: GRIN, CODEX, ENGLER, LOGAN (picture), MAGNESS
11. Production Map: EPA Crop Production Regions 3 and 10.
12. Plant Codes:
a. Bayer Code: CIDGR

496

1. Pumpkin
(Calabaza, Connecticut field pumpkin, Howden pumpkin, Jack-o-lantern, Mammouth pumpkin, Mimi pumpkin, West Indian pumpkin, Cuban squash, Toadback, Japanese pie pumpkin, Cheese pumpkin, Cushaw, Potiron, Calabaza comun)
Cucurbitaceae
Cucurbita argyrosperma C. Huber
C. argyrosperma C. Huber ssp. *argyrosperma* (syn: *C. mixta* Pangalo)
C. argyrosperma var. *callicarpa* L. Merrick & D.M. Bates
C. maxima Duchesne ex Lam.
C. maxima Duchesne ex Lam. ssp. *maxima* (syn: *C. maxima* var. *turbaniformis* (Roem.) Alef.)
C. moschata (Duchesne ex Lam.) Duchesne ex Poir.
C. pepo L.
C. pepo L. var. *pepo*
C. pepo var. *ovifera* (L.) L.H. Bailey
2. As indicated, varieties of several species of *Cucurbita* carry the name pumpkin. There is much confusion also between the terms pumpkin and squash. Pumpkin, as here described, is the edible fruit of cucurbits used for feed or food when ripe, and having somewhat coarse, strongly flavored flesh. Winter squash has finer textured and less strongly flavored flesh. Pumpkins are produced on trailing annual plants, having large, generally 5-pointed leaves. Fruits are variable in size, color and shape, ranging from about 3 to 100 pounds in weight. With a large diversity in size, shape and color, there are 4 basic types: small types (4-6 pounds) used mostly for cooking and pies; naked seeded types (no seed coat) used for roasted snack seeds; intermediate (8-15 pounds) and large (15-25 pounds) cultivars for cooking and jack-o-lantern;

jumbo types (50-100 pounds) for exhibitions; and miniature cultivars for specialty markets. They have a moderately hard rind, with a thick, edible flesh below, and a central seed cavity. Calabaza is also called pumpkin, West Indian pumpkin, Cuban squash and Tadback. There are 200 seeds per ounce.

3. Season, seeding to mature: 3 to 5 months.

4. Production in U.S.: 63,260 acres, 1992 CENSUS. In 1992, the top 10 states are Illinois (8,297 acres), California (5,552), New York (4,574), Pennsylvania (4,023), Texas (3,465), Ohio (3,345), New Jersey (3,314), Michigan (2,976), Minnesota (2,406) and Indiana (2,197) for 31,852 acres or 50 percent of the total U.S. acreage (1992 CENSUS). In 1992, Puerto Rico reported 3,085 acres and 6,528 tons of calabaza (pumpkin) (1992 CENSUS).

5. Other production regions: In 1995, Canada reported 2,988 acres (ROTHWELL 1996a).

6. Use: As food, culinary frozen and canned commercially. Also stock feed.

7. Part(s) of plant consumed: Internal flesh as food; whole fruit as feed. Small amounts of seeds are consumed as nuts

8. Portion analyzed/sampled: Fruit

9. Classifications:
   a. Authors Class: Cucurbit vegetables
   b. EPA Crop Group (Group & Subgroup): Cucurbit vegetables (9B)
   c. Codex Group: 011 (VC 0429) Cucurbits fruiting vegetables
   d. EPA Crop Definition: Squash = Pumpkin; Also proposing that winter squash = pumpkins

10. References: GRIN, CODEX, LOGAN, MAGNESS, RICHE, US EPA 1994, US EPA 1995a, CROPS

11. Production Map: EPA Crop Production Regions 1, 2, 5, 6 and 10 for 86 percent of total U.S. acreage.

12. Plant Codes:
    a. Bayer Code: CUUMA (*C. maxima*), CUUMO (*C. moschata*), CUUPE (*C. pepo*)

497

1. Purple prairieclover

Fabaceae (Leguminosae)
Dalea purpurea Vent.

2. Purple prairieflower is a erect warm season perennial forb that grows between 20 to 90 cm tall and native to the U.S. It flowers June through August and seeds mature July through September. Native Americans used purple prairieclover to treat wounds and the leaves were also boiled and used as a leafy vegetable. Its habitat includes prairies, plains and hills and on a number of soil types. It reproduces from seeds and rootstocks.

3. Season: Warm season perennial forb that make good hay and grazing from spring through the summer.

4. Production in U.S.: No specific data for Purple prairieclover however, pasture and rangelands are produced on more than 410 million A (U.S. CENSUS, 1992).

5. Other production regions: No entry.

6. Use: Grazing and for hay.

7. Part(s) of plant consumed: Leaves and stems

8. Portion analyzed/sampled: Forage and hay

9. Classifications:
   a. Authors Class: Nongrass Animal Feeds
   b. EPA Crop Group (Group & Subgroup): Miscellaneous

c. Codex Group: No specific citation although the general class 050 (AL 0157) Legume Animal Feeds could be used.
   d. EPA Crop Definition: None

10. References: GRIN, US EPA 1994, US EPA 1995, CODEX, HOLZWORTH, STUBBIENDIECK (picture), RICHE

11. Production Map: EPA Crop Production Region are 5, 7, 8, and 9.

12. Plant Codes:
    a. Bayer Code: No specific entry

498

1. Purslane, Garden
(Pusley, Fatweed, Kitchen purslane, Common purslane, Verdolaga, Pourpier commun)

Portulacaceae
Portulaca oleracea L.

2. Purslane is a common warm season annual that is a persistent, trailing weed with fleshy, succulent stems, and widely distributed. Improved strains are also cultivated as potherbs. Leaves are small, spatulate or narrow obovate, thick and green or red in color. The wild plant is gathered for a potherb in some areas. Cultivated forms are more upright growing, larger and of better flavor. Plants may be grown from seed or from cuttings. Rarely cultivated in the U.S

3. Season, seed to first harvest: About 3 weeks.

4. Production in U.S.: No data, negligible. Most abundant in the eastern states and least common in Pacific Northwest (STEPHENS 1988).

5. Other production regions: Purslane is a popular vegetable in France, other European countries and Eqypt and Sudan (RUBATZKY 1997).

6. Use: As potherb or salad.

7. Part(s) of plant consumed: Leaves and young stems. Also seeds have been eaten raw or ground and made into bread. (One plant can produce 50,000 small black edible seeds)

8. Portion analyzed/sampled: Leaves and young stems

9. Classifications:
   a. Authors Class: Leafy vegetables (except *Brassica* vegetables)
   b. EPA Crop Group (Group & Subgroup): Leafy vegetables (except *Brassica* vegetables) (4A)
   c. Codex Group: 013 (VL 0492) Leafy vegetables (including *Brassica* leafy vegetables)
   d. EPA Crop Definition: None

10. References: GRIN, DUKE(c)(picture), FACCIOLA, MAGNESS, STEPHENS 1988 (picture), US EPA 1995a, RUBATZKY

11. Production Map: EPA Crop Production Regions 1,2 and 3.

12. Plant Codes:
    a. Bayer Code: POROL

499

1. Purslane, Winter
(Cuban spinach, Miner's lettuce, Springbeauty)

Portulacaceae
Claytonia perfoliata Donn ex Willd. (syn: *Montia perfoliata* (Donn ex Willd.) Howell)

2. Winter purslane is a short-lived annual with opposite, somewhat fleshy leaves mostly rising from the root, generally ovate in shape and up to 3 inches long. The plant is mainly

grown in winter in warm countries, and the leaves are used as salad and potherbs. Growth characteristics are similar to spinach.

3. Season, seeding to harvest: About 1 month.
4. Production in U.S.: No data. Cultivated in North America (FACCIOLA 1990). Herb native to North America (MARTIN 1979).
5. Other production regions: No entry.
6. Use: Salad and potherb.
7. Part(s) of plant consumed: Leaves
8. Portion analyzed/sampled: Leaves
9. Classifications:
   a. Authors Class: Leafy vegetables (except *Brassica* vegetables)
   b. EPA Crop Group (Group & Subgroup): Leafy vegetables (except *Brassica* vegetables) (4A)
   c. Codex Group: 013 (VL 0493) Leafy vegetables (including *Brassica* leafy vegetables)
   d. EPA Crop Definition: None
10. References: GRIN, BAILEY 1976, CODEX, FACCIOLA, MAGNESS, MARTIN 1979, US EPA 1995a
11. Production Map: No entry.
12. Plant Codes:
    a. Bayer Code: CLAPE

500

1. Quackgrass

(Couchgrass, Doggrass, Quickgrass, Twitchgrass)

Poaceae (Gramineae)

Elytrigia repens (L.) Desv. ex Nevski (syn: *Agropyron repens* (L.) Beauv., *E. repens* var. *repens* (L.) Desv. ex B.D. Jackson)

2. Quackgrass is native to Europe and Asia and is best known as a troublesome noxious weedy grass of croplands, orchards and lawns. It spreads both by creeping rootstocks and seed. It is believed to have been introduced from Eurasia but is now established over most temperate regions of the world. It grows aggressively and seeds abundantly, and is persistent in both cultivated and abandoned fields. Stems grow up to 3 feet. Leaf blades are thin, flat, up to 6 inches long; and less than $1/2$ inch wide, and sparsely pubescent on the upper surface. Quackgrass is tolerant of saline and alkaline soils and helps bind soil in areas with a rainfall of >330 mm. It has good forage qualities and is useful for pasture, hay and silage. It is used as forage in humid temperate regions and under irrigated western areas. It is more susceptible to drought than other wheatgrasses. It may be harvested from volunteer established areas but is not planted agriculturally. It has some limited use on the rough areas of golf courses. Newer hybrids are less rhizomatous than their parents, and have potential for use on irrigated pasture areas in combination with legumes.
3. Season: Cool season perennial with aggressive rhizomes, and is most productive in the early spring. It can be grazed or cut for hay or silage.
4. Production in U.S.: No data for quackgrass production, however, pasture and rangeland are produced on over 410 million A (U.S. CENSUS, 1992). Adapted to plant hardiness zones 3, 4, 5 and 6.
5. Other production regions: See above.
6. Use: Grazing, forage, hay and silage. Rough areas of golf course.

7. Part(s) of plant consumed: Leaves, stems and seedheads
8. Portion analyzed/sampled: Forage and hay
9. Classifications:
   a. Authors Class: See Forage grass
   b. EPA Crop Group (Group & Subgroup): See Forage grass
   c. Codex Group: See Forage grass
   d. EPA Crop Definition: None
10. References: GRIN, MAGNESS, US EPA 1994, US EPA 1995, CODEX, BARNES, ALDERSON, HOLZWORTH, MOSLER, BALL, RICHE
11. Production Map: EPA Crop Production Regions 7, 9 and 11.
12. Plant Codes:
    a. Bayer Code: AGRRE

501

1. Quince

(Golden apple, Membrillo)

Rosaceae

Cydonia oblonga Mill. (syn: *C. vulgaris* Pers; *Pyrus cydonia* L.)

2. Quinces are fruits closely related to apples and pears, but are of lesser economic importance. The plants are deciduous thornless shrubs or small trees. Generally they are not over 20 feet high. Fruits are mostly 2 inches up in diameter, covered with pubescence. Fruit color is generally yellow. Fruit flesh is rather hard, with a bitter acid taste. The most common cultivars/varieties include 'Champion', 'Pineapple', 'Smyrna' and 'Van Deman'.
3. Season: Bloom to harvest in about 150 days.
4. Production in U.S.: California is key production area with over 200 acres. No available statistical information from other areas.
5. Other production regions: Mediterranean, Argentina and the Middle East.
6. Use: Rarely eaten fresh, mostly utilized for jelly and other preserves. The 'Pineapple' quince must be cooked before using. The flowering quince (*Chaenomeles* spp.*),* widely grown as an ornamental, often sets some fruit that also may be utilized for jelly.
7. Part(s) of plant consumed: Internal flesh, but all cooked prior to pressing for jelly
8. Portion analyzed/sampled: Whole fruit
9. Classifications:
   a. Authors Class: Pome fruits
   b. EPA Crop Group (Group & Subgroup): Crop Group 11: Pome Fruits Group
   c. Codex Group: 002 (FP 0231) Pome fruits
   d. EPA Crop Definition: None
10. References: GRIN, CODEX, LOGAN, MAGNESS, REICH, SCHREIBER, US EPA 1995, WESTPORT, WHEALY, IVES, REHM
11. Production Map: EPA Crop Production Region 10.
12. Plant Codes:
    a. Bayer Code: CYDOB

502

1. Quinoa (Keen-wah)

(Arroz del Peru, Quinua, Vegetable Caviar, Inca rice, Trigo inca, Petty rice)

Chenopodiaceae

Chenopodium quinoa Willd.

2. Staple ancient Incan grain rich in protein. It's an annual herb, $1^1/_2$ to $6^1/_2$ feet tall, native to the Andes Mountains. Quinoa means "mother grain" in the Incan language. Quinoa is a pseudocereal similar to buckwheat and amaranth. The seed coats are covered with bitter saponin compounds that must be removed before human consumption.

Quinoa has a thick, erect woody stalk that made be branched or unbranched and the leaves on young plants are green. Seeded 4 to 6 inches apart in 20 inch rows, then thin to 6 to 8 inches apart. Yields range from 1200 to 1800 lb/A.

3. Season, seeding to harvest: 80 to 190 days. Planted late March to early June and harvested after frost.

4. Production in U.S.: In 1987, about 250 tons. Southern Colorado and Central California (OELKE 1992). Potential production of 6000 acres in Colorado at elevation >7,000 feet.

5. Other production regions: Along the Coast to Canada (OELKE 1992). Western South America (FACCIOLA 1990).

6. Use: Grain used in soup, ground flour for bread, pasta and cereal. Also used as feed. In the U.S. sold as grain and cooked as rice. The leaves can be eaten as a vegetable like spinach.

7. Part(s) of plant consumed: Grain, leaves

8. Portion analyzed/sampled: Grain

9. Classifications:
   a. Authors Class: Cereal grains
   b. EPA Crop Group (Group & Subgroup): Miscellaneous
   c. Codex Group: 020 (GC 0648) Cereal grains
   d. EPA Crop Definition: None

10. References: GRIN, CODEX, OELKE(d), JACOBSEN, FACCIOLA, JOHNSON(a)

11. Production Map: EPA Crop Production Regions 9 and 10.

12. Plant Codes:
   a. Bayer Code: CHEQU

503

1. Radish

(Rabano, Spring radish, Common radish, Small radish, Ravonet, Rabanillo, Garden radish)

Brassicaceae (Cruciferae)

Raphanus sativus L. var. *sativus*

2. The radish commonly grown in the U.S., found in nearly all home gardens and an important market crop, is a rapidly developing annual which produces a tender, spicy, enlarged fleshy taproot. The edible roots are nearly globular or much elongated. They must be harvested while tender, before they become pithy and too strong flavored. The leaves form a rosette from the top of the root. Including the petiole they are up to a foot long, and rough to the touch. The top of the enlarged root is about even with the soil surface. Bunching is done for the local market, but most radishes are topped and packed for the retail market. There are 3,100 seeds per ounce.

3. Season, seeding to harvest: 3 to 6 weeks. Successive planting can be made every 10-14 days.

4. Production in U.S.: 33,217 acres, 1959 CENSUS; 29,893 acres (1992 CENSUS). In 1992, the top 9 states reported as Florida (17,177 acres), Michigan (2,967), California (2,675), Ohio (2,547), Minnesota (1,492), New York (897), Oregon (399), Washington (305), and New Jersey (292) (1992 CENSUS).

5. Other production regions: In 1995, Canada reported 1,717 acres (ROTHWELL 1996a).

6. Use: Mainly as raw in salad for roots. Leaves can be eaten. U.S. per capita consumption of roots in 1994 is 0.4 pounds (PUTNAM 1996).

7. Part(s) of plant consumed: Whole root and top

8. Portion analyzed/sampled: Root and top (leaves). Analyze separately

9. Classifications:
   a. Authors Class: Root and tuber vegetables; Leaves of root and tuber vegetables
   b. EPA Crop Group (Group & Subgroup): Root and tuber vegetables (1A and 1B); Leaves of Root and tuber vegetables (2). (Representative crop)
   c. Codex Group: 016 (VR 0494) Root and tuber vegetables; 013 (VL 0494) Leafy vegetables (including *Brassica* leafy vegetables)
   d. EPA Crop Definition: None

10. References: GRIN, CODEX, LOGAN (picture), MAGNESS, RICHE, ROTHWELL 1996a, SCHREIBER, SCHNEIDER 1985, US EPA 1996a, US EPA 1995a, US EPA 1994, WARE, SWIADER, ZANDSTRA, PUTNAM 1996

11. Production Map: EPA Crop Production Regions 1, 3, 5 and 10.

12. Plant Codes:
   a. Bayer Code: RAPSR

504

1. Radish, Oriental

(Tsumamina, Lohbaak, Clover radish, Chinese white winter radish, Chinese rose winter radish, Daikon, Japanese radish, Lobok, Chinese radish, Winter radish, Lopak, Lorbark, Mingho, Sakurajima, Kaiware, Black radish, Chinese winter radish, Asian radish)

Brassicaceae (Cruciferae)

Raphanus sativus var. *caudatus* (L.) L.H. Bailey
(Mougri, Rat-tail radish)

R. sativus var. *niger* J.Kern. (syn: *R. sativus* L. var. *longipinnatus* L.H. Bailey)
(Daikon)

2. These radishes differ from the common radish in being larger, more pungent, slower in development and producing roots that remain crisp and tender much longer. They usually are seeded in midsummer or later, so the roots reach marketable size in cool weather. Roots can be stored like carrots. The leaf crown becomes larger than that of the common or spring type, so they require more space. The roots become large, 2 or more inches in diameter up to 16 inches or more long. Radishes have been grown in the Orient, developing very large roots, reportedly up to 50 pounds or more, and with leaf top spreads of more than 2 feet. They require a long growing season for such development. Roots are used mainly as cooked vegetables and are a major food in the Orient. They are also preserved by salting as in making sauerkraut. These types are grown in the U.S., mainly for use in Oriental dishes. Culture and exposure of edible parts is similar to those for other root crops. Normally, tops are removed before the roots are put into storage. The normal commercial root weight ranges from about 2 to 8 pounds. The Black radishes ('Long black Spanish' and 'Round black Spanish' cultivars) weigh 2 to 16 ounces and are used as garnish, or cooked. The tops are discarded and the roots are tailed, scrubbed, trimmed and

peeled (SCHNEIDER 1985). Daikon seeds can be used for sprouts, called Kaiware, Tsumamina, Clover radish and Spicy sprouts. In Washington, 90 to 100 percent of the seed produced is exported to Asian countries to produce Kaiware which is used like alfalfa sprouts. In 1993, Washington reported 227 acres for seed in Region 11. Sprouts grow in greenhouses in the U.S. and take about 1 week to produce.

3. Season, seeding to harvest: About 2 to 3 months for best usable size root. Some take about 6 months to reach full size.

4. Production in U.S.: 367 acres of Daikon reported (1992 CENSUS). Hawaii reported 233 acres and California 130 acres of daikon (1992 CENSUS). Also, Oriental radishes are grown on the East Coast for the Oriental market. Washington reported 227 acres for Daikon seed and limited acreage for roots (SCHREIBER 1995).

5. Other production regions: In 1995, Canada reported growing Black radish and Chinese radish. Specific acreage not reported but Chinese vegetables reported as 220 acres (ROTHWELL 1996a).

6. Use: Roots mainly as cooked vegetable, some as salad, or salted. Also leaves can be eaten, normally pickled. Preparation: rinse and peel root. Kimchee is made with Oriental radish. Also sprouts are used for garnish, sandwich or salad. The young seed pods of the rat-tail radish are eaten fresh, cooked or pickled.

7. Part(s) of plant consumed: Fleshy root, usually after peeling, and tops. Seed sprouts. Pods

8. Portion analyzed/sampled: Root and top (leaves). Analyze separately

9. Classifications:
   a. Authors Class: Root and tuber vegetables: Leaves of root and tuber vegetables
   b. EPA Crop Group (Group & Subgroup): Root and tuber vegetables (1A and 1B); Leaves of root and tuber vegetables (2)
   c. Codex Group: 016 (VR 0591) Root and tuber vegetables for Japanese radish and 016 (VR 0590) Root and tuber vegetables for Black radish
   d. EPA Crop Definition: Oriental radish (root and top) = Chinese or Japanese radish (both white and red), winter radish, daikon, lobok, lopak and other cultivars and/or hybrids of these

10. References: GRIN, CODEX, LOGAN (picture), MAGNESS, MYERS, RICHE, ROTHWELL 1996a, SCHREIBER, SCHNEIDER 1985 (picture), STEPHENS 1988 (picture), US EPA 1995a, WARE, YAMAGUCHI 1983, RUBATZKY

11. Production Map: EPA Crop Production Regions 10, 11, 12 and 13.

12. Plant Codes:
   a. Bayer Code: RAPSA (Japanese radish), RAPSN (*R. sativa* var. *niger*)

505

1. Rambutan
(Shaotzu, Ramboutan)
Sapindaceae

Nephelium lappaceum L. (syn: *Euphoria nephelium* DC., *Dimocarpus crinita* Lour.)

2. Grows in hot, wet tropical lowlands. Rambutan is related to lychee, but is more tropical and the trees are smaller, only

reaching 40 ft. The oval fruits are borne in loose terminal or axillary clusters, up to 2 inches long, and covered with soft, fleshy spines about $\frac{1}{2}$ inch in length. The red skin is thin and leathery. In eating, the basal part is cut or torn away, and with slight pressure the pulpy, translucent flesh is pressed out. The flavor is tart and pleasant.

3. Season: Bloom to harvest: About 100 to 150 days. Generally fruits twice a year.

4. Production in U.S.: Hawaii (60 acres-1993) (KAWATE 1995).

5. Other production regions: Tropical southeast Asia, Central & South America(MORTON 1987).

6. Use: Pulp eaten fresh or stewed, canned or preserved into jams and jellies.

7. Part(s) of plant consumed: Fruit pulp

8. Portion analyzed/sampled: Whole fruit

9. Classifications:
   a. Authors Class: Tropical and subtropical fruits – inedible peel
   b. EPA Crop Group (Group & Subgroup): Miscellaneous
   c. Codex Group: 006 (FI 0358) Assorted tropical and subtropical fruits – inedible peel
   d. EPA Crop Definition: See Lychee

10. References: GRIN, ANON(a), CODEX, LOGAN, MAGNESS, MARTIN 1987, US EPA 1994, KAWATE 1995, MORTON (picture)

11. Production map: EPA Crop Production Region 13.

12. Plant Codes:
   a. Bayer Code: NEELA

506

1. Rampion
(Bellflower, Little turnip)
Campanulaceae

Campanula rapunculus L.

2. Rampion is a biennial plant to 3 feet tall, but in cultivation is grown as an annual. Both roots and leaves are eaten, mainly in salads. Leaves are entire, obovate to linear lanceolate in shape, 6 inches or more in length. They form a rosette at the root crown. The roots are long, up to 1 foot, slender, and white. For young roots, the plant resembles radish in culture and exposure. Older roots are used as turnips. Roots can be stored for winter use.

3. Season, seeding to harvest: Up to 5 months.

4. Production in U.S.: No data. Minor. Grown in Florida gardens (STEPHENS 1988). Naturalized in North America (FACCIOLA 1990).

5. Other production regions: Eurasia (FACCIOLA 1990).

6. Use: Mainly in raw salads, both young leaves and roots. Older roots are used as turnips.

7. Part(s) of plant consumed: Mainly root, but leaves also

8. Portion analyzed/sampled: Roots; Leaves

9. Classifications:
   a. Authors Class: Root and tuber vegetables: Leaves of root and tuber vegetables
   b. EPA Crop Group (Group & Subgroup): Miscellaneous
   c. Codex Group: 016 (VR 0592) Root and tuber vegetables
   d. EPA Crop Definition: None

10. References: GRIN, BAILEY 1976, CODEX, FACCIOLA, MAGNESS, STEPHENS 1988 (picture), USDA NRCS

11. Production Map: No entry

12. Plant Codes:
   a. Bayer Code: CMPRP

507

1. Rape
(Rape greens, Colpa, Chou oleifere, Bird rape, Forage rape)
1. Rapeseed oil
(Canola, Colza, Colsat)
Brassicaceae (Cruciferae)
Brassica napus L. var. napus
1. Field mustard seed
(Brown sarson, Toria, Rapeseed oil, Indian rape, Spring turnip rape, Indian colza, Yellow sarson)
   B. rapa ssp. *trilocularis* (Roxb.) Hanelt (syn: *B. rapa* ssp. *sarson* (Prain) Denford; *B. campestris* var. *sarson* Prain)
   (Yellow sarson, Indian colza)
   B. rapa ssp. *dichotoma* (Roxb.) Hanelt (syn: *B. rapa* var. *dichotoma* (Roxb.) Kitam; *B. campestris* var. *toria* Duthie & Fuller)
   (Toria, Indian rape, Brown sarson, Canola, Spring turnip rape)
   B. rapa ssp. *oleifera* Metzg.
   (Canola, Natane, Winter turnip rape, Yu tsai)
2. Rape is grown sparingly as a potherb, more generally for livestock feed and as an oil source. As a potherb the leaves of young plants are used. Leaves are generally lobed, 4 to 12 inches long, half as wide, near glabrous, but with scattered hairs. Flower stems are much branched, up to 3 feet. In exposure of edible parts young rape plants used as potherbs are similar to spinach. Rapeseed oil is obtained from the seeds primarily of the species *B. napus* and *B. rapa* and the oil from different species is not distinguished on the market, since all have similar properties. Production in the U.S. is limited, but rapeseed oil is of major importance in Europe, Canada and Asia. The small, near globular seeds are borne in elongated, closed capsules. They contain 30 to 45 percent of a semi-drying oil. The oil is separated either by solvent extraction or by cold or hot pressing. The term "colza" refers to refined oil. Low glucosinolate and erucic acid levels are now defined as "double low" or canola quality, which is less than 2 percent erucic acid and having less than 30 micromoles of aliphatic glucosinolates per gram of defatted meal and are called low erucic acid rapeseed (LEAR). Rapeseed oil (non-canola type) is not usable as edible oils.
3. Season, growing: Both spring and winter annuals.
4. Production in U.S.: Canola and other rapeseed about 90,000 acres in 1992 CENSUS. Primarily grown in Idaho, North Dakota, Washington, Minnesota, Michigan and Georgia (1992 CENSUS). In 1995, U.S. planted 445,000 acres of canola, 2,500 acres of rapeseed, and 22,900 acres of mustard seed. Canola production reported as 546,984,000 pounds, rapeseed 3,012,000 pounds, and mustard seed 18,304,000 pounds (USDA 1996c).
5. Other production regions: About 10,040,000 acres of canola in Canada for 1995 (ROTHWELL 1996a).
6. Use: Primarily industrial and edible oils. Also used for potherb and grazing.
7. Part(s) of plant consumed: Seeds and leaves (green)
8. Portion analyzed/sampled: Oil use: Seeds and its processed commodities meal and refined oil. Foliage use: Leaves (green). For rapeseed meal, residue data are not needed for non-canola type rapeseed oil since it is produced for industrial uses and is not an edible oil. The edible oil is only produced from canola. Meal is required for both but it is recommended to conduct the trials on canola to include both. For rape greens, commodity is listed in Crop Group 05; *Brassica* (Cole) Leafy Vegetable Group under 40 CFR 180.41
9. Classifications:
   a. Authors Class: Oilseed; *Brassica* (cole) leafy vegetables
   b. EPA Crop Group (Group & Subgroup): Miscellaneous for oil; *Brassica* (cole) leafy vegetables for rape greens
   c. Codex Group: 023 (SO 0495) Oilseed; 067 (OC 0495) Crude vegetable oils; 068 (OR 0495) Edible vegetable oils; 013 (VL 0495) Leafy vegetables (including *Brassica* leafy vegetables) for rape greens; 023 (SO 0694) Oilseed for field mustard seed
   d. EPA Crop Definition: Rapeseed = *B. napus*, *B. campestris* and *Crambe absyssinica* (oilseed-producing varieties only which include canola and crambe)
10. References: GRIN, CODEX, MAGNESS, RICHE, ROBBELEN 1989 (picture), ROTHWELL 1996a, SCHREIBER, US EPA 1995a, US EPA 1995b, USDA 1996c, HANNAWAY(b)
11. Production Map: EPA Crop Production Regions 2, 5, 7 and 11 (mainly oil use).
12. Plant Codes:
   a. Bayer Code: BRSRO (*B. rapa* ssp. *oleifera*), BRSNN (*B. napus* var. *napus*)

508

1. Raspberry
(Red raspberry, Black raspberry, Caneberry, Mayberry, Thimbleberry, Blackcap, Himbeere, Framboise, Frambueso, Yellow raspberry, Bababerry, Keriberry, Tulameen)
Rosaceae
Rubus spp.
1. Red raspberry
(European red raspberry, Frambueso rojas, Chordon)
   R. idaeus L. var. *idaeus*
1. American red raspberry
(Framboisier rouge)
   R. strigosus Michx. (syn: *R. idaeus* var. *strigosus* (Michx.) Maxim.)
1. Black raspberry
(Blackcap, Framboisier de virginia, Frambueso nigro)
   R. occidentalis L.
1. Hill raspberry
(Mysore raspberry)
   R. niveus Thunb.
1. Purple raspberry
   R. occidentalis L. x *R. idaeus* L.
1. Western thimbleberry
(Salmonberry, Thimbleberry)
   R. parviflorus Nutt.
1. Mauritius raspberry
   R. rosifolius Sm.
2. There are three main types of raspberry grown commericaly:
Red raspberry – *R. idaeus* L. and *R. strigosus* Michx., the most common form of raspberry cultivated (over 75 percent of US production).
Black raspberry or blackcaps – *R. occidentalis* L.

Purple raspberry – a cross between red and black types.

Some varieties of red raspberry are yellow to gold in color and sometimes referred to as yellow raspberry. The crown and root system of raspberry are perennial, while the canes are biennial. During the first year the canes usually do not flower and are called primocanes. Following a dormant period the canes flower and set fruit. Plants are propagated vegetatively. Red types tend to be best propagated from root cuttings or suckers with attached roots. Black and purple types are usually propagated by tip layering. Some varieties, noticeably 'Heritage', produce a fall crop on the terminals of current season canes. Canes are stiff and may attain height of 10 feet or more in red and purple raspberries, and 4 or 5 feet in black. Commercially, they are usually supported by parallel wires along each side of the row, and headed to not over 6 feet high. The aggregate fruits, which are borne in loose clusters, consist of numerous seeds, each surrounded by fleshy pulp. The fruits are attached to a receptacle until ripe, but when harvested the receptacle remains on the plant (with blackberry the receptacle stays attached to the picked fruit). The harvested fruit is hemispherical, either rounded or conic and open at the center. Red fruit is $1/2$ to near 1 inch in diameter and similar in length; black fruit is somewhat smaller and flatter; purple fruit is as large or larger than red. Important cultivars include 'Willamette', 'Sweet Briar', 'Meeker', 'Amity' and 'Heritage'.

3. Season: Typical bloom to harvest is 80 to 90 days.
4. Production in U.S.: 15,000 domestic acres in 1994. In 1995 some key production states included Washington (5,900 acres), Oregon (5,200) and California (1,900).
5. Other production regions: Significant imported fruit from Canada and Chile. Additional commercial production in Columbia, Guatemala, Australia, New Zealand and Russia. In 1995, Canada grew 8,031 acres (ROTHWELL 1996a).
6. Use: Fruit are used fresh or processed into desserts, canned fruit, frozen fruit, preserves, ice cream and jam.
7. Part(s) of plant consumed: Fruit
8. Portion analyzed/sampled: Whole Fruit
9. Classifications:
   a. Authors Class: Berries
   b. EPA Crop Group (Group & Subgroup): Crop Group 13: Berries Group as well as this group's Caneberry Subgroup (Representative crop). Based on the above, concerning the retention of the receptacle on the blackberry fruit and not the raspberry; it is recommended to use the raspberry fruit for sampling in residue testing for the caneberry subgroup; thereby providing a cleaner frozen sample for analysis
   c. Codex Group: 004 (FB 0272) Berries and other Small Fruit
   d. EPA Crop Definition: Caneberry = *Rubus* spp.
10. References: GRIN, CODEX, CRANDALL, LOGAN, MAGNESS, SHOEMAKER, SCHREIBER, US EPA 1994, US EPA 1995, WHEALY, ROTHWELL 1996a, REHM, USDA 1996f, CRANDALL, PRITTS
11. Production Map: EPA Crop Production Regions 1, 5 and 12.
12. Plant Codes:
   a. Bayer Code: RUBID (*R. idaeus*), RUBSG (*R. strigosus*), RUBOC (*R. occidentalis*), RUBPA (*R. parviflorus*), RUBSS (*R.* spp.)

509

1. Redtop

(Redtop bentgrass, Black bentgrass, Herdsgrass, Fiorin, Fine bentgrass, Bentgrass-redtop)

Poaceae (Gramineae)

Agrostis gigantea Roth (syn: *A. alba* L., *A. alba* var. *gigantea* Roth (G. Meyer)

2. Redtop is a creeping perennial, which speads by rhizomes and is related to the other bentgrasses. It's a member of the grass tribe *Aveneae*. It is now found over much of the Northern U.S. in meadows, although it is native to Europe. The stems are slender. Leaves are narrow, about $1/4$ inch, and grows to a height ranging from 20 to 150 cm. The panicle is loose and pyramidal in shape, and reddish in color, which accounts for the name. Redtop is used in pasture mixtures under humid conditions and for hay. Also used in lawns and golf greens in the Southeast. The distribution of redtop is similar to Kentucky bluegrass, but is better adapted to acidic (pH 5.5), low fertility and clay type soils. It tolerates wet soils and is used to revegetate strip mining areas and roadbanks. It is planted at a rate of 3 to 5 lb/acre alone. It is low in palatability, but grows and forms a sod quickly, and so protects from soil erosion until slower growing grasses become established. It is rarely seeded alone. However, since the 1940s its use in pasture has declined due to the increase in use of orchardgrass and tall fescue. There are 4,990,000 to 5,740,000 seeds/lb.
3. Season: Cool season, spring and fall growth. Seeded in the fall or spring. Harvest for hay at the full bloom stage in June or for seed in July. Good for cattle and horses, yields ≤1 ton per acre.
4. Production in U.S.: There is 43,089 lb redtop seed produced on 378 acres in Idaho and Illinois (U.S. CENSUS, 1992).
5. Other production regions: See above.
6. Use: Pasture, hay, turfgrass, soil building and for erosion control, revegetating areas such as roadbanks and strip mining fields.
7. Part(s) of plant consumed: Leaves and stems
8. Portion analyzed/sampled: Forage and hay
9. Classifications:
   a. Authors Class: See Forage grass
   b. EPA Crop Group (Group & Subgroup): See Forage grass
   c. Codex Group: See Forage grass
   d. EPA Crop Definition: None
10. References: GRIN, MAGNESS, US EPA 1994, US EPA 1995, CODEX, BARNES, ALDERSON, MOSLER (picture), RICHE
11. Production Map: EPA Production Regions 1, 2, 3, 4, 5, 7, 8, 9 and 11.
12. Plant Codes:
   a. Bayer Code: AGSGI

510

1. Reedgrass

Poaceae (Gramineae)

Calamagrostis spp.

1. Bluejoint

(Canada reedgrass, Bluejoint reedgrass)

Calamagrostis canadensis (Michx.) P. Beauv.

1. Pine reedgrass

(Pinegrass)

C. rubescens Buckley

2. There are nearly 30 reedgrass species (*Calamagrostis* spp.) cool season, rhizomatous, perennial grasses native to North America. Bluejoint and pine reedgrass are two of the most important forage grasses, and are members of the grass tribe, *Aveneae*. Bluejoint is abundant from Alaska through Canada and the northern two-thirds of the U.S., and is used on burned-over forest lands and forage areas and prefers wet meadow areas. Bluejoint produces good quality forage for pasture and hay, and is the main source of wild hay in Wisconsin and Minnesota, and ranges in height from 60 to 150 cm. Pine reedgrass is restricted to the Rocky Mountain, North Central, and prairie and rangeland areas of the U.S. and Canada. It ranges in height from 0.4 to 1 m tall, with flat blades, 8 to 30 cm long and 2 to 5 cm wide. It is adapted to moderately dry soils. While forage quality is considered to be fair for sheep, cattle and horses it is important where it is abundant. Forage quality is lowest when the grass is mature, and does not persist under full sun.

3. Season: Cool season, growth is in the spring and fall. Most palatable in the spring. Forage quality is good until the plant matures. Bluejoint has higher forage quality than does Pine reedgrass.

4. Production in U.S.: No data for either Bluejoint or Pine reedgrass production, however, pasture and rangeland are produced on more than 410 million A (U.S. CENSUS, 1992). See above.

5. Other production regions: Both are common in Canada.

6. Use: Pasture hay, and for erosion control.

7. Part(s) of plant consumed: Leaves and stems

8. Portion analyzed/sampled: Forage and hay

9. Classifications:
   a. Authors Class: See Forage grass
   b. EPA Crop Group (Group & Subgroup): See Forage grass
   c. Codex Group: See Forage grass
   d. EPA Crop Definition: None

10. References: GRIN, US EPA 1994, US EPA 1995, CODEX, HOLZWORTH, STUBBENDIECK(a) (picture), MOSLER, RICHE

11. Production Map: EPA Production Regions for bluejoint are 5, 7, 8, 11 and 12; and for pine reedgrass are 10, 11 and 12.

12. Plant Codes:
    a. Bayer Code: CLMCD (*C. canadensis*)

511

1. Rhodesgrass

(Pasto rhodes, Windmill grass)

Poaceae (Gramineae)

Chloris gayana Kunth

2. This grass is native to South Africa and was brought to the U.S. in 1902. It is a fine-stemmed, leafy warm season subtropical and tropical perennial, growing up to 3 feet high. It produces abundant seed, and also spreads by running stolons which may reach 6 feet in length and produce a plant at each node. The plants are not winter hardy, being killed by temperatures below about 15 degrees F. Consequently, it is adapted in this country only to near the Gulf Coast from Florida to southern Texas and southern Arizona and California. While fairly drought resistant, ample moisture is needed for good production. Under favorable conditions of moisture and soil fertility, heavy yields of high quality pasturage and

hay are secured. Stands are established by seeding. There are 1,300,000 seeds/lb.

3. Season: warm season perennial grass. Hay cut before flowering. Seed matures about 25 days after flowering. Can be grazed 4 to 6 months after planting.

4. Production in U.S.: No specific data, See Forage grass. Southern tier of states.

5. Other production regions: South Africa and worldwide in tropics and warm temperate regions (GOULD 1975).

6. Use: Pasture, hay and erosion control.

7. Part(s) of plant consumed: Foliage

8. Portion analyzed/sampled: Forage and hay

9. Classifications:
   a. Authors Class: See Forage grass
   b. EPA Crop Group (Group & Subgroup): See Forage grass
   c. Codex Group: See Forage grass
   d. EPA Crop Definition: None

10. References: GRIN, MAGNESS, ALDERSON, SKERMAN, GOULD

11. Production Map: EPA Crop Production Regions 2, 3, 4, 6, 8 and 10.

12. Plant Codes:
    a. Bayer Code: CHRGA

512

1. Rhubarb

(Pieplant, Ruibarbo, Wineplant) (others: Chinese rhubarb, Turkey rhubarb, Da-huang)

Polygonaceae

Rheum x *hybridum* Murray (syn: *Rheum rhabarbarum* auct.; *R. rhaponticum* auct.)

2. Rhubarb is a cool season perennial plant which forms large fleshy rhizomes and large leaves with long, thick petioles, the edible portion. Rhizomes and the crown persist for many years. The leaves grow from the crown in early spring. Leaf blades, which are poisonous, are up to a foot or more in width and length. The edible petioles are up to 18 inches long, 1 to 2 inches in diameter, generally somewhat hemispherical in cross section. Roots are also taken up and bedded in cellars or houses in winter, forcing growth in darkness to produce etiolated leaf-stems, which are much prized. A stand of rhubarb is most productive beginning its third year and lasting through its sixth year. Chinese rhubarb is *R. officinale* Baill., *R. palmatum* L., and *R. tanguticum* Maxim. ex Balf. *R. tanguticum* is also called Turkey rhubard and Da-huang.

3. Season, start of spring growth to first harvest: About 3-4 weeks with 2 harvest cycles.

4. Production in U.S.: 3,010 acres reported 1959 CENSUS. 861 acres reported (1992 CENSUS); also in many home gardens. Processed rhubarb yields about 6 tons per acre, and fresh rhubarb yields about 4 tons per acre (SCHREIBER 1995). Strawberry rhubarb is grown in greenhouses and available from Washington, January to April (LOGAN 1996). In 1992, the top 6 states reported, Washington (361 acres), Oregon (307), Michigan (93), California (27), Massachusetts (17), and Wisconsin (11) (1992 CENSUS).

5. Other production regions: In 1995, Canada reported 512 acres (ROTHWELL 1996a).

6. Use: Fresh as sauce or culinary; also frozen commercially.

7. Part(s) of plant consumed: Large fleshy leaf petioles only

8. Portion analyzed/sampled: Petioles without leaf blades
9. Classifications:
   a. Authors Class: Leafy vegetables (except *Brassica* vegetables)
   b. EPA Crop Group (Group & Subgroup): Leafy vegetables (except *Brassica* vegetables) (4B)
   c. Codex Group: 017 (VS 0627) Stalk and stem vegetables
   d. EPA Crop Definition: None
10. References: GRIN, CODEX, LOGAN, MAGNESS, RICHE, SCHREIBER, STEPHENS 1988 (picture), US EPA 1994, US EPA 1996a, US EPA 1995a, SWIADER
11. Production Map: EPA Crop Production Regions 5 and 12.
12. Plant Codes:
    a. Bayer Code: RHERH

513

1. Rice
(Upland rice, Arroz, Riz, Paddy rice, Rough rice, Wet paddy rice, Dryland rice, Lowland rice, Common rice, Sweet rice, Waxy rice, Nomi, Sake)

Poaceae (Gramineae)

Oryza sativa L.

2. Rice is the principal food crop in practically all the tropical regions of the world as well as in most subtropical areas and in some temperate zone areas having a relatively long growing season. Probably half of the world population depends on rice for its major food source. Rice is grown on around 200 million acres outside of mainland China. Probably an additional 100 million acres are grown there. In the U.S., rice acreage (1966-67) was just under 2 million, mainly in California, Arkansas, Louisiana and Texas. Acreage (1995) was just over 3 million in Arkansas, California, Louisiana, Mississippi, Missouri and Texas. Rice has been cultivated since antiquity. Seeding of rice was a religious rite in China nearly 5000 years ago. The cultivated plant probably originated in Southeast Asia where wild types still persist. Rice cultivation in Europe began around 700 A.D. Rice was brought to America by the earliest colonists. All rice grown in the U.S. and most of that cultivated in other countries is of the species *Oryza sativa* L. This is really a cultivar species, not found in the wild. Some 20 to 25 species of *Oryza* are known, but the ancestry of the cultivated types is uncertain. The species *O. glaberrima* Steud. is cultivated in Africa. More than 8,000 variety names for rice are known, but many of these may be local names given to similar or identical varieties. However, many hundreds of varieties distinct in characteristics or in adaptation are known. These are roughly classed in three groups as follows:

Japonica **Group.** In general, varieties in this group have short kernels. Stems are stiff, short and upright. Leaves are short, dark green and the second leaf forms a narrow angle with the stem. Plants are pubescent and form many tillers. Panicles are numerous, short, dense and heavy. Spikelets are awnless. This group is generally grown in more northern climates as Japan, Korea, Northern China, Europe, as well as California and Arkansas in this country (short grain).

Indica **Group.** This group is more tropical in adaptation than *Japonica* varieties and includes the kinds grown in Southern Asia, the Philippines, and the South-Central states and California in the U.S. These varieties are characterized by long kernels, long, light green leaves, tall, somewhat spreading stems, much less stiff than stems in *Japonica* varieties and tend to lodge. Panicles are numerous, long, light in weight, medium in density. Spikelets are awnless (long grain).

Bulu **or** *Javanica* **Group.** This group is of minor importance compared with *Japonica* and *Indica*. Varieties classed here are grown mainly on the islands off Southeast Asia. In the U.S., medium grain rice is grown in Arkansas, California, Louisiana, Missouri and Texas. They are somewhat intermediate in characteristics. The kernels are large, stems are tall, stiff and upright. Panicles are few in number, of medium length and density, but heavy. Awns are numerous (medium grain).

Rice Culture

The rice plant differs from most grains in that it thrives best when grown with the soil surface covered with water. Some rice is grown on soils that are not flooded; but soils well supplied with moisture are necessary and even then yields are much less than with flooded rice. This so-called "upland" rice is not grown in this country.

Upland or dryland rice is direct seeded in a dry seedbed and grows to maturity without natural or artifical prolonged flooding conditions. It is generally restricted to tropical areas.

Lowland or wet paddy rice is grown in the U.S. It grows under artifical flood that may be continuous from seeding until drained for harvest.

In Arkansas, long grain types mature in about 110 to 130 days, while medium grain types mature in 135 days. These grains are flooded 66 to 90 days during the heading period. Some crop rotations for rice include oat, lespedeza, soybean, fish, and grain sorghum.

Rice seed requires a high soil moisture content to germinate. It germinates and grows readily when seeded on the soil surface under water. Much of the seeding in this country is on flooded sites, the seed being spread from airplanes. It may also be drilled in prepared soil, followed by flooding after growth starts. In Oriental countries where land is scarce, rice is often started in nurseries and transplanted to the field. This is an economical use of land, but does not result in higher yields than seeding in place. For highest yields, fields are kept flooded from time of seeding or transplanting until shortly before harvest, with water 4 to 8 inches deep. Water should not become stagnant. If possible, a slow inflow and outflow of water should be maintained. An alternative practice is to draw off the old water and flood with fresh water at intervals.

In the U.S. nearly all rice is harvested with combines. Water is withdrawn 2 to 3 weeks before harvest so as to dry the soil enough for combine operation. Rice so harvested generally contains too much moisture to keep in storage and must be dried artificially prior to storage. An alternative method, no longer used in this country, is to mow the rice and let it lie in swaths to dry prior to threshing. Some modification of this method is widely used in most countries where rice is grown. In very humid climates the mowed rice may need to be removed from the field and kept off the ground to permit drying prior to threshing.

The Rice Plant

Rice is cultivated as an annual. In areas where freezing does not occur, plants will persist for more than one season through rooting of tillers. Whether under water or in moist soil the central stem first emerges. Within a few days tiller or branch plants grow from buds in leaf axils near the soil level. The number of such tillers varies with the density of the plant stand and with the variety. Usually not more than half a dozen stems develop from a single seed, but up to 50 tillers may form on some kinds if very widely spaced. Only tillers that form early produce panicles and ripen grain.

Stem height ranges from less than 20 inches to 6 feet or more. Stems are hollow between nodes. Tall growing kinds are subject to lodging or falling over, particularly if fertilized. The shorter- and stiffer-stemmed varieties are therefore more satisfactory as use of fertilizers generally results in greatly increased yields. Recent development of short, stiff-stemmed varieties adapted to the tropics, coupled with the increased use of fertilizer, is resulting in greatly increased production in South Asian countries.

The panicle or seed bearing head roughly resembles that of oats but is more compact and tends to droop more. The panicle is usually 4 to 10 inches long with branches that rise singly or in whorls. Each branch bears several spikelets, each with a single flower. The panicles usually contain from 75 to 150 spikelets, but the number may be greater. The panicle is initially enclosed in a sheath. It emerges from the sheath about 75 to 80 days after seeding in the earliest varieties and in 125 to 130 days in late varieties. The period from panicle emergence to full seed ripeness is about 35 days. The rice flower consists of two small, sterile lemmas which partially cover the developing seed; floral bracts, the lemma and palea; and within these the stamens and pistil, or sex organs. As in most oat and barley varieties, the lemma and palea adhere to the developing seed or kernel. The lemma may be awned, but is awnless, or near awnless, in most cultivated varieties. The adhering lemma and palea constitute the "hull" of the threshed grain or so-called "rough" rice. After the hull is removed in milling the rice is termed "brown" or "hulled". The hull comprises about 20 percent of the kernel weight. Further milling removes the bran (5 to 8 percent), or seed coat, the germ (4 percent), and some of the endosperm (68 percent). This results in about an additional 10 percent removal by weight.

Milling and Uses of Rice

The primary use of rice is for human food. Primary culinary variants include stickiness and aroma. For food use, the hull, present on the threshed or "rough" rice, is first removed. For home use in Oriental countries, this may be done by use of a hollow block and wooden pounder, which leaves the bran layer and the germ on the rice. In the U.S., other European countries, and to a considerable extent in the Orient, rice is machine milled. This removes the hull, bran, germ and some of the endosperm.

The first step after cleaning the rice to remove any chaff and foreign matter, is to pass the kernels through a sheller which may consist of paired rubber rollers revolving at different speeds. Hulling stones may also be used. One horizontal stone is stationary and a second one revolves. When properly spaced they loosen the hulls with minimum kernel breakage. The product from this step is brown rice, the bran layer and germ being still present on the grain. Brown rice is generally further milled in the U.S. and Europe to remove the bran and germ. This is accomplished by rubbing or scouring. A final step is termed brushing. It consists of passing the rice through a rapidly revolving vertical cylinder covered with overlapping pieces of cowhide or pigskin. This results in a polished surface on the grains.

A step termed parboiling, preliminary to hulling, is often used in Oriental countries and to some extent in the U.S. The rough rice is first soaked, then steamed either under pressure or without pressure, and subsequently dried before removing the hulls. This results in a toughening of the endosperm and less breakage. Also, minerals and vitamins present in the bran and embryo move into the grain proper to some extent, so parboiled rice is higher in these important nutrients than rice hulled without such treatment.

In the U.S. a process having a similar effect is sometimes used. The rough rice is first put under vacuum to remove air, then is soaked under pressure and finally steamed under pressure. This process shortens the time of treatment compared with parboiling and results in similar reduced breakage of the endosperm and enhanced nutritive value.

Milled rice is used mainly for direct consumption, usually after cooking by boiling. It may be sold as precooked or partially precooked rice. Rice is also used extensively as breakfast foods, as puffed rice, flakes or rice crispies. Broken rice may be used as food or in manufacture of alcoholic beverages. Rice flour, a product of final milling, is used in various mixes. The bran is used mainly as livestock feed. Rice hulls are used for fuel, insulation, mulch and in certain manufacturing processes. About 80 percent of the rice in the U.S. is used for human food and 16 percent for brewers use. Rice oil can be extracted from brown rice and used for salad dressing, cooking oil and soaps. Glutinous rice, also called nomi, waxy or sweet rice, is used for desserts.

3. Season, seeding to harvest: About 110-130 days for early varieties and 160-195 for late varieties.

4. Production in U.S.: In 1995, all rice 3,121,000 acres planted with 8,693,550 tons (USDA 1996c). For all rice in 1995, the six states reporting: Arkansas (1,350,000 planted acres), Louisana (575,000), California (467,000), Texas (320,000), Mississippi (290,000), and Missouri (119,000). Most production was long grain with 2,335,000 acres planted, next is medium grain with 774,000 acres and short grain at 12,000 acres. The top state for medium and short grains is California. The top state for long grain is Arkansas (USDA 1996c).

5. Other production regions: Oriental countries, tropical Asia. Also see above.

6. Use: See above; Human food, bran, livestock feed and flour. Per capita consumption in 1994 is 18.4 pounds. A fermented drink is called sake.

7. Part(s) of plant consumed: Grain and straw

8. Portion analyzed/sampled: Grain and straw, and grain processed commodities polished rice, hulls and bran. Grain is kernel plus hull. For rice straw, stubble (basal portion of the stems) left standing after harvesting the grain

9. Classifications:

   a. Authors Class: Cereal grains; Forage, fodder and straw of cereal grains

   b. EPA Crop Group (Group & Subgroup): Cereal grains (15) (Representative crop); Forage, fodder and straw of cereal grain

   c. Codex Group: 020 (GC 0649) Cereal grain; 065 (CF 0649) Processed rice bran; 058 (CM 1206) Unprocessed rice bran, 051 (AS 0649) Dry straw and fodder, 058 (CM 0649) Husked rice, 058 (CM 1205) Polished rice

   d. EPA Crop Definition: None

10. References: GRIN, CODEX, FACCIOLA, MAGNESS, USDA 1996c, US EPA 1996a, US EPA 1995a, US EPA 1994, IVES, MARSHALL 1993, SMITH 1995, ADAIR, HELM, HILL

11. Production Map: EPA Crop Production Regions 4, 5, 6 and 10 with most in 4.

12. Plant Codes:

a. Bayer Code: ORYSA, ORYSP (Paddy rice), ORYSW (Seeded paddy rice), ORYSD (Seeded upland rice), ORYSI (Dry seeded upland rice, followed by irrigation)

514

1. Rose apple
(Pomarrosa, Jambos, Pommier rose, Malabar plum)
Myrtaceae
Syzygium jambos (L.) Alston (syn: *Eugenia jambos* L.)
2. Tropical evergreen tree to 30 feet in height. The fruits are ovoid or globular up to 2 inches in length and are white or pale yellow. The fruit has a thin layer of pale yellow flesh and 1 to 3 brown seeds, which lie loose in a large, hollow seed cavity. The fruit is sometimes eaten out of hand but is more generally used to make jelly and jams. It also can be stewed or preserved in syrup.
3. Season, harvest: In Florida, the main season is May through July. Tree fruits sporadically all year.
4. Production in U.S.: No data. Florida and Hawaii (MORTON 1987).
5. Other production regions: West Indies.
6. Use: Fresh, jams and jelly and candied.
7. Part(s) of plant consumed: Fruit
8. Portion analyzed/sampled: Whole fruit
9. Classifications:
   a. Authors Class: Tropical and subtropical fruits – edible peel
   b. EPA Crop Group (Group & Subgroup): Miscellaneous
   c. Codex Group: 005 (FT 0309) Assorted tropical and subtropical fruits – edible peel
   d. EPA Crop Definition: None
10. References: GRIN, CODEX, MAGNESS, MORTON (picture)
11. Production Map: EPA Crop Production Region 13.
12. Plant Codes:
    a. Bayer Code: SYZJA

515

1. Roselle
(Florida cranberry, Agrio de Guinea, Jamaica sorrel, Indian sorrel, Red sorrel, Vina, Sorrel)
Malvaceae
Hibiscus sabdariffa L.
2. The plant is a strong annual, 5 to 7 feet in height. The leaves are lobed. The calices of the flowers are red and thick or fleshy, and are the principal edible portion. When cooked, they make a sauce or jelly somewhat like cranberry. The juice is also used as an acid drink. The culture of Roselle is similar to eggplant or peppers. Plants are started in beds, then transplanted to the field. While the bolls are still green they are picked and the fleshy calices removed. Roselle is grown rather extensively in the tropics, and to some extent in warmer sections of the U.S. Exposure of edible parts is similar to that of small fruited tomatoes. Culture is similar to eggplants and okra.
3. Season, seeding to first harvest: About 3 to 5 months.
4. Production in U.S.: No data. Limited. Florida (STEPHENS 1988).
5. Other production regions: Tropics.
6. Use: Jellies, jams, culinary, acid drink, wines, potherb.
7. Part(s) of plant consumed: Thick, fleshy sepals, or calyx, which envelop seed boll. Also, dried flowers for tea and leaves as greens

8. Portion analyzed/sampled: Flower head (fresh and dry); Leaves
9. Classifications:
   a. Authors Class: Fruiting vegetables (except cucurbits)
   b. EPA Crop Group (Group & Subgroup): Miscellaneous
   c. Codex Group: 012 (VO 0446) Fruiting vegetables other than cucurbits; 013 (VL 0446) Leafy vegetables (including *Brassica* leafy vegetables); 066 (DT 0446) Teas
   d. EPA Crop Definition: None
10. References: GRIN, CODEX, KENNARD, MAGNESS, STEPHENS 1988 (picture), RUBATZKY, DUKE(a)
11. Production Map: EPA Crop Production Regions 3 and 13.
12. Plant Codes:
    a. Bayer Code: HIBSA

516

1. Rosemary
(Alercim, Romarin, Romero)
Lamiaceae (Labiatae)
Rosmarinus officinalis L.
2. The plant is a hardy, evergreen perennial shrub, 2 to 4 feet high, native to the Mediterranean region. The needle-like leaves are narrow and entire. Leaves are used in cooking meats, fish and stews. Such use is more extensive in Europe than in the U.S. Oil of rosemary is obtained by distillation from the leaves and flowering tops.
3. Season, transplant to harvest: Harvested on continuing basis for fresh herbs; dried-leaf production, harvest 1 to 2 times during growing season once it is 2 years old.
4. Production in U.S.: No data. California (FOSTER 1993), Florida (STEPHENS 1988).
5. Other production regions: See above.
6. Use: Seasoning for meat, salads, baked products, confections and tea. Also grown as an ornamental hedge.
7. Part(s) of plant consumed: Mainly leaves
8. Portion analyzed/sampled: Leaves (fresh and dried)
9. Classifications:
   a. Authors Class: Herbs and spices
   b. EPA Crop Group (Group & Subgroup): Herbs and spices (19A)
   c. Codex Group: 027 (HH 0741) Herbs; 057 (DH 0741) Dried herbs
   d. EPA Crop Definition: None
10. References: GRIN, CODEX, FACCIOLA, FARRELL (picture), FOSTER (picture), LEWIS, MAGNESS, MARTIN 1979, REHM, STEPHENS 1988, US EPA 1995a
11. Production Map: EPA Crop Production Regions 3 and 10.
12. Plant Codes:
    a. Bayer Code: RMSOF

517

1. Rough pea
(Singletary pea, Wild winter pea, Caley pea)
Fabaceae (Leguminosae)
Lathyrus hirsutus L.
1. Flat pea
(Flat peavine)
L. sylvestris L.
2. Rough pea, native to the Mediterranean region, is a winter annual adapted to the southern third of the U.S. The stems

are weak and trailing up to 3 feet long except in dense stands. The leaves have one pair of long, narrow leaflets and terminate in a coiled tendril. The seed pods are rough and hairy with 5 to 10 seeds each. Seeds are usually sown in the fall. Plantings are used for pasture, hay or turning under for soil improvement. Livestock may be injured from grazing plants with mature seeds or feeding hay from such plants, but prior to seed ripening good pasturage and hay are produced. Growing rough pea has declined greatly in recent years as Austrian winter peas and hairy vetch make more growth. Acreage harvested for seed was 34,631 in 1949 and only 8,109 in 1959, according to CENSUS figures. Important winter legume and soil improvement crop in southern U.S.

Flat pea is a long-lived leafy legume useful for revegetation of roadbanks. It is adapted to low pH soils and fed with other forages. Often seeded with Tall fescue, Perennial ryegrass or Redtop.

3. Season, grazing March-May. Discontinue grazing late spring to let it naturally reseed.
4. Production in U.S.: No recent data. Southern U.S.
5. Other production regions: Canada grows lathyrus, see vetch, chickling.
6. Use: Pasture, hay.
7. Part(s) of plant consumed: Foliage
8. Portion analyzed/sampled: Forage and hay
9. Classifications:
   a. Authors Class: Nongrass animal feeds
   b. EPA Crop Group (Group & Subgroup): Miscellanous
   c. Codex Group: No specific entry
   d. EPA Crop Definition: None
10. References: GRIN, MAGNESS, DUKE(h) (picture), BALL (picture), BARNES
11. Production Map: EPA Crop Production Regions 2, 3 and 4.
12. Plant Codes:
   a. Bayer Code: LTHHI (*L. hirsutus*), LTHSY (*L. sylvestris*)

518

1. Roundleaf cassia
(Roundleaf sensitive pea)
Fabaceae (Leguminosae)
Chamaecrista rotundifolia (Pers.) Greene (syn: *Cassia rotundifolia* Pers.)

2. Roundleaf cassia is a herbaceous short-lived perennial that grows prostrate on stems 30 to 110 cm long, and is native to Florida. It grows on light textured soils in Mexico, Central America, the Caribbean, northern South America and Australia. It produces pods 1.5 to 4 cm long and 3 to 5 cm wide, and is grown in areas receiving 500 to 600 mm rainfall/season. Roundleaf cassia produces up to 7000 kg DM/ha/season and has seed yields ranging from 200 to 800 kg/ha. There are 200,000 to 470,000 seeds/kg.
3. Season: Warm season short-lived perennial forage that can be managed as an annual crop. It is used for pasture.
4. Production in U.S.: No specific data for Roundleaf cassia production, however, pasture and rangelands are produced on more than 410 million A (U.S. CENSUS, 1992). See above.
5. Other production regions: See above.
6. Use: Pasture grazing, hay, and revegetation of eroded areas.
7. Part(s) of plant consumed: Leaves and stems

8. Portion analyzed/sampled: Forage and hay
9. Classifications:
   a. Authors Class: Nongrass Animal Feeds
   b. EPA Crop Group (Group & Subgroup): Miscellaneous
   c. Codex Group: No specific citation although the general class 050 (AL 0157) Legume Animal Feeds could be used
   d. EPA Crop Definition: None
10. References: GRIN, US EPA 1994, US EPA 1995, CODEX, HOLZWORTH, SKERMAN(a) (picture), RICHE
11. Production Map: EPA Crop Production Region are 3 and 13.
12. Plant Codes:
   a. Bayer Code: CASRO

519

1. Rue
(Common rue, Herb-of-Grace, Garden rue, Ruda)
Rutaceae
Ruta graveolens L.

2. The plant is a perennial herb, woody at the base, up to 3 feet in height. The fern-like leaves are much divided and have a very strong odor. In ancient times, rue was much used as a flavoring and for medicinal purposes. It is now used to some extent by people who like bitter flavors in cookery and beverages. The volatile oil, distilled from the whole plant, is used in aromatic vinegar and toilet preparations.
3. Season, seeding to harvest: Harvest as needed before bloom.
4. Production in U.S.: No data. It is hardy in southern U.S. (FOSTER 1993).
5. Other production regions: Mediterranean region (FACCIOLA 1990).
6. Use: Culinary. Leaves added to salads and flavoring component in pickles.
7. Part(s) of plant consumed: Leaves
8. Portion analyzed/sampled: Leaves (fresh and dried)
9. Classifications:
   a. Authors Class: Herbs and spices
   b. EPA Crop Group (Group & Subgroup): Herbs and spices (19A)
   c. Codex Group: 027 (HH 0742) Herbs; 057 (DH 0742) Dried herbs
   d. EPA Crop Definition: None
10. References: GRIN, BAILEY 1976, CODEX, FACCIOLA, FOSTER (picture), MAGNESS, NEAL, US EPA 1995a
11. Production Map: No entry
12. Plant Codes:
   a. Bayer Code: RUAGR

520

1. Rutabaga
(Swede, Swedish turnip, Turnip-rooted cabbage, Laurentian turnip, Russian turnip, Nabo)
Brassicaceae (Cruciferae)
Brassica napus var. *napobrassica* (L.) Rchb.

2. The rutabaga is a root crop, similar to turnip, but differing from turnip in having a denser-textured root which bears more side roots. The leaves are glabrous with a bluish bloom, instead of hairy as in turnip. Roots are harvested at a more advanced stage than turnip, so the growing season is longer. Roots may be stored for winter use. In culture and

exposure of plant parts, rutabaga is similar to beets. Rutabagas are better adapted to northern regions than turnips and are biennials grown as an annual. Rutabagas are harvested at 3-5 inches in diameter, generally yellow-fleshed, and are longer and rounder than turnips. Leaves are larger and fleshier with a stronger flavor than turnips. There are 11,000 seeds per ounce.

3. Season, seeding to harvest: 3 months or more.

4. Production in U.S.: No data, normally combined with turnip roots. Grown in Washington (SCHREIBER 1995). U.S. production is concentrated in Wisconsin, Minnesota and Washington.

5. Other production regions: Canada.

6. Use: Root is cooked like turnip. Eaten raw or cooked, peeled before use.

7. Part(s) of plant consumed: Fleshly root only. Leaves are edible, but not highly regarded as a cooking green

8. Portion analyzed/sampled: Root, (40 CFR 180.1 (J)(6) states that rutabaga tops are removed and discarded before analyzing roots)

9. Classifications:
   a. Authors Class: Root and tuber vegetables; Leaves of root and tuber vegetables
   b. EPA Crop Group (Group & Subgroup): Root and tuber vegetables (1A and 1B); Leaves of root and tuber vegetables (2)
   c. Codex Group: 016 (VR 0497) Root and tuber vegetables for roots; 013 (VL 0497) Leafy vegetables (including *Brassica* leafy vegetables)
   d. EPA Crop Definition: None

10. References: GRIN, CODEX, LOGAN (picture), MAGNESS, SCHREIBER, STEPHENS 1988 (picture), US EPA 1996a, US EPA 1995a, REHM, SWIADER, FORTIN

11. Production Map: EPA Crop Production Regions 5, 8, 10, 11 and 12.

12. Plant Codes:
    a. Bayer Code: BRSNA

521

1. Rye

(Roggen, Seigle, Centeno)

Poaceae (Gramineae)

Secale cereale L. ssp. *cereale*

2. Rye in the U.S. was planted on about 3.5 million acres (1966-67), but only about one-third of that acreage was harvested for grain. The rest was used for hay, silage, pasture, erosion control or plowed under for soil improvement. Production of grain for the two years averaged 25.9 million bushels. In 1994-95, production averaged over 10 million bushels. Most of the rye for grain is produced in the southern and central states. The rye plant grows rapidly and vigorously from seed, resulting in a rapid cover valuable for erosion control or early pasture. Selected varieties are hardier to cold than other cereal grains, so rye as a winter crop can be grown in areas too cold for winter wheat. Also, rye will produce better on light, sandy soils and on soils of low fertility than other small grains. Because rye develops rapidly, especially in early spring, it can be plowed-in early and still give a good volume of organic matter for soil improvement. Rye as pasture or hay is less palatable than other small grains or legumes but is readily grazed if other grazing is not available. Rye as a grain

crop is similar in most respects to wheat. Practically all the rye for grain is sown in the fall and harvested in early summer. It is earlier maturing than wheat. Stems reach 3 to 5 feet in height. The spikes are 3 to 5 inches long, slender and awned. Spikelets generally contain two fertile flowers. Seeds are enclosed in the palea and lemma, as in wheat, but tend to protrude when near ripe, so are less completely enclosed than in wheat. The seed threshes free of the palea and lemma. The seeds tend to shatter or fall out when ripe. For this reason, coupled with earlier ripening, rye may be a bad weed in wheat fields. In the Midwest, rye is primarily grown for grain and occasionally for hay or pasture. Rye makes excellent forage, especially when combined with red or crimson clovers and ryegrass. Rye fits into crop rotations and can be grazed followed by warm season crops, such as corn, sorghum and an oilseed crop. The rye kernel is composed of 12 to 17 percent pericarp, 80 to 85 percent endosperm, and 2 to 4 percent germ. About 125 pounds of rye grain gives a milling yield of 100 pounds of rye flour. Uses of rye grain: Rye is second only to wheat for flour production. Milling of rye is essentially similar to wheat. Baked goods made with rye flour have a distinctive flavor. As feed, rye is not relished by livestock, so rye grain is usually fed in mixtures with other cereals. In nutritive value, rye is a little lower than wheat. Substantial quantities of rye are also used for making distilled alcoholic beverages. In 1994, the per capita consumption of rye was 0.61 lb.

3. Season, seeding to harvest: About 3 months.

4. Production in U.S.: In 1995, 1,612,000 acres planted with 9,928,000 bushels of grain (USDA 1996c). In 1995, the top six states reported as Georgia (300,000 acres planted), Oklahoma (190,000), Texas (150,000), North Carolina (100,000), Michigan (90,000) and Virginia (90,000) (USDA 1996c).

5. Other production regions: Canada planted about 479,000 acres in 1996, mainly Saskatchewan and Alberta (ROTHWELL 1997).

6. Use: Food grain, bread, distilled alcohol, animal feed, pasture, hay, silage and cover crop.

7. Part(s) of plant consumed: Grain and foliage

8. Portion analyzed/sampled: Grain, forage and straw. The grain processed commodities are flour and bran. Grain is kernel with hull removed. For rye forage, cut sample at 6 to 8 inch stage to stem elongation (jointing) stage, at approximately 30 percent dry matter. For rye straw, cut plant residue (dried stalks or stems with leaves) left after the grain has been harvested (threshed)

9. Classifications:
   a. Authors Class: Cereal grains; Forage, fodder and straw of cereal grains
   b. EPA Crop Group (Group & Subgroup): Cereal grains (15); Forage, fodder and straw of cereal grain (16)
   c. Codex Group: 020 (GC 0650) Cereal grain; 065 (CF 0650) Processed rye bran, 058 (CM 0650) Unprocessed rye bran; 078 (CP 1250) Rye bread, 065 (CF 1250) Rye flour, 051 (AF 0650) Green fodder, 051 (AS 0650) Dry straw and fodder; 065 (CF 1251) Rye wholemeal
   d. EPA Crop Definition: None

10. References: GRIN, CODEX, MAGNESS, USDA 1996c, US EPA 1996a, US EPA 1995a, US EPA 1994, IVES, REHM, SCHNEIDER 1996a, OELKE(f), ROTHWELL 1997

11. Production Map: EPA Crop Production Regions 1, 2, 3, 5, 7 and 8 with most in 3, 5 and 7.

12. Plant Codes:

a. Bayer Code: SECCE (*S. cereale*), SECCS (Spring rye), SECCW (Winter rye)

522

1. Ryegrass, Italian
(Annual ryegrass, Australian ryegrass, Raigras italiano, Nezumimugi, Darnel, Dokumugi, Rabillo)

Poaceae (Gramineae)

Lolium multiflorum Lam.

2. This is an annual grass from Europe, grown for hay in Oregon and Washington west of the Cascade Mountains. It grows to a height of 2 to 3 feet and is adapted to poorly drained soils. Ryegrass is considered the most important pasture forage/turf grass in the world. Over 90 percent of the annual ryegrass is utilized as winter pasture and 80 percent of these pastures are overseeded into warm season perennial grasses to extend the grazing season. In the southern states it is grown as a winter annual for pasture, hay, silage and cover crop. The plant develops rapidly from seed, making a quick cover suitable for early grazing. It is nutritious and palatable. Many seed sources contain both Italian and perennial ryegrass seed. There are 227,000 seeds/lb. An additional short-lived annual *Lolium* sp. is Darnel, *L. temulentum* L., also called Dokumugi and Rabillo. For perennial ryegrass, which see.

3. Season, cool season grass. Newly seeded pasture can be grazed within 2 months at heights of 6 to 8 inches.

4. Production in U.S.: No specific data, See Forage grass. In 1992, 142,190 tons of ryegrass seed were produced on 216,171 acres (1992 CENSUS). Southern states and Oregon and Washington. In 1992 and 1987, the top state for ryegrass seed production was Oregon (213,873 acres).

5. Other production regions: See above.

6. Use: Hay, pasture, silage and cover crop. Also used as turf grass cover, for roadbanks, golf courses and lawns.

7. Part(s) of plant consumed: Foliage

8. Portion analyzed/sampled: Forage and hay

9. Classifications:
   a. Authors Class: See Forage grass
   b. EPA Crop Group (Group & Subgroup): See Forage grass
   c. Codex Group: See Forage grass; 051 (AS 5251) Darnel
   d. EPA Crop Definition: None

10. References: GRIN, MAGNESS, ALDERSON, STUBBENDIECK, BARNES, MOSLER (picture), BALL, BAYER, BAILEY 1976

11. Production Map: EPA Crop Production Regions 2, 3, 4, 5, 6, 10 and 12 for hay and winter pasture. Seed crop use in regions 11 and 12.

12. Plant Codes:
   a. Bayer Code: LOLMU (*L. multiflorum* Italian), LOLTE (*L. temulentum*), LOLPS (*L. persicum*), LOLMG (*L. multiflorum* annual)

523

1. Ryegrass, Perennial
(English ryegrass, Lymegrass, Terrellgrass, Strand wheat, Hosomugi, Raigras ingles, Ivraie vivace)

Poaceae (Gramineae)

Lolium perenne L.

2. This is an important cool-season short-lived perennial bunchgrass, introduced from Europe, where it has been grown under culture for more than 3 centuries. The numerous leaves are long and slender, and the plant spreads by stolons. It may reach a height of 12 inches. In the U.S. it is best adapted to the Pacific Northwest. The major use is in permanent pastures where it is usually seeded in mixtures with other grasses. It can be seeded at a rate of 16 to 20 kg/ha, April through May and also early August. It thrives best in cool, moist regions with mild winters. It is nutritious and palatable both for pasturage and hay. It is more compatible with alfalfa than most other cool season perennial grasses, but it is more sensitive to temperature extremes and drought than annual ryegrasses. Its lack of persistence has limited its use for U.S. pastures. There are 550,000 seeds/kg.

3. Season: Cool season grass.

4. Production in U.S.: No specific data, See Forage grass. Also see Italian ryegrass for seed crop data. Pacific Northwest and Northeast.

5. Other production regions: British Columbia, Canada.

6. Use: Pasture, hay, turf, silage.

7. Part(s) of plant consumed: Foliage

8. Portion analyzed/sampled: Forage and hay

9. Classifications:
   a. Authors Class: See Forage grass
   b. EPA Crop Group (Group & Subgroup): See Forage grass
   c. Codex Group: See Forage grass
   d. EPA Crop Definition: None

10. References: GRIN, MAGNESS, RICHE, SCHREIBER, MOSLER, BAILEY 1976, BARNES(a)

11. Production Map: EPA Crop Production Regions 1, 5, 10 and 12.

12. Plant Codes:
   a. Bayer Code: LOLPE

524

1. Safflower
(False Saffron, Cartamo)

Asteraceae (Compositae)

Carthamus tinctorius L.

2. The plant is an annual grown from seed, up to 3 feet high, with glabrous, ovate, spiny leaves, which are almost as broad as long. It develops a flower head which tapers upward. The yellow to orange flowers are used to color and flavor foods. The flower heads yield a dyestuff, somewhat used on fabrics, especially silk. The seeds yield a valuable food oil. Seeds are borne partially exposed in globular heads, with 15 to 50 seeds per head and 1 to 5 heads per plant. Seeds are elongated, $1/4$ to $1/3$ inch long and a third as much in diameter. The seed contains 32 to 40 percent oil. The seed coats are fibrous, so seeds are decorticated before pressing or put through expellers to obtain the oil. Most of the oil is used for edible purposes, but it is also comparable to linseed oil for industrial use. The press cake is a valuable high protein feed supplement for cattle, sheep and poultry.

3. Season, seeding to harvest: Most cultivars require a minimum growing period of 120 days. In central California the growing period extends to 150 days.

4. Production in U.S.: 193,490 tons of seed on 264,837 acres reported in 1992 CENSUS of Agriculture. 1992 CENSUS

reports the top four states as California, North Dakota, Montana and South Dakota. In 1994, California harvested 163,458 acres (MELNICOE 1996e).

5. Other production regions: Canada – 500 tons of seed on 2,719 acres in 1995 (ROTHWELL 1996a).

6. Use: Flower heads to color and flavor foods; the primary use is for the edible oil from the seed.

7. Part(s) of plant consumed: Primarily as oil and meal (press cake); secondarily the flower head

8. Portion analyzed/sampled: Seeds/meal/refined oil for oilseed use, and flower heads (fresh and dry) for spice use

9. Classifications:
   a. Authors Class: Oilseed; Herbs and spices
   b. EPA Crop Group (Group & Subgroup): Miscellaneous
   c. Codex Group: 023 (SO 0699) Oilseed; 067 (OC 0699) Vegetable oils crude; and 068 (OR 0699) Vegetable oils edible
   d. EPA Crop Definition: None

10. References: GRIN, CODEX, MAGNESS, RICHE, ROBBELEN 1989, ROTHWELL 1996a, US EPA 1995a, US EPA 1995b, US EPA 1994, MELNICOE 1996e, IVES

11. Production Map: EPA Crop Production Regions 7 and 10 which accounts for 97 percent of U.S. Production.

12. Plant Codes:
    a. Bayer Code: CAUTI

525

1. Saffron crocus
(Saffron, Zafran, Azafran)
Iridaceae
Crocus sativus L.

2. Similar to the garden crocus and is propagated by planting corms in the fall or spring. The leaves are narrow and grass-shaped and grow to 3 to 18 inches tall. It is grown for its flower part called the stigma. Approximately 35,000 flowers (60,000 stigmas) are handpicked to make 1 pound of saffron. The stigmas are extracted by drying the flowers. Saffron is available ground or whole threads.

3. Season: Blooms in autumn.

4. Production in U.S.: No data.

5. Other production regions: Grows in Mediterranean area, Iran and Latin America. About 3.2 metric tons were imported into the U.S. in 1992 (USDA 1993a).

6. Use: Spice, flavoring and coloring agent in sauces, cakes, rice and fish and chicken casseroles.

7. Part(s) of plant consumed: Dried stigmas of flower

8. Portion analyzed/sampled: Stigmas

9. Classifications:
   a. Authors Class: Herbs and spices
   b. EPA Crop Group (Group & Subgroup): Herbs and spices (19B)
   c. Codex Group: No specific entry
   d. EPA Crop Definition: None

10. References: GRIN, BAILEY 1976, FARRELL (picture), SCHNEIDER 1994a, USDA 1993a

11. Production Map: No entry

12. Plant Codes:
    a. Bayer Code: CVOSA

526

1. Sage
(Common sage, Garden sage, Salvia real, Broadleaf sage)
Lamiaceae (Labiatae)
Salvia officinalis L.

2. Sage is a shrub-like perennial, up to 35 inches high, widely cultivated in moderate climates for the leaves used in flavoring, especially of meats and dressings. The leaves are entire and oblong, 1 to 1½ inches long, densely pubescent. The young leaves may be used fresh, or after careful curing and drying. Cuttings of leaves and succulent stems may be made up to 3 times per season. A volatile oil from sage is used in perfumery. Commercial production is very limited. About 2,656 tons of dried sage are imported annually (1992), mainly from Mediterranean countries. Imports have tripled in 24 years. *S. fruticosa* mill. (syn: *S. triloba* L.f.) is called "Greek sage" in commerce and represents over 50 percent of the imported sage.

3. Season, seeding to harvest: 1 year, then up to 3 harvests each year prior to flowering. Sage can be transplanted. The top 6-8 inches of growth are harvested.

4. Production in U.S.: 300 acres in California (ELMORE 1985).

5. Other production regions: Mediterranean countries.

6. Use: Flavoring of meats and dressings. Also flowers eaten raw in salads.

7. Part(s) of plant consumed: Leaves and flowers

8. Portion analyzed/sampled: Leaves (fresh and dried)

9. Classifications:
   a. Authors Class: Herbs and spices
   b. EPA Crop Group (Group & Subgroup): Herbs and spices (19A)
   c. Codex Group: 027 (HH 0743) Herbs; 057 (DH 0743) Dried herbs. Also clary
   d. EPA Crop Definition: None

10. References: GRIN, BAILEY 1976, CODEX, ELMORE 1985, FOSTER (picture), MAGNESS, USDA 1993a, US EPA 1995a, SPLITTSTOESSER

11. Production Map: EPA Crop Production Region 10.

12. Plant Codes:
    a. Bayer Code: SALOF

527

1. Sainfoin
(Esparcet, Holy clover, Esparceta)
Fabaceae (Leguminosae)
Onobrychis viciifolia Scop. (syn: *O. sativa* Lam., *O. viciaefolia*)

2. Sainfoin was introduced from Turkey and is promising as a hay and pasture crop for the Northern Plains, particularly Montana, Idaho and western North Dakota. It is a perennial legume, growing to 3 to 4 feet, with pinnate-compound leaves. Under tests in the above area it has outyielded alfalfa and has not caused bloat in cattle. It appears to be about equal to alfalfa in palatability and nutritive value. It does not compete well with weeds as regrowth after mowing is limited. It is not a major forage in the U.S. because of disease problems, but is adapted to drylands in Western U.S. and adjacent Canada. Productive for 3 to 5 years.

3. Season, dryland pasture or hay. For hay cut at 50 percent flower stage, flowers June to August.

4. Production in U.S.: No data. Montana, Idaho and western North Dakota.
5. Other production regions: In 1995, Canada reported 35 acres (ROTHWELL 1996a).
6. Use: Hay and pasture.
7. Part(s) of plant consumed: Foliage
8. Portion analyzed/sampled: Forage and hay
9. Classifications:
   a. Authors Class: Nongrass animal feeds
   b. EPA Crop Group (Group & Subgroup): Nongrass animal feeds (18)
   c. Codex Group: 050 (AL 1027) Legume animal feeds
   d. EPA Crop Definition: Alfalfa = Sainfoin
10. References: GRIN, CODEX, MAGNESS, ROTHWELL 1996a, US EPA 1995a, DUKE(h), BARNES
11. Production Map: EPA Crop Production Regions 7 and 9.
12. Plant Codes:
    a. Bayer Code: ONBVI

528

1. **Salal**

(Shallon, Lemonleaf)

Ericaceae

Gaultheria shallon Pursh

2. Evergreen, erect shrub to 6 feet tall. Branchlets are pubescent and leaves are shiny oblong or ovate to 4 inches long and finely serrated. Flowers are in racemes and fruit is purple to black in color. Grows best in moist sandy or peaty soils. Berries ripen in August and often remain on the bushes into October.
3. Season, planting: Propagated by seeds, layers, suckers, division or cuttings.
4. Production in U.S.: No data. Grows from southern Alaska to southern California (BAILEY 1976).
5. Other production regions: Western North American (FACCIOLA 1990).
6. Use: Sweet, juicy fruits are eaten fresh, cooked, dried or used in pies, jellies and drinks. Dried fruit used like raisins. When used by florist, known as lemonleaf.
7. Part(s) of plant consumed: Fruit
8. Portion analyzed/sampled: Whole fruit (fresh and dried)
9. Classifications:
   a. Authors Class: Berries
   b. EPA Crop Group (Group & Subgroup): Miscellaneous
   c. Codex Group: No specific entry
   d. EPA Crop Definition: None
10. References: GRIN, BAILEY 1976, FACCIOLA, KUHNLEIN (picture)
11. Production Map: No entry
12. Plant Codes:
    a. Bayer Code: GAHSH

529

1. **Salsify**

(Oyster plant, Vegetable oyster, Kalapi, White salsify, Salsifi blanco, Goats beard)

Asteraceae (Compositae)

Tragopogon porrifolius L.

2. Salsify is grown for its edible roots, which when cooked have a flavor resembling that of oysters. Although naturally a biennial, the cool season crop is produced as an annual. Grows to over 3 feet tall. The leaves are smooth and very long, slender and grasslike. The roots are white, long and slender, about 1 inch diameter at the crown, 10 to 12 inches long, and tapering. Cultural conditions are similar to parsnips and carrots, but plants have a smaller leaf surface and require a longer growing season than carrots. There are 2,000 seeds per ounce.
3. Season, seeding to harvest: 4-5 months or more for common varieties. Roots are harvested in the fall after a freeze.
4. Production in U.S.: 34 acres reported in 1954 CENSUS. More in home gardens. In the northern U.S. it is planted in the spring while in Florida it is planted in the fall (STEPHENS 1988).
5. Other production regions: Canada (ROTHWELL 1996a). Mediterranean region (FACCIOLA 1990).
6. Use: Mainly as cooked vegetable but can be used raw in salads. The leaves can be used in salads or cooked.
7. Part(s) of plant consumed: Mainly roots, but it is marketed with 2 to 3 inches or more of the leaves which are edible. Sometimes the young shoots that develop in the second year can be used in salads or cooked. Also sprouted seeds can be used in salads. Preparation includes rinsing the root and peeling (SCHNEIDER 1985)
8. Portion analyzed/sampled: Root and top (leaves)
9. Classifications:
   a. Authors Class: Root and tuber vegetables; Leaves of root and tuber vegetables
   b. EPA Crop Group (Group & Subgroup): Root and tuber vegetables (1A and 1B)
   c. Codex Group: 016 (VR 0498) Root and tuber vegetables; 013 (VL 0498) Leafy vegetables (including *Brassica* leafy vegetables)
   d. EPA Crop Definition: None
10. References: GRIN, CODEX, FACCIOLA, LORENZ, MAGNESS, MYERS (picture), NEAL, STEPHENS 1988 (picture), SCHNEIDER 1985, ROTHWELL 1996a, US EPA 1995a, YAMAGUCHI 1983, RUBATZKY
11. Production Map: No entry.
12. Plant Codes:
    a. Bayer Code: TROPS

530

1. **Salsify, Black**

(Scorzonera, Coconut root, Mock oyster, Black oyster plant, Serpent root, Viper grass, Oysterplant, Escorzonera, Spanish salsify)

Asteraceae (Compositae)

Scorzonera hispanica L.

2. The plant has a long, fleshy tap root to about 1 foot long, similar to salsify, but black in surface color, with white flesh. The leaves are entire and grasslike, but wider than grass. The plant reaches 2 feet in height. The plant is perennial, but is cultivated as an annual. Culture is similar to that of other root crops, as salsify or carrots.
3. Season, seeding to harvest: About 5-6 months. A new cultivar 'Flandria' can be harvested in about 3 months (FACCIOLA 1990).
4. Production in U.S.: No data, very limited. Naturalized in U.S. as summer or winter crop (MYERS 1991).

5. Other production regions: Canada (ROTHWELL 1996a). Mediterranean region (FACCIOLA 1990).
6. Use: Roots as cooked vegetable. Leaves sometimes used as salad. Preparation includes rinsing the root and peeling (SCHNEIDER 1985).
7. Part(s) of plant consumed: Mainly roots, leaves sometimes
8. Portion analyzed/sampled: Root and tops (leaves)
9. Classifications:
   a. Authors Class: Root and tuber vegetables; Leaves of root and tuber vegetables
   b. EPA Crop Group (Group & Subgroup): Root and tuber vegetables (1A and 1B); Leaves of root and tuber vegetables (2)
   c. Codex Group: 016 (VR 0594) Root and tuber vegetables
   d. EPA Crop Definition: None
10. References: GRIN, CODEX, FACCIOLA, LOGAN (picture), LORENZ, MAGNESS, MYERS, ROTHWELL 1996a, SCHNEIDER 1985, STEPHENS 1988, US EPA 1996a, US EPA 1995a
11. Production Map: No entry.
12. Plant Codes:
   a. Bayer Code: SCVHI

531

1. Salsify, Spanish
(Spanish oyster plant, Scolymus, Golden thistle, Sunnariah, Cardo amarillo, Epine jaune)
1. Spanish oyster
Asteraceae (Compositae)
Scolymus hispanicus L.

2. This plant makes a root much like salsify, which see, but roots are larger in size and hence more production is attained than with salsify. The leaves are narrow and long, and very prickly or spiny. The plant grows 2 to 2½ feet in height. Roots, the edible part, are up to 10 inches long and 1 inch diameter. Flavor of the cooked roots is milder than salsify. This is a vegetable of minor importance, mainly limited to a few home gardens in the U.S. Culture is similar to that of other root crops, as carrots.
3. Season, seeding to harvest: About 5-6 months for common varieties.
4. Production in U.S.: No data, very limited. Naturalized in U.S. as summer or winter crop (MYERS 1991).
5. Other production regions: Mediterranean region (FACCIOLA 1990).
6. Use: As a cooked vegetable.
7. Part(s) of plant consumed: Mainly roots. Young leaves can be blanched and used in salads locally
8. Portion analyzed/sampled: Root
9. Classifications:
   a. Authors Class: Root and tuber vegetables
   b. EPA Crop Group (Group & Subgroup): Root and tuber vegetables (1A and 1B)
   c. Codex Group: 016 (VR 0593) Root and tuber vegetables
   d. EPA Crop Definition: None
10. References: GRIN, CODEX, FACCIOLA, LORENZ, MAGNESS, MYERS, US EPA 1995a
11. Production Map: No entry
12. Plant Codes:
   a. Bayer Code: SCYHI

532

1. Sand bluestem
(Bluestem-sand)
Poaceae (Gramineae)
Andropogon hallii Hack. (syn: *Andropogon gerardii* var. *paucipilus* (Nash) Fern.)

2. This is a warm season, sod-forming grass native from North Dakota and Montana south to Texas and Arizona, and belongs to the grass tribe *Andropogoneae*. It generally resembles big bluestem and is considered a ecotype or subspecies of it. It differs from big bluestem in having a hairy panicle (the seed head) and more vigorous rhizomes, resulting in more rapid and extensive lateral spread. Stems reach up to 7 feet under the best conditions. It can readily hybridize with big bluestem. It is found mainly on deep, sandy soils and is a valuable range grass on such soils and is useful as a revegetating species. Sand bluestem is seeded in the spring at a rate of 3 to 5 lb/A or 1 to 6 lb/A in mixtures and contains from 20,000 to 45,000 seeds/lb.
3. Season: Warm season forage and grows rapidly from midspring through the early fall. Seeds mature in October.
4. Production in U.S.: No data for Sand bluestem production, however, pasture and rangeland are produced on more than 410 million A (U.S. CENSUS, 1992). See above.
5. Other production regions: No entry.
6. Use: Pasture and rangeland, hay, rangeland reseeding, and for erosion control.
7. Part(s) of plant consumed: Leaves, stems, and seeds
8. Portion analyzed/sampled: Forage and hay
9. Classifications:
   a. Authors Class: See Forage grass
   b. EPA Crop Group (Group & Subgroup): See Forage grass
   c. Codex Group: See Forage grass
   d. EPA Crop Definition: None
10. References: GRIN, MAGNESS, US EPA 1994, US EPA 1995, CODEX, BARNES, ALDERSON, RICHE
11. Production Map: EPA Production Regions 5, 6, 7, 8 and 9.
12. Plant Codes:
   a. Bayer Code: No specific entry, genus entry ANOSS

533

1. Sand dropseed
(Sand fallsame, Zacaton arenoso)
Poaceae (Gramineae)
Sporobolus cryptandrus (Torr.) A. Gray

2. This is a tufted, native bunchgrass abundant in the Southern Plains and from Idaho and Oregon southward. It is most prevalent on sandy soil. Plants are 2 to 3 feet tall, with solid stems and rather numerous leaves up to 12 inches long and ¼ inch wide. Roots are coarse and deep penetrating. The plants produce a fairly large amount of foliage which is eaten readily by livestock while green, but sparingly when ripe. If not overgrazed, stands tend to increase in density. It produces seed in abundance and is useful for reseeding depleted range land. Its relatively low palatability, however, limits its overall usefulness. There are 5 million seeds/lb.
3. Season: Starts growth early spring (perennial). Mature June-August.

4. Production in U.S.: No specific data, See Forage grass. Southern Plains and from Idaho and Oregon south. Not reported in the southeastern U.S.
5. Other production regions: Canada and Mexico (GOULD 1975).
6. Use: Rangeland.
7. Part(s) of plant consumed: Foliage
8. Portion analyzed/sampled: Forage and hay
9. Classifications:
   a. Authors Class: See Forage grass
   b. EPA Crop Group (Group & Subgroup): See Forage grass
   c. Codex Group: See Forage grass
   d. EPA Crop Definition: None
10. References: GRIN, MAGNESS, STUBBENDIECK(a), GOULD
11. Production Map: EPA Crop Production Regions 7, 8, 9, 10 and 11.
12. Plant Codes:
    a. Bayer Code: SPZCR

534

1. Sapodilla
(Naseberry, Nispero, Chicozapote, Chiku, Chico zapote, Chicle, Zapotle, Sapota, Dilly, Sapodilla plum)
Sapotaceae
Manilkara zapota (L.) P. Royen (syn: *M. achras* (Mill.) Fosberg; *M. zapotilla* (Jacq.) Gilly; *Arachas zapota* L.)

2. The tree is a tropical evergreen, native to the American tropics. Grows in hot tropical lowlands with low to high rainfall and is not tolerant of frost. Sapodilla is propagated by seed and grafting with first fruit crop within 3-5 years after the graft. The tree becomes large, with smooth, thick, shiny leaves. The bark contains a milky latex, which is obtained by tapping. Fruits are up to 6 inches in diameter, flattened at the stem end, globose-conic in shape. The skin is thin and scurfy. Flesh is honey-brown, tender, granular and with a flavor similar to maple sugar. Fruit structure is somewhat similar to apple, with a central core of 10 to 12 cells.
3. Season: Bloom to harvest: 4 to 10 months. Flowering July-October and fruits February-June in Florida. Fruit at all stages of maturity can be found on the tree at the same time.
4. Production in U.S.: Minor production in South Florida (30 acres-1995) (CRANE 1996a), Guam, Hawaii (2 growers, 11 trees and 1 acre) (KAWATE 1996b), and Puerto Rico.
5. Other production regions: Mexico, India, Guatemala, Venezuela.
6. Use: Ripe fruit pulp eaten fresh or used for desserts. Latex from the tapped tree is used for chicle gum. Also grown as an ornamental.
7. Part(s) of plant consumed: Fruit pulp and latex
8. Portion analyzed/sampled: Whole fruit
9. Classifications:
   a. Authors Class: Tropical and subtropical fruits – inedible peel
   b. EPA Crop Group (Group & Subgroup): Miscellaneous
   c. Codex Group: 006 (FI 0359) Assorted tropical and subtropical fruits – inedible peel
   d. EPA Crop Definition: See Avocado and Papaya
10. References: GRIN, CODEX, KAWATE 1996b, KNIGHT(a), LAKSHMINARAYANA, MAGNESS, MARTIN 1987, US

EPA 1994, MORTON (picture), CRANE 1996a, CAMPBELL (e), KUTE
11. Production map: EPA Crop Production Region 13.
12. Plant Codes:
    a. Bayer Code: MNKZA

535

1. Sapote, Black
(Black persimmon, Sapote negro, Black sapote)
Ebenaceae
Diospyros digyna Jacq.

2. Tropical evergreen tree up to 80 feet in height. The black sapote is native to Mexico and Central America with glossy, leathery leaves 4 to 12 inches long. The fruit is smooth, thin skin with glossy brown to black pulp.
3. Season, bloom to harvest: Fruit matures in 7 to 10 months. Flowers March-May with harvest December-March in Florida.
4. Production in U.S.: Minor Production in Florida (1 acre). (CRANE 1996a)
5. Other production regions: See above.
6. Use: Fresh eating, dessert. Also grown as an ornamental.
7. Part(s) of plant consumed: Pulp only
8. Portion analyzed/sampled: Whole fruit
9. Classifications:
   a. Authors Class: Tropical and subtropical fruits – inedible peel
   b. EPA Crop Group (Group & Subgroup): Miscellaneous
   c. Codex Group: 006 (FI 0360) Assorted tropical and subtropical fruits – inedible peel
   d. EPA Crop Definition: See Avocado and Papaya
10. References: GRIN, CODEX, CRANE 1996a, CRANE 1995, MARTIN 1987, MORTON (picture), BALERDI
11. Production map: EPA Crop Prodution Region 13.
12. Plant Codes:
    a. Bayer Code: No specific entry

536

1. Sapote, Green
(Injerto, Raxtul, Faisan)
Sapotaceae
Pouteria viridis (Pittier) Cronquist (syn: *Calocarpum viride* Pittier, *Achradelpha viridis* O.F. Cook)

2. The green sapote tree is a large, up to 80 feet, tropical evergreen with leaves up to 10 inches long by 2 inches wide. Fruits are ovoid or eliptical, 3 to 5 inches long, with usually one large seed. The fruit peel is thin, scurfy and roughened. Flesh is red or reddish brown, firm and somewhat granular, with a rich, sweet flavor. The green sapote tree is similar to mamey, but with smaller leaves. Fruits are similar in size and other characteristics to the mamey sapote.
3. Season, bloom to maturity: 6 to 8 months.
4. Production in U.S.: No data. Dooryard trees only.
5. Other production regions: Central America (MORTON 1987).
6. Use: Fresh eating, preserves.
7. Part(s) of plant consumed: Inner pulp
8. Portion analyzed/sampled: Fruit
9. Classifications:
   a. Authors Class: Tropical and subtropical fruits – inedible peel
   b. EPA Crop Group (Group & Subgroup): Miscellaneous

---

c. Codex Group: 006 (FI 0361) Assorted tropical and subtropical fruits – inedible peel

d. EPA Crop Definition: None

10. References: GRIN, CODEX, MAGNESS, MORTON

11. Production Map: EPA Crop Production Region 13.

12. Plant Codes:

a. Bayer Code: No specific entry

537

1. Sapote, Mamey

(Mamey, Mamey zapote, Chachaas, Mammee sapote, Sapote, Zapote, Mamey colorado, Marmalade fruit, Marmalade plum)

Sapotaceae

Pouteria sapota (Jacq.) H.E. Moore & Stearn (syn: *Calocarpum sapota* (Jacq.) Merr.)

2. Grows in hot tropical lowlands. Propagated by seed and grafting with first fruit harvest 4-5 years after graft. The plant is a tropical evergreen tree that grows to 60 feet in height. The evergreen leaves cluster at the branch tips. The large fruit is round to elliptic usually weighing 1 to 3 lbs. with a scurfy, thick, woody and brown colored skin with usually one large seed. Flesh is pale salmon to tropical red, firm and somewhat granular, with a rich, sweet flavor.

3. Season: Bloom to harvest period may vary from 12 to 24 months. In Florida, most flower February-July but some bloom all year. Trees may have flowers, immature fruits and mature fruits at the same time.

4. Production in U.S.: In 1995, 318 acres in South Florida (CRANE 1996a). Small plantings in Puerto Rico. (MONTALVO-ZAPATA 1995).

5. Other production regions: Mexico, Cuba and Central America (LOGAN 1996).

6. Use: Fresh eating, beverages and desserts.

7. Part(s) of plant consumed: Mostly the pulp, however the seed can be milled to prepare a bitter chocolate

8. Portion analyzed/sampled: Whole fruit

9. Classifications:

a. Authors Class: Tropical and subtropical fruits – inedible peel

b. EPA Crop Group (Group & Subgroup): Miscellaneous

c. Codex Group: 006 (FI 0362) Assorted tropical and subtropical fruits – inedible peel

d. EPA Crop Definition: See Avocado and Papaya

10. References: GRIN, CODEX, CRANE 1996a, CRANE 1995, LOGAN, MAGNESS, MARTIN 1987, US EPA 1994, MORTON (picture), MONTALVO-ZAPATA 1995, PHILLIPS(f)

11. Production Map: EPA Crop Production Region 13.

12. Plant Codes:

a. Bayer Code: No specific entry

538

1. Sapote, White

(Casimiroa, Matasano, Sapote blanco, Zapote, Mexican-apple, Zapote-blanco, Woolly-leaf white sapote)

Rutaceae

Casimiroa edulis La Llave & Lex.

C. tetrameria Millsp.

(Matasano, Woolly-leaf white sapote)

2. This fruit is distantly related to citrus. The tree is an evergreen, native to Central America, about as hardy as lemon. Grows in subtropical or tropical highlands with medium rainfall. Propagated by seed, cutting, layering and grafting. Fruits are near globose, 3 to 4 inches in diameter, with a thin, nearly smooth skin that is usually green in color with a yellow blush when mature. The white colored pulp is soft and juicy. It has a sweet mild flavor that resembles peaches, lemons, mangoes, coconut or vanilla, depending on the variety.

3. Season: Trees bloom and set fruit in both spring and fall. Bloom to maturity is about 5 months from spring bloom, 7 to 8 months from fall bloom. Fruit is available August through November and April through June.

4. Production in U.S.: Production in California, Florida and a few trees in Hawaii.

5. Other production regions: Mexico, Central America, South America, West Indies, Mediterranean region, East Indies and India (MORTON 1987).

6. Use: Mostly fresh eating and preserves.

7. Part(s) of plant consumed: Inner pulp, after skin and seed are removed

8. Portion analyzed/sampled: Whole Fruit

9. Classifications:

a. Authors Class: Citrus fruits

b. EPA Crop Group (Group & Subgroup): Miscellaneous

c. Codex Group: 006 (FI 0363) Assorted tropical and subtropical fruits – inedible peel

d. EPA Crop Definition: Proposing – Citrus = White sapote

10. References: GRIN, CODEX, LOGAN, KAWATE 1995, MAGNESS, MARTIN 1987, US EPA 1994, MORTON (picture), SAULS(a)

11. Production Map: EPA Crop Production Regions 10 and 13.

12. Plant Codes:

a. Bayer Code: No specific entry

539

1. Savory, Summer

(Ajedrea comun, Sarriette)

Lamiaceae (Labiatae)

Satureja hortensis L.

1. Savory, Winter

(Ajedra, Tomillo real, Savory)

S. montana L.

2. Savories are plants of the mint family, native in the Mediterranean area. Summer savory is an annual, pubescent herb about 18 inches high, with oblong-linear leaves. Winter savory is perennial, 6 to 12 inches high, less pubescent than summer savory. Both kinds are sometimes used as potherbs, but now are more commonly used as spices or condiments to flavor meat dishes, sauces, stews, etc. For this purpose, summer savory is more used than winter. Commercial production in the U.S. is slight.

3. Season, seeding to harvests: About 90 days.

4. Production in U.S.: 300 acres of summer savory in California (ELMORE 1980).

5. Other production regions: 15 acres of summer savory in Canada (ROTHWELL 1996a).

6. Use: Flavoring in drinks, baked goods and condiments. One of the spices in "Fines Herbes".

7. Part(s) of plant consumed: Leaves and flowering tops

8. Portion analyzed/sampled: Leaves and flowering tops (fresh and dried)
9. Classifications:
   a. Authors Class: Herbs and spices
   b. EPA Crop Group (Group & Subgroup): Herbs and spices (19A)
   c. Codex Group: 027 (HH 0745) Herbs; 057 (DH 0745) Dried herbs
   d. EPA Crop Definition: None
10. References: GRIN, CODEX, ELMORE 1980, FARRELL (picture), FOSTER (picture), MAGNESS, MARTIN 1979, ROTHWELL 1996a, STEPHENS 1988, US EPA 1995a
11. Production Map: EPA Crop Production Region 10.
12. Plant Codes:
    a. Bayer Code: STIHO (*S. hortensis*)

540

1. **Seagrape**
(Seaside grape, Uva de playa, Platterleaf, Jamaican kino, Kino)
Polygonaceae
Coccoloba uvifera (L.) L.
2. The sea grape is native to sandy shores of the American tropics. The plant is a small tree, sometimes reaching to 30 feet with stiff near-round leaves up to 8 inches across. Fruits are globose to pyriform, about $3/4$ inch diameter and borne in clusters. The skin is pubescent, enclosing an edible pulp and single seed. The pulp is eaten directly and makes an excellent jelly. Ripening is mainly in midsummer. The sea grape appears not to be grown commercially, but some fruit from native plants is harvested.
3. Season, to harvest:. Ripens mainly in midsummer.
4. Production in U.S.: No data. Home gardens.
5. Other production regions: Tropical America, warm tropical lowlands and coastal areas (MARTIN 1987).
6. Use: Fresh and jelly. Also grown as an ornamental.
7. Part(s) of plant consumed: Fruit
8. Portion analyzed/sampled: Whole fruit
9. Classifications:
   a. Authors Class: Berries
   b. EPA Crop Group (Group & Subgroup): Miscellaneous
   c. Codex Group: 005 (FT 0310) Assorted tropical and subtropical fruits – edible peel
   d. EPA Crop Definition: None
10. References: GRIN, CODEX, MAGNESS, MARTIN 1987, PHILLIPS(h)
11. Production Map: EPA Crop Production Region 13.
12. Plant Codes:
    a. Bayer Code: CODUV

541

1. **Sentul**
(Santol, Wild mangosteen, Santor, Kechapi, Katul, Sayai)
Meliaceae
Sandoricum koetjape (Burm. f.) Merr. (syn: *S. indicum* Cav.)
2. The sentul is a fast growing, straight-trunked, pale-barked tree 50 to 150 feet tall, branched close to the ground with young branchlets densely hairy. Primarily evergreen with spirally-arranged compound leaves that are 4 to 10 inches long. The fruit is globose or oblate with wrinkles extending a short distance from the base. The rind is downy and edible.

The juicy pulp (aril) is white and translucent. The fruit is $1^{1}/_{2}$ to 3 inches wide and yellow. The seeds are inedible. Called Santor or Wild mangosteen in Guam.
3. Season, harvest: Fruit ripens in Florida, August and September.
4. Production in U.S.: No data. Grows well in south Florida (MORTON 1987).
5. Other production regions: Native to Indochina and naturalized in India and the Philippines (MORTON).
6. Use: The fruit is usually consumed raw without peeling. With the seeds removed, the fruit is made into jam or jelly. Also with the peel and seed removed, the pulp is commercially preserved in syrup. This marmalade, called Santol, is exported from the Philippines to Oriental food dealers in the U.S.
7. Part(s) of plant consumed: Fruit
8. Portion analyzed/sampled: Whole fruit
9. Classifications:
   a. Authors Class:Tropical and subtropical fruits – edible peel
   b. EPA Crop Group (Group & Subgroup): Miscellaneous
   c. Codex Group: 006 (FI 0364) Assorted tropical and subtropical fruits – inedible peel
   d. EPA Crop Definition: None
10. References: GRIN, CODEX, MORTON (picture)
11. Production Map: EPA Crop Production Region 13.
12. Plant Codes:
    a. Bayer Code: No specific entry

542

1. **Serviceberry**
(Mountain ash)
Rosaceae
Sorbus spp.
1. **Servicetree**
(Sorbapple)
   S. domestica L.
1. **Wild servicetree**
(Checkertree, Chequers)
   S. torminalis (L.) Crantz
2. Deciduous trees from 30 to 60 feet tall of the Northern Hemisphere. Fruit is a small pome about 1 inch in diameter.
3. Season, harvest: Fruit harvested after frost as food.
4. Production in U.S.: No data. USDA Plant Hardiness Zone 6 (BAILEY 1976).
5. Other production regions: Mediterranean region.
6. Use: Fruit is eaten fresh after exposed to frost or when overripe like the Medlar. Also fermented into wine, dried like prunes or processed into jellies. Bark used for tanning leather. Also grown as an ornamental.
7. Part(s) of plant consumed: Fruit
8. Portion analyzed/sampled: Whole fruit
9. Classifications:
   a. Authors Class: Berries
   b. EPA Crop Group (Group & Subgroup): Miscellaneous
   c. Codex Group: 004 (FB 0274) Berries and other small fruits
   d. EPA Crop Definition: None
10. References: GRIN, BAILEY 1976, CODEX, FACCIOLA
11. Production Map: No entry
12. Plant Codes:
    a. Bayer Code: SOUSS (*S.* spp.)

543

1. Sesame

(Benneseed, Beniseed, Simsim, Sesamo, Sesame seed, Sesame oil, Sesam, Ajonjoli)

Pedaliaceae

Sesamum indicum L. (syn: *S. orientale* L.)

2. The sesame plant is an erect-growing annual, 3 to 4 feet high, grown for the small, obovate, flattened seeds, which are widely used on bread, rolls and other culinary items, and are also extracted for the oil. The leaves are entire, 3 to 5 inches long, oblong or lanceolate, and somewhat roughened. Seeds are produced in pods. The plant requires a fairly long, warm growing season. Only the seeds are utilized. The small, flattened seeds are borne in two-valved pods, and have a content of semi-drying oil of 50 percent or more. About 15,000 acres of sesame for seed were reported in the U.S. in the 1959 CENSUS. Sesame for oil is a major crop in Asia and Mexico. The oil is usually expressed in Europe and Asia in three stages. The first is a cold press, followed by hot presses. Hot pressed oil is refined before being suitable for edible use. Sesame oil is used mainly as a salad and cooking oil, and in the manufacture of margarine and shortening. Culture and exposure are similar to that of soybean. U.S. imported about 40,616 tons of seed in 1993 with an averaged of about half from Mexico since 1990 (USDA 1993a).

3. Season, seeding to harvest: 4 to 5 months. Non-shattering types can be combined directly.

4. Production in U.S.: 15,087 acres reported in 1959 CENSUS. About 25,000 acres in 1996 (MITCHELL 1996). Arizona, Oklahoma and Texas (CRISWELL 1996 and MITCHELL 1996). Also grown in southeast Colorado (JOHNSON 1986).

5. Other production regions: In 1985, about 470,000 acres in Mexico (ROBBELEN 1989). Also see above.

6. Use: Seeds used for flavoring of baked breads. Fixed oil used as salad oil and in other foods.

7. Part(s) of plant consumed: Seeds. Meal can be used in limited quantities in feed

8. Portion analyzed/sampled: Seed and oil

9. Classifications:
   a. Authors Class: Oilseed; Herbs and spices (seed)
   b. EPA Crop Group (Group & Subgroup): Miscellaneous
   c. Codex Group: 023 (SO 0700) Oilseed; 028 (HS 0700) Spices; 067 (OC 0700) Crude vegetable oil; 068 (OR 0700) Edible (refined) vegetable oil
   d. EPA Crop Definition: None

10. References: GRIN, CODEX, CRISWELL 1996, FARRELL (picture), JOHNSON 1986, MAGNESS, MITCHELL 1996, REHM, ROBBELEN 1989 (picture), USDA 1994a, USDA 1993a, US EPA 1995a, IVES

11. Production Map: EPA Crop Production Regions 6, 8 and 10.

12. Plant Codes:
    a. Bayer Code: SEGIN

544

1. Sesbania

(Sesban, Colorado river hemp, Peatree, Coffeebean, Indigobean, Hemp sesbania)

Fabaceae (Leguminosae)

Sesbania exaltata (Raf.) Rydb. ex A.W. Hill (syn: *S. macrocarpa* Muhl.)

2. This is an upright-growing annual legume reaching to 8 feet, grown only for turning under as a soil-improving crop. It produces a heavy tonnage for this purpose. On irrigated lands in the Southwest, it is extensively grown as a summer crop and turned-under prior to planting winter vegetables or in citrus orchards. It is grown mainly in areas of high summer temperatures. Little value as a forage crop. It is grazed or used as green manure crop.

3. Season: Summer crop. Flowers April-October.

4. Production in U.S.: No data. Southwest U.S.

5. Other production regions: No entry.

6. Use: Primarily, cover crop. Also, the flowers, leaves and immature fruits are cooked and used as food.

7. Part(s) of plant consumed: Foliage

8. Portion analyzed/sampled: Forage and hay

9. Classifications:
   a. Authors Class: Nongrass animal feeds
   b. EPA Crop Group (Group & Subgroup): Miscellanous
   c. Codex Group: No specific entry
   d. EPA Crop Definition: None

10. References: GRIN, DUKE(h) (picture), MAGNESS, SKERMAN(a)

11. Production Map: EPA Crop Production Regions 10 and 13.

12. Plant Codes:
    a. Bayer Code: SEBEX

545

1. Shallot

(Eschalot, Scallion, Cebollin, Green shallots, Dried shallots, Cipollina)

Liliaceae (Amaryllidaceae)

Allium cepa var. *aggregatum* G. Don (syn: *A. ascalonicum* auct.)

2. Shallots produce a cluster of bulbs from a single planted bulb. Otherwise, they are similar to the common onion. Commercially they are grown mainly for marketing as green onions, mainly in the South. The mother or "seed" bulbs are planted in late summer or fall. As daughter bulbs and plants develop, soil is pushed around them to blanch the lower portion. Daughter plants are pulled at suitable size, the outer skin removed from the bulb and base of leaves, and the small bulb and green leaves are marketed as green onions. The hollow, rounded leaves are to 24 inches long and bulbs are $^3/_4$ to $2^1/_2$ inches in diameter and clusters of bulbs may contain up to 15 bulbs. Some varieties are 'Louisiana Pearl', 'Bayou Pearl' and 'Wilmington'. Shallots are also grown for the dry bulbs, which are milder flavored than most onions. Culture for dry bulbs is essentially like that for onions. Scallions are generally considered young green onions.

3. Season, planting to harvest: About 2-3 months for green; 3-6 months for dry bulbs. Shallots grown for green onion are pulled when their tops are 6-8 inches long.

4. Production in U.S.: About 50 acres in Washington state for dried shallots. Washington state for dry and Louisiana and other Southern states for green shallots (STEPHENS 1988). Estimated 1,000 acres in the U.S. (SWIADER 1992).

5. Other production regions: Europe.

6. Use: Fresh, in salads, and culinary cookery.

7. Part(s) of plant consumed: Inner bulb and leaves for green onions; bulb with scales removed for dry

8. Portion analyzed/sampled: Green shallot – Whole plant without roots; Dried shallot – bulb
9. Classifications:
   a. Authors Class: Bulb vegetables
   b. EPA Crop Group (Group & Subgroup): Bulb vegetables
   c. Codex Group: 009 (VA 0388) Bulb vegetables
   d. EPA Crop Definition: Dry bulb onions = Shallots (dry bulb only); Green onions = Green shallots or Green eschalots
10. References: GRIN, CODEX, MAGNESS, SCHREIBER, STEPHENS 1988, US EPA 1995a, US EPA 1995a, YAMAGUCHI 1983, RUBATZKY, SWIADER, FACCIOLA
11. Production Map: EPA Crop Production Regions 4, 11 and 12.
12. Plant Codes:
    a. Bayer Code: ALLAS (*A. ascalonicum*)

546

1. Shea butter tree
(Bambuk butter, Butterseed, Shea nut, Sheatree, Shea, Karite nut)
Sapotaceae
Vitellaria paradoxa C.F. Gaertn. (syn: *Butyrospermum parkii* (G. Don) Kotschy, *B. paradoxum* (C.F. Gaertn.) Hepper; *B. paradoxum* ssp. *parkii* (G. Don) Hepper)
2. This fat is obtained from nuts of the above species, which is a large tree native to West Africa. The dried fruit consists of a thin shell, enclosing an egg-shaped seed. The seeds weigh about 3 grams. The kernels contain about 50 percent of a non-drying fat. Both nuts and the fat are exported to Europe as well as used locally. In Europe, the fat is used as a cooking fat, in the manufacture of margarine, and as a substitute for cacao butter. The press cake or extracted meal is fed to cattle.
3. Season, Bloom to harvest: About 90 days.
4. Production in U.S.: No data.
5. Other production regions: Hot tropical lowlands with low rainfall and definite dry season, primarily tropical west Africa.
6. Use: Oil.
7. Part(s) of plant consumed: Seed
8. Portion analyzed/sampled: Seed and its processed commodity edible oil
9. Classifications:
   a. Authors Class: Oilseed
   b. EPA Crop Group (Group & Subgroup): Miscellaneous
   c. Codex Group: 023 (SO 0701) Oilseed
   d. EPA Crop Definition: None
10. References: GRIN, CODEX, MAGNESS, MARTIN 1983
11. Production Map: No entry
12. Plant Codes:
    a. Bayer Code: No specific entry

547

1. Silky bluegrass
(Queensland bluegrass)
Poaceae (Gramineae)
Dichanthium sericeum (R. Br.) A. Camus
2. This is a perennial warm season bunchgrass and is useful for pasture. It is adapted to heavy clay soils and grows to 1 foot tall.

3. Season: Warm season pasture, growing spring through autumn.
4. Production in U.S.: See Forage grass. Texas (SKERMAN 1989).
5. Other production regions: Native to Australia (GOULD 1975).
6. Use: Pasture, hay and rangeland.
7. Part(s) of plant consumed: Foliage (leaves and stems)
8. Portion analyzed/sampled: Forage and hay, See Forage grass
9. Classifications:
   a. Authors Class: See Forage grass
   b. EPA Crop Group (Group & Subgroup): See Forage grass
   c. Codex Group: See Forage grass
   d. EPA Crop Definition: None
10. References: GRIN, SKERMAN, BAYER, GOULD (picture)
11. Production Map: EPA Crop Production Regions 6 and 8.
12. Plant Codes:
    a. Bayer Code: DIHSE

548

1. Silver bluestem
(Silver beardgrass)
Poaceae (Gramineae)
Bothriochloa lagaroides ssp. *torreyana* (Steud.) Allred & Gould (syn: *Bothriochloa saccharoides* (Sw.) Rydb. var. *torreyana* (Steud.) Gould; *B. saccharoides* auct. nonn.)
2. This is a native warm season bunchgrass. It grows in the spring and begins to flower 3 to 4 weeks later and produces abundant seeds. It grows erect to 1.3 meter tall and is adapted to prairies and rocky slopes and a broad range of soil types. It has fair forage quality for all livestock and withstands light grazing when mature.
3. Season: Warm season pasture, spring to summer.
4. Production in U.S.: See Forage grass. Adapted to Arizona, California, Kansas, New Mexico, Oklahoma and Texas (STUBBENDIECK 1992).
5. Other production regions: Northern Mexico (GOULD 1975).
6. Use: Forage and pasture.
7. Part(s) of plant consumed: Foliage (leaves and stems)
8. Portion analyzed/sampled: Forage and hay, See Forage grass
9. Classifications:
   a. Authors Class: See Forage grass
   b. EPA Crop Group (Group & Subgroup): See Forage grass
   c. Codex Group: See Forage grass
   d. EPA Crop Definition: None
10. References: GRIN, BAYER, STUBBENDIECK, GOULD (picture)
11. Production Map: EPA Crop Production Regions 4, 5, 6 and 10.
12. Plant Codes:
    a. Bayer Code: No specific entry

549

1. Sixweeks threeawn
(Common needlegrass)
Poaceae (Gramineae)
Aristida adscensionis L. (syn: *A. submucronata* Schumach; *A. fasciculata* Torr.)
2. This is a annual, warm season grass native to tropical Africa and introduced into the U.S. It belongs to the grass tribe *Aris-*

tideae. It occurs on poor grassland soils and is most common on sandy soils. Grows erect to a height of 90 cm, in area with rainfall ≤250 mm and altitudes varying from sea level to 2250 m. It is a poorer quality forage than most perennials, however, it provides early forage in wasteland, rocky areas and abused rangeland areas. It has a short life and quality declines as it matures. It can set seed in Arizona.

3. Season: Warm season forage. Grows rapidly from mid-spring through the early fall. Must be grazed before flowering.
4. Production in U.S.: No data for Sixweeks threeawn production, however, pasture and rangeland are produced on more than 410 million A (U.S. CENSUS, 1992). Western Missouri, Kansas and Texas to southern Nevada and California (GOULD 1975).
5. Other production regions: Mexico, South America and Africa (GOULD 1975).
6. Use: Grazing rangeland and for erosion control.
7. Part(s) of plant consumed: Leaves and stems
8. Portion analyzed/sampled: Forage and hay
9. Classifications:
   a. Authors Class: See Forage grass
   b. EPA Crop Group (Group & Subgroup): See Forage grass
   c. Codex Group: See Forage grass
   d. EPA Crop Definition: None
10. References: GRIN, US EPA 1994, US EPA 1995, CODEX, SKERMAN (picture), RICHE, GOULD (picture)
11. Production Map: EPA Production Regions 5, 6, 8, 9 and 10.
12. Plant Codes:
   a. Bayer Code: ARKAD

550

1. Skirret
(Skirrit, Skirwort, Sugar root, Escaravia, Sisaro, Shen quin, Berle a sucre)
Apiaceae (Umbelliferae)
Sium sisarum L.

2. Skirret is a vegetable grown for its edible roots. The plant grows to 3 to 4 feet high and has compound, pinnate leaves. The roots grow in clusters, like sweet potatoes, but are longer and somewhat cylindrical and jointed. They have a sweet taste and are tender if well grown, but have a woody, nonedible core. Plants are usually grown from seed, which may be started in beds and transplanted to the field, or may be field sown.
3. Season, seeding to harvest: 6 to 8 months.
4. Production in U.S.: No data, very minor. In Florida, planted in fall. Mainly garden (STEPHENS 1988).
5. Other production regions: Still used widely in China and Japan (STEPHENS 1988).
6. Use: As cooked vegetable. The nonedible root core is removed before cooking.
7. Part(s) of plant consumed: Fleshy, tuberous roots with nonedible core
8. Portion analyzed/sampled: Root
9. Classifications:
   a. Authors Class: Root and tuber vegetables
   b. EPA Crop Group (Group & Subgroup): Root and tuber vegetables (1A and 1B)
   c. Codex Group: 016 (VR 0595) Root and tuber vegetables
   d. EPA Crop Definition: None

10. References: GRIN, CODEX, KAYS, MAGNESS, STEPHENS 1988 (picture), US EPA 1995a
11. Production Map: EPA Crop Production Region: many regions for the garden use, including 3.
12. Plant Codes:
   a. Bayer Code: No specific entry

551

1. Sloughgrass
(American sloughgrass)
Poaceae (Gramineae)
Beckmannia syzigachne (Steud.) Fernald (syn: *B. eruciformis* Auct.)

2. This is a leafy perennial bunchgrass important to the U.S. and prairie provinces. Sloughgrass grows in wet areas near lake shores and tolerates moderate levels of salinity in the soil. Suitable for both cultivated and irrigated lands. It produces a good quality forage and is readily eaten by cattle.
3. Season: Cool season, early spring growth through early summer. Grazing in the spring and fall and stored for hay in winter.
4. Production in U.S.: No data for sloughgrass production, however, pasture and rangeland are produced on more than 410 million A (U.S. CENSUS, 1992).
5. Other production regions: In Canada, it is grown in the prairie provinces.
6. Use: Rangeland grazing, forage, and hay, and cover crop.
7. Part(s) of plant consumed: Leaves, stems, and seedheads
8. Portion analyzed/sampled: Forage and hay
9. Classifications:
   a. Authors Class: See Forage grass
   b. EPA Crop Group (Group & Subgroup): See Forage grass
   c. Codex Group: See Forage grass
   d. EPA Crop Definition: None
10. References: GRIN, US EPA 1994, US EPA 1995, CODEX, BARNES, MOSLER, RICHE
11. Production Map: EPA Crop Production Regions 7, 8, 9 and 11.
12. Plant Codes:
   a. Bayer Code: BECSY

552

1. Smilograss
(Smilo, Rice millet)
Poaceae (Gramineae)
Piptatherum miliaceum (L.) Coss. (syn: *Oryzopsis miliacea* (L.) Benth. & Hook. f. ex Asch. & Schweinf.)

2. This is a perennial cool season grass introduced from the Mediterranean region to the U.S. and is distributed only to southern California. It's a member of the grass tribe *Stipeae*. It is an erect bunchgrass 0.6 to 1.5 m tall growing, flat blades 20 to 50 cm long and 3 to 10 mm wide and has a panicle type inflorescence (15 to 30 cm long) in dryland fields, chaparral and converted wasteland areas. It produces good forage for livestock, especially when immature because if allowed to mature the grass is too rank and coarse for grazing. Smilograss reproduces from seeds and tillers and are used to revegetate chaparral areas after wildfires. There are 1,949,220 seeds/kg.

3. Season: Cool season grass that starts growing in early spring and produces heavy vegetation. Stands should not be grazed too late in the season to assure good seed set.
4. Production in U.S.: There is no production data for Smilograss, however, pasture and rangeland are produced on more than 410 million A (U.S. CENSUS, 1992). Southern California.
5. Other production regions: Mediterranean region.
6. Use: Pasture grazing and revegetation of chaparral areas after wildfires.
7. Part(s) of plant consumed: Leaves and stems
8. Portion analyzed/sampled: Forage and hay
9. Classifications:
   a. Authors Class: See Forage grass
   b. EPA Crop Group (Group & Subgroup): See Forage grass
   c. Codex Group: See Forage grass
   d. EPA Crop Definition: None
10. References: GRIN, US EPA 1994, US EPA 1995, CODEX, HOLZWORTH, MOSLER, STUBBENDIECK (picture), RICHE
11. Production Map: EPA Crop Production Region 10.
12. Plant Codes:
    a. Bayer Code: ORZMI

553

1. Snakegourd
(Culebra, Patole, Guada bean, Serpent cucumber, Serpent gourd, Viper's gourd, Serpent vegetal)
Cucurbitaceae
Trichosanthes cucumerina var. *anguina* (L.) Haines (syn: *T. anguina* L.)

2. Annual vine with branched tendrils. Leaves are heart shaped. Fruits are edible when green and are 1 to 6 feet long and 1¹/₂ to 4 inches wide at the largest diameter. Plants are grown on trellises about 7 feet high.
3. Season, seeding to harvest: About 3 months, immature fruits up to 2¹/₂ feet can be harvested.
4. Production in U.S.: Grows in Hawaii (NEAL 1965).
5. Other production regions: Tropics (YAMAGUCHI 1983).
6. Use: Primarily young fruit, but leaves also used as a vegetable.
7. Part(s) of plant consumed: Fruit
8. Portion analyzed/sampled: Whole fruit
9. Classifications:
   a. Authors Class:Cucurbit vegetables
   b. EPA Crop Group (Group & Subgroup): Miscellaneous
   c. Codex Group: 011 (VC 0430) Cucurbits fruiting vegetables
   d. EPA Crop Definition: None
10. References: GRIN, CODEX, MARTIN 1983, NEAL, YAMAGUCHI 1983
11. Production Map: EPA Crop Production Region 13.
12. Plant Codes:
    a. Bayer Code: TTHCA

554

1. Sorghum
(Hirse)
Poaceae (Gramineae)
Sorghum spp.
1. Sorghums, Forage
(Broomcorn, Sorgo, Great millet)
S. bicolor (L.) Moench (syn: *S. vulgare* Pers.)

1. Johnsongrass
(Canota, Sorgo, Sorgho d'Alep)
S. halepense (L.) Pers.
1. Sorghum almum
(Columbusgrass, Sorgrass, Sorghum hybrid, Sorghumgrass, Sorgo negro, maicillo, Argentinegrass, Garavi Almum grass)
S. x *almum* Parodi
1. Sudangrass
(Chicken corn, Sorgrass, Sorghum hybrid)
S. x *drummondii* (Nees ex Steud.) Millsp. & Chase (syn: *S. sudanense* (Piper) Stapf.)
1. Sorgrass
(Sorghum hybrids)
S. bicolor x *S. halepense* (L.) Pers.
S. bicolor x *S. arundinaceum* (Desv.) Stapf

2. Sorghums are grown for four principal uses: grain, forage, syrup or sugar, and industrial use of the stems and fibers. The plants are tall annuals or perennials with flat leaf blades. The grain sorghums are listed under grain sorghum, and syrup or sugar sorghums under sweet sorghum. Industrial sorghums include broomcorn, *S. bicolor* (L.) Moench, the tough stems and panicles of which are used for brooms, listing under grain sorghum. These classes are not distinct, except industrial sorghum. Thus both grain and sugar sorghums may be used for pasturage or hay. *Sorghum* species grown primarily for pasture, silage or hay are listed as follows.

Forage sorghums are similar to the grain sorghums but differ from the latter in having sweet or slightly sweet and juicy stems and are more leafy. The leaves are broad and coarse. The stems vary from 2 feet up to 15 feet in height depending on variety and growing conditions. They are mainly summer annuals, usually seeded in rows like corn. They have been grown since prehistoric times in Asia and Africa and a great many varieties have been obtained through natural selection. Forage sorghums were introduced into the U.S. about 1850, and numerous varieties have developed here. They tolerate heat and limited moisture and are valuable for hay or silage, especially in the central and southern Plains. Many of the forage sorghums are dangerous to livestock while green because of the prussic acid (hydrocyanic acid) content. This disappears as the fodder is cured. They produce more dry matter tonnage than grain sorghum. The forage sorghums are used mostly for silage.

Johnsongrass differs from other sorghum species in being a perennial that spreads by vigorous rootstocks. For this reason it is difficult to eradicate where not wanted and may be a troublesome weed. It is, however, a valuable livestock feed in many sections of the South. It is not winter hardy in cold climates. Its area of adaptation is roughly the Cotton Belt. Johnsongrass was brought to this country from Turkey about 1830. Stems are about ¹/₄ inch in diameter, up to 4 or 5 feet tall. Leaves are numerous, long and slender. Growth is very vigorous and two or three crops of hay may be harvested in a season. It is palatable and nutritious both as hay and as pasture. It may contain small amounts of prussic acid, but rarely enough to poison livestock.

In recent years many varieties that are hybrids of *Sorghum* spp. have been developed and become commercial. These represent both grain and forage types, or dual purpose types. Also, methods of producing hybrid seed of *Sorghum* spp. have been developed commercially so a number of hybrid forage sorghum varieties are now available. In general, these forage types resemble *Sudangrass* but may vary in vigor, coarseness of plant, sweetness and time from seeding to maturity.

Sorghum almum grass was introduced from South African, Australian, and New Zealand sources. It is possibly a hybrid between Johnsongrass, *S. halepense*, and a cultivated sorghum, *S. bicolor*. It resembles Johnsongrass but is coarser and taller growing with larger stems and leaves. The rhizomes are short and stout in contrast to the long ones of Johnsongrass, and it does not become a troublesome weed. Its adaptation is similar to that of Johnsongrass, or roughly to the Cotton Belt.

Chicken corn was apparently introduced by chance and became naturalized on black soils of Alabama around the middle of the last century. Later it largely disappeared as a naturalized plant. A selection has been increased recently, primarily for use in wildlife plantings. It is a sweet sorghum of medium size. The seed shatters in late summer, remains dormant over winter and germinates the following spring. Seed is in limited commercial production.

Sudangrass was introduced into the U.S. in 1909 from Africa and is now one of the most valuable summer annual forage grasses. It is widely adapted, drought resistant, and grows rapidly from late seeding. Usually ready for grazing 5 to 6 weeks after planting. It is usually seeded alone in low-rainfall areas but is often combined with soybeans in more humid areas. The grass stems reach up to 7 feet under the most favorable conditions. They are slender, usually only about $1/4$-inch across. Several stems rise from a single clump. Leaves are numerous, long and narrow. Sudangrass is valuable for hay, silage or pasture. If growth is short and stunted the prussic acid content may be high enough to make pasturing hazardous to livestock, but it is safe to use as hay. The prussic acid content is lower than in the other forage sorghums. Seed of many improved varieties are available in commerce. Acreage planted annually to sudangrass in the U.S. is estimated at around 4,000,000. Sudangrass is harvested for hay when 30 inches tall. Forage sorghums are used primarily as silage while sudangrass and sorghum-sudangrass hybrids are grazed by livestock or fed as green chop or hay. Forage sorghums are harvested July-August.

3. Season: Summer annual for midsummer pasture, silage or hay.
4. Production in U.S.: No specific data, See Forage grass. in 1992, 4,257 tons of sudangrass seeds were produced on 4,248 acres (1992 CENSUS). In 1992, the top 3 seed producing states for sudangrass were California (2,760 acres), Texas (1,193) and Washington (169) (1992 CENSUS). Central and Southern Plains and Cotton Belt.
5. Other production regions: Common throughout the temperate and warmer regions of the world (GOULD 1975).
6. Use: Pasture, silage, hay and green chop. Also as cover crop following potatoes.
7. Part(s) of plant consumed: Foliage
8. Portion analyzed/sampled: Forage and hay. Also see Forage grass
9. Classifications:
   a. Authors Class: See Forage grass
   b. EPA Crop Group (Group & Subgroup): See Forage grass
   c. Codex Group: See Forage grass
   d. EPA Crop Definition: Sorghum (grain) = Sorghum (grain), Sudangrass (seed crop) and hybrids of these grown for its seed. Sorghum (fodder, forage) = Sorghum (fodder, for-

age), Sudangrass, and hybrids of these grown for fodder and/or forage
10. References: GRIN, MAGNESS, RICHE, SCHREIBER, US EPA 1995a, BALL, BARNES, SKERMAN, UNDER-SANDER(b), GOULD
11. Production Map: EPA Crop Production Regions 2, 3, 4, 5, 6, 7 and 8. Seed crop in Regions 8, 10 and 11.
12. Plant Codes:
   a. Bayer Code: SORSS (*S. spp.*), SORAL (*S. almum*), SORVU (*S. bicolor*), SORHA (*S. halepense*)

555

1. Sorghum, Grain

(Milo, Durra, Kaffir-corn, Indian millet, Great millet, Grand millet, Kaoliang, Chinese sorghum, Shattercane, Guineacorn, Sorgo comun)

Poaceae (Gramineae)

Sorghum bicolor (L.) Moench (syn: *S. caffrorum* (Thunb.) P. Beauv.; *S. dochna* (Forssk.) Snowden; *S. guineense* Stapf; *S. nervosum* Besser ex Schult. & Schult.f.; *S. vulgare* Pers.)

2. Grain sorghum was grown on 13,902,000 acres in the U.S. (average for 1966-67), mainly in the Central and Southern Plains states. Yields for the two years averaged 53.2 bushels per acre for a total average production of about 740 million bushels (1960s data retained for comparsion below). In 1981 worldwide, grain sorghum was grown on about 100 million acres. China, India and Africa grow large quantities. In the U.S. most of the grain sorghum is used as livestock feed, but in the Orient and Africa most is used as food.

Sorghum culture goes back to antiquity with Egypt being an early area. Grain sorghums grown in this country mainly trace to African origins. Although they were brought here during early colonial days they did not become important crops until farming developed in drier sections of the U.S. They generally out-yielded other grains under conditions of limited moisture.

Grain sorghum plants are coarse annual grasses. Nearly all of the varieties grown in the U.S. are so-called dwarf types, with stems under 5 feet in height and suitable for harvesting with combines. In other countries many taller-stemmed kinds are grown. Leaves are relatively broad, have numerous but small stomata, and are covered with a waxy bloom. They tend to roll along the midrib under moisture stress. Thus the plant is more drought resistant than most other grains and requires less water per pound of dry matter produced. Flowers and seeds are borne in relatively dense panicles that vary from 3 to 20 inches in length and up to 3 inches in width. The panicle is enclosed in a rather strong sheath until just before the first flowers open. Branch stems in the panicle rise in whorls and may be few or many. They are also variable in length, resulting in variable density of the panicle. Spikelets are partially enclosed in two rather short, thick glumes. Each spikelet contains two flowers, only one of which is usually fertile and sets a seed. The fertile flower consists of a thin lemma and thin palea, and inside these the stamens and pistil, the latter developing into the kernel. The lemma may be awned or awnless.

When threshed, the seed separates from the floral bracts as in wheat. The kernels are small, averaging about two-thirds the weight of wheat grains. Weight of 1000 sorghum grains is mostly between 20 and 30 grams. Kernels are generally near round to broad-conic in shape. The grain consists of about 6 percent bran, the pericarp or surface layers; 10 percent germ; and 84 percent endosperm, which is largely starch. In protein content, sorghum is higher than corn and about equal to wheat. In fat content it is lower than corn but higher than wheat.

Grain Sorghum Groups

Grain sorghum varieties are classed in seven agronomic groups, as follows:

Kafir sorghums, originally from South Africa, have thick, juicy stems, large leaves and awnless cylindrical-shaped panicles. Seeds may be white, pink or red and are medium in size.

Milo sorghums, originally from East Africa, have stems that are less juicy than in Kafir. Leaf blades are wavy with a yellow midrib. Heads are bearded or awned, compact, oval in shape. Seeds are large, pale pink to cream in color. Plants tend to be more tolerant to heat and drought than the Kafirs.

Feterita sorghums, came from Sudan. Leaves are sparse in number. Stems are slender and dry. Panicles are compact and oval in shape. Seeds are very large for sorghum, chalky white in color.

Durra sorghums, are from the Mediterranean area, the Near East, and Middle East. Stems are dry. Panicles are bearded and hairy and may be compact or open. Seeds are large and flattened.

Shallu sorghums, are from India with tall, slender, dry stems. Heads are loose. Seeds are pearly white in color and late maturing, thus requiring a relatively long growing season.

Koaliang sorghums, typical of those mainly grown in China, Manchuria and Japan, have slender, dry, woody stems with sparse leaves. Panicles are wiry and semicompact. Seeds are brown and bitter in taste.

Hegari sorghums, from Sudan are somewhat similar to Kafirs but have more nearly oval panicles, and plants that tiller profusely. Seeds are chalky white.

In the U.S. most varieties have been derived from crosses involving Kafir and Milo. Other groups have also entered into some varieties, so the varieties now grown are generally not typical of any specific group. White sorghum without tannins is preferred for human food. Most sorghums now being grown in this country are from hybrid seeds, made possible by the finding and isolation of male sterile strains. When the male-sterile line is planted alongside suitable lines with fertile pollen all the seed produced on the male-sterile line is hybrid. Use of such seed, coupled with improved agronomic practices, have resulted in recent average yields which are more than double those being obtained even as late as from 1952 to 1956. The average yield for 1993-95 is about 10 bushels more per acre over 1966-67.

Uses of Grain Sorghum

Nearly all (98 percent) the sorghum grain consumed in the U.S. is used for livestock feed for cattle, swine and poultry. Of the 1994 supply about one-third was exported, mainly to Japan, India and Europe. Most of that exported was probably used as food. For food use, the grain may be roughly ground and made into bread-like preparations, used after grinding and stewing as a mush or porridge, or made into flour for mixing with wheat flour for breads. Varieties with waxy endosperms are a source of starch having properties similar to tapioca. The grain is also a source of native beers, particularly in Africa. For feed use, sorghum grain should be ground for most classes of livestock, since the grains are small and relatively hard. In feeding value it is almost equal to kernel corn.

Some quantities of grain sorghums go into industrial uses in this country. Starch is manufactured by a wet-milling process similar to that used for corn starch (See under corn). The starch is then made into dextrose for use in foods. Starch from waxy sorghums is used in adhesives and for sizing paper and fabrics, also in the "mud" used in drilling for oil. The grain is also a source of grain and butyl alcohol. Like the forage sorghums, the green grain sorghum plants contain the glucoside dhurrin, which converts to prussic acid (HCN) and is poisonous to livestock. For this reason grain sorghums are not suitable for pasturage. Grain sorghum fits well into a number of crop rotations and fits into a double crop production system. Some rotated crops are small grains, soybeans and cotton.

3. Season, seeding to harvest: About 95 to 120 days or more. Days to flowering vary from 62 to 90.

4. Production in U.S.: In 1995, 8,278,000 acres harvested for grain with 460,373,000 bushels and as average yield of 55.6 bushels per acre. The 3-year average for 1993-95 was 8,704,000 acres harvested for grain sorghum with 547,917,000 bushels and an average yield of 62.77 bushels per acre (USDA 1996c). In 1995, the top 10 states were Kansas (3,100,000 acres), Texas (2,400,000), Nebraska (980,000), Missouri (490,000), Oklahoma (320,000), Arkansas (185,000), Illinois (170,000), Colorado (165,000), New Mexico (130,000) and South Dakota (120,000) for 97 percent of the U.S. acreage harvested for grain sorghum (USDA 1996c).

5. Other production regions: In 1981, Mexico grew 4,364,490 acres of grain sorghum with about 9 percent of the world production as compared to the U.S. with about 31 percent and India 16 percent (MARTIN 1983).

6. Use: In U.S., mainly used for feed. Other domestic uses are for food, alcohol and seed. In the rest of the world, the grain is largely used as cereal, ground into flour or used in preparation of beer.

7. Part(s) of plant consumed: Grain and stalks

8. Portion analyzed/sampled: Grain, forage, stover, aspirated grain fractions and processed commodity, flour. For sorghum forage, cut whole aerial portion of the plant at soft to hard dough stage. For sorghum stover, use mature dried stalks from which the grain has been removed with about 85 percent dry matter. Aspirated grain fractions are grain dust collected at grain elevators. Sorghum flour residue data are not needed at this time since the flour is used exclusively in the U.S. as a component for drywall, and not as food or feed. However, 50 percent of the worldwide sorghum production goes toward human consumption, data may be needed on flour at a later date.

9. Classifications:
   a. Authors Class: Cereal grains; Forage, fodder and straw of cereal grains
   b. EPA Crop Group (Group & Subgroup): Cereal grains (15) and Forage, fodder and straw of cereal grains (16) (Representative crop)
   c. Codex Group: 020 (GC 0651) Cereal grains; 051 (AS 0651 and AF 0651) Straw, fodder and forage of cereal grains and grasses
   d. EPA Crop Definition: Sorghum (fodder, forage) = *Sorghum* spp. (Sorghum fodder and forage, sudangrass and hybrids of these grown for fodder and/or forage). Sorghum (grain) = *Sorghum* spp. (Sorghum grain and sudangrass seed crops) and hybrids of these grown for its seed

10. References: GRIN, CODEX, FACCIOLA, MAGNESS, MARTIN 1983, USDA 1996a, USDA 1996c, US EPA 1996a, US EPA 1995a, SMITH 1995
11. Production Map: EPA Crop Production Regions 2, 4, 5, 6, 7 and 8.
12. Plant Codes:
    a. Bayer Code: SORVU

556

1. Sorghum, Sweet
(Sorgo, Sorgo dulce, Zuckerhirse, Sorgho doux)
Poaceae (Gramineae)

Sorghum bicolor (L.) Moench (syn: *S. vulgare* Pers.; *S. nervosum* Besser ex Schult. & Schult.f.; *S. guineense* Stapf.; *S. durra* (Forssk.) Stapf; *S. dochna* (Forssk.) Snowden; *S. caffrorum* (Thunb.) P. Beauv.)

2. Sorgo or sweet sorghums are plants grown primarily as a source of syrup (sirup) although they also may be used as silage or for hay. They are closely related to sorghums grown for silage, hay or grain but are characterized by containing an abundance of sweet juice in the stems. The plants are perennial in warm climates but in continental U.S. they winter kill in most areas, so are grown as annuals. Some sorgo for syrup is grown in most states except the far north, but major production is in areas adapted to cotton growing. Sorgo is of Old World origin and has been long cultivated in Africa and southern Asia. The culture is quite similar to that of corn. Seeding should be delayed until the soil is warm. Then the seeds are planted in well prepared soil in rows about 3 feet apart. The initial stem forms branches or tillers at underground nodes so a clump of several stems forms from a single seed. Stems in different varieties reach from 8 to 12 feet or more in height under good growing conditions and an inch or more in diameter. A long, slender leaf rises from each node, with the leaf base or sheath encircling the stem. Stalks terminate in a panicle containing the flowers and later the seeds. Harvesting for syrup production is best done as soon as the seed becomes hard and ripe. The interval from planting to harvest may range from around 100 days up to near 200, depending on such factors as variety, earliness of planting and growing season temperature. Most sorghum for syrup is grown on small acreages, generally less than an acre per farm, and hand methods are used in harvesting. Leaves are hand stripped from the standing stalks by beating or by cutting with a cane knife. If the leaves remain, delay milling for 3 to 5 days for the leaves to dry out and the stalks to lose some water, and natural enzymes within the stalk to invert some of the sucrose. Heads are cut off and the stalks cut at ground level. Crushing the stems for extraction of the juice, and preparation of the syrup from the juice are the same as described for cane syrup production, which see. Under normal conditions, an efficient mill will deliver 50 to 55 pounds of juice from 100 pounds of clean stalks. It takes from 6 to 12 gallons of raw juice to finish 1 gallon of syrup. The finished syrup will weigh about 11¹/₂ pounds per gallon. Recent data on sorghum syrup production are not available. Sweet sorghum yields 100 to 300 gallons of syrup per acre. All is used as food. Leaves and tops removed from the stems may be fed to livestock on the farm, but do not enter commerce. The crushed stalks are termed bagasse, pomace or chews.

3. Season, seeding to harvest: About 100 to 200 days.

4. Production in U.S.: In 1995, grain and sweet sorghum for all uses, including syrup, planted to 9,454,000 acres (USDA 1996a). This is the lowest planting since 1930 (USDA 1996c). In 1987, about 1,000 acres reported as harvested for syrup only with 1,088,289 pounds of syrup produced (1987 CENSUS). The top 5 states as reported in 1987, Tennessee (266 acres), Kentucky (177), Missouri (54), Georgia (49) and Indiana (29). Southeastern states (MASK 1991).
5. Other production regions: China and South Africa (BAILEY 1976).
6. Use: Syrup production and forage (silage).
7. Part(s) of plant consumed: Stalks
8. Portion analyzed/sampled: Stalk and processed commodity syrup. The seed and forage can be covered by sorghum grain tolerances
9. Classifications:
    a. Authors Class: Grasses for sugar or syrup
    b. EPA Crop Group (Group & Subgroup): Miscellaneous
    c. Codex Group: 021 (GS 0658) Grasses for sugar or syrup production; 020 (GC 0651) Cereal grains; 069 (DM 0658) Miscellaneous derived products of plant origin for molasses
    d. EPA Crop Definition: None
10. References: GRIN, CODEX, MAGNESS, MASK, PAUTLER (1989 CENSUS), USDA 1996a, USDA 1996c, US EPA 1996a, BAILEY 1976, BITZER 1987a, BITZER 1987b
11. Production Map: EPA Crop Production Regions 2, 4 and 5.
12. Plant Codes:
    a. Bayer Code: SORVS

557

1. Sorrel, French
(Round sorrel, Suan mo, Bucklerleaved sorrel, Dock, Rumex a ecusson, Acedera romana, Yuam ye juan mo)
Polygonaceae

Rumex scutatus L.

2. French sorrel differs from garden sorrel, which see, in that the stems are more branched and less upright. The leaves are somewhat fleshy. Rosette leaves have long petioles and are somewhat heart-shaped, while stem leaves are more pointed on short petioles.
3. Season, seeding to first harvest: About 2 months, but harvested throughout the summer.
4. Production in U.S.: No data, very limited.
5. Other production regions: Native to Europe and Asia (BAILEY 1976). See Garden sorrel.
6. Use: As potherb and use in salad for flavoring and in soups.
7. Part(s) of plant consumed: Leaves only
8. Portion analyzed/sampled: Leaves
9. Classifications:
    a. Authors Class: Leafy vegetables (except *Brassica* vegetables); Herbs and spices
    b. EPA Crop Group (Group & Subgroup): Leafy vegetables (except *Brassica* vegetables) (4A)
    c. Codex Group: 027 (HH 0746) Herbs (Common and related *Rumex*)
    d. EPA Crop Definition: None
10. References: GRIN, BAILEY 1976, CODEX, FACCIOLA, KAYS, LORENZ, MAGNESS, SCHNEIDER 1985 (picture), US EPA 1995a
11. Production Map: No entry.

12. Plant Codes:
  a. Bayer Code: No specific entry

558

1. Sorrel, Garden

(Sourgrass, Greensauce, Broadleaved sorrel, Sorrel, Common sorrel, Dock, Sour dock, Acedera, Grand oseille, Sorelu, Vinagrera)
Polygonaceae
  Rumex acetosa L.
2. Garden sorrel is a perennial plant, the roots of which persist for several years. It is grown to a limited extent for the leaves, gathered in early spring and used as greens or pot-herbs. The stems are erect, reaching 3 or more feet in height. Rosette leaves are thin, light green and oblong. Stem leaves are narrow and pointed. All leaves are glabrous and have a tart flavor.
3. Season, start of growth to first harvest: About 4 weeks. Seeding to first harvest about 2 months.
4. Production in U.S.: No data; very limited. Mostly home gardens. Grows well in Florida gardens (STEPHENS 1988). Naturalized in U.S. (GARLAND 1993). Sorrel appears to be available from greenhouses in U.S. (SCHNEIDER 1985).
5. Other production regions: Grows in Canada (ROTHWELL 1996a). Sorrel is an important fresh herb crop in Germany (PALLUTT 1996).
6. Use: As potherb and in salads. Sorrel is a main ingredient in a relish served with meat and fowl, called Greensauce (SCHNEIDER 1985). Preparation: clean sorrel by dunking in water and draining.
7. Part(s) of plant consumed: Young leaves. Flowers and seeds are edible (FACCIOLA 1990)
8. Portion analyzed/sampled: Leaves
9. Classifications:
  a. Authors Class: Leafy vegetables (except *Brassica* vegetables); Herbs and spices
  b. EPA Crop Group (Group & Subgroup): Leafy vegetables (except *Brassica* vegetables) (4A)
  c. Codex Group: 027 (HH 0746) Herbs (Common and related *Rumex*)
  d. EPA Crop Definition: None
10. References: GRIN, CODEX, FACCIOLA, GARLAND, KAYS, LORENZ, MAGNESS, PALLUTT, ROTHWELL 1996a, SCHNEIDER 1985, STEPHENS 1988 (picture), US EPA 1995a
11. Production Map: No entry
12. Plant Codes:
  a. Bayer Code: RUMAC (*R. acetosa*), RUMAH (*R. acetosa* var. *hortensis*)

559

1. Soursop

(Guanabana, Currosol, Graviola, Sapote agrio, Coracao de raihna, Suirsaak, Zuursaak, Chachiman-epineux, Prickly custard apple, Mundla)
Annonaceae
  Annona muricata L.
2. This tropical tree is a small evergreen, up to 15 to 20 feet, with leathery, obovate leaves, native to tropical America. The plant likes hot tropical lowlands with high rainfall. It is very susceptible to frost. The plant is propagated by seeds or grafting. The large flowers are hermaphrodite, thick and fleshy. Poor fruit set is a common problem. The fruit grow on small auxiliary stems, often directly on the trunk of the tree. There is great variability in size and weight of fruit. They can be very large, up to 12 pounds, the largest fruit among the Annonaceae. The oval, heart-shaped or oblong fruit has a deep green skin with many short fleshy spines. The interior flesh is white, juicy, aromatic with a tart flavor. Its texture is somewhat cotton-like and contains many seeds.
3. Season: Bloom to maturity in 70 to 120 days.
4. Production in U.S.: It is an important fruit in Puerto Rico (115 tons from 50 acres in 1992 CENSUS). It is too tender for culture in continental U.S., except in the warmest parts of Florida.
5. Other production regions: South and Central America including Mexico, Brazil, Venezuela.
6. Use: Pulp is eaten fresh, made into beverages (carato), jelly, ice cream or other desserts.
7. Part(s) of plant consumed: Fruit pulp, which has the flavor like pineapple and mango
8. Portion analyzed/sampled: Whole fruit
9. Classifications:
  a. Authors Class: Tropical and subtropical fruits – inedible peel
  b. EPA Crop Group (Group & Subgroup): Miscellaneous
  c. Codex Group: 006 (FI 0365) Assorted tropical and subtropical fruits – inedible peel
  d. EPA Crop Definition: Proposing – Sugar apple = Soursop
10. References: GRIN, BUESO, CODEX, KNIGHT(a), LOGAN, MAGNESS, MARTIN 1987, US EPA 1994, PUROHIT, RICHE
11. Production Map: EPA Crop Production Region 13.
12. Plant Codes:
  a. Bayer Code: ANUMU

560

1. South African bluestem

(Tambookie grass, Coolataigrass, Common thatching grass)
Poaceae (Gramineae)
  Hyparrhenia hirta (L.) Stapf (syn: *Andropogon hirtus* L.)
2. This is a perennial warm season grass native to the Mediterranean region, and introduced into the southern U.S. It is an erect grass up to 90 cm tall with its inflorescence a panicle. South African bluestem is adapted to areas 1200 to 2500 m with rainfall >500 mm, and will not survive below freezing temperatures. It is not real palatable except when young. It can be grazed heavily to prevent seeding and it will provide fair quality hay and silage. In the U.S., it has been useful as a soil conservation grass on stony and severely eroded soils. There are 1,320,000 seeds/kg.
3. Season: Warm season tropical grass that starts growing in the spring and is productive in the summer. It can be grazed heavily to prevent seed production and further reduction in forage quality.
4. Production in U.S.: There is no production data for South African bluestem, however, pasture and rangeland are produced on more than 410 million A (U.S. CENSUS, 1992). See above.
5. Other production regions: Mediterranean region.
6. Use: Pasture grazing, hay, silage, and soil conservation crop.

7. Part(s) of plant consumed: Leaves and stems
8. Portion analyzed/sampled: Forage and hay
9. Classifications:
   a. Authors Class: See Forage grass
   b. EPA Crop Group (Group & Subgroup): See Forage grass
   c. Codex Group: See Forage grass
   d. EPA Crop Definition: None
10. References: GRIN, US EPA 1994, US EPA 1995, CODEX, HOLZWORTH, SKERMAN (picture), RICHE
11. Production Map: EPA Crop Production Regions 3, 4, 6 and 13.
12. Plant Codes:
    a. Bayer Code: HYRHI

561

1. Southernwood
(Old Man, Southern wormwood, Slovenwood)
Asteraceae (Compositae)
Artemisia abrotanum L.

1. Mugwort
(Felon herb, Sage brush, St. John's plant, Carlinethistle)
A. vulgaris L.

1. Roman wormwood
A. pontica L.

2. The plants are shrubby perennials, 2 to 5 feet high, with green, glabrous, finely divided leaves. Grown from seed or from cuttings. The young, aromatic leaves and shoots are sometimes used for flavoring cakes and other culinary preparations. The plants are of minor importance as food, but are more common as an ornamental. Codex Alimentarius Commission listed *A. vulgaris* and *A. abrotanum* in the edible wormwood group.
3. Season, harvest: The herb is harvested two times a year when in flower and produces up to 10 years. They bloom in late summer.
4. Production in U.S.: No data.
5. Other production regions: Europe.
6. Use: Fresh and dried herbs.
7. Part(s) of plant consumed: Leaves and stems
8. Portion analyzed/sampled: Tops (fresh and dried)
9. Classifications:
   a. Authors Class: Herbs and spices
   b. EPA Crop Group (Group & Subgroup): Miscellaneous
   c. Codex Group: 027 (HH 0754) Herbs; 057 (DH 0754) Dried herbs. (Except *A. pontica*)
   d. EPA Crop Definition: None
10. References: GRIN, CODEX, FOSTER (picture), GARLAND (picture), MAGNESS, USDA NRCS
11. Production Map: No entry.
12. Plant Codes:
    a. Bayer Code: ARTAT (*A. abrotanum*), ARTVU (*A. vulgaris*)

562

1. Soybean
(Soya, Tofu, Temph, Miso, Soya bean, Soybean oil, Soja, Chinese pea, Manchurian bean, Japan pea, Japan bean, Japanese fodder plant)
Fabaceae (Leguminosae)
Glycine max (L.) Merr. (syn: *G. hispida* (Moench) Maxim.; *Soja max* (L.) Piper)

2. The soybean has become the most important source of vegetable oil in the U.S. Quantity of beans crushed for oil averaged about 1.270 billion bushels, 1991-93, producing 7,005,000 tons of oil. Production of oil has almost tripled in the past 30 years. The plant is a bushy, hairy annual herb up to 3 feet. The hairy pods grow in small clusters, maxillary along the stem, and each contains 2 to 4 seeds. Seeds are variable in size, generally about $1/4$ inch long. Pods are closed until seeds are threshed out. The seeds contain up to 25 percent of drying oil. The oil is extracted either with solvents, hydraulic presses or expellers, the latter two methods involving heating. Much soybean oil is used as salad and cooking oil and for the manufacture of margarine. Large quantities are also used in industry. The press cake is a high protein feed. The beans and plants are also important livestock feeds. There are many cultivars of soybeans with yellow-seeded beans generally used for food, green-seeded beans preferred for sprouts, and black-seeded beans used for oil, horse feed or fermented. Soybeans are generally double-cropped in middle Atlantic and Southeastern U.S. with barley and wheat. Also, soybeans are cultivated similar to corn or cotton and fits well into rotations with corn, cotton or rice. There are several maturity dates for soybeans from 75 to 200 days to reach maturity, generally about 80 to 100 days. Most cultivars are developed for oil and not forage. Some forage soybean cultivars are 'Donegal' and 'Derry'. There are 4,500 seeds per pound.
3. Season, seeding to harvest: About 4 to 6 months.
4. Production in U.S.: 60,135,000 acres planted in 1993 (U.S.). Top 9 states that accounted for 74 percent of acres planted in 1993 are Illinois, Iowa, Minnesota, Indiana, Missouri, Ohio, Arkansas, Nebraska and Mississippi.
5. Other production regions: Canada reported 1,778,173 acres in 1995. Brazil and China are large producers (RUBATZKY 1997).
6. Use: Immature seeds as green vegetable. Mature seeds for confectionery products; sprouts; soymilk which is the base for tofu; fermented seeds for soysauce or shoya, and miso and temph; and the major product oil which is used as salad oil, shortening, margarine and many specialty oil products. The defatted meal is used mostly as cattle feed. Some nonfat meal is used for human and pet consumption. Foliage can be used for livestock feed. Food, feed and industrial uses; and used whole, meal and for oil.
7. Part(s) of plant consumed: Seed and foliage
8. Portion analyzed/sampled: Seed, forage, hay, aspirated grain fractions; and the seed processed commodities meal, hulls and refined oil. For soybean forage, cut samples at 6 to 8 inches tall (sixth node) to beginning pod formation, at approximately 35 percent dry matter. For soybean hay, cut samples at mid-to-full bloom stage and before bottom leaves begin to fall or when pods are approximately 50 percent developed. Hay should be field-dried to a moisture content of 10 to 20 percent. Label restrictions against feeding forage hay may be allowed; e.g. *Do not feed green immature growing plants to livestock, or Do not harvest for livestock feed*. For soybean silage, residue data on silage are optional. Harvest sample when pods are one-half to fully mature (full pod stage). In the absence of silage data, residues in forage will be used for silage, with correction for dry matter. For aspirated grain fractions (previously called grain dust), dust collected at grain elevators for environmental and safety

reasons. Residue data should be provided for any postharvest use on soybeans. For a preharvest use after the reproduction stage begins and seed heads are formed, data are needed unless residues in the grain are less then the limit of quantitation of the analytical method. For a preharvest use during the vegetative stage (before the reproduction stage begins), data will not normally be needed unless the plant metabolism or processing study shows a concentration of residues of regulatory concern in an outer seed coat (e.g. soybean hulls)

9. Classifications:
   a. Authors Class: Oilseed; Legume vegetables (succulent and dried); Foliage of legume vegetables
   b. EPA Crop Group (Group & Subgroup): Soybeans: Legume vegetables (succulent or dried); and Foliage of legume vegetables. (Representative commodities). Soybeans (immature seed) Legume vegetables (succulent and dried) (6A)
   c. Codex Group: 014 (VP 0541) Legume vegetables (immature seed); 015 (VD 0541) Pulses; 050 (AL 0541 and 1265) Legume animal feeds; 067 (OC 0541) Vegetable crude oils; 068 (OR 0541) Vegetable refined oils
   d. EPA Crop Definition: None
10. References: GRIN, BAILEY 1976, CODEX, MAGNESS, MARTIN, ROBBELEN 1989 (picture), ROTHWELL 1996a, USDA 1994a, US EPA 1995a, US EPA 1995b, US EPA 1994, SMITH 1995, BALL, BARNES(a), DUKE(h), RUBATZKY, SMITH 1978
11. Production Map: EPA Crop Production Regions 2, 4 and 5 for dried soybean production.
12. Plant Codes:
    a. Bayer Code: GLXMA

563

1. Soybean, Vegetable

(An-ing, Habichuela soya, Coffee bean, Coffee berry, Edamame, Japan bean, Stock pea, Edamane, Edible soybean, Soya, Garden soybean, Soybean Immature seed)

Leguminosae

Glycine max (L.) Merr. (syn: *G. hispida* (Moench) Maxim; *G. gracilis* Skvortzov; *Soja max* (L.) Piper; *Dolichos soja* L.)

2. In the U.S., soybeans are not normally grown commercially as a vegetable, although tremendous acreages are grown for forage, feed and for crushing the seed for oil. In the Orient, soybeans are important food crops and are used as beans and peas. As vegetables, they are used to some extent as green pods, like snap beans, or for the unripe seeds, like garden peas or green lima beans; or as ripe seeds. Culture and exposure of edible parts are similar to that of lima beans, which see. The following refers to soybeans as a food vegetable.
3. Season, seeding to harvest: About 3 to 6 months, depending on type and usage.
4. Production in U.S.: No data. Limited acreage. Mainly a garden vegetable (STEPHENS 1988). New crop in Washington (SCHREIBER 1995).
5. Other production regions: Orient.
6. Use: As cooked vegetables, stir-fry; seed as snack food. Mainly the seeds are used. The traditional Asian uses of soybean as a food crop are based on the processed dry beans, which see Soybean, but boiled green seeds are eaten as vegetables. Viable seeds can be used for edible sprouts or planting. Some vegetable varieties in the U.S. are, 'Bansei',

'Fuji', 'Verde' and 'Seminole'. One use as a vegetable is to boil the whole pod, then shell and eat the seeds.
7. Part(s) of plant consumed: Whole pods or seeds. Foliage is an insignificant feed item
8. Portion analyzed/sampled: Succulent seed without pod and succulent seed with pod
9. Classifications:
   a. Authors Class: Legume vegetables (succulent or dried)
   b. EPA Crop Group (Group & Subgroup): Legume vegetables (succulent or dried) (6A)
   c. Codex Group: 014 (VP 0541) Legume vegetables (immature seed); 050 (AL 0541) Legume animal feeds
   d. EPA Crop Definition: None
10. References: GRIN, CODEX, MAGNESS, MARTIN 1983, SCHREIBER, STEPHENS 1988 (picture), US EPA 1996a, US EPA 1995a, YAMAGUCHI 1983, DUKE(h)
11. Production Map: EPA Crop Production Region 11. Many regions for garden use, including 3.
12. Plant Codes:
    a. Bayer Code: GLXMA

564

1. Spanish fennel

(Love-in-a-mist, Fennel flower, Wild fennel)

Ranunculaceae

Nigella hispanica L.
N. damascena L.

2. An annual herb to almost 2 feet tall from southern Europe. It is grown for its fernlike leaves, 1 to 2 inches long.
3. Season: No data.
4. Production in U.S.: No data.
5. Other production regions: Native to Mediterranean region (FACCIOLA 1990).
6. Use: Condiment.
7. Part(s) of plant consumed: Leaves and seeds
8. Portion analyzed/sampled: Leaves (fresh and dried); Seeds
9. Classifications:
   a. Authors Class: Herbs and spices
   b. EPA Crop Group (Group & Subgroup): Miscellaneous
   c. Codex Group: No specific entry
   d. EPA Crop Definition: None
10. References: GRIN, FACCIOLA, NEAL
11. Production Map: No entry
12. Plant Codes:
    a. Bayer Code: NIGHA (*N. hispanica*)

565

1. Spanish lime

(Genip, Mamoncillo, Quenepa, Macao, Limoncillo, Genipe, Honeyberry, Guenepa, Guayo, Mamon)

Sapindaceae

Melicoccus bijugatus Jacq. (syn: *Melicocca bijuga* L.)

2. This tree, native to the American tropics, is widely cultivated in the West Indies. It is not related to citrus limes. The tree is medium to large. Fruits are borne in clusters, like grapes. Individual fruits are ovoid in shape, an inch or more long, and have a thin, brittle skin. Inside is a thin layer of tart-to-sweet pulp and a large seed. Fruit ripens in late summer. The edible portion is the juicy pulp. The seeds are edible after roasting.

3. Season: In Florida, fruits ripen from June to September. Bloom to harvest: 90-150 days.
4. Production in U.S.: No data. Puerto Rico and Florida.
5. Other production regions: West Indies and Central America (MORTON 1987).
6. Use: Fresh, jam, jelly. For cold drink, the peeled fruit is boiled.
7. Part(s) of plant consumed: Fruit pulp
8. Portion analyzed/sampled: Fruit
9. Classifications:
   a. Authors Class: Tropical and subtropical fruits – inedible peel
   b. EPA Crop Group (Group & Subgroup): Miscellaneous
   c. Codex Group: 006 (FI 0366) Assorted tropical and subtropical fruits – inedible peel
   d. EPA Crop Definition: Proposing – Lychee = Spanish lime
10. References: GRIN, CODEX, MAGNESS, MARTIN 1987, MORTON (picture)
11. Production Map: EPA Crop Production Region 13.
12. Plant Codes:
    a. Bayer Code: No specific entry

566

1. Spike bentgrass
(Western redtop, Spike redtop, Bentgrass-spike)
Poaceae (Gramineae)
Agrostis exarata Trin.

2. Spike bentgrass is a cool season, perennial, erect or decumbent bunchgrass that grows from 20 to 120 cm tall with a spike-like panicle (5 to 30 cm long), and spreads by short rhizomes. It is an important rangegrass species that is native to the U.S. and grows in the western North America rangeland, and is used for cattle and horse grazing. It grows along streams and meadows and is adapted from sea level to 3000 m elevation. It grows from spring until the fall.
3. Season: Cool season grass with spring and fall growth. It flowers in July or August and matures in early September. Grazing spring through summer, makes good hay.
4. Production in U.S.: No data for Spike bentgrass production, however, pasture and rangeland are produced on more than 410 million A (U.S. CENSUS, 1992). See above.
5. Other production regions: In Canada, Spike bentgrass is grown in British Columbia.
6. Use: Pasture and rangeland grazing.
7. Part(s) of plant consumed: Leaves and stems
8. Portion analyzed/sampled: Forage and hay
9. Classifications:
   a. Authors Class: See Forage grass
   b. EPA Crop Group (Group & Subgroup): See Forage grass
   c. Codex Group: See Forage grass
   d. EPA Crop Definition: None.
10. References: GRIN, US EPA 1994, US EPA 1995, CODEX, ALDERSON, STUBBENDIECK(a) (picture), MOSLER, RICHE
11. Production Map: EPA Production Regions 7, 8, 9, 10 and 11.
12. Plant Codes:
    a. Bayer Code: AGSEN

567

1. Spike trisetum
Poaceae (Gramineae)
Trisetum spicatum (L.) K. Richt.

2. This is a cool season perennial bunchgrass native to the U.S. and a member of the grass tribe *Aveneae*. It is an erect bunchgrass growing from 10 to 50 cm tall that spreads by tillers, and has a panicle seedhead (2 to 15 cm long). Spike trisetum is adapted to bottomland soils, alpine meadows and slopes and is distributed from Canada, northeast and northwest U.S. and south into Mexico. It has good forage quality for all livestock and wildlife throughout the growing season. It is one of the most important forage grasses grown in the mountains. It reproduces from seeds and tillers.
3. Season: Warm season with spring growth. Remains green until August.
4. Production in U.S.: There is no production data for Spike trisetum, however, pasture and rangeland are produced on more than 410 million A (U.S. CENSUS, 1992). See above.
5. Other production regions: Most of Canada. See above.
6. Use: Grazing on pasture.
7. Part(s) of plant consumed: Leaves and stems
8. Portion analyzed/sampled: Forage and hay
9. Classifications:
   a. Authors Class: See Forage grass
   b. EPA Crop Group (Group & Subgroup): See Forage grass
   c. Codex Group: See Forage grass
   d. EPA Crop Definition: None
10. References: GRIN, US EPA 1994, US EPA 1995, CODEX, BARNES, HOLZWORTH, SKERMAN, STUBBENDIECK (picture), RICHE
11. Production Map: EPA Crop Production Regions 1, 2, 5, 9, 10, 11, 12 and 13.
12. Plant Codes:
    a. Bayer Code: No specific entry

568

1. Spikeoat
Poaceae (Gramineae)
Helictotrichon hookeri (Scribn.) Henrard

2. This is a cool season perennial bunchgrass native to North America, and is a member of the grass tribe *Aveneae*. It occurs naturally in plant communities of the Rocky Mountains and Canadian prairies. Spike oat grows erect from 10 to 75 cm tall, and has a panicle 5 to 12 cm long at maturity. It is found growing on hillsides, prairies and mountain tops on all soil types. It starts growing early in the spring, flowers June to July and seeds mature July through August. Spikeoat reproduces by seeds or tillers. It has good forage value for cattle and horses, and fair for sheep and wildlife, but forage quality declines rapidly at maturity.
3. Season: Cool season, early spring growth through early summer. Grazing in the spring and fall.
4. Production in U.S.: No data for Spikeoat production, however, pasture and rangeland are produced on over 410 million A (U.S. CENSUS, 1992). See above.
5. Other production regions: In Canada it is grown in the prairie provinces.
6. Use: Pasture grazing, hay and for erosion control.
7. Part(s) of plant consumed: Leaves and stems

8. Portion analyzed/sampled: Forage and hay
9. Classifications:
   a. Authors Class: See Forage grass
   b. EPA Crop Group (Group & Subgroup): See Forage grass
   c. Codex Group: See Forage grass
   d. EPA Crop Definition: None
10. References: GRIN, US EPA, 1994, US EPA 1995, CODEX, BARNES, MOSLER, RICHE
11. Production Map: EPA Crop Production Regions 7 and 8.
12. Plant Codes:
    a. Bayer Code: No specific entry

569

1. Spinach
(Gemuese spinat, epinard, Espinaca)
Chenopodiaceae
Spinacia oleracea L.

2. Next to cabbage, spinach is the most important of the cool season vegetables grown for greens or potherbs in the U.S. The cool season annual plants are grown from seed (12-20 pounds of seed per acre) and harvested while young and tender. Varieties differ in leaf shape from smooth and broad arrow-shaped to savoyed or wrinkled, but all leaves are non-hairy. Plants form a rosette of leaves on a very short stem, which later grows into a seed stalk. For greens, whole plants are harvested by cutting near the ground level prior to appreciable elongation of the stem. Leaves are largely separate and individually exposed during development. Savoy and semi-savoy are used for fresh market, and flat-leafed and semi-savoy for processing. Most of the commercial crop is machine-harvested by cutting the plants about 1 inch above the soil surface. There are 2,900 seeds per ounce.
3. Season, seeding to harvest: Usually 35 to 70 days depending on season.
4. Production in U.S.: Fresh market 17,200 acres with 97,100 tons in 1995; Processing 22,220 acres with 149,680 tons in 1995 (USDA 1996b). In 1995, processed spinach from Texas (8,200 acres) with the remaining states: Arkansas, California, Oklahoma, Tennessee and Wisconsin (14,020). Fresh market spinach reported for six states: California (6,700 acres), Colorado (3,000), Texas (2,700), New Jersey (2,600), Maryland (1,600) and Virgina (600) (USDA 1996b). In 1993, Washington produced 75 percent of the U.S. spinach seed production on 3,000 to 4,000 acres (SCHREIBER 1995).
5. Other production regions: In 1995, Canada reported 1,065 acres (ROTHWELL 1996a).
6. Use: As potherbs and salads; shipped fresh, canned, frozen.
7. Part(s) of plant consumed: Leaves
8. Portion analyzed/sampled: Leaves
9. Classifications:
   a. Authors Class: Leafy vegetables (except *Brassica* vegetables)
   b. EPA Crop Group (Group & Subgroup): Leafy vegetables (except *Brassica* vegetables) (4A) (Representative crop)
   c. Codex Group: 013 (VL 0502) Leafy vegetables (including *Brassica* leafy vegetables)
   d. EPA Crop Definition: None
10. References: GRIN, CODEX, LOGAN (picture), MAGNESS, ROTHWELL 1996a, SCHREIBER, USDA 1996b,

US EPA 1996a, US EPA 1995a, US EPA 1994, IVES, YAMAGUCH 1983, SWIADER
11. Production Map: EPA Crop Production Regions 1, 2, 4, 5, 6, 9 and 10.
12. Plant Codes:
    a. Bayer Code: SPQOL

570

1. Spinach, New Zealand
(Tsuru-na, Warrigal cabbage)
Aizoaceae
Tetragonia tetragonoides (Pall.) Kuntze (syn: *T. expansa* Murray)

2. New Zealand spinach is used in the same manner as spinach, but the plant is very different. It reaches a height of 1 to 2 feet, and is much branched, spreading to 3 to 4 feet across. In home gardens, tender shoots, tips and leaves are cut and used throughout the summer. Commercially, whole plants are usually cut above the ground when small. New growth from the cut stem base will produce a later crop. The plant is adapted to warmer growing conditions than spinach, and will produce summer greens where spinach will not. Leaves are generally similar in appearance to spinach. The dark green leaves have a slight fuzz.
3. Season, seeding to first harvest: About 2 months.
4. Production in U.S.: No data. Minor as compared to spinach, Mainly in home gardens. Available year-round from California (LOGAN 1996).
5. Other production regions: Native to New Zealand and naturalized in Europe (RUBATZKY 1997).
6. Use: As potherb or greens in salads.
7. Part(s) of plant consumed: Young leaves and stem tips
8. Portion analyzed/sampled: Tops (leaves)
9. Classifications:
   a. Authors Class: Leafy vegetables (except *Brassica* vegetables)
   b. EPA Crop Group (Group & Subgroup): Leafy vegetables (except *Brassica* vegetables) (4A)
   c. Codex Group: 013 (VL 0486) Leafy vegetables (including *Brassica* leafy vegetables)
   d. EPA Crop Definition: None
10. References: GRIN, CODEX, LOGAN, MAGNESS, STEPHENS 1988 (picture), US EPA 1995a, YAMAGUCHI 1983 (picture), RUBATZKY (picture)
11. Production Map: EPA Crop Production Region 10.
12. Plant Codes:
    a. Bayer Code: TEATE

571

1. Spinach, Vine
(Indian spinach, Malabar nightshade, Ceylon spinach, Libato, Gui, Red vine spinach, Malabar spinach, Country spinach, Basella)
Basellaceae
Basella alba L. (syn: *B. rubra* L.)

2. This is a plant sparingly grown in the U.S. as a potherb, but more important in the tropics. As a vegetable in the north it may be started under glass, set in the field after the soil warms, and provides a potherb to follow spinach. Leaves are succulent and of various shapes. Culture and leaf exposure

are comparable to those of spinach, but growth is slower. Grown as a perennial vine in the Caribbean and tropical South America (MARTIN 1979).

3. Season, seeding to harvest: 55 to 80 days. Normally, after about 3 months, the established plants may be cut or pruned on a weekly basis. Propagation is by direct seeding, transplants or stem cuttings.
4. Production in U.S.: No data.
5. Other production regions: Tropics, see above.
6. Use: As potherb or salad. The leaves have a mucilagious substance useful to thicken soups.
7. Part(s) of plant consumed: Leaves
8. Portion analyzed/sampled: Leaves
9. Classifications:
   a. Authors Class: Leafy vegetables (except *Brassica* vegetables)
   b. EPA Crop Group (Group & Subgroup): Leafy vegetables (except *Brassica* vegetables) (4A)
   c. Codex Group: 013 (VL 0503) Leafy vegetables (including *Brassica* leafy vegetables)
   d. EPA Crop Definition: None
10. References: GRIN, MAGNESS, MARTIN 1979 (picture), STEPHENS 1988 (picture), US EPA 1995a
11. Production Map: EPA Crop Production Region 13.
12. Plant Codes:
    a. Bayer Code: BADAL

572

1. Spiny hopsage
(Grays saltbush, Spiny sage)
Chenopodiaceae
Grayia spinosa (Hook.) Moq.

2. Spiny hopsage is a warm season perennial shrub that grows erect to 1.5 m tall and is native to the U.S. It flowers April through July and reproduces from seeds. Its habitat includes mesa flats and valleys and on alkaline soils. Native Americans ground the parched seeds to make pinole flour. Its spines may cause minor injury, but its forage is of excellent quality for all classes of livestock, browsed in the fall, winter and spring.
3. Season: Warm season perennial forage that can be used for grazing in the fall, winter and spring.
4. Production in U.S.: No specific data for Spiny hopsage production, however, pasture and rangelands are produced on more than 410 million A (U.S. CENSUS, 1992). See above.
5. Other production regions: No entry
6. Use: Grazing and seeds for flour.
7. Part(s) of plant consumed: Leaves, stems, seeds and flowers
8. Portion analyzed/sampled: Forage and hay
9. Classifications:
   a. Authors Class: Nongrass Animal Feeds
   b. EPA Crop Group (Group & Subgroup): Miscellaneous
   c. Codex Group: No specific citation although the general class 052 (AM 0165) Miscellaneous Fodder and Forage Crops could be used
   d. EPA Crop Definition: None
10. References: GRIN, US EPA 1994, US EPA 1995, CODEX, HOLZWORTH, STUBBENDIECK (picture), RICHE
11. Production Map: EPA Crop Production Region are 9 and 11.
12. Plant Codes:
    a. Bayer Code: No specific entry

573

1. Squash, Summer
(Cizelle, Pattypan, Straightneck squash, Vegetable marrow, Zucchini, Crookneck squash, Scallop squash, White-bush squash, Courgette, Marrow, Patisson, Zucchetti, Cocozelle, Calabaza de verano, Butterblossom)
Cucurbitaceae
Cucurbita pepo L.
C. pepo L. var. *pepo*
C. pepo var. *ovifera* (L.) L.H. Bailey

2. Summer squashes are varieties of *C. pepo* that are commonly harvested while the rinds on the fruit are soft and tender. The plants are essentially similar to those of winter squash and pumpkin. All have large leaves. The stems may be long and trailing or shorter and more upright in the "bush" forms. Fruits have smooth skin and vary greatly in form, from flat scallop-shape to long, slender kinds up to 2 feet or more long and 3 inches or more in diameter. This latter type is commonly termed 'Zucchini'. The soft-shelled summer squash types include 'Zucchini', 'Cizelle', 'Scallopini', 'Yellow crookneck', 'Yellow straightneck', 'Cucuzza', 'Sunburst', 'Marrow', 'Pattypan', and Chayote (which see). 'Butterblossom' is a squash cultivar grown especially for producing squash blossoms to be eaten. The plants produce a large amount of male flowers harvested at 8-10 inches long. Flowers from any pumpkin or squash plant can be harvested and used. Squash blossoms are dipped in flour and fried.
3. Season, seeding to first harvest: 40 to 50 days. Usually hand harvested.
4. Production in U.S.: No separate data for summer squash. 47,782 acres reported for all squash, 1959 CENSUS. In 1992, no separate data for summer squash. 69,029 acres reported for all squash, 1992 CENSUS. The 1959 CENSUS data set was retained for comparsion only. In 1992, the top 15 states for all squash are Florida (13,292 acres), California (8,374), Georgia (8,339), Michigan (4,277), New Jersey (3,951), Texas (2,833), New York (2,586), North Carolina (2,578), Oregon (2,286), South Carolina (1,117), Colorado (1,115), Maryland (1,109), Washington (1,106), Minnesota (1,060) and Arkansas (1,004) with a total of 55,027 acres or 80 percent of total U.S. acreage (1992 CENSUS).
5. Other production regions: In 1995, Canada reported 7,507 acres for all squash (ROTHWELL 1996a). Large imports to the U.S. from Mexico and Central America for all squash (LOGAN 1996).
6. Use: As cooked vegetable, or for fresh salads. Flowers are fried.
7. Part(s) of plant consumed: Primarily whole fruit. Flowers are edible
8. Portion analyzed/sampled: Whole fruit
9. Classifications:
   a. Authors Class: Cucurbit vegetables
   b. EPA Crop Group (Group & Subgroup): Cucurbit vegetables (9B) (Representative crop)
   c. Codex Group: 011 (VC 0431) Cucurbits fruiting vegetables
   d. EPA Crop Definition: Summer squash = Gourd fruits consumed when immature (whole fruit edible) including *C. pepo*, *Lageneria* spp., *Momordica* spp., *Sechium edule* (chayote) and other varieties and/or hybrids of these. Squash = pumpkins, summer, and winter squash

10. References: GRIN, CODEX, LOGAN, MAGNESS, RICHE, ROTHWELL 1996a, US EPA 1996a, US EPA 1995a, US EPA 1995c, SWIADER, SPLITTSTOESSER
11. Production Map: EPA Crop Production Regions for squash (summer and winter) are 1, 2, 3, 4, 5, 6, 10 and 11 for 95 percent of the U.S. acreage.
12. Plant Codes:
    a. Bayer Code: CUUPE (*C. pepo*)

574

1. Squash, Winter

(Pumpkin, Calabaza, Japanese pumpkin, Kabocha squash, Cuban pumpkin, Butternut squash, Marrow, Cuban squash, Turban squash, Hubbard squash, Cushaws, Spaghetti squash, Acorn squash, Table queen squash, Carnival squash, Delicata squash, Sweet dumpling squash, Golden nugget squash, Buttercup squash, Orange marrow, Banana squash, Australian blue squash, Sweet meat squash, Mediterranean squash, Silverseed gourd, Butter cup squash, Winter crookneck squash, Fordhook squash, Citrouille, Calabaza de invierno)

Cucurbitaceae

Cucurbita argyrosperma C. Huber

C. argyrosperma C. Huber ssp. *argyrosperma* (syn: *C. mixta* Pangalo)

C. argyrosperma var. *callicarpa* L. Merrick & D.M. Bates

C. maxima Duchesne ex Lam.

C. maxima Duchesne ex Lam. ssp. *maxima* (syn: *C. maxima* var. *turbaniformis* (Roem.) Alef.)

C. moschata (Duchesne ex Lam) Duchesne ex Poir.

C. pepo L.

C. pepo L. var *pepo*

C. pepo var. *ovifera* (L.) L.H. Bailey

2. The winter squashes are varieties of *Cucurbita* species that are harvested when the fruits are fully mature and the rinds are hard. The flesh is usually finer grained and milder flavored than pumpkins, otherwise they are similar. Fruits have smooth hard rinds. They vary in shape from flattened to heart shaped to very oblong, and in size from 4 to 5 inches in length and diameter to 2 feet long and 18 inches in diameter. The hard-shelled mature small winter squash types include 'Acorn', 'White table queen', 'Green table queen', 'Gold table queen', 'Carnival', 'Turban', 'Delicata' ('Sweetpotato'), 'Butternut', 'Sweet dumpling', 'Kabocha' squash, 'Golden nugget' and 'Buttercup'. The hard-shelled mature large winter squash include 'Spaghetti' squash, 'Orange marrow', 'Hubbard', 'Banana' squash, 'Australian blue', 'Sweet meat', 'Mediterranean' squash and 'Cababaza'. Pumpkin, which see. 'Hubbard' squash can be stored for up to 6 months.
3. Season, seeding to harvest: 3 to 4 months.
4. Production in U.S.: 47,782 acres for all squash reported, 1959 CENSUS. 69,029 acres for all squash reported, 1992 CENSUS. The 1959 CENSUS data set retained for comparsion only.
5. Other production regions: See squash, summer, and pumpkin. No separate data for winter squash.
6. Use: Fresh as cooked vegetable, also canned, pickled and frozen.
7. Part(s) of plant consumed: Inner flesh, after rind and seeds removed. Flowers also edible
8. Portion analyzed/sampled: Fruit
9. Classifications:
    a. Authors Class: Cucurbit vegetables
    b. EPA Crop Group (Group & Subgroup): Cucurbit vegetables (9B)
    c. Codex Group: 011 (VC 0433) Winter squash; 011 (VC 0429) Pumpkin, Cucurbits fruiting vegetables
    d. EPA Crop Definition: Squash = Pumpkins, summer, and winter squash; Proposing, Winter squash = Mature gourd fruit
10. References: GRIN, CODEX, LOGAN, MAGNESS, RICHE, US EPA 1996a, US EPA 1995a, SWIADER, SPLITTS-TOESSER
11. Production Map: EPA Crop Production Regions for squash (summer and winter) are 1, 2, 3, 4, 5, 6, 10 and 11 for 95 percent of the U.S. acreage. Pumpkins are 1, 2, 5, 6 and 10 for 86 percent of the U.S. acreage.
12. Plant Codes:
    a. Bayer Code: CUUMA (*C. maxima*), CUUMO (*C. moschata*), CUUPE (*C. pepo*)

575

1. Squirreltail

(Bottlebrush squirreltail, Desert squirreltail, Cola de zorra, Zacate triguillo)

Poaceae (Gramineae)

Elymus elymoides (Raf.) Swezey (syn: *Sitanion hystrix* (Nutt.) J.G. Smith)

2. This is a cool season perennial native to the U.S. and a member of the grass tribe *Triticeae*. It is an erect to spreading bunchgrass from 10 to 60 cm tall that spreads by tillers and has a spike seedhead (2 to 15 cm long). Squirreltail is adapted to dry hills, plains, open woods and rocky slopes of deserts. It has fair forage quality for cattle and horses, poor for sheep. It reproduces from seeds and tillers.
3. Season: Warm season, spring growth, flowers in late spring and may regrow to flower again. Forage can be grazed until late summer to fall.
4. Production in U.S.: There is no production data for Squirreltail, however, pasture and rangeland are produced on more than 410 million A (U.S. CENSUS, 1992).
5. Other production regions: Grows in British Columbia, Canada.
6. Use: Grazing and for control of soil erosion.
7. Part(s) of plant consumed: Leaves, stems and seed
8. Portion analyzed/sampled: Forage and hay
9. Classifications:
    a. Authors Class: See Forage grass
    b. EPA Crop Group (Group & Subgroup): See Forage grass
    c. Codex Group: See Forage grass
    d. EPA Crop Definition: None
10. References: GRIN, US EPA 1994, US EPA 1995, CODEX, BARNES, HOLZWORTH, STUBBENDIECK (picture), RICHE
11. Production Map: EPA Crop Production Regions 7, 8, 9, 10, 11 and 12.
12. Plant Codes:
    a. Bayer Code: SITHY

576

1. St. Augustine grass

(Buffalograss, Ramsammygrass, Penbagrass, Yerba de San Agustin, Chiendent de boeuf)

Poaceae (Gramineae)

Stenotaphrum secundatum (Walter) Kuntze

2. This grass, native to the West Indies, is planted along the southeastern Coastal Plains. It is an extensively creeping perennial with stolons that have rather long internodes. These stolons send up branches at the nodes, which are not over 12 inches high. Leaf blades are 4 to 6 inches long, glabrous and blunt. It will withstand salt water spray so can be used along sea coasts. In the area of adaptation it is used for lawns, golf fairways and pastures. It forms dense sods which stand tramping well and is a widespread lawn grass in the South. As pasture it is grown mainly on muck soils in the Florida Everglades. It is not hardy north of central Georgia and Alabama. It forms little seed and is propagated by planting the rooted runners. Ample fertilizer must be used to secure good growth.

3. Season: Warm season sod grass.

4. Production in U.S.: No specific data, See Forage grass. Southeastern Coastal Plains.

5. Other production regions: West Indies.

6. Use: Pasture, Turfgrass

7. Part(s) of plant consumed: Foliage

8. Portion analyzed/sampled: Forage and hay

9. Classifications:
   a. Authors Class: See Forage grass
   b. EPA Crop Group (Group & Subgroup): See Forage grass
   c. Codex Group: See Forage grass
   d. EPA Crop Definition: None

10. References: GRIN, MAGNESS, ALDERSON, SKERMAN, STUBBENDIECK, BARNES

11. Production Map: EPA Crop Production Regions 2, 3, 4, 6 and 13.

12. Plant Codes:
    a. Bayer Code: STPSE

577

1. Star apple

(Caimito, Cainit, Guayabillo, Star plum, Goldenleaf tree)

Sapotaceae

Chrysophyllum cainito L. (syn: *C. bicolor* Poir.)

2. The tree, native or naturalized in tropical America, is a thick-headed tropical evergreen, up to 50 feet, with oval to oblong, stiff leaves up to 6 inches long, silky pubescent on the underside. Fruit is nearly globose, smooth and hard, up to 3 inches in diameter. In a cross-section, the core is star-shaped – hence the name. The flesh is tender, mild and sweet in flavor. The skin is smooth and waxy. The tree is cultivated to some extent in the American tropics. Trees are propagated by seed, cutting and budding.

3. Season, bloom to harvest: Approximately 180 days and must be hand-picked by clipping the stem.

4. Production in U.S.: 1 acre in Florida (CRANE 1996a). Minor production in Hawaii and south Florida.

5. Other production regions: West Indies, Central America, Mexico and southeastern Asia (MORTON 1987).

6. Use: Fresh eating. The ripe fruit is purple or pale green.

7. Part(s) of plant consumed: Pulp only

8. Portion analyzed/sampled: Whole fruit

9. Classifications:
   a. Authors Class: Tropical and subtropical fruits – inedible peel

   b. EPA Crop Group (Group & Subgroup): Miscellaneous
   c. Codex Group: 006 (FI 0368) Assorted tropical and subtropical fruits – inedible peel
   d. EPA Crop Definition: See Avocado and Papaya

10. References: GRIN, CODEX, CRANE 1996a, CRANE 1995, KAWATE 1996b, MAGNESS, MARTIN 1987, MORTON (picture), USDA NRCS

11. Production map: EPA Crop Production Region 13.

12. Plant Codes:
    a. Bayer Code: CSFCA

578

1. Starfruit

(Carambola, Belimbing manis, Jalea, Caramba, Blimbing, Country gooseberry)

Oxalidaceae

Averrhoa carambola L.

2. The tree attains a height of 35 feet. It likes hot, wet tropical lowlands. It is tolerant of a variety of soils if they are well drained and mildly acidic. Will grow well in warm subtropical areas. Plants are propagated by seed, layering and grafting. Its fruits are star-shaped when cut across, ovoid to ellipsoid, yellow, 4 to 5 inches in length and are acutely 5-angled. The fruit is crisp, juicy and aromatic. There are distinct variety types which are based on fruit taste. The sweet varieties ('Arkin', 'Fwang Tung', 'Thai Knight', 'Maha') are eaten fresh. The tart varieties ('Golden Star', 'Thayer', 'Newcombe', 'Star King') are mainly used for cooking.

3. Season: Bloom to maturity in about 90 days.

4. Production in U.S.: The majority of domestic production is from Florida. Fruit from Hawaii (35 acres in 1994) must be treated prior to shipment to the U.S. mainland. Approximately 1 ton of fruit harvested from Guam in 1992.

5. Other production regions: Taiwan and southeast Asia.

6. Use: Sweet varieties are eaten fresh or used in salads or juices. Tart varieties are used for cooking, jams, jellies and garnish. Unripe fruit has high oxalate content and can be used to remove rust and clean metals.

7. Part(s) of plant consumed: Whole fruit

8. Portion analyzed/sampled: Whole fruit

9. Classifications:
   a. Authors Class: Tropical and subtropical fruits – edible peel
   b. EPA Crop Group (Group & Subgroup): Miscellaneous
   c. Codex Group: 005 (FT 0289) Assorted tropical and subtropical fruits – edible peel
   d. EPA Crop Definition: Proposing – Guava = Starfruit

10. References: GRIN, ANON (a), CODEX, LOGAN, MAGNESS, MARTIN 1987, RICHE, US EPA 1994, ROY, CAMPBELL(i), CRANE 1992

11. Production Map: EPA Crop Production Region 13.

12. Plant Codes:
    a. Bayer Code: AVRCA

579

1. Stevia (steve-e-ya)

(Paraguayan grass, Kahee, Honeygrass, Kaa-he-e, Sugarleaf, Sweet herb of Paraguay, Yerba dulce)

Asteraceae (Compositae)

Stevia rebaudiana (Bertoni) Bertoni

2. Stevia is a small tropical perennial herb that can grow in hot and cold climates to 4 feet tall. In the spring, plug transplants are planted in 30 inch rows. In the fall, the plants are harvested for leaves with the waste stem material spread back on the field. Plants will overwinter in Davis, California, and may be grown as a perennial. Also, stevia can be grown as a seed crop in California for export to Canada. Canada grows the plants from transplants and they are harvested with a modified peanut harvester. The crop is dried with conventional peanut drying technology before the leaves are threshed from the stems. Dried leaves are used in Paraguay as a local sweetner and herbal tea. Also grown in the south of Brazil. Dried leaves can be extracted with methanol and water for the sweetner.

3. Season, transplanting to harvest: About 120 days for leaves. Stevia can live 5 to 10 years and be commercially productive for 4 to 5 years.

4. Production in U.S.: Seed crop use in California (MELNICOE 1997a) and production potential for sugar crop (SHOCK 1982). In 1997 up to 7 acres in California.

5. Other production regions: South America, Japan, Korea and Vietnam (MELNICOE 1997a, SHOCK 1982 and LOVERING 1997). Canada planted about 49 acres in Ontario and British Columbia in 1996 (ROTHWELL 1997).

6. Use: Sugar substitute crop. Stevioside diterpene glycosides, extracted from the leaves, is about 300 times sweeter than sucrose sugar. Also, dried leaves are sold in health stores, as is the powdered form. Non-nutritive sweetening agent.

7. Part(s) of plant consumed: Leaves which contain 3 to 10 percent stevioside by leaf dry weight

8. Portion analyzed/sampled: Commercially, leaves (dried) and extract as a sweetner

9. Classifications:
   a. Authors Class: Herbs and spices; Teas
   b. EPA Crop Group (Group & Subgroup): Miscellaneous
   c. Codex Group: No specific entry. Teas are 066 (DT 0171) Teas (Tea and Herb Tea)
   d. EPA Crop Definition: None

10. References: GRIN, ARKCOLL 1990, CODEX, FACCIOLA, GRIN, LOVERING, MELNICOE 1997a, SHOCK, MERCK, ROTHWELL

11. Production Map: EPA Crop Production Region 10.

12. Plant Codes:
    a. Bayer Code: No specific entry

580

1. Strawberry
(Garden strawberry, Hybrid strawberry, Fresa, Cultivated strawberry)
Rosaceae
Fragaria x *ananassa* Duchesne (*F. chiloensis* (L.) Duchesne x *F. virginiana* Duchesne)

1. Chilean strawberry
(Beach strawberry, Chiloe strawberry)
F. chiloensis (L.) Duchesne

1. Wild strawberry
(Virginia strawberry)
F. virginiana Duchesne

1. European strawberry
(Sowteat strawberry, Woodland strawberry)
F. vesca L.

2. Modern cultivated strawberry varieties apparently have originated from crosses of the above top two species. Plants are perennial herbs, evergreen, with crowns near ground level and leaves reaching upward to as much as a foot during early summer. Flowers do not appear at once. In fact, some varieties are considered "ever bearing" with flowering/fruiting over an extended period of time. Fruit is borne on trusses, with 3 or more fruits per truss, generally somewhat below the leaf canopy. Fruit consists mainly of a fleshy receptacle, with numerous seeds imbedded on the surface. Fruits $1/2$ inch to $1 1/2$ inches in diameter, generally conic, some nearly round. Commercial fields are either planted as matted row or as annual production. California, Florida and other southern areas generally follow annual plantings. Here, crowns are planted in the fall and fruit is harvested in spring. The strawberry fields are abandoned after last harvest. Most other parts of the country use the matted row system. Crowns are planted in the spring. The first crop is not harvested until the following year. The strawberry plant is maintained as a perennial, usually for 3 or more years. After harvest, matted row fields are mowed down and renovated. Important varieties include 'Pjaro', 'Chandler' and 'Driscoll's' (major California grower) proprietary varieties.

3. Season: Bloom to harvest in about 30 days. Often blooms and fruits of various stages of ripeness are on the same plant at once.

4. Production in U.S.: In 1995 there were 48,430 acres harvested. California accounts for almost 50 percent of acres (23,600) and 85 percent of the fruit volume. Other major states include Florida (6,000 acres), Oregon (5,700), New York and North Carolina (both at 2,400), Michigan (1,800) Pennsylvania (1,400) and Washington (1,300). The fruit is widely planted in gardens and "Pick Your Own" operations.

5. Other production regions: U.S. imports significant fruit from Mexico. Also imported from Canada, New Zealand and Guatemala. In 1995, Canada grew 14,214 acres (ROTHWELL 1996a).

6. Use: Fresh, frozen, preserves, culinary.

7. Part(s) of plant consumed: Entire fruit without cap

8. Portion analyzed/sampled: Whole fruit without cap

9. Classifications:
   a. Authors Class: Small fruits
   b. EPA Crop Group (Group & Subgroup): Miscellaneous, use to be part of Small fruits and berry group
   c. Codex Group: 004 (FB 0275) Berries and other small fruit
   d. EPA Crop Definition: None

10. References: GRIN, CODEX, LOGAN, MAGNESS, ROTHWELL 1996a, SCHREIBER, US EPA 1994, US EPA 1995, USDA NRCS, IVES, USDA 1996f, COURTER

11. Production Map: EPA Crop Production Regions 1, 2, 3, 5, 10 and 12.

12. Plant Codes:
    a. Bayer Code: FRAAN (*F. ananassa*), FRACH (*F. chiloensis*), FRAVE (*F. vesca*), FRAVI (*F. virginiana*)

581

1. Strawberrypear
(Pitaya, Pitajaya, Pitahaya, Nightblooming cactus, Nightblooming cereus)
Cactaceae
Hylocereus undatus (Haw.) Britton & Rose (syn: *Cereus undatus* Haw.)

2. This climbing cactus may be terrestial or epiphytic and believed to be native to southern Mexico. Its heavy 3-sided, green, fleshy, much-branched stems with flat wavy wings, may reach 20 feet in length. The night-blooming, bell-shaped, white flowers are up to 14 inches long and 9 inches wide. The non-spiny fruit is oblong-oval to 4 inches long and $2^1/_2$ inches thick, coated with red or yellow ovate bases of scales similar to the outside of a globe artichoke. Related species are *H. ocamponis* Britton and Rose, and *Cereus peruvianus* Mill. (**Apple cactus**).

3. Season, bloom to harvest: It blooms and fruits mainly in August and September.

4. Production in U.S.: No data. Southern Florida (MORTON 1987).

5. Other production regions: Commonly cultivated and natural-ized throughout tropical American lowlands, West Indies, Bahamas, and Bermuda (MORTON 1987).

6. Use: Fruit flesh eaten fresh. Also the juice as a cool drink. A syrup made of the whole fruit is used to color pastries and candy. The unopened flowerbud can be cooked and eaten as a vegetable.

7. Part(s) of plant consumed: Fruit and unopened flowerbud

8. Portion analyzed/sampled: Whole fruit

9. Classifications:
   a. Authors Class: Tropical and subtropical fruits – inedible peel
   b. EPA Crop Group (Group & Subgroup): Miscellaneous
   c. Codex Group: No specific entry
   d. EPA Crop Definition: None

10. References: GRIN, MORTON (picture), USDA NRCS

11. Production Map: EPA Crop Production Region 13.

12. Plant Codes:
    a. Bayer Code: No specific entry

582

1. Sugar apple
(Sweetsop, Anon, Rinon, Anona blanca, Corazon, Custard apple, Ate, Ata, Sitaphal, Sharifa)
Annonaceae
   Annona squamosa L.
2. The tree is related to cherimoya, but is smaller. They can grow up to 20 feet tall. It has narrower leaves, and is a little less hardy than the cherimoya. Tree requires hot tropical lowlands or subtropical climate, with medium rainfall. Prop-agated by seed or grafting. Flowers and fruits are borne in clusters of 2 to 4 on short axillary branches. Flowers open and fruits set over several months, so fruits in various stages of development are on the tree at one time. The fruit resem-bles the soursop in appearance and the cherimoya in compo-sition.The fruit is generally heart-shaped, up to 3 inches across, and is composed of many loosely coherent carpels, with rounded tips. The surface of the fruit is very uneven, a result of depressions between the carpel terminals. The skin is yellow-green in color. Carpel tips may separate in ripe fruit, exposing some of the white pulp below. The fruit is sensitive to chilling injury.
3. Season: Bloom to maturity in 120 to 150 days.
4. Production in U.S.: Increasing production in South Florida, 30 acres in 1994 (CRANE 1995). Guam harvested 1,010 pounds in 1992 (1992 CENSUS). Of some importance in Puerto Rico.

5. Other production regions: India and tropical America. Sugar apples are important fruits in many tropical areas.

6. Use: The fruit is generally consumed fresh or used in ice cream, sherbets or other desserts.

7. Part(s) of plant consumed: Fruit pulp, which is custard like, sweet and contains numerous seeds

8. Portion analyzed/sampled: Whole fruit

9. Classifications:
   a. Authors Class: Tropical and subtropical fruits – inedible peel
   b. EPA Crop Group (Group & Subgroup): Miscellaneous
   c. Codex Group: 006 (FI 0368 and FI 4151) Assorted tropical and subtropical fruits – inedible peel
   d. EPA Crop Definition: Sugar apple = Atemoya, and custard apple. Also proposing Sugar apple = Cherimoya, Ilama, Soursop and Biriba.

10. References: GRIN, CODEX, LOGAN, MAGNESS, MAR-TIN 1987, US EPA 1994, PUROHIT, CRANE 1995, US EPA 1995a, RICHE, MORTON (picture), PHILLIPS(a), SANEWSKI

11. Production Map: EPA Crop Production Region 13

12. Plant Codes:
    a. Bayer Code: ANUSQ

583

1. Sugar beet
(Sugarbeet, Mangel [sugar beet 9d], Remolacha azucarera, Betterave sucriere, Mangel beet)
Chenopodiaceae
   Beta vulgaris L. ssp *vulgaris* (syn: *B. vulgaris* L. var. *sacharifera* and *altissima*).
2. Sugar beets are the principal source of domestic sugar for the temperate areas of the world. Worldwide production of sugar beets was about 19.8 million acres (average for 1992-95) with 1,420,250 acres in the U.S. Sugar beet production in the U.S. for 1992 and 1993 averaged 4,217,00 tons of raw sugar or 3,946,000 tons of refined sugar (calculated on the basis that 1.07 tons of raw sugar is required to produce 1 ton of refined sugar) with about 3 tons of refined sugar per acre of beets grown.

Cultivated beets are believed to have been derived from *B. vul-garis* L., native to Mediterranean coastal areas of Europe. Though used much earlier as vegetable and feed crops, beets as a sugar source have been developed only during the past 190 years. In 1811, following knowledge that some kinds of beets were rich in sugar, Napoleon ordered extensive pro-duction of such beets and the construction of plants to extract the sugar in France. Although this was partially suc-cessful and provided some sugar during the Napoleonic wars, the industry faded out following his defeat at Waterloo and the opening of the country to imports of cane sugar from tropical areas. However, around the middle of that century a substantial industry developed in Germany and France based on higher sugar content beets and improved techniques of sugar extraction. Although sporadic attempts to produce sugar from beets in the U.S. occurred from 1830 onward, the first successful mill was built in 1879 at Alvarado, Califor-nia. A factory on the site operated until 1968.

Today sugar beet growing and sugar manufacture are major industries from the Great Lakes west to the Pacific. Major production centers are irrigated lands of the Rocky Mountain states and westward, but substantial production also occurs in the North Central and Plains states, and limited quantities are produced in New York and Maine.

The beet plant is an overwintering biennial that produces a rosette of leaves and enlarged root one year and a fruiting stalk the following season. Except for seed production, however, it is grown as an annual, seeded in spring and harvested in the fall. The leaves during that year rise from the crown of the enlarged root. Largest leaves may reach 18 inches or more in overall length. Half or more of that length is petiole and the rest leaf blade. The blade is generally elongated oval to arrow shaped but irregular and somewhat roughened. Up to 75 leaves may rise on a vigorous plant. The first-formed leaves die after about a month, but leaves formed in midsummer persist until harvest. Roots are generally broadly conic in shape and average about 4 inches diameter at the top, and near twice as long. Under favorable growing conditions, roots of high sugar content varieties may contain 20 percent sugar by fresh weight at harvest. Beets are planted in rows 24 to 28 inches or more apart. After growth has started plants are thinned in the row to leave them 8 to 12 inches apart.

Maximum yields are obtained with a long growing season so planting is done as early in spring as the soil can be worked. Harvest is usually as late as possible before the ground freezes. For later harvest, California can overwinter the roots in the ground from a spring planted crop. Harvesting in the U.S. is highly mechanized. Machines dig the plants, cut off the top of the root with adhering leaves and deliver the roots into trucks. The leaves with crowns are used for livestock feed either directly or as ensilage. The roots are delivered to the sugar mills where they may be stored in piles before the mill can take them. In very large piles, aeration of the beets may be necessary to prevent heating.

Sugar Extraction Process

Roots are thoroughly washed, then cut into thin strips termed cossettes or chips. Sugar is removed from these by diffusion with hot water through a series of compartments, the fresh hot water first reaching the cossettes from which most of the sugar has been removed and progressively passing on to cossettes containing more sugar. This hot water, moving counter to the sugar content of the cossettes, emerges as "raw juice" with a sugar content of 10 to 15 percent. This is treated with lime to precipitate nonsugars; then with CO_2 gas and filtered, emerging as so-called "thin juice". This is run through a series of five steam-heated vacuum evaporators. The final super-saturated solution is "seeded" with sugar crystals to promote crystallization of the sugar. Crystals are removed by centrifuge. The separated molasses is reboiled and centrifuged to remove additional sugar. Finally the molasses is treated with lime and mixed with raw juice to extract still more sugar. In addition to the beet tops (used directly as animal feed, or after drying or ensiling), the pulp (after sugar extraction) is excellent feed for cattle and sheep. It is commonly dried at the sugar mill but may be fed as wet pulp. The molasses may be added to the pulp for feed or may be sold separately for mixing with other feeds. It is also used in industry for fermentation.

3. Season, seeding to harvest: 80 to 120 days

4. Production in U.S.: In 1995, 1,442,500 acres planted, producing 27,954,000 tons (USDA 1996c). In 1995, the top 10 states as Minnesota (426,000 acres), North Dakota (206,000), Idaho (198,000), Michigan (190,000), California (117,000), Nebraska (75,900), Wyoming (63,000), Montana

(55,700), Colorado (42,800) and Texas (20,200) for 97 percent of the U.S. acreage (USDA 1996c).

5. Other production regions: In 1995, Canada reported 54,857 acres (ROTHWELL 1996a).

6. Use: Root for sugar and feed pulp; Leaves for feed.

7. Part(s) of plant consumed: Roots and tops

8. Portion analyzed/sampled: Roots and tops (leaves) and processed commodities, refined sugar, dried pulp and molasses. Residue data may be supplied for raw or refined sugar or both.

9. Classifications:
   a. Authors Class: Root and tuber vegetables; Leaves of root and tuber vegetables
   b. EPA Crop Group (Group & Subgroup): Root and tuber vegetables (1A); Leaves of root and tuber vegetables (2). (Representative crop)
   c. Codex Group: 016 (VR 0596) Root and tuber vegetables; 052 (AV 0596) Miscellaneous fodder and forage crops for tops; 069 (DM 0596) Miscellaneous derived products of plant origin edible for molasses; 071 (AB 0596 for dry pulp; AB 1201 for wet pulp). By-products used for animal feeding purposes derived from fruit and vegetable processing
   d. EPA Crop Definition: None. We have included the term, Mangel, with sugar beets for its cultural and large root similarities as compare to table beets. Mangel roots weigh 15 to 30 pounds with a growing season of about 100 days and are used mainly as feed. Mangels belong to *B. vulgaris*

10. References: GRIN, CODEX, FACCIOLA, MAGNESS, ROTHWELL 1996a, USDA 1996c, USDA 1996a, US EPA 1996a, US EPA 1995a, US EPA 1994, IVES, REHM

11. Production Map: EPA Crop Production Regions 5, 7, 8, 9, 10 and 11.

12. Plant Codes:
    a. Bayer Code: BEAVA

584

1. Sugar maple
(Hard maple, Rock maple)
Aceraceae
Acer saccharum Marshall ssp. *saccharum*

1. Black sugar maple
(Hard maple, Rock maple, Black maple)
   A. saccharum ssp. *nigrum* (F. Michx.) Desmarais (syn: *A. nigrum* F. Michx.)

2. Maple sugar and syrup are obtained from the sap of these two species and are solely products of the U.S. and Canada. Native Americans were making crude syrups and sugar from maple sap before the coming of the Europeans. The preparation of maple sugar and syrup is strictly a farm industry, occurring from Kentucky northwest to Iowa, northeast to Maine and north into Canada. Native stands of these maple species are tapped to obtain the dilute juice or sap. The trees are not a cultivated crop, although competing useless trees may be removed and maple stands may be thinned to promote better growth and sugar yield.

Only a small proportion of the available trees of these species are actually tapped. The tapping is done by boring a small hole (under $1/2$-inch diameter) horizontally into the tree so as to penetrate through the outer or sap wood. On large trees up to four such taps may be made at one time. Tapered spouts (hollow tubes) are driven into the holes to fit tightly, and the sap flows through this tube and is collected in sap buckets. It is important to protect the buckets and contents from rain water. Tapping is done in late winter, before bud break. During periods when temperatures are above freezing, sap flow is quite abundant. A tap hole usually produces 5 to 15 gallons of sap, though much more than that is sometimes obtained. Sugar content of the sap also varies widely, from 1 degree to 3 degrees Brix or higher.

Portable tanks of various types are used to collect the sap, which is poured into the tank through strainers. An alternative method is to use pipelines to carry the sap to the evaporation equipment. Originally a single open kettle over a fire was used to evaporate the excess water in the sap to produce syrup. Now multiple evaporators are mainly used, the syrup being transferred as it becomes more dense. Usually 2 or 3 transfers are made. Modern evaporating pans have flues in them through which the heat from the fuel passes to speed the process and conserve fuel. For standard-density syrup, concentration is to 65.5 Brix, which is about 85 percent solids by weight. If the sap tests 2.4 Brix, 34 gallons would be required to produce one gallon of syrup. Slow evaporation, or longer heating time, in the final stages of concentration result in a darker colored syrup. More rapid evaporation at this stage gives a lighter colored, higher grade syrup. Sensitive thermometers are used to determine when the syrup is concentrated to the standard of 65.5 Brix.

The completed syrup contains solid granules, mainly calcium malate, termed sugar sand. For table syrup these must be removed. On the farm they may be allowed to settle out or are removed by filtering. Centrifuging is efficient if available. To produce various types of maple sugar, the syrup is further heated and additional water driven off. If heated to a boiling point of 230 degrees F. and cooled rapidly without stirring, a solid cake is formed. Stirring during cooling results in crystal formation. For fine crystals, the highly supersaturated solution is seeded with fine crystals and stirred rapidly, which results in rapid formation of great numbers of fine crystals. Numerous products, as maple cream, or butters, soft-sugar candies, maple spread and candies utilize maple syrup or sugar. From 1961-66, total maple syrup production in the U.S. averaged approximately 1,400,000 gallons, inclusive, as compared to 1995 with 1,096,000 gallons. This includes that made into sugar.

3. Season: Sap flow in late winter.
4. Production in U.S.: In 1995, 1,096,000 gallons of maple syrup (USDA 1996c). In 1995, the top 8 states in maple syrup production reported as Vermont (365,000 gallons), New York (208,000), Maine (162,000), Wisconsin (98,000), Ohio (65,000), New Hampshire (64,000), Michigan (55,000), Pennsylvania (43,000) (USDA 1996c).
5. Other production regions: 3,127,000 gallons of maple syrup were imported into the U.S. in 1995.
6. Use: Maple syrups for candy and sweetener. Also grown as an ornamental tree.
7. Part(s) of plant consumed: Sap
8. Portion analyzed/sampled: Sap and processed commodity, syrup
9. Classifications:
   a. Authors Class: Forestry
   b. EPA Crop Group (Group & Subgroup): Miscellaneous
   c. Codex Group: No specific entry

d. EPA Crop Definition: None
10. References: GRIN, MAGNESS, USDA 1996c
11. Production Map: EPA Crop Production Regions 1 and 5.
12. Plant Codes:
   a. Bayer Code: ACRSC (*Acer saccharum* Marshall spp. *saccharum*)

585

1. Sugarcane
(Turbinado, Cana de azucar, Sugarcane batons, Ko, Bagasse, Canne a sucre, Zuckerrohr)
Poaceae (Gramineae)
Saccharum officinarum L.

2. Sugarcane is a perennial grass and the source of sugar in all tropical and subtropical countries of the world. Estimates for 1992 to 1995 indicate world production of sugarcane averaged about 807 million tons. Production in the U.S., excluding Puerto Rico, averaged 23,859,000 tons during those years, from 876,850 acres of cane in Hawaii, Florida, Texas and Louisiana. Sugarcane production in Puerto Rico was 926,088 tons for 1992.

Several species of *Saccharum* are found in Southeast Asia and neighboring islands, and from these, cultivated cane probably originated. The sweet juice and crystallized sugar were known in China and India some 2500 years ago. Sugarcane reached the Mediterranean countries in the eighth century A.D., and reached the Americas in early colonial times. The cane plant is a coarse growing member of the grass family with juice or sap high in sugar content.

It is tender to cold, the tops being killed by temperatures a little below freezing. In the continental U.S., where freezing may occur during the winter, it is mainly planted in late summer or early fall and harvested a year later. In tropical countries it may be planted at almost any time of the year since the plant does not have a rest period. The season of active growth in the continental U.S. is 7 to 8 months while in tropical countries growth is near continuous until harvest. This results in heavier yields of cane and sugar under tropical conditions. For example, yields of cane and sugar per acre in Hawaii, where the cane is grown for about 2 years before harvesting, are 3 to 4 times higher than yields in Louisiana and Florida from one season's growth.

Sugarcane plants are propagated by planting sections of the stem. The mature stems may vary from 4 to 12 feet or more in height, and in commercial varieties are from $3/_4$ to 2 inches in diameter. The stem has joints or nodes as in other grasses. These range from 4 to 10 inches apart along the above-ground section of the stem. At each node a broad leaf rises which consists of a sheaf or base and the leaf blade. The sheaf is attached to the stem at the node and at that point entirely surrounds the stem with edges overlapping. The sheath from one node encircles the stem up to the next node above and may overlap the base of the leaf on the next higher node. The leaf blade is very long and narrow, varying in width from 1 to 3 inches and up to 5 feet or more in length. Also, at each node along the stem is a bud, protected under the leaf sheath. When stem sections are planted by laying them horizontally and covering with soil a new stem grows from the bud, and roots grow from the base of the new stem. The stem branches below ground so several may rise as a clump from the growth of the bud at a node. In planting cane fields, mature cane stalks are cut into sections and laid horizontally in furrows. In the continental U.S., sections with several nodes are laid while in tropical countries sections with 2 or 3 nodes are commonly used, since temperatures for growth are more favorable. Usually only one node on a stem piece develops a new plant because of polarity along the stem piece.

Planting is in rows about 6 feet apart to make possible cultivation and use of herbicides for early weed control. As plants become tall lower leaves along the stems are shaded and die. These ultimately drop off, so only leaves toward the top remain green and active. Between the nodes, the stems have a hard, thin, outer tissue or rind and a softer center. The high-sugar-containing juice is in this center. Complete shading of the ground occurs about 5 to 8 months after planting and is referred to as close-in or lay-by. More than one crop may be harvested from a planting. After the first crop is removed two or more so-called stubble or ratoon crops may be obtained. These result from growth of new stalks from the bases of stalks cut near the ground level in harvesting. Generally, continental U.S. and Puerto Rico sugarcane production areas utilize up to two ratoon crops which are harvested at 8 to 10 month intervals while a small percentage of the Hawaiian crop is ratooned for another 24 month crop. Hawaii is experimenting with annually harvested cane in 1997 (KAWATE 1997).

Harvesting

Harvesting of cane in Hawaii and Louisiana is highly mechanized. Machines top the canes at a uniform height, cut them off at ground level, and deposit them in rows. In Florida, cane is mainly cut by hand. Leaves and trash are burned from the cane in the rows by use of flame thrower type machines. An alternate method is to burn the leaves from the standing cane, after which it is cut and taken directly to the mill. Delay between cutting and milling in either case should be as short as possible since delay results in loss of sugar content. Machines are under development that will cut, clean and load the cane so it can be taken directly to the mill. In the continental U.S., where winter freezing is a hazard, cane harvest must start earlier than is desirable for maximum yields. When plants are killed by freezing, sugar loss occurs rapidly. While such plants are suitable for sugar extraction if harvested promptly after freezing, this may not be possible when large acreages are involved. In nonmechanized areas cane is still cut and the leaves stripped off by using cane knives. This is arduous and time consuming work.

Sugar Manufacture

Sugar is obtained from the cane at mills located near centers of production. The cane first goes through a washer, then is cut into small pieces by revolving knives. These cut pieces may then be shredded or may move to crushers directly. The crushers consist of two large grooved rollers mounted horizontally, one above the other. The crushed, macerated cane then goes through three or more roller mills which consist of grooved rollers with heavy hydrolic pressure maintained on the upper roller. Water, equal to about 20 percent, is added before the mixture is passed through each set of rollers except the last one. Efficient mills extract at least 90 percent of the sugar in the cane.

The mixture of plant sap and water, collected from the roller mills with the sugar in solution, is slightly acidic in reaction with a pH of 5 to 5.5. It is neutralized with lime, which precipitates some of the colloids and other nonsugars and also stops conversion of sucrose to reducing sugars. The limed juice is then heated

to boiling, which results in further formation of precipitates that settle to the bottom of the tanks. These are drawn off and filtered to remove more juice. The nearly clear juice is continuously drawn off from the top of the tank and goes to the evaporators. The evaporators are a set of three vacuum pans or "bodies" arranged in series, with each successive pan maintained under higher vacuum. The juice enters the first pan at 16 to 18 degrees Brix and leaves the third at 55 to 75 degrees Brix. It then goes to high-vacuum boiling pans, about 25 inches of mercury, there it is further concentrated to 96 degrees Brix and contains sugar crystals. It then is centrifuged to remove most of the liquid or molasses. The remaining raw or brown sugar is then ready for final refining. Much of the imported sugar enters this country as raw sugar and is further refined here before being marketed. The final refining steps include melting the raw sugar, decoloring by passing through carbon filters, recrystallizing in vacuum boiling pans, and drying by centrifuging.

A hundred pounds of raw sugar produces about 94 pounds of refined. A ton of cane yields from less than 170 pounds to more than 225 pounds of raw sugar, depending on such factors as variety, maturity when harvested, promptness of milling and incidence of diseases on the cane in the field. Average per acre cane yields in 1993 to 1995 were 24.2 tons in Louisiana, 33.9 tons in Florida, 85.6 tons in Hawaii and Texas 32.5 tons. The molasses obtained in milling totals around 194 million gallons in the U.S. for 1993. It is used as an additive in livestock feed, in the manufacture of alcohol and alcoholic beverages, as rum, and to some extent in foods. The final molasses in the refining process is blackstrap molasses or "C" strike molasses. "A" strike molasses is termed "Mother Liquor". Blackstrap molasses is a viscous liquid residual fraction from which crystalline sugar cannot be further obtained by the repeated crystallization of cane syrup. Mainland U.S. production of refiners' blackstrap molasses was 125,400 tons in 1994 for cane and beet. Mainland U.S. cane produced 906,898 tons of molasses and beets produced 1,320,000 tons of molasses. In shipments of molasses from Hawaii to mainland U.S. was 166,289 tons in 1994. U.S. imported 206,528 tons of molasses from Mexico in 1994. The fibrous plant residue, called "bagasse" from the roller mills may be used as fuel at the mill, biomass fuel to produce electricity, made into paper or insulating board, or used as plant mulches, bedding for livestock or feed.

Sugarcane Syrup and Liquid Sweetners

From 50 to 60 pounds of juice should be obtained from 100 pounds of cane. Open-type, continuous flow evaporators are generally used to concentrate the juice. The cold juice enters the lower end of the evaporator which is heated by fire beneath or, in larger installations, by steam coils. When the juice is heated, proteins and some other nonsugar constituents coagulate, float on the surface, and are skimmed off at the upper end. In manufactured apparatus a final finishing or evaporating vat may be used. Proper density of the finished product is determined by using a hydrometer (35-36 degrees Baume), or determining the boiling point with a thermometer (226-228 degrees F.). The finished syrup is then filtered and placed in containers while hot.

Production of cane syrup has fluctuated widely, reaching more than 28 million gallons in 1945 when sugar was scarce because of World War II to not being reported beginning 1990 by USDA. Presently, corn syrups have taken the liquid sweetner market from sugarcane. High-fructose syrups including grades 55 and 42, glucose syrup and dextrose are all derived from corn. In 1994, domestic consumption of corn syrups was 2,312,546,000 gallons with the U.S. producing 2,307,980,000 gallons of this. Some liquid cane sugar is produced for local markets, *eg.* Hawaii. Also with the increased use of low or no-caloric sweeteners, significant growth of liquid cane sweeteners appear to be limited with the existing corn syrups competition that started in the early 1980s.

Sugarcane Processing

The cane is crushed with at least two runs to remove the juice. This is followed by a series of strikes (A, B and C) or crystallizations. The result is raw sugar and liquid molasses. The liquid molasses is recrystallized in the "B" and "C" strikes for more raw sugar. The final strike or "C" strike results in the final or blackstrap molasses and the final crystallized raw sugar in the syrup. Refinery or raw sugar is further processed to refined white sugar by filtration, char-decolorization, crystallization and centrifuging. White sugar is purified and crystallized sucrose (saccharose). Soft sugar means fine-grain purified moist sugar. Turbinado is raw sugar.

3. Season, planting to first harvest: One-year crop of 10 to 15 months in continental U.S. depending on location to a two-year crop in tropical areas. Ratoon crops are produced chiefly in continental U.S. and Puerto Rico.

4. Production in U.S.: In 1995, 882,300 acres harvested with 29,386,000 tons (USDA 1996c). Sugar production from sugarcane for 1993 reported as 3,482,000 tons for raw and 3,255,000 tons for refined sugar. Raw value is the equivalent of 96 degrees sugar, as defined in the Sugar Act of 1948. The refined value is calculated on the basis that 100 pounds of raw sugar is required to produce 93.46 pounds of refined sugar (USDA 1996a). In 1995, the top four states for sugar production are Florida (427,000 acres), Louisiana (368,000), Hawaii (46,000), and Texas (41,300) for 100 percent of the U.S. acreage. In 1992, Puerto Rico had 3,693 acres of Fall cane with 135,455 tons; 6,375 acres of Spring cane with 191,466 tons; 27,383 acres of Ratoon cane with 599,167 tons, and Sojourn cane with 963 acres for seed or feed use. Totals for Puerto Rico in 1992 are 38,414 acres with 926,088 tons (1992 CENSUS). In 1990, Hawaii's total caneland acreage was 161,991 acres of which about half (71,998 acres) were harvested with 819,631 tons of raw sugar which averaged 11.38 tons per acre or 5.7 tons on an annual basis since it is a 24 month crop (HAWAIIAN SUGAR PLANTERS' ASSOCIATION 1991).

5. Other production regions: In North America, Mexico is the largest producer with about 1.3 million acres with 44,000,000 tons in 1994 to 1995 (USDA 1996a).

6. Use: Sugarcane batons are produced in Hawaii and California. Peel the outer bark and eat out of hand or cut into fine strips (LOGAN 1996). Mainly used as sugar source.

7. Part(s) of plant consumed: Cane (stalk). Bagasse is an insignificant feed item

8. Portion analyzed/sampled: Cane and processed commodities molasses and refined sugar. Residue data for refined sugar may be supplied for raw or refined or both. Residue data are also needed for blackstrap molasses. For sugarcane bagasse, information indicates that sugarcane bagasse is mainly used for fuel. Residue data will not be needed at this time, but may be needed at a later date

9. Classifications:
   a. Authors Class: Grasses for sugar or syrup
   b. EPA Crop Group (Group & Subgroup): Miscellaneous
   c. Codex Group: 021 (GS 0659) Grasses for sugar or syrup production; 052 (AM 0659) Miscellaneous fodder and forage crops for fodder; 052 (AV 0659) Miscellaneous fodder and forage crops for forage
   d. EPA Crop Definition: None

10. References: GRIN, CODEX, HAWAIIAN SUGAR PLANTERS' ASSOCIATION 1991, KUNKEL, LOGAN, MAGNESS, NEAL, RICHE, USDA 1996c, USDA 1996a, US EPA 1996a, US EPA 1994, KAWATE 1997

11. Production Map: EPA Crop Production Regions 3, 4, 6 and 13.

12. Plant Codes:
    a. Bayer Code: SACOF

586

1. Sumac, Smooth
(Scarlet sumac, Vinegartree)
Anacardiaceae
Rhus glabra L.

2. The perennial shrub grows to about 60 inches tall the first year which is cut back for the second year growth and harvests. The shrub is erect, dioecious and glabrous. The leaves are compound with 11-31 leaflets to 5 inches long. The tannin (polyphenol) is found in the leaves. Japan wax (Japan tallow, Sumac wax) is prepared from hot pressing of immature fruits of the Oriental sumac (*R. succedanea, R. vernicifera,* and *R. trichocarpa*). Japan wax is used in food-contact cotton bags.

3. Season, harvest: Leaves harvested twice a season. First harvest is toward the end of May and second harvest in July in the southern U.S. Harvest starts the second year and plants can produce for 8 years.

4. Production in U.S.: No data. Southern U.S.

5. Other production regions: Eastern temperate North America (BAILEY 1976).

6. Use: Primarily, leaf tannin used to flavor beer, similar to wood chips. The tannins are extracted primarily from the leaves which are sun dried to 10 percent moisture before extracted. Also grown as an ornamental.

7. Part(s) of plant consumed: Leaves. Also a refreshing pink lemonade-like beverage is prepared from the fruit and the peeled young shoots and peeled roots can be eaten raw

8. Portion analyzed/sampled: Leaves (dried) and extract

9. Classifications:
   a. Authors Class: Herbs and spices
   b. EPA Crop Group (Group & Subgroup): Miscellaneous
   c. Codex Group: No specific entry
   d. EPA Crop Definition: None

10. References: GRIN, BAILEY 1976, BAYER, FACCIOLA

11. Production Map: EPA Crop Production Regions 2, 4 and 6.

12. Plant Codes:
    a. Bayer Code: RHUGL

587

1. Sunflower
(Sunflower seed oil, Sunflower seed, Hopi sunflower, Girasol, Polocote, Girasole)
Asteraceae (Compositae)
Helianthus annuus L.

2. Sunflower is native to the western U.S., but principal commercial production of the seed for oil is in other countries, especially Russia. World production of oil averaged near

8,026,150 tons from 1992 to 1994, and sunflower oil production in the U.S. has increased since 1966. Grown from southern Canada south to Mexico and Argentine. The plant is a large, rough annual, with a stiff stem up to 10 or 12 feet tall. Leaves are cordate to ovate, rough, up to 12 inches long. The angular seeds are up to $\frac{1}{2}$ inch long, and are densely packed in the flat, terminal heads, which may be more than a foot across. Seeds are exposed in the head during growth. Seeds normally contain about 25 percent of the semi-drying oil, but this has been increased by breeding to above 40 percent. Oil is usually expressed by an initial cold press, followed by hot pressing. The cold-press oil is used as a salad and cooking oil and for margarine, the hot-press mainly in industry. The meal is a very valuable animal feed. Hulls can be used as a roughage. Seeds are also consumed as nuts with the non-oil varieties for human food and feeding birds. Industrial uses of sunflower oil are for paints, plastics, soaps, detergents and as a pesticide carrier, surfactant and lubricant. Use as a snack food has grown over the last 17 years, with the non-oil varieties with larger seeds used. Sunflowers can also be used as silage crop and can be double cropped in areas with small grains or vegetables and where the season is too short to produce mature corn for silage.

3. Season, planting to harvest: About 120 days.
4. Production in U.S.: 2,757,000 acres planted in 1993 with the non-oil varieties included at 460,000 acres. (USDA 1994a). 94 percent of the oil varieties are produced in North Dakota, South Dakota, Minnesota and Kansas. 76 percent of the non-oil varieties are produced in North Dakota, Minnesota and Kansas in 1993.
5. Other production regions: Canada reported 190,024 acres in 1995. (ROTHWELL 1996a).
6. Use: Vegetable oil, snack food, birdseed, livestock feed, forage, silage and industrial use of the oil. Also grown as an ornamental.
7. Part(s) of plant consumed: Seed primarily
8. Portion analyzed/sampled: Seed and its processed commodities meal and refined oil
9. Classifications:
   a. Authors Class: Oilseed
   b. EPA Crop Group (Group & Subgroup): Miscellaneous
   c. Codex Group: 023 (SO 0702) Oilseed; 067 (OC 0702) Vegetable crude oil; 068 (OR 0702) Vegetable edible oil
   d. EPA Crop Definition: None
10. References: GRIN, CODEX, FOLLETT, MAGNESS, ROBBELEN 1989 (picture), ROTHWELL 1996a, USDA 1994a, US EPA 1995a, US EPA 1995b, US EPA 1994, IVES, REHM, PUTNAM 1990, ADAMS
11. Production Map: EPA Crop Production Regions 5, 7 and 8.
12. Plant Codes:
   a. Bayer Code: HELAN

588

1. Sunolgrass
Poaceae (Gramineae)
Phalaris coerulescens Desf.
1. Koleagrass
(Bulbous canarygrass)
P. elongata Braun-Blanq. (syn: *P. tuberosa* var. *hirtiglumis* Trab.)

2. Sunolgrass, introduced from Australia in 1935, and Koleagrass, introduced from Morocco in 1955, are not being grown commercially at present in the U.S. Both are bunchgrasses that have round bulb-like enlargements at the base of the stems. Both are rapid growers. Sunolgrass is a poor seed producer. Koleagrass, which produces ample seed, is of more promise. Adaptation of both is limited to areas having mild winters.
3. Season: Warm season grass.
4. Production in U.S.: No specific data, See Forage grass. Southeast U.S.
5. Other production regions: See above.
6. Use: Pasture.
7. Part(s) of plant consumed: Foliage
8. Portion analyzed/sampled: Forage and hay
9. Classifications:
   a. Authors Class: See Forage grass
   b. EPA Crop Group (Group & Subgroup): See Forage grass
   c. Codex Group: See Forage grass
   d. EPA Crop Definition: None
10. References: GRIN, MAGNESS
11. Production Map: EPA Crop Production Region 2.
12. Plant Codes:
   a. Bayer Code: PHACO (*P. coerulescens*)

589

1. Surinam cherry
(Pitanga, Cerezade Cayena, Grumichama, Brazil cherry, Cayenne cherry)
Myrtaceae
Eugenia uniflora L.

2. The tree is small, up to 20 feet, sufficiently hardy to grow in central Florida and southern California. It is often grown as a bush or hedge. Leaves are entire, oblong, up to 3 inches, pointed at the terminal and glabrous. The fruits are 1 inch or more in diameter, near globose to oblate in shape, glabrous, but generally ridged. Each fruit contains 1 or 2 seeds. Flavor is quite tart, but not quite as tart as Tart cherry. Fruits are seen on the markets in tropical countries, where two crops per year may be produced. In areas of the continental U.S., where grown, only one crop is produced, and production is not commercial. Trees can be propagated from seed and cutting.
3. Season, bloom to harvest: About 3 weeks. Normally, fruits mature in 30-50 days.
4. Production in U.S.: No data. Hawaii and Florida (MORTON 1987).
5. Other production regions: West Indies and South America (MORTON 1987).
6. Use: Fresh, jams, jelly, relish, pickles or sherbet. Also grown as an ornamental.
7. Part(s) of plant consumed: Fruit
8. Portion analyzed/sampled: Whole fruit
9. Classifications:
   a. Authors Class: Tropical and subtropical fruits – edible peel
   b. EPA Crop Group (Group & Subgroup): Miscellaneous
   c. Codex Group: 005 (FT 0311) Assorted tropical and subtropical fruit – edible peel
   d. EPA Crop Definition: None
10. References: GRIN, CODEX, MAGNESS, MARTIN 1987, MORTON (picture)

11. Production Map: EPA Crop Production Region 13.
12. Plant Codes:
    a. Bayer Code: No specific entry

590

1. Swamp leaf
(Rau Ngo, Rau Om, Ngo Om, Beremi, Kerak kerak, Rumput jari, Sebueh)
Scrophulariaceae
Limnophila chinensis spp. *aromatica* (Lam.) T.Yamaz. (syn: *L. aromatica* (Lam.) Merr.)
   (Rau Ngo')
L. chinensis (Osbeck) Merr. ssp. *chinensis*
   (Rau Om, Ngo' Om, Beremi, Kerak kerak, Rumput jari, Sebueh)
2. Aquatic plant that prefers mud and a film of standing water. Usually sold in bunches in Oriental groceries and is easily recognized by the whorls of three leaves.
3. Season: No data.
4. Production in U.S.: No data. Grown in the southern U.S. near Vietnamese communities (KUEBEL 1988).
5. Other production regions: Southeast Asia (MARTIN 1979).
6. Use: Herb used in sweet and sour dishes and soups.
7. Part(s) of plant consumed: Leaves
8. Portion analyzed/sampled: Leaves (fresh)
9. Classifications:
    a. Authors Class: Herbs and spices
    b. EPA Crop Group (Group & Subgroup): Miscellaneous
    c. Codex Group: No specific entry
    d. EPA Crop Definition: None
10. References: GRIN, KUEBEL, MARTIN 1979, USDA NRCS
11. Production Map: No entry
12. Plant Codes:
    a. Bayer Code: No specific entry

591

1. Sweet bay
(Laurel, Bay leaf, Grecian laurel, Bay laurel, Bay)
Lauraceae
Laurus nobilis L.
2. The sweet bay or laurel is a small evergreen tree, native to the Mediterranean region, with thick, simple, dull-green, lanceolate leaves. It is widely grown as an ornamental, both in the open and in tubs. It is the laurel of history and poetry, long a symbol of victory. The leaves have a pleasant odor and a bitter, aromatic taste. They are used both fresh and dried in cookery to impart flavor. No data are available as to U.S. production, but it is limited. The bay-rum or bayberry of commerce is from the leaves of *Pimenta racemosa* Mill.
3. Season, harvest: In late summer or autumn.
4. Production in U.S.: No data. Grown to some extent in southern California. (FOSTER 1993).
5. Other production regions: Mediterranean region.
6. Use: Flavoring in cookery.
7. Part(s) of plant consumed: Leaves, mainly dried for commerce. Takes about 2 weeks to dry
8. Portion analyzed/sampled: Leaves (Fresh and dried)
9. Classifications:

a. Authors: Class: Herbs and spices
b. EPA Crop Group (Group & Subgroup): Herbs and spices group (19A)
c. Codex Group: 027 (HH 0723) Herbs; 057 (DH 0723) Dried herbs
d. EPA Crop Definition: None
10. References: GRIN, CODEX, FARRELL, FOSTER (picture), LEWIS, MAGNESS, US EPA 1995a
11. Production Map: EPA Crop Production Region 10.
12. Plant Codes:
    a. Bayer Code: LURNO

592

1. Sweet cicely
(Garden myrrh, Myrrh, Anise, Great chervil, Sweet chervil)
Apiaceae (Umbelliferae)
Myrrhis odorata (L.) Scop.
2. The myrrh plant is a pubescent perennial, up to 3 feet in height, with thin, soft, pinnate leaves. Leaflets are lanceolate. Leaves, tender stems and roots are sweet scented. The plants persist for years. Formerly myrrh was used for flavoring salads. Now apparently little grown, except as an ornamental. Taste is similar to celery and sweet anise. The root is eaten after steaming and the seeds are used as a flavoring.
3. Season: The fernlike leaves can be harvested as they develop. Propagated by seedling transplants or plant divisions.
4. Production in U.S.: No data.
5. Other production regions: Native to the mountain areas of Europe (FOSTER 1993).
6. Use: Herb; the young shoots can be eaten raw or cooked. See above.
7. Part(s) of plant consumed: Leaves primarily; Dried seed as spice; Fresh root in salad or as vegetable
8. Portion analyzed/sampled: Leaves (fresh and dried)
9. Classifications:
    a. Authors Class: Herbs and spices
    b. EPA Crop Group (Group & Subgroup): Miscellaneous
    c. Codex Group: 027 (HH 0747) Herbs; 057 (DH 0747) Dried herbs
    d. EPA Crop Definition: None
10. References: GRIN, CODEX, GARLAND (picture), FOSTER (picture), MAGNESS
11. Production Map: No entry
12. Plant Codes:
    a. Bayer Code: No specific entry

593

1. Sweet clover
(Melilot, Clover-sweet)
Fabaceae (Leguminosae)
Melilotus spp.
1. White sweet clover
(White melilot, Hubam, White flowered hubam, Bukhara clover)
M. alba Medik.
1. Yellow sweet clover
(Yellow-flowered sweet clover, Yellow melilot, Melist)
M. officinalis Lam.
1. Tall yellow sweet clover
M. altissima Thuill.

1. Sour clover
(Senji, Indian sweet clover)

M. indicus (L.) All.

2. The sweet clovers are native to temperate Europe and Asia. Although reported as found in Virginia as early as 1739, it was not until the present century that their great value for soil improvement, pasture, hay and silage became recognized. During the decade 1948-57, seed production in the U.S. averaged about 46,000,000 pounds annually with near 16,000,000 additional pounds imported, sufficient to seed more than 5,000,000 acres. Since 1960, however, seed production has declined in this country from an annual average of about 21,600,000 pounds to about 1 million pounds in 1987 and 1992. Most of the sweet clovers grown here are biennial, although some annual kinds are grown. They are adapted only to soils that are near neutral in acidity. Major production is in a belt from the Great Lakes west to Montana and south to the Gulf of Mexico in areas having 17 inches or more annual precipitation. All the biennial sweet clovers fix large amounts of atmospheric nitrogen. The deep-penetrating tap roots decompose after the second year, so the crops are very useful for opening up subsoil. They are therefore probably the best of the crops for soil improvement. The sweet clovers have a high content of coumarin which reduces palatability. More important, in hay spoiled due to excess moisture when stored or in improperly prepared silage, dicoumarol, which reduces the clotting of blood, is formed. Animals fed such hay or silage are subject to excessive external or internal bleeding. The species of *Melilotus* of most value agriculturally are described as follows.

In general, the **white-flowered forms of sweet clover** are somewhat ranker growing, heavier yielding, and have coarser stems than the yellow flowered. They are later maturing so are generally preferred for pastures in areas of ample moisture. The more vigorous growth and heavier yields make them somewhat superior for soil improvement. In growth habit and appearance the two are similar except for flower color. Most of the *M. alba* grown is so-called common white. Two selected varieties in addition to common are in the trade. 'Spanish' is leafier and somewhat more productive than common and is recommended for higher rainfall grain areas of the Pacific Northwest, as well as the Great Plains. 'Evergreen' is late maturing, providing long grazing and heavy forage yields. It is generally adapted in the Corn Belt. 'Penta' is a low coumarin variety, bred in Wisconsin. Annual forms of *M. alba* are also in the trade. 'Hubam', 'Floranna' and 'Israel' are such kinds. They are most useful in the central and southern regions where the growing season is relatively long.

The **yellow-flowered sweet clovers** grown in the U.S. are all biennial. As compared to white-flowered *M. alba*, the yellow-flowered is finer stemmed, matures earlier in summer, is more tolerant to drought and competition with companion crops, and gives a better quality but lower yield of hay. Because of better drought tolerance the yellow-flowered is better adapted to the Great Plains. The first season a central much-branched stem is produced with a deep taproot which becomes fleshy in the fall. The second year, crown buds start growth early with vigorous, rather coarse stems. Leaves are trifoliate, the leaflets being long-oval in shape. For hay, the second season crop should be cut early. Several varieties are in the trade. 'Madrid' makes strong seedling growth, is leafy

and later maturing the second year, and well adapted to the Great Plains. 'Gold Top' is vigorous and late maturing, giving longer pasture. 'Erector', developed in Canada, makes good growth in the northeastern Great Plains.

Sour clover is a very important cover crop in California.

3. Season: Production May to August. Seeded spring or autumn at 10 to 15 lbs/acre.
4. Production in U.S.: In 1992, sweet clover seed produced on 380 acres with 48 tons. In 1987, 8,601 acres with 912 tons (1992 CENSUS). In 1992 and 1987, seed production from Ohio and Minnesota (1992 CENSUS). Grown as forage legume in Washington (SCHREIBER 1995).
5. Other production regions: In 1995, Canada reported 5,812 acres (ROTHWELL 1996a).
6. Use: Pasture, hay, silage, cover crop and soil improvement.
7. Part(s) of plant consumed: Foliage
8. Portion analyzed/sampled: Forage and hay
9. Classifications:
   a. Authors Class: Nongrass animal feeds
   b. EPA Crop Group (Group & Subgroup): Nongrass animal feed (18)
   c. Codex Group: 050 (AL 1023) Legume animal feed
   d. EPA Crop Definition: None
10. References: GRIN, CODEX, MAGNESS, RICHE, ROTHWELL 1996a, SCHREIBER, US EPA 1996a, US EPA 1995a, BARNES, DUKE (h), BALL
11. Production Map: EPA Crop Production Region 5 for seed production. EPA Crop Production Regions 3, 5, 6 and 7 for forage crop.
12. Plant Codes:
    a. Bayer Code: MEUAL (*M. albus*), MEUAT (*M. altissimus*), MEUOF (*M. officinalis*)

594

1. Sweet potato
(Batata, Yam, Cuban sweetpotato, Boniato, Camote, Sweet potato vine)
1. Sweetpotato

Convolvulaceae

Ipomoea batatas (L.) Lam. var. *batatas*

2. The sweetpotato is a major food crop in all tropical and subtropical countries, and in warmer parts of the U.S. It is grown for the enlarged fleshy tuberous roots. The plant is naturally a perennial, but under cultivation is grown as an annual. The plant is a trailing vine. Moist fleshed varieties in the U.S. are often erroneously called yams. See yam. In the U.S., sweet potato roots of the previous crop are laid in beds, which in cooler climates are heated, then covered. Sprouts (slips) growing from these are pulled free and field planted. Stem cuttings from field plantings may also be planted in the field in areas of long season. Leaves vary in shape, but are commonly heart shaped, and form a canopy over the soil by the time the roots are becoming tuberous. Edible roots develop entirely underground. Boniato or Cuban sweetpotato has a distinctive white interior. Sweetpotatoes are called Batatas in Puerto Rico. Food types of sweetpotatoes are either soft-fleshed or firm-fleshed cultivars. The soft flesh sweetpotatoes or yams are sweeter, softer and used for baking. The firm or dry types are used for frying, candied or boiling. Depending upon cultivars, they can be stored from

4-12 months. Sweetpotatoes are native to Central and South America.

3. Season, field setting to harvest: About 4 to 5 months. Boniatos grown in Dade Country take 4 to 6 months.
4. Production in U.S.: 644,150 tons from 88,600 acres in 1995 (USDA 1996c). Guam reported 21 acres in 1992 and Puerto Rico reported 1,092 acres in 1992 (1992 CENSUS). Dade County in Florida produced 5,000 acres of Boniato in 1983 (STEPHENS 1988). In 1995, the top 6 states are: North Carolina (35,000 acres), Louisiana (22,000), California (8,300), Mississippi (6,200), Texas (5,600), and Alabama (4,400) which comprise 92 percent of the total U.S. acreage (USDA 1996a).
5. Other production regions: China is the largest producer (RUBATZKY 1997).
6. Use: Culinary, as cooked vegetable, also frozen and canned, and used for chips, flour and noodles. U.S. per capita consumption in 1994 is 4.7 pounds (PUTNAM 1996). Also grown as an ornamental.
7. Part(s) of plant consumed: Tuberous roots, generally with peel removed. Non-marketable roots are used as livestock feed
8. Portion analyzed/sampled: Root
9. Classifications:
   a. Authors Class: Root and tuber vegetables
   b. EPA Crop Group (Group & Subgroup): Root and tuber vegetables (1C and 1D); Leaves of root and tuber vegetables (2). Representative crop for 1D
   c. Codex Group: 016 (VR 0508) Root and tuber vegetables; 013 (VL 0508) Leafy vegetables (including *Brassica* leafy vegetables)
   d. EPA Crop Definition: Sweetpotatoes = yams
10. References: GRIN, CODEX, MAGNESS, RICHE, STEPHENS 1988 (picture), USDA 1996a, US EPA 1996a, US EPA 1995a, US EPA 1994, YAMAGUCHI 1983 (picture), IVES, REHM, RUBATZKY, SWIADER, EDMOND, PUTNAM 1996
11. Production Map: EPA Crop Production Region 2, 3, 4, 6 and 10 which comprise 99 percent of the U.S. acreage. Boniato is represented by Region 13.
12. Plant Codes:
   a. Bayer Code: IPOBA

595

1. Swiss chard
(Chard, Palang sag, Palak, Sea kale beet, Nikon, Leaf beet, Indian spinach, Acelga cardo, Sicilian broadrib beet, Mangold, Silver beet, White beet)
Chenopodiaceae
   Beta vulgaris var. *flavescens* (Lam.) Lam. & DC.

1. Spinach beet
(Foliage beet, Kwoon taat tsoi, Paak tim tsoi, Tojisa)
   B. vulgaris var. *cicla* L. (syn: *B. cicla* (L.) L.)
2. Swiss chard is a foliage beet, developed for its large fleshy leaf petioles and broad, crisp leaf blades. Plants may be started in beds and transplanted to the field; but more commonly direct field seeding, followed by plant thinning, is practiced. Plantings made in the spring will produce leaves for greens in about 60 days and will continue to produce "greens" or leaves until frost if the growing point is not injured. The edible leaf stems and blades are exposed as in spinach and turnip greens. Leaf blades which can measure 6 inches across are prepared like

spinach and midribs or petioles are cooked like asparagus. There are 1,500 seeds per ounce.
3. Season, planting to first harvest: 50 to 60 days.
4. Production in U.S.: In 1993, Washington state grew 150 acres of Swiss chard seed (SCHREIBER 1995). Grows in Florida both as a winter vegetable and a summer cooking green (STEPHENS 1988). In 1994, California harvested 85 acres of chard (MELNICOE 1996e).
5. Other production regions: Normally locally grown and marketed. In 1995, 5 acres were reported in Canada (ROTHWELL 1996a).
6. Use: Mostly marketed fresh; cooked before consumed.
7. Part(s) of plant consumed: Leaf blades and petioles
8. Portion analyzed/sampled: Petioles (includes leaves)
9. Classifications:
   a. Authors Class: Leafy vegetables (except *Brassica* vegetables)
   b. EPA Crop Group (Group & Subgroup): Leafy vegetables (except *Brassica* vegetables) (4B)
   c. Codex Group: 013 (VL 0464) Leafy vegetables (including *Brassica* leafy vegetables) as Chard
   d. EPA Crop Definition: None
10. References: GRIN, CODEX, FACCIOLA, MAGNESS, SCHREIBER, STEPHENS 1988 (picture), US EPA 1995a, US EPA 1995b, YAMAGUCHI 1983, MELNICOE 1996e, SWIADER, SPITTSTOESSER
11. Production Map: EPA Crop Production Region 10.
12. Plant Codes:
   a. Bayer Code: BEAVF (*B. vulgaris* var. *flavescens*)

596

1. Swordbean
(Haba de burro, Pois sabre, Sword jackbean)
Fabaceae (Leguminosae)
   Canavalia gladiata (Jacq.) DC.
2. Similar to jackbean, which see. The swordbean plant is a viney perennial. Pods range from 6 to 16 inches long and contain 5 to 10 red seeds. The ratio of pod length to width is used to identify Jackbean from Swordbean which is smaller. Like Jackbean, mature seeds must be boiled to remove toxins before eating.
3. Season, seeding to first harvest: 3-4 months. Mature seed in 5-10 months.
4. Production in U.S.: Grown in Florida gardens (STEPHENS 1988).
5. Other production regions: Tropical and subtropical plant (YAMAGUCHI 1983). India and humid Africa as secondary grain legume (JANICK 1990).
6. Use: Used as snap beans (edible-podded) usually boiled. Fodder for feed.
7. Part(s) of plant consumed: Primarily young pods about 6 inches long and immature seeds. Sometimes flowers and young leaves are used as potherbs. Also mature seeds but to a much lesser extent
8. Portion analyzed/sampled: Young pods and immature seeds (edible-podded)
9. Classifications:
   a. Authors Class: Legume vegetables (succulent or dried)
   b. EPA Crop Group (Group & Subgroup): Legume vegetables (succulent or dried) (6A)
   c. Codex Group: 014 (VP 0542) Legume vegetables

d. EPA Crop Definition: None
10. References: GRIN, CODEX, JANICK 1990, MARTIN 1983, NEAL, STEPHENS 1988 (picture), US EPA 1995a, YAMAGUCHI 1983, USDA NRCS, RUBATZKY
11. Production Map: EPA Crop Production Regions 3 and 13.
12. Plant Codes:
a. Bayer Code: CNAGL

597

1. Tall dropseed
(Rough dropseed, Zacaton)
Poaceae (Gramineae)
Sporobolus compositus (Poir.) Merr. (syn: *Sporobolus asper* (Michx.) Kunth)
2. This is a perennial warm season grass native to the U.S. and is a member of the grass tribe *Eragrosteae*. It is an erect bunchgrass from 0.6 to 1.2 m tall with leaf blades 10 to 70 cm long, and its inflorescence is a panicle (5 to 30 cm long). It is adapted to prairies, foothills and on soils that are intermittently wet and dry. Tall dropseed produces fair forage for livestock. It reproduces by seeds, tillers or rhizomes. It is seeded at a rate of 4 to 7 kg/ha. There are 1,109,115 seeds/kg.
3. Season: Warm season grass that starts growing in the late spring and flowers in August. Its palatability declines rapidly as it matures.
4. Production in U.S.: There are no production data for Tall dropseed, however, pasture and rangeland are produced on more than 410 million A (U.S. CENSUS, 1992). Vermont to eastern Washington, south to Mississippi, Louisiana, Texas and Arizona (GOULD 1975).
5. Other production regions: No entry.
6. Use: Grazing and to prevent soil erosion.
7. Part(s) of plant consumed: Leaves and stems
8. Portion analyzed/sampled: Forage and hay
9. Classifications:
a. Authors Class: See Forage grass
b. EPA Crop Group (Group & Subgroup): See Forage grass
c. Codex Group: See Forage grass
d. EPA Crop Definition: None.
10. References: GRIN, US EPA 1994, US EPA 1995, CODEX, HOLZWORTH, STUBBENDIECK (picture), RICHE, GOULD
11. Production Map: EPA Crop Production Regions 1, 4, 5, 6, 7, 8 and 9.
12. Plant Codes:
a. Bayer Code: No specific entry, genus entry is SPZSS

598

1. Tall oatgrass
(Falseoat, French-rye, Oatgrass-tall)
Poaceae (Gramineae)
Arrhenatherum elatius (L.) P. Beauv. ex J. Presl & C. Presl ssp. *elatius*
2. Tall oatgrass, native to Europe, was brought to the U.S. early in the last century, and it is a member of the grass tribe *Aveneae*. It is grown widely in the central and northern states, and in the Pacific Northwest. It is a hardy, upright cool season, short-lived perennial bunchgrass reaching to 6 feet, with many leaves. The seed head resembles that of

oats, hence the name. It tends to grow in bunches and is well adapted to well-drained light-textured soils with a pH of 5.5 to 7.0. It is suitable for pastures and yields a good palatable hay with yields of 1 to 2 tons/A. It is frequently seeded in combination with other grasses and legumes as sweet and red clover. It is shorter lived than most other bunchgrasses. It is seeded at a rate of 40 to 50 lb/A. Tall oatgrass contains 150,000 seeds/lb. It is a poor seed producer, but can produce 100 to 300 lb/A.
3. Season: Cool season forage. Grows rapidly from mid-spring through the early fall. Harvest for hay before heading. Seed either in spring or fall.
4. Production in U.S.: No data for Tall oatgrass production, however, pasture and rangeland are produced on more than 410 million A (U.S. CENSUS, 1992). See above.
5. Other production regions: See above.
6. Use: Pasture, hay, and for erosion control.
7. Part(s) of plant consumed: Leaves and stems
8. Portion analyzed/sampled: Forage and hay
9. Classifications:
a. Authors Class: See Forage grass
b. EPA Crop Group (Group & Subgroup): See Forage grass
c. Codex Group: See Forage grass
d. EPA Crop Definition: None
10. References: GRIN, MAGNESS, US EPA 1994, US EPA 1995, CODEX, BARNES, ALDERSON, MOSLER (picture), RICHE
11. Production Map: EPA Production Regions 1, 5, 7, 9, 11 and 12.
12. Plant Codes:
a. Bayer Code: ARREL

599

1. Tamarind
(Indian date, Tamarindo, Indian tamarind)
Fabaceae (Leguminosae)
Tamarindus indica L.
2. Grows in hot tropical lowlands. Does not fruit well in climates of high rainfall all year. Tree is propagated by seed, layering and grafting. First fruit production within 3-4 years from vegetative propagation. The tree is a slow growing legume. Large pinnate leaves are graceful. Tamarind trees can become quite large (75 ft.). The fruit is a pod, 3 to 8 inches long, an inch wide, containing up to 6 or more seeds. The pulp is very acid. It is a good source of calcium, phosphorus and riboflavin. The shell or skin is thin and brittle, and readily removed from the edible pulp.
3. Season: Fruit matures within 300-360 days from flowering.
4. Production in U.S.: New Mexico and Florida produces some fruit. For the most part, tamarind production is of minor importance in the U.S.
5. Other production regions: India and other tropical and subtropical countries. Domestic imports from Jamaica, Haiti, Grenada and Mexico.
6. Use: The young seedlings, tender leaves and flowers can be used as a salad crop, immature pods can be used as a seasoning, the pulp from seeds and shell is an important ingredient in Worcestershire and barbecue sauces. The pulp can also be used in drinks, in preserves and in meat sauces. It is an excellent tropical ornamental tree.

7. Part(s) of plant consumed: Every part of the tree can be utilized, however and pulp and seeds are most often used parts
8. Portion analyzed/sampled: Whole fruit
9. Classifications:
   a. Authors Class: Tropical and subtropical fruits – inedible peel
   b. EPA Crop Group (Group & Subgroup): Miscellaneous
   c. Codex Group: 006 (FI 0369) Assorted tropical and subtropical fruits – inedible peel
   d. EPA Crop Definition: None
10. References: GRIN, BUESO, CODEX, LOGAN, MAGNESS, MARTIN 1987, US EPA 1987, MORTON (picture), SAULS(e)
11. Production Map: EPA Crop Production Region 13.
12. Plant Codes:
    a. Bayer Code: No specific entry

600

1. Tangelo
(Small sized tangelo, Medium sized tangelo, K-early citrus, Ugli, Ugli fruit, Unique fruit, Uniq)
Rutaceae
   Citrus x *tangelo* J.W. Ingram & H.E. Moore (syn: *C. paradisi* x *C. reticulata*)

2. See also citrus fruits and citrus hybrids. Crosses of grapefruit and tangerine oranges have resulted in several varieties, termed tangelos, which are increasing in production and popularity in the U.S. These hybrids are intermediate in character between the parents. Fruits are similar in size to sweet orange, but with a tendency to be necked at the stem end. They are extremely juicy and intermediate in acidity between the parents. Peel is relatively thin and tender. General characteristics are similar to sweet orange, except for more tender peel and juicier, more tart flesh. Some cultivars are 'Minneola' ('Honeybell'), 'Orlando', 'Nova', 'Early K' ('Sunrise' tangelo), and 'Sampon'. 'Ugli' tangelo of Jamaica is believed to be a chance hybrid between mandarin orange and grapefruit (MORTON 1987).
3. Season, bloom to harvest: 8 to 14 months.
4. Production in U.S.: In 1995-96, 12,700 bearing acres with 110,000 tons of tangelo. K-early citrus was 300 acres and 7,000 tons (USDA 1996d). In 1995-96, Florida was the only state reporting tangelos and K-early citrus (USDA 1996d).
5. Other production regions: No entry.
6. Use: Mainly juice, some fresh.
7. Part(s) of plant consumed: Interior vesicles
8. Portion analyzed/sampled: Whole fruit
9. Classifications:
   a. Authors Class: See citrus hybrids
   b. EPA Crop Group (Group & Subgroup): See citrus hybrids
   c. Codex Group: 001 (FC 0003) Citrus fruits
   d. EPA Crop Definition: See citrus hybrids
10. References: GRIN, CODEX, LOGAN (picture), MAGNESS, MORTON (picture of 'Ugli'), USDA 1996d, US EPA 1994
11. Production Map: EPA Crop Production Region 3 and some in 10.
12. Plant Codes:
    a. Bayer Code: CIDRP

601

1. Tangerine
(Mandarin, Dancy, Clementine, Mandarin orange, Tangerino, Culate mandarin, Ponkan, Mandarina, Naranjita, Kid-glove orange)
Rutaceae
   Citrus reticulata Blanco (syn: *C. nobilis* Andrews, *C. ponnensis* Hort., *C. chrysocarpa* Lush., *C. tangerina* Hort., *C. reshni* Hort.)

1. Satsuma orange
(Owari satsuma, Unshu)
   C. unshiu Marcov.

2. Also see citrus fruits. Tangerines are characterized by a loose skin which separates readily from the pulp, and by segments which separate readily from each other. Fruit is generally oblate and smaller than sweet or round oranges. Diameter is mostly $2^1/_2$ inches or less, though some varieties average larger. Tangerines are grown in all citrus areas of the U.S. Two main types are grown in the U.S.: The Satsuma group, characterized by a small tree hardier than other citrus, and early ripening fruit of yellow or light orange color; and the Tangerine group, characterized by deep orange color and later ripening. Some Satsuma plantings are along the Gulf states, in areas too cold for other citrus. 'Dancy' is the leading variety of Tangerine and 'Owari' of Satsuma. About 10 other varieties occur occasionally. 'Clementine' is a cultivar of *C. reticulata*. Tangerine cultivars are 'Dancy' 'Algerian', 'Fairchild', 'Honey' ('Murcott'), 'Robinson', and 'Sunburst'. Mandarin cultivars are 'Royal mandarin', 'Honey', and 'Clementine'. Satsuma cultivars are 'Owari', 'Wase' and 'Kara'.
3. Season, bloom to harvest: 6 to 10 months.
4. Production in U.S.: In 1995-96, 348,000 tons from 38,700 bearing acres (USDA 1996d). In 1995-96, the top 3 states reported as Florida (24,300 bearing acres), California (8,700), and Arizona (5,700) (USDA 1996d).
5. Other production regions: Japan is a leading producer of Mandarins (KALE).
6. Use: Mainly fresh, but juice also frozen.
7. Part(s) of plant consumed: Internal segments
8. Portion analyzed/sampled: Whole fruit
9. Classifications:
   a. Authors Class: See citrus fruits
   b. EPA Crop Group (Group & Subgroup): See citrus fruits
   c. Codex Group: 001 (FC 0206) Citrus fruits
   d. EPA Crop Definition: See citrus fruits
10. References: GRIN, CODEX, LOGAN (picture), MAGNESS, USDA 1996d, US EPA 1994, IVES, KALE
11. Production Map: EPA Crop Production Regions 3 and 10 with 66 percent of the acreage in 3.
12. Plant Codes:
    a. Bayer Code: CIDRE (*C. reticulata*); CIDUN (*C. unshiu*)

602

1. Tanglehead
(Black speargrass, Bunch speargrass, Pili grass, Zacate colorado, Herbe polisson, Speargrass)
Poaceae (Gramineae)
   Heteropogon contortus (L.) P. Beauv. ex Roem. & Schult. (syn: *Andropogon contortus* L.; *H. hirtus* Pers.)

2. This is a hardy perennial warm season grass native to the southwest U.S. It is a member of the grass tribe *Andropogoneae*. It is an erect bunchgrass (20 to 80 cm tall) that flowers with a raceme 3 to 8 cm long. It is common to rocky hills and canyons in sandy soils with a pH 5 to 6, and is most abundant on heavily grazed soils. It grows from areas at sea level to 2000 m, and in Hawaii it grows at areas <300 m and with rainfall varying 500 to 1500 mm. Tanglehead flowers from June through November, and reproduces from seeds and tillers. It produces a fair to good quality forage for cattle and horses if cut for hay before flowering. Its palatability declines rapidly at maturity, and its awns may be irritating to grazing livestock and can reduce the quality of the wool of sheep. Tanglehead is also useful to control soil erosion.
3. Season: Warm season, growth starts in spring and it matures in August. It can be cut for hay before it sets seed.
4. Production in U.S.: No data for Tanglehead production, however, pasture and rangeland are produced on over 410 million A (U.S. CENSUS, 1992). See above.
5. Other production regions: Widespread in Mexico.
6. Use: Rangeland grazing, hay, and reseeding rangeland for erosion control.
7. Part(s) of plant consumed: Leaves and stems
8. Portion analyzed/sampled: Forage and hay
9. Classifications:
   a. Authors Class: See Forage grass
   b. EPA Crop Group (Group & Subgroup): See Forage grass
   c. Codex Group: See Forage grass
   d. EPA Crop Definition: None
10. References: GRIN, US EPA 1994, US EPA 1995, CODEX, HOLZWORTH, SKERMAN (picture), STUBBENDIECK, RICHE
11. Production Map: EPA Crop Production Regions 9, 10 and 13.
12. Plant Codes:
    a. Bayer Code: HTOCO

603

1. Tanier spinach
(Tahitian taro, Belembe, Tahitian spinach, Calalou, Malanga, Tahitian spinach)
Araceae
Xanthosoma brasiliense (Desf.) Engl.
2. This perennial plant to about 2 feet tall develops from a rather insignificant corm. The leaves are sagittate to trilobed, glabrous, dark green and succulent. The petioles are long and succulent with no above-ground stem produced. Leaves and petioles are eaten from very young to older leaves. The leaves may be harvested for food anytime. The whole, mature leaves and petioles are cut and bundled under commercial practices.
3. Season, planting with corms to harvest: Suitable leaves for harvest are produced in 2 to 3 weeks, but 6 weeks are needed for mature leaves. Single leaves cut weekly or all leaves every 6 to 8 weeks.
4. Production in U.S.: No data. Limited. It is cultivated in tropical America and some Pacific islands (NEAL 1965).
5. Other production regions: Tropics (MARTIN 1979).
6. Use: Tender leaves and petioles are used as a potherb like spinach. Before cooking, the leaves and petioles are cut up. They are boiled for 10-15 minutes.

7. Part(s) of plant consumed: Leaves and petioles. Tubers are edible when cooked but too small to be used for food
8. Portion analyzed/sampled: Leaves and petioles
9. Classifications:
   a. Authors Class: Leaves of root and tuber vegetables
   b. EPA Crop Group (Group & Subgroup): Miscellaneous
   c. Codex Group: No specific entry
   d. EPA Crop Definition: None
10. References: GRIN, MARTIN 1979 (picture), NEAL (picture), YAMAGUCHI 1983, RUBATZKY (picture)
11. Production Map: EPA Crop Production Region 13.
12. Plant Codes:
    a. Bayer Code: XATBR

604

1. Tansy
(Common tansy, Rainfarn, Tanaceto, Tanaisie commune, Golden buttons, Tanarida)
Asteraceae (Compositae)
Tanacetum vulgare L. (syn: *Chrysanthemum vulgare* (L.) Bernh.)
2. The plant is a coarse-growing herbaceous perennial 2 to 3 feet in height with finely divided leaves. The bitter, aromatic leaves were formerly much used for flavoring, especially puddings and omelets, also medicinally. Use in both categories has declined in recent years. In the U.S., 8 acres were reported in the 1949 CENSUS, with no later report.
3. Season, seeding to harvest: Harvested at full bloom from late July to September.
4. Production in U.S.: No data. Grows from Maine to North Carolina, west to Minnesota and Missouri, and the Pacific Northwest (FOSTER 1993).
5. Other production regions: Eurasia (FACCIOLA 1990).
6. Use: Flavoring, condiment.
7. Part(s) of plant consumed: Leaves
8. Portion analyzed/sampled: Leaves (fresh and dried)
9. Classifications:
   a. Authors Class: Herbs and spices
   b. EPA Crop Group (Group & Subgroup): Herbs and spices (19A)
   c. Codex Group: 027 (HH 0748) Herbs; 057 (DH 0748) Dried herbs
   d. EPA Crop Definition: None
10. References: GRIN, CODEX, FOSTER (picture), KUHNLEIN, MAGNESS, REHM, US EPA 1995a, FACCIOLA
11. Production Map: EPA Crop Production Regions 1, 2, 5 and 12.
12. Plant Codes:
    a. Bayer Code: CHYVU

605

1. Tapertip hawksbeard
(Longleaf hawksbeard, Mountain hawksbeard)
Asteraceae (Compositae)
Crepis acuminata Nutt.
2. Tapertip hawksbeard is a cool season perennial forb that grows 20 to 70 cm tall. Native to the U.S. It belongs to the tribe *Cichoriceae*. It flowers in May through August and reproduces from seeds. Its habitat includes prairies, hillsides and is most abundant in well-drained soils on open areas. As

a forage, tapertip hawksbeard is most palatable late spring to early summer, and is of good to fair quality for cattle, sheep, and horses. It's preferred by sheep.

3. Season: Cool season perennial forage that can be used for grazing in the spring and early summer.
4. Production in U.S.: No specific data for Tapertip hawksbeard production, however, pasture and rangelands are produced on more than 410 million A (U.S. CENSUS, 1992).
5. Other production regions: No entry.
6. Use: Grazing.
7. Part(s) of plant consumed: Leaves and stems
8. Portion analyzed/sampled: Forage and hay
9. Classifications:
   a. Authors Class: Nongrass Animal Feeds
   b. EPA Crop Group (Group & Subgroup): Miscellaneous
   c. Codex Group: No specific citation although the general class 052 (AM 0165) Miscellaneous Fodder and Forage Crops could be used
   d. EPA Crop Definition: None
10. References: GRIN, US EPA 1994, US EPA 1995, CODEX, HOLZWORTH, STUBBENDIECK (picture), RICHE
11. Production Map: EPA Crop Production Regions 9, 10 and 11.
12. Plant Codes:
   a. Bayer Code: No specific entry

606

1. Taro
(Kalo, Taro de chine, Chinese potato, Malanga, Sato imo, Poi, Luau, See coo, Wetland taro, Old cocoyam, Eddoe, Eddo, Upland taro, Yu tau, Woo chai, Gabi, Elephants' ear, Cocoyam, Yu)

Araceae

Colocasia esculenta (L.) Schott (syn: *Caladium esculentum* Vent.)

2. Taro has long been an important perennial herb crop in the tropics. Grows to 6 feet tall. One type of Taro, the Dasheen, which see, is nearly free of crystals of calcium oxalate, which are abundant in the roots of other taros, and give them an acrid taste when raw. This disappears with cooking. Taro is the base of the popular Hawaiian dish "poi". While the corm is the main food, the young, unopened leaves are also used as greens, termed "luau" or "See coo". Some varieties are grown in water, others in upland culture. See Dasheen for cultural data. A Chinese form (bunlong) is grown for table taro and the leaves furnish most of the luau sold in Hawaii. The taro "mother" corm which is to 8 inches high and to 5 inches in diameter produces 2 to 15 side tubers or cormels up to 6 inches in diameter. Corm and cormels are used as food. A few kinds of wetland taros are used for poi, other upland kinds for the table. Propagation is usually corms or cormels. In Hawaii the cut stem portion with petiole attached called huli or set are obtained at harvest.
3. Season, planting to harvest: About 7 to 18 months (13 months normal). Harvested when tops are dried. Upland taro about 8-9 months.
4. Production in U.S.: Hawaii produced 4,570 tons in 1968 and 3,645 tons from 495 acres in 1992. Guam reported 9 acres and about 10 tons. Total U.S. 496 acres and 3,653 tons (1992 CENSUS). The 1994 Agricultural Statistics reported an average of 492 acres for 5 years (1989 to 1993). Yields range

from 2-8 pounds per plant. Hawaii and Guam, mainly the Pacific area.

5. Other production regions: Tropics.
6. Use: As cooked vegetable. Also grown as an ornamental.
7. Part(s) of plant consumed: Corms and young leaves prior to unfolding
8. Portion analyzed/sampled: Corm and foliage
9. Classifications:
   a. Authors Class: Root and tuber vegetables; Leaves of root and tuber vegetables
   b. EPA Crop Group (Group & Subgroup): Root and tuber vegetables (1C and 1D); Leaves of roots and tuber vegetables (2)
   c. Codex Group: 016 (VR 0505) Root and tuber vegetables; 013 (VL 0505) Leafy vegetables (including *Brassica* leafy vegetables)
   d. EPA Crop Definition: None
10. References: GRIN, CODEX, KAYS, MAGNESS, MYERS, NEAL (picture), RICHE, USDA 1994a, US EPA 1995a, US EPA 1995b, YAMAGUCHI 1983, SPLITTSTOESSER
11. Production Map: EPA Crop Production Region 13.
12. Plant Codes:
   a. Bayer Code: CXSES

607

1. Tarragon
(Estragon, French tarragon)

Asteraceae (Compositae)

Artemisia dracunculus L.

2. The plant is an herbaceous perennial, with erect, branched stems, up to 3 feet in height. The small leaves are lanceolate and entire. The green leaves and succulent stems may be used fresh or dried for flavoring salads, vinegar, pickles, meats, etc. Tarragon oil is obtained by distillation of the green parts, while tarragon vinegar is an infusion of the green parts in vinegar. It is propagated by root cuttings or division into 2-3 new plants in early spring. Root clusters are planted 12-20 inches apart and split into new plants every 3-4 years. Tarragon is the special ingredient in Dijon mustard from France. The oil of tarragon is used in beverages, vinegar, mustard and baked products. It has an anise-licorice like odor.
3. Season, harvest: Tender tops of the plants are harvested at intervals and may be dried. The plants are replaced every 3 to 4 years. Harvested 2-3 times per season to encourage stem branching.
4. Production in U.S.: Limited production in U.S. 250 acres in California (ELMORE 1985).
5. Other production regions: Europe.
6. Use: Flavoring of vinegar, poultry, fish, sauces, cheese, vegetable products and salads.
7. Part(s) of plant consumed: Tops, mainly dried leaves
8. Portion analyzed/sampled: Tops (Fresh and dried)
9. Classifications:
   a. Authors Class: Herbs and spices
   b. EPA Crop Group (Group & Subgroup): Herbs and spices (19A)
   c. Codex Group: 027 (HH 0749) Herbs
   d. EPA Crop Definition: None

10. References: GRIN, CODEX, ELMORE 1985, FARRELL, FOSTER (picture), LEWIS, MAGNESS, US EPA 1995a, FORTIN, SPLITTSTOESSER
11. Production Map: EPA Crop Production Region 10.
12. Plant Codes:
    a. Bayer Code: ARTDR

608

1. Tea

(Teaplant, China tea, Black tea, Green tea, Assam tea, Oolong tea)

Theaceae

Camellia sinensis (L.) Kuntze var. *sinensis* (syn: *C. thea* Link)
C. sinensis var. *assamica* (J. Mast.) Kitam. (syn: *C. theifera* Griff.)

(Assamtea)

C. sinensis (L.) Kuntze

(Teaplant)

2. The tea plant is an evergreen shrub or small tree that may reach 30 feet if unpruned. It is adapted to sub-tropical areas, and is widely grown for its leaves, which are dried and constitute the tea of commerce. The plants tolerate some frost, and are grown in southern U.S., primarily South Carolina. The leaves are lanceolate, glabrous, but sometimes pubescent on lower surface, and 2 to 5 inches long. The harvested portions are the succulent short tips and young leaves. Including older leaves reduces the quality of the tea. Leaves are harvested at intervals of 2 weeks or less. For green tea, leaves are heated quickly after harvest to inactivate enzymes. For black tea, leaves are wilted and partially dried in shallow layers, then are rolled by twisting or wringing. A short oxidation or fermentation period is followed by heating at 160 degrees F or above to stop oxidation. Oolong tea combines flavors of green and black tea. Leaves are fermented before drying. The U.S. imported about 93,382 tons of tea primarily from Asia and South America in 1993.
3. Season, harvest: Biweekly May through October. Machine harvested.
4. Production in U.S.: Limited, about 130 acres in production with 30 acres of mature and harvestable plants (LEIDNER 1992). South Carolina with 20,000 pounds per acre of leaves and buds per year (1992).
5. Other production regions: See above.
6. Use: Beverage. For every 4 pounds of green shoots about 1 pound of dried tea is produced. Also grown as an ornamental hedge.
7. Part(s) of plant consumed: Leaves
8. Portion analyzed/sampled: Plucked (or fresh picked) leaves and processed commodities dried tea and instant tea
9. Classifications:
   a. Authors Class: Dried edible plant tops (stimulants)
   b. EPA Crop Group (Group & Subgroup): Miscellaneous
   c. Codex Group: 066 (DT 1114) Teas
   d. EPA Crop Definition: None
10. References: GRIN, CODEX, LEIDNER 1992, MAGNESS, USDA 1994a, US EPA 1995b, US EPA 1994, USDA NRCS, DUKE(a)(picture)
11. Production Map: EPA Crop Production Region 2.
12. Plant Codes:
    a. Bayer Code: CAHSI

609

1. Teff

(Tef, Williams lovegrass, Chimanganga, Teffgrass, Taf, Tafi)

Poaceae (Gramineae)

Eragrostis tef (Zuccagni) Trotter (syn: *E. abyssinica* (Jacq.) Link)

2. A staple cereal grain crop of Ethiopia. The plant is a slender upright tufted annual that is drought resistant. It has a brown panicle with red, white or brown seed with yields averaging 0.9 tons/ha and improved cultivars at 1.7 to 2.25 tons/ha. Hay yields are 3200 to 6000 lb/A. Teff is an intermediate grass between tropical and temperate grasses. It provides more than two-thirds of the human nutrition in Ethiopia. In the U.S., teff is considered a health food. The grain contains no gluten.
3. Season, seeding to harvest: 45 to 160 days depending on cultivar. Planted early June.
4. Production in U.S.: No data. Limited. Northwest U.S. and Montana, Oklahoma and South Dakota.
5. Other production regions: Ethiopia.
6. Use: Grain used in breads, pancakes, soups and alcoholic beverages.
7. Part(s) of plant consumed: Grain and foliage
8. Portion analyzed/sampled: Grain, straw and hay
9. Classifications:
   a. Authors Class: Cereal grains; Forage, fodder and straw of cereal grains
   b. EPA Crop Group (Group & Subgroup): Miscellaneous
   c. Codex Group: 020 (GC 0652) Cereal grains
   d. EPA Crop Definition: None
10. References: GRIN, CODEX, CHEVERTON, STALLKNECHT(a), BOE
11. Production Map: EPA Crop Production Regions 7, 9 and 11.
12. Plant Codes:
    a. Bayer Code: ERATF

610

1. Thorn mimosa

(Egyptian mimosa, Bubul, Gum arabic tree, Egyptian acacia, India gum-arabic-tree, Babul acacia, Thorny acacia, Babul, Suntwood, Egyptian thorn)

Fabaceae (Leguminosae)

Acacia nilotica (L.) Willd. ex Delile (syn: *Mimosa nilotica* L.)

2. Thorn mimosa is a perennial warm season small tropical tree 2.5 to 14 m tall that is native to the tropics and is grown mostly in the Sudan and west Africa. The tender pods, shoots and seed are used as a vegetable crop and as a forage for sheep, goats and camels. It is grown on soils with a pH 5 to 8. Thorn mimosa flowers from October through December and fruits in March through June. Thorn mimosa seeds can be roasted before they are eaten. Its bark, gum, leaves and pods are used medicinally in west Africa. Its inner bark contains 18 to 23 percent tannin and it is used in tanning leather. It is propagated by seeds.
3. Season: Warm season perennial trees, leaves, seeds, shoots, and pods are readily grazed by livestock.
4. Production in U.S.: No specific data for Thorn mimosa production, however, pasture and rangelands are produced on more than 410 million A (U.S. CENSUS, 1992).
5. Other production regions: See above.

6. Use: Grazing, forage, leather tanning and as a human food (seeds, pods, shoots). Also grown as an ornamental.
7. Part(s) of plant consumed: Leaves, stems and seeds
8. Portion analyzed/sampled: Forage, hay and seeds
9. Classifications:
   a. Authors Class: Nongrass Animal Feeds
   b. EPA Crop Group (Group & Subgroup): Miscellaneous
   c. Codex Group: No specific citation although the general class 050 (AL 0157) Legume Animal Feeds could be used
   d. EPA Crop Definition: None
10. References: GRIN, US EPA 1994, US EPA 1995, CODEX, HOLZWORTH, DUKE(h), SKERMAN(a) (picture), RICHE
11. Production Map: EPA Crop Production Region are 3 and 13.
12. Plant Codes:
    a. Bayer Code: ACANL

611

1. Threadleaf sedge
(Blackroot)
Cyperaceae
Carex filifolia Nutt.
2. Threadleaf sedge is a cool season perennial monecious grass-like herb that grows from 5 to 30 cm tall. It is native to the U.S. It begins growth in early spring and flowers in April and May and reproduces by seeds and tillers. Its leaves are grass-like blades 3 to 20 cm long and its inflorescence is a spike 5 to 20 mm long and 2 to 6 mm wide. Its habitat includes prairies, hills and valleys and is most abundant on dry soils. Native Americans ate the culm bases of threadleaf sedge during times of famine. Its forage is of excellent quality for sheep, horses and wildlife, and good quality for cattle early in the spring. Maintains its palatability during the season.
3. Season: Cool season. Highly palatable perennial forage that can be used for grazing in the early spring.
4. Production in U.S.: No specific data for Threadleaf sedge production, however, pasture and rangelands are produced on more than 410 million A (U.S. CENSUS, 1992). See above.
5. Other production regions: Distributed in Canada.
6. Use: Grazing.
7. Part(s) of plant consumed: Leaves and stems
8. Portion analyzed/sampled: Forage and hay
9. Classifications:
   a. Authors Class: Nongrass Animal Feeds
   b. EPA Crop Group (Group & Subgroup): Miscellaneous
   c. Codex Group: No specific citation although the general class 052 (AM 0165) Miscellaneous Fodder and Forage Crops could be used
   d. EPA Crop Definition: None
10. References: GRIN,US EPA 1994, US EPA 1995, CODEX, HOLZWORTH, STUBBENDIECK (picture), RICHE
11. Production Map: EPA Crop Production Regions are 7, 9, 10 and 11.
12. Plant Codes:
    a. Bayer Code: No specific entry, genus entry is CRXSS

612

1. Thyme
(Tomillo)
Lamiaceae (Labiatae)
Thymus spp.

1. Common thyme
(Garden thyme, English thyme)
T. vulgaris L.
1. Creeping thyme
(Mother-of-thyme, Wild thyme, Serpol, Lemon thyme, Serpolet)
T. serpyllum L. (syn: *T. praecox* ssp. *arcticus* (Dur.) Jalas)
1. Lemon thyme
T. x *citriodorus* (Pers.) Schreb. ex Schweigg. & Korte
2. These species of *Thymus* are grown under the general name Thyme. Common or Garden thyme is *T. vulgaris*, a sub-shrub up to 16 inches high with stiff branches, linear or lanceolate leaves. *T. serpyllum* is Mother-of-thyme or Creeping thyme, and has prostrate, creeping stems with ascending shoots, elliptic to oblong leaves, 1/2 inch long. Leaves of both kinds are used for seasoning soups, stews, meats, etc. They may be used either fresh or dried. Culture is similar to that of mint, to which thyme is related. Oil distilled from the plants is used in perfume. The chief part of the oil is thymol which is used in cough drops and baked goods.
3. Season, transplanting to harvest: Harvest when in bloom.
4. Production in U.S.: 150 acres in California (ELMORE 1985).
5. Other production regions: Mediterranean region (FACCIOLA 1990).
6. Use: Seasoning soups, stews, meats, and in pickles and vinegar.
7. Part(s) of plant consumed: Leaves and flowering tops
8. Portion analyzed/sampled: Leaves (fresh and dried)
9. Classifications:
   a. Authors Class: Herbs and spices
   b. EPA Crop Group (Group & Subgroup): Herbs and spices (19A)
   c. Codex Group: 027 (HH 0750) Herbs; 057 (DH 0750) Dried herbs
   d. EPA Crop Definition: None
10. References: GRIN, CODEX, ELMORE 1985, FARRELL (picture), FOSTER (picture), GARLAND, LEWIS, MAGNESS, STEPHENS 1988, USDA NRCS, US EPA 1995a, FACCIOLA, PHILLIPS
11. Production Map: EPA Crop Production Region 10.
12. Plant Codes:
    a. Bayer Code: THYSS (*T.* spp.) THYSE (*T. serpyllum*), THYVU (*T. vulgaris*)

613

1. Ti
(Ki, Oke, Hawaiian good-luckplant, Good-luck-plant, Bongbush, Tiplant, Ti Palm, Lai, Tree-of-kings, Andong)
Agavaceae
Cordyline fruticosa (L.) A. Chev. (syn: *Cordyline terminalis* (L.) Kunth)
2. This monocotyledon is a low, slender tree-like plant, 6 to 12 feet tall with leaves up to 30 inches long and 5 inches wide, native in the East Indies. There it is cultivated for the large, tuberous roots, normally up to 14 pounds but as much as 300 pounds, which contain much sugar. Roots are usually baked or roasted. Baked, macerated roots are also fermented in water and an intoxicating liquor is obtained by distillation. So far as known, Ti is not grown as a food plant in the U.S. outside of Hawaii. The leaves have long served as plates or wrappers for food and as a fodder for horses and cattle.

3. Season, propagation: By cuttings from stems.
4. Production in U.S.: No data. Hawaii. Grown as a ornamental in Florida and Hawaii (FRANK 1997).
5. Other production regions: Tropical Asia.
6. Use: A brandy called okolehao (Oke) is made from the baked fermented mash of the root.
7. Part(s) of plant consumed: Root. Leaves are insignificant feed item, but leaves are used in Hawaiian cooking as pot-herb to impart flavor without consumption of the leaves
8. Portion analyzed/sampled: Roots and leaves
9. Classifications:
   a. Authors Class: Root and tuber vegetables; Leaves of root and tuber vegetables
   b. EPA Crop Group (Group & Subgroup): Miscellaneous
   c. Codex Group: No specific entry
   d. EPA Crop Definition: None
10. References: GRIN, MAGNESS, NEAL, USDA NRCS, GRIN, BAILEY 1976, FACCIOLA, FRANK, KAWATE 1997a
11. Production Map: EPA Crop Production Region 13.
12. Plant Codes:
   a. Bayer Code: CDLFR

614

1. Timothy
(Herdgrass, Horsegrass, Meadow cat's-tail, Cat's-tail, Cola de topo, Common timothy)
Poaceae (Gramineae)
Phleum pratense L. (syn: *P. nodosum* L.)
2. Timothy is of European origin but was first cultivated in the U.S. It became the most important hay-grass in this country by the start of the last century. It is still of major importance in cool sections. The plant is a perennial bunchgrass, long-lived and winter hardy in cool, moist areas. Stems are leafy, and reach to 40 inches under the best conditions. Timothy is palatable and nutritious both for hay and for grazing. It does not tolerate drought, high temperatures or close grazing. Over 70 cultivars are available. It is often seeded in mixtures with other grasses. Timothy is the standard hay grass in the northeast U.S., cool humid regions of the northwest U.S., and widely used across the U.S. and Canada. Cut at stem elongation, stands will persist for 4 years. It does not persist as long as perennial ryegrass or tall fescue. Normally seeded at rates of 8 to 10 pounds per acre alone or 6 to 8 pounds per acre with clover, alfalfa or trefoil.
3. Season: Normally 1-2 cuttings per year for hay. Most productive spring and early summer. Can be harvested under optimum conditions up to 4 times a season. Hay cut at boot to early bloom stage and for silage harvested at early heading stage.
4. Production in U.S.: No specific data, See Forage grass. In 1992, 2,522 tons of seed were produced from 15,768 acres (1992 CENSUS). In 1992, the top 5 states for seed production reported as Minnesota (11,181 acres), Idaho (1,758), New York (941), Pennsylvania (744) and Ohio (402) (1992 CENSUS). See above.
5. Other production regions: In 1995, Canada reported 87,258 acres (ROTHWELL 1996a).
6. Use: Hay, pasture, silage.
7. Part(s) of plant consumed: Foliage
8. Portion analyzed/sampled: Forage and hay

9. Classifications:
   a. Authors Class: See Forage grass
   b. EPA Crop Group (Group & Subgroup): See Forage grass
   c. Codex Group: See Forage grass
   d. EPA Crop Definition: None
10. References: GRIN, MAGNESS, ALDERSON, MOSLER (picture), STUBBENDIECK(a), BALL, BARNES, RICHE, ROTHWELL 1996a
11. Production Map: EPA Crop Production Regions 1, 5, 9, 11 and 12. Seed crop regions by order of importance, regions 5, 11 and 1.
12. Plant Codes:
   a. Bayer Code: PHLPR

615

1. Timothy, Alpine
(Mountain timothy, Wild timothy, Timothy alpino, Alpine cat's-tail)
Poaceae (Gramineae)
Phleum alpinum L.
2. This is a perennial cool season grass native to the U.S. and is distributed as far north as Alaska. It's a member of the grass tribe *Aveneae*. It is an erect bunchgrass 15 to 60 cm tall, flat blades 2 to 15 cm long and 3 to 8 mm wide and has a panicle-type inflorescence (15 to 30 cm long) in mountain areas and meadows with elevations >1250 m. Alpine timothy provides a good to excellent quality forage for all livestock. Reproduces by seeds and tillers.
3. Season: Cool season grass that starts growing in early spring, flowers in June through August.
4. Production in U.S.: There is no production data for Alpine timothy, however, pasture and rangeland are produced on more than 410 million A (U.S. CENSUS, 1992). See above
5. Other production regions: Canada.
6. Use: Pasture grazing and hay.
7. Part(s) of plant consumed: Leaves and stems
8. Portion analyzed/sampled: Forage and hay
9. Classifications:
   a. Authors Class: See Forage grass
   b. EPA Crop Group (Group & Subgroup): See Forage grass
   c. Codex Group: See Forage grass
   d. EPA Crop Definition: None
10. References: GRIN, US EPA 1994, US EPA 1995, CODEX, HOLZWORTH, MOSLER, STUBBENDIECK (picture), RICHE
11. Production Map: EPA Crop Production Regions 1, 5, 7, 9, 11 and 12.
12. Plant Codes:
   a. Bayer Code: No specific entry, genus entry is PHLSS

616

1. Tobacco
(Tabaco)
Solanaceae
Nicotiana tabacum L.
2. The tobacco plant is a thick-stemmed annual bearing large leaves with short petioles or leaf stems. Leaf blades are often more than 20 inches long and half as wide. They rise in a spiral along the stem. Stems grow 4 to 6 feet tall and terminate in a cluster of flowers if not topped. Except for seed production,

plant terminals are usually removed when flowering begins in order to increase size and thickness of leaves – the marketable portion. Plants are usually started in beds under cloth cover in early spring and moved to the field after all hazard of frost is past. Growth in the field from setting to harvest covers 3 to 5 months. Harvesting may consist of removing most mature leaves by hand at about weekly intervals or cutting the whole stem. The former method is more generally used as it gives higher leaf yield and better quality. Leaves are then dried by one of several processes. In flue-cured tobacco, heat is applied in such a way that no smoke reaches the leaf hung in racks. In fire-curing, open fires are used and the smoke results in a darker colored, distinctly flavored leaf. In air-cured tobacco no heat is added except as necessary to prevent mold during humid periods. Kinds of curing depends on type of tobacco grown and ultimate use. Cigar wrapper tobacco, grown mainly in Wisconsin and Connecticut, is produced under partial shade – resulting in thinner leaves and less damage to the leaves. In the final products from tobacco (cigarettes, pipe tobacco, cigars, chewing tobacco and snuff) the leaf midribs and larger veins are largely removed. They may be processed to obtain nicotine insecticides or used as mulching material. They are not used as feed. In 1993, tobacco was grown on 746,405 acres in the U.S.. About 54 percent of this was flue-cured and nearly 43 percent air cured. Less than 3 percent was fire cured. Around 13,200 acres were devoted to cigar tobacco.

3. Season, transplanting to harvest: 3 to 5 months.
4. Production in U.S.: In 1993, 746,405 acres harvested in U.S. 93 percent of the U.S. production is in North Carolina (36 percent), Kentucky (28 percent), Tennessee (10 percent) and the remainder (19 percent) in South Carolina, Virginia, and Georgia (1994 AGRICULTURAL STATISTICS). Also grown in other states, e.g., Connecticut, Florida, Indiana, Maryland, Massachusetts, Missouri, Ohio, Pennsylvania, West Virginia and Wisconsin.
5. Other production regions: 74,409 acres in Canada; and 91,983 acres in Mexico (USDA 1994a).
6. Use: Smoking, chewing and snuff.
7. Part(s) of plant consumed: Leaves and stems
8. Portion analyzed/sampled: Green tobacco leaves. If residues >0.1 ppm, also need cured tobacco and pyrolysis study (OPPTS 860.1000)
9. Classifications:
   a. Authors Class: Dried edible plant tops (stimulants)
   b. EPA Crop Group (Group & Subgroup): Miscellaneous
   c. Codex Group: No specific entry
   d. EPA Crop Definition: None
10. References: GRIN, MAGNESS, USDA 1994a, IVES, US EPA 1996a
11. Production Map: EPA Crop Production Regions 2 and 5.
12. Plant Codes:
   a. Bayer Code: NIOTA

617

1. Tomato
(Pumate, Pomme d'amour, Pomodoro, Baby tear drop tomato, Tomate, Kamako, 'Ohi'a-lomi, Plum tomato, Cherry tomato, Pear tomato, Roma tomato, Gold apple, Love apple)
Solanaceae
Lycopersicon esculentum Mill. var. *esculentum* (syn: *L. lycopersicum* (L.) Karsten; *Solanum lycopersicum* L.)

1. Cherry tomato
(Salad tomato)
L. esculentum var. *cerasiforme* (Dunal) A. Gray (syn: *L. lycopersicum* var. *cerasiforme* (Dunal) Alef.)

1. Pear tomato
(Baby tear drop tomato, Salad tomato)
L. esculentum Mill. var. *esculentum* (syn: *L. lycopersicum* var. *pyriforme* (Dunal) Alef.)

2. Tomatoes are among the most important food crops in all parts of the world except areas of cool, short growing seasons. The plant is an annual, grown from seed. They are usually started in beds and moved to the field at 6 to 8 weeks, but may be seeded in place. When grown without support the plants of most varieties are much branched and sprawling. If supported on stakes or trellis it attains heights up to 6 to 8 feet. Leaves are compound and pinnate, a foot or more in length. Fruits have a smooth peel. They vary in size from an inch up to 4 inches in diameter. In shape they are generally oblate, but some varieties are globose or pyriform or conic. Tomato plant types are determinate (bush), indeterminate and semideterminate. Chief cultivar groups are for home use, fresh market, greenhouse and processing.

Some specialty tomatoes: Pear tomatoes are red or yellow, 1 to 2 inches across and a favorite for preserves. Plum and cherry tomatoes are about 1 inch across. Roma tomatoes are mainly used for paste and canning. Dried tomatoes are called pumate. Small-fruited tomatoes are ideal for serving whole in salads, termed salad tomatoes.

Greenhouse vegetables and fruits include tomato, cucumber, mesclun mix (fresh greens), melon, pepper, eggplant and strawberry. Acreage of U.S. greenhouse vegetables and fruits is about 550 to 650 acres (NAEGELY 1997) with predictions that larger greenhouse facilities will be in the southwest U.S. and Mexico. Currently, Leamington (Ontario, Canada) has the biggest concentration of greenhouse vegetables in North America. The top 10 states for greenhouse tomato acreage are Colorado (94 acres), Texas (72), Pennsylvania (56), Arizona (44), New York (35), California (30), Ohio (20), Tennessee (20), Mississippi (16) and New Jersey (15). (NAEGELY 1997). There are 10,000 seeds per ounce.

3. Season, seed planting to first ripe fruit: About 4 months. Transplanting to harvest: About 2-3 months. Usually ripen 5-8 weeks after fruit set. Processing tomatoes are mechanically harvested.
4. Production in U.S.: In 1995, commercial about 13 million tons. Also in most home gardens. In 1995, commercial U.S. acreage reported as 135,910 for fresh market and 359,080 for processed tomatoes with total acreage of 494,990. Production reported as 1,642,000 tons for fresh and 11,276,090 tons for processed (USDA 1996b). In 1995, the top 10 states for fresh market are Florida (47,300 acres), California (38,000), Georgia (5,000), New Jersey (4,800), Pennsylvania (4,400), Tennessee (4,100), South Carolina (4,000), Alabama (3,900), Texas (3,800) and Virginia (3,800) for 88 percent of U.S. acreage. The top 6 states for processed tomatoes are California (331,000 acres), Ohio (11,000), Indiana (6,900), Michigan (4,200), Pennsylvania (1,500) and Colorado (220) for 99 percent of U.S. acreage (USDA 1996b).
5. Other production regions: In 1995, Canada reported 27,528 acres of field tomatoes and 353 acres of greenhouse tomatoes (ROTHWELL 1996a).

6. Use: Fresh in salads, cooked, juice. Commercially canned and canned juice, unripened fruits in relishes. Tomato fruit can be baked, stewed, fried, dried, pickled, juiced, pureed and processed into catsup and sauces. U.S. per capita consumption in 1994 was 89.7 pounds (PUTNAM 1996).
7. Part(s) of plant consumed: Generally whole fruit, except peel. Peel is generally removed after steaming or immersion in hot water. Also whole fruit is consumed fresh. Fresh tomatoes are also dried. Also an edible oil is obtained from the seeds
8. Portion analyzed/sampled: Fruit and its processed products paste and puree. Residue data on paste covers processed products (e.g. sauce, juice, catsup) except puree which covers canned tomatoes
9. Classifications:
   a. Authors Class: Fruiting vegetables (except cucurbits)
   b. EPA Crop Group (Group & Subgroup): Fruiting vegetables (except cucurbits)(8) (Representative crop)
   c. Codex Group: 012 (VO 0448) Fruiting vegetables (other than cucurbits); 070 (JF 0448) Fruit juices
   d. EPA Crop Definition: Tomatoes = Tomatillos
10. References: GRIN, CODEX, FACCIOLA, KAYS, LOGAN (picture), MAGNESS, NEAL, USDA 1996b, US EPA 1996a, US EPA 1994, WARE, SWIADER, NAEGELY, PUTNAM 1996
11. Production Map: EPA Crop Production Regions 1, 2, 3, 5 and 10 for 97 percent of the acreage.
12. Plant Codes:
   a. Bayer Code: LYPES, LYPXS (direct seeded), LYPXP (transplants)

618

1. Tomato, Currant
('Ohi'a-ma-kanahele, Cocktail tomato, German raisin tomato, Tomatillo)
Solanaceae
Lycopersicon pimpinellifolium (L.) Mill.

2. The currant tomato is normally seen in gardens. Because of its resistance to several diseases it has been used in breeding more disease resistant tomato varieties. The plant is more slender in growth than tomato. The grape-like fruit is borne in rather long, loose racemes. Individual fruits are small, generally not more than $1/3$- to $1/2$-inch in diameter, with smooth skin.
3. Season, seeding to first ripe fruit: 60 to 70 days.
4. Production in U.S.: As novelty in some home gardens. At least 5 cultivars are known.
5. Other production regions: Native to Andean South America (FACCIOLA 1990)
6. Use: Fruit edible. Sometimes used as food novelty, either fresh, cooked or dried and used in salads and soups.
7. Part(s) of plant consumed: Whole fruit
8. Portion analyzed/sampled: Whole fruit
9. Classifications:
   a. Authors Class: Fruiting vegetables (except cucurbits)
   b. EPA Crop Group (Group & Subgroup): Miscellaneous
   c. Codex Group: No specific entry
   d. EPA Crop Definition: None
10. References: GRIN, FACCIOLA, MAGNESS, NEAL
11. Production Map: No entry.
12. Plant Codes:
   a. Bayer Code: LYPPI

619

1. Tomato, Tree
(Palo de tomate, Tamarillo, Tomate de Arbol)
Solanaceae
Cyphomandra betacea (Cav.) Sendtn. (syn: *C. crassifolia* Kuntze)

2. The tree tomato is a perennial shrub 6 to 10 feet high, with large, soft pubescent heart-shaped leaves. It can be grown in the open only in frost free locations. The fruit is 2 to 3 inches long, oval in shape, smooth and many seeded, borne on a long stem. Fruit resembles a tomato in appearance, use and flavor.
3. Season, bloom to mature fruit: About 3-6 months. It begins blooming 2 years after seeding and can continue for 10 years but usually replaced after 5-6 years of production. In the West Indies, the fruit ripens from October to January.
4. Production in U.S.: No data, limited. Cultivated in California and available October to January (LOGAN 1996). Also, Florida and Puerto Rico.
5. Other production regions: Widely grown in South America (STEPHENS 1988). Also cultivated in Central America (SCHNEIDER 1985). New Zealand (RUBATZKY 1997).
6. Use: Fresh and culinary, like tomato. Preparation – fruit must be peeled before eating, whether it's raw or cooked (SCHNEIDER 1985).
7. Part(s) of plant consumed: Whole fruit
8. Portion analyzed/sampled: Whole fruit
9. Classifications:
   a. Authors Class: Fruiting vegetables (except cucurbits)
   b. EPA Crop Group (Group & Subgroup): Miscellaneous
   c. Codex Group: 005 (FT 0312) Assorted tropical and subtropical fruits, edible peel
   d. EPA Crop Definition: None
10. References: GRIN, CODEX, KENNARD (picture), LOGAN, MAGNESS, SCHNEIDER 1985, STEPHENS 1988 (picture), DUKE (a), RUBATZKY
11. Production Map: EPA Crop Production Regions 10 and 13.
12. Plant Codes:
   a. Bayer Code: CYJBE

620

1. Tonka bean
(Tonka bean oil, Cumaru, Tonga, Tonquin)
Fabaceae (Leguminosae)
Dipteryx odorata (Aubl.) Willd.
Taralea oppositifolia Aubl. (syn: *D. oppositifolia* (Aubl.) Willd.)
(English tonka bean)

2. The oil is obtained from the seed of the above species, trees native to Central and South America, but now cultivated to some extent in other tropical areas. The fruit is a pod about 2 inches long, containing a single fragrant seed. The seed after curing is used chiefly for scenting tobacco and snuff. The non-drying oil is used in flavoring.
3. Season: Flowers March-May and fruits from June-July. Fallen pods are harvested from January to March. Propagation is usually by seed but can be propagated by cuttings.
4. Production in U.S.: No data.
5. Other production regions: Tropical areas. Native to Central and South America.

6. Use: Flavoring.
7. Part(s) of plant consumed: Seed
8. Portion analyzed/sampled: Seed and its oil
9. Classifications:
   a. Authors Class: Herbs and spices
   b. EPA Crop Group (Group & Subgroup): Miscellaneous
   c. Codex Group: 006 (FI 0370) Assorted tropical and subtropical fruits – inedible peel; 028 (HS 0370) Spices
   d. EPA Crop Definition: None
10. References: GRIN, CODEX, MAGNESS, NEAL, DUKE(a)
11. Production Map: EPA Crop Production Region 13.
12. Plant Codes:
    a. Bayer Code: No specific entry

621

1. Trefoil
Fabaceae (Leguminosae)
Lotus spp.
1. Birdsfoot trefoil
(Cuernecillo)
L. corniculatus L. var. *corniculatus*
1. Narrowleaf trefoil
(Slender trefoil)
L. glaber Mill. (syn: *L. corniculatus* L. var. *tenuifolius* L.; *L. tenuis* Waldst. & Kit. ex Willd.)
1. Big trefoil
(Greater birdsfoot trefoil)
L. uliginosus Schkuhr

2. Trefoils are widely distributed throughout the world, and are native to Europe, North Africa and Asia. Birdsfoot trefoil is a perennial, fine-stemmed, leafy legume that has become of increased importance in American agriculture in recent years. Introduced by chance from Europe, strains selected in this country are now of major importance as pasture and hay crops. There are two types, Empire and European, that are grown in U.S. and Canada. The Empire or New York type is finer stemmed, more prostrate, indeterminate and more winter hardy than the European types. It is hardy and adapted to areas of ample moisture supply from the Ohio and Potomac Rivers north into Canada and west to the edge of the Great Plains, also in humid parts of the Pacific states. The leaves are sessile each with 5 linear to oval leaflets. Stems are decumbent unless in fairly dense stands, reaching 20 to 40 or more inches in length. The plant has a deep, branched root system and tolerates both wet and moderately dry conditions. It is unusual among legumes in that it does not cause bloat in cattle. Both as pasture and as hay is highly palatable and nutritious. Harvested seed increased six-fold from 1949 to 1959, and in the latter year was sufficient to plant about 300,000 acres, according to 1960 CENSUS data. The 1992 CENSUS data for seed production is half that of the 1987 seed acreage, which was sufficient to plant 200,000 acres. Plant 4 to 6 pounds of seed per acre. There are 370,000 seeds per pound.

In general appearance Big trefoil resembles Birdsfoot trefoil, but its range of adaptation is quite different. It is much less winter hardy and adapted only to humid areas with mild winters. It is grown mostly in western Oregon but also is promising for the southeastern states. It tolerates submergence and grows well on wet, poorly drained soil. It spreads by underground stems. The root system is shallow. It can be grown in combination with sod-forming grasses, competing well with them. It is high in palatability, both as pasturage and as hay. Seed inoculation is important in establishing plantings. While less important nationally than birdsfoot trefoil, it is a valuable crop for special areas.

3. Season: Harvest for hay or silage in early June. Two to three hay cuttings per season. Graze up to early September. Seeded late August to September.
4. Production in U.S.: In 1992, 7,375 acres of seed produced with 260 tons (1992 CENSUS). In 1992, the top 3 seed producing states reported as Michigan (3,463 acres), Wisconsin (3,083), and Minnesota (644) (1992 CENSUS). For forage and hay, Ohio and Potomac Rivers north into Canada and west to the edge of the Great Plains. Also western Oregon.
5. Other production regions: In 1995, Canada reported 9,983 acres (ROTHWELL 1996a).
6. Use: Pasture, hay, silage, soil improvement and roadbank stabilization.
7. Part(s) of plant consumed: Foliage
8. Portion analyzed/sampled: Forage and hay. For trefoil forage, cut sample at 5 to 10 inch or early bloom stage, at approximately 30 percent dry matter. For trefoil hay, cut at first flower to full bloom. Hay should be field-dried to a moisture content of 10 to 20 percent
9. Classifications:
   a. Authors Class: Nongrass animal feeds
   b. EPA Crop Group (Group & Subgroup): Nongrass animal feeds (18)
   c. Codex Group: 050 (AL 1028) Legume animal feeds
   d. EPA Crop Definition: Alfalfa = Birdsfoot trefoil
10. References: GRIN, CODEX, MAGNESS, RICHE, ROTHWELL 1996a, US EPA 1996a, US EPA 1995a, BARNES
11. Production Map: EPA Crop Production Region 5 for seed production. Regions 1, 2 and 5 for forage.
12. Plant Codes:
    a. Bayer Code: LOTCO (*L. corniculatus*), LOTTE (*L. glaber*), LOTUL (*L. uliginosus*)

622

1. Triticale (trit-i-kay-lee)
Poaceae (Gramineae)
x *Triticosecale* sp. (syn: x *Triticosecale rimpaui* Wittm.; *Triticum aestivum* x *Secale cereale*)

2. Triticale is a small grain cross between wheat and rye. It is the first man-made commercial crop. It has the advantage of out yielding wheat and rye in many marginal crop production areas. The grain is mainly used for feed and the foliage as a forage crop. The grain is also used in bread, noodles, cakes and crackers. There are both spring and winter varieties. It has major developmental programs in Canada and Mexico. The lemma and palea thresh free from the grain. The grain weighs 45 to 50 pounds per bushel.
3. Season, seeding to harvest: About 100 to 120 days.
4. Production in U.S.: In 1992, 22,188 acres harvested with 639,818 bushels (1992 CENSUS). Most production as a grain is in the Sestern states. In Southern states, winter types are grazed in the fall. In 1992, the top 11 states reported as Washington (3,688 acres), Texas (3,587), Kansas (3,100), Wyoming (1,940), Montana (1,760), New Mexico (1,747), Colorado (917), North Dakota (828), Nebraska (807), South

Carolina (610) and Oregon (567) for 19,551 acres or 88 percent of the total acreage (1992 CENSUS).

5. Other production regions: Triticale is a major crop in Europe (PALLUTT 1996). In 1995, Canada reported 19,702 acres (ROTHWELL 1996a). Reported 4 million acres worldwide (MATZ 1991).

6. Use: In U.S., mainly used as feed. Also used in certain bread products and cereals. In Europe, the grain is also processed into flour for baked goods and cereal. In South America used for malt.

7. Part(s) of plant consumed: Seed and foliage

8. Portion analyzed/sampled: Grain, forage, straw; and its processed commodities; flour and bran, if grown for food purposes

9. Classifications:
   a. Authors Class: Cereal grains; Forage, fodder and straw of cereal grains
   b. EPA Crop Group (Group & Subgroup): Cereal grains (15); Forage, fodder, and straw of cereal grains (16)
   c. Codex Group: 020 (GC 0653) Cereal grains
   d. EPA Crop Definition: Wheat = Triticale

10. References: GRIN, CODEX, FACCIOLA, HARDIN, PALLUTT, RICHE, ROTHWELL 1996a, SCHREIBER, US EPA 1995a, MATZ, OELKE(c), VARUGHESE

11. Production Map: EPA Crop Production Regions 2, 7, 8, 9 and 11.

12. Plant Codes:
   a. Bayer Code: TTLSS, TTLSO (Summer triticale), TTLWI (Winter triticale)

623

1. Tufted hairgrass
(Salt and pepper grass)
Poaceae (Gramineae)
Deschampsia cespitosa (L.) P. Beauv.

2. This is a perennial cool season grass native to the U.S., and is an important widespread forage. There are 12 hairgrass species in North America. It is a member of the grass tribe *Aveneae*. It is distributed from Alaska through most of the northeast and western U.S. and throughout Canada. It is an erect bunchgrass (20 to 160 cm tall) with a panicle 10 to 25 cm long that is adapted to many soil types and habitats, including boggy areas, prairies, stream banks and wet meadows. Growth starts early in the spring and flowers July to September. Seed matures in August through September. Tufted hairgrass produces good quality pasture for all livestock and withstands close grazing. It is widely used in Alaska for revegetation, forage and a low maintenance ground cover.

3. Season: Cool season, growth starts in early spring, and matures in September. It is very palatable and can be grazed closely or cut for hay.

4. Production in U.S.: No data for Tufted hairgrass productions, however, pasture and rangeland are produced on more than 410 million A (U.S. CENSUS, 1992). See above.

5. Other production regions: Grown throughout Canada.

6. Use: Rangeland grazing, hay, pasture, revegetation and ground cover for erosion control.

7. Part(s) of plant consumed: Leaves and stems

8. Portion analyzed/sampled: Forage and hay

9. Classifications:

   a. Authors Class: See Forage grass
   b. EPA Crop Group (Group & Subgroup): See Forage grass
   c. Codex Group: See Forage grass
   d. EPA Crop Definition: None

10. References: GRIN, US EPA 1994, US EPA 1995, CODEX, HOLZWORTH, STUBBENDIECK (PICTURE), MOSLER, RICHE

11. Production Map: EPA Crop Production Regions 1, 5, 9, 10, 11 and 12.

12. Plant Codes:
   a. Bayer Code: DECCA

624

1. Turmeric
(Turmerico, Indian saffron)
Zingiberaceae
Curcuma longa L. (syn: *C. domestica* Valeton)

2. The plant is a large-leaved herb, closely related to ginger, which see. It is cultivated in tropical countries for the thick, rounded, underground rhizomes, which constitute the spice, turmeric. Turmeric contains an oil, which consists in part of curcumin, which on oxidation is changed into vanillin, the active principle in vanilla. The rootstocks of turmeric, both fresh and dried, are also used as flavoring in curries and other cookery. Turmeric is a perennial crop grown as an annual.

3. Season, planting to harvest: 7-10 months.

4. Production in U.S.: No data.

5. Other production regions: Turmeric grows best in a tropical, humid climate. It is grown in Haiti, Jamaica and India.

6. Use: Natural coloring agent in processed cheese, an essential ingredient of all types of curry. Also a flavoring of poultry and rice dishes. Yellow coloring agent.

7. Part(s) of plant consumed: Root

8. Portion analyzed/sampled: Root (Fresh and dried)

9. Classifications:
   a. Authors Class: Root and tuber vegetables; Herbs and spices
   b. EPA Crop Group (Group & Subgroup): Root and tuber vegetables (1C, 1D)
   c. Codex Group: No specific entry
   d. EPA Crop Definition: None

10. References: GRIN, FARRELL (picture), LEWIS (picture), MAGNESS, US EPA 1995a, USDA NRCS, DUKE(a)

11. Production Map: EPA Crop Production Region 13.

12. Plant Codes:
   a. Bayer Code: CURLO

625

1. Turnip
(Garden turnip, Rappina, Rappone, Namenia, Tendergreen, Turnip greens, Nabo forrajero, Nabo-colza, Chou rave, Nabo, Rabano)
Brassicaceae (Cruciferae)
Brassica rapa L. ssp. *rapa* (syn: *B. campestris* L. (Rapifera group); *B. campestris* ssp. *rapifera* (Metzger) Sinsk.)

1. Turnip, Seven-top
(Japanese greens, Turnip greens, Rappone, Namenia, Rapini, Turnip tops)
Brassica rapa L. ssp. *rapa* (syn: *B. septiceps* (L.H. Bailey) L.H. Bailey; *B. rapa* var. *septiceps* L.H. Bailey)

2. The turnip is grown from seed as an annual, used both for the enlarged fleshy root, and for the leaves as potherbs. The plant first forms a rosette of thin, hairy leaves on slender petioles. The root soon enlarges to a globular or generally flattened tuberous tissue, which is tender but later becomes tougher and somewhat fibrous. Roots are generally harvested when 3 inches or less in diameter. Certain varieties, such as 'Seven-top', 'Topper' and 'Italian kale', are grown primarily for the leaves. Plants should grow rapidly for best quality, both for greens and roots. Turnip varieties grown for tops only can be harvested 1 to 3 times per season with first cutting about 1 month after seeding. There are 14,000 seeds per ounce.

3. Season, seeding to harvest: For greens about 4 to 7 weeks. For roots about 8 to 14 weeks.

4. Production in U.S.: 9,256 acres for roots; 10,034 acres for greens (1992 CENSUS). In 1992, turnip roots reported from the top 10 states as Georgia (1,127 acres), California (1,117), North Carolina (777), Texas (679), New Jersery (662), Oregon (443), South Carolina (403), Michigan (338), Tennessee (319) and Indiana (255) with about 66 percent of the U.S. acreage. Turnip greens reported from the top 10 states as Tennessee (2,037 acres), Georgia (1,622), South Carolina (807), Oklahoma (669), Alabama (661), Texas (630), Arkansas (546), North Carolina (410), Illinois (364) and Ohio (318) with about 80 percent of the U.S. acreage (1992 CENSUS).

5. Other production regions: In 1995, Canada grew turnips (ROTHWELL 1996a).

6. Use: Roots as cooked vegetable; leaves as potherbs or fresh greens. Leaves also canned and frozen commercially. Turnip forage is grazed for both the tops and roots; Washington grows 2,000 to 4,000 acres of forage on the average (SCHREIBER 1995). Preparation: Peel and cube the root before cooking or serve raw in salads (LOGAN 1996).

7. Part(s) of plant consumed: Roots, usually peel removed; whole leaves

8. Portion analyzed/sampled: Roots and tops (leaves). Analyze separately

9. Classifications:
   a. Authors Class: Root and tuber vegetables; Leaves of root and tuber vegetables. Also, *Brassica* (cole) leafy vegetables for tops
   b. EPA Crop Group (Group & Subgroup): Root and tuber vegetables (1A and 1B); Leaves of root and tuber vegetables (2) (Tops are representative crop)
   c. Codex Group: 013 (VL 0506) Leafy vegetables (including *Brassica* leafy vegetables) for tops; 016 (VR 0506) Root and tuber vegetables for roots; 052 (AM 0506) Fodder; 052 (AV 0506) Forage
   d. EPA Crop Definition: Turnip tops or greens = Broccoli raab (raab, raab salad), hanover salad

10. References: GRIN, CODEX, LOGAN (picture), MAGNESS, RICHE, ROTHWELL 1996a, SCHREIBER, US EPA 1996a, US EPA 1995a, US EPA 1994, YAMAGUCHI 1983, USDA NRCS, IVES, REHM, SWIADER, CRAMER

11. Production Map: EPA Crop Production Regions: Turnip roots: 1, 2, 3, 4, 5, 6, 10, 11 and 12; Turnip tops: 2, 3, 4, 5, 6, 8 and 10.

12. Plant Codes:
    a. Bayer Code: BRSRR (Root), BRSRE (Top)

626

1. Tyfon
(Holland greens, *Brassica* forage, Colbaga (see text))
Brassicaceae (Cruciferae)
   Brassica rapa L. 'Tyfon'

2. A cross between Chinese cabbage and stubble turnip. Fast growing crop, can be grown in all parts of U.S. and very winter hardy. Professor David W. Koch, University of New Hampshire agronomist, has experimented with tyfon. Another similar cross is colbaga which is a cross between Chinese cabbage and rutabaga and grown and stored like rutabagas. Developed by Professor E.M. Meader at the University of New Hampshire. In 1991, tyfon was listed by Gurney Seed and Nursery Co., Yankton, SD 57079.

3. Season, seeding to harvest: 40 days with additional cuttings at 30 to 40 day intervals.

4. Production in U.S.: Grown in New Hampshire as a forage crop (IR-4 REQUEST DATABASE 1996).

5. Other production regions: No entry.

6. Use: Leaves can be used in salads or as a cooked vegetable like spinach. Leaves and roots can be used as forage, especially for sheep. As a forage *Brassica*, it is used as a fall grazing season extender, and in some cases during the summer. Other forage *Brassica* are turnips, kale, rape greens and rutabagas.

7. Part(s) of plant consumed: Leaves and roots

8. Portion analyzed/sampled: Leaves and roots

9. Classifications:
   a. Authors Class: Root and tuber vegetables; Leaves of root and tuber vegetables
   b. EPA Crop Group (Group & Subgroup): Miscellaneous
   c. Codex Group: No specific entry
   d. EPA Crop Definition: None

10. References: GRIN, BOWMAN, FACCIOLA, IR-4 REQUEST DATABASE 1996

11. Production Map: EPA Crop Production Region 1 (and other parts of the U.S. in the future).

12. Plant Codes:
    a. Bayer Code: No specific entry

627

1. Udo
(Spikenard, Oudo, Japanese asparagus, Tu-huo)
Araliaceae
   Aralia cordata Thunb. (syn: *A. edulis* Siebold & Zucc.)

2. Udo is a vegetable grown for its tender, etiolated spring shoots, somewhat like asparagus. It is grown in Japan and to a limited extent by Oriental gardeners in the U.S. The plant is a strong-growing perennial, producing the edible shoots from the roots each spring. The summer growth reaches 4 to 8 feet in height, with large, compound pinnate leaves. In culture, the roots are established in beds or rows, like asparagus. As the young shoots start in spring, they are kept covered with soil for complete blanching. The shoots harvested are up to 18 inches long and $1^1/_2$ inches diameter at the base. Prior to use, shoots are boiled in salt water, or are sliced and held in cold water, to remove a turpentine-like resin. They are then eaten raw as a salad or cooked.

3. Season, from growth start in spring to harvest: 2 to 4 weeks. Propagation is by stem or root cuttings or by seed.

4. Production in U.S.: No data, very limited. (BAILEY 1976).
5. Other production regions: Grown in Japan (BAILEY 1976). Native to Asia.
6. Use: As raw salad or cooked vegetable. Prepare as asparagus or for soups.
7. Part(s) of plant consumed: Young etiolated (blanched) stems or shoots only
8. Portion analyzed/sampled: Young stems
9. Classifications:
   a. Authors Class: Stalk and stem vegetables
   b. EPA Crop Group (Group & Subgroup): Miscellaneous
   c. Codex Group: No specific entry
   d. EPA Crop Definition: None
10. References: GRIN, BAILEY 1976, FACCIOLA, MAGNESS, RUBATZKY
11. Production Map: No entry.
12. Plant Codes:
    a. Bayer Code: ARLCO

628

1. Ulluco (oo-yoo-koh)
(Ullucu, Papalisa, Olluco, Melloco, Ulluca, Ullugue, Ruba)
Basellaceae

Ullucus tuberosus Caldas (syn: *U. tuberosis* Caldas, *U. kunthii* Moq., *Basella tuberosa* (Caldas) Kunth, *Melloca tuberosa* (Caldas) Lindl., *M. peruviana* Lindl.)

2. Ullucu is a low growing herb. Alternate heart-shaped leaves are borne on long petioles. Normally propagated by small tubers. The tubers are dug by hand and up to 3 inches long.
3. Season, planting to harvest: As short as 5 months, but normally 6 to 8 months.
4. Production in U.S.: No data.
5. Other production regions: Andes of South America (NATIONAL RESEARCH COUNCIL 1989).
6. Use: Tubers cooked like potatoes. The green leaves are nutritious and used in salad or as a potherb. Ullucu is related to malabar spinach.
7. Part(s) of plant consumed: Tuber and leaves
8. Portion analyzed/sampled: Tuber and leaves
9. Classifications:
   a. Authors Class: Root and tuber vegetables; Leaves of root and tuber vegetables
   b. EPA Crop Group (Group & Subgroup): Miscellaneous
   c. Codex Group: 016 (VR 0599) Root and tuber vegetables
   d. EPA Crop Definition: None
10. References: GRIN, CODEX, NATIONAL RESEARCH COUNCIL 1989 (picture), YAMAGUCHI 1983, RUBATZKY
11. Production Map: No entry.
12. Plant Codes:
    a. Bayer Code: ULLTU

629

1. Vanilla
(Vanilla bean, Vainilla, Vanille, Tahiti bean)
Orchidaceae

Vanilla planifolia Jacks. (syn: *V. fragrans* (Salisb.) Ames)

2. Vanilla is a tall, climbing plant, with thick, oblong-lanceolate leaves, belonging to the orchid family. Vanilla is native to the American tropics, but is now grown in most tropical areas. The vines are trained on supports. The spice is obtained from the pods, which are harvested while green then dried and fermented for 3 to 5 weeks. The vanillin crystallizes on the surface of the pods which are 5 to 6 inches long. Importance of natural vanilla for flavoring foods has decreased greatly, due to development of a far cheaper synthetic product. U.S. imported about 1,388 tons of vanilla beans in 1992 with the majority of it coming from Indonesia and Madagascar (USDA 1993a).

3. Season, planting to first blossoms: About 1½ years. Blossom to harvest: Almost 9 months. It continues to bear annually for 30 to 40 years.
4. Production in U.S.: Limited acreage in Hawaii (Tahiti bean or Tahitian vanilla, *V. tahitensis* J.W. Moore) (NEAL 1995) (FACCIOLA 1990).
5. Other production regions: Grows in Mexico (NEAL 1965) (BAILEY 1976). Also see above.
6. Use: Spice.
7. Part(s) of plant consumed: Whole bean and extract
8. Portion analyzed/sampled: Whole bean or pod
9. Classifications:
   a. Authors Class: Herbs and spices
   b. EPA Crop Group (Group & Subgroup): Herbs and spices (19B)
   c. Codex Group: 028 (HS 0795) Spices
   d. EPA Crop Definition: None
10. References: GRIN, BAILEY 1976, CODEX, FACCIOLA, MAGNESS, NEAL (picture), PENZEY, USDA 1993a, US EPA 1995a, IVES
11. Production Map: EPA Crop Production Region 13.
12. Plant Codes:
    a. Bayer Code: VANPL

630

1. Vaseygrass
(Giant paspalum, Upright paspalum)
Poaceae (Gramineae)

Paspalum urvillei Steud.

2. This is a warm season perennial bunchgrass quite similar to Dallisgrass and introduced from South America before 1880. It is now common in the humid Southeast, especially the Gulf Coast. It thrives both on wet sites and on well-drained soils. The erect stems may reach to 6 feet, with leaf blades up to 15 inches long and half an inch wide. Vaseygrass is seldom planted but where growing is utilized for pasture and hay. It has good hay quality until mature. A minor crop in the U.S., it can withstand drought and tolerates wet soils. It is easily overpastured. As hay, palatability is good and yields heavy with up to four cuttings per year possible. There are 970,000 seeds/kg.

3. Season, warm season pasture.
4. Production in U.S.: No specific data, See Forage grass. Southeast, especially Gulf Coast.
5. Other production regions: South America.
6. Use: Pasture, hay.
7. Part(s) of plant consumed: Foliage
8. Portion analyzed/sampled: Forage and hay
9. Classifications:
   a. Authors Class: See Forage grass
   b. EPA Crop Group (Group & Subgroup): See Forage grass
   c. Codex Group: See Forage grass

d. EPA Crop Definition: None
10. References: GRIN, MAGNESS, SKERMAN, ALDERSON, STUBBENDIECK (picture)
11. Production Map: EPA Crop Production Regions 2, 3, 4, 6 and 13.
12. Plant Codes:
   a. Bayer Code: PASUR

631

1. Velvet bean
(Cowage velvetbean, Mauritius velvetbean, Bengal velvetbean, Florida velvetbean, Cowitch, Benguk, Lyon bean, Yokohama velvetbean, Florida velvetbean, Bengal bean)
Fabaceae (Leguminosae)
  Mucuna pruriens var. *utilis* (Wall. ex Wight) Baker ex. Burck (syn: *M. aterrima* (Piper & Tracy) Holland, *M. deeringiana* (Bort.) Merr., *Stizolobium deeringianum* Bort.)
2. This is a strong-growing annual plant native to the tropics. Most of the several varieties grown in the U.S. are *M. pruriens* although some kinds are of other *Mucuna* species or interspecies hybrids. The slender stems may grow to 30 feet in some kinds. They are mainly grown with a support crop, usually corn, on which they climb and are partially supported. The leaves are trifoliate, with large ovate leaflets. Pods are pubescent, up to 6 inches long, with three to six seeds per pod. Velvet beans are well adapted to sandy soils and require a long growing season to produce much pasturage. They are grown mainly in the southeastern Coastal Plain for late summer pasturage and soil improvement. All parts of the plant are nutritious and palatable to livestock.
3. Season, long season to produce pasture, late summer.
4. Production in U.S.: No data. Southeastern Coastal Plain.
5. Other production regions: Tropics.
6. Use: Pasture. Also the young pods can be eaten as snap beans. The plant has showy flowers and also is grown as an ornamental.
7. Part(s) of plant consumed: Foliage
8. Portion analyzed/sampled: Forage and hay
9. Classifications:
   a. Authors Class: Nongrass animal feeds
   b. EPA Crop Group (Group & Subgroup): Nongrass animal feeds (18)
   c. Codex Group: 050 (AL 1022) Legume animal feeds
   d. EPA Crop Definition: None
10. References: GRIN, CODEX, FACCIOLA, MAGNESS, US EPA 1995a, DUKE(h)
11. Production Map: EPA Crop Production Region 2.
12. Plant Codes:
   a. Bayer Code: MUCAT

632

1. Velvetgrass
Poaceae (Gramineae)
  Holcus spp.
1. Velvetgrass
(Yorkshire fog)
  H. lanatus L.
1. German velvetgrass
(Creeping velvetgrass, Creeping softgrass)
  H. mollis L.

2. Velvetgrasses are cool season perennial bunchgrasses that are widely distributed over the western and eastern U.S. They range in height from 25 to 100 cm, and are usually an indication of poorly managed pastures. They have relatively low palatability, but can produce two cuttings of hay per year.
3. Season: Cool season, growth is in the early spring, and matures in August. They are usually only a fair quality forage.
4. Production in U.S.: No production data for any of the velvetgrasses is available, however, pasture and rangeland are produced on more than 410 million acres (U.S. CENSUS, 1992). See above.
5. Other production regions: Europe (GOULD 1975).
6. Use: Pasture, hay, and for erosion control.
7. Part(s) of plant consumed: Leaves and stems
8. Portion analyzed/sampled: Forage and hay
9. Classifications:
   a. Authors Class: See Forage grass
   b. EPA Crop Group (Group & Subgroup): See Forage grass
   c. Codex Group: See Forage grass
   d. EPA Crop Definition: None
10. References: GRIN, US EPA 1994, US EPA 1995, CODEX, HOLZWORTH, MOSLER, HITCHCOCK (picture), RICHE, GOULD
11. Production Map: EPA Production Regions for Velvetgrasses are 2, 3, 4, 5, 6 and 8.
12. Plant Codes:
   a. Bayer Code: HOLLA (*H. lanatus*); HOLMO (*H. mollis*); HOLSS (*H.* spp.)

633

1. Vernonia
(Ironweed)
Asteraceae (Compositae)
  Vernonia galamensis (Cass.) Less. (syn: *V. pauciflora* (Willd.) Less.)
2. Perennial herbs in North America. Leaves are alternate. Flowers are usually purple and in terminal clusters in late summer.
3. Season, planting: Easily cultivated and propagated by division, seeds or cuttings.
4. Production in U.S.: Limited acreage in Puerto Rico and Arizona (SENFT 1994).
5. Other production regions: Native to equatorial Africa (SENFT 1994). Grown in tropical Africa and Central America (1992 YEARBOOK OF AGRICULTURE).
6. Use: Oil used as drying agent in paints and plastics. Further research is needed on using the defatted meal for livestock feed.
7. Part(s) of plant consumed: Seed (potential with feed use)
8. Portion analyzed/sampled: Seed and its meal, if it is used for feed
9. Classifications:
   a. Authors Class: Oilseed
   b. EPA Crop Group (Group & Subgroup): Miscellaneous
   c. Codex Group: No specific entry
   d. EPA Crop Definition: None
10. References: GRIN, BAILEY 1976, SENFT 1994 (picture), ROBBELEN 1989, YEARBOOK OF AGRICULTURE 1992 (picture)
11. Production Map: No entry.

12. Plant Codes:
   a. Bayer Code: No specific entry, genus entry is VERSS

634

1. Vetch
(Wicke, Vesce)
Fabaceae (Leguminosae)
Vicia spp.
1. Purple vetch
(Reddish tufted vetch)
V. benghalensis L. (syn: *V. atropurpurea* Desf.)
1. Hungarian vetch
V. pannonica Crantz
1. Common vetch
(Tare, Veza forrajera, Spring vetch, Oregon vetch, White vetch)
V. sativa L. ssp. *sativa*
1. Hairy vetch
(Smooth vetch, Madison vetch, Winter vetch, Russian vetch, Downy vetch, Fodder vetch, Vesce velue, Sand vetch)
V. villosa Roth ssp. *villosa*

2. The vetches are weak-stemmed, semi-vining plants with pinnate leaves terminating in tendrils. Some 150 species are known, about 25 of which are native in the U.S. However, the species grown agriculturally here are all introduced, being native to Europe or western Asia. The vetches are extensively used as green manure for soil improvement, hay, and in the South for winter pasture. Vetch seed is harvested from more than 100,000 acres annually (112,956 acres in 1959, CENSUS figure) in the U.S., sufficient to seed around 1,000,000 acres. In 1992, vetch seed was harvested from 15,334 acres with 6,593,950 pounds of seed (1992 CENSUS). Vetches may become troublesome weeds in grain fields but are readily controlled with herbicides. Tares as mentioned in the Bible are believed to have been common vetch.

In general **purple vetch** resembles hairy vetch, with pubescent stems and pods. It is, however, the least winter hardy of the commonly grown vetches, and temperatures below 20 degrees F. generally cause injury. It is grown mainly in milder sections of California where it is a hay crop and a major cover crop. Stems are up to 30 inches or more in length, prostrate or climbing. Leaves consist of five to eight pairs of linear to oblong-linear leaflets. It is grown but little in colder climates where spring seeding would be necessary.

Hungarian vetch is intermediate in winter hardiness between hairy and common. Its special virtue is tolerance to wet soils. It is mainly grown as a hay crop on heavy soils in western Oregon where winter precipitation is heavy. Plants are pubescent. Leaves consist of four to eight pairs of linear-to-oblong leaflets. In Oregon it is commonly fall seeded with a grain crop for partial support. It produces a palatable and nutritious hay on soils poorly adapted for other vetches.

Common vetch is less winter hardy than hairy. When fall seeded, winter injury often occurs at temperatures below 10 degrees F. The plants are sparingly pubescent with procumbent stems up to 3 feet or more. Leaves consist of up to seven pairs of elliptic or oblong leaflets. Common vetch is grown both as a soil improvement crop to prevent erosion and for hay. Common vetch also used in California as a cover crop and is being replaced mostly by Purple vetch. It is cut for hay in early bloom stage and can be seeded with small grains. For hay or seed production it is usually planted with a grain crop, as oats or wheat, to support the vetch plants. Seeding is in the fall in mild climates, in spring in cold areas. Common vetch produces a palatable hay, especially for cattle. Good winter and spring grazing is provided in mild climates. 'Williamette', 'Warrior' and 'Doark' are important varieties of common vetch.

Hairy vetch is the most winter hardy of the vetches, enduring below 0 degrees F. winter temperatures. It is generally grown as a winter annual. Each leaf consists of about 10 pairs of elliptic-oblong leaflets. The weak stems reach to 4 feet and are procumbent unless supported. Stems and pods of common hairy vetch are pubescent. Smooth vetch is a strain of hairy with stems and pods smooth or nearly so. Although a little less hardy than the pubescent strain, it is the type now largely grown. 'Madison' is a strain well suited to the Midwest as well as the South. Hairy vetch, including strains, is used mainly as a green manure crop, being seeded in the fall and turned into the soil in the spring. Such use is general throughout the Corn Belt, in the Southeast, and to some extent in Western orchards. In the South, winter pasturing is often practiced. Little hairy vetch is harvested for hay as other species are more suitable. Hairy vetch is grazed as a pasture in May and June and often grown with small grains for forage. It can be cut for hay when pods are in the soft dough stage.

3. Season: In southern U.S. vetch flowers April-May and seeds mature in late May to June.

4. Production in U.S.: In 1992, 15,334 acres of seed crop with 6,593,950 pounds (1992 CENSUS). In 1992, the top five seed-producing states were Oregon (6,682 acres), California (1,565), Nebraska (1,246), Minnesota (1,090) and Pennsylvania (1,061) (1992 CENSUS). As a pasture, hay or cover crop it is grown primarily in southern, midwest and western coastal states.

5. Other production regions: See above.

6. Use: See above; Pasture, hay, cover crop, silage, green manure crop. Also grown as an ornamental.

7. Part(s) of plant consumed: Foliage

8. Portion analyzed/sampled: Forage and hay. For vetch forage, cut sample at 6 inch to prebloom stage, at approximately 30 percent dry matter. For vetch hay, cut at early bloom stage to when seeds in the lower half of the plant are approximately 50 percent developed. Hay should be field-dried to a moisture content of 10 to 20 percent. Vetch does not include crownvetch

9. Classifications:
   a. Authors Class: Nongrass animal feeds
   b. EPA Crop Group (Group & Subgroup): Nongrass animal feeds (18)
   c. Codex Group: 050 (AL 1029) Legume animal feeds
   d. EPA Crop Definition: None

10. References: GRIN, CODEX, MAGNESS, RICHE, US EPA 1996a, US EPA 1995a, BALL, DUKE(h), UNDERSANDER, BARNES

11. Production Map: EPA Crop Production Regions 2, 5, 10 and 12 for seed crop.

12. Plant Codes:
   a. Bayer Code: VICSS (*V.* spp.), VICBE (*V. benghalensis*), VICPA (*V. pannonica*), VICSA (*V. sativa*), VICVI (*V. villosa*)

635

1. Vetch, Chickling
(Grass pea, Chickling pea, Riga pea, White peavine, Chickling vetchling, Dogtooth pea, Khesari, Wedge peavine, Grass peavine, Vetchling, Wild pea, Lathyrus)
Fabaceae (Leguminosae)
Lathyrus sativus L.
2. Grows in generally poor soil conditions. The plant is much branched, herbaceous annual. Native to southern Europe and southwest Asia and introduced to southern U.S. Adapted to cotton producing areas of the South and better on clay soils with a pH of 5.5 to 8.2 than many other annual legumes.
3. Season, seeding to harvest: No data.
4. Production in U.S.: Cultivated in Alabama, Louisiana and Mississippi.
5. Other production regions: Grown in Canada (ROTHWELL 1996a).
6. Use: Seed for human food. Leaves for potherb. Mainly for feed.
7. Part(s) of plant consumed: Seed, leaves and plant for livestock feed which is not a significant feed item
8. Portion analyzed/sampled: Forage and hay. Forage cut at prebloom with approximately 30 percent dry matter. Hay cut a bloom stage with seeds 50 percent developed on lower half of plants and field dry to 80 to 90 percent dry matter
9. Classifications:
   a. Authors Class: Nongrass animal feeds
   b. EPA Crop Group (Group & Subgroup): Miscellaneous
   c. Codex Group: 050 (AL 1029) Legume animal feeds as vetch; 050 (AL 5235) as chickling vetch
   d. EPA Crop Definition: None
10. References: GRIN, CODEX, ROTHWELL 1996a, US EPA 1996a, YAMAGUCHI 1983, USDA NRCS, BARNES, PUTNAM 1993
11. Production Map: EPA Crop Production Regions 2 and 4.
12. Plant Codes:
    a. Bayer Code: LTHSA

636

1. Vetch, Milk
(Chickpea milkvetch, Garbanzo-silvestre)
1. Cicer milkvetch
Fabaceae (Leguminosae)
Astragalus cicer L.
2. Cicer milkvetch is a hardy, herbaceous long-lived cool season perennial that spreads by rhizomes up to 1.3 m long, that is native to southern Europe and was introduced into the U.S. in the 1920s for use as a pasture, hay or soil conservation crop. It is adapted to dry meadows, orchards and woodlands receiving 40 cm annual precipitation in the Great Plains, western U.S. and Canada. It is a valuable forage for range and pasture plantings and for erosion control. Yields about 75 to 80 percent of that for alfalfa. Cicer milkvetch reproduces by seeds or rhizomes, and seed yields are 500 kg/ha. There are 260,000 seeds/kg. Another promising related species is Sicklepod milkvetch, *A. falcatus* Lam.
3. Season: Cool season long-lived perennial forage that can be used as an pasture and soil conservation crop.
4. Production in U.S.: No specific data for Cicer milkvetch production, however, pasture and rangelands are produced on

more than 410 million acres (U.S. CENSUS 1992). See above.
5. Other production regions: Temperate region of Northern Hemisphere (BAILEY 1976). In 1995 Canada reported 128 acres (ROTHWELL 1996a).
6. Use: Pasture and rangeland grazing, hay, and revegetation of eroded areas, soil conservation crop.
7. Part(s) of plant consumed: Leaves and stems (foliage)
8. Portion analyzed/sampled: Forage and hay
9. Classifications:
   a. Authors Class: Nongrass animal feeds
   b. EPA Crop Group (Group & Subgroup): Nongrass animal feeds (forage, fodder, straw and hay) Group 18
   c. Codex Group: 050 (AL 1029) Legume animal feeds
   d. EPA Crop Definition: None
10. References: GRIN, US EPA 1994, US EPA 1995, CODEX, HOLZWORTH, DUKE(h) (picture), BARNES, ROTHWELL 1996a, RICHE, BAILEY 1976, JANICK 1990
11. Production Map: EPA Crop Production Regions 7, 8, 9 and 11.
12. Plant Codes:
    a. Bayer Code: ASACI

637

1. Vetch, Minor
Fabaceae (Leguminosae)
Vicia spp.
1. Monantha vetch
(Single-flowered vetch, One-flowered vetch)
V. articulata Hornem.
1. Bard vetch
(Barn vetch)
V. monantha Retz.
1. Narrowleaf vetch
(Blackpod vetch, Spring vetch)
V. sativa ssp. *nigra* (L.) Ehrh. (syn: *V. augustifolia* L.)
1. Horsebean
(Field bean, Haba cabalar)
V. faba var. *equina* Pers.
1. Woolypod vetch
(Winter vetch, Hairy vetch, Large Russian vetch)
V. villosa ssp. *varia* (Host) Corb. (syn: *V. dasycarpa* Ten.)
2. Additional vetch species are grown to a limited extent in the U.S. Monantha vetch, *V. articulata* can be grown under the same conditions as common vetch. Plants are smooth or nearly so. The light lavender colored flowers are borne singly. It is distinguishable mainly by flat seeds. Bard vetch, *V. monantha* is much restricted in adaptation. It is grown only in irrigated areas of southern California and Arizona. It is very similar to monantha vetch in appearance, but seeds are oval to round. Narrowleaf vetch, *V. sativa* is adapted to soil and climatic conditions like those of common vetch. It is early maturing, ripening seed from spring plants in the north and is characterized by the narrow leaflets. It is suitable for pasturage throughout the Cotton Belt and used as a winter cover and green manure crop. It can be a high-quality forage and grows 4 to 20 inches tall. It flowers in March-October and fruits May-November. It may be a troublesome weed in spring wheat. Horse bean, *V. faba*, is grown as a feed crop in the U.S. only in coastal valleys in California, and now much

less than previously. The upright-growing, near glabrous plants have large leaves with broad, oval leaflets. Weevils infesting the seeds are responsible for reduced usage of horsebean as a feed crop. Woolypod vetch, *V. villosa* ssp. *varia*, is quite similar to hairy or smooth vetch, *V. villosa* ssp. *villosa*, but is slightly less winter hardy and grows at slightly lower winter temperatures. It is a valuable winter vetch for the Cotton Belt. Also grown in California as cover crops.

3. Season, See Vetch.
4. Production in U.S.: No data, See Vetch.
5. Other production regions: See Vetch.
6. Use: See above; cover crop, pasture, hay.
7. Part(s) of plant consumed: Foliage
8. Portion analyzed/sampled: Forage and hay
9. Classifications:
   a. Authors Class: Nongrass animal feeds
   b. EPA Crop Group (Group & Subgroup): Nongrass animal feeds (18)
   c. Codex Group: 050 (AL 1029) Legume animal feeds
   d. EPA Crop Definition: None
10. References: GRIN, CODEX, MAGNESS, US EPA 1995a, BALL, DUKE(h). BARNES
11. Production Map: EPA Crop Production Regions 2, 5, 10 and 12.
12. Plant Codes:
    a. Bayer Code: VICSS (*V.* spp.), VICAR (*V. articulatea* and *V. monantha*), VICFE (*V. faba*), VICAN (*V. sativa*), VICVV (*V. villosa*)

638

1. Vietnamese coriander
(Rau Ram, Coriander-Vietnamese, Marshpepper smartweed)
Polygonaceae
Polygonum odoratum Lour. (syn: *Persicaria odorata* (Lour.) Sojak)

2. A tender perennial herb with red stems and green lanceolate leaves marked with red and closely resembling the European water-pepper or marshpepper smartweed (*P. hydropiper* L.).
3. Season, planting: Propagated by cuttings.
4. Production in U.S.: No data. Grown in southern U.S. near Vietnamese communities (KUEBEL 1988).
5. Other production regions: Southeast Asia (KUEBEL 1988).
6. Use: To garnish meat dishes, especially fowl. Ingredient of Du'a Can, a pickled dish resembling sauerkraut.
7. Part(s) of plant consumed: Leaves
8. Portion analyzed/sampled: Leaves (fresh and dried)
9. Classifications:
   a. Authors Class: Herbs and spices
   b. EPA Crop Group (Group & Subgroup): Miscellaneous
   c. Codex Group: No specific entry
   d. EPA Crop Definition: None
10. References: GRIN, KUEBEL, MARTIN 1979 (*P. hydropiper*)
11. Production Map: No entry
12. Plant Codes:
    a. Bayer Code: No specific entry

639

1. Violet
Violaceae
Viola spp.

1. Sweet violet
(English violet, Waioleka, Garden violet, Florist violet)
V. odorata L.
1. Pansy
(Wild violet, Johnny-jump-up, Heartsease, Wild pansy, Field pansy)
V. tricolor L.

2. Perennial herbs to about 6 inches tall. Leaves cordate-ovate to reniform and toothed. Flowers are fragrant. *V. tricolor* is a short-lived perennial to 12 inches tall.
3. Season: Propagated by seed.
4. Production in U.S.: No data.
5. Other production regions: Naturalized in North America (GARLAND 1993).
6. Use: Flower extract used for flavoring in cream puddings, jellies, ices, and liqueurs. Flowers used in salads and to make a tea. Leaves are dried to make a tea and young leaves in salads. Also grown as an ornamental bedding plant.
7. Part(s) of plant consumed: Mainly flowers; leaves
8. Portion analyzed/sampled: Flowers (fresh and dried); Leaves (fresh and dried)
9. Classifications:
   a. Authors Class: Herbs and spices
   b. EPA Crop Group (Group & Subgroup): Miscellaneous
   c. Codex Group: No specific entry
   d. EPA Crop Definition: None
10. References: GRIN, BAILEY 1976, FACCIOLA, GARLAND (picture), NEAL, SCHNEIDER 1994c
11. Production Map: No entry.
12. Plant Codes:
    a. Bayer Code: VIOSS (*V.* spp.), VIOOD (*V. odorata*), VIOTR (*V. tricolor* – Wild pansy), VIOAR (*V. tricolor* – Field pansy)

640

1. Walnut, Black
Juglandaceae
Juglans spp.
1. Walnut, American black
(Eastern black walnut)
J. nigra L.
1. Walnut, Northern California black
(Hinds' black walnut, Hand's black walnut, Northern California walnut)
J. hindsii Jeps. ex R.E. Sm. (syn: *J. californica* var. *hindsii* Jeps.)
1. Walnut, Texas black
(Little walnut, Nogal, River walnut)
J. microcarpa var. *microcarpa* Berland

2. Black walnuts in the U.S. are widely distributed and are of several species, not all of which are listed. The trees vary from small, up to 20 feet, to very large, near 100 feet in height. All have long, compound leaves, with up to 20 or more leaflets. Leaflets are generally oblong-lanceolate, and smooth. In all, the nuts are encased in a semi-pulpy husk, which does not separate from the nut readily. The nut shell is thick and very hard. The kernel does not separate from the shell readily. The American or Eastern black walnut is in limited commercial cultivation. Substantial quantities are gathered from native trees and marketed, mainly after shelling. About 80 percent by weight of nut is shell. Grounded

shells are used in industry as an abrasive, and in oil and gas drillings.

3. Season, harvest: Nuts ripen in September and October.
4. Production in U.S.: Most are harvested from the wild in the eastern and central U.S. As of 1993, Hammons Products Co. in Stockton, Missouri was the only company to buy and process eastern black walnuts. In 1994, California reported 117 acres of Black walnuts (MELNICOE 1996e).
5. Other production regions: No entry.
6. Use: Primarily used in bakery products, ice cream and candies. Also used as an ornamental tree and for its fine-grain wood as a timber product.
7. Part(s) of plant consumed: Nutmeat and extracted oil
8. Portion analyzed/sampled: Nutmeat
9. Classifications:
   a. Authors Class: Tree nuts
   b. EPA Crop Group (Group & Subgroup): Tree nuts
   c. Codex Group: 022 (TN 0678) Tree nuts (*J. nigra*)
   d. EPA Crop Definition: None
10. References: GRIN, FACCIOLA, GREAT LAKES FRUIT GROWERS NEWS 1993, MAGNESS, USDA NRCS, US EPA 1995a, US EPA 1995b, US EPA 1994, WOODROOF(b) (picture), MELNICOE 1996e
11. Production Map: EPA Crop Production Regions 5 and 10. (EPA reports walnuts as Black walnuts and English walnuts with 98 percent grown in Region 10).
12. Plant Codes:
   a. Bayer Code: IUGMI (*J. microcarpa*), IUGNI (*J. nigra*)

641

1. Walnut, English

(California walnut, Persian walnut, Madeira nut, Nogal, Carpathian walnut, Nogal comum, Hu-tao)

Juglandaceae

Juglans regia L.

2. This species is the principal walnut of commerce. The trees attain heights above 50 feet unless restricted by pruning. Leaves are compound, with up to a dozen oblong-ovate, near glabrous leaflets. Fruits are near globular to oblong, with a smooth surface, borne at the terminals of shoots. The husks surrounding the nuts are semi-fleshy and generally dehisce at maturity, allowing nut to drop out or to be readily separated. The shell is hard but generally thin, and the kernel separates from it readily. About 50 percent by weight of nut is shell.
3. Season, bloom to maturity: About 5 months.
4. Production in U.S.: About 260,00 tons from 175,000 acres (USDA 1994a). The 1992 CENSUS reported California with 211,541 acres and Oregon with 2,178 acres. California with about 99 percent of the U.S. acreage.
5. Other production regions: Eurasia.
6. Use: Direct eating and in confections, and cookery. Also used as an ornamental tree and for its fine-grain wood as a timber product.
7. Part(s) of plant consumed: Internal kernels only
8. Portion analyzed/sampled: Nutmeat
9. Classifications:
   a. Authors Class: Tree nuts
   b. EPA Crop Group (Group & Subgroup): Tree nuts
   c. Codex Group: 022 (TN 0678) Tree nuts
   d. EPA Crop Definition: None

10. References: GRIN, BAILEY 1976, CODEX, MAGNESS, RICHE, USDA 1994a, US EPA 1995a, US EPA 1995b, US EPA 1994, WOODROOF(b), IVES, REHM, MENNINGER, RAMOS
11. Production Map: EPA Crop Production Region 10.
12. Plant Codes:
   a. Bayer Code: IUGRE

642

1. Wasabi

(Japanese horseradish, Hata)

Brassicaceae (Cruciferae)

Wasabia japonica (Miq.) Matsum. (syn: *Eutrema wasabi* Maxim.)

2. Wasabi is a perennial herb grown for its large, fleshy rhizomes used to add spicy flavor to Japanese dishes such as sushi. Grows in cool, wet, shaded climates. Fields are semi-flooded with a constant water flow like watercress. Fleshy rhizomes are grated and prepared into a fresh green paste. An upland form, known as 'Hata' is also grown.
3. Season, transplanting to harvest: About 1 1/2 -2 years. Large stem cuttings (8-12 inches) are used for propagation.
4. Production in U.S.: 10 acres in Washington state and Oregon (SCHREIBER 1995).
5. Other production regions: Japan.
6. Use: Flavoring.
7. Part(s) of plant consumed: Rhizomes (fresh) for flavoring and leaves for greens. Rhizomes are also dehydrated and ground into a powder for use after rehydration
8. Portion analyzed/sampled: Rhizomes (fresh)
9. Classifications:
   a. Authors Class: Root and tuber vegetables
   b. EPA Crop Group (Group & Subgroup): Miscellaneous
   c. Codex Group: No specific entry
   d. EPA Crop Definition: None
10. References: GRIN, FACCIOLA, SCHREIBER, RUBATZKY (picture)
11. Production Map: EPA Crop Production Region 12.
12. Plant Codes:
   a. Bayer Code: ETMWA

643

1. Water bamboo

(Manchurian wildrice, Coba, Kuw-sun, Kwo-bai, Jiao-bai, Makomo dake, Manchurian waterrice, Gau sun, Water oats, Wild rice)

Poaceae (Gramineae)

Zizania latifolia (Griseb.) Turcz. ex Stapf

2. This aquatic plant is closely related to wild rice (*Z. palustris*) of North America. A perennial that grows in stagnant ponds and in poorly drained soils. The plant grows from 4 to 7 feet tall with leaves from 12 to 24 inches long. The enlarged stems are harvested, upper leaves cut off and only the stem with husk-like wrapper leaves sent to market. The edible part is the succulent stem after the husks are removed. There are three types: Green stem (early maturing), White stem (mid-late-season), and Pink or Red stem (mid-late-season). Preparation of the field is similar to paddy rice.

3. Season, planting to harvest: About 5 months (150 days for Green type, 170 days for White or Pink types) after transplanting.
4. Production in U.S.: No data. Would do well in area in southern U.S. where paddy rice is grown (JANICK 1990, YAMAGUCHI).
5. Other production regions: Eastern Asia as well as Japan and Taiwan (RUBATZKY 1997).
6. Use: Swollen stems are sliced and eaten raw or cooked. Remains crisp when stir-fried.
7. Part(s) of plant consumed: Stem with husks removed. Dried leaves have been used as animal feed
8. Portion analyzed/sampled: Stems (with and without husks)
9. Classifications:
   a. Authors Class: Stalk and stem vegetables
   b. EPA Crop Group (Group & Subgroup): Miscellaneous
   c. Codex Group: No specific entry
   d. EPA Crop Definition: None
10. References: GRIN, JANICK 1990, YAMAGUCHI, RUBATZKY (picture)
11. Production Map: EPA Crop Production Regions for rice (which see).
12. Plant Codes:
    a. Bayer Code: ZIZLA

644

1. Water dropwort
(Rau Can, Ghora-ajowan, Seri, Shui qin, Shelum, Piopo, Water celery, Chinese celery, Javan water dropwort, Oenanthe)
Apiaceae (Umbelliferae)
Oenanthe javanica (Blume) DC.

2. Stoloniferous cool season perennial is a substitute for celery which it closely resembles. Prefers a wet situation and a cool season. In Japan, seedlings are planted after the rice crop has been harvested in the early fall. The crop is grown in flooded culture into the winter and harvested in January and February when plants are about 12 inches tall. Water dropwort petioles are more numerous, slender, rounded and hollow as compared to celery.
3. Season: Propagation is primarily by stolons but seeds may be used. Harvested 2-3 months after planting when plants are about 12 inches tall.
4. Production in U.S.: Cultivated in Hawaii (KUEBEL 1988). Grown in Florida in 1977 (STEPHENS 1988).
5. Other production regions: Can be maintained the year round under greenhouse conditions (KUEBEL 1988). Native of Southeast Asia (YAMAGUCHI 1983).
6. Use: Used as a salad herb and eaten raw or steamed with rice.
7. Part(s) of plant consumed: Leaves and petioles
8. Portion analyzed/sampled: Tops (fresh)
9. Classifications:
   a. Authors Class: Leafy-vegetables (except *Brassica* vegetables)
   b. EPA Crop Group (Group & Subgroup): Miscellaneous
   c. Codex Group: No specific entry
   d. EPA Crop Definition: None
10. References: GRIN, KUEBEL, STEPHENS 1988, YAMAGUCHI 1983
11. Production Map: EPA Crop Production Region 13.
12. Plant Codes:
    a. Bayer Code: OENJA

645

1. Water spinach
(Swamp cabbage, Kangkong, Weng cai, Rau muong, Green engtsai, Kancon, Chinese convolvulus, Water convolvulus, Water sweet potato, Potatovine, Tangkong, Swamp moringglory, Kangkung, Chingquat, Pakquat)
Convolvulaceae
Ipomoea aquatica Forssk. (syn: *I. reptans* Poir)

2. A trailing perennial vine that spreads rapidly by rooting at the nodes. Vertical branches arise from the leaf axils. It is glabrous with sagittate, alternate leaves. The foliage is succulent, especially the wetland form. There are several varieties divided between dry (upland) forms and wet (paddy) forms. It is grown for its long tender shoots about 16 inches which are used as green vegetable. Water spinach has hollow stems. 'Ching Quat' (green stem) and 'Pak Quat' (white stem) are cultivated.
The plant is closely related to sweet potato, but mainly grown in water. Roots are cooked and eaten by natives in the tropics, but are inferior to sweet potato. Plants are grown around pools or in tanks, mainly for the succulent leaves which are used as a potherb resembling spinach. Plants, foliage and roots resemble sweet potato.
3. Season, seeding to harvest: Upland culture ('Ching Quat') – 50 to 60 days. In aquatic culture ('Pak Quat') transplant to harvest is 30 days with three or more cuttings in a year.
4. Production in U.S.: Limited data. Under paddy culture, Hawaii grew 4 acres in 1985 (HYLIN 1985). Interest in Florida to grow the crop (MEISTER 1996a).
5. Other production regions: Tropical and subtropical areas and greenhouse (YAMAGUCHI 1983).
6. Use: Potherb, cooked as spinach. Young succulent tips are eaten fresh in salads.
7. Part(s) of plant consumed: Practically all parts of the young plant are eaten but mainly the tops
8. Portion analyzed/sampled: Tops (fresh)
9. Classifications:
   a. Authors Class: Leafy vegetables (except *Brassica* vegetables)
   b. EPA Crop Group (Group & Subgroup): Miscellaneous
   c. Codex Group: 013 (VL 0507) Leafy vegetables (including *Brassica* leafy vegetables) (CODEX uses 'Kangkung' as commodity name)
   d. EPA Crop Definition: None
10. References: GRIN, CODEX, HYLIN, KUEBEL, MARTIN 1979 (picture), MAGNESS, MEISTER 1996a, YAMAGUCHI 1983, USDA NRCS
11. Production Map: EPA Crop Production Region 13.
12. Plant Codes:
    a. Bayer Code: IPOAQ

646

1. Waterchestnut, Chinese
(Waternut, Horse's hoof, Matai, Hon matai, Kweilin matai, Pi chi, Pi tsi, Suimatai, Kuro-kuwai, Ling, Ground chestnut, Matai)
Cyperaceae
Eleocharis dulcis (Burm. f.) Trin. ex Hensch. (syn: *E. tuberosa* (Roxb.) Schult.)

1. Waterchestnut

(Caltrop, Water caltrop, Jesuit nut, Ling-kok, European waterchestnut, Risotto)

Trapaceae

Trapa natans L. var *natans*

T. bicornis Osbeck

(Horn nut)

2. Primarily two types of aquatic plants are grown under the name waterchestnut: One, *E. dulcis*, is a rush-like plant grown extensively in China for its near-round turnip-shaped tubers. They are grown in ponds, and the tubers are harvested by scooping them off the bottom with forks. This is the chinese waterchestnut or "ling" widely used in Chinese foods. The other plant, also called waterchestnut, or Jesuit nut, or Water caltrops, is *T. natans*, a water plant with large leaves that floats on the water surface. It is grown to some extent in southern Europe and Asia. The edible part is the nutlike fruit, 1 to 2 inches in diameter, with 4 spined angles, which grows below the leaf blade. It is roasted and eaten like chestnuts. Neither type of waterchestnut is produced commercially in the U.S., although there has been some effort with *E. dulcis*. *E. dulcis* is usually grown in rotation with paddy rice in the Orient. Corms are planted in a nursery in early spring and once the plants are about 8 inches tall, they are transplanted into paddy fields. *T. bicornis* is the same as *T. natans* except the nut has two horns and *T. natans* has four horns.

3. Season: Perennial plants, with a crop harvested annually. Water is drained 30 days before harvest for *E. dulcis*. Waterchestnut fruit harvest begins about 9 months after planting and continues at 8-10 day intervals for about 3 months.

4. Production in U.S.: No data. Negligible. *E. dulcis* seldom grown in Florida, California and Hawaii (STEPHEN 1988). *E. dulcis* is endemic to the tropics (YAMAGUCHI 1983), and has been grown in Georgia, Florida, Alabama and California (SCHNEIDER 1985). *T. natans* naturalized in North America (FACCIOLA 1990).

5. Other production regions: Eurasia and Africa for *T. bicornis*, and tropical eastern Asia for *E. dulcis* (RUBATZKY 1997).

6. Use: Cooked, eaten out of hand or in other foods. *T. natans* seeds are boiled, roasted and used as the main ingredient in "risotto". Preparation for *E. dulcis* includes scrubbing, rinsing and peeling.

7. Part(s) of plant consumed: Tuberous bulbs (corm) in *E. dulcis*; nut-like fruits in *T. natans* (seed kernel).

8. Portion analyzed/sampled: Tubers for *E. dulcis*; Seeds for *Trapa* spp.

9. Classifications:
   a. Authors Class: Root and tuber vegetable for *E. dulcis*; Herbs and spices for *Trapa* spp.
   b. EPA Crop Group (Group & Subgroup): Miscellaneous
   c. Codex Group: No specific entry
   d. EPA Crop Definition: None

10. References: GRIN, FACCIOLA, KAYS, MAGNESS, NEAL, SCHNEIDER 1985, STEPHENS 1988 (picture), YAMAGUCHI 1983 (picture), RUBATZKY

11. Production Map: No entry

12. Plant Codes:
   a. Bayer Code: ELODU (*E. dulcis*), TRPNA (*T. natans*)

647

1. Watercress

(Crestles, Berro, Eker, Biller, Bilure, Ribcress, Brown Cress, Wellgrass, Leko, Teng tongue, Long tails, Berro de agua, Cresson de fontaine, Upland watercress)

Brassicaceae (Cruciferae)

Rorippa nasturtium-aquaticum (L.) Hayek (syn: *Nasturtium officinale* R. Br.; *N. nasturtium-aquaticum* (L.) Karsten)

2. Watercress is a perennial plant grown for the pungent leaves and young stems, which are widely used for garnishing and in salads. Commercially, watercress is normally grown in pools of gently flowing water. It also grows naturally in many streams. The leaves are smooth and compound, with three to a dozen nearly round leaflets. Plants can be propagated from seeds or cuttings. The leaves and stems are partially submerged during growth. Springs are usually the water source, since any contamination in the water would contaminate the cress.

3. Season: Cutting continuous throughout year. About three weeks after the seedlings appear, the plants are ready to harvest or four to six weeks after cuttings are transplanted. Harvested leaves are from 4 to 12 inches. From seeding to harvest about six months.

4. Production in U.S.: 505 acres reported (1992 CENSUS). In 1992, Florida harvested 329 acres, California (90), Hawaii (38) and Maryland-Virginia area with about 3 acres (1992 CENSUS). In 1996, about 4 acres were grown under upland culture in Hawaii (KAWATE 1997a). In other watercress growing areas, we understand that the growers can keep the water off the beds with the use of overhead irrigation when pest control materials are applied, therefore, there is no standing or flowing water on the beds during application (MEISTER 1997).

5. Other production regions: In 1995, Canada reported that watercress is a Canadian crop (ROTHWELL 1996a).

6. Use: Mainly for garnishing, in salads and sandwiches.

7. Part(s) of plant consumed: Leaves and young stems (about the top 6 inches)

8. Portion analyzed/sampled: Leaves and stems

9. Classifications:
   a. Authors Class: Leafy vegetables (except *Brassica* vegetables)
   b. EPA Crop Group (Group & Subgroup): Miscellaneous
   c. Codex Group: 013 (VL 0473) Leafy vegetables (including *Brassica* leafy vegetables)
   d. EPA Crop Definition: None, but proposing for upland watercress – Upland cress = Upland watercress

10. References: GRIN, CODEX, LOGAN, MAGNESS, NEAL, RICHE, ROTHWELL 1996a, STEPHENS 1988 (picture), US EPA 1996a, YAMAGUCHI 1983, KAWATE 1997a, MEISTER 1997, MCHUGH, RUBATZKY

11. Production Map: EPA Crop Production Regions 1, 2, 3, 10 and 13.

12. Plant Codes:
   a. Bayer Code: NAAOF

648

1. Watermelon

(Melon, Sandia, Patilla)

Cucurbitaceae

Citrullus lanatus (Thunb.) Matsum. & Nakai var. *lanatus* (syn: *C. vulgaris* Schrad.; *Colocynthis citrullus* (L.) O. Ktze.)

2. Watermelons are large, smooth skinned fruits, grown on prostrate, vinelike stems. Stems grow to 10 or more feet in length. Leaves are fairly large, a foot or more in length, and deeply notched. Fruits generally spherical to pronounced oblong oval. On different varieties, they range from 6 to 50 pounds or more in weight. Fruit consists of a firm outer rind, a layer of white inner rind flesh $1/2$ to 1 inch thick, and an interior colored edible pulp in which the seeds are imbedded.
3. Season, seeding to harvest: $2^{1}/_{2}$ to 4 months.
4. Production in U.S.: Commercial 1,999,300 tons in 1994 (USDA-NASS 1995a). In 1994, 234,600 acres planted in U.S. with the top 10 states as Texas (56,000 acres), Florida (40,000), Georgia (37,000), California (16,700), South Carolina (12,500), Alabama (12,100), Oklahoma (11,000), North Carolina (9,500), Mississippi (8,500) and Arizona (6,800) with 90 percent (USDA-NASS 1995a).
5. Other production regions: Asia, Europe and Africa with China having 23 percent of the world's watermelon production (RUBATZKY 1997).
6. Use: Fresh eating. Some inner rind pulp preserved.
7. Part(s) of plant consumed: Mainly interior colored pulp only. Some preserving of white inner rind flesh
8. Portion analyzed/sampled: Whole fruit or representative samples
9. Classifications:
   a. Authors Class: Cucurbit vegetables
   b. EPA Crop Group (Group & Subgroup): Cucurbit vegetables (9A)
   c. Codex Group: 011 (VC 0432) Fruiting vegetables, cucurbits
   d. EPA Crop Definition: Melons = Watermelon
10. References: GRIN, CODEX, MAGNESS, USDA-NASS 1995a, US EPA 1995b, US EPA 1995a, US EPA 1994, RUBATZKY, DOOLITTLE
11. Production Map: EPA Crop Production Regions 2, 3, 4, 5, 6, 8 and 10.
12. Plant Codes:
    a. Bayer Code: CITLA

649

1. Wax jambu
(Java apple, Samarang roseapple, Wax apple, Water apple, Jambosa, Jumrool)
Myrtaceae

Syzygium samarangense (Blume) Merr. & L.M. Perry (syn: *Eugenia javanica* Lam.)

2. The tree is 16 to 50 feet tall with a short trunk to 1 foot thick. The fruits are pear-shaped, waxy, about 2 inches long and 2 inches wide with flesh that is spongy and very bland in flavor. Cultivars are called 'Pink' and 'Srinark'. 'Srinark' has a large copper colored fruit with white flesh, usually with a single seed. 'Srinark' is thought to be the best flavored wax jambu.
3. Season, bloom to harvest: Flowers April-May and fruits June-July in Florida (CRANE 1995).
4. Production in U.S.: Florida (FACCIOLA 1990). 1 acre in Florida (CRANE 1995).
5. Other production regions: Tropical Asia (FACCIOLA 1990).
6. Use: Eaten raw or cooked as a sauce or in a stew with true apples. Also grown as an ornamental tree.

7. Part(s) of plant consumed: Fruit pulp
8. Portion analyzed/sampled: Whole fruit
9. Classifications:
   a. Authors Class: Tropical and subtropical fruits – inedible peel
   b. EPA Crop Group (Group & Subgroup): Miscellaneous
   c. Codex Group: 006 (FI 0340) Assorted tropical and subtropical fruits – inedible peel (CODEX principal vernacular is Java apple)
   d. EPA Crop Definition: Proposing – Guava = Wax jambu
10. References: GRIN, CODEX, CRANE 1995a, CRANE 1995, FACCIOLA, MORTON (picture)
11. Production Map: EPA Crop Production Region 13.
12. Plant Codes:
    a. Bayer Code: No specific entry

650

1. Waxgourd
1. Chinese squash
(Mokwa, Chinese fuzzy gourd, Summer squash, Zitkwa, Petha, Tao-tue, Maoqwa, Tsitgwa, Tankoy, Bi-dao, Moqua, Fuzzy melon, Fuzzy gourd, Zit-kwa)
1. Chinese waxgourd
(Chinese wintermelon, Chinese preserving melon, White gourd, Ash gourd, Ton kwa, Dung kwa, Winter squash, Ash pumpkin, Calabaza blanca, Tunka, Chinese watermelon, Tocan, Christmas melon, White pumpkin, Tallowgourd, Winter melon, Tougan, Doongua, Chamkwa, Tankoy)
Cucurbitaceae

Benincasa hispida (Thunb.) Cogn. (syn: *B. cerifera* Savi)

2. The plant is an annual; running squash-like vine with large, soft, hairy leaves. Fruits are large, oblong, 10 to 16 inches in length, hairy, with a waxy bloom when ripe, solid white flesh and cucumber-like seeds. In culture it is similar to cucumber or cantaloupe. Mature fruit is used for preserves and sweet pickles, also said to be eaten raw in tropics. Fruit said to be much esteemed in Asia, particularly China. The small unripe fruit weighing about 3 pounds and 4 to 6 inches long is called zitkwa, Chinese squash, fuzzy melon or gourd, Chinese fuzzy gourd, petha, tao-tue, maogwa, tsitgwa, tankoy, bi-dao, mokwa and moqua. When it matures, the melon that weighs about 40 pounds is called waxgourd, winter melon, preserving melon, white gourd, ash gourd, dung-kwa, tung quo, ton kwa, dong quo, white pumpkin, tallow gourd, Chinese watermelon, Christmas melon, tocan, tougan, doongua, chamkwa, tankoy and tunka. Chinese waxgourd can be kept in storage for up to 6 months.
3. Season, seeding to harvest: As summer squash about 67 to 87 days. As winter squash about 90 to 150 days. Flowering begins 60-80 days after planting and young immature fruit are harvested 7-8 days later. Fully mature fruit is harvested 60-70 days after flowering.
4. Production in U.S.: About 50 acres in California (MELNICOE 1997a). Grown as spring and fall crops in southern California or as a summer crop in the rest of the state (MYERS 1991). In Florida, planted up to 3 times (July, December and March) in southern Florida, while the rest of the state is planted in March and July (STEPHEN 1988).
5. Other production regions: In 1985, 99 percent of the total U.S. imports of gourds which included *B. hispida* came from

the Dominican Republic (LAMBERTS 1990). Grown in tropical and subtropical regions (YAMAGUCHI 1983).

6. Use: Preserves, pickling, cooked and fresh eating. Ripe fruits, with rind and seeds discarded, are normally cooked like other vegetables, especially in soups, but also for fresh eating like sliced cucumbers and preserves. Immature fruits are used as summer squash.

7. Part(s) of plant consumed: Interior of fruit when mature like winter squash. Whole fruit when unripe like summer squash. Seeds, young leaves and flowers are also edible

8. Portion analyzed/sampled: Whole fruit

9. Classifications:
   a. Authors Class: Cucurbit vegetables
   b. EPA Crop Group (Group & Subgroup): Cucurbit vegetables (9B)
   c. Codex Group: 011 (VC 4255) Fruiting vegetables, cucurbits (as waxgourd)
   d. EPA Crop Definition: None

10. References: GRIN, CODEX, LAMBERTS(a), MAGNESS, MARTIN 1984, NEAL, REHM, US EPA 1995a, YAMAGUCHI 1983 (picture), USDA NRCS, MYERS (picture), STEPHENS 1988, MELNICOE 1997a, BAYER, RUBATZKY (picture)

11. Production Map: EPA Crop Production Regions 3, 10 and 13.

12. Plant Codes:
   a. Bayer Code: BNCHI

651

1. Wheat
(Farine, Harina, Weizen, Froment, Trigo, Ble, Tarwe)
Poaceae (Gramineae)
Triticum spp.

2. Wheat is the most important food grain of the temperate zones, both north and south, with over 549 million acres worldwide in 1993. Acreage in wheat is near 60 million acres in the U.S. Production in the U.S. was 1,524,340,000 bushels in 1967. In 1993-95, acreage averaged over 70 million acres in the U.S. with over 2 billion bushels. Wheat has been a food crop for mankind since the beginning of agriculture. Carbonized grains dating to at least as early as 8,000 B.C. have been found in Iraq and Syria, and many other findings in Eastern Mediterranean countries are nearly as old. The Middle East is probably the area of origin, and wheat apparently spread throughout Europe not later than the Stone Age.

Wheat is essentially a cool season crop that thrives best at preharvest temperatures averaging around 60 degrees F. The minimum frost-free growing season is about 100 days. In continental U.S. wheat is grown in every state although production in New England is minor. From 15 to 20 or more inches of precipitation, are necessary for annual cropping. In some areas with not more than 10 to 15 inches of precipitation wheat is grown once in two years, with the land kept free of vegetation one of the years to accumulate moisture in the soil. The wheat plant is an annual grass. It is mainly grown as a winter annual in milder climates, with seeding in the fall and harvest from June through August depending on the length of the winter. In areas with rigorous winter climates it is mainly spring seeded. Planting is as early as soil can be worked, and harvest is in late summer and early fall.

Winter wheat is seeded in dryland areas at 60 to 120 lbs/acre, irrigated areas 60 to 150 lbs/acre, and for grazing at 75 to 120 lbs/acre.

In early growth stages the wheat plant consists of a much compressed stem or crown and numerous narrowly linear or linear-lanceolate leaves. Leaves are mainly near glabrous. Buds in the leaf axils below the soil surface grow into lateral branches termed tillers. From both the main crown and the tillers, elongated stems develop later and terminate in a spike or head in which the flowers, and finally the seed or grain, develop. In fall-seeded wheat the plant usually remains in the rosette stage throughout the fall and winter, sending up the elongated stems in late spring. In spring-seeded wheat the rosette period may be short, and tillering is usually much less than in fall plantings. During late fall and early spring, fall-seeded wheat can be lightly pastured without greatly reducing grain yields, and this is frequently done. The pasturage at this stage is nutritious and highly palatable. Stems of wheat reach from 18 inches to 4 or more feet in height depending on kind and growing conditions. The spike or head may be from less than 2 inches to 4 or 5 inches long. Both stems and spikes from the latest-formed tillers are usually somewhat smaller than those of the early formed tillers. Spikelets develop at nodes in the spike and in these spikelets the flowers and seeds develop. Spikelets vary in number from 10 to 30 per spike and each develops from 1 to 5 flowers and seeds, depending on the kind. Density of spikelets and overall shape of the spike also vary. The flower consists of two outer membranous tissues, one termed lemma and one palea, which enclose the stamens and the single ovary which develops into the seed or kernel. Together these are termed the glumes, and they continue to enclose and completely cover the developing seed.

In awned varieties, the awns or beards are at the terminal of the lemma. The palea is membranous and awnless. In most wheats the lemmas and paleas are separated from the kernels in threshing, forming the chaff. In spelt and emmer, formerly grown to a very limited extent for feed in this country, they continue to adhere to the seed following threshing. The wheat grain or kernel is roughly ovate or egg shaped and from $1/6$ to $2/5$ inch in length depending on kind. The dorsal surface is generally smooth and rounded but the ventral surface is creased. At the apex a brush consisting of short hairs is generally present. Color of the kernel varies from dark red through light brown – classed commercially as red wheat, to white, cream or yellow – classed commercially as white wheat, or amber, in durum wheat.

The wheat kernel is made up of three main parts: (1) The outer covering consists of several distinct cell layers and is the bran, separated from the flour during most milling processes. It comprises about 12 percent of the kernel weight. (2) The endosperm consists mainly of starch and makes up about 85 to 86 percent of the kernel. It is the portion present in white flour. (3) The germ, or embryo, expands into the new plant at and following germination. It makes up only about 2.5 percent of the kernel and is also separated out in most milling processes.

Botanical Classification of Wheat
In USDA Technical Bulletin 1287, titled "Classification of *Triticum* species and Wheat Varieties Grown in the U.S.", wheat is classified into 10 species of *Triticum*. Six of these are cultivated and four are noncultivated, or rarely so. Winter wheat includes

hard red, soft red and white. Spring wheat includes hard red, durum and white. Wheats are classified as "hard" or "soft" depending upon the endosperm granularity. Hard wheats have flinty, translucent grains while soft wheats have starchy, opaque looking grain.

Uses of Wheat

Wheat is used mainly for food (about 70 percent), but substantial quantities are also used as feed for livestock (about 22 percent). Some wheat is cut for hay. Wheat grown for the grain crop may also be used for pasture before the stems begin to elongate. The straw following threshing formerly was an important sustenance feed for livestock. Since nearly all wheat is now combined, with the straw scattered in the field, this use is now decreased. As temporary pasturage wheat is nutritious and palatable. As a feed grain, wheat is fed to livestock either whole or after coarse grinding. In either case the feed includes the entire kernels. The bran from flour milling is also an important livestock feed, and the germ is a valuable addition to feed concentrate. For food, most of the wheat is made into flour, the base of most baked foods as breads, cakes, etc. Macaroni is made from durum wheat. Durum wheat is harder than bread wheats, and its cultivars have spring growth habits in the U.S. Most of the flour used in this country is white. In making white flour the bran and germ are removed mechanically and the resulting product consists essentially of the ground endosperm. Whole wheat flour is also an important food. Some of the bran and germ separated out in milling also are used as food. In addition to food and feed uses, some wheat is used as a source of starch and in the making of alcoholic beverages. Approximately 100 pounds of wheat are processed into 72 to 75 pounds of flour and 25 to 28 pounds of wheat mill feeds. The per capita consumption of wheat flour is 139 lbs and cereal is 4.9 lbs (1992-93).

3. Season, seeding to harvest: 90 to 260 days depending on variety and time of planting. Harvesting of winter wheat may begin before 1 June in southeastern U.S. and late summer in Montana. Harvesting of spring wheat begins mid-July in Midwest and Pacific Northwest, while beginning after 1 August in northern states.

4. Production in U.S.: In 1995, 69,177,000 acres planted with 2,185,539,000 bushels and winter wheat accounted for 70 percent of the acreage (USDA 1996c). In 1995, the top 10 states reported for all wheat as Kansas (11,700,000 acres planted), North Dakota (11,290,000), Oklahoma (6,900,000), Texas (5,800,000), Montana (5,720,000), Colorado (2,940,000), South Dakota (2,883,000), Washington (2,700,000), Minnesota (2,298,000), and Nebraska (2,150,000). Winter wheat planted on 48,726,000 acres; durum wheat on 3,436,000 acres and other spring wheat on 17,015,000 acres (USDA 1996c). Most durum wheat is produced in North Dakota or about 86 percent. The other spring wheat, about 93 percent is grown in North Dakota, Montana, Minnesota and South Dakota (USDA 1996c).

5. Other production regions: Canada harvested about 31,200,000 acres of all wheat in 1995 with spring wheat about 86 percent of total (ROTHWELL 1996a).

6. Use: See above; Wheat products for food include; whole-grain wheat, wheat bran, flours, cracked wheat, wheat flakes, grits, wheat meal, puffed wheat, shredded whole wheat, wheat germ and bulgur wheat.

7. Part(s) of plant consumed: Grain and foliage

8. Portion analyzed/sampled: Grain, forage, hay, straw and aspirated grain fractions. The grain processed commodities bran, flour, middlings, shorts and germ. Grain is kernel with

hull removed. Aspirated grain fractions (previously called "grain dust") are collected at grain elevators during the moving/handling of grains/oilseeds for environmental and safety reasons. Residue data should be provided for any postharvest use on corn, sorghum, soybeans or wheat. For a preharvest use after the reproduction stage begins and seed heads are formed, data are needed unless residues in the grain are less than the limit of quantitation of analytical method. For a preharvest use during the vegetative stage (before the reproduction stage begins), data will not normally be needed unless the plant metabolism or processing study shows a concentration of residues of regulatory concern in an outer seed coat (e.g., wheat bran). Wheat forage: Cut sample at 6 to 8 inch stage to stem elongation (jointing) stage, at approximately 25 percent DM. Wheat hay: Cut samples at early flower (boot) to soft dough stage. Hay should be field dried to a moisture content of 10 to 20 percent. Wheat straw: Cut plant residue (dried stalks or stems with leaves) left after the grain has been harvested (threshed). Wheat: Includes emmer wheat and triticale. No processing study is needed for a specific tolerance on emmer wheat. Wheat milled byproducts: Use highest value for wheat middlings, bran and shorts.

9. Classifications:
   a. Authors Class: Cereal grains; Forage, fodder and straw of cereal grains
   b. EPA Crop Group (Group & Subgroup): Cereal grains (15); Forage, fodder, and straw of cereal grains (16) (Representative commodity for Crop Groups 15 and 16)
   c. Codex Group: 020 (GC 0654) Cereal grains; 065 (CF 0654) Processed bran; 058 (CM 0654) Unprocessed bran; 065 (CF 1210) Wheat germ; 051 (AS 0654) Dry straw and fodder; 065 (CF 1211) Wheat flour; 065 (CF 1212) Wheat wholemeal; 078 (CP 1211) White bread
   d. EPA Crop Definition: Wheat = Triticale

10. References: GRIN, CODEX, FACCIOLA, MAGNESS, USDA 1996c, US EPA 1996a, US EPA 1995a, US EPA 1994, IVES, ROTHWELL 1996a, SMITH 1995, MARTIN (a), BAKER, SCHNEIDER 1996a

11. Production Map: EPA Crop Production Regions 1, 2, 3, 4, 5, 6, 7, 8, 9, 10, 11 and 12 with most in 5, 7 and 8.

12. Plant Codes:
    a. Bayer Code: TRZSS

652

1. Wheat, Club
(Cluster wheat, Dwarf wheat, Hedgehog wheat)
Poaceae (Gramineae)

Triticum compactum Host (syn: *T. aestivum* ssp. *compactum* (Host) Mackey)

2. Cultivars of this species may be either of winter or of spring type. Stems vary in height but are generally stiff. Spikes are short, usually under 2.5 inches in length, very compact and flattened. Spikelets usually contain five flowers and spread at near right angles to the rachis or stem. Spikelets are generally awnless, but sometimes awned. Kernels are small, flattened, have a very shallow, narrow crease and a short brush. Principal use is flour manufacture for crackers, cookies and cakes.

3. Season, See Wheat.

4. Production in U.S.: See Wheat.

5. Other production regions: See Wheat.

6. Use: See Wheat.
7. Part(s) of plant consumed: See Wheat
8. Portion analyzed/sampled: See Wheat
9. Classifications:
   a. Authors Class: See Wheat
   b. EPA Crop Group (Group & Subgroup): See Wheat
   c. Codex Group: See Wheat
   d. EPA Crop Definition: See Wheat
10. References: GRIN, See Wheat
11. Production Map: EPA Crop Production Region: See Wheat; Mostly Region 11.
12. Plant Codes:
   a. Bayer Code: No specific entry

653

1. Wheat, Common
(Bread wheat)
Poaceae (Gramineae)
Triticum aestivum L. (syn: *T. sativum* Lam., *T. vulgare* Vill.)
2. This species has a long, slender spike which is somewhat flattened. Spikelets are two to five flowered, relatively far apart on the stem and nearly erect. Awns are either lacking or less than ¹/₂ inch long. Stem centers are generally hollow but may be pithy. Leaves are more narrow than in some other wheats. Kernels may be red or white, hard or soft and the grain threshes free of glumes. This is the source of most of the wheat varieties cultivated in the U.S. Over 200 such varieties have been described, with nearly 100 now cultivated. They may be either spring or winter type and comprise nearly 95 percent of the wheat grown in this country. Principle use is for flour for leavened breads and pastries.
3. Season, See Wheat.
4. Production in U.S.: See Wheat.
5. Other production regions: See Wheat.
6. Use: See Wheat.
7. Part(s) of plant consumed: See Wheat
8. Portion analyzed/sampled: See Wheat
9. Classifications:
   a. Authors Class: See Wheat
   b. EPA Crop Group (Group & Subgroup): See Wheat
   c. Codex Group: See Wheat
   d. EPA Crop Definition: See Wheat
10. References: GRIN, See Wheat
11. Production Map: EPA Crop Production Region: See Wheat; Mostly Regions 2, 4, 5, 6, 7, 8 and 11.
12. Plant Codes:
   a. Bayer Code: TRZAX, TRZAS (Spring wheat), TRZAW (Winter wheat)

654

1. Wheat, Durum
(Hard wheat)
Poaceae (Gramineae)
Triticum durum Desf. (syn: *T. pyramidale* Percival)
2. Cultivars of this species grown in the U.S. are all spring wheats. Stems generally are pithy internally and leaves are relatively broad. Spikes are intermediate in length and flattened. Awns are nearly always present and are long and coarse, white, yellow or black in color. Kernels are white or red, usually long and pointed, very hard and translucent with

angular sides and a short brush. Durum wheat is used mainly for the manufacture of semolina which is made into macaroni, spaghetti and related products. About eight cultivars are grown on more than 3 million acres in this country, mainly in North Dakota and neighboring states.
3. Season: See Wheat.
4. Production in U.S.: See Wheat.
5. Other production regions: See Wheat.
6. Use: See Wheat.
7. Part(s) of plant consumed: See Wheat
8. Portion analyzed/sampled: See Wheat
9. Classifications:
   a. Authors Class: See Wheat
   b. EPA Crop Group (Group & Subgroup): See Wheat
   c. Codex Group: See Wheat
   d. EPA Crop Definition: See Wheat
10. References: GRIN, See Wheat
11. Production Map: EPA Crop Production Region: See Wheat; Mostly Regions 5 and 7.
12. Plant Codes:
   a. Bayer Code: TRZDU

655

1. Wheat, Einkorn
(Small spelt)
Poaceae (Gramineae)
Triticum monococcum L.
2. Einkorn or one-grained wheat is a primitive kind, the cultivation of which goes back to prehistoric times. It is grown in isolated areas in the Middle East and southern Europe. It is used as flour for making dark breads and as cattle or horse feed. It is also finding a market as a health food. Both winter and spring forms occur. Spikes are awned, slender, narrow, flattened and fragile. Spikelets contain only a single fertile flower and thus produce only one seed. Seeds are pale red, slender, flattened, almost without crease and remain in the spikelets after threshing. Einkorn is little grown at present and not at all in the U.S.
3. Season: See Wheat.
4. Production in U.S.: See Wheat.
5. Other production regions: See Wheat.
6. Use: See Wheat.
7. Part(s) of plant consumed: See Wheat
8. Portion analyzed/sampled: See Wheat
9. Classifications:
   a. Authors Class: See Wheat
   b. EPA Crop Group (Group & Subgroup): See Wheat
   c. Codex Group: See Wheat
   d. EPA Crop Definition: See Wheat
10. References: GRIN, See Wheat
11. Production Map: EPA Crop Production Region: See Wheat.
12. Plant Codes:
   a. Bayer Code: No specific entry

656

1. Wheat, Emmer
(Amidonnier, Escana mayor)
Poaceae (Gramineae)
Triticum dicoccon Schrank (syn: *T. dicoccum* Schrank, *T. dicoccum* Schuebl.)

2. Emmer is one of the most ancient of cultivated cereals. It was a major cereal until durum wheat was cultivated. In the last 40 years, its production has increased in Ethiopia and is sometimes ground for flour and baked into special bread and porridge. It may be either winter or spring in habit. Leaves generally are pubescent. Spikes are very dense and flattened laterally. Spikelets generally contain two flowers and generally are awned. The red or white kernels remain enclosed in the glumes after threshing. They are slender and acute at both ends. Emmer was formerly grown in the U.S. for feed on a limited acreage but now has substantially disappeared from cultivation. In 1992, Emmer and Spelt wheats reported with 8,942 acres from the top 4 states of Ohio (3,674 acres), Montana (1,696), Pennsylvania (1,677), and Michigan (1,150) (1992 CENSUS).
3. Season: See Wheat.
4. Production in U.S.: See Wheat.
5. Other production regions: See Wheat.
6. Use: See Wheat.
7. Part(s) of plant consumed: See Wheat
8. Portion analyzed/sampled: See Wheat
9. Classifications:
   a. Authors Class: See Wheat
   b. EPA Crop Group (Group & Subgroup): See Wheat
   c. Codex Group: See Wheat; 020 (GC 4625) Cereal grains
   d. EPA Crop Definition: See Wheat
10. References: GRIN, See Wheat
11. Production Map: EPA Crop Production Region: See Wheat. Mainly 1, 5 and 7.
12. Plant Codes:
   a. Bayer Code: TRZDI

657

1. Wheat, Macha
(Trigo maka)
Poaceae (Gramineae)
Triticum macha Dekapr. & A.M. Menabde
2. This is a late-maturing winter wheat with tall, hollow stems. Spikes vary in density from open to dense, with short awns. Kernels remain in the spikelets after threshing. They are elliptical, red and intermediate in hardness. Macha wheat is grown in the Republic of Georgia and southern Russia, but not commercially in the U.S.
3. Season: See Wheat.
4. Production in U.S.: See Wheat.
5. Other production regions: See Wheat.
6. Use: See Wheat.
7. Part(s) of plant consumed: See Wheat
8. Portion analyzed/sampled: See Wheat
9. Classifications:
   a. Authors Class: See Wheat
   b. EPA Crop Group (Group & Subgroup): See Wheat
   c. Codex Group: See Wheat
   d. EPA Crop Definition: See Wheat
10. References: GRIN, See Wheat
11. Production Map: EPA Crop Production Region: See Wheat.
12. Plant Codes:
   a. Bayer Code: No specific entry

658

1. Wheat, Oriental
(Khorassan wheat)
Poaceae (Gramineae)
Triticum turanicum Jakubz. (syn: *T. orientale* Percival)
2. This is a spring wheat, early in maturity with narrow, pubescent leaves. Spikes are long, loose and almost square in cross-section. Awns are long and often black. Spikelets produce two or three kernels which are long, narrow, white and hard. This wheat is grown in the Mediterranean area and the Near East, but not in the U.S.
3. Season: See Wheat.
4. Production in U.S.: See Wheat.
5. Other production regions: See Wheat.
6. Use: See Wheat.
7. Part(s) of plant consumed: See Wheat
8. Portion analyzed/sampled: See Wheat
9. Classifications:
   a. Authors Class: See Wheat
   b. EPA Crop Group (Group & Subgroup): See Wheat
   c. Codex Group: See Wheat
   d. EPA Crop Definition: See Wheat
10. References: GRIN, See Wheat
11. Production Map: EPA Crop Production Region: See Wheat.
12. Plant Codes:
   a. Bayer Code: No specific entry

659

1. Wheat, Persian
(Persian black wheat, Trigo de Persia)
Poaceae (Gramineae)
Triticum carthlicum Nevski
2. Persian wheat is of spring habit, early maturing and somewhat resistant to fungus diseases. It has strong yellow to light-red stems. Spikes are flexible, tending to lean over. While several flowers are present in each spikelet only three usually develop kernels. Kernels are free-threshing, flinty, almost always red. Persian wheat is grown in the eastern Mediterranean area, including southern Russia, but not commercially in the U.S.
3. Season, See Wheat.
4. Production in U.S.: See Wheat.
5. Other production regions: See Wheat.
6. Use: See Wheat.
7. Part(s) of plant consumed: See Wheat
8. Portion analyzed/sampled: See Wheat
9. Classifications:
   a. Authors Class: See Wheat
   b. EPA Crop Group (Group & Subgroup): See Wheat
   c. Codex Group: See Wheat
   d. EPA Crop Definition: See Wheat
10. References: GRIN, See Wheat
11. Production Map: EPA Crop Production Region: See Wheat.
12. Plant Codes:
   a. Bayer Code: No specific entry

660

1. Wheat, Polish
(Trigo polaco)
Poaceae (Gramineae)

Triticum polonicum L.

2. Polish wheat varieties are spring wheats with tall stems. Spikes are large, up to 0.5 inches long, open or dense, awned, and square or rectangular in cross-section. Kernels are very long, narrow and hard, similar to rye. They thresh free of the glumes. While grown extensively in Mediterranean countries, Polish wheat has proven inferior in the U.S. both in yield and in quality for bread or macaroni products. For these reasons it has substantially disappeared from commercial production.
3. Season: See Wheat.
4. Production in U.S.: See Wheat.
5. Other production regions: See Wheat.
6. Use: See Wheat.
7. Part(s) of plant consumed: See Wheat
8. Portion analyzed/sampled: See Wheat
9. Classifications:
  a. Authors Class: See Wheat
  b. EPA Crop Group (Group & Subgroup): See Wheat
  c. Codex Group: See Wheat
  d. EPA Crop Definition: See Wheat
10. References: GRIN, See Wheat
11. Production Map: EPA Crop Production Region: See Wheat.
12. Plant Codes:
  a. Bayer Code: No specific entry

661

1. Wheat, Poulard
(Cone wheat, Rivet wheat, Turgid wheat, Wild emmer, Trigo poulard)
Poaceae (Gramineae)

Triticum turgidum L.

2. Poulard wheats may be winter or spring in habit. Stems are usually tall, thick and solid or pithy. Leaves are broad. Spikes are long and dense, sometimes compound or branched. They are near square in cross-section, with long awns. Kernels are short, ovate and humped in shape. Poulard wheat is closely related to durum but is somewhat inferior in this country both in production and in macaroni-making quality, so has practically disappeared from cultivation. It is grown quite extensively in Mediterranean countries.
3. Season: See Wheat.
4. Production in U.S.: See Wheat.
5. Other production regions: See Wheat.
6. Use: See Wheat.
7. Part(s) of plant consumed: See Wheat
8. Portion analyzed/sampled: See Wheat
9. Classifications:
  a. Authors Class: See Wheat
  b. EPA Crop Group (Group & Subgroup): See Wheat
  c. Codex Group: See Wheat
  d. EPA Crop Definition: See Wheat
10. References: GRIN, See Wheat
11. Production Map: EPA Crop Production Region: See Wheat.
12. Plant Codes:
  a. Bayer Code: TRZTU

662

1. Wheat, Shot
(Indian dwarf wheat, Trigo indio)

Poaceae (Gramineae)

Triticum sphaerococcum Percival

2. This is an early maturing spring wheat with short, stiff stems that tiller profusely. Spikes are awnless or short-awned and dense. They appear square in cross-section. Spikelets contain six or seven flowers and develop four or five kernels. Kernels are short and almost spherical, unique among wheats, and thresh free and differ from other wheats by being shorter with smaller heads. Kernels are red or white in color. Shot wheat is grown in northwest India, but not commercially in the U.S.
3. Season, See Wheat.
4. Production in U.S.: See Wheat.
5. Other production regions: See Wheat.
6. Use: See Wheat.
7. Part(s) of plant consumed: See Wheat
8. Portion analyzed/sampled: See Wheat
9. Classifications:
  a. Authors Class: See Wheat
  b. EPA Crop Group (Group & Subgroup): See Wheat
  c. Codex Group: See Wheat
  d. EPA Crop Definition: See Wheat
10. References: GRIN, See Wheat
11. Production Map: EPA Crop Production Region: See Wheat.
12. Plant Codes:
  a. Bayer Code: TRZSC

663

1. Wheat, Spelt
(Dinkel wheat, Epeautre, Escanda, Spetz)
Poaceae (Gramineae)

Triticum spelta L.

2. Spelt may be either winter or spring in habit and awned or awnless. The spike is long and narrow. Spikelets are two-kerneled and upright, closely pressed to the rachis or central stem. Kernels are red, long, flattened, with a sharp tip and a narrow, shallow crease. They remain enclosed in the glumes after threshing. Spelt has higher proteins and minerals than other wheats. Spelt flour has a nutty taste. Spelt has been grown in Europe over 300 years and introduced into the U.S. in 1890s. The spelt hull must be removed before it can be used for food. Hulls comprise 20 to 30 percent of the grain weight. Spelt can be used in flour and baked goods to replace soft red winter wheat. Spelt is more winter hardy than most soft red winter wheat but less than most hard red winter wheat. Spelt was formerly grown in the U.S. on a small acreage for livestock feed but has now almost disappeared from cultivation. In 1992, Emmer and Spelt wheats reported with 8,942 acres from the top four states of Ohio (3,674 acres), Montana (1,696), Pennsylvania (1,677) and Michigan (1,150) (1992 CENSUS).
3. Season, See Wheat; planted same as winter wheat, mid-September in upper midwest.
4. Production in U.S.: See Wheat.
5. Other production regions: See Wheat.
6. Use: See Wheat.
7. Part(s) of plant consumed: See Wheat
8. Portion analyzed/sampled: See Wheat
9. Classifications:
  a. Authors Class: See Wheat
  b. EPA Crop Group (Group & Subgroup): See Wheat

c. Codex Group: See Wheat; 020 (GC 4673) Cereal grains
d. EPA Crop Definition: See Wheat
10. References: GRIN, OPLINGER(a), See Wheat
11. Production Map: EPA Crop Production Region: See Wheat and Emmer.
12. Plant Codes:
   a. Bayer Code: TRZSP

664

1. Wheat, Timopheevi
(Trigo de Georgia)
Poaceae (Gramineae)
   Triticum timopheevii (Zhuk.) Zhuk. var. *timopheevii*
2. This is a late maturing spring wheat with leaf blades that are pubescent on both sides. Spikes are very compact, rather short, somewhat pyramidal in shape with soft, thin, rather short awns. Spikelets usually contain two kernels. Kernels are medium long, slender and hard or flinty. The species occurs in the Republic of Georgia and southern Russia. It is not grown in the U.S.
3. Season: See Wheat.
4. Production in U.S.: See Wheat.
5. Other production regions: See Wheat.
6. Use: See Wheat.
7. Part(s) of plant consumed: See Wheat
8. Portion analyzed/sampled: See Wheat
9. Classifications:
   a. Authors Class: See Wheat
   b. EPA Crop Group (Group & Subgroup): See Wheat
   c. Codex Group: See Wheat
   d. EPA Crop Definition: See Wheat
10. References: GRIN, See Wheat
11. Production Map: EPA Crop Production Region: See Wheat.
12. Plant Codes:
   a. Bayer Code: No specific entry

665

1. Wheat, Vavilovi
Poaceae (Gramineae)
   Triticum vavilovii Jakubz.
2. This a winter type wheat, mid-season in maturity with thick, strong, stems that resembles Spelt. Spikes are medium dense to loose, and awned. Kernels remain in the spikelets after threshing. They are ovate, white and hard. This wheat is grown somewhat in Russia but not commercially in the U.S.
3. Season: See Wheat.
4. Production in U.S.: See Wheat.
5. Other production regions: See Wheat.
6. Use: See Wheat.
7. Part(s) of plant consumed: See Wheat
8. Portion analyzed/sampled: See Wheat
9. Classifications:
   a. Authors Class: See Wheat
   b. EPA Crop Group (Group & Subgroup): See Wheat
   c. Codex Group: See Wheat
   d. EPA Crop Definition: See Wheat
10. References: GRIN, See Wheat
11. Production Map: EPA Crop Production Region: See Wheat.
12. Plant Codes:
   a. Bayer Code: No specific entry

666

1. Wheat, Wild einkorn
Poaceae (Gramineae)
   Triticum boeoticum Boiss. (syn: *T. aegilopoides* (Link) Balansa ex Korn.)
2. It grows as a native grass in the Balkans and Anatolia. Differs little from einkorn.
3. Season: See Wheat.
4. Production in U.S.: See Wheat.
5. Other production regions: See Wheat.
6. Use: See Wheat.
7. Part(s) of plant consumed: See Wheat
8. Portion analyzed/sampled: See Wheat
9. Classifications:
   a. Authors Class: See Wheat
   b. EPA Crop Group (Group & Subgroup): See Wheat
   c. Codex Group: See Wheat
   d. EPA Crop Definition: See Wheat
10. References: GRIN, See Wheat
11. Production Map: EPA Crop Production Region: See Wheat.
12. Plant Codes:
   a. Bayer Code: No specific entry

667

1. Wheat, Wild emmer
Poaceae (Gramineae)
   Triticum dicoccoides (Korn. ex Asch. & Graebn.) Aarons.
2. This plant grows in the area from Palestine to Russia. It grows in North and South Dakota. It is a winter annual with loose, flattened spikes bearing long, stiff awns. Spikelets fall from the fragile spike at maturity. Spikelets are large, usually with three flowers but developing only two kernels. Wild emmer appears not to be cultivated. Kernels remain enclosed in the glumes after threshing and the grain is red or white in color.
3. Season: See Wheat.
4. Production in U.S.: See Wheat.
5. Other production regions: See Wheat.
6. Use: See Wheat.
7. Part(s) of plant consumed: See Wheat
8. Portion analyzed/sampled: See Wheat
9. Classifications:
   a. Authors Class: See Wheat
   b. EPA Crop Group (Group & Subgroup): See Wheat
   c. Codex Group: See Wheat
   d. EPA Crop Definition: See Wheat
10. References: GRIN, See Wheat
11. Production Map: EPA Crop Production Region: See Wheat.
12. Plant Codes:
   a. Bayer Code: No specific entry

668

1. Wheatgrass
Poaceae (Gramineae)
   Agropyron spp.; *Elymus* spp.; *Elytrigia* spp.; *Pascopyrum* spp.; *Pseudoroegneria* spp.
2. The wheatgrasses are hardy, mainly perennial, erect grasses, important especially in the northern Great Plains of the U.S. and in Canada. The seed heads resemble wheat heads, hence

the name. They may form sods or grow in bunches. They are suitable feed for all classes of livestock. They are often drought-resistant perennial species suitable for livestock and wildlife and valued for soil stabilization and watershed management. In North America, wheatgrasses are most prevalent in the northern Great Plains as well as semiarid to arid rangeland areas of the Intermountain Great Basin regions. The wheatgrasses are members of the *Triticeae* grass tribe which also includes wheat, and hybridization occurs frequently among the perennial members of this tribe. They produce growth early in the spring. Recent taxonomic realignment of the wheatgrasses and wildryes based on genomic or biological relationships, as well as plant morphology has been completed, and is reflected in this book. The new taxonomic realignment limits *Agropyron* to the crested wheatgrass complex, while slender and bearded wheatgrass are included in the genus *Elymus*, intermediate and tall wheatgrass are included in the genus *Elytrigia*, bluebunch wheatgrass is included in the genus *Pseudoroegneria*, western wheatgrass is included in the genus *Pascopyrum*, and quackgrass in the genus *Elytrigia*. Around 150 species of wheatgrass (*Agropyron*) are known in the temperate regions of the world, about 30 in North America. The species most important in agriculture in the U.S. follow.

3. Season: Cool season crop, generally spring through summer.
4. Production in U.S.: Approximately 2,423,506 lbs. of wheatgrass seed was produced in the U.S. on 15,719 acres (U.S. CENSUS, 1992). More than 84 percent of the wheatgrass seed was produced in Montana, Utah, Colorado, Kansas, South Dakota and Washington. Primarily Great Plains area.
5. Other production regions: Important in British Columbia, Alberta and southern Saskatchewan, Canada.
6. Use: Rangeland grazing, forage, hay and silage. Also wheatgrass is used in health food preparations.
7. Part(s) of plant consumed: Leaves, stems and seedheads
8. Portion analyzed/sampled: Forage and hay
9. Classifications:
   a. Authors Class: See Forage grass
   b. EPA Crop Group (Group & Subgroup): See Forage grass
   c. Codex Group: See Forage grass
   d. EPA Crop Definition: None.
10. References: GRIN, MAGNESS, US EPA 1994, US EPA 1995, CODEX, BARNES (picture), ALDERSON, MOSLER, RICHE
11. Production Map: EPA Crop Production Regions 7, 8, 9 and 11.
12. Plant Codes:
   a. Bayer Code: See specific wheatgrass

669

1. **Wheatgrass, Bluebunch**

Poaceae (Gramineae)

Pseudoroegneria spicata (Pursh) A. Love (syn: *Agropyron spicatum* Pursh; *P. spicata* ssp. *spicata* (Pursh) A. Love)

1. **Beardless bluebunch wheatgrass**
(Beardless wheatgrass)

P. spicata (Pursh) A. Love (syn: *P. spicata* ssp. *inermis* (Scribn. & J.G. Sm.) A. Love; *Agropyron spicatum f. inerme* (Scribn. & J.G. Sm.) Beetle; *A. inerme* (Scribn. & J.G. Sm.) Rydb.)

2. Bluebunch wheatgrass is a perennial cool-season drought-resistant bunchgrass native to the dry areas of the Western

states. It is a dominant species in the Pacific Northwest and Intermountain states, and north into Canada. Both taxa are very similar, except the beardless lacks the awns, making the plants more palatable in late stages of growth. Plant growth is vigorous, starting early in spring. Plants may reach a height of 4 feet. Leaves are up to 10 inches long and $^1/_2$ inch wide, flat and tending to droop. They often grow in association with sagebrushes (*Artemisa* spp.). They are valuable native grasses for the intermountain region, and are adapted to the same soil and climatic conditions as crested wheatgrass with elevations less than 855m and rainfall less than 16 inches. It is considered one of the most valuable rangeland forage grasses. Leaves remain green throughout the summer and are palatable even when dry. Stands may be depleted under heavy grazing pressure and livestock must be removed before the boot stage is reached. Good animal feed for horses. Propagation is by seed. For seed production row spacing is 24 inches in irrigated areas and 36 inches or wider in dryland areas with 25 to 30 seeds/linear foot. There are 140,000 seeds/lb for bluebunch wheatgrass, and 125,000 seeds/lb for beardless bluebunch wheatgrass.

3. Season: Cool season, spring growth through early summer and also good forage in the fall and early winter, and for seed production harvest by mid-July. For seed production, stands persist for 3 years. Grazing spring and early summer, stored for hay in winter. Grazing wheatgrass lightly or cutting for hay in late fall after the last killing frost is recommended. Grazing livestock recommended until the late boot stage in the spring and summer.
4. Production in U.S.: No data for Bluebunch wheatgrass production, however, pasture and rangeland are produced on over 410 million acres (U.S. CENSUS, 1992). Adapted to USDA Plant Hardiness Zone 5 for Beardless bluebunch wheatgrass, and zones 5 and 6 for Bluebunch wheatgrass.
5. Other production regions: In Canada, these wheatgrasses are grown in British Columbia.
6. Use: Rangeland grazing, forage, and hay.
7. Part(s) of plant consumed: Leaves, stems and seedheads
8. Portion analyzed/sampled: Forage and hay
9. Classifications:
   a. Authors Class: See Forage grass
   b. EPA Crop Group (Group & Subgroup): See Forage grass
   c. Codex Group: See Forage grass
   d. EPA Crop Definition: None
10. References: GRIN, MAGNESS, US EPA 1994, US EPA 1995, CODEX, BARNES, ALDERSON, HOLZWORTH, STUBBENDIECK (picture), MOSLER, RICHE
11. Production Map: EPA Crop Production Regions 7, 8, 9, 11 and 12 for Bluebunch wheatgrass, and Region 9 for Beardless bluebunch wheatgrass.
12. Plant Codes:
   a. Bayer Code: No specific entry

670

1. **Wheatgrass, Crested**

(Desert wheatgrass, Standard crested wheatgrass, Desert crested wheatgrass)

Poaceae (Gramineae)

Agropyron desertorum (Fisch. ex Link) Schult. (syn. *Agropyron cristatum* (L.) Garetn. ssp. *desertorum* (Fisch. ex Link) A. Love)

2. This is a hardy cool-season, perennial bunchgrass, originally introduced from eastern Russia, western Siberia, and Central Asia. Leaves are abundant, both at the base and along the stems. It is a very important rangeland grass, and useful for revegetation of the western rangelands. Leaves are 6 to 10 inches long, about $1/4$ inch wide, flat and slightly hairy on the upper surface. Stems are slender, 2 to 3 feet in height, growing in dense clumps. It is deep-rooted and drought resistant, well adapted to the Northern Plains and higher elevations in the Rocky Mountains and the arid intermountain regions, and the prairie provinces of Canada. Adapted to areas with 8 to 18 inches rainfall. Growth starts early in spring. Crested wheatgrass is credited with salvaging vast areas of deteriorated rangelands during the depression and dust bowls in the 1930s. If cut early, excellent-quality hay is produced. This is a highly valuable grass in its area of adaptation. It is also good for soil stabilization and as a biological suppressant of noxious weeds. Once established, stands will persist for many years (>30 years). Propagation is by seeding. For seed production, row spacing is 25 to 30 inches in irrigated areas and 36 inches in dryland areas. There are 188,000 seeds/lb.
3. Season: Cool season, very productive in the spring and harvest in August for seed production. Grazing spring, early summer, and in the fall it is cut for hay.
4. Production in U.S.: There are more than 12.6 million acres seeded in North America to crested wheatgrass (BARNES 1995). Adapted to USDA Plant Hardiness Zones 2-4.
5. Other production regions: In Canada, there are 15,951 acres (ROTHWELL 1996a).
6. Use: Rangeland grazing, forage, hay and silage.
7. Part(s) of plant consumed: Leaves, stems and seedheads
8. Portion analyzed/sampled: Forage and hay
9. Classifications:
   a. Authors Class: See Forage grass
   b. EPA Crop Group (Group & Subgroup): See Forage grass
   c. Codex Group: See Forage grass
   d. EPA Crop Definition: None
10. References: GRIN, MAGNESS, US EPA 1994, US EPA 1995, CODEX, BARNES (picture), ALDERSON, HOLZWORTH, ROTHWELL 1996a, MOSLER, RICHE
11. Production Map: EPA Crop Production Regions 7, 8, 9 and 11.
12. Plant Codes:
   a. Bayer Code: No specific spp. entry, genus entry AGRSS

671

1. **Wheatgrass, Fairway**
(Fairway crested wheatgrass, Crested wheatgrass)
Poaceae (Gramineae)
Agropyron cristatum (L.) Gaertn.
2. Fairway is a cool season bunch wheatgrass that spreads by rhizomes and is of Siberian origin more widely grown in Canada than in the U.S. It is considered a complex of crested wheatgrass subspecies and generally resembles crested wheatgrass but is finer stemmed and shorter (0.2 to 1 m tall) than crested and yields less forage. It is especially suited for dryland lawns and other turf planting because of its dense growth and relatively fine texture. It is used extensively to reseed sagebrush areas. It is also used extensively for rangeland, pasture and hay in western Canada and to a more limited extent in the northern Great Plains and Intermountain

Region in the U.S. Propagation is by seeds. For seed production, row spacing is 24 inches in irrigated areas and 36 inches or wider in dryland areas with 25 to 30 seeds/linear foot. There are 200,000 seeds/lb.
3. Season: Cool season and is most productive in the early spring. It can be grazed or cut for hay, and for seed production harvest in August.
4. Production in U.S.: No data for Fairway wheatgrass production, however, pasture and rangeland are produced on over 410 million acres (U.S. CENSUS, 1992). Adapted to USDA Plant Hardiness Zones 4, 5, 6 and 7.
5. Other production regions: In Canada it is grown in western areas.
6. Use: Rangeland grazing, forage, and hay and reseeding.
7. Part(s) of plant consumed: Leaves, stems and seedheads
8. Portion analyzed/sampled: Forage and hay
9. Classifications:
   a. Authors Class: See Forage grass
   b. EPA Crop Group (Group & Subgroup): See Forage grass
   c. Codex Group: See Forage grass
   d. EPA Crop Definition: None
10. References: GRIN, MAGNESS, US EPA 1994, US EPA 1995, CODEX, BARNES, ALDERSON, USDA 1995, HOLZWORTH, MOSLER, RICHE
11. Production Map: EPA Crop Production Regions 8, 9 and 11.
12. Plant Codes:
   a. Bayer Code: AGRCR

672

1. **Wheatgrass, Intermediate**
(Crested intermediate wheatgrass)
Poaceae (Gramineae)
Elytrigia intermedia ssp. *intermedia* (Host) Nevski (syn: *Thinopyrum intermedium* (Host) Barkworth & D.R. Dewey; *Agropyron intermedium* (Host) P. Beauv.)
2. This is a perennial cool season sod-forming wheatgrass, introduced from Eurasia. It has proved well adapted to the Northern and Central Great Plains, the Pacific Northwest and Canada. It is a little less hardy and drought resistant than Crested wheatgrass. It is adapted to areas with rainfall >350 mm and elevations up to 3000 m. It is suited to be grown with alfalfa under dryland conditions. Plant growth is vigorous, and is relished by all classes of livestock. Stems reach a height of 3 to 4 feet. For both pasture and hay production, this is a valuable grass for its area of adaptation, as well as good for soil erosion control. Forage quality declines rapidly at advanced stages of maturity. It is readily established by seeding, and is increasing in importance on marginal lands where orchardgrass and bromegrass are not well adapted. Often grown with alfalfa for hay or pasture in areas with a shortage of water. For seed production, row spacing is 24 inches in irrigated areas and 36 inches in dryland areas with 25 to 30 seeds/linear foot. There are 79,000 seeds/lb. It reproduces by seed, tiller and rhizome.
3. Season: Cool season, spring growth and matures June-August, and for seed production harvest by mid-August. For seed production stands persist for up to 5 years. Grazing spring, early summer and can also fall graze or cut for hay.
4. Production in U.S.: No data for Intermediate wheatgrass production, however, pastures and rangelands are produced on

more than 410 million acres (U.S. CENSUS, 1992). Adapted to USDA Plant Hardiness Zones 2-6.

5. Other production regions: In Canada, Intermediate wheatgrass grows in Manitoba, Saskatchewan, Alberta, and British Columbia. In 1995, Canada planted about 16,000 acres (ROTHWELL 1996a).

6. Use: Rangeland grazing, forage, and hay.

7. Part(s) of plant consumed: Leaves, stems and seedheads

8. Portion analyzed/sampled: Forage and hay

9. Classifications:
   a. Authors Class: See Forage grass
   b. EPA Crop Group (Group & Subgroup): See Forage grass
   c. Codex Group: See Forage grass
   d. EPA Crop Definition: None

10. References: GRIN, MAGNESS, US EPA 1994, US EPA 1995, CODEX, SCHREIBER, BARNES, ALDERSON, HOLZWORTH, ROTHWELL 1996a, STUBBENDIECK, MOSLER (picture), RICHE

11. Production Map: EPA Crop Production Regions 5, 6, 7, 8, 9, 11 and 12.

12. Plant Codes:
    a. Bayer Code: No specific entry

673

1. Wheatgrass, Pubescent
(Intermediate pubescent wheatgrass)

Poaceae (Gramineae)

Elytrigia intermedia ssp. *intermedia* (Host) Nevski (syn: *Thinopyrum intermedium* ssp. *barbulatum* (Schur) Barkworth & D.R. Dewey; *Agropyron intermedium* var. *trichophorum* (Link) Halac.)

2. This is a cool season, sod-forming grass with creeping rhizomes native to Eurasia. It's closely related to Intermediate wheatgrass and has the same general range of adaptation. It is better adapted to the southern part of the region. The heads and seeds of Pubescent wheatgrass are covered with short, stiff hairs. A named cultivar of Pubescent, 'Topar', developed in 1964, is better adapted to low-fertility and saline soils and is more drought resistant than Intermediate wheatgrass. It is adapted to areas with rainfall >380 mm and elevation <3050 m. In palatability and general appearance, Pubescent wheatgrass is very similar to Intermediate wheatgrass. Propagation is by seeds. For seed production, row spacing is 24 inches in irrigated areas and 36 inches in dryland areas with 25 to 30 seeds/linear foot. There are 80,000 seeds/lb.

3. Season: Cool season. Is most productive in the early spring. It can be grazed or cut for hay in the spring and fall, and for seed production harvest in August. Stands are productive for 7 to 10 years.

4. Production in U.S.: No data for Pubescent wheatgrass production, however, pasture and rangeland are produced on over 410 million acres (U.S. CENSUS, 1992). Adapted to USDA Plant Hardiness Zones 2, 3, 4 and 5.

5. Other production regions: In Canada, it is grown in Manitoba.

6. Use: Rangeland grazing, forage and hay and reseeding.

7. Part(s) of plant consumed: Leaves, stems and seedheads

8. Portion analyzed/sampled: Forage and hay

9. Classifications:
   a. Authors Class: See Forage grass

b. EPA Crop Group (Group & Subgroup): See Forage grass
c. Codex Group: See Forage grass
d. EPA Crop Definition: None

10. References: GRIN, MAGNESS, US EPA 1994, US EPA 1995, CODEX, BARNES, ALDERSON, HOLZWORTH, RICHE

11. Production Map: EPA Crop Production Regions 6, 7, 9, 10 and 11.

12. Plant Codes:
    a. Bayer Code: No specific entry

674

1. Wheatgrass, Siberian
(Siberian crested wheatgrass)

Poacece (Gramineae)

Agropyron fragile ssp. *sibiricum* (Willd.) Melderis (syn: *Agropyron sibiricum* (Willd.) P. Beauv.)

2. This is a drought-resistant cool season bunchgrass introduced from Russia in 1934. It is generally similar to Crested wheatgrass in use but appears to do better than Crested on poor sites or under other adverse conditions. For this reason it is replacing Crested wheatgrass in some areas. The stems are finer and heads more narrow than in Crested wheatgrass. Nutritive value and palatability are similar to those of Crested wheatgrass. It is also useful for reseeding.

3. Season: Cool season. Is most productive in the early spring. It can be grazed or cut for hay, and for seed production harvest in August.

4. Production in U.S.: No data for Siberian wheatgrass production, however, pasture and rangeland are produced on over 410 million acres (U.S. CENSUS, 1992). Adapted to USDA Plant Hardiness Zones 5.

5. Other production regions: See above.

6. Use: Rangeland grazing, forage and hay and range reseeding.

7. Part(s) of plant consumed: Leaves, stems and seedheads

8. Portion analyzed/sampled: Forage and hay

9. Classifications:
   a. Authors Class: See Forage grass
   b. EPA Crop Group (Group & Subgroup): See Forage grass
   c. Codex Group: See Forage grass
   d. EPA Crop Definition: None

10. References: GRIN, MAGNESS, US EPA 1994, US EPA 1995, CODEX, BARNES, ALDERSON, HOLZWORTH, RICHE

11. Production Map: EPA Crop Production Regions 9 and 11.

12. Plant Codes:
    a. Bayer Code: No specific sp. entry, genus entry AGRSS

675

1. Wheatgrass, Slender
(Agropiro delgado)

Poaceae (Gramineae)

Elymus trachycaulus ssp. *trachycaulus* (Link) Gould ex Shinners (syn: *Agropyron trachycaulum* (Link) Malte ex H. F. Lewis)

2. This is a native bunchgrass generally distributed throughout the U.S., except in the southeastern and south central regions. It is prevalent in the Northern Great Plains and the Rocky Mountain states, as far east as North Carolina and as

far north as Alaska. It is a short-lived perennial (2 to 5 years) that grows to 3 feet in dense leafy clumps or bunches, a foot or more in diameter. The flowering stems are erect and rather coarse. Most of the leaves are basal. They are up to 1 foot long and $^1/_2$ inch wide. A great deal of morphological variation is found in slender wheatgrass types such as having either smooth or hairy leaves. This grass furnishes abundant pasture and a nutritious hay for cattle and sheep if harvested early. Adapted to areas receiving at least 35 cm of annual precipitation, and an altitude <2000m. The forage also matures well on the ground, so furnishes winter grazing. Slender wheatgrass is less drought tolerant than most wheatgrasses, including crested and bluebunch wheatgrass. Stands may easily be overgrazed, and are not as persistent as the sod-forming wheatgrasses, and is less drought tolerant than most wheatgrasses. Its major uses are for revegetation programs because it establishes rapidly or as a green manure crop in the western U.S. and Canada. Propagation is by seeds. For seed production, row spacing is 25 to 30 inches in irrigated areas and 36 inches or wider in dryland areas with 25 to 30 seeds/linear foot. There are 140,000 seeds/lb.

3. Season: Cool season, spring growth through early summer, seed in late August or early spring, and for seed production harvest July and August. For seed production stands persist for 3 years. Grazing spring, early summer, and fall and store for hay in winter. Grazing livestock recommended until the late boot stage.

4. Production in U.S.: No data for Slender wheatgrass production, however, pasture and rangeland are produced on over 410 million acres (U.S. CENSUS, 1992). Adapted to USDA Plant Hardiness Zones 3 and 5.

5. Other production regions: In Canada it is grown in Saskatchewan and western provinces and as far north as Alaska.

6. Use: Rangeland grazing, forage, and hay, cover crop and soil stabilization.

7. Part(s) of plant consumed: Leaves, stems and seedheads

8. Portion analyzed/sampled: Forage and hay

9. Classifications:
   a. Authors Class: See Forage grass
   b. EPA Crop Group (Group & Subgroup): See Forage grass
   c. Codex Group: See Forage grass
   d. EPA Crop Definition: None

10. References: GRIN, MAGNESS, US EPA 1994, US EPA 1995, CODEX, BARNES, ALDERSON, HOLZWORTH, STUBBENDIECK (picture), MOSLER, RICHE

11. Production Map: EPA Crop Production Regions 7, 8, 9, 10 and 11.

12. Plant Codes:
   a. Bayer Code: AGRTR

676

1. Wheatgrass, Streambank

Poaceae (Gramineae)
   Elymus lanceolatus ssp. *lanceolatus* (Scribn. & J.G. Sm.) Gould (syn: *Agropyron riparium* Scribn. & J.G. Sm.)

2. This cool season grass is native from Alberta and British Columbia in Canada south to Nevada and Colorado, resembles Thickspike wheatgrass except that leaf blades are more narrow. It develops vigorous rhizomes, resulting in dense sods. It is drought tolerant and especially valuable for erosion control. Top growth is short and it produces less forage

than some other wheatgrasses. Its greatest value is for sods on airports, road banks and irrigation canal banks and rangeland reseeding. Propagation is by seeds. For seed production, row spacing is 36 inches in irrigated and dryland areas with 25 to 30 seeds/linear foot. There are 152,000 seeds/lb.

3. Season: Cool season and is most productive in the early spring. It can also be fall grazed or cut for hay, and for seed production harvest in mid-August. Stands are productive for up to five years.

4. Production in U.S.: No data for Streambank wheatgrass production, however, pasture and rangeland are produced on over 410 million acres (U.S. CENSUS, 1992). Adapted to USDA Plant Hardiness Zones 4 and 5. See above.

5. Other production regions: In Canada it is grown from Alberta to British Columbia.

6. Use: Rangeland grazing, forage, and hay and reseeding.

7. Part(s) of plant consumed: Leaves, stems and seedheads

8. Portion analyzed/sampled: Forage and hay

9. Classifications:
   a. Authors Class: See Forage grass
   b. EPA Crop Group (Group & Subgroup): See Forage grass
   c. Codex Group: See Forage grass
   d. EPA Crop Definition: None

10. References: GRIN, MAGNESS, US EPA 1994, US EPA 1995, CODEX, BARNES, ALDERSON, HOLZWORTH, RICHE

11. Production Map: EPA Crop Production Regions 9 and 11.

12. Plant Codes:
   a. Bayer Code: No specific entry

677

1. Wheatgrass, Tall

Poaceae (Gramineae)
   Elytrigia elongata (Host) Nevski (syn: *Agropyron elongatum* (Host) Beauv.)

2. Tall wheatgrass is a cool season winter hardy, tall-growing, coarse-textured bunchgrass native to the eastern Mediterranean region, introduced from Turkey in 1909. It is adapted to growing on wet, alkaline and saline soils and has been used extensively for seeding such sites in the Northern Plains and Intermountain regions with rainfall in the range of 35 to 40 cm annually, and elevations <1700 m. Leaves are erect and blue-green and Tall wheatgrass is the latest maturing wheatgrass adapted to the temperate western growing regions. Clumps reach up to 6 feet and yield heavily with sufficient available moisture. It is less drought resistant and less palatable than crested wheatgrass, but is useful for both hay and pasturage on soils not suitable for the other wheatgrasses, as well as an excellent wind barrier grass for soil stabilization. Tall wheatgrass remains green 3 to 4 weeks longer than other rangegrasses, and is a valuable forage late in the season. Tall wheatgrass is related to the intermediate wheatgrass complex and is being used in breeding programs with wheat as a source of resistance to drought and salinity. Seeding rate is 22 kg/ha. There are 79,000 seeds/lb.

3. Season: Cool season. Is a good source of pasture and hay in the late summer, and is also used for silage. It should be left with a stubble height of 20 cm at the end of the season to prevent overgrazing. It is usually recommended for cattle and sheep.

4. Production in U.S.: No data for Tall wheatgrass production, however, pasture and rangeland are produced on over 410 million acres (U.S. CENSUS, 1992). Adapted to USDA Plant Hardiness Zones 3, 4 and 5. See above.
5. Other production regions: Reported in Canada (ROTHWELL 1996a).
6. Use: Rangeland grazing, pasture, forage, hay and silage.
7. Part(s) of plant consumed: Leaves, stems and seedheads
8. Portion analyzed/sampled: Forage and hay
9. Classifications:
   a. Authors Class: See Forage grass
   b. EPA Crop Group (Group & Subgroup): See Forage grass
   c. Codex Group: See Forage grass
   d. EPA Crop Definition: None
10. References: GRIN, MAGNESS, US EPA 1994, US EPA 1995, CODEX, BARNES, ALDERSON, HOLZWORTH (picture), RICHE, ROTHWELL 1996a
11. Production Map: EPA Crop Production Regions 5, 6, 7, 8 and 9.
12. Plant Codes:
    a. Bayer Code: AGREL

678

1. Wheatgrass, Thickspike
(Northern wheatgrass)
Poaceae (Gramineae)
   Elymus lanceolatus ssp. *lanceolatus* (Scribn. & J.G. Sm.) Gould (syn: *Agropyron dasystachym* (Hook.) Scribn.)
2. Species of the genus *Elymus* are the most widely distributed and diverse of the *Triticeae* grass tribe, however, only Thickspike and Slender wheatgrass are agronomically important. Thickspike wheatgrass is a sod-forming, cross-pollinating perennial wheatgrass native from the Hudson Bay area and Alaska south to Nevada and Colorado. It is closely related to Streambank wheatgrass and is the smooth (glabrous) form. Stems reach to 3 feet. The creeping rhizomes result in formation of fairly dense drought resistant sod. It does well on light textured soils and is very hardy. There are two basic growth types: one is more procumbent and is valuable as a low maintenance soil stabilization ground cover crop, while the upright type is the preferred forage type. It is more drought resistant than western wheatgrass. Growth starts early and provides good pasturage while succulent, but becomes tough and wiry late in the season. It is also good for soil stabilization in areas up to 915 m elevation with 305 mm rainfall. It is rarely seeded but is an important native grass over a wide area. Propagation is by seeds. For seed production, row spacing is 24 inches in irrigated areas and 36 inches or wider in dryland areas with 25 to 30 seeds/linear foot. There are 145,000 seeds/lb.
3. Season: Cool season, most productive in the early summer and can be grazed or cut for hay, and for seed production harvest July and August. For seed production stands persist for up to 5 years. It will not persist well under heavy grazing pressure.
4. Production in U.S.: No data for Thickspike wheatgrass production, however, pasture and rangeland are produced on over 410 million acres (U.S. CENSUS, 1992). Adapted to USDA Plant Hardiness Zones 3, 4, 5 and 6.
5. Other production regions: In Canada, it is mainly grown in British Columbia.

6. Use: Rangeland grazing, forage and hay and soil stabilization crop.
7. Part(s) of plant consumed: Leaves, stems and seedheads
8. Portion analyzed/sampled: Forage and hay
9. Classifications:
   a. Authors Class: See Forage grass
   b. EPA Crop Group (Group & Subgroup): See Forage grass
   c. Codex Group: See Forage grass
   d. EPA Crop Definition: None
10. References: GRIN, MAGNESS, US EPA 1994, US EPA 1995, CODEX, BARNES, ALDERSON, HOLZWORTH, MOSLER, RICHE
11. Production Map: EPA Crop Production Regions 7, 9, 11 and 12.
12. Plant Codes:
    a. Bayer Code: No specific entry

679

1. Wheatgrass, Western
(Bluestem, Wyoming wheatgrass, Colorado bluestem, Bluestem wheatgrass, Bluejoint, Agropiro del oeste)
Poaceae (Gramineae)
   Pascopyrum smithii (Rydb.) A. Love (syn: *Agropyron smithii* Rydb.; *Elymus smithii* (Rydb.) Gould)
2. Western wheatgrass is a long-lived perennial, cross pollinated, and sod-forming grass by rhizomes, native to most parts of the U.S. except the humid Southeast. It is a dominant species in the central and northern Great Plains, southeast U.S., and in Canada from Ontario to Alberta. Plant growth is vigorous, reaching 2 to 3 feet in height. Leaves are up to 12 inches long, $^1/_4$ inch wide, rather stiff and erect. The whole plant is covered with a grayish bloom, and the leaves are typically bluegreen in color. It thrives best on rather heavy soil, but is adapted to a wide range of soil types, including alkaline soil, and can withstand period of drought. It grows up to elevations of 1,530 m and annual rainfall less than 400 mm. It is relished by all classes of livestock both as pasturage and when cut for hay while still succulent. The plants are usually grown from seed, but spread from underground rhizomes to form dense sods. Western wheatgrass is often confused with Thickspike wheatgrass. Western wheatgrass is commonly associated with Blue grama and Needlegrass in the Great Plains and with Bluebunch wheatgrass and Thickspiked bluegrass in intermountain regions. This is a very valuable grass, both for feed and for wind erosion control. It is recommended for revegetating saline soils and surface mining areas. It is hard to establish by seeding but can be vegetatively propagated in small areas to stabilize the soil. For seed production, row spacing is 36 inches in irrigated areas and 36 inches or wider in dryland areas with 25 to 30 seeds/linear foot. There are 114,000 seeds/lb.
3. Season: Cool season, spring growth through early summer and also good forage in the fall and early winter up to the last killing frost, and for seed production harvest by mid-August. For seed production, stands persist for 3 years. Grazing spring and early summer, stored for hay in winter.
4. Production in U.S.: No data for Western wheatgrass production, however, pasture and rangeland are produced on more than 410 million acres (U.S. CENSUS, 1992). Adapted to USDA Plant Hardiness Zones 2-5.

5. Other production regions: In Canada, Western wheatgrass grows from Ontario west through Alberta.
6. Use: Rangeland grazing, forage, and hay.
7. Part(s) of plant consumed: Leaves, stems and seedheads
8. Portion analyzed/sampled: Forage and hay
9. Classifications:
   a. Authors Class: See Forage grass
   b. EPA Crop Group (Group & Subgroup): See Forage grass
   c. Codex Group: See Forage grass
   d. EPA Crop Definition: None
10. References: GRIN, MAGNESS, US EPA 1994, US EPA 1995, CODEX, BARNES, ALDERSON, HOLZWORTH, STUBBENDIECK (picture), MOSLER, RICHE
11. Production Map: EPA Crop Production Regions 6, 7, 8, 9 and 11.
12. Plant Codes:
    a. Bayer Code: AGRSM

680

1. **Wild leek**
(Ramp, Wood leek)
Liliaceae (Amaryllidaceae)
Allium tricoccum Aiton

2. While the wild leek is not a cultivated crop, it is harvested in some areas. It is a strongly scented and flavored onion relative, indigenous from New England to Wisconsin and south to the Carolinas and neighboring states. In the mountains of West Virginia residents celebrate each year the coming of the "ramp" with "ramp" festivals which feature ramp cooked with Canadian bacon fat. Both the small bulbs and the leaves are eaten.
3. Season, harvest: Harvested from wild in late summer.
4. Production in U.S.: No data. Native from New England to Wisconsin and south to the Carolinas. Also grows well in Florida (STEPHENS 1988).
5. Other production regions: No entry.
6. Use: Mainly as flavoring.
7. Part(s) of plant consumed: Leaves and bulbs
8. Portion analyzed/sampled: Whole plant without roots
9. Classifications:
   a. Authors Class: Bulb vegetables
   b. EPA Crop Group (Group & Subgroup): Bulb vegetables
   c. Codex Group: No specific entry
   d. EPA Crop Definition: Onions = green onions
10. References: GRIN, BAILEY 1976, FACCIOLA, MAGNESS, STEPHENS 1988, US EPA 1995a
11. Production Map: EPA Crop Production Regions 1, 2, 3 and 5.
12. Plant Codes:
    a. Bayer Code: ALLTC

681

1. **Wild rice**
(Indian Rice, Water Oats, Northern wildrice, American wildrice, Manomin, Canadian rice, Cultivated wildrice, Squaw rice, Lake rice, Folle avoine, Rice wild, Tame wildrice)
Poaceae (Gramineae)
Zizania palustris L.
1. **Wildrice, Eastern**
(Southern wildrice, Annual wildrice, Riz d'eau)
Z. aquatica L.

2. Wild rice is not closely related to cultivated rice, belonging to a different genera and even a different tribe of the grass family. The plants are native to North America, growing in shallow lakes and ponds from southern Canada south to Florida. While much variation exists in size of plant, size of seed and other characteristics in plants from different locations, all are included in *Z. aquatica* and *Z. palustris*. The plant is an aquatic annual grass with stems hollow except at the nodes. It grows only in water, preferably in lakes or ponds that are not stagnant yet have very little current. Growth is usually in water 3 feet or less in depth, with a mud bottom. Stems may reach 3 to 6 feet above the water surface and are terminated by large branched flower panicles. The basal part of the panicles bears only staminate flowers, while the terminal part bears only pistillate flowers and sets the seed crop. The seed is tightly enclosed in an outer palea and inner lemma. The lemma terminates in a stiff, twisted awn. In the wild, seed can be harvested from boats by bending the stems over the boat and beating the heads with a stick. In commercial production, harvesting machines are in use. The harvested seed is enclosed in the palea and lemma and is quite moist. Seed does not mature uniformly and when mature, drops into the water. Therefore, successive harvests are necessary to obtain a maximum yield from a location. Sufficient seed is always lost to maintain a stand under suitable growing conditions. The seed is composed of about 14 percent hull, 77 percent endosperm, 4 percent pericarp and 4 percent embryo.

Following harvest the moist seed must be quickly dried to avoid spoilage. For food use it is dried by parching, which also loosens the hulls. Parching is done by heating in a rotating drum or in an open kettle with constant stirring to prevent burning. Hulls are separated from the grains by tramping or pounding in a pit and putting through a fanning mill. In California, the green rice (35 to 45 percent moisture) is processed by cleaning, curing, parboiling, drying and shelling (dehulling). The hulls go to landfills and the straw is usually left in the field to decompose or is burned. The separated kernels are near cylindrical, 12 to 20 mm long by 2 mm or less in thickness. They are highly esteemed as food by the Native Americans and by others familiar with the taste. The seed is also an excellent wildlife feed.

When not a cultivated crop, favorable sites are often seeded in hunting preserves for wildlife feed and also for grain harvest.

Primarily, wild rice for food is harvested annually in the north central states, California and Canada, much by Native Americans. Wild rice is unusually high in protein content and low in fat compared with other cereals. It normally sells at a price 2 to 3 times that of cultivated rice. Eastern wildrice is *Z. aquatica* which is not cultivated, while *Z. palustris* is cultivated. In 1963, a cultivar with reduced grain shattering was developed and allowed the use of harvesting equipment similar to cultivated rice. Cultivated wildrice is planted on dry ground and flooded to 6 inches with air seeding at 30 to 45 lbs per acre required the first year. The paddy will reseed itself and produce for 3 to 4 years in Minnesota. Fields are drained 3 weeks before harvest. Yields range from 50 to 1,750 lbs per acre. Approximately, 2.5 lbs of grain are processed into 1 lb of dry processed grain. Crops rotated with cultivated wildrice are buckwheat, rye, wheat, mustards and forage grasses for seed production.

3. Season, seeding to harvest: Commercial crop, similar culture to rice. About 106 to 130 days to mature after seeding. Seed in fall in Minnesota and in the spring in California.
4. Production in U.S.: In 1992, 34,437 acres with 11,604 tons (1992 CENSUS). In 1992, the top three states reported as Minnesota (21,717 acres), California (11,739) and Idaho (665). In production, California accounted for over half (7,340 tons) and Minnesota second at 4,200 tons (1992 CENSUS).
5. Other production regions: Canada.
6. Use: See above. Edible grain cooked like rice, and used for flour, wildlife feed and shelter.
7. Part(s) of plant consumed: Grain and straw. Straw is insignificant feed item, since plant residues are normally left in the field
8. Portion analyzed/sampled: Grain. See above for straw, if ever a significant feed item, the straw would be sampled
9. Classifications:
   a. Authors Class: Cereal grains; Forage, fodder, and straw of cereal grains
   b. EPA Crop Group (Group & Subgroup): Cereal grain (15), Forage, fodder and straw of cereal grains (16)
   c. Codex Group: 020 (GC 0655) Cereal grains
   d. EPA Crop Definition: None
10. References: GRIN, CODEX, MAGNESS, RICHE, US EPA 1995a, US EPA 1994, OELKE (a & i)
11. Production Map: EPA Crop Production Regions 5 and 10.
12. Plant Codes:
   a. Bayer Code: ZIZAQ (*Z. aquatica*), ZIZPA (*Z. palustris*)

682

1. Wildrye grass
Poaceae (Gramineae)
Elymus spp.
Leymus spp.
Psathyrostachys spp.
1. Canada wildrye
(Centeno silvestre)
E. canadensis L.
1. Giant wildrye
L. condensatus (J. Presl) A. Love (syn: *E. condensatus* J. Presl)
1. Blue wildrye
E. glaucus Buckley
1. Russian wildrye
Psathyrostachys juncea (Fisch.) Nevski (syn: *E. junceus* Fisch.)
1. Basin wildrye
L. cinereus (Scribn. & Merr.) A. Love (syn: *E. cinereus* Scribn. & Merr.)

2. Wildrye species are widely distributed among the native cool season grasses of the Western states. Most forage species are perennial bunch growers, but some form sods. These grasses are coarse and of low palatability for livestock. Because of their vigor and ease of establishment they form a quick cover and are useful in mixtures with slower growing kinds. Wildrye grasses are susceptible to ergot fungus, which replaces the seed kernel and is highly toxic to livestock. A recent taxonomic realignment of wildrye grasses separated them into three genera. The most useful wildrye species are described as follows.

Canada wildrye grass is a native bunchgrass in the Plains, Rocky Mountain and Pacific Northwest states. Seed heads may reach to 5 feet. Leaf blades are broad, flat and rough, up to 12 inches long and $1/2$ inch or more broad. Growth begins later in spring than most grasses, but continues throughout summer if moisture is available. Palatability while succulent is fair, but poor when the plants become woody. Good quality hay can be obtained with early mowing. It is usually seeded in combination with other slower-growing grasses in order to obtain a quick cover.

Giant wildrye grass, native throughout the Western states, is the most robust of the native rye grasses, reaching to 10 feet. It is a perennial bunchgrass, forming large clumps. Leaves are large, up to 2 feet long and $3/4$ inch wide. It grows abundantly on wet and saline soils, but also occurs on moderately dry soils. Giant wildrye is suitable for grazing while succulent. The ripened clumps provide winter sustenance feed for cattle and horses. Propagation is by sprigs.

Blue wildrye is a perennial bunchgrass native throughout the Western states. It grows in small tufts, reaching up to 5 feet. Leaves are broad and flat, up to 12 inches long. It is abundant on moist soils but will tolerate drought. The coarse leaves are relished by cattle, particularly while succulent. It is shade tolerant and gives high yields of pasturage or of good hay if cut early. It is propagated by seeds but has not been widely planted.

Russian wildrye is a cool season bunch grass introduced in 1926 from Russia. It has proved adapted to the northern Great Plains, intermountain regions and Canadian prairie, where it is used primarily for pasture. Growth starts early in spring. Plants are leafy and nutritious, with dense basal leaves. Seedling vigor is low, but once established, plants are deep rooted, drought resistant and salt tolerant.

Basin wildrye is a cool season, slightly spreading grass found throughout the Western states and Canada, particularly on alkaline soils. It is a tall, coarse grower, sometimes reaching to 10 feet. It is relatively low in palatability and is not planted commercially but affords emergency summer or winter pasturage where present. Propagation is by seed and tiller.

3. Season: Growth starts early in the spring. Cool season pasture, used fall and early winter for grazing. Cut at boot stage for hay.
4. Production in U.S.: No specific data, See Forage grass. Primarly the Western states and northern Great Plains.
5. Other production regions: Canada.
6. Use: Pasture, erosion control, rangegrass, cover crop, silage.
7. Part(s) of plant consumed: Foliage
8. Portion analyzed/sampled: Forage and hay
9. Classifications:
   a. Authors Class: See Forage grass
   b. EPA Crop Group (Group & Subgroup): See Forage grass
   c. Codex Group: See Forage grass
   d. EPA Crop Definition: None
10. References: GRIN, MAGNESS, MOSLER (picture), STUBBENDIECK (picture), BARNES, ALDERSON
11. Production Map: General EPA Crop Production Regions 5, 7, 9, 10, 11 and 12. Specific Regions for Canada wildrye: 7, 9 and 11; Giant wildrye: 5, 7, 9 and 11; Blue wildrye: 9, 11 and 12; Russian wildrye: 7, 9 and 11; and Basin wildrye: 7, 9, 10 and 11.
12. Plant Codes:
   a. Bayer Code: ELYCA (*E. canadensis*), ELYJU (*P. juncea*)

683

1. Wintergreen
(Boxberry, Checkerberry, Partridge berry, Teaberry, Mountain tea, Ivry-leaves, Creeping wintergreen)
Ericaceae
Gaultheria procumbens L.

2. Evergreen perennial plant with creeping stems, which send up upright branches about 5 inches high, bearing dark green, oval glabrous leaves and small, scarlet fruit. Oil of wintergreen is distilled from the leaves. The plant is native throughout eastern U.S., from Georgia northward, in woodland areas. The oil is used in flavoring and medicine. The plant is not cultivated in the U.S., as a cheaper source of a similar oil is birch (*Betula lenta* L.), or a synthetic product.
3. Season, start of growth to harvest: About 2 months. Flowering June to August and fruits in the fall.
4. Production in U.S.: Not cultivated in the U.S. Plant is native throughout eastern U.S. from Georgia north.
5. Other production regions: No entry.
6. Use: Flavoring and medicine.
7. Part(s) of plant consumed: Oil distilled from leaves
8. Portion analyzed/sampled: Leaves (Fresh and dried)
9. Classifications:
   a. Authors: Class: Herbs and spices
   b. EPA Crop Group (Group & Subgroup): Herbs and spices (19A)
   c. Codex Group: 027 (HH 0752) Herbs; 057 (DH 0752) Dried herbs
   d. EPA Crop Definition: None
10. References: GRIN, BAILEY 1976, CODEX, DUKE(c) (picture), GARLAND (picture), MAGNESS, US EPA 1995a
11. Production Map: No entry.
12. Plant Codes:
    a. Bayer Code: GAHPR

684

1. Woodruff
(Sweet woodruff, Waldmeister)
Rubiaceae
Galium odoratum (L.) Scop. (syn: *Asperula odorata* L.)

2. The plant is a perennial herb, 8 inches tall. Leaves are in whorls, lanceolate and generally smooth. The leaves are used in some wines and in summer drinks in Europe. Woodruff is little known in the U.S. except as a rock garden plant for shady places.
3. Season, harvest: Harvest the herb just before flowering in May or June.
4. Production in U.S.: No data.
5. Other production regions: Europe.
6. Use: Flavoring in May wine and in summer drinks.
7. Part(s) of plant consumed: Leaves
8. Portion analyzed/sampled: Leaves (Fresh and dried)
9. Classifications:
   a. Authors Class: Herbs and spices
   b. EPA Crop Group (Group & Subgroup): Herbs and spices (19A)
   c. Codex Group: 027 (HH 0753) Herbs; 057 (DH 0753) Dried herbs
   d. EPA Crop Definition: None

10. References: GRIN, CODEX, FOSTER (picture), MAGNESS, US EPA 1995a
11. Production Map: No entry.
12. Plant Codes:
    a. Bayer Code: GALOD

685

1. Wormwood
(Absinthium, Absinthe, Absinthe wormwood, Common wormwood)
Asteraceae (Compositae)
Artemisia absinthium L.

2. The plant is a spreading perennial herb, hardy in the northern U.S., and grown commercially in Michigan and Wisconsin. The top, grown from overwintering roots, attains a height of 2 to 4 feet, and is much branched. Leaves are parted into 2 to 3 long lobes and pubescent. Flower heads are small and numerous. Tops are gathered in midsummer and dried. The extract was formerly much used in medicine and liquors. At present, its use is mainly as the principal ingredient in absinthe, in which it is combined with other aromatic oils. Absinthe is a dark green bitter-tasting oil. It is related to sagebrush.
3. Season, start of growth to harvest: About 3 months. It is propagated from plant divisions.
4. Production in U.S.: No data. Michigan and Wisconsin.
5. Other production regions: No entry.
6. Use: Extracted for oil or dried herb, Leaves used sparingly as a seasoning. Also grown as an ornamental.
7. Part(s) of plant consumed: Leaves, stems and flower heads
8. Portion analyzed/sampled: Tops (fresh & dried)
9. Classifications:
   a. Authors Class: Herbs and spices
   b. EPA Crop Group (Group & Subgroup): Herbs and spices (19A)
   c. Codex Group: 027 (HH 0754) Herbs; 057 (DH 0754) Dried herbs
   d. EPA Crop Definition: None
10. References: GRIN, CODEX, FOSTER, MAGNESS, US EPA 1995a, SPLITTSTOESSER
11. Production Map: EPA Crop Production Region 5.
12. Plant Codes:
    a. Bayer Code: ARTAB

686

1. Yacon
(Yacon strawberry, Jacon, Llacon, Llamon, Arboloco, Puhe, Poir de terre cochet, Bolivian sunroot, Aricoma, Jiquima, Jiquimilla, Jicama, Chancaca, Apple of the earth)
Asteraceae (Compositae)
Smallanthus sonchifolius (Poepp. & Endl.) H. Rob. (syn: *Polymnia sonchifolia* Poepp. & Endl.)

2. Yacon is a compact, herbaceous plant with dark-green celery-like leaves. The aerial stems to 6 feet tall are hairy with purple markings. Small daisy-like flowers, yellow or orange, are packed together at the top of the plants. The tubers usually weigh about $1/2$ to 1 pound each, but can reach 4 pounds. Tubers are fused to the swollen stem in bunches of four to five and up to 20. The tubers resemble a brown potato with the texture of jicama. It is called jicama in Ecuador but it is

not the Jicama of commerce, which see. Plants are grown as annuals for it leaves and as a perennial for its underground tuberous root.

3. Season, planting to harvest: About six to seven months. Propagated by plantlets or divided tubers. Harvested after flowering and top die-back.

4. Production in U.S.: Limited. Grows in California (MELNICOE 1996d).

5. Other production regions: South America, mainly Columbia and Ecuador (NATIONAL RESEARCH COUNCIL 1989).

6. Use: Tubers are used for flavoring and eaten raw like a fruit. The sweet tuber is added to salads, or consumed boiled or baked. Also grated and squeezed to make a drink or sugar called *chancaca*. The main stem can be used as a cooked vegetable. Also the tops can be used as forage.

7. Part(s) of plant consumed: Tubers and main stem for food, and tops for feed

8. Portion analyzed/sampled: Tubers and tops (the tuber is the food of commerce)

9. Classifications:
   a. Authors Class: Root and tuber vegetables; Leaves of root and tuber vegetables
   b. EPA Crop Group (Group & Subgroup): Miscellaneous
   c. Codex Group: No specific entry
   d. EPA Crop Definition: None

10. References: GRIN, FACCIOLA, NATIONAL RESEARCH COUNCIL 1989 (picture)

11. Production Map: EPA Crop Production Region 10.

12. Plant Codes:
   a. Bayer Code: No specific entry

687

1. Yam bean
(Potato bean, Jicama)
Fabaceae (Leguminosae)
Pachyrhizus spp.

1. Jicama (hic-a-ma)
(Mexican potato, Sargott, Chopsui potato, Mexican turnip, Manoic pea, Potato yam, Dou shu, Sincamas)
P. erosus (L.) Urb. (syn: *P. angulatus* Rich.ex DC.; *P. bulbosus* Kurz; *Dolichos erosus* L.)

1. Potato bean
(Ajipo, Da di gua)
P. tuberosus (Lam.) Spreng.

1. Ahipa (a-hee-pa)
(Ajipa)
P. ahipa (Wedd.) Parodi (syn: *Dolichos ahipa* Wedd.)

2. These are perennial climbing herbs, with very long and large tuberous roots, ultimately up to 6 or 8 feet long and weighing 50 or more pounds. The vining tops reach 10 to 25 feet. Leaves are compound. Pods are 8 to 12 inches long, and near an inch wide. Young roots are eaten both raw and boiled and are a source of starch. The name Jicama is also used in Spanish for any edible root. Mature seeds and pods contain rotenone.

3. Season, seed to harvest: About 5 months. Whole small roots can be used to yield mature roots in about 3 months. Commercial roots weigh 2-5 pounds.

4. Production in U.S.: No data. *P. erosus* grown in South Florida in home gardens and commercially in such tropical regions as Puerto Rico, Hawaii and Mexico (STEPHENS 1988) for good quality roots.

5. Other production regions: Native to Central and South America (YAMAGUCHI 1983) and Mexico for *P. erosus* (MYERS 1991).

6. Use: Roots eaten raw or cooked, and source of starch. In preparation, the fibrous brown outer tissue of the root is peeled and the crisp white flesh is cut for salads, dips or cooked like potatoes (MYERS 1991).

7. Part(s) of plant consumed: Young tuberous roots. Young pods are edible like snap beans and used locally

8. Portion analyzed/sampled: Root

9. Classifications:
   a. Authors Class: Root and tuber vegetables
   b. EPA Crop Group (Group & Subgroup): Root and tuber vegetables (1C and 1D)
   c. Codex Group: *P. erosus* 016 (VR 0601) Root and tuber vegetables
   d. EPA Crop Definition: None

10. References: GRIN, CODEX, KAYS, LOGAN (picture), MAGNESS, MYERS, NATIONAL RESEARCH COUNCIL 1989, NEAL, STEPHENS 1988, US EPA 1995a, YAMAGUCHI 1983, SPLITTSTOESSER

11. Production Map: EPA Crop Production Region 13.

12. Plant Codes:
   a. Bayer Code: PACER (*P. erosus*)

688

1. Yam, True
(Ñame, Igname, Yam, Fufu, Yam shoot)
Dioscoreaceae
Dioscorea spp.

1. Greater yam
(Wateryam, Winged yam, Uhi, Ten months yam, Pei tsao, Lisbon yam, Bak chiu, Agua yam, Guyana arrowroot, White yam, Name-de-agu, Ubi yam)
D. alata L.

1. Chinese yam
(Cinnamon vine, Chinese potato, Japanese yam, Nago imo, Shan yao, Shan yuek)
D. batatas Decne.

1. Lesser yam
(Gado, Asiatic yam, Spiny yam, Chinese yam, Potato yam)
D. esculenta (Lour.) Burkill

1. White yam
(White Guinea yam, Guinea yam, Eboe yam)
D. rotundata Poir.

1. Mapuey
(Cushcush yam, Aja, Yampi, Maona, Napi, Cara doce, Yampee)
D. trifida L.f.

1. Yellow yam
(Yellow Guinea yam, Twelve months yam, Lagos yam)
D. cayenensis Lam.

2. Monocot tropical herb plants with long trailing vines reaching 10 to 30 feet with cordate, shiny leaves. Depending on the species, from one to several tubers are produced per plant. The vines need support (poles or stakes) for better yields. It is grown for the edible underground tubers which vary in size and shape averaging 3 to 8 pounds. The tubers develop deep in the ground and are difficult to dig. Time to

first harvest depends on species, the size of the seed tuber and if presprouted; normally 6 to 10 months after emergence (MARTIN 1983). In the U.S., moist varieties of sweet potatoes are often erroneously termed yams. The true yams are grown primarily in the tropics.

3. Season, planting to harvest: Plant small tubers or portions of tubers or presprout tubers before planting. About 10 months. In Florida, tubers are planted in March-April and harvested 10-11 months later (STEPHENS 1988).

4. Production in U.S.: In 1992, Guam reported 10 acres and 19,890 pounds; Virgin Islands reported 4 acres and 4,693 pounds; and Puerto Rico reported 2,230 acres with 7,592,200 pounds produced (1992 CENSUS). Also grown in south Florida (SCHNEIDER 1985).

5. Other production regions: Tropical and semitropical regions (YAMAGUCHI 1983).

6. Use: After harvest, the tubers are dried (curing) for a few hours, then stored under shade or in a building called a "yam barn". Yams can be stored for several months. Most tubers are marketed fresh. The most common method of preparation is boiling. Once boiled the tuber can be mashed and eaten as fufu. Also, tubers can be baked or fried. In preparation, scrub the tuber with a brush while rinsing, remove skin and underlayer, rinse and place in bowl of cold water before cooking.

7. Part(s) of plant consumed: Primarily the tuber. The shoots of some wild species are boiled and eaten as greens. Tubers can be used for starch, and poultry and livestock feed. Tubers are not a significant feed item

8. Portion analyzed/sampled: Tubers

9. Classifications:
   a. Authors Class: Root and tuber vegetables; Leaves of root and tuber vegetables
   b. EPA Crop Group (Group & Subgroup): Root and tuber vegetables; Leaves of root and tuber vegetables (2).
   c. Codex Group: 016 (VR 0600) Root and tuber vegetables
   d. EPA Crop Definition: Sweet potatoes = True yams

10. References: GRIN, CODEX, STEPHENS 1988 (picture), MAGNESS, MARTIN 1983, NEAL (*D. alata* picture), SCHNEIDER 1985, US EPA 1995a, US EPA 1996a, YAMAGUCHI 1983 (*D. alata* and *batatas* pictures)

11. Production Map: EPA Crop Production Region 13.

12. Plant Codes:
   a. Bayer Code: DIUAL (*D. alata*), DIUBA (*D. batatas*)

689

1. Yarrow
(Milfoil, Common yarrow, Sanquinary, Thousand seal, Nose-bleed)
Asteraceae (Compositae)
Achillea millefolium L.

2. Aromatic perennial growing to 3 feet tall. Finely divided feathery leaves are about 2 to 8 inches long. Each flower head about ¹/₄ inch across. It blooms from June to September. Common yarrow native to eastern North America is *A. millefolium* var. *occidentalis* DC. Yarrow in western North America (California to Alaska) may be *A. millefolium* ssp. *borealis* (Bong.) Breitung (FOSTER 1993).

3. Season, seeding to harvest: Harvest plants as they come into bloom.

4. Production in U.S.: No data. See above.

5. Other production regions: Grows throughout the Northern Hemisphere.

6. Use: Food-flavoring ingredient for beverages. Also grown as an ornamental in rock gardens.

7. Part(s) of plant consumed: Leaves

8. Portion analyzed/sampled: Leaves (fresh and dried)

9. Classifications:
   a. Authors Class: Herbs and spices
   b. EPA Crop Group (Group & Subgroup): Miscellaneous
   c. Codex Group: No specific entry
   d. EPA Crop Definition: None

10. References: GRIN, FOSTER (picture), GARLAND (picture), USDA NRCS

11. Production Map: No entry

12. Plant Codes:
   a. Bayer Code: ACHMI

690

1. Yautia
(New cocoyam)
1. Tanier
(New cocoyam)
Araceae
Xanthosoma spp.

1. Tannia
(Malanga, Ocumo, Cuban dasheen, Tarrier, Taya, Yellow yautia, Tannie, Mangaras, Tanyah, Tiquisque)
X. sagittifolium (L.) Schott (syn: *X. edule* (Mey) Schott; *X. xanthorrhizon* (Jacq.) Koch; *Arum sagittaefolium* L.)

1. Blue ape
(Oto, Badu)
X. violaceum Schott

2. Tropical plants to 5 feet tall related to Taro but without peltate leaves, extensively cultivated in the West Indies and other tropical countries for the edible corm and cormels. On certain varieties the young leaves and main stems (petioles) (madre) are also used as potherbs. The leaves are large, 1 to 2 feet long, broad-arrow in shape, borne on long petioles radiating from the "mother" or large corm. The top of the corm may be at or above ground level. In general exposure of the corm, the Yautia is similar to turnip. Propagation is by planting small cormels or the cut-off top of the "mother" corm with some of the petioles still remaining. Yautia blanca corm is white-fleshed and Yautia amarilla is yellow-fleshed. Growing Yautias do not require as much soil moisture as does Dasheen, which see. Yautias are sometimes referred to as *Alocasia* spp.

3. Season, planting to harvest: 6 to 12 months. In Florida, spring planting to harvest takes 9-10 months. Crop is usually hand harvested.

4. Production in U.S.: In 1992, Puerto Rico reported 1,563 acres with 1,121 tons. Virgin Islands reported 2 acres with 1,740 pounds (1992 CENSUS).

5. Other production regions: Tropics.

6. Use: Corms are washed and peeled before cooking and used like potatoes. Leaves and petioles (peeled) used as potherb. They are boiled 10-15 minutes and served like spinach.

7. Part(s) of plant consumed: Corms and leaves

8. Portion analyzed/sampled: Corms and leaves

9. Classifications:

a. Authors Class: Root and tuber vegetables; Leaves of root and tuber vegetables
b. EPA Crop Group (Group & Subgroup): Root and tuber vegetables (1C and 1D); Leaves of root and tuber vegetables for *X. sagittifolium* only
c. Codex Group: 013 (VL 0504) Leafy vegetables (including *Brassica* leafy vegetables); 016 (VR 0504) Root and tuber vegetables
d. EPA Crop Definition: None
10. References: GRIN, CODEX, MAGNESS, MARTIN 1979, STEPHENS 1988 (picture), US EPA 1995a, US EPA 1994, YAMAGUCHI 1983, O'HAIR, RUBATZKY
11. Production Map: EPA Crop Production Regions 3 and 13.
12. Plant Codes:
a. Bayer Code: XATSA (*X. sagittifolium*)

691

1. Zoysia grass
(Flawn)
Poaceae (Gramineae)
Zoysia spp.
1. Manilagrass
(Japanese carpetgrass, Korean grass)
Z. matrella (L.) Merr.
1. Japanese lawngrass
(Common zoysia, Shiba, Noshiba, Korean lawngrass)
Z. japonica Steud.
1. Mascarene grass
(Korean velvetgrass)
Z. tenuifolia Willd. ex Thiele
2. All of these grasses are native to tropical or Eastern Asia. In this country they are used exclusively for lawns, golf courses and occasionally for erosion control. They form excellent green turf, green in summer but becoming brown during the winter months. Mascarene grass is the smallest, finest-leaved and least hardy of the three. It grows only 2 inches high, is shallow rooted. It is grown somewhat in southern areas. Japanese lawngrass has a broad, coarse leaf and makes a dense sod. It is winter hardy as far north as Boston. Japanese lawngrasses are faster growing and form a less dense sod than the other types and some are seeded cultivars. It is tough, harsh, unpalatable to livestock. Once established, it is very persistent. Propagation is by inserting plugs of sod. Cultivars are available, of which 'Meyer' is best known. Manilagrass has a general limit of hardiness at about 40 degrees N. It appears best adapted to rather heavy soil but will thrive on other types. It is propagated by planting sod pieces and is rather slow to become established, but makes a dense, persistent sod. There are also many seeded zoysia grasses, such as Japanese lawngrass. The seeded types are coarser texture. Zoysia grasses are slow to establish.
3. Season: Warm season, vegetative and seeded propagation.
4. Production in U.S.: No specific data, See Forage grass. Southern states.
5. Other production regions: See above.
6. Use: Turf, pasture.
7. Part(s) of plant consumed: Foliage
8. Portion analyzed/sampled: Forage and hay
9. Classifications:
a. Authors Class: See Forage grass
b. EPA Crop Group (Group & Subgroup): See Forage grass
c. Codex Group: See Forage grass
d. EPA Crop Definition: None
10. References: GRIN, MAGNESS, TURGEON, EMMONS
11. Production Map: EPA Crop Production Regions 2 and 4.
12. Plant Codes:
a. Bayer Code: ZOYSS (*Z.* spp.), ZOYJA (*Z. japonica*), ZOYMA (*Z. matrella*), ZOYTE (*Z. tenuifolia*)

BIBLIOGRAPHY

This bibliography provides both the references for this book and a listing of source documents for food and feed commodities.

ADAIR: Adair, C.R. 1973. Rice Varieties in the United States. Varieties and Products. USDA Agricultural Research Service. Agriculture Handbook, No. 289.

ADAMS: Adams, J. 1982. Sunflower. National Sunflower Association. Bismarck, North Dakota. 40 pp.

ADSULE: Adsule, R.N. and S.S. Kadam. 1995. Guava. In Handbook of Fruit Science and Technology, D.K. Salunkhe and S.S. Kadam, editors. Marcel Dekker, Inc., New York, NY.

ADSULE(a): Adsule, R.N. and N.B. Patil. 1995. Pomegranate. In Handbook of Fruit Science and Technology, D.K. Salunkhe and S.S. Kadam, editors. Marcel Dekker, Inc., New York, NY.

AHMED: Ahmed, E.M. and C.R. Barmore. 1980. Avocado. In Tropical and Subtropical Fruits, S. Nagy and P. Shaw, editors. AVI Publishing Company, Inc. Westport, CT.

AHMEDULLAHH: Ahmedullahh, M. and D.G. Himelrick. 1990. Grape Management, pp. 383-471. In Small Fruit Crop Management, G.J. Galletta and D.G. Himelrick, editors. Prentice-Hall, Englewood Cliffs, NJ.

AIKENS: Aikens, C.G. 1991. Green Grocer's Guide to the Harvest. Peachtree Publishing, Atlanta, GA. 176 pp.

ALDERMAN: Alderman, D.C. and W.B. Davis. 1981. Native Edible Fruits, Nuts, Vegetables, Herbs, Spices, and Grasses of California. I. Fruit Trees and Nuts. University of California Division of Agricultural Sciences, Cooperative Extension Service Leaflet 2226.

ALDERMAN(a): Alderman, D.C. and W.B. Davis. 1981. Native Edible Fruits, Nuts, Vegetables, Herbs, Spices, and Grasses of California. III. Vegetables. University of California Division of Agricultural Sciences, Cooperative Extension Service Leaflet 2705.

ALDERMAN(b): Alderman, D.C. and W.B. Davis. 1981. Native Edible Fruits, Nuts, Vegetables, Herbs, Spices, and Grasses of California. II. Small or Bush Fruits. University of California Division of Agricultural Sciences Cooperative Extension Service Leaflet 2278.

ALDERSON: Alderson, J. and W.C. Sharp. 1995. Grass Varieties in the United States. USDA. CRC Inc., Lewis Publishing, Boca Raton, FL. 295 pp.

ALKOFAHI: Alkofahi, A., J.K. Rupprecht, J.E. Anderson, J.L. McLaughlin, K.L. Milolajczak, and B.A. Scott. 1989. Search for new pesticides from higher plants, pp 25-43. In J.T. Arnason, B.J.R. Philogene and P. Morand (eds.). Amer. Chem. Soc. Sym. Ser:2:387.

ALLEN: Allen, P. 1986. Glossary of Oriental Vegetables. University of California Cooperative Extension Service. Small Farm Center. Family Farm Series. 8pp.

ALMEYDA: Almeyda, N. and F.W. Martin. 1976. Cultivation of Neglected Tropical Fruits with Promise. Part 2. The Mamey Sapote. USDA. Agricultural Research Service. ARS-S-156. 13 pp.

ANDERSON: Anderson, K.N. and L.E. Anderson. 1993. The International Dictionary of Food and Nutrition. John Wiley & Sons, NY. 330 pp.

ANDREWS: Andrews, D.J., J.F. Rajewski, and K.A. Kumar. 1993. Pearl Millet: New Feed Grain Crop, pp. 198-208. In New Crops. J. Janick and J. Simon. John Wiley & Sons, NY.

ANOCHILI: Anochili, B.C. 1984. Tropical Agricultural Handbook. Food Crop Production. MacMillan Publishing Co., London. 60 pp.

ANON(a): Anon(a). 1994. Tropical Special Fruit: Number of farms, area, number of trees, State of Hawaii, 1991-1994. In Statistics of Hawaiian Agriculture. Hawaii Agric. Rptg. Serv., Honolulu, HI.

ANON(b): Anon(b). 1989. Acreages of minor fruit crop production in Hawaii. Hawaii Tropical Fruit Growers Industry Analysis. Honolulu, HI.

ANON(c): Anon(c). 1916 Where are the best pawpaws? J. Hered. 7:291-296.

ANON(d): Anon(d). 1996. Turfgrass. Grounds Maintenance. 31(8):10, 51.

ANTHONY: Anthony, K.R.M., J. Meadley, and G. Robbelen. 1993. New Crops for Temperate Regions. Chapman and Hall, NY. 245 pp.

ARKCOLL: Arkcoll, D. 1990. New Crops from Brazil. In Advances in New Crops by J. Janick and J.E. Simon, editors. Timber Press, Portland, OR. pp. 367-371.

ARRIOLA: Arriola, M.C. de, J.F. Calzada, J.F. Menchu, C. Rolz, and R. Garcia. 1980. Papaya. In Tropical and Subtropical Fruits, S. Nagy and P. Shaw, editors. AVI Publishing Co., Inc., Westport, CT.

ASENJO: Asenjo, C.F. 1980. Acerola. In Tropical and Subtropical Fruits, S. Nagy and P. Shaw, editors. AVI Publishing Co., Inc., Westport, CT.

ATLAS: Atlas, N. 1994. The Rebirth of the Ancient Grains. Vegetarian Times. 199: 46-55.

ATLAS(a): Atlas, N. 1989. The Essential Guide to Grains. Vegetarian Times. 144: 6-9.

ATLAS(b): Atlas, N. 1989. The Essential Guide to Beans. Vegetarian Times. 144: 16-21.

BAILEY 1976: Bailey, L.H. and E.Z. Bailey. 1976. Hortus Third, A Concise Dictionary of Plants Cultivated in the United

States and Canada. MacMillan Publishing Company, New York, NY. 1290 pp.

BAILEY(a): Bailey, L.H. 1969. Manual of Cultivated Plants. MacMillan Publishing. Co., New York, NY. 1116 pp.

BAILEY(b): Bailey, L.H. 1953. The Standard Cyclopedia of Horticulture. Macmillan Publishing Co., New York, NY.

BAILEY(c): Bailey, L.H. 1948. Some Recent Chinese Vegetables. Cornell University Agricultural Experimental Station Bulletin No. 67.

BAKER: Baker, A.E. and P. A. Riley. 1994. Feed Situation and Outlook Yearbook. USDA Economic Research Service. FDS-330. 76 pp.

BALDER: Balder, B., S. Ramanajam, and H. Jain. 1988. Pulse Crops (Grain Legumes). Oxford and IBH Publishing Co., New Delhi, India.

BALERDI: Balerdi, C. and M. Misitis. 1994. The Black Sapote. University of Florida Cooperative Extension Service. Institute of Food and Agricultural Services: Fact Sheet, 2pp.

BALL: Ball, D.M., C.S. Hoveland, and G.D.Lacefield. 1991. Southern Forages. Potash, Phosphate Institute and Foundation. Agronomy Research. Atlanta, GA.

BALL(a): Ball, D.M., C.S. Hoveland. 1978. Alkaloid Levels in *Phalaris aquatica* L. is Affected by Environment. Agronomy Journal 70: 977-981.

BARNES: Barnes, R.F., D.A. Miller, and C.J. Nelson. 1995. Forages, Volume I: An Introduction to Grassland Agriculture. Fifth Edition, Iowa State University Press. Ames, IA. 516 pp.

BARNES(a): Barnes, R.F., D.A. Miller and C.J. Nelson. 1995. Forages, Volume II: The Science of Grassland Agriculture. Iowa State University Press. Ames, IA. 357 pp.

BARRETT: Barrett, R.P. 1990. Legume Species as Leaf Vegetables, pp. 391-396. In Advances in New Crops. Janick, J. and J. Simon, Editors. Timber Press. Portland, OR.

BARRETT(a): Barrett, M. 1986. The New Old Grains, Vegetarian Life and Times 101: 28-31, 51.

BATES: Bates, D.M., R.W. Robinson, C. Jeffrey. 1990. Biology and Utilization of the Cucurbitaceae. Cornell University Press, Ithaca, NY. 486 pp.

BAYER: Bayer AG. 1992. Important Crops of the World and their Weeds (2nd Edition). Scientific and Common Names, Synonyms and WSSA/WSSJ Approved Computer Codes. Bayer AG Business Group Crop Protection. Leverkusen, Germany. 1682 pp.

BEATTIE: Beattie, W.R. 1937. Production of Chicory and Endive. USDA Leaflet No. 133.

BEBEE 1991: Bebee, C.N. 1991. The Protection of Nut Crops, 1979-April, 1991. USDA National Agricultural Library-EPA. BLA No. 109.

BEBEE 1990: Bebee, C.N. 1990. The Protection of Tropical and Subtropical Fruits, 1979-April, 1990. USDA National Agricultural Library-EPA. BLA No. 97.

BHUTANI: Bhutani, V.P. and V.K. Joshi. 1995. Plum. In Handbook of Fruit Science and Technology, D.K. Salunkhe and S.S. Kadam, editors. Marcel Dekker, Inc., New York, NY.

BIANCHINI: Bianchini, F. and F. Corbetta. 1985. The Complete Book of Health Plants. Crescent Books, NY. 242 pp.

BITZER 1987a: Bitzer, M.J. 1987. Production of Sweet Sorghum for Sirup in Kentucky. Part 1. University of Kentucky, Cooperative Extension Service. AGR-122. 4 pp.

BITZER 1987b: Bitzer, M.J. and J.D. Fox. Processing Sweet Sorghum Sirup. Part 2. University of Kentucky, Cooperative Extension Service. AGR-123. 6 pp.

BLACK: Black, L.L. 1988. Personal Communications *Lagenaria* in Louisiana. Dept. of Plant Pathology and Crop Physiology, Louisiana State University, Baton Rouge, 19 Oct. 88.

BLACKBURNE: Blackburne, P. 1992. Fungal Feast. J. Royal Horticultural Society. 117 (12): 590-592.

BLACKWELL: Blackwell, W.H. 1990. Poisonous and Edible Plants. Prentice-Hall. Englewood Cliffs, NJ. 329 pp.

BOE: Boe, A., E.K. Twidwell, and D.P. Casper. 1991. Forage Potential of Teff. Proceedings Forage Grassland Conference. American Forage and Grassland Council. pp. 236-239.

BONAR: Bonar, A. 1985. Macmillan Treasury of Herbs. Macmillan Publishing Co., New York, NY. 114 pp.

BONAVIA: Bonavia, E. 1973. The Cultivated Oranges and Lemons of India and Ceylon. Volumes I and II. W.H. Allen, London. 385 pp.

BOOTE: Boote, K. 1982. Growth Stages of the Peanut (*Arachis hypogoea* L.). Peanut Science 9 :35-40.

BOSISIO 1989: Bosisio, M. 1989. Meadowfoam: Pretty Flowers, Pretty Possibilities. Agricultural Research. Agricultural Research Service, USDA, Washington, D.C. February, 1989: 10-11.

BOWMAN: Bowman, J.S. 1984. Personal Communications on Tyfon. Cooperative Extension, University of New Hampshire, Durham. 25 Jan 84.

BOXER: Boxer, A. and P. Bock. 1980. The Herb Book. Octopus Book Ltd., New York, NY. 274 pp.

BREENE: Breene, W.M. 1991. Food Uses of Grain Amaranth.

Cereal Foods World 36 (5): 426-437.

BREVOORT: Brevoort, P. 1995. The U.S. Botanical Market – An Overview. Herbalgram 36: 49-57.

BROUK: Brouk, B. 1975. Plants Consumed By Man. Academic Press. New York, NY. 480 pp.

BROWN: Brown, W.L., M.S. Zuber, L.L. Darrah, and D.V. Glover. 1984. Origin, Adaptation, and Types of Corn. National Corn Handbook. Purdue University Cooperative Extension Service. NCH-10. 6 pp.

BRYANT: Bryant, E. and K. Yumkella. 1994. Michigan Dry Bean Digest 18 (2): 5-9.

BUESO: Bueso, C.E. 1980. Soursop, Tamarind and Chironja. In Tropical and Subtropical Fruits, S. Nagy and P. Shaw, editors. AVI Publishing Company, Inc. Westport, CT.

BURR: Burr, F. 1990. The Field and Garden Vegetables of America. The American Botanist Booksellers. Chillicothe, IL. 667 pp.

CALLAWAY: Callaway, M.B. 1992. Current research for the commercial development of pawpaws (*Asimina triloba* (L) Dunal). Hort. Sci. Vol 27 (2).

CAMPBELL 1985: Campbell, C. S. 1985. The Subfamilies and Tribes of Gramineae (Poaceae) in the Southeastern United States. Journal of the Arnold Arboretum 66(2): 123-199.

CAMPBELL(a): Campbell, C.W. 1994. Handling of Florida-grown and Imported Tropical Fruits and Vegetables. Hort Science 29 (9): 975-978.

CAMPBELL(b): Campbell, C.W. and J.W. Sauls. 1991. Spondias In Florida, University of Florida. Florida Cooperative Extension Service. Institute of Food and Agricultural Services. Fruit Crops Fact Sheet FC-63, 3 pp.

CAMPBELL(c): Campbell, C.W. and R.L. Phillips. 1990. The Atemoya. University of Florida. Florida Cooperative Extension Service. Florida Institute of Food and Agricultural Science. Fruit Crops Fact Sheet FC-64, 4 pp.

CAMPBELL(d): Campbell, C.W. and S.E. Malo. 1990. The Loquat. University of Florida. Florida Cooperative Extension Service. Florida Institute of Food and Agricultural Science. Fruit Crops Fact Sheet FC-5, 2 pp.

CAMPBELL(e): Campbell, C.W., S.E. Malo and S. Goldwater. 1990. The Sapodilla. University of Florida. Florida Cooperative Extension Service. Florida Institute of Food and Agricultural Science. Fruit Crops Fact Sheet FC-1, 2 pp.

CAMPBELL(f): Campbell, C.W. and S.E. Malo. 1986. The Mango. University of Florida. Florida Cooperative Extension Service. Florida Institute of Food and Agricultural Science. Fruit Crops Fact Sheet FC-2, 4 pp.

CAMPBELL(g): Campbell, C.W. and S.E. Malo. 1979. The Lychee. University of Florida. Florida Cooperative Extension Service. Florida Institute of Food and Agricultural Science. Fruit Crops Fact Sheet FC-6, 2 pp.

CAMPBELL(h): Campbell, C.W. and S.E. Malo. 1984. University of Florida. Florida Cooperative Extension Service. Florida Institute of Food and Agricultural Science. Fruit Crops Fact Sheet FC-12, 2 p.

CAMPBELL(i): Campbell, C.W. 1965. The Golden Star Carambola. University of Florida. Agricultural Experiment Station Circular S-173.

CAMPBELL-PLATT: Campbell-Platt, G. 1987. Fermented Foods of the World. A Dictionary and Guide. Butterworths. London. 291 pp.

CARTER 1978: Carter, J.F. 1978. Sunflower Science and Technology. American Society of Agronomy Monograph No. 19. Madison, WI. 505 pp.

CARTER 1989: Carter, P.R., D.R. Hicks, J.D. Doll, E.E. Schulte, and B. Holmes. 1989. Popcorn. Alternative Field Crops Manual. University of Wisconsin and University of Minnesota. 6 pp.

CHAMBERS 1996: Chambers, C. 1996. Personal Communications. Cilantro Grower and Manager Crop R & D. Duda Valley Onions. McAllen, TX. 15 JAN 96.

CHAPMAN: Chapman, G.P. 1996. The Biology of Grasses. First Ed. CAB International, UK. 273 pp.

CHAPMAN(a): Chapman, G.P. 1992. An Introduction to the Grasses. First Ed. CAB International, UK. 111 pp.

CHAPMAN(b): Chapman, G.P. 1992. Grass Evolution and Domestication. First Ed. Cambridge University Press, NY. 390 pp.

CHASE: Chase, L. 1992. The Plumcot Joins the Ranks of Summer Fruit. California Summer Fruits, The Packer. May 9.

CHATTERTON 1996: Chatterton, N.J. 1996. Personal Communications on Dr. M.D. Rumbaugh's Releases of Globe-mallow. USDA-ARS, Forage and Research Laboratory, Utah State University, Logan. 6 SEP 96.

CHAVAN: Chavan, J.K. and S.J. Jadhav. 1995. Coconut. In Handbook of Fruit Science and Technology, D.K. Salunkhe and S.S. Kadam, editors. Marcel Dekker, Inc., New York, NY.

CHAVAN(a): Chavan, U.D. and S.S. Kadam. 1995. Passion Fruit. In Handbook of Fruit Science and Technology, D.K. Salunkhe and S.S. Kadam, editors. Marcel Dekker, Inc., New York, NY.

CHEVERTON: Cheverton, M.R. and G.P. Chapman. 1989. Teff. In New Crops for Food and Industry. Wickens, G.E., editor. Chapman and Hall, NY. pp. 235-238.

CHILDERS: Childers, N.F. 1983. Modern Fruit Science. 9th Edition. N.F. Childers Publisher, Gainesville, FL. 583 pp.

CHILDERS(a): Childers, N.F. 1969. Modern Fruit Science. 4th Edition. Chapter 15: Culture of Plums pp. 388-412. Horticultural Publications/Rutgers University, New Brunswick, NJ.

CHRISTIE: Christie, B.R. and A.A. Hanson. 1987. CRC Handbook of Plant Science in Agriculture. Vol. I. CRC Press. Boca Raton, FL. 190 pp.

CHRISTOPHER: Christopher, E.P. 1958. Introductory Horticulture. McGraw-Hill Book Co., NY. 482 pp.

CLARK: Clark, L.G. and P.W. Pohl. 1996. Agnes Chases' First Book of Grasses. Fourth Ed. Smithsonian Institute, Washington, D.C.

CLAYTON: Clayton, W.D. and F.R. Richardson. 1972. The Tribe *Zoysieae* Miq. Studies in the Gramineae: XXXII. Kew Bulletin 27: 37-49.

CLEMENT: Clement, C.R., R.M. Manshardt, J. DeFrank, F. Zee, and P. Ito. 1991. Introduction and Evaluation of Pejbaye (*Bactris gasipaes*) for Palm Heart Production in Hawaii. In J. Janick, and J.E. Simon, editors, 2nd National Symposium on New Crops – Exploration, Research, and Commercialization, 6-9 October 1991. John Wiley & Sons, Inc. c 1993: 465-472.

CODEX: Codex Alimentarius. 1993. Pesticide Residues in Food. Section 2. Codex Classification of Foods and Animal Feeds. FAO/WHO, Rome, Italy. Vol. 2: 218 pp.

CODEX(a): Codex Alimentarius. 1996. Pesticide Residues on Food – Maximum Residue Limits. FAO/WHO, Rome, Italy. Vol. 2B, 301 pp.

CODEX(b): Codex Alimentarius. 1989. Guide to Codex Recommendations Concerning Pesticide Residues, part 4. Codex Classification of Foods and Animal Feeds (Second Edition). Food and Agriculture Organization of the United Nations World Health Organization, Rome. CAC/PR4-1989. 139 pp.

COFFMAN: Coffman,F.G. 1977. Oat History, Identification and Classification. USDA. ARS. Washington, D.C. Technical Bulletin, No. 1516. 356 pp.

COLLINS: Collins, J.L. 1960. The Pineapple. Interscience, New York, NY.

COLLINS 1996: Collins, R.A. 1996. Personal Communications on Ondeev. California Vegetable Specialties, Inc. 15 Poppy House Road, Rio Vista, CA. 10 June 1996.

CONSIDINE: Considine, D.M. and G.D. Considine. 1982. Foods and Food Production Encyclopedia. Van Nostrand Reinhold Co. NY. 2305 pp.

COOK: Cook, J. 1990. Allium Underground. Organic Gardening 37 (8): 28-35.

COOKE: Cooke, L. and D.A. Konstant. 1991. What's New in Oilseeds? Check Out Crambe! Agricultural Research. Agricultural Research Service, USDA, Washington, D.C. March Issue: 16-17.

COURTER: Courter, J.W. and G.B. Holcomb. 1991. Strawberries. A Small Scale Agriculture Alternative. USDA Cooperative State Research Service. Washington, D.C. USGPO: 282-933: 40115/CSRS. 2 pp.

COYLE: Coyle, L.P. 1982. The World Encyclopedia of Food. Facts on File Publishing, NY. 730 pp.

CRAMER: Cramer, C. 1993. Turned on to Turnips. The New Farm. 15(6): 20-23.

CRANDALL: Crandall, P.C. and H.A. Daubeny. 1990. Raspberry Management, p. 157-213. In Small Fruit Crop Management, G.J. Galletta and D.G. Himelrick, editors. Prentice Hall, Englewood Cliffs, NJ.

CRANE 1996a: Crane, J.H. 1996a. Personal Communications. Review of Monographs. Tropical Research and Education Center. University of Florida, Homestead. 5 March 1996.

CRANE 1995: Crane, J.H. 1995. Personal Communications. Tropical Crops. Tropical Research and Education Center. University of Florida, Homestead. 7 September 1995.

CRANE 1995a: Crane, J.H. 1995. Personal Communications on Tropical Crops. Tropical Research and Education Center, University of Florida, Homestead. February 1995.

CRANE 1994: Crane, J.H. 1994. Personal Communications. Tropical Fruit Trees. Tropical Research and Education Center, University of Florida, Homestead. 15 September 1994.

CRANE 1992: Crane, J.H. 1992. The Carambola (Star Fruit) FC-12. Fruit Crop Fact Sheet: Tropical Research and Education Center, University of Florida, Homestead. 4 pp.

CRANE(a): Crane, J.H., C.F. Balerdi and C.W. Campbell. 1991. The Avocado. University of Florida. Florida Cooperative Extension Service. Institute of Food and Agricultural Science. Fruit Crops Fact Sheet FC-3, 7 pp.

CRANE(b): Crane, J.H. and C.W. Campbell. 1991. The Mango. University of Florida. Florida Cooperative Extension Service. Institute of Food and Agricultural Science. Fruit Crops Fact Sheet FC-2, 5 pp.

CRANE(c): Crane, J.H. 1989. Acreage and Plant Densities of Commercial Carambola, Mamey Sapote, Lychee, Longan, Sugar Apple, Atemoya, and Passion Fruit Plantings in South Florida. Proceedings Florida State Horticultural Society 102: 239-242.

CRISWELL 1996: Criswell, J.T. 1996. Personal Communications. Sesame Production. Oklahoma State University, Stillwater. April.

CROCKER(a): Crocker, T.E. 1994. Peaches and Nectarines in Florida. University of FLorida. Florida Cooperative Extension Service. IFAS. Circular 299-D. 9 pp.

CROCKER(b): Crocker, T.E. 1993. The Pecan. University of Florida. Florida Cooperative Extension Service. Institute of Food and Agricultural Science. Fruit Crops Fact Sheet FC-43. 4 pp.

CROCKER(c): Crocker, T.E. and J.A. Mortensen. 1988. The Muscadine Grape. University of Florida. Florida Cooperative Extension Service. Institute of Food and Agricultural Science. Fruit Crops Fact Sheet FC-16A. 4 pp.

CROCKER(d): Crocker,T.E. and W.B. Sherman. 1984. The Apple. University of Florida. Florida Cooperative Extension Service. Institute of Food and Agricultural Science. Fruit Crops Fact Sheet FC-14. 3 pp.

CROCKER(e): Crocker, T.E. and J.A. Mortensen. 1983. The Bunch Grape. University of Florida. Florida Cooperative Extension Service. Institute of Food and Agricultural Science. Fruit Crops Fact Sheet FC-17A. 4 pp.

CROCKER(f): Crocker, T.E. and C.P. Andrews. 1980. The Persimmon. University of Florida. Florida Cooperative Extension Service. Institute of Food and Agricultural Science. Fruit Crops Fact Sheet FC-20. 4 pp.

CROCKER(g): Crocker, T.E. and C.P. Andrews. 1979. Pears for Florida. University of Florida. Florida Cooperative Extension Service. Institute of Food and Agricultural Science. Fruit Crops Fact Sheet FC-29. 3 pp.

CROPS: Crops Research Division. 1963. Growing Pumpkins and Squash. USDA Farmer's Bulletin No. 2086. US GPO. Washington, D.C. 27 pp.

CUOMO: Cuomo, G.J., D.C. Blouin, and J.F. Beatty. 1996. Forage Potential of Dwarf Napiergrass and a Pearl Millet X Napiergrass Hybrid. Agronomy Journal 88 (3) 434-438.

DAHLEN: Dahlen, M. 1983. A Popular Guide to Chinese Vegetables. Crown Publishing, NY. First Edition. 113 pp.

DANA: Dana, M.N. and J.E. Simon. 1989. Chinese Vegetables. Purdue University Cooperative Extension Service. Bulletin HO-187.

DAVIES: Davies, F.S. 1990. The Naval Orange. University of Florida. Florida Cooperative Extension Service. Institute of Food and Agricultural Science. Fruit Crops Fact Sheet FC-83. 4 pp.

DAVIS: Davis, A.R., M. Tosiano, S. Slayton, and R. Helrich. 1996. Maine wild blueberries. New England Agriculture Statistics Service, October 15, 1996.

DAVIS(a): Davis, E. W., E.A. Oelke, E.S. Oplinger, J. Doll, C.V. Houson, and D.H. Putnam. 1992. Cowpea. In Alternative Field Crops Manual. Universities of Wisconsin and Minnesota Cooperative Extension Service. 4 pp.

DEAN: Dean, L. (Editor). 1997. North Carolina Specialty Crops School for February 21, Truffles. The Great Lakes Vegetable Growers News, Sparta, MI. p. 47.

DESAI: Desai, B.B. and D.K. Salunkhe. 1995. Cherry. In Handbook of Fruit Science and Technology, D.K. Salunkhe and S.S. Kadam, editors. Marcel Dekker, Inc., New York, NY.

DESAI(a): Desai, U.T. and P.M. Kotecha. 1995. Fig. In Handbook of Fruit Science and Technology, D.K. Salunkhe and S.S. Kadam, editors. Marcel Dekker, Inc., New York, NY.

DEWITT: Dewitt, D. and N. Gerlach. 1994. A Garden of Fiery Delights. Chili Pepper 8 (2): 28-31.

DEWITT(a): Dewitt, D. and N. Gerlach. 1990. The Whole Chili Pepper Book. Little Brown & Co., Boston. 373 pp.

DEWOLF: Dewolf, G.P. 1987. Taylor's Guide to Vegetables and Herbs. Chanticleer Press. NY.

DOOLITTLE: Doolittle, S.P., A.L. Taylor, L.L. Danielson, and L.B. Reed. 1962. Commercial Watermelon Growing. USDA ARS Agriculture Bulletin, No. 259. US GPO. Washington, D.C. 31 pp.

DORSCHNER: Dorschner, K.W. 1997. Hop Update Personal Communications. Rutgers, the State University of New Jersey, New Brunswick, NJ. 1 pp.

DOTY: Doty, W.L. 1980. All About Vegetables. Ortho Books. San Francisco, CA. 112 pp.

DOWNING: Downing, J. 1997. Personal Communications Cranberry Production. Cranberry Institute, East Wareham, MA. 20 FEB 97.

DUKE(a): Duke, J.A. and J.L. duCellier. 1993. CRC Handbook of Alternative Cash Crops. CRC Press, Boca Raton, FL. 537 pp.

DUKE(b): Duke, J.A. 1992. Handbook of Phytochemical Constituents of GRAS Herbs and Other Economic Plants. CRC Publishing Co., Boca Raton, FL. 654 pp.

DUKE(c): Duke, J.A. 1992. Handbook of Edible Weeds. CRC Publishing Co., Boca Raton, FL. 246 pp.

DUKE(d): Duke, J.A. 1989. Ginseng: A Concise Handbook. Reference Publications, Inc. Algonac, MI. 273 pp.

DUKE(e): Duke, J.A. 1986. Handbook of Proximate Analysis Tables of Higher Plants. CRC Publishing Co., Boca Raton, FL. 389 pp.

DUKE(f): Duke, J.A. 1985. Culinary Herbs: A Potpourri. Trado-

Medisa Books, NY. 195 pp.

DUKE(g): Duke, J.A. 1985. Handbook of Nuts. CRC Publishing Co., Boca Raton, FL. 343 pp.

DUKE(h): Duke, J.A. 1981. Handbook of Legumes of World Economic Importance. Plenum Press, NY. 345 pp.

DUPONT: DuPont, J., E. M. Osman. 1987. Cereals and Legumes in the Food Supply. Iowa State University Press. Ames, IA. 360 pp.

EDMINSTEN: Edminsten, K.L., A.C. York, and J.M. Ferguson. 1993. 1993 Cotton Information. North Carolina Cooperative Extension Service Bulletin AG-417. 145 pp.

EDMOND: Edmond, J.B. and G.R. Ammerman. 1971. Sweet Potatoes: Production, Processing, Marketing. AVI Publishing Co., Westport, CT. 334 pp.

EHRENSING 1994: Ehrensing, D. 1994. Personal Communications Concerning Meadowfoam. Department of Crop and Soil Science, Oregon State University, Corvallis. 19 July, 1994.

ELLIOTT: Elliott, D.B. 1976. Roots: An Underground Botany and Foragers Guide. Chatham Press, CT. 128 pp.

ELMORE 1985: Elmore, C.L. 1985. Personal Communications. Herbs and Spices. University of California, Davis. August

ELMORE 1980: Elmore, C.L. 1980. Personal Communications. Herbs and Spices. University of California, Davis. July.

ELVING: Elving, P. 1987. Sunset Fresh Produce. Lane Publishing Co. Menlo Park, CA. 128 pp.

EMMONS: Emmons, R.D. 1995. Turfgrass Science and Management. Second ed. Delmar Publ., Albany, NY. 512 pp.

ENGLER: Engler, P. 1995. Personal Communications on Citrus. California Citrus Quality Council. Diamond Bar, California. February & April, 1995.

ENSMINGER(a): Ensminger, M.E. 1992. The Stockman's Handbook. 7th ed. Interstate Publ., Inc. Danville, IL. 1030 pp.

ENSMINGER(b): Ensminger, M.E., J.E. Oldfield, and W.W. Heinemann. 1990. Feeds and Nutrition Digest. 2nd ed. Regus Publ. Co., Clovis, CA 795 pp.

ENSMINGER(c): Ensminger, A.H., M.E. Ensminger, J.E. Konlande, and J.R.K. Robson. 1983. Food for Health. First ed. Regus Press. Clovis, CA. 1178 pp.

ENSMINGER(d): Ensminger, A.H., M.E. Ensminger, J.E. Konlande, and J.R.K. Robson. 1983. Foods and Nutrient Encyclopedia. Vol. I and II. First ed. Regus Press. Clovis, CA. 2410 pp.

ERICHSEN: Erichsen-Brown, C. 1979. Use of Plants for the Past 500 Years. Breezy Creeks Press. Ontario. 510 pp.

ERICKSON: Erickson, K. 1987. If You Don't Know Beans About Beans Or Even If You Do. 3rd ed. Mineral Nutrition Program Publ. St. Paul, MN. 64 pp.

ESHBAUGH: Eshbaugh, W.H. 1980. Personal Communications for Peppers. Department of Botany, Miami University, Oxford, Ohio. 6 November, 1980.

ESSER: Esser, W.L. 1983. Dictionary of Natural Foods. Natural Hygiene Press. Bridgeport, CT. 166 pp.

FABER 1996a: Faber, B. 1996a. Personal Communications. Tropicals in California. County Advisor, University of California, Ventura. 2 February 1996.

FACCIOLA: Facciola, S. 1990. Cornucopia: A Source Book of Edible Plants. Kampong Publ. Vista, CA. 677 pp.

FARRELL: Farrell, K.T. 1985. Spices, Condiments, and Seasonings. First ed. AVI Book, Van Nostrand Reinhold Company, NY. 414 pp.

FARRELL(a): Farrell, K.T. 1990. Spices, Condiments, and Seasonings. Second ed. AVI Book, Van Nostrand Reinhold Company, NY. 415 pp.

FELKER 1993: Felker, P. 1993. Personal Communications Concerning Leucaena. Center for Semi-Arid Forest Resources. Texas A & I University, Kingsville. 24 August 1993.

FERGUSON: Ferguson, L., G.S. Sibbet, and G.C. Martin. 1994. Olive Production Manual. University of California, Division of Agricultural and Natural Resources Publication 3353. 160 pp.

FERGUSON(a): Ferguson, L. and M. Arparia. 1990. New Subtropical Tree Crops in California. Pp. 331-337 in Advances in New Crops. Janick, J. and J. Simon, editors. Timber Press. Portland, OR.

FERNALD: Fernald, M.L. and A.C. Kinsey. 1943. Edible Wild Plants of the Eastern North America. Idlewild Press, NY.

FICK: Fick, G.W. and S.C. Mueller. 1989. Alfalfa Quality, Maturity and Mean Stage of Development. Cornell University, Ithaca, NY. Cornell Information Bulletin 217.

FITZGIBBON: Fitzgibbon, T. 1976. The Food of the Western World. Hutchinson & Publ., Co. London. 529 pp.

FLATH: Flath, R.A. 1980. Pineapple. In Tropical and Subtropical Fruits. S. Nagy and P. Shaw, editors. AVI Publishing Company, Inc. Westport, CT.

FLAX COUNCIL OF CANADA 1995: The Flax Council of Canada. 1995. Flax Focus. 465-467. Lombard Ave., Winnipeg Manitoba, Canada. Spring 1995. Vol. 8, No. 2.

FOHNER 1986: Fohner, G. 1986. Personal Communications Concerning Marigolds. Goldsmith Seeds, Inc. Gilroy, CA. 10 March 1986.

FOLLETT: Follett, R.H. and J.W. Echols. 1982. Sunflower Production in Colorado. Service in Action Publication No. 102. Colorado State University. Fort Collins. 3 pp.

FORSYTH: Forsyth, W.G.C. 1980. Banana and Plantain. In Tropical and Subtropical Fruits, S. Nagy and P. Shaw, editors. AVI Publishing Company, Inc. Westport, CT.

FORTIN: Fortin, F. 1996. The Visual Food Encyclopedia. First ed. Macmillan, NY. 685 pp.

FOSTER: Foster, S. 1993. Herbal Renaissance. Gibbs-Smith Publishers, Salt Lake City. 234 pp.

FRANK: Frank, J.R. 1997. Personal Communications on Book Review. USDA-ARS (Retired), Fort Detrick, Frederick, MD.

FREITUS: Freitus, J. 1975. 160 Edible Plants Commonly Found in the Eastern United States. Stonewall Press. Lexington, MA. 85 pp.

FRISCH: Frisch, R.E. 1966. Plants that Feed the World. D. VanNostrand Co., NJ. 104 pp.

FUKUDA: Fukuda, S.K. 1987. Personal Communications which included Marie C. Neal's "In Gardens of Hawaii" article. Cooperative Extension Service, University of Hawaii at Manoa. 2 pp.

GARLAND: Garland, S. 1993. The Complete Book of Herbs and Spices. The Reader's Digest Association, Inc. Pleasantville, NY. 288 pp.

GARRISON: Garrison, S.A. 1994. Personal Communications for Asparagus Production. Cooperative Extension, Cook College, Rutgers, the State University, Bridgeton, NJ. 11 August, 1994.

GENTRY: Gentry, H.S., H. Mittleman, and P.R. McCrohan. 1992. Introduction of Chia and Gum Tragacanth, New Crops for the U.S. Diversity 8 (1): 28-31.

GHORPADE: Ghorpade, V.M., M.A. Hanna and S.S. Kadam. 1995 Apricot. In Handbook of Fruit Science and Technology, D.K. Salunkhe and S.S.Kadam, editors. Marcel Dekker, Inc., New York, NY.

GIBBON: Gibbon, T.F. 1976. The Food of the Modern World. Hutchinson of London. 529 pp.

GIBBONS: Gibbons, E. 1979. Handbook of Edible Plants. Donning Press, VA. 319 pp.

GLEASON: Gleason, H.A. 1958. New Britton and Brown Illustrated Flora of the Northeastern U.S. and Adjacent Canada. Volumes I-III. NY Botanical Garden, Lancaster Press, PA.

GOFORTH: Goforth, C.E. 1994. A Summary of Kenaf Production and Product Development Research 1989-1993. Mississippi State University Bulletin 1011. 33 pp.

GOULD: Gould, F.W. 1975. The Grasses of Texas. Texas A&M University Press, College Station, TX. 653 pp.

GRANBERRY: Granberry, D.M., P. Colditz, and W.J. McLaurin. 1986. Southern Pea. Commercial Vegetable Production. University of Georgia. College of Agriculture. Circular 485. 4 pp.

GREAT LAKES VEGETABLES GROWERS NEWS 1993: Editor. 1993. Black Walnuts: One Company Supports the Whole Industry. March. p. 28.

GREAT LAKES VEGETABLES GROWERS NEWS 1990: Editor. 1990. It's Horseradish Heaven at Huntsinger Farms of Eau Claire, WI. The Great Lakes Vegetables Growers News, Great American Publishing, Inc. Sparta, MI. November. p 9.

GREEN: Green, R.J. 1985. Peppermint and Spearmint Production in the Midwest. The Herb, Spice, and Medicinal Plant Digest 3 (1): 1-2,5.

GRIN: Germplasm Resources Information Network. 1997. Plant Taxonomy. USDA-ARS, National Germplasm Resources Laboratory. Beltsville, MD. Database.

GUENAULT: Guenault, B. 1985. New Plants and Plant Products as Food. Proceedings Nutr. Soc. 94:31-35.

GUPTA: Gupta, O.P. and S.S. Kadam. 1995. Ber (Jujube). In Handbook of Fruit Science and Technology. D.K. Salunkhe and S.S. Kadam, editors. Marcel Dekker, Inc., New York, NY.

HACKETT: Hackett, C. and J. Carollane. 1982. Edible Horticultural Crops: A Compendium of Information on Fruit, Vegetables, Spices and Nuts. Academic Press, NY. 673 pp.

HALBROOKS: Halbrooks, M.C. 1987. Bunch Grapes: Another Fruit Crop for Florida, University of Florida. Florida Cooperative Extension Service. IFAS. Fruit Crops Fact Sheet. FC-78. 4 pp.

HALPIN: Halpin, A.M. 1982. Organic Gardener's Complete Guide to Vegetables and Fruits. Rodale Press. Emmaus, PA. 510 pp.

HALPIN(a): Halpin, A.M. 1978. Unusual Vegetables. Rodale Press. Emmaus, PA. Pp. 327-331.

HAMASAKI: Hamasaki, R.T., H.R. Valenzuela, D.M. Tsuda and J.Y. Uchida. 1994. "Fresh Basil Production Guidelines for Hawaii," College of Tropical Agriculture and Human Resources. Series 154. University of Hawaii. 9 pp.

HANNAWAY: Hannaway, D.B., W.S. McGuire, and H.W. Youngberg. 1983. Growing Turnips for Forage. Oregon State University Extension Service. FS 296. 2 pp.

HANNAWAY(a): Hannaway, D.B. and W. S. McGuire. 1982.

Growing Field Peas for Forage. Oregon State University Extension Service Fact Sheet. FS 288. 2 pp.

HANNAWAY(b): Hannaway, D.B., H.W. Youngberg, and W.S. McGuire. 1982. Growing Rape and Kale for Forage. Oregon State University Extension Service. FS 287.

HANNON: Hannon, J. (Editor) 1997. *Connexions*. Cook College, NJAES, Office of Communications and Public Affairs. Rutgers, the State University, New Brunswick, NJ. 20 pp.

HANSON: Hanson, C.V., E.A. Oelke, D.H. Putnam, and E.S. Oplinger. 1992. Psyllium. Alternative Field Crops Manual. University of Minnesota and University of Wisconsin Extension Service. 3 pp.

HANSON(a): Hanson, C.V., D.K. Wildung, V.A. Fritz, and L. Waters. 1991. Production of Belgian Endive (Witloof) in Minnesota. Center for Alternative Plant and Animal Products. University of Minnesota. 25 pp.

HARDIN: Hardin, L.S. 1990. International Repercussions of New Crops. In Advances in New Crops, edited by J. Janick and J.E. Simon. Timber Press, Inc. Portland, OR. pp. 47-51.

HARRINGTON: Harrington, G. 1984. Grow Your Own Chinese Vegetables. Garden Way Publ. Book, Pownal, VT. 268 pp.

HARRIS: Harris, B.C. 1955. Eat the Weeds. 1955 Worchester, Mass. 147 pp.

HARRISON: Harrison, S.G. 1989. The Oxford Book of Food Crops. Oxford University Press. London. 206 pp.

HARTMANN: Hartmann, R.W., Y. Nakagawa, R. Sakuoka. 1978. Mustard Cabbage. HI Cooperative Home Gardening Vegetable Series. No. 10.

HASSY: Hassy, J.K., R.S. Johnson, J.A. Grant, and W.O. Reil. 1994. Kiwifruit: Growing and Handling. University of California Publication 3344, ANR Publications, Oakland, CA.

HAWAIIAN SUGAR PLANTERS' ASSOCIATION: Hawaiian Sugar Planters' Association. 1991. Hawaiian Sugar Manual. HSPA, Aiea, HI. 25 pp.

HEATH: Heath, M.E., R.F. Barnes, and D.S. Metcalfe. 1985. Forages: The Science of Grassland Agricultures. Fourth Ed. Iowa State University Press, Iowa. 643 pp.

HEATON: Heaton, D.D. 1997. A Produce Reference Guide to Fruits and Vegetables from Around the World. Nature's Harvest. Food Product Press, NY. 244 pp.

HEDRICK: Hedrick, V.P., Editor. 1972. Sturdivant's Edible Plants of the World – 1919. Dover Publ. Co., NY. 686 pp.

HEISER: Heiser, C.B. 1979. The Gourd Book. University of Oklahoma Press, OK. 248 pp.

HELM: Helm, R.S. 1993. Rice Production Handbook. University of Arkansas Cooperative Extension Service. MP 192.

HENNINGER: Henninger, M. 1996 Personal Communications on Potatoes. NJAES, Cook College, Rutgers, the State University, New Brunswick, NJ. 19 December 1996.

HENSON: Henson, P.R. and J.L. Stephens. 1958. Lupines. USDA Farmer's Bulletin 2114. 12 pp.

HERBST: Herbst, S.T. 1991. Food Lovers' Companion. Barrons, NY. 582 pp.

HEYWOOD: Heywood, V.H. 1982. Popular Encyclopedia of Plants. Cambridge University Press, NY. Pp. 336-340.

HILL: Hill, J.E., S.R. Roberts, D.M. Brandon, S.C. Scardaci, J.F. Williams, C.M. Wick, W.M. Canevari, and B.L. Weir. 1992. Rice Production in California. University of California Cooperative Extension Division of Agriculture and Natural Resources. Publication 21498. 22 pp.

HITCHCOCK: Hitchcock, A.S. 1971. Manual of the Grasses of the United States. USDA Misc. Publ. 200, Revised.

HOFMANN: Hofmann, L., A.M. Decker, and J.L. Newcomer. 1993. Hay and Pasture Seedlings for Maryland, University of Maryland Cooperative Extension Service Bulletin MEP 299. 6 pp.

HOFSTETTER: Hofstetter, B. 1988. Cover Crop Guide. The New Farm 5: 17-31.

HOLCOMB: Holcomb, G.B. 1989. Exotic Fruits: A Small-Scale Agriculture Alternative. USDA Cooperative State Research Service. Washington, D.C. 2 pp.

HOLCOMB(a): Holcomb, G.B. 1988. Specialty Vegetables. A Small-Scale Agriculture Alternative. USDA Cooperative State Research Service. Washington, D.C.

HOLLAND: Holland, J. et al. 1991. The Composition of Food. 55th ed. McCance & Widdowsons, Royal Chemistry Publ., Cambridge. 462 pp.

HOLLE: Holle, M. 1994. ART: The Biodiversity of Andean Root and Tuber Program. Diversity 10(4): 32-33.

HOLT: Holt, C. 1991. Complete Book of Herbs. Conran Octopus Ltd., London. 220 pp.

HOLZWORTH: Holzworth, L.K., L.E. Wiesner, and H.F. Bowman. 1990. Grass and Legume Seed Production in Montana and Wyoming. Montana and Wyoming Soil Conservation District Special Report No. 12. 31 pp.

HOSENEY: Hoseney, R.C. 1994. Principles of Cereal Science and Technology. Second ed. American Association of Cereal Chemists. St. Paul, MN. 378 pp.

HOSENEY(a): Hoseney, R.C. 1986. Principles of Cereal Science and Technology. First Ed. American Association of Cereal Chemists, Inc. St. Paul, MN. 327 pp.

HOWES: Howes, F.N. 1975. A Dictionary of Useful and Everyday Plants and Their Common Names. Cambridge University Press, London. 290 pp.

HU: Hu, S.T., D.B. Hannaway, and H.W. Youngberg. 1992 Forage Resources of China. Pudoc Wageningen, Netherlands. 327 pp.

HUMPHREY: Humphrey, S.W. 1973. Spices, Seasonings, and Herbs. Macmillan Co., NY. 370 pp.

HYLIN: Hylin, J.W. 1985. Personal Communications Concerning Swamp Cabbage. University of Hawaii, Honolulu. 18 November, 1985.

HYLTON: Hylton, W.H. 1993. The Rodale Herb Book. Rodale Press, Emmaus, PA. 650 pp.

HYMONITZ: Hymonitz, T. 1990. Grain Legumes. Pp. 154-158 in Advances in New Crops. Janick, J. and J. Simon, editors. Timber Press. Portland, OR.

IR-4 REQUEST DATABASE 1996: IR-4 State Request Database. 1996. IR-4 Food-Use Requests by State Provided by D.K. Infante, IR-4, New Jersey Agricultural Experiment Station, Cook College, Rutgers, the State University of New Jersey, New Brunswick, NJ. 30 April 1996.

IVES: Ives, F. 1997a. Unpublished Data on Crop Vernaculars for Mexico. OPP, US EPA, Washington, D.C. 3 pp.

JACKSON: Jackson, L.K. 1991. Citrus Growing in Florida. Third Ed. University of Florida Press. Gainesville, FL. 293 pp.

JACKSON(a): Jackson, L.K. and J.J. Ferguson. 1986. Tangerines and Tangerine Hybrids. University of Florida. Florida Cooperative Extension Service. IFAS. Fruit Crops Fact Sheet FC-33A. 4 pp.

JACKSON(b): Jackson, L.K. and J.W. Sauls. 1986. Grapefruit. University of Florida. Florida Cooperative Extension Service. IFAS. Fruit Crops Fact Sheet FC- 35. 4 pp.

JACOBSEN: Jacobsen, S.E. and O. Stolen. 1993. Quinoa - Morphology, Phenology, and Prospects for Its Production as a New Crop in Europe. European Journal Agronomy 2 (1): 19-29.

JANICK 1996: Janick, J. (Editor). 1996. Progress in New Crops. ASHS Press, 600 Cameron Street, Alexandria, VA. 660 pp.

JANICK 1991: Janick, J. and J.E. Simon. 1991 Proceedings Second National Symposium New Crops. John Wiley & Sons, Inc. 710 pp.

JANICK 1990: Janick, J. and J.E. Simon. 1990. Advances in New Crops. Proceedings of the First National Symposium. New Crops: Research, Development, Economics. Timber Press, Inc. Portland, OR. 560 pp.

JANICK 1981: Janick, J., R.W. Schenz, F. Woods, and W.W. Ruttan. 1981. Plant Science: Introduction to World Crops. 3rd ed. W.H. Freeman & Co. 868 pp.

JOHNSON 1986: Johnson, D.L. and R.L. Croissant. 1986. Sesame Production. Services in Action Publication No. 100. Colorado State University Cooperative Extension. 3/86. 2 pp.

JOHNSON(a): Johnson, D.L. and S.M. Ward. 1993. Quinoa. Pp. 222-227. In New Crops. J.Janick and J.E. Simon, editors. John Wiley & Sons. NY.

JOHNSON(b): Johnson, D.L. and M.N. Jha. 1993. Blue Corn. Pp. 228-230. In New Crops. J. Janick and J.E. Simon, editors. John Wiley & Sons, NY.

JOHNSON(c): Johnson, D.L. 1990. New Grains and Pseudo-grains. Pp. 122-127 in Advances in New Crops. Janick, J. and J.E. Simon, editors. Timber Press. Portland, OR.

JOSHI: Joshi, V.K. and V.P. Bhutani. 1995. Peach and Nectarine. In Handbook of Fruit Science and Technology, D.K. Salunkhe and S.S. Kadam, editors. Marcel Dekker, Inc. New York, NY.

KADAM: Kadam, S.S. and D.K. Salunkhe. 1995. Avocado. In Handbook of Fruit Science and Technology, D.K. Salunkhe and S.S. Kadam, editors, Marcel Dekker, Inc. New York, NY.

KADAM(a): Kadam, S.S. and S.S. Deshpande. 1995. Lychee. In Handbook of Fruit Science and Technology, D.K. Salunkhe and S.S. Kadam, editors. Marcel Dekker, Inc., New York, NY.

KADAM(b): Kadam, P.Y., S.A. Dhumal, and N.N. Shinde. 1995. Pear. In Handbook of Fruit Science and Technology, D.K. Salunkhe and S.S. Kadam, editors. Marcel Dekker, Inc., New York, NY.

KADANS: Kadans, J.M. 1970. Modern Encyclopedia of Herbs. W. Parker Publ. Co., New York. 257 pp.

KALE: Kale, P.N. and P. G. Adsule. 1995. Citrus. In Handbook of Fruit Science and Technology, D.K. Salunkhe and S.S. Kadam, editors. Marcel Dekker, Inc., New York, NY.

KALRA: Kalra, S.K., D.K. Tandon, and B.P. Singh. 1995. Mango. In Handbook of Fruit Science and Technology, D.K. Salunkhe and S.S. Kadam, editors. Marcel Dekker, Inc., New York, NY.

KAUFFMAN: Kauffman, C.S. and L.E. Weber. 1990. Grain Amaranth. Pp. 127-139. in Advances in New Crops. Janick, J. and J.E. Simon, editors. Timber Press. Portland, OR.

KAUSHAL: Kaushal, B.B. Lal, and P.C. Sharma. 1995. Apple. In Handbook of Fruit Science and Technology, D.K. Salunkhe and S.S. Kadam, editors. Marcel Dekker, Inc. New York, NY.

KAWATE 1997: Kawate, M. 1997. Personal Communications concerning Sugarcane. University of Hawaii at Manoa, Honolulu, HI. 11 June, 1997.

KAWATE 1997a: Kawate, M. 1997a. Personal Communications concerning Ti. University of Hawaii at Manoa, Honolulu, HI. 9 June, 1997.

KAWATE 1997b: Kawate, M. 1997b. Personal Communications concerning Peach Palm. University of Hawaii at Manoa, Honolulu, HI. 1 August, 1997.

KAWATE 1997c: Kawate, M. 1997c. Personal Communications concerning Upland Watercress. University of Hawaii at Manoa, Honolulu, HI. 31 January, 1997.

KAWATE 1996a: Kawate, M. 1996a. Personal Communications. Tropical fruits. University of Hawaii at Manoa, Honolulu, HI. 11 March, 1996.

KAWATE 1996b: Kawate, M. 1996b. Personal Communications. Specialty tropical fruits. University of Hawaii at Manoa, Honolulu, HI. 30 April, 1996.

KAWATE 1996c: Kawate, M. 1996c. Personal Communications. Cilantro, Parsley. University of Hawaii at Manoa, Honolulu, HI. 29 February, 1996.

KAWATE 1996d: Kawate, M. 1996d. Personal Communication, letter sent 30 April, 1996 to G. Markle on Specialty Tropical Fruit.

KAWATE 1995: Kawate, M. 1995. Personal Communications on Minor Crops in Hawaii. University of Hawaii at Manoa, Honolulu, HI. 6 December, 1995.

KAWATE 1995a: Kawate, M. 1995a. Personal Communications on Poha. University of Hawaii at Manoa, Honolulu, HI. 11 May, 1995.

KAYE: Kaye, G.C. 1986. Wild and Exotic Mushroom Cultivation in North America. 2nd ed. Farlow Ref. Library. Harvard University. Cambridge, MA. 59 pp.

KAYS: Kays, S.J. and J.C. Silva Dias. 1995. Common Names of Commercially Cultivated Vegetables of the World in 15 Languages. Economic Botany 49(2) 115-152.

KAYS 1996: Kays, S.J. and J.C. Silva Dias. 1996. Cultivated Vegetables of the World. Exon Press. P.O. Box 80803, Athens, GA 30608-0803. 170 pp.

KENNARD: Kennard, W.C. and H.F. Winters. 1960. Some Fruits and Nuts of the Tropics. USDA/ARS Misc. Publication No. 801. Superintendent of Documents, Washington, D.C. 135 pp.

KEPHART: Kephart, K.D., G.A. Murray, and D. Auld. 1990. Alternate Crops for Dryland Production Systems in Northern Idaho. Pp. 62-67 in Advances in New Crops. Janick, J. and J.E. Simon, editors. Timber Press. Portland, OR.

KIRK: Kirk, D.R. 1970. Wild Edible Plants of the Western United States Including Almost All of Southwestern Canada and Northwestern Mexico. Naturegraph Publ., CA. 307 pp.

KLEIN: Klein, M.B. 1985. All About Citrus and Subtropical Fruits. Ortho Books. San Francisco, CA. 96 pp.

KNIGHT: Knight, Jr., R.J. and J.W. Sauls. 1986. The Passion Fruit. University of Florida. Florida Cooperative Extension Service. Institute of Food and Agricultural Services. Fruit Crop Fact Sheet FC-60. 4 pp.

KNIGHT(a): Knight, Jr., R.J. 1980. Origin and World Importance of Tropical and Subtropical Fruit Crops. In Tropical and Subtropical Fruits, S. Nagy and P. Shaw, editors. AVI Publishing Company, Inc. Westport, CT.

KOTCH: Kotch, R.S., J.H.Murphy, M.D. Orzolek, and P.A. Ferretti. 1995. Factors Affecting the Production of Baby Corn. Journal Vegetable Crop Production 1(1):19-28.

KOTECHA: Kotecha, P.M. and B.B. Desai. 1995. Banana. In Handbook of Fruit Science and Technology, D.K. Salunkhe and S.S. Kadam, editors. Marcel Dekker, Inc., New York, NY.

KOTECHA(a): Kotecha, P.M. and D.L. Madhavi. 1995. Berries. In Handbook of Fruit Science and Technology, D.K. Salunkhe and S.S. Kadam, editors. Marcel Dekker, Inc. New York, NY.

KOWALCHIK: Kowalchik, C. and W. Hylton. 1987. Rodale Illustrated Encyclopedia of Herbs. Rodale Press, Emmaus, PA. 545 pp.

KRAUS: Kraus, J.E. 1940. Chinese Cabbage Varieties, Their Classification, Description, and Culture in the Central Great Plains. USDA Circular No. 571.

KREZDORN: Krezdorn, A.H. and M.L. DuBois. 1986. The Fig. University of Florida. Florida Cooperative Extension Service. IFAS. Fruit Crops Fact Sheet FC-27. 2 pp.

KUEBEL: Kuebel, K.R. and A.O. Tucker. 1988. Vietnamese Culinary Herbs in the United States. Economic Botany, 43 (3). pp. 413-419.

KUGLER: Kugler, D.E. 1988. Kenaf Newsprint. U.S. Department of Agriculture, Cooperative State Research Service, Washington, D.C. 13 pp.

KUHNLEIN: Kuhnlein, H.V. and N.J. Turner. 1991. Traditional Plant Foods of Canadian Indigenous Peoples Nutrition, Botany and Use. Gordon and Breach Science Publishers, Philadelphia. Volume 8. 602 pp.

KUNKEL: Kunkel, D. 1993. Personal Communications on Sugarcane Processing. Cook College, Rutgers, the State University, New Brunswick, NJ. 22 June, 1993.

KUNKEL(a): Kunkel, G. 1984. Plants for Human Consumption. Koeltz, Scientific Books, W. Germany. 394 pp.

KURTZ 1995: Kurtz, E.A. 1995. Personal Communications. Parsley. EAK AG, Inc. Salinas, CA. 21 December, 1995.

KUTE: Kute, L.S. and M.B. Shete. 1995. Sapota (Sapodilla). In Handbook of Fruit Science and Technology, D.K. Salunkhe and S.S.Kadam, editors. Marcel Dekker, Inc., New York, NY.

LAIDIG: Laidig, G.L., E.G. Knox, and R. Buchanan. 1985. Underexploited Crops. Macmillan Publ. Co., NY.

LAKSHMINARAYANA: Lakshminarayana, S. 1980. Mango. In Tropical and Subtropical Fruits, S. Magy and P. Shaw, editors. AVI Publishing Company, Inc. Westport, CT.

LAKSHMINARAYANA(a): Lakshminarayana, S. 1980. Sapodilla and Prickly Pear. In Tropical and Subtropical Fruits, S. Nagy and P. Shaw, editors. AVI Publishing Company, Inc. Westport, CT.

LAMBERTS: Lamberts, M. and J. Crane. 1990. Tropical Fruits. Pp. 337-355 in Advances in New Crops. Janick, J. and J.E. Simon, editors. Timber Press. Portland, OR.

LAMBERTS(a): Lamberts, M. 1990. Latin American Vegetables. Pp. 378-387 in Advances in New Crops. Janick, J. and J.E. Simon, editors. Timber Press. Portland, OR.

LANGER: Langer, R.H.M. and G.D. Hill. 1991. Agricultural Plants. Second Ed. Cambridge University Press, NY. 387 pp.

LARKCOM: Larkcom, J. 1991. Oriental Vegetables. Kodansha International, NY. 232 pp.

LARKCOM(a): Larkcom, J. 1984. The Salad Garden. Viking Press. NY. 168 pp.

LEIDNER 1992: Leidner, J. 1992. All the Tea in Charleston. Progressive Farmer. January, 1992.

LEONARD: Leonard, W.H. and J.H. Martin. 1963. Cereal Crops, Macmillan Publishing, NY. 824 pp.

LEWIS: Lewis, Y.S. 1984. Spices and Herbs for the Food Industry. Food Trade Press. Orpington, England. 208 pp.

LIVINGSTON: Livingston, A.D. and H. Livingston. 1993. Edible Plants and Animals: Unusual Foods from Aardvark to Zamia. Facts on File. 293 pp.

LOGAN: Logan, M. 1996. The Packer 1996 Produce Availability and Merchandising Guide. Vance Publishing Corporation. Lincolnshire, IL.

LONGBRAKE: Longbrake, T.D., M.L. Baker, S. Cotner, J. Parsons, R. Roberts, and L. Stein. 1993. Specialty Vegetables in Texas. Texas Agricultural Extension Service. B-1613. 11 pp.

LORENZ: Lorenz, O.A. and D.N. Maynard. 1988. Knott's Handbook for Vegetable Growers. 3rd Edition. John Wiley & Sons. 456 pp.

LOTSCHERT: Lotschert, W. and G. Beese. 1988. Collins Guide to Tropical Plants. William Collins Sons & Co., London. 256 pp.

LOVERING: Lovering, N. 1997. Personal Communications on Stevia. Royal Sweet International Technologies Ltd. P.O. Box 186, Delhi, Ontario, Canada. 27 February and 24 March, 1997.

LUMPKIN: Lumpkin, T.A., J.C. Konovsky, K.J. Larson, and D.C. McClary. Potential New Specialty Crops from Asia: Adzuki Bean, Edamane Soybean, and Astragalus. Pp. 45-52 In Janick, J. and J.E. Simon. 1991. Proceedings of Second National Symposium on New Crops. John Wiley & Sons, Inc.

LUNDY: Lundy, R. 1997. Personal Communications on Mint Production. Mint Industry Research Council, Washington.

LYRENE: Lyrene, P.M. and T.E. Crocker. 1990. The Blueberry. University of Florida. Florida Cooperative Extension Service. Institute of Food and Agricultural Science. Fruit Crops Fact Sheet FC-46A. 4 pp.

LYRENE(a): Lyrene, P.M. and T.E. Crocker. 1984. The Chinese Jujube. University of Florida. Florida Cooperative Extension Service. Institute of Food and Agricultural Science. Fruit Crops Fact Sheet FC-50. 3 pp.

MABBERLY: Mabberly, D.J. 1987. The Plant Book. Cambridge University Press, NY. 706 pp.

MABEY: Mabey, R. 1978. Plants with a Purpose. Collins Publ., NY. 176 pp.

MACGREGOR: MacGregor, A.W. and R.S. Bhatty. 1993. Barley: Chemistry and Technology. American Association of Cereal Chemists. 486 pp.

MACKIN: Mackin, J. 1993. The Cornell Book of Herbs and Edible Flowers. Cornell Cooperative Extension. 104 pp.

MADLENER: Madlener, J.C. 1977. The Sea Vegetable Book. Clarkson N. Potter, Inc., Publ. NY. 288 pp.

MAGNESS: Magness, J.R., G.M. Markle, and C.C. Compton. 1971. Food and Feed Crops of the United States. NJAES Bulletin 828, Rutgers University, New Brunswick, NJ. 255 pp.

MALKUS: Malkus, D.J. 1984. Evaluation of Amaranth as a Potential Greens Crop in the Mid-South. HortScience 19 (6): 881-883.

MALO(a): Malo, S.E. and C.W. Campbell. 1991. The Guava. Florida Cooperative Extension Service. Institute of Food and Agricultural Science. Fruit Crops Fact Sheet FC-4. 2 pp.

MALO(b): Malo, S.E. and C.W. Campbell. 1991. The

Macadamia. University of Florida. Florida Cooperative Extension Service. Institute of Food and Agricultural Science. Fruit Crops Fact Sheet FC-9. 3 pp.

MALO(c): Malo, S.E. and C.W. Campbell. 1991. The Papaya. University of Florida. Florida State Cooperative Extension Services. Institute of Food and Agricultural Science. Fruit Crops Fact Sheet FC-11. Revised. 3 pp.

MALO(d): Malo, S.E. and C.W. Campbell. 1991 The Tahiti Lime. University of Florida. Florida Cooperative Extension Service. Institute of Food and Agricultural Science. Fruit Crops Fact Sheet FC-8. 3 pp.

MALO(e): Malo, S.E. and C.W. Campbell. 1990. The Pineapple. University of Florida. Florida Cooperative Extension Service. Institute of Food and Agricultural Science. Fruit Crops Fact Sheet FC-7. 3 pp.

MALO(f): Malo, S.E. and C. W. Campbell. 1986. The Papaya. University of Florida, Florida Cooperative Extension Service. Institute of Food and Agricultural Science. Fruit Crops Fact Sheet FC-11. 3 pp.

MALO(g): Malo, S.E. and C.W. Campbell. 1989. The Banana. University of Florida. Florida Cooperative Extension Service. Institute of Food and Agricultural Science. Fruit Crops Fact Sheet FC-10. 4 pp.

MALO(h): Malo, S.E. and C.W. Campbell. 1983. The Avocado. University of Florida. Florida Cooperative Extension Service. Institute of Food and Agricultural Science. Fruit Crops Fact Sheet FC-3. 4 pp.

MANOS 1986: Manos, C.G., P.J. Schrynemeeckers, D.E. Hogue, J.N. Telford, G.S. Stoewsand, D.H. Beerman, J.G. Babish, J.T. Blue, B.S.Shane, and D.J. Lisk. 1986. Toxicologic Studies with Lambs Fed Jojoba Meal Supplemented Rations. Journal Agricultural Food Chemistry. 34, 801-805.

MARGEN: Margen, S. 1992. The Wellness Encyclopedia of Food and Nutrition. First Ed. Rebus, NY. 512 pp.

MARR: Marr, C.W. and W.J. Lamont. 1991. Farming a Few Acres of Vegetables. Kansas State University Cooperative Extension Service. Manhattan, KS. Bulletin 1113.

MARSH: Marsh, D.B. 1991. Ethnic Crop Production: An Overview and Implications for Missouri. HortSci. 26(9):1133-1135.

MARSHALL 1993: Marshall, W.E. and J.I. Wadsworth. 1993. Rice Science and Technology, First Ed. 470 pp.

MARSHALL 1992: Marshall, H.G. and M. E. Sorrello. 1992. Oat Science and Technology. American Society of Agronomy. Crop Science Society of America. Madison, WI. No. 33. 846 pp.

MARTIN 1987: Martin, F.W., C.W. Campbell, and R.M. Ru-

berte. 1987. Perennial Edible Fruits of the Tropics: An Inventory. U.S. Department of Agriculture. Agricultural Handbook No. 642. 252 pp (Illus.)

MARTIN 1983: Martin, F.W. 1983. Handbook of Tropical Food Crops. CRC Press, Inc. Boca Raton, FL. 296 pp.

MARTIN 1980: Martin, F.W. 1980. Durian and Mangosteen. In Tropical and Subtropical Fruits, S. Nagy and P. Shaw, editors. AVI Publishing Company, Inc. Westport, CT.

MARTIN 1979: Martin, F.W. and R.M. Ruberte. 1979. Edible Leaves of the Tropics (2nd Edition). Antillian College Press, Mayaguez, Puerto Rico. 234 pp.

MARTIN 1975: Martin, F.W. and R.M. Ruberte. 1975. Edible Leaves of the Tropics. Antillian College Press. Mayaguez, Puerto Rico. 235 pp.

MARTIN(a): Martin, J.H., W.H. Leonard, and D.L. Stamp. 1976. Principles of Field Crop Production. Macmillan Publ. Co., Inc. NY. First Ed. 1118 pp.

MARTIN(b): Martin, J.H. and W.H. Leonard. 1967. Principles of Field Crop Production. Macmillan Publ. Co., Inc. NY. 1044 pp.

MARTINI: Martini, J.H. 1992. Unpublished Data. Weight to weight ratio study for Kohlrabi leaves and enlarged stem. Department of Food Sciences and Technology. NYAES, Cornell University, Geneva, NY. 3 pp.

MASK: Mask, P.L. and W.C. Morris. 1991. Sweet Sorghum Culture and Syrup Production. Circular ANR 625. Alabama Cooperative Extension, Auburn University, Alabama. 12 pp.

MATZ: Matz, S.A. 1991. The Chemistry and Technology of Cereals as a Food and Feed. Second ed. AVI Book Van Nostrand Reinhold, NY. 751 pp.

MCCALLUM: McCallum, M. 1996. New dried carrot pieces could spawn other vegetable products. Great Lakes Vegetable Grower, No. 10, p. 1-3, October, 1996.

MCCLURE: McClure, S. 1995. The Herb Gardener. Storey Communications, Inc. Schoolhouse Road, Pownal, VT. 236 pp.

MCDONALD: McDonald, M.B. and L. Copeland. 1997. Seed Production: Principles and Practices. Chapman and Hall, NY. Second Ed. 749 pp.

MCGEE: McGee, H. 1984. On Food and Cooking. Charles Scribner's Sons. NY. 685 pp.

MCHUGH: McHugh, J.J., Jr., S.K. Fukuda, and K.Y. Takeda. 1987. Hawaii Watercress Production. University of Hawaii College of Tropical Agriculture and Human Resources. Research Extension Series - 088.

MCMILLIN: McMillin, B. 1997. Personal Communications in

Horseradish Production. Illinois Horseradish Growers (Tri-County Vegetable Growers), 8490 Forest Blvd., Caseyville, IL. 31 January, 1997.

MCRAE: McRae, B.A. 1992. Hops. The Herb Companion 4 (4): 60-64.

MEDVED: Medved, E. 1981. The World of Food. Ginn & Co. Lexington, MA. 599 pp.

MEISTER 1997: Meister, C.W. 1997. Personal Communications concerning watercress. University of Florida, Gainesville, October 1987.

MEISTER 1996a: Meister, C.W. 1996a. Personal Communications concerning Swamp Cabbage. University of Florida, Gainesville. June, 1996.

MELNICOE 1997a: Melnicoe, R.S. 1997a. Unpublished data on Stevia and Waxgourd. IR-4, University of California, Davis. 10 March and 22 April, 1997

MELNICOE 1997b: Melnicoe, R.S. 1997b. Unpublished data. California Crop List. IR-4, University of California, Davis. 8 pp.

MELNICOE 1996a: Melnicoe, R.S. 1996a. Personal Communications on Vegetable Crops in California. University of California, Davis. 22 April, 1996.

MELNICOE 1996b: Melnicoe, R.S. 1996b. Personal Communications on Vegetable Crops in California (Epazote). University of California, Davis. 14 June, 1996.

MELNICOE 1996c: Melnicoe, R.S. 1996c. Personal Communications on Jojoba. University of California, Davis. 27 August, 1996.

MELNICOE 1996d: Melnicoe, R.S. 1996d. Personal Communications on Feijoa, Pepino, Yacon, and Babaco. University of California, Davis. 22 August, 1996.

MELNICOE 1996e: Melnicoe, R.S. 1996e. Personal Communications on Review for Crop Acreages in California. IR-4, University of California, Davis. November, 1996.

MELNICOE 1995: Melnicoe, R.S. 1995. Personal Communications. Crops grown on Guam. University of California, Davis. 7 April, 1995.

MENNINGER: Menninger, E.A. 1977. Edible Nuts of the World. Horticultural Books. Stuart, FL. 175 pp.

MENZEL: Menzel, C. 1989. The Subtropical Longan: An Australian View. California Grower 13 (7): 20,22-23.

MERCK: Budavari, S. 1996. The Merck Index. Merck Research Laboratories Division of Merck & Co., Inc. Whitehouse Station, NJ. 1741 pp.

MERONUCK: Meronuck, R.A., H. Meredith, and D.H. Putnam. 1991. Lupin Production and Utilization Guide. Center for Alternative Plant and Animal Production. University of Massachussetts. 27 pp.

METCALF: Metcalf, D.S. and D.M. Elkins. 1980. Crop Production. Macmillan Publ. NY. 4th Ed. Pp. 366-387.

MILLER: Miller, P.R., W.L. Graves, W. A. William, and B.A. Madison. 1989. Covercrops for California Agriculture. University of California Division of Agriculture and Natural Resources. Publication 21471. 24 pp.

MILLER(a): Miller, D.A. 1984. Forage Crops. First Ed. McGraw-Hill, Inc. NY. 530 pp.

MITCHELL 1996: Mitchell, T. 1996. Personal Communications. Sesame Production. Texas Department of Agriculture. April and August.

MONTAGNE: Montagne, P. 1977. The New Larousse Gastronomique. Crown Publ., Inc. NY. 1064 pp.

MONTALVO-ZAPATA 1996a: Montalvo-Zapata, R.1996a. Personal Communications. Coffee. University of Puerto Rico, Mayaquez. 8 August, 1996.

MONTALVO-ZAPATA 1996b: Montalvo-Zapata, R. Personal Communications. Corianders. University of Puerto Rico, Mayaguez. 3 September, 1996.

MONTALVO-ZAPATA 1995: Montalvo-Zapata, R. 1995. Personal Communications. Tropical Fruits. University of Puerto Rico, Mayaguez. 16 March, 1995.

MONTALVO-ZAPATA 1988: Montalvo-Zapata, R. 1988. Personal Communications on Peppers, et. al. University of Puerto Rico, Mayaguez. 14 April, 1988

MOORE: Moore, J.N. and R. M. Skirvin. 1990. Blackberry Management, p. 214-244. In Small Fruit Crop Management, G.J. Galletta and D.G. Himelrick, editors. Prentice Hall, Englewood Cliffs, NJ.

MOORE(a): Moore, K.J., L.E. Moser, K.P. Vogel, S.S. Waller, B.E. Johnson, and J.F. Pederson. 1991. Describing and Quantifying Growth Stages of Perennial Forage Grasses. Agronomy Journal 83: 1073-1077.

MORTENSEN: Mortensen, E. and E.T. Bullard. 1970. Handbook of Tropical and Subtropical Horticulture, U.S. Department of State, Agency for International Development. Washington, D.C. 186 pp.

MORTON: Morton, J.F. 1987. Fruits of Warm Climates. Media Inc. First Ed. Greensboro, NC. 505 pp.

MORTON(a): Morton, J.F. 1976. The Pigeon Pea. (*Cajanus cajan* Millsp.), A High-protein, Tropical Bush Legume. HortScience 11 (1): 11-18.

MOSLER: Mosler, L.E., D.R. Buxton, and M.D. Asler. 1996. Cool-Season Forage Grasses. American Society of Agronomy Publ. Agronomy Series 34. Madison, WI. 841 pp.

MOTT: Mott, J. 1969. The Market for Passion Fruit Juice. Tropical Products Institute Report G38.

MOWRY: Mowry, H., R. Toy,. and H. Wolfe. 1967. Miscellaneous Tropical and Subtropical Florida Fruit. University of Florida. Food and Agricultural Science ARC. Bulletin 156A.

MULHERIN: Mulherin, J. 1988. The MacMillan Treasury of Spices and Natural Flavorings. MacMillan Publ., Co. NY.

MULHLBAUER: Mulhlbauer, F.J., R.W. Short, R.J. Summerfield, K.J. Morrison and P.G. Swan. 1981. Description and Culture of Lentils. Washington State University Cooperative Extension Bulletin EB0957. 8 pp.

MULLINS: Mullins, C. A. 1997. Personal Communications on Upland Cress Production. Plateau Experimental Station. The University of Tennessee, Crossville, TN. 29 May, 1997.

MUSHROOMS 1995 STATS: Mushrooms. 1995. National Agricultural Statistics Service, USDA. Washington, D.C. VG. 2-1-2 (8-95) 15 pp.

MYERS: Myers, C. 1991. Specialty and Minor Crops Handbook. Small Farm Center. University of California. Publication 3346.

NAEGELY: Naegely, S.K. 1997. Greenhouse Vegetables - Business is Booming (Includes comments from Drs. Merle Jensen (AZ) and Richard Snyder (MS). Greenhouse Grower. Meister Publishing Co., Willoughby, OH. June: pp 14-8.

NAGY: Nagy, S., P.E. Shaw, and W.F. Wardowski. 1990. Fruits of Tropical and Subtropical Origin-Composition, Properties, and Uses. Florida Science Source. Lake Alfred, FL. 391 pp.

NAJ: Naj, A. 1992. Peppers. Alfred A. Knopf Publishers. NY. 245 pp.

NALEWAJA 1996: Nalewaja, J.D. 1996. Personal Communications on Flaxseed. North Dakota State University, Fargo. 14 August, 1996.

NANTACHAI: Nantachai, S. 1994. Durian. Asian Food Handling Bureau, Thailand. 156 pp.

NATIONAL ACADEMY OF SCIENCE: National Academy of Science. 1984. Underexploited Tropical Plants with Promising Economic Value. National Academy of Science. Washington, D.C. 190 pp.

NATIONAL RESEARCH COUNCIL 1989: National Research Council. 1989. Lost Crops of the Incas. National Academy of Sciences Press. Washington, D.C. 415 pp.

NATIONAL RESEARCH COUNCIL 1984: National Research Council. 1984. Leucaena: Promising Forage and Tree Crop for the Tropics. Second Ed. National Academy of Science, Washington, D.C. 100 pp.

NATIONAL RESEARCH COUNCIL 1975: National Research Council. 1975. Products from Jojoba: A Promising New Crop for Arid Lands. National Academy of Sciences, Washington, D.C. 30 pp.

NEAL: Neal, M.C. 1965. In Gardens of Hawaii. Bishop Museum Press, Honolulu. 924 pp.

NELSON: Nelson, K.A. and S.J. Eilertson. 1996. Cereal Grain Specialty Products: A Brief Overview. Cereal Foods World. 41(5): 383-385.

NELSON(a): Nelson, T. 1993. Adzuki Beans. Small Farm News. Jan./Feb. Pp. 3.

NENE: Nene, Y.L., S.D. Hall, and V.K. Sheila. 1990. The Pigeonpea. CAB International Crops Research Institute for Semi-arid Tropicals. UK. 490 pp.

NEVE: Neve, R.A. 1991. Hops. Chapman & Hall. NY. 265 pp.

NISHINA: Nishina, M.S., D.M. Sato, W.T. Nishijima, and R.F.L. Mau. 1992. Ginger Root Production in Hawaii. Commodity Fact Sheet GIN-3(A) Rhizome. Hawaii Institute of Tropical Agriculture and Human Resources, University of Hawaii at Manoa. 4 pp.

NORTON: Norton, R.A. 1980. The Pistachio Nut - A New Crop for California. University of Florida. Cooperative Extension Service. Institute of Food and Agricultural Science. 3 pp.

OELKE(a): Oelke, E.A. 1993. Wild Rice; Domestication of a Native North American Genus. Pp. 235-243. In New Crops, J.Janick and J.E. Simon, editors. John Wiley & Sons, NY.

OELKE(b): Oelke, E.A., E.S. Oplinger, C.V. Hanson, D. Davis, D. Putnam, and C. Rosen. 1992. Dry Field Pea. In Alternative Field Crops Manual. Universities of Wisconsin and Minnesota Cooperative Extension Service. 4 pp.

OELKE(c): Oelke, E.A., E.S. Oplinger, and M.A. Brinkman. 1992. Triticale. In Alternative Field Crops Manual. Universities of Wisconsin and Minnesota Cooperative Extension Service. 8 pp.

OELKE(d): Oelke, E.A., D.H. Putnam, T. M. Teynor, and E.S. Oplinger. 1992. Quinoa. Alternative Field Crops Manual. Universities of Wisconsin and Minnesota Cooperative Extension Service. 6 pp.

OELKE(e): Oelke, E.A., E.S. Oplinger, C.V. Hanson, and K.A. Kelling. 1990. Meadowfoam. Alternative Field Crops Manual. Universities of Wisconsin and Minnesota Cooperative Extension Service. 4 pp.

OELKE(f): Oelke, E.A., E.S. Oplinger, H. Bahri, B.R. Durgan, D.H. Putnam, J.D. Doll, and K.A. Kelling. 1990. Rye. Alternative Field Crops Manual. Universities of Wisconsin and Minnesota Cooperative Extension Service. 6 pp.

OELKE(g): Oelke, E.A., E.S. Oplinger, D.H. Putnam, B.R. Durgan, J.D. Doll and D.J. Undersander. 1990. Millets. Alternative Field Crops Manual. Universities of Wisconsin and Minnesota Cooperative Extension Service. 5 pp.

OELKE(h): Oelke, E.A., E.S. Oplinger, and M.A. Brinkman. 1989. Triticale. Alternative Field Crops Manual. Universities of Wisconsin and Minnesota Cooperative Extension Service.

OELKE(i): Oelke, E.A., J. Grava, D. Noetzel, D. Barron, J. Percich, C. Schertz, J. Strait, and R. Stucker. 1982. Wild Rice Production in Minnesota. University of Minnesota Agricultural Extension Service Bulletin 464.

O'HAIR: O'Hair, S.K. 1990. Tropical Root and Tuber Crops. Pp. 424-428 in Advances in New Crops. Janick,J. and J.E. Simon, editors. Timber Press. Portland, OR.

O'HAIR(a): O'Hair, S.K. 1990. Tropical Root and Tuber Crops. Horticultural Reviews 12: 157-196.

OLDFIELD: Oldfield, J.E. 1986. Forage: Resources for the Future. Council for Agricultural Science and Technology (CAST) Report No. 108. 50 pp.

OLSZACK 1995: Olszack, R. 1995. Personal Communications on Tropical Fruit Crops in the Trade. Brooks Tropicals. Homestead, FL. September, 1995.

OPLINGER: Oplinger, E., L. Hardman, A. Kaminski, S. Comb, and J. Doll. 1992. Mungbean in Alternative Field Crops Manual. Universities of Wisconsin and Minnesota Cooperative Extension Service. 4 pp.

OPLINGER(a): Oplinger, E.S., E.A. Oelke, A.R. Kaminski, K.A. Kelling, J.D. Doll, B.R. Durgan, and R.T. Schuler. 1990. Spelt. Alternative Field Crops Manual. Universities of Wisconsin and Minnesota Cooperative Extension Service. 3 pp.

OPLINGER(b): Oplinger, E.S., E.A. Oelke, M.A. Brinkman, and K.A. Kelling. 1989. Buckwheat. Alternative Field Crops Manual. Universities of Wisconsin and Minnesota Cooperative Extension Service.

PALLUTT: Pallutt, W. 1996. Personal Communications. Important Crops in Germany and Crop Grouping. Biologische Bundesanstalt Fur Land. Institut fur Integrierten Pflenzenschutz Stahnsdorfer Damm 81, D-14532. Kleinmachnow, Germany. July, 1996.

PARKE: Parke, J.L. 1995. Personal Communications. Ginseng. University of Wisconsin, Madison. 23 March, 1995.

PARKER: Parker, S.P. (Editor). 1982. Synopsis and Classification of Living Organisms. McGraw-Hill Book Company, New York, NY. 1232 pp.

PATAKY: Pataky, J.K. 1991. Production of Cuitlacoche [*Ustilago maydis* (DS) Corda] on Sweet Corn, HortScience 26 (11): 1374-1377.

PATIL: Patil, V.K., V.R. Chakrawar, P.R. Narwadkar, and G.S. Shinde. 1995. Grape. In Handbook of Fruit Science and Technology, D.K. Salunkhe and S.S. Kadam, editors. Marcel Dekker, Inc., New York, NY.

PATTEE: Pattee, H.E. and C.T. Young. 1982. Peanut Science and Technology, American Peanut and Research & Ed. Yoakum, TX. 825 pp.

PAUTLER (1987 Census): Pautler, C.P. 1987. 1987 Census of Agriculture. Bureau of Census. U.S. Dept. of Commerce. Superintendent of Documents, U.S. Gov't. Printing Office, Washington, D.C. 1 (Part 51).

PAYNE: Payne, J.A. and G.W. Krewer. 1990. Mayhaw: A New Fruit Crop for the South. In Advances in New Crops, edited by J. Janick and J.E. Simon. Timber Press. Portland, OR. p. 317-321.

PEMBERTON: Pemberton, R.W. and N.S. Lee. 1996. Wild Food Plants in South Korea. Market Presence, New Crops and Exports to the United States. Economic Botany 50 (1): 57-90.

PENZEY: Penzey, W. 1995. Merchants of Quality Spices. Penzeys Ltd. Catalogue, P.O. Box 1448, Waukesha, WI. 34 pp.

PETERSON: Peterson, R.N. 1991. Pawpaw (Asminia), p. 567-600. In Genetic resources of Temperate Fruit and Nut Crops. Acta Horticulturae, Number 290.

PHELPS: Phelps, L. 1997a. Personal Communications and Review on Mushrooms. American Mushroom Institute. Washington, D.C. 27 February, 1997.

PHILLIPS: Phillips, H.F. 1989. What Thyme is it: A Guide to Thyme Taxa Cultivation in the U.S. Proceedings of the 4th National Herb Growing and Marketing Conference – Herbs 89: 44-50.

PHILLIPS(a): Phillips, R.L. and C.W. Campbell. 1993. The Sugar Apple. University of Florida. Florida Cooperative Extension Service. Institute of Food and Agricultural Sciences. Fruit Crops Fact Sheet FC-38. 3 pp.

PHILLIPS(b): Phillips, R.L. and S. Goldweke. 1993. The Jaboticaba. University of Florida. Florida Cooperative Extension Service. Institute of Food and Agricultural Sciences. Fruit Crops Fact Sheet FC-39. 2 pp.

PHILLIPS(c): Phillips, R.L. 1991. The Coconut. University of Florida. Florida Cooperative Extension Service. Institute of Food and Agricultural Sciences. Fruit Crops Fact Sheet FC-40. 3 pp.

PHILLIPS(d): Phillips, R.L. 1991. Barbados Cherry. University of Florida. Florida Cooperative Extension Service. Institute of Food and Agricultural Sciences. Fruit Crops Fact Sheet FC-28. 3 pp.

PHILLIPS(e): Phillips, R.L., C.W. Campbell, and S.E. Malo. 1991. The Longan. University of Florida. Florida Cooperative Extension Service. Institute of Food and Agricultural Sciences. Fruit Crops Fact Sheet FC-49. 2 pp.

PHILLIPS(f): Phillips, R.L., S.E. Malo and C.W. Campbell. 1986. The Mamey Sapote. University of Florida. Florida Cooperative Extension Service. Institute of Food and Agricultural Sciences. Fruit Crops Fact Sheet FC-30.

PHILLIPS(g): Phillips, R.L., S. Goldwater and C.W. Campbell. 1983. Key Lime. University of Florida. Florida Cooperative Extension Service. Institute of Food and Agricultural Sciences. Fruit Crops Fact Sheet FC-19. 3 pp.

PHILLIPS(h): Phillips, R.L. and G. Joyner. 1978. The Sea Grape. University of Florida. Florida Cooperative Extension Service. Institute of Food and Agricultural Sciences. Fruit Crops Fact Sheet FC-37. 2 pp.

PHILLIPS(i): Phillips, S. 1992. Vintage Vegetables. The Garden 117 (11): 507-511.

PITKANEN: Pitkanen, A.L. 1970. Tropical Fruits, Herbs, and Spices. 2nd Ed. E.T. Prevost Publ. Lemon Grove, CA. 300 pp.

PONCAVAGE: Poncavage, J. 1992. Native Nutmeats. Organic Gardening 39 (8): 45-50.

POPE: Pope, D.D. and S.M. McCarter. 1992. Evaluation of Inoculation Methods for Inducing Common Smut on Corn Ears. Phytopathology 82 (9): 950-955.

POPENONE: Popenone, W. 1920. Manual of Tropical and Subtropical Fruits. Macmillan, New York, NY.

PRAKASH: Prakash, V. 1991. Leafy Spices. First Edition. CRC Press. Boca Raton, FL. Pp. 1-113.

PRASHAR: Prashar, P. 1980. Edible Beans. South Dakota University Agricultural Experiment Station. Bulletin No. 671.

PREVOST: Prevost, R. 1970. Tropical Fruits. Second Ed. Lemon Grove Publ., CA. 300 pp.

PRITTS: Pritts, M.P. and G.B. Holcomb. 1991. Brambles. USDA Cooperative Extension Service Office Small-Scale Agriculture. 2 pp.

PROCTOR: Proctor, J.T.A. and W.G. Bailey. 1987. Ginseng: Industry, Botany, and Culture. Horticultural Reviews 9:187-236.

PROHENS: Prohens, J., J.J. Ruil, and F. Nuez. 1996. The Pepino (*Solanum muricatum*, Solanceae): A New Crop with a History. Economic Botany 50 (4): 355-368.

PUROHIT: Purohit, A.G. 1995. Annonaceous Fruits. In Handbook of Fruit Science and Technology, D.K. Salunkhe and S.S. Kadam, editors. Marcel Dekker, Inc., New York, NY.

PURSEGLOVE: Purseglove, J.W. 1977. Tropical Crops: Dicotyledons. 3rd Edition. Longman Group, London. 719 pp.

PURSEGLOVE(a): Purseglove, J.W. 1972. Tropical Crops: Monocotyledons. 1st Edition. John Wiley & Sons, NY. 334 pp.

PUTNAM 1993: Putnam, D., W. Breeze, and D. Somers. 1993. Vetching Vetch. BioOptions 4 (3): 8.

PUTNAM 1992: Putnam, D.H., E.S. Oplinger, J. Doll, and E.E. Schulte. 1992. Amaranth In Alternative Field Crops Manual. Universities of Wisconsin and Minnesota Cooperative Extension Service. 4 pp.

PUTNAM 1990: Putnam, D.H., E.S. Oplinger, D.R. Hicks, B.R. Durgan, D.M. Noetzel, R.A. Meronuck, J.D. Doll, and E.E. Schulte. 1990. Sunflower. Alternative Field Crops Manual. Universities of Wisconsin and Minnesota Cooperative Extension Service. 8 pp.

PUTNAM 1996: Putnam, J.J. and J.E. Allshouse. 1996. Food Consumption, Prices and Expenditures, 1996. Annual Data 1970-1994. USDA Economic Research Service (ERS) Statistical Bulletin Number 928.

QUAIL MOUNTAIN: Quail Mountain Herbs®. 1996. Picture Poster of Edible Herbs and Lettuces. P.O. Box 1049. Watsonville, CA.

RABINOWITCH: Rabinowitch, H.D. and J.L. Brewster. 1990. Onions and Allied Crops. Vol. I. Botany, Physiology, and Genetics. CRC Press. Boca Raton, FL. 273 pp.

RABINOWITCH(a): Rabinowitch, H.D. and J. L. Brewster. 1990. Onions and Allied Crops. Vol. II. Agronomy, Biotic, Pathology, and Crop Protection. CRC Press. Boca Raton, FL. 320 pp.

RACHIE: Rachie, K.O. 1984. Tropical Legumes: Resources for the Future. National Academy of Science. Washington, D.C. 326 pp.

RAINA: Raina, B.L. 1995. Olive. In Handbook of Fruit Science and Technology, D.K. Salunkhe and S.S. Kadam, editors. Marcel Dekker, Inc., New York, NY.

RAMOS: Ramos, D.E. 1985. Walnut Orchard Management. University of California Cooperative Extension Service. Publication No. 21410. 178 pp.

RATTO 1996a: Ratto, R. 1996. Personal Communications. Cilantro. California Grower. 6 March, 1996.

RATTO 1988: Ratto, R. 1988. Personal Communications on Celeriac. Ratto Bros., Modesto, CA. 25 February 1988.

RAY: Ray, G. and L. Waldheim. 1981. Citrus. Horticultural Publ. Co., Inc. Tucson, AZ.

REHM: Rehm, S. 1994. Multilingual Dictionary of Agronomic Plants. Kluwer Academic Publishers, 101 Philip Drive, Norwell, MA. 286 pp.

REHM(a): Rehm, S. and S. Espig. 1991. The Cultivated Plants of the Tropics and Subtropics. Verlag Josef Margraf Publ., West Germany. First Ed. 552 pp.

REICH: Reich, L. 1991. Uncommon Fruits worthy of Attention: A gardener's guide. Addison-Wesley Publishing.

REUTHER: Reuther, W. 1973. The Citrus Industry. Vol. III. Production Technology. University of California, Davis, CA. 528 pp.

REUTHER(a): Reuther, W., L.D. Batchelor, and H.J. Weber. 1968. The Citrus Industry, Vol. II Anatomy, Physiology, Genetics, and Reproduction. University of California, Davis, CA. 398 pp.

REUTHER(b): Reuther, W., H.J. Weber, and L.D. Batchelor. 1967. The Citrus Industry. Vol. I. History, World, Distribution, Botany, and Varieties. University of California, Davis, CA. 611 pp.

RICE: Rice, R.P., L.W. Rice, and H.D. Tindall. 1991. Fruit and Vegetable Production in Warm Climates. Macmillan Education Ltd., London. 486 pp.

RICHE: Riche, M. 1994. 1992 Census of Agriculture. Volume I, Geographic Area Series, Part 51. United States Summary and State Data. U.S. Department of Commerce. Bureau of the Census. 463 pp.

RINZLER: Rinzler, C.A. 1990. The Complete Book of Herbs, Spices and Condiments. Facts on File, Inc. NY 199 pp.

RINZLER(a): Rinzler, C.A. 1987. The Complete Book of Food. World Almanac. NY. 426 pp.

RIOTTE: Riotte, L. 1993. The Complete Guide to Growing Nuts. Taylor Publishing Co. Dallas, TX. 162 pp.

ROBBELEN 1989: Robbelen, G., R.K. Downey, and A. Ashri. 1989. Oil Crops of the World. McGraw-Hill Publishing Company. New York. 554 pp.

ROBINSON: Robinson, R.G. 1986. Amaranth, Quinoa, Ragi, Tef, and Niger: Tiny Seeds of Ancient History and Modern Interest. University of Minnesota Agricultural Experiment Station Bulletin AD-SB-2949.

RODGERS: Rodgers, E.G. 1974. Needed: A Federal Noxious Weed Act. Weeds Today. Weed Science Society of America. December, 1974, pp. 9-10.

ROECKLEIN: Roecklein, J.C. and P.S. Leung. 1987. A Profile of Economic Plants. Transaction Books. New Brunswick, GA. Pp. 487.

ROETHELI 1991: Roetheli, J.C. 1991. Lesquerella as a Source of Hydroxy Fatty Acids for Industrial Products. Growing Industrial Materials Series. USDA-CSRS, Office of Agricultural Materials. Washington, D.C. 46 pp.

ROETHELI 1991a: Roetheli, J.C. 1991a. Personal Communications Concerning Crambe. USDA-CSRS, Office of Agricultural Materials, Washington, D.C. 8 April, 1991.

ROSENGARTEN: Rosengarten, F. 1984. The Book of Edible Nuts. First Edition. Walker & Co., NY. 384 pp.

ROSENGARTEN(a): Rosengarten, F. 1969. The Book of Spices. First Edition. Livingston Publishing Co., Philadelphia, PA. 497 pp.

ROTHWELL 1997: Rothwell, J.D. and L. D'Costa. 1997. Area & Dollar Statistics on Grain Crops in Canada (1991-1996) and Value to Area of Vegetables in Canada 1996 Summary. Unpublished data. Pest Management Regulatory Agency. Health Canada. Ontario.

ROTHWELL 1996a: Rothwell, J.D. and M. Beaith. 1996a. Personal Communications. Comprehensive Canadian Crop List. Health Canada. Ontario. 8 February, 1996.

ROY: Roy, S.K., D.P. Waskar, and D.S. Khurdiya. 1995. Other Tropical Fruits. In Handbook of Fruit Science and Technology, D.K. Salunkhe and S.S. Kadam, editors. Marcel Dekker, Inc., New York, NY.

ROY(a): Roy, S.K. and G.D. Joshi. 1995. Minor Fruits – Tropical. In Handbook of Fruit Science and Technology, D.K. Salunkhe and S.S. Kadam, editors. Marcel Dekker, Inc., New York, NY.

ROYSE 1995: Royse, D.J. 1995. Specialty Mushrooms. Mushroom News. The American Mushroom Institute. Washington, D.C. (May, 1995). pp. 4-19.

RUBATZKY: Rubatzky, V.E. and M. Yamaguchi. 1997. World Vegetables. Second Ed. Chapman & Hall, NY. 843 pp.

RUELE: Ruele, G. and L. Bruce. 1956. Mango Growing in Florida. University of Florida Agricultural Experiment Station Bulletin. No. 574.

RUPPRECHT: Rupprecht, J.K. and C.J. Chang, J.M. Cassady and J.L. McLaughlin. 1986. Asimicin, a new cytyotoxic and pesticidal acetogenin from the pawpaw, *Asimini triloba* (Annonaceae). Heterocycles 24:1197-1201.

RYDER 1984: Ryder, E.J. 1984. Leafy Salad Vegetables. Pp. 171-194. AVI Publishing Co., Inc. Westport, CT.

RYDER 1979: Ryder, E.J. 1979. Leafy Salad Vegetables. AVI Publishing Co., Inc. Westport, CT. 266 pp.

SALUNKHE: Salunkhe, D.K. and S.S. Kadam. 1995.

Handbook of Fruit Science and Technology. Marcel Dekker, Inc., New York, NY. 610 pp.

SALUNKHE(a): Salunkhe, D.K. and S.S. Despande. 1991. Foods of Plant Origin. AVI Book Van Nostrand Reinhold Publishing. NY. 502 pp.

SALVI: Salvi, M.J. and J.C. Rajput. 1995. Pineapple. In Handbook of Fruit Science and Technology, D.K. Salunkhe and S.S. Kadam, editors. Marcel Dekker, Inc. New York, NY.

SAMSON: Samson, J.A. 1986. Tropical Fruits. 2nd Ed. Longman Scientific and Technical Publishing. Singapore. 336 pp.

SANEWSKI: Sanewski, G. 1992. Custard Apples: Cultivation and Crop Protection. Queensland Department of Primary Industries, Second Edition. Brisbane. Information Science QI90031. 103 pp.

SARRACINO 1988: Sarracino, R. 1988. Personal Communications: Section 18 Emergency Exemption No. 88-6. Pesticide Registration. Department of Food and Agriculture, State of California, Sacramento. 4 pp.

SAULS(a): Sauls, J.W. and C.W. Campbell. 1991. White Sapote. University of Florida. Florida Cooperative Extension Service. Institute of Food and Agricultural Science. Fruit Crops Fact Sheet FC-59. 2 pp.

SAULS(b): Sauls, J.W. and C.W. Campbell. 1986. Mango Propagation. University of Florida. Florida Cooperative Extension Service. Institute of Food and Agricultural Science. Fruit Crops Fact Sheet FC-58. 4 pp.

SAULS(c): Sauls, J.W. and C.W. Campbell. 1983. The Canistel. University of Florida. Florida Cooperative Extension Service. Institute of Food and Agricultural Science. Fruit Crops Fact Sheet FC-61. 2 pp.

SAULS(d): Sauls, J.W. and L.J. Jackson. 1983. Lemons, Limes and Other Acid Citrus. University of Florida. Florida Cooperative Extension Service. Institute of Food and Agricultural Science. Fruit Crops Fact Sheet FC-42. 4 pp.

SAULS(e): Sauls, J.W. and C.W. Campbell. 1980. The Tamarind. University of Florida. Florida Cooperative Extension Service. Institute of Food and Agricultural Science. Fruit Crops Fact Sheet FC-62. 2 pp.

SAUNDERS: Saunders, C.F. 1989. Edible and Useful Wild Plants in the U.S. and Canada. Dover Publ. NY. 275 pp.

SAUNT: Saunt, J. 1990. An Illustrated Guide to Citrus Varieties of the World. Sinclair International Limited. 126 pp.

SCARPA: Scarpa, J. 1993. Grains: Amaranth, Tef, Spelt, Kamut, Quinoa, and Triticale are Moving onto New-Wave Menus. Restaurant Business Magazine 92 (14): 162.

SCHERY 1972: Schery, R.W. 1972. Plants for Man. Second Edition. Prentice-Hall, Inc. Englewood Cliffs, NJ. 657 pp.

SCHNEIDER 1996a: Schneider, B.A. 1996a. Uses of Brans of Wheat, Oats, Barley and Rye as Foods. Unpublished Data D 220744, U.S. EPA, Washington, D.C. 14 February 1996. 32 pp.

SCHNEIDER 1996b: Schneider, B.A. 1996b. Response to Comments from DuPont Agriculture Products that are not Normally Rotated for Residue Chemistry Guidelines. D226438. May 15. US EPA. 6 pp. (Unpublished Data).

SCHNEIDER 1995: Schneider, B.A. and R.A. Loranger. 1995. Evaluation of Washington State Department of Agriculture Report for Nonfood/Nonfeed Status for Small-Seeded Vegetable Crops. US-EPA. 25 February, 1995. 14 pp.

SCHNEIDER 1994: Schneider, B.A. 1994. Response to Public Comments on Pesticide Tolerances; Crop Grouping Regulation: Proposed Rule Part VII. Evaluation of U-Toy to Add to Crop Group. CB No. 14421. DP: D207721. US-EPA. 7 pp.

SCHNEIDER 1994a: Schneider, B.A. 1994a. Response to Commodities, Subcommodities, and Commodity Groups Not Yet Included in the New DRES Vocabulary. Part VIII: S, Entries 664-784. US - EPA. (Unpublished)

SCHNEIDER 1994b: Schneider, B.A. 1994b. Response to Commodities, Subcommodities, and Commodity Groups Not Yet Included in the new DRES Vocabulary. Part IX: T, Entries 785-833. US-EPA. (Unpublished).

SCHNEIDER 1994c: Schneider, B.A. 1994c. Response to Commodities, Subcommodities, and Commodity Groups Not Yet Included in the new DRES Vocabulary. Part X: U-Z, Entries 834-920. US-EPA. (Unpublished).

SCHNEIDER 1993a: Schneider, B.A. 1993a. Response to Commodities, Subcommodities, and Commodity Groups Not Yet Included in the New DRES Vocabulary. Parts I-V: A-R, Entries 1-663. US-EPA (Unpublished). 20 July, 20 August, 14 September and 17 September, 1993.

SCHNEIDER 1993b: Schneider, B.A. 1993b. Leucanena Cultural Practices New Crop. ID# 93TX0023 and 24. Oxyfluorfen and metolachlor, CB# 11974.D191863 and 191868. US-EPA, June 16. 5 pp. (Unpublished)

SCHNEIDER 1992: Schneider, B.A. 1992. Crop Groupings, Part I: Evaluation of the Crop Grouping Scheme (40 CFR 180.34) and Response to Workshop Questions CB# 8059, 8205, 8482, 8486, 10388, 10389. US-EPA OPPTS, HED, CBTS. September 17. 55 pp.

SCHNEIDER 1992a: Schneider, B.A. 1992. Information Concerning Use of Marigolds as a Poultry Feeding Supplement. Hexakis. CB# 8838,8839. US-EPA. 11 February, 1992. 14 pp.

SCHNEIDER 1985: Schneider, Elizabeth. 1985. Uncommon Fruits & Vegetables: A Common Sense Guide. Harper & Row Publishers, Inc. New York, NY. 546 pp.

SCHNEIDER 1970: Schneider, B.A. 1970. Agronomy I - Field Crop Production. University of Maryland. College Park, MD. 273 pp.

SCHOLAR: Scholar, R. and L. Edwards. 1986. Mungbean Production in Oklahoma. Oklahoma State University. Cooperative Extension Facts No. 2060.

SCHREIBER: Schreiber, A. and L. Ritchie. 1995. Washington Minor Crops. Food and Environmental Quality Lab, Washington State University. 325 pp.

SCHULTER: Schulter, R.E. et al. 1975. The Winged Bean. A High-Profit Crop for the Tropics. National Academy of Sciences. Washington, D.C. 35 pp.

SCOTT: Scott, C.L. 1987. The National Gardening Book of Lettuce and Greens. National Gardening Association. Burlington, VT. 34 pp.

SEELIG: Seelig, R.A. and M.C. Bing. 1990. Encyclopedia of Produce. United Fresh Fruit and Vegetable Association. Arlington, VA.

SEELIG(a): Seelig, R.A. 1978. Fruit and Vegetable Series. A-Z. United Fresh Fruit and Vegetable Association. Arlington, VA.

SENFT 1994: Senft, D. 1994. Onward Vernonia! Agricultural Research. Agricultural Research Service. USDA, Washington, D.C. September, 1994: 16-17.

SERNA-SALDIVAR: Serna-Saldivar, S.O., M.H. Gomez, and L.W. Rooney. 1994. Food Uses of Regular and Specialty Corns and Their Dry-Milled Fractions. Chapter 9 Pp. 263-298. In Specialty Corns. CRC Press. Boca Raton, FL.

SHEETS: Sheets, M.D. and M.L. DuBois. 1991. The Pomegranate. University of Florida. Florida Cooperative Extension Service. Institute of Food and Agricultural Science. Fruit Crops Fact Sheet FC-44. 2 pp.

SHOCK: Shock, C.C. 1982. Experimental Cultivation of Rebaudi's Stevia in California. Agronomy Progress Report No. 122. University of California (Davis), April 1982. 7 pp.

SHOEMAKER: Shoemaker, J.S. 1983. Small Fruits Culture. 5th Edition. AVI Publishing Co. Westport, CT. 385 pp.

SHOEMAKER(a): Shoemaker, J.S. 1978. Small Fruit Culture, 4th Edition. AVI Publishing Co. Westport, CT.

SHULER: Shuler, K.D. 1994. Personal Communications on Chinese Celery. Cooperative Extension, Institute of Fruit and Agricultural Science. University of Florida, West Palm Beach, FL. 25 February, 1994.

SIMON: Simon, J.E. 1990. Essential Oils and Culinary Herbs. Pp. 472-483. In Advances in New Crops. Janick, J. and J.E. Simon, editors. Timber Press. Portland, OR.

SIMON(a): Simon, J.E., E. Cebert, and D. Reiss-Bubenheim. 1989. Cultivar Evaluation of Culinary Herbs. Proceedings of the 4th National Herb Growing and Marketing Conference. Herbs 89: 12-21.

SIMON(b): Simon, J.E. 1985. Sweet Basil: A Production Guide. HO-189. Cooperative Extension Service, Purdue University, West Lafayette, IN. 2 pp.

SINGH: Singh, U. and B. Singh. 1992. Tropical Grain Legume as Important Human Foods. Economic Botany 46 (3): 310-321.

SISSON: Sisson, M. and E. Sorensen. 1989. Edible Flowers: A Brief History. Proceedings of the 4th National Herb Growing and Marketing Conference. Herbs 89: 96-99.

SKERMAN: Skerman, P.J. and F. Riveros. 1989. Tropical Grasses. Food & Agricultural Organization of the United Nations. Vol. 23. FAO. Rome. 832 pp.

SKERMAN(a): Skerman, P.J., D.G. Cameron and F. Riveros. 1988. Tropical Forage Legumes. Food & Agricultural Organization of the United Nations. Plant Protection and Protection Series. No 2. Second Ed. Rome. 692 pp.

SMALL: Small, E. and M. Jomphe. 1988. A Synopsis of the Genus *Medicago* (*Leguminosae*). Canada Journal of Botany 67: 3260-3294.

SMITH 1978: Smith, A.K. and J.J. Circle. 1978. Soybeans: Chemistry and Technology. Vol. I. Proteins. 2nd Edition. AVI Publishing Co. Westport, CT. 470 pp.

SMITH 1995: Smith, C.W. 1995. Crop Production: Evolution, History, and Technology. First Edition. John Wiley & Sons, NY. 467 pp.

SMITH 1997: Smith, D. 1997. The Small Acreage Crops of Texas: An Inventory and Implications for Crop Production, Communication. Texas Agricultural Experiment Station.

SMITH 1989: Smith, J.D. and W.H. Isom. 1989. Common Dry Bean Production in California. University of California Cooperative Extension Service Publication 21468.

SOMMER: Sommer, R. 1995. Why I Will Continue to Eat Corn Smut. Natural History 18, 20, 22.

SPICES: Spices, etc. 1997. Fall-Winter, 1997-1998. Herb and Spice Catalogue, 48 pp. Charlottesville, VA.

SPLITTSTOESSER: Splittstoesser, W.E. 1990. Vegetable Growing Handbook: Organic and Traditional Methods. Third Edition. AVI Publishing Co., Inc. NY. 362 pp.

STALLKNECHT: Stallknecht, G.F., K.M. Gilbertson, and J.L. Eckhoff. 1993. Teff: Food Crop for Humans and Animals. Pp. 228-233. In New Crops. J.Janick and J.E. Simon, editors. John Wiley & Sons, NY.

STALLKNECHT(a): Stallknecht, G.F., K.M. Gilbertson, and J. Eckhoff. 1991. Teff: Food Crop for Humans and Animals. Pp. 231-234. In Janick,J. and J.E. Simon 1991. Proceedings Second National Symposium New Crops. John Wiley & Sons, Inc.

STAMETS 1983: Stamets, P. and J.S. Chilton. 1983. The Mushroom Cultivator. Agarikon Press. Olympia, WA. 415 pp.

STANG: Stang, E.J. 1990. Elderberry, Highbush Cranberry, and Juneberry Management, p. 363-382 In Small Fruit Crop Management, G.J. Galletta and D. G. Himelrick, editors. Prentice-Hall, Englewood Cliffs, NJ.

STEPHENS 1988: Stephens, J.M. 1988. Manual of Minor Vegetables. Florida Cooperative Extension Service. Bulletin SP-40. University of Florida, Gainesville. 123 pp.

STERRETT: Sterrett, S.B. and C.P. Savage. 1986. Chinese Cabbage Varieties. The Vegetable Growers News 41 (2): 1-3.

STOBART: Stobart, P. 1970. Herbs, Spices and Flavorings. McGraw-Hill Book, Co. NY. 262 pp.

STUART: Stuart, M. 1981. The Encyclopedia of Herbs and Herbalism. Gossett and Dunlap. Orbis Publ. Co. NY. 304 pp.

STUBBENDIECK: Stubbendieck. J., S.L. Hatch, and C.H. Butterfield. 1992. North American Range Plants. Fourth Edition. University of Nebraska Press, Lincoln. 493 pp.

STUBBENDIECK(a): Stubbendieck, J., S.L. Hatch, and C.H. Butterfield. 1997. North American Range Plants. Fifth Ed. University of Nebraska, Lincoln. 501 pp.

STURTEVANT: Sturtevant, E.L. 1972. Sturtevant's Notes on Plants of the World. U.P. Hedrick, Editor. Dover Publ. NY. 696 pp.

SULLIVAN: Sullivan, G.A. 1996. Peanut Information. North Carolina Cooperative Extension Service. North Carolina State University Bulletin. AG-331. 83 pp.

SUMMERFIELD: Summerfield, R.J. and A.H. Bunting. 1980. Advances in Legume Science. Royal Botanical Gardens. Kew, England. 667 pp.

SWEET: Sweet, C. 1990. Mango Origins and Culture. California Grower 12 (14): 38-40.

SWEET(a): Sweet, C. 1986. Thin Sapote Skin Cuts Its Market Potential. California Avocado Grower 10 (1): 34-35.

SWIADER: Swiader, J.M., G.W. Ware, and J.P. McCollum. 1992. Producing Vegetable Crops. Fourth Ed. Interstate Publ. Co. Danville, IL. 626 pp.

TAINTER: Tainter, D.R. and A.T. Grenis. 1993. Spices and Seasonings. A Food Technology Handbook. VCH Publication. Des Moines, IA.

TANAKA: Tanaka, T. 1976. Tanaka's Cyclopedia of Edible Plants of the World. First Ed. 924 pp. Keigaku Publ. Co. Japan.

TANNER: Tanner, G.W., J.J. Mullahey and D. Maehr. 1996. Saw-palmetto: An Ecologically and Economically Important Native Palm. Cooperative Extension Circular WEC-109. University of Florida, Gainesville. 4 pp.

TAYLOR: Taylor, N.L. 1990. The True Clovers. In Advances in New Crops edited by J.Janick and J.E. Simon. Timber Press, OR. pp. 177-182.

TAYLOR(a): Taylor, N.L. and D. Henry. 1989. Kura Clover for Kentucky. University of Kentucky College of Agriculture Cooperative Extension Services. Bulletin AGR-141. 2 pp.

TAYLOR(b): Taylor, N.L. and J.N.N. Campbell. 1989. Native Kentucky Clovers. Buffalo Clover. University of Kentucky College Agriculture Cooperative Extension Service. Bulletin AGR-142. 2 pp.

TAYLOR(c): Taylor, N.L. 1985. Clover Science and Technology. American Society of Agronomy. Series No. 25. Madison, WI. 616 pp.

TEERI: Teeri, J.A. and L.E. Stowe. 1976. Climatic Patterns and the Distribution of C4 Grasses in North American. Oecologia (Berl.) 23:1-12.

TEMPLETON: Templeton, E. 1985. Enjoying Florida Mangoes. University of Florida. Florida Cooperative Extension Service. Institute of Food and Agricultural Science. Food & Nutrition Fact Sheet HFS-871. 4 pp.

TERRELL: Terrell, E.E., S.R. Hill, J.H. Wiersema, and W.E. Rice. 1986. Scientific Names of Plants: A Checklist of Names for 3000 Vascular Plants of Economic Importance. USDA Agricultural Research Service. Agricultural Handbook No. 505. 244 pp.

TESKEY: Teskey, B.J.E. and J.S. Shoemaker. 1972. Tree Fruit Production. Chapter 5: Cherries Pp. 238-293. The AVI Publishing Co. Westport, Ct.

THOMPSON: Thompson, D.A. 1986. *Leucaena leucocephala* and Other Fast Growing Trees and Their Cultivation and Uses. Organization of American States. Caribbean Region Leucaena Project. 34 pp.

TUCKER: Tucker, A.O. and T. DeBaggio. 1992. Cilantro Around the World. The Herb Companion 4 (4): 36-41.

TULL: Tull, D. 1987. A Practical Guide to Edible and Useful

Plants. Texas Monthly Press. Austin, TX. 518 pp.

TURGEON: Turgeon, A.J. 1996. Turfgrass Management. Fourth Edition. Prentice-Hall, NJ. 406 pp.

UEBERSAX: Uebersax, M.A. and L.G. Occena. 1993. Suitability of Dry Edible Beans for World Food Relief Programs: Applications and Nutritive Value. Michigan Dry Bean Digest 18 (1): 2-20.

UNDERSANDER: Undersander, D.J., N.J. Ehlke, A.R. Kaminski, J.D. Doll, and K.A. Kelling. 1990. Hairy Vetch. Universities of Wisconsin and Minnesota Cooperative Extension Service. Alternative Field Crops Manual. 4 pp.

UNDERSANDER(a): Undersander, D.J., B.R. Durgan, A.R. Kaminski, J.D. Doll, G.L. Worf, and E.E. Schulte. 1990. Kochia. Universities of Wisconsin and Minnesota Cooperative Extension Service. Alternative Field Crops Manual. 3 pp.

UNDERSANDER(b): Undersander, D.J., L.H. Smith, A.R. Kaminski, K.A. Kelling, and J.D. Doll. 1990. Sorghum Forage. Universities of Wisconsin and Minnesota Cooperative Extension Service. Alternative Field Crops Manual. 5 pp.

USDA 1997a: USDA. 1997a. Noncitrus Fruits and Nuts. 1996. Preliminary Summary, ERS-National Agricultural Statistics Service, USDA. Washington, D.C. January, 1997. 72 pp.

USDA 1997b: USDA. 1997b. Agricultural Statistics. National Agricultural Statistics Service. United States Government Printing Office. Washington, D.C.

USDA 1996a: USDA. 1996a. Agricultural Statistics. National Agricultural Statistics Service, USDA. U.S. Government Printing Office. Washington, D.C. 1995-1996.

USDA 1996b: USDA. 1996b. Vegetables 1995 Summary. National Agricultural Statistics Service. USDA. Washington, D.C. January 1996.

USDA 1996c: USDA. 1996c. Crop Production 1995 Summary. National Agricultural Statistics Service. USDA. Washington, D.C. January, 1996.

USDA 1996d: USDA. 1996d. Citrus Fruits 1996 Summary. National Agricultural Statistics Service. USDA. Washington, D.C. September, 1996.

USDA 1996e: USDA. 1996e. Fresh Fruit and Vegetable Shipments by Commodities, States and Months, Calendar Year 1995. Agricultural Marketing Service, Fruit and Vegetable Division. Market News Branch. USDA. Washington, D.C. FVAS-4, Issued April 1996. 56 pp.

USDA 1996f: USDA. 1996f. Noncitrus Fruits and Nuts: 1995 Summary. National Agricultural Statistics Service. USDA. Washington, D.C.

USDA 1995: USDA. 1995. Ginseng Developments, pp. 27-29. In

Tropical Products: World Markets and Trade. USDA Foreign Agricultural Service. Circular Series FTROP 3-95.

USDA 1994a: USDA 1994a. Agricultural Statistics. National Agricultural Statistics Service. US Government Printing Office. Washington, D.C. 486 pp.

USDA 1994b: USDA. 1994b. American Ginseng. A Small-Scale Agriculture Alternative. Office for Small-Scale Agriculture, CSREES. Washington, D.C. 2 pp.

USDA 1993a: USDA. 1993a. U.S. Spice Trade. United States Department of Agriculture (USDA) Foreign Agricultural Service. Circular Series FTEA 1-93. Prepared by Rex E.T. Dull HTPD. 49 pp.

USDA 1993b: USDA 1993b. Agricultural Research. Agricultural Research Service, U.S. Department of Agriculture. Washington, D.C. 41 (6), p. 19.

USDA 1970: USDA. 1970. Commercial Growing of Horseradish. USDA-ARS Leaflet No. 547. Superintendent of Documents, U.S. Government Printing Office. Washington, D.C. 6 pp.

USDA-NASS 1996: USDA. 1996. Bean Statistics. National Agricultural Statistics Service. Washington, D.C. 1 August, 1996.

USDA-NASS 1995a: USDA. 1995a. Vegetables 1994 Summary. National Agricultural Statistics Service, ASB, USDA. Washington, D.C. January 1995. Vg. 1-2 (95)

USDA, NRCS: USDA, NRCS. 1995. The Plants Database. Natural Resources Conservation Service (formerly Soil Conservation Service). National Plant Data Center, USDA. Baton Rouge, LA. 1008 pp.

USDI 1997: USDI. 1997. Endangered and Threatened Wildlife and Plants; Proposed Endangered Status for 10 Plant Taxa from Maui Nui, Hawaii. Fish and Wildlife Service, U.S. Dept. of Interior. Washington, D.C. (Federal Register 62:94:26757-70).

US EPA 1997: US EPA. 1997. The FIFRA and FFDCA as amended by the Food Quality Protection Act (FQPA) of August 3, 1996. OPP, EPA. March, 1997. Pamphlet 730L97001. 189 pp.

US EPA 1996: US EPA. 1996. Title 40 - Protection of Environment Parts 150-189, CFR, Office of Federal Register. U.S. Government Printing Office. Washington, D.C. 707 pp.

US EPA 1996a: US EPA 1996a. Residue Chemistry Test Guidelines, OPPTS 860 1000 to 1900. U.S. Government Printing Office, Washington, D.C. August, 1996.

US EPA 1995: US EPA. 1995. Title 40-Protection of Environment. Parts 150-189. Code of Federal Regulations. Office of the Federal Register. U.S. Government Printing Office, Washington, D.C. 720 pp.

US EPA 1995a: US EPA. 1995a. Table II (September 1995): Raw Agricultural and Processed Commodities and Feedstuffs Derived from Field Crops. Subdivision O, Residue Chemistry, Pesticide Assessment Guidelines OPP/HED/EPA, Washington, D.C. 22 pp.

US EPA 1995b: US EPA. 1995. Summer Squash; Definitions and Interpretations, 40 CFR 180.1(h), Chayote. Federal Register, U.S. Government Printing Office, Superintendent of Documents. Washington, D.C. (16 August, 1995) 60:158:42447-9.

US EPA 1994: US EPA. 1994. Pesticide Reregistration Rejection Rate Analysis Residue Chemistry. Follow-up Guidance for: Updated Livestock Feeds Tables, and Number and Location of Domestic Crop Field Trials. U.S. Government Printing Office. EPA 738-K-94-001.

USHER: Usher, G. 1974. A Dictionary of Plants Used by Man. Constable, London. 619 pp.

UPHOF: Uphof, J.C.T. 1968. Dictionary of Economic Plants. 2nd ed. Verlag Publ., NY. 591 pp.

VALVERDE: Valverde, M.E., O. Paredes-Lopez, J.K. Pataky, and R. Guevara-Lara. 1995. Huitlacoche (Ustilago maydis) As a Food Source – Biology, Composition, and Production. Critical Review in Food Science and Nutrition 35 (3): 191-229.

VANDERMAESEN: Vandermaesen, L. and S. Somaatmadja, Editors. 1989. Plant Resources of South-East Asia. Prosea. No. 1. Pulses. Pudoc Wageningen. 105 pp.

VANDYNE: Van Dyne, D.L., M.G. Blase and K.D. Carlson. 1990. Industrial Feedstocks and Products from High Erucic Acid Oil. Crambe and Industrial Rapeseed. Printing Services, University of Missouri - Columbia. 29 pp.

VANVRANKEN 1996: VanVranken, R.W. 1996. Personal Communications on Designer Asian and Leafy Vegetables and its Mix called Mesclun Greens. Cook College, Rutgers University, New Brunswick, NJ. September, 1996.

VARUGHESE: Varughese, G., W.H. Pfeiffer, and R.J. Pena. 1996. Triticale: A Successful Alternative Crop. Part I. Cereal Foods World 41 (6): 474-482.

VARUGHESE(a): Varughese, G. W.H. Pfeiffer, and R.J. Pena. 1996. Triticale: A Successful Alternative Crop. Part II. Cereal Foods World 41 (7): 635-645.

VAVRINA: Vavrina, C.S. 1991. Chinese Cabbage Production Guide for Florida. Florida Institute of Food and Agricultural Science Publication SP-100.

VAVRINA(a): Vavrina, C.S. ,T.A. Obreza, and J. Cornell. 1993. Response of Chinese Cabbage to N. Rate and Source in Sequential Planting. HortScience 23(12): 1164-1165.

VERHEIG: Verheig, E. and R. Coronel. 1991. Plant Resources of South-East Asia. Vol.2. Edible Plants and Nuts. Pudoc Wageningen. 447 pp.

VIETMEYER: Vietmeyer, N.D. 1996. Lost Crops of Africa. Volume I. Grains. National Academy of Science Press. Washington, D.C. 381 pp.

WALLACE 1992: Wallace, R.D. and L.G. Spinella. 1992. World Chestnut Industry Conference. Chestnut Marketing Association. Alachua, FL. 160 pp.

WARE: Ware, G.W. and J.P. McCollum. 1980. Producing Vegetable Crops. The Interstate Printers & Publishers, Inc. Danville, IL. 607 pp.

WATSON: Watson, S.A. and P.E. Ramstad. 1987. Corn: Chemistry and Technology. American Association of Cereal Chemists, Inc. St. Paul, MN. 605 pp.

WEBSTER: Webster, F.H. 1986. Oats: Chemistry and Technology, American Association of Cereal Chemists. 433 pp.

WEINEMANN: Weinemann, S. 1991. The Packer 1991 Produce Availability and Merchandising Guide. 20th Edition. Vance Publishing Co., San Francisco, CA. 444 pp.

WELBAUM: Welbaum, G. 1991. Lost and Found: Food Crops from the Tropics. The Virginia Farmer 10 (4):3.

WELCH: Welch, D.F. and S.D. Cotner. 1991. The Texas Master Gardener Handbook. Texas Agricultural Experiment Service. Texas A & M University, College Station, TX. 452 pp.

WEISS: Weiss, R. 1992. A Curious Fruit Medlar. J. Royal Horticultural Society 117 (11): 538-539.

WESTPORT: Westport, M.N. 1978. Temperate-Zone Pomology. W.H. Freeman Co., NY. 428 pp.

WHEALY: Whealy, K. and S. Demuth. 1993. Fruit, Berry and Nut Inventory. Second Edition. Seed Savers Publications. Decorah, IA. 518 pp.

WICKENS: Wickens, G.E., H. Haq, and P. Day. New Crops for Food and Industry. Chapman & Hall. London. 444 pp.

WILLIAMS: Williams, J.T. 1993. Underutilized Crops. Pulses and Vegetables. First Edition. Chapman and Hall, NY. 250 pp.

WILLIAMS(a): Williams, L. and J.A. Duke. 1978. Growing Ginseng. Farmers' Bulletin Number 2201. USDA, Washington, D.C. 8 pp.

WILLIAMSON: Williamson, J.G. and L.K. Jackson. 1989. The Sweet Orange. University of Florida. Florida Cooperative Extension Service. Institute of Food and Agricultural Science. Fruit Crops Fact Sheet FC-25A. 5pp.

WILSON: Wilson, C.W. 1980. Guava. In Tropical and Subtropical Fruits, S. Nagy and P. Shaw, editors. AVI Publishing Co., Inc. Westport, CT.

WININGER 1991: Wininger, W. 1991. The Citizen Forester. Fen's Rim Publications, Inc. Elk Rapids, MI. 20 pp.

WISEMAN: Wiseman, A.J.L., H.J.S. Finch, and A.M. Samuel. 1993. Crop Husbandry Including Grassland. 7th Edition. Pergamon Press. NY. 316 pp.

WOODROOF: Woodroof, J.G. 1986. Commercial Fruit Processing. 2nd Edition. AVI Publishing Co. Westport, CT. 678 pp.

WOODROOF(a): Woodroof, J.G. 1982. Tree Nuts. 2nd Edition. AVI Publishing Co. Westport, CT. 731 pp.

WOODROOF(b): Woodroof, J.G. 1967. Tree Nuts Production Processing Products. The AVI Publishing Co. Volumes One and Two.

WOOLLEY 1986: Woolley, D.G. 1986. Personal Communications on Crambe. Department of Agronomy, Iowa State University, Ames. September, 1986. 2 pp.

YAMAGUCHI: Yamaguchi, M. 1990. Asian Vegetables. Pp. 387-390 in Advances in New Crops. Janick, J. and J.E. Simon, Editors. Timber Press. Portland, OR.

YAMAGUCHI 1983: Yamaguchi, M. 1983. World Vegetables. Van Nostrand Reinhold Company Ltd., New York, NY. 415 pp.

YAMAGUCHI 1973: Yamaguchi, M. 1973. Production of Oriental Vegetables in the U.S. Hort.Sci. 8 (5): 362-370.

YANOVSKY: Yanovsky, E. 1936. Food Plants of North American Indians. USDA Miscellaneous Publications No. 237. U.S. Government Printing Office. Washington, D.C.

YEARBOOK OF AGRICULTURE 1992: Yearbook of Agriculture. 1992. New Crops, New Uses, New Markets. Office of Publishing and Visual Communications, USDA. Washington, D.C. 302 pp.

ZANDSTRA: Zandstra, B.H. and D.W. Warncke. 1989. Radish, Rutabaga and Turnip. Ag. Facts. Michigan State University Cooperative Extension Service Ext. Bulletin E-2207.

ZHU: Zhu, Y., C.C. Sheaffer, and D.K. Barnes. 1996. Forage Yield and Quality of Six Annual Medicago Species in the North Central USA. Agronomy Journal 88 (6): 955-960.

Crop Production Regions and Map as accepted by U.S. EPA

Contains U.S. map of growing regions for trial distribution

Border Definitions of Regions

I. ME, NH, VT, MA, RI, CT, NY, PA
   NJ   N of Rt. 1
   MD NW of I-95
   VA   N of I-64 and W of I-81
   WV N of I-64 and E of I-77
   OH   E of I-77

II. NC, SC, GA, DE
   VA   E of I-81 or S of I-64
   MD SE of I-95
   NJ   S of Rt. 1
   WV S of I-64
   KY   S of I-64 and S of BGP and E of I-65
   TN   E of I-65
   AL   Except Mobile and Baldwin Counties

III. FL, AL: Mobile and Baldwin Counties
   (see Region XIII for South Florida)

IV. LA, AR, MS
   TN   W of I-65
   MO E of Rt. 67 and S of Rt. 60

V. MI, IN, IL, WI, MN, IA
   OH   W of I-77
   WV N of I-64 and W of I-77
   KY   N of I-64 or N of BGP or W of I-65
   MO W of Rt. 67 or N of Rt. 60
   KS   E of Rt. 281
   NE   E of Rt. 281
   SD   E of Rt. 281
   ND   E of Rt. 281

VI. OK   E of Rt. 281/183
   TX   E of Rt. 283 and SE of Rt. 377

VII. MT   E of Rt. 87 or E of I-15
   WY E of I-25 or N of I-90
   ND   W of Rt. 281
   SD   W of Rt. 281
   NE   W of Rt. 281

VIII. KS   W of Rt. 281
   CO   E of I-25
   NM   E of I-25
   TX   W of Rt. 283 and NW of Rt. 377
   OK   W of Rt. 281/183

IX. UT, NV
   NM   W of I-25 and N of I-10
   CO   W of I-25
   WY W of I-25 and W of I-90
   MT   W of Rt. 87 and W of I-15
   AZ   NE of Rt. 89/93 and N of I-10

X. CA   Except Medocino, Humboldt, Trinity, Del Norte and
   Siskiyou Counties
   AZ   SW of Rt. 89/93 or S of I-10
   NM   S of I-10

XI. ID
   OR   E of Cascades
   WA   E of Cascades

XII. CA   Counties excluded from Region X
   OR   W of Cascades
   WA   W of Cascades

XIII. HI, PR
   South FL   S of Rt. 41

Notes
• Crop Regions: I through XIII, also equate to 1 through 13 in the crop monographs.
• States: ME, etc., are state abbreviations.
• Directions: N – North; S – South; E – East; W – West, etc.
• Roads: I-95 equates to Interstate Highway 95, etc.

Source
US EPA Guidelines OPPTS 860.1500 dated August 1996 and US EPA Opinion letter dated 12 DEC 94 noting South FL as part of Region XIII. Both sources included in this book as Appendix I, page 324, and Appendix VII.

U.S. Crop Production Regions

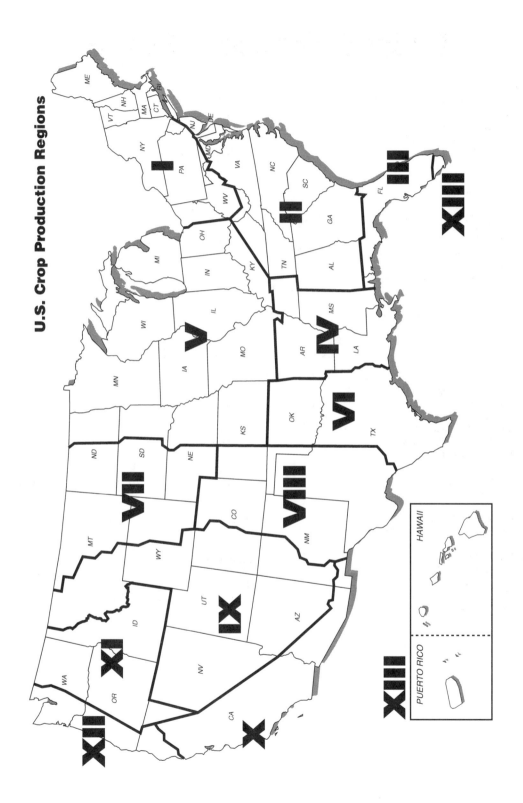

UNITED STATES ENVIRONMENTAL PROTECTION AGENCY
WASHINGTON, D.C. 20460

December 12, 1994

OFFICE OF
PREVENTION, PESTICIDES AND
TOXIC SUBSTANCES

Charles W. Meister, Ph.D.
IR-4 Southern Region Coordinator
Institute of Food and Agricultural Sciences
Food Science and Human Nutrition Department
Pesticide Research Laboratory
University of Florida
Gainesville, FL 32611-0720

Dear Dr. Meister:

Thank you for your letter of October 20, 1994 in which you proposed that the tropical region of South Florida be included in Region XIII, which now includes Puerto Rico and Hawaii. You stated that tropical fruits and vegetables are commercially grown in Monroe and Dade counties south of State Road 41 in South Florida.

Your proposal has been accepted and will be included in a future revision of our guidance document concerning the number and location of domestic crop field trials. In the meantime we suggest that you include a copy of this letter with any relevant submission to the Agency.

Sincerely,

/s/

Richard A. Loranger, Ph.D.
Acting Chief
Chemistry Branch I
Health Effects Division
(7509C)

/s/

Edward Zager
Chief
Chemistry Branch II
Health Effects Division
(7509C)

EPA Crop Grouping Scheme

EPA Crop Grouping Abstract
Development by authors and Hoyt Jamerson, EPA Minor Use Officer

This unit contains an alphabetical index to the crops in all the crop groups, giving the Crop Group Number. The index will be included in Title 40 of the Code of Federal Regulations as a finding aid after its publication in the Federal Register.

Crop Group (Subgroup) Number and Name	Representative Commodities	Commodities
1. Root and Tuber Vegetables	Carrot, potato, radish and sugar beet	Arracacha; arrowroot; artichoke, Chinese; artichoke, Jerusalem; beet, garden; beet, sugar; burdock, edible; canna, edible; carrot; cassava, bitter and sweet; celeriac; chayote (root); chervil, turnip-rooted; chicory; chufa; dasheen (taro); ginger; ginseng; horseradish; leren; parsley, turnip-rooted; parsnip; potato; radish; radish, oriental; rutabaga; salsify; salsify, black; salsify, Spanish; skirret; sweet potato; tanier; turmeric; turnip; yam bean; yam, true.
1A. Root vegetables subgroup	Carrot, radish and sugar beet	Beet, garden; beet, sugar; burdock, edible; carrot; celeriac; chervil, turnip-rooted; chicory; ginseng; horseradish; parsley, turnip-rooted; parsnip; radish; radish, oriental; rutabaga; salsify; salsify, black; salsify, Spanish; skirret turnip
1B. Root vegetables (except sugar beet) subgroup	Carrot and radish	Beet, garden; burdock, edible; carrot; celeriac; chervil, turnip-rooted; chicory; ginseng; horseradish; parsley, turnip-rooted; parsnip; radish; radish, oriental; rutabaga; salsify; salsify, black; salsify, Spanish; skirret; turnip
1C. Tuberous and corm vegetables subgroup	Potato	Arracacha; arrowroot; artichoke, Chinese; artichoke, Jerusalem; canna, edible; cassava, bitter and sweet; chayote (root); chufa; dasheen (taro); ginger; leren; potato; sweet potato; tanier; turmeric; yam bean; yam, true
1D. Tuberous and corm vegetables (except potato) subgroup	Sweet potato	Arracacha; arrowroot; artichoke, Chinese; artichoke, Jerusalem; canna, edible. cassava, bitter and sweet; chayote (root); chufa; dasheen (taro); ginger; leren; sweet potato; tanier; turmeric; yam bean; yam, true
2 Leaves of Root and Tuber Vegetables (Human Food or Animal Feed)	Turnip and garden beet or sugar beet	Beet, garden; beet, sugar; burdock, edible; carrot; cassava, bitter and sweet; celeriac; chervil, turnip-rooted; chicory; dasheen (taro); parsnip; radish; radish, oriental (daikon); rutabaga; salsify, black; sweet potato; tanier; turnip; yam, true
3. Bulb Vegetables	Onion, green; and onion, dry bulb	Garlic; garlic, great-headed; leek; onion, dry bulb and green; onion, Welsh; shallot

Crop Group / Subgroup	Representative Commodities	Commodities
4. Leafy vegetables (except *Brassica* vegetables)	Celery, head lettuce, leaf lettuce and spinach	Amaranth (Chinese spinach); arugula (roquette); cardoon; celery; celery, Chinese; celtuce; chervil; chrysanthemum, edible-leaved; chrysanthemum, garland; corn salad; cress, garden; cress, upland; dandelion; dock (sorrel); endive (escarole); fennel, Florence; lettuce, head and leaf; orach; parsley; purslane, garden; purslane, winter; radicchio (red chicory); rhubarb; spinach; spinach, New Zealand; spinach, vine; Swiss chard
4A Leafy greens subgroup	Head lettuce and leaf lettuce, and spinach	Amaranth; arugula; chervil; chrysanthemum, edible-leaved; chrysanthemum, garland; corn salad; cress, garden; cress, upland; dandelion; dock; endive; lettuce; orach; parsley; purslane, garden; purslane, winter; radicchio; spinach; spinach, New Zealand; spinach, vine
4B Leaf petioles subgroup	Celery	Cardoon; celery; celery, Chinese; celtuce; fennel, Florence; rhubarb; Swiss chard
5. *Brassica* (Cole) Leafy Vegetables	Broccoli or cauliflower; cabbage; and mustard greens	Broccoli; broccoli, Chinese (gai lon); broccoli raab (rapini); Brussels sprouts; cabbage; cabbage, Chinese (bok choy); cabbage, Chinese (napa); cabbage, Chinese mustard (gai choy); cauliflower; cavalo broccolo; collards; kale; kohlrabi; mizuna; mustard greens; mustard spinach; rape greens
5A. Head & Stem *Brassica* subgroup	Broccoli or cauliflower and cabbage	Broccoli; broccoli, Chinese; brussles sprouts; cabbage; cabbage, Chinese (napa); cabbage, Chinese mustard; cauliflower; cavalo broccolo; kohlrabi
5B. Leafy *Brassica* greens subgroup	Mustard greens	Broccoli raab; cabbage, Chinese (bok choy); collards; kale; mizuna; mustard greens; mustard spinach; rape greens
6. Legume vegetables (Succulent or dried)	Bean (*Phaseolus*), (succulent & dried), pea (*Pisum*) (succulent & dried) and soybean	Bean (*Lupinus*) (includes grain lupin, sweet lupin, white lupin, and white sweet lupin); bean (*Phaseolus*) (includes field bean, kidney bean, lima bean, navy bean, pinto bean, runner bean, snap bean, tepary bean, wax bean); bean (*Vigna*) (includes adzuki bean, asparagus bean, blackeyed pea, catjang, Chinese longbean, cowpea, crowder pea, moth bean, mung bean, rice bean, southern pea, urd bean, yardlong bean); broad bean (fava); chickpea (garbanzo); guar; jackbean; lablab bean; lentil; pea (*Pisum*) (includes dwarf pea, edible-podded pea, English pea, field pea, garden pea, green pea, snowpea, sugar snap pea); pigeon pea; soybean; soybean (immature seed); sword bean
6A. Edible-podded legume vegetables subgroup	Any one succulent cultivar of edible-podded bean (*Phaseolus*) and any one succulent cultivar of edible-podded pea (*Pisum*)	Bean (*Phaseolus*) (includes runner bean, snap bean, wax bean); bean (*Vigna*) (includes asparagus bean, Chinese longbean, moth bean, yardlong bean); jackbean; pea (*Pisum*) (includes dwarf pea, edible-podded pea, snow pea, sugar snap pea); pigeon pea; soybean (immature seed); sword bean

6B. Succulent shelled pea and bean subgroup	Any succulent shelled cultivar of bean (*Phaseolus*) and garden pea (*Pisum*)	Bean (*Phaseolus*) (includes lima bean, green; broad bean, succulent; bean (*Vigna*) (includes backeyed pea, cowpea, southern pea); pea (*Pisum*) (includes English pea, garden pea, green pea); pigeon pea
6C. Dried shelled pea and bean (except soybean) subgroup	Any one dried cultivar of bean (*Phaseolus*) and any one dried cultivar of pea (*Pisum*)	Dried cultivars of bean (*Lupinus*); bean (*Phaseolus*) (includes field bean, kidney bean, lima bean (dry), navy bean, pinto bean, tepary bean); bean (*Vigna*) (includes adzuki bean, blackeyed pea, catjang, cowpea, crowder pea moth bean, mung bean, rice bean, southern pea, urd bean); broad bean (dry); chickpea; guar; lablab bean; lentil; pea (*Pisum*) (includes field pea); pigeon pea
7. Foliage of Legume Vegetables	Any cultivar of bean (*Phaseolus*), field pea (*Pisum*) and soybean	Plant parts of any legume vegetable included in the legume vegetables that will be used as animal feed
7A. Foliage of legume vegetables (except soybeans) subgroup	Any cultivar of bean (*Phaseolus*) and field pea (*Pisum*)	Plant parts of any legume vegetable (except soybeans) included in the legume vegetables group that will be used as animal feed.
8. Fruiting Vegetables (except cucurbits)	Tomato, bell pepper, and one cultivar of non-bell pepper	Eggplant; groundcherry (*Physalis* spp.); pepino; pepper (includes bell pepper chili pepper, cooking pepper, pimento, sweet pepper); tomatillo; tomato
9. Cucurbit Vegetables	Cucumber, muskmelon, and summer squash	Chayote (fruit); Chinese waxgourd (Chinese preserving melon); citron melon; cucumber; gherkin; gourd, edible (includes hyotan, cucuzza, hechima, Chinese okra); *Momordica* spp. (includes balsam apple, balsam pear, bittermelon, Chinese cucumber); muskmelon (includes cantaloupe); pumpkin squash, summer; squash, winter (includes butternut squash, calabaza, hubbard squash, acorn squash, spaghetti squash); watermelon
9A. Melon subgroup	Cantaloupe	Citron melon; muskmelon; watermelon
9B. Squash/Cucumber subgroup	One cultivar of summer squash and cucumber	Chayote (fruit); Chinese waxgourd; cucumber; gherkin; gourd, edible; *Momordica* spp.; pumpkin; squash, summer; squash, winter
10. Citrus Fruits	Sweet orange; lemon and grapefruit	Calamondin; citrus citron; citrus hybrids (includes chironja, tangelo, tangor); grapefruit; kumquat; lemon; lime; mandarin (tangerine); orange, sour; orange, sweet; pummelo; Satsuma mandarin
11. Pome Fruits	Apple and pear	Apple; crabapple; loquat; mayhaw; pear; pear, oriental; quince

12. Stone Fruits	Sweet or tart cherry; peach; and plum or fresh prune	Apricot; cherry, sweet; cherry, tart; nectarine; peach; plum; plum, Chickasaw; plum, Damson; plum, Japanese; plumcot; prune (fresh)
13. Berries	Any one blackberry or any one raspberry; and blueberry	Blackberry (including bingleberry, boysenberry; dewberry; lowberry, marionberry, olallieberry, youngberry); blueberry; currant; elderberry; gooseberry; huckleberry; loganberry; raspberry, black and red
13A. Caneberry (blackberry and raspberry) subgroup	Any one blackberry or any one raspberry	Blackberry; loganberry; red and black raspberry; cultivars and/or hybrids of these
13B. Bushberry subgroup	Blueberry, highbush	Blueberry, highbush and lowbush; currant; elderberry; gooseberry; huckleberry
14. Tree Nuts	Almond and pecan	Almond; beech nut; Brazil nut; butternut; cashew; chestnut; chinquapin; filbert (hazelnut); hickory nut; macadamia nut; pecan; walnut, black and English
15. Cereal Grains	Corn (sweet and field), rice, sorghum, and wheat	Barley; buckwheat; corn; millet, pearl; millet, proso; oats; popcorn; rice; rye; sorghum (milo); teosinte; triticale; wheat; wild rice
16. Forage, Fodder and Straw of Cereal Grains	Corn, wheat, and any other cereal grain crop	Forage, fodder, and straw of all commodities included in the cereal grains group
17. Grass Forage, Fodder, and Hay Group	Bermuda grass; bluegrass; and bromegrass or fescue	Any grass, Gramineae family (either green or cured) except sugarcane and those included in the cereal grains group, that will be fed to or grazed by livestock, all pasture and range grasses and grasses grown for hay or silage
18. Nongrass Animal Feeds (Forage, Fodder, Straw, and Hay)	Alfalfa and clover (*Trifolium*)	Alfalfa; bean, velvet; clover (*Trifolium, Melilotus*); kudzu; lespedeza; lupin; sainfoin; trefoil; vetch; vetch, crown; vetch, milk

19. Herbs and Spices	Basil (fresh & dried); black pepper; chive; and celery seed or dill seed	Allspice; angelica; anise; anise, star; annatto (seed); balm; basil; borage; burnet; camomile; caper buds; caraway; caraway, black; cardamom; cassia bark; cassia buds; catnip; celery seed; chervil (dried); chive; chive, Chinese; cinnamon; clary; clove buds; coriander leaf (cilantro or Chinese parsley); coriander seed (cilantro); costmary; culantro (leaf); culantro (seed); cumin; curry (leaf); dill (dillweed); dill (seed); fennel (common); fennel, Florence (seed); fenugreek; grains of paradise, horehound; hyssop; juniper berry; lavender; lemongrass; lovage (leaf); lovage (seed); mace; marigold, marjoram; mustard (seed); nasturtium; nutmeg; parsley (dried); pennyroyal; pepper, black; pepper, white; poppy (seed); rosemary; rue; saffron; sage; savory, summer and winter; sweet bay; tansy; tarragon; thyme; vanilla; wintergreen; woodruff; wormwood
19A. Herb subgroup	Basil (fresh & dried) and chive	Angelica; balm; basil; borage; burnet; camomile; catnip; chervil (dried); chive; chive, Chinese; clary; coriander (leaf); costmary; culantro (leaf); curry (leaf); dillweed; horehound; hyssop; lavender; lemongrass; lovage (leaf); marigold; marjoram; nasturtium; parsley (dried); pennyroyal; rosemary; rue; sage; savory, summer and winter; sweet bay; tansy; tarragon; thyme; wintergreen; woodruff; wormwood
19B. Spice subgroup	Black pepper; and celery seed or dill seed	Allspice; anise (seed); anise, star; annatto (seed); caper (buds); caraway; caraway, black; cardamom; cassia (bark); cassia (buds); celery (seed); cinnamon; clove (buds); coriander (seed); culantro (seed); cumin; dill (seed); fennel, common; fennel, Florence (seed); fenugreek; grains of paradise; juniper (berry); lovage (seed); mace; mustard (seed); nutmeg; pepper, black pepper, white; poppy (seed); saffron; vanilla

Published as regulation in 17 May 95 *Federal Register*. 40 CFR 180.41

EPA CROP GROUPING
17 MAY, 1995

Federal Register
Rule Publication
Authored by H.L. Jamerson and B.A. Schneider of EPA

Pesticide Tolerances; Revision of Crop Groups

[Federal Register: May 17, 1995
(Volume 60, Number 95)]
[Rules and Regulations]
From the Federal Register Online via GPO Access
[wais.access.gpo.gov]

Part VIII

Environmental Protection Agency

---

40 CFR Part 180

Pesticide Tolerances; Revision of Crop Groups; Final Rule
===

ENVIRONMENTAL PROTECTION AGENCY

40 CFR Part 180
[OPP-300269A; FRL-4939-9]
RIN 2070-AB78

Pesticide Tolerances; Revision of Crop Groups

AGENCY: Environmental Protection Agency (EPA).
ACTION: Final rule.

Summary: EPA is revising pesticide tolerance crop-grouping regulations to create new crop subgroups, expand existing crop groups by adding new commodities, and revise the representative crops in some groups. EPA expects these revisions to promote greater use of crop grouping for tolerance-setting purposes and to facilitate availability of pesticides for minor crop uses. EPA initiated these regulations.

Effective Date: This regulation becomes effective May 17, 1995.

Addresses: Written objections and hearing requests, identified by the document control number, [OPP-300269A], may be submitted to: Hearing Clerk (1900), Environmental Protection Agency, Rm. M3708, 401 M St., SW., Washington, DC 20460. Fees accompanying objections and hearing requests shall be labeled "Tolerance Petition Fees" and forwarded to: EPA Headquarters Accounting Operations Branch, OPP (Tolerance Fees), P.O. Box 360277M, Pittsburgh, PA 15251. A copy of any objections and hearing requests filed with the Hearing Clerk should be identified by the document control number and submitted to: Public Response and Program Resources Branch, Field Operations Division (7506C), Office of Pesticide Programs, Environmental Protection Agency, 401 M St., SW., Washington, DC 20460. In person, bring copy of objections and hearing request to Rm. 1132, CM #2, 1921 Jefferson Davis Hwy., Arlington, VA 22202.

A copy of objections and hearing requests filed with the Hearing Clerk may also be submitted electronically by sending electronic mail (e-mail) to: opp-docket@epamail.epa.gov. Copies of objections and hearing requests must be submitted as an ASCII file avoiding the use of special characters and any form of encryption. Copies of objections and hearing requests will also be accepted on disks in WordPerfect in 5.1 file format or ASCII file format. All copies of objections and hearing requests in electronic form must be identified by the docket number [OPP-300269A]. No Confidential Business Information (CBI) should be submitted through e-mail. Electronic copies of objections and hearing requests on this rule may be filed online at many Federal Depository Libraries. Additional information on electronic submissions can be found in unit VIII of this document.

For Further Information Contact: By mail: Hoyt Jamerson, Registration Support Branch, Registration Division (7505C), Office of Pesticide Programs, Environmental Protection Agency, 401 M St., SW., Washington, DC 20460. Office telephone number: (703)-308-9368; e-mail: jamerson.hoyt@epamail.epa.gov.

Supplementary Information
I. Background

The crop grouping regulations currently in 40 CFR 180.34(f) enable the establishment of tolerances for a group of crops based on residue data for certain crops that are representative of the group. EPA issued a proposed rule, published in the Federal Register of August 25, 1993 (58 FR 44990), under the provisions of the Federal Food, Drug, and Cosmetic Act (FFDCA), which proposed to revise the crop grouping regulations primarily by adding subgroups to 8 of the 19 existing crop groups. Each subgroup is a smaller and more closely related grouping of the commodities included in the "parent" crop group, and the representative commodities for each subgroup are also a smaller subset of those for the parent group. In addition, EPA proposed to add new commodities to expand some of the existing crop groups, and to revise representative crops for some crop groups to provide petitioners more flexibility in obtaining supporting residue data. EPA also proposed to add an alphabetical listing of commodities with cross-references to the assigned crop groups as an Index to Commodities in the Finding Aids section at the end of title 40 of the Code of Federal

Regulations (CFR), parts 150 to 189. This action is intended to promote more extensive use of crop group tolerances as part of the EPA's efforts to improve utilization of existing and new residue data. Written comments were solicited and were received from more than 22 interested parties and groups in response to the proposal. Comments were received from the pesticide industry, State pesticide regulatory authorities, agricultural grower and marketing organizations, and the Interregional Research Project No. 4 (IR-4). All of these comments have been reviewed and are on file with the Agency in the Public Response and Program Resources Branch at the address provided above. All of the comments were supportive of the proposal in concept, but some comments wanted modifications to the proposal. Most comments were substantially satisfied by editorial changes and deletions from or additions to the proposal. Comments of significance and changes to the rule, as previously proposed, are discussed by topic in succeeding units of this preamble.

II. General Revisions

A. Requirements for New Residue Data When There Are Existing Tolerances

A commenter recommended that, where representative crops are removed from some crop groups by this revision, a way should be found to maintain established crop group tolerances without requiring additional residue data. The Agency expects that residue data from the remaining representative commodities should provide sufficient support for the existing crop group tolerances. Also, available data which previously supported the removed representative crop can still be considered in support of the group tolerance, whether or not that commodity is currently included as a representative commodity. However, all existing tolerances will be subject to reassessment as part of the reregistration program.

The crop group most affected by the removal of representative commodities is the previous small fruits and berries crop group which has been amended to become the new berries crop group, with the removal of cranberries, grapes, and strawberries from the group. Cranberries, grapes, and strawberries have been removed from the crop group since their cultural practices and residue chemistry concerns are distinct enough from the other small fruits and berries to have been an impediment to registrants who might have sought a crop group tolerance. Residue data will still be required to support tolerances for any of these three commodities, which have been included in the listing of miscellaneous crops in 40 CFR 180.41(b).

Only one tolerance for the small fruits and berries group has been established since 1983. However, some tolerances were established for a small fruits crop group which existed before the small fruits and berries crop group was established in 1983, and before specific representative commodities were named in the crop grouping regulations. There are also a substantial number of tolerances for pre-1983

crop groups other than small fruits. All of the existing crop group tolerances will continue in effect until the pesticides undergo the reregistration process or a petition is submitted requesting conversion to a new crop group or subgroup. At that time, consideration will be given to setting individual tolerances for any commodities covered by the old crop group tolerance that are not supportable under the new regulations.

B. Addition of Crops

EPA has accepted several suggestions to add certain commodities to the crop groupings, which were proposed in the Federal Register of August 25, 1993. Comments requesting additions to the crop groupings and revisions to the crop group tables, as were previously proposed, are discussed in unit III of this preamble under specific crop group headings.

Future changes to the crop group tables or other portions of Sec. 180.40 or Sec. 180.41 will be subject to notice-and-comment rulemaking procedures, except for technical amendments to the tables, e.g., to update the scientific nomenclature, or to add a new cultivar of a commodity that is already listed. Minor technical amendments to Secs. 180.40 and 180.41 will be made by publication of a final rule.

C. Representative Commodities

There are no changes to the representative commodities, as proposed, with the exception of editorial revisions to several crop groupings to clearly identify the commodities for which residue data are required. Several commenters suggested the deletion or substitution of representative commodities for certain crop groupings. These comments and the editorial revisions to the representative commodities are discussed in unit III of this preamble under specific crop group headings.

D. Regional/Common Names for Commodities

In response to a request that efforts should be made to incorporate additional regional commodity names in the crop groupings, a number of common names have been added to the Index to Commodities, with references to the commodity name as it is listed in the crop group.

Additional regional/common names will be added to the Index as warranted with references to the commodity as it is named in the crop group tables. In order not to lengthen the crop group tables unnecessarily, new common names will be added only to the Index.

E. Miscellaneous Commodities

Some suggestions were made to list in the commodity index crops not included in a crop grouping. However, the commodity index is intended to complement the crop group tables, rather than be a comprehensive listing of all commodities with tolerances. Some of the ungrouped crops may be considered for inclusion in a crop grouping at a future date, at which time they will be added to the index. Crops that were intentionally not included in any groups were listed previously in Sec. 180.34(f)(7); such miscellaneous commodities are now listed in Sec. 180.41(b).

III. Specific Revisions to Crop Groups

1. **Crop Group 1.** Root and tuber vegetables group. In the crop group listing, the designation edible canna (Queensland arrowroot) replaces purple arrowroot, oriental radish replaces Japanese radish, and yam bean has been expanded to include jicama and manioc pea.

In response to a request, chayote root has been added to the root and tuber vegetables group and to subgroups 1-C (tuberous and corm vegetables) and 1-D (tuberous and corm vegetables, except potato).

Chicory, grown for its roots and leaves, has not been expanded to include witloof chicory (or the common names French endive and Belgian endive) because of the likelihood of confusion due to chicory root being included in crop group 1, chicory leaves being included in crop group 2, and endive being included in crop group 4. Cultural practices associated with witloof chicory production also differ from the production of chicory leaves and endive. Witloof chicory is produced from chicory roots, which are transplanted from the field to indoor growth chambers, where the edible compact head of blanched leaves is "forced" from the root.

2. **Crop Group 2.** Leaves of root and tuber vegetables (human food or animal feed) group. In the crop group listing, the designation Japanese radish was changed to Oriental radish.

A commenter requested that the common names French endive, Belgian endive, and witloof be added to the chicory entry in the leaves of root and tuber vegetables crop group. For the reasons given for crop group 1 above, these common names have not been added to the crop group or to the Index to Commodities.

3. **Crop Group 3.** Bulb vegetables (*Allium* spp.) group. The representative commodities for the bulb vegetables (*Allium* spp.) group are listed clearly as two separate commodities—onion, green and onion, dry bulb—to clarify that residue data are required for both green and dry bulb onions.

4. **Crop Group 4.** Leafy vegetables (except *Brassica* vegetables) group.

The representative commodities for leafy vegetables (except *Brassica* vegetables) group have been editorially modified to clarify that residue data in support of crop group 4 and subgroup 4-A (leafy greens) tolerances are required for both head lettuce and leaf lettuce.

Cardoon (*Cynara cardunculus*) and Chinese celery (*Apium graveolens* var. *secalinum*) have been added to crop group 4 and subgroup 4-B, leaf petioles.

Florence fennel has been expanded to include the name finocchio.

The common or loose-leaf chicory (asparagus chicory, radichetta, or green chicory) was considered but not added to crop group 4 at this time because of the potential confusion with chicory leaves, which are in crop group 2.

5. **Crop Group 5.** *Brassica* (cole) leafy vegetables group. Mizuna and mustard spinach have been added to the crop group and to subgroup 5-B, leafy *Brassica* greens subgroup.

6. **Crop Group 6.** Legume vegetables (succulent or dried) group. The representative commodities for subgroup 6-A have been clarified as being succulent cultivars. Adzuki bean, moth bean, mung bean and rice bean have been moved from the bean (*Phaseolus* spp.) listing and are now included with the bean (*Vigna* spp.) to reflect recent taxonomic changes.

Pea (*Pisum* spp.) has been expanded to include sugar snap pea and snow pea. Sugar pea was deleted, since snow pea is the preferred common name.

As requested by a commenter, pigeon pea (*Cajanus cajan*) has been added to subgroup 6-B, succulent shelled pea and bean, of crop group 6, legume vegetables, because it is used extensively in the Caribbean and Central and South American countries as a fresh green pea removed from its pod.

7. **Crop Group 7.** Foliage of legume vegetables group. No comments were submitted; no changes have been made.

8. **Crop Group 8.** Fruiting vegetables (except cucurbits) group. The representative commodities for fruiting vegetables (except cucurbits) group have been modified to clarify that residue data in support of crop group 8 are required for both bell pepper and a nonbell pepper, as well as for tomatoes. This is not a change in policy; however, the data requirement was not articulated in the regulation previously.

9. **Crop Group 9.** Cucurbit vegetables group. A commenter questioned the listing of cantaloupe, a specific type of muskmelon, as representative commodity for subgroup 9-A; whereas for the parent crop group, any muskmelon is a suitable representative commodity. Muskmelon of any type is considered acceptable for the parent crop group because there are two other representative commodities, a cucumber and a summer squash, to balance out the group. However, as the only representative commodity for the subgroup, cantaloupe would be the best because its finely-ridged rough surface would result in higher surface residues compared to the smooth-skinned melons like the honeydew melon. Therefore, in the final rule, cantaloupe has been retained as the representative commodity for subgroup 9-A.

A commenter requested that EPA reconsider adding a third subgroup "winter squash and pumpkins" under cucurbits or, alternatively, place winter squash and pumpkins with "melons" since they all have inedible rinds. The commenter is concerned that inappropriately high tolerances might be set otherwise, based on residue in summer squash and cucumbers, which could utilize more of the Reference Dose than would be necessary. A review of established tolerances for the proposed subgroup 9-B, squash/cucumber, shows that tolerances for summer squash, cucumber, winter squash, and pumpkins are the same or fall within the 5X limitation for tolerance levels in the same subgroup. Therefore, EPA has retained winter squash and pumpkins in subgroup 9-B.

In response to comments, chayote (*Sechium edule*) fruit has been added to the cucurbit vegetables group and the squash/cucumber subgroup.

10. **Crop Group 10.** Citrus fruits group. A commenter recommended that the representative commodities for the citrus fruits group should be reduced to two, to include sweet orange and a choice between lemon and grapefruit rather than both of the latter. An alternative recommendation was to delete grapefruit as a representative commodity on the basis that there is no difference between residues in sweet oranges and those in grapefruit. However, EPA has not reduced the number of representative commodities because of the importance of citrus in the diet; the consumption of combined citrus exceeds that of any commodity in the general population, and for infants the consumption of citrus is second to apples. Therefore, EPA believes it is important to require residue data for three representative commodities (grapefruit, lemon, and sweet orange) in support of tolerances for the citrus fruits group.

11. **Crop Group 11.** Pome fruits group. The only change to this group is in the scientific name for apple to reflect current nomenclature.

12. **Crop Group 12.** Stone fruits group. A commenter recommended that sweet cherries, rather than sour cherries, should be a representative commodity for the stone fruits group. The commenter explained that, at harvest, sour cherries are always flushed with water while sweet cherries are usually handled dry. Thus, higher pesticide residues would be expected on the sweet cherries, indicating they should be the preferred representative commodity of the two types of cherries. However, EPA prefers to allow the option of either sour or sweet cherries as representative commodity, provided that sour cherries should be analyzed for residues in their unwashed state. In addition, the term "tart cherry" will replace "sour cherry."

A request was made to include pomegranates in the stone fruits group. Since this commodity is not similar to other members of the stone fruits group, it was not included.

Several commenters requested that olives be added to the stone fruits group. A major problem with grouping olives with stone fruits is the need for processing studies to determine the concentration of residues in olive oil. Because residue studies for olives, including processing studies, would be required for a stone fruit group that includes olives, olives would have to be a representative crop, which would reduce the usefulness of the group and negate any benefit to olives from being in the group. EPA may reevaluate pomegranate and olive as tropical/subtropical fruits, when a tropical/subtropical fruit crop group is researched in the future.

13. **Crop Group 13.** Berries group. A commenter requested clarification of the discussion of the bushberry subgroup 13-B in Section III of the proposed rule. The bushberry subgroup includes woody shrubs and bushes that produce fruit in clusters, including the blueberry. Blackberries are included in subgroup 13-A with other caneberries. Youngberry has been added to blackberry since it is a blackberry-raspberry hybrid similar to boysenberry and marionberry, which are included with blackberry.

14. **Crop Group 14.** Tree nuts group. Several commenters requested that representative commodities for the tree nuts group should be revised to allow a choice between pecan and English walnut as a representative crop in addition to almond. Previously all three commodities were required representative crops. EPA proposed deletion of English walnuts as a representative commodity to streamline the tree nut crop group data requirements by requesting field residue data for only the minimum number of representative commodities that will enable EPA to adequately evaluate the residue data for establishing a tolerance. Almost all English walnuts are produced in California while pecans, which are the major tree nut crop produced in the U.S., are distributed throughout the U.S., particularly in the southeastern region. Residue data on almonds and pecans are needed to obtain geographically representative residue data for the tree nuts. Almonds and pecans have been retained as the representative crops for this group. However, EPA will be flexible on using residue data already developed for English walnuts and such data will be useful in establishing tolerances for tree nuts or supporting reregistration actions.

Two commenters requested that pistachios be added to the tree nut group, stating that residue data are now available that demonstrate that pesticide residue levels on pistachios are "at levels similar to, if not lower than other nuts in the grouping." Pistachios have been requested and considered for inclusion in the crop group previously, but not included because pistachio shells split and hence would be expected to permit greater residues on the edible portion of the nut than other nut crops. Since the commenters did not submit any comparative field residue data between pistachios and other tree nuts, EPA has not added pistachios to the tree nuts group.

15. **Crop Group 15.** Cereal grains group. No comments were submitted, and no changes have been made.

16. **Crop Group 16.** Forage, fodder and straw of cereal grains group. No comments were submitted, and no changes have been made.

17. **Crop Group 17.** Grass forage, fodder, and hay group. No comments were submitted, and no changes have been made.

18. **Crop Group 18.** Non-grass animal feeds (forage, fodder, straw and hay) group. No comments were submitted, and no changes have been made.

19. **Crop Group 19.** Herbs and spices group. In response to various comments, a number of commodities have been added to crop group 19 and its subgroups: star anise, annatto seed, black caraway, cardamom, chervil (dried), Chinese chive, culantro, grains of paradise, mustard seed, white pepper, and poppy seed. Common fennel and Florence fennel (seed) have replaced Italian and sweet fennel.

The commodities *capsicum* (peppers), ginger, paprika, peppermint leaves, sesame seed, spearmint leaves, and turmeric were considered, but not added to, crop group 19 because they are members of other crop groups, they have other processed commodities that would require them to be representative commodities, or their cultural practices and pest problems are too dissimilar.

A commenter requested that there be only one representative commodity in the herb subgroup and one in the spice subgroup, only two for the total crop group, contending that the cost of generating residue data on even two commodities for a subgroup would greatly exceed the sales value of the crops themselves. Another commenter requested that additional herbs and spices be included in the crop group, contending that the EPA-proposed list does not include many spices that meet the definitions of spice issued by the U.S. Food and Drug Administration, the U.S. Department of Agriculture, and the American Spice Trade Association.

EPA cannot further reduce the number of representative commodities at this time because of the great diversity in plant classification between the numerous herbs and spices, the wide variation in cultural practices and pest problems between the various commodities, the wide differences in plant parts that are the raw agricultural commodity, and the lack of comparative field residue trial data for many of the commodities. As the field residue database on herbs and spices becomes more extensive, then the possibility exists for further reducing and/or changing some of the representative commodities, or further subdividing the current subgroups, so that the number of representative commodities might be reduced.

IV. Addition of New Crop Groups

Several commenters requested the addition of new crop groups, as follows: Oil seed crops, to include sunflower, rape, canola, crambe, flax, safflower, jojoba, and Lesquerella; tropical fruits to include banana, mango, papaya, passion fruit, etc.; and subtropical fruits to include avocado, kiwifruit, persimmon, cherimoya, guava, mango, and pomegranate. The development of new crop groups for tropical and subtropical fruits and for oil seed crops is beyond the scope of this rulemaking, but may be considered for future rulemaking.

As indicated in the proposed rule, any future recommended changes to the crop groups or subgroups should be presented in a form which includes all necessary background and supporting information, such as a list of all commodities to be included, accompanied by scientific names, naming all representative commodities and providing a rationale for selecting the particular commodities and representative commodities to be included. EPA welcomes an opportunity to evaluate crop group/subgroup proposals, when they are submitted from interested parties, and/or to work with such parties on the types of information and data necessary to evaluate a new crop group.

V. Other Comments

A commenter suggested defining cotton to include kenaf, said to be a related fiber crop also in the family Malvaceae. Adding or amending crop definitions in 40 CFR 180.1(h) is beyond the scope of EPA's current efforts to revise the crop groupings. A request to establish a commodity definition in Sec. 180.1(h) may be submitted to EPA for review as a separate amendment. The amendment should include rationale for change, comparative cultural practices including pest problems, application timing, food/feed uses, and geographical distribution for commodity production, as well as processing food items.

At this time EPA has no plans to set tolerances on a crop group or subgroup basis for pesticide residues in processed food or animal feed commodities, even when the parent raw agricultural commodity is a member of a crop group. Generally the processed forms of commodities are very different from their raw forms and, within a crop group, also different from each other's processed forms, including in terms of expected residues. Also, processed commodities may have incurred pesticide residues from direct or indirect application of pesticides to the processed food as well as application to the raw form from which the processed form is derived. This would present a problem of too much variability in expected residues in the various processed commodities. In addition, some chemicals have a tendency to concentrate as a result of processing whereas others may remain constant or dissipate during processing; this lack of consistency in resulting residues would also make it difficult to set a crop group tolerance to cover several dissimilar processed commodities.

VI. Implementation

Petitions pending at the time this final rule is published will continue to be processed based on the previous regulation, except they will be given the benefit of any appropriate revised or reduced residue data requirements if needed. Likewise, residue studies which are currently underway should not be adversely affected by this new rule.

Residue data requirements imposed by these regulations for a crop group tolerance are substantially the same as those that were imposed by Sec. 180.34(f), except that for a number of crop groups, fewer representative commodities are required or a choice of representative commodities is allowed. For the bulb vegetables group, the tree nuts group, and the herbs and spices group, the number of representative commodities is now fewer than before. For the leaves of root and tuber vegetables group, *Brassica* (cole) leafy vegetables group, and herbs and spices group, there is some choice allowed in terms of representative commodities.

Because of a major change to the former small fruits and berries crop group—deletion of cranberries, grapes, and strawberries as group members and as representative commodities, resulting in the new berries crop group—any petition for a tolerance for the small fruits and berries crop group that is currently pending or submitted within 30 days after publication of this rule will be processed as if it were a petition for tolerances for the berries crop group and the individual commodities cranberries, grapes, and strawberries. No additional fee will be imposed because of this action, although any other amendment to such a petition would be subject to the usual fees.

A pending petition for a crop group tolerance which is found deficient in terms of residue data may be reconsidered by EPA as a petition for one or more related crop subgroup tolerances. EPA's response to the petitioner will indicate whether such subgroup tolerances can be supported. Similarly, crop group tolerances being reassessed for reregistration that are determined not to be supported by the available data will be evaluated to determine whether the data might support one or more related crop subgroup tolerances or one or more individual crop tolerances.

A petition for a crop group or crop subgroup tolerance which relies on existing individual tolerances for all or some of the representative crops for the crop group or subgroup will be subject to reassessment of all available data in support of the individual tolerances to determine if such data are currently considered adequate to support the crop group or subgroup tolerance.

All existing crop group tolerances will continue in effect until the pesticides undergo the reregistration process or a petition is submitted requesting conversion to a new crop group or subgroup. At that time, consideration will be given to setting individual tolerances for any commodities covered by the old crop group tolerance that are not supportable under the new regulations.

Fees imposed by 40 CFR 180.33(h) for petitions for crop group tolerances will apply to petitions for subgroup tolerances as well. For fee purposes, each request for a crop subgroup tolerance will be considered as if it were a request for a single commodity tolerance.

VII. Index to Commodities

This unit contains an alphabetical index to the crops in all the crop groups, giving the Crop Group number. The index will be included in Title 40 of the Code of Federal Regulations as a finding aid after its publication in the *Federal Register*.

Calabaza (see squash, winter) 9
Calaloo (see amaranth) 4
Calamondin. 10
Calilu (see amaranth) 4
Camomile.. 19
Canna, edible 1
Cantaloupe (see muskmelon). 9
Cape gooseberry (see groundcherry) 8
Caper buds 19
Caraway 19
Cardoni (see cardoon) 4
Cardoon 4
Cardamom. 19
Carrot 1
Carrot (foliage) 2
Casaba (see muskmelon). 9
Cashew 14
Cassava, bitter and sweet. 1
Cassava, bitter and sweet (foliage). 2
Cassia bark 19
Cassia buds 19
Catjang (see bean (*Vigna* spp.)) 6
Catmint (see catnip) 19
Catnip 19
Cauliflower 5
Cavalo broccolo 5
Celeriac. 1
Celeriac (foliage). 2
Celery 4
Celery cabbage (see Chinese cabbage (napa)) 5
Celery mustard (see Chinese cabbage (bok choy)). 5
Celery root (see celeriac). 1
Celery seed 19
Celtuce 4
Ceylon spinach (see vine spinach). 4
Chayote (fruit). 9
Chayote (root). 1
Cherokee blackberry (see blackberry) 13
Cherry, sweet 12
Cherry, tart 12
Chervil 4
Chervil (dried). 19
Chervil, turnip-rooted 1
Chervil, turnip-rooted (foliage) 2
Chesterberry (see blackberry) 13
Chestnut 14
Cheyenne blackberry (see blackberry) 13
Chickasaw plum 12
Chickpea. 6
Chickpea (foliage). 7
Chicory 1
Chicory (foliage). 2
Chihili cabbage (see Chinese cabbage (napa)) 5
Chili pepper (see pepper (*Capsicum* spp.)) 8
China pea (see pea (*Pisum* spp.) (snow pea)). 6
China star anise (see star anise). 19
Chinese artichoke 1
Chinese broccoli 5
Chinese cabbage (bok choy) 5

Chinese cabbage (napa). 5
Chinese celery. 4
Chinese celery cabbage (see Chinese cabbage (napa)). 5
Chinese chive 19
Chinese cucumber (see *Momordica* spp.). 9
Chinese green mustard (see Chinese mustard cabbage) 5
Chinese green mustard cabbage (see Chinese mustard cabbage). . 5
Chinese kale (see Chinese broccoli) 5
Chinese lantern plant (see tomatillo) 8
Chinese leek (see Chinese chive). 19
Chinese longbean (see bean (*Vigna* spp.)) 6
Chinese mustard (see mustard greens). 5
Chinese mustard cabbage 5
Chinese okra (see edible gourd). 9
Chinese parsley (see coriander) 19
Chinese pea (see pea (*Pisum* spp.) (snow pea)) 6
Chinese pear (see Oriental pear) 11
Chinese preserving melon (see Chinese waxgourd) 9
Chinese radish (see Oriental radish) 1
Chinese spinach (see amaranth) 4
Chinese squash (see Chinese waxgourd) 9
Chinese turnip (see Oriental radish). 1
Chinese waxgourd. 9
Chinese white cabbage (see Chinese cabbage (bok choy)). . . . 5
Chinquapin 14
Chironja (see citrus hybrids) 10
Chive. 19
Choi sum (see Chinese cabbage (bok choy)) 5
Chopsuey greens (see chrysanthemum, edible-leaved). 4
Choy sum (see Chinese cabbage (bok choy)). 5
Chrysanthemum, edible-leaved 4
Chrysanthemum, garland. 4
Chufa 1
Ciboule (see Welsh onion). 3
Cilantro (see coriander) 19
Cilantro del monte (see culantro). 19
Cinnamon 19
Citrus citron 10
Citron melon 9
Citrus hybrids (*Citrus* spp.) 10
Clary 19
Clove buds 19
Clover (forage, fodder, straw, hay). 18
Cluster bean (see guar) 6
Cocoyam (see tanier). 1
Cocoyam (foliage). 2
Collards 5
Common bean (see bean (*Phaseolus* spp.) (kidney bean)) 6
Common millet (see proso millet) 15
Common vetch (see vetch) 18
Congo pea (see pigeon pea). 6
Cooking pepper (see pepper (*Capsicum* spp.)). 8
Coriander (leaf and seed). 19
Corn 15
Corn (forage, fodder). 16
Corn salad. 4
Coryberry (see blackberry) 13
Costmary. 19
Courgette (see squash, summer) 9

VIII. Electronic Copies of Objections and Hearing Requests

A record has been established for this rulemaking under docket number [OPP-300269A] (including any objections and hearing requests submitted electronically as described below). A public version of this record, including printed, paper versions of electronic comments, which does not include any information claimed as CBI, is available for inspection from 8 a.m. to 4:30 p.m., Monday through Friday, excluding legal holidays. The public record is located in Room 1132 of the Public Response and Program Resources Branch, Field Operations Division (7506C), Office of Pesticide Programs, Environmental Protection Agency, Crystal Mall #2, 1921 Jefferson Davis Highway, Arlington, VA.

Written objections and hearing requests, identified by the document control number [OPP-300269A], may be submitted to the Hearing Clerk (1900), Environmental Protection Agency, Rm. 3708, 401 M St., SW., Washington, DC 20460.

A copy of electronic objections and hearing requests filed with the Hearing Clerk can be sent directly to EPA at: opp-Docket@epamail.epa.gov

A copy of electronic objections and hearing requests filed with the Hearing Clerk must be submitted as an ASCII file avoiding the use of special characters and any form of encryption.

The official record for this rulemaking, as well as the public version, as described above will be kept in paper form. Accordingly, EPA will transfer any objections and hearing requests received electronically into printed, paper form as they are received and will place the paper copies in the official rulemaking record which will also include all objections and hearing requests submitted directly in writing. The official rulemaking record is the paper record maintained at the address in "ADDRESSES" at the beginning of this document.

IX. Regulatory Requirements

A. Executive Order 12866

Under Executive Order 12866 (58 FR 51735, Oct. 4, 1993), the Agency must determine whether the regulatory action is "significant" and therefore subject to all the requirements of the Executive Order (i.e., Regulatory Impact Analysis, review by the Office of Management and Budget (OMB)). Under section 3(f), the order defines "significant" as those actions likely to lead to a rule

(1) having an annual effect on the economy of $100 million or more, or adversely and materially affecting a sector of the economy, productivity, competition, jobs, the environment, public health or safety, or State, local, or tribal governments or communities (also known as "economically significant");

(2) creating serious inconsistency or otherwise interfering with an action taken or planned by another agency;

(3) materially altering the budgetary impacts of entitlement, grants, user fees, or loan programs; or

(4) raising novel legal or policy issues arising out of legal mandates, the President's priorities, or the principles set forth in this Executive Order.

Pursuant to the terms of this Executive Order, EPA has determined that this rule is not "significant" and is therefore not subject to OMB review.

B. Regulatory Flexibility Act

This regulatory action has been reviewed under the provisions of section 3(a) of the Regulatory Flexibility Act, and EPA has determined that it will not have a significant adverse economic impact on a substantial number of small businesses, small governments, or small organizations.

As this regulatory action is intended to simplify established policy, it is expected that no adverse economic impact will occur on any small entity.

Accordingly, EPA certifies that this regulatory action does not require a separate regulatory flexibility analysis under the Regulatory Flexibility Act.

C. Paperwork Reduction Act

This rule contains no information collection requirements subject to the Paperwork Reduction Act.

List of Subjects in 40 CFR Part 180

Environmental protection, Administrative practice and procedures, Agricultural commodities, Pesticides and pests, Reporting and recordkeeping requirements.

Dated: May 5, 1995.

Lynn R. Goldman,

Assistant Administrator for Prevention, Pesticides and Toxic Substances

Therefore, 40 CFR part 180 is amended as follows:

PART 180—[AMENDED]

1. The authority citation for part 180 continues to read as follows:

Authority: 21 U.S.C. 346a and 371.

2. In Sec. 180.1, by revising paragraph (g), to read as follows:

Sec. 180.1 Definitions and interpretations.

* * * * *

(g) For the purpose of computing fees as required by Sec. 180.33, each group of related crops listed in Sec. 180.34(e) and each crop group or subgroup listed in Sec. 180.41 is counted as a single raw agricultural commodity in a petition or request for tolerances or exemption from the requirement of a tolerance.

Sec. 180.34 [Amended]

3. By amending Sec. 180.34 Tests on the amount of residue remaining by removing paragraph (f).

4. By adding new Sec. 180.40, to read as follows:

Sec. 180.40 Tolerances for crop groups.

(a) Group or subgroup tolerances may be established as a result of:

(1) A petition from a person who has submitted an application for the registration of a pesticide under the Federal Insecticide, Fungicide, and Rodenticide Act.

(2) On the initiative of the Administrator.

(3) A petition by an interested person.

(b) The tables in Sec. 180.41 are to be used in conjunction with this section for the establishment of crop group tolerances. Each table in Sec. 180.41 lists a group of raw agricultural commodities that are considered to be related for the purposes of this section. Refer also to Sec. 180.1(h) for a listing of commodities for which established tolerances may be applied to certain other related and similar commodities.

(c) When there is an established or proposed tolerance for all of the representative commodities for a specific group or subgroup of related commodities, a tolerance may be established for all commodities in the associated group or subgroup. Tolerances may be established for a crop group or, alternatively, tolerances may be established for one or more of the subgroups of a crop group.

(d) The representative crops are given as an indication of the minimum residue chemistry data base acceptable to the Agency for the purposes of establishing a group tolerance. The Agency may, at its discretion, allow group tolerances when data on suitable substitutes for the representative crops are available (e.g., limes instead of lemons).

(e) Since a group tolerance reflects maximum residues likely to occur on all individual crops within a group, the proposed or registered patterns of use for all crops in the group or subgroup must be similar before a group tolerance is established. The pattern of use consists of the amount of pesticide applied, the number of times applied, the timing of the first application, the interval between applications, and the interval between the last application and harvest. The pattern of use will also include the type of application; for example, soil or foliar application, or application by ground or aerial equipment.

(f) When the crop grouping contains commodities or byproducts that are utilized for animal feed, any needed tolerance or exemption from a tolerance for the pesticide in meat, milk, poultry and/or eggs must be established before a tolerance will be granted for the group as a whole. The representative crops include all crops in the group that could be processed such that residues may concentrate in processed food and/or feed. Processing data will be required prior to establishment of a group tolerance, and food additive tolerances will not be granted on a group basis.

(g) If maximum residues (tolerances) for the representative crops vary by more than a factor of 5 from the maximum value observed for any crop in the group, a group or subgroup tolerance will ordinarily not be established. In this case individual crop tolerances, rather than group tolerances, will normally be established.

(h) Alternatively, a commodity with a residue level significantly higher or lower than the other commodities in a group may be excluded from the group tolerance (e.g., cereal grains, except corn). In this case an individual tolerance at the appropriate level for the unique commodity would be established, if necessary. The alternative approach of excluding a commodity with a significantly higher or lower residue level will not be used to establish a tolerance for a commodity subgroup. Most subgroups have only two representative commodities; to exclude one such commodity and its related residue data would likely provide insufficient residue information to support the remainder of the subgroup. Residue data from crops addi-

tional to those representative crops in a grouping may be required for systemic pesticides.

(i) The commodities included in the groups will be updated periodically either at the initiative of the Agency or at the request of an interested party. Persons interested in updating this section should contact the Registration Division of the Office of Pesticide Programs.

(j) Establishment of a tolerance does not substitute for the additional need to register the pesticide under a companion law, the Federal Insecticide, Fungicide, and Rodenticide Act. The Registration Division of the Office of Pesticide Programs should be contacted concerning procedures for registration of new uses of a pesticide.

5. By adding new Sec. 180.41, to read as follows:

Sec. 180.41 Crop group tables.

(a) The tables in this section are to be used in conjunction with Sec. 180.40 to establish crop group tolerances.

(b) Commodities not listed are not considered as included in the groups for the purposes of this paragraph, and individual tolerances must be established. Miscellaneous commodities intentionally not included in any group include asparagus, avocado, banana, cranberry, fig, globe artichoke, grape, hops, kiwifruit, mango, mushroom, okra, papaya, pawpaw, peanut, persimmon, pineapple, strawberry, water chestnut, and watercress.

(c) Each group is identified by a group name and consists of a list of representative commodities followed by a list of all commodity members for the group. If the group includes subgroups, each subgroup lists the subgroup name, the representative commodity or commodities, and the member commodities for the subgroup. Subgroups, which are a subset of their associated crop group, are established for some but not all crops groups.

(1) Crop Group 1: Root and Tuber Vegetables Group.

(i) Representative commodities. Carrot, potato, radish and sugar beet.

(ii) Table. The following Table 1 lists all the commodities included in Crop Group 1 and identifies the related crop subgroups.

Table 1—Crop Group 1: Root and Tuber Vegetables

Commodities	Related crop subgroups
Arracacha (*Arracacia xanthorrhiza*)	1-C, 1-D
Arrowroot (*Maranta arundinacea*)	1-C, 1-D
Artichoke, Chinese (*Stachys affinis*)	1-C, 1-D
Artichoke, Jerusalem (*Helianthus tuberosus*)	1-C, 1-D
Beet, garden (*Beta vulgaris*)	1-A, 1-B
Beet, sugar (*Beta vulgaris*)	1-A
Burdock, edible (*Arctium lappa*)	1-A, 1-B
Canna, edible (Queensland arrowroot) (*Canna indica*)	1-C, 1-D
Carrot (*Daucus carota*)	1-A, 1-B
Cassava, bitter and sweet (*Manihot esculenta*)	1-C, 1-D
Celeriac (celery root) (*Apium graveolens* var. *rapaceum*)	1-A, 1-B
Chayote (root) (*Sechium edule*)	1-C, 1-D
Chervil, turnip-rooted (*Chaerophyllum bulbosum*)	1-A, 1-B
Chicory (*Cichorium intybus*)	1-A, 1-B

Chufa (*Cyperus esculentus*) 1-C, 1-D
Dasheen (taro) (*Colocasia esculenta*) 1-C, 1-D
Ginger (*Zingiber officinale*) 1-C, 1-D
Ginseng (*Panax quinquefolius*) 1-A, 1-B
Horseradish (*Armoracia rusticana*) 1-A, 1-B
Leren (*Calathea allouia*)..1-C, 1-D
Parsley, turnip-rooted (*Petroselinum crispum*
   var. *tuberosum*)...1-A, 1-B
Parsnip (*Pastinaca sativa*)...1-A, 1-B
Potato (*Solanum tuberosum*) 1-C
Radish (*Raphanus sativus*)..1-A, 1-B
Radish, oriental (daikon) (*Raphanus sativus*
   subvar. *longipinnatus*)...1-A, 1-B
Rutabaga (*Brassica campestris* var. *napobrassica*). 1-A, 1-B
Salsify (oyster plant) (*Tragopogon porrifolius*)..............1-A, 1-B
Salsify, black (*Scorzonera hispanica*)...........................1-A, 1-B
Salsify, Spanish (*Scolymus hispanicus*) 1-A, 1-B
Skirret (*Sium sisarum*)...1-A, 1-B
Sweet potato (*Ipomoea batatas*) 1-C, 1-D
Tanier (cocoyam) (*Xanthosoma sagittifolium*)................1-C, 1-D
Turmeric (*Curcuma longa*)...1-C, 1-D
Turnip (*Brassica rapa* var. *rapa*) 1-A, 1-B
Yam bean (jicama, manoic pea) (*Pachyrhizus* spp.) 1-C, 1-D
Yam, true (*Dioscorea* spp.) 1-C, 1-D

(iii) Table. The following Table 2 identifies the crop subgroups for Crop Group 1, specifies the representative commodity(ies) for each subgroup, and lists all the commodities included in each subgroup.

Table 2—Crop Group 1 Subgroup Listing

Representative commodities	Commodities
Crop Subgroup 1-A. Root vegetables subgroup.	
Carrot, radish, and sugar beet	Beet, garden; beet, sugar; burdock, edible; carrot; celeriac; chervil, turnip-rooted; chicory; ginseng; horseradish; parsley, turnip-rooted; parsnip; radish; radish, oriental; rutabaga; salsify; salsify, black; salsify, Spanish; skirret; turnip.
Crop Subgroup 1-B. Root vegetables (except sugar beet) subgroup.	
Carrot and radish	Beet, garden; burdock, edible; carrot; celeriac; chervil, turnip-rooted; chicory; ginseng; horseradish; parsley, turnip-rooted; parsnip; radish; radish, oriental; rutabaga; salsify; salsify, black; salsify, Spanish; skirret; turnip.
Crop Subgroup 1-C. Tuberous and corm vegetables subgroup.	

Potato	Arracacha; arrowroot; artichoke, Chinese; artichoke, Jerusalem; canna, edible; cassava, bitter and sweet; chayote (root); chufa; dasheen; ginger; leren; potato; sweet potato; tanier; turmeric; yam bean; yam, true.
Crop Subgroup 1-D. Tuberous and corm vegetables (except potato) subgroup.	
Sweet potato	Arracacha; arrowroot; artichoke, Chinese; artichoke, Jerusalem; canna, edible; cassava, bitter and sweet; chayote (root); chufa; dasheen; ginger; leren; sweet potato; tanier; turmeric; yam bean; yam, true.

----

(2) Crop Group 2. Leaves of Root and Tuber Vegetables (Human Food or Animal Feed) Group (Human Food or Animal Feed) Group.

    (i) Representative commodities. Turnip and garden beet or sugar beet.

    (ii) Commodities. The following is a list of all the commodities included in Crop Group 2:

Crop Group 2: Leaves of Root and Tuber Vegetables (Human Food or Animal Feed) Group—Commodities

Beet, garden (*Beta vulgaris*)
Beet, sugar (*Beta vulgaris*)
Burdock, edible (*Arctium lappa*)
Carrot (*Daucus carota*)
Cassava, bitter and sweet (*Manihot esculenta*)
Celeriac (celery root) (*Apium graveolens* var. *rapaceum*)
Chervil, turnip-rooted (*Chaerophyllum bulbosum*)
Chicory (*Cichorium intybus*)
Dasheen (taro) (*Colocasia esculenta*)
Parsnip (*Pastinaca sativa*)
Radish (*Raphanus sativus*)
Radish, Oriental (daikon) (*Raphanus sativus* subvar. *longipinnatus*)
Rutabaga (*Brassica campestris* var. *napobrassica*)
Salsify, black (*Scorzonera hispanica*)
Sweet potato (*Ipomoea batatas*)
Tanier (cocoyam) (*Xanthosoma sagittifolium*)
Turnip (*Brassica rapa* var. *rapa*)
Yam, true (*Dioscorea* spp.)

(3) Crop Group 3. Bulb Vegetables (*Allium* spp.) Group.

    (i) Representative commodities. Onion, green; and onion, dry bulb.

    (ii) Commodities. The following is a list of all the commodities in Crop Group 3:

Crop Group 3: Bulb Vegetables (*Allium* spp.) Group Commodities

Garlic (*Allium sativum*)
Garlic, great-headed (elephant) (*Allium ampeloprasum* var. *ampeloprasum*)

Leek (*Allium ampeloprasum, A. porrum, A. tricoccum*)
Onion, dry bulb and green (*Allium cepa, A. fistulosum*)
Onion, Welch (*Allium fistulosum*)
Shallot (*Allium cepa* var. *cepa*)

(4) Crop Group 4. Leafy Vegetables (Except *Brassica* Vegetables) Group.

(i) Representative commodities. Celery, head lettuce, leaf lettuce, and spinach (*Spinacia oleracea*).

(ii) Table. The following Table 1 lists all the commodities included in Crop Group 4 and identifies the related crop subgroups.

Table 1—Crop Group 4: Leafy Vegetables (Except *BRASSICA* Vegetables) Group

Commodities	Related crop subgroups
Amaranth (leafy amaranth, Chinese spinach, tampala) (*Amaranthus* spp.)	4-A
Arugula (Roquette) (*Eruca sativa*)	4-A
Cardoon (*Cynara cardunculus*)	4-B
Celery (*Apium graveolens* var. *dulce*)	4-B
Celery, Chinese (*Apium graveolens* var. *secalinum*)	4-B
Celtuce (*Lactuca sativa* var. *angustana*)	4-B
Chervil (*Anthriscus cerefolium*)	4-A
Chrysanthemum, edible-leaved (*Chrysanthemum coronarium* var. *coronarium*)	4-A
Chrysanthemum, garland (*Chrysanthemum coronarium* var. *spatiosum*)	4-A
Corn salad (*Valerianella locusta*)	4-A
Cress, garden (*Lepidium sativum*)	4-A
Cress, upland (yellow rocket, winter cress) (*Barbarea vulgaris*)	4-A
Dandelion (*Taraxacum officinale*)	4-A
Dock (sorrel) (*Rumex* spp.)	4-A
Endive (escarole) (*Cichorium endivia*)	4-A
Fennel, Florence (finochio) (*Foeniculum vulgare* Azoricum Group)	4-B
Lettuce, head and leaf (*Lactuca sativa*)	4-A
Orach (*Atriplex hortensis*)	4-A
Parsley (*Petroselinum crispum*)	4-A
Purslane, garden (*Portulaca oleracea*)	4-A
Purslane, winter (*Montia perfoliata*)	4-A
Radicchio (red chicory) (*Cichorium intybus*)	4-A
Rhubarb (*Rheum rhabarbarum*)	4-B
Spinach (*Spinacia oleracea*)	4-A
Spinach, New Zealand (*Tetragonia tetragonioides, T. expansa*)	4-A
Spinach, vine (Malabar spinach, Indian spinach) (*Basella alba*)	4-A
Swiss chard (*Beta vulgaris* var. *cicla*)	4-B

(iii) Table. The following Table 2 identifies the crop subgroups for Crop Group 4, specifies the representative commodities for each subgroup, and lists all the commodities included in each subgroup.

Table 2—Crop Group 4 Subgroup Listing

Representative commodities	Commodities
Crop Subgroup 4-A. Leafy greens subgroup. Head lettuce and leaf lettuce, and spinach (*Spinacia oleracea*)	Amaranth; arugula; chervil; chrysanthemum, edible-leaved; chrysanthemum, garland; corn salad; cress, garden; cress, upland; dandelion; dock; endive; lettuce; orach; parsley; purslane, garden; purslane, winter; radicchio (red chicory); spinach; spinach, New Zealand; spinach, vine.
Crop Subgroup 4-B. Leaf petioles subgroup. Celery	Cardoon; celery; celery, Chinese; celtuce; fennel, Florence; rhubarb; Swiss chard.

(5) Crop Group 5. *Brassica* (Cole) Leafy Vegetables Group.

(i) Representative commodities. Broccoli or cauliflower; cabbage; and mustard greens.

(ii) Table. The following Table 1 lists all the commodities included in Crop Group 5 and identifies the related crop subgroups.

Table 1—Crop Group 5: *Brassica* (Cole) Leafy Vegetables

Commodities	Related crop subgroups
Broccoli (*Brassica oleracea* var. *botrytis*)	5-A
Broccoli, Chinese (gai lon) (*Brassica alboglabra*)	5-A
Broccoli raab (rapini) (*Brassica campestris*)	5-B
Brussels sprouts (*Brassica oleracea* var. *gemmifera*)	5-A
Cabbage (*Brassica oleracea*)	5-A
Cabbage, Chinese (bok choy) (*Brassica chinensis*)	5-B
Cabbage, Chinese (napa) (*Brassica pekinensis*)	5-A
Cabbage, Chinese mustard (gai choy) (*Brassica campestris*)	5-A
Cauliflower (*Brassica oleracea* var. *botrytis*)	5-A
Cavalo broccolo (*Brassica oleracea* var. *botrytis*)	5-A
Collards (*Brassica oleracea* var. *acephala*)	5-B
Kale (*Brassica oleracea* var. *acephala*)	5-B
Kohlrabi (*Brassica oleracea* var. *gongylodes*)	5-A
Mizuna (*Brassica rapa* Japonica Group)	5-B
Mustard greens (*Brassica juncea*)	5-B
Mustard spinach (*Brassica rapa* Perviridis Group)	5-B
Rape greens (*Brassica napus*)	5-B

(iii) Table. The following Table 2 identifies the crop subgroups for Crop Group 5, specifies the representative commodity(ies) for each subgroup, and lists all the commodities included in each subgroup.

Table 2—Crop Group 5 Subgroup Listing

Representative commodities	Commodities
Crop Subgroup 5-A. Head and stem *Brassica* subgroup Broccoli or cauliflower; and cabbage	Broccoli; broccoli, Chinese; brussels sprouts; cabbage; cabbage, Chinese (napa); cabbage, Chinese mustard; cauliflower; cavalo broccolo; kohlrabi
Crop Subgroup 5-B. Leafy *Brassica* greens subgroup Mustard greens	Broccoli raab; cabbage, Chinese (bok choy); collards; kale; mizuna; mustard greens; mustard spinach; rape greens

(6) Crop Group 6. Legume Vegetables (Succulent or Dried) Group.

(i) Representative commodities. Bean (*Phaseolus* spp.; one succulent cultivar and one dried cultivar); pea (*Pisum* spp.; one succulent cultivar and one dried cultivar); and soybean.

(ii) Table. The following Table 1 lists all the commodities included in Crop Group 6 and identifies the related crop subgroups.

Table 1—Crop Group 6: Legume Vegetables (Succulent or Dried)

Commodities	Related crop subgroups
Bean (*Lupinus* spp.) (includes grain lupin, sweet lupin, white lupin, and white sweet lupin)	6-C
Bean (*Phaseolus* spp.) (includes field bean, kidney bean, lima bean, navy bean, pinto bean, runner bean, snap bean, tepary bean, wax bean)	6-A, 6-B, 6-C
Bean (*Vigna* spp.) (includes adzuki bean, asparagus bean, black-eyed pea, catjang, Chinese longbean, cowpea, Crowder pea, moth bean, mung bean, rice bean, southern pea, urd bean, yardlong bean)	6-A, 6-B, 6-C
Broad bean (fava bean) (*Vicia faba*)	6-B, 6-C
Chickpea (garbanzo bean) (*Cicer arietinum*)	6-C
Guar (*Cyamopsis tetragonoloba*)	6-C
Jackbean (*Canavalia ensiformis*)	6-A
Lablab bean (hyacinth bean) (*Lablab purpureus*)	6-A
Lentil (*Lens esculenta*)	6-C
Pea (*Pisum* spp.) (includes dwarf pea, edible-pod pea, English pea, field pea, garden pea, green pea, snow pea, sugar snap pea)	6-A, 6-B, 6-C
Pigeon pea (*Cajanus cajan*)	6-A, 6-B, 6-C
Soybean (*Glycine max*)	N/A
Soybean (immature seed) (*Glycine max*)	6-A
Sword bean (*Canavalia gladiata*)	6-A

(iii) Table. The following Table 2 identifies the crop subgroups for Crop Group 6, specifies the representative commodities for each subgroup, and lists all the commodities included in each subgroup.

Table 2—Crop Group 6 Subgroup Listing

Representative commodities	Commodities
Crop Subgroup 6-A. Edible-podded legume vegetables subgroup. Any one succulent cultivar of edible-podded bean (*Phaseolus* spp.) and any one succulent cultivar of edible podded pea (*Pisum* spp.)	Bean (*Phaseolus* spp.) (includes runner bean, snap bean, wax bean); bean (*Vigna* spp.) (includes asparagus bean, Chinese longbean, moth bean, yardlong bean); jackbean; pea (*Pisum* spp.) (includes dwarf pea, edible-pod pea, snow pea, sugar snap pea); pigeon pea; soybean (immature seed); sword bean.
Crop Subgroup 6-B. Succulent pea and bean subgroup. Any succulent shelled cultivar of bean (*Phaseolus* spp.) and garden pea (*Pisum* spp.)	Bean (*Phaseolus* spp.) (includes lima bean (green)); broad bean (succulent); bean (*Vigna* spp.) (includes black-eyed pea, cowpea, southern pea); pea (*Pisum* spp.) (includes English pea, garden pea, green pea); pigeon pea.
Crop Subgroup 6-C. Dried shelled pea and bean (except soybean) subgroup. Any one dried cultivar of bean (*Phaseolus* spp.); and any one dried cultivar of pea (*Pisum* spp.)	Dried cultivars of bean (*Lupinus* spp.) (includes grain lupin, sweet lupin, white lupin, and white sweet lupin); (*Phaseolus* spp.) (includes field bean, kidney bean, lima bean (dry), navy bean, pinto bean; tepary bean; bean (*Vigna* spp.) (includes adzuki bean, blackeyed pea, catjang, cowpea, Crowder pea, moth bean, mung bean, rice bean, southern pea, urd bean); broad bean (dry); chickpea; guar; lablab bean; lentil; pea (*Pisum* spp.) (includes field pea); pigeon pea.

(7) Crop Group 7. Foliage of Legume Vegetables Group.

(i) Representative commodities. Any cultivar of bean (*Phaseolus* spp.), field pea (*Pisum* spp.), and soybean.

(ii) Table. The following Table 1 lists the commodities included in Crop Group 7.

Table 1—Crop Group 7: Foliage of Legume Vegetables Group

Representative commodities	Commodities
Any cultivar of bean (*Phaseolus* spp.) and field pea (*Pisum* spp.), and soybean (*Glycine max*)	Plant parts of any legume vegetable included in the legume vegetables that will be used as animal feed

(iii) Table. The following Table 2 identifies the crop subgroup for Crop Group 7 and specifies the representative commodities for the subgroup, and lists all the commodities included in the subgroup.

Table 2—Crop Group 7 Subgroup Listing

Representative commodities	Commodities
Crop Subgroup 7-A. Foliage of legume vegetables (except soybeans) subgroup	
Any cultivar of bean (*Phaseolus* spp.), and field pea (*Pisum* spp.)	Plant parts of any legume vegetable (except soybeans) included in the legume vegetables group that will be used as animal feed

(8) Crop Group 8. Fruiting Vegetables (Except Cucurbits) Group.

(i) Representative commodities. Tomato, bell pepper, and one cultivar of non-bell pepper.

(ii) Commodities. The following is a list of all the commodities included in Crop Group 8:

Crop Group 8: Fruiting Vegetables (Except Cucurbits)—Commodities

Eggplant (*Solanum melongena*)
Groundcherry (*Physalis* spp.)
Pepino (*Solanum muricatum*)
Pepper (*Capsicum* spp.) (includes bell pepper, chili pepper, cooking pepper, pimento, sweet pepper)
Tomatillo (*Physalis ixocarpa*)
Tomato (*Lycopersicon esculentum*)

(9) Crop Group 9. Cucurbit Vegetables Group.

(i) Representative commodities. Cucumber, muskmelon, and summer squash.

(ii) Table. The following Table 1 lists all the commodities included in Crop Group 9 and identifies the related subgroups.

Table 1—Crop Group 9: Cucurbit Vegetables

Commodities	Related crop subgroups
Chayote (fruit) (*Sechium edule*)	9-B
Chinese waxgourd (Chinese preserving melon) (*Benincasa hispida*)	9-B
Citron melon (*Citrullus lanatus* var. *citroides*)	9-A
Cucumber (*Cucumis sativus*)	9-B
Gherkin (*Cucumis anguria*)	9-B
Gourd, edible (*Lagenaria* spp.) (includes hyotan, cucuzza); (*Luffa acutangula*, L. *cylindrica*) (includes hechima, Chinese okra)	9-B
Momordica spp. (includes balsam apple, balsam pear, bitter melon, Chinese cucumber)	9-B
Muskmelon (hybrids and/or cultivars of *Cucumis melo*) (includes true cantaloupe, cantaloupe, casaba, crenshaw melon, golden pershaw melon, honeydew melon, honey balls, mango melon, Persian melon, pineapple melon, Santa Claus melon, and snake melon)	9-A
Pumpkin (*Cucurbita* spp.)	9-B
Squash, summer (*Cucurbita pepo* var. *melopepo*) (includes crookneck squash, scallop squash, straightneck squash, vegetable marrow, zucchini)	9-B
Squash, winter (*Cucurbita maxima*; C. *moschata*) (includes butternut squash, calabaza, hubbard squash); (C. *mixta*; C. *pepo*) (includes acorn squash, spaghetti squash)	9-B
Watermelon (includes hybrids and/or varieties of *Citrullus lanatus*)	9-A

(iii) Table. The following Table 2 identifies the crop subgroups for Crop Group 9, specifies the representative commodities for each subgroup, and lists all the commodities included in each subgroup.

Table 2—Crop Group 9 Subgroup Listing

Representative commodities	Commodities
Crop Subgroup 9-A. Melon subgroup. Cantaloupes	Citron melon; muskmelon; watermelon
Crop Subgroup 9-B. Squash/cucumber subgroup. One cultivar of summer squash and cucumber	Chayote (fruit); Chinese wax gourd; cucumber; gherkin; gourd, edible; *Momordica* spp.; pumpkin; squash, summer; squash, winter.

(10) Crop Group 10. Citrus Fruits (*Citrus* spp., *Fortunella* spp.) Group.

(i) Representative commodities. Sweet orange; lemon and grapefruit.

(ii) Commodities. The following is a list of all the commodities in Crop Group 10:

Crop Group 10: Citrus Fruits (*Citrus* spp., *Fortunella* spp.) Group—Commodities

Calamondin (*Citrus mitis* X *Citrofortunella mitis*)
Citrus citron (*Citrus medica*)
Citrus hybrids (*Citrus* spp.) (includes chironja, tangelo, tangor)
Grapefruit (*Citrus paradisi*)
Kumquat (*Fortunella* spp.)
Lemon (*Citrus jambhiri, Citrus limon*)
Lime (*Citrus aurantiifolia*)
Mandarin (tangerine) (*Citrus reticulata*)
Orange, sour (*Citrus aurantium*)
Orange, sweet (*Citrus sinensis*)
Pummelo (*Citrus grandis, Citrus maxima*)
Satsuma mandarin (*Citrus unshiu*)

(11) Crop Group 11: Pome Fruits Group.
(i) Representative commodities. Apple and pear.
(ii) Commodities. The following is a list of all the commodities included in Crop Group 11:

Crop Group 11: Pome Fruits Group—Commodities

Apple (*Malus domestica*)
Crabapple (*Malus* spp.)
Loquat (*Eriobotrya japonica*)
Mayhaw (*Crataegus aestivalis, C. opaca*, and *C. rufula*)
Pear (*Pyrus communis*)
Pear, oriental (*Pyrus pyrifolia*)
Quince (*Cydonia oblonga*)

(12) Crop Group 12. Stone Fruits Group.
(i) Representative commodities. Sweet cherry or tart cherry; peach; and plum or fresh prune (*Prunus domestica, Prunus* spp.)
(ii) Commodities. The following is a list of all the commodities included in Crop Group 12:

Crop Group 12: Stone Fruits Group—Commodities

Apricot (*Prunus armeniaca*)
Cherry, sweet (*Prunus avium*)
Cherry, tart (*Prunus cerasus*)
Nectarine (*Prunus persica*)
Peach (*Prunus persica*)
Plum (*Prunus domestica, Prunus* spp.)
Plum, Chickasaw (*Prunus angustifolia*)
Plum, Damson (*Prunus domestica* spp. *insititia*)
Plum, Japanese (*Prunus salicina*)
Plumcot (*Prunus armeniaca* x *P. domestica*)
Prune (fresh) (*Prunus domestica, Prunus* spp.)

(13) Crop Group 13. Berries Group.
(i) Representative commodities. Any one blackberry or any one raspberry; and blueberry.
(ii) Table. The following Table 1 lists all the commodities included in Crop Group 13 and identifies the related subgroups.

Table 1—Crop Group 13: Berries Group

Commodities	Related crop subgroups
Blackberry (*Rubus eubatus*) (including bingleberry, black satin berry, boysenberry, Cherokee blackberry, Chesterberry, Cheyenne blackberry, coryberry, darrowberry, dewberry, Dirksen thornless berry, Himalayaberry, hullberry, Lavacaberry, lowberry, Lucretiaberry, mammoth blackberry, marionberry, nectarberry, olallieberry, Oregon evergreen berry, phenomenalberry, rangeberry, ravenberry, rossberry, Shawnee blackberry, youngberry, and varieties and/or hybrids of these)	13-A
Blueberry (*Vaccinium* spp.)	13-B
Currant (*Ribes* spp.)	13-B
Elderberry (*Sambucus* spp.)	13-B
Gooseberry (*Ribes* spp.)	13-B
Huckleberry (*Gaylussacia* spp.)	13-B
Loganberry (*Rubus loganobaccus*)	13-A
Raspberry, black and red (*Rubus occidentalis, Rubus strigosus, Rubus idaeus*)	13-A

(iii) Table. The following Table 2 identifies the crop subgroups for Crop Group 13, specifies the representative commodities for each subgroup, and lists all the commodities included in each subgroup.

Table 2—Crop Group 13 Subgroups Listing

Representative commodities	Commodities
Crop Subgroup 13-A. Caneberry (blackberry and raspberry) subgroup.	
Any one blackberry or any one raspberry	Blackberry; loganberry; red and black raspberry; cultivars and/or hybrids of these
Crop Subgroup 13-B. Bushberry subgroup.	
Blueberry, highbush	Blueberry, highbush and lowbush; currant; elderberry; gooseberry; huckleberry

(14) Crop Group 14. Tree Nuts Group.
(i) Representative commodities. Almond and pecan.
(ii) Commodities. The following is a list of all the commodities included in Crop Group 14:

Crop Group 14: Tree Nuts—Commodities

Almond (*Prunus dulcis*)
Beech nut (*Fagus* spp.)
Brazil nut (*Bertholletia excelsa*)
Butternut (*Juglans cinerea*)
Cashew (*Anacardium occidentale*)
Chestnut (*Castanea* spp.)
Chinquapin (*Castanea pumila*)
Filbert (hazelnut) (*Corylus* spp.)
Hickory nut (*Carya* spp.)

Macadamia nut (bush nut) (*Macadamia* spp.)
Pecan (*Carya illinoensis*)
Walnut, black and English (Persian) (*Juglans* spp.)

(15) Crop Group 15. Cereal Grains Group.

(i) Representative commodities. Corn (fresh sweet corn and dried field corn), rice, sorghum, and wheat.

(ii) Commodities. The following is a list of all the commodities included in Crop Group 15:

Crop Group 15: Cereal Grains—Commodities
Barley (*Hordeum* spp.)
Buckwheat (*Fagopyrum esculentum*)
Corn (*Zea mays*)
Millet, pearl (*Pennisetum glaucum*)
Millet, proso (*Panicum milliaceum*)
Oats (*Avena* spp.)
Popcorn (*Zea mays* var. *everta*)
Rice (*Oryza sativa*)
Rye (*Secale cereale*)
Sorghum (milo) (*Sorghum* spp.)
Teosinte (*Euchlaena mexicana*)
Triticale (*Triticum-Secale* hybrids)
Wheat (*Triticum* spp.)
Wild rice (*Zizania aquatica*)

(16) Crop Group 16. Forage, Fodder and Straw of Cereal Grains Group.

(i) Representative commodities. Corn, wheat, and any other cereal grain crop.

(ii) Commodities. The commodities included in Crop Group 16 are: Forage, fodder, and straw of all commodities included in the group cereal grains group.

(17) Crop Group 17. Grass Forage, Fodder, and Hay Group.

(i) Representative commodities. Bermuda grass; bluegrass; and bromegrass or fescue.

(ii) Commodities. The commodities included in Crop Group 17 are: Any grass, Gramineae family (either green or cured) except sugarcane and those included in the cereal grains group, that will be fed to or grazed by livestock, all pasture and range grasses and grasses grown for hay or silage.

(18) Crop Group 18. Nongrass Animal Feeds (Forage, Fodder, Straw, and Hay) Group.

(i) Representative commodities. Alfalfa and clover (*Trifolium* spp.)

(ii) Commodities. The following is a list of all the commodities included in Crop Group 18:

Crop Group 18: Nongrass Animal Feeds (Forage, Fodder, Straw, and Hay) Group—Commodities
Alfalfa (*Medicago sativa* subsp. *sativa*)
Bean, velvet (*Mucuna pruriens* var. *utilis*)
Clover (*Trifolium* spp., *Melilotus* spp.)
Kudzu (*Pueraria lobata*)
Lespedeza (*Lespedeza* spp.)
Lupin (*Lupinus* spp.)

Sainfoin (*Onobrychis viciifolia*)
Trefoil (*Lotus* spp.)
Vetch (*Vicia* spp.)
Vetch, crown (*Coronilla varia*)
Vetch, milk (*Astragalus* spp).

(19) Crop Group 19. Herbs and Spices Group.

(i) Representative commodities. Basil (fresh and dried); black pepper; chive; and celery seed or dill seed.

(ii) Table. The following Table 1 lists all the commodities included in Crop Group 19 and identifies the related subgroups.

Table 1—Crop Group 19: Herbs and Spices Group

Commodities	Related crop subgroups
Allspice (*Pimenta dioica*)	19-B
Angelica (*Angelica archangelica*)	19-A
Anise (anise seed) (*Pimpinella anisum*)	19-B
Anise, star (*Illicium verum*)	19-B
Annatto (seed)	19-B
Balm (lemon balm) (*Melissa officinalis*)	19-A
Basil (*Ocimum basilicum*)	19-A
Borage (*Borago officinalis*)	19-A
Burnet (*Sanguisorba minor*)	19-A
Camomile (*Anthemis nobilis*)	19-A
Caper buds (*Capparis spinosa*)	19-B
Caraway (*Carum carvi*)	19-B
Caraway, black (*Nigella sativa*)	19-B
Cardamom (*Elettaria cardamomum*)	19-B
Cassia bark (*Cinnamomum aromaticum*)	19-B
Cassia buds (*Cinnamomum aromaticum*)	19-B
Catnip (*Nepeta cataria*)	19-A
Celery seed (*Apicum graveolens*)	19-B
Chervil (dried) (*Anthriscus cerefolium*)	19-A
Chive (*Allium schoenoprasum*)	19-A
Chive, Chinese (*Allium tuberosum*)	19-A
Cinnamon (*Cinnamomum verum*)	19-B
Clary (*Salvia sclarea*)	19-A
Clove buds (*Eugenia caryophyllata*)	19-B
Coriander (cilantro or Chinese parsley) (leaf) (*Coriandrum sativum*)	19-A
Coriander (cilantro) (seed) (*Coriandrum sativum*)	19-B
Costmary (*Chrysanthemum balsamita*)	19-A
Culantro (leaf) (*Eryngium foetidum*)	19-A
Culantro (seed) (*Eryngium foetidum*)	19-B
Cumin (*Cuminum cyminum*)	19-B
Curry (leaf) (*Murraya koenigii*)	19-A
Dill (dillweed) (*Anethum graveolens*)	19-A
Dill (seed) (*Anethum graveolens*)	19-B
Fennel (common) (*Foeniculum vulgare*)	19-B
Fennel, Florence (seed) (*Foeniculum vulgare* Azoricum Group)	19-B
Fenugreek (*Trigonella foenumgraecum*)	19-B
Grains of paradise (*Aframomum melegueta*)	19-B
Horehound (*Marrubium vulgare*)	19-A
Hyssop (*Hyssopus officinalis*)	19-A
Juniper berry (*Juniperus communis*)	19-B

Lavender (*Lavandula officinalis*). 19-A
Lemongrass (*Cymbopogon citratus*) 19-A
Lovage (leaf) (*Levisticum officinale*) 19-A
Lovage (seed) (*Levisticum officinale*). 19-B
Mace (*Myristica fragrans*). 19-B
Marigold (*Calendula officinalis*) 19-A
Marjoram (*Origanum* spp.) (includes sweet or annual marjoram,
    wild marjoram or oregano, and pot marjoram) 19-A
Mustard (seed) (*Brassica juncea, B. hirta, B. nigra*) 19-B
Nasturtium (*Tropaeolum majus*) 19-A
Nutmeg (*Myristica fragrans*). 19-B
Parsley (dried) (*Petroselinum crispum*) 19-A
Pennyroyal (*Mentha pulegium*) 19-A
Pepper, black (*Piper nigrum*). 19-B
Pepper, white. 19-B
Poppy (seed) (*Papaver somniferum*) 19-B
Rosemary (*Rosemarinus officinalis*) 19-A
Rue (*Ruta graveolens*). 19-A
Saffron (*Crocus sativus*) 19-B
Sage (*Salvia officinalis*). 19-A
Savory, summer and winter (*Satureja* spp.) 19-A
Sweet bay (bay leaf) (*Laurus nobilis*) 19-A
Tansy (*Tanacetum vulgare*) 19-A
Tarragon (*Artemisia dracunculus*) 19-A
Thyme (*Thymus* spp.) 19-A
Vanilla (*Vanilla planifolia*) 19-B
Wintergreen (*Gaultheria procumbens*). 19-A
Woodruff (*Galium odorata*). 19-A
Wormwood (*Artemisia absinthium*). 19-A

(iii) Table. The following Table 2 identifies the crop subgroups for Crop Group 19, specifies the representative commodities for each subgroup, and lists all the commodities included in each subgroup.

Table 2—Crop Group 19 Subgroups

Representative commodities	Commodities
Crop Subgroup 19-A. Herb subgroup. Basil (fresh and dried) and chive	Angelica; balm; basil; borage; burnet; camomile; catnip; chervil (dried); chive; chive, Chinese, clary; coriander (leaf); costmary; culantro (leaf); curry (leaf); dillweed; horehound; hyssop; lavender; lemongrass; lovage (leaf); marigold; marjoram (*Origanum* spp.); nasturtium; parsley (dried); pennyroyal; rosemary; rue; sage; savory, summer and winter; sweet bay; tansy; tarragon; thyme; wintergreen; woodruff; and wormwood.
Crop Subgroup 19-B. Spice subgroup Black pepper; and celery seed or dill seed	Allspice; anise (seed); anise, star; annatto (seed); caper (buds); caraway; caraway, black; cardamom; cassia (buds); celery (seed); cinnamon; clove (buds); coriander (seed); culantro (seed); cumin; dill (seed); fennel, common; fennel, Florence (seed); fenugreek; grains of paradise; juniper (berry); lovage (seed); mace; mustard (seed); nutmeg; pepper, black; pepper, white; poppy (seed); saffron; and vanilla

Codex Alimentarius Commission

Classification of Foods and Animal Feeds

Food and Agriculture Organization of the United Nations

World Health Organization
24 June, 1996

Index of Classes, Types and Groups of Commodities

Class A Primary Food Commodities of Plant Origin

Type	No.	Group	Group Letter Code
01 Fruits	001	Citrus fruits	FC
	002	Pome fruits	FP
	003	Stone fruits	FS
	004	Berries and other small fruits	FB
	005	Assorted tropical and sub-tropical fruits – edible peel	FT
	006	Assorted tropical and sub-tropical fruits – inedible peel	FI
	007	Reserved	
	008	Reserved	
02 Vegetables	009	Bulb vegetables	VA
	010	Brassica (cole or cabbage) vegetables, Head cabbages, Flowerhead brassicas	VB
	011	Fruiting vegetables, Cucurbits	VC
	012	Fruiting vegetables, other than Cucurbits	VO
	013	Leafy vegetables (including Brassica leafy vegetables)	VL
	014	Legume vegetables	VP
	015	Pulses	VD
	016	Root and tuber vegetables	VR
	017	Stalk and stem vegetables	VS
	018	Reserved	
	019	Reserved	
03 Grasses	020	Cereal grains	GC
	021	Grasses, for sugar or syrup production	GS
04 Nuts and Seeds	022	Tree nuts	TN
	023	Oilseed	SO
	024	Seed for beverages and sweets	SB
	025	Reserved	
	026	Reserved	
05 Herbs and Spices	027	Herbs	HH
	028	Spices	HS
	029	Reserved	

Class B Primary Food Commodities of Animal Origin

Type	No.	Group	Group Letter Code
06 Mammalian products	030	Meat (from mammals other than marine mammals)	MM
	031	Mammalian fats	MF
	032	Edible offal (mammalian)	MO
	033	Milks	ML
	034	Reserved	
	035	Reserved	
07 Poultry products	036	Poultry meat (including Pigeon meat)	PM
	037	Poultry fats	PF
	038	Poultry, Edible offal of	PO
	039	Eggs	PE
08 Aquatic animal products	040	Freshwater fish	WF
	041	Diadromous fish	WD
	042	Marine fish	WS
	043	Fish roe (including milt = soft roe) and edible offal of fish	roe WR, offal WL
	044	Marine mammals	WH
	045	Crustaceans	WC
	046	Reserved	
	047	Reserved	
09 Amphibians and reptiles	048	Frogs, lizards, snakes and turtles	AR
10 Invertebrate animals	049	Molluscs (including Cephalopods) and other invertebrate animals	IM

Class C Primary Animal Feed Commodities

Type	No.	Group	Group Letter Code
11 Primary feed commodities of plant origin	050	Legume animal feeds	AL
	051	Straw, fodder and forage of cereal grains and grasses (including buckwheat fodder) (forage)	AF

Type	No.	Group	Group Letter Code
	051	Straw, fodder and forage of cereal grains and grasses (including buckwheat fodder) (straws and fodder dry)	AS
	052	Miscellaneous Fodder and Forage crops (fodder)	AM
	052	Miscellaneous Fodder and Forage crops (forage)	AV
	053	Reserved	
	054	Reserved	

Class D Processed Foods of Plant Origin

Type	No.	Group	Group Letter Code
12 Secondary food commodities of plant origin	055	Dried fruits	DF
	056	Dried vegetables	DV
	057	Dried herbs	DH
	058	Milled cereal products (early milling stages)	CM
	059	Miscellaneous secondary food commodities of plant origin	SM
	060	Reserved	
	061	Reserved	
	062	Reserved	
	063	Reserved	
	064	Reserved	
13 Derived products of plant origin	065	Cereal grain milling fractions	CF
	066	Teas	DT
	067	Vegetable oils, crude	OC
	068	Vegetable oils, edible (or refined)	OR
	069	Miscellaneous derived edible products of plant origin	DM
	070	Fruit juices	JF
	071	By-products, used for animal feeding purposes, derived from fruit and vegetable processing	AB
	072	Reserved	

Type	No.	Group	Group Letter Code
	073	Reserved	
	074	Reserved	
14 Manufactured foods (single-ingredient) of plant origin	075	Reserved	
	076	Reserved	
	077	Reserved	
15 Manufactured foods (multi-ingredient) of plant origin	078	Manufactured multi-ingredient cereal products	CP
	079	Reserved	

Class E Processed Foods of Animal Origin

Type	No.	Group	Group Letter Code
16 Secondary food commodities of animal origin	080	Dried meat and fish products	MD
	081	Reserved	
	082	Secondary milk products	LS
	083	Reserved	
17 Derived edible products of animal origin	084	Crustaceans, processed	SC
	085	Animal fats, processed	FA
	086	Milk fats	FM
	087	Derived milk products	LD
	088	Reserved	
	089	Reserved	
18 Manufactured food (single-ingredient) of animal origin	090	Manufactured milk products (single-ingredient)	LI
	091	Reserved	
19 Manufactured food (multi-ingredient) of animal origin	092	Manufactured milk products (multi-ingredient)	LM
	093	Reserved	

Codex Classification of Foods and Feeds
Foods and Feeds of Plant Origin

AB 0001	Citrus pulp, dry		AM 5255	Mangel or Mangold, see Fodder beet
AB 0226	Apple pomace, dry		AS 0081	Straw and fodder (dry) of cereal grains
AB 0269	Grape pomace, dry		AS 0061	Straw, fodder (dry) and hay of cereal
AB 0596	Sugar beet pulp, dry			grains and other
AB 1201	Sugar beet pulp, wet		AS 0162	Hay or fodder (dry) of grasses
AF 0645	Maize forage		AS 0640	Barley straw and fodder, dry
AF 0647	Oat forage (green)		AS 0641	Buckwheat fodder
AF 0650	Rye forage (green)		AS 0645	Maize fodder
AF 0651	Sorghum forage (green)		AS 0646	Millet fodder, dry
AF 5249	Corn forage		AS 0647	Oat straw and fodder, dry
AL 0061	Bean fodder		AS 0649	Rice straw and fodder, dry
AL 0072	Pea hay or Pea fodder (dry)		AS 0650	Rye Straw and fodder, dry
AL 0157	Legume animal feeds		AS 0651	Sorghum straw and fodder, dry
AL 0524	Chick-pea fodder		AS 0654	Wheat straw and fodder, dry
AL 0528	Pea vines (green)		AS 0657	Teosinte fodder
AL 0541	Soya bean fodder		AS 5241	Bermuda grass
AL 0545	Lupin, forage		AS 5243	Bluegrass
AL 0697	Peanut fodder		AS 5245	Brome grass
AL 1020	Alfalfa fodder		AS 5247	Corn fodder
AL 1021	Alfalfa forage (green)		AS 5251	Darnel
AL 1022	Bean, Velvet		AS 5253	Fescue
AL 1023	Clover		AV 0353	Pineapple Forage
AL 1024	Kudzu		AV 0480	Kale Forage
AL 1025	Lespedeza		AV 0506	Turnip leaves or tops
AL 1027	Sainfoin		AV 0596	Sugar beet leaves or tops
AL 1028	Trefoil		AV 0659	Sugar cane forage
AL 1029	Vetch		AV 1050	Cow cabbage
AL 1030	Bean forage (green)		AV 1051	Fodder beet leaves or tops
AL 1031	Clover hay or fodder		AV 1052	Marrow-stem cabbage or Marrow-stem kale
AL 1265	Soya bean forage (green)		CF 0081	Cereal brans, processed
AL 1270	Peanut forage (green)		CF 0645	Maize meal
AL 5217	Chickling Vetch, see Vetch, Chickling		CF 0649	Rice bran, processed
AL 5219	Grass pea, see Vetch, Chickling		CF 0650	Rye bran, processed
AL 5221	Kudzu, Tropical, see Kudzu		CF 0654	Wheat bran, processed
AL 5223	Melilot, see Clovers		CF 1210	Wheat germ
AL 5225	One extra no. for alfalfa products: AL 1021		CF 1211	Wheat flour
	(alfalfa forage, green)		CF 1212	Wheat wholemeal
AL 5227	Puero, see Kudzu, Tropical		CF 1250	Rye flour
AL 5229	Sericea, see Lespedeza		CF 1251	Rye wholemeal
AL 5231	Tropical kudzu, see Kudzu, Tropical		CF 1255	Maize flour
AL 5233	Velvet bean, see Bean, Velvet		CF 5273	Corn flour, see Maize flour
AL 5235	Vetch, Chickling, see Vetch		CF 5275	Corn meal, see Maize meal
AL 5237	Vetch, Crown, see Vetch		CM 0081	Bran, unprocessed of cereal grain
AL 5239	Vetch, Milk, see Vetch		CM 0649	Rice, husked
AM 0165	Miscellaneous fodder and forage crops		CM 0650	Rye bran, unprocessed
AM 0353	Pineapple fodder		CM 0654	Wheat bran, unprocessed
AM 0497	Swedish turnip or Swede fodder		CM 1205	Rice, polished
AM 0506	Turnip fodder		CM 1206	Rice bran, unprocessed
AM 0659	Sugar cane fodder		CP 0179	Bread and other cooked cereal products
AM 0660	Almond hulls		CP 0645	Maize bread
AM 0691	Cotton fodder, dry		CP 1211	White bread
AM 0738	Mint hay		CP 1212	Wholemeal bread
AM 1051	Fodder beet		CP 1250	Rye bread

CP 5295	Corn bread, see Maize bread	DT 5279	Camomile, Roman or Noble, see Camomile
DF 0014	Prunes	DT 5281	Mayweed, Scented, see Camomile, German
DF 0167	Dried fruits	DT 5283	Paraguay tea, see Maté
DF 0226	Apples, dried	DT 5285	Peppermint tea (succulent or dry leaves)
DF 0240	Apricots, dried		see Peppermint
DF 0269	Dried grapes (= Currants, Raisins and Sultanas)	DT 5287	*T. bohea* L.; *T. viridis* L.
DF 0295	Dates, dried or dried and candied	DV 0168	Dried vegetables
DF 0297	Figs, dried or dried and candied	FB 0018	Berries and other small fruits
DF 5257	Currants	FB 0019	Vaccinium berries, including bearberry
DF 5259	Dried vine fruits, see Dried grapes	FB 0020	Blueberries
DF 5261	Muscatel, see Dried grapes	FB 0021	Currants, Black, Red, White
DF 5263	Raisins (seedless white grape var., partially	FB 0260	Bearberry
	dried) see Dried grapes	FB 0261	Bilberry
DF 5265	Sultanas, see Dried grapes	FB 0262	Bilberry, Bog
DH 0007	Bay leaves, dry	FB 0263	Bilberry, Red
DH 0170	Dried herbs	FB 0264	Blackberries
DH 0624	Celery leaves, dry	FB 0265	Cranberry
DH 0720	Angelica, including Garden Angelica, dry	FB 0266	Dewberries (including Boysenberry
DH 0721	Balm leaves, dry		and Loganberry)
DH 0722	Basil, dry	FB 0267	Elderberries
DH 0724	Borage, dry	FB 0268	Gooseberry
DH 0726	Catmint, dry	FB 0269	Grapes
DH 0728	Burning bush, dry	FB 0270	Juneberries
DH 0731	Fennel, dry	FB 0271	Mulberries
DH 0732	Horehound, dry	FB 0272	Raspberries, Red, Black
DH 0733	Hyssop, dry	FB 0273	Rose hips
DH 0734	Lavender, dry	FB 0274	Service berries
DH 0736	Marjoram, dry	FB 0275	Strawberry
DH 0738	Mints, dry	FB 0276	Strawberries, Wild
DH 0741	Rosemary, dry	FB 0277	Cloudberry
DH 0742	Rue, dry	FB 0279	Currant, Red, White, see also Currants,
DH 0743	Sage, dry		Black, Red, White
DH 0745	Savory, Summer; Winter, dry	FB 1235	Table-grapes
DH 0747	Sweet cicely, dry	FB 4073	Blueberry, Highbush, see Blueberries
DH 0748	Tansy and related species, dry	FB 4075	Blueberry, Lowbush, see Blueberries
DH 0750	Thyme, dry	FB 4077	Blueberry, Rabbiteye, see Blueberries
DH 0752	Wintergreen leaves, dry	FB 4079	Boysenberry, see Dewberries
DH 0753	Woodruff, dry	FB 4081	Cowberry, see Bilberry, Red
DH 0754	Wormwoods, dry	FB 4083	Huckleberries
DH 0755	Lovage, dry	FB 4085	Loganberry, see Dewberries
DH 1100	Hops, dry	FB 4087	Olallieberry, see Dewberries
DH 5269	Cretan Dittany, dry, see Burning bush, dry	FB 4089	Service Berries
DH 5271	Oregano (= Wild Marjoram), dry, see Marjoram	FB 4091	Strawberry, Musky, see Strawberries, Wild
DM 0001	Citrus Molasses	FB 4093	Whortleberry, Red, see Bilberry, Red
DM 0305	Olives, processed	FC 0001	Citrus Fruit
DM 0596	Sugar beet molasses	FC 0002	Lemons and Limes (including Citron)
DM 0658	Sorghum molasses	FC 0003	Mandarins (including Mandarin-like hybrids)
DM 0659	Sugar cane molasses	FC 0004	Oranges, Sweet, Sour (including
DM 0715	Cocoa powder		Orange-like hybrids)
DM 1215	Cocoa butter	FC 0005	Shaddocks or Pomelos (including
DM 1216	Cocoa mass		Shaddock-like hybrids)
DT 0171	Teas (Tea and Herb teas)	FC 0201	Calamondin, see also Subgroup 0003 Mandarins
DT 0446	Roselle (calyx and flowers), dry	FC 0202	Citron, see also Subgroup 0002, Lemons and Limes
DT 1110	Camomile or Chamomile	FC 0203	Grapefruit, see also Subgroup 0005
DT 1111	Lemon verbena (dry leaves)		Shaddocks or Pomelos
DT 1112	Lime blossoms	FC 0204	Lemon, see also Subgroup 0002 Lemons and Limes
DT 1113	Maté (dry leaves)	FB 0278	Currant, Black, see also Currants,
DT 1114	Tea, Green, Black (black, fermented and dried)		Black, Red, White
DT 5277	Camomile, German or Scented, see Camomile	FC 0205	Lime, see also Subgroup 0002 Lemons and Limes

FC 0206	Mandarin, see also Subgroup 0003 Mandarins	FI 0345	Mango
FC 0207	Orange, Sour, see also Subgroup 0004 Oranges, Sweet, Sour	FI 0346	Mangostan
		FI 0347	Marmaladedos
FC 0208	Orange, Sweet, see also Subgroup 0004 Oranges, Sweet, Sour	FI 0348	Mombin, Yellow
		FI 0349	Naranjilla
FC 0209	Shaddock, see also Subgroup 0005 Shaddocks or Pomelos	FI 0350	Papaya
		FI 0351	Passion Fruit
FC 4000	Bigarade, see Orange, Sour	FI 0352	Persimmon, American
FC 4001	Blood orange, see Orange, Sweet	FI 0353	Pineapple
FC 4002	Chinotto, see Orange, Sour	FI 0354	Plantain
FC 4003	Chironja, see Subgroup Oranges, Sweet, Sour	FI 0355	Pomegranate
FC 4005	Clementine, see Mandarin	FI 0356	Prickly pear
FC 4006	Cleopatra mandarin, see Subgroup 0003 Mandarins	FI 0357	Pulasan
FC 4007	Dancy or Dancy mandarin, see Subgroup 0003 Mandarins	FI 0358	Rambutan
		FI 0359	Sapodilla
FC 4008	King Mandarin, see Subgroup 0003 Mandarins	FI 0360	Sapote, Black
FC 4011	Malta orange, see Blood Orange	FI 0361	Sapote, Green
FC 4014	Mediterranean mandarin, see Subgroup 0003 Mandarins	FI 0362	Sapote, Mammey
		FI 0363	Sapote, White
FC 4016	Myrtle-leaf orange, see Chinotto	FI 0364	Sentul
FC 4018	Natsudaidai, see Subgroup 0005 Shaddocks or Pomelos	FI 0365	Soursop
		FI 0366	Spanish lime
FC 4019	Orange, Bitter, see Orange, Sour	FI 0367	Star apple
FC 4020	Pomelo, see Shaddocks or Pomelos	FI 0068	Sugar apple
FC 4022	Satsuma or Satsuma mandarin, see Subgroup 0003 Mandarin	FI 0369	Tamarind
		FI 0371	Elephant apple
FC 4024	Seville Orange, see Orange, Sour	FI 4127	Chinese gooseberry, see Kiwifruit
FC 4027	Tangerine, see Subgroup 0003 Mandarins	FI 4129	Egg fruit, see Canistel
FC 4029	Tangelo, large-sized cultivars, see Subgroup 0005	FI 4131	Genip, see Marmaladebox
FC 4031	Tangelo, small and medium sized cultivars, see Subgroup 0003	FI 4133	Indian Fig, see Prickly pear
		FI 4135	Lulo, see Naranjilla
FC 4033	Tangelolo, see Subgroup 0005 Shaddocks or Pomelos	FI 4137	Mangosteen, see Mangostan
		FI 4139	Papaw or Pawpaw, see Papaya
FC 4035	Tangors, see Subgroup 0003 Mandarins	FI 4141	Persimmon, Japanese
FC 4037	Tankan mandarin, see Subgroup 0003 Mandarins	FI 4143	Pineapple guava, see Feijoa
FC 4039	Ugli, see Subgroup 0005 Shaddocks or Pomelos	FI 4145	Quito orange, see Naranjilla
FC 4041	Willowleaf mandarin, see Mediterranean mandarin	FI 4147	Sesso vegetal, see Akee apple
FI 0030	Assorted tropical and subtropical Fruits – inedible peel	FI 4149	Strawberry peach, see Kiwifruit
		FI 4151	Sweetsop, see Sugar apple
FI 0325	Akee apple	FP 0009	Pome fruits
FI 0326	Avocado	FP 0226	Apple
FI 0327	Banana	FP 0227	Crab-apple
FI 0328	Banana, dwarf	FP 0228	Loquat
FI 0329	Breadfruit	FP 0229	Medlar
FI 0330	Canistel	FP 0230	Pear
FI 0331	Cherimoya	FP 0231	Quince
FI 0332	Custard Apple	FP 4044	Japanese medlar, see Loquat
FI 0333	Doum or Dum palm	FP 4047	Nashi pear, see Pear, Oriental
FI 0334	Durian	FP 4049	Pear, Oriental, see Pear
FI 0335	Feijoa	FP 4051	Sand pear, see Pear, Oriental
FI 0336	Guava	FS 0012	Stone fruits
FI 0337	Ilama	FS 0013	Cherries
FI 0338	Jackfruit	FS 0014	Plums (including Prunes)
FI 0339	Jambolan	FS 0240	Apricot
FI 0340	Java Apple	FS 0241	Bullace
FI 0341	Kiwifruit	FS 0242	Cherry plum
FI 0342	Longan	FS 0243	Cherry, Sour
FI 0343	Litchi	FS 0244	Cherry, Sweet
FI 0344	Mammey Apple	FS 0245	Nectarine

FS 0246	Morello		FT 4125	Tree strawberry, see Arbutus berry
FS 0247	Peach		GC 0080	Cereal grains
FS 0248	Plum, Chickasaw		GC 0081	Cereal grains, except Buckwheat,
FS 0249	Sloe			Canihun and Quinoa
FS 4053	Chicksaw plum, see Plum, Chickasaw		GC 0640	Barley
FS 4055	Damsons (Damsons plums) see Plum, Damson		GC 0641	Buckwheat
FS 4057	Mirabelle, see Plum, Mirabelle		GC 0642	Canihua
FS 4059	Myrobolan plum, see Cherry plum		GC 0643	Hungry rice
FS 4061	Plum, American, see Sloe		GC 0644	Job's tears
FS 4063	Plum, Damason, see Bullace		GC 0645	Maize
FS 4065	Plum, Greengage, see Plums		GC 0646	Millet
FS 4067	*Prunus insititia* L. var. *italica* (Borkh.) L.M. Neum.		GC 0647	Oats
FS 4069	Plum, Japanese, see Plums		GC 0648	Quinoa
FS 4071	Plum, Mirabelle, see Bullace		GC 0649	Rice
FT 0026	Assorted tropical and sub-tropical		GC 0650	Rye
	fruits – edible peel		GC 0651	Sorghum
FT 0285	Ambarella		GC 0652	Teff or Tef
FT 0286	Arbutus berry		GC 0653	Triticale
FT 0287	Barbados cherry		GC 0654	Wheat
FT 0288	Bilimbi		GC 0655	Wild rice
FT 0289	Carambola		GC 0656	Popcorn
FT 0290	Caranda		GC 0657	Teosinte
FT 0291	Carob		GC 4597	Acha, see Hungry Rice
FT 0292	Cashew apple		GC 4599	Adlay, see Job's tears
FT 0293	Chinese olive, Black, White		GC 4601	African millet, see Millet, Finger
FT 0294	Coco plum		GC 4603	Brown-corn millet, see Millet, Common
FT 0295	Date		GC 4605	Extra no. for commodities
FT 0296	Desert date			derived from cereal grains
FT 0297	Fig		GC 4607	Bulrush millet, see Millet, Bulrush
FT 0298	Grumichama		GC 4609	Cat-tail millet, see Millet, Bulrush
FT 0299	Hog plum		GC 4611	Chicken corn, see Sorghum
FT 0300	Jaboticaba		GC 4613	Corn, see Maize
FT 0301	Jujube, Indian		GC 4615	Corn-on-the-cob, (Codex Stand. 133-1981)
FT 0302	Jujube, Chinese		GC 4617	Corn, whole kernel (Codex Stand. 132-1981)
FT 0303	Kumquats		GC 4619	Dari see, see Sorghum
FT 0304	Natal Plum		GC 4621	Durra, see Sorghum
FT 0305	Olives		GC 4623	Durum wheat, see Wheat
FT 0306	Otaheite gooseberry		GC 4625	Emmer, see Wheat
FT 0307	Persimmon, Japanese		GC 4627	Feterita, see Sorghum
FT 0308	Pomerac		GC 4629	Finger millet, see Millet, Finger
FT 0309	Rose apple		GC 4631	Fonio, see Hungry Rice
FT 0310	Sea grape		GC 4633	Foxtail millet, see Millet, Foxtail
FT 0311	Surinam cherry		GC 4635	Fundi, see Hungry Rice
FT 0312	Tree tomato		GC 4637	Guinea corn, see Sorghum
FT 4095	Acerola, see Barbados cherry		GC 4639	Hog millet, see Millet, Common
FT 4097	Aonla, see Otaheite gooseberry		GC 4641	Kaffir corn, see Sorghum
FT 4099	Brazilian cherry, see Grumichana		GC 4643	Kaoliang, see Sorghum
FT 4101	Icaco plum, see Coco plum		GC 4645	Millet, Barnyard, see Millet
FT 4103	Java almond		GC 4647	Millet, Bulrush, see Millet
FT 4105	Kaki or Kaki fruit, see Persimmon, Japanese		GC 4649	Millet, Common, see Millet
FT 4107	Kumquat, Marumi, see Kumquats		GC 4651	Millet, Finger, see Millet
FT 4109	Kumquat, Nagami, see Kumquats		GC 4653	Millet, Foxtail, see Millet
FT 4111	Locust tree, see Carob		GC 4655	Millet, Little, see Millet
FT 4113	Persimmon Chinese, see Persimmon, Japanese		GC 4657	Milo, see Sorghum
FT 4115	Pitanga, see Surinam cherry		GC 4659	Oat, Red, see Oats
FT 4117	Pomarrosa, see Rose apple		GC 4661	Pearl millet, see Millet, Bulrush
FT 4119	Pomarrosa, Malay, see Pomerac		GC 4663	Extra no. for processed maize products: 1255
FT 4121	St. John's bread, see Carob		GC 4665	Proso millet, see Millet, Common
FT 4123	Tamarillo, see Tree tomato		GC 4667	Russian millet, see Millet, Common

GC 4669	Shallu, see Sorghum
GC 4671	Sorgo, see Sorghum
GC 4673	Spelt, see Wheat
GC 4675	Spiked millet, see Millet, Bulrush
GC 4677	Extra nos. for processed rice products
GS 0658	Sorgo or Sorghum, Sweet
GS 0659	Sugar cane
GS 4679	Extra no. for processed sugar cane products
HH 0624	Celery leaves
HH 0720	Angelica, including Garden Angelica
HH 0721	Balm leaves
HH 0722	Basil
HH 0723	Bay leaves
HH 0724	Borage
HH 0725	Burnet, Great
HH 0726	Herbs
HH 0727	Chives
HH 0729	Curry leaves
HH 0730	Dill
HH 0731	Fennel
HH 0732	Horehound
HH 0733	Hyssop
HH 0734	Lavender
HH 0735	Lovage
HH 0736	Marjoram
HH 0737	Marigold flowers
HH 0738	Mints
HH 0739	Nasturtium, Garden, leaves
HH 0740	Parsley
HH 0741	Rosemary
HH 0742	Rue
HH 0743	Sage and related Salvia species
HH 0744	Sassafras leaves
HH 0745	Savory, Summer; Winter
HH 0746	Sorrel, Common, and related Rumex species
HH 0747	Sweet Cicely
HH 0748	Tansy and related species
HH 0749	Tarragon
HH 0751	Winter cress, Common; American
HH 0752	Wintergreen leaves
HH 0753	Woodruff
HH 0754	Wormwoods
HH 4731	Burnet, Salad, see Burnet, Great
HH 4733	Catnip, see Catmint
HH 4735	Chervil, see Group 013: Leafy vegetables
HH 4737	Chives, Chinese, see Chives
HH 4739	Clary, see Sage (and related Salvia species)
HH 4741	Costmary, see Tansy (and related species)
HH 4743	Cretan Dittany, see Burning Bush
HH 4745	Estragon, see Tarragon
HH 4747	Fennel, Bulb, see Group 009: Bulb vegetables, No. VA 0380
HH 4749	Marjoram, Sweet, see Marjoram
HH 4751	Marjoram, Wild, see Marjoram
HH 4753	Mugwort, see Wormwoods
HH 4755	Myrrh, see Sweet Cicely
HH 4757	Oregano, see Marjoram
HH 4759	Pennyroyal, see Mints
HH 4761	Peppermint, see Mints

HH 4763	Southernwood, see Wormwoods
HH 4765	Spearmint, see Mints
HH 4767	Watercress, see Group 013: Leafy vegetables
HS 0093	Spices
HS 0370	Tonka bean, see also Group 006: Assorted tropical and sub-tropical
HS 0624	Celery seed
HS 0720	Angelica seed
HS 0730	Dill, seed
HS 0735	Lovage, seed
HS 0739	Nasturtium pods
HS 0771	Anise seed
HS 0772	Calamus, root
HS 0773	Caper buds
HS 0774	Caraway seed
HS 0775	Cardamom seed
HS 0776	Cassia buds
HS 0777	Cinnamon bark (including Cinnamon, Chinese bark)
HS 0778	Cloves, buds
HS 0779	Coriander, seed
HS 0780	Cumin seed
HS 0781	Elecampane, root
HS 0782	Fenugreek, seed
HS 0783	Galangal, rhizomes
HS 0784	Ginger, root
HS 0785	Grains of paradise
HS 0786	Juniper berries
HS 0787	Liquorice, roots
HS 0788	Mace
HS 0789	Nutmeg
HS 0790	Pepper, Black; White
HS 0791	Pepper, Long
HS 0792	Pimento, fruit
HS 0794	Turmeric, root
HS 0795	Vanilla, beans
HS 4769	Allspice fruit, see Pimento
HS 4771	Angelica, root, stem and leaves, see Group 027: Herbs
HS 4773	Aniseed, see Anise seed
HS 4775	Cassia bark, see Cinnamon bark (including Cinnamon, Chinese)
HS 4777	Fennel, seed
HS 4779	Horseradish, see VR 0583, Group 016: Root and Tuber
HS 4781	Licorice, see Liquorice
HS 4783	Poppy seed, see Group 023: Oilseed
HS 4785	Sesame seed, see Group 023: Oilseed
HS 4787	Tamarind, see Group 006: Assorted tropical and subtropical
JF 0001	Citrus juice
JF 0004	Orange juice
JF 0175	Fruit juices
JF 0203	Grapefruit juice
JF 0226	Apple juice
JF 0269	Grape juice
JF 0341	Pineapple juice
JF 0448	Tomato juice
JF 1140	Black currant juice

JF 5293	Cassis, see Black currant juice
OC 0172	Vegetable oils, crude
OC 0305	Olive oil, crude
OC 0495	Rape seed oil, crude
OC 0541	Soya bean oil, crude
OC 0645	Maize oil, crude
OC 0665	Coconut oil, crude
OC 0691	Cotton seed oil, crude
OC 0696	Palm oil, crude
OC 0697	Peanut oil, crude
OC 0699	Safflower seed oil, crude
OC 0700	Sesame seed oil, crude
OC 0702	Sunflower seed oil, crude
OC 1240	Palm kernel oil, crude
OC 5289	Corn oil, crude, see Maize oil, crude
OR 0172	Vegetable oils, edible
OR 0305	Olive oil, refined
OR 0495	Rape seed oil, refined
OR 0541	Soya bean oil, refined
OR 0645	Maize oil, edible
OR 0665	Coconut oil, refined
OR 0691	Cotton seed oil, edible
OR 0696	Palm oil, edible
OR 0697	Peanut oil, edible
OR 0699	Safflower seed oil, edible
OR 0700	Sesame seed oil, edible
OR 0702	Sunflower seed oil, edible
OR 1240	Palm kernel oil, edible
OR 5291	Corn oil, edible, see Maize oil, edible
SB 0091	Seed for beverages
SB 0715	Cacao beans
SB 0716	Coffee beans
SB 0717	Cola nuts
SB 4727	Kola, see Cola nuts
SM 0716	Coffee beans, roasted
SO 0088	Oilseed
SO 0089	Oilseed except peanut
SO 0090	Mustard seeds
SO 0478	Mustard seed, Indian
SO 0485	Mustard seed
SO 0495	Rape seed
SO 0690	Ben Moringa seed
SO 0691	Cotton seed
SO 0692	Kapok
SO 0693	Linseed
SO 0694	Mustard seed, Field
SO 0695	Niger seed
SO 0696	Pala nut
SO 0697	Peanut
SO 0698	Poppy seed
SO 0699	Safflower seed
SO 0700	Sesame seed
SO 0701	Shea nuts
SO 0702	Sunflower seed
SO 0703	Peanut, whole
SO 4701	Coconut, see Group 022: Tree nuts
SO 4703	Colza, see Rape seed
SO 4705	Colza, Indian, see Mustard seed, Field
SO 4707	Desert date, see Group 005: Assorted tropical

	and sub-tropical fruits – edible peel
SO 4709	Drumstick tree seed, see Ben Moringa seed
SO 4711	Flax-seed, see Linseed
SO 4713	Groundnut, see Peanut
SO 4715	Horseradish tree seed, see Ben Moringa seed
SO 4717	Extra nos. for commodities derived from oilseed
SO 4719	Olive, see Group 005: Assorted tropical and subtropical
SO 4721	Rape seed, Indian, see Mustard seed, Field
SO 4723	Soya bean (dry), see group 015: Pulses
SO 4725	Extra no. for products derived from palm nut
TN 0085	Tree nuts
TN 0295	Cashew nuts
TN 0660	Almonds
TN 0661	Beech nuts
TN 0662	Brazil nuts
TN 0663	Butter nuts
TN 0664	Chestnuts
TN 0665	Coconuts
TN 0666	Hazelnuts
TN 0667	Hickory nuts
TN 0668	Japanese horse-chestnuts
TN 0669	Macadamia nuts
TN 0670	Pachira nuts
TN 0671	Paradise nuts, see Sapucaia nuts
TN 0672	Pecans
TN 0673	Pine nuts
TN 0674	Pili nuts
TN 0675	Pistachio nuts
TN 0676	Sapucaia nuts
TN 0677	Tropical Almonds
TN 0678	Walnuts
TN 4681	Bush nuts, see Macadamia nuts
TN 4683	Chinquapin, see Chestnuts
TN 4685	Filberts, see Hazelnuts
TN 4687	Java almonds, see Pili nuts
TN 4689	Pignolia or Pignoli, see Pine nuts
TN 4691	Pinocchi, see Pine nuts
TN 4693	Pinon nut, see Pine nuts
TN 4695	Queensland nuts, see Macadamia nuts
TN 4697	Walnuts, Black, see Walnuts
TN 4699	Walnuts, English; Walnuts, Persian, see Walnuts
VA 0035	Bulb vegetables
VA 0036	Bulb vegetables, except Fennel, Bulb
VA 0380	Fennel, Bulb
VA 0381	Garlic
VA 0382	Garlic, Great-headed
VA 0383	Kurrat
VA 0384	Leek
VA 0385	Onion, bulb
VA 0386	Onion, Chinese
VA 0387	Onion, Welsh
VA 0388	Shallot
VA 0389	Spring onion
VA 0390	Silverskin onion
VA 0391	Tree onion
VA 4153	Carosella, see Fennel, Italian
VA 4155	Chives, see Group 027: Herbs
VA 4157	Chives, Chinese, see Group 027: Herbs

VA 4159	Fennel, Italian, see Fennel, bulb
VA 4161	Fennel, Roman, see Fennel, bulb
VA 4163	Fennel, Sweet, see Fennel, Roman
VA 4165	Japanese bunching onion, see Welsh onion
VA 4167	Multiplying onion, see Welsh onion
VA 4169	Onion, Egyptian, see Tree onion
VA 4171	Rakkyo, see Onion, Chinese
VB 0040	Brassica (cole or cabbage) vegetables, Head cabbages
VB 0041	Cabbages, head
VB 0042	Flowerhead brassicas (includes Broccoli: Broccoli, Chinese)
VB 0400	Broccoli
VB 0401	Broccoli, Chinese
VB 0402	Brussels sprouts
VB 0403	Cabbage, Savoy, see also Cabbages, Head
VB 0404	Cauliflower, see also Flowerhead brassicas
VB 0405	Kohlrabi
VB 4173	Broccoli, Sprouting, see Broccoli
VB 4175	Cabbage, see Cabbages, Head
VB 4177	Cabbage, Green, see Cabbage, Savoy
VB 4179	Cabbage, Red, see Cabbages, Head
VB 4181	Cabbage, Oxhead, see Cabbages, Head
VB 4183	Cabbage, Pointed, see Cabbage, Oxhead
VB 4185	Cabbage, White, see Cabbage, Head
VB 4187	Cabbage, Yellow, see Cabbage, Savoy
VB 4189	Cauliflower, Green, see Cauliflower
VB 4191	Kailan, see Broccoli, Chinese
VC 0045	Fruiting vegetables, Cucurbits
VC 0046	Melons, except Watermelon
VC 0420	Balsam apple
VC 0421	Balsam pear
VC 0422	Bottle gourd
VC 0423	Chayote
VC 0424	Cucumber
VC 0425	Gherkin
VC 0426	Gherkin, West Indian
VC 0427	Loofah, Angled
VC 0428	Loofah, Smooth
VC 0429	Pumpkins
VC 0430	Snake gourd
VC 0431	Squash, Summer
VC 0432	Watermelon
VC 0433	Winter Squash, see also Pumpkins
VC 4193	Bitter cucumber, see Balsam pear
VC 4195	Bitter gourd, see Balsam pear
VC 4197	Bitter melon, see Balsam pear
VC 4199	Cantaloupe, see Melons
VC 4201	Casaba or Casaba melon, see Subgroup Melons except Watermelon
VC 4203	Christophine, see Chayote
VC 4205	Citron melon, see Watermelon
VC 4207	Courgette, see Squash, Summer
VC 4209	Cucuzzi, see Bottle Gourd
VC 4211	Cushaws, see Pumpkins
VC 4213	Marrow, see Squash, Summer
VC 4215	Melon, Crenshaw, see VC 0046
VC 4217	Melon, Honey Ball, see VC 0046
VC 4219	Melon, Honeydew, see VC 0046

VC 4221	Melon, Mango, see VC 0046
VC 4223	Melon, Netted, see VC 0046
VC 4225	Melon, Oriental Pickling, see VC 0046
VC 4227	Melon, Persian, see VC 0046
VC 4229	Melon, Pomegranate, see VC 0046
VC 4231	Melons, Serpent, see VC 0046
VC 4233	Melon, Snake, see VC 0046
VC 4235	Melon, White-skinned, see VC 0046
VC 4237	Melon, Winter, see VC 0046
VC 4239	Muskmelon, see VC 0046
VC 4241	Patisson, see Squash, White Bush
VC 4243	Sinkwa or Sinkwa towel gourd, see Loofah, Angled
VC 4245	Sponge gourd, see Loofah, Smooth
VC 4247	Squash, see Squash, Summer and Winter squash
VC 4249	Squash, White bush, see Squash, Summer
VC 4251	Vegetable spaghetti, see Pumpkins
VC 4253	Vegetable sponge, see Loofah, Smooth
VC 4255	Wax gourd
VC 4257	West Indian gherkin, see Gherkin, West Indian
VC 4259	Winter Melon, see Melon, Winter
VC 4261	Zucchetti, Squash, Summer
VC 4263	Zucchini, see Squash Summer
VD 0070	Pulses
VD 0071	Beans (dry)
VD 0072	Peas (dry)
VD 0520	Bambara groundnut (dry seed)
VD 0521	Black gram (dry)
VD 0523	Broad bean (dry)
VD 0524	Chick-pea (dry)
VD 0526	Common bean (dry)
VD 0527	Cowpea (dry)
VD 0531	Hyacinth bean (dry)
VD 0533	Lentil (dry)
VD 0534	Lima bean (dry)
VD 0535	Mat bean (dry)
VD 0536	Mung bean (dry)
VD 0537	Pigeon pea (dry)
VD 0539	Rice bean (dry)
VD 0541	Soya bean (dry)
VD 0545	Lupin (dry)
VD 0560	Adzuki bean (dry)
VD 0561	Field pea (dry)
VD 0562	Horse gram
VD 0563	Kersting's groundnut
VD 0564	Tepary bean (dry)
VD 4465	Angola pea, see Pigeon pea
VD 4467	Black-eyed pea, see Cowpea
VD 4469	Bonavist bean, see Hyacinth bean
VD 4471	Cajan pea, see Pigeon pea
VD 4473	Dwarf bean (dry), see Common bean (dry)
VD 4475	Fava bean (dry), see Broad bean (dry)
VD 4477	Field bean (dry), see Common bean (dry)
VD 4479	Flageolet (dry), see Common bean (dry)
VD 4481	French bean, see Group 014: Legume vegetables
VD 4483	Geocarpa groundnut or Geocarpa bean, see Kersting's groundnut
VD 4485	Garden pea, see Group 014: Legume vegetables
VD 4487	Goa bean, see Group 014: Legume vegetables
VD 4489	Gram (dry), see Chick-pea (dry)

VD 4491	Green beans, see Group 014: Legume Vegetables
VD 4493	Green gram (dry), see Mung bean (dry)
VD 4495	Groundnut, see Peanut, Group 023: Oilseed
VD 4497	Haricot bean, see Common bean, Group 014: Legume vegetables
VD 4499	Horse bean (dry), see Broad bean (dry)
VD 4501	Jack bean, see Group 014: Legume vegetables
VD 4503	Kidney bean (dry), see Common bean (dry)
VD 4505	Lablab (dry), see Hyacinth bean (dry)
VD 4507	Moth bean (dry), see Mat bean (dry)
VD 4509	Navy bean (dry), see Common bean (dry)
VD 4511	Pea (dry), see Field pea (dry)
VD 4513	Red gram (dry), see Pigeon pea (dry)
VD 4515	Runner bean, see Common bean, Group 014: Legume vegetables
VD 4517	Scarlet runner bean, see Group 014: Legume vegetables
VD 4519	Sieva bean (dry), see Lima bean (dry)
VD 4521	Soybean (dry), see Soya bean (dry)
VD 4523	Urd bean (dry), see Black gram (dry)
VD 4525	Wrinkled pea (dry), see Field pea (dry)
VL 0053	Leafy vegetables
VL 0054	Brassica leafy vegetables
VL 0269	Grape leaves
VL 0337	Papaya leaves
VL 0421	Balsam pear leaves
VL 0446	Roselle leaves
VL 0460	Amaranth
VL 0461	Betel leaves
VL 0462	Box thorn
VL 0463	Cassava leaves
VL 0464	Chard
VL 0465	Chervil
VL 0466	Pak-choi or Paksoi
VL 0467	Chinese cabbage, (type Pe-tsai)
VL 0468	Choisum
VL 0469	Chicory leaves (green and red cultivars)
VL 0470	Corn salad
VL 0471	Marsh marigold
VL 0472	Cress, Garden
VL 0473	Watercress
VL 0474	Dandelion
VL 0475	Dock
VL 0476	Endive
VL 0477	Goosefoot
VL 0478	Indian mustard
VL 0479	Japanese greens, various species
VL 0480	Kale (including among others: Collards, Curly Kale)
VL 0481	Komatsuma
VL 0482	Lettuce, Head
VL 0483	Lettuce, Leaf
VL 0484	Mallow
VL 0485	Mustard greens
VL 0486	New Zealand spinach
VL 0487	Nightshade, Black
VL 0488	Orach
VL 0489	Pepper leaves
VL 0490	Plantain leaves
VL 0491	Pokeweed
VL 0492	Purslane
VL 0493	Purslane, Winter
VL 0494	Radish leaves (including Radish tops)
VL 0495	Rape greens
VL 0496	Rucola
VL 0497	Rutabaga greens
VL 0498	Salsify leaves
VL 0499	Sea kale
VL 0500	Senna leaves
VL 0501	Sowthistle
VL 0502	Spinach
VL 0503	Spinach, Indian
VL 0504	Tannia leaves
VL 0505	Taro leaves
VL 0506	Turnip greens
VL 0507	Kangkung
VL 0508	Sweet potato, leaves
VL 0510	Cos lettuce
VL 4313	Amsoi, see Indian Mustard
VL 4315	Arrugula, see Rucola
VL 4317	Beet leaves, see Chard
VL 4319	Bitter cucumber leaves, see Balsam pear leaves
VL 4321	Blackjack
VL 4323	Bledo, see Amaranth
VL 4325	Borecole, see Kale, curly
VL 4327	Broccoli raab
VL 4329	Celery cabbage, see Chinese cabbage
VL 4331	Celery mustard, see Pak-choi
VL 4333	Cowslip, (American english) see Marsh marigold
VL 4335	Crisphead lettuce, see Lettuce, Head
VL 4337	Curly kale, see Kale, curly
VL 4339	Cutting lettuce, see Lettuce, Leaf
VL 4341	Endive, broad or plain leaved, see Endive
VL 4343	Endive, curled, see Endive
VL 4345	Fennel, see Group 027 Herbs
VL 4347	Fennel, Bulb, see Group 009 Bulb vegetables
VL 4349	Garden cress, see Cress, Garden
VL 4351	Gow Kee, see Box thorn
VL 4353	Jamaican sorrel, see Roselle leaves
VL 4355	Kale, curly, see Kale
VL 4357	Lambs lettuce, see Corn salad
VL 4359	Lettuce, Red, see Lettuce, Head
VL 4361	Matrimony vine, see Box Thorn
VL 4363	Mustard, Indian, see Indian Mustard
VL 4365	Namenia, see Turnip greens
VL 4367	Pak-tsai, see Chinese cabbage, (type Pe-tsai)
VL 4369	Poke-berry leaves, see Pokeweed
VL 4371	Red leaved chicory, see Chicory leaves
VL 4373	Silver beet, see Chard
VL 4375	Spinach beet, see Chard
VL 4377	Sugar loaf, see Chicory leaves
VL 4379	Swiss chard, see Chard
VL 4381	Tendergreen, see Turnip greens
VL 4383	Tsai shim, see Choisum
VL 4385	Tsoi sum, see Choisum
VL 4387	Vine spinach, see Spinach, Indian
VL 4389	Water Spinach, see Kangkung
VL 4391	Yautia leaves, see Tannia leaves

VO 0050	Fruiting vegetables, other than Cucurbits		VP 0533	Lentil (young pods)
VO 0051	Peppers		VP 0534	Lima bean (young pods and/or immature beans)
VO 0440	Egg plant		VP 0535	Mat bean (green pods, mature, fresh seeds)
VO 0441	Ground cherries		VP 0536	Mung bean (green pods)
VO 0442	Okra		VP 0537	Pigeon pea (green pods and/or young green seeds)
VO 0443	Pepino		VP 0538	Podded pea (young pods)
VO 0444	Peppers, Chili		VP 0539	Rice bean (young pods)
VO 0445	Peppers, Sweet (including pimento or pimiento)		VP 0540	Scarlet runner bean (pods & seeds)
VO 0446	Roselle		VP 0541	Soya bean (immature seeds)
VO 0447	Sweet corn, (corn-on-the-cob)		VP 0542	Sword bean (young pods & beans)
VO 0448	Tomato		VP 0543	Winged pea (young pods)
VO 0449	Fungi, Edible (not including mushrooms)		VP 0544	Yard-long bean (pods)
VO 0450	Mushrooms		VP 0545	Lupin
VO 1275	Sweet corn (kernels), see definition in Codex Standard 132-1981		VP 4393	Angola pea (immature seed), see Pigeon pea
			VP 4395	Asparagus bean (pods), see Yardlong bean
VO 4265	Alkekengi, see Ground cherries		VP 4397	Asparagus pea (pods), see Goa bean
VO 4267	Aubergine, see Eggplant		VP 4399	Bonavist bean (young pods and immature seeds)
VO 4269	Bell pepper, see Peppers, Sweet		VP 4401	Butter bean (immature pods), see Lima bean
VO 4271	Cape gooseberry, see Ground cherries		VP 4403	Dwarf bean (immature pods and/or seeds), see Common bean
VO 4273	Cherry pepper, see Peppers, Chili		
VO 4275	Cherry tomato, see Ground cherries		VP 4405	Edible-podded pea, see Podded pea
VO 4277	Chili peppers, see Peppers, Chili		VP 4407	Fava bean (green pods and immature beans), see Broad bean
VO 4279	Chinese lantern plant, see Ground cherries		
VO 4281	Cluster pepper, see Peppers, Chili		VP 4409	Field bean (green pods), see Common bean
VO 4283	Cone pepper, see peppers, Chili		VP 4411	Flageolet (fresh beans), see Common bean
VO 4285	Corn-on-the-cob, see Sweet corn (corn-on-the-cob)		VP 4413	Four-angled bean (immature pods), see Goa bean
VO 4287	Fungus "Chanterelle," see Fungi, Edible		VP 4415	French bean (immature pods and seeds), see Common bean
VO 4289	Golden berry, see Ground cherries		
VO 4291	Husk tomato, see Ground cherries		VP 4417	Garbanzos, see Chick-pea
VO 4293	Lady's finger, see Okra		VP 4419	Gram (green pods), see Chick-pea
VO 4295	Melon pear, see Pepino		VP 4421	Green bean (green pods and immature seeds), see Common bean
VO 4297	Naranjilla, see Group 006 Assorted tropical and subtropical fruits inedible peel		
			VP 4423	Green gram (green pods), see Mung bean
VO 4299	Paprika, see Peppers, Sweet		VP 4425	Guar (young pods), see Cluster bean
VO 4301	Peppers, Long, see Peppers, Sweet		VP 4427	Haricot bean (green pods, and/or immature seeds)
VO 4303	Pimento or Pimiento, see Peppers, Sweet		VP 4429	Horse bean (green pods and/or immature seeds)
VO 4305	Quito Orange, see Naranjilla		VP 4431	Kidney bean (pods and/or immature seeds), see Common bean
VO 4307	Strawberry tomato, see Ground cherries		
VO 4309	Tomatillo, see Ground cherries		VP 4433	Lablab (young pods; immature seeds), see Hyacinth bean
VO 4311	Tree melon, see Pepino		
VP 0060	Legume vegetables		VP 4435	Mangetout or Mangetout pea, see Podded pea
VP 0061	Beans, except broad bean and soya bean		VP 4437	Moth bean, see Mat bean
VP 0062	Beans, shelled		VP 4439	Navy bean (young pods and/or immature seeds)
VP 0063	Peas (pods and succulent = immature seeds)		VP 4441	Pea, see Garden pea
VP 0064	Peas, shelled (succulent seeds)		VP 4443	Pigeon bean, (green pods and immature seeds)
VP 0520	Bambara groundnut (immature seeds)		VP 4445	Vicia faba L., subsp. eu-faba, var. minor Beck, see Broad bean
VP 0521	Black gram (green pods)		
VP 0522	Broad bean (green pods and immature seeds)		VP 4447	Red gram (green pods and/or young green seeds), see Pigeon Pea
VP 0523	Broad bean, shelled (succulent) (= immature seeds)		
VP 0524	Chick-pea (green pods)		VP 4449	Runner bean (green pods & seeds), see Common bean
VP 0525	Cluster bean (yound pods)		
VP 0526	Common bean (pods and/or immature seeds)		VP 4451	Sieva bean (young pods and/or green fresh beans), see Lima
VP 0527	Cowpea (immature pods)		
VP 0528	Garden pea (young pods) (=succulent, immature seeds)		VP 4453	Snap bean (young pods), see Common bean
			VP 4455	Soybean, see Soya bean
VP 0529	Garden pea, shelled (succulent seeds)		VP 4457	Sugar pea (young pods), see Podded pea
VP 0530	Goa bean (immature pods)		VP 4459	Urd bean (green pods), see Black gram
VP 0531	Hyacinth bean (young pods, immature seeds)		VP 4461	Winged bean (immature pods), see Goa bean
VP 0532	Jack bean (young pods, immature seeds)		VP 4463	Wrinkled pea, see Garden pea

368

VR 0075	Root and tuber vegetables		VR 4531	Cassava, Bitter, see Cassava
VR 0423	Chayote root		VR 4533	Cassava, Sweet, see Cassava
VR 0463	Cassava		VR 4535	Chinese radish, see Radish, Japanese
VR 0469	Chicory, roots		VR 4537	Christophine, see Chayote root
VR 0494	Radish		VR 4539	Cocoyam, see Tannia and Taro
VR 0497	Swede		VR 4541	Dasheen, see Taro
VR 0498	Salsify		VR 4543	Daikon, see Radish, Japanese
VR 0504	Tannia		VR 4545	Eddoe, see Taro
VR 0505	Taro		VR 4547	Globe artichoke, see Group 017; Artichoke Globe
VR 0506	Turnip, Garden		VR 4549	Gruya, see Canna, edible
VR 0508	Sweet potato		VR 4551	Jicama, see Yam bean
VR 0530	Goa bean root		VR 4553	Leren, see Topee Tambu
VR 0570	Alocasia		VR 4555	Manioc, see Cassava, bitter
VR 0571	Arracacha		VR 4557	Oyster plant, see Salsify
VR 0572	Arrowhead		VR 4559	Potato yam, see Yam bean
VR 0573	Arrowroot		VR 4561	Queensland arrowroot, see Canna, edible
VR 0574	Beetroot		VR 4563	Rutabaga, see Swede
VR 0575	Burdock, greater or edible		VR 4565	Salsify, Black, see Scorzonera
VR 0576	Canna, edible		VR 4567	Tanier, see Tannia
VR 0577	Carrot		VR 4569	Tapioca, see Cassava
VR 0578	Celeriac		VR 4571	Turnip, see Swede
VR 0579	Chervil, Turnip-rooted		VR 4573	Turnip, Swedish, see Swede
VR 0580	Chufa, see Tiger nut		VR 4575	Yam, Cush-cush, see Yams
VR 0581	Galangal, Greater		VR 4577	Yam, Eight-months, see Yam, White Guinea
VR 0582	Galangal, Lesser		VR 4579	Yam, Greater, see Yams
VR 0583	Horseradish		VR 4581	Extra nos. for processed sugar beet products
VR 0584	Japanese artichoke		VR 4583	Yam, Twelve-months, see Yam Yellow Guinea
VR 0585	Jerusalem artichoke		VR 4585	Yam, White, see Yam, White Guinea
VR 0586	Oca		VR 4587	Yam, White Guinea, see Yams
VR 0587	Parsley, Turnip-rooted		VR 4589	Yam, Yellow, see Yam, Yellow Guinea
VR 0588	Parsnip		VR 4591	Yam, Yellow Guinea, see Yams
VR 0589	Potato		VR 4593	Yautia, see Tannia
VR 0590	Radish, Black		VS 0078	Stalk and stem vegetables
VR 0591	Radish, Japanese		VS 0469	Witloof chicory (sprouts)
VR 0592	Rampion roots		VS 0620	Artichoke, Globe
VR 0593	Salsify, Spanish		VS 0621	Asparagus
VR 0594	Scorzonera		VS 0622	Bamboo shoots
VR 0595	Skirrit or Skirret		VS 0623	Cardoon
VR 0596	Sugar beet		VS 0624	Celery
VR 0599	Ullucu		VS 0625	Celtuce
VR 0600	Yams		VS 0626	Palm hearts
VR 0601	Yam bean		VS 0627	Rhubarb
VR 4527	Achira, see Canna, edible		VS 4595	Celery leaves, see Group 027: Herbs
VR 4529	Black salsify, see Scorzonera			

APPENDIX IV

Authors
Classification
For Food and Feed Crops

Authors Crop Grouping Scheme

The first edition (1971) of *Food and Feed Crops of the United States* "Green Book" outlined the general plant grouping scheme with the primary utilization of the edible part of the plant highlighted, e.g. vegetable, fruit, grass, grain, etc. This included plant growth characteristics, part(s) consumed, botanical relationships, commercial importance, geographical distribution, livestock feed items and processed products. In the second edition, we have added the EPA and Codex commodity classifications for harmonization including representative commodities for the current U.S. EPA crop grouping scheme as well as revising the crop listings.

The crop groupings scheme proposed by the authors of the "Green Book" in 1971 is just as valid now as then. It has been used as an outline for regulatory crop grouping schemes, both national and international. We might use different names or numbers but the concept is the same. The "Green Book" might well be considered the father of the present day crop grouping schemes. It provided the necessary seed for crop groupings, let alone a common vocabulary.

The concept of utilizing crop grouping to estimate maximum pesticide residues in each commodity within a group of related commodities has been used since publication of crop grouping regulations in the *Federal Register* of 6 December 1962 (27FR12100), 29 June 1983 (48FR29855) and 17 May 1995 (60FR26626). From the very restrictive and little used groups that were established without group names and still utilizing individual crop names in 1962, the revised crop groups of 1983 and 1995 were expanded to include crop group names with many minor crops that increased usage. The 1983 rule revised the regulations with a list of representative crops for each group, these crops were most likely to have the highest residue and the most economically important crops in each group. The 1995 rule revised the regulations to include additional crops and subgroups. The subgroups are smaller and more closely related groupings of commodities with their own representative commodity(s) which were chosen primarily from the representative commodities for the parent group. The crop grouping schemes were developed, in part, to minimize the burden of developing data in support of commodity-specific tolerances for minor crops. A number of minor crops can be included in a group tolerance based upon residue data for the representative crops. The concept of crop grouping is legally, scientifically and economically sound.

The EPA 1995 Crop Grouping (including the previous one in 1983) followed the *Food and Feed Crops of the United States* plant grouping outline in establishing the crop groupings and helped to further define the crops relationship, including commodity definitions which can be considered subgroups in many cases. As the regulatory agencies, and food safety and pest management researchers (public and private) become more comfortable with using established crop groupings, additional ones will be added, e.g. oilseed, tropical fruits, etc. The authors continue to focus on this efficient approach to developing food safety research data.

The primary authors of the present crop grouping scheme for the U.S. EPA are Dr. Bernard A. Schneider and the EPA Minor Use Officer, Hoyt L. Jamerson. Additionally, the Minor Use Coordinator, J. Douglas Rothwell, for the Pest Management Regulatory Agency of Health Canada helped harmonize the Canadian Crop Grouping Scheme with the U.S. These scientists deserve much of the credit in recognizing the importance of crop groupings for the minor/specialty crops in the regulatory agencies. In the university and USDA arenas, scientists George Markle, Charles Compton, John Magness, Bailey Pepper, Ed Swift and Jerry Baron deserve much of the credit in initiating and following through on the crop grouping concept of including minor / specialty crops in a scientific / regulatory compatible scheme. To place minor crops in perspective, the Food Quality Protection Act of 1996 (FQPA), established the minor crop definition as being less than 300,000 acres, thereby, some 30 commodities are now considered major crops. Listed in this book are over 400 minor crops that represent a large part of a healthy diet. The second edition of the "Green Book" serves as an extensive literature reference on food and feed commodities.

In the future, more pest management programs for minor crop registrations will be based on crop grouping research data which will help address similarities in crop metabolism and common modes of actions in related species. The authors crop grouping classification (crop monographs, which see) continues to represent harmonization with the U.S. EPA, Health Canada and Codex, as follows by having eleven vegetable crop groups; eight fruit crop groups; one tree nut crop group; two cereal grain crop groups; one grass, forage and pasture crop group; one legume, nongrass, forage and pasture crop group; one condiment, spice and essential oil plants crop group; one edible food oil crop group; three stimulant plants crop groups; and two sugar and syrup plants crop groups.

The Authors classification of crops lists them by type of crop, such as vegetable crops followed by the specific crop groups. The specific crop groups exhibit specific reference to U.S., Canadian, Codex and Authors crop groups. The 1995 U.S. crop groups (1 through 19) are exhibited in Appendix II. The Canadian crop groups (1 through 19) are similar to the U.S., except for the additional Canadian crop group 20, which is their oilseed crop group. The Codex crop groups are shown in Appendix III. The Authors classification references the specific crop monographs under each category which under most conditions will include more than one species.

Crop Groupings
Food and Feed Crops of the United States 2nd Edition (Revised)
By Markle, Baron, and Schneider
Authors Classification

Vegetable Crops List
- Root and tuber vegetables
- Leaves of root and tuber vegetables
- Bulb vegetables
- Leafy vegetables (except Brassica vegetables)
- Brassica (Cole) leafy vegetables
- Legume vegetables (succulent or dried)
- Foliage of legume vegetables
- Fruiting vegetables (except cucurbits)
- Cucurbit vegetables
- Stalk and stem vegetables
- Edible fungi

Fruit Crops List
- Citrus fruits
- Pome fruits
- Stone fruits
- Berries
- Small fruits
- Tropical and subtropical fruits
- Tropical and subtropical fruits — edible peel
- Tropical and subtropical fruits — inedible peel

Tree Nut Crops List
- Tree nuts

Cereal Grain Crops List
- Cereal grains
- Forage, fodder and straw of cereal grains

Grass Forage and Pasture Crops List
- Grass forage, fodder and hay

Legume (Nongrass) Forage and Pasture Crops List
- Nongrass animal feeds

Condiment, Spice and Essential Oil Plants List
- Herbs and spices

Edible Food Oil Crops List
- Oilseed

Stimulant Crops List
- Dried edible plant tops (stimulants)
- Teas
- Tropical and subtropical trees with edible seeds for beverages and sweets

Sugar and Syrup Plants List
- Grasses for sugar or syrup
- Forestry

Authors Classification

Vegetable Crops List

Vegetables constitute a major portion of the diet of most people of the world. In the United States, average per capita consumption of commercially produced vegetables in 1994 was estimated at 271 pounds, of which about 146 pounds was marketed fresh and 125 pounds was in processed form (USDA 1997b). For comparison in 1967, it was estimated at 211 pounds, of which almost 100 pounds was marketed fresh and 111 pounds was in processed form (MAGNESS). Most vegetables grown in the U.S. are grown as annual plants, asparagus being one of the most important exceptions. The parts of the plant consumed may be in different forms such as the leaves, stems, fruiting parts, roots, tubers, or bulbs. Therefore, exposure of edible parts to direct contact with applied pesticides varies greatly in the various crops. Vegetable crops grown in the U.S., including tropical areas, are briefly described in the crop monographs. A few crops not believed to be grown domestically, but important in other countries are also listed.

Root and tuber vegetables

Crop Groups
- U.S. – 1
- Canada – 1
- Codex – VR

Authors listing by crop monograph —

Arracacha	Ginger	Rampion
Arrowhead	Ginseng	Rutabaga
Arrowroot	Honewort	Salsify
Artichoke, Chinese	Horseradish	Salsify, Black
Artichoke, Jerusalem	Leren	Salsify, Spanish
Bean, Goa	Lotus root	Skirret
Beet, Fodder	Maca	Sugar beet
Beet, Garden	Mashua	Sweet potato
Burdock, Edible	Mauka	Taro
Canna, Edible	Mustard, Tuberous rooted Chinese	Ti
Carrot	Oca	Turmeric
Cassava	Parsley, Turnip-rooted	Turnip
Celeriac	Parsnip	Tyfon
Chayote (root)	Polynesian arrowroot	Ulluco
Chervil, Turnip-rooted	Potato	Wasabi
Chicory	Potato, Specialty	Waterchestnut, Chinese
Chufa	Radish	Yacon
Dahlia, Garden	Radish, Oriental	Yam bean
Dasheen		Yam, True
Evening primrose, Common		Yautia

Leaves of root and tuber vegetables

Crop Groups
- U.S. – 2
- Canada – 2
- Codex – VL and AV

Authors Listing by crop monograph —

Arracacha	Dasheen	Salsify
Artichoke, Jerusalem	Honewort	Salsify, Black
Beet, Fodder	Maca	Sugar beet
Beet, Garden	Mashua	Tanier spinach
Burdock, Edible	Mauka	Taro
Cassava	Parsley, Turnip-rooted	Ti
Celeriac	Radish	Turnip (Tops)
Chervil, Turnip-rooted	Radish, Oriental	Tyfon
Chicory	Rampion	Ulluco
	Rutabaga	Yacon
		Yautia

Bulb vegetables

Crop Groups
- U.S. – 3
- Canada – 3
- Codex – VA

Authors listing by crop monograph —

Chive	Leek	Onion, Tree
Chive, Chinese	Onion	Onion, Welsh
Daylily	Onion, Beltsville bunching	Shallot
Garlic	Onion, Chinese	Wild leek
Garlic, Great-headed	Onion, Potato	

Leafy vegetables (except *Brassica* vegetables)

Crop Groups
- U.S. – 4
- Canada – 4
- Codex – VL

Authors listing by crop monograph —

Amaranth, Chinese	Comfrey	Fennel, Florence
Amaranth, Leafy	Coriander (Cilantro)	Fenugreek
Arugula	Coriander, False	Fern, Edible
Cardoon	Corn salad	Good-King-Henry
Celery	Cress, Garden	Gow Kee
Celery, Chinese	Cress, Upland	Iceplant
Chervil	Dandelion	Japanese honewort
Chicory	Dock	Jute, Nalta
Chrysanthemum, Edible-leaved	Endive (Escarole)	Kale, Sea
	Fameflower	Lettuce

Lettuce, Celery
Lettuce, Head
Lettuce, Leaf
Marshmarigold
Orach
Parsley
Plantain

Pokeweed
Purslane, Garden
Purslane, Winter
Rhubarb
Sorrel, French
Sorrel, Garden
Spinach

Spinach, New
Zealand
Spinach, Vine
Swiss chard
Water dropwort
Water spinach
Watercress

Brassica (Cole) leafy vegetables:

Crop Groups
U.S. – 5
Canada – 5
Codex – VL and VB
Authors listing by crop monograph -

Abyssinian cabbage
Bok choy
Broccoli
Broccoli, Chinese
Broccoli raab
Brussels sprouts

Cabbage
Cabbage, Chinese
Cabbage, Seakale
Cauliflower
Collards
Hanover salad

Kale, Common
Kohlrabi
Mustard, Wild
Mustard greens
Rape
Turnip (Tops)

Legume vegetables (succulent or dried):

Crop Groups
U.S. – 6
Canada – 6
Codex – VD and VP
Authors listing by crop monograph —

Bambara groundnut
Bean
Bean, Adzuki
Bean, Broad
Bean, Dry common
Bean, Goa
Bean, Hyacinth
Bean, Lima
Bean, Moth
Bean, Mung
Bean, Rice

Bean, Scarlet runner
Bean, Succulent
common
Bean, Tepary
Bean, Urd
Bean, Yardlong
Catjang
Chickpea
Guar
Jackbean
Lentil

Lupin
Pea, Edible-podded
Pea, Field
Pea, Garden
Pea, Pigeon
Pea, Southern
Pea, Winged
Peanut
Soybean
Soybean, Vegetable
Swordbean

Foliage of legume vegetables

Crop Groups
U.S. – 7
Canada – 7
Codex – AL
Authors listing by crop monograph —

Catjang
Guar
Lupin
Pea, Field

Pea, Southern
Peanut
Soybean

Fruiting vegetables (except cucurbits):

Crop Groups
U.S. – 8
Canada – 8
Codex – VO
Authors listing by crop monograph —

Eggplant
Groundcherry
Huckleberry, Garden
Martynia

Okra
Pepino
Peppers
Peppers, Hot
Peppers, Sweet

Roselle
Tomato
Tomato, Currant
Tomato, Tree

Cucurbit vegetables:

Crop groups
U.S. – 9
Canada – 9
Codex – VC
Authors listing by crop monograph —

African horned
cucumber
Balsam pear
Cantaloupe
Chayote (fruit)
Citron melon
Cucumber
Cucumber,
Armenian

Cucumber, Wild
Gherkin, West
Indian
Gourd, Edible
Melon, Garden
Melon, Oriental
pickling
Melon, Winter

Muskmelon
Pumpkin
Snakegourd
Squash, Summer
Squash, Winter
Watermelon
Waxgourd

Stalks and stem vegetables:

Crop Groups
U.S. – Miscellaneous
Canada – None
Codex – VS
Authors listing by crop monograph —

Airpotato
Artichoke, Globe
Asparagus

Bamboo
Japanese knotweed
Palm heart

Udo
Water bamboo

Edible Fungi:

Crop Groups
U.S. – Miscellaneous
Canada – None
Codex – VO
Authors listing by crop monograph —

Mushroom,
Agaricus

Mushroom,
Specialty

Fruit Crops List

Fruit crops, as listed, are those grown for the edible pulpy, juicy tissues, associated with the seeds, produced on perennial plants. Fruits grown on annual plants, as tomatoes and melons, are listed under vegetables. Fruits comprise a major part of the diet of people in the United States. Total fruit production in the United States averaged about 32 million tons annually, 1993 to 1995, inclusive, or almost 200 pounds per capita. Actual annual consumption per capital for 1995 was estimated at 101 pounds fresh fruit, 91 pounds as canned fruit and juice, 3 pounds as dried fruit and 3.4 pounds as frozen fruit (USDA 1997b). Types of dried fruits include apple, apricot, date, fig, peach, pear, prune and raisin.

Citrus fruits:

Crop Groups
    U.S. – 10
    Canada – 10
    Codex – FC

Authors listing by crop monograph —

Calamondin	Lemon	Orange, Trifoliate
Citron, Citrus	Lime	Pummelo
Citrus fruits	Lime, Sweet	Sapote, White
Citrus hybrids	Orange, Sour	Tangelo
Grapefruit	Orange, Sweet	Tangerine
Kumquat		

Pome fruits:

Crop Groups
    U.S. – 11
    Canada – 11
    Codex – FP

Authors listing by crop monograph —

Apple	Mayhaw	Pear
Crabapple	Medlar	Quince
Loquat		

Stone fruits:

Crop Groups
    U.S. – 12
    Canada – 12
    Codex – FS

Authors listing by crop monograph —

Apricot	Peach	Plum, Chickasaw
Cherry, Black	Plum	Plum, Damson
Cherry, Nanking	Plum, American	Plum, Japanese
Cherry, Sweet	Plum, Cherry	Plumcot
Cherry, Tart		

Berries:

Crop Groups
    U.S. – 13
    Canada – 13
    Codex – FB

Authors listing by crop monograph —

Aronia berry	Currant, Black	Juneberry
Bearberry	Currant, Red	Lingonberry
Bilberry	Dewberry	Mulberry
Blackberry	Elderberry	Partridgeberry
Blueberry	Gooseberry	Raspberry
Blueberry, Lowbush	Highbush cranberry	Salal
Caneberries	Huckleberry	Seagrape
Cloudberry	Jostaberry	Serviceberry

Small fruits:

Crop Groups
    U.S. – Miscellaneous
    Canada – None
    Codex – FB

Authors listing by crop monographs —

Cranberry	Maypop
Grape	Strawberry
Kiwifruit	

Tropical and subtropical fruits:

Crop Groups
    U.S. – Miscellaneous
    Canada – None
    Codex – FT and FI

Authors listing by crop monograph —
Reserved as a super crop group to include edible and inedible peel tropical and subtropical fruits, as appropriate.

Tropical and subtropical fruits – edible peel:

Crop Groups
    U.S. – Miscellaneous
    Canada – None
    Codex – FT

Authors listing by crop monograph —

Acerola	Imbu	Papaya, Mountain
Ambarella	Jaboticaba	Persimmon
Blimbe	Jujube	Pomerac
Carob bean	Jujube, Indian	Rose apple
Cashew	Natal plum	Sentul
Date	Olive	Starfruit
Fig	Otaheite gooseberry	Surinam cherry
Guava		

Tropical and subtropical fruits – inedible peel:

Crop Groups
    U.S. – Miscellaneous
    Canada – None
    Codex – FI

Authors listing by crop monograph —

Abiu	Jackfruit	Prickly pear
Akee apple	Longan	Pulasan
Atemoya	Lychee	Rambutan
Avocado	Mamey apple	Sapodilla
Banana	Mango	Sapote, Black
Biriba	Mangosteen	Sapote, Green
Breadfruit	Marmaladebox	Sapote, Mamey
Canistel	Monstera	Soursop
Cherimoya	Naranjilla	Spanish lime
Custard apple	Papaya	Star apple
Durian	Passionfruit	Strawberrypear
Feijoa	Pawpaw	Sugar apple
Governor's plum	Pineapple	Tamarind
Ilama	Pomegranate	Wax jambu
Imbe		

Tree Nut Crops List

The tree nut crops listed here are the edible, oily seeds of trees or large shrubs, which are encased in fleshy or fibrous husks, or are essentially covered by such tissues, during growth. The edible portions are additionally enclosed in woody or semi-woody shells, protected from direct contact with any pesticides. Only systemic materials could be present in the edible parts of the plants. Therefore only one classification of these crops from the standpoint of exposure of edible parts to applied pesticides has been made, since all are similar, and represent the minimum of such exposure. Per capita consumption of tree nuts for 1994/1995 is 2.25 pounds (USDA 1997b).

A number of other edible crops have the term "nut" in the name. Some are produced on annual plants, as peanut, others on herbaceous plants, as water chestnut. These are listed under vegetables and/or oilseed.

Tree nuts:
Crop Groups
U.S. – 14
Canada – 14
Codex – TN

Authors listing by crop monograph —

Almond	Chestnut	Oak
Almond, Tropical	Coconut	Pecan
Beechnut	Ginkgo	Pili nut
Betelnut	Hazelnut	Pine nut
Brazil nut	Heartnut	Pistachio
Butternut	Hickory nut	Walnut, Black
Cashew	Macadamia nut	Walnut, English

Cereal Grain Crops List

Cereal grain crops listed, except for grain amaranth, buckwheat, psyllium and quinoa, are members of the grass family Poaceae (Gramineae), grown primarily for their edible starch-containing seeds. Other types of seed crops as dry beans, field peas and soybeans, are listed in other categories. All of the cereal grain crops are used to some extent for human food and animal feed. Most may also be used at times for pasture, hay or silage –

where the whole of the above-ground plant is utilized.

All cereal grain crops are annual plants. They may be winter annuals, seeded in the fall and harvested the following summer; or summer annuals, seeded in the spring and harvested during the same growing season. In total, they provide the main carbohydrate foods for humankind as well as the main concentrated feeds for livestock.

In most cereal grains, except in oats and barley, the kernels or seeds are enclosed in papery coverings until harvested. These usually are removed during threshing, which in this country is commonly a part of the harvesting process. During threshing the seeds are in contact with these coverings as well as with the stems and leaves of the plant prior to separation into grain and plant refuse. This refuse consists of straw (stems and leaves), chaff (mainly the papery coverings that enclosed the seeds) and aspirated grain fractions. The straw or stover from the harvested grains may also be used for livestock feed.

Seeds are borne in various ways. They may be closely packed on a base as in corn, in compact spike or head as in wheat and barley, or in rather compact or loose panicles as in oats, rice and sorghums.

Cereal grain crops (wheat, rye, rice, corn, oat and barley) are grown on about 170 million acres in the U.S. Worldwide acreage in these grain crops is about 860 million (USDA 1997b).

Cereal grains:
Crop Groups
U.S. – 15
Canada – 15
Codex – GC, CF, CP and CM

Authors listing by crop monograph —

Amaranth, Grain	Oat, Sand	Wheat, Emmer
Barley	Psyllium	Wheat, Macha
Buckwheat	Quinoa	Wheat, Oriental
Canarygrass, Annual	Rice	Wheat, Persian
Corn	Rye	Wheat, Polish
Millet, Foxtail	Sorghum, Grain	Wheat, Poulard
Millet, Pearl	Teff	Wheat, Short
Millet, Proso	Triticale	Wheat, Spelt
Oat	Wheat	Wheat, Timopheevi
Oat, Abyssinian	Wheat, Club	Wheat, Vavilovi
Oat, Animated	Wheat, Common	Wheat, Wild einkorn
Oat, Common	Wheat, Durum	Wheat, Wild emmer
Oat, Naked	Wheat, Einkorn	Wild rice

Forage, fodder and straw of cereal grains:
Crop Groups
U.S. – 16
Canada – 16
Codex – AF and AS

Authors listing by crop monograph —

Amaranth, Grain	Corn	Oat, Animated
Barley	Millet, Pearl	Oat, Common
Buckwheat	Millet, Proso	Oat, Naked
Canarygrass, Annual	Oat	Oat, Sand
	Oat, Abyssinian	Rice

Rye	Wheat, Einkorn	Wheat, Shot
Sorghum, Grain	Wheat, Emmer	Wheat, Spelt
Teff	Wheat, Macha	Wheat, Timopheevi
Triticale	Wheat, Oriental	Wheat, Vavilovi
Wheat	Wheat, Persian	Wheat, Wild einkorn
Wheat, Club	Wheat, Polish	Wheat, Wild emmer
Wheat, Common	Wheat, Poulard	Wild rice
Wheat, Durum		

Grass Forage and Pasture Crops List

Grasses constitute the most important food and feed crops of the world. The cereal grain crops, produced primarily for the seeds, are primarily grasses but here are treated under the heading "Cereal Grain Crops." Under the heading "Grasses," kinds grown primarily for pasture, hay and silage are considered

Botanically, grasses are herbaceous — or sometimes woody — monocotyledonous plants having jointed stems. Stems are closed at the joints, mostly hollow but sometimes solid between joints. Leaves are generally long and narrow, with veins parallel to the long axis. The leaf base, or sheath, is clasping to the stem. Flowers are mostly perfect (with both male and female organs present) although imperfect flowers (one-sexed) occur in some species. Flowers are generally borne in small clusters. Each flower contains a single one-celled ovary which, following fertilization with the male pollen, develops into a single seed.

Native grasses occur in all parts of the world except the more extreme arctic regions. Many of these are important pasturage and in some areas are also harvested for hay. In addition, grasses are extensively seeded on farm and range lands for pasturage, hay and silage. According to the 1992 Census, about 600 million acres of land in the United States were used for pasturage and grazing. Most of this was grassland.

Around 1500 species of grasses occur in the United States. Only the more important types are discussed in the crop monographs.

No differential among grasses from the standpoint of pesticide residues would be expected and therefore grass forage and pasture crops are treated as one class.

Grass forage, fodder and hay:

Crop Groups
U.S. – 17
Canada – 17
Codex – AS

Authors listing by crop monograph —

Alkali sacaton	Bluestem, Big	Buffalograss
Alkaligrass	Bluestem,	Buffelgrass
Arizona cottontop	Caucasian	Canarygrass, Reed
Bahiagrass	Bluestem, Diaz	Caribgrass
Beachgrass	Bluestem, Yellow	Carpetgrass
Bentgrass	Broadleaf carpet	Centipedegrass
Bermudagrass	grass	Crabgrass
Blowoutgrass	Bromegrass	Creeping foxtail
Bluegrass	Bromegrass, Minor	Curly mesquite
Bluestem,	annual	Dallisgrass
Australian	Bromesedge	Eastern gamagrass

Feather fingergrass	Pangolagrass	Tall dropseed
Fescue grass	Panicgrass	Tall outgrass
Forage grass	Panicgrass,	Tanglehead
Galleta grass	Introduced	Timothy
Gaint cane	Paspalum	Timothy, Alpine
Grama grass	Perennial veldtgrass	Tufted hairgrass
Green sprangletop	Pine dropseed	Vaseygrass
Hardinggrass	Plains bristlegrass	Velvetgrass
Hooded windmill-	Polargrass	Wheatgrass
grass	Prairie sandreed	Wheatgrass,
Indian ricegrass	Quackgrass	Bluebunch
Indiangrass	Redtop	Wheatgrass,
Junegrass	Reedgrass	Crested
Limpograss	Rhodesgrass	Wheatgrass,
Little bluestem	Ryegrass, Italian	Fairway
Lovegrass	Ryegrass, Perennial	Wheatgrass,
Maidencane	Sand bluestem	Intermediate
Mannagrass	Sand dropseed	Wheatgrass,
Marshhay cord-	Silky bluegrass	Pubescent
grass	Silver bluestem	Wheatgrass,
Meadow foxtail	Sixweeks threeawn	Siberian
Millet, Foxtail	Sloughgrass	Wheatgrass,
Millet, Japanese	Smilograss	Slender
Molassesgrass	Sorghum	Wheatgrass,
Muhly grass	South African	Streambank
Napiergrass	bluestem	Wheatgrass, Tall
Needlegrass	Spike bentgrass	Wheatgrass,
Oat, Slender	Spike trisetum	Thickspike
Oat, Wild	Spikeoat	Wheatgrass,
Oatgrass	Squirreltail	Western
Oniongrass	St. Augustine grass	Wildrye grass
Orchardgrass	Sunolgrass	Zoysia grass

Legume (Nongrass) Forage and Pasture Crops List

Legume plants, member of the family Fabaceae (Leguminosae), are second only in importance to the grasses as pasture and hay crops for livestock feed. The seeds are borne in pods, with one to many seeds per pod. In some kinds, as red and white clover, many short pods are borne in a head. In others, such as alfalfa and peas, the pods are in racemes. Leaves are generally compound, either pinnately as in vetches, or palmate as in clover.

Legumes in general are very high in palatability and are highly nutritious. All plant parts tend to be higher in protein than comparable parts of nonlegume plants. They also contain large amounts of calcium and a moderate amount of phosphorus, as well as being an excellent source of vitamins A and D. For these reasons legumes, both as pasture and as hay, are highly valuable livestock feeds.

Legumes are also highly valuable agriculturally because they have the ability, in symbiotic relationship with Rhizobia bacteria, to use nitrogen from the air. The bacteria develop nodules on the roots of the legumes and take nitrogen from the air as they develop and multiply. This "fixed" nitrogen is in turn available to the legume plant. In many legume species it is desirable to inoculate the seed or soil with cultures of the particular kind of *Rhizobia* symbiotic with the legume species, particularly when the legume is planted for the first time on a site.

In some legumes the flowers are self-pollinating, the parts being so arranged that the stigma is in contact with the pollen bearing anthers. In others the stigma and anthers must be released from the keel, after which pollen from the anthers escapes and can fall on the stigma. Some kinds of legumes, as red clover, are self-incompatible, and cross pollination from neighboring seedling plants is necessary for seed set.

Legumes have been know throughout history. Alfalfa was one of the earliest cultivated crops. In China the soybean is referred to in earliest writings. Tares, mentioned in the Bible, are believed to be common vetch.

The entire top of the plant is consumed by livestock, either as pasturage, hay or silage. There is no significant difference between various kinds from the standpoint of the residue that would be expected following pesticide applications.

A number of legumes, grown for feed, are also valuable as food and edible oil crops. These are listed in both categories. This group includes legumes and other nongrass feeds.

Nongrass animal feeds (forage, fodder, straw and hay):

Crop Groups
U.S. – 18
Canada – 18
Codex – AL
Authors listing by crop monograph —

Alfalfa	Clover, Striata	Mulga
Arrowleaf balsam-	Clover, Sub	Multiflower false-
root	Clover, White	rhodesgrass
Black wattle	Clover, Whitetip	Perennial peanut
Burclover	Clovers, True	Purple prairieclover
Calopo	Crotalaria	Rough pea
Camwood	Crown vetch	Roundleaf cassia
Clover, Alsike	Globemallow	Sainfoin
Clover, Alyce	Guajillo	Sesbania
Clover, Arrowleaf	Gumweed, Great	Spring hopsage
Clover, Ball	valley	Sweet clover
Clover, Berseem	Huisache	Tapertip hawks-
Clover, Bigflower	Kenaf	beard
Clover, Crimson	Kidney vetch	Thorn mimosa
Clover, Hop	Koa	Threadleaf sedge
Clover, Lappa	Kochia	Trefoil
Clover, Persian	Kudzu	Velvet bean
Clover, Red	Leadplant	Vetch
Clover, Rose	Lespedeza	Vetch, Chickling
Clover, Seaside	Leucaena	Vetch, Milk
Clover, Strawberry	Lupin	Vetch, Minor

Condiment, Spice, and Essential Oil Plants List

Plants in this grouping are those that are used to impart special flavors to foods, including confections. Therefore, the quantity of any of them in the diet is extremely limited since very small amounts are required to produce the desired flavors. Some are more important as vegetables than as condiment plants, in which case they are described under vegetables. Sources of essential oils used in cosmetics are not included unless they are also used in foods.

Herbs and Spices
Crop Groups
U.S. – 19
Canada – 19
Codex – HH, HS and DH
Authors listing by crop monograph —

Agrimony	Elecampane	Parsley
Allspice	Epazote	Pennyroyal
Angelica	Evening Primrose,	Pennyroyal,
Anise	Common	American
Anise hyssop	Fennel	Pepper
Annatto	Fennel, Florence	Pepper, Long
Asafetida	Fenugreek	Pepper leaf
Balm	Flower, Edible	Perilla
Basil	Galangal	Poppy seed
Black bread weed	Geranium, Scented	Rosemary
Borage	Ginger, White	Rue
Burnet, Salad	Ginkgo	Safflower
Calamus-root	Grains of Paradise	Saffron crocus
Camomile	Gumweed, Curlycup	Sage
Caper	Honewort	Savory, Summer
Caraway	Hop	Savory, Winter
Caraway, Black	Horehound	Sesame (seed)
Cardamon	Horseradish	Sorrel, French
Cardamon-	Hyssop	Sorrel, Garden
amomum	Indian borage	Southernwood
Cassia	Juniper	Spanish fennel
Catnip	Laurel, Cherry	Stevia
Celery (Celery seed)	Lavender	Sumac, Smooth
Chaya	Lemon verbena	Swamp leaf
Chervil	Lemongrass	Sweet bay
Chia	Licorice	Sweet cicely
Chive	Lovage	Tansy
Chive, Chinese	Marigold	Tarragon
Cinnamon	Marigold, Pot	Thyme
Clary	Marjoram	Tonka bean
Clove	Mexican mint	Turmeric
Coneflower, Purple	marigold	Vanilla
Coriander (seed)	Mexican oregano	Vietnamese
Coriander, False	Mint	coriander
Costmary	Mioga	Violet
Cumin	Monarda	Waterchestnut
Curry	Mountainmint	Wintergreen
Daylily	Mustard seed	Woodruff
Dill	Nasturtium	Wormwood
Dock	Nutmeg	Yarrow
Dokudami		

Edible Food Oil Crops List

Vegetable oils are of two major kinds. The plants that are sources of volatile or essential oils used mainly as flavoring in cooking or in confections are listed under condiment, spice and essential oil plants, since these oils are used mainly for their flavoring and aromatic qualities. Essential oils from many plant sources are important constituents of perfumes and cosmetics. These are not listed since they are not used in foods. The second group of edible vegetable oils are non-volatile except at very high temperatures. They are widely used as butter substitutes, in salad dressings and as cooking fats. This group can be further divided into non-drying, semi-drying, and drying oils or fats. Sources of these oils are described in the crop monographs.

Practically all seeds contain oil, which can be extracted either by heating and pressing or by solvent extraction. Also, the fleshy tissues which surround the seeds of some fruits contain an extractable oil.

Plants which are important sources of edible food oil in the U.S., or which are significant in international trade, are listed. Per capita consumption of edible food oils in the U.S. for 1994 is about 60 pounds (USDA 1997b).

Oilseed

Crop Groups
   U.S. – Miscellaneous
   Canada – 20
   Codex – SO, OR and OC
Authors listing by crop monograph —

Ben moringa seed	Flax	Peanut
Borage	Jojoba	Poppy seed
Buffalo gourd	Kapok oil	Rape (seed)
Castor oil plant	Lesquerella	Safflower
Coconut	Meadowfoam	Sesame
Cottonseed	Mustard seed	Shea butter tree
Crambe	Niger seed	Soybean
Cuphea	Olive	Sunflower
Euphorbia	Palm oil	Veronia
Evening primrose, Common		

Stimulant Crops List

This includes a small group of crops used in mildly stimulating drinks (coffee, tea, cola) and tobacco. Cacao, the source of chocolate, is also listed. Although not strictly a stimulant, it appears more compatible to this group than to any other of those used.

Dried edible plant tops (stimulants):

Crop Groups
   U.S. – Miscellaneous
   Canada – None
   Codex – DT
Authors listing by crop monograph —
Tea
Tobacco

Teas:

Crop Groups
   U.S. – Miscellaneous
   Canada – None
   Codex – DT
Authors listing by crop monograph —

Comfrey	Stevia
Lime blossom	

Tropical and Subtropical trees with edible seeds for beverages and sweets:

Crop Groups
   U.S. – Miscellaneous
   Canada – None
   Codex – SB and SM
Authors listing by crop monograph —

Cacao bean	Cola
Coffee	Guarana

Sugar and Syrup Plants List

Sugar is obtained primarily from two sources in the United States, sugarcane and sugar beet. A small quantity of maple sugar is produced from natural stands of the sugar or rock maple, *Acer saccharum*, and some corn sugar is produced by conversion of starch to reducing sugar. Sugarcane, *Saccharum officinarum*, is the main source of sugar in tropical and subtropical countries while the sugar beet, *Beta vulgaris*, is the principal source in temperate regions. About 9 million tons of sugar are consumed annually in the United States. In 1994 about 37 percent of this came from domestic sugar beet production, 50 percent from domestic sugarcane (including Puerto Rico) and about 13 percent was imported — practically all of which was from cane (USDA 1997b). Sugar beets are listed under root and tuber vegetables.

Syrups may be made from the same sources used for sugar, but additional sources are also used. These include the sweet sorghums and grains, particularly corn. Maple syrup is an esteemed product. Syrups are essentially partially refined sugars in rather concentrated solutions. Lack of complete refinement results in characteristic flavors. Sugar and syrup crops and preparation are described in the crop monographs.

Grasses for sugar or syrup:

Crop Groups
   U.S. – Miscellaneous
   Canada – None
   Codex – GS
Authors listing by crop monograph —
Sorghum, Sweet
Sugarcane

Forestry:

Crop Groups
   U.S. – Miscellaneous
   Canada – None
   Codex – None
Authors listing by crop monograph —
Sugar maple

EPA Residue Chemistry Test Guidelines

OPPTS 860.1000
Background
(Includes Table 1)
(Updated)

Introduction

This guideline is one of a series of test guidelines that have been developed by the Office of Prevention, Pesticides and Toxic Substances, United States Environmental Protection Agency for use in the testing of pesticides and toxic substances, and the development of test data that must be submitted to the Agency for review under Federal regulations.

The Office of Prevention, Pesticides and Toxic Substances (OPPTS) has developed this guideline through a process of harmonization that blended the testing guidance and requirements that existed in the Office of Pollution Prevention and Toxics (OPPT) and appeared in Title 40, Chapter I, Subchapter R of the Code of Federal Regulations (CFR), the Office of Pesticide Programs (OPP) which appeared in publications of the National Technical Information Service (NTIS) and the guidelines published by the Organization for Economic Cooperation and Development (OECD).

The purpose of harmonizing these guidelines into a single set of OPPTS guidelines is to minimize variations among the testing procedures that must be performed to meet the data requirements of the U. S. Environmental Protection Agency under the Toxic Substances Control Act (15 U.S.C. 2601) and the Federal Insecticide, Fungicide and Rodenticide Act (7 U.S.C. 136, *et seq.*).

Final Guideline Release: This guideline is available from the U.S. Government Printing Office, Washington, DC 20402 on *The Federal Bulletin Board*. By modem dial 202-512-1387, telnet and ftp: fedbbs.access.gpo.gov (IP 162.140.64.19), internet: http://fedbbs.access.gpo.gov, or call 202-512-0132 for disks or paper copies. This guideline is also available electronically in Adobe format from the EPA Public Access webpage: http://www.epa.gov/docs/OPPTS_Harmonized/860_Residue_Chemistry_Test_Guidelines

OPPTS 860.1000 Background

(a) **Scope**

(1) **Applicability.** This guideline is intended to meet testing requirements of both the Federal Insecticide, Fungicide, and Rodenticide Act (FIFRA) (7U.S.C. 136, *et seq.*) and the Federal Food, Drug, and Cosmetic Act (FFDCA) (21 U.S.C. 301, *et seq.*).

(2) **Background.** The source materials used in developing this harmonized OPPTS test guideline are OPP test 170–1 Scope of Data Requirements, OPP 171–1 List of Requirements, OPP 171–14, Special Considerations for Temporary Tolerance Petitions, and OPP 171–5 Presentation of Residue Data (Pesticide Assessment Guidelines, Subdivision O: Residue Chemistry, EPA Report 540/9–82–023, October 1982).

(b) **Purpose and scope of data requirements**

(1) **General.**

(i) This guideline provides general information and overall guidance for the 860 series on residue chemistry and registrants should use it in conjunction with the other listed guidelines to assure compliance with registration requirements. Topics addressed in this guideline include: Purpose and scope of data requirements, regulatory authority, minor change in use pattern, food use/nonfood use determinations, tobacco use, aquatic uses, special considerations for temporary tolerance petitions, data requirements for temporary tolerances, presentation of residue data, guidance on submittal of raw data, and references. Also included (under paragraph (m) of this guideline) is a table "Raw Agricultural and Processed Commodities and Livestock Feeds Derived from Field Crops."

(ii) Sections of this series describe the residue chemistry data requirements specified by 40 CFR part 158, the Federal Food, Drug and Cosmetic Act (FFDCA), and some other information needed for pesticide uses that may result in residues in food, feed, or tobacco. Residue chemistry data are used by the Agency to estimate the exposure of the general population to pesticide residues in food, and for setting and enforcing tolerances for pesticide residues in food or feed.

(iii) Information on the chemical identity and composition of the pesticide product, the amounts, frequency and time of pesticide application, and results of test on the amount of residues remaining on or in the treated food or feed, are needed to support a finding as to the magnitude and identity of the residues which result in food or animal feed as a consequence of a proposed pesticide usage.

(iv) Residue chemistry data are also needed to support the adequacy of one or more methods for the enforcement of the tolerance, and to support practicable methods for removing residues that exceed any proposed tolerance.

(2) **Petitions for tolerance.** Residue chemistry data for a new use of a pesticide are generally submitted to the Agency in a petition for tolerance as required under section 408 or 409 of the FFDCA. The format, 2 procedures and fees associated with petitions for tolerance are included in sections 408 and 409 of the FFDCA, and 40 CFR 180.1 to 180.35.

(c) **Regulatory authority.** The Agency regulates pesticides under two acts, FFDCA and FIFRA. The FFDCA gives the Agency the authority to set legally enforceable limits, or tolerances, for pesticides in foods. The Agency sets tolerances for pesticide residues remaining in raw agricultural commodities (RACs) under section 408 of the FFDCA. The Agency sets food additive tolerances under section 409 of the FFDCA for pesticide residues which concentrate in processed foods above raw food tolerances, or which are the result of pesticide application during or after food processing. In some cases maximum residue limits are established under section 701 of the FFDCA for residues in processed commodities. Under FIFRA, all pesticides must be registered with the Agency before they may be sold or distributed in commerce. FIFRA sets an overall risk/benefit standard for pesticide registration, requiring that pesticides perform their intended function, when used according to labeling directions, without posing unreasonable risks of adverse effects to human health or the environment.

(d) **Minor changes in use pattern.**

(1) If a minor change in the use pattern or formulation of a currently registered pesticide is requested, the registrant may have to submit additional residue data to demonstrate that the change will not result in residues exceeding the established tolerance. Examples of changes in pesticide use patterns that are

likely to require residue studies, and possibly another petition for a new tolerance, include:

(i) Significant changes in preharvest interval and/or in postharvest treatment.

(ii) Extension of use patterns to include low volume or ultralow volume (ULV) aerial as well as ground application.

(iii) Addition of a sticker or extender to the formulation.

(iv) Conversion to a slow-release formulation.

(v) Use in additional climatic regions.

(vi) An increase in the application rate.

(vii) An increase in the number of applications allowed.

(2) Examples of minor changes that may not require residue studies include:

(i) A change in surfactant concentration.

(ii) Substitution of a new but similar surfactant.

(iii) Substitution of one clay diluent for another.

(iv) Exceptions must be made individually based on a thorough knowledge of the chemistry of the ingredients involved.

(3) Residue data in support of a changed use pattern are normally submitted to the Agency under FIFRA in a form described in 40 CFR part 162.

(e) **Food use/nonfood use determinations**

(1) **Definitions.** The term "food" is defined in section 201(f) of the FFDCA as articles used for food or drink for man or other animals, chewing gum, and articles used for components of any such articles. If a pesticide use is likely to result in residues on food, the use is a food use, a petition for tolerance or exemption is required, and appropriate residue chemistry considerations apply. Nonfood uses are those uses that are not likely to yield residues in food. Uses that could result in residues in meat, milk, poultry, or eggs are also considered to be food uses.

(2) **Food use/nonfood use determination data requirements.**

(i) In some cases residue chemistry data are needed to determine whether a proposed use is a food use or a nonfood use. The general criteria for food use/nonfood use determinations is that if residues could occur in foods or feed, the use is a food use and a petition for tolerance/exemption from tolerance is required. In some cases this determination can be made based on the nature of the site to which the pesticide is to be applied. Thus application to land other than cropland is considered a nonfood use. In other cases the distinction is not as clear. For example, baiting with a rodenticide around the borders of cropland or in a tamper-resistant bait box within cropland would be considered a nonfood use, but applying the bait directly to the crop would be considered a food use.

(ii) For the following types of uses, the food use/nonfood use determination will be based on the results of the data described in each section. Registration for these types of uses will not be granted until the necessary data have been submitted to the Agency and found acceptable.

(A) Seed treatments.

(1) This includes cases where seed is treated either by the seed company (and dyed according to 40 CFR 153.155) or by the farmer (planting box or hopper treatments).

(2) In order for a seed treatment to be considered a nonfood use, data from a radiotracer study must be available showing no uptake of residues (radioactivity) from treated seed into the aerial portion of the growing crop.

(3) If residues occur in the aerial portion of the plant, or if there is no data available to make this determination, seed treatments are considered to be food uses requiring tolerances.

(B) Crops grown for seed only.

(1) These crops may qualify as nonfood uses provided there is no likelihood of residues in crops grown from the harvested seeds and the seeds or other parts of the treated crop are not diverted to food/feed use. Factors affecting this include the level of residues on the harvested seed, the half-life of seed residues, the weight of the seed in relation to that of the subsequent crop, and the amount of residue uptake from the seed into the aerial portion of the crop. More details are provided in paragraphs (e)(3) and (e)(4) of this guideline.

(2) On the other hand, uses on crops where the seed itself is a major RAC (such as corn, sorghum, soybeans, small grains, and sunflowers) are considered to be food uses. In these cases, seeds from treated crops could not be distinguished from untreated crops and could be diverted to human and animal consumption.

(3) With the exception of certain FIFRA section 24(c) registrations (also referred to as special local needs (SLNs)), as discussed in paragraph (e)(2) of this guideline, alfalfa and clover grown for seed cannot be considered a nonfood use because of the economic importance of alfalfa and clover hay. Subsequent cuttings for hay could be taken regardless of label restrictions. Also, because of the increasing importance of alfalfa sprouts as a human food item, use on alfalfa grown for seed only cannot be considered a nonfood use.

(4) Desiccation uses for clover grown for seed will be considered nonfood uses because the desiccation renders the hay unfit for consumption by livestock.

(C) Fallow land. Use of a pesticide on fallow land requires data indicating whether residues persist in soil long enough for uptake by crops. Fallow land uses must include a time limitation on planting to food/feed crops or tobacco. Twelve months is the longest time interval deemed practical for a fallow land use restriction. If residues persist in soil and are taken up by food/ feed crops for the length of the time of planting limitation, or 12 months (whichever is shorter), a petition for tolerance for all crops which could be planted on the fallow land will be required. Additonal guidance is provided in OPPTS 860.1850 and 860.1900.

(D) Nonbearing crop uses. Nonbearing crops are perennial crops that will not produce a harvestable RAC during the season of application. Application of a pesticide to a nonbearing food or feed crop will be considered a nonfood use only if data are available to demonstrate that no detectable residues occur in the crop at the first harvest. However, provided the label contains a restriction against harvesting food/feed within 1 year of application, the Agency will consider such uses to be nonfood unless the pesticide is known to be persistent and systemic as shown, for example, by rotational crop data. If residues are detected in the crop at first harvest, a petition for tolerance with the full range of residue chemistry data requirements will be needed before a tolerance and registration will be granted.

(3) **Additional considerations.**

(i) One area in which numerous questions have arisen over the years is whether uses of pesticides on crops grown for seed are nonfood uses, which do not require tolerances. The Agency's concern has been that the crop may be diverted to food use before seeds are produced or that the seeds themselves may be used for food or feed, regardless of label statements prohibiting such practices.

(ii) In recent years the Agency has accepted some uses on crops grown for seed as nonfood uses. If a *State* is willing to provide assurance that seed crops will not be diverted to food or feed and information on acreage and cultural practices is provided, the Agency will consider pesticidal treatment of crops grown for seed as a nonfood use under FIFRA section 24(c) registrations. The information on cultural practices should emphasize those practices which distinguish the seed crop from the corresponding food crop and which would render the seed crops commercially unviable for food. Examples of such practices might be smaller crop spacing preventing adequate root formation and planting in a different season to encourage bolting versus head formation. Information would also be needed on possible residues in followup crops grown from the harvested seeds as described in paragraph (e)(4)(i)(B) of this guideline.

(iii) The Agency will consider these uses on a case-by-case basis for FIFRA section 24(c) registrations once the assurances and cultural practice information outlining a rationale as to why these are nonfood uses have been submitted by a State. Most cole, leafy vegetable, and root crops have a good chance of obtaining pesticidal treatment of their seed crops on a nonfood basis. Crops where the seeds themselves are major RACs, such as grains, beans, and peas, have very little chance of such registrations on a nonfood basis. Alfalfa, clover, and grass grown for seed are generally not considered nonfood uses since they can be cut for hay prior to going to seed. However, the Agency has accepted use on alfalfa grown for seed in certain states, including Washington State, based on rules issued by the States governing the use of the seeds, seed screenings, and other crop parts. Recently, the Washington State rule was expanded to cover more than 30 small-seeded leafy and root vegetables.

(4) **Seed use/nonfood use special local needs.**

(i) Seed use may refer to direct application of a pesticide to seeds before planting (seed treatments) or application in the field to crops grown for seed. This section clarifies

the type of information that the Agency needs so that a FIFRA section 24(c) or SLN for seed use may be deemed to be nonfood. If the Agency has a legitimate concern about carryover to subsequent crops, residue information will be needed.

(A) Seed treatments. Seed treatment uses can be considered nonfood only if a radiotracer study shows no uptake of radioactivity into the aerial portion of the crop (or into the underground portion of root crops). Our experience with such studies is that it is quite unlikely a seed treatment will be considered a nonfood use. In most cases a tolerance is established at the quantitation limit of the analytical method on the crop grown from the treated seeds.

(B) Applications to crops grown for seed. Applications to crops grown for seed can be considered nonfood uses if the following two conditions are met:

(1) Subsequent to treatment no parts of the crop will be diverted to use as human food or livestock feed.

(2) There is no likelihood of residues in crops grown from the harvested seed.

(3) These conditions are discussed in more detail under paragraphs (e)(4)(ii) and (e)(4)(iii) of this guideline.

(ii)

(A) In some instances the first condition may be met by the timing of the application. In other words, the condition is met if the pesticide is applied to the seed crop at a point when it is no longer fit for consumption. An example would be use of a desiccant on carrots or radishes near the time of seed harvest. At this time the roots would no longer be desirable as a food.

(B) In those cases where the application timing does not satisfy the first condition, two other possible means of meeting that condition exist:

(1) The State in which the registration is sought provides assurance through some regulatory process that the seed crop will not be diverted to food or feed. This assurance must include all crop parts that could be consumed by humans or livestock. Crops of special concern are alfalfa, clover, and grass, which may be cut for hay. The first example of a State using this procedure was Washington State for registration of pesticides on alfalfa grown for seed.

(2) Cultural practices information is submitted showing how the seed crop may be distinguished from the corresponding food crop and how the seed crop is commercially unviable as a food crop. Examples of such cultural practices might be smaller crop spacing preventing adequate root formation or planting in a different season to encourage bolting versus head formation.

(C) Some general statements can be made at a crop grouping level with regard to the chances of the first condition being met. Most cole, leafy vegetable, and root crops grown for seed have a good chance of meet-

Food and Feed Crops of the United States

ing this requirement. On the other hand, crops where the seeds themselves are major RACs such as grains, beans, and peas have very little chance of nonfood registrations. Cucurbits and fruiting vegetables are probably not eligible for nonfood uses since the fruit is still edible at the stage when the seeds have formed.

(iii) The second condition to be met for a nonfood use will be addressed:

(A) In many cases this condition may be met without actual residue data on the harvested seed or the crop grown from that seed. The registrant and/or State should consider data or information on the following factors when calculating a theoretical residue in the crop grown from the harvested seeds.

(1) Weight of seed.

(2) Weight of edible portion of following crop.

(3) Total weight of following crop.

(4) The weights could be expressed in terms of an individual seed/plant or on a per acre basis. In either case the figures would allow an estimate of the dilution of residues due to growth of the plant.

(5) Tolerances on other crops with similar rate and preharvest interval.

(6) Information as to whether the seeds are directly exposed to the pesticide spray. (The above two factors can be used to estimate maximum likely residues on the harvested seeds.)

(7) Seed treatment data for the pesticide on other crops.

(8) Degree to which the pesticide translocates (how systemic?).

(9) Half-life of the pesticide on other crops. (The above three factors can be used to estimate how much the pesticide might move from seed to the growing crop.)

(B) [Reserved]

(iv) If a calculation using reasonable assumptions and taking into account the above factors indicates residues in the RAC grown from the harvested seeds will be well below (for example, 1 order of magnitude) the detection limit of the analytical method, condition two would be met. On the other hand, if the calculation shows residues close to or above the detection limit, actual residue data on the harvested seeds and/or following crop will be required to show that the use can be considered nonfood.

(f) **Tobacco uses.** Use of pesticides on tobacco does not require a tolerance or an exemption from a tolerance. Nonetheless, data are needed to assess the exposure of humans to residues on tobacco. The following residue chemistry data for tobacco uses should be submitted to the Agency using the procedures for submitting data under FIFRA referred to above.

(1) **Nature of the residue study.** A tobacco metabolism study is required. This study is similar to plant metabolism studies required in support of tolerances on RACs and discussed in OPPTS 860.1300. The commodity of interest is green tobacco leaves. If total toxic residues (parent plus metabolites of potential risk concern, as determined by the Agency) in green tobacco are ≤0.1 ppm, no further residue chemistry data are required. (If the Nature of the Residue in plants is considered adequately under-

stood based on plant metabolism studies accepted by the Agency for uses on food crops, the registrant may choose not to perform a tobacco metabolism study, but may move immediately to generation of field trial data. However, it should be noted that if total toxic residues ≤0.1 ppm are found in a tobacco metabolism study, field trial data would not be required.)

(2) **Field Trials.**

(A) If metabolism data show that the total toxic residue (TTR) is ≤0.1 ppm in green tobacco, a validated analytical method (see OPPTS 860.1340 for basic requirements) and field trial data are required. Field trials should be conducted, as with food crops, using the maximum application rate and number of applications, minimum intervals between applications, and minimum preharvest interval. Samples of green tobacco should be analyzed for the total toxic residue identified in the metabolism study.

(B) Adequate geographical representation is required. OPPTS 860.1500 indicates that three field trials are required (which should all be performed in Region 2). Other applicable requirements in that guideline such as number of samples per site should also be met.

(C) If the maximum residue in all individual composite samples of green tobacco is ≤0.1 ppm, no additional residue chemistry data are required. If the maximum residue in any individual composite sample of green tobacco is ≤0.1 ppm, residues in samples of cured tobacco should be determined (three locations may be the same as those for green tobacco). If the maximum residue in all individual composite samples of cured tobacco is ≤0.1 ppm, no additional residue chemistry data are required.

(3) **Pyrolysis study.** If the maximum residue in any individual composite sample of cured tobacco is >0.1 ppm, a pyrolysis study is required. Pyrolysis products resulting from the total toxic residue must be identified/ characterized in a manner similar to that required for plant metabolism studies. Information from the pyrolysis study will be presented to the Toxicology Branches and/or the Health Effects Division (HED) Metabolism Committee for determination of their human health significance.

(g) **Aquatic uses.** When a pesticide is applied directly to water, data on potential residues in water, fish, shellfish, and irrigated crops will be required. Tolerances are not established for potable water under FFDCA, thus data on residues in water are submitted under FIFRA. Residue limits for potable water will be established under the auspices of the Safe Water Drinking Act (SWDA) when appropriate. Tolerances for fish, shellfish, and/or irrigated crops should be submitted in the form of a petition for tolerance or exemption from tolerance under the FFDCA.

(h) **Special considerations for temporary tolerance petitions.**

(1) A temporary tolerance may be established in conjunction with an Experimental Use Permit (EUP). A petition for a temporary tolerance should specify the amount of pesticide which will be used under the conditions of the experimental permit, the crop acreage which will be treated, and the geographical areas where the treatments will be made.

(2) The chemistry data requirements for permanent tolerances will in general apply to petitions for temporary tolerances except that the metabolism and residue data need not be extensive.

Whether the latter exception is to be made will depend on the toxicity of the pesticide or possible degradation products and/or metabolites, the amount of acreage to be treated, the importance of the food or feed commodity, and similar considerations. It is assumed that data will be obtained while the EUP is in effect and will be made available to the Registration Division of OPP.

(i) **Data requirements for temporary tolerances.**

(1) Temporary tolerances are established to cover residues of pesticides that result from EUPs. These permits are used to gather additional data on efficacy, phytotoxicity, and pesticide residues. The locations, acreages, amounts of pesticide to be applied, and the names of cooperators must be specified. In addition, these EUPs allow use of the pesticide on only a very small percentage of the total national acreage of a crop. For these reasons, the data requirements for a temporary tolerance are not as stringent as those for a permanent tolerance. This guideline provides more details as to what is meant by the phrase "need not be as extensive".

(2) With regard to metabolism studies, characterization and/or identification of the residue need not be as thorough as required for permanent tolerances. If the parent pesticide is the predominant residue in that portion of the residue which has been identified, it is usually acceptable to regulate only the parent compound. However, additional characterization and/or identification of the residue will be required for permanent tolerances. It may also be acceptable to defer regulation of major metabolites (e.g. those comprising more than 10 percent of the total residue) to the permanent tolerance stage. Such decisions will be made via the temporary tolerance review and consultation with the HED Metabolism Committee.

(3) For plant metabolism, the usual requirement for studies on three diverse crops need not be met for temporary tolerance petitions. In those instances where no detectable residues are found on feed items (assuming detection limits for feed items are ≤0.1 ppm), livestock metabolism studies may be delayed until the permanent tolerance stage. In addition, it may be possible to delay poultry metabolism studies to the permanent tolerance stage when low residues are found in feed items and a ruminant metabolism or feeding study indicates there is little transfer of residues to tissues or milk.

(4) Petitioners will be required to submit enforcement methods for temporary tolerances. The methods must be accompanied by independent laboratory validations as directed by PR Notice 96–1 (see paragraph (n)(5) of this guideline). EPA laboratory validations will also be initiated as part of the temporary tolerance process. However, if no other data deficiencies exist, recommendation for a temporary tolerance should not be held up pending the completion of EPA's method validation. If the EPA method trial fails for reasons of enforcement practicality (e.g., availability of glassware) but shows the method is valid for collection of residue data, approval of temporary tolerances may proceed, but the petitioner should be informed to correct the problem prior to establishing permanent tolerances. On the other hand, if a method fails and is not valid for the collection of residue data (due to low or erratic recoveries, for example), the validity of the residue data needs to be determined. Any future tolerances (temporary or permanent, including extensions of temporary tolerances about to expire) depending on that method may not receive a favorable recommendation.

(5) Data showing whether the FDA multiresidue procedures

recover a pesticide may be delayed until submission of a permanent tolerance petition.

(6) The number and/or geographic representation of crop field trials need not be as complete for temporary tolerance petitions. Translation of residue data from similar crops may also be more liberal than that employed in review of permanent tolerances. For example, if use patterns are essentially the same, field trials on apples may be used to supplement pear data for establishing a temporary tolerance on the latter. Storage stability data can also be translated from different crop groups at this stage.

(7) Since temporary food and feed additive tolerances may be established, processing studies are normally required to determine whether residues concentrate in the processed commodities listed in Table 1 under paragraph (m) of this guideline. In some instances it may be possible to waive the processing study if no detectable residues are found in the RAC following application of exaggerated rates (see discussion of processed food/feed under OPPTS 860.1520).

(8) Certain label restrictions may be considered practical for an EUP that would never be considered such for full registration. For example, if an extremely small acreage of grapes were to be treated, it might be practical to restrict the grapes to the fresh market only. Another case may be restricting the feeding of wheat straw to livestock. If such restrictions can be deemed practical for a particular EUP, data on certain RACs or processed commodities (e.g., grape juice, grape pomace, and raisins in the first case cited above) may be waived for the temporary tolerance approval. However, the petitioner should be informed that data are needed to remove such restrictions for a permanent tolerance.

(9) With regard to product chemistry data, information on beginning materials, the manufacturing process, and some analyses of the technical grade active ingredient are required. This may be provided on batches produced by a pilot plant. Data on physical/chemical properties may be deferred to permanent tolerance petitions.

(10) The Agency normally does not object to the extension of EUPs and the accompanying temporary tolerances for a year or two, provided a significant expansion of acreage has not been requested.

(j) **Presentation of residue data.**

(1) Individual analyses, not average results, should be reported. The data should include blank values and uncorrected values for the treated samples. If applicable, it should be indicated how corrections have been made for blanks and recoveries. When analytical methods that rely upon retention time for chemical identification (e.g. gas-liquid chromatography (GLC) and high performance liquid chromatography (HPLC)) are used, corrections should not be made for blanks due to discrete peaks (at the retention time of the pesticide sought) in the chromatograms of untreated controls.

(2) It is preferable to summarize the data in tables showing crop, residue found, dosage, intervals between treatments and interval from final treatment to harvest (PHI), number of applications made, and formulation used. However, the tabulated data should be keyed for ready reference to the raw data and sample history sheets. The sample history sheets should note rainfall, sample treatment (washed, brushed, trimmed, etc.), sample collection and analysis dates, storage conditions, and other factors which might affect the residue levels.

(3) Standard curves, optical absorbance readings, or copies of appropriately labeled chromatograms should be included in the raw data. It is desirable that at least some of the chromatograms show the analyst's sample dilution factors, sample equivalent injected, and the way in which peaks were quantitated. Photographs of thin-layer chromatography plates, paper chromatograms, or radioautographs of plants treated with labeled pesticides should be furnished when such evidence is necessary for the evaluation of the data. More detail on raw data appears in paragraphs (k) and (l) of this guideline.

(4) A statistical treatment of data may be used to express the precision and accuracy of the analytical results when sufficient data make such treatment valid. The null hypothesis technique is useful in calculating the confidence level at which a set of values from control samples do not differ significantly from a set of values from treated samples. A graphical representation of residues versus time (decline or dissipation rate curves) is also desirable. This is usually plotted on semilog coordinates with time (days after treatment) as the linear function.

(k) **Guidance on submittal of raw data.**

(1) The distinction between summary tables and raw data is very important. According to the Good Laboratory Practice (GLP) standards under 40 CFR 160.3, raw data are defined as:

* * * any laboratory worksheets, records, memoranda, notes, or exact copies thereof, that are the result of original observations and activities of a study and are necessary for the reconstruction and evaluation of the report of that study.

* * * "Raw data" may include, but is not limited to, photographs, microfilm or microfiche copies, computer printouts, magnetic media, including dictated observations, and recorded data from automated instruments.

(2) A study report is designed to provide specific information in order to fulfill an Agency requirement. It must include a description of the methods and materials used, the results of the experiment, a discussion of those results, and a conclusion should be made regarding the objective of the study. Calculations or statistical evaluations should be clearly presented and sufficient raw data must be submitted with the report so that the Agency can completely understand how the study was conducted and how the results were derived. Registrants may also want to consider submitting additional information with some studies if they feel it would enhance the review process.

(3) There is a considerable amount of information which is not necessarily considered to be raw data, but which can be crucial for evaluation of a study. Therefore the Agency has decided to include not only requirements for raw data, but also any critical information which is necessary for evaluation of a study and may not be discussed in other guidelines.

(4) The first portion of paragraph (l) of this guideline describes general raw data requirements which apply to virtually all residue chemistry studies. Individual types of studies (i.e. metabolism, analytical method, field trials, etc.) are discussed. Examples are provided whenever specific problems have been noted or special circumstances may exist. Original raw data should not be sent; copies are sufficient. Transcription of data may be adequate for some situations, particularly if the original data are illegible.

(5) Other considerations.

(i) When preparing reports and organizing the raw data associated with the final report the registrant should ensure that information is provided in a manner which is complete, legible, and logical. The data should be well-organized so that the Agency can easily find the desired information. Each sample value listed in a summary table should be linked to a field report, chromatogram, and quantitative data worksheet. Sample storage information and dates should be readily available so the Agency does not need to look at many different places in a report to find the dates of application, harvest, shipping, sample transfer, extraction, and analysis.

(ii) The registrant may want to consider tabulating much of the data into a spreadsheet or other appropriate format. For example, residue data could be organized in a spreadsheet which includes all information necessary for calculation of the residues such as sample weight, volume of final extract, peak height/area, etc. so the Agency can verify the reported results. It would not be necessary to submit each and every calculation sheet if the results were tabulated in such a manner.

(iii) Chromatograms should be completely labeled. A time scale should be on the chromatogram along with the complete sample identification. The retention times of the peaks of interest should also be noted on the chromatogram.

(iv) Data are often submitted which are illegible after copying once or twice and cannot be reproduced readily on microfiche. If the original raw data will not be legible after copying, the data should be transcribed, a statement should be made that the data have been transcribed (or included in a worksheet or spreadsheet), and the transcriptions should be submitted. Raw data are not useful to the Agency if the data cannot be understood. Thin-layer chromatograms (from metabolism studies) are particularly difficult to read when photocopied. Extra attention should be paid to reproductions of these data.

(v) A complete table of contents listing any data appended should be in the report. If a checklist of raw data is appropriate, the registrant may want to consider including such a list.

(vi) A summary of data from associated studies (e.g. storage stability, bridging formulations, etc.) would be helpful, along with a specific reference, including Master Record Identification (MRID) number. The Agency frequently spends a considerable amount of time tracking down such information. While this is not raw data, inclusion of such a table could substantially reduce review time.

(l) **Requirements for residue chemistry studies.**

(1) **General.** Most residue chemistry studies have several common elements: The pesticide is applied to a crop or animal, samples are taken from crops or animals, samples are stored prior to analysis, and the samples are subjected to an analysis. The following list of raw data requirements will apply to most types of residue chemistry studies.

(i) Field notes and/or reports on application, harvest, and plot maintenance.

(ii) Calibration of application equipment (for confirmation of application rate).

(iii) A specific description (which may be in the field

notes) as to how and where (within the plot) the sample was taken. What was done to ensure that a sample is representative of the test plot?

(iv) All dates (not just intervals) associated with the study: Planting (if applicable), application, harvest, shipping, processing, sample preparation, extraction, and analysis. Tabulation in a single table along with the calculated intervals would be helpful.

(v) Information on reference substance characterization as required by GLP standards.

(vi) Techniques used for preparation of the subsamples used for analysis.

(vii) If automated data calculation methods are used, some discussion/ examples on the calculation techniques, such as the spreadsheet formulas, calculation parameters from a commercial system, etc.

(viii) When reports reference registrant/laboratory standard operating procedures (SOPs) and standard methods. The registrant must ensure that descriptions of any critical referenced procedures must be available to the Agency **at the time of the review**. If these SOPs have been previously submitted to the Agency, the registrant may reference the documents by the MRID number or accession number.

(ix) Data from analyses aborted due to instrument failure or other circumstances should not be included. GLP regulations require that such data be kept, but it need not be submitted with the final report.

(2) **Requirements for plant and animal metabolism studies.**
(i) The purpose of these studies is to define the chemical nature of the pesticide residue in plants and animals. Because all other residue chemistry studies depend on the residue definition, these are the most important residue chemistry studies considered. Review of metabolism studies is time consuming because they are so closely scrutinized. Generally more raw data are required for metabolism studies than any other type of study. It can be discerned from the list below that all the raw data associated with any analysis must be submitted. Ancillary raw data such as freezer temperature logs, standard preparation records, and instrument maintenance logs need not be submitted.

(ii) It must be emphasized that **raw** data must be submitted. Corrections for losses should not be made. Requirements for raw data are outlined in the following list. These items should be considered in addition to the general list provided under paragraphs (l)(1)(i) through (l)(1)(ix) of this guideline.

(A) All quantitative data which are used to assess the amount of radioactivity in all samples and extracts. This would include, but is not limited to, specific activity of the radiolabeled pesticide and quantitative details of its dilution (how much labeled and how much not labeled) in the final formulation, counting data (in disintegrations per minute), sample weights, counting efficiencies, and other appropriate measurements so that the concentration of radioactivity (test chemical) may be verified. Data should be provided for control and treated samples. Background data for solvents must be included as well.

(B) All chromatograms, mass spectra, NMR spectra, infrared spectra, UV/vis spectra, and any other data used to make a determination of the characterization and identification of degradation products. This would include chromatograms or spectra which demonstrate negative results (i.e. proposed metabolites are not found).

(C) Quantitative data associated with chromatograms, autoradiograms, spectra, etc., so that quantitative assessment can be made of metabolite identification of degradation products. This would include the amount of radioactivity (in disintegration per minute) used in the analysis (e.g. applied to the TLC plate) and the amount recovered. A complete set of sample calculations should be included with formulas and variables defined in generally understood terms, using data from the submitted report.

(D) Chromatograms/spectra of standards and controls.

(E) A flow chart is a useful format for presenting data on distintegrations per minute for extractions.

(F) Information on sample and extract storage including conditions, temperatures, intervals, etc.

(G) The route of exposure to the animals should be specifically described. This is particularly crucial for dermal treatments; location on the animal's body should be described.

(H) Vital statistics of the test animals throughout the study including body weights, egg or milk production, and feed consumption. Animal housing should be described as well.

(3) **Requirements for residue analytical methods studies.**
Once the pesticide residues of concern have been determined from the metabolism studies, analytical methods must be developed so the residues of concern in the commodities of interest can be quantified. There are three types of studies which are submitted under this category: Residue methods specific for the residues of concern which may be used for collection of residue data and enforcement of pesticide tolerances, independent laboratory validation of the tolerance enforcement method, and testing of the pesticide and its residues of concern through the FDA multiresidue protocols. Requirements for raw data are outlined in the following list. These items should be considered in addition to the general list provided under paragraphs (l)(1)(i) through (l)(1)(ix) of this guideline.

(i) All validation data (recoveries) associated with all methods submitted for all commodities of interest. This would include not only final results, but also sample weights, extraction volumes, final volume of extracts, peak heights/areas, injection volumes, and any other data which would allow the Agency to reproduce the calculations. These data can easily be summarized in a table, and may be reported along with analytical data for treated samples.

(ii) Chromatograms from all commodities of interest at all spiking levels, including the claimed limit of quantitation (LOQ). A minimum of 10 chromatograms is suggested for each commodity including both split and control samples, particularly in those instances where tolerances are proposed at or near the limit of quantitation. There is a need for more control and method blank chromatograms at this time than is generally submitted so the Agency can assess the reported limits of detection and quantitation. (If the method has been previously submitted to the Agency for other commodities, it is not necessary to resubmit data, and it may be referenced by MRID number).

(iii) If multiple commodities are tested by the independent laboratory, data (including chromatograms) for **all** tested commodities should be submitted.

(iv) Notes on all technical communications between the study sponsor and the independent laboratory validating the method. This could be summarized by the analysis in a report rather than being notes from a laboratory notebook.

(v) Analyst notes on the difficulties of the method and modifications made to facilitate method implementation.

(vi) All appropriate manufacturer and lot numbers for chromatography columns, chemicals, and equipment.

(4) **Requirements for storage stability.** Most samples collected for later analysis in conjunction with a metabolism or magnitude of residue study are stored for an extended period (more than 1 month) prior to analysis. Therefore registrants must assure the Agency that the residues are stable in the commodity during the storage period under the conditions actually used for storage of field samples or demonstrate the degree of loss of residues during storage. Storage stability studies are designed to provide this information. Requirements for raw data are outlined in the following list. These items should be considered in addition to the general list provided under paragraphs (l)(1)(i) through (l)(1)(ix) of this guideline. Separate guidance on storage stability studies is available under OPPTS guideline 860.1380.

(i) Storage intervals, locations, temperatures, containers.

(ii) Preparation of samples prior to placement into storage. The report should state whether the samples were stored intact or homogenized.

(iii) Dates of spiking, extraction, and analysis.

(iv) If samples with field-weathered residues are used, a summary of the field trials used to generate the samples (refer to the section on crop field trials). Registrants may choose to reference this information (MRID number) if it has been submitted to the Agency in a separate report.

(v) All quantitative data so that the Agency can reproduce the calculations performed by the registrant. This would include, but not necessarily be limited to, sample weights, extraction volumes, aliquot volumes/ weights, injection volumes, final extract volumes, and peak heights/areas. This may be summarized in a table or spreadsheet.

(vi) Representative chromatograms of control, spiked, and treated (if applicable) samples, including those used to demonstrate storage stability in metabolism studies.

(5) **Requirements for magnitude of the residue studies.**

(i) Several types of residue chemistry studies are used to obtain a quantitative assessment of the residues of concern in/on commodities consumed by humans and/or livestock. Table 1, under paragraph (m) of this guideline, identifies all significant food commodities and feed stuffs, both raw and processed, for which residue data are

collected and tolerances are set. Table 1, titled "Raw Agricultural and Processed Commodities and Livestock Feeds Derived from Field Crops," was Table II in the earlier OPP guidelines. In addition, the table provides the following information for feed stuffs: A maximum percent of the diet on a dry weight or as-fed basis for beef and dairy cattle, poultry, and swine, and guidance on the acceptability of label restriction prohibiting acommodity's use as a livestock feed. The Agency recognizes all plant commodities can be, and are, fed to livestock. However, the Agency requires livestock metabolism and feeding studies, and residue data only on those commodities considered to be "significant."

(ii) Most raw data requirements are common to each type of magnitude-of-the-residue study. Therefore general requirements are described followed by requirements which would be unique to each type of study. These items should be considered in addition to the general list provided under paragraphs (l)(1)(i) through (l)(1)(ix) of this guideline.

(A) Specific information on sample storage including conditions, sample form (intact or homogenized), and sample container. Often a range of temperatures is provided, which account for some temperature spikes, but a statement as to the sample condition throughout the storage period is generally not provided. For example a power outage may have driven the temperature (of the freezer) up to 5°C from –20°C, which is duly reported, but the Agency does not know how long the samples were stored at 5°C or whether they remained frozen. Magnitude of residue study reports should detail how samples are handled and stored prior to receipt by the lab. In particular, the interval between sample harvest and refrigeration of the sample (or placing it on Dry Ice) should also be reported.

(B) A sufficiently wide range of chromatograms must be submitted so that the Agency can make an assessment of the reproducibility of the method, potential interferences, variations in signal-to-noise ratios, and so on. If a study consists of 10 samples or fewer, chromatograms from all samples should be provided. If more than 10 samples are analyzed, a minimum of 10 treated-sample chromatograms should be provided. Chromatograms of samples with unusual or inconsistent results should also be included.

(C) A minimum of three chromatograms each of control and fortified control samples for each RAC and processed fraction. These chromatograms are necessary to assess the limit of quantitation.

(D) A sample calculation should be presented in which a treated sample and fortified control are taken through the entire calculation procedure. The Agency should be able to tie this calculation to a chromatogram and quantitative data included in the report.

(E) All quantitative data associated with all samples should be provided so that the Agency can independently calculate the results. This would include, but not necessarily be limited to: Sample weights, extraction volumes, aliquot volumes/weights, injection volumes, final extract volumes, and peak heights/areas. Dates of each step should be specified so the treated sample data can be associated with control samples, spike recovery samples, and standards. These data can be summarized in a spreadsheet.

(F) All data associated with the calibration standards must be submitted. If a single point external standard quantitation method is used, the data for each standard should be presented so the Agency is able to calculate the residues in/on treated samples. If a calibration curve is calculated by linear regression or some other method, information for each standard should be included, along with the regression data. The Agency should have sufficient information regarding the curve calculation techniques to reproduce the standard curve. If a calibration curve is manually drawn, each calibration curve should be provided.

(G) All quantitative data associated with spike recovery samples must be submitted. This is particularly crucial if residues in/on treated samples are corrected for spike recoveries. It must be explicitly stated in the report whether results are corrected for an **average** recovery or the set recovery. If corrected values are reported for treated samples, apparent values should be reported as well.

(6) **Requirements for crop field trial studies.** Crop field trials are used to assess the magnitude of the pesticide residue in/on a commodity at the time of harvest. These studies usually involve application of the pesticide to a crop in accordance with label directions in a manner which would "* * * expose the crop to the maximum legal amount of the pesticide." The information provided is used to set pesticide tolerances. Requirements for raw data are outlined in the following list. These items should be considered in addition to the general list provided under paragraphs (l)(1)(i) through (l)(1)(ix) of this guideline.

(i) A summary of the weather conditions for the growing season, stating whether the weather conditions were normal, should be provided. For example, the report could state that the crop was subjected to normal temperatures with higher than average rainfall. More detailed weather data should be provided if unusual circumstances exist such as drought or hurricane which could help to explain any unusual residues found. The Agency should have sufficient information to ascertain if a field was irrigated, and the type of irrigation system used. Daily rain, wind, and temperature data need not be provided on a routine basis.

(ii) Actual analysis for moisture content of potential animal feed commodities may be required by the Agency if the samples do not appear to reflect the appropriate crop stage.

(7) **Requirements for processing studies.** Processing studies are used to assess the effect of commodity processing on the pesticide residue, i.e. if residues concentrate or are diluted. Concentration of residues may necessitate establishment of food or feed additive tolerances or maximum residue limits under section 701 of the FFDCA. Because crop field trials are an integral part of these studies, the requirements for raw data listed below should be considered **in addition** to those outlined in the general list provided under paragraphs (l)(1)(i) through (l)(1)(ix) of this guideline, and the list for crop field trials.

(i) Identification of the source of the commodity which was processed (the place from which the commodity being processed was obtained).

(ii) Detailed description of processing procedures actually used and a comparison to commercial processing.

(iii) Weights of sample processed and processed products. Material balances of the processes are very helpful, particularly if presented in a flow chart format.

(iv) If any phytotoxicity concerns are raised in the processing study report (precluding the use of exaggerated rates which may be necessary to produce detectable residues in/on the RAC), all data which support any statements made must be provided.

(v) Moisture content of potential animal feed commodities (e.g., wet apple pomace, pineapple process residue, processed potato waste).

(8) **Requirements for magnitude of residue—meat, milk, poultry, and eggs studies.** The Agency is concerned about pesticide residues in animal products primarily as a result of two means of exposure: Consumption of treated feeds by livestock and direct application of the pesticide to animals and their premises. Requirements for raw data are outlined in the following list. These items should be considered in addition to the general list provided under paragraphs (l)(1)(i) through (l)(1)(ix) of this guideline.

(i) Vital statistics such as feed consumption, and milk/egg production.

(ii) A brief description of compositing procedures, such as combining milk samples from morning and afternoon milking or combining tissues from hens within a treatment group. Preparation techniques should also be specified.

(iii) If the animals are exposed to the pesticide through dermal or premise treatment, the application techniques (from applicator notes, etc.) should be provided. The Agency should have a complete understanding of the exposure.

(iv) Moisture content of feeds used in feeding studies, as well as any data on the amount of feed consumed daily by the animal. The Agency should be able to calculate the animal's exposure on dry matter and as-fed bases.

(9) **Requirements for other residue chemistry studies.** There are other types of residue chemistry studies which are less frequently submitted. The raw data requirements described in the general list provided under paragraphs (l)(1)(i) through (l)(1)(ix) of this guideline and in crop field trial lists typically apply to these studies as well and should be included. Listed below are some additional factors to consider for these individual studies, most of which are concerned with application of the pesticide.

(i) Magnitude of residue—water, under OPPTS 860.1400:

(A) Specific description of the location and technique of the pesticide application should be provided. The description should be related to maps of the area with regard to introduction of the water (containing pesticide residue) into water resources.

(B) Sampling locations (as related to the maps) should be very specifically described. Sampling technique and timing is crucial as well.

(C) Storage stability considerations for water samples are often neglected; this information should be included.

(ii) Magnitude of residue—fish, under OPPTS 860.1400:

(A) Application of the pesticide to the water and introduction of the fish into the water must be thoroughly described in the field notes submitted.

(B) A complete description of the test system (static vs. dynamic, etc.) is very important in the assessment of the study.

(iii) Magnitude of residue—irrigated crops, under OPPTS 860.1400:

(A) Application of the pesticide to the test system must be explicitly described.

(B) Include introduction of the pesticide into the irrigation water and frequency and method of crop irrigation.

(iv) Magnitude of residue—food handling establishments, under OPPTS 860.1460:

(A) The test location and setup must be thoroughly described. Placement of the food or commodities in the treated area should be explicitly stated. Maps of the treated areas have been very helpful to the Agency. The type of packaging and/or covers should be described.

(B) The application of the pesticide must be described in detail. This would include a description of the site of application in relation to the location of the food.

(m) **Raw Agricultural and Processed Commodities and Livestock Feeds Derived from Field Crops.**

(1) Table 1 provides a listing of all significant food and feed commodities, both raw and processed, for which residue data are collected and tolerances may be set. In addition, for feed commodities (commonly called feedstuffs) the table provides the maximum percent in the diet for beef and dairy cattle, poultry and swine and guidance on the acceptability of label restrictions prohibiting use as a feedstuff.

(2) According to the information available, most agricultural crops and their corresponding raw agricultural and processed commodities can be, and are, fed to livestock. However, for regulatory purposes, the Agency requires residue data and livestock metabolism and feeding studies only for those feedstuffs considered to be "significant". In some cases, the criteria under paragraph (m)(3)(i) of this guideline were also applied to byproducts of food processing to determine if there were "significant" feedstuffs produced and fed to livestock.

(3) The following criteria were used to decide what feedstuffs are considered "significant":

(i) The annual production of the crop (RAC) in the United States is more than 250,000 tons.

(ii) The maximum amount in the livestock diet is more than 10 percent.

(iii) The commodity is grown mainly as a feedstuff.

(iv) The amount of a commodity (raw agricultural or processed) produced or diverted for use as a feedstuff is at least 0.04 percent of the total estimated annual tonnage of all feedstuffs available for livestock utilization in the United States. This amount is equivalent to approximately 270,000 tons on a dry matter basis.

(v) Feedstuffs less than 0.04 percent of the total estimated annual tonnage of all feedstuffs available were included if one or more of the following three criteria are met:

(A) The feedstuff is listed and routinely traded on the commodities exchange markets.

(B) There is regional production, seasonal considerations, or an incident history for use of the feedstuff.

(C) The feedstuff is grown exclusively for livestock feeding in quantities greater than 10,000 tons on a dry matter basis (0.0015 percent of the total estimated annual tonnage of all feedstuffs available).

(4) Using these criteria for inclusion of feedstuffs in Table 1, the Agency expects to account for more than 99 percent of the available annual tonnage (on a dry-matter basis) of feedstuffs used in the domestic production of more than 95 percent of beef and dairy cattle, poultry, swine, milk, and eggs.

(5)

(i) Historically, a label restriction has precluded the need for residue data, tolerances or consideration of selected commodities in the livestock diet. In reviewing the data collected on feedstuffs and public comments the Agency also reevaluated the policy of allowing as a substitute for data, a label restriction prohibiting the use (or sale) of a commodity for feeding purposes. The Agency has developed three criteria to determine whether a label restriction of a commodity from use as a feedstuff could be allowed in lieu of data.

(ii) Criteria developed by the Agency are :

(A) The feedstuff must remain under the control of the grower. (For example, byproducts of processing would usually not be under the control of the grower.)

(B) The crop must not be grown primarily as a feedstuff.

(C) A label restriction should cause no economic hardship.

(D) The Agency's view is that there are only three cases where a label restriction should be allowed: peanut hay, soybean forage, and soybean hay.

(E) Finally, the Agency has added a statement to the table (endnote (3)) on percent dry matter requesting that the percent moisture be reported for representative samples of raw agricultural and processed commodities which are beef or dairy cattle feedstuffs. This is needed since the dietary burdens for these animals are calculated on a dry matter basis. This information will also allow the Agency to determine whether samples were harvested at the proper crop growth stage or dried to the appropriate moisture level after harvest or during processing. In the case of commodities which may have widely varying levels of moisture (e.g., apple pomace, processed potato waste), this information may be used to adjust residue levels for the establishment of tolerances.

(5) A blank space in the processed commodity, feedstuff, or percent of livestock diet columns of Table 1 for a specific crop does not necessarily mean that such items are not produced from this crop, and/or used as human foods or feedstuffs, but that the Agency, at this time, does not consider these items significant in the daily dietary risk assessment of the population of the United States from pesticide use on that crop.

Table 1.—Raw Agricultural and Processed Commodities and Feedstuffs Derived From Crops

Crop	Raw Agricultural Commodity	Processed Commodity	Feedstuff	% DM (3)	Beef Cattle	Dairy Cattle	Poultry	Swine
			Feedstuff		Percent of Livestock Diet (1, 2)			
Alfalfa (4)	forage	....	forage	35	70	60	NU (6)	NU
	hay	....	hay	89	70	60	NU	NU
	seed (5)	....	meal (7)	89	25	50	10	10
		....	silage (8)	40	70	60	NU	NU
Almond	nutmeat hulls	....	hulls	90	10	10	NU	NU
Apple	fruit	pomace, wet juice	pomace, wet	40	40	20	NU	NU
Apricot	fruit (9)	....	....	....	....	....	....	....
Artichoke, Globe	flower head	....	....	....	....	....	....	....
Asparagus	spears (stems)	....	....	....	....	....	....	....
Avocado	fruit (9)	....	....	....	....	....	....	....
Banana (10)	whole fruit	....	....	....	....	....	....	....
Barley (11)	grain (12)	pearled barley	grain (12)	88	50	40	75	80
	hay	flour	hay	88	25	60	NU	NU
	straw	bran	straw	89	10	10	NU	NU
Bean (13)	bean, succulent seed	....	....	....	....	....	....	....
Beet, garden	root tops (leaves)	....	....	....	....	....	....	....
Beet, sugar	root	sugar, refined (14)	tops (leaves)	23	20	10	NU	NU
	tops (leaves)	pulp, dried	pulp, dried	88	20	20	NU	NU
		molasses	molasses	75	10	10	NU	NU
Blackberry (15)	berry	....	....	....	....	....	....	....
Blueberry	berry	....	....	....	....	....	....	....

Footnotes begin on page 400.

Crop	Raw Agricultural Commodity	Processed Commodity	Feedstuff		Percent of Livestock Diet (1, 2)			
			Feedstuff	% DM (3)	Beef Cattle	Dairy Cattle	Poultry	Swine
Broccoli	flower head and stem	....	....	....	....	....	....	....
Brussels sprouts	leaf sprouts	....	....	....	....	....	....	....
Buckwheat	grain (16)	flour	....	....	....	....	....	....
Cabbage	fresh, w/wrapper leaves (17)	....	....	....	....	....	....	....
Cacao bean	bean	roasted bean cocoa powder chocolate	....	....	....	....	....	....
Canola	seed	meal oil, refined	meal	88	15	15	15	15
Carob bean	bean	....	....	....	....	....	....	....
Carrot	root	....	culls (18)	12	25	25	NU	10
Cauliflower	flower head and stem	....	....	....	....	....	....	....
Celery	untrimmed leaf stalk (petiole)	....	....	....	....	....	....	....
Cherry, sweet	fruit (9)	....	....	....	....	....	....	....
Cherry, tart (sour)	fruit (9)	....	....	....	....	....	....	....
Chicory	root tops (leaves)	....	....	....	....	....	....	....
Citrus	fruit, whole	pulp, dried oil juice	pulp, dried	91	20	20	NU	NU
Clover (19)	forage	....	forage	30	30	60	NU	NU
	hay	....	hay	89	30	60	NU	NU
		....	silage (20)	30	30	60	NU	NU
Coconut	coconut (meat and liquid combined)	copra (dried meat) oil	....	....	....	....	....	....
Coffee (21)	bean, green	bean, roasted instant	....	....	....	....	....	....
Collards	greens	....	....	....	....	....	....	....

Footnotes begin on page 400.

| Crop | Raw Agricultural Commodity | Processed Commodity | Feedstuff | | Percent of Livestock Diet (1, 2) | | | |
			Feedstuff	% DM (3)	Beef Cattle	Dairy Cattle	Poultry	Swine
Corn, field	grain starch (25)	wet milling: oil, refined	grain	88	80	40	80	80
	forage (22)		forage (22)	40	40	50	NU	NU
	stover (23) grits flour	dry milling: meal oil, refined	stover (23)	83	25	15	NU	NU
	aspirated grain fractions (24)		aspirated grain fractions (24)	85	20	20	NU	20
			milled byproducts (26)	85	50	25	60	75
Corn, pop	grain	....	grain	88	80	40	80	80
	stover (23)	....	stover (23)	85	25	15	NU	NU
Corn, sweet (27)	sweet corn, K + CWHR (28)	....	forage (29)	48	40	50	NU	NU
	forage (29)	....	stover (23)	83	25	15	NU	NU
	stover (23)	....	cannery waste (30)	30	35	20	NU	NU
Cotton	undelinted seed	meal	undelinted seed	88	25	25	NU	NU
	cotton gin byproducts (31)	hulls	cotton gin byproducts (31)	90	20	20	NU	NU
		oil, refined	meal	89	15	15	20	15
			hulls	90	20	15	NU	NU
Cowpea (32)	seed	....	seed	88	20	20	10	50
	hay	....	hay	86	40	40	NU	NU
	forage	....	forage	30	40	40	NU	15
Crabapple	fruit	....	....	.....	.....	.....	.....	.....
Cranberry	berry	....	....	.....	.....	.....	.....	.....
Crownvetch (33)	forage	....	forage	30	20	60	NU	NU
	hay	....	hay	90	20	60	NU	NU
Cucumber	fruit	....	....	.....	.....	.....	.....	.....
Currant	fruit	....	....	.....	.....	.....	.....	.....

Footnotes begin on page 400.

Crop	Raw Agricultural Commodity	Processed Commodity	Feedstuff		Percent of Livestock Diet (1, 2)			
			Feedstuff	%DM (3)	Beef Cattle	Dairy Cattle	Poultry	Swine
Date	fruit, dried (9)	....	....	....	....	....	....	....
Dewberry	berry	....	....	....	....	....	....	....
Eggplant	fruit	....	....	....	....	....	....	....
Elderberry	berry	....	....	....	....	....	....	....
Endive/Escarole	leaves	....	....	....	....	....	....	....
Fig	fruit	dried	....	....	....	....	....	....
Flax	seed	meal	meal	88	10	10	30	10
Garlic	bulb	....	....	....	....	....	....	....
Ginseng	root, dried	....	....	....	....	....	....	....
Gooseberry	berry	....	....	....	....	....	....	....
Grape	fruit	raisin juice	....	....	....	....	....	....
Grass (pasture & range-land (34)	forage	....	forage	25	60	60	NU	NU
	hay	....	hay	88	60	60	NU	NU
		....	silage (35)	40	60	60	NU	NU
	straw	....	straw	88	10	10	NU	NU
	seed screenings	....	seed screenings	88	25	30	NU	NU
Herbs (36)	fresh	dried	....	....	....	....	....	....
Hops	hops cones, dried (37)	....	....	....	....	....	....	....
Horseradish	root	....	....	....	....	....	....	....
Huckleberry	berry	....	....	....	....	....	....	....
Jerusalem artichoke	tuber	....	....	....	....	....	....	....
Kale	leaves	....	....	....	....	....	....	....
Kiwifruit	fruit	....	....	....	....	....	....	....
Kohlrabi	bulbous stem and leaves	....	....	....	....	....	....	....
Kumquat	fruit	....	....	....	....	....	....	....
Leek	whole plant	....	....	....	....	....	....	....
Lentil	seed	....	....	....	....	....	....	....
Lespedeza (38)	forage	....	forage	22	20	60	NU	NU
	hay	....	hay	88	20	60	NU	NU

Footnotes begin on page 400.

Crop	Raw Agricultural Commodity	Processed Commodity	Feedstuff		Percent of Livestock Diet (1, 2)			
			Feedstuff	% DM (3)	Beef Cattle	Dairy Cattle	Poultry	Swine
Lettuce, head	fresh, w/wrapper leaves (39)	....	....	....	....	....	....	....
Lettuce, leaf	leaves (40)	....	....	....	....	....	....	....
Loganberry	berry	....	....	....	....	....	....	....
Lupin	seed	....	seed	88	20	20	15	20
Mango	fruit (9)	....	....	....	....	....	....	....
Millet (41)	grain (42)	flour (44)	grain (42)	.88	50	40	70	75
	forage		forage	30	25	60	NU	NU
	hay		hay	85	25	60	NU	NU
	straw (43)		straw (43)	90	10	10	NU	NU
Mung bean	bean bean sprouts (45)	....	....	....	....	....	....	....
Mushroom	cap and stem	....	....	....	....	....	....	....
Muskmelon (46)	fruit	....	....	....	....	....	....	....
Mustard greens	greens (leaves)	....	....	....	....	....	....	....
Nectarine	fruit (9)	....	....	....	....	....	....	....
Nuts (47)	nutmeat	....	....	....	....	....	....	....
Oats (48)	grain (12)	flour	grain (12)	89	50	40	80	80
	forage	groats/rolled oats	forage	30	25	60	NU	NU
	hay		hay	90	25	60	NU	NU
	straw		straw	90	10	10	NU	NU
Okra	fruit (pods)	....	....	....	....	....	....	....
Olives	fruit (9)	oil	....	....	....	....	....	....
Onion, bulb	bulb	....	....	....	....	....	....	....
Onion, green	whole plant, w/o roots	....	....	....	....	....	....	....

Footnotes begin on page 400.

Table 1.—Raw Agricultural and Processed Commodities and Feedstuffs Derived From Crops—Continued

Crop	Raw Agricultural Commodity	Processed Commodity	Feedstuff	% DM (3)	Percent of Livestock Diet (1, 2) Beef Cattle	Dairy Cattle	Poultry	Swine
Papaya	fruit	....	....	....	....	....	....	....
Parsley (49)	leaves, fresh	dried	....	....	....	....	....	....
Parsnip	root	....	....	....	....	....	....	....
Passion fruit	fruit	....	....	....	....	....	....	....
Pawpaw	fruit	....	....	....	....	....	....	....
Pea (50)	pea, succulent (51) seed (52)	....	....	....	....	....	....	....
Pea, field (53)	seed	....	seed	90	20	20	20	20
	vines	....	vines	25	25	50	NU	NU
	hay	....	hay	88	25	50	NU	NU
		....	silage (54)	40	25	50	NU	NU
Peach	fruit (9)	....	....	....	....	....	....	....
Peanut	nutmeat	meal	meal	85	15	15	25	15
	hay (55)	oil, refined	hay (55) (R) (56)	85	25	50	NU	NU
Pear	fruit	....	....	....	....	....	....	....
Pepper, bell and non-bell (57)	fruit	....	....	....	....	....	....	....
Peppermint	tops (leaves and stems)	oil	....	....	....	....	....	....
Pimento (58)	fruit	....	....	....	....	....	....	....
Pineapple	fruit	process residue (59) juice	process residue (59)	25	30	20	NU	NU
Plantain (60)	whole fruit	....	....	....	....	....	....	....
Plum	fruit (9)	prune	....	....	....	....	....	....
Potato	tuber	granules/ flakes (61)	culls	20	75	40	NU	50
		chips peel, wet	processed potato waste (62)	15	75	40	NU	NU
Pumpkin	fruit	....	....	....	....	....	....	....
Quince	fruit	....	....	....	....	....	....	....
Radicchio (red chicory)	leaves, fresh	....	....	....	....	....	....	....

Footnotes begin on page 400.

Table 1.—Raw Agricultural and Processed Commodities and Feedstuffs Derived From Crops—Continued

Crop	Raw Agricultural Commodity	Processed Commodity	Feedstuff	% DM (3)	Beef Cattle	Dairy Cattle	Poultry	Swine
Radish	root	....	....	......	....	....	....	......
	tops (leaves)							
Rape	seed	meal (63)	meal	88	15	15	15	15
	forage		forage	30	30	30	NU	NU
Rape greens (64)	greens (leaves)	....	....	......	....	....	....	......
Raspberry, black and red	berry	....	....	......	....	....	....	......
Rhubarb	petioles	....	....	......	....	....	....	......
Rice (65)	grain (12)	polished rice	grain (12)	88	40	40	60	65
	straw	hulls	straw	90	10	10	NU	NU
		bran	hulls	90	10	10	15	NU
			bran	90	15	15	25	15
Rutabaga	root	....	....	......	....	....	....	......
Rye (66)	grain (67)	flour	grain (67)	88	40	40	50	80
	forage	bran	forage	30	25	60	NU	NU
	straw		straw	88	10	10	NU	NU
Safflower	seed	meal oil, refined	meal	91	10	10	25	25
Salsify	root	....	....	......	....	....	....	......
	tops (leaves)							
Sesame	seed	oil	....	......	....	....	....	......
Shallot	bulb	....	....	......	....	....	....	......
Sorghum, grain	grain	flour (68)	grain	86	40	40	80	90
	forage (22)		forage (22)	35	40	50	NU	NU
	stover (23)		stover (23)	88	25	15	NU	NU
	aspirated grain fractions (24)		aspirated grain fractions (24)	85	20	20	NU	20
Sorghum, sweet (69)	stalk	syrup	....	......	....	....	....	......
Sorghum forages, Sudan grass	(See Grass)	....	....	......	....	....	....	......

Footnotes begin on page 400.

Table 1.—Raw Agricultural and Processed Commodities and Feedstuffs Derived From Crops—Continued

Crop	Raw Agricultural Commodity	Processed Commodity	Feedstuff	% DM (3)	Beef Cattle	Dairy Cattle	Poultry	Swine
					Percent of Livestock Diet (1, 2)			
Soybean (70)	seed	meal	seed	89	15	15	20	25
	forage	hulls	forage (R) (56)	35	30	30	NU	NU
	hay	oil, refined	hay (R) (56)	85	30	30	NU	NU
	aspirated grain fractions (24)		aspirated grain fractions (24)	85	20	20	NU	20
			meal	92	15	15	40	25
			hulls	90	20	20	20	NU
			silage (71)	30	30	30	NU	NU
Spearmint	tops (leaves and stems)	oil	....	....	....	....	....	....
Spices (72)	fresh	dried	....	....	....	....	....	....
Spinach	leaves	....	....	....	....	....	....	....
Squash	fruit	....	....	....	....	....	....	....
Strawberry	berry	....	....	....	....	....	....	....
Sugarcane (73)	cane	molasses (74) sugar, refined (14)	molasses (74)	75	10	10	NU	NU
Sunflower	seed	meal oil (refined)	meal	92	15	15	30	20
Sweet potato	root	....	....	....	....	....	....	....
Swiss chard	petioles	....	....	....	....	....	....	....
Taro	corm foliage	....	....	....	....	....	....	....
Tea (75)	plucked leaves	dried instant	....	....	....	....	....	....
Tomato	fruit	paste (76) puree	....	....	....	....	....	....
Trefoil (77)	forage hay	....	forage	30	20	60	NU	10
		....	hay	85	20	60	NU	NU
Turnip	root tops (leaves)	....	root	15	75	20	NU	40
		....	tops (leaves)	30	50	30	NU	NU

Footnotes begin on page 400.

Crop	Raw Agricultural Commodity	Processed Commodity	Feedstuff		Percent of Livestock Diet (1, 2)			
			Feedstuff	% DM (3)	Beef Cattle	Dairy Cattle	Poultry	Swine
Vetch (78)	forage hay	....	forage	30	20	60	NU	NU
		....	hay	85	20	60	NU	NU
Watercress	leaves and stems	....	....	....	....	....	....	....
Watermelon	fruit	....	....	....	....	....	....	....
Wheat (79) (80)	grain (67)	bran	grain (67)	89	50	40	80	80
	forage	flour	forage	25	25	60	NU	NU
	hay	middlings	hay	88	25	60	NU	NU
	straw	shorts	straw	88	10	10	NU	NU
	aspirated grain fractions (24)	germ	aspirated grain fractions (24)	85	20	20	NU	NU
			milled byproducts (81)	88	40	50	50	50
Yam	tuber	....	....	....	....	....	....	....

(ii) Table notes. The following notes are referenced in the table.

(1) **Percent of Livestock Diet.** For percentages of feedstuffs in livestock diets other than those listed here, contact one of the Chemistry Branches, Health Effects Division, Mail Code 7509C, Office of Pesticide Programs, Environmental Protection Agency, 401 M. St. SW., Washington, DC 20460.

(2) **Percent of Livestock Diet.** Maximum percent of diet on a dry weight basis for finishing beef and lactating dairy cattle, and on an as-fed basis for poultry and finishing swine (hogs).

(3) **% DM (percent dry matter)** For beef and dairy feedstuffs, the percent moisture should be reported for representative samples of raw agricultural and processed commodities.

(4) **Alfalfa.** Residue data are needed from a minimum of three cuttings, unless climatic conditions restrict the number of cuttings. Cut sample at late bud to early bloom stage (first cut), and/or at early (one-tenth) bloom stage (later cuts).

(5) **Alfalfa seed.** For registered uses on alfalfa grown for seed, residue data should be provided on seed, forage, and hay; for all other uses data should only be provided on forage and hay.

(6) **NU.** Not used or a minor feedstuff (less than 10 percent of livestock diet).

(7) **Alfalfa meal.** Residue data are not needed for meal; however, the meal should be included in the livestock diet, using the hay tolerance level. Hay should be field-dried to a moisture content of 10 to 20 percent.

(8) **Alfalfa silage.** Residue data on silage are optional, but are desirable for assessment of dietary exposure. Cut at late bud to one-tenth bloom stage for alfalfa, allow to wilt to approximately 60 percent moisture, then chop fine, pack tight, and allow to ferment for three weeks maximum in an airtight environment until it reaches pH 4. This applies to both silage and haylage. In the absence of silage data, residues in forage will be used for silage, with correction for dry matter.

(9) **Fruit.** Fruit should be analyzed after removing and discarding stem, and stone or pit.

(10) **Banana.** Field residue data on both bagged and unbagged bananas should be provided. The required number of field trials may be split between bagged and unbagged bananas. Alternatively, one sample each of bagged and unbagged bananas may be taken from each site. Data are required on the whole commodity (including peel after removing and discarding crown tissue and stalk) for establishing tolerances. At the petitioner's discretion, residue data on just the banana pulp may be provided for purposes of dietary risk assessment.

(11) **Barley hay.** Cut when the grain is in the milk to soft dough stage. Hay should be field-dried to a moisture content of 10 to 20 percent. Barley straw. Plant residue (dried stalks or stems with leaves) left after the grain has been harvested (threshed).

(12) **Barley grain, oat grain, or rice grain.** Kernel (caryopsis) plus hull (lemma and palea).

(13) **Bean.** See Crop Group 6: Legume Vegetables under 40 CFR 180.41 for cultivars of beans. **Bean seed.** Dried seed for

uses on dried shelled beans; succulent seed without pod for uses on succulent shelled beans (e.g. lima beans); succulent seed with pod for edible-podded beans (e.g. snap beans). Cowpea is the only bean crop considered for livestock feeding. (See cowpea). Residue data for forage and hay are required only for cowpea.

(14) **Beet, sugar.** Residue data may be supplied for raw sugar or refined sugar, or both raw and refined. **Sugarcane.** Residue data may be supplied in the same manner.

(15) **Blackberry.** See Crop Group 13: Berries under 40 CFR 180.41 for cultivars of blackberries.

(16) **Buckwheat grain.** Seed (achene) plus hull.

(17) **Cabbage fresh, with wrapper leaves.** Entire cabbage head with obviously decomposed or withered leaves removed. In addition, residue data on cabbage head, without wrapper leaves, are desirable particularly when a more accurate assessment of dietary exposure is necessary.

(18) **Carrot culls.** Data for raw agricultural commodities will cover residues on culls.

(19) **Clover forage.** Cut sample at the 4 to 8 inch to prebloom stage, at approximately 30 percent DM. **Clover hay.** Cut at early to full bloom stage. Hay should be field-dried to a moisture content of 10 to 20 percent. Residue data for clover seeds are not needed.

(20) **Clover silage.** Residue data on silage are optional, but are desirable for assessment of dietary exposure. Cut sample at early to one-fourth bloom for clover, allow to wilt to approximately 60 percent moisture, then chop fine, pack tight, and allow to ferment for 3 weeks maximum in an air-tight environment until it reaches pH 4. This applies to both silage and haylage. In the absence of silage data, residues in forage will be used for silage, with correction for dry matter.

(21) **Coffee.** Residue data are required on the green bean, the roasted bean, and on instant coffee. Tolerances on the green bean will be established under section 408 of the FFDCA. Maximum residue limits on the roasted bean and instant coffee will be established under section 701 of the FFDCA if residues exceed those on the green bean. The green bean is the dried seed of the coffee bean.

(22) **Field corn forage.** Cut sample (whole aerial portion of the plant) at late dough/early dent stage (black ring/layer stage for corn only). **Sorghum forage.** Cut sample (whole aerial portion of the plant) at soft dough to hard dough stage). Forage samples should be analyzed as is, or may be analyzed after ensiling for 3 weeks maximum, and reaching pH 5 or less, with correction for dry matter.

(23) **Corn stover.** Mature dried stalks from which the grain or whole ear (cob + grain) have been removed; containing 80 to 85 percent DM. **Sorghum stover.** Mature dried stalks from which the grain has been removed; containing approximately 85 percent DM.

(24) **Aspirated grain fractions** (previously called **grain dust**). Dust collected at grain elevators for environmental and safety reasons. Residue data should be provided for any postharvest use on corn, sorghum, soybeans, or wheat). For a preharvest use after the reproduction stage begins and seed heads are formed, data are needed unless residues in the grain are less than the limit of quantitation of the analytical method. For a preharvest use during the vegetative stage (before the reproduction stage begins), data will not normally be needed unless the plant

metabolism or processing study shows a concentration of residues of regulatory concern in an outer seed coat (e.g. wheat bran, soybean hulls).

(25) **Corn starch.** Residue data for starch will be used for corn syrup. Petitioners may also provide data on syrup for a more accurate assessment of dietary exposure.

(26) **Corn milled byproducts.** Use residue data for corn dry-milled processed commodities having the highest residues, excluding oils.

(27) **Sweet corn.** Residue data on early sampled field corn should suffice to provide residue data on sweet corn, provided the residue data are generated at the milk stage on kernel plus cob with husk removed and there are adequate numbers of trials and geographical representation from the sweet corn growing regions.

(28) **Sweet corn (K + CWHR).** Kernels plus cob with husks removed.

(29) **Sweet corn forage.** Samples should be taken when sweet corn is normally harvested for fresh market, and may or may not include the ears. Petitioners may analyze the freshly cut samples, or may analyze the ensiled samples after ensiling for 3 weeks maximum, and reaching pH 5 or less, with correction for percent dry matter.

(30) **Sweet corn cannery waste.** Includes husks, leaves, cobs, and kernels). Residue data for forage will be used for sweet corn cannery waste.

(31) **Cotton gin byproducts** (commonly called **gin trash**). Include the plant residues from ginning cotton, and consist of burrs, leaves, stems, lint, immature seeds, and sand and/or dirt. Cotton must be harvested by commercial equipment (stripper and mechanical picker) to provide an adequate representation of plant residue for the ginning process. At least three field trials for each type of harvesting (stripper and picker) are needed, for a total of six field trials.

(32) **Cowpea forage.** Cut sample at 6 inch to prebloom stage, at approximately 30 percent DM. **Cowpea hay.** Cut when pods are one-half to fully mature. Hay should be field-dried to a moisture content of 10 to 20 percent.

(33) **Crownvetch forage.** Cut sample at 6 inch to prebloom stage, at approximately 30 percent DM. **Crown vetch hay.** Cut at full bloom stage. Hay should be field-dried to a moisture content of 10 to 20 percent.

(34) **Grass.** Zero day crop field residue data for grasses cut for forage should be provided unless it is not feasible, e.g. preplant/preemergent pesticide uses. A reasonable interval before cutting for hay is allowed. **Grass forage.** Cut sample at 6 to 8 inch to boot stage, at approximately 25 percent DM. **Grass hay.** Cut in boot to early head stage. Hay should be field-dried to a moisture content of 10 to 20 percent. Grasses include barnyardgrass, bentgrass, Bermudagrass, Kentucky bluegrass, big bluestem, smooth bromegrass, buffalograss, reed canarygrass, crabgrass, cupgrass, dallisgrass, sand dropseed, meadow foxtail, eastern gramagrass, side-oats grama, guineagrass, Indiangrass, Johnsongrass, lovegrass, napiergrass, oatgrass, orchardgrass, pangolagrass, redtop, Italian ryegrass, sprangletop, squirreltail-grass, stargrass, switchgrass, timothy, crested wheatgrass, and wildryegrass. Also included are sudangrass and sorghum forages and their hybrids. For grasses grown for seed only: "Residue data for grass straw (plant material remaining in field after harvest of

seeds) and seed screenings should be provided only for uses on grass grown for seed. A label restriction against the feeding or grazing of directly treated forage is considered practical for uses on grass grown for seed. In such cases, PGIs (pregrazing intervals) and PHIs (preharvest intervals) should be included in the use directions and the residue data for forage and hay may be based on the regrowth after the seed crop has been harvested. If a pesticide is to be used on pasture/range grass in addition to grass grown for seed, the forage (usually zero day) and hay data for the former will cover these two commodities for grass grown for seed (provided the application rates on the latter are not higher). In such cases, only residue data for straw and seed screenings should be provided for the use on grass grown for seed."

(35) **Grass silage.** Residue data on silage are optional, but are desirable for assessment of dietary exposure. Cut sample at boot to early head stage, allow to wilt to 55 to 65 percent moisture, then chop fine, pack tight, and allow to ferment for 3 weeks maximum in an air-tight environment until it reaches pH 4. In the absence of silage data, residues in forage will be used for silage, with correction for dry matter.

(36) **Herbs.** Consist primarily of leaves, stems, and flowers and are marketed fresh (succulent) or dried. See Crop Subgroup 19–A under 40 CFR 180.41 for listing of herbs.

(37) **Hops, cones, dried.** According to PR Notice 93–12 (December 23, 1993), dried hops will be considered as a raw agricultural commodity for regulatory purposes. Residue data are needed for dried hops only.

(38) **Lespedeza forage.** Cut sample at 4 to 6 inch to prebloom stage, at 20 to 25 percent DM. **Lespedeza hay: Annual/Korean.** Cut at early blossom to full bloom stage. **Sericea.** Cut when 12 to 15 inches tall. Hay should be field-dried to a moisture content of 10 to 20 percent.

(39) **Lettuce, fresh, with wrapper leaves.** Entire lettuce head with obviously decomposed or withered leaves removed. In addition, residue data on lettuce head, without wrapper leaves, are desirable particularly when more accurate assessment of dietary exposure is necessary.

(40) **Lettuce, leaf.** Residue data should be on samples with obviously decomposed or withered leaves removed.

(41) **Millet forage.** Cut sample at 10 inches to early boot stage, at approximately 30 percent DM. **Millet hay.** Cut at early boot stage or approximately 40 inches tall, whichever is reached first. Hay should be field-dried to a moisture content of 10 to 20 percent. Millet includes pearl millet.

(42) **Millet grain.** Kernel plus hull (lemma and palea). **Pearl millet** grain. Kernel with hull (lemma and palea) removed.

(43) **Millet straw.** Data are required for proso millet only. **Proso millet straw.** Plant residue (dried stalks or stems with leaves) left after the grain has been harvested.

(44) **Millet flour.** Not produced significantly in the United States for human consumption. Residue data are not needed at this time.

(45) **Mung bean.** Data on mung bean covers sprouts except when the pesticide is used on the sprouts per se.

(46) **Muskmelon.** Includes cantaloupe, casaba, crenshaw, etc. See Crop Group 9: Cucurbit Vegetables under 40 CFR 180.41 for other cultivars of muskmelons.

(47) **Nuts.** Includes Crop Group 14: Tree Nuts under 40 CFR 180.41, except for almonds. Pistachio is under consideration to

be added to Crop Group 14. Residue data for tree nuts may be used to support uses on pistachio. See Crop Group 14 for a listing of nuts. Also see almonds. Almond hulls are considered a significant feedstuff. Hulls from other tree nuts are not considered significant feedstuffs.

(48) **Oats forage.** Cut sample between tillering to stem elongation (jointing) stage. **Oats hay.** Cut sample from early flower to soft dough stage. Hay should be field-dried to a moisture content of 10 to 20 percent. **Oats straw.** Cut plant residue (dried stalks or stems with leaves) left after the grain has been harvested (threshed).

(49) **Parsley.** Fresh parsley is included in Crop Group 4: Leafy Vegetables under 40 CFR 180.41. Dried parsley is included in Crop Subgroup 19A: Herbs under 40 CFR 180.41.

(50) **Pea.** Residue data for forage and hay are required for cowpea. (See cowpea). Residue data for vines and hay are required for field peas only. (See pea, field).

(51) **Pea, succulent.** Succulent seed with pod for edible-podded peas (e.g. snow peas); succulent seed without pod for uses on succulent shelled peas (e.g. English peas).

(52) **Pea seed.** Mature dried seed for uses on dried shelled peas.

(53) **Pea, field.** Does not include the canning field pea cultivars used for human food) Includes cultivars grown for livestock feeding only such as **Austrian winter pea. Field pea vines.** Cut sample anytime after pods begin to form, at approximately 25 percent DM) **Field pea hay.** Succulent plant cut from full bloom through pod formation). Hay should be field-dried to a moisture content of 10 to 20 percent.

(54) **Pea, field, silage.** Use field pea vine residue data for field pea silage with correction for dry matter.

(55) **Peanut hay.** Peanut hay consists of the dried vines and leaves left after the mechanical harvesting of peanuts from vines that have been sundried to a moisture content of 10 to 20 percent.

(56) **(R):** Label restrictions against feeding may be allowed; e.g. Do not feed green immature growing plants to livestock, or Do not harvest for livestock feed.

(57) **Pepper.** Nonbell pepper includes chili pepper.

(58) **Pimento.** The official name adopted by the Georgia Pimento Growers Association.

(59) **Pineapple process residue** (also known as wet bran). A wet waste byproduct from the fresh-cut product line that includes pineapple tops (minus crown), bottoms, peels, any trimmings with peel cut up, and the pulp (left after squeezing for juice); it can include culls.

(60) **Plantain.** Banana tolerance will cover plantain.

(61) **Potato granules/flakes.** Residue data may be provided for either.

(62) **Processed potato waste.** Tolerance levels for wet peel should be used for dietary burden calculations. Residue data may be provided from actual processed potato waste generated using a pilot or commercial scale process that gives the highest percentage of wet peel in the waste.

(63) **Rapeseed meal.** Residue data are not needed for rapeseed oil since it is produced for industrial uses and is not an edible oil. The edible oil is only produced from canola. (See canola.)

(64) **Rape greens.** A commodity listed in Crop Group 5: Brassica (Cole) Leafy Vegetable Group under 40 CFR 180.41.

(65) **Rice straw.** Stubble (basal portion of the stems) left standing after harvesting the grain.

(66) **Rye forage.** Cut sample at 6 to 8 inch stage to stem elongation (jointing) stage, at approximately 30 percent DM. **Rye straw.** Cut plant residue (dried stalks or stems with leaves) left after the grain has been harvested (threshed).

(67) **Rye grain** or **wheat grain.** Kernel (caryopsis) with hull (lemma and palea) removed.

(68) **Sorghum flour.** Residue data are not needed at this time since sorghum flour is used exclusively in the United States as a component for drywall, and not as either a human food or a feedstuff. However, because 50 percent of the worldwide sorghum production goes toward human consumption, data may be needed at a later date.

(69) **Sorghum, sweet.** Sweet sorghum commodities (i.e., seed and forage) will be covered by the sorghum grain tolerances.

(70) **Soybean forage.** Cut samples at 6 to 8 inches tall (sixth node) to beginning pod formation, at approximately 35 percent DM. **Soybean hay.** Cut samples at mid-to-full bloom stage and before bottom leaves begin to fall or when pods are approximately 50 percent developed. Hay should be field-dried to a moisture content of 10 to 20 percent.

(71) **Soybean silage.** Residue data on silage are optional. Harvest sample when pods are one-half to fully mature (full pod stage). In the absence of silage data, residues in forage will be used for silage, with correction for dry matter.

(72) **Spices.** Include aromatic seeds, buds, bark, berries, pods, and roots consumed and marketed primarily in their dried form. See Crop Subgroup 19–B under 40 CFR 180.41 for listing of spices.

(73) **Sugarcane bagasse.** Information indicates that sugarcane bagasse is mainly used for fuel. Residue data will not be needed at this time, but may be needed at a later date.

(74) **Sugarcane molasses.** Residue data are needed for blackstrap molasses.

(75) **Tea.** Residue data are required on plucked (or freshly picked leaves, dried tea, and instant tea. Tolerances on plucked tea leaves will be established under section 408 of FFDCA. Maximum residue limits on dried tea and instant tea will be established under section 701 of the FFDCA if residues exceed those on the plucked tea.

(76) **Tomato paste.** Residue data on tomato paste cover tomato processed products (e.g. sauce, juice, catsup), except tomato puree which covers canned tomatoes.

(77) **Trefoil forage.** Cut sample at 5 to 10 inch or early bloom stage, at approximately 30 percent DM. **Trefoil hay.** Cut at first flower to full bloom. Hay should be field-dried to a moisture content of 10 to 20 percent.

(78) **Vetch forage.** Cut sample at 6 inch to prebloom stage, at approximately 30 percent DM. **Vetch hay.** Cut at early bloom stage to when seeds in the lower half of the plant are approximately 50 percent developed. Hay should be field-dried to a moisture content of 10 to 20 percent. Vetch does not include crownvetch.

(79) **Wheat forage.** Cut sample at 6 to 8 inch stage to stem elongation (jointing) stage, at approximately 25 percent DM. **Wheat hay.** Cut samples at early flower (boot) to soft dough stage. Hay should be field-dried to a moisture content of 10 to 20 percent. **Wheat straw.** Cut plant residue (dried stalks or stems with leaves) left after the grain has been harvested (threshed).

(80) **Wheat.** Includes emmer wheat and triticale. No processing study is needed for a specific tolerance on emmer wheat.

(81) **Wheat milled byproducts.** Use highest value for wheat middlings, bran, and shorts.

(n) References. The following references should be consulted for additional background material on this test guideline.

(1) Environmental Protection Agency. Pesticide Reregistration Rejection Rate Analysis—Residue Chemistry; Follow-up Guidance for: Generating Storage Stability Data; Submission of Raw Data; Maximum Theoretical Concentration Factors; Flowchart Diagrams. EPA Report No. 737–R–93–001, February 1993.

(2) Environmental Protection Agency. Pesticide Reregistration Rejection Rate Analysis—Residue Chemistry; Follow-up Guidance for: Updated Livestock Feeds Tables; Aspirated Grain Fractions (Grain Dust); A Tolerance Perspective; Calculating Livestock Dietary Exposure; Number and Location of Domestic Crop Field Trials. EPA Report No. 737–K– 94–001, June 1994.

(3) Environmental Protection Agency. Pesticide Reregistration Rejection Rate Analysis—Residue Chemistry. EPA Report No. 738–R–92–001, June 1992.

(4) Environmental Protection Agency. FIFRA Accelerated Reregistration—Phase 3 Technical Guidance. EPA Report No. 540/09–90–078, December 1989.

(5) Environmental Protection Agency. Pesticide Regulation (PR) Notice 96–1, Tolerance Enforcement Methods—Independent Laboratory Confirmation by Petitioner, February 7, 1996.

EPA
Residue Chemistry
Test Guidelines

OPPTS 860.1500
Crop Field Trials
(Updated)

INTRODUCTION

This guideline is one of a series of test guidelines that have been developed by the Office of Prevention, Pesticides and Toxic Substances, United States Environmental Protection Agency for use in the testing of pesticides and toxic substances, and the development of test data that must be submitted to the Agency for review under Federal regulations.

The Office of Prevention, Pesticides and Toxic Substances (OPPTS) has developed this guideline through a process of harmonization that blended the testing guidance and requirements that existed in the Office of Pollution Prevention and Toxics (OPPT) and appeared in Title 40, Chapter I, Subchapter R of the Code of Federal Regulations (CFR), the Office of Pesticide Programs (OPP) which appeared in publications of the National Technical Information Service (NTIS) and the guidelines published by the Organization for Economic Cooperation and Development (OECD).

The purpose of harmonizing these guidelines into a single set of OPPTS is to minimize variations among the testing procedures that must be performed to meet the data requirements of the U. S. Environmental Protection Agency under the Toxic Substances Control Act (15 U.S.C. 2601) and the Federal Insecticide, Fungicide and Rodenticide Act (7 U.S.C. 136, *et seq.*).

Final Guideline Release: This guideline is available from the U.S. Government Printing Office, Washington, DC 20402 on *The Federal Bulletin Board*. By modem dial 202-512-1387, telnet and ftp: fedbbs.access.gpo.gov (IP 162.140.64.19), internet: http://fedbbs.access.gpo.gov, or call 202-512-0132 for disks or paper copies. This guideline is also available electronically in Adobe format from the EPA Public Access webpage: http://www.epa.gov/docs/OPPTS_Harmonized/860_Residue_Chemistry_Test_Guidelines.

OPPTS 860.1500 Crop Field Trials

(a) **Scope**

(1) **Applicability.** This guideline is intended to meet testing requirements of both the Federal Insecticide, Fungicide, and Rodenticide Act (FIFRA) (7 U.S.C 135 et seq.) and the Federal Food, Drug, and Cosmetic Act (FFDCA) (21 U.S.C. 301 *et seq.*).

(2) **Background.** The source materials used in developing this harmonized OPPTS test guideline are OPP 171-4 Results of Tests on the Amount of Residue Remaining, Including A Description of the Analytical Methods Used and OPP 171-12 Food Use/Nonfood Use Determination Data Requirements (Pesticide Assessment Guidelines, Subdivision O: Residue Chemistry, EPA Report 540/9-82-023, October 1982). This guideline should be used in conjuction with OPPTS 860.1000, Background.

(b) **Purpose.** Crop field trials are conducted to determine the magnitude of the pesticide residue in or on raw agricultural commodities (RACs) and to reflect pesticide use patterns that could lead to the highest possible residues. The pesticide must be applied at known application rates and in a manner similar to the use directions intended for the pesticide label. Data are normally required for each crop or the representative crops of each crop group for which a tolerance and registration is requested and for each RAC derived from the crop.

(c) **General considerations.** The residue field experiments consist of examination of RACs for residues of the pesticide chemical after treatment corresponding to the uses proposed in Section B of a petition. Residue investigations should be specifically designed so as to circumscribe the total residue situation. Data should be available to show whether residues occur on any plant parts that may be used in foods or feeds. Use on rice may necessitate residue data for water, shellfish, fish, and irrigated crops when water from the rice field is diverted to such uses.

(1) **Residues determined.** The purpose of field experiments is to quantify by chemical analyses the terminal residues of concern that have been previously demonstrated in the plant metabolism studies. The total toxic residue (TTR), as defined in OPPTS 860.1300 Nature of the Residue – Plants, Livestock, should be determined by the analytical method of choice. In some cases, it may be necessary to employ more than one analytical method to determine the TTR. Terminal residues in some commodities may differ from those found in other commodities, e.g. residues in meat or milk may differ from those in plants. In cases where determination of bound residues or minor metabolites require separate analysis or unduly increase the cost of processing residue samples, not all samples must be analyzed for these components. However, enough of the samples must have been analyzed for these components to allow estimation of the ratio of these components to the parent compound. As an alternative to separate analyses for such residues, consideration will also be given to the use of a compound (often the parent pesticide) as a marker or indicator of the TTR. Based on the nature of the residue studies, a ratio of the TTR to the marker compound can often be determined. Petitioners contemplating use of an indicator or marker compound in analytical methods (either for tolerance enforcement or for collecting data in studies such as crop field trials) are advised to contact the Agency about the acceptability of this approach.

(2) **Sampling.**

(i) The samples taken should be of the whole raw agricultural commodity (RAC) as it moves in interstate commerce. For some crops there may be more than one RAC derived from the crop. For example the RACs for field corn include the seed, stover, and forage. Table 1 of OPPTS 860.1000 (page 392) contains a list of the RACs derived from each crop. The sample should not be brushed, stripped, trimmed, or washed except to the extent that these are commercial practices before shipment, or to the extent allowable in 40 CFR 180.1(j) or the Pesticide Analytical Manual (PAM) (see References, paragraph (L)(12) of this guideline). Petitioners are also advised to consult the proposed regulation (40 CFR 180.45) on portion of food commodities to be analyzed (see paragraph (l)(8) of this guideline). In the enforcement program, produce is examined for residues on an as-is basis, regardless of whether it meets any Federal or State quality grading standards with respect to washing, brushing, or number of wrapper leaves retained. Because certain crops (cabbage, celery, and lettuce) may be shipped without having been stripped or trimmed, samples of these crops should be untrimmed for determining tolerances and only obviously decomposed outer leaves should be removed. Data on trimmed and/or washed samples may be generated at petitioner's option for use in risk assessment. The preparation of each sample prior to analysis should be indicated.

(ii) Samples should be collected to reflect the various raw agricultural commodities that might be marketed separately, consumed or fed at various times. For example, in an early post-planting use on winter wheat, the green plant should be sampled at the time it might be foraged and cut for hay, the mature wheat grain should be sampled, and the dried straw should be sampled. Table 1 of OPPTS 860.1000 (page 392) includes a list of the various RACs that should be analyzed.

(iii) The sample taken from a field should be representative of all portions of the crop from the field. Thus, there should be a valid statistical basis for sampling. Standardized procedures, such as the use of the Latin squares for a forage crop, selection of tree fruits from the upper, middle, and lower levels of opposing quadrants of the tree, the use of grain triers for the taking of core samples of commodities in bulk quantities, and sample reduction by quartering of replicate samples from a field are desirable. It is preferable to have additional field site data rather than replicate data from within a field if only a limited number of samples are analyzed. More detail on numbers of samples are provided in paragraph (e)(2)(iv) of this guideline.

(iv) Accepted procedures for maintaining sample integrity should be followed after taking the sample. Normally samples should be frozen as soon as possible and kept frozen until analyzed. Information should be furnished on how samples are shipped and stored until analyzed. If samples are likely to be held in storage, storage stability data should be obtained (see OPPTS 860.1380).

(d) **Design of residue experiments.**

(1) **Field studies.**

(i) The field experiments should be designed specifically to yield residue data, not merely as an adjunct to field performance (efficacy) tests. Field experiments must reflect the proposed use with respect to the rate and mode of application, number and timing of applications, and formulations proposed. Because of differences observed in residue levels resulting from ultra low volume (ULV) and aerial applications, these too may need to be represented unless the proposed label specifically prohibits such application methods. Some crops may be grown under conditions of little or no rainfall, such as lettuce grown in the Imperial Valley of California. Such areas need to be included in field trial programs. More details on number and location of trials to be conducted are provided in paragraph (e)(2)(iv) of this guideline.

(ii) For significant food/feed crops, the field experiments may be required to provide for residue dissipation or decline studies in which samples are taken at intervals during the period from the last application of the pesticide to normal harvest. The data obtained should indicate the pattern of uptake of the pesticide and its decline. When presented graphically, these data are useful in determining a preharvest interval (PHI) if one is needed. More details on residue decline data appear in paragraph

(d) of this guideline.

(2) **Fumigation areas.** In addition to fumigation treatments at the proposed use conditions, treatment at exaggerated rates is desirable. The studies should adequately represent those commodities which might be treated, including oily foods (peanuts, butter), and high surface area foods (flour), and types of packaging allowable under the directions for use. The studies should reflect the effect of parameters such as times of exposure, dosage, temperature, pressure, geometry and airtightness of the container upon residue levels. The effect of aeration time and procedure upon residue reduction should be demonstrated.

(3) **Slow-release encapsulated formulation uses.** The use of slow-release encapsulated formulations may lead to higher residues than conventional formulations. Thus, if use of a slow-release formulation is proposed, residue data reflecting this formulation will be required. Data showing that the analytical method detects any active ingredient remaining in the encapsulating material at the time of analysis is required. The registrant should consult the Agency chemists and toxicologists concerning whether residue data on the encapsulating wall material are needed. The general criteria used by the Agency is that if the encapsulating material is an inert polymer and is not absorbed from the gut, residue data are not required. For polymers not previously cleared, this requires a radiolabeled encapsulating material feeding study with rats showing essentially 100 percent excretion of activity, with no residual activity in tissues. Data on the residue levels of the encapsulating material will not be required for uses involving application before edible parts form.

(e) **Number and location of domestic crop field trials for establishment of pesticide residue tolerances.**

(1) **Summary**

(i) **General.** The number of trials required for a crop takes into account not only its production acreage, but also its dietary significance. All field trial programs – initiated in 1995 or later – should adhere to this guidance. If fewer trials than required in this section are submitted or if a certain region is not represented in trials conducted prior to 1995, judgment will be used to determine whether any additional data are needed. Factors such as the level of residues (i.e. how close to the limit of quantition (LOQ)) and the variability of residues in those trials that were conducted, and the available data for the pesticide on related crops will be considered. When the data fall seriously short, it may not be possible to establish a tolerance until the missing data are provided. In those cases where the data come close to the requirements in this document (e.g. one too few field trials, lack of data from a region of relatively low production) or where the data sufficiently represent residues likely to be seen based on data for similar uses and professional judgment, the tolerance may be granted on a conditional basis.

(ii) **Definitions.** The following definitions apply to this guideline.

A *pesticide field trial* entails one or more applications per growing season of a formulated pesticide product to a specified crop (or the soil) at one site following actual or simulated cultural practices. Such applications are usually in accordance with registered or proposed uses (or a fraction or multiple thereof in some cases) to provide treated commodity samples for estimating pesticide tolerances and/or dietary exposure to pesticides.

A *pesticide field trial site* is a geographically defined address/location within a country/region/state of a field, space, water body, or other area in or on which a pesticide field trial is conducted. (In most cases this definition boils down to a site being one farm.) A site typically consists of several *plots* (areas of ground with defined boundaries on which a crop is grown), each of which receives a specified pesticide application regimen.

Sample is a defined amount of individual agricultural commodity units (e.g. specific number of fruits or tubers, a set weight of grain, etc.) randomly selected from a plot which may be composited for pesticide analysis. (NOTE: As discussed in paragraph (e)(1)(v)(H) of this guideline, tolerances will continue to be based on analyses of composite samples. In the future EPA may also require analyses of individual servings (e.g. one apple, one potato) to assess the dietary risk from acutely toxic pesticides. (This possible requirement will not be discussed further in this guideline.)

(iii) **Numbers of field trials.**

(A) The actual numbers of field trials that will be required for a large number of crops are shown in Table 1 (page 431) and Attachment 7 (page 452, "Required Number of Field Trials" column) of Appendix A of this guideline. The required numbers of trials range from 1 to 20. Crops having large acreage and high consumption for the general population or infants will need up to 20 trials, whereas crops of <200 total U.S. acres will need only one trial. In each case these represent the number of acceptable trials reflecting the label use pattern (maximum rate, etc.) producing the highest residue. Trials which reflect other use patterns or which for some reason do not generate viable samples (e.g. crop failure) will not be counted. For the purposes of standardizing the number of field trials, it should be emphasized that in most cases (see paragraph (e)(1)(iv) of this guideline), these numbers represent the minimum that will be accepted to establish a tolerance (with the exception of crop group tolerances or uses resulting in no quantifiable residues). Additional trials are always welcome and even encouraged by the Agency.

(B) EPA has taken into consideration several major factors to determine the necessary numbers of trials and believes these numbers will be applicable in most cases. However, in limited circumstances the Agency may require additional trials or accept fewer trials than specified in Table 1 of Appendix A of this guideline. An example of fewer trials being required might be a soil fumigant which is so phytotoxic that planting of crops must be delayed until soil residues dissipate and the plant metabolism study shows no uptake

of residues of concern for such a use. Any petitioner believing that fewer trials are adequate for a given crop will need to provide a convincing rationale. In such cases the Agency strongly advises petitioners to submit a protocol and rationale before initiating such trials. Likewise, any residue chemistry reviews requesting additional trials will include a justification as to the need for such data.

(iv) **Residue decline data.** This guideline also gives more specific guidance for current residue decline requirements. Residue decline data will be required for uses where:

(A) The pesticide is applied when the edible portion of the crop has formed.

(B) It is clear that residues may occur on the food or feed commodities at, or close to, the earliest harvest time. The number of decline studies needed is one for crops requiring 5 to 12 total trials and two for crops requiring 16 to 20 total trials. These studies are included in the 5 to 12 or 16 to 20 total trials (i.e. not in addition to these numbers of trials). For a given pesticide additional decline studies will not be required crop by crop if studies on representative crops (tree fruit, root crop, leafy vegetable, grain, and fruiting vegetable) indicate residues do not increase with longer PHIs.

(v) **Compositing of samples.** Two independently composited samples of treated commodity should be collected and separately analyzed in each field trial. These two samples may be taken from the same plot. An exception is met with crops requiring only 1 to 2 trials; for these crops four samples (one each from four separate plots (two at 1× rate, two at 2× rate)) will be needed for each trial. In all cases Codex guidelines on minimum sample sizes should be followed (see Attachment 8 (page 456) of Appendix A of this guideline). A control crop sample should also be collected from each crop field trial site for analysis. For commodities not specifically listed in the Codex guidelines, petitioners are advised to use the Codex guidance on minimum sample size for a crop part having a similar form (e.g. another seed, leafy material, root/tuber).

(vi) **Crop grouping.**

(A) The number of trials in Table 1 (page 431) and Attachment 7 (page 452) of Appendix A of this guideline are based upon each crop being the only one within its crop group for which a tolerance is requested. In the case of crop group tolerances for which there are at least two representative crops, the number of trials can be reduced by 25 percent for those representative commodities that need eight or more trials when requested individually (i.e. 20-15, 16-12, 12-9, 8-6). Table 2 of Appendix A (page 434) of this guideline shows the resulting numbers of trials needed for all crop groups in 40 CFR 180.41.

(B) Since the Agency has also created subgroups within the existing crop grouping scheme, guidance on the number of field trials needed for the representative commodities in these subgroups is provided in Table 3 of Appendix A (page 437) of this guideline. These numbers of trials were determined on a case-by-case basis looking at the acreages and consumption of the representative commodities and of the whole subgroup. Similar principles were applied to crop groups established in 40 CFR 180.1(h) as specified in Table 4 of Appendix A (page 441) of this guideline.

(vii) **Nonquantifiable residues.**

(A) If metabolism data (on the crops of interest or related crops) or field trial data on related crops indicate quantifiable residues are not likely, a petitioner may elect to conduct 25 percent fewer trials for crops normally requiring eight or more trials. However, if all of these trials do not show residues below the LOQ of the method, additional trials will normally be required to bring the total number conducted up to the standard requirement. In addition to residues being below the LOQ, the following two conditions must be met for 25 percent fewer trials to be accepted:

(1) The method has a sufficiently low LOQ (usually ≤0.01-0.05 ppm).

(2) The trials still represent all significant regions of production.

(B) The application of both 25 percent reductions discussed in paragraph (e)(1)(vi)(A) of this guideline (crop group and residues less than the LOQ) to a given crop will not be acceptable. In addition, neither 25 percent reduction will be applied to crops requiring five or more trials.

(C) The numbers of trials in Table 1 (page 431) and Attachment 7 (page 452) of Appendix A of this guideline are also predicated upon only one formulation type being requested for use on each crop. If additional types of formulation are desired, additional data such as side-by-side bridging studies may be needed as discussed in paragraph (e)(2)(x) of this guideline.

(D) Some special considerations are also provided in this document for early season uses on annual crops and spray volumes – ground versus aerial equipment.

(viii) **Amended registrations.** For amended registrations that involve significant changes in application rate or PHI, the number of field trials required will normally be 25 percent less than that needed to establish an original tolerance, provided that the tolerance level is shown by the reduced number of trials to be adequate to cover the new use. However, if the reduced number of trials indicates that the original tolerance is inadequate, or if the original number of trials was five or less or already included a 25 percent reduction (crop group or less than the LOQ), the number of trials needed for an amended registration will be the same as that for the original tolerance. On a case-by-case basis the Agency may require less additional data than described for an amended registration.

(ix) **Location of field trials.** With regard to the location of trials, the Agency has agreed to the division of the country into 13 regions as proposed by NACA/IR-4 (see map, page 325, and Attachment 10, page 459) of Appendix A of this guideline. This will allow greater flexibility in data collection for minor uses. For crops requiring more than three total trials, Table 5 (page 443) of Appendix A of this guideline shows suggested distributions of trials among these 13 regions. These distributions were developed using the concept that the number of trials per region should generally correlate with the percentage of the crop grown in that region. However, where possible, at least one trial should be included in each region having ≥2 percent of the national production.

(x) **Other considerations.**

(A) The distributions of trials in Table 5 of Appendix A of this guideline are not intended to be absolute requirements. Petitioners may wish to contact EPA regarding the suitability of alternative distributions of trials. To aid petitioners in determining distribution of trials, the production of numerous crops by region is specified in Table 6 (page 447) of Appendix A of this guideline.

(B) For crops requiring three or fewer trials, the data should represent to the extent possible a balance of the highest production areas, different geographic/climatic conditions, and/or major differences in varieties of the crop. At least one trial should be conducted in the region of highest production.

(C) With respect to the distribution of multiple trials within a region, this should generally follow the relative production in the individual growing areas (states or counties as appropriate) of the region. However, the sites should also be sufficiently separated to reflect the diversity of the growing region.

(D) To aid the Agency's review process, petitioners are requested to include a copy of the map on page 325 showing the locations of all sites of acceptable trials in the volume of field trial reports for each crop.

(E) Finally, separate guidance has been provided in Attachment 11 (page 459) to address requirements for tolerances with geographically restricted registrations and for FIFRA section 24(c) registrations.

(2) **Detailed discussion**

(i) Background.

(A) In 1992 EPA conducted an analysis of residue chemistry studies that had been submitted in support of the reregistration of pesticides to determine the factors that led to rejection of certain studies (i.e. classified as unacceptable). This analysis included active participation by representatives of the pesticide industry (American Crop Protection Association (ACPA), formerly known as National Agricultural Chemicals Association or NACA) and the IR-4 program, the two major groups which generate residue chemistry data. A frequent reason for rejection of crop field trials was insufficient geographical representation. This could be due to either an insufficient number of trials being conducted or to the trials not being conducted in all areas of significant production for a given crop.

(B) As a result of this analysis, the document entitled "NACA Recommendations for Residue Site Selection and Number of Field Trials" (hereafter referred to as the "ACPA proposal", see paragraph (l)(13) of this guideline) was prepared by members of ACPA, USDA and IR-4 and submitted to EPA in September 1992.

(ii) **Summary of ACPA proposal.**

(A) ACPA/IR-4/USDA proposed dividing crops into three groups (based on total acres) for purposes of defining the number and location of crop field trials:

(1) Major crops involve more than 2 million acres.

(2) Major-Minor crops involve more than 300,000 acres but less than 2 million acres.

(3) Minor crops involve 300,000 acres or less.

(B) The number of trials suggested for major crops (20 trials) and minor crops (3-6 trials) did not take into account factors such as dietary significance or the geographical distribution of production. Such factors were considered for the major-minor crops (8-12 trials).

(C) With regard to location of trials, the ACPA proposal divided the country into 13 regions based on natural geography and climatic boundaries. For distribution of trials it was stated that "* * * The number of trials per region should generally correlate with the percent of the crop grown in that area * * *."

(iii) **EPA analysis of ACPA proposal.**

(A) The ACPA proposal was studied in detail by the Agency and it was decided that it was a useful starting point. However, there were two major concerns. First, there was no consideration given to the dietary significance of the crops that ACPA had placed in the "minor" crop category (three to six trials). It was concluded that more trials were necessary for a significant number of the fruits and vegetables categorized as minor crops. On the other hand, fewer trials were considered necessary for those minor crops with very low dietary intake. The second major concern was that the ACPA proposal did not address the definition of a site, how samples should be collected, and the number of samples per site. The proposal was revised to take into account these two concerns.

(B) Crop field trial topics such as definitions of site, numbers of trials, sampling, and distribution of trials are discussed in more detail in paragraph (e)(2)(iv) of this guideline. Tables are also included to specify the numbers of trials for many crops, crop groups, and crop subgroups, the percentages of crop production by region, and suggested distribution of trials in each region for numerous crops.

(iv) **Sampling requirements.**

(A) With respect to how samples should be collected, the Agency will continue to base tolerances on composite samples. As to the number or weight of agricultural commodity that should be collected for each composite sample, petitioners should follow the Codex "Guidelines on Minimum Sample Sizes for Agricultural Commodities from Supervised Field Trials for Residue Analysis", ALINORM 87/24A (1987)(see Attachment 8 (page 456) of Appendix A of this guideline). For commodities not specifically listed in the Codex guidelines, petitioners are advised to use the Codex guidance on minimum sample size for a crop part having a similar form (e.g. another seed, leafy material, root/tuber). In each field trial report the petitioner should indicate whether or not these guidelines were followed. If they were not, an explanation should be provided along with details on how the sampling deviated from the Codex recommendations. Petitioners should also include in the field trial report the number of agricultural commodity units making a composite as well as the weight of the composite sample.

(B) With regard to the number of samples per site, the Agency has decided that more than one treated sample is needed to provide some estimate of variability, but that three or more samples are unlikely to result in much additional information since compositing will tend to mask much of the variability. Therefore, the Agency has concluded that two independently composited samples should be collected at each site (i.e. for each field trial – with the exception of crops needing only one or two trials as described under paragraph (e)(2)(iv)(D) of this guideline). The treated samples may be taken from two separate plots or from the same plot. In addition, at least one control (untreated) sample should be collected and analyzed at each site.

(C) In those cases where the two treated samples are obtained from the same plot, it needs to be emphasized that the samples be collected by two separate runs through the plot following the aforementioned Codex guidelines. Splitting one sample from a plot or conducting two analyses on one sample will not be an acceptable alternative to separately collecting and analyzing two samples. In other words, multiple analyses of a single sample or of subsamples constitute the equivalent of only one data point. (However, if such multiple analyses are conducted, each value should be reported and clearly indicated as to which sample it represents.)

(D) For crops which require only one or two field trials (≤ 200 and >200-2000 acres, respectively), at least one composite sample should be collected from each of the four separate treated plots (plus the control plot) at each site. It is strongly suggested that more than one sample be collected from each plot. Two plots should be treated at the proposed or registered application rate (1×) and two plots at a 2× rate. Furthermore, each plot should receive independently prepared applications of the pesticide. The same tank mixture should not be used to treat more than one plot. This will allow some assessment of variability due to factors such as preparation of the tank mix. (NOTE: As discussed later in this guideline, petitioners always have the option of conducting three or more field trials at the 1× rate (with two treated samples per trial) instead of the one or two trials with at least four treated samples per trial and plots reflecting both 1× and 2× rates.)

(E) With regard to the handling of samples at the residue analysis stage, petitioners should follow the guidance in section 142 of PAM, Volume I (see paragraph (l)(12) of this guideline on sample compositing and communiting. Multiple analyses of a sample are not required, but are advised as a check in those cases where the residue values from the two composite samples are significantly different.

(F) In all field trial reports petitioners need to indicate clearly whether each reported residue refers to a separate sample or a second analysis of the same sample. In either case, all analyses should be reported-petitioners should not average multiple analyses of a single sample or the results of multiple samples in a trial. See also OPPTS 860.1000 for a discussion of submission of raw data concerning this point.

(v) **Number of trials for individual crops.**

(A) The required number of trials for a crop can be found in either Table 1 (page 431) or Attachment 7 (page 452) (column "Required Number of Field Trials") (see Appendix A of this guideline). Table 1 is an alphabetical list of crops with the minimum number of trials and treated samples. Attachment 7 of Appendix A of this guideline lists the crops in order of number of required field trials, but does not specify numbers of samples. However, Attachment 7 of Appendix A of this guideline does include the acreages and consumptions of crops that were used to determine the number of trials as discussed in paragraphs (e)(2)(v)(b) through (E)(2)(v)(M) of this guideline. Although the list of crops is not all inclusive, an attempt was made to include all crops for which acreage and/or consumption information was available. With regard to names for crops, the Index to Commodities as published in the "Pesticide Tolerances; Revision of Crop Groups" rule (see paragraph (l)(7) of this guideline) was used.

(B) The Agency believed that dietary significance needed to be a greater factor than in the ACPA proposal for determining the amount of residue data required for each crop. First, criteria were developed to assign a base number of field trials dependent solely on total U.S. acreage of the crop. Acreage was used instead of production by weight since the former is more consistent from year-to-year. The primary sources used for acreage information were USDA's Agricultural Statistics (see paragraph (l)(11) of this guideline) and the Census of Agriculture (Department of Commerce, see paragraph (l)(14) of this guideline). IR-4 also provided information on some low acreage crops that are not included in the afore-mentioned publications. When acreage figures varied between sources, the highest figure was used. Acreage from

Puerto Rico was included for coffee and bananas since such production was greater than or comparable to that in the 50 States. The base numbers of field trials as a function of acreage are delineated in Attachment 7 of Appendix A. For simplicity the base numbers of trials are limited to 16, 12, 8, 5, 3, 2, and 1.

(C) Next, criteria were developed to adjust the number of trials based on dietary importance of the commodity. The figures contained in the Agency's Dietary Risk Evaluation System (DRES) for the general population were used to make a first cut. The diets of non-nursing infants and children aged 1-6 were examined to adjust upward the number of trials on any commodities that had significantly higher consumption by these groups than by the general population. The consumption percentages used are those of the whole diet (i.e. food plus water consumption) and are shown along with the acreages of crops in Attachment 7.

(D) For crops having 8-16 base trials (>300,000 acres), it was decided that the number of trials could be increased or decreased based on human consumption. Crops which comprise >0.4 percent of the general population diet had the number of trials increased by one level (e.g. from 8 to 12, from 12 to 16). For those crops having 16 base trials, the number of trials was increased to 20 if they comprise >0.4 percent of the diet. In addition, any crop with >300,000 acres and comprising >1.0 percent of general population consumption requires at least 16 field trials. This particular criterion results in an increased number of trials for apples, oranges, and tomatoes. On the other hand, crops with >300,000 acres accounting for <0.1 percent of consumption had their number of trials decreased by one level. The crops affected by this criterion are primarily or exclusively animal feeds: Alfalfa, clover, cotton, grasses, and sorghum.

(E) For crops ≤300,000 acres the Agency has concluded that due to the small number of base trials (≤5) for such crops, it would not be appropriate to decrease the number of trials based on low consumption. However, any such crops comprising ≥0.02 percent of the general population diet had their number of trials increased by one level (e.g. 3 to 5, from 5 to 8). This criterion affected a significant number of fruits and vegetables such as broccoli, carrots, grapefruit, lettuce, peaches, pears, and snap beans.

(F) Addressing concerns raised in the recent National Academy of Sciences (NAS) report "Pesticides in the Diets of Infants and Children" (see paragraph (l)(15) of this guideline), the Agency also looked at the contribution of crops to the diets of nonnursing infants and children 1-6 years of age. In most cases, crops that are significant in these diets are also important in the diet of the general population. However, rice and oats were found to exceed the 0.4 percent of the diet criterion for large acreage crops using the infant diet, but not when using the diet of the general population. Therefore, the number of trials for these two crops was increased from 12 to 16. In addition, peaches comprise a much higher percentage (1.12 percent) of the nonnursing infant diet than of the general population diet (0.366 percent). Therefore, the number of trials required for peaches was increased from 8 (number based on general population) to 12. (Based on the relatively low acreage of peaches, it was decided not to increase the number of peach trials to 16, the number of trials required for crops having >300,000 acres and comprising >1.0 percent of the diet.)

(G) No information could be located for a number of crops as to total acreage. The acreage for such crops is "0.00" in Attachment 7 (page 452) of Appendix A of this guideline. While most of these are almost certainly very minor crops, for such crops, a minimum of three field trials will be required unless documentation of national acreage can be provided to show fewer trials are an acceptable number.

(H) In addition to total acreage and percentage of the diet, one other factor was considered in determining the number of trials for crops. The Agency believes that the number of trials can be reduced if most of a crop is grown in one region. Therefore, for most crops which have ≥90 percent of their production in one region the number of trials has been reduced one level (e.g. from 8 to 5, from 5 to 3). Crops subject to this reduction include avocados, olives, and pistachios. It should be noted, however, that for some crops having >90 percent of production in one region the number of trials was not reduced due to the dietary significance of these commodities. In the case of crops which only require three trials based on total acreage but have ≥90 percent of production in one region, registrants/petitioners will have the option of conducting three trials with two treated samples per trial or two trials with four treated samples each (four plots per trial – two at 1× rate and two at 2× rate). Some of the crops having this option include globe artichokes, brussels sprouts, figs, mangoes, and parsley. For crops which require two or fewer trials based on total acreage, there will be no reduction based on production being primarily in one region.

(I) The effect of the 90 percent production being in one region can be ascertained by comparing the "Required Number of Field Trials W/O 90 percent" AND "Required Number of Field Trials" columns in Attachment 7 of this guideline. Those crops which have a smaller number of trials in the "Required Number of Field Trials" have received a reduction due to ≥90 percent of production being in one region. The "Required Number of Field Trials" column agrees with the "Minimum No. of Trials" column in Table 1 (page 431) of Appendix A of this guideline.

(J) For the purposes of standardizing the number of required field trials, it should be emphasized that in most cases the number of trials based on paragraphs (e)(2)(v)(A) through (e)(2)(v)(I) of this guideline criteria and listed in Table 1 and Attachment 7 (Required Number of Field Trials) of Appendix A of this guideline represent the minimum number of trials that will be accepted (with the exception of crop group tolerances or uses resulting in no quantifiable residues as described in paragraphs (e)(2)(vii) and (e)(2)(viii) of this guideline). Additional trials are always welcome and, in fact, encouraged in the sense that more data points provide greater certainty of expected residue levels. The Agency has taken into consideration several major factors to determine the necessary numbers of trials and believes these numbers will be applicable in most cases. However, in limited circumstances the Agency may require additional trials or accept fewer trials than specified in Table 1 of Appendix A of this guideline. Any registrant/petitioner believing that fewer trials are adequate for a given crop will need to provide a convincing rationale. In such cases the Agency strongly advises registrants/petitioners to submit a protocol (outlining number and locations of trials) and rationale before initiating such trials. Likewise, any residue chemistry reviews requesting additional trials will include a justification as to the need for such data.

(K) The numbers of trials in Table 1 and Attachment 7 of Appendix A of this guideline represent how many acceptable trials are required reflecting the label use pattern producing the highest residue. In most cases such trials include the maximum rate per application per season, the minimum intervals between applications, and the minimum PHI. Trials which reflect other use patterns will not be counted unless the difference in use is insignificant (e.g. application rate 5 percent higher; PHI of 23 days versus 21 days). In those cases where multiple use patterns are desired and it is not clear which would result in the highest residue (e.g. different PHIs as a function of application rate), the full number of trials will be needed for each use unless side-by-side studies consistently show higher residues from one use pattern. (Additional guidance on this subject for early season uses appears in a in paragraph (e)(2)(ix) of this guideline.) Registrants/petitioners should also be aware that trials which for some other reason do not generate viable samples reflecting the proposed use will not be counted. Possible causes of the absence of such samples are crop failure, mislabeling of samples, contamination, and insufficient documentation of sample integrity from collection to analysis. For these reasons it would be prudent for petitioners to conduct at least the field portions of a greater number of trials than the minimum listed in Table 1 of Appendix A of this guideline.

(L) The Agency believes that one or two trials are adequate for very low acreage crops (≤200 and >200-≤2,000 acres, respectively). A greater uncertainty in residue levels is tolerable for these crops based on their extremely low contribution to the diet. However, if considerable variability is encountered between plots or between trials for such crops, the Agency may set the tolerance noticeably higher than the highest observed residue. In such scenarios registrants/petitioners have the option of conducting additional field trials to attempt to show that a lower tolerance level would suffice. In fact, registrants/petitioners always have the option of conducting three or more field trials at the 1× rate (with two treated samples per trial) instead of the one or two trials with at least four treated samples per trial and plots reflecting both 1× and 2× rates.

(M) Additional points need to be made with regard to the numbers of trials listed in Table 1 (page 430) and Attachment 7 (page 452) of Appendix A of this guideline:

(1) Residue decline studies are included for many uses on crops needing ≥5 trials.

(2) These numbers are based upon each crop being the only one within its crop group for which a tolerance is requested. Refer to the Crop Group Tolerances for how many trials are needed for uses on crop groups.

(3) Fewer trials may be accepted for uses that do not yield quantifiable residues.

(4) The numbers are also predicated upon only one formulation type being requested for use on each crop. Refer to paragraph (e)(2)(x) of this guideline for data requirements for additional types of formulations.

(5) The spray volumes specified for certain uses, especially ultralow volume (ULV) and orchard uses, can affect the number of required trials. This is discussed in more detail later in this guideline (see paragraph (e)(2)(xi) of this guideline).

(6) Fewer trials will be needed for an amended registration provided the existing tolerance is shown to be adequate. Refer to the appropriate section later in this guideline for more details (see paragraph (e)(2)(xii) of this guideline).

(7) Table 1 of Appendix A of this guideline addresses only national registration of terrestrial uses on domestic crops. Import tolerances are not covered. Refer to Attachment 11 (page 459) of Appendix A for guidance on crop field trials to support regional and FIFRA section 24(c) registrations.

(8) The numbers represent trials required for permanent tolerances. With the exception of the small acreage crops, fewer trials will normally be accepted for temporary tolerances (experimental use permits).

(9) Validated analytical methodology, appropriate storage stability data, and documentation on sample handling, shipping, and storage intervals and conditions from sampling to analysis are needed to support all field trials.

(10) Sampling and analysis of treated and control samples for each RAC of a crop as specified in Table 1 of OPPTS 860.1000 (page 392) (e.g. corn grain, forage and stover) should be included in all field trials unless a livestock feeding restriction is considered practical in Table 1 of OPPTS 860.1000 (page 392) and is placed on the pesticide label for a commodity. One exception to this is cotton gin byproducts, for which six trials (three each for "stripper cotton" and "picker cotton") are required as opposed to twelve trials being needed for cottonseed.

(11) Commercially important varieties of a crop as well as seasonal variations (e.g. winter wheat vs. spring wheat) should be covered by the field trials. Data on different varieties are especially important if there are significant differences in size and/or length of growing season. Residue data from more than one year are desirable, but not required for national registration. (NOTE: Data from more than one year will be required for regional registration of crops which require ≥8 trials for national registration as detailed in Attachment 11.)

(12) The numbers of trials are intended to cover terrestrial food uses on growing crops. They do not address postharvest applications to commodities such as fruit or stored grain. These will continue to be handled on a case-by-case basis.

(13) Unless radiolabeled data show a seed treatment to be a nonfood use, it will not be treated differently than any other food use. However, in many cases such uses may be eligible for the 25 percent reduction in the number of trials due to residues being below the method's LOQ.

(14) If the label of a product recommends addition of another ingredient such as crop oil or a specific class of surfactants, the field trials should reflect the use of that additive.

(vi) **Residue decline studies.**

(A) Residue decline studies are required. Such data will be needed for uses where the pesticide is applied when the edible portion of the crop has formed or it is clear that quantifiable residues may occur on the food or feed commodities at, or close to, the earliest harvest time. The primary purpose of these studies is to determine if residues are higher at longer PHIs than requested and the approximate half-life of the residues. In addition, such studies are frequently of great value for determining an appropriate tolerance when a use pattern is changed. The number of decline studies needed is one for crops requiring 5 to 12 total trials and two for crops requiring 16 to 20 total trials. These studies are included in the 5 to 12 or 16 to 20 trials (i.e. not in addition to these numbers of trials). Decline studies will not be required for crops needing three or fewer total trials.

(B) The design of the decline studies should include 3 to 5 sampling times in addition to the requested PHI. The sampling times should all fall within the crop stage when harvesting could reasonably be expected to occur. The time points should be approximately equally spaced and, where possible, represent both shorter and longer PHIs than that requested. Of course, shorter PHIs cannot be examined in the case of a use with a zero-day PHI. In addition, for an at-plant/preplant use, the PHI is usually predetermined by the length of the growing season of the crop. Therefore, for such uses that result in quantifiable residues, petitioners should attempt to stretch the harvest period by sampling immature fruit, tubers, etc. if necessary.

(C) Only one composite sample will be required for each time point in a decline study. However, petitioners are advised to take two or more samples to prevent method and sampling variability from masking or appearing to create residue changes with time.

(D) It is anticipated that for most pesticides residue decline studies will not be necessary for all crops. For a given pesticide additional decline studies will not be required if studies on representative crops indicate residues do not increase with longer PHIs. This will provide some assurance that the tolerances represent the maximum residues that will occur from proposed or registered uses of a pesticide. The representative crop approach to be used is similar to that described in OPPTS 860.1380. If a pesticide is to be applied to all types of crops, it is recommended that decline data be obtained on the following five representative commodities: A tree fruit, a root crop, a leafy vegetable, a grain, and a fruiting vegetable. Some flexibility in the choice of crops will be permitted. For example, a legume vegetable could be substituted for the fruiting vegetable.

(vii) **Crop group tolerances.**

(A) The number of trials in Table 1 (page 431) of Appendix A of this guideline are based upon each crop being the only one within its crop group for which a tolerance is requested. In the case of crop group tolerances for which there is more than one representative crop, the number of trials can be reduced based on the reasonable assumption that residues in the representative commodities should reflect residues on all crops in the group. The reduction in the number of trials is 25 percent (i.e. 20 to 15, 16 to 12, 12 to 9, 8 to 6) for those representative commodities that need eight or more trials when requested individually. Crops which require five or fewer field trials will not receive any reduction when used as a representative commodity. Table 2 (page 434) of Appendix A of this guideline shows the resulting numbers of trials needed for all crop groups in 40 CFR 180.41.

(B) As stated in 40 CFR 180.41, if maximum residues for the representative crops vary by more than a factor of 5 from the maximum value observed for any crop in the group, a group tolerance will ordinarily not be established. In this case individual crop toler-

ances will normally be established and the 25 percent reduction in the number of trials will not apply. Petitioners should keep this in mind when planning crop field trials for crop group tolerances.

(C) It should be noted that a similar 25 percent reduction in the number of trials may be applied to uses that do not yield quantifiable residues. However, both of these 25 percent reductions may not be applied to the same crop. In other words, the number of trials can not be reduced 50 percent for a representative commodity that does not contain quantifiable residues.

(D) The Agency has also created subgroups within the existing crop grouping scheme (see paragraph (l)(7) of this guideline). Guidance on the number of field trials needed for the representative commodities in these subgroups is also provided in Table 3 (page 437) of Appendix A of this guideline. The number of trials were determined on a case-by-case basis looking at the acreages and consumption of the representative commodities and the whole subgroup. Refer to the footnotes of Table 3 of Appendix A of this guideline for more details.

(E) In effect, some crop groups have been established in 40 CFR 180.1(h). For example, a tolerance on onions applies to dry bulb onions, green onions, and garlic. To determine the number of trials required for the groups in 40 CFR 180.1(h), refer to Table 4 (page 439) of Appendix A of this guideline.

(F) Although there is no crop group for "small grains" in CFR 180.41, for data generation purposes wheat, barley, oats, and rye may be treated as a group. Provided use patterns and resulting residues are similar, the numbers of trials for wheat, barley, and oats may be reduced to 15, 9, and 12, respectively. Five trials are still needed for rye. The tolerances will be established on the individual crops due to the lack of an official small grain crop group.

(viii) **Uses resulting in no quantifiable residues.**

(A) If metabolism data or field trial data on related crops indicate quantifiable residues are not likely, a petitioner may elect to conduct 25 percent fewer trials for crops normally requiring eight or more trials. However, if all of these trials do not show residues below the LOQ of the method, additional trials will normally be required to bring the total number conducted up to the standard requirement. Thus, registrants/petitioners could risk a delay in obtaining a tolerance if this option is chosen. In addition to residues being below the LOQ, two other conditions must be met for the 25 percent fewer trials to be acceptable. First, the method must have a sufficiently low LOQ both from an analytical chemistry standpoint and for risk assessment purposes. This means the LOQ will need to be in the ≤0.01-0.05 ppm range in most cases. Second, the trials still need to represent all significant regions of production. Distribution of trials across regions is discussed in more detail in paragraph (e)(2)(xiii) of this guideline.

(B) The 25 percent reduction in the number of field trials for residues below the LOQ cannot be applied to representative commodities being used to establish crop group tolerances. The reduction is also not applicable to crops that require five or fewer field trials.

(C) For crops which have more than one raw agricultural commodity (RAC), the 25 percent reduction for residues below the LOQ may be applied to one commodity even if the others have quantifiable residues. For example, if a pesticide is applied to an early stage of corn, it is possible residues are found on forage and fodder, but not in the grain. In this case, 16 trials may be acceptable for grain, even though 20 are needed for the forage and fodder. This is not meant to imply that separate trials are to be conducted for different crop parts. In other words, corn grain, forage, and fodder should be collected from each trial site. If no residues are found on grain from a minimum of 16 geographically representative sites, the grain collected at other sites need not be analyzed.

(D) To take advantage of this option, registrants/petitioners should be certain to submit adequate recovery data and chromatograms establishing the limit of quantitation of the method. The petitioner should describe how the LOQ was calculated and cite any appropriate references.

(ix) **Additional considerations for early season uses on annual crops.**

(A) Many pesticide labels give options for applications made prior to crop emergence, such as allowing the use to be preplant, at-plant, or preemergence. Because the Agency has concluded that these types of application can be grouped for the purposes of determining the total number of field trials, the trials for a specific crop can be divided among these three applications at the registrant/petitioner's discretion. For example, the twelve trials for a particular pesticide on cotton could consist of three preplant, three at-plant, and six preemergence applications (plus the maximum rate and number of any proposed postemergence applications – see paragraph (e)(2)(ix)(D) of this guideline).

(B) If the label gives a choice for surface application versus incorporation into the soil, data reflecting both of these modes of application will be required. There are two options for conducting and determining the number of trials in this instance. The preferred option is for each trial to include both applications on side-by-side plots. Only one composite treated sample would be required for each plot. The minimum number of trials should be as designated in Table 1 (page 431) of Appendix A of this guideline. This means that the total number of samples would be equivalent to that required for most other uses on the same crop. Again using cotton as the example, at least 12 trials would be needed with each having two samples (one for surface applied and one for soil incorporated). The 12 trials can be divided among preplant, at-plant and preemergence applications if all these appear on the label.

(C) The alternative option is to divide the total number of trials in Table 1 of Appendix A of this guideline roughly equally between those having only the surface treatment and those reflecting only soil incorporation. Two composite treated samples are needed in each trial. Since the trials for each mode of application will need to have adequate geographic representation, this option may result in a greater number of trials for those crops which have a regions normally needing only one trial. Using the cotton example, the result would be at least two additional trials (14 total) since regions 6 and 2 (representing 10 percent and 8 percent of production, respectively) would each need to have two trials (one for surface and one for incorporation). If the side-by-side option in paragraph (e)(2)(ix)(B) of this guideline were chosen, only one trial would be required in each of those regions.

(D) Particularly in the case of herbicides, the label may permit pre-and/or postemergence applications. If both are allowed, all field trials should include both applications. If the choice is limited to one or the other, the full number of trials as specified in Table 1 of Appendix A of this guideline should be conducted for each type of application. However, fewer total trials will be accepted if some side-by-side studies show a consistent pattern between the residues from the pre- and postemergence uses. In this instance the full number of trials will be needed only for the mode of application consistently resulting in higher residues. (NOTE: The discussion in this paragraph refers to before or after the emergence of the food/feed crop. Occasionally, labels specify application timing in terms of before or after weeds emerge. The critical factor for purposes of this discussion is whether or not the food/feed crop has emerged.)

(x) **Formulations.**

(A) It is stated under paragraph (e)(2)(v) of this guideline that the numbers are based upon only one formulation type being requested for use on each crop. The number of trials needed to register additional formulation types or classes will be addressed on a case-by-case basis. In some instances the full number of trials will also be needed for a new type of formulation, whereas other formulation classes may be registered with a few bridging studies or perhaps no field trials at all. The decision depends upon how similar the formulations are in composition and physical form, the mode of application, and the timing of the application.

(B) One type of formulation which will normally require a full set of field trials is the microencapsulated or controlled release formulation. Since these are designed to control the release rate of the active ingredient, the same number of field trials is needed as to obtain an original tolerance regardless of the timing and mode of its application and the amount of data available on other formulations classes.

(C) Most of the remaining types of formulations can be divided into two groups – those which are diluted with water prior to application and those which are applied intact. Granules and dusts are the most common examples of the latter. Granular formulations will generally require the full number of field trials regardless of what data are already available for other formulation classes. This is based on several observed cases of residue uptake being quite different for granules versus other types of formulations of the same active ingredient. No residue data will be required for dusts if data are available at the same application rate and PHI for a formulation applied as a wetting spray (e.g. emulsifiable concentrates (EC), wettable powder (WP)).

(D) The most common formulation types which are diluted in water prior to application include EC, WP, water dispersible granules (WDG, WG) or dry flowables (DF), flowable concentrates (FlC), and soluble concentrates (liquid or solid) (SC, SL). Residue data may be translated among these classes of formulations for applications that are made prior to crop emergence (i.e. preplant, at-plant, and preemergence applications) or just after crop emergence. Data may also be translated among these formulation classes for applications directed to the soil (as opposed to foliar treatments).

(E) For mid-to-late season foliar applications of formulation types listed, two options are available. The new type of formulation could be treated similarly to an amended registration – 25 percent fewer trials would be required than were required for the formulation class used to obtain the original tolerance. Alternatively, side-by-side studies (often referred to as bridging data) could be conducted. These involve applications of the registered formulation (the type used to obtain the tolerance) and the new type of formulation to side-by-side plots using the same rates and preharvest intervals. If residues from the new formulation are comparable to or less than those from the registered formulation, the new formulation can be registered. However, if residues are higher from the new formulation in the side-by-side comparison, the full number of trials specified in Table 1 (page 431) of Appendix A of this guideline will be required for that formulation to determine the higher tolerance level needed to cover its registration.

(F) The exact number of side-by-side studies required will be decided on a case-by-case basis. A "representative crops" approach may be used if the new formulation is requested for use on numerous crops. Submission of protocols outlining the crops and sites to be used in these bridging studies is encouraged. The most common questions from registrants/petitioners in this area have involved use of EC data to support registrations of wettable powders. It is EPA's understanding that the American Crop Protection Association (ACPA) is compiling data from its members that compare residues from ECs and WPs. If a sufficient number of such studies are available, it is possible a conclusion could be made in the future that no additional crop field trials are required to register a WP if data for an EC reflecting the same use pattern are available.

(G) If registrants/petitioners wish to register two or more formulation classes when obtaining the initial tolerance and registration, the same basic concepts would apply. That is, a complete set of trials as specified in Table 1 of Appendix A of this guideline should be conducted on one type of formulation and the additional formulation classes handled like an amended registration (25 percent fewer trials than the primary formulation) or compared to the primary type of formulation using side-by-side studies.

(H) A few other statements can be made concerning data requirements for formulations. Dry flowable or water dispersible granular formulations are sufficiently similar to wettable powders to allow translation of residue data between them. Placing a formulation (typically WP) in a water soluble bag does not require additional residue data provided adequate data are available for the unbagged product.

(I) Some pesticides (e.g. phenoxy herbicides) can be applied as one or more salts and/or esters. Generally, different salts or esters of an active ingredient can be treated as new formulations of that active ingredient for purposes of determining the number of crop field trials. Thus, a new salt could be treated like an amended registration (25 percent fewer trials than the original salt or form of the active ingredient) or compared to the registered form of the active ingredient using side-by-side studies.

(xi) **Spray volumes – ground versus aerial equipment.**
(A) The subjects of spray volumes and aerial versus ground equipment are often interconnected and were addressed under paragraph (l)(9) of this guideline. The notice stated the following:

"* * * Provided that the pesticide product label specifies that aerial applications are to be made in a minimum of 2 gallons water per acre (or 10 gallons per acre in the case of tree or orchard crops), crop field trials reflecting aerial application will be waived in those cases where adequate data are available from use of ground equipment reflecting the same application rate, number of applications, and PHI. This data waiver does not apply to aerial applications using diluents other than water (e.g. vegetable oils). In addition, the Agency reserves the right to require aerial data if special circumstances warrant it."

(B) Based on paragraph (e)(2)(xi)(A) of this guideline, there are only a few instances where the number of field trials will be affected by the spray volumes or type of equipment (at least for aerial versus ground) specified on the label. However, the following two

Food and Feed Crops of the United States

exceptions should be kept in mind:

(1) Ultra-low volume uses (<2 gallons spray per acre; <10 gallons per acre for orchards) in mid- to late-season will be treated as separate use patterns regardless of the nature of the diluent (water, vegetable oil, etc.). If the ULV application is the first use on the crop (i.e. no tolerance established), the minimum number of field trials specified in Table 1 (page 431) or Attachment 7 (page 452) of Appendix A of this guideline is required. If data are already available reflecting higher spray volumes, the ULV application can be handled similarly to an amended registration (i.e. 25 percent fewer trials than specified in Table 1 of Appendix A of this guideline providing these trials show the existing tolerance is adequate – see amended registrations in paragraph (e)(2)(xii) of this guideline). Alternatively, it would be acceptable for registrants/ petitioners to demonstrate using side-by-side studies that residues from the ULV applications are comparable to or lower than those from higher spray volumes. However, if residues are higher from the ULV application in these side-by-side studies, the full number of trials specified in Table 1 of Appendix A of this guideline will be required for this use.

(2) For treatment of orchards, dilute sprays (typically 100-400 gallons per acre) and concentrate sprays (typically 20-100 GPA) will be treated as separate uses. The number of trials will depend upon which of two options is chosen, analogous to the discussion earlier in this guideline for surface applied versus soil incorporation (see *Additional Considerations for Early Season Uses on Annual Crops* in paragraph (e)(2)(ix) of this guideline). If side-by-side plots (dilute vs. concentrate) are included at all sites (the preferred option), the numbers of trials in Table 1 of Appendix A of this guideline will apply and one treated sample from each plot (instead of the normally required two) will be acceptable. Alternatively, the trials could be divided roughly equally between dilute and concentrate sprays with adequate geographic representation required for each type of spray. In this case, two treated samples are needed at each site and the total number of required trials may exceed that in Table 1 of Appendix A of this guideline if one or more regions require only one study. Refer to the example for cotton in the section on Early Season Uses.

(3) If either dilute or concentrate sprays are already approved for use on an orchard crop, the request to add the other type of spray to the label will be treated as an amended registration requiring 25 percent fewer trials than specified in Table 1 of Appendix A of this guideline (see amended registrations in paragraph (e)(2)(xii) of this guideline) or a number of side-by-side studies estab-

lishing that residues from the requested type of spray are not higher than those from the registered one. The exact number of side-by-side studies required will be determined on a case-by-case basis. Submission of protocols outlining the locations and numbers of sites is encouraged.

(C) One final comment on spray volumes concerns chemigation – the application of pesticides by injection into irrigation water. The Agency views this as a type of ground application using very large spray volumes. Provided that data are available for typical ground spray volumes, data reflecting chemigation are not required.

(xii) **Amended registrations.** For amended registration requests that involve a significant change in application rate (either individual or seasonal), interval between applications, or preharvest interval (PHI), the number of field trials required will normally be 25 percent less than that needed to establish an original tolerance, provided that the latter is shown by the reduced number of trials to be adequate to cover the new use. However, if the reduced number of trials indicates that the original tolerance is inadequate, or if the original number of trials was ≤5 or already included a 25 percent reduction (crop group or residues <LOQ), the number of trials for an amended registration is the same as that for the original tolerance. On a case-by-case basis the Agency may require less additional data than described above in paragraph (e)(2)(V)(J) of this guideline for an amended registration. This could be particularly true when residue decline studies are available reflecting a proposed change in a PHI. In some instances, no additional data may be necessary. An example would be a request to reduce the application rate for a use that already does not produce quantifiable residues.

(xiii) **Location of Trials.**

(A) The Agency divided the United States into 13 regions based on growing conditions as proposed by ACPA (see map on page 325). The dividing lines reflect natural geography or climatic boundaries and, therefore, in many cases do not coincide with state lines. The exact definitions of the regions are specified by states, counties, highways, or mountain ranges in Attachment 10 (page 459) of Apppendix A. The Agency has decided that Puerto Rico is more similar to Hawaii (Region 13) than Florida (Region 3) in terms of climate and geography. Therefore, Puerto Rico should be considered to be combined with Hawaii to form Region 13. The production figures in Table 6 (page 447) of Appendix A of this guideline and distributions of trials in Table 5 (page 443) of Appendix A of this guideline have been developed on this basis. Also, as noted in paragraph (e)(2)(xiii)(D) of this guideline, it may be acceptable for trials in the southern extreme of Florida to represent Region 13.

(B) Using crop production figures the Agency has developed suggested distributions of trials among the 13 regions for crops requiring >3 trials. These distributions are delineated in Table 5 (page 443) of Appendix A of this guideline and were developed using the following general criteria. The number of trials per region should generally correlate with the percentage of the crop grown in that region. However, where possible, at least one trial should be included in each region having ≥2 percent of the national production. The latter criterion can be met in most, if not all, cases for crops requiring ≥12 trials. However, for some crops needing 5-8 trials, trying to satisfy this criterion would result in regions with a high percentage of the production having too few trials. For example, in the case of sweet cherries the Agency has not suggested that trials be conducted in Regions 1 and 9 (3 percent each of national production) since this would leave too few trials in the major regions of production (5, 10, 11).

(C) The distributions of trials in Table 5 of Appendix A (page 443) of this guideline are not intended to be absolute requirements, but "suggested" designs for these studies. There are likely to be several acceptable alternatives for most crops. Registrants/petitioners may wish to contact the Agency regarding the suitability of alternative distributions of trials.

(D) It should also be noted that the regional borders specified in Attachment 10 (page 459, map on page 325) of Appendix A of this guideline are not absolute lines; rather, they have been drawn as rough approximations of climatically similar areas. Field trials conducted within reasonable distances of regional borders can be acceptable for fulfilling requirements for the region on either side of the border as long as weather conditions and cropping practices are representative of either region. Therefore, if it can be demonstrated that a site is representative of two regions, crop field trials for both regions can be performed at that site. For example, a site in northern Florida (part of region 3) may be acceptable for a crop grown in the southern part of region 2. Similarly, the southern extreme of Florida can be regarded as bordering region 13 (Puerto Rico plus Hawaii). Field trials in southern Florida can thus be used toward satisfying the requirements for assorted tropical fruits. However, in any of these cases where a site near a border may represent two regions, the total number of crop field trials required for the two regions will not change. Therefore, if more than one trial is required from the two regions, sites in addition to the one near the regional border will be needed in the two regions or trials will be required for more than one year at the site near the border.

(E) For crops requiring ≤3 trials, it is more difficult to develop guidance on distribution of trials since the number of growing regions is often comparable to or even greater than the total number of trials. In these cases the data should represent to the extent possible a balance of the highest production areas, different geographic/climatic conditions, and/or major differences in varieties of the crop. At least one trial should

be conducted in the region of highest production.

(F) To aid registrants/petitioners in determining distribution of trials for crops not listed in Table 5 of Appendix A of this guideline or alternative distributions of trials for crops that are in that table, the production of numerous crops by region is specified in Table 6 (page 447) of Appendix A of this guideline. Most of these figures were obtained using acreage information from USDA's *Agricultural Statistics* (1991) and the 1987 *Census of Agriculture* (Dept. of Commerce). These publications list production by state instead of region. Since numerous states fall into more than one region, the distribution of acreage within these states had to be estimated to calculate regional production. Numerous crops (primarily minor crops such as spices, herbs, and unusual berries) are not listed at all in this table since no regional production figures for them were available. As can also be seen in Table 6 of Appendix A of this guideline, the total accountability of production is <100 percent for a considerable number of crops. However, the Agency believes sufficient percentages of production (most are >85 percent) are accounted for to determine the distribution of trials.

(G) A special comment needs to be made concerning distribution of trials for crop group or crop subgroup tolerances for legume vegetables. The regulation specifies that the representative commodities include one cultivar of succulent bean, one cultivar of dried bean, one cultivar of dried pea, etc. depending upon the crop group or subgroup. If possible, the cultivar chosen should be one that is grown in all significant areas of production for that class of bean or pea. If this cannot be done, a combination of cultivars may be used to encompass all regions of production. As an example, it will not be acceptable to provide data from only one region for a certain cultivar of dried bean even if that dried bean is grown only in that region. The data need to reflect all significant regions of production for all dried beans if a crop group or subgroup tolerance is desired.

(H) The discussion in paragraphs (e)(2)(xiii)(A) through (e)(2)(xiii)(G) of these guidelines focuses on the distribution of trials among regions. With respect to the distribution of multiple trials within a region, this should generally follow the relative production in the individual growing areas (states or counties as appropriate) of the region. However, the sites should also be sufficiently separated to reflect the diversity of the growing region including soil types. In other words, if production is scattered throughout much of a region, the trials should not be clustered in one small portion of that region.

(I) To aid the Agency's review process with regard to the distribution of trials among and within regions, registrants/petitioners are requested to include a copy of the map on page 325 showing the locations of all sites of acceptable trials (i.e. those reflecting the proposed use and generating viable samples) in the volume of field trial reports for each crop.

(xiv) **Requirements for tolerances with geographically restricted registrations and for FIFRA section 24(c) registrations.** The preceding discussion in this guideline on determining the number of crop field trials addresses national registration of pesticides. Since regional registration is accepted by the Agency under certain circumstances, separate guidance has been developed as detailed in Attachment 11 (page 459) of Appendix A of this guideline. This attachment also addresses field trial requirements for FIFRA section 24(c) or Special Local Needs registrations. In summary, the basic concept described in Attachment 11 is that the number of trials for a regional registration should be determined by multiplying the number of field trials required for national registration by the proportion of the crop (on an acreage basis) grown in the region in which registration is sought.

(f) **Aspirated grain fractions: A tolerance perspective.**

(1) **Background.**

(i) When cereal grains or oilseeds (e.g. corn, wheat, sorghum, barley, oats, rye, and rice, and soybeans) are moved into, transferred within, or shipped from U.S. grain handling facilities, dust is generated as the grain moves through a transfer point, e.g. bucket elevator, one belt to another, etc. This dust escapes as an air pollutant and is potentially damaging to workers if inhaled, and because of its flammability when it becomes airborne, it is also a highly explosive dust which creates a hazardous work environment. The grain elevator industry refer to this elevator dust as "grain dust". However, the livestock feed manufacturers for aesthetical reasons commonly refer to this material as "aspirated grain fractions". Since the Agency interest is related to livestock feeding "aspirated grain fractions" is the preferred term in this discussion of elevator "grain dust".

(ii) Aspirated grain fractions are a nutritious livestock feed somewhat comparable to the whole grain. Since various pesticides are applied either preharvest to growing grains or postharvest to stored grains, the harvested/stored grains could have pesticide surface residues which could concentrate in the aspirated dust. This concentration occurs from postharvest treatment because pesticide residues are absorbed onto the large surface areas of the dust particles on the grain. The particle sizes of the dust can range from <1 μm to 2500 μm, with as much as 50 percent being <100 μm. Incorporation of this fine dust into animal feeds can cause increased exposure of pesticide residues to animals, and these residues could be transferred into the human food chain through livestock meat, milk, or eggs. Concentration of residues in aspirated dust can also potentially occur if measurable surface residues of pesticides are found on harvested grain/oilseeds even from preharvest treatment.

(iii) The incorporation of "aspirated grain fractions" into animal feeds would fall under the auspices of the Federal Food, Drugs, and Cosmetic Act (Amended January 1980) if a tolerance for pesticide residues is needed as a result

of moving cereal grains and oil seeds through commerce.

(2) **Definitions/characteristics.**

(i) The 1993 Official Publication of the Association of American Feed Control Officials (AAFCO) (see paragraph (l)(16)) defines "grain dust" (Section 60.43) as "aspirated grain fractions" ((IFN 4-12-208) Cereals – oil seeds grain and seed fractions aspirated.): "Aspirated grain fractions" are obtained during the normal aspiration of cereal grains and/or oil seeds for the purpose of environmental control and safety within a grain handling facility. It should consist primarily of seed parts and may not contain more than 15 percent ash. It should not contain aspirations from medicated feeds." (Note: Medicated feeds refer to those treated with animal drugs; Ash is defined as the mineral residue remaining after combustion in air.). (International Feed Numbers and Names (IFN) were developed and provided by the Feed Composition Data Bank, USDA National Agricultural Library, Beltsville, MD.).

(ii) A related grain byproduct is called "chaff and/or dust". This material is collected in grain processing plants solely to clean the grain, whereas "aspirated grain fractions" are collected at grain elevators for environmental and safety reasons. The AAFCO defines "chaff and/or dust" (IFN 4-02-149 Cereals-legumes chaff and/or dust (Section 81.3, Screenings)) as follows: "Chaff and/or dust" is material that is separated from grains or seeds in the usual commercial cleaning processes. It may include hulls, joints, straw, mill or elevator dust, sweepings, sand, dirt, grains, seeds. It must be labelled, "chaff and/or dust". If it contains more than 15 percent ash the words 'sand' and 'dirt' must appear on the label." "Chaff and/or dust" is normally recombined with unprocessed broken grain pieces and/or bran before being used in animal feeds. Any pesticide residues in regard to tolerance needs would be considered in grain byproducts from the grain processing.

(iii) Therefore, only the residue data requirements of the tolerance setting process for "aspirated grain fractions" need to be considered in this guideline.

(iv) Although "aspirated grain fractions" can be defined in general by IFN 4-12-208, more specific characteristics for this dust are not as easily defined. First, the dust collection systems are designed to achieve safety and air quality, and not to isolate the dust by particle size or content, i.e. dust, and/or chaff, bran, other light materials. There are no specific guidelines or industrial standards of dust collection equipment for grain handling facilities. Second, the large variability of the dust composition is governed by the location, time of year, and crop condition at harvest, as most elevators handle grains on a seasonal basis, e.g. wheat in the summer, and corn, sorghum, and soybeans in early fall. Third, the dust is not normally segregated by an individual grain or seed commodity as it is collected, but is trapped in a common container or bin. Normally this dust will be recombined with other transient grain at the elevator site. Thus, a composite of this aspirated dust will probably be found at inland and export terminal elevators. In general, aspirated dust from one commodity will only be found at country elevators.

(3) **Utilization in animal feedstuffs.**

(i) Based upon the feed industry uses of "aspirated grain fractions", the estimate of 20 percent of the diet is to be used for all livestock, although some research has shown that the dust can be fed up to 50 percent to cattle and swine. "Aspirated grain fractions" is normally mixed with other feedstuffs (e.g. molasses as a binding agent), or it can be pelleted by mixing with alfalfa meal at 50 percent to produce "range cubes" which are fed possibly at 20-30 percent in addition to other feedstuffs, e.g. grasses, hay, etc. Dairy farmers and processors of dairy feeds also tend not to use "aspirated grain fractions" in feeds because of the possibility of pesticide contamination of milk. Leading U.S. poultry producers have stated that the current poultry production practices prevent the use of "aspirated grain fractions" in their feed mixes because of the possible presence of high pesticide concentrations in the feed which can result in a lower weight gain for broiler and/or a drop in egg production with laying hens. Thus, the inclusion of "aspirated grain fractions" in poultry diets should not be considered. It also appears that much more of the dust may be used for beef cattle, than for other livestock.

(ii) Based upon the U.S. export volumes it appears that corn, wheat, sorghum, and soybeans are the major grains/oil seed that will generate significant volumes of elevator dust. Barley, oats, and rye would make up a very small percentage (<2 percent) of the total "aspirated grain fractions" available for animal feeds. In addition, rice grain dust is not used in animal feeds because of a high silica content of >30 percent.

(iii) Because of different growing patterns, i.e. difference in the grain exposure because of protective glumes around the kernels in several crops, and possibly different application patterns of the pesticide, individual data will be required for corn, wheat, sorghum, and soybeans, and should not be translated from one to the other to support a proposed or registered use.

(iv) Presently the grains of corn, wheat, and sorghum, and the seed of soybeans are considered RAC's. When the grain is harvested and stored some dust is present on the grain. "Aspirated grain fractions" from these crops are removed by aspiration methods for environmental and safety reasons as the grain and seed are moved through commerce. This dust is normally added back to the whole grain/seed as it travels through country and inland elevators, with final removal, in many cases, occurring at the export elevators. Removal and/or addition of this aspirated dust does not change the RAC. There is no processing per se involved in its removal or its addition. Therefore, for consistency, "aspirated grain fractions", which is only a portion of the whole grain or seed at harvest and storage, should also be considered a RAC.

(v) According to the grain elevator industry, "aspirated grain fractions" is normally a composite of more than one grain. The collected dust from the grain being moved

through the elevator is added to a common dust bin, meaning that the dust from corn can be added to dust from wheat, the dust from sorghum can be added to corn, etc. Therefore, a tolerance for "aspirated grain fractions" should be established for the pesticide, and this tolerance should consider the use of a pesticide on corn, wheat, sorghum, and/or soybeans. For example, if the pesticide is used only on one grain/oil seed, the tolerance should be established assuming this crop will represent 100 percent of the dust. If the pesticide is used on several crops, the RAC with the highest residues in the dust will be used to establish the tolerance.

(g) **Test method.**

(1) Presently residue data for "aspirated grain fractions" are required for all postharvest applications of pesticides for corn, wheat, sorghum, soybeans, and on some preharvest applications for these crops with a zero day or short PHI whose seed heads are formed at the time of application.

(2) Residue data should be submitted in support of all postharvest uses. Data needs for a preharvest use follow the discussions on postharvest uses.

(3) For a postharvest use the following can be used as a reference to help design a laboratory experiment to measure residue levels in "aspirated grain fractions" from transient grains in elevator operations. Only one residue study is needed for each grain (corn, wheat, sorghum, soybeans) that is treated postharvest (or has a preharvest use resulting in quantifiable residues as described in paragraph (g)(4) of this guideline).

(4) If the pesticide is currently registered for a postharvest use, treated grain from a commercial operation can be used. The treated grain should be analyzed for residues of the pesticide under investigation, cleaned by an aspirated method identical or similar to a commercial elevator operation to trap the dust. For each 100 lb of grain, the amount of dust should be approximately 200 g. Depending upon the pesticide residue levels, this may or may not be a sufficient amount for fractionation and analyses; larger quantities of grain may be utilized. Next, the cleaned grain and the dust should be analyzed for the pesticide residues, and the level of pesticide residue concentration determined. However, before analysis of the dust, it should be fractionated into four or five different ranges, e.g, under 400 μm, 400 to 800 μm, 800 to 1200 μm, 1200 to 2000 μm, and 2000 to 2500 μm, or any other similar sieve sizes to determine the particle size distribution. The purpose of this distribution data is to show that the aspirated dust sample typifies a sample of commercial elevator "aspirated grain fractions"; normally, at least 50 percent of the elevator "aspirated grain fractions" have a particle size of <400 μm. But, for purposes of residue analysis, the pesticide treated dust should be recombined since this reconstituted dust sample would be more representative of "aspirated grain fractions" used in commercial feed production and/or feeding practices. In addition, since "grain fractions" are defined according to the American Feed Control Association to contain ash at less than 15 percent, the ash content of the combined dust fractions should also be determined. The elevator dust sample should be analyzed using methodology for the pesticide under investigation which does not exhibit interference problems from residues of other registered cereal/oilseed pesticides that might be present from prior appli-

cations. It is recommended that triplicate samples be taken. Duplicate analyses of pesticide residue levels should be performed on all samples.

(5) An alternative procedure for either a currently registered postharvest pesticide, or a proposed registration of a newly developed postharvest pesticide could be as follows:

(i) First, "aspirated grain fractions" that has been collected by a commercial elevator aspiration system should be acquired. A particle size distribution of the aspirated dust should be measured from 2500 (or 2000) μm to under 400 μm (using 4 or 5 sieve sizes to cover this range as described in paragraph (g)(4) of this guideline), and the ash content determined. Analysis of the untreated dust sample as the control will indicate any problems if other pesticides are present from prior applications. A sample of the grain should be cleaned by aspiration, using a method identical or similar to commercial operations. Next, using the unfractionated aspirated dust sample that was aquired as described in paragraph (g)(4) of this guideline from a commercial grain elevator, apply the dust to the cleaned grain at a rate of 0.2 percent (by weight), and mix to distribute the dust evenly over the grain. Apply the pesticide at its maximum allowable label rate, and after the solvent has dried, a portion of the grain, which is now covered with the aspirated dust and the pesticide, should be sampled for residue analysis. Remove the treated dust from the grain by an aspiration method, and analyze this cleaned grain and the treated dust for pesticide residues. It is recommended that triplicate samples be taken. Duplicate analyses of pesticide residue levels should also be performed on all samples. The concen-

tration factor should be determined for the pesticide using average residue levels in the dust (aspirated grain fractions) and in the grain after application of the pesticide but before cleaning. The tolerance level for the aspirated grain fractions is calculated using this concentration factor and the tolerance for the grain.

(ii) Storage stability data on the whole grain will adequately support storage of "aspirated grain fractions" samples prior to residue analysis. Refer to OPPTS 860.1380 for guidance on generating storage stability data.

(iii) One application at the maximum allowable label rate, followed by the collection and the analysis of the "aspirated grain fractions" for the pesticide immediately after application, should provide sufficient data to adequately determine the expected level of a pesticide in commercial elevator "aspirated grain fractions". The collection and analysis of this dust should follow in paragraph (g)(4) of this guideline suggestions for the gathering of "aspirated grain fractions" data from a postharvest pesticide use. The Agency reserves the right to change this data requirement if actual commercial practices change to require additional applications.

(iv) For a preharvest use on wheat, corn soybeans or sorghum after the reproduction stage begins and seed heads are formed, the following flowchart can be used to determine if residue data are required on aspirated grain fractions.

(v) Residue data for "aspirated grain fractions" will not normally be needed if the pesticide is applied during the vegetative stage and before the reproduction stage begins

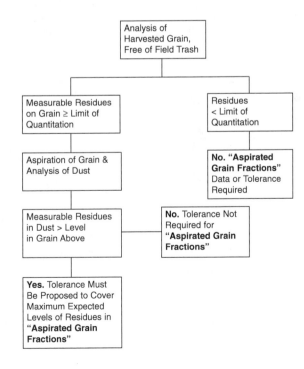

and seed heads are formed, unless the plant metabolism and/or processing study shows a concentration of residues of regulatory concern in outer seed coat (e.g. wheat bran, soybean hulls).

(h) **Data reporting – crop field trials.**

(1) **Purpose.** Crop field trials provide residue chemistry data on the magnitude of the residue in or on RACs to support registration of any pesticide intended for use on a food or feed crop. Residue chemistry data on RACs are used by the Agency to estimate the exposure of the general population to pesticide residues in food, and for setting and enforcing tolerances for pesticide residues in or on raw agricultural foods or feeds.

(2) **Objective.**

(i) This guideline is designed to aid the petitioner/registrant in generating reports which are compatible with the Agency's review process. While following this guidance is not mandatory, data submitters are encouraged to submit complete reports which can be efficiently reviewed by the Agency.

(ii) The Agency recognizes there are sections in the guideline which do not apply in all cases. Therefore, registrants should exercise scientific judgement in deciding which portions are germane to a specific data submission.

(iii) This guideline is intended to organize the submission of data to facilitate the review process.

(iv) The petitioner/registrant's report on crop field trials on a RAC should include all information necessary to provide a complete and accurate description of field trial treatments and procedures; sampling (harvesting), handling, shipping, and storage of the RAC; storage stability validation (or reference thereto) of the test chemical (and metabolites of toxicological concern in a plant matrix; residue analyses of field samples for the TTR and for individual components of toxicological concern); validation (recovery studies) of the residue analytical methodology; reporting of the data and statistical analyses; and, quality control measures/precautions taken to ensure the fidelity of these operations. The following is a suggested format for the report.

(3) **Format of the data report.** The following describes a suggested order and format for a study report item by item. However, other formats are also acceptable provided the information described in this paragraph is included.

(i) Master cover page. Title page and additional documentation requirements (i.e. requirements for data submission and procedures for claims of confidentiality of data) if relevant to the study report should precede the contents of the study formatted below. These requirements are described in PR Notice 86-5 (see paragraph (l)(10).

(ii) Table of contents. The table of contents should indicate the overall organization of the study, including tables and figures.

(iii) Summary/introduction.

(A) Purpose of studies

(B) Results (including explanations for apparently aberrant, atypical values, or outliers; discussion of geographical representation (major growing areas),

seasonal variation (summer/winter, wet/dry, etc.) and representativeness of types and varieties of the RAC

(C) Field procedures

(D) Analytical procedures/instrumentation

(E) Method recovery validation data

(F) Storage stability

(G) Discussion (including Quality Control measures taken; statistical treatments of data; and information on the levels of the TTR (including any individual components of the residue of special concern) in or on the RAC (specific plant parts) arising from the use of the pesticide formulated product on the test crop under specific use conditions. Results should also be correlated to the storage stability study).

(H) Conclusions

(iv) Data tables and other graphic representations.

(A) Summary map (U.S.A. with regions as shown in map on page 325. Include outside USA, if applicable) of crop field study sites (by crop)

(B) Summary tables of residue results of individual field trials

(C) Graphic representations (e.g. residue decline, figures, flowcharts, etc.)

(D) Summary tables of recovery data via the analytical methodology

(E) Summary tables of storage stability validation data

(v) Information/raw data on individual field trials (specifically, each individual field trial report should include the following information):

(A) Test substance (pesticide).

(1) Identification of the test pesticide active ingredient (a.i.), including CAS or IUPAC chemical name, common name (ANSI, BSI, ISO), and company developmental/experimental name.

(2) Identification of the pesticide formulated products used in the field trial, including trade name, type (EC, WP, G, etc.), and amount of active ingredient per gallon, pound, etc., EPA registration number (if available), and manufacturer.

(3) Information on other relevant parameters, as pertinent, (e.g. tank mates, spray additives, carrier (encapsulating polymer, etc.)).

(4) Other. Any and all additional information the registrant/petitioner considers appropriate and relevant to provide a complete and thorough description of the test chemical.

(B) Test commodity (RAC).

(1) Identification of the RAC, including type/variety and crop group classification (40 CFR 180.41).

(2) Identification of specific crop parts harvested; used in residue analytical methodology validations; and subjected to residue analysis for a determination of the TTR.

(3) The developmental stages, general condition (immature/mature, green/ripe, fresh/dry, etc.) and sizes of the RAC at time of pesticide applications

and at harvestings.

(4) Other. Any and all additional information the registrant/petitioner considers appropriate and relevant to provide a complete and thorough description of the RAC.

(vi) Test procedures.

(A) A detailed description of the experimental design and procedures followed in the growing of the RAC, applications of the pesticide formulated products, and harvestings of samples. The information provided, which may be presented on standardized field sheets, should include (in addition to a description of the test substance and test commodity):

(1) Trial identification number.

(2) Cooperator (name and address), test location (region number as shown in map on page 325 and Attachment 10 (page 459) of Appendix A of this guideline, county and state-country, if outside USA), and year.

(3) Field trial layout (e.g. size and number of control and experimental plots; number of plants per plot/unit area, number of rows per plot, length of rows and row spacing).

(4) Cultural treatments-farming practice (cultivation, irrigation, etc.) and cropping system.

(5) Soil characteristics (name/designation of the soil type. If application rate of the pesticide is dependent on any soil properties such as percent of organic matter, these should also be described).

(6) Methods of application (air or ground) of the pesticide formulated products, description of the application equipment, type of application (band/ broadcast, soil/foliar/ directed, ULV/concentrate/ dilute, other), and, calibration of pesticide application equipment, including methods and dates.

(7) Dose rates (amount of active ingredient and formulated product per acre, row, volume, etc.) and spray volumes per acre).

(8) Number and timing of applications (total number, during dormancy, preplant, preemergence, prebloom, etc., between-application-intervals, and treatment-to-sampling intervals (also known as the preharvest interval (PHI)).

(9) Other pesticides applied (identity (name and type of formulated products, active ingredients), rates, dates, tank-mate or separate, purpose of use).

(10) Climatological data (record of temperature and rainfall during the growing season from the nearest weather station, and wind speed during application). (See guidance on raw data in OPPTS 860.1000, page 387.)

(11) Dates (planting/sowing/transplanting, as applicable, other significant dates in the growing of the crop (e.g. husk split for tree crops), pesticide applications, harvests).

(12) Harvest procedures (method of harvesting (mechanical/hand, from the plant/ground/flota-tion, etc,), type equipment used, number/weight of samples collected per replication and number of replications per treatment level, statistical nature of sampling (e.g. fruit taken from upper, middle, and lower portions of tree exterior and interior), sample coding (cross-referenced to sample history), etc.).

(13) Quality control (control measures/precautions followed to ensure the fidelity of the crop field test).

(14) Other. Any and all additional information the registrant/petitioner considers appropriate and relevant to provide a complete and thorough description of the growing of the RAC, applications of the pesticide formulated products, and harvesting of samples.

(B) A detailed description of the handling, preshipping storage, and shipping procedures for harvested RAC samples. The information provided, which may be presented on a standardized form, should include (in addition to a description of the test substance and the test commodity):

(1) Sample identification (means of labeling/coding);

(2) Conditions (temperatures, container types/ sizes, sample sizes, form (e.g., whole commodity; chopped), etc.) and duration of storage before shipping.

(3) Methods of packaging for shipment (container types/sizes, sample sizes, ambient/iced, labeling/coding, etc.)

(4) Means of transport from the field to the laboratory.

(5) Dates (harvest, preshipping storage, shipping, and receipt in the laboratory).

(6) Quality control (control measures/precautions followed to ensure the integrity of harvested samples during handling, preshipping storage, and shipping operations).

(7) Other. Any and all additional information the registrant/petitioner considers appropriate and relevant to provide a complete and thorough description of the handling, preshipping storage, and shipping procedures for harvested samples.

(C) A detailed description of the conditions and length of storage of harvested RAC samples following their receipt in the laboratory.

(D) A detailed description of the residue analyses used in determining the TTR in RAC field trial and storage stability samples. If the specified information is provided elsewhere within the overall data submission package, it need not be reiterated here. In that case, a reference to the relevant analytical methodology would be sufficient.

(E) Method recovery validation studies should be run concurrently with the residue analyses of crop field trial samples from each individual field trial in order to provide information on the recovery levels of the test compounds from the test substrates at various fortification levels using the residue analytical methods, and to establish a validated limit of quantification. The following information specific to the method validations, which may be presented on a standardized form, should include:

(1) Experimental design: Identity of test substrates (specific plant parts) and test compounds (parent/specific metabolites). Number and magnitude of fortification levels, number of replicate samples per test compound per fortification level, sample coding, control samples, etc.

(2) Fortification procedure: Detail the preparation of the test compounds and test substrates and the manner in which the test compounds was/were introduced to the test substrates.

(3) Dates: Test sample preparation (maceration/extraction/etc.), test compounds preparation (standard solutions of known concentration), residue analyses.

(4) Residue results: Raw data, ppm found uncorrected (corrected values may also be reported but the basis of correction should be explained), procedures for calculating percent recoveries, recovery levels (range), and limits of quantitation and detection.

(5) Other. Any and all additional information the registrant/petitioner considers appropriate and relevant to provide a complete and thorough description of analytical methodology validation procedures.

(vii) Organization of data tables and forms.

(A) Tables of residue assay data for specific plant parts analyzed. Residue levels should be reported uncorrected. Corrected values may also be presented but the procedure needs to be explained.

(B) Tables on residue recovery values.

(C) Graphs, as pertinent (e.g. residue decline).

(D) Forms containing field trial history information.

(E) Forms containing harvesting, shipping, storage information.

(F) Tables of weather data if unusual conditions claimed to result in aberrant residues. See raw data guidance in OPPTS 860.1000 on page 387.

(viii) Certification. A signed and dated certification of authenticity by, and identifying information (typed name, title, affiliation, address, telephone number) of the personnel responsible for the report.

(ix) References.

(x) Appendices.

(A) Representative chromatograms, spectra, etc. of reagent blanks, solvent blanks, reference standards, controls, field samples, fortified samples, etc. (cross-referenced to individual field trial study reports).

(B) Reprints of published and unpublished literature, company reports, letters, analytical methodology, etc. cited (or used) by the petitioner/ registrant (unless physically located elsewhere in the overall data report, in which case cross-referencing will suffice).

(C) Other. Any relevant material not fitting in any of the other sections of this report.

(i) **Data reporting – specialty applications.**

(1) **Foreword.** This data reporting section of specialty applications is divided into three parts:

(i) Classification of seed treatments and treatment of crops grown for seed use only as nonfood or food uses.

(ii) Postharvest fumigation of crops and processed foods and feeds.

(iii) Postharvest treatment (except fumigation) of crops and processed foods and feeds.

Each part gives the format/outline recommended by the Agency to be used by the petitioner/registrant for reports on the particular specialty application study.

(2) **Format of the data report – seed treatments.** For seed treatments to be classified as a nonfood use, data from a radiotracer study are needed demonstrating no uptake of radioactivity to the aerial portion and edible root (both human and livestock consumption) portion of the crop. If the radiotracer study demonstrates that the particular seed treatment is a nonfood use, no further studies are needed. If the seed treatment is classified as a food use, data as given in the appropriate OPPTS 860 series guidelines are required (e.g. plant metabolism, crop field trials). The following guidance is a suggested format/outline for reporting the radiotracer study determining whether the seed treatment results in uptake of radioactivity to the aerial edible and root portions of the crop.

(i) Master cover page. Title page and additional documentation requirements (i.e. requirements for data submission and procedures for claims of confidentiality of data) if relevant to the study report should precede the contents of the study formatted below. These requirements are described in PR Notice 86-5.

(ii) Table of contents. The table of contents should indicate the overall organization of the study, including tables and figures.

(iii) Introduction.

(A) Background and historical information on the pesticide.

(1) Brief summary of nature of the residue in plants, including the structures of the parent and residues considered to be of toxicological concern.

(2) [Reserved].

(B) Purpose of study.

(C) Abstract of study.

(1) Brief summary of application and field procedures.

(2) Results, including unexpected problems.

(3) Conclusions.

(iv) Materials and methods.

(A) Test substance.

(1) Identification of the test pesticide active ingredient (a.i.), including CAS or IUPAC chemical name, common name (ANSI, BSI, ISO), registrant developmental/experimental name and chemical structure.

(2) Description of the radiolabeled test material. Identify the radiolabel and the site of the label. A rationale should be provided for selection of a radiolabel other than 14 C and for the site of the label (where possible the ring position should be labeled). The purity, specific activity in Curries/ mole and disintegrations per minute per gram (dpm/g) should be reported here.

(3) Identification of the pesticide formulated products in which the radiolabeled pesticide active ingredient was applied, including trade name, type (EC, WP, G, etc.), pounds of active ingredient per gallon, percent a.i. by weight, EPA registration number, and manufacturer.

(4) Physical state and nature of the solvent, carrier, bait, adjuvant or other matrix in which the pesticide was applied.

(B) Test crop.

(1) Identification of the test crop including variety.

(2) Identification of specific crop parts harvested and subjected to analysis for radioactivity.

(3) Developmental stages, general condition (immature/mature, green/ripe, fresh/dry, etc.), sizes of the test crop at time of harvest.

(C) Test site.

(1) Description of test site. Overall testing environment (outdoor test plots, greenhouse, plant growth chamber); environmental conditions (temperature, rainfall, sunlight); soil type.

(2) Location (county, state).

(3) Cooperator.

(D) Field trial methods.

(1) Detailed description of application of radiolabeled pesticide to seeds. Information to be reported includes dose rate, pounds active ingredient and formulated product per pounds of seed, concentration of treatment solution, volume of application solution per pounds of seed, formulation, physical state in which pesticide is applied, diluent, additives, etc., method of application (hopper box, commercial equipment). The pesticide should be applied at the maximum proposed application rate.

(2) Field trial layout. Information to be reported includes size of plots/pots, number of plants per plot/pot, number of plots/pots, number of plants per unit area, length of rows and row spacing.

(3) Farming practice. Information on practices such as cultivation, irrigation, and treatments with other pesticides should be included here.

(4) Harvest procedures, including the number of days between planting and harvesting.

(E) Sampling, handling and storage.

(1) Dates of sampling, shipping, storage, and analyses.

(2) Description of sampling procedure and size of samples.

(3) Handling, preshipping, shipping, post-shipping storage conditions, including storage times.

(F) Analytical procedures/instrumentation.

(1) Description of sample preparation (i.e. dissection, grinding, lyophilization, number of plants contained in one sample, etc.) prior to analyses of radioactivity.

(2) Details of analytical method to measure radioactivity, including descriptions of equipment and instrument parameters.

(G) Quality control. Description of control measures and precautions followed to ensure the fidelity of the field tests, samples and measurement of the residue.

(H) Other pertinent information on materials and methods.

(v) Results and conclusions.

(A) Brief summary of study procedures. The summary of the study procedures should include the number of field trials, descriptions of the application of the radiolabeled pesticide to the seed (dose rate, method, formulation), the site (greenhouse, outdoors, plant growth chamber), number of days between planting and harvest, number of plants sampled, part of the plant analyzed for radioactivity, and the method of detection.

(B) Results.

(1) Total recovered (i.e. combustible) radioactivity on seeds at time of planting, if measured: The radioactivity should be reported as:

(i) disintegrations per minute (dpm)

(ii) dpm/mg

(iii) ppm equivalents (expressed as parent compound). A sample calculation of ppm from radioactive counts should be provided, especially if other units (i.e. not dpm) are used.

(2) The distribution of radioactivity in the treated crop at the time of harvest or sampling. The data to be reported are the total recovered (i.e. combustible) radioactivity remaining at time of sampling or harvest on the plant's parts of interest as defined in Table 1 of OPPTS 860.1000 (page 392) (i.e. the aerial and edible root portions of the plant). The radioactivity for those plant parts should be reported in tabular format as:

(i) dpm

(ii) dpm/mg

(iii) ppm equivalents (expressed as parent compound).

(3) Graphs and figures of the results: Graphs, if provided, should be accompanied by tables of actual values from which graphs were constructed.

(4) Narrative of results. Narrative should include a discussion of the quantitative accountability for a majority of the total radioactivity recovered from the aerial and edible root portions of the plant and a discussion of unexpected problems, the way in which they were resolved, and explanations for

apparently aberrant, atypical values.

(C) Conclusions. The petitioner/registrant's conclusion on whether the results of this study and any other relevant studies support a nonfood use classification for the seed treatment in question should be given.

(vi) Raw data and information on individual field trials.

(A) Details of radioactive counting data for selected representative samples. Details should include counting time, total counts recorded, corrected counts, counting efficiencies, other raw data (sample sizes, ppm equivalents found, sensitivity, limit of detection) and other pertinent information needed to check the registrant's calculations.

(B) Description of calculations, including examples.

(C) Description of statistical tests, including examples.

(D) Representative raw data figures. As applicable, printout sheets, chromatograms, spectra, etc.

(E) Other. Any additional information the registrant considers appropriate and relevant to provide a complete and thorough description of the study.

(vii) Certification.

Certification of authenticity by the Study Director (including signature, typed name, title, affiliation, address, telephone number and date).

(viii) References.

(ix) Appendices.

(A) Reprints of published and unpublished literature, company reports, letters, etc., not expected to be in OPP files, but which the petitioner/registrant feels will aid the review of the study.

(B) Other pertinent information which does not fit in any other section of this outline.

(j) **Data reporting – postharvest fumigations.**

(1) **Foreword.**

(i) Fumigation may be defined as the act of releasing and dispersing a toxic chemical so that it reaches the organism wholly or primarily in the gaseous or vapor state. Both the RACs and their processed products may be treated postharvest by fumigation.

(ii) The report for a study on the postharvest fumigation of raw crops and processed foods should include all information necessary to provide a complete and accurate description of the study.

(2) **Format of the data report – fumigation.**

(i) Master cover page. Title page and additional documentation requirements (i.e. requirements for data submission and procedures for claims of confidentiality of data) if relevant to the study report should precede the contents of the study formatted below. These requirements are described in PR Notice 86-5.

(ii) Table of contents. The table of contents should indicate the overall organization of the study, including tables and figures.

(iii) Introduction.

(A) Background and historical information on the pesticide.

(1) Brief summary of nature of the residue in plants, including the structures of the parent and residues considered to be of toxicological concern.

(2) [Reserved].

(B) Purpose of study.

(C) Abstract of study.

(1) Brief summary of application procedures.

(2) Results, including unexpected problems.

(3) Conclusions.

(iv) Materials and methods.

(A) Test Substance

(1) Identification of the test pesticide active ingredient (a.i.), including chemical name, common name (ANSI, BSI, ISO), registrant developmental/experimental name and chemical structure.

(2) Identification of the pesticide formulated products in which the pesticide active ingredient was applied, including trade name, formulation type, pounds of active ingredient per gallon, percent a.i. by weight, EPA registration number and manufacturer.

(3) Information on the matrix in which the formulated pesticide was applied and on any additives.

(4) Physical/chemical parameters on the test substance.

(B) Test raw or processed commodity.

(1) Identification of the raw or processed test commodity, including variety.

(2) Identification of specific crop parts harvested.

(3) Developmental stages, general condition (immature/mature, green/ripe, fresh/dry, etc.), sizes of the test commodity at time of fumigation.

(4) Size and kind of containers holding the commodity (e.g. wood, burlap, etc.).

(5) Information on whether the raw or processed commodity, or its storage container, had been treated prior to the test postharvest treatment, including application rates, PHIs, and the residue prior to the test postharvest treatment.

(C) Test site.

(1) Description of fumigation chamber. Information to be reported includes:

(i) Type of fumigation chamber (grain elevator and flat storage, tarpaulin covering, shophold, fumigation vault, vacuum chamber, etc.).

(ii) Size and geometry of fumigation chamber.

(iii) Measures taken to seal the fumigation chamber (e.g. covering surfaces with asphalt paper or plastic tarpaulins, sealing of vents, windows, cracks, etc.).

(iv) Temperature inside the chamber.

(v) The relative size of the chamber as compared to the commodity load.

(2) Location of fumigation chamber. Information to be reported includes:

(i) County and state.

(ii) Enviromental conditions, if applicable (temperature, wind, humidity) *(iii)* Cooperator.

(D) Application of the pesticide.

(1) Type of fumigant dispensing system and method of fumigant volatilization.

(2) Measures taken to hasten gas circulation.

(3) Dose rate, exposure time, temperature, and pressure;

(4) Layout of the fumigation chamber (i.e. discharge points and positioning of circulating fans/blowers in relation to arrangement of commodities, size of stacks of commodities, etc.).

(5) Number and dates of applications.

(6) Formulation.

(E) Aeration of the commodities.

(1) The aeration time and the dates of the aeration.

(2) Description of aeration procedures inside (e.g. removal of seals and covers, opening of doors and windows, use of exhaust fans and air suction system) and outside the fumigation chamber.

(3) Description of any aeration following sampling.

(F) Sampling, handling, and storage.

(1) Dates of sampling, shipping, storage, and analyses.

(2) Description of sampling procedure, including the location of the sampling (e.g. top, bottom or side outer layer or center of stack; side or middle of chamber), size of the samples, and measures taken to prevent desorption of the fumigant during sampling.

(3) Handling, preshipping, shipping, and postshipping storage conditions, including storage times, special measures taken to prevent desorption of the fumigant during the time between sampling and analysis, and description of sample containers and storage temperature.

(G) Analytical procedures/instrumentation.

(1) Description of sample preparation (compositing, subsampling, grinding, extraction, etc.) and measures taken to prevent desorption of the fumigant during sample preparation.

(2) Details of analytical method to measure residue, including descriptions of equipment/instrumentation and instrument parameters.

(H) Quality control. Description of control measures and precautions to ensure the integrity of the test, samples and measurement of the residue.

(I) Any other pertinent information on material and methods.

(v) Results and conclusions.

(A) Brief summary of the study procedures. The summary of the study procedures should include the number of trials, the commodities, whether the commodities had been previously treated with the test active ingedient, descriptions of the fumigations and fumigation chambers, the fomulation, aeration time, and the method of detection.

(B) Results of analyses of treated and control samples and fortified samples.

(1) Tables of the results. Residue data should be given in a tabular format, providing the following information:

(i) Commodity.

(ii) Plant part.

(iii) Type of fumigation chamber.

(iv) Dose.

(v) Exposure time.

(vi) Temperature.

(vii) Aeration time.

(viii) Residue. Residue testing should extend beyond sampling immediately after the label-specified aeration to include studies to follow the rate of residue decline that could be expected under various shipping and storage conditions and temperature.

(2) Graphs and figures of the results. Graphs, if provided, should be accompanied by tables of actual values from which graphs were constructed.

(3) Narrative on the results. Narrative should include a discussion of problems and ways in which they were resolved and explanations for apparently aberrant, atypical values.

(C) Conclusions on the appropriate tolerances for the proposed uses.

(vi) Raw data and information on individual trials.

(A) Raw data tables for residue analyses of treated, control and fortification recovery samples and standards.

(B) Representative raw data figures.

(1) As applicable, printouts, spectra, chromatograms of treated samples, control samples, fortified samples and standards, etc.

(2) Calibration curves.

(C) Description of calculations, including examples.

(D) Description of statistical tests, including examples.

(E) Other. Any additional information the petitioner/registrant considers appropriate and relevant to provide a complete and thorough description of the study.

(vii) Certification. Certification of authenticity by the Study Director (including signature, typed name, title, affiliation, address, telephone number, and date).

(viii) References.

(ix) Appendices.

(A) Reprints of published and unpublished literature, company reports, letters, etc., not expected to be in OPP files, but which the petitioner/ registrant feels will aid the review of the study.

(B) Other pertinent information which does not fit in any other section of this outline.

(k) Data reporting – postharvest treatment (except fumigation).

(1) **Foreword.** Postharvest treatments of foods and feeds are applied by various means, including dips, drenches, mechanical foamers, and spray and brush applicators. The pesticide may be applied directly (to the commodity) or indirectly (to the storage bin). Often, the application of a wax coating on the commodity is

involved. Both the RAC and its processed product may be treated postharvest. The report for a study on the postharvest treatment of raw crops and processed foods and feeds should include all information necessary to provide a complete and accurate description of the study.

(2) **Format of the data report.**

(i) Master cover page. Title page and additional information requirements (i.e. requirements for data submission and procedure for claims of confidentiality of data) if relevant to the study report should precede the content of the study formatted below. These requirements are described in PR Notice 86-5.

(ii) Table of contents. The table of contents should indicate the overall organization of the study, including tables and figures.

(iii) Introduction.

(A) Background and historical information on the pesticide.

(1) Brief summary of nature of the residue in plants, including the structures of the parent and the residues considered to be of toxicological significance.

(2) [Reserved].

(B) Purpose of study.

(C) Abstract of study.

(1) Brief summary of application procedures.

(2) Results, including unexpected problems.

(3) Conclusions.

(iv) Materials and methods.

(A) Test Substance.

(1) Identification of the test pesticide active ingredient (a.i.), including chemical name, common name (ANSI, BSI, ISO), registrant developmental/experimental name and chemical structure.

(2) Identification of the pesticide formulated products in which the pesticide active ingredient was applied, including trade name, type (EC, WP, G, etc.), pounds of active ingredient per gallon, percent a.i. by weight, EPA registration number and manufacturer.

(3) Information on the matrix (e.g. water, wax) in which the formulated pesticide was applied and on any additives.

(B) Test raw or processed commodity.

(1) Identification of the raw or processed test commodity, including variety.

(2) Identification of specific crop parts treated and analyzed.

(3) Developmental stages, general condition (mature/immature, green/ ripe, fresh/dry, etc.), sizes of the test commodity at time of treatment.

(4) Information on whether the commodity or storage container had been treated with the test active ingredient prior to the test postharvest treatment, including application rates, PHIs, and the residue prior to the test postharvest treatment.

(C) Test site.

(1) Description of test site. Overall testing environment (outdoor, indoor, climate controlled packinghouse, etc.), temperature;

(2) Location (county, state).

(3) Cooperator.

(D) Application of the pesticide.

(1) Physical state in which the pesticide was applied.

(2) Description of method/equipment for pesticide application e.g. direct (applied to commodity) or indirect (applied to storage container), dips, drenches, mechanical towers, spray applicators, brush applicators, wax applicators.

(3) Pounds active ingredient and formulation per pounds treated commodity, concentration of treatment solution, volume of treatment per pounds treated commodity, exposure time, number of treatments, and temperature of solution.

(4) Description of postharvest practices accompanying the postharvest treatment such as application of wax coatings after treatment, detergent washes, and rinses, including number, timing, and volume.

(5) Dates of applications.

(6) Formulation.

(F) Sampling, handling, and storage.

(1) Dates of sampling, shipping, storage, and analyses.

(2) Description of sampling procedure and size of the samples.

(3) Handling, preshipping, shipping, and postshipping storage conditions, including storage time.

(G) Analytical procedures/instrumentation.

(1) Description of sample preparation (compositing, subsampling, grinding, extraction, etc.).

(2) Details of analytical method to measure residue, including descriptions of equipment/ instrumentation and instrument parameters.

(H) Quality control. Description of control measures and precautions to ensure the fidelity of the field test, samples and measurement of the residue.

(I) Any other pertinent information on materials and methods.

(v) Results and conclusions.

(A) Brief summary of study procedures. The summary of the study procedures should include the number of trials, the commodities, whether the commodities had been previously treated with the test active ingredient, description of the postharvest treatment (e.g. concentration, exposure time, temperature), the formulation, and the method of detection.

(B) Results of analyses of treated and control samples and fortified samples.

(1) Tables of the results. Residue data should be given in a tabular format, providing the following information, as applicable:

(i) Commodity.

(ii) Plant part.

(iii) Method/equipment for pesticide application.

(iv) Pounds active ingredient per pounds commodity.

(v) Concentration of treatment solution.

(vi) Volume treatment solution per pounds commodity.

(vii) Exposure time.

(viii) Number of treatments.

(ix) Other pertinent information affecting the level of residue (e.g. use of wax, rinse, volume and time of rinse).

(x) Formulation.

(xi) Residue. Residue testing should provide information on the rate of residue decline that could be expected under various shipping and storage conditions and temperature.

(2) Graphs and figures of the results: Graphs, if provided, should be accompanied by tables of actual values from which graphs were constructed.

(3) Narrative on the results: Narrative should include a discussion of problems and ways in which they were resolved and explanations for apparently aberrant, atypical values.

(C) Conclusions on the appropriate tolerances for the proposed uses.

(vi) Raw data and information on individual trials.

(A) Raw data tables for residue analyses of treated, control and fortification recovery samples, and standards.

(B) Representative raw data figures.

(1) As applicable, printouts, spectra, chromatograms of treated samples, control samples, fortified samples and standards, etc.

(2) Calibration curves.

(C) Description of calculations, including examples.

(D) Description of statistical tests, including examples.

(E) Other. Any additional information the petitioner/registrant considers appropriate and relevant to provide a complete and thorough description of the study.

(vii) Certification. Certification of authenticity by the Study Director (including signature, typed name, title, affiliation, address, telephone number and date).

(viii) References.

(ix) Appendices.

(A) Reprints of published and unpublished literature, company reports, letters, etc., not expected to be in OPP files, but which the registrant feels will aid the review of the study.

(B) Other pertinent information which does not fit in any other section of this outline.

(L) References

The following references should be consulted for additional background material on this test guideline.

1. Environmental Protection Agency, Pesticide Reregistration Rejection Rate Analysis – Residue Chemistry; Followup Guidance for: Generating Storage Stability Data; Submission of Raw Data; Maximum Theoretical Concentration Factors; Flowchart Diagrams. EPA Report 737-R-93-001, February 1993.

2. Environmental Protection Agency, Pesticide Reregistration Rejection Rate Analysis – Residue Chemistry; Follow-up Guidance for: Updated Livestock Feeds Tables; Aspirated Grain Fractions (Grain Dust); A Tolerance Perspective; Calculating Livestock Dietary Exposure; Number and Location of Domestic Crop Field Trials. EPA Report 737-K-94-001, June 1994.

3. Environmental Protection Agency, Pesticide Reregistration Rejection Rate Analysis – Residue Chemistry; EPA Report 738-R-92-001, June 1992.

4. Environmental Protection Agency, FIFRA Accelerated Reregistration – Phase 3 Technical Guidance. EPA Report 540/09-90-078, December 1989.

5. Environmental Protection Agency, 1986, Pesticide Assessment Guidelines, Subdivision O, Residue Chemistry, Series 171-4; Addendum No. 2 on Data Reporting, Magnitude of the Residue: Crop Field Trials, EPA Report 540/09-86-151.

6. Environmental Protection Agency, 1988, Pesticide Assessment Guidelines, Subdivision O, Residue Chemistry, Series 171-4; Addendum No. 5 on Data Reporting, Specialty Applications, EPA Report 540/09-88-008.

7. FEDERAL REGISTER Notice, 60 FR 26625-26643, May 17, 1995: Pesticide Tolerances; Revision of Crop Groups.

8. FEDERAL REGISTER Notice, 58 FR 50888-50893, September 29, 1993.

9. Environmental Protection Agency, Pesticide Regulation Notice, PR 93-2, Waiver of Crop Field Trial Data for Aerial Applications, February 11, 1993.

10. Environmental Protection Agency, Pesticide Regulation Notice PR 86-5, Standard Format for Data Submitted under the FIFRA and Certain Provisions of the Federal Food, Drug, and Cosmetic Act (FFDCA), May 3, 1986.

The following references provide additional background information.

11. Department of Agriculture, Agricultural Statistics, 1991.

12. Department of Agriculture, Pesticide Analytical Methods (PAM), Vols. I and II, 1994. (Available from National Technical Information Service (NTIS), Springfield, VA).

13. American Crop Protection Association (ACPA); NACA Recommendations for Residue Site Selection and Number of Field Trials, September 1992.

14. Department of Commerce, Census of Agriculture, 1987.

15. National Academy of Sciences, Pesticides in the Diets of Infants and Children, 1993, National Research Council, Washington, D.C.

16. Association of American Feed Control Officials (AAFCO), 1993 Official Publication, 1993.

Appendix A – Tables and Attachments for Guidance on Number and Location of Domestic Crop Field Trials

Table 1. – Minimum Numbers of Crop Field Trials and Treated Samples for Tolerances in Individual Crops

Table 2. – Required Numbers of Field Trials for Crop Groups (40 CFR 180.41)

Table 3. – Required Numbers of Field Trials for Crop Subgroups (40 CFR 180.41)

Table 4. – Required Numbers of Field Trials for Crop "Groups" (40 CFR 180.1(h))

Table 5. – Suggested Distribution of Field Trials by Region for Crops Requiring >3 Trials

Table 6. – Regional Distribution of Crop Production

Attachment 7. – Methodology for Determining Number of Field Trials

Attachment 8. – Codex Guidelines on Minimum Sample Sizes for Agricultural Commodities from Supervised Field Trials for Residue Analysis

Attachment 10. – Border Definitions of Regions

Attachment 11. – Number of Field Trials Required for Tolerances with Geographically Restricted Registration and for FIFRA Section 24(c) Special Local Needs Registrations

Table 1 – Minimum Numbers of Crop Field Trials and Treated Samples for Tolerances on Individual Crops

Following the procedure explained in the body of this guideline and in Attachment 7 of this appendix, this table specifies the minimum numbers of field trials and treated samples required to obtain tolerances on individual crops. For those crops requiring ≥8 trials in this table, a 25 percent reduction in the number of trials is acceptable for uses resulting in no quantifiable residues providing certain criteria are met (see OPPTS 860.1500 for details). The same reduction is acceptable for representative commodities that are being used to obtain crop group tolerances (see Table 2 of this appendix) and some crop subgroup tolerances (see Table 3 of this appendix A).

(NOTE: Application of both 25 percent reductions (residues <LOQ and crop group) to a given crop will not be acceptable.)

The numbers in this table represent the minimum number of acceptable trials reflecting the label use pattern producing the highest residue. Trials reflecting other use patterns or which for some reason do not generate viable samples will not be counted. In addition, these numbers of trials are predicated upon only one formulation type being requested for use on the crop. If additional types of formulations are desired, additional data may be needed as discussed in the Formulations section of this guideline.

A minimum of two treated samples is required from each field trial for crops requiring ≥3 trials. For crops requiring <3 trials, a minimum of four treated samples from four independently treated plots is required for each trial – two samples reflecting the maximum proposed application rate (1×) and two reflecting a 2× rate. As discussed in the Sampling Requirements section of this guideline, each composite sample should be collected by a separate run through a treated plot. Splitting one sample from a plot or conducting two analyses on one sample will not be an acceptable alternative to separately collecting and analyzing two samples. Multiple analyses of a single sample or of subsamples constitute the equivalent of only one data point.

Table 1.—Minimum Numbers of Crop Field Trials and Treated Samples for Tolerances on Individual Crops

Crop	Minimum No. of Trials	Minimum No. of Treated Samples
Acerola	1	4
Alfalfa	12	24
Almond	5	10
Apple, Sugar	2	8
Apple	16	32
Apricot	5	10
Arracacha	2	8
Artichoke, Globe	3 or 2*	6 or 8*
Artichoke, Jerusalem	3	6
Asparagus	8	16
Atemoya	1	4
Avocado	5	10
Banana	5	10
Barley	12	24
Bean, Dried[1]	12	24
Bean, Edible Podded[1]	8	16
Bean, Lima, Dried	3	6
Bean, Lima, Green	8	16
Bean, Mung	3 or 2*	6 or 8*
Bean, Snap	8	16
Bean, Succulent Shelled[1]	8	16
Beet, Garden	5	10
Blackberry	3[2]	6[2]
Blueberry	8	16
Bok choi	2	8
Boysenberry	2	8
Broccoli	8	16
Broccoli, Chinese (gai lon)	2	8
Brussels Sprouts	3 or 2*	6 or 8*
Buckwheat	5	10
Cabbage	8	16
Cabbage, Chinese	3	6
Cacao Bean (cocoa)	3	6
Calabaza	2	8
Calamondin	1	4
Canola	8	16
Cantaloupe	8	16
Carambola	2	8
Carob	3	6
Carrot	8	16
Cassava, bitter or sweet	2	8
Cauliflower	8	16
Celery	8	16
Cherry, Tart (Sour)	8	16
Cherry, Sweet	8	16
Chestnut	3	6
Chickpea (garbanzo bean)	3	6
Chicory	2	8
Clover	12	24
Coconut	5	10
Coffee	5	10
Collards	5	10
Corn, Field	20	40
Corn, Pop	3	6
Corn, Sweet	12	24
Cotton	12	24

Footnotes are found on page 434.

Table 1.—Minimum Numbers of Crop Field Trials and Treated Samples for Tolerances on Individual Crops— Continued

Crop	Minimum No. of Trials	Minimum No. of Treated Samples
Cowpea (dried shelled bean)	5	10
Cowpea (succulent shelled bean)	3	6
Cowpea (forage/hay)	3	6
Crabapple	3	6
Cranberry	5	10
Cress, Upland	1	4
Cucumber	8	16
Currant	2	8
Dandelion	1	4
Dasheen (taro)	2	8
Date	3 or 2*	6 or 8*
Dill (dill seed, dillweed)	2	8
Eggplant	3	6
Elderberry	3	6
Endive (escarole)	3	6
Fig	3 or 2*	6 or 8*
Filbert (hazelnut)	3 or 2*	6 or 8*
Flax	5	10
Garlic	3	6
Genip	1	4
Ginger	2	8
Ginseng	3 or 2	6 or 8
Gooseberry	3	6
Grapefruit	8	16
Grape	12	24
Grasses (crop group) (also see Table 2)	12	24
Guar	3 or 2*	6 or 8*
Guava	2	8
Hops	3	6
Horseradish	3	6
Huckleberry	3	6
Kale	3	6
Kiwi fruit	3 or 2*	6 or 8*
Kohlrabi	3	6
Kumquat	1	4
Leek	3	6
Lemon	5	10
Lentil	3	6
Lettuce, Head	8	16
Lettuce, Leaf	8	16
Lime	3	6
Loganberry	2	8
Longan	1	4
Lotus Root	1	4
Lychee	1	4
Macadamia Nut	3 or 2*	6 or 8*
Mamey Sapote	2	8
Mandarin (tangerine)	5	10
Mango	3 or 2*	6 or 8*
Melon, Casaba	3	6
Melon, Crenshaw	3	6
Melon, Honeydew	5	10
Millet, Proso	5	10
Mint[3]	5	10
Mulberry	3	6

Footnotes are found on page 434.

Crop	Minimum No. of Trials	Minimum No. of Treated Samples
Mushrooms	3	6
Muskmelons[4]	8	16
Mustard, Chinese	2	8
Mustard Greens	5[5]	10[5]
Nectarine	8	16
Oat	16	32
Okra	5	10
Olive	3	6
Onion, Dry Bulb	8	16
Onion, Green	3	6
Orange, Sour and Sweet	16	32
Papaya	3 or 2*	6 or 8*
Parsley	3	6
Parsnip	3	6
Passion Fruit	2	8
Pawpaw	3 or 2*	6 or 8*
Peach	12	24
Peanut	12	24
Peanut, Perennial	3	6
Pear	8	16
Pea, Field (Austrian Winter) (forage/hay)[6]	3	6
Pea, Chinese	1	4
Pea, Dried Shelled[1,7]	5	10
Pea, Edible Podded[1]	3	6
Pea, Succulent Shelled[1] (Pea, Garden, Succulent)	8	16
Pecan	5	10
Pepper, Bell	8	16
Pepper, (other than bell)	3	6
Persimmon	3 or 2*	6 or 8*
Pimento	2	8
Pineapple	8	16
Pistachio	3	6
Plantain	3 or 2*	6 or 8*
Plum	8	16
Pomegranate	3 or 2*	6 or 8*
Potato	16	32
Pumpkin	5	10
Quince	3 or 2*	6 or 8*
Radish, Oriental (daikon)	2	8
Radish	5	10
Rapeseed	3	6
Raspberry, Black and Red	3[2]	6[2]
Rhubarb	2	8
Rice	16	32
Rice, Wild	5	10
Rutabaga	3	6
Rye	5	10
Safflower	5	10
Sainfoin	3	6
Salsify	3	6
Sesame	3	6
Shallot	1	4
Sorghum, Grain	12	24
Soybean (dried)	20	40
Spinach	8	16

Footnotes are found on page 434.

Crop	Minimum No. of Trials	Minimum No. of Treated Samples
Squash, Summer	5	10
Squash, Winter	5	10
Strawberry	8	16
Sugar Beet	12	24
Sugarcane	8	16
Sunflower	8	16
Sweet Potato	8	16
Swiss Chard	3	6
Tangelo	3	6
Tanier (cocoyam)	2	8
Tobacco	3	6
Tomato	16	32
Turnip, root	5	10
Turnip, tops (leaves)	5	10
Walnut, Black and English	3	6
Watercress	2	8
Watermelon	8	16
Wheat	20	40
Yam, True	3	6

* For these crops registrants/petitioners have the option of doing 3 trials with two treated samples (1x rate) per trial or 2 trials with four treated samples (two at 1x rate, two at 2x rate) per trial.

[1] These bean/pea commodities include more than one type of bean/pea. The specific commodities included in each of these groups are shown below. The specific representative commodity for which field trials should be run in each case are those representative commodities provided in crop subgroup in 40 CFR 180.41. *bean, edible podded:* include those commodities listed in the subgroup 6-A as *Phaseolus* spp., *Vigna* spp., jackbean, soybean (immature seed) and sword bean. *pea, edible podded:* include those commodities listed in the subgroup 6-A as *Pisum* spp. and pigeon pea. *bean, succulent shelled:* include those commodities listed in the subgroup 6-B as *Phaseolus* spp., *Vigna* spp. and broad bean. *pea, succulent shelled:* include those commodities listed in subgroup 6-B as *Pisum* spp. and pigeon pea. *bean, dried shelled (except soybean):* include those commodities listed in subgroup 6-C as Lupinus spp., *Phaseolus* spp., *Vigna* spp., guar and lablab bean. *pea, dried shelled:* include those commodities listed in subgroup 6-C as *Pisum* spp., lentil and pigeon pea.

[2] A minimum of five trials (and 10 samples) is required on any one blackberry or any one raspberry if a tolerance is sought on "caneberries" (see Table 3 or Table 4 of this appendix). A minimum of three trials (and six samples) is required if a tolerance is sought only on blackberries or only on raspberries.

[3] A tolerance for mint may be obtained using residue data for spearmint and/or peppermint. If a tolerance is sought for either spearmint or peppermint separately, five trials are still required.

[4] A tolerance for muskmelons may be obtained using residue data for cantaloupes.

[5] A minimum of eight trials (and 16 samples) is required on mustard greens if a tolerance is sought on the crop subgroup leafy *Brassica* greens (see Table 3 of this appendix).

[6] A minimum of three trials is required for field pea forage and hay with Austrian winter pea the preferred cultivar. Field pea seeds will be considered dried shelled peas and require a minimum of five trials.

[7] The number of trials required for dried shelled pea is based on combined acreage and consumption of dried garden pea (*Pisum* spp.) and lentil.

Table 2.—Required Numbers of Field Trials For Crop Groups (40 CFR 180.41)

Crop Group	Representative Commodities	Number of Field Trials for Commodity if Not Part of Crop Group	Number of Field Trials for Commodity as Part of Crop Group
(1) Root and Tuber Vegetables	carrot	8	6
	potato	16	12
	radish	5	5
	sugar beet	12	9
			Total = 32

434

Table 2.—Required Numbers of Field Trials For Crop Groups (40 CFR 180.41)—Continued

Crop Group	Representative Commodities	Number of Field Trials for Commodity if Not Part of Crop Group	Number of Field Trials for Commodity as Part of Crop Group
(2) Leaves of Root and Tuber Vegetables (Human Food or Animal Feed)	turnip	5	5
	sugar beet or garden beet	12	9
			Total = 14
(3) Bulb Vegetables (*Allium* spp.)	green onion	3	3
	dry bulb onion	8	6
			Total = 9
(4) Leafy Vegetables (Except *Brassica* Vegetables)	leaf lettuce	8	6
	head lettuce	8	6
	celery	8	6
	spinach	8	6
			Total = 24
(5) *Brassica* (Cole) Leafy Vegetables	broccoli or cauliflower	8	6
	cabbage	8	6
	mustard greens	5	5
			Total = 17
(6) Legume Vegetables (Succulent or Dried)	bean (*Phaseolus* spp.), succulent	16	12[1]
	bean (*Phaseolus* spp.), dried	12	9
	pea (*Pisum* spp.), succulent	11	9[2]
	pea (*Pisum* spp.), dried	5	5
	soybean	20	15
			Total = 50
(7) Foliage of Legume Vegetables	bean[3]	3	3
	field pea (*Pisum* spp.)	3	3
	soybean	20	15
			Total = 21
(8) Fruiting Vegetables (Except Cucurbits)	tomato	16	12
	pepper (bell + one cultivar other than bell)	11 (8 + 3)	9 (6 + 3)
			Total = 21
(9) Cucurbit Vegetables	cucumber	8	6

Footnotes are found on page 437.

Crop Group	Representative Commodities	Number of Field Trials for Commodity if Not Part of Crop Group	Number of Field Trials for Commodity as Part of Crop Group
	muskmelon	8	6
	summer squash	5	5
			Total = 17
(10) Citrus Fruits (*Citrus* spp., *Fortunella* spp.)	orange, sweet	16	12
	lemon	5	5
	grapefruit	8	6
			Total = 23
(11) Pome Fruits	apple	16	12
	pear	8	6
			Total = 18
(12) Stone Fruits	sweet or tart (sour) cherry	8	6
	peach	12	9
	plum (or fresh prune)	8	6
			Total = 21
(13) Berries	any one blackberry or any one raspberry	3	3
	blueberry, highbush	8	6
			Total = 9
(14) Tree Nuts	almond	5	5
	pecan	5	5
			Total = 10
(15) Cereal Grains	fresh sweet corn	12	9
	dried field corn	20	15
	rice	16	12
	sorghum	12	9
	wheat	20	15
			Total = 60
(16) Forage, Fodder and Straw of Cereal Grains	corn	20	15
	wheat	20	15
	any other cereal grain	16	12
			Total = 42
(17) Grass Forage, Fodder, and Hay	Bermuda grass, bluegrass, and bromegrass or fescue	12 (4 trials for each cultivar)	Total = 12

Footnotes are found on page 437.

Crop Group	Representative Commodities	Number of Field Trials for Commodity if Not Part of Crop Group	Number of Field Trials for Commodity as Part of Crop Group
(18) Nongrass Animal Feeds (Forage, Fodder, Straw, and Hay)	alfalfa	12	9
	clover	12	9
			Total = 18
(19) Herbs and Spices	basil (fresh and dried)	3	3
	chive	3	3
	dill seed or celery seed	2	3
	black pepper	3	3
			Total = 12

[1] Twelve total field trials are required, 6 for an edible podded bean, and 6 for a succulent shelled bean.
[2] Nine total field trials are required, 3 for an edible podded pea, and 6 for a succulent shelled pea.
[3] Cowpea is the suggested representative commodity.

Table 3.—Required Numbers of Field Trials for Crop Subgroups (40 CFR 180.41)

The number of field trials required for crop groups is provided in Table 2. For crop groups, the required number of field trials shown in Table 1 should be done for each representative commodity, except that 25 percent fewer trials are required for representative commodities normally requiring 8 or more trials. This procedure does not necessarily apply to the crop subgroups shown in the Table below since there are fewer representative commodities in many cases (see 40 CFR 180.41). The table below and the corresponding footnotes describe the required numbers of field trials for crop subgroups.

Crop Subgroup	Representative Commodities	Other Commodities [1]	Production Acres[8] (x1000)	% Consumption	×Field Trials[2]
1A. Root Vegetables[3]	carrot		98	0.322	6
	radish		46	0.003	5
	sugar beet		1350	0.617	9
		Total	1494	0.942	20
1B. Root Vegetables Except Sugar Beets[3]	carrot		98	0.322	6
	radish		46	0.003	5
		beet, garden	13	0.042	
		turnip	20	0.043	
		Total	177	0.410	11
1C. Tuberous and Corm Vegetables[4]	potato		1310	2.091	16
		sweet potato[8]	90.5	0.072	
		Total	1400	2.163	16
1D. Tuberous and Corm Vegetables (Except Potato)[4]	sweet potato[8]		90.5	0.072	8

Footnotes are found on page 440.

Table 3.—Required Numbers of Field Trials for Crop Subgroups (40 CFR 180.41)—Continued

The number of field trials required for crop groups is provided in Table 2. For crop groups, the required number of field trials shown in Table 1 should be done for each representative commodity, except that 25 percent fewer trials are required for representative commodities normally requiring 8 or more trials. This procedure does not necessarily apply to the crop subgroups shown in the Table below since there are fewer representative commodities in many cases (see 40 CFR 180.41). The table below and the corresponding footnotes describe the required numbers of field trials for crop subgroups.

Crop Subgroup	Representative Commodities	Other Commodities [1]	Production Acres [8] (x1000)	% Consumption	xField Trials [2]
4A. Leafy Greens [3]	lettuce, head		240	0.394	6
	lettuce, leaf		51	0.025	6
	spinach		36	0.081	6
		Total	327	0.500	18
4B. Leaf Petioles [4]	celery [8]		36	0.114	8
5A. Head and Stem Brassica [3]	cabbage [8]		98.7	0.182	6
	cauliflower or (broccoli)		65 (115)	0.029 (0.091)	6
		Total	278.7	0.302	12
5B. Leafy Brassica Greens [7]	mustard greens		9.7	0.027	5 [7]
		cabbage, Chinese (bok choy)	8.7	0.007	
		collards	15	0.035	
		kale	6.2	0.003	
		Total	39.6	0.072	8 [7]
6A. Edible-Podded Legume Vegetables [3]	one succulent cultivar of edible podded bean (Phaseolus ssp.)		289	0.372	6
	one succulent cultivar of edible podded pea (Pisum spp.)		unknown	unknown	3
		Total	289	0.372	9
6B. Succulent, Shelled Pea and Bean [3]	one succulent shelled cultivar of bean (Phaseolus spp.)		51	0.048	6
	one garden pea (Pisum spp.)		314	0.319	6
		Total	365	0.367	12
6C. Dried, Shelled Pea and Bean (Except Soybean) [3]	one dried cultivar of bean (Phaseolus spp.)		1750	0.267	9
	one dried cultivar of pea (Pisum spp.)		395	0.005	5

Footnotes are found on page 440.

The number of field trials required for crop groups is provided in Table 2. For crop groups, the required number of field trials shown in Table 1 should be done for each representative commodity, except that 25 percent fewer trials are required for representative commodities normally requiring 8 or more trials. This procedure does not necessarily apply to the crop subgroups shown in the Table below since there are fewer representative commodities in many cases (see 40 CFR 180.41). The table below and the corresponding footnotes describe the required numbers of field trials for crop subgroups.

Crop Subgroup	Representative Commodities	Other Commodities[1]	Production Acres[8] (x1000)	% Consumption	xField Trials[2]
		Total	2145	0.272	14
7A. Foliage of Legume Vegetables Except Soybeans[3]	any cultivar of bean (*Phaseolus* spp.)		2090	0	6
	field pea (*Pisum* spp.)		709	0	5
		Total	2799	0	11
9A. Melon[4]	cantaloupe		130	0.083	8
		watermelon	193	0.142	
		melon, honeydew	29	0.034	
		Total	352	0.259	8
9B. Squash/Cucumber[3]	one cultivar of summer squash		29	0.059	5
	cucumber		130	0.134	6
		pumpkin	41	0.008	
		winter squash	29	0.060	
		Total	229	0.261	11
13A. Caneberry (Blackberry and Raspberry)[5]	any one blackberry[8] (or any one raspberry)		7.9 (15)	0.018 (0.006)	3 (3)
		Total	22.9	0.024	5
13B. Bushberry[4]	blueberry, highbush		59	0.017 (consumption for non-nursing infants = 0.043 percent)	8
19A. Herb[6]	basil, fresh and dried				3
	chive				3
		Total	2.75	0.014	6
19B. Spice[6]	black pepper				3
	celery seed or dill seed				3

Footnotes are found on page 440.

Table 3.—Required Numbers of Field Trials for Crop Subgroups (40 CFR 180.41)—Continued

The number of field trials required for crop groups is provided in Table 2. For crop groups, the required number of field trials shown in Table 1 should be done for each representative commodity, except that 25 percent fewer trials are required for representative commodities normally requiring 8 or more trials. This procedure does not necessarily apply to the crop subgroups shown in the Table below since there are fewer representative commodities in many cases (see 40 CFR 180.41). The table below and the corresponding footnotes describe the required numbers of field trials for crop subgroups.

Crop Subgroup	Representative Commodities	Other Commodities[1]	Production Acres[8] (x1000)	% Consumption	xField Trials[2]
		Total	2.75	0.014	6

[1] The column "other commodities" only includes commodities which account for >5 percent of the acreage estimates for the representative commodities.

[2] A minimum of 3 field trials is required for any representative commodity.

[3] The number of required field trials for these crops was determined in the same manner as for crop groups.

[4] For each of these crop subgroups, the normal number of field trials required for the representative commodity is required for the crop subgroup.

[5] The required number (five) of field trials for Caneberries was determined using the total acreage and consumption estimates for blackberries and raspberries, and applying the same criteria as used for determining the number of required field trials for individual commodities. A minimum of three field trials is required if a tolerance is sought for either blackberries or raspberries separately.

[6] For the subgroups "Herb" and "Spice", the minimum number of required field trials (3) was required for each representative commodity.

[7] The required number of field trials for Leafy *Brassica* Greens was determined using the total acreage and consumption estimates for the major commodities in the subgroup (since mustard greens represents a relatively small fraction of this total), and applying the same criteria as used for determining the number of required field trials for individual commodities. Therefore, a minimum of eight trials is required if a tolerance is sought on Leafy *Brassica* Greens. If a tolerance on only "mustard greens" is desired, a minimum of five trials is required (see Table 1).

[8] Acreage information (given in thousands of acres) and consumption for the following commodities include values for both the commodity itself, and the acreages and consumptions of other commodities for which the tolerance would apply as defined in 40 CFR 180.1(h): blackberries: blackberries (6.7), boysenberries (1.2); cabbage: cabbage (90), Chinese cabbage (napa) (8.7); celery: celery (36), fennel (only consumption data available); sweet potatoes: sweet potatoes (87), yams (3.5).

Table 4. – Required Numbers of Field Trials for Crop "Groups" CFR 180.1(H))

The Code of Federal Regulations (40 CFR 180.1(h)) states the following:

"Tolerances and exemptions established for pesticide chemicals in or on the general category of RACs listed in column A apply to the corresponding specific RACs listed in column B. However, a tolerance or exemption for a specific commodity in column B does not apply to the general category in Column A." This section of the CFR addresses two distinct situations. In the first situation, a specific commodity is included in both columns A and B. Residue data for that commodity support a registration or tolerance for itself as well as for the additional items listed in column B. These include the following column A commodities: alfalfa, bananas, blackberries, broccoli, cabbage, celery, endive, lettuce (head), lettuce (leaf), marjoram, muskmelons, onions (dry bulb only), onions (green), peaches, sorghum (grain, stover, forage) sugar apple, summer squash, sweet potatoes, tangerines, tomatoes, turnip tops or turnip greens, and wheat. The required number and distribution of field trials for items in column A support items in column B for this situation. The minimum numbers of field trials for these commodities are specified in Table 1 (page 429) or Attachment 7 (page 450) of this appendix. (Note: Although "muskmelons", "Oriental radish", "rapeseed" and "summer squash" do not appear by name in column B next to their entry in column A, for practical purposes these entries are treated as falling under the situation described above with the numbers of field trials specified in Table 1 of this appendix.)

The second situation occurs in cases where the item in column A is a term identifying a group of commodities in column B. These include the following column A commodities: beans, beans (dry), beans (succulent), caneberries, cherries, citrus fruits, lettuce, melons, onions, peas, peas (dry), peas (succulent), peppers, and squash. Since these column A commodities are in essence crop "groups", the number of field trials required for these "commodities" will be determined in a similar manner as for crop subgroups (or crop groups in the case of citrus). Listed in the following Table 4 are the field trial requirements to support a tolerance for these column A commodities. In each case, one or more representative commodities from column B are shown for which field trial data are required to support the "commodity" in column A. The required number of field trials for each representative commodity is also provided. Since these are treated similar to crop groups and/or subgroups, a 25 percent reduction in the required number of field trials for commodities typically requiring 8 or more field trials was employed in those case where there is more than one representative commodity.

Table 4.—Required Numbers of Field Trials for Crop "Groups" (40 CFR 180.1(h))

Column A Commodities	Representative Column B Commodities	Acres (x1000)	percent Consumption	Number of Field Trials if Not in Crop Group	Number of Field Trials if Part of Crop "Group" in Column A	Total Number of Required Field Trials for Tolerance on Crop "Group" in Column A
Beans	one succulent cultivar of edible podded bean (*Phaseolus* spp.)	289	0.372	8	6	
	one succulent shelled cultivar of bean (*Phaseolus* spp.)	51	0.048	8	6	
	one dried cultivar of shelled bean (*Phaseolus* spp.)	1750	0.267	12	9	
						21
Beans, dry	one dried cultivar of shelled bean (*Phaseolus* spp.)	1750	0.267	12	12	12
Beans, succulent	one succulent edible cultivar of podded bean (*Phaseolus* spp.)	289	0.372	8	6	

Column A Commodities	Representative Column B Commodities	Acres (x1000)	percent Consumption	Number of Field Trials if Not in Crop Group	Number of Field Trials if Part of Crop "Group" in Column A	Total Number of Required Field Trials for Tolerance on Crop "Group" in Column A
	one succulent shelled cultivar of bean (*Phaseolus* spp.)	51	0.048	8	6	
						12
Caneberries	any one blackberry or any one raspberry	23	0.024	3	5	5
Cherries	tart (sour) cherry	68.4	0.035	8	6	
	sweet cherry	60.5	0.031	8	6	
						12
Citrus fruits	See Table 2					23
Lettuce	lettuce, head	240	0.394	8	6	
	lettuce, leaf	51	0.025	8	6	
						12
Melons	cantaloupe	130	0.083	8	8	8
Onions	dry bulb onions	246	0.199	8	6	
	green onions	18.1	0.004	3	3	
						9
Peas	one succulent cultivar of edible podded pea (*Pisum* spp.)	unknown	unknown	3	3	
	one succulent cultivar of shelled pea (*Pisum* spp.)	314	0.319	8	6	
	one dried cultivar of shelled pea (*Pisum* spp.)	395	0.005	5	5	
						14
Peas (dry)	one dried cultivar of shelled pea (*Pisum* spp.)	395	0.005	5	5	5
Peas (succulent)	one succulent cultivar of edible podded pea (*Pisum* spp.)	unknown	unknown	3	3	

Column A Commodities	Representative Column B Commodities	Acres (x1000)	percent Consumption	Number of Field Trials if Not in Crop Group	Number of Field Trials if Part of Crop "Group" in Column A	Total Number of Required Field Trials for Tolerance on Crop "Group" in Column A
	one succulent cultivar of shelled pea (*Pisum* spp.)	314	0.319	8	6	
						9
Peppers	peppers, bell	70.6	0.040	8	6	
	peppers, other than bell	27.7	0.016	3	3	
						9
Squash[1]	one cultivar of summer squash	29.0	0.055	5	8	8

[1] To be consistent with the proposed squash/cucumber subgroup (see Table 3) of this appendix, one variety of summer squash was chosen as representative of all squash and pumpkins. However, since the combined acreage and consumption for all these commodities far exceeds that for the representative commodity, summer squash (combined acreage = 99,000 including 58,000 for summer and winter squash, and 41,000 for pumpkins; combined consumptions are 0.118 percent and 0.2 percent for the general population and nonnursing infants, respectively), the required number of field trials for the latter was increased one level from 5 to 8. Alternatively, five trials each could be conducted on summer squash and winter squash to obtain a tolerance on "squash".

Table 5.—Suggested Distribution of Field Trials By Region For Crops Requiring >3 Trials

Crop	Total No. of Trials[1]	1	2	3	4	5	6	7	8	9	10	11	12	13
Alfalfa	12	1	1			6		1		1	1	1		
	9	1				4		1		1	1	1		
Almond	5										5			
Apple	16	4	2			3				1	1	5		
	12	3	1			2				1	1	4		
Apricot	5										4	1		
Asparagus	8		1			2					3	2		
	6		1			2					2	1		
Avocado	5			1							4			
Banana	5			1										4
Barley	12	1[2]	1[2]			3		4		1	1	2		
	9	1[2]	1[2]			2		3		1	1	1		
Bean, Dried	12	1				5		2	1	1	1	1		
	9					4		1	1	1	1	1		

Footnotes are found on page 447.

Crop	Total No. of Trials[1]	1	2	3	4	5	6	7	8	9	10	11	12	13
						Number of Trials in Region								
Bean, Lima, Green	8		4			1					2	1		
	6		3			1					1	1		
Bean, Snap	8	1	1	1		3					1	1		
	6	1	1	1		2						1		
Beet, Garden	5	1				2	1						1	
Blackberry[3]	5		1			1							3	
Blueberry	8	1	3			3							1	
	6	1	2			2							1	
Broccoli	8						1				6		1	
	6						1				4		1	
Buckwheat	5	1				1		3						
Cabbage	8	2	1	1		1	1		1		1			
	6	1	1	1		1	1				1			
Canola	8		1			2		2				3		
	6		1			2		1				2		
Cantaloupe	8		1			1	2				4			
	6		1			1	1				3			
Carrot	8			1		1	1				4	1		
	6			1		1	1				3			
Cauliflower	8	1				1					5		1	
	6	1				1					3		1	
Celery	8			2		1					5			
	6			1		1					4			
Cherry, Sour (tart)	8	1				5				1		1[2]	1[2]	
	6	1				4				1				
Cherry, Sweet	8					2					2	3	1	
	6					2					2	2		
Clover	12	1	1		1	3	1	1	1	1	1	1		
	9	1	1		1	2	1	1	1		1[2]	1[2]		
Coconut	5													5
Coffee	5													5
Collards	5		2	1			1				1			
Corn, Field	20	1	1			17	1							
	15	1	1			12	1							

Footnotes are found on page 447.

Crop	Total No. of Trials[1]	Number of Trials in Region												
		1	2	3	4	5	6	7	8	9	10	11	12	13
Corn, Sweet	12	2	1	1		5					1	1	1	
	9	1	1	1		3					1	1	1	
Cotton	12		1		3		1		4		3			
	9		1		2		1		3		2			
Cowpea (dried shelled bean)	5			1		1	3							
Cowpea[6] (forage/hay) (succulent shelled bean)	3		1		1		1							
Cranberry	5	2				2							1	
Cucumber	8		3	1		2	1				1			
	6		2	1		2	1							
Flax	5					2		3						
Grapefruit	8		5			1					2			
	6		3			1					2			
Grape	12	2									8	1	1	
	9	2									5	1	1	
Grasses (All Areas Across the Country)	12/9													
Lemon	5			1							4			
Lettuce, Head	8	1[2]	1[2]	1							6			
	6	1[2]	1[2]	1							4			
Lettuce, Leaf	8	1[2]	1[2]	1							6			
	6	1[2]	1[2]	1							4			
Mandarin (tangerine)	5			3							2			
Melon, Honeydew	5						1				4			
Millet, Proso	5					1		2	2					
Mint	5					2						3		
Mustard Greens[4]	8		2	1	1	1	1				2			
	5		1		1	1	1				1			
Nectarine	8	1	1								5	1		
	6	1[2]	1[2]								4	1		
Oat	16	1	1			9	1	3	1					
	12	1	1			6	1	2	1					
Okra	5		1	1	1		2							
Onion, Dry Bulb	8	1				1	1		1		2	1	1	
	6	1					1		1		2	1		

Footnotes are found on page 447.

Table 5.—Suggested Distribution of Field Trials By Region For Crops Requiring >3 Trials—Continued

Crop	Total No. of Trials[1]	Number of Trials in Region												
		1	2	3	4	5	6	7	8	9	10	11	12	13
Orange, Sour and Sweet	16			11			1				4			
	12			8			1				3			
Peache	12	1	4		1	1	1				4			
	9	1	3			1	1				3			
Peanut	12		8	1			2		1					
	9		5	1			2		1					
Pear	8	1									3	4		
	6	1									2	3		
Pea, Garden, Succulent	8	1[2]	1[2]			4						2	1	
	6	1[2]	1[2]			3						1	1	
Pecan	5		2		1		1		1					
Pepper, Bell	8		2	2		1	1				2			
	6		1	1		1	1				2			
Pineapple	8													8
	6													6
Plum	8					1					5	1	1	
	6					1					4		1	
Potato	16	2	1	1		4				1	1	6		
	12	2	1	1		2				1	1	4		
Pumpkin	5	1	1			1	1				1			
Radish	5	1		2		1					1			
Rapeseed[6]	3							1			2			
Raspberry, Black and Red[3]	5	1				1							3	
Rice	16			11	1	2					2			
	12			7	1	2					2			
Rice, Wild	5					4					1			
Rye	5		1			2		2						
Safflower	5							2			3			
Sorghum, Grain	12		1		1	4	2	1	3					
	9				1	3	2	1	2					
Soybean (dried)	20		2		3	15								
	15		2		2	11								
Spinach	8	1	2			2				1	2			
	6	1	1			1				1	2			

Footnotes are found on page 447.

Table 5.—Suggested Distribution of Field Trials By Region For Crops Requiring >3 Trials—Continued

Crop	Total No. of Trials[1]	1	2	3	4	5	6	7	8	9	10	11	12	13
Squash, Summer[5]	8	1	2	1		1	1				1	1		
	5	1	1	1		1					1			
Squash, Winter	5	1	1	1		1					1			
Strawberry	8	1	1	1		1					3		1	
	6	1		1		1					2		1	
Sugar Beet	12					5		1	1	1	2	2		
	9					5		1	1		1	1		
Sugarcane	8		3	3		1								1
	6		3	2										1
Sunflower	8			3		4	1							
	6			2		3	1							
Sweet Potato	8		4	1	1		1				1			
	6		3		1		1				1			
Tomato	16	1	1	2		1					11			
	12	1	1	2		1					7			
Turnip Root	5		2		1	1					1			
Turnip Tops (leaves)	5		2		1	1	1							
Watermelon	8		2	2		1	2				1			
	6		2	1			2				1			
Wheat	20		1		1	5	1	5	6			1		
	15		1		1	3	1	4	4			1		

[1] Where two entries are provided for a crop (with the exception of mustard greens and summer squash as explained below), the second is for situations where a 25 percent reduction in the number of trials is possible due to the crop being a representative commodity used to obtain a crop group tolerance or due to the pesticidal use resulting in no quantifiable residues.

[2] Either region is acceptable.

[3] A minimum of five trials is required on any one blackberry or any one raspberry if a tolerance is sought on "caneberries" (see Table 3 or Table 4 of this appendix). A minimum of three trials is needed if a tolerance is sought on only blackberries or only on raspberries.

[4] A minimum of eight trials is required on mustard greens if a tolerance is sought on the crop subgroup leafy *Brassica* greens (see Table 3 of this appendix). A minimum of five trials is required if a tolerance is sought on only mustard greens.

[5] A minimum of five trials is required for a tolerance on "summer squash". If a tolerance is sought on "squash", at least 8 trials are required on summer squash as a representative commodity (see Table 4 of this appendix). Alternatively, five trials each could be conducted on summer squash and winter squash to obtain a tolerance on "squash".

[6] Although regional production figures are not provided in Table 6 of this appendix for rapeseed, cowpea forage/hay, and cowpea succulent shelled bean, sufficient information is available for determining suggested regions for these crops

Table 6.—Regional Distribution of Crop Production

Crop	Total percent Production Accounted	1	2	3	4	5	6	7	8	9	10	11	12	13
Alfalfa	99	8	3			51		14		13	4	6		
Almond	100										100			
Apple	97	27	11			20				3	6	30		
Apricot	96										89	7		

Table 6.—Regional Distribution of Crop Production—Continued

Crop	Total percent Production Accounted	Percentage of Crop Production (Acreage Basis) in Region												
		1	2	3	4	5	6	7	8	9	10	11	12	13
Artichoke, globe	100										100			
Asparagus	97		3			28					38	28		
Avocado	100			9							91			
Banana	99			<10										>90
Barley	99	2	2			29	36	2	6		3	19		
Bean, Dried	99	2				45	17	11	3		10	11		
Bean, Lima, Dried	99										97	2		
Bean, Lima, Green	97		46			12					28	11		
Bean, Mung	95						95							
Bean, Snap	97	14	16	9		45					3	10		
Beet, Garden	97	28	2			45	6				5		11	
Blackberry	95		7		3	6	6						73	
Blueberry[1]	94	11	36			40							7	
Bok choy	99		13	40							39			7
Boysenberry	99												99	
Broccoli	100						5				92		3	
Brussels sprouts	97	2									95			
Buckwheat	96	15				15		66						
Cabbage	93	21	16	11		18	12		3		12			
Cabbage, Chinese	97	5	8	34							46			4
Cacao Bean	100													100
Canola[2]	90		15			25	20					30		
Cantaloupe	95	2	5			6	23				59			
Carrot	98			10		13	9				59	5	2	
Cauliflower	97	4		2		4					77		10	
Celery	99			23		9	4				63			
Cherry, Tart (Sour)	100	11				75			9			3	2	
Cherry, Sweet	98	4				20			4		22	39	9	
Coconut	100													100
Coffee	100													100
Collards	99	4	60	8	4	6	7				10			
Corn, Field	97	3	6			86	2							
Corn, Pop	95					91			4					

Footnotes are found on page 451.

Table 6.—Regional Distribution of Crop Production—Continued

Crop	Total percent Production Accounted	Percentage of Crop Production (Acreage Basis) in Region												
		1	2	3	4	5	6	7	8	9	10	11	12	13
Corn, Sweet	96	13	4	8		50					3	11	7	
Cotton	97		8		26		11		37		15			
Cowpea (dried shelled bean)	89			18		10	61							
Cranberry	88	45				33							10	
Cucumber	94	3	36	10		27	10				5		3	
Currant	98												98	
Date	100										100			
Eggplant	94	5	35	35	2	5					10			2
Endive (escarole)	96	5	13	66							12			
Fig	99										99			
Filbert/hazelnut	100												100	
Garlic	100									7	82	11		
Ginger	100													100
Grapefruit	100			73			13				14			
Grape	96	5									86	5		
Guar	94								94					
Hops	94											94		
Kale	96	8	44	9	4	11	10				10			
Kiwifruit	99										99			
Kumquat	100			42							58			
Lemon	100			3							97			
Lentil	99							4				95		
Lettuce (head + leaf)	94	2[3]	2[3]	4							88			
Lime	99			80							19			
Loganberry	97												97	
Macadamia nut	100										3			97
Mandarin[4] (tangerine)	99			66							33			
Melon, honeydew	98						17				81			
Millet, Proso	99					29		35	35					
Mint	99					30						69		
Mustard Greens	97		24	5	12	15	20				21			
Nectarine	98	3	3								88	4		
Oat	97	8	2			61	3	20	3					

Footnotes are found on page 451.

Table 6.—Regional Distribution of Crop Production—Continued

Crop	Total percent Production Accounted	Percentage of Crop Production (Acreage Basis) in Region												
		1	2	3	4	5	6	7	8	9	10	11	12	13
Okra	97		28	8	9		46				6			
Olive	100										100			
Onion, Dry Bulb	96	10	2			9	15		10	5	21	16	8	
Onion, Green	97	3	4			8	18		3		54	2	5	
Orange, Sour and Sweet	100			72			2				26			
Papaya	96													96
Parsley	99	3	20	20		6	15				33			2
Pawpaw	100		100											
Peach	97	7	39		3	9	6			2	29	2		
Peanut	100		72	5			16		7					
Pear	95	7				2					33	53		
Pea, Field (Austrian Winter)	100											100		
Pea, Garden, Dried	97											97		
Pea, Garden, Succulent	92	5	4			49					3	22	9	
Pecan	100		35	3	8	2	34		10	2	6			
Pepper, Bell	92	4	20	25		10	8		2		23			
Pepper, other than bell	94		4	3		4			50	15	18			
Persimmon	93			3							90			
Pimento	92		86							6				
Pineapple	100													100
Pistachio	100										100			
Plantain	100													100
Plum	98					3					90	2	3	
Pomegranate	99										99			
Potato	95	11	4	3		27				7	4	39		
Pumpkin	86	20	12			39	5				10			
Quince	100										100			
Radish	96	2		67		21					6			
Raspberry, Black and Red	97	8				15							74	
Rhubarb	94					22							72	
Rice	100				70	3	13				15			
Rice, Wild	100					79					21			

Footnotes are found on page 451.

Table 6.—Regional Distribution of Crop Production—Continued

Crop	Total percent Production Accounted	Percentage of Crop Production (Acreage Basis) in Region												
		1	2	3	4	5	6	7	8	9	10	11	12	13
Rye	97	4	13	2		32		42	4					
Safflower	97							44			53			
Sorghum, Grain	100		2		6	34	17	12	29					
Soybean (dried)	99		9		15	75								
Spinach	96	7	17		3	3	26			9	31			
Squash, Summer or Winter[5]	95	10	23	19	3	12	8				14	6		
Strawberry	99	9	6	11	2	10					43		18	
Sugar Beet	100					52		7	8	5	14	14		
Sugarcane	100			51	35		4							10
Sunflower	100					35		58	7					
Sweet Potato	99		66	2	17		7				7			
Tangelo	100			61							39			
Tobacco	98		77			21								
Tomato	97	3	6	12		8					68			
Turnip root	94	5	35	5	3	13	13				14	2	4	
Turnip tops (leaves)	98		40	3	22	14	10		2		7			
Walnuts, Black and English	98										98			
Watermelon	94		21	19	5	8	21		7		13			
Wheat	94		4		4	23	6	26	26			5		

[1] Distribution of blueberry production based on tame (highbush) blueberries only.
[2] Approximate distribution of canola production based on 1993 acreage information provided by Intermountain Canola.
[3] Represents Regions 1 (NY) and 2 (NJ).
[4] Distribution of mandarin (tangerine) production based on acreage information for "honey tangerines" plus "other tangerines".
[5] Distributions of both summer squash and winter squash based on acreage information for "squash".

Attachment 7. – Methodology for Determining Number of Field Trials

(1) Assign a base number of field trials to each commodity based on acreage as follows:

Number of Acres	Number of Field Trials
> 10,000,000	16
> 1,000,000 - ≤ 10,000,000	12
> 300,000 - ≤ 1,000,000	8
> 30,000 - ≤ 300,000	5
> 2000 - ≤ 30,000	3
> 200 - ≤ 2000	2
≤ 200	1

(2) For commodities with acreage >300,000 A, increase the number of required field trials one level (e.g. 5 to 8, or 8 to 12) if consumption ≥0.4% of total consumption (general population, children 1 to 6, or non-nursing infants).

(3) For commodities with acreage >300,000 A, decrease the number of required field trials one level if consumption <0.1% of total consumption.

(4) For commodities with acreage >300,000 A, increase the number of required field trials one level if consumption ≥0.02% of total consumption.

(5) For the column "Required No. of Field Trials", the number of required field trials was decreased one level for commodities with greater than 90% production in one region unless indicated otherwise by a footnote.

(6) A minimum of 16 field trials is required for commodities with production greater than 300,000 acres and comprising greater than 1% of dietary consumption for the U.S. general population, non-nursing infants, or children 1 to 6 (except sugarcane, see footnote 3).

(7) A minimum of 12 field trials is required for commodities with production less than 300,000 acres and comprising greater than 1% of dietary consumption for the U.S. general population, non-nursing infants, or children 1 to 6.

Attachment 7 – Methodology for Determining Number of Field Trials

Commodity notes	Best acreage estimate (x1000)	Percent consumption		Required No. of field trials W/O 90%	Required No. of field trials	Footnotes
		General population	Non-nursing infants			
Corn, field	63300.00	0.530	0.402	20	20	
soybean (dried)	59200.00	0.631	0.875	20	20	
wheat	61700.00	2.620	0.554	20	20	11
apple	601.00	1.260	3.391	16	16	
oat	6120.00	0.153	0.402	16	16	2
orange (sour and sweet)	791.00	2.313	1.439	16	16	
potato	1310.00	2.091	0.601	16	16	
rice	2800.00	0.294	0.710	16	16	2
tomato	455.00	1.484	0.318	16	16	
alfalfa	26000.00	0.000	0.000	12	12	
barley	9180.00	0.106	0.075	12	12	
bean, dried, shelled (except soybean)	1750.00	0.267	0.054	12	12	15
clover	37300.00	0.000	0.000	12	12	
corn, sweet	671.00	0.440	0.219	12	12	
cotton	11000.00	0.038	0.003	12	12	
grape	833.00	0.437	0.127	12	12	
grasses (crop group 180.41)	475000.00	0.000	0.000	12	12	
peach	273.00	0.424	1.120	12	12	11
peanut	1690.00	0.139	0.016	12	12	
sorghum, grain (milo)	11200.00	0.044	0.003	12	12	
sugarbeet	1350.00	0.617	0.207	12	12	3
asparagus	97.00	0.024	0.002	8	8	
bean, edible podded	289.00	0.372	0.440	8	8	15
bean, lima, green	51.00	0.048	0.022	8	8	
bean, snap	289.00	0.372	0.440	8	8	
bean, succulent shelled	51.00	0.048	0.022	8	8	15
blueberry	59.00	0.017	0.043	8	8	
broccoli	115.00	0.091	0.014	8	8	9, 11
cabbage	98.70	0.182	0.010	8	8	11
canola	278.00	0.000	0.000	8	8	7
cantaloupes	130.00	0.083	0.004	8	8	
carrots	98.00	0.322	0.786	8	8	
cauliflower	65.00	0.029	0.001	8	8	
celery	36.00	0.114	0.013	8	8	11

Footnotes are found on page 455.

Commodity notes	Best acreage estimate (x1000)	Percent consumption				Footnotes
		General population	Non-nursing infants	Required No. of field trials W/O 90%	Required No. of field trials	
cherry, tart (sour)	68.40	0.035	0.027	8	8	4
cherry, sweet	60.50	0.031	0.024	8	8	4
cucumber	130.00	0.134	0.014	8	8	
grapefruit	189.00	0.271	0.059	8	8	
lettuce, head	240.00	0.394	0.000	8	8	
lettuce, leaf	51.00	0.025	0.002	8	8	16
muskmelon	159.00	0.118	0.007	8	8	11, 18
nectarine	33.00	0.024	0.000	8	8	
onion, dry bulb	246.00	0.199	0.022	8	8	11
pear	84.00	0.228	0.848	8	8	
pea, garden, succulent (pea, succulent shelled)	314.00	0.319	0.288	8	8	15
pepper, bell	70.60	0.040	0.007	8	8	
pineapple	36.00	0.126	0.175	8	8	9
plum	151.00	0.083	0.251	8	8	9
spinach	36.00	0.081	0.099	8	8	
strawberry	53.00	0.064	0.015	8	8	
sugarcane	830.00	1.386	0.463	8	8	3
sunflower	1980.00	0.008	0.000	8	8	
sweet potato	90.50	0.072	0.154	8	8	11
watermelon	193.00	0.142	0.008	8	8	
almond	428.00	0.005	0.001	5	5	10
apricot	24.00	0.063	0.207	5	5	
avocado	88.00	0.023	0.001	8	5	
banana	24.20	0.426	0.555	5	5	5, 11
beet, garden	13.00	0.042	0.102	5	5	
buckwheat	81.00	0.002	0.000	5	5	
coconut	0.00	0.050	0.660	5	5	
coffee	75.00	0.086	0.000	8	5	
collards	15.00	0.035	0.009	5	5	
cowpea (dried shell bean)	36.8	0.006	0.001	5	5	
cranberry	27.00	0.060	0.008	5	5	
flax	430.00	0.000	0.000	5	5	
lemon	69.00	0.040	0.002	8	5	
mandarin (tangerines)	28.00	0.021	0.001	5	5	11
melon, honeydew	29.00	0.034	0.003	5	5	
millet, proso	292.00	0.000	0.000	5	5	
mint	122.00	0.001	0.000	5	5	19
mustard greens	9.70	0.027	0.000	5	5	12
okra	5.70	0.027	0.000	5	5	
pea, dried	394.00	0.005	0.016	5	5	9, 15
pecan	453.00	0.010	0.000	5	5	
pumpkin	41.00	0.008	0.000	5	5	
radish	46.00	0.003	0.000	5	5	
rice, wild	31.40	0.000	0.000	5	5	
rye	546.00	0.008	0.000	5	5	
safflower	211.00	0.003	0.000	5	5	
squash, summer	29.00	0.059	0.003	5	5	13
squash, winter	29.00	0.060	0.198	5	5	13
turnip	20.00	0.043	0.000	5	5	
artichoke, jerusalem	0.00	0.000	0.000	3	3	
bean, lima, dried	44.40	0.015	0.000	5	3	
blackberry	7.90	0.018	0.000	3	3	11
cabbage, chinese	8.70	0.007	0.000	3	3	17
cacao bean (cocoa)	0.00	0.077	0.008	5	3	
carob	0.00	0.000	0.000	3	3	
chestnut	2.32	0.000	0.000	3	3	
chickpea (garbanzo)	4.00	0.001	0.000	3	3	
corn, pop	268.00	0.013	0.001	5	3	
cowpea (succulent shelled bean)	20.2	0.019	0.009	3	3	
cowpea (forage/hay)	16.4	0.000	0.000	3	3	
crabapple	0.00	0.001	0.000	3	3	
eggplant	5.30	0.011	0.000	3	3	
elderberry	0.00	0.000	0.000	3	3	
endive (escarole)	6.00	0.002	0.000	3	3	
garlic	14.80	0.001	0.000	3	3	
gooseberry	0.00	0.001	0.000	3	3	
hops	34.00	0.007	0.000	5	3	
horseradish	0.00	0.000	0.000	3	3	
huckleberry	0.00	0.000	0.000	3	3	
kale	6.20	0.003	0.000	3	3	
kohlrabi	0.00	0.000	0.000	3	3	
leeks	0.00	0.000	0.000	3	3	
lentil	162.00	0.002	0.000	5	3	
lime	6.60	0.006	0.000	3	3	
melon, casaba	0.00	0.001	0.000	3	3	
melon, crenshaw	0.00	0.000	0.000	3	3	

Footnotes are found on page 455.

Commodity notes	Best acreage estimate (x1000)	Percent consumption				Footnotes
		General population	Non-nursing infants	Required No. of field trials W/O 90%	Required No. of field trials	
mulberry	0.00	0.000	0.000	3	3	
mushroom	3.00	0.040	0.002	3	3	6
olive	33.30	0.017	0.003	5	3	
onion, green	18.10	0.004	0.000	3	3	11
parsley	5.10	0.007	0.001	3	3	
parsnip	0.00	0.001	0.000	3	3	
peanut, perennial	6.00	0.000	0.000	3	3	
peas, field (austrian winter)(forage/hay)	36.40	0.000	0.000	5	3	
pea, edible podded	0.00	0.000	0.000	3	3	15
pea, garden, dried	232.00	0.003	0.016	5	3	
pepper, non-bell	27.70	0.016	0.000	3	3	
pistachio	52.00	0.000	0.000	5	3	
rapeseed	7.7	0.000	0.000	3	3	
raspberry, black and red	15.00	0.006	0.000	3	3	
rutabaga	0.00	0.005	0.000	3	3	
sainfoin	0.00	0.000	0.000	3	3	
salsify	0.00	0.000	0.000	3	3	
sesame	0.00	0.001	0.000	3	3	
swiss chard	0.00	0.003	0.000	3	3	
tangelo	20.00	0.005	0.000	3	3	
tobacco	681.00	0.000	0.000	3	3	8
walnut, black and english	214.00	0.009	0.005	5	3	
yam, true	3.50	0.003	0.006	3	3	
apple, sugar	0.30	0.000	0.000	2	2	11
arracacha	0.30	0.000	0.000	2	2	
artichoke, globe	12.00	0.006	0.000	3	2	
bean, mung	15.00	0.012	0.000	3	2	
bok choy	1.70	0.001	0.000	2	2	17
boysenberry	1.20	0.001	0.000	2	2	
broccoli, chinese (gai lon)	0.60	0.000	0.000	2	2	
brussels sprouts	4.40	0.013	0.007	3	2	
calabaza	2.00	0.000	0.000	2	2	
carambola	0.50	0.000	0.000	2	2	
cassava, bitter or sweet	0.50	0.000	0.000	2	2	
chicory	0.42	0.001	0.000	2	2	
currant	0.34	0.001	0.000	2	2	
dasheen (taro)	1.70	0.000	0.000	2	2	
date	6.80	0.001	0.000	3	2	
dill (dill seed, dillweed)	0.30	0.000	0.000	2	2	
fig	17.00	0.005	0.001	3	2	
filbert/hazelnut	28.70	0.001	0.000	3	2	
ginger	0.30	0.002	0.000	2	2	
ginseng	5.00	0.000	0.000	3	2	
guar	6.70	0.000	0.000	3	2	
guava	1.20	0.000	0.000	2	2	
kiwifruit	9.00	0.000	0.000	3	2	
loganberry	0.24	0.000	0.000	2	2	
macadamia nut	24.00	0.000	0.000	3	2	
mamey sapote	0.30	0.000	0.000	2	2	
mango	3.16	0.001	0.000	3	2	
mustard, chinese	0.40	0.000	0.000	2	2	
papaya	3.90	0.010	0.000	3	2	
passion fruit	0.27	0.000	0.000	2	2	
pawpaw	0.00	0.000	0.000	3	2	
persimmon	2.60	0.001	0.000	3	2	
pimento	1.80	0.004	0.000	2	2	
plantain	12.50	0.003	0.000	3	2	
pomegranate	3.50	0.000	0.000	3	2	
quinces	0.00	0.000	0.000	3	2	
radish, oriental (daikon)	0.30	0.000	0.000	2	2	
rhubarb	0.90	0.007	0.000	2	2	
tanier (cocoyam)	2.00	0.000	0.000	2	2	
watercress	0.46	0.001	0.000	2	2	
acerola	0.05	0.000	0.000	1	1	
atemoya	0.10	0.000	0.000	1	1	
calamondin	0.01	0.000	0.000	1	1	
cress, upland	0.10	0.000	0.000	1	1	
dandelion	0.20	0.001	0.000	1	1	
genip	0.05	0.000	0.000	1	1	
kumquat	0.10	0.000	0.000	1	1	
longan fruit	0.08	0.000	0.000	1	1	
lotus root	0.03	0.000	0.000	1	1	
lychee	0.20	0.000	0.000	1	1	
pea, snow	0.10	0.000	0.000	1	1	
shallot	0.10	0.001	0.000	1	1	

Footnotes are found on page 455.

Footnotes:

1. Acreage values in this table are presented to three significant figures.

2. The number of field trials required for these commodities (oats and rice) was increased 1 level from 12 to 16 due to high consumption by non-nursing infants.

3. The numbers of field trials required for sugar beets and sugarcane were not increased one level based on comprising ≥0.4% consumption for the general population (or increased based on >1% consumption as discussed in criterion (6) under "methodology") because (a) the major human food derived from these two commodities (sugar) is highly refined, and (b) neither are major national animal feeds.

4. Individual consumption estimates for sweet and tart (sour) cherries were obtained by weighting the total consumption of cherries by the production (acreage) of each.

5. Limited geographical production was not used as a criterion to reduce the number of field trials for bananas because of the small number of trials required and the large consumption of this commodity by infants and children.

6. The number of field trials required for mushrooms was decreased since mushrooms are generally grown indoors under relatively constant growing conditions likely leading to little residue variability.

7. Due to the expanding production and consumption of canola products, eight field trials will be required for this commodity.

8. The number of field trials for tobacco was reduced because of the limited importance of this commodity for dietary exposure assessment.

9. Limited geographical production was not used as a criterion to reduce the number of field trials for broccoli, pineapples, and plums because of the widespread high consumption of these commodities, and was not used to reduce the number of field trials required for peas, dry because of the high acreage and the large variety of peas represented by this commodity.

10. Since the number of field trials for almonds was already reduced based on low dietary consumption, a further reduction due to >90% production being in one region was not made.

11. Acreage information (given in thousands of acres) and consumption for the following commodities include values for both the commodity itself, and the acreages and consumptions of other commodities for which the tolerance would apply as defined in 40 CFR 180.1(h): *apples, sugar:* apple, sugar (0.2), atemoya (0.1) *bananas:* bananas (11.7), plantains (12.5) *blackberries:* blackberries (6.7), boysenberries (1.2) *broccoli:* broccoli (114), broccoli, chinese (gai lon, 0.6) *cabbage:* cabbage (90), chinese cabbage (8.7) *celery:* celery (36), fennel (only consumption data available) *mandarins (tangerines):* tangerines (15), tangelos (13) *muskmelons:* cantaloupe (130), honeydew (29); consumption information is also available for casaba, crenshaw and persian melons *onions, dry bulb:* onions, dry bulb (231), garlic (14.6) *onions, green:* green onions (18), shallots (0.1) *peaches:* peaches (240), nectarines (33) *sweet potatoes:* sweet potatoes (87), yams (3.5) *wheat:* wheat (61577), triticale (152).

12. Eight field trials are needed if mustard greens are used as a representative commodity of the brassica leafy greens crop subgroup (see Table 3 of the appendix, page 434).

13. Acreage estimates for all squash were divided by 2 to estimate acreage for summer and winter squash separately.

14. Acreage estimates are given as 0.000 when acreage information is not available for a commodity.

15. These bean/pea commodities include more than one type of bean/pea. The specific commodities included in each of these groups are shown below. The specific representative commodity for which field trials should be run in each case are those representative commodities provided in 40 CFR 180.41: *bean, edible podded:* include those commodities listed in crop subgroup 6-A as *Phaseolus* spp., *Vigna* spp., jackbean, immature soybean seeds, and swordbean. *pea, edible podded:* include those commodities listed in subgroup 6-A as *Pisum* spp., and pigeon pea. *bean, succulent shelled:* include those commodities listed in crop subgroup 6-B as *Phaseolus* spp., *Vigna* spp. and broad bean. *pea, succulent shelled:* include those commodities listed in crop subgroup 6-B as *Pisum* spp. and pigeon pea. *bean, dried, shelled (except soybean):* include those commodities listed in crop subgroup 6-C as *Lupinus* spp., *Phaseolus* spp., *Vigna* spp., guar and lablab bean. *pea, dried shelled:* include those commodities listed in crop subgroup 6-C as *Pisum* spp., lentil and pigeon pea.

16. Consumption estimates for leaf lettuce include "lettuce-leafy varieties" and "lettuce-unspecified."

17. Individual consumption estimates for chinese cabbage and bok choy were obtained by weighting the total consumption of the two commodities by the relative production (acreage) of each.

18. A tolerance for muskmelons may be obtained using residue data for cantaloupes.

19. If a tolerance is sought for either spearmint or peppermint separately, five field trials are still required.

Attachment 8. – Codex Guidelines On Minimum Sample Sizes for Agricultural Commodities From Supervised Field Trials for Residues Analysis

CCPR 1987 ALINORM 87/24A APPENDIX IV ANNEX I ["Provisionally Adopted" by 19th CCPR ALINORM 87/24A para 251, 1987]

The "Guidelines on Pesticide Residue Trials to Provide Data for the Registration of Pesticides and the Establishment of Maximum Residues Limits" include a section entitled "Guide to Sampling" in which minimum sample sizes are recommended for a number of crops, selected as examples. Practical experience in sampling in recent years has indicated the need to reconsider the recommendations in the guidelines for the sample sizes and the ad hoc Working Group on the Development of Residues Data and Sampling recommends that the Annex II which follows this Annex I replaces the relevant section in the trials guidelines.

The major changes are the results of adopting a general principle that, with certain exceptions, such as very small items like berries, nuts, grain, and immature vegetables, it is more appropriate to recommend taking a number of crop units rather than a minimum weight. In many cases, the recommendation is to take 12 units for large items or 24 units for smaller items. The choice of 12 units permits easier planning of composite samples, for example, 3 units from each of 4 replicates (6 units for smaller items). It is useful, too, in sampling tree fruits, where 6 fruits from each of 4 trees is recommended. The principle of taking 12 units is readily extended to crops such as cereals, fodders, or grain where a minimum sample weight is proposed with sampling from 12 areas of the plot.

A number of crops can be harvested mechanically and in these cases 12 primary samples from the harvester as it proceeds through the treated plot is recommended.

Although it is not normally recommended it may sometimes be necessary to subsample bulky or heavy items before shipment to the residue laboratory. This practice must be limited to special sampling problems identified in Annex II always bearing in mind the importance of maintaining a fully representative subsample and avoiding any possible contamination or deterioration of the material. It is essential that it should only be done if a clean area is available and if the personnel involved have received specific instruction or training in this respect.

The ad hoc Working party emphasized that the recommendations for minimum sizes are for samples of crops at the stage of growth at which they would be harvested for consumption when taken from supervised trials, which frequently involve relatively small plots. It may be necessary to take larger sample in certain circumstances, especially if larger plots or fields are being sampled. Larger samples of some crops may also be needed if particularly low limits of determination are involved (thus possibly requiring larger analytical samples) or for multi-residue determinations (requiring larger, or multiple, analytical samples). The small sample size required by most analytical methods is not the major factor in deciding the size of field samples – obtaining representative material must be the priority in the field.

Alternative considerations may apply when deciding on the quantities of immature crops required from residue dissipation trials.

Attachment 8. – Codex Guidelines on Minimum Sample Sizes for Agricultural Commodities

Sample Type	Codex Code No.	Previous Recommendation	New Recommendation
Fodder and sugar beets	VR 0596 AM 1051	5 kg (min 5 plants)	12 plants
Potatoes	VR 0589	5 kg or 5 items	24 tubers or 12 of very large from at least 6 plants
Other root crops, e.g., carrots, red beet, Jerusalem artichoke, sweet potato, celeriac, turnip, swede, parsnip, horseradish, salsify, chicory, radish, scorzonera.	Group 016	5 kg (large) 2 kg (small items).	12 large roots or 24 (or more) small for minimum sample weight of 2 kg
Leeks	VA 0384	2 kg	12 plants
Spring Onions	VA 0389	2 kg	24 plants (or more) for a minimum sample weight of 2 kg
Garlic, shallots	VA 0381 VA 0388	2 kg	24 bulbs from at least 12 plants
Small-leaf salad crops, e.g., cress, dandelion, corn salad	Group 013	2 kg	0.5 kg from at least 12 plants (or sites in plot)
Spinach, chicory leaves	VL 0469 VL 0502 VL 0503	2 kg	1 kg from at least 12 plants

Sample Type	Codex Code No.	Previous Rec-ommendation	New Recommendation
Lettuce	VL 0482 VL 0483	2 kg	12 plants or 1 kg from at least 12 plants if individual leaves are collected
Endive	VL 0476	2 kg	12 plants
Kale forage, kale	AV 0480 VL 0480	5 kg	2 kg from at least 12 plants sampled from at least 2 levels on the plant
Green cruciferous e.g., fodder crops, rape mustard, green oil poppy.	Group 023		2 kg from at least 12 separate areas of plot[b]
Large *Brassica* crops e.g., cauliflower, cabbage	Group 010	5 kg or 5 items	12 plants
Brussels sprouts, Broccoli	Group 010	2 kg	1 kg from at least 12 plants and for Brussels sprouts sampled from at least 2 levels on the plant
Kohlrabi	VB 0405	5 kg or 5 items	12 plants
Celery	VS 0624	2 kg	12 plants
Rhubarb	VS 0627	2 kg	12 sticks (or more) from at least 12 separate plants for minimum sample weight of 2 kg
Asparagus	VS 0621	2 kg	24 sticks (or more) from at least 24 separate plants for minimum sample weight of 2 kg
Globe artichoke	VS 0620		12 heads
Soybeans	VS 0541	1 kg	1 kg from at least 12 separate areas of plot
Peas, *Phaseolus* beans (French, Kidney, Runner etc.), broad beans, field beans, lentils.	Group 014 Group 015	2 kg 2 kg	1 kg (fresh green or dry seed as appropriate)
Tomatoes, green peppers	Group 012	2 kg	24 fruits or 12 from large fruiting varieties from at least 12 plants (more if necessary for a mini-mum sample weight of 2 kg)
Aubergines (eggplants)	VO 0440	5 kg or 5 items	12 fruits from 12 separate plants
Cucumbers	VO 0424	5 kg or 5 items	12 fruits from 12 separate plants
Gherkins, courgettes squash	Group 011	2 kg	24 fruits from at least 12 plants (more if necessary to make a minimum weight of 2 kg)
Melons, gourds, pumpkins, watermelons	Group 011	5 kg or 5 items	12 fruits from 12 separate plants
Sweet corn	VO 0447	2 kg	12 ears (more if necessary to make a minimum weight of 2 kg

Fruit
  Citrus fruit e.g., orange, lemon, clementine, mandarin, pomelo, grapefruit, tangelo, tangerine.
  Pome fruit e.g., apples, pears, quinces, medlars
  Large stone fruit, e.g., apricots, nectarines, peaches, plums.
  Small stone fruit
	
	Group 001
	
	Group 002
	Group 003
	
	Group 003
		
		5 kg
		
		5 kg
		5 kg (2 kg for plums)
		
		2 kg
			24 fruit from several places on at least 4 individual trees (more if necessary for a minimum sam-ple weight of 2 kg
			
			1 kg from several places on at least 4 trees

Grapes	FB 0269	2 kg	12 bunches, or parts of 12 bunches from separate vines to give at least 1 kg
Currants, raspberries and other small berries	Group 004	2 kg	0.5 kg from at least 12 separate areas of bushes
Strawberries, gooseberries	FB 0268 FB 0275 FB 0276	2 kg	1 kg from at least 12 separate areas of bushes
Miscellaneous, small fruits, e.g., olives, dates, figs	Group 005	2 kg	1 kg from several places on at least 4 trees

Sample Type	Codex Code No.	Previous Recommendation	New Recommendation
Bananas	FI 0327	5 kg or 4 fruits from each of 5 bunches.	24 fruits from at least 6 bunches from separate trees and from several places of each of the bunches
Miscellaneous fruit e.g., avocados, guavas, mangoes, pawpaws, pomegranates, persimmons, kiwifruit, lychee.	Group 006	5 kg	24 fruits from at least 4 separate trees or plants (more fruit if necessary for a minimum sample weight of 2 kg)
Pineapples	FI 0353	5 kg or 5 items	12 fruit
Grain of wheat, barley, oats, rye, triticale and other small grain cereals; maize (off the cob), rice, sorghum.	Group 020	1 kg (2 kg maize)	1 kg from at least 12 separate areas of a plot or treatment lot (applies to both field and post-harvest trials)
Straw of the above crops	Group 051	1 kg	0.5 kg from at least 12 separate areas of a plot[b]
Maize, straw/stover/fodder (mature plants excluding cobs)	AF 0645	5 plants	12 plants[a]
Green or silage maize		5 plants	12 plants[a]
Green forage/silage crops of alfalfa, clover, fodder peas and beans, vetch, sainfoin, lotus, fodder soybeans, ryegrass, fodder cereals, sorghum.	Group 050	1 kg (smaller leaves) 2 kg (larger leaves).	1 kg from at least 12 separate areas of a plot
Dry hay of the above crops	Group 050	1-2 kg	0.5 kg from at least 12 separate areas of a plot[b]
Peanuts	SO 0697	1 kg (2 kg with fibre)	1 kg from at least 24 plants
Treenuts, Walnuts, chestnuts, almonds, etc	Group 022	1 kg	1 kg (with or without shells)
Coconut	TN 0665	5 kg or 5 items	12 nuts
Rapeseed, flax and wild mustard	Group 023	1 kg	0.5 kg from at least 12 separate areas of a plot[b]
Sunflower, safflower	SO 0702	1 kg	12 heads or 1 kg from 12 separate areas of a plot[b]
Cottonseed	SO 0691	1 kg	1 kg with or without fibre
Coffee, cocoa	Group 024	2 kg	1 kg (fresh or dry)
Garden herbs and medicinal plants, e.g., parsley, thyme	Group 027 Group 028 Group 057.		0.5 kg fresh 0.2 kg dry
Tea (dry leaves)	Group 066	1 kg	0.2 kg
Mushrooms	VO 0450		12 items or more with a minimum sample weight of 0.5 kg
Sugarcane	GS 0659	5 kg (20 cm of stem)	12 × 20 cms lengths of stem from 12 areas of the plot[a]
Hops (dry cones)	DH 1100		0.5 kg
Beer, wine, cider, fruit juices	Group 070		1 litre

[a] Divide each stem with leaves attached into 3 equal lengths. Take top, middle and bottom portions respectively from each of three groups of four stems ensuring that parts of all 12 stems are included in the sample.

[b] Crops which are harvested mechanically can be sampled from the harvester as it proceeds through the crop.

Attachment 10. – Border Definitions of Regions
See map on page 323

I. ME, NH, VT, MA, RI, CT, NY, PA
NJ - N of Rt. 1
MD - NW of I-95
DVA - N of I-64 and W of I-81
WV - N of I-64 and E of I-77
OH - E of I-77
II. NC, SC, GA, DE,
VA - E of I-81 or S of I-64
MD - SE of I-95
NJ — S of Rt. 1
WV — S of I-64
KY — S of I-64 and S of BGP and E of I-65
TN — E of I-65
AL — Except Mobile and Baldwin Co.'s
III. FL, AL, — Mobile and Baldwin Co.'s
IV. LA, AR, MS
TN — W of I-65
MO — E of Rt. 67 and S of Rt. 60
V. MI, IN, IL, WI, MN, IA,
OH — W of I-77
WV — N of I-64 and W of I-77
KY — N of I-64 or N of BGP or W of I-65
MO — W of Rt. 67 or N of Rt. 60
KS, NE, SD, ND — all E of Rt. 281
VI. OK — E of Rt. 281/183

TX — E of Rt. 283 and SE of Rt. 377
VII. MT — E of Rt. 87 or E of I-15
WY — E of I-25 or N of I-90
ND, SD, NE — all W of Rt. 281
VIII. KS — W of Rt. 281
CO — E of I-25
NM — E of I-25
TX — W of Rt 283 and NW of Rt. 377
OK — W of Rt. 281/183
IX. UT, NV,
NM — W of I-25 and N of I-10
CO — W of I-25
WY — W of I-25 and S of I-90
MT — W of Rt. 87 and W of I-15
AZ — NE of Rt. 89/93 and N of I-10
X. CA — Except Mendocino, Humboldt, Trinity, Del Norte, and Siskiyou counties
AZ — SW of Rt. 89/93 or S of I-10
NM — S of I-10
XI. ID
OR & WA — E of Cascades
XII. CA — Counties excluded from Region X
OR & WA — W of Cascades
XIII. HI, PR
South FL — S of Rt. 41 (Added)

Note: Tables 5 and 6 in this Appendix still reference all of Florida as Region III.

Attachment 11. – Number of Field Trials Required for Tolerances With Geographically Restricted Registration, and for FIFRA Section 24(C) Special Local Needs Registrations

Throughout this guideline, guidance has been provided regarding field trial data requirements for tolerances with national registrations. This attachment provides guidance concerning the number of field trials required for tolerances with geographically restricted registrations, and FIFRA section 24(C) Special Local Needs (SLN) registrations. Sampling requirements and other criteria presented elsewhere in this guideline also apply to the data requirements discussed in this attachment. A flow chart follows the text below to facilitate determination of field trial data requirements.

Tolerances with geographically restricted registrations may be established for minor agricultural uses (1990 Farm Bill). Specifically (see 7/7/93 memorandum, Victor J. Kimm, Acting Assistant Administrator, OPPTS, to Honorable Bob Graham, U.S. Senate),

The Administrator will not require a person to submit, in relation to registration or reregistration of a pesticide for minor agricultural use under this Act, any field residue data from a geographic area where the pesticide will not be registered for such use.

Comments below address the data requirements for both tolerances with geographically restricted registration, and the additional state-specific data required for section 24(C) SLN registrations. When discussing the number of required field trials below, the term "geographically restricted region" will apply to either of these situations.

The number of field trials required for a tolerance with geographically restricted registration is equal to the number of field trials required for the commodity for a national tolerance or registration, multiplied by the proportion (by acres) of the crop grown in that region. However, regardless of the acreage in the specific region for which the regional registration is requested, at least 2 field trials will be required (except in the case of very minor crops as specified elsewhere in this guideline which require only 1 field trial for national registration). Two composite samples per field trial are generally required. However, when 3 or fewer field trials are required for any registration, the registrant may choose to (a) obtain samples from 1× and 2× application rates from separate plots in each of 2 field trials (i.e. one composite sample taken from each of two 1× and two 2× separately treated plots, resulting in 4 total samples per field trial), or (b) perform 3 field trials in different locations at the 1× rate (2 composite samples obtained from each plot).

Field trial locations must be representative of growing conditions throughout the region covered by the regional registration.

For section 24(C) SLN registrations requested for two neighboring states, data from one state will be accepted for a section 24(C) use in a neighboring state only if:

1. The states, or pertinent parts thereof, are in the same geographical region as defined in this guideline.

2. A sufficient number of field trials are available from the state to fulfill the requirements of the paragraph above for the acreage of commodity grown in both states.

3. Field trials are performed in sufficiently diverse areas such that conditions likely to be found in both states are represented in the field trials.

For crops requiring 8 or more field trials nationally, regional and section 24(C) registrations will require multiple year field trial data. Multiple year data are required to account for variability due

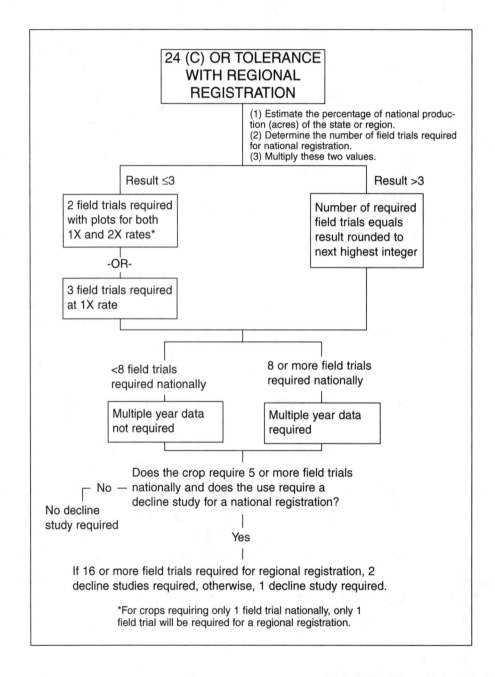

to varying climatic conditions and other factors which would normally be expected to be seen by obtaining field trial data from more diverse regions, but would not be seen for regional registrations since field trial data are obtained from more limited geographical areas. The total required number of field trials must be performed over at least 2 different years (e.g. if 4 total field trials are required, 2 would be performed in one year, and 2 in the next year). Multiple year data will not be required if sufficient nationally representative or multiple year data are available for other pesticide formulations of the same active ingredient or similar uses from which the Agency can estimate likely variability.

For crops normally requiring 5 or more field trials for a national registration, *and uses requiring a decline study* (discussed elsewhere in this guidance), one or more decline studies will be required for a section 24(C) or regional registration. The number of decline studies required for a use will not exceed the number required for a national tolerance/registration for that commodity. See the flow chart for further details.

Examples

Example 1: A section 24(C) SLN is desired for use of a pesticide on apples in Washington (WA). Since WA accounts for approximately 27 percent of national apple production, and since 16 field trials are required for apples nationally, 5 field trials will be required from WA for this use (0.27 x 16 = 4.3 or 5 field trials). Since greater than 8 field trials are required nationally (16), multiple year data will be required (3 field trials the first year, 2 the second year). Finally, if the use were one requiring a decline

study, 1 decline study would also be required for this section 24(C) use.

Example 2: A section 24(C) SLN is desired for use of a pesticide on alfalfa in Iowa (IA). Since IA accounts for approximately 8 percent of alfalfa grown nationally, and since 12 field trials are required for alfalfa nationally, 2 field trials will be required from IA to support this registration (0.08 x 12 = 0.96, however, except for crops requiring only 1 field trial nationally, at least 2 trials are required for any regional registration; therefore, 2 field trials are required). Since greater than 8 field trials are required nationally (12), the two required field trials would have to be distributed over two years (one field trial in each of two years). Since alfalfa requires greater than 5 field trials for a national registration (12), one of these studies would have to be a decline study if the use pattern requires a decline study. For the other study, one sample from each of 4 separately treated plots (two at 1× and two at 2× rates) would be required.

Example 3: A tolerance with regional registration is requested for application of a pesticide to peanuts in the Southeastern U.S. (GA, 42 percent, AL, 14 percent, NC, 9 percent, FL, 5 percent, SC, 1 percent, total 71 percent of U.S. peanut production). Since 12 field trial are required for peanuts nationally, 9 field trials would be required for this use (0.71 x 12 = 8.5 or 9 field trials required). Since 12 field trials are required nationally, the required field trials would have to be distributed over at least two years (preferably 5 the first year, four the second). If the use pattern was one requiring a decline study, a single decline study, would be required.

EPA Residue Chemistry Test Guidelines

OPPTS 860.1520
Processed Food/Feed

INTRODUCTION

This guideline is one of a series of test guidelines that have been developed by the Office of Prevention, Pesticides and Toxic Substances, United States Environmental Protection Agency for use in the testing of pesticides and toxic substances, and the development of test data that must be submitted to the Agency for review under Federal regulations.

The Office of Prevention, Pesticides and Toxic Substances (OPPTS) has developed this guideline through a process of harmonization that blended the testing guidance and requirements that existed in the Office of Pollution Prevention and Toxics (OPPT) and appeared in Title 40, Chapter I, Subchapter R of the Code of Federal Regulations (CFR), the Office of Pesticide Programs (OPP) which appeared in publications of the National Technical Information Service (NTIS) and the guidelines published by the Organization for Economic Cooperation and Development (OECD).

The purpose of harmonizing these guidelines into a single set of OPPTS guidelines is to minimize variations among the testing procedures that must be performed to meet the data requirements of the U. S. Environmental Protection Agency under the Toxic Substances Control Act (15 U.S.C. 2601) and the Federal Insecticide, Fungicide and Rodenticide Act (7 U.S.C. 136, et seq.).

Final Guideline Release: This guideline is available from the U.S. Government Printing Office, Washington, DC 20402 on The Federal Bulletin Board. By modem dial 202-512-1387, telnet and ftp: fedbbs.access.gpo.gov (IP 162.140.64.19), internet: http://fedbbs.access.gpo.gov, or call 202-512-0132 for disks or paper copies. This guideline is also available electronically in Adobe format EPA Public Access webpage: http://www.epa.gov/docs/OPPTS_Harmonized/860_Residue_Chemistry_ Test_Guidelines

OPPTS 860.1520
Processed food/feed.

(a) **Scope.**

(1) Applicability. This guideline is intended to meet testing requirements of both the Federal Insecticide, Fungicide, and Rodenticide Act (FIFRA) (7 U.S.C. 136, *et seq.*) and the Federal Food, Drug, and Cosmetic Act (FFDCA) (21 U.S.C. 301, *et seq.*).

(2) Background. The source material used in developing this harmonized OPPTS test guideline is OPP 171-4, Results of Tests on the Amount of Residue Remaining, Including a Description of the Analytical Methods Used (Pesticide Assessment Guidelines, Subdivision O: Residue Chemistry., EPA Report 540/9-82-023, October 1982). This guideline should be used in conjunction with OPPTS 860.1000, Background (page 381).

(b) **Purpose.** Processing studies are required to determine whether residues in raw commodities may be expected to degrade or concentrate during food processing. If residues do concentrate in a processed commodity, a food or feed additive tolerance must be established under section 409 of the FFDCA (or a section 701 Maximum Residue Limit (MRL) in some cases). However, if residues do not concentrate in processed commodities, the tolerance for the raw agricultural commodity (RAC) itself applies to all processed food or feed derived from it.

(c) **Concentration of residues on processing.**

(1) Whenever there is a possibility of residue levels in processed foods/feeds exceeding the level in a RAC, processing data are required. Examples of processed foods/feeds in which residues may concentrate are apple juice and apple pomace, the hulls, meal, and refined oil from cottonseed, or the sugar, dried pulp and molasses from sugar beet roots. A list of processed byproducts is contained in Table 1 of OPPTS 860.1000, Background (page 392).

(2) Processing studies should simulate commercial practices as closely as possible. RAC samples used in processing studies should contain field-treated quantifiable residues, preferably at or near the proposed tolerance level, so that concentration factors for the various byproducts can be determined. As discussed in paragraph (f)(3)

of this guideline, this may require field treatment at exaggerated application rates to obtain sufficient residue levels for processing studies. Processing studies utilizing spiked samples are not acceptable, unless it can be demonstrated that the RAC residue consists entirely of a surface residue.

(3) Only one processing study is required for each crop in Table 1 of OPPTS 860.1000 having a processed commodity. However, it is advisable to have multiple samples of the RAC and processed commodities in the study. As stated in paragraph (f)(2) of this guideline, if multiple processing studies are available for a given crop, the Agency will use the average concentration factor obtained across these studies. In some cases the requirement for a processing study may be waived based on field trial data for the RAC reflecting exaggerated application rates. This is discussed in more detail in paragraph (f)(3) of this guideline.

(4) The total toxic residue should be measured in the raw agricultural commodity at the time processing is initiated and in all processed commodities of the crop listed in Table 1 of OPPTS guideline 860.1000. With the exception of the small grains, the Agency will not normally translate data between crops. In the case of small grains, a processing study on wheat satisfies the requirement for studies on barley, buckwheat, millet, oats and rye if the pesticide is applied to all these crops in a similar manner and comparable residue levels occur in the grains.

(5) Unless the processed commodities are analyzed within 30 days of their production, data demonstrating the stability of residues in representative processed commodities during storage are required as described in OPPTS guideline 860.1380.

(6) If the processing studies indicate that residues concentrate on processing, then a food additive petition, including a food additive regulation proposal, is normally required as specified by section 409 of the FFDCA. However, for a processed food or feed that is not ready-to-eat, an MRL may need to be proposed under section 701 of the FFDCA. This is explained in more detail in paragraph (f)(5) of this guideline.

(d) **Reduction of the residue level on processing.** In those cases where the assumption of tolerance level residues occurring in commodities results in unacceptable exposure, then the petitioner has the option of submitting data on food prepared for consumption. The Agency will take into account data on washing, trimming, cooking, peeling or processing to the extent that these procedures are used on specific commodities. Although the lower levels of residues resulting from such processes may be used in the risk assessment, the tolerance will still be set on the commodity as it travels in interstate commerce. Of course, if these data indicate that residues concentrate in some fractions while decreasing in others, both the higher and lower residue levels will be used in the risk assessment. The Agency will also take into account the wide variation in techniques used to prepare food. For example, if cooking completely destroys the residue on a vegetable, the Agency will use, at a maximum, the limit of quantitation (LOQ) in the cooking study as the residue level for cooked vegetables. The Agency will also use the consumption of uncooked vegetables and the tolerance level to estimate the exposure from uncooked vegetables.

(e) **Maximum theoretical concentration factors.** (1) This paragraph addresses maximum theoretical concentration factors for use in determination of the exaggerated application rate needed for field trials on commodities which can be processed. The use of exaggerated rate studies is discussed in more detail in paragraph (f)(3) of this guideline. The following Table 1 provides a listing of maximum theoretical concentration factors.

A secondary use of this list could be for worst case dietary exposure assessment, when experimental processing data are unavailable.

Table 1.—Maximum Theoretical Concentration Factors by Crop

Crop	Maximum Concentration Factor
Apples	>14×*
Barley	8×
Beets, sugar	>20×*
Canola	3×
Citrus	1000×
Coconut	3×
Coffee	4.5×
Corn	25×
Cottonseed	6×
Figs	3.5×
Grapes	5×*
Mint (peppermint, spearmint)	330×
Oats	8×**
Olive	10×
Peanuts	3×
Pineapple	4×
Potatoes	5×
Plums (prunes)	3.5×
Rapeseed	2×
Rice	8×
Rye	10×
Safflower	9×
Soybeans	12×
Sugarcane	>20×*
Sunflower	4.5×
Tomatoes	5.5×*
Wheat	8×

*Experimental factor
**Based on factor for wheat

(2) The list is not all inclusive as factors are not available for all processed commodities listed in Table 1 of OPPTS 860.1000 (page 392). In addition, some processed commodities may have greater potential for concentration than those processed commodities for which factors were calculated. For those commodities for which higher concentrations are expected, the Agency has tabulated some experimental concentration factors, by comparing proposed and established food/feed additive tolerances to the proposed and established tolerances for the RAC. Additional factors may be added or updated in the future as further information becomes available.

(3) There are two types of processes for which maximum theoretical concentration factors can easily be calculated. The first type is where the concentration is based on the loss of water during processing. In this case, the theoretical concentration factor is the ratio of the percent of dry matter in the processed commodity to the percent of dry matter in the RAC. For example, grapes contain 18 percent dry matter while raisins contain 85 percent dry matter. The theoretical concentration factor for the processing of grapes into raisins is 85/18 or 4.7×. The second type of process is that in which a RAC is separated into components, such as the processing of corn grain into corn oil. In this case, the theoretical concentration factor is 100 percent divided by the percentage of the processed commodity in the RAC. Corn grain may contain as little as 4 percent corn oil. The theoretical concentration factor for processing of corn into oil then is 100/4, or 25×.

(4) In cases where a crop had multiple processed fractions, only the fraction having the highest maximum the-

oretical concentration factor is listed in Table 1 (see paragraph (e)(1) of this guideline). In some cases, only typical yields were available for a particular RAC, particularly for grains. A factor was still calculated, but may not actually be the maximum theoretical concentration factor. The following Table 2 shows calculations for those commodities where concentration is based on loss of water,

Table 3 shows calculations for those commodities where concentration is based on separation into components, and Table 4 is a tabulation of experimentally determined factors obtained by comparing proposed and established food/feed additive tolerances to the proposed and established tolerances for the RAC. A bibliography for the tables is given in paragraph (e)(5) of this guideline.

Table 2.—Theoretical Concentration Factors Based on Loss of Water

Crop	percent dry matter	Factor	Reference
Figs	22		PAM I, sec. 202.12
dry figs	76	3.5	PAM I sec., 202.12
Grapes	18		Harris Guide
raisins	85	4.7	Harris Guide
Potatoes	20		USDA
dried (flakes, granules)	93	4.7	USDA
Plums	21		PAM I, sec. 202.12
prunes	72	3.4	PAM I, sec. 202.12
Tomatoes	6		p. 311, Commercial Vegetable Processing, 2nd Ed.
puree	8.5	1.4	p. 272, Commercial Vegetable Processing, 2nd Ed.
paste	33	5.5	p. 277, Commercial Vegetable Processing, 2nd Ed.

Table 3.—Theoretical Concentration Factors Based on Separation into Components

Crop	Minimum percent of whole	Factor	Reference
Barley grain			
bran	13	7.7	based on wheat bran
pearled	82	1.2	p. 426, Principles of Field Crop Production
Beets, sugar			
sugar	8	12.5	Advances in Sugar Beet Production
molasses			
dried pulp			
Canola			
meal	52	1.9	p. 259, by difference, CRC Handbook of Processing and Utilization in Agriculture
oil	33	3.0	p. 259, CRC Handbook of Processing and Utilization in Agriculture

Crop	Mini-mum percent of whole	Factor	Reference
Citrus			
peel	30	3.3	p. 1391, Considine Foods and Food Production Encyclopedia
oil	0.1	1000	PAM I, sec. 202.12
pulp, dehydrated			
juice	50	2	p. 1387, Considine Foods and Food Production Encyclopedia
Coconut			
oil	35	2.9	PAM I, sec. 202.15
copra (dried meal)		2.1	DRES (from USDA Handbook No. 102)
Coffee			
roasted bean	1.2	18 percent loss in weight in roasting	p. 459 Considine
instant		4.4	PP no. 0E3875-based on weights in processing study
Corn grain			
oil	4	25.0	p. 243, Corn, Culture, Processing, Products
Cottonseed			
hulls	26	3.8	p. 187, CRC Handbook of Processing and Utilization in Agriculture
meal	45	2.2	p. 187, CRC Handbook of Processing and Utilization in Agriculture
oil	16	6.3	p. 187, CRC Handbook of Processing and Utilization in Agriculture
Grapes			
juice	82	1.2	Harris Guide
Mint			
oil	0.3	333	15 mL oil from 10 lb hay
Oats			
flour			
rolled oats	70	1.4	p. 577-8, Cereal Crops
Olive oil			
oil	10	108	p. 1372, Considine Foods and Food Production Encyclopedia
Peanuts			
meal	46	2.2	p. 139, by difference, see p. 293, Peanuts:....
oil	36	2.8	PAM I, sec. 202.25
Pineapple			PP no. 6F0482
process residue	26	3.8	
juice			
Potatoes			
processed waste	25	4.0	Northwest Food Processors Association
Rapeseed			
meal	52	1.9	p. 259, by difference, CRC Handbook of Processing and Utilization in Agriculture
Rice grain (rough rice) ...			
hulls	20	5.0	pp. 649, 652, Cereal Crops
bran	13	7.7	pp. 649, 652, Cereal Crops
Rye grain			
bran	10	10.0	pp. 244-5, CRC Handbook of Processing and Utilization in Agriculture
flour			
Safflower			
meal	11	9.1	p. 114, CRC Handbook of Processing and Utilization in Agriculture

Table 3.—Theoretical Concentration Factors Based on Separation into Components—Continued

Crop	Minimum percent of whole	Factor	Reference
oil (safflower)	30	3.3	p. 114, CRC Handbook of Processing and Utilization in Agriculture
Soybeans			
hulls	9	11.3	MRID No. 424482-03, Appendix B, p67
meal	46	2.2	CBRS No. 10541, D. Miller, 1/29/93
oil	8	12.0	CBRS No. 10541, D. Miller, 1/29/93
Sugarcane			
molasses			
sugar	8.5	11.8	p. 426, Principles of Field Crop Production
Sunflower			
meal	22	4.5	p. 146, by difference, CRC Hand book of Processing and Utilization in Agriculture
oil	40	2.5	p. 146, CRC Hand book of Processing and Utilization in Agriculture
Tomatoes			
juice	70	1.4	p. 303, Commercial Vegetable Processing, 2nd. Ed.
Wheat grain			
bran	13	7.7	p. 2125, Considine
flour	72	1.4	pp. 295-6, Cereal Crops
shorts	12	8.3	pp. 295-6, Cereal Crops

Table 4.—Maximum Observed (Experimental) Concentration Factors

Crop	Maximum Concentration Factor [1]
apple pomace	14×
sugar beet pulp, dry	20×
sugarcane molasses	20×

[1] These factors are based on a comparison of proposed and established food additive tolerances to the proposed and established tolerances on raw agricultural commodities.

(5) **The following is a bibliography for Tables 1 through 4.**

Pesticide Analytical Manual, Volume I (PAM I), 1994, Food and Drug Administration.

Agriculture Handbook No. 8, Composition of Foods: Raw, Processed, prepared, U. S. Department of Agriculture, Agricultural Research Service, B. K. Watt, and A. L. Merrill, December, 1963.

CRC Handbook of Processing and Utilization in Agriculture, Volume II, Part 2 Plant Products, I. A. Wolff, ed., CRC Press, Boca Raton, FL 1983.

Foods and Food Production Encyclopedia, D. M. Considine, and G. D. Considine, eds., Van Nostrand Reinhold, New York, 1982.

Commercial Vegetable Processing, 2nd Edition, ed. B. S. Luk, and J. G. Woodroof, Avi/Van Nostrand Reinhold, New York, 1988.

Peanuts: Production, Processing, Products, 2nd Edition, J. G. Woodroof, Avi Publishing, Westport, CT, 1973.

Corn: Culture, Processing, Products, Ed. G. E. Inglett, Avi Publishing, Westport, CT, 1970.

Oats: Chemistry and Technology, ed. F. H. Webster, American Association of Cereal Chemists, Inc., St. Paul, MN 1986.

Advances in Sugar Beet Production: Principles and Practices, eds., R. T. Johnson, et. al., Iowa State University Press, Ames, IA 1971.

Harris Guide.

Feeds & Nutrition – Complete, First Edition, Ensminger, M. E., and C. G. Olentine, Jr., Ensminger Publishing Co., Clovis, CA 1978.

Cereal Crops, Leonard, W. H., and J. H. Martin, Macmillan Co., New York, 1963.

Principles of Field Crop Production. 3rd. Edition, Martin, J. H., W. H. Leonard, and D. L. Stamp, Macmillan, New York, 1976.

(f) **Determining the need for food/feed additive tolerances.**

(1) **RAC residue value.**

(i) The Agency will consider using some average residue value from field trials if it can be determined that there is sufficient mixing during processing such that variation among individual samples from a field will be substantially evened out. It has been stated that, "* * * * the most relevant 'average' residue value from crop field trials is the highest average residue value from the series of individual field trials * * * *."

This value is sometimes referred to as the HAFT (highest average field trial). Other average values (e.g. average of all field trials) may be considered if the circumstances involved in processing of the crop warrant. Such an example would be where processing is likely to involve blending of crop from across a regional or national market.

(ii) As a result of this policy, it is necessary to determine the HAFT for each RAC for which a processing study has shown concentration of residues. For each field trial reflecting the maximum residue use (i.e. maximum number and rate of application, minimum preharvest interval) and considered acceptable for determining the section 408 of FFDCA tolerance (i.e. values discarded for reasons such as contamination should not be included), residue values for all samples at that site reflecting that use should be averaged.

(NOTE: If residues were corrected for low method recoveries or for losses during storage in order to determine the tolerance, the corrected values should also be used in this exercise.) The highest such average value is the HAFT and is to be used to calculate the maximum expected residue in processed commodities. For field trials in which only one sample per site reflects the maximum residue use no averaging can be done and the highest individual residue value becomes the HAFT.

(2) **Multiple processing studies.**

(i) Whenever more than one processing study has been conducted for a particular pesticide on a given RAC, the average concentration factor should be used for each processed commodity when determining the need for section 409 tolerances (or section 701 MRLs under paragraph (c)(6) of this guideline). Similarly, if multiple samples or subsamples are analyzed within a processing study, the average residue value should be used for each commodity as opposed to using the lowest value from the RAC samples and the highest value for the processed fraction samples, which would result in the highest concentration factors. When averaging concentration factors across studies, factors which exceed the theoretical maximum should be lowered to the latter for averaging purposes. In no instance should a section 409 tolerance (or section 701 MRL) be based on a concentration factor greater than the theoretical maximum. If only one processing study has been conducted and the theoretical concentration factor has been exceeded, the section 409 or section 701 residue level should be based on the fac-

tor (if available) listed in Tables 1 through 4 of this guideline.

(ii) As stated in paragraph (c)(2) of this guideline, processing studies should reflect actual commercial practices. If several studies are available and a step (e.g. washing) that is routinely used in the processing of that RAC is omitted, it may be inappropriate to include that study in the calculation of the average concentration factor.

(3) **Use of exaggerated rate studies.**

(i) The Agency encourages use of field trials with exaggerated application rates in cases where residues near or below the analytical method's LOQ are expected in the RAC from the maximum registered rate (1×). For purposes of this discussion, pesticide uses can be divided into those which result in quantifiable residues in the RAC and those which do not. The former would have section 408 tolerances set above the LOQ, while the latter would usually have tolerances set at the LOQ. In either case, if possible, processing studies should use RAC samples which contain quantifiable residues.

(ii) For uses which result in quantifiable residues in the RAC from the registered application rate, exaggerated rate applications are not needed to generate RAC samples for processing if all field trials lead to residues well above the LOQ. However, if residues below or near the LOQ are observed in some field trials, it is advisable for an exaggerated application rate to be used to generate RAC samples for the processing study. Regardless of whether exaggerated application rates are used, if a section 408 tolerance is based on the presence of quantifiable residues and concentration of residues is observed in a processed commodity, that concentration factor will be used in conjunction with the HAFT or other applicable average value and other relevant factors (e.g. variability of the analytical method) to determine the need for a section 409 tolerance (or section 701 MRL). In other words, the concentration factor will not be adjusted for the use of exaggerated rates in cases where quantifiable residues are observed in the RAC from the registered use.

(iii) In those cases where all RAC samples from the field trials show residues below the LOQ and the residue data cover all significant growing regions for the crop as delineated in OPPTS 860.1500 (page 405), it may be possible to waive the processing study and conclude that section 409 tolerances (or section 701 MRLs) are not needed based on the results of field trials conducted at exaggerated application rates. With the exception of mint and citrus, if exaggerated rate data are available and these field trials result in no quantifiable residues in the RAC, no processing study and section 409 tolerances are required provided that the rate was exaggerated by at least the highest theoretical concentration factor among all the processed commodities derived from that crop or 5×, whichever is less. Processing studies will be needed for citrus and mint in virtually all cases due to the extremely high potential concentration factors for citrus oil (1,000×) and mint oil (330×).

(iv) If no quantifiable residues are found in the RAC from the maximum registered rate, but the exaggerated rate does produce quantifiable residues, the latter samples should be processed and residues measured in the appropriate commodities. Any residues still above the LOQ in the processed commodities should be adjusted for the degree of exaggeration. These adjusted residues should then be compared to the LOQ for the RAC. If the adjusted residues are greater than or equal to twice the LOQ, a section 409 tolerance (or section 701 MRL) is needed. Due to the variability associated with an analytical method near its LOQ, a food additive tolerance (or section 701 MRL) will not normally be established for residues less than twice the LOQ. For example, consider a field corn RAC tolerance set at 0.05 ppm (LOQ) and residues of 0.08 ppm being found in the RAC and 0.30 ppm in the oil following a 5× application rate. Adjusting for the 5× rate, oil residues would be 0.06 ppm, which is less than twice the LOQ. Therefore, a section 409 tolerance is not necessary. However, if the oil residues were 1.0 ppm, a section 409 tolerance (or perhaps section 701 MRL) at 0.20 ppm (1.0 ppm/5) would be necessary.

(v) One additional scenario needs to be discussed regarding use of exaggerated rates. In some cases no quantifiable residues may be found in the RAC, but the exaggerated rate is less than the maximum theoretical concentration factor (or 5×, whichever is less) due to phytotoxicity limitations. In these instances a decision will be made case-by-case as to the need for a processing study. If a processing study is deemed necessary, any quantifiable residues in processed fractions would be adjusted for the degree of exaggeration as explained in the previous paragraph. Some of the factors to consider when determining if the processing study is needed include how close the degree of exaggeration comes to the theoretical factor (or 5×, whichever is less) and whether detectable residues (i.e. greater than limit of detection but less than LOQ) are found in any RAC samples. Another consideration would be whether the pesticide is likely to be present on a specific portion of the RAC based on when it is applied and/or its ability to translocate. For example, a pesticide applied late in the growing season would be more likely to be on the surface of a fruit and have greater potential to concentrate in pomace than one applied only at the bloom stage or earlier

(4) **Impact of Ready-to-Eat (RTE).**

(i) The classification of a processed food as RTE or not RTE will determine whether or not the possibility of setting a section 701 MRL needs to be explored as discussed under paragraph (f)(5) of this guideline. Until recently, the Agency has considered any food available for sale as being ready-to-eat. The Agency now holds that RTE food has a common sense meaning of food which is consumed without further preparation and will apply this interpretation in future actions. Therefore, food should now be considered "ready-to-eat" if it consumed "as-is" or is added to other RTE foods (e.g. condiments).

(ii) The Agency also realizes that application of this definition of RTE may be difficult in many instances. The following processed foods are examples of not-ready-to-eat: mint oil, citrus oil, guar gum, and dried tea. Examples of clearly RTE foods are raisins, olives, and potato chips. Vegetable oils are an example of foods not so easily characterized under this RTE standard. The Agency is presently analyzing information on food consumption and mixing of livestock feeds in order to classify processed commodities with respect to RTE. As such decisions are made, they will be made available to the public.

(5) **Determining the need for section 409 tolerances or section 701 MRL's.**

(i) The Agency will establish food/feed additive tolerances (FATs) under section 409 of the FFDCA for processed foods or feeds that are classified as RTE if residues in those processed commodities are likely to exceed the corresponding section 408 tolerances. Therefore, for an RTE food such as raisins, the concentration factor (taking into account multiple processing studies and exaggerated rates, if applicable) should be multiplied by the HAFT (or other applicable average value) and that value compared to the RAC tolerance. If that number is appreciably higher than the section 408 tolerance, a food/feed additive (section 409) tolerance will be needed. The judgment as to "appreciably higher" will need to take into account how close the residue level is to the LOQ of the analytical method. If residues in the processed food are less than twice the LOQ, a section 409 tolerance is normally not needed. On the other hand, when residues in the processed food (i.e. concentration factor times HAFT) are significantly above the LOQ, a section 409 tolerance will normally be needed if those residues are approximately 1.5× the section 408 tolerance (or higher). For situations in which the processed food/feed residues are close to that level (e.g. 1.3 to 1.7× those in the RAC), all relevant information including variability in recovery data will be considered by the Agency when assessing the need for food/feed additive tolerances.

(ii) The procedure is more complex for processed foods or feeds that are not RTE (nRTE). If residues in an nRTE processed food exceed the section 408 tolerance, residues in the RTE forms of those foods/feeds will need to be determined and then compared to the section 408 tolerance. If the residues in the RTE (i.e. mixed/diluted) form do not exceed the RAC tolerance, the Agency will establish an MRL on the nRTE processed commodity under section 701 of the FFDCA. On the other hand, if residues in the RTE (mixed/diluted) form still appreciably exceed those in the RAC, a food/feed additive tolerance will be established for the nRTE processed commodity under section 409 of the FFDCA.

(iii) In order to determine whether residues in the

RTE (mixed/diluted) forms of nRTE processed foods/feeds exceed those in the RAC, the Agency will develop dilution factors. These will be based on the least amount of dilution that may occur for the nRTE food. For example, flour, assuming it is classified as nRTE, is likely to have a relatively low dilution factor based on its use in preparation of commodities such as crackers, bagels, and tortillas. Dried tea, on the other hand, is likely to have a large dilution factor based on the relative weight of water used to brew tea. At this time there is no list of dilution factors. As these factors are derived, the Agency will announce them to the public periodically.

(iv) The procedure for assessing nRTE processed commodities is as follows. The concentration factor (accounting for multiple processing studies and exaggerated rates, if necessary) is multiplied by the HAFT (or other applicable average value) to determine residues in the nRTE processed food. If the residue in the nRTE food does not appreciably exceed the section 408 tolerance, neither a section 409 tolerance nor section 701 MRL is needed. If the residue in the nRTE processed food does appreciably exceed the RAC tolerance, that residue should be divided by the dilution factor to determine the residue level in the RTE form. If the residue in the RTE (mixed/diluted) food is basically equal to or less than the section 408 tolerance, a section 701 MRL is needed for the nRTE processed commodity. If the residue in the RTE (mixed/diluted) food still appreciably exceeds the section 408 tolerance, a section 409 (i.e. food or feed additive) tolerance is needed for the nRTE processed commodity.

(v) This procedure can be illustrated by some examples using mint and the nRTE food mint oil. The assumption is made that for three different pesticides that the HAFT value is 8.0 ppm and that the RAC tolerance is 10 ppm. The assumption is also made that the dilution factor for mint oil is 160 for its use in preparation of RTE foods. Pesticide A is observed to concentrate 1.3× in mint oil. The concentration factor times the HAFT is equal to 10.4 ppm, which is not appreciably higher than the RAC tolerance of 10 ppm. Neither a section 409 tolerance nor section 701 MRL is needed for the mint oil. Pesticide B is found to concentrate 40× in mint oil. The concentration factor (40) times the HAFT (8.0 ppm) is equal to 320 ppm, well above the RAC tolerance of 10 ppm. The residues in the RTE (mixed/diluted) food are then calculated to be 2 ppm by dividing the mint oil residue of 320 ppm by the dilution factor of 160. The 2 ppm residue in the RTE food is below the 10 ppm RAC tolerance. Therefore, a section 701 MRL of 320 ppm should be established for the nRTE mint oil. Pesticide C is found to concentrate 320× in mint oil. The concentration factor (320) times the HAFT (8.0 ppm) is 2,560 ppm, which is well above the RAC tolerance of 10 ppm. The residues in the RTE food are then calculated to be 16 ppm by dividing the mint oil residue of 2,560 ppm by the dilution factor of 160. The 16 ppm

in the RTE (mixed/diluted) food appreciably exceeds the 10 ppm RAC tolerance. Therefore, a section 409 or food additive tolerance is needed for mint oil at 2,560 ppm (or more likely at 2,500 ppm considering significant figures).

(g) **Data report format.**

The following describes a suggested format for a study report, item by item. However, other formats are also acceptable, provided that the information described in this paragraph is included.

(1) *Title/cover page.* Title page and additional documentation requirements (i.e. requirements for data submission and procedures for claims of confidentiality of data if relevant to the study report) should precede the content of the study. These requirements are described in PR Notice 86–5 (see paragraph (h)(5) of this guideline).

(2) *Table of contents.*

(3) *Summary/introduction.*

(4) *Materials.*

(i) *Test substance.*

(A) Identification of the pesticide formulated product used in the field trial from which the RAC used in the processing study was derived, including the active ingredient therein, or if fortified RAC samples were used in the processing study, identity of the fortifying substances.

(B) Identification and amount of residues in experimentally treated RAC samples at the time the processing study is initiated.

(C) Any and all additional information petitioners consider appropriate and relevant to provide a complete and thorough description and identification of the test substances used in the processing study.

(ii) *Test commodity.*

(A) Identification of the RACs (crop/type/variety) and the specific crop parts used in the processing study.

(B) Sample identification (source of samples, field trial identification number; control or weathered residue sample, coding and labeling information (should be the same as or cross-referenced to the sample coding/labeling assigned at harvest)).

(C) Treatment histories (pesticides used, rates, number of applications, preharvest intervals (PHIs), etc.) of the RAC samples used in the processing study.

(D) The developmental stages, general condition (immature/mature, green/ripe, fresh/dry, etc.) and sizes of the RAC samples used in the processing study.

(E) Any and all additional information the petitioner considers appropriate and relevant to provide a complete and thorough description of the RACs used in the processing study.

(5) *Methods.*

(i) *Experimental design.* For example:

(A) Number of test/control samples.

(B) Number of replicates.

(C) Residue levels in the RACs to be used.

(D) Representativeness of test commodities to the matrices of concern, etc.

(ii) *Test procedures.*

(A) Fortification (spiking) procedure, if used (detail the manner in which the test compounds were introduced to the RACs).

(B) A description of the processing procedure used and how closely it simulates commercial practice. Quantities of starting RAC and of resulting processed commodities.

(C) A description of the methods of residue analysis (see OPPTS 860.1340, Residue analytical method).

(D) A description of the means of validating the methods of residue analysis (see OPPTS 860.1340).

(E) A description of any storage stability validation studies that may have been performed (see OPPTS 860.1380, Storage stability data).

(6) *Results/discussion.*

(i) *Residue results.*

(A) Raw data and correction factors applied, if any.

(B) Recovery levels.

(C) Storage stability levels, if applicable.

(D) Direct comparison of residues in the RAC with those in each processed product or processing fraction derived from that sample, etc.

(ii) *Statistical treatments.* Describe tests applied to the raw data.

(iii) *Quality control.* Include if not covered elsewhere. Describe control measures/precautions followed to ensure the fidelity of the processing study.

(iv) *Other.* Constituting any and all additional information the petitioner considers appropriate and relevant to provide a complete and thorough description of the processing study or studies.

(7) *Conclusions.* Discuss conclusions that may be drawn concerning the concentration/reduction of the test compounds in the test matrices as a function of the standard commercial processing procedure, and the need for food/feed additive tolerances or section 701 MRLs.

(8) *Certification.* Certification of authenticity by the Study Director (including signature, typed name, title, affiliation, address, telephone number, date).

(9) *Tables/figures.*

(i) Tables of raw data from the processing study, method recovery data, storage stability recovery data (if applicable); etc.

(ii) Graphs, figures, flowcharts, etc. (as relevant—include the processing procedure with weights of RAC and processed fractions).

(10) *Appendixes.*

(i) Representative chromatograms, spectra, etc. (as applicable).

(ii) Reprints of methods and other studies (unless physically located elsewhere in the overall data submission, in which case cross-referencing will suffice) which will support the registrant's conclusions.

(iii) Other, comprising any relevant material not fitting in any of the other portions of this report.

(h) *References.* The following references should be consulted for additional background material on this test guideline.

(1) Environmental Protection Agency, Pesticide Reregistration Rejection Rate Analysis – Residue Chemistry; Follow-up Guidance for: Generating Storage Stability Data; Submission of Raw Data; Maximum Theoretical Concentration Factors; Flowchart Diagrams. EPA Report No. 737-R-93-001, February, 1993.

(2) Environmental Protection Agency, Pesticide Reregistration Rejection Rate Analysis – Residue Chemistry; Follow-up Guidance for: Updated Livestock Feeds Tables; Aspirated Grain Fractions (Grain Dust); A Tolerance Perspective; Calculating Livestock Dietary Exposure; Number and Location of Domestic Crop Field Trials. EPA Report No. 737-K-94-001, June, 1994.

(3) Environmental Protection Agency, Pesticide Reregistration Rejection Rate Analysis – Residue Chemistry; EPA Report No. 738-R-92-001, June, 1992.

(4) Environmental Protection Agency, FIFRA Accelerated Reregistration – Phase 3 Technical Guidance. EPA Report No. 540/09-90-078. (Available from National Technical Information Service, Springfield, VA).

(5) Environmental Protection Agency, Pesticide Registration Notice PR 86-5, Standard Format for Data Submitted under the FIFRA and Certain Provisions of the Federal Food, Drug, and Cosmetic Act (FFDCA), May 3, 1986.

Metric Equivalents and Approximate Conversions

Metric Equivalents and Approximate Conversions

Equivalents, Length

1 inch	=	25.4 millimeters
1 millimeter	=	0.04 inch
1 foot	=	0.3 meter
1 meter	=	3.3 feet
1 yard	=	0.9 meter
1 meter	=	1.1 yards
1 mile	=	1.6 kilometers
1 kilometer	=	0.6 mile

Equivalents, Area

1 sq. inch	=	6.45 sq. centimeter
1 sq. centimeter	=	0.15 sq. inch
1 sq. foot	=	0.09 sq. meter
1 sq. meter	=	10.8 sq. feet
1 sq. yard	=	0.8 sq. meter
1 sq. meter	=	1.2 sq. yards
1 acre	=	0.4 hectare
1 hectare	=	2.47 acres

Equivalents, Weight (Mass)

1 ounce	=	28.35 grams
1 gram	=	0.035 ounce
1 pound	=	453.59 grams
1 kilogram	=	2.2 pounds
1 pound	=	0.454 kilogram
1 metric ton (tonne)	=	1.1 tons
1 ton	=	2,000 pounds

Conversions, Length

inches x 25.4	=	millimeters
millimeters x 0.04	=	inches
feet x 0.3	=	meters
meters x 3.3	=	feet
yards x 0.9	=	meters
meters x 1.1	=	yards
miles x 1.6	=	kilometers
kilometers x 0.6	=	miles

Conversions, Area

sq. inches x 6.45	=	sq. centimeters
sq. centimeters x 0.16	=	sq. inches
sq. feet x 0.09	=	sq. meters
sq. meters x 10.8	=	sq. feet
sq. yards x 0.8	=	sq. meters
sq. meters x 1.2	=	sq. yard
acres x 0.4	=	hectares
hectares x 2.5	=	acres

Conversions, Weight (Mass)

1 ounce x 28.35	=	grams
1 gram x 0.035	=	ounces
1 pound x 0.45	=	kilograms
1 kilogram x 2.2	=	pounds
1 ton (short) x 0.9	=	metric ton (tonne)
1 metric ton x 1.1	=	tons (short)

References:

Keller, J.J. 1974. Metric Manual. J.J. Keller & Associates, Inc. Nennak, WI. 362 pp.

Pernezny, K. *et. al.* 1995. Florida Tomato Scouting Guide, Second Edition. IFAS, Cooperative Extension Service SP22 , University of Florida, Gainesville. 44 pp.

Indices

Includes Scientific Names Index and Common Names Index listed by crop monograph number(s) for each name

General Index
Plant Scientific Names
Referenced by Crop Monograph Number

Plants' Common Names,
Referenced by Crop Monograph Number

Abata cola 187
Abiu 001
Abricot 023
Abricotier 023
Absinthe 685
Absinthe wormwood 685
Absinthium 685
Abyssinian cabbage 002
Abyssinian kale 312
Abyssinian mustard 002
Acedera 558
Acedera romana 557
Aceituna 418
Acelga 064
Acelga cardo 595
Acerola 003
Achicoria 153
Achiote 021
Achira 118
Achita 015, 016
Achoccha 208
Acom 006
Acopate 192
Acore odorant 110
Acorn 406
Acorn squash 574
Acuyo 461
Adanka bean 048
Aerial yam 006
African cabbage 002
African fan palm 432
African foxtailgrass 100
African horned cucumber 004
African horned melon 004
African jointgrass 342
African mangosteen 293
African marigold 359
African mustard 396
African oil palm 433
African sandalwood 113
Agarista 068
Agastache 020
Agrimony 005
Agrio de Guinea 515
Agropiro del oeste 679
Agropiro delgado 675
Agua yam 688
Ahipa 687
Airella rouge 343
Airelle 077
Airpotato 006
Aja 688

Ajankuiri (a-han-hwee-ri) 490
Ajawiri 490
Ajedra 539
Ajedrea comun 539
Aji 462
Aji amarillo 463
Aji dulce 464
Aji picante 463
Ajibravo 463
Ajipa 687
Ajipo 687
Ajo 249
Ajo porro 325
Ajonjoli 543
Akee 007
Akee apple 007
Aki 007
'Ala 'ala 006
Alachofa tuberosa 030
Alaea 021
Alaska oniongrass 425
Albahaca 045
Albaricoque 023
Alcachofa 031
Alecost 195
Alegria 015
Alercim 516
Alfalfa 008
Algarroba 127
Algazul 291
Algodonera 196
Alho 249
Alholva 240
Aliipoe 118
Alkali bluegrass 079
Alkali sacaton 009
Alkaligrass 010
Alkekengi 269
Alko chines 155
Allegheny blackberry 074
Allegheny chinkapin 150
Allegheny serviceberry 307
Allgood 259
Alligator pear 037, 040, 142
Alligator pepper 264
Allouia 330
Allouya 330
Allspice 011
Almendro 012, 013
Almond 012
Almond oil 012
Almond, Tropical 013

Alpine cat's-tail 615
Alpine cranberry 343
Alpiste 114
Alta fescue 242
Amandier 012
Amarante caudee 015
Amarante elegante 015
Amaranth 014
Amaranth, Chinese 014
Amaranth, Grain 015
Amaranth, Leafy 016
Amaranto 014, 015
Amatungula 402
Ambarella 017
Amberique 055, 060
American basil 045
American basswood 341
American beachgrass 046
American beech 063
American bird pepper 463
American blueberry 077
American Bunch 266
American chestnut 150
American cotton 196
American cowslip 364
American crab 197
American cranberry 200
American cranberrybush 281
American dewberry 074
American eggplant 228
American elder 229
American elderberry 229
American falsepennyroyal 457
American filbert 278
American fountain grass 380
American ginseng 257
American hazelnut 278
American lime 341
American linden 341
American mannagrass 358
American marigold 359
American nightshade 483
American oil palm 433
American parsley 439
American persimmon 468
American pima cotton 196
American pinon 471
American plum 477
American red raspberry 508
American red elder 229
American red plum 477
American sloughgrass 551

Common Name Index

Common Name Index

Food And Feed Crops Of The United States